# ENGINEERING MECHANICS

## (APPLIED MECHANICS)

*[For the Students of B.E./B.Tech., B.Arch., Diploma in Engineering, U.P.S.C. (Engg and Civil Services), AMIE, GATE, and other Competitive Examinations]*

AF889482

**J.K. Gupta**
**S.K. Gupta**

CENGAGE

Australia • Brazil • Canada • India • Mexico • Singapore • United Kingdom • United States

CENGAGE

*Engineering Mechanics*
*(Applied Mechanics)*

J.K. Gupta

S.K. Gupta

Cover Image: © Thinkstock

**© 2021 Cengage Learning India Pvt. Ltd.**

ALL RIGHTS RESERVED. No part of this work covered by the copyright herein may be reproduced or distributed in any form or by any means without the prior written permission of the copyright owner.

For product information and technology assistance, contact us at
**asia.infoindia@cengage.com**

For permission to use material from this text or product, submit all requests online at
**www.cengage.com/permissions**

**ISBN-13:** 978-93-5350-524-0
**ISBN-10:** 93-5350-524-0

**Cengage Learning India Pvt. Ltd.**
**T-303, 3rd Floor**
**DAV Complex, Plot 3**
**Mayur Vihar Phase 1**
**Delhi-110091**

Cengage is a leading provider of customized learning solutions with employees residing in nearly 40 different countries and sales in more than 125 countries around the world. Find your local representative at **www.cengage.com**

Cengage is committed to creating inclusive and diverse products, including perspectives that represent the diversity of people across all backgrounds, identities, opinions, and experiences. If you have any feedback related to this, please contact us at **support.cengage.com**

To learn more about Cengage products, platforms and services, visit at **www.cengage.co.in**

**Printed by:**
**Rajiv Book Binding House**
D-1518, DSIIDC Industrial Complex, Narela, New Delhi 110040

Second Impression 2022

# Preface

It gives us immense pleasure and profound satisfaction to present this textbook titled ***Engineering Mechanics*** to the students and teaching faculty. This book covers the complete syllabi of all Indian and Foreign Technical Universities.

The subject of engineering mechanics deals with the behaviour of bodies or systems subjected to external forces. The subject matter has been arranged systematically in **21 chapters**. Each chapter focuses on the basic concepts and explains the fundamental principles, methods, and step-by-step approach to solve problems in engineering mechanics.

While writing this book, we have constantly kept in mind the latest as well as the changing trends in the examinations. To make it more useful at all levels, the book has been written in simple and easy-to-understand style with well-graded solved examples placed appropriately throughout the text.

At the end of each chapter, *Highlights, Numerical Problems with Answers, Review Questions, and Objective-type Questions with Answers* have been provided to present a complete and comprehensive textbook on this subject. In fact, the book will also be useful as a reference text for post-graduate students as well as the industry professionals.

Even though all effort has been made to prevent errors of any kind, it is difficult to claim perfection as human error is plausible. It would be highly appreciated if any error, omission, and/or suggestion for further improvement of this book is brought to our notice. The same would be gratefully acknowledged and incorporated in the next edition.

## Acknowledgements

We express our sincere gratitude to numerous teachers and students who have contributed significantly to this text through their valuable suggestions and comments shared during the preparation of the manuscript. We also extend our gratitude to the editorial and production staff at Cengage Learning India, our publishers, for taking praiseworthy initiative to improve, design, and timely publish this book.

**J.K. Gupta**
**S.K. Gupta**

# Contents

# 1

# BASIC CONCEPTS



## 1.1 INTRODUCTION

*Engineering Mechanics* is a branch of science which deals with the behaviour of bodies under the action of forces. It deals with the practical application of mechanics in the field of engineering and technology and is divided into the following two main classes:

**1.** Statics; and **2.** Dynamics.

*Statics* deals with the forces and their effects when the body is at rest. *Dynamics* deals with the forces and their effects while the body is in motion. Dynamics is further divided into the following two groups:

**1.** Kinematics; and **2.** Kinetics.

*Kinematics* deals with the bodies in motion without any reference to forces responsible for the motion. In other words, it deals with the relative motion.

*Kinetics* deals with the bodies in motion due to direct application of forces. In other words, it deals with the inertia forces which arise from the combined effect of mass and motion of the body.

## 1.2 FUNDAMENTAL AND DERIVED UNITS

The measurement of physical quantities is one of the most important operations in engineering. Every quantity is measured in terms of some arbitrary, but internationally accepted units, called *fundamental units.*

Some units are expressed in terms of other units, which are derived from fundamental units, are known as *derived units*, *e.g.* the unit of area, velocity, acceleration, pressure, etc.

## 1.3 SYSTEMS OF UNITS

There are four systems of units *i.e.* C.G.S. units; F.P.S. units; M.K.S. units; and S.I. units. Since our course of studies are conducted in S.I. system of units, therefore we shall only discuss this system of units.

## 1.4 S.I. UNITS

This is an international system of units, which is widely used in many countries including India.

In this system of units, there are seven fundamental units and two supplementary units, which cover the entire field of science and engineering. These units are shown in Table 1.1.

**Table 1.1** Fundamental and supplementary units

| *S. No.* | *Physical quantity* | *Unit* |
|---|---|---|
| | *Fundamental units* | |
| **1.** | Length ($l$) | Metre (m) |
| **2.** | Mass ($m$) | Kilogram (kg) |
| **3.** | Time ($t$) | Second (s) |
| **4.** | Temperature ($T$) | Kelvin (K) |
| **5.** | Electric current ($I$) | Ampere (A) |
| **6.** | Luminous intensity ($I_v$) | Candela (cd) |
| **7.** | Amount of substance ($n$) | Mole (mol) |
| | *Supplementary units* | |
| **8.** | Plane angle ($\alpha$, $\beta$, $\theta$, $\phi$) | Radian (rad) |
| **9.** | Solid angle ($\Omega$) | Steradian (sr) |

The derived units, which will be commonly used in this book, are given in Table 1.2.

**Table 1.2** Derived units

| *S. No.* | *Quantity* | *Symbol* | *Unit* |
|---|---|---|---|
| **1.** | Linear velocity | $V$ | m/s |
| **2.** | Linear acceleration | $a$ | m/s$^2$ |
| **3.** | Angular velocity | $\omega$ | rad/s |
| **4.** | Angular acceleration | $\alpha$ | rad/s$^2$ |
| **5.** | Mass density | $\rho$ | kg/m$^3$ |
| **6.** | Force or Weight | $F$, $W$ | N; 1 N = 1 kg-m/s$^2$ |
| **7.** | Pressure | $p$ | N/m$^2$ |
| **8.** | Work, Energy, Enthalpy | $W, E, H$ | J; 1 J = N-m |
| **9.** | Power | $P$ | W; 1 W = 1 J/s |
| **10.** | Absolute or dynamic viscosity | $\mu$ | N-s/m |
| **11.** | Kinematic viscosity | $v$ | m$^2$/s |
| **12.** | Frequency | $f$ | Hz; 1 Hz = 1 cycle/s |
| **13.** | Gas constant | $R$ | J/kg K |
| **14.** | Thermal conductance | $h$ | W/m$^2$ K |
| **15.** | Thermal conductivity | $k$ | W/m K |
| **16.** | Specific heat | $c$ | J/kg K |
| **17.** | Molar mass or Molecular mass | $M$ | kg/kg-mole |

## 1.5 METRE, KILOGRAM, AND SECOND

The *metre* is defined as the length equal to 1 650 763.73 wavelengths in vacuum of the radiation corresponding to the transition between the levels $2p_{10}$ and $5d_5$ of the Krypton 86 atom.

The *kilogram* is defined as the mass of the international prototype (standard block of platinum iridium alloy) of the kilogram, kept at the International Bureau of Weights and Measures at Sevres, near Paris.

The *second* is defined as the duration of 9 192 631 770 periods of the radiation corresponding to the transition between the two hyperfine levels of the ground state of the cesium 133 atom.

## 1.6 RULES FOR S.I. UNITS

The eleventh General Conference of Weights and Measures recommended only the fundamental and derived units for S.I. system, but it did not elaborate the rules for the usage of the units. Later on, many scientists and engineers held a number of meetings for the style and usage of S.I. units. Some of the outcomes of the meetings are as follows:

1. For numbers having five on more digits, the digits should be placed in groups of three separated by spaces (instead of commas) counting both to the left and right to the decimal point.
2. In a four-digit number, the space is not required unless the four-digit number is used in a column of numbers with five or more digits.
3. A dash is to be used to separate units that are multiplied together. For example, newton metre is written as N-m. It should not be confused with mN, which stands for millinewton.
4. Plurals are never used with symbols. For example, metre or metres are written as m.
5. All symbols are written in small letters except the symbols derived from the proper names. For example, N for newton and W for watt.
6. The units with names of scientists should not start with capital letters when written in full. For example, 90 newton and not 90 Newton.

## 1.7 IMPORTANT TERMS USED IN ENGINEERING MECHANICS

The following terms are widely used in engineering mechanics:

1. ***Matter*** A matter is anything that occupies space, possesses mass and offers resistance to any external force.
2. ***Particle*** A particle is a body that occupies negligible space but has a mass which is considered to be concentrated at a point.
3. ***Space*** The space is the geometric region occupied by the body.
4. ***Rigid body*** A rigid body consists of a system of innumerable particles. It does not change its shape and size under the effect of forces acting over it. Strictly speaking, all the solid bodies are not perfectly rigid bodies. However, they may be considered as a rigid body, because the shape and size of the body (*i.e.* deformation) does not change under the effect of the forces acting on it.

**Note:** An elastic body undergoes deformation but regains its shape after the external forces are removed.

## 1.8 SCALAR AND VECTOR QUANTITIES

A *scalar quantity* is one which has magnitude only *e.g.* mass, time, volume, density, etc.

A *vector quantity* is one which has magnitude as well as direction *e.g.* velocity, acceleration, force, etc.

A vector quantity is represented by a vector which is a straight line of a certain length possessing a starting point and a terminal point at which it carries an arrow head. The length of the vector represents the magnitude to some scale while the arrow head of the vector represents the direction of the vector quantity.

The addition and subtraction of vector quantities may be done as follows:

Consider two vector quantities $P$ and $Q$, which are required to be added or subtracted, as shown in Fig. 1.1 (*a*). In order to add the vectors, take any point $A$ and draw a line $AB$ parallel and equal in magnitude to the vector $P$, to some suitable scale. Through $B$, draw $BC$ parallel and equal in magnitude to the vector $Q$. Join $AC$, which gives the required sum of the two vectors $P$ and $Q$, as shown in Fig. 1.1 (*b*).

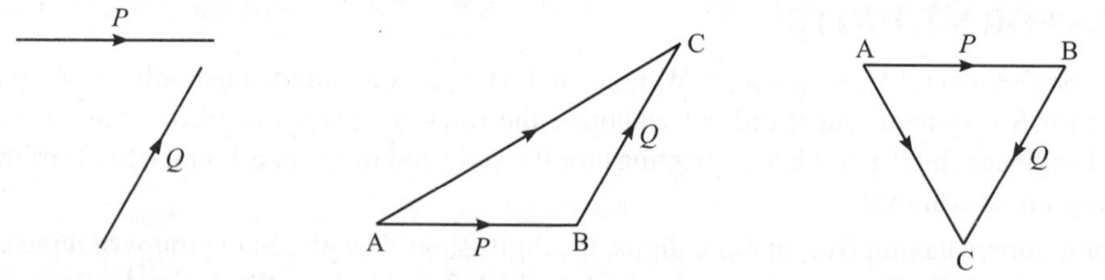


(*a*) Vector quantities (*b*) Addition of vectors (*c*) Subtraction of vectors

**Fig. 1.1**

Now in order to subtract the vectors, draw a line $AB$ parallel and equal in magnitude to the vector $P$, to some suitable scale. Through $B$, draw $BC$ parallel and equal in magnitude to the vector $Q$, but in *opposite* direction. Join $AC$, which gives the required difference of the vectors $P$ and $Q$, as shown in Fig. 1.1 (*c*).

## 1.9 USEFUL MATHEMATICAL FORMULAE

The following useful mathematical formulae are important from the subject point of view:

**1. *Algebra***

**(*a*)** When the bases are same and powers are different, then in multiplication, the powers are added, *e.g.*

$$x^m \times x^n = x^{m+n}$$

and in division, the powers are subtracted, *e.g.*

$$\frac{x^m}{x^n} = x^{m-n}$$

**(*b*)** In a quadratic equation $ax^2 + bx + c = 0$, the value of $x$ is given by

$$x = \frac{-b \pm \sqrt{b^2 - 4ac}}{2a}$$

where $a$ = Coefficient of $x^2$,
$b$ = Coefficient of $x$, and
$c$ = Constant term.

**(*c*)** Anything raised to the power zero is one, *i.e.*

$$x^0 = 1$$

**(*d*)** If $x^a = y$, then $x = y^{1/a}$

**(*e*)** The Binomial theorem is given by

$$(x+y)^n = {}^nC_0\, x^n y^0 + {}^nC_1\, x^{n-1} y^1 + {}^nC_2\, x^{n-2} y^2 + {}^nC_3\, x^{n-3} y^3 + {}^nC_4\, x^{n-4} y^4 + {}^nC_5\, x^{n-5} y^5 + \ldots + {}^nC_n\, x^{n-n} y^n$$

$$= x^n + \frac{\underline{\lfloor n}}{\underline{\lfloor (n-1)} \times \underline{\lfloor 1}} \times x^{n-1} y + \frac{\underline{\lfloor n}}{\underline{\lfloor (n-2)} \times \underline{\lfloor 2}} \times x^{n-2} y^2 + \frac{\underline{\lfloor n}}{\underline{\lfloor (n-3)} \times \underline{\lfloor 3}} \times x^{n-3} y^3$$

$$+ \frac{\underline{\lfloor n}}{\underline{\lfloor (n-4)} \times \underline{\lfloor 4}} \times x^{n-4} y^4 + \frac{\underline{\lfloor n}}{\underline{\lfloor (n-5)} \times \underline{\lfloor 5}} \times x^{n-5} y^5 + \ldots + y^n$$

It may be noted that

$$ {}^{n}C_{0} = \frac{\underline{|n}}{\underline{|n-0} \times \underline{|0}} = \frac{\underline{|n}}{\underline{|n} \times \underline{|0}} = 1 $$

$$ {}^{n}C_{1} = \frac{\underline{|n}}{\underline{|n-1} \times \underline{|1}} = \frac{\underline{|n}}{\underline{|n-1} \times 1} $$

$$ {}^{n}C_{2} = \frac{\underline{|n}}{\underline{|n-2} \times \underline{|2}} = \frac{\underline{|n}}{\underline{|n-2} \times 1 \times 2} = \frac{\underline{|n}}{\underline{|n-2} \times 2} $$

$$ {}^{n}C_{3} = \frac{\underline{|n}}{\underline{|n-3} \times \underline{|3}} = \frac{\underline{|n}}{\underline{|n-3} \times 1 \times 2 \times 3} = \frac{\underline{|n}}{\underline{|n-3} \times 6} $$

Similarly,

$$ {}^{n}C_{n} = \frac{\underline{|n}}{\underline{|n-n} \times \underline{|n}} = \frac{\underline{|n}}{\underline{|0} \times \underline{|n}} = 1 $$

**2. *Co-ordinate geometry***

**(*a*)** The equation of a straight line is given by

$$ y = mx + c $$

**(*b*)** The equation of a parabola is given by

$$ y^2 = 4ax $$

**(*c*)** The equation for a circle is given by

$$ x^2 + y^2 = a^2 $$

where $a$ = Radius of the circle.

**(*d*)** The equation for an ellipse is given by

$$ \frac{x^2}{a^2} + \frac{y^2}{b^2} = 1 $$

where $a$ = Semi-major axis, and $b$ = Semi-minor axis.

**3. *Differential calculus***

**(*a*)** $\frac{d}{dx}(C) = 0$, where $C$ is a constant

**(*b*)** $\frac{d}{dx}(x^n) = n\,x^{n-1}$

**(*c*)** $\frac{d}{dx}(u.v) = u \times \frac{dv}{dx} + v \times \frac{du}{dx}$

**(*d*)** $\frac{d}{dx}\left(\frac{u}{v}\right) = \frac{v \times \frac{du}{dx} - u \times \frac{dv}{dx}}{v^2}$

**(*e*)** $\frac{d}{dx}(\sin x) = \cos x$; $\frac{d}{dx}(\text{cosec } x) = -\text{cosec } x.\cot x$

**(*f*)** $\frac{d}{dx}(\cos x) = -\sin x$; $\frac{d}{dx}(\sec x) = \sec x \,.\, \tan x$

**(*g*)** $\frac{d}{dx}(\tan x) = \sec^2 x$; $\frac{d}{dx}(\cot x) = -\text{cosec}^2\, x$

**4. *Integral calculus***

**(*a*)** $\int a\, dx = ax + C$

(*b*) $\int x^n \, dx = \frac{x^{n+1}}{n+1}$ (Increase the power by one divided by the same increased power)

(*c*) $\int (ax+b)^n \, dx = \frac{(ax+b)^{n+1}}{(n+1)a}$

(*d*) $\int a^x \, dx = \frac{a^x}{\log_e a} + C$

(*e*) $\int a^x \log_e a \, dx = a^x + C$

**5. *Trigonometry***

(***a***) In a right angled triangle *ABC*, as shown in Fig. 1.2,

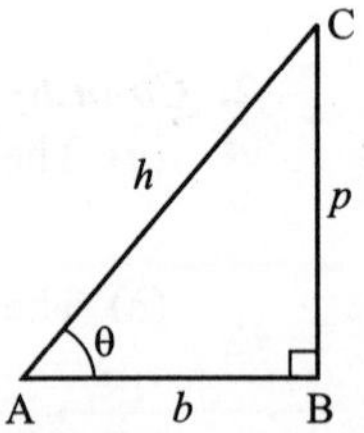


**Fig. 1.2**

$$\sin\theta = \frac{BC}{AC} = \frac{\text{Perpendicular }(p)}{\text{Hypotenuse }(h)}$$

$$\cos\theta = \frac{AB}{AC} = \frac{\text{Base }(b)}{\text{Hypotenuse }(h)}$$

$$\tan\theta = \frac{\sin\theta}{\cos\theta} = \frac{BC}{AB} = \frac{\text{Perpendicular}(p)}{\text{Base }(b)}$$

$$\text{cosec}\,\theta = \frac{1}{\sin\theta} = \frac{h}{p}$$

$$\sec\theta = \frac{1}{\cos\theta} = \frac{h}{b}$$

and $$\cot\theta = \frac{1}{\tan\theta} = \frac{\cos\theta}{\sin\theta} = \frac{b}{p}$$

(***b***) The values of different trigonometric functions are given as follows:

| *Angle* | 0° | 30° | 45° | 60° | 90° | 180° | 270° | 360° |
|---|---|---|---|---|---|---|---|---|
| sin θ | 0 | $\frac{1}{2}$ | $\frac{1}{\sqrt{2}}$ | $\frac{\sqrt{3}}{2}$ | 1 | 0 | –1 | 0 |
| cos θ | 1 | $\frac{\sqrt{3}}{2}$ | $\frac{1}{\sqrt{2}}$ | $\frac{1}{2}$ | 0 | –1 | 0 | 1 |
| tan θ | 0 | $\frac{1}{\sqrt{3}}$ | 1 | $\sqrt{3}$ | ∞ | 0 | ∞ | 0 |
| cosec θ | ∞ | 2 | $\sqrt{2}$ | $\frac{2}{\sqrt{3}}$ | 1 | ∞ | –1 | ∞ |
| sec θ | 1 | $\frac{2}{\sqrt{3}}$ | $\sqrt{2}$ | 2 | ∞ | –1 | ∞ | 1 |
| cot θ | ∞ | $\sqrt{3}$ | 1 | $\frac{1}{\sqrt{3}}$ | 0 | ∞ | 0 | ∞ |

It may be noted that

***(i)*** From 0° to 90° (*i.e.* in the first quadrant), *all* trigonometrical ratios are *positive.*

***(ii)*** From 90° to 180° (*i.e.* in the second quadrant), only sin θ and cosec θ are *positive.*

***(iii)*** From 180° to 270° (*i.e.* in the third quadrant), only tan θ and cot θ are *positive.*

***(iv)*** From 270° to 360° (*i.e.* in the fourth quadrant), only cos θ and sec θ are *positive.*

| Remember | Add | Sugar | To | Coffee |
|---|---|---|---|---|
| | ↓ | ↓ | ↓ | ↓ |
| | All | sin | tan | cos |
| | Quadrant I | Quadrant II | Quadrant III | Quadrant IV |

***(c)*** The various trigonometric relations are as follows:

$$\sin^2\theta + \cos^2\theta = 1$$

$$1 + \tan^2\theta = \sec^2\theta$$

$$1 + \cot^2\theta = \text{cosec}^2\,\theta$$

***(d) Sum and difference of two angles A and B***

$$\sin(A+B) = \sin A\cos B + \cos A\sin B$$

$$\sin(A-B) = \sin A\cos B - \cos A\sin B$$

$$\cos(A+B) = \cos A\cos B - \sin A\sin B$$

$$\cos(A-B) = \cos A\cos B + \sin A\sin B$$

$$\tan(A+B) = \frac{\tan A + \tan B}{1 - \tan A.\tan B}$$

$$\tan(A-B) = \frac{\tan A - \tan B}{1 + \tan A.\tan B}$$

$$\sin(A+B) - \sin(A-B) = 2\cos A\sin B$$

***(e) Functions of multiple and half of angles***

$$\sin A = 2\sin\left(\frac{A}{2}\right)\cos\left(\frac{A}{2}\right)$$

$$\cos A = \cos^2\left(\frac{A}{2}\right) - \sin^2\left(\frac{A}{2}\right)$$

$$= 2\cos^2\left(\frac{A}{2}\right) - 1$$

$$= 1 - 2\sin^2\left(\frac{A}{2}\right)$$

$$\tan A = \frac{2\tan\left(\frac{A}{2}\right)}{1 - \tan^2\left(\frac{A}{2}\right)}$$

$$\sin 2A = 2\sin A\cos A$$

$$\cos 2A = \cos^2 A - \sin^2 A = 2\cos^2 A - 1 = 1 - 2\sin^2 A$$

$$\tan 2A = \frac{2\tan A}{1 - \tan^2 A}$$

$$\sin 3A = 3\sin A - 4\sin^3 A$$

$$\cos 3A = 4\cos^3 A - 3\cos A$$

$$\sin^2 \frac{A}{2} = \frac{1-\cos A}{2}$$

$$\cos^2 \frac{A}{2} = \frac{1+\cos A}{2}$$

***(f) Products and powers of functions***

$$2\sin A \sin B = \cos(A-B) - \cos(A+B)$$

$$2\cos A \cos B = \cos(A+B) + \cos(A-B)$$

$$2\sin A \cos B = \sin(A+B) + \sin(A-B)$$

$$2\cos A \sin B = \sin(A+B) - \sin(A-B)$$

$$\sin^2 A = \frac{1}{2}(1-\cos 2A)$$

$$\cos^2 A = \frac{1}{2}(1+\cos 2A)$$

$$\sin^3 A = \frac{1}{4}(3\sin A - \sin 3A)$$

$$\cos^3 A = \frac{1}{4}(3\cos A + \cos 3A)$$

***(g) Sum and difference of functions***

$$\sin A + \sin B = 2\sin\left(\frac{A+B}{2}\right)\cos\left(\frac{A-B}{2}\right)$$

$$\sin A - \sin B = 2\cos\left(\frac{A+B}{2}\right)\sin\left(\frac{A-B}{2}\right)$$

$$\cos A + \cos B = 2\cos\left(\frac{A+B}{2}\right)\cos\left(\frac{A-B}{2}\right)$$

$$\cos A - \cos B = -2\sin\left(\frac{A+B}{2}\right)\sin\left(\frac{A-B}{2}\right)$$

$$\tan A + \tan B = \frac{\sin(A+B)}{\cos A \cos B}$$

$$\tan A - \tan B = \frac{\sin(A-B)}{\cos A \cos B}$$

$$\sin^2 A - \sin^2 B = \sin(A+B)\sin(A-B)$$

$$\cos^2 A - \cos^2 B = -\sin(A+B)\sin(A-B)$$

$$\cos^2 A - \sin^2 B = \cos(A+B)\cos(A-B)$$

***(h) Rule for change of trigonometric ratios***

$\sin(-\theta) = -\sin\theta$ ; $\text{cosec}(-\theta) = -\text{cosec}\,\theta$

$\cos(-\theta) = \cos\theta$ ; $\sec(-\theta) = \sec\theta$

$\tan(-\theta) = -\tan\theta$ ; $\cot(-\theta) = -\cot\theta$

$\sin(90° - \theta) = \cos\theta$ ; $\text{cosec}(90° - \theta) = \sec\theta$

$\cos(90° - \theta) = \sin\theta$ ; $\sec(90° - \theta) = \text{cosec}\,\theta$

$\tan(90° - \theta) = \cot\theta$ ; $\cot(90° - \theta) = \tan\theta$

$\sin(90° + \theta) = \cos\theta$ ; $\text{cosec}(90° + \theta) = \sec\theta$

$\cos(90° + \theta) = -\sin\theta$ ; $\sec(90° + \theta) = -\text{cosec}\,\theta$

$\tan(90° + \theta) = -\cot\theta$ ; $\cot(90° + \theta) = -\tan\theta$

$\sin(180° - \theta) = \sin\theta$ ; $\text{cosec}(180° - \theta) = \text{cosec}\,\theta$

$\cos(180° - \theta) = -\cos\theta$ ; $\sec(180° - \theta) = -\sec\theta$

$\tan(180° - \theta) = -\tan\theta$ ; $\cot(180° - \theta) = -\cot\theta$

$\sin(180° + \theta) = -\sin\theta$ ; $\text{cosec}(180° + \theta) = -\text{cosec}\,\theta$

$\cos(180° + \theta) = -\cos\theta$ ; $\sec(180° + \theta) = -\sec\theta$

$\tan(180° + \theta) = \tan\theta$ ; $\cot(180° + \theta) = \cot\theta$

(*i*) In any triangle, having sides $a$, $b$ and $c$, as shown in Fig. 1.3, the sine formula is given by

$$\frac{a}{\sin A} = \frac{b}{\sin B} = \frac{c}{\sin C}$$

and the cosine formula is $a^2 = b^2 + c^2 - 2bc\cos A$

or $b^2 = a^2 + c^2 - 2ac\cos B$

and $c^2 = a^2 + b^2 - 2ab\cos C$

where $A$, $B$ and $C$ are opposite angles of sides $a$, $b$ and $c$, respectively.

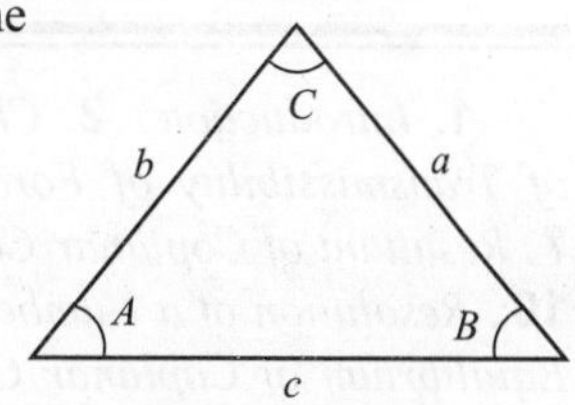


Fig. 1.3

(*j*) In a right angled triangle, as shown in Fig. 1.4,

$$(AC)^2 = (AB)^2 + (BC)^2$$

*i.e.* $(\text{Hypotenuse})^2 = (\text{Base})^2 + (\text{Perpendicular})^2$

or $h^2 = b^2 + p^2$

This is known as *Pythagoras Theorem.*

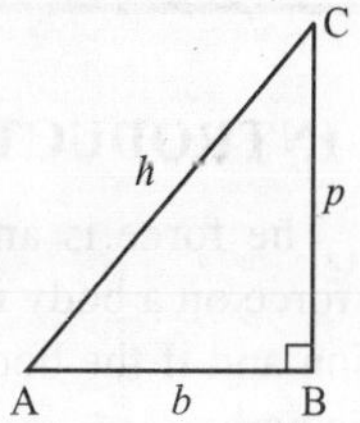


Fig. 1.4

# 2

# COPLANAR CONCURRENT FORCES

**1.** *Introduction* **2.** *Characteristics of a Force* **3.** *Classification of a Force System* **4.** *Principle of Transmissibility of Forces* **5.** *Principle of Superposition of Forces* **6.** *Resultant and Equilibrant* **7.** *Resultant of Coplanar Concurrent Forces* **8.** *Parallelogram Law of Forces* **9.** *Resolution of a Force* **10.** *Resolution of a Number of Coplanar Concurrent Forces (Principle of Resolution)* **11.** *Condition for Equilibrium of Coplanar Concurrent System of Forces* **12.** *Free Body Diagram* **13.** *Triangle Law of Forces* **14.** *Polygon Law of Forces* **15.** *Graphical Method of Determining Resultant of a Number of Coplanar Concurrent Forces* **16.** *Lami's Theorem*

## 2.1 INTRODUCTION

The force is an agent which produces or tends to produce, destroy or tends to destroy motion. The force on a body may change the motion of the body, *i.e.* if the body is at rest, the force may set it in motion and if the body is already in motion, it may accelerate it. The force may also retard the motion of the body.

## 2.2 CHARACTERISTICS OF A FORCE

In order to determine the effects of a force acting on a body, we must know the following characteristics of a force:

**(*a*)** Magnitude of the force (*i.e.* 50 N, 100 N etc.),

**(*b*)** Direction or line of action of force,

**(*c*)** Sense or nature of force (*i.e.* push or pull), and

**(*d*)** Point of application of the force.

## 2.3 CLASSIFICATION OF A FORCE SYSTEM

When several forces of different magnitude and direction act upon a body, they are called to form a *system of forces*.

The forces, according to the plane in which forces are applied and the relative positions of the lines of action, may be classified as follows:

1. ***Collinear forces*** When the lines of action of all the forces lie along the same line, then they are known as *collinear forces*.
2. ***Coplanar forces*** When the lines of action of all the forces lie on the same plane, then they are known as *coplanar forces*.
3. ***Concurrent forces*** When the lines of action of all the forces meet at a single point, then they are known as *concurrent forces*.
4. ***Coplanar concurrent forces*** When the lines of action of all the forces lie in the same plane and pass through a common point, then they are known as *coplanar concurrent forces*.
5. ***Coplanar non-concurrent forces*** When the lines of action of all the forces lie in the same plane but do not pass through a common point, then they are known as *coplanar non-concurrent forces*.
6. ***Non-coplanar concurrent forces*** When all the forces do not lie in the same plane but their lines of action pass through a common point, then they are known as *non-coplanar concurrent forces*.
7. ***Non-coplanar non-concurrent forces*** When all the forces do not lie in the same plane and their lines of action also do not pass through a common point, then they are known as *non-coplanar non-concurrent forces*.

## 2.4 PRINCIPLE OF TRANSMISSIBILITY OF FORCES

It states that '*the state of rest or motion of the rigid body is unaltered if a force acting on the body is replaced by another force of the same magnitude and direction, but acting anywhere on the body in the line of action of the replaced force*'.

In Fig. 2.1, $F$ is the force acting on a rigid body at point $A$. According to the principle of transmissibility of forces, if the point of application of this force is shifted to another point $B$, which is on the line of action of force $F$, the state of rest or the steady motion of the body is still unaltered.

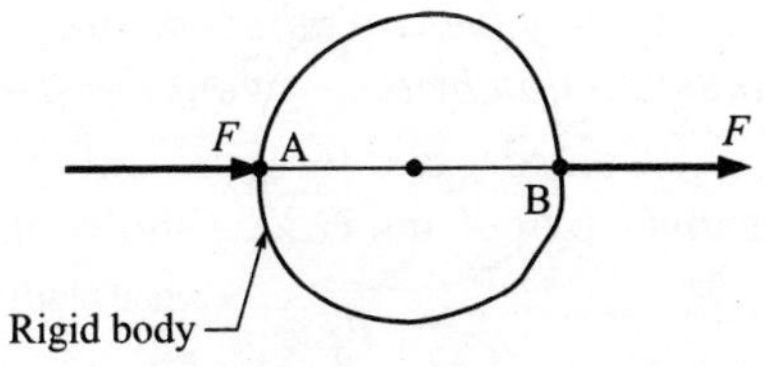


**Fig. 2.1**

**Notes: 1.** The nature of force can be changed, *i.e.* pull to push or push to pull. In Fig. 2.1, we see that force $F$ acting at point $A$ is push, while the force acting at $B$ is pull.

**2.** The principle of transmissibility of forces is only applicable to rigid bodies. It is not valid for deformable bodies.

## 2.5 PRINCIPLE OF SUPERPOSITION OF FORCES

It states that '*the combined effect of the force system acting on a particle of a rigid body is the vector sum of effects of individual forces*'.

Consider two forces $P$ and $Q$ acting at $A$ on a boat, as shown in Fig. 2.2. Let $R$ be the resultant of these two forces $P$ and $Q$. According to Newton's second law of motion, the boat will move in the direction of the resultant force $R$ with acceleration proportional to $R$. The same motion can be obtained when $P$ and $Q$ are applied simultaneously.

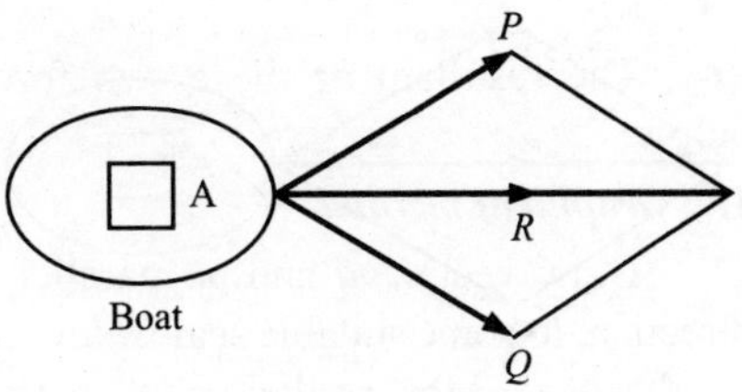


**Fig. 2.2**

## 2.6 RESULTANT AND EQUILIBRANT

When a number of forces act on a body in such a way that the body remains in a state of rest or of uniform motion, then the system of forces is said to be in *equilibrium*. A single force which produces the same effect as the given system of forces acting on the body, is known as *resultant*.

The resultant of given system of forces may change the state of rest or of uniform motion of the body, *i.e.* the body is not in equilibrium. In order to bring the body in the equilibrium state, a single force, is to be applied to keep the body in equilibrium. This single force is known as *equilibrant* and it is equal and opposite to the resultant of the given system of forces acting on the body.

## 2.7 RESULTANT OF COPLANAR CONCURRENT FORCES

When a number of concurrent coplanar forces act upon a body to keep the body in equilibrium, then the resultant of these forces is the vector sum of all these forces.

The process of determining the resultant force of a number of forces acting on a body, is called *composition of forces.* The various laws used for finding the resultant force are as follows:

1. Parallelogram law of forces;
2. Triangle law of forces; and
3. Polygon law of forces.

These laws are discussed, in detail, in the following pages:

## 2.8 PARALLELOGRAM LAW OF FORCES

It states that '*if two forces acting simultaneously on a particle, be represented in magnitude and direction by the two adjacent sides of a parallelogram, then their resultant is represented in magnitude and direction by the diagonal of the parallelogram passing through their point of intersection*'.

Consider two forces $P$ and $Q$ making an angle $\theta$, acting at a point $O$. The magnitude and direction is given by $OA$ and $OB$, as shown in Fig. 2.3 (*a*).

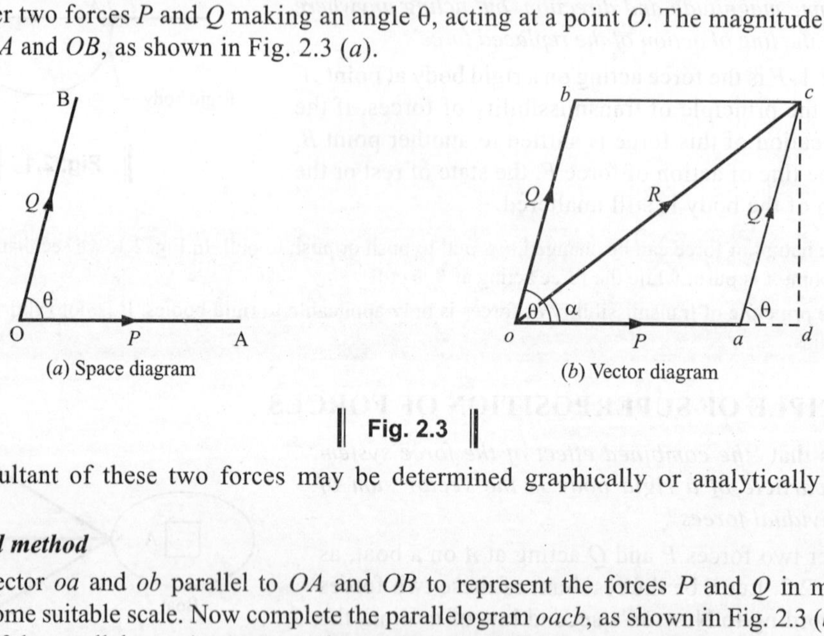


**Fig. 2.3**

The resultant of these two forces may be determined graphically or analytically as discussed below:

**1. *Graphical method***

Draw vector $oa$ and $ob$ parallel to $OA$ and $OB$ to represent the forces $P$ and $Q$ in magnitude and direction, to some suitable scale. Now complete the parallelogram $oacb$, as shown in Fig. 2.3 (*b*). The vector $oc$ (diagonal of the parallelogram) represents the resultant $R$ of the two forces $P$ and $Q$.

**2. *Analytical method***

Draw perpendicular from $c$ meeting the line $oa$ produced at $d$. Since $ac$ is parallel to $ob$, therefore angle $cad$ = angle $aob = \theta$. Now from the right angled triangle $acd$,

$$\frac{cd}{ac} = \sin\theta \quad \text{or} \quad cd = ac\sin\theta = Q\sin\theta \quad ...(\because ac = ob = Q)$$

and
$$\frac{ad}{ac} = \cos\theta \quad \text{or} \quad ad = ac\cos\theta = Q\cos\theta$$

From right angled triangle *odc*,

$$(oc)^2 = (od)^2 + (cd)^2 = (oa + ad)^2 + (cd)^2$$

$$\therefore \quad R^2 = (P + Q\cos\theta)^2 + (Q\sin\theta)^2$$

$$= P^2 + Q^2\cos^2\theta + 2PQ\cos\theta + Q^2\sin^2\theta$$

$$= P^2 + Q^2(\cos^2\theta + \sin^2\theta) + 2PQ\cos\theta$$

$$= P^2 + Q^2 + 2PQ\cos\theta \quad ...(\because \sin^2\theta + \cos^2\theta = 1)$$

or
$$R = \sqrt{P^2 + Q^2 + 2PQ\cos\theta} \quad ...(i)$$

The inclination of the line of action of the resultant $R$ with the force $P$ is given by $\alpha$.

Now from right angled triangle *odc*,

$$\tan\alpha = \frac{cd}{od} = \frac{cd}{oa + ad} = \frac{Q\sin\theta}{P + Q\cos\theta}$$

or
$$\alpha = \tan^{-1}\left[\frac{Q\sin\theta}{P + Q\cos\theta}\right] \quad ...(ii)$$

It may be noted that

(*i*) When the two forces are equal, *i.e.* $P = Q$, then resultant,

$$R = \sqrt{P^2 + P^2 + 2P \times P\cos\theta} = \sqrt{2P^2 + 2P^2\cos\theta}$$

$$= \sqrt{2P^2(1 + \cos\theta)} = \sqrt{2P^2 \times 2\cos^2\left(\frac{\theta}{2}\right)} = 2P\cos\left(\frac{\theta}{2}\right)$$

$$...\left[\because 1 + \cos\theta = 2\cos^2\left(\frac{\theta}{2}\right)\right]$$

and
$$\alpha = \tan^{-1}\left[\frac{P\sin\theta}{P + P\cos\theta}\right] = \tan^{-1}\left[\frac{\sin\theta}{1 + \cos\theta}\right]$$

$$= \tan^{-1}\left[\frac{2\sin\left(\frac{\theta}{2}\right)\cos\left(\frac{\theta}{2}\right)}{2\cos^2\left(\frac{\theta}{2}\right)}\right] = \tan^{-1}\left[\frac{\sin\left(\frac{\theta}{2}\right)}{\cos\left(\frac{\theta}{2}\right)}\right]$$

$$...\left[\because \sin\theta = 2\sin\left(\frac{\theta}{2}\right)\cos\left(\frac{\theta}{2}\right)\right]$$

$$= \tan^{-1}\left[\tan\left(\frac{\theta}{2}\right)\right] = \frac{\theta}{2}$$

This shows that when the two forces $P$ and $Q$ are equal, then the resultant $R$ bisects the angle between the two forces.

(*ii*) When the two forces are at right angles, *i.e.* when $\theta = 90°$, then resultant,

$$R = \sqrt{P^2 + Q^2 + 2PQ\cos 90°} = \sqrt{P^2 + Q^2} \quad ...(\because \cos 90° = 0)$$

and
$$\alpha = \tan^{-1}\left[\frac{Q\sin\theta}{P + Q\cos\theta}\right] = \tan^{-1}\left[\frac{Q\sin 90°}{P + Q\cos 90°}\right]$$

$$= \tan^{-1}\left(\frac{Q}{P}\right) \quad ...(\because \sin 90° = 1)$$

**(iii)** When the two forces $P$ and $Q$ act in the same straight line and in the same direction, *i.e.* when $P$ and $Q$ are collinear, then $\theta = 0°$

$$\therefore \qquad R = \sqrt{P^2 + Q^2 + 2PQ \cos 0°} = \sqrt{P^2 + Q^2 + 2PQ} = P + Q$$

This is the maximum value of the resultant $R$. In other words, the resultant of the forces $P$ and $Q$ is maximum when $\cos \theta$ is maximum, *i.e.* when $\cos \theta = 1$ or $\theta = 0°$.

**(iv)** When two forces $P$ and $Q$ act in the same straight line but in opposite direction, then $\theta = 180°$.

$$\therefore \qquad R = \sqrt{P^2 + Q^2 + 2PQ \cos 180°} = \sqrt{P^2 + Q^2 - 2PQ} = P - Q$$

$$\text{...}(\because \cos 180° = -1)$$

The resultant $R$ will act in the direction of the greater force. This is the minimum value of the resultant $R$. In other words, the resultant of the forces $P$ and $Q$ is minimum when $\cos \theta$ is minimum, *i.e.* when $\cos \theta = -1$ or $\theta = 180°$.

**Example 2.1** *Find the magnitude and direction of the resultant of two forces 50 N and 30 N acting at a point with an included angle of 60° between them.*

***Solution*** Given : $P = 50$ N ; $Q = 30$ N ; $\theta = 60°$

***Magnitude of the resultant***

We know that resultant,

$$R = \sqrt{P^2 + Q^2 + 2PQ \cos \theta}$$

$$= \sqrt{(50)^2 + (30)^2 + 2 \times 50 \times 30 \times \cos 60°}$$

$$= \sqrt{2500 + 900 + 1500} = \sqrt{4900} = 70 \text{ N} \quad \textbf{Ans.}$$

***Direction of the resultant***

Let $\alpha$ = Angle of inclination with the force $P$ (*i.e.* 50 N).

We know that

$$\tan \alpha = \frac{Q \sin \theta}{P + Q \cos \theta}$$

$$= \frac{30 \sin 60°}{50 + 30 \cos 60°} = \frac{30 \times 0.866}{50 + 30 \times 0.5} = 0.3997$$

$$\therefore \qquad \alpha = \tan^{-1}(0.3997) = 21.786° \quad \textbf{Ans.}$$

**Example 2.2** *The resultant of two forces at 60° angle is 14 N. If the same forces are applied on right angle, then the resultant is $\sqrt{136}$ N. Determine the magnitude of both the forces.*

***Solution*** Given : $\theta_1 = 60°$ ; $R_1 = 14$ N ; $\theta_2 = 90°$ ; $R_2 = \sqrt{136}$ N

Let $P$ and $Q$ be the two forces, as shown in Fig. 2.4. In the first case, from the parallelogram law of forces,

$$R_1 = \sqrt{P^2 + Q^2 + 2PQ \cos \theta_1}$$

or

$$R_1^2 = P^2 + Q^2 + 2PQ \cos \theta_1$$

$$(14)^2 = P^2 + Q^2 + 2PQ \cos 60°$$

$$196 = P^2 + Q^2 + 2PQ \times \frac{1}{2} = P^2 + Q^2 + PQ \qquad \text{...}(i)$$

**Fig. 2.4**

In the second case, $R_2^2 = P^2 + Q^2 + 2PQ\cos\theta_2$

$$(\sqrt{136})^2 = P^2 + Q^2 + 2PQ\cos 90°$$

$$136 = P^2 + Q^2 \quad ...(\because \cos 90° = 0) \quad ...(ii)$$

Subtracting equation (*ii*) from equation (*i*), we get

$$196 - 136 = PQ$$

or $$PQ = 60 \quad ...(iii)$$

We know that $(P+Q)^2 = P^2 + Q^2 + 2PQ = 136 + 2 \times 60 = 256$

$$\therefore \quad P + Q = \sqrt{256} = 16 \text{ N} \quad ...(iv)$$

We also know that $(P-Q)^2 = P^2 + Q^2 - 2PQ = 136 - 2 \times 60 = 16$

$$\therefore \quad P - Q = \sqrt{16} = 4 \quad ...(v)$$

Adding equations (*iv*) and (*v*), we have

$$2P = 16 + 4 = 20$$

$$\therefore \quad P = \frac{20}{2} = 10 \text{ N } \textbf{Ans.}$$

and from equation (*v*), $Q = P - 4 = 10 - 4 = 6$ N **Ans.**

**Example 2.3** *Two equal forces are acting at a point with an included angle of 60°. If the resultant force is 35 N, find the magnitude of each force.*

***Solution*** Given : $\theta = 60°$ ; $R = 35$ N

Let $P$ = Magnitude of each force.

We know that resultant for equal forces,

$$R = 2P\cos\left(\frac{\theta}{2}\right)$$

or $$35 = 2P\cos\left(\frac{60°}{2}\right) = 2P\cos 30° = 2P \times 0.866$$

$$\therefore \quad P = \frac{35}{2 \times 0.866} = 20.2 \text{ N } \textbf{Ans.}$$

**Example 2.4** *A push of 180 N and a pull of 350 N act simultaneously at a point. If the angle between them is 135°, find the resultant of the forces.*

***Solution*** Given : Push, $P = 180$ N ; Pull, $Q = 350$ N ; $\theta = 180° - 135° = 45°$

The forces of 180 N and 350 N acting at point $O$ is shown in Fig. 2.5.

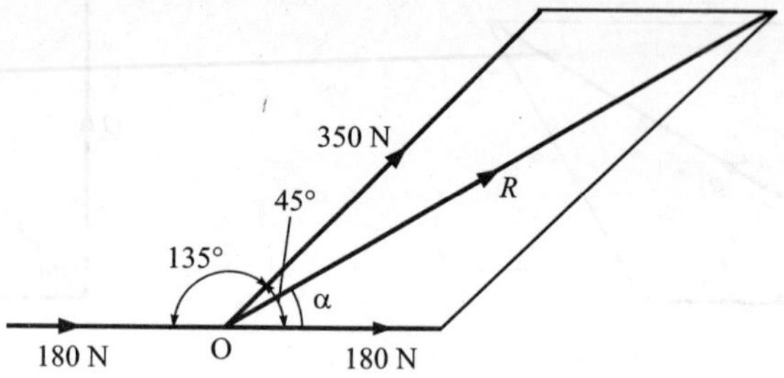


**Fig. 2.5**

We know that resultant, $R = \sqrt{P^2 + Q^2 + 2PQ\cos\theta}$

$= \sqrt{(180)^2 + (350)^2 + 2 \times 180 \times 350 \cos 45°}$

$= \sqrt{32\,400 + 122\,500 + 89\,082} = 494$ N **Ans.**

Let $\alpha$ = Angle of inclination with the force 180 N.

We know that $\tan\alpha = \dfrac{Q\sin\theta}{P + Q\cos\theta} = \dfrac{350\sin 45°}{180 + 350\cos 45°}$

$= \dfrac{350 \times 0.707}{180 + 350 \times 0.707} = \dfrac{247.45}{427.45} = 0.5789$

$\therefore$ $\alpha = \tan^{-1}(0.5789) = 30.06°$ **Ans.**

**Example 2.5** *The resultant of two concurrent forces is 1220 N and the angle between the forces is 90°. The resultant makes an angle of 35° with one of the force. Find the magnitude of each force.*

***Solution*** Given : $R$ = 1220 N ; $\theta$ = 90° ; $\alpha$ = 35°

Let $P$ and $Q$ be the two concurrent forces, as shown in Fig. 2.6.

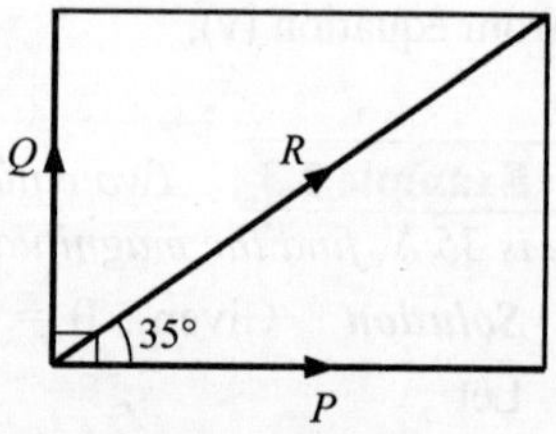


**Fig. 2.6**

We know that $\tan\alpha = \dfrac{Q\sin\theta}{P + Q\cos\theta}$

$\tan 35° = \dfrac{Q\sin 90°}{P + Q\cos 90°} = \dfrac{Q}{P}$

$0.7 = \dfrac{Q}{P}$

or $Q = 0.7\,P$ ...(*i*)

We also know that resultant of the two forces at right angles is

$R = \sqrt{P^2 + Q^2}$

$1220 = \sqrt{P^2 + (0.7P)^2} = \sqrt{1.49P^2} = 1.22\,P$

$\therefore$ $P = \dfrac{1220}{1.22} = 1000$ N **Ans.**

and $Q = 0.7P = 0.7 \times 1000 = 700$ N **Ans.** ...[From equation (*i*)]

**Example 2.6** *The maximum and minimum resultant of two forces acting at a point are 350 N and 50 N, respectively. Determine the value of the two forces. Also find the angle between them if the resultant is to be 250 N.*

**Solution** Given : $R_{max} = 350$ N ; $R_{min} = 50$ N

***Value of the two forces***

Let $P$ and $Q$ be the two forces.

We know that the resultant is maximum when the angle between the two forces is 0°, *i.e.* when the forces are collinear. In such a case,

$$R_{max} = P + Q \quad \text{or} \quad 350 = P + Q \qquad ...(i)$$

Also the resultant is minimum when the angle between the two forces is 180°, *i.e.* when the forces act in opposite direction. In such a case,

$$R_{min} = P - Q \quad \text{or} \quad 50 = P - Q \qquad ...(ii)$$

From equations (*i*) and (*ii*), we have

$$P = 200 \text{ N}; \text{ and } Q = 150 \text{ N} \quad \textbf{Ans.}$$

***Angle between the two forces***

Let $\theta$ = Angle between the two forces if $R$ = 250 N.

We know that $R^2 = P^2 + Q^2 + 2PQ\cos\theta$

$$(250)^2 = (200)^2 + (150)^2 + 2 \times 200 \times 150 \cos\theta$$

$$62\,500 = 40\,000 + 22\,500 + 60\,000 \cos\theta$$

or $$60\,000\cos\theta = 0 \quad \text{or} \quad \cos\theta = 0$$

$\therefore$ $$\theta = 90° \quad \textbf{Ans.}$$

**Example 2.7** *The sum of two concurrent forces P and Q is 270 N and their resultant is 180 N. The angle between the force P and the resultant R is 90°. Find the magnitude of each force and the angle between them.*

**Solution** Given : $P + Q = 270$ N ; $R = 180$ N ; $\alpha = 90°$

***Magnitude of each force*** (Refer Fig. 2.7)

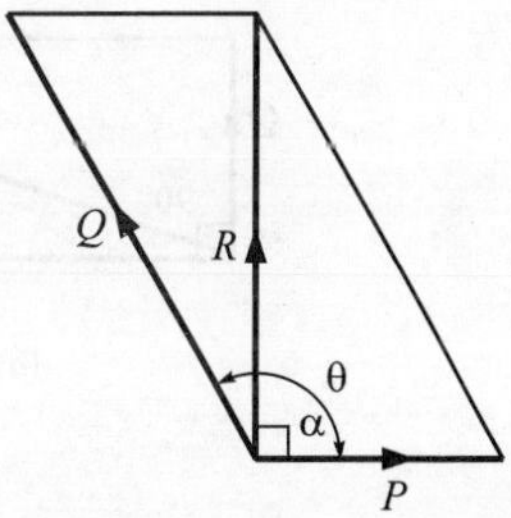

**Fig. 2.7**

We know that angle between the resultant $R$ and the force $P$, is given by

$$\tan\alpha = \frac{Q\sin\theta}{P + Q\cos\theta}$$

or $$\tan 90° = \frac{Q\sin\theta}{P + Q\cos\theta}$$

Since tan 90° = ∞ (infinity), therefore $P + Q\cos\theta$ must be equal to zero (because anything divided by zero is infinity).

$\therefore$ $$P + Q\cos\theta = 0$$

or $$Q\cos\theta = -P \qquad ...(i)$$

We also know that $R^2 = P^2 + Q^2 + 2PQ\cos\theta$

$$(180)^2 = P^2 + Q^2 + 2P(-P) = P^2 + Q^2 - 2P^2 - Q^2 - P^2$$

$$= (Q + P)(Q - P) = 270\,(Q - P)$$

$\therefore$ $$Q - P = \frac{(180)^2}{270} = 120 \qquad ...(ii)$$

We know that $$Q + P = 270 \qquad ...(\text{Given}) \; ...(iii)$$

From equations (*ii*) and (*iii*), we have

$$Q = 195 \text{ N}; \text{ and } P = 75 \text{ N} \quad \textbf{Ans.}$$

***Value of angle between P and Q (angle θ)***

Substituting the value of $P$ and $Q$ in equation (*i*), we have

$$Q \cos\theta = -P \quad \text{or} \quad 195 \cos\theta = -75$$

$$\therefore \qquad \cos\theta = \frac{-75}{195} = -0.3846 \quad \text{or} \quad \theta = 112.619° \ \textbf{Ans.}$$

**Example 2.8** *Find the magnitude of two forces such that, when they act at right angle, their resultant is $\sqrt{10}$ N, but when they act at 60°, the resultant is $\sqrt{13}$ N.*

***Solution*** Given : When $\theta = 90°$, $R = \sqrt{10}$ N ; When $\theta = 60°$, $R = \sqrt{13}$ N

Let $P$ and $Q$ be the two forces.

When $\theta = 90°$ [Fig. 2.8 (*a*)], resultant

$$R = \sqrt{P^2 + Q^2}$$

or $$\sqrt{10} = \sqrt{P^2 + Q^2} \quad \text{or} \quad 10 = P^2 + Q^2 \qquad ...(i)$$

When $\theta = 60°$, [Fig. 2.8 (*b*)], resultant

$$R = \sqrt{P^2 + Q^2 + 2PQ\cos\theta}$$

or $$\sqrt{13} = \sqrt{P^2 + Q^2 + 2PQ\cos 60°}$$

or $$13 = P^2 + Q^2 + 2PQ \times 0.5 = P^2 + Q^2 + PQ \qquad ...(ii)$$

Subtracting equation (*i*) from equation (*ii*), we have

$$13 - 10 = P^2 + Q^2 + PQ - (P^2 + Q^2) \quad \text{or} \quad 3 = PQ$$

$$\therefore \qquad P = \frac{3}{Q} \quad \text{or} \quad P^2 = \frac{9}{Q^2} \qquad ...(iii)$$

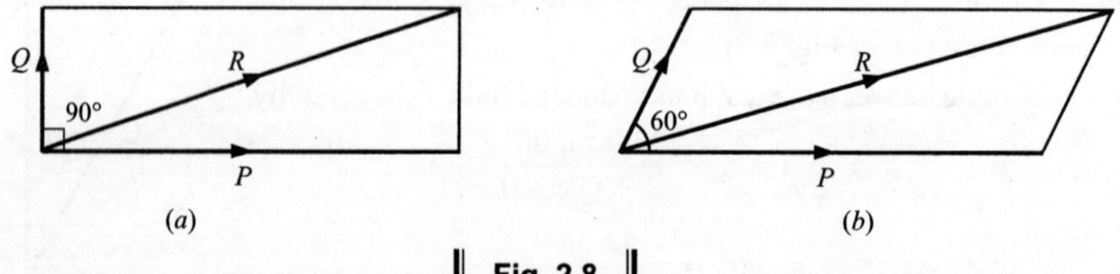


**Fig. 2.8**

Substituting the value of $P^2$ in equation (*i*), we have

$$10 = \frac{9}{Q^2} + Q^2 \quad \text{or} \quad 10Q^2 = 9 + Q^4$$

or $$Q^4 - 10Q^2 + 9 = 0$$

Solving this as a *quadratic equation, we have

$$Q^2 = \frac{10 \pm \sqrt{(10)^2 - 4 \times 9 \times 1}}{2} = \frac{10 \pm \sqrt{100 - 36}}{2} = \frac{10 \pm 8}{2}$$

$$\therefore \qquad Q^2 = \frac{10 \pm 8}{2} = 9 \text{ and } 1$$

*Solution of a quadratic equation, $ax^2 + bx + c = 0$ is given as: $x = \dfrac{-b \pm \sqrt{b^2 - 4ac}}{2a}$

When $Q^2 = 9$, then $Q = \pm 3$

When $Q^2 = 1$, then $Q = \pm 1$

For $Q = +3$, $P = \frac{3}{Q} = \frac{3}{3} = 1\text{ N}$ ...[From equation (*iii*)]

For $Q = -3$, $P = \frac{3}{Q} = \frac{3}{-3} = -1\text{ N}$

For $Q = +1$, $P = \frac{3}{Q} = \frac{3}{1} = 3\text{ N}$

For $Q = -1$, $P = \frac{3}{Q} = \frac{3}{-1} = -3\text{ N}$

Hence the two forces $P$ and $Q$ are (3 and 1) or (–3 and –1) **Ans.**

## 2.9 RESOLUTION OF A FORCE

The process of finding the components of a given force in two given directions is called *resolution of a force* and each component is called a *resolved part*. These components will have the same effect on the body as the given single force.

Let a given force $R$ is making an angle $\theta$ with the horizontal or $X$-axis, as shown in Fig. 2.9. It is required to find the components of the force $R$ along the horizontal (*i.e.* $X$-axis) and vertical (*i.e.* $Y$-axis) directions. From Fig. 2.9, we find that

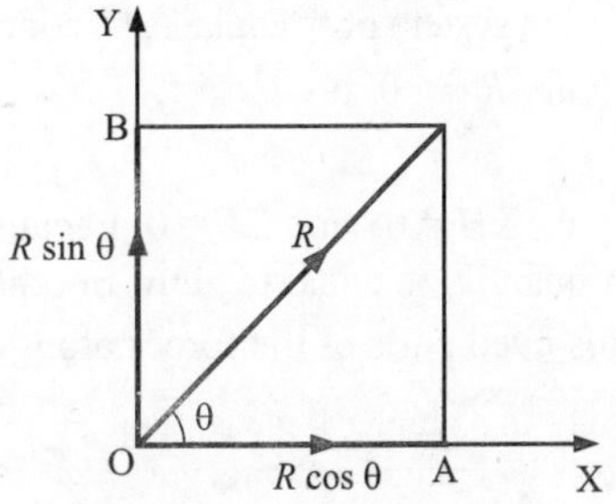


**Fig. 2.9**

Component of $R$ along $X$-axis or in the horizontal direction

$$= R\cos\theta$$

and component of $R$ along $Y$-axis or in the vertical direction

$$= R\sin\theta$$

## 2.10 RESOLUTION OF A NUMBER OF COPLANAR CONCURRENT FORCES (PRINCIPLE OF RESOLUTION)

When a number of coplanar concurrent forces are acting on a body, then the algebraic sum of the resolved parts of the number of forces in a given direction, is equal to the resolved part of their resultant in the same direction.

Consider a number of coplanar concurrent forces $P_1$, $P_2$, $P_3$ and $P_4$ acting at a point $O$, making an angle $\theta_1$, $\theta_2$, $\theta_3$ and $\theta_4$ with $OX$, as shown in Fig. 2.10.

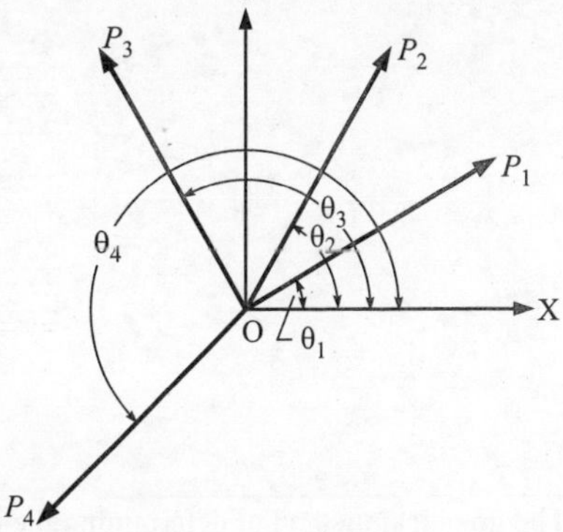


**Fig. 2.10**

Each force may be resolved into two components, one along $X$-axis (in the horizontal direction) and the other along $Y$-axis (in the vertical direction).

We know that algebraic sum of all the horizontal components,

$$\Sigma H = P_1\cos\theta_1 + P_2\cos\theta_2 + P_3\cos\theta_3 + P_4\cos\theta_4$$

and algebraic sum of all the vertical components,

$$\Sigma V = P_1\sin\theta_1 + P_2\sin\theta_2 + P_3\sin\theta_3 + P_4\sin\theta_4$$

Now the resultant of all the forces is given by

$$*R = \sqrt{(\Sigma H)^2 + (\Sigma V)^2}$$

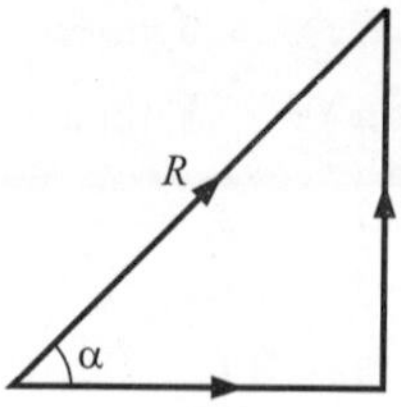


Fig. 2.11

and its inclination with the horizontal, as shown in Fig. 2.11, is given by

$$\tan \alpha = \frac{\Sigma V}{\Sigma H}$$

**Notes:** **1.** When $\Sigma H$ and $\Sigma V$ are both *positive*, then resultant $R$ lies in the first quadrant (*i.e.* between 0° and 90°).
**2.** When $\Sigma H$ is *negative* and $\Sigma V$ is *positive*, then resultant $R$ lies in the second quadrant (*i.e.* between 90° and 180°).
**3.** When $\Sigma H$ and $\Sigma V$ are both *negative*, then resultant $R$ lies in the third quadrant (*i.e.* between 180° and 270°).
**4.** When $\Sigma H$ is *positive* and $\Sigma V$ is *negative*, then resultant $R$ lies in the fourth quadrant (*i.e.* between 270° and 360°).

## 2.11 CONDITION FOR EQUILIBRIUM OF COPLANAR CONCURRENT SYSTEM OF FORCES

We have already discussed that coplanar concurrent forces pass through a common point.

If a system of **coplanar** concurrent forces are in equilibrium, then its resultant ($R$) must be equal to zero *i.e.* $R = 0$, or

$$0 = (\Sigma H)^2 + (\Sigma V)^2$$

$\therefore\ \Sigma H = 0$, and $\Sigma V = 0$, because the sum of the squares of the two real quantities is zero only when each quantity is equal to zero. In other words, for the equilibrium of system of forces, the algebraic sum of the resolved parts of the forces along any two mutually perpendicular directions is separately equal to zero.

**Note:** In a graphical method, the system of forces will be in equilibrium, when these forces are represented, both in magnitude and direction, by the sides of a polygon taken in order.

## 2.12 FREE BODY DIAGRAM

A drawing of a body, a structure or a portion of a structure showing all the forces acting on it, is called a *free body diagram* (FBD). Such a diagram is essential for applying the equations of equilibrium for which we must account for all the known and unknown forces acting on the body.

**Example 2.9** *A boat is moved uniformly along a canal at a given speed by two ropes P and Q, as shown in Fig. 2.12. If the resultant of the forces in ropes, P and Q be R = 700 N, find the values of the forces P and Q.*

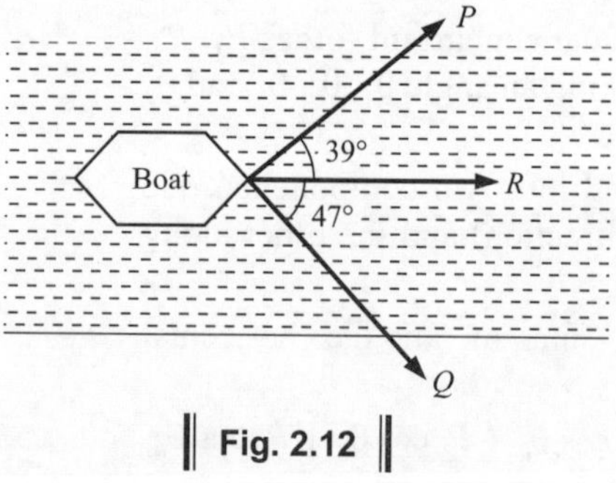


Fig. 2.12

* The graphical method of determining the resultant is discussed in Sec. 2.15.

** The condition of equilibrium for coplanar non-concurrent forces is discussed in the next chapter.

***Solution*** Given : Resultant, $R = 700$ N

Resolving the forces $P$ and $Q$ horizontally and vertically, as shown in Fig. 2.13.

Resolving the forces horizontally,

$$P \cos 39° + Q \cos 47° = R$$

or $$P \times 0.7771 + Q \times 0.682 = 700 \qquad ...(i)$$

Now resolving the forces vertically,

$$P \sin 39° = Q \sin 47°$$

$$P \times 0.6293 = Q \times 0.7313$$

$$\therefore \quad \frac{P}{Q} = \frac{0.7313}{0.6293} = 1.162$$

or $$P = 1.162\, Q$$

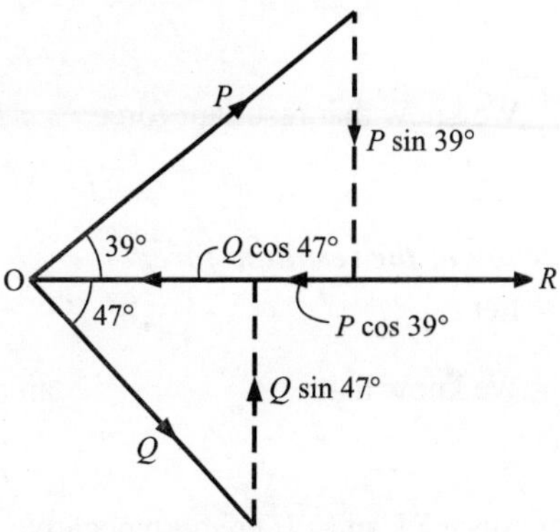


**Fig. 2.13**

Substituting this value of $P$ in equation (*i*), we have

$$1.162 \times 0.7771Q + Q \times 0.682 = 700$$

$$0.903Q + 0.682Q = 700$$

$$1.585Q = 700$$

$$\therefore \quad Q = \frac{700}{1.585} = 441.64 \text{ N} \quad \textbf{Ans.}$$

and $$P = 1.162Q = 1.162 \times 441.64 = 513.186 \text{ N} \quad \textbf{Ans.}$$

**Example 2.10** *Three forces of magnitude 15 kN, 20 kN and 40 kN are acting at a point O, as shown in Fig. 2.14. The angles made by 15 kN, 20 kN and 40 kN forces are 60°, 120° and 240°, respectively, with the X-axis. Find the magnitude and direction of the resultant force.*

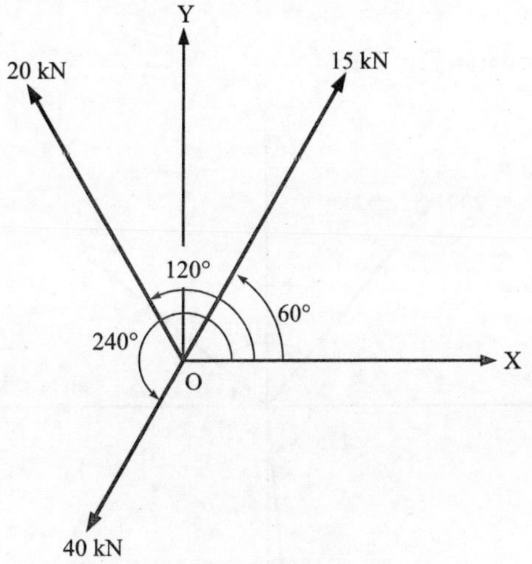


**Fig. 2.14**

***Solution*** Given : $P_1 = 15$ kN ; $P_2 = 20$ kN ; $P_3 = 40$ kN ; $\theta_1 = 60°$ ; $\theta_2 = 120°$ ; $\theta_3 = 240°$

***Magnitude of the resultant force***

Let $R$ = Resultant force.

Resolving all the forces horizontally, *i.e.* along $OX$,

$$\Sigma H = 15 \cos 60° + 20 \cos 120° + 40 \cos 240°$$

$$= 15 \times 0.5 + 20(-0.5) + 40(-0.5) = 7.5 - 10 - 20 = -22.5 \text{ kN}$$

Now resolving all the forces vertically, *i.e.* along $OY$,

$$\Sigma V = 15 \sin 60° + 20 \sin 120° + 40 \sin 240°$$
$$= 15 \times 0.866 + 20 \times 0.866 + 40\,(-0.866)$$
$$= 13 + 17.32 - 34.64 = -4.32 \text{ kN}$$

We know that resultant force, $R = \sqrt{(\Sigma H)^2 + (\Sigma V)^2} = \sqrt{(-22.5)^2 + (-4.32)^2}$

$$= \sqrt{506.25 + 18.66} = 22.9 \text{ kN} \quad \textbf{Ans.}$$

***Direction of the resultant force***

Let $\alpha$ = Angle which the resultant $R$ makes with horizontal.

We know that $\tan \alpha = \dfrac{\Sigma V}{\Sigma H} = \dfrac{-4.32}{-22.5} = 0.192$

$\therefore \quad \alpha = \tan^{-1}(0.192) = 10.87°$

Since $\Sigma V$ and $\Sigma H$ are both negative, therefore the resultant $R$ lies in the third quadrant, *i.e.* between 180° and 270°. Then actual angle of the resultant $R$

$$= 180° + 10.87°$$
$$= 190.87° \text{ in the anticlockwise direction from } OX \quad \textbf{Ans.}$$

**Example 2.11** *Four forces of magnitude 40 N, 50 N, 20 N and 60 N act at a point O along straight lines OA, OB, OC and OD, respectively, such that ∠AOB = 40°, ∠BOC = 100°, ∠COD = 125° and ∠DOA = 95°. Determine the magnitude and direction of the resultant.*

***Solution*** Given : $P_1 = 40$ N ; $P_2 = 50$ N ; $P_3 = 20$ N ; $P_4 = 60$ N ; $\angle AOB = 40°$ ; $\angle BOC = 100°$ ; $\angle COD = 125°$ ; $\angle DOA = 95°$

***Magnitude of the resultant***

The system of forces is shown in Fig. 2.15.

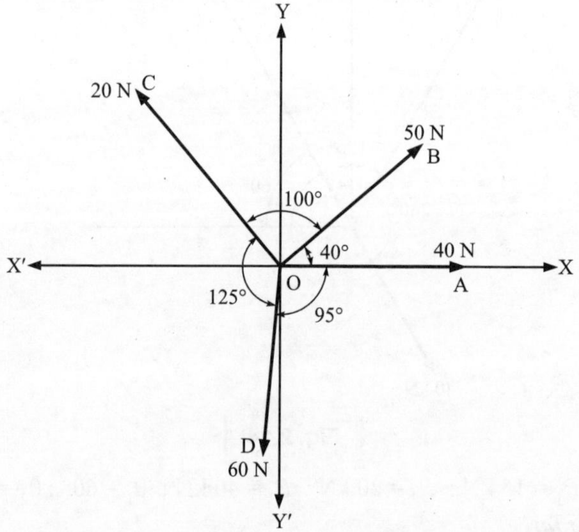


**Fig. 2.15**

From the figure, we see that

$\theta_1 = 0°$; $\theta_2 = 40°$; $\theta_3 = 40 + 100 = 140°$; and $\theta_4 = 40 + 100 + 125 = 265°$

Resolving all the forces horizontally, *i.e.* along *X*-axis or *OA*,

$$\Sigma H = 40 \cos 0° + 50 \cos 40° + 20 \cos 140° + 60 \cos 265°$$
$$= 40 + 50 \times 0.766 - 20 \times 0.766 - 60 \times 0.087$$
$$= 40 + 38.3 - 15.32 - 5.22 = 78.3 - 20.54 = 57.76 \text{ N}$$

Now resolving all the forces vertically, *i.e.* along *Y*-axis or *OY*,

$$\Sigma V = 40 \sin 0° + 50 \sin 40° + 20 \sin 140° + 60 \sin 265°$$
$$= 0 + 50 \times 0.643 + 20 \times 0.643 - 60 \times 0.9962$$
$$= 0 + 32.15 + 12.86 - 59.77 = -14.76 \text{ N} \quad \textbf{Ans.}$$

We know that resultant force, $R = \sqrt{(\Sigma H)^2 + (\Sigma V)^2} = \sqrt{(57.76)^2 + (-14.76)^2}$

$$= \sqrt{3336.2 + 217.85} = 59.6 \text{ N} \quad \textbf{Ans.}$$

***Direction of the resultant***

Let $\alpha$ = Inclination of the resultant force with the force 40 N, *i.e.* along *OA*.

We know that $\tan \alpha = \dfrac{\Sigma V}{\Sigma H} = \dfrac{-14.76}{57.76} = -0.2555$

$$\therefore \quad \alpha = \tan^{-1}(-0.2555) = -14.33°$$

Since $\Sigma V$ is –ve and $\Sigma H$ is +ve, therefore the resultant *R* lies in the fourth quadrant (*i.e.* between 270° and 360°). Thus actual angle of the resultant = 360° – 14.33° = 345.67° in the anticlockwise direction from *OA*. **Ans.**

**Note:** The *negative* sign indicates that the resultant force makes an angle of 14.33° in the clockwise direction (or 360° – 14.33° = 345.67° in the anticlockwise direction) from *OA*. **Ans.**

**Example 2.12** *Determine the magnitude and direction of the resultant force, for the system of forces acting at the corners of a rectangular block, as shown in Fig. 2.16.*

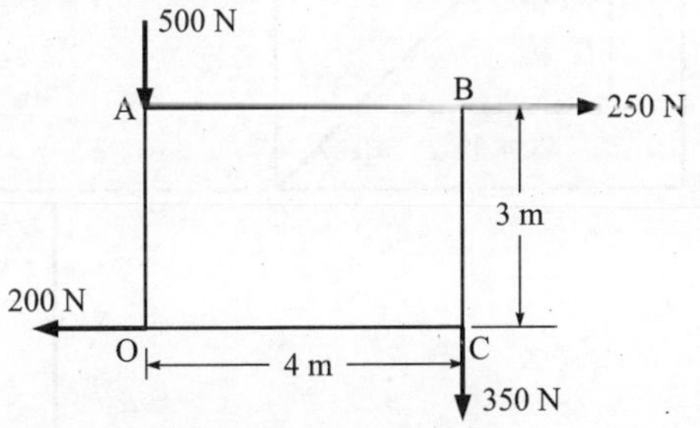


**Fig. 2.16**

***Solution*** Given : $P_1 = 500$ N ; $P_2 = 200$ N ; $P_3 = 250$ N ; $P_4 = 350$ N

***Magnitude of the resultant force***

Let *R* = Resultant force.

Resolving all the forces horizontally,

$$\Sigma H = 250 - 200 = 50 \text{ N}$$

and resolving all the forces vertically, $\Sigma V = -500 - 350 = -850$ N ...(Taking downward forces as negative)

We know that magnitude of the resultant force,

$$R = \sqrt{(\Sigma H)^2 + (\Sigma V)^2} = \sqrt{(50)^2 + (-850)^2} = 851.47 \text{ N} \quad \textbf{Ans.}$$

***Direction of the resultant force***

Let $\alpha$ = Angle of inclination of the resultant $R$ with the horizontal *i.e.* along $\Sigma H$, as shown in Fig. 2.17.

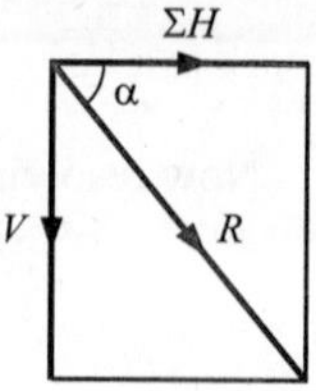


Fig. 2.17

We know that $\tan\alpha = \dfrac{\Sigma V}{\Sigma H} = \dfrac{-850}{50} = -17$

$\therefore$ $\alpha = \tan^{-1}(-17) = -86.63°$

Since $\Sigma H$ is +ve and $\Sigma V$ is –ve, therefore resultant $R$ lies in the fourth quadrant (*i.e.* between 270° and 360°). Thus actual angle of the resultant = 360° – 86.63 = 273.37° in the anticlockwise direction with the horizontal. **Ans.**

**Note:** The –ve sign indicates that the resultant $R$ makes an angle of 86.63° in clockwise direction with the horizontal.

**Example 2.13** *ABCD is a square. Forces of 10 N, 8 N and 4 N act at A in the directions AD, AC and AB, respectively. Using analytical method, determine:*

1. *The resultant in magnitude and direction.*
2. *The magnitude and sense of the two forces along directions AJ and AH, where J and H are midpoints of CD and BC, respectively, which together will balance the above resultant.*

***Solution*** Given : Force in $AD$ = 10 N ; Force in $AC$ = 8 N ; Force in $AB$ = 4 N

**1. *Magnitude of the resultant***

The forces acting on a square $ABCD$ are shown in Fig. 2.18.

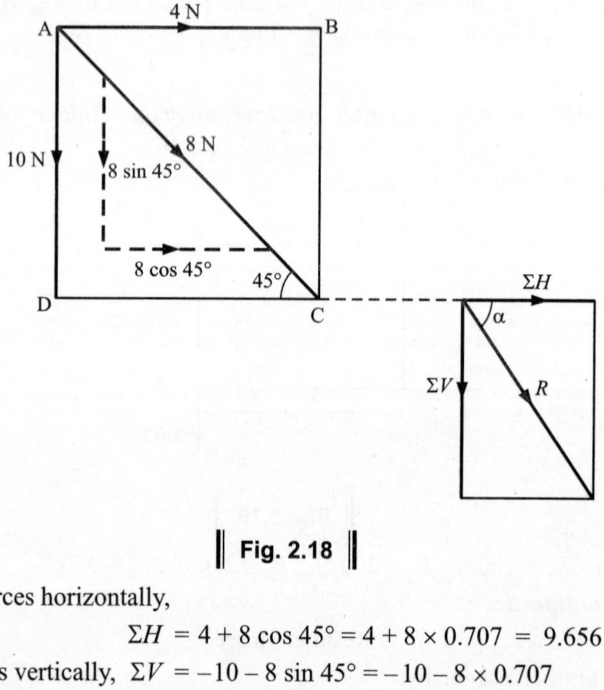


Fig. 2.18

Resolving all the forces horizontally,

$$\Sigma H = 4 + 8\cos 45° = 4 + 8 \times 0.707 = 9.656 \text{ N}$$

and resolving all the forces vertically, $\Sigma V = -10 - 8\sin 45° = -10 - 8 \times 0.707$

$$= -15.656 \text{ N} \quad \text{...(Taking downward forces as negative)}$$

We know that magnitude of the resultant,

$$R = \sqrt{(\Sigma H)^2 + (\Sigma V)^2} = \sqrt{(9.656)^2 + (-15.656)^2}$$

$$= \sqrt{93.24 + 245.11} = 18.394 \text{ N} \quad \textbf{Ans.}$$

***Direction of the resultant***

Let $\alpha$ = Angle of inclination of the resultant $R$ with the horizontal (*i.e.* with $CD$ or $\Sigma H$)

We know that $$\tan \alpha = \frac{\Sigma V}{\Sigma H} = \frac{-15.656}{9.656} = -1.6214$$

$\therefore$ $$\alpha = \tan^{-1}(-1.6214) = -58.336°$$

The –ve sign indicates that the resultant $R$ makes an angle of 58.336° in the clockwise direction with the horizontal. **Ans.**

**2. *Magnitude and sense of force along AJ and AH***

Let $F_1$ = Force along $AJ$ making an angle $\theta_1$ with the vertical force $AD$, and

$F_2$ = Force along $AH$ making an angle $\theta_2$ with the horizontal force $AB$.

In order to balance the resultant $R$, the sum of the forces $F_1$ and $F_2$ should be equal and their directions should be from $J$ to $A$ and from $H$ to $A$, as shown in Fig. 2.19.

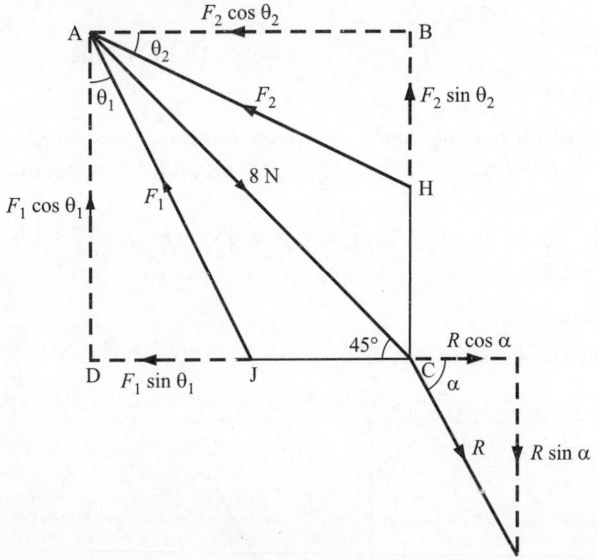


**Fig. 2.19**

Now resolving the forces $F_1$, $F_2$ and $R$ into horizontal and vertical components and equating the same. First of all, equating the horizontal components, we have

$$F_1 \sin \theta_1 + F_2 \cos \theta_2 = R \cos \alpha \quad ...(i)$$

Similarly, equating the vertical components,

$$F_1 \cos \theta_1 + F_2 \sin \theta_2 = R \sin \alpha \quad ...(ii)$$

If $a$ is the side of a square, then $DJ = BH = \frac{a}{2}$

$\therefore$ $$\tan \theta_1 = \frac{DJ}{AD} = \frac{\frac{a}{2}}{a} = 0.5$$

or $$\theta_1 = \tan^{-1}(0.5) = 26.565°$$

Similarly, $$\tan \theta_2 = \frac{BH}{AB} = \frac{\frac{a}{2}}{a} = 0.5$$

$\therefore$ $$\theta_2 = \tan^{-1}(0.5) = 26.565°$$

Now substituting the values of $\theta_1$, $\theta_2$ and $\alpha$ in equations (*i*) and (*ii*), we have

$$F_1 \sin 26.565° + F_2 \cos 26.565° = 18.394 \cos 58.336°$$

$$F_1 \times 0.447 + F_2 \times 0.894 = 18.394 \times 0.525$$

or $$F_1 + 2F_2 = 21.6 \quad \text{...(Dividing throughout by 0.447)} \quad ...(iii)$$

Similarly equation (*ii*) becomes,

$$F_1 \cos 26.565° + F_2 \sin 26.565° = 18.394 \sin 58.336°$$

$$F_1 \times 0.894 + F_2 \times 0.447 = 18.394 \times 0.851$$

or $$F_1 + 0.5F_2 = 17.5 \quad \text{...(Dividing throughout by 0.894)} \quad ...(iv)$$

Subtracting equation (*iv*) from equation (*iii*), we have

$$2F_2 - 0.5F_2 = 21.6 - 17.5 = 4.1$$

$$\therefore \quad F_2 = \frac{4.1}{1.5} = 2.73 \text{ N} \quad \textbf{Ans.}$$

Substituting the value of $F_2$ in equation (*iii*), we have

$$F_1 + 2 \times 2.73 = 21.6$$

$$\therefore \quad F_1 = 21.6 - 5.46 = 16.14 \text{ N} \quad \textbf{Ans.}$$

**Example 2.14** *ABCDEF is a regular hexagon. Forces of magnitude 2N, $4\sqrt{3}$ N, 8 N, $2\sqrt{3}$ N and 4 N are acting at A in the directions of AB, AC, AD, AE and AF, respectively. Find the magnitude and direction of the resultant force.*

***Solution*** Given : $P_1 = 2$ N ; $P_2 = 4\sqrt{3}$ N ; $P_3 = 8$ N ; $P_4 = 2\sqrt{3}$ N ; $P_5 = 4$ N ; $\theta_1 = 0°$ ; $\theta_2 = 30°$ ; $\theta_3 = 60°$ ; $\theta_4 = 90°$ ; $\theta_5 = 120°$

***Magnitude of the resultant force***

The system of forces is shown in Fig. 2.20.

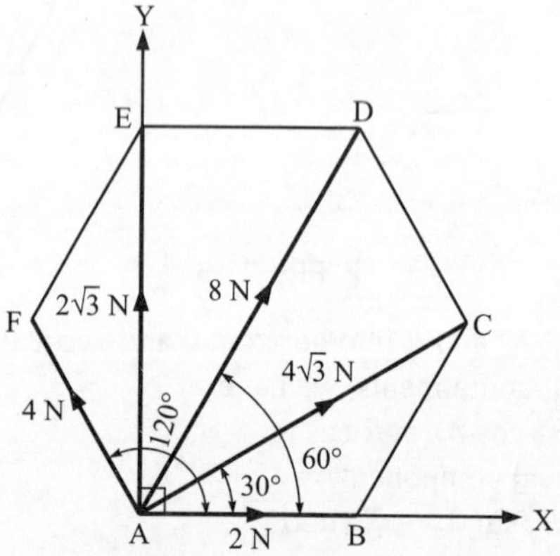


**Fig. 2.20**

Resolving all the forces horizontally *i.e.* along *X*-axis (or *AB*),

$$\Sigma H = 2 \cos 0° + 4\sqrt{3} \cos 30° + 8 \cos 60° + 2\sqrt{3} \cos 90° + 4 \cos 120°$$

$$= 2 + 4\sqrt{3} \times \frac{\sqrt{3}}{2} + 8 \times 0.5 + 0 + 4(-0.5)$$

$$= 2 + 6 + 4 - 2 = 10 \text{ N}$$

Now resolving all the forces vertically *i.e.* along *Y*-axis (or *AE*),

$$\Sigma V = 2 \sin 0° + 4\sqrt{3} \sin 30° + 8 \sin 60° + 2\sqrt{3} \sin 90° + 4 \sin 120°$$

$$= 0 + 4\sqrt{3} \times 0.5 + 8 \times \frac{\sqrt{3}}{2} + 2\sqrt{3} \times 1 + 4 \times \frac{\sqrt{3}}{2}$$

$$= 2\sqrt{3} + 4\sqrt{3} + 2\sqrt{3} + 2\sqrt{3} = 10\sqrt{3} \text{ N}$$

We know that resultant force, $R = \sqrt{(\Sigma H)^2 + (\Sigma V)^2} = \sqrt{(10)^2 + (10\sqrt{3})^2} = 20$ N **Ans.**

***Direction of the resultant force***

Let $\alpha$ = Inclination of the resultant force with the force 2 N *i.e.* along *AB*.

We know that $\tan \alpha = \dfrac{\Sigma V}{\Sigma H} = \dfrac{10\sqrt{3}}{10} = \sqrt{3}$

$\therefore$ $\alpha = 60°$ **Ans.**

Since $\Sigma V$ and $\Sigma H$ are both positive, therefore the resultant force lies in the first quadrant.

**Example 2.15** *In a regular hexagon ABCDEF, as shown in Fig. 2.21, five forces of magnitude 20 N, P, Q, 40 N and 30 N act along AB, CA, AD, AE and FA, respectively. Find the value of the forces P and Q, for the system to be in equilibrium.*

***Solution*** Given : $P_1 = 20$ N ; $P_2 = -P$ (Push) ; $P_3 = Q$ ; $P_4 = 40$ N ; $P_5 = -30$ N (push) ; $\theta_1 = 0°$ ; $\theta_2 = 30°$ ; $\theta_3 = 60°$ ; $\theta_4 = 90°$ ; $\theta_5 = 120°$

***Value of the forces P and Q***

The angular position of the given forces acting at *A* is shown in Fig. 2.22.

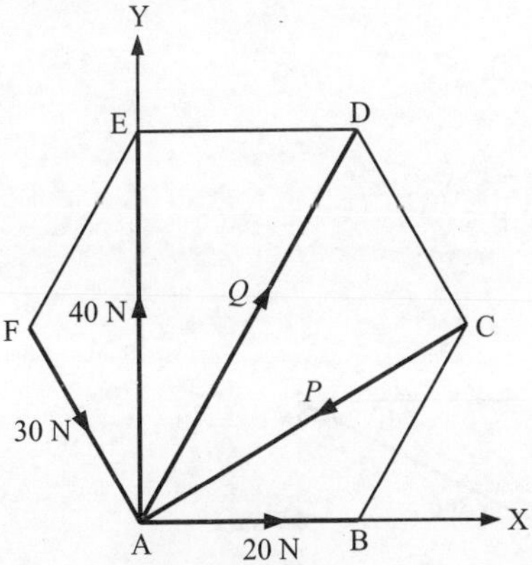


**Fig. 2.21**

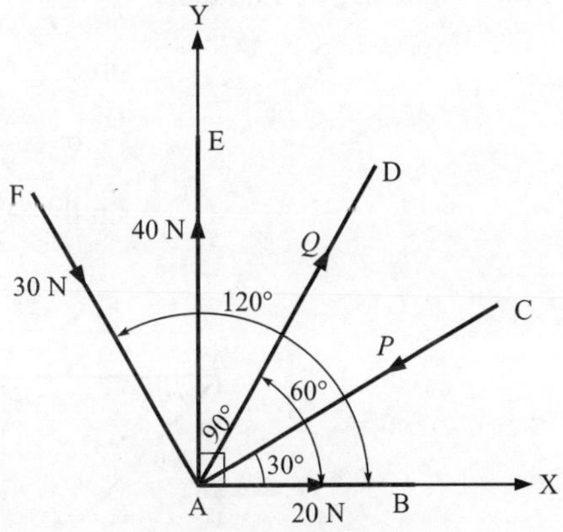


**Fig. 2.22**

Resolving all the forces horizontally *i.e.* along *X*-axis (or *AB*),

$$\Sigma H = 20 \cos 0° - P \cos 30° + Q \cos 60° + 40 \cos 90° - 30 \cos 120°$$

$$= 20 - P \times 0.866 + Q \times 0.5 + 40 \times 0 - 30\,(-0.5)$$

$$= 20 - 0.866\,P + 0.5Q + 15$$

Resolving all the forces vertically *i.e.* along *Y*-axis (or *AE*),

$$\Sigma V = 20 \sin 0° - P \sin 30° + Q \sin 60° + 40 \sin 90° - 30 \sin 120°$$

$$= 0 - P \times 0.5 + Q \times 0.866 + 40 \times 1 - 30 \times 0.866$$

$$= -0.5\,P + 0.866\,Q + 14.02$$

For the system to be in equilibrium, $\Sigma H = 0$ and $\Sigma V = 0$.

Equating $\Sigma H = 0$, we have

$$20 - 0.866P + 0.5Q + 15 = 0$$

or
$$35 - 0.866P + 0.5Q = 0$$

$$\therefore \quad Q = \frac{0.866P - 35}{0.5} = 1.732P - 70 \quad \text{...}(i)$$

Now equating $\Sigma V = 0$, we have

$$-0.5P + 0.866Q + 14.02 = 0$$

Substituting the value of $Q$ from equation ($i$),

$$-0.5P + 0.866(1.732P - 70) + 14.02 = 0$$

$$-0.5P + 1.5P - 60.62 + 14.02 = 0$$

or
$P = 46.6$ N **Ans.**

and
$Q = 1.732P - 70$ ...[From equation ($i$)]

$= 1.732 \times 46.6 - 70 = 10.71$ N **Ans.**

**Example 2.16** *A string ABC, as shown in Fig. 2.23, is tied at two points A and C. A smooth ring of weight 500 N which can slide along the string at B is 200 mm away from A along the string and pulled by a horizontal force P as shown in the figure. If the tensions in the string on both sides of B are same, find the magnitude of the force P.*

***Solution*** Given : $AB = 200$ mm ; $BC = 300$ mm ; $BD = 150$ mm ; $W = 500$ N

***Magnitude of the force P***

From Fig. 2.23, we find that

$$\sin \alpha = \frac{BD}{AB} = \frac{150}{200} = 0.75$$

or
$$\alpha = 48.6°$$

and
$$\sin \beta = \frac{BD}{BC} = \frac{150}{300} = 0.5$$

or
$$\beta = 30°$$

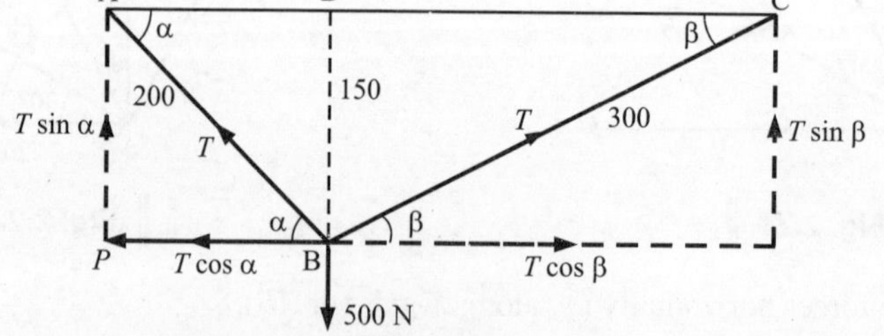

All dimensions in mm.

**Fig. 2.23**

Let $T$ = Tension in the strings $BA$ and $BC$.

Resolving the forces horizontally, we have

$$T \cos \beta = P + T \cos \alpha$$

$$T\cos 30° = P + T\cos 48.6°$$

$$T \times 0.866 = P + T \times 0.661$$

or $$P = 0.205\,T \qquad ...(i)$$

Now resolving the forces vertically, we have

$$T\sin\alpha + T\sin\beta = 500$$

$$T\sin 48.6° + T\sin 30° = 500$$

$$T \times 0.75 + T \times 0.5 = 500 \quad \text{or} \quad 1.25T = 500$$

$$\therefore \qquad T = \frac{500}{1.25} = 400 \text{ N} \quad \textbf{Ans.}$$

and from equation (*i*), $$P = 0.205T = 0.205 \times 400 = 82 \text{ N} \quad \textbf{Ans.}$$

**Example 2.17** *A particle is acted upon by three forces equal to 5 N, 10 N and 13 N, along the three sides of an equilateral triangle, taken in order. Find the magnitude and direction of the resultant force.*

***Solution*** Given : Forces 5 N, 10 N and 13 N acting on three sides *AB*, *BC* and *CA* of an equilateral triangle *ABC*, taken in order are shown in Fig. 2.24 (*a*)

The free body diagram of the given system of forces is shown in Fig. 2.24 (*b*).

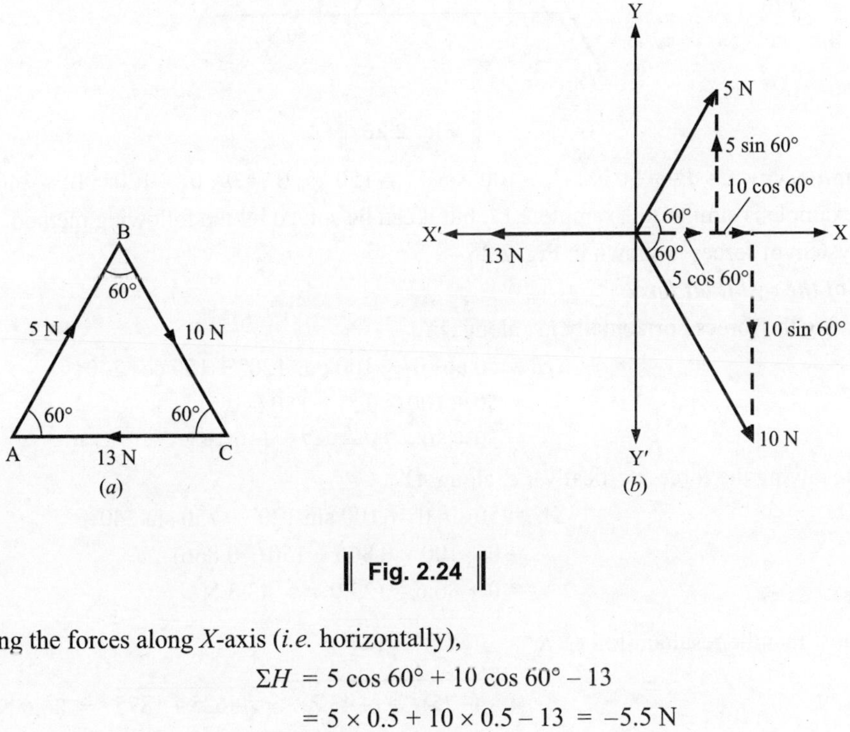


**Fig. 2.24**

Resolving the forces along *X*-axis (*i.e.* horizontally),

$$\Sigma H = 5\cos 60° + 10\cos 60° - 13$$

$$= 5 \times 0.5 + 10 \times 0.5 - 13 = -5.5 \text{ N}$$

Now resolving the forces along *Y*-axis (*i.e.* vertically),

$$\Sigma V = 5\sin 60° - 10\sin 60°$$

$$= 5 \times 0.866 - 10 \times 0.866 = -4.33 \text{ N}$$

Let *R* be the magnitude and α be the direction of the resultant force *R* along *X*-axis.

From Fig. 2.25, we see that $R = \sqrt{(\Sigma H)^2 + (\Sigma V)^2}$

$$= \sqrt{(5.5)^2 + (4.33)^2}$$

$$= \sqrt{30.25 + 18.75} = 7 \text{ N } \textbf{Ans.}$$

and $\tan \alpha = \dfrac{\Sigma V}{\Sigma H} = \dfrac{4.33}{5.5} = 0.7873$

$\therefore$ $\alpha = \tan^{-1}(0.7873) = 38.212° \textbf{ Ans.}$

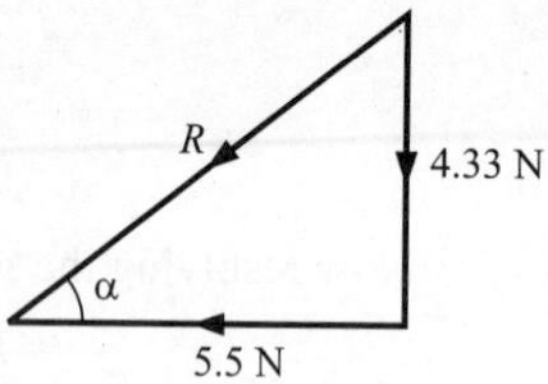


**Fig. 2.25**

**Example 2.18** *A particle is acted upon by three forces equal to 50 N, 100 N and 150 N along the three sides of an equilateral triangle, taken in order, as shown in Fig. 2.26. Find the magnitude and direction of the resultant force.*

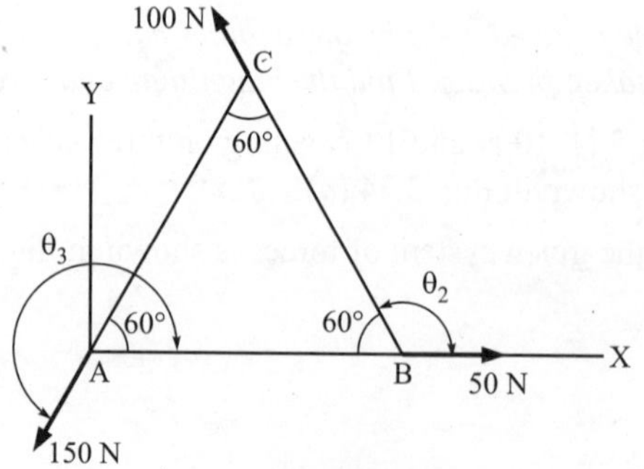


**Fig. 2.26**

***Solution*** Given : $P_1 = 50$ N ; $P_2 = 100$ N ; $P_3 = 150$ N ; $\theta_1 = 0°$, $\theta_2 = 120°$ ; $\theta_3 = 240°$

This example is similar to Example 2.17, but it can be solved by the following method.

The system of forces is shown in Fig. 2.26.

***Magnitude of the resultant force***

Resolving the forces horizontally *i.e.* along *AX*,

$$\Sigma H = 50 \cos 0° + 100 \cos 120° + 150 \cos 240°$$
$$= 50 + 100(-0.5) + 150(-0.5)$$
$$= 50 - 50 - 75 = -75 \text{ N}$$

Now resolving the forces vertically, *i.e.* along *AY*,

$$\Sigma V = 50 \sin 0° + 100 \sin 120° + 150 \sin 240°$$
$$= 0 + 100 \times 0.866 + 150(-0.866)$$
$$= 0 + 86.6 - 129.9 = -43.3 \text{ N}$$

We know that the resultant force, $R = \sqrt{(\Sigma H)^2 + (\Sigma V)^2}$

$$= \sqrt{(-75)^2 + (-43.3)^2} = \sqrt{5625 + 1875} = 86.6 \text{ N}$$

***Direction of the resultant force***

Let $\alpha$ = Angle of inclination of the resultant force.

We know that $\tan \alpha = \dfrac{\Sigma V}{\Sigma H} = \dfrac{-43.3}{-75} = 0.5773$

$\therefore$ $\alpha = \tan^{-1}(0.5773) = 30°$

Since $\Sigma H$ and $\Sigma V$ are both negative, therefore resultant lies in the third quadrant.

∴ Angle of inclination of the resultant force from $AX$ in the anticlockwise direction

$$= 180° + 30° = 210° \quad \textbf{Ans.}$$

## 2.13 TRIANGLE LAW OF FORCES

It states that '*if two forces acting simultaneously on a particle, be represented in magnitude and direction by the two sides of a triangle taken in order, then their resultant is represented in magnitude and direction by the third side of the triangle, taken in opposite order*'.

Consider two forces $P$ and $Q$ making an angle θ, acting at a point $O$. The magnitude and direction is given by $OA$ and $OB$, as shown in Fig. 2.27 (*a*). The resultant of these two forces may be determined graphically or analytically as discussed below:

**1. *Graphical method***

Draw vector $oa$ parallel to $OA$ and vector $ab$ parallel to $OB$ to represent forces $P$ and $Q$ in magnitude and direction. Join $ob$ which represent the resultant $R$ in magnitude but in opposite direction, as shown in Fig. 2.27 (*b*).

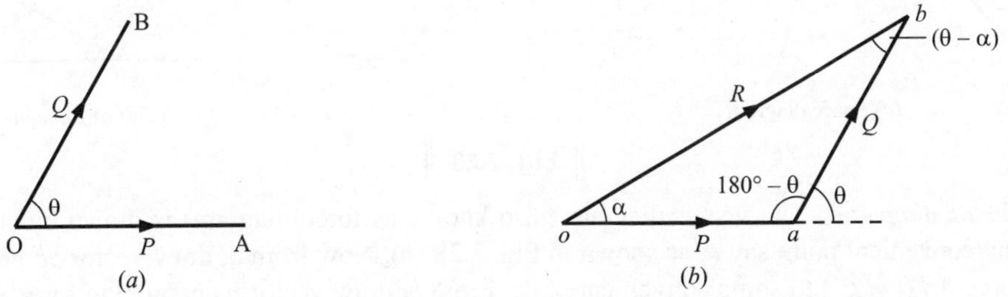


**Fig. 2.27**

**2. *Analytical method***

According to the geometry of triangle $oab$,

$$\frac{oa}{\sin(\theta - \alpha)} = \frac{ab}{\sin \alpha} = \frac{ob}{\sin(180° - \theta)}$$

or

$$\frac{P}{\sin(\theta - \alpha)} = \frac{Q}{\sin \alpha} = \frac{R}{\sin \theta} \quad \text{...[∵ } \sin(180° - \theta) = \sin\theta]$$

## 2.14 POLYGON LAW OF FORCES

It states that '*if a number of forces acting simultaneously on a body are represented in magnitude and direction by the sides of a polygon taken in order, then their resultant may be represented in magnitude and direction by the closing side of a polygon, taken in opposite order*'.

## 2.15 GRAPHICAL METHOD OF DETERMINING RESULTANT OF A NUMBER OF COPLANAR CONCURRENT FORCES

We have already discussed the analytical method (principle of resolution) of determining the resultant of a number of coplanar concurrent forces. The magnitude and direction of the resultant force of a number of coplanar concurrent forces may be determined graphically as discussed below:

1. ***Space diagram*** First of all, draw the space diagram. The construction of a diagram showing the various forces along with their magnitude and direction, as shown in Fig. 2.28 (*a*), is called space diagram.

2. ***Bow's notation*** All the forces in the space diagram are named by using Bow's notation, *i.e.* each force is represented by two capital letters placed in space on either side of the force, as shown in Fig. 2.28 (*a*). For example, force $P_1$ is represented by *AB*, force $P_2$ is represented by *BC* and so on.

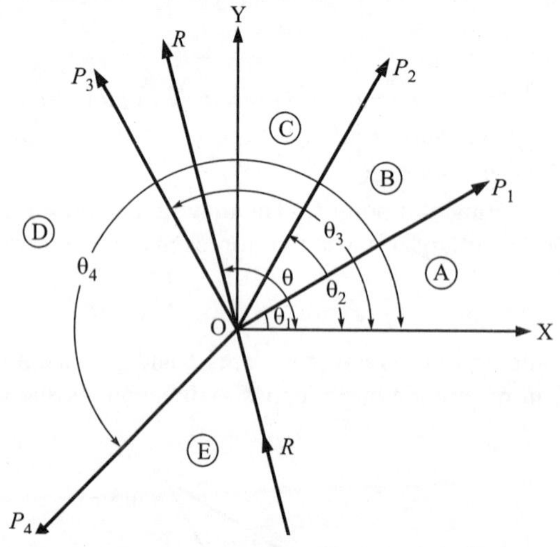


(*a*) Space diagram

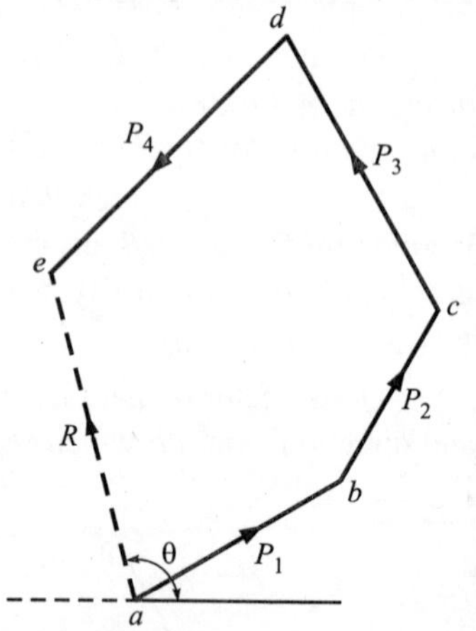


(*b*) Vector or force diagram

**Fig. 2.28**

3. ***Vector diagram*** The vector diagram (also known as force diagram) is drawn by starting from any convenient point say *a*, as shown in Fig. 2.28 (*b*). Now from *a*, draw vector *ab* parallel to the force *AB* (*i.e.* $P_1$) to some convenient scale. From *b*, draw vector *bc* parallel to force *BC* (*i.e.* $P_2$). Similarly, draw vectors *cd* and *de* parallel to forces *CD* (*i.e.* $P_3$) and *DE* (*i.e.* $P_4$), respectively. Now the closing side *ea* represents the magnitude and direction of the resultant *R*, in the opposite direction, as shown in the figure. In the space diagram, draw a line parallel to vector *ae* and measure angle θ which gives the required inclination of the resultant *R* with the horizontal.

## 2.16 LAMI'S THEOREM

It states that '*if three forces acting at a point are in equilibrium, then each force is proportional to the sine of the angle between the other two forces*'.

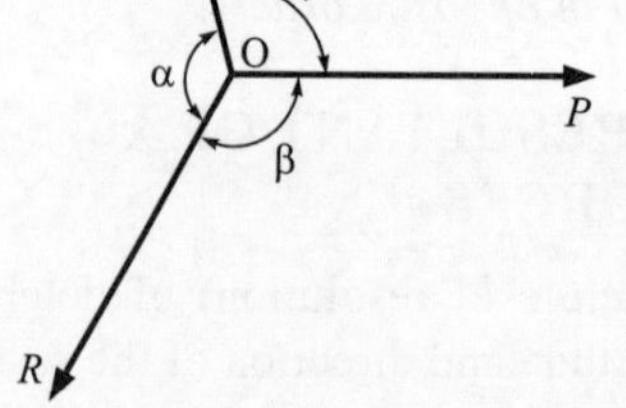


**Fig. 2.29**

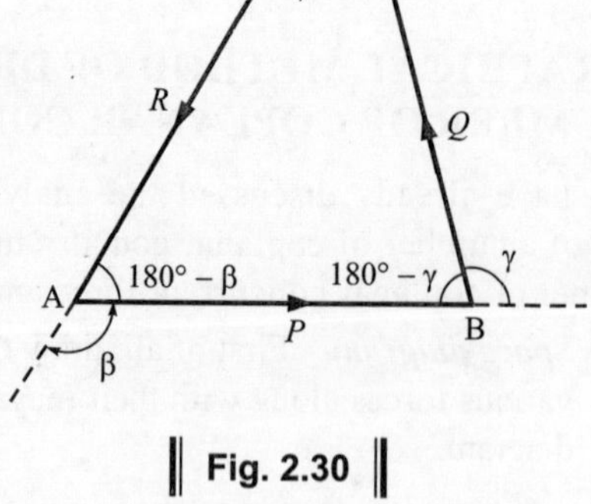


**Fig. 2.30**

Let $P$, $Q$ and $R$ be the three forces acting at a point $O$ are in equilibrium, as shown in Fig. 2.29. Now according to Lami's theorem,

$$\frac{P}{\sin\alpha} = \frac{Q}{\sin\beta} = \frac{R}{\sin\gamma}$$

***Proof*** Since the three forces $P$, $Q$ and $R$ are in equilibrium, therefore they can be the represented by the three sides of a triangle $ABC$, taken in order, as shown in Fig. 2.30. The sides $AB$, $BC$ and $CA$ represents the forces $P$, $Q$ and $R$, respectively.

According to the sine rule for the triangle $ABC$,

$$\frac{AB}{\sin(180^\circ - \alpha)} = \frac{BC}{\sin(180^\circ - \beta)} = \frac{CA}{\sin(180^\circ - \gamma)}$$

or

$$\frac{P}{\sin\alpha} = \frac{Q}{\sin\beta} = \frac{R}{\sin\gamma}$$

**Example 2.19** *The following forces act at a point:*

*(a) 500 N towards north-east,*

*(b) 400 N towards east,*

*(c) 250 N inclined at 30° west of north, and*

*(d) 300 N inclined at 60° south of west.*

*Determine the magnitude and direction of the resultant by graphical and analytical methods.*

***Solution*** Given : $P_1 = 400$ N ; $\theta_1 = 0°$ ; $P_2 = 500$ N ; $\theta_2 = 45°$ ; $P_3 = 250$ N ; $\theta_3 = 90° + 30° = 120°$ ; $P_4 = 300$ N ; $\theta_4 = 180° + 60° = 240°$

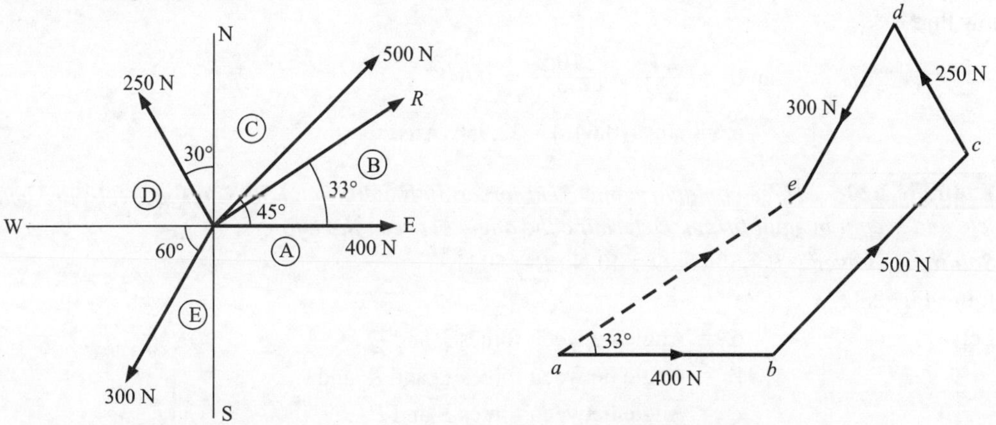


(*a*) Space diagram (*b*) Vector or force diagram

**Fig. 2.31**

***Graphical method***

The magnitude and direction of the resultant for the given system of forces by graphical method is obtained as discussed below:

1. First of all, draw the space diagram as shown in Fig. 2.31 (*a*). The space diagram means the construction of a diagram showing the various forces along with their magnitude and direction.
2. All the forces in the space diagram are named by using Bow's notation. The naming of each force by capital letters on either side is termed as Bow's notation.

3. Now draw vector or force diagram as shown in Fig. 2.31 (*b*). Draw vector *ab* parallel to force *AB* (*i.e.* 400 N) to some suitable scale. From point *b*, draw vector *bc* parallel to force *BC* (*i.e.* 500 N). Similarly draw vector *cd* parallel to force *CD* (*i.e.* 250 N) and draw *de* parallel to force *DE* (*i.e.* 300 N).
4. Join *ea* which gives the magnitude and direction of the resultant force.

By measurement, we find that magnitude of the resultant force is equal to 570 N and its inclination is 33° with the 400 N force. **Ans.**

***Analytical method***

Resolving all the forces horizontally *i.e.* along the east-west line,

$$\Sigma H = 400 \cos 0° + 500 \cos 45° + 250 \cos 120° + 300 \cos 240°$$
$$= 400 \times 1 + 500 \times 0.707 + 250\,(-0.5) + 300\,(-0.5)$$
$$= 400 + 353.5 - 125 - 150 = 478.5 \text{ N}$$

Now resolving all the forces vertically *i.e.* north-south line,

$$\Sigma V = 400 \sin 0° + 500 \sin 45° + 250 \sin 120° + 300 \sin 240°$$
$$= 0 + 500 \times 0.707 + 250 \times 0.866 + 300\,(-0.866)$$
$$= 353.5 + 216.5 - 259.8 = 310.2 \text{ N}$$

We know that resultant force,

$$R = \sqrt{(\Sigma H)^2 + (\Sigma V)^2} = \sqrt{(478.5)^2 + (310.2)^2}$$
$$= \sqrt{228\,962 + 96\,224} = 570.25 \text{ N} \quad \textbf{Ans.}$$

Let α be the inclination of the resultant force makes with the 400 N force *i.e.* along the east line. We know that

$$\tan \alpha = \frac{\Sigma V}{\Sigma H} = \frac{310.2}{478.5} = 0.648$$

$$\therefore \quad \alpha = \tan^{-1}(0.648) = 32.95° \quad \textbf{Ans.}$$

**Example 2.20** *Three coplanar concurrent forces 30 N ; 40 N and 50 N, respectively, are acting on a particle and keep it in equilibrium. Determine the angle between three forces.*

***Solution*** Given : $P = 30$ N ; $Q = 40$ N ; $R = 50$ N

Refer Fig. 2.32.

Let
$$\alpha = \text{Angle between forces } P \text{ and } Q,$$
$$\beta = \text{Angle between forces } Q \text{ and } R, \text{ and}$$
$$\gamma = \text{Angle between forces } R \text{ and } P.$$

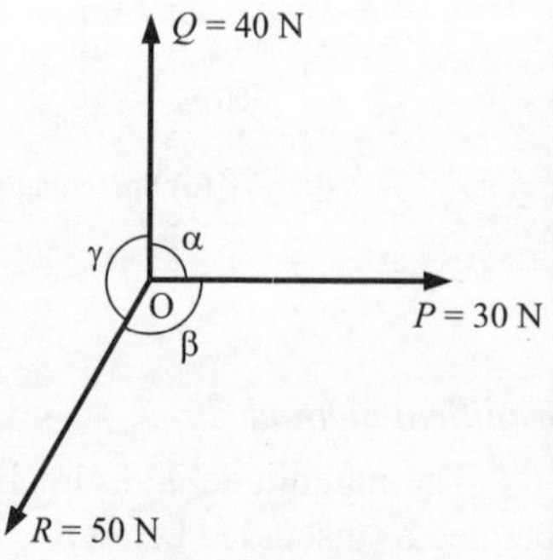


**Fig. 2.32**

When three forces are acting on a particle and keep it in equilibrium, then each force is equal to the resultant of the other two forces.

First of all, let us consider that the force $R = 50$ N is the equilibrium force (*i.e.* resultant) of the remaining two forces 30 N and 40 N.

$$\therefore \quad R = \sqrt{P^2 + Q^2 + 2PQ\cos\alpha}$$

or
$$R^2 = P^2 + Q^2 + 2PQ\cos\alpha$$
$$(50)^2 = (30)^2 + (40)^2 + 2 \times 30 \times 40 \cos\alpha$$
$$2500 = 900 + 1600 + 2400 \cos\alpha$$

or
$$2400 \cos\alpha = 2500 - 2500 = 0$$

$$\therefore \quad \cos\alpha = 0 \text{ or } \alpha = 90° \quad \textbf{Ans.}$$

Now considering that the force $Q = 40$ N is the equilibrium force of the remaining two forces 30 N and 50 N. In the similar way as above,

$$(40)^2 = (30)^2 + (50)^2 + 2 \times 30 \times 50 \cos \beta$$
$$1600 = 900 + 2500 + 3000 \cos \beta$$

or
$$3000 \cos \beta = 1600 - 900 - 2500 = -1800$$

$\therefore$
$$\cos \beta = \frac{-1800}{3000} = -0.6 \quad \text{or} \quad \beta = 126.87° \quad \textbf{Ans.}$$

Again, considering that the force $P = 30$ N is the equilibrium force of the remaining two forces 40 N and 50 N. Thus

$$(30)^2 = (40)^2 + (50)^2 + 2 \times 40 \times 50 \cos \gamma$$
$$900 = 1600 + 2500 + 4000 \cos \gamma$$

or
$$4000 \cos \gamma = 900 - 1600 - 2500 = -3200$$

$\therefore$
$$\cos \gamma = \frac{-3200}{4000} = -0.8 \quad \text{or} \quad \gamma = 143.13° \quad \textbf{Ans.}$$

**Note:** The value of $\gamma$ may be obtained as follows:

We know that
$$\alpha + \beta + \gamma = 360°$$
$$90° + 126.87° + \gamma = 360°$$

$\therefore$
$$\gamma = 360° - 90° - 126.87° = 143.13° \quad \textbf{Ans.}$$

**Example 2.21** *A force P, as shown in Fig. 2.33, is applied at O to the string LOM. If the tension in each part of the string is 100 N, find the magnitude and direction of the force P for equilibrium conditions.*

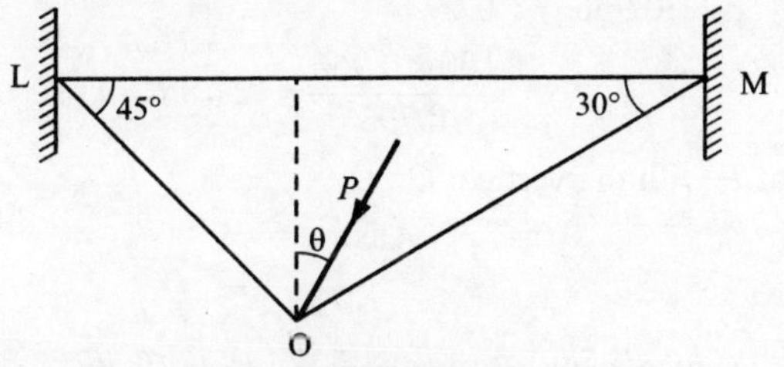


**Fig. 2.33**

***Solution*** Given : Tension in string *OL* and *OM* each = 100 N

From the geometry of the Fig. 2.34, we find that

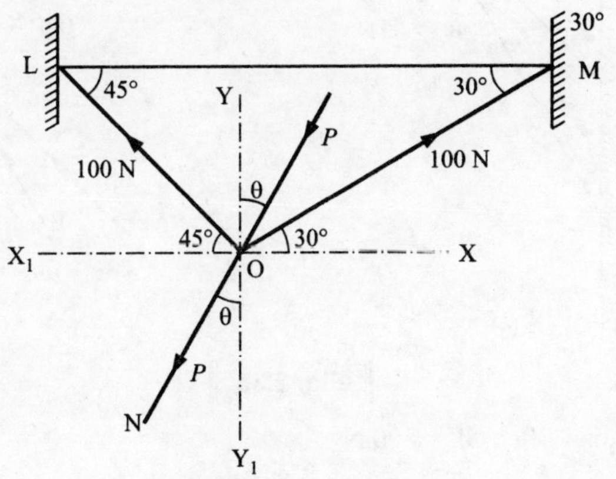


**Fig. 2.34**

$$\angle LON = 45° + 90° - \theta = 135° - \theta$$
$$\angle NOM = 30° + 90° + \theta = 120° + \theta$$
$$\angle LOM = 180° - (45° + 30°) = 105°$$

For equilibrium conditions, applying Lami's theorem, we have

$$\frac{P}{\sin \angle LOM} = \frac{100}{\sin \angle LON} = \frac{100}{\sin \angle NOM}$$

$$\frac{P}{\sin 105°} = \frac{100}{\sin (135° - \theta)} = \frac{100}{\sin (120° + \theta)} \quad ...(i)$$

From above, we see that

$$\frac{100}{\sin (135° - \theta)} = \frac{100}{\sin (120° + \theta)}$$

or $$\sin (135° - \theta) = \sin (120° + \theta)$$

$$135° - \theta = 120° + \theta$$

∴ $$135° - 120° = \theta + \theta = 2\theta$$

or $$\theta = \frac{135° - 120°}{2} = 7.5°$$

Now $$\frac{P}{\sin 105°} = \frac{100}{\sin (135° - 7.5°)} = \frac{100}{\sin 127.5°} \quad ...[\text{From equation } (i)]$$

$$\frac{P}{0.966} = \frac{100}{0.793}$$

∴ $$P = \frac{100 \times 0.966}{0.793} = 121.8 \text{ N} \quad \textbf{Ans.}$$

and the angle of inclination of $P$ with the vertical $OY$ is

$$\theta = 7.5° \quad \textbf{Ans.}$$

**Example 2.22** *A jib crane, as shown in Fig. 2.35 (a), is required to lift a load of 500 N. Find the forces in the jib and tie.*

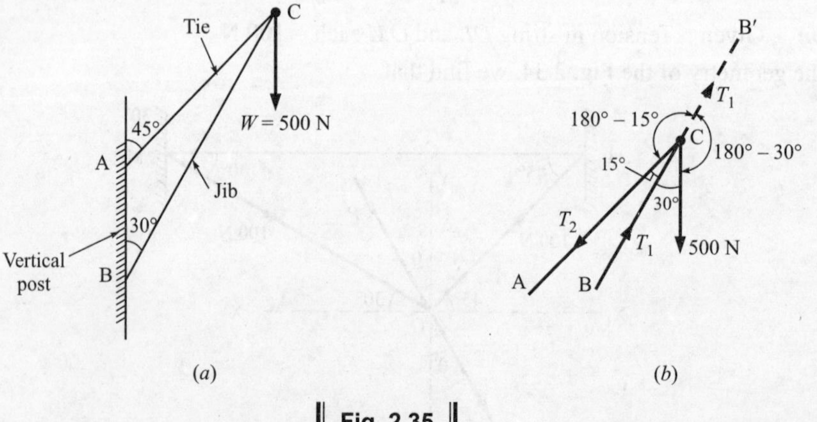


**Fig. 2.35**

***Solution*** Given : Load lifted, $W = 500$ N

Let $T_1$ = Force in the jib, and

$T_2$ = Force in the tie.

It may be noted that in a jib crane, the tie rod is always in tension and the jib is under compression.

In Fig. 2.35 (*b*), the line of $T_1$ *i.e.* $BC$ is produced to $B'$ to have the same direction of $T_1$, $T_2$ and $W$.

Now for the equilibrium of forces, applying Lami's theorem at point $C$, as shown in Fig. 2.35 (*b*). We have,

$$\frac{T_1}{\sin 45°} = \frac{T_2}{\sin (180° - 30°)} = \frac{500}{\sin (180° - 15°)}$$

$$\frac{T_1}{\sin 45°} = \frac{T_2}{\sin 30°} = \frac{500}{\sin 15°}$$

$$\therefore \quad T_1 = \frac{500}{\sin 15°} \times \sin 45° = \frac{500}{0.2588} \times 0.707$$

$$= 1366 \text{ N (compressive)} \quad \textbf{Ans.}$$

and $$T_2 = \frac{500}{\sin 15°} \times \sin 30° = \frac{500}{0.2688} \times 0.5 = 930 \text{ N (tensile)} \quad \textbf{Ans.}$$

**Example 2.23** *Two bars AC and BC are hinged together at C. Their other ends are hinged to a vertical wall at A and B, as shown in Fig. 2.36. Find the axial force in the bars.*

***Solution*** Given : Load at $C$ = 5000 N

The free body diagram of the given system is shown in Fig. 2.37 (*a*). Let $T_1$ be the tension in the bar $AC$ and $T_2$ be the compression in the bar $BC$. The load of 5000 N is suspended vertically downwards at $C$.

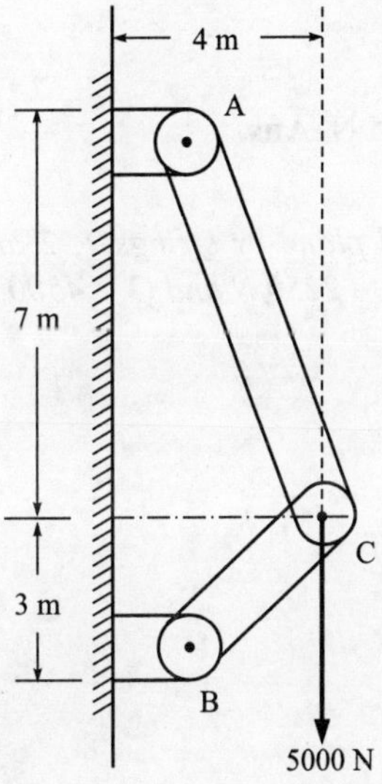


Fig. 2.36

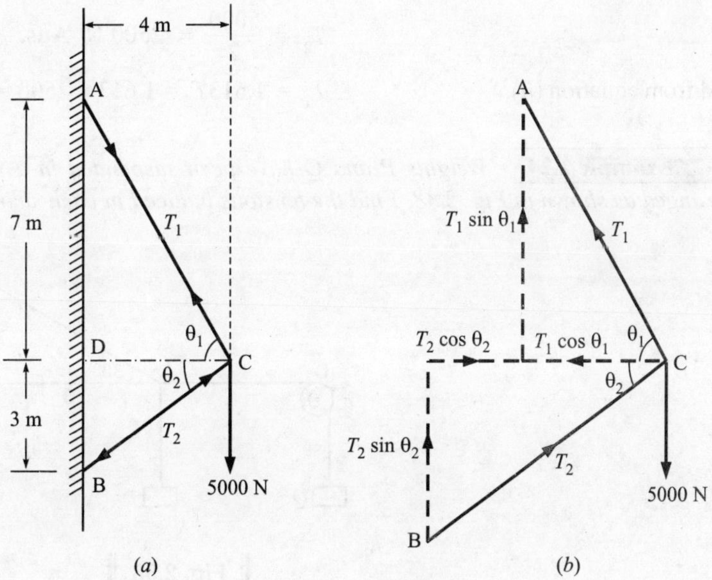


Fig. 2.37

From triangle $ACD$, $$\tan \theta_1 = \frac{AD}{CD} = \frac{7}{4} = 1.75$$

$$\therefore \quad \theta_1 = \tan^{-1} (1.75) = 60.255°$$

and from triangle $BCD$, $$\tan \theta_2 = \frac{BD}{CD} = \frac{3}{4} = 0.75$$

$$\therefore \quad \theta_2 = \tan^{-1} (0.75) = 36.87°$$

The joint $C$ is in equilibrium under the action of the following forces [See Fig. 2.37 (*b*)]:

1. Tension $T_1$ in bar $AC$,
2. Compression $T_2$ in bar $BC$, and
3. Load 5000 N acting vertically downwards at $C$

Resolving the forces vertically and equating $\Sigma V = 0$, we have

$$T_1 \sin \theta_1 + T_2 \sin \theta_2 = 5000$$

$$T_1 \sin (60.255°) + T_2 \sin (36.87°) = 5000$$

$$T_1 \times 0.868 + T_2 \times 0.6 = 5000$$

$$0.868\, T_1 + 0.6 T_2 = 5000 \quad ...(i)$$

Now resolving the forces horizontally and equating $\Sigma H = 0$, we have

$$T_1 \cos \theta_1 = T_2 \cos \theta_2$$

$$T_1 \cos (60.255°) = T_2 \cos (36.87°)$$

$$T_1 \times 0.496 = T_2 \times 0.8$$

$$0.496 T_1 = 0.8 T_2$$

or $$T_1 = \frac{0.8 T_2}{0.496} = 1.613 T_2 \quad ...(ii)$$

Substituting the value of $T_1$ in equation (*i*), we have

$$0.868 \times 1.613 T_2 + 0.6 T_2 = 5000$$

$$1.4 T_2 + 0.6 T_2 = 5000 \quad \text{or} \quad 2T_2 = 5000$$

$\therefore$ $$T_2 = \frac{5000}{2} = 2500 \text{ N} \quad \textbf{Ans.}$$

and from equation (*ii*), $$T_1 = 1.613 T_2 = 1.613 \times 2500 = 4032.5 \text{ N} \quad \textbf{Ans.}$$

**Example 2.24** *Weights P and Q have been suspended in a vertical plane by strings 1, 2 and 3, arranged as shown in Fig. 2.38. Find the tensions induced in each string, if P = 2250 N and Q = 4500 N.*

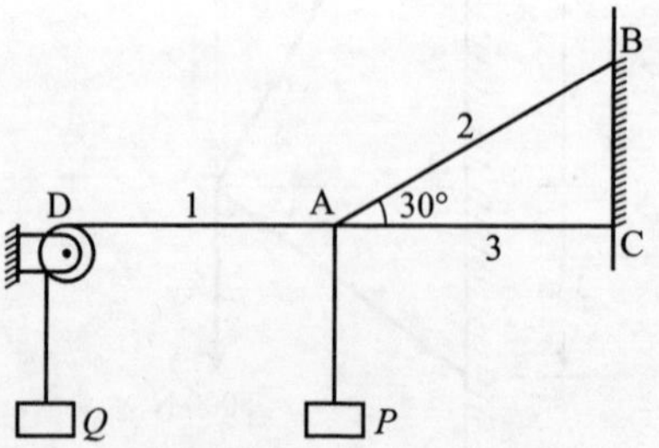


**Fig. 2.38**

***Solution*** Given : $P = 2250$ N ; $Q = 4500$ N

The free body diagram of the weight $Q = 4500$ N suspended by a string 1 passing over a pulley $D$ is shown in Fig. 2.39 (*a*). Since the pulley $D$ is frictionless, therefore tension in the string 1 must be equal to $T$ *i.e.* $T_1 = T = 4500$ N.

Now consider the free body diagram of the joint $A$, as shown in Fig. 2.39 (*b*). The joint $A$ is in equilibrium under the action of the following forces:

1. Tension $T_1 = 4500$ N, in string 1,
2. Tension $T'$ in the string carrying the load $P = 2250$ N,

3. Tension $T_2$ in the string 2, and
4. Tension $T_3$ in the string 3.

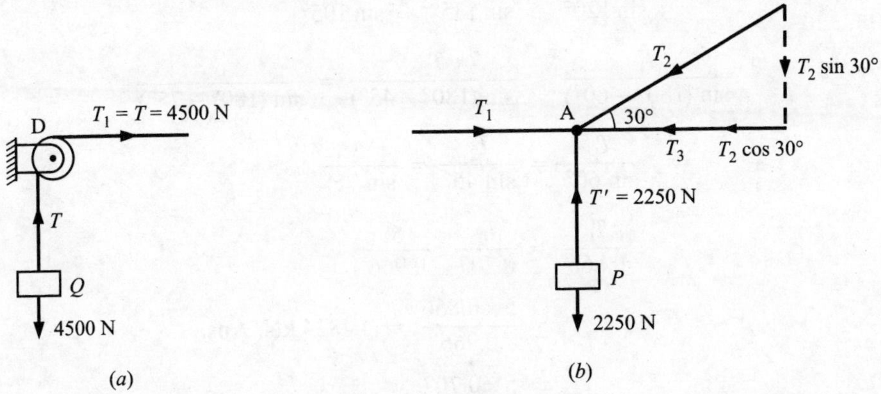


**Fig. 2.39**

Resolving the forces vertically, and equating $\Sigma V = 0$, we have

$$T_2 \sin 30° = T' = 2250$$

$$\therefore \quad T_2 = \frac{2250}{\sin 30°} = \frac{2250}{0.5} = 4500 \text{ N} \quad \textbf{Ans.}$$

Now resolving the forces horizontally and equating $\Sigma H = 0$, we have

$$T_3 = T_1 - T_2 \cos 30° = 4500 - 4500 \times 0.866$$
$$= 4500 - 3897 = 603 \text{ N} \quad \textbf{Ans.}$$

**Example 2.25** *A machine weighing 5 kN is supported by two chains attached to some point on the machine. One chain goes to the hook in the ceiling and has an inclination of 45° with the horizontal. The other chain goes to the eye bolt in the wall and is inclined at 30° to the horizontal. Make calculations for the tension induced in the chain.*

***Solution*** Given : Weight of the machine = 5 kN

The machine supported by two chains *OA* and *OB* attached to some point *O* on the machine is shown in Fig. 2.40. The point *A* represents the hook in the ceiling and has an inclination of 45° with the horizontal. The point *B* represents the eyebolt in the wall and has an inclination of 30° with the horizontal.

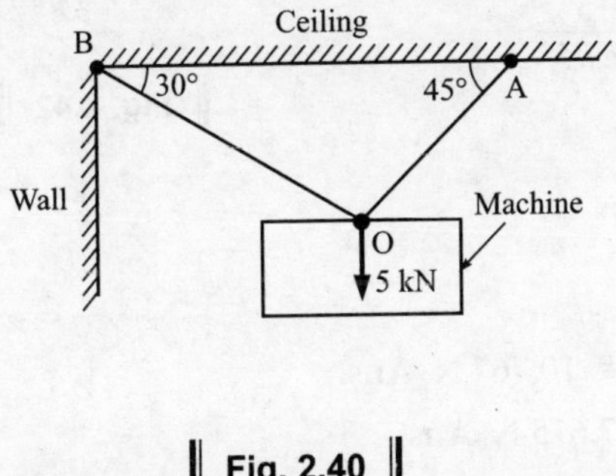


**Fig. 2.40**

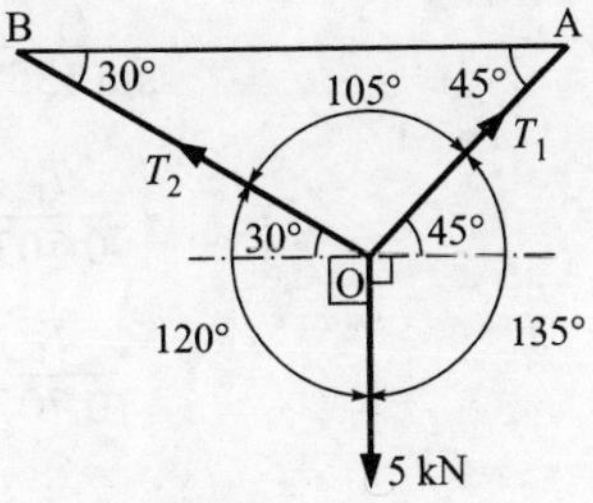


**Fig. 2.41**

Let $T_1$ and $T_2$ be the tensions in the chains *OA* and *OB*, respectively, as shown in Fig. 2.41. Now the point *O* is in equilibrium under the action of three forces *i.e.* $T_1$, $T_2$ and 5 kN. The angle between these forces is shown in Fig. 2.41.

Applying Lami's theorem,

$$\frac{T_1}{\sin 120°} = \frac{T_2}{\sin 135°} = \frac{5}{\sin 105°}$$

$$\frac{T_1}{\sin (180° - 60°)} = \frac{T_2}{\sin (180° - 45°)} = \frac{5}{\sin (180° - 75°)}$$

$$\frac{T_1}{\sin 60°} = \frac{T_2}{\sin 45°} = \frac{5}{\sin 75°}$$

$$\frac{T_1}{0.866} = \frac{T_2}{0.707} = \frac{5}{0.966}$$

$\therefore$ $$T_1 = \frac{5 \times 0.866}{0.966} = 4.4824 \text{ kN} \quad \textbf{Ans.}$$

and $$T_2 = \frac{5 \times 0.707}{0.966} = 3.6594 \text{ kN} \quad \textbf{Ans.}$$

**Example 2.26** *An electric light fixture weighing 15 N hangs from a point C by two strings AC and BC. The string AC is inclined at 60° to the horizontal and BC at 45° to the vertical. Using Lami's theorem or otherwise, determine the forces in the strings AC and BC.*

***Solution*** Given : Weight at $C$ = 15 N

Let $T_1$ = Force in the string $AC$, and

$T_2$ = Force in the string $BC$.

The system of forces is shown in Fig. 2.42.

From the figure, we find that

$$\angle ACB = 180° - (60° + 45°) = 75°$$

$$\angle ACE = 60° + 90° = 150°$$

$$\angle BCE = 45° + 90° = 135°$$

Using Lami's theorem,

$$\frac{T_1}{\sin \angle BCE} = \frac{T_2}{\sin \angle ACE} = \frac{15}{\sin \angle ACB}$$

$$\frac{T_1}{\sin 135°} = \frac{T_2}{\sin 150°} = \frac{15}{\sin 75°}$$

$$\frac{T_1}{0.707} = \frac{T_2}{0.5} = \frac{15}{0.966}$$

or $$\frac{T_1}{0.707} = \frac{T_2}{0.5} = 15.23$$

$\therefore$ $$T_1 = 0.707 \times 15.23 = 10.767 \text{ N} \quad \textbf{Ans.}$$

and $$T_2 = 0.5 \times 15.23 = 7.615 \text{ N} \quad \textbf{Ans.}$$

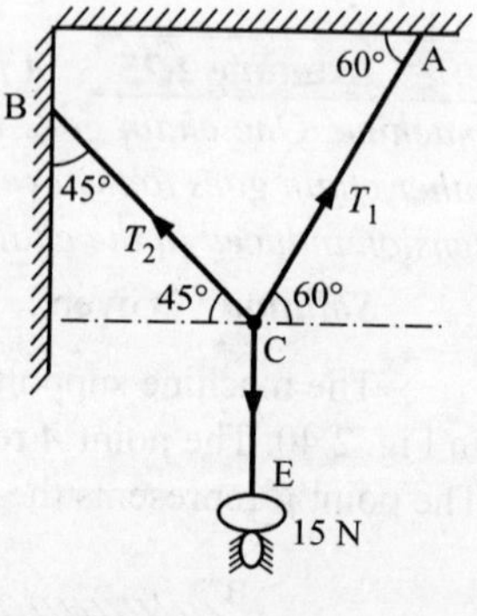


Fig. 2.42

**Example 2.27** *A fine light string ABCDE whose extremity A is fixed, has weights W and $W_1$ attached to it at B and C. It passes round a small smooth peg at D carrying a weight of 400 N at the free* *end E, as shown in Fig. 2.43. If in the position of equilibrium, BC is horizontal and AB and CD make 150° and 120° with BC. Find: 1. the tensions in the portions AB, BC and CD of the string; and 2. the magnitude of W and $W_1$.*

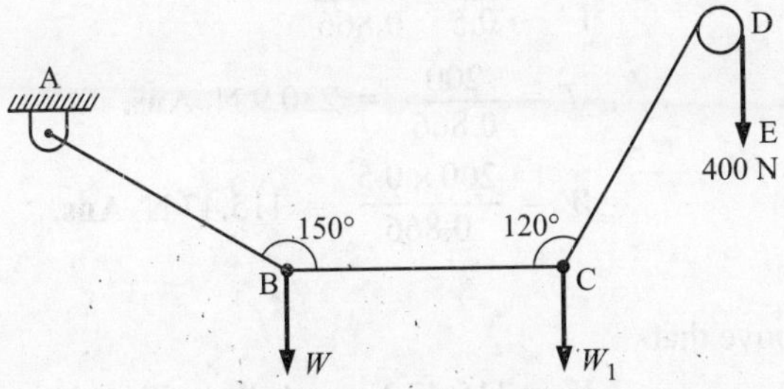


**Fig. 2.43**

***Solution*** Given : Weight at $B = W$ ; Weight at $C = W_1$ ; Weight at $E = 400$ N

**1. *Tensions in the portions AB, BC and CD of the string***

First of all consider the equilibrium of point *C*, as shown in Fig. 2.44. At this point, the following three forces are acting:

(*a*) Tension in the, string *CD* ($T_1$) which is equal to tension in the string *DE i.e.* 400 N,

(*b*) Tension in the string *CB* ($T_2$), and

(*c*) Weight at *C* ($W_1$).

Since $\angle BCD = 120°$; $\angle BCF = 90°$, therefore

$$\angle DCF = 360° - (120° + 90°) = 150°$$

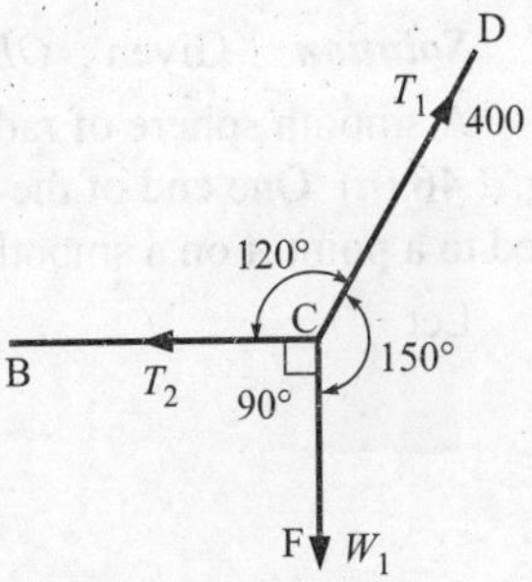


**Fig. 2.44**

Now according to Lami's theorem,

$$\frac{W_1}{\sin \angle BCD} = \frac{T_2}{\sin \angle DCF} = \frac{T_1}{\sin \angle BCF}$$

$$\frac{W_1}{\sin 120°} = \frac{T_2}{\sin 150°} = \frac{T_1}{\sin 90°}$$

$$\frac{W_1}{0.866} = \frac{T_2}{0.5} = 400$$

$\therefore$ $$W_1 = 400 \times 0.866 = 346.4 \text{ N} \quad \textbf{Ans.}$$

and $$T_2 = 400 \times 0.5 = 200 \text{ N} \quad \textbf{Ans.}$$

Considering the equilibrium of point *B*, as shown in Fig. 2.45. At this point, the following three forces are acting:

(*a*) Tension in the string *BA* (*T*),

(*b*) Tension in the string *BC* ($T_2 - 200$ N), and

(*c*) Weight at *B* (*W*).

Since $\angle ABC = 150°$, and $\angle CBG = 90°$, therefore

$$\angle ABG = 360° - (150° + 90°) = 120°$$

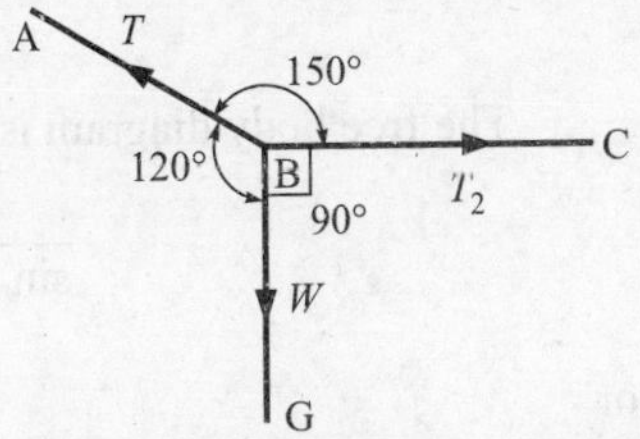


**Fig. 2.45**

Now according to Lami's theorem,

$$\frac{T}{\sin \angle CBG} = \frac{W}{\sin \angle ABC} = \frac{T_2}{\sin \angle ABG}$$

$$\frac{T}{\sin 90°} = \frac{W}{\sin 150°} = \frac{T_2}{\sin 120°}$$

$$\frac{T}{1} = \frac{W}{0.5} = \frac{200}{0.866}$$

$\therefore$ $$T = \frac{200}{0.866} = 230.9 \text{ N} \quad \textbf{Ans.}$$

and $$W = \frac{200 \times 0.5}{0.866} = 115.47 \text{ N} \quad \textbf{Ans.}$$

**2. *Magnitude of W and $W_1$***

We have calculated above that

$$W = 115.47 \text{ N; and } W_1 = 346.4 \text{ N} \quad \textbf{Ans.}$$

**Example 2.28** *A smooth sphere of radius r and weight W hangs by a light string of length l. One end of the string is fastened to a point on the surface of the sphere while the other end is fixed to a point on a smooth vertical wall. Determine the reaction of the wall and the tension in the string.*

***Solution*** Given : $OB = OC = r$ ; Weight = $W$ ; $CA = l$

A smooth sphere of radius $OB = r$ and weight $W$ is hung by a light string $CA$ of length $l$, as shown in Fig. 2.46 (*a*). One end of the string is fastened to a point $C$ on the surface of the sphere and the other end is fixed to a point $A$ on a smooth vertical wall.

Let $R_B$ = Reaction of the wall, and

$T$ = Tension in the string.

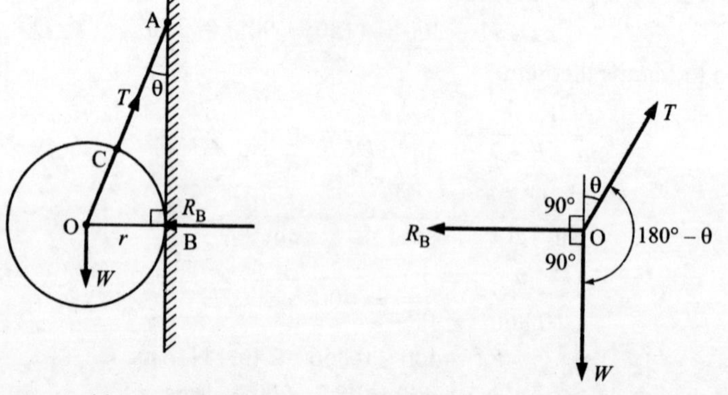


**Fig. 2.46**

The free body diagram is shown in Fig. 2.46 (*b*). Applying Lami's theorem, we have

$$\frac{R_B}{\sin(180° - \theta)} = \frac{W}{\sin(90° + \theta)} = \frac{T}{\sin 90°}$$

or $$\frac{R_B}{\sin \theta} = \frac{W}{\cos \theta} = \frac{T}{1}$$

$\therefore$ $$R_B = T \sin \theta \qquad ...(i)$$

and $$W = T\cos\theta \text{ or } T = \frac{W}{\cos\theta} \qquad ...(ii)$$

Now from Fig. 2.46 (*a*), $$\sin\theta = \frac{OB}{OA} = \frac{OB}{OC + CA} = \frac{r}{r+l}$$

and $$\cos\theta = \sqrt{1-\sin^2\theta} = \sqrt{1-\left(\frac{r}{r+l}\right)^2}$$

$$= \sqrt{\frac{(r+l)^2 - r^2}{(r+l)^2}} = \sqrt{\frac{r^2 + l^2 + 2rl - r^2}{(r+l)^2}}$$

$$= \sqrt{\frac{l^2 + 2rl}{(r+l)^2}} = \frac{\sqrt{l^2+2rl}}{r+l}$$

Substituting the value of sin θ and cos θ in equations (*i*) and (*ii*), we have

$$R_B = T\sin\theta = \frac{W}{\cos\theta}\times\sin\theta$$

$$= W \times \frac{r}{(r+l)} \times \frac{(r+l)}{\sqrt{l^2+2rl}} = \frac{W.r}{\sqrt{l^2+2rl}} \quad \textbf{Ans.}$$

and $$T = \frac{W}{\cos\theta} = \frac{W(r+l)}{\sqrt{l^2+2rl}} \quad \textbf{Ans.}$$

**Example 2.29** *A spherical ball weighing 200 N is supported by a string AB of length 150 mm. The ball is in contact with the smooth vertical wall at C as shown in Fig. 2.47 (a). The radius of the sphere is equal to the length of string AB. Find the tension in the string and the reaction of the wall.*

***Solution*** Given : $W = 200$ N ; $AB = 150$ mm ; $r = OB = AB = 150$ mm

Let $T$ = Tension in the string *BA*, and

$R$ = Reaction of the wall.

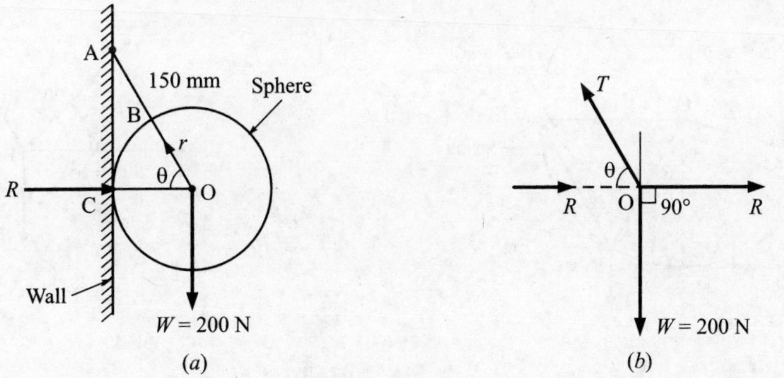


**Fig. 2.47**

From the triangle *AOC*, $$\cos\theta = \frac{OC}{OA} = \frac{150}{150+150} = 0.5$$

∴ $$\theta = 60°$$

The sphere is in equilibrium under the action of the following three forces:

(*a*) Weight of 200 N acting vertically downwards at *O*,

(*b*) Tension in the string (*T*), and

(*c*) Reaction *R* of the wall at the point of contact *C*.

From Fig. 2.47 (*b*), we find that

$$\text{Angle between } T \text{ and } R = 180° - \theta = 180° - 60° = 120°$$

$$\text{Angle between } T \text{ and } W = 90° + \theta = 90° + 60° = 150°$$

and $$\text{Angle between } R \text{ and } W = 90°$$

Now applying Lami's theorem, we have

$$\frac{T}{\sin 90°} = \frac{R}{\sin 150°} = \frac{200}{\sin 120°}$$

or $$\frac{T}{1} = \frac{R}{0.5} = \frac{200}{0.866}$$

$\therefore$ $$T = \frac{200}{0.866} = 230.9 \text{ N} \quad \textbf{Ans.}$$

and $$R = \frac{200}{0.866} \times 0.5 = 115.45 \text{ N} \quad \textbf{Ans.}$$

**Example 2.30** *A small ring is free to slide on a smooth circular wire of radius r, fixed to a vertical plane. It is attached by a string of length l* $(< \sqrt{2}r)$ *to a point of the wire at the same height as the centre. If W be the weight of the ring, show that tension in the string is* $\dfrac{(2r^2 - l^2)W}{r\sqrt{4r^2 - l^2}}$

***Solution*** Given: $AB = l (< \sqrt{2}r)$; $OB = r$

The ring shall be at rest at point *B*, which is higher than the lowest point of wire, because length *l* is less than * $\sqrt{2}r$ (Refer Fig. 2.48).

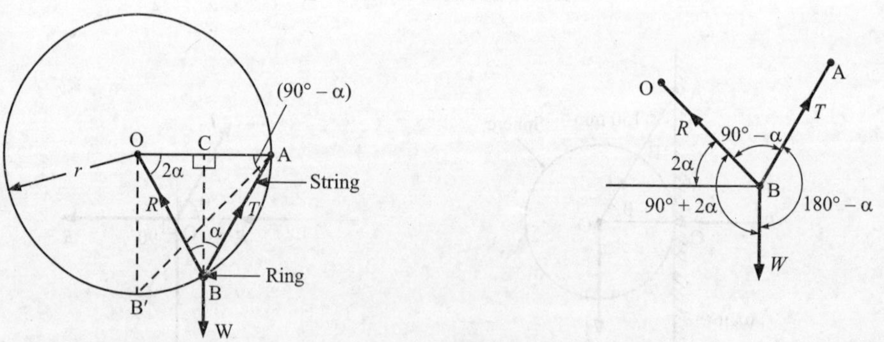


**Fig. 2.48**

Let *R* be the reaction of the wire on the ring and *T* be the tension in the string *AB*. Let $\angle AOB = 2\alpha$. We know that in triangle *AOB*,

$$(AB)^2 = (OA)^2 + (OB)^2 - 2OA \times OB \cos 2\alpha \quad \text{...(Cosine formula)}$$

* $AB' = \sqrt{(OA)^2 + (OB')^2} = \sqrt{r^2 + r^2} = \sqrt{2r^2} = \sqrt{2}r$. Thus *AB* is less than *AB*′ or $\sqrt{2}r$.

or $$l^2 = r^2 + r^2 - 2r \times r \cos 2\alpha = 2r^2 - 2r^2 \cos 2\alpha$$

$\therefore$ $$\cos 2\alpha = \frac{2r^2 - l^2}{2r^2}$$

$$\angle OAB = \angle OBA = \frac{180° - 2\alpha}{2} = 90° - \alpha$$

$$\angle OBC = 180° - (90° + 2\alpha) = 90° - 2\alpha$$

Applying Lami's theorem at point $B$, we have

$$\frac{W}{\sin (90° - \alpha)} = \frac{T}{\sin (90° + 2\alpha)}$$

or $$\frac{W}{\cos \alpha} = \frac{T}{\cos 2\alpha}$$

$\therefore$ $$T = \frac{W \cos 2\alpha}{\cos \alpha} \quad ...(i)$$

We know that $$\cos 2\alpha = 2 \cos^2 \alpha - 1$$

or $$2 \cos^2 \alpha = 1 + \cos 2\alpha$$

$\therefore$ $$\cos \alpha = \sqrt{\frac{1 + \cos 2\alpha}{2}}$$

Now substituting the value of $\cos 2\alpha$, we have

$$\cos \alpha = \sqrt{\frac{1 + \left[\dfrac{2r^2 - l^2}{2r^2}\right]}{2}} = \sqrt{\frac{4r^2 - l^2}{4r^2}} = \frac{\sqrt{4r^2 - l^2}}{2r}$$

Substituting the values of $\cos 2\alpha$ and $\cos \alpha$ in equation $(i)$, we have

$$T = W\left[\left(\frac{2r^2 - l^2}{2r^2}\right) \times \frac{2r}{\sqrt{4r^2 - l^2}}\right] = \frac{(2r^2 - l^2)W}{r\sqrt{4r^2 - l^2}}$$

**Example 2.31** *A roller, as shown in Fig. 2.49, is of weight 20 kN. What force P is necessary to start the roller over the block A and reaction at the contact point C.*

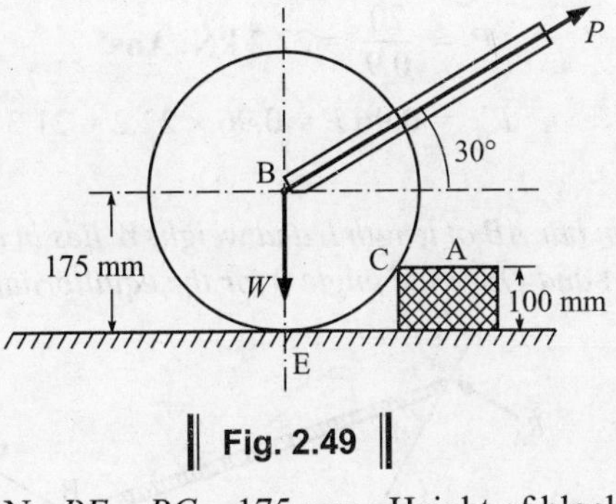


**Fig. 2.49**

***Solution*** Given : $W = 20$ kN ; $BE = BC = 175$ mm ; Height of block = 100 mm

From the geometry of the Fig. 2.50, we find that

$$\sin \theta = \frac{BD}{BC} = \frac{75}{175} = 0.4285$$

$\therefore$ $$\theta = \sin^{-1} (0.4285) = 25.37°$$

The roller is in equilibrium under the action of the following three forces (Refer Fig. 2.51):

1. Weight ($W = 20$ kN) acting vertically downwards at $B$,
2. Force $P$ acting at 30° to the horizontal, and
3. Reaction $R_C$ acting at the point of contact $C$.

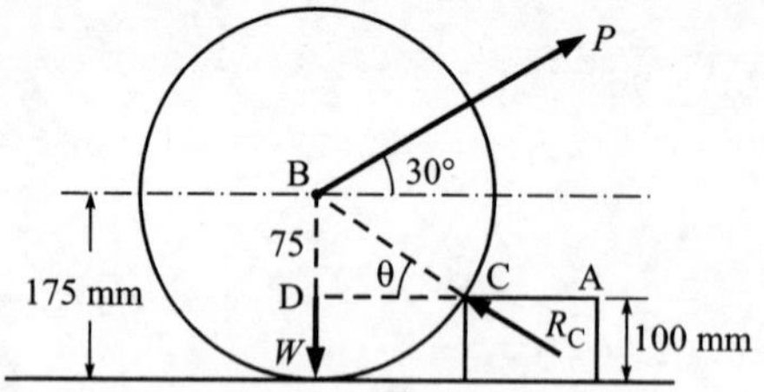


Fig. 2.50

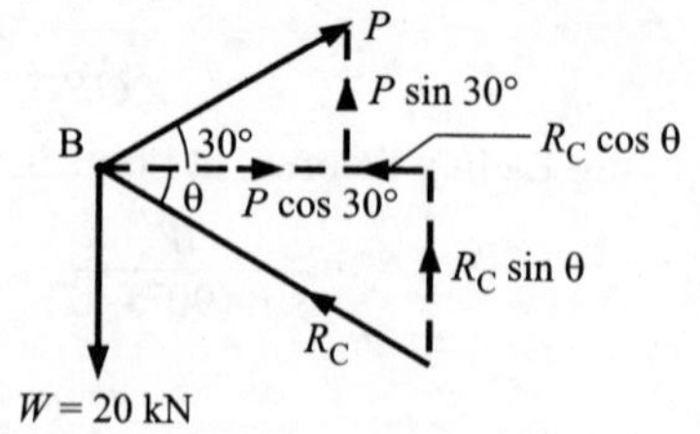


Fig. 2.51

For the equilibrium, these forces must pass through a common point $B$. Resolving the forces $P$ and $R_C$ into vertical and horizontal components and equating the same.

Equating the vertical forces,

$$P \sin 30° + R_C \sin \theta = W$$

$$P \times 0.5 + R_C \sin 25.37° = 20$$

$$P \times 0.5 + R_C \times 0.4285 = 20$$

or $$0.5P + 0.4285 R_C = 20 \quad ...(i)$$

Now equating horizontal forces,

$$P \cos 30° = R_C \cos \theta = R_C \cos 25.37°$$

$$P \times 0.866 = R_C \times 0.9035$$

or $$R_C = \frac{P \times 0.866}{0.9035} = 0.96\,P \quad ...(ii)$$

Substituting the value of $R_C$ in equation (*i*), we have

$$0.5P + 0.4285 \times 0.96P = 20$$

$$0.5P + 0.4P = 20$$

$$0.9P = 20$$

∴ $$P = \frac{20}{0.9} = 22.2 \text{ kN} \quad \textbf{Ans.}$$

and from equation (*ii*), $$R_C = 0.96\,P = 0.96 \times 22.2 = 21.3 \text{ kN} \quad \textbf{Ans.}$$

**Example 2.32** *A uniform bar AB of length L and weight W lies in a vertical plane with its ends resting on two smooth inclined planes OA and OB. Find angle θ for the equilibrium of the bar, as shown in Fig. 2.52.*

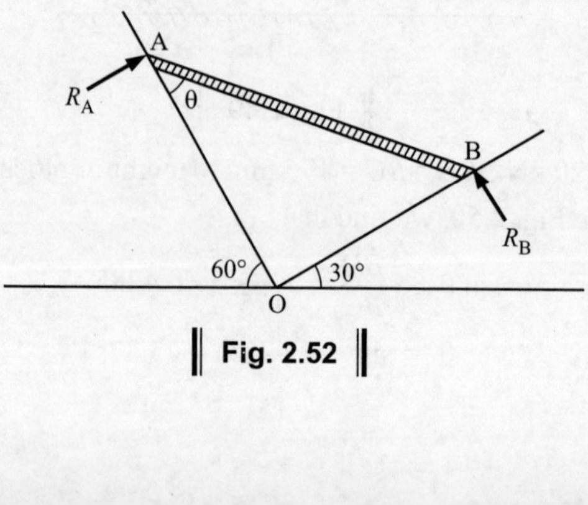


Fig. 2.52

***Solution*** Given : $AB = L$

For the equilibrium of the rod, the weight of the rod ($W$) and the reactions at $A$ and $B$ (*i.e.* $R_A$ and $R_B$) must meet at the same point $C$, as shown in Fig. 2.53.

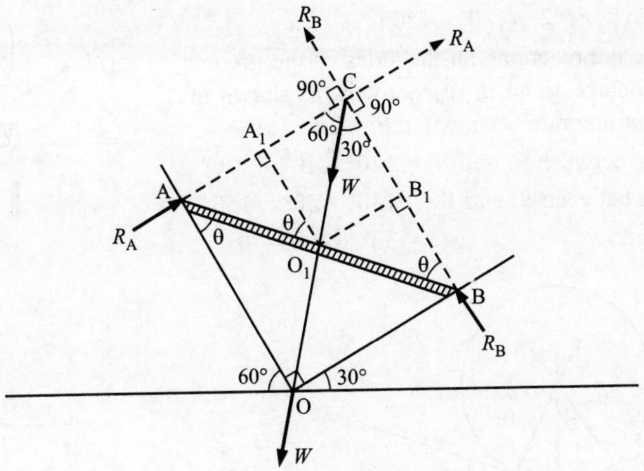


**Fig. 2.53**

Now applying Lami's theorem at point $C$,

$$\frac{R_A}{\sin(90° + 60°)} = \frac{R_B}{\sin(90° + 30°)} = \frac{W}{\sin 90°}$$

or
$$\frac{R_A}{0.5} = \frac{R_B}{0.866} = W$$

∴
$$R_A = 0.5\,W \quad ...(i)$$

and
$$R_B = 0.866\,W \quad ...(ii)$$

Since $O_1A = O_1B = \frac{L}{2}$, therefore

$$O_1A_1 = O_1A\cos\theta = \frac{L}{2}\cos\theta$$

and
$$O_1B_1 = O_1B\sin\theta = \frac{L}{2}\sin\theta$$

Now taking moments about the centre of the rod *i.e.* at point $O_1$, we have

$$R_A \times O_1A_1 = R_B \times O_1B_1$$

$$0.5\,W \times \frac{L}{2}\cos\theta = 0.866 \quad \times - \sin\theta$$

or
$$\frac{\sin\theta}{\cos\theta} = \frac{0.5}{0.866} = 0.577$$

$$\tan\theta = 0.577$$

∴
$$\theta = \tan^{-1}(0.577) = 30° \textbf{ Ans.}$$

**Example 2.33** *A smooth circular cylinder of radius 2 metres is lying in a triangular groove, one side of which makes an angle of 10° and the other an angle of 30° with the horizontal, as shown in Fig. 2.54. Find the reactions at the surface of contact, if there is no friction and the cylinder weighs 1500 N.*

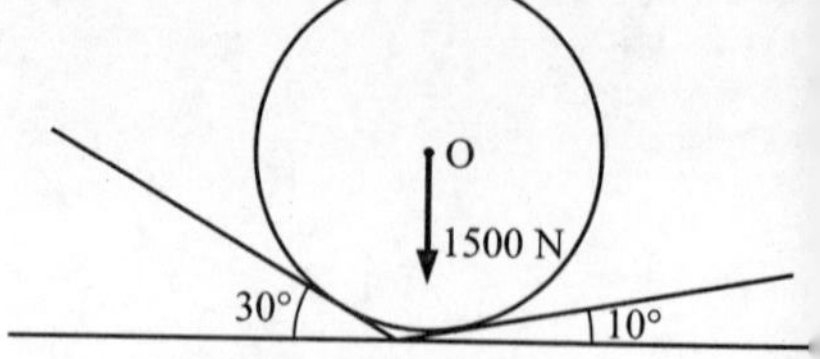


**Fig. 2.54**

***Solution*** Given : $r = 2$ m ; $W = 1500$ N

Let $R_1$ and $R_2$ be the reactions on the sides of the groove *i.e.* at the surface of contact *A* and *B*, respectively, as shown in Fig. 2.55 (*a*). The vector diagram is shown in Fig. 2.55 (*b*).

We see that angle between $R_1$ and $W = 180° - 10° = 170°$

Angle between $R_2$ and $W = 180° - 30° = 150°$

and angle between $R_1$ and $R_2$ $= 30° + 10° = 40°$

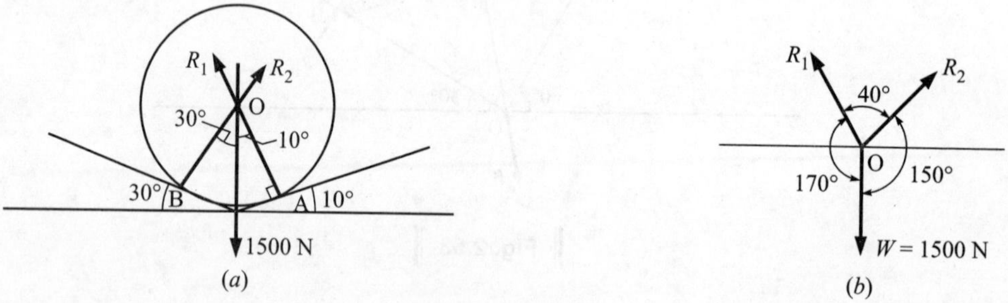


**Fig. 2.55**

Now applying Lami's theorem, we have

$$\frac{R_1}{\sin 150°} = \frac{R_2}{\sin 170°} = \frac{1500}{\sin 40°}$$

or

$$\frac{R_1}{0.5} = \frac{R_2}{0.1736} = \frac{1500}{0.6428}$$

$\therefore$

$$R_1 = \frac{1500}{0.6428} \times 0.5 = 1166.8 \text{ N} \quad \textbf{Ans.}$$

and

$$R_2 = \frac{1500}{0.6428} \times 0.1736 = 405.1 \text{ N} \quad \textbf{Ans.}$$

**Example 2.34** *Two smooth spheres of weight W and radius r each are in equilibrium in a horizontal channel of width b (< 4r) and vertical sides, as shown in Fig. 2.56. Find the three reactions from the sides of the channel which are all smooth. Find also the force exerted by each sphere on the other. Calculate these values, if r = 250 mm ; b = 900 mm ; and W = 1000 N.*

***Solution*** Given : $r = 250$ mm ; $b = 900$ mm ; $W = 1000$ N

Since the spheres are smooth, therefore the pressure at various points of contact will be normal to the surface. It may also be noted that at the point of contact between the two spheres the forces will act along the line joining their centres, as shown in Fig. 2.57.

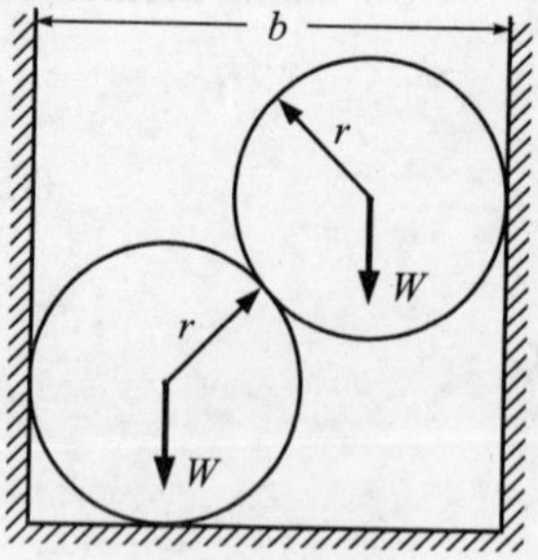


**Fig. 2.56**

Let $R_A$ = Reaction of the vertical wall on the bottom sphere,

$R_B$ = Reaction of the vertical wall on the top sphere,

$R_C$ = Reaction of the horizontal wall on the bottom sphere,

$F_1$ = Force exerted by the bottom sphere on the top sphere,

$F_2$ = Force exerted by the top sphere on the bottom sphere, and

$\alpha$ = Angle made by the line $OO_1$ with the vertical line $O_1D$.

From triangle $OO_1D$, as shown in Fig. 2.57,

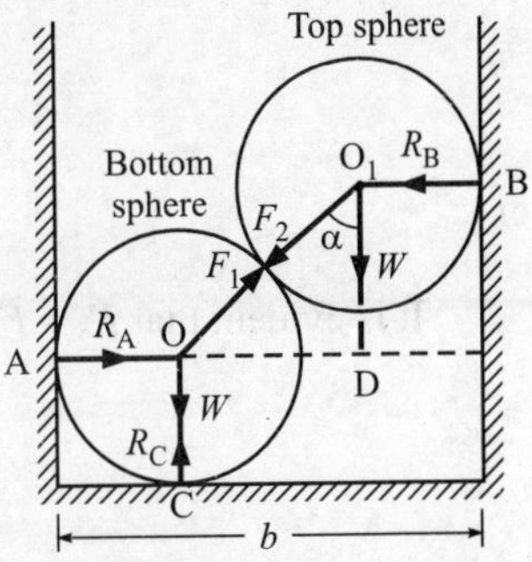


**Fig. 2.57**

$$\sin\alpha = \frac{OD}{OO_1} = \frac{b-2r}{2r}$$

and

$$\cos\alpha = \sqrt{1-\sin^2\alpha} = \sqrt{1-\left(\frac{b-2r}{2r}\right)^2}$$

$$= \frac{1}{2r}\sqrt{(2r)^2-(b-2r)^2}$$

$$= \frac{1}{2r}\sqrt{4br-b^2} \qquad ...(ii)$$

In order to determine the reactions, let us consider the equilibrium of each sphere separately. First of all, let us consider the equilibrium of the top sphere, as shown in Fig. 2.58 (*a*) and (*b*).

**Fig. 2.58**

Applying Lami's theorem, we have

$$\frac{F_1}{\sin 90°} = \frac{}{\sin(90°+\alpha)} \qquad \frac{}{\sin(180°-\alpha)}$$

or

$$F_1 = \frac{W}{\cos\alpha} = \frac{R_B}{\sin\alpha} \qquad ...(iii)$$

$\therefore$

$$F_1 = \frac{W}{\cos\alpha} = \frac{W\times 2r}{\sqrt{4br-b^2}} \qquad \text{...[From equation (ii)]}$$

From equation (*iii*),

$$R_B = \frac{W}{\cos\alpha}\times\sin\alpha = \frac{W\times 2r}{\sqrt{4br-b^2}}\times\frac{(b-2r)}{2r} = \frac{W(b-2r)}{\sqrt{4br-b^2}}$$

Now considering the equilibrium of the bottom sphere, as shown in Fig. 2.59 (*a*), (*b*) and (*c*).

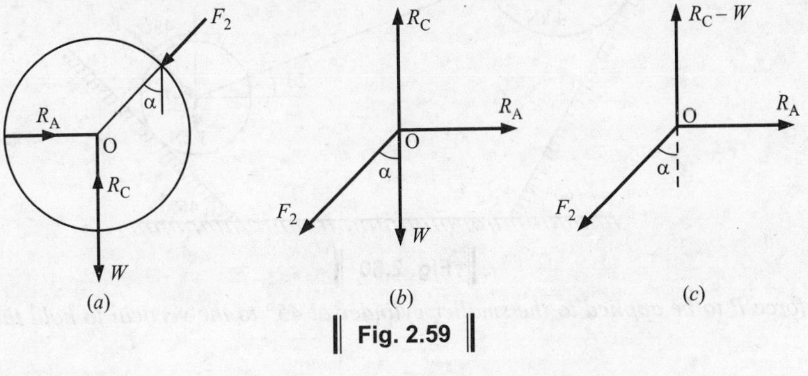


**Fig. 2.59**

Applying Lami's theorem, we have

$$\frac{F_2}{\sin 90^\circ} = \frac{R_A}{\sin(180^\circ - \alpha)} = \frac{R_C - W}{\sin(90^\circ + \alpha)}$$

or

$$F_2 = \frac{R_A}{\sin\alpha} = \frac{R_C - W}{\cos\alpha}$$

It is evident that $F_2 = F_1$.

$$\therefore \quad F_2 = F_1 = \frac{W \times 2r}{\sqrt{4br - b^2}}$$

and

$$\frac{W \times 2r}{\sqrt{4br - b^2}} = \frac{R_A}{\sin\alpha} = \frac{R_C - W}{\cos\alpha}$$

$$\therefore \quad R_A = \frac{W \times 2r}{\sqrt{4br - b^2}} \times \sin\alpha = \frac{W \times 2r}{\sqrt{4br - b^2}} \times \frac{b - 2r}{2r} = \frac{W(b - 2r)}{\sqrt{4br - b^2}}$$

and

$$R_C - W = F_2 \cos\alpha = F_1 \cos\alpha = \frac{W \times 2r}{\sqrt{4br - b^2}} \times \frac{1}{2r}\sqrt{4br - b^2} = W$$

$$\therefore \quad R_C = W + W = 2W$$

Since we are given that $r = 250$ mm; $b = 900$ mm and $W = 1000$ N; therefore

$$R_A = \frac{W(b - 2r)}{\sqrt{4br - b^2}} = \frac{1000(900 - 2 \times 250)}{\sqrt{4 \times 900 \times 250 - (900)^2}} = 1333.33 \text{ N} \quad \textbf{Ans.}$$

$$R_B = \frac{W(b - 2r)}{\sqrt{4br - b^2}} = R_A = 1333.33 \text{ N} \quad \textbf{Ans.}$$

$$R_C = 2W = 2 \times 1000 = 2000 \text{ N} \quad \textbf{Ans.}$$

$$F_1 = F_2 = \frac{W \times 2r}{\sqrt{4br - b^2}} = \frac{1000 \times 2 \times 250}{\sqrt{4 \times 900 \times 250 - (900)^2}} = 1666.67 \text{ N} \quad \textbf{Ans.}$$

**Example 2.35** *Two cylinders A and B weighing 4 kN and 3 kN, respectively, rest on smooth inclined plane, as shown in Fig. 2.60. They are connected by a bar of negligible weight hinged to each cylinder at its geometric centre by smooth pins.*

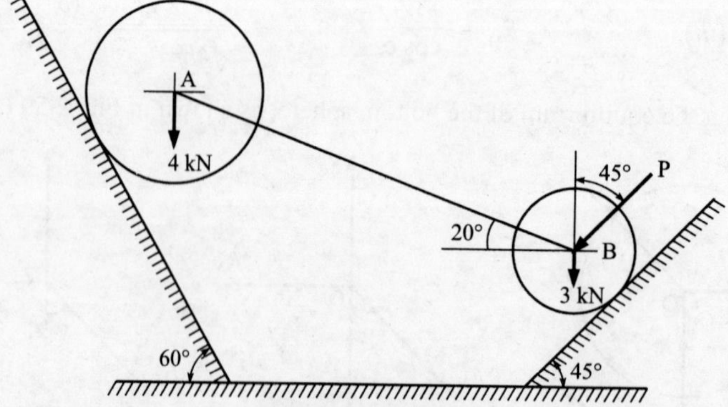


**Fig. 2.60**

*Find the force P to be applied to the smaller cylinder at 45° to the vertical to hold the system in the given position.*

***Solution*** Given : $W_A = 4$ kN = 4000 N ; $W_B = 3$ kN = 3000 N

The various forces acting on the given system is shown in Fig. 2.61.

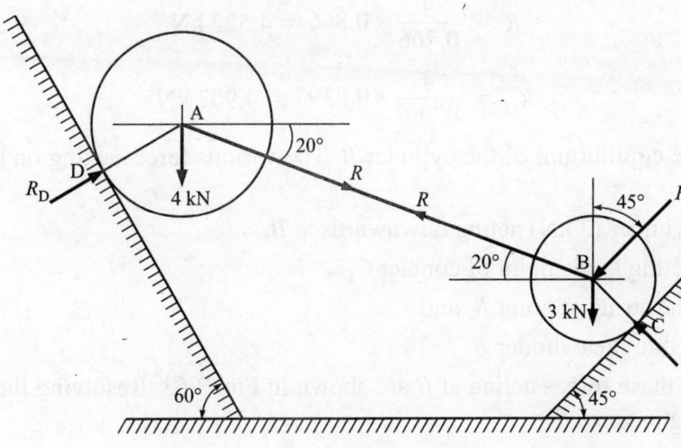


**Fig. 2.61**

First of all, let us consider the equilibrium of the cylinder $A$. The various forces acting on the cylinder $A$ are as follows:

1. Weight of the cylinder (4 kN) acting downwards at $A$,
2. Reaction ($R_D$) acting at the point of contact $D$, and
3. The tension ($R$) in the string at point $A$.

The angles between these forces acting at $A$ are shown in Fig. 2.62.

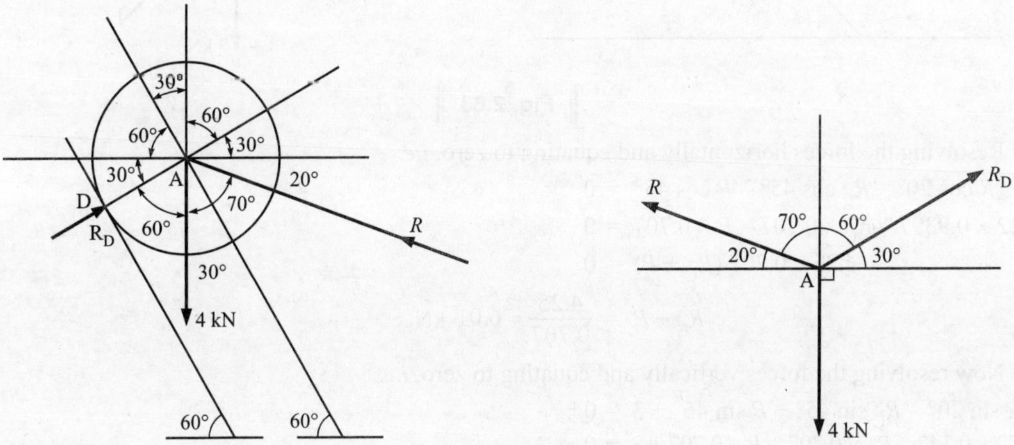


**Fig. 2.62**

Applying Lami's theorem at $A$,

$$\frac{R}{\sin(90° + 30°)} = \frac{R_D}{\sin(90° + 20°)} = \frac{4}{\sin(70° + 60°)}$$

$$\frac{R}{\sin 120°} = \frac{R_D}{\sin 110°} = \frac{4}{\sin 130°}$$

$$\frac{R}{0.866} = \frac{R_D}{0.9397} = \frac{4}{0.766}$$

$$\therefore \qquad R = \frac{4}{0.766} \times 0.866 = 4.522 \text{ kN}$$

and
$$R_D = \frac{4}{0.766} \times 0.9397 = 4.907 \text{ kN}$$

Now considering the equilibrium of the cylinder *B*. The various forces acting on the cylinder *B* are as follows:

(*a*) Weight of the cylinder (3 kN) acting downwards at *B*,

(*b*) Reaction ($R_C$) acting at the point of contact *C*,

(*c*) Tension (*R*) in the string at point *B*, and

(*d*) Force (*P*) applied to the cylinder *B*.

The angles between these forces acting at *B* are shown in Fig. 2.63. Resolving the forces horizontally and vertically and equating to zero.

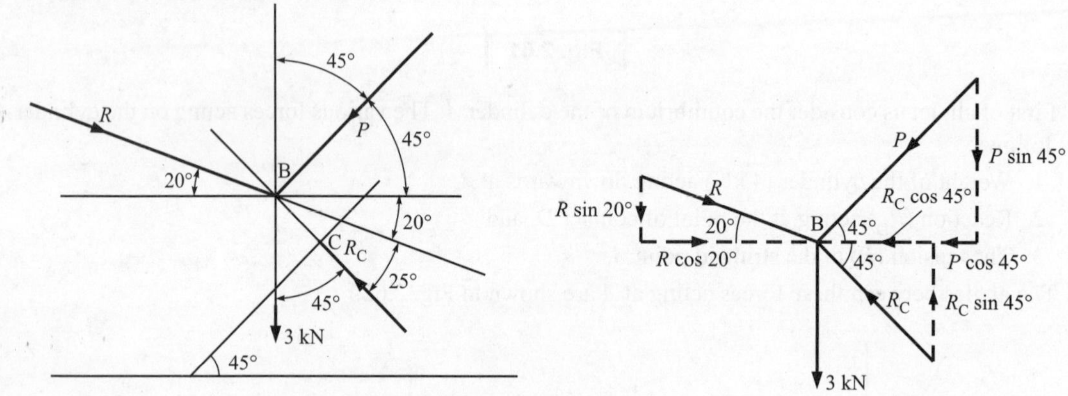


**Fig. 2.63**

Resolving the forces horizontally and equating to zero, *i.e.*

$$R \cos 20° - R_C \cos 45° - P \cos 45° = 0$$

$$4.522 \times 0.9397 - R_C \times 0.707 - P \times 0.707 = 0$$

$$4.25 - 0.707\,(R_C + P) = 0$$

or
$$R_C + P = \frac{4.25}{0.707} = 6.01 \text{ kN} \qquad ...(i)$$

Now resolving the forces vertically and equating to zero, *i.e.*

$$R \sin 20° - R_C \sin 45° + P \sin 45° + 3 = 0$$

$$4.522 \times 0.342 - R_C \times 0.707 + P \times 0.707 + 3 = 0$$

$$1.546 - 0.707\,(R_C - P) + 3 = 0$$

$$-0.707\,(R_C - P) + 4.546 = 0$$

or
$$R_C - P = \frac{4.546}{0.707} = 6.43 \text{ kN} \qquad ...(ii)$$

Adding equations (*i*) and (*ii*), we have

$$2R_C = 6.01 + 6.43 = 12.44 \text{ kN}$$

or
$$R_C = \frac{12.44}{2} = 6.22 \text{ kN}$$

$\therefore \quad P = R_C - 6.43$ ...[From equation (*ii*)]

$= 6.22 - 6.43 = -0.21$ kN $= -210$ N **Ans.**

The minus sign indicates that $P$ will act in the opposite direction.

## HIGHLIGHTS

1. A force is an agent which produces or tends to produce, destroy or tends to destroy the motion.
2. In order to determine the effects of a force acting on a body, we must know the following characteristics of a force:

   **(*a*)** Magnitude of the force (*i.e.* 50 N, 100 N etc.); **(*b*)** Direction or line of action of force; **(*c*)** Sense or nature of force (*i.e.* push or pull); and **(*d*)** Point of application of the force.
3. When the lines of action of all the forces lie along the same line, then the forces are said to be *collinear forces.*
4. When the lines of action of all the forces lie on the same plane, then the forces are known as *coplanar forces.*
5. When the lines of action of all the forces meet at a single point, then the forces are known as *concurrent forces.*
6. When the lines of action of all the forces lie in the same plane and pass through a common point, then the forces are said to be *coplanar concurrent forces.*
7. When the lines of action of all the forces lie in the same plane but do not pass through a common point, then the forces are known as *coplanar non-concurrent forces.*
8. When all the forces do not lie in the same plane, but their lines of action pass through a common point, then the forces are said to be *non-coplanar concurrent forces.*
9. When all the forces do not lie in the same plane, and also their lines of action do not pass through a common point, then the forces are known as *non-coplanar non-concurrent forces.*
10. The principle of transmissibility states that '*the state of rest or motion of a rigid body is unaltered if a force acting on the body is replaced by another force of the same magnitude and direction, but acting anywhere on the body in the line of action of the replaced force*'.
11. The principle of superposition of forces states that '*the combined effect of the force system acting on a particle or a rigid body is the vector sum of effects of individual forces*'.
12. A *resultant* is a single force which produces the same effect as the given system of forces acting on the body.
13. A single force, equal and opposite to the resultant, is known as *equilibrant.* This force keeps the body in equilibrium.
14. Parallelogram law of forces states that '*if two forces acting simultaneously on a body be represented in magnitude and direction by the two adjacent sides of a parallelogram, then their resultant is represented in magnitude and direction by the diagonal of the parallelogram passing through their point of intersection*'.
15. The process of finding the components of a given force in two given directions is called *resolution of a force,* and each component is called *resolved part.*
16. When a number of coplanar concurrent forces are acting on a body, then the algebraic sum of the resolved parts of the number of forces in a given direction is equal to the resolved part of their resultant in the same direction.
17. If a system of coplanar concurrent forces is in equilibrium, then its resultant ($R$) must be equal to zero, *i.e.* $R = 0$ or $0 = (\Sigma H)^2 + (\Sigma V)^2$.

**18.** Triangle law of forces states that '*if two forces acting simultaneously on a particle, be represented in magnitude and direction by the two sides of a triangle taken in order, then their resultant is represented, in magnitude and direction by the third side of the triangle, taken in opposite order*'.

**19.** Polygon law of forces states that '*if a number of forces acting simultaneously on a body, are represented in magnitude and direction by the sides of a polygon taken in order, then their resultant may be represented in magnitude and direction by the closing side of a polygon, taken in opposite order*'.

**20.** Lami's theorem states that '*if three forces acting at a point are in equilibrium, then each force is proportional to the sine of the angle between the other two forces.*' Mathematically,

$$\frac{P}{\sin \alpha} = \frac{Q}{\sin \beta} = \frac{R}{\sin \gamma}$$

where $P$, $Q$ and $R$ are the three forces acting at a point and $\alpha$, $\beta$ and $\gamma$ are the angles between the three forces.

## NUMERICAL PROBLEMS

**1.** Find the resultant of two forces 100 N and 150 N acting at a point. The angle between the forces is 60°. **[Ans.** 218 N]

**2.** Two forces act at an angle of 120°. The larger force is 40 N and the resultant is perpendicular to the smaller force. Find the magnitude of the smaller force. **[Ans.** 20 N]

**3.** The resultant of two forces $P$ and $Q$ is $R$. If $Q$ is doubled, the new resultant is perpendicular to $P$. Show that $Q = R$.

**4.** Two forces equal to $2P$ and $P$ act on a particle. If the first force is doubled and second force is increased by 12 N, the direction of the resultant remains same. Find the value of $P$. **[Ans.** 12 N]

**5.** The resultant of the two forces, when they act at an angle of 60° is 14 N. If the same forces are acting at right angles, their resultant is $\sqrt{136}$ N. Find the magnitude of the two forces. **[Ans.** 10 N; 6 N]

**6.** The following forces are acting on a body:

(*a*) 200 N inclined at 30° with east towards north,

(*b*) 250 N towards north,

(*c*) 300 N towards north-west, and

(*d*) 350 N inclined at 40° with west towards south.

Determine the magnitude and direction of the resultant force.

**[Ans.** 456 N; 132.3° along $OX$ in the anticlockwise direction]

**7.** $ABCDEF$ is a regular hexagon. Forces of magnitude 2 N, $\sqrt{3}$ N, 5 N, $\sqrt{3}$ N and 2 N act at $A$ in the directions of $AB$, $AC$, $AD$, $AE$ and $AF$, respectively. Find the magnitude and direction of the resultant force.

**[Ans.** 10 N; 60° with the horizontal]

**8.** Four forces equal to $P$, $2P$, $3P$ and $4P$ act along the sides $AB$, $BC$, $CD$ and $DA$ of a square $ABCD$. Find in magnitude and direction of the resultant. **[Ans.** $2\sqrt{2}P$; 45°]

**9.** A weight of 2 kN is supported by two chains $AC$ and $BC$, as shown in Fig. 2.64. Find the tension in each chain. **[Ans.** 1 kN; 1.732 kN]

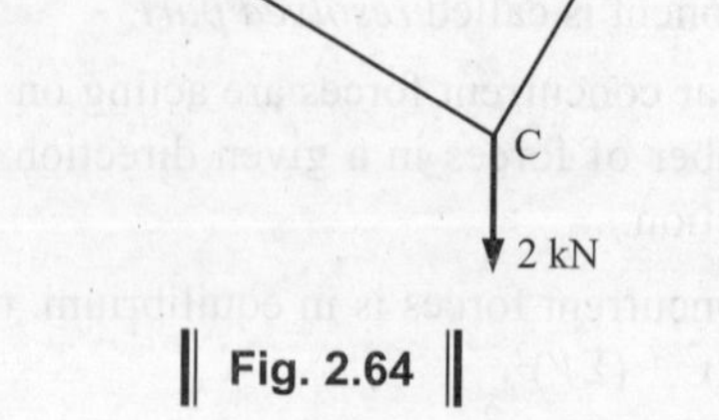


**Fig. 2.64**

**10.** A particle is acted upon the following coplanar concurrent forces:

(*a*) 150 N pull towards east,

(*b*) 250 N pull towards north-east,

(*c*) 200 N push towards north,

(*d*) 100 N push inclined at 30° with south towards west (S 30°W), and

(*e*) 300 N push inclined at 45° with east towards south (E 45°S).

Find the magnitude and direction of the resultant force. **[Ans.** 320.85 N; 59.15°]

**11.** Five strings are tied at a point and are pulled in all directions equally spaced from one another in the same plane. If the magnitude of the pulls on three consecutive strings are 10 kN, 14 kN and 12 kN, respectively, find the magnitude of the pulls on other two strings, for the system to be in equilibrium position.

**[Ans.** 11.236 kN; 14.476 kN]

**[Hint:** Angle between two consecutive strings = 360/5 = 72°]

**12.** The system of forces, all in newtons, is shown in Fig. 2.65. Determine the magnitude and direction of the resultant force. **[Ans.** 45.6 N; 132.3° with *OE*]

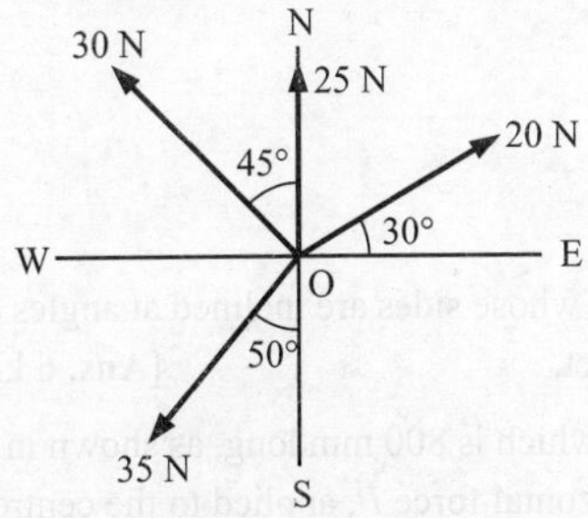


**Fig. 2.65**

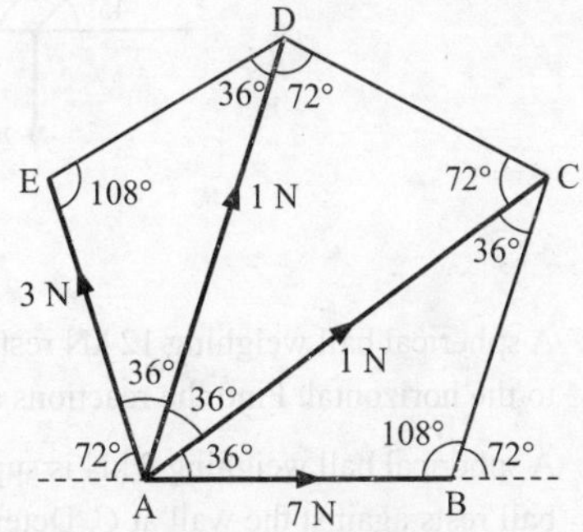


**Fig. 2.66**

**13.** Forces of 7 N, 1 N, 1 N and 3 N, respectively act at one angular point of regular pentagon towards the other four points, taken in order, are shown in Fig. 2.66. Find the magnitude and direction of the resultant force.

**[Ans.** 8.426 N; 31.42°]

**14.** *ABCDEF* is a regular hexagon. Forces of 8 N (pull), $P_1$ (push), $P_2$ (pull), 16 N (pull) and 12 N (push) act along the sides *AB*, *CA*, *AD*, *AE* and *FA*, respectively, are in equilibrium. Find the magnitude of $P_1$ and $P_2$.

**[Ans.** 18.64 N; 4.28 N]

**15.** A weight of 200 N is supported by two strings, as shown in Fig. 2.67. The supports are 80 mm apart. Find the tension in the strings. **[Ans.** $T_1 = T_2$ = 155.56 N]

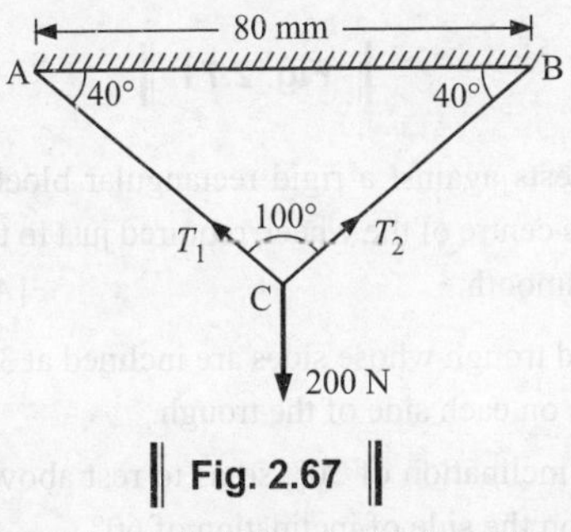


**Fig. 2.67**

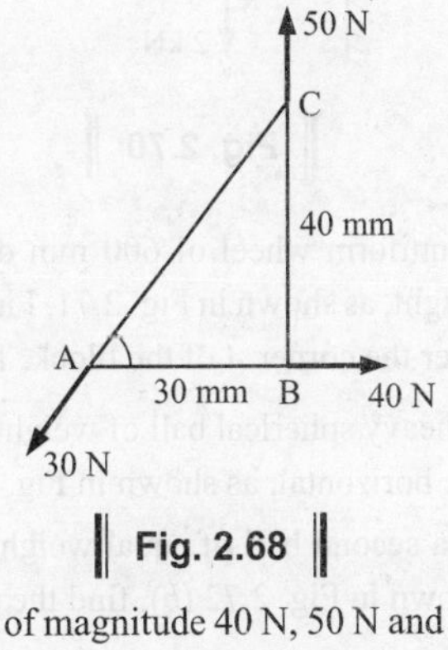


**Fig. 2.68**

**16.** In a right angled triangle *ABC*, as shown in Fig. 2.68, three forces of magnitude 40 N, 50 N and 30 N act along the sides *AB*, *BC* and *CA*, respectively. Find the magnitude and direction of the resultant force.

**[Ans.** 34 N; 49.76°]

17. *ABCD* is a square. Forces of 10 N, 8 N and 4 N act at *A* in the directions *AD*, *AC* and *AB*, respectively. Using the analytical method, determine the resultant in magnitude and direction. [**Ans.** 18.394 N; 58.31°]

18. A body weighing 800 N is supported at the end of a rod and a chain. The other end of the rod and chain are fixed in a wall. The rod is kept horizontal and the chain makes an angle of 30° with the rod. Determine the magnitude and direction of the forces exerted by the chain and the rod. [**Ans.** 400 N; 346.4 N]

19. A string is connected between two points at the same level. It carries two weights of 200 N and *W*. The middle portion of the string is kept horizontal where the weights of 200 N and *W* are suspended. The other portions of the string makes angles of 45° and 30° with the horizontal, as shown in Fig. 2.69. Determine the tension *T* in the middle portion of the string and the unknown weight *W*. [**Ans.** 200 N; 115.47 N]

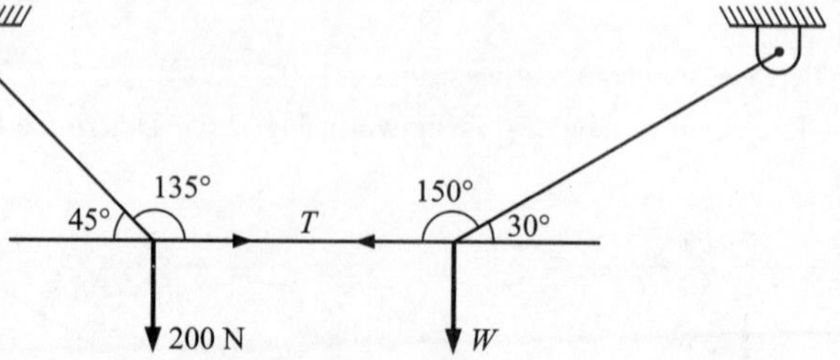


**Fig. 2.69**

20. A spherical ball weighing 12 kN rests in a triangular groove whose sides are inclined at angles of 30° and 45° to the horizontal. Find the reactions at the surfaces of contact. [**Ans.** 6 kN; 10.392 kN]

21. A spherical ball weighing 2 kN is supported by a string *AB* which is 800 mm long, as shown in Fig. 2.70. The ball rests against the wall at *C*. Determine the value of horizontal force *P*, applied to the centre of the ball so that the reaction of the wall at the point *C* is zero. The radius of the ball is equal to the length of the string *AB*. Find also the tension in the string. [**Ans.** 1.155 kN; 2.31 kN]

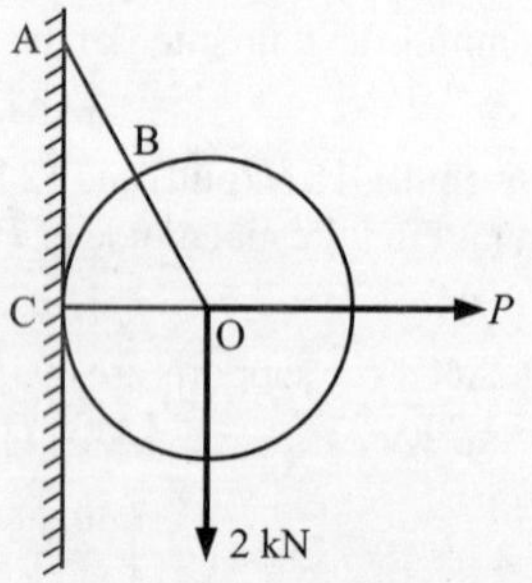


**Fig. 2.70**

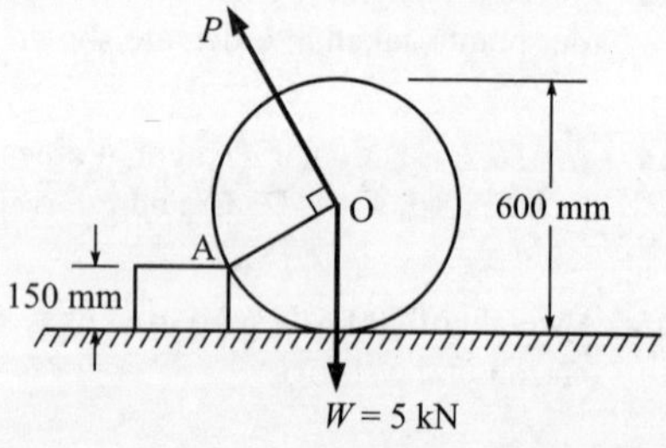


**Fig. 2.71**

22. A uniform wheel of 600 mm diameter, weighing 5 kN rests against a rigid rectangular block of 150 mm height, as shown in Fig. 2.71. Find the least pull through the centre of the wheel, required just to turn the wheel over the corner *A* of the block. Take all the surfaces to be smooth. [**Ans.** 4.33 kN]

23. A heavy spherical ball of weight 1000 N rests in a *V* shaped trough whose sides are inclined at 30° and 60° to the horizontal, as shown in Fig. 2.72 (*a*). Find the pressure on each side of the trough.

If a second ball of equal weight be placed on the side of inclination of 30°, so as to rest above the first, as shown in Fig. 2.72 (*b*), find the pressure of the lower ball on the side of inclination of 60°.

[**Ans.** 866 N, 500 N; 1000 N]

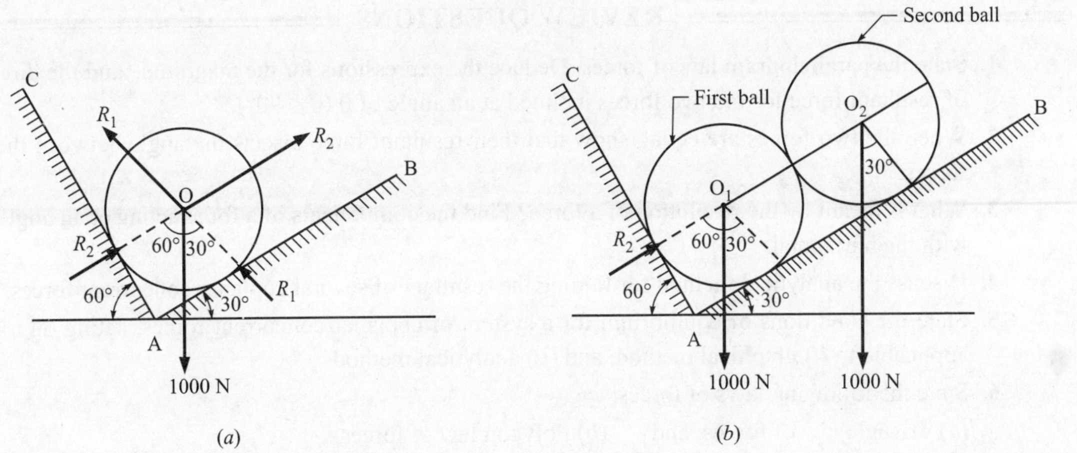


**Fig. 2.72**

**24.** Two smooth spheres rest in a horizontal channel, as shown in Fig. 2.73. The larger sphere has a radius of 180 mm and a weight of 500 N. The smaller sphere is of 100 mm radius and 200 N weight. The channel is 360 mm wide at the bottom with one side vertical. The other side is inclined at 60° with the horizontal. Find the reactions at the contact points *i.e.* *Q*, *P*, *L* and *N*.

[**Ans.** $R_Q$ = 240.84 N, $R_P$ = 134.18 N; $R_L$ = 154.93 N, $R_N$ = 622.5 N]

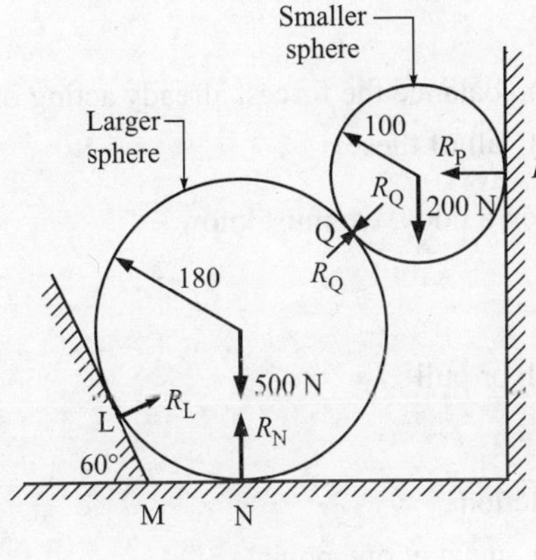


**Fig. 2.73**

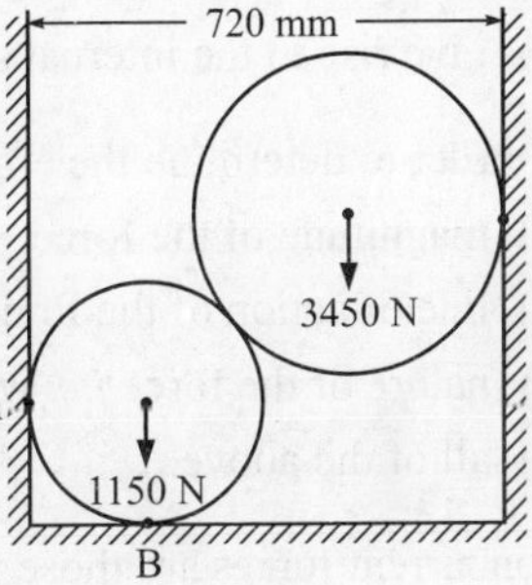


**Fig. 2.74**

**25.** Two spherical balls rest between two vertical walls, as shown in Fig. 2.74. The radius of smaller ball is 160 mm and weight is 1150 N. The radius of the larger ball is 240 mm and its weight is 3450 N. The distance between the walls is 720 mm. Assuming the contact surfaces to be smooth, determine the reactions at *A*, *B* and *C*. [**Ans.** $R_A = R_B = R_C$ = 4600 N]

**26.** Two cylinders *P* and *Q* rest in a channel, as shown in Fig. 2.75. The cylinder *P* has diameter of 100 mm and weight 200 N, whereas the cylinder *Q* has diameter of 180 mm and weight 600 N.

If the bottom width of the box is 180 mm with one side vertical and the other inclined at 60°, determine the pressures at all the four points of contact.

[**Ans.** $R_A$ = 134.2 N; $R_B$ = 240.8 N; $R_C$ = 154.9 N; $R_D$ = 722.5 N]

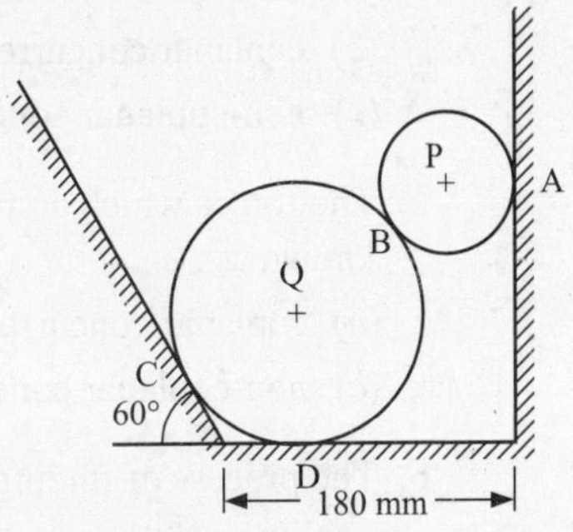


**Fig. 2.75**

## REVIEW QUESTIONS

1. State the parallelogram law of forces. Deduce the expressions for the magnitude and the direction of resultant force for the two forces inclined at an angle of θ ($\theta < 90°$).
2. When the two forces are equal, show that their resultant force bisects the angle between the two forces.
3. What is meant by the resolution of a force? Find the components of a force acting at an angle of θ with the horizontal.
4. Discuss the analytical method for finding the resultant of several coplanar concurrent forces.
5. State the conditions of equilibrium for a system of coplanar concurrent forces, acting on a body applicable to (*i*) graphical method; and (*ii*) analytical method.
6. State the following laws of forces:

   (*a*) Triangle law of forces; and (*b*) Polygon law of forces.
7. In the context of concurrent coplanar forces, explain the terms resultant and equilibrant.
8. How a resultant force is obtained by graphical method when a number of coplanar concurrent forces are acting at a point?
9. State and prove Lami's theorem.

## OBJECTIVE TYPE QUESTIONS

1. A force while acting on a body may

   (*a*) change its motion (*b*) balance the forces, already acting on it
   (*c*) give rise to the internal stresses in it (*d*) all of these

2. In order to determine the effects of a force, acting on a body, we must know

   (*a*) magnitude of the force
   (*b*) line of action of the force
   (*c*) nature of the force *i.e.* whether the force is push or pull
   (*d*) all of the above

3. Concurrent forces are those forces whose lines of action

   (*a*) lie on the same line (*b*) meet at one point
   (*c*) meet on the same plane (*d*) none of these

4. The forces, which meet at one point and their lines of action also lie on the same plane, are known as

   (*a*) coplanar concurrent forces (*b*) coplanar non-concurrent forces
   (*c*) non-coplanar concurrent forces (*d*) non-coplanar non-concurrent forces

5. The forces which do not meet at one point and their lines of action do not lie on the same plane are known as

   (*a*) coplanar concurrent forces (*b*) coplanar non-concurrent forces
   (*c*) non-coplanar concurrent forces (*d*) none of these

6. The process of finding out the resultant force is called ........... of forces.

   (*a*) composition (*b*) resolution

**7.** The algebraic sum of the resolved parts of a number of forces in a given direction is equal to the resolved part of their resultant in the same direction. This is known as

(*a*) principle of independence of forces

(*b*) principle of resolution of forces

(*c*) principle of transmissibility of forces

(*d*) none of these

**8.** The resultant of two forces $P$ and $Q$ acting at an angle $\theta$ is

(*a*) $\sqrt{P^2 + Q^2 + 2PQ \sin \theta}$ (*b*) $\sqrt{P^2 + Q^2 + 2PQ \cos \theta}$

(*c*) $\sqrt{P^2 + Q^2 - 2PQ \cos \theta}$ (*d*) $\sqrt{P^2 + Q^2 - 2PQ \tan \theta}$

**9.** If the resultant of two forces $P$ and $Q$ acting at an angle $\theta$, makes an angle $\alpha$ with the force $P$, then

(*a*) $\tan \alpha = \dfrac{P \sin \theta}{P + Q \cos \theta}$ (*b*) $\tan \alpha = \dfrac{P \cos \theta}{P + Q \cos \theta}$

(*c*) $\tan \alpha = \dfrac{Q \sin \theta}{P + Q \cos \theta}$ (*d*) $\tan \alpha = \dfrac{Q \cos \theta}{P + Q \sin \theta}$

**10.** The resultant of two forces $P$ and $Q$ (such that $P > Q$) acting along the same straight line, but in opposite direction, is given by

(*a*) $P + Q$ (*b*) $P - Q$ (*c*) $P/Q$ (*d*) $Q/P$

**11.** The resultant of two equal forces $P$ making an angle $\theta$, is given by

(*a*) $2P \sin \dfrac{\theta}{2}$ (*b*) $2P \cos \dfrac{\theta}{2}$ (*c*) $2P \tan \dfrac{\theta}{2}$ (*d*) $2P \cot \dfrac{\theta}{2}$

**12.** The resultant of two forces each equal to $P$ and acting at right angles is

(*a*) $\dfrac{P}{\sqrt{2}}$ (*b*) $\dfrac{P}{2}$ (*c*) $\dfrac{P}{2\sqrt{2}}$ (*d*) $\sqrt{2}P$

**13.** The angle between two forces when the resultant is maximum and minimum, respectively, are

(*a*) 0° and 180° (*b*) 180° and 0° (*c*) 90° and 180° (*d*) 90° and 0°

**14.** If the resultant of two equal forces has the same magnitude as either of the forces, then the angle between the two forces is

(*a*) 30° (*b*) 60° (*c*) 90° (*d*) 120°

**15.** The resultant of the two forces $P$ and $Q$ is $R$. If $Q$ is doubled, the new resultant is perpendicular to $P$. Then

(*a*) $P = Q$ (*b*) $Q = R$ (*c*) $Q = 2R$ (*d*) none of these

**16.** Two forces are acting at an angle of 120°. The bigger force is 40 N and the resultant is perpendicular to the smaller one. The smaller force is

(*a*) 20 N (*b*) 40 N (*c*) 80 N (*d*) none of these

**17.** A number of forces acting at a point will be in equilibrium, if

(*a*) all the forces are equally inclined
(*b*) sum of all the forces is zero
(*c*) sum of resolved parts in the vertical direction is zero (*i.e.* $\Sigma V = 0$)
(*d*) sum of resolved parts in the horizontal direction is zero (*i.e.* $\Sigma H = 0$)

**18.** If a number of forces are acting at a point, their resultant is given by

(*a*) $(\Sigma V)^2 + (\Sigma H)^2$ (*b*) $\sqrt{(\Sigma V)^2 + (\Sigma H)^2}$

(*c*) $(\Sigma V)^2 + (\Sigma H)^2 + 2(\Sigma V) + (\Sigma H)$ (*d*) $\sqrt{(\Sigma V)^2 + (\Sigma H)^2 + 2(\Sigma V) + (\Sigma H)}$

**19.** If a number of forces are acting at a point, their resultant will be inclined at an angle θ with the horizontal, such that

(*a*) $\tan\theta = \dfrac{\Sigma H}{\Sigma V}$ (*b*) $\tan\theta = \dfrac{\Sigma V}{\Sigma H}$

(*c*) $\tan\theta = \Sigma V \times \Sigma H$ (*d*) $\tan\theta = \sqrt{\Sigma V + \Sigma H}$

**20.** According to Lami's theorem

(*a*) the three forces must be equal
(*b*) the three forces must be at 120° to each other
(*c*) the three forces must be in equilibrium
(*d*) if the three forces acting at a point are in equilibrium, then each force is proportional to the sine of the angle between the other two.

***Answers to Objective Type Questions***

| | | | | |
|---|---|---|---|---|
| **1.** (*d*) | **2.** (*d*) | **3.** (*b*) | **4.** (*a*) | **5.** (*d*) |
| **6.** (*a*) | **7.** (*b*) | **8.** (*b*) | **9.** (*c*) | **10.** (*b*) |
| **11.** (*b*) | **12.** (*d*) | **13.** (*a*) | **14.** (*d*) | **15.** (*b*) |
| **16.** (*a*) | **17.** (*c*), (*d*) | **18.** (*b*) | **19.** (*b*) | **20.** (*d*) |

# 3

# COPLANAR NON-CONCURRENT FORCES



## 3.1 INTRODUCTION

We have already discussed in the previous chapter (Sec. 2.3) that the coplanar non-concurrent forces are those forces which do not meet at one point, but their lines of action lie in the same plane. In this chapter, we shall first discuss the moment of a force and parallel forces.

## 3.2 MOMENT OF A FORCE

The *moment* of a force is the turning effect produced by a force, on the body, on which it acts. It is equal to the product of the force and the perpendicular distance of the point about which the moment is required and the line of action of the force. Mathematically,

$$\text{Moment of a force} = F \times x$$

where $F$ = Force acting on the body, and

$x$ = Perpendicular distance of the point and the line of action of the force, as shown in Fig. 3.1.

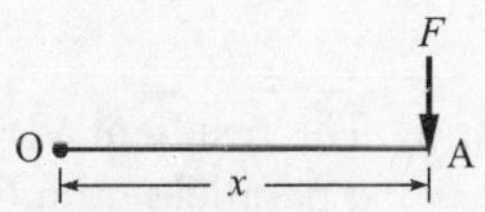


**Fig. 3.1**

The unit of moment depends upon the unit of force and the unit of perpendicular distance. If the force is in newton and the perpendicular distance is in metres, then the unit of moment will be newton metre (briefly written as N-m).

**Note:** The moment of a force is a vector quantity, as it has both magnitude as well as direction.

## 3.3 TYPES OF MOMENTS

The moments are of the following two types:

**1.** Clockwise moment; and **2.** Anticlockwise moment.

**Fig. 3.2**

When the moment of a force is such that it rotates or turns the body in the same direction in which the hands of the clock move, then it is said to be *clockwise moment*, as shown in Fig. 3.2 (*a*).

Similarly, when the moment of a force is such that it rotates or turns the body in the direction opposite to the movement of the hands of a clock, it is then said to be *anticlockwise moment*, as shown in Fig. 3.2 (*b*).

**Note:** It is a general practice that clockwise moments are taken as *positive* while anticlockwise moments are taken as *negative*.

## 3.4 VARIGNON'S THEOREM OF MOMENTS

It states that '*when a number of coplanar forces are acting about a point in their plane, then the algebraic sum of the moments of all the forces about a point is equal to the moment of their resultant force about the same point*'.

The Varignon's theorem of moments may be proved as discussed below:

**1.** Let the two coplanar concurrent forces *P* and *Q* acting at point *O* are represented in magnitude and direction by *OA* and *OB*, as shown in Fig. 3.3.

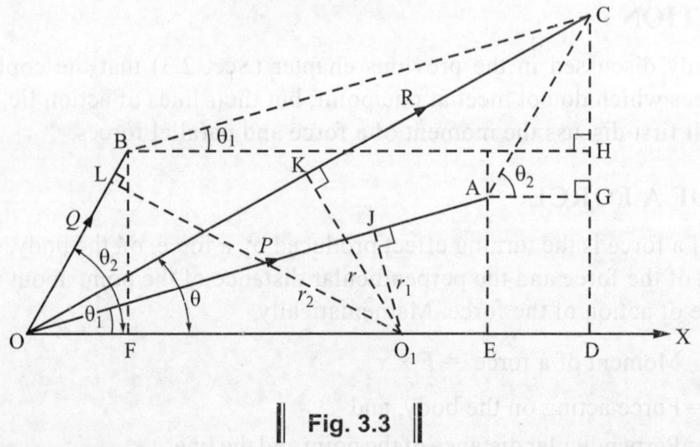


**Fig. 3.3**

**2.** The resultant (*R*) of *P* and *Q* is represented in magnitude and direction by *OC*, the diagonal of a parallelogram *OACB*.

**3.** Take any point $O_1$ on the horizontal axis *OX*. From $O_1$, draw perpendiculars $O_1J$, $O_1L$ and $O_1K$ on the forces *P*(*OA*), *Q*(*OB*) and resultant *R*(*OC*), respectively.

**4.** From points *A*, *B* and *C*, draw perpendiculars *AE*, *BF* and *CD*, respectively, on the horizontal axis *OX*.

**5.** From points *A* and *B*, draw perpendiculars *AG* and *BH*, respectively, on the vertical line *CD*.

Let $O_1J = r_1$; $O_1L = r_2$; and $O_1K = r$

$\angle AOX = \theta_1$; $\angle BOX = \theta_2$; and $\angle COX = \theta$

From Fig. 3.3, we see that $OA$ is parallel to $BC$ and $OA = BC$. Therefore, projection from $A$ and $B$ on the same vertical line $CD$ is equal, *i.e.* $DG = CH$.

Similarly, $OB$ is parallel to $AC$ and $OB = AC$. Therefore, projection from $A$ and $B$ on the horizontal line $OX$ will be equal, *i.e.* $OF = ED$.

For better understanding, the forces $P$, $Q$ and $R$ are shown separately in Fig. 3.4 (*a*), (*b*) and (*c*), respectively. Now resolving these forces in the horizontal and vertical directions, we have

From Fig. 3.4 (*a*), $OE = P\cos\theta_1$; and $AE = GD = CH = P\sin\theta_1$

From Fig. 3.4 (*b*), $OF = ED = Q\cos\theta_2$; and $BF = HD = Q\sin\theta_2$

Similarly, from Fig. 3.4 (*c*), $OD = R\cos\theta$; and $CD = R\sin\theta$

Let length of point $O_1$ from $O$, *i.e.* $OO_1 = x$.

$\therefore$ $O_1J = r_1 = x\sin\theta_1$

$O_1L = r_2 = x\sin\theta_2$

and $O_1K = r = x\sin\theta$

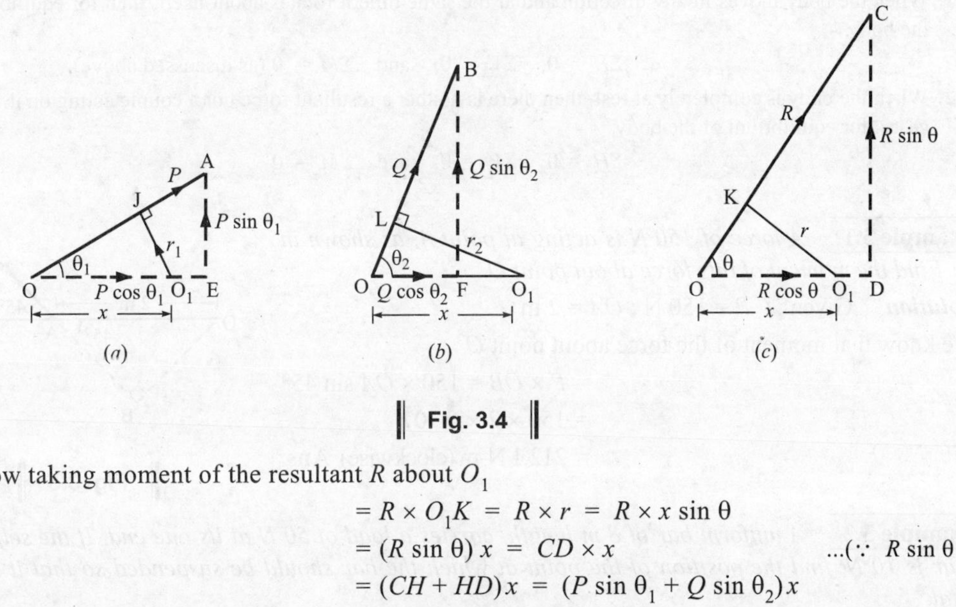


**Fig. 3.4**

Now taking moment of the resultant $R$ about $O_1$

$$= R \times O_1K = R \times r = R \times x\sin\theta$$

$$= (R\sin\theta)\,x = CD \times x \qquad ...(\because R\sin\theta = CD)$$

$$= (CH + HD)x = (P\sin\theta_1 + Q\sin\theta_2)x$$

$$= P \times x\sin\theta_1 + Q \times x\sin\theta_2$$

$$= P \times r_1 + Q \times r_2$$

$= $ Moment of $P$ about $O_1$ + Moment of $Q$ about $O_1$

Hence, it is proved that the moment of resultant $R$ about any point in the plane is equal to the algebraic sum of the moments of $P$ and $Q$ about the same point.

This theorem may be applied to any coplanar force system whether concurrent, or non-concurrent or parallel force system.

## 3.5 PRINCIPLE OF MOMENTS

When a body acted upon by a system of coplanar forces, is in equilibrium, then the algebraic sum of the moments of all the forces about any point in their plane, is equal to zero. Mathematically, $\Sigma M = 0$.

In other words, the sum of all the clockwise moments must be equal to the sum of all the anticlockwise moments, *i.e.*

Sum of all clockwise moments = Sum of all anticlockwise moments

**Note:** The principle of moments finds wide application in determining the reactions of the beams and the forces in the various members of the frame.

## 3.6 CONDITIONS OF EQUILIBRIUM FOR COPLANAR NON-CONCURRENT FORCES

We have already discussed in the previous chapter (Sec. 2.11), the conditions of equilibrium for coplanar concurrent forces. When a number of *coplanar non-concurrent forces are acting upon a body, and the body moves in any direction, then the resultant forces causing motion of a body, must be equal to zero. In other words, the algebraic sum of the horizontal and vertical components of all the forces must be equal to zero, *i.e.* $\Sigma H = 0$; and $\Sigma V = 0$.

Similarly, when the body rotates about itself, without moving, then the body is subjected to a single resultant couple without any resultant force. In such a case, the moment of the couple causing the rotation of the body must be equal to zero, *i.e.* $\Sigma M = 0$.

**Notes: 1.** When the body moves in any direction and at the same time it rotates about itself, then for equilibrium of the body,

$$\Sigma H = 0; \quad \Sigma V = 0; \quad \text{and} \quad \Sigma M = 0 \text{ (as discussed above)}$$

**2.** When the body is completely at rest, then there is neither a resultant force nor a couple acting on it. In such cases, for equilibrium of the body,

$$\Sigma H = 0; \quad \Sigma V = 0; \quad \text{and} \quad \Sigma M = 0$$

**Example 3.1** *A force of 150 N is acting at point A, as shown in Fig. 3.5. Find the moment of this force about point O.*

***Solution*** Given : $P = 150$ N ; $OA = 2$ m

We know that moment of the force about point $O$

$$= P \times OB = 150 \times OA \sin 45°$$

$$= 150 \times 2 \times 0.707$$

$$= 212.1 \text{ N-m (clockwise)} \textbf{ Ans.}$$

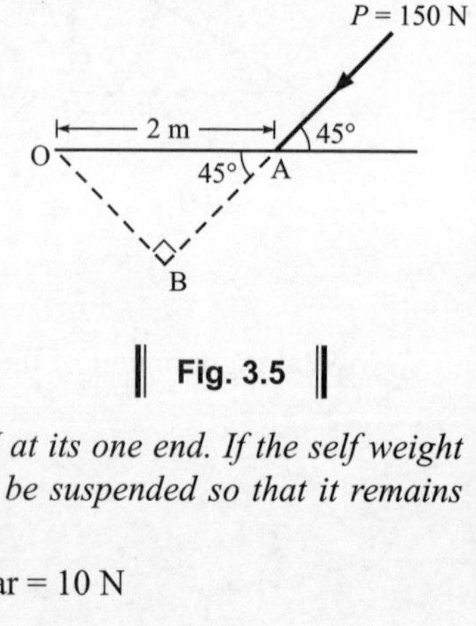


Fig. 3.5

**Example 3.2** *A uniform bar of 8 m length, carries a load of 50 N at its one end. If the self weight of the bar is 10 N, find the position of the point at which the bar should be suspended so that it remains horizontal.*

***Solution*** Given : $AB = 8$ m ; Load at $B = 50$ N ; Self weight of bar = 10 N

The bar is shown in Fig. 3.6.

The self weight of the bar will act at the centre of the bar, *i.e.*

$$AC = CB = 4 \text{ m}$$

Let $D$ is the point where it should be hanged to keep the bar horizontal.

Now taking moments about point $D$ and equating clockwise moments equal to anticlockwise moments, we have

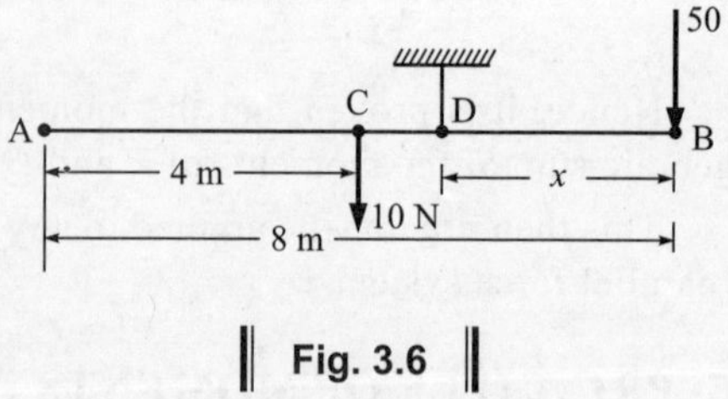


Fig. 3.6

* We have already discussed that coplanar non-concurrent forces do not meet at one point.

$$50 \times x = 10 \times CD = 10\,(CB - BD)$$
$$= 10\,(4 - x) = 40 - 10x$$

or $$50x + 10x = 40$$

∴ $$x = \frac{40}{60} = \frac{2}{3} = 0.667 \text{ m } \textbf{Ans.}$$

**Example 3.3** *A simply supported beam AB of span 6 m is carrying point loads of 5 kN, 10 kN and 15 kN at 1.5 m, 3 m and 4.5 m, respectively, from the support A. Calculate the reactions at the supports A and B.*

***Solution*** Given : $AB = 6$ m ; Load at $C = 5$ kN ; Load at $D = 10$ kN ; Load at $E = 15$ kN

The *simply supported beam is shown in Fig. 3.7.

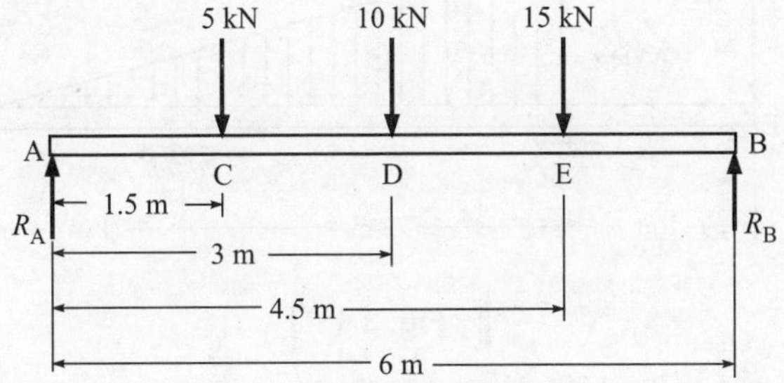


**Fig. 3.7**

Let $R_A$ and $R_B$ = Reactions at supports $A$ and $B$, respectively.

Taking moments about $B$ and equating clockwise moments equal to anticlockwise moments, we have

$$R_A \times 6 = 5 \times 4.5 + 10 \times 3 + 15 \times 1.5 = 22.5 + 30 + 22.5 = 75$$

∴ $$R_A = \frac{75}{6} = 12.5 \text{ kN } \textbf{Ans.}$$

We know that $$R_A + R_B = 5 + 10 + 15 = 30 \text{ kN}$$

∴ $$R_B = 30 - R_A = 30 - 12.5 = 17.5 \text{ kN } \textbf{Ans.}$$

**Example 3.4** *A horizontal beam of length 3 m is loaded at C and D, as shown in Fig. 3.8. Find the reactions at A and B.*

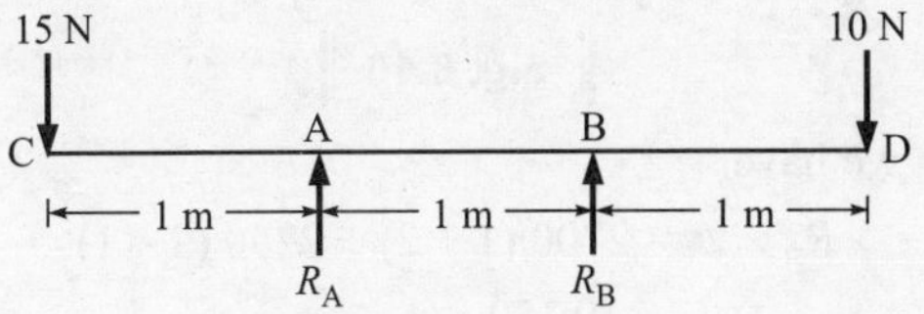


**Fig. 3.8**

***Solution*** Given : Load at $C = 15$ N ; Load at $D = 10$ N ; $CA = AB = BD = 1$ m

Let $R_A$ and $R_B$ = Reactions at $A$ and $B$, respectively.

*A beam which has supports at its both ends is called *simply supported beam.*

Taking moments about $B$ and equating clockwise moments equal to anticlockwise moments, we have

$$R_A \times 1 + 10 \times 1 = 15 \times 2$$

$\therefore$ $$R_A = 30 - 10 = 20 \text{ N} \quad \textbf{Ans.}$$

We know that $$R_A + R_B = 15 + 10 = 25$$

$\therefore$ $$R_B = 25 - R_A = 25 - 20 = 5 \text{ N} \quad \textbf{Ans.}$$

**Example 3.5** *Calculate the reactions $R_A$ and $R_B$ for the beam loaded, as shown in Fig. 3.9. Neglect the weight of the beam.*

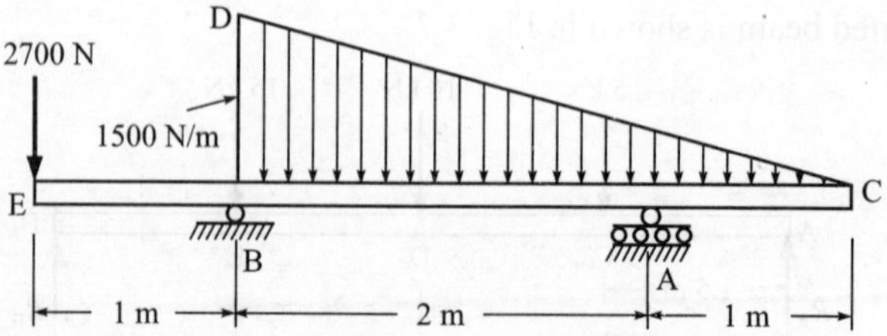


Fig. 3.9

***Solution*** Given : Load at $B$ = 1500 N/m

We know that load on $BCD$ = Area of triangle $BCD$

$$= \frac{1}{2} BD \times BC = \frac{1}{2} \times 1500(2+1) = 2250 \text{ kN}$$

This load will act at the centre of gravity ($G$) of the triangle $BCD$ *i.e.* at a distance of $\frac{BC}{3} = \frac{2+1}{3} = 1$ m from $B$. Now the beam may be considered as shown in Fig. 3.10.

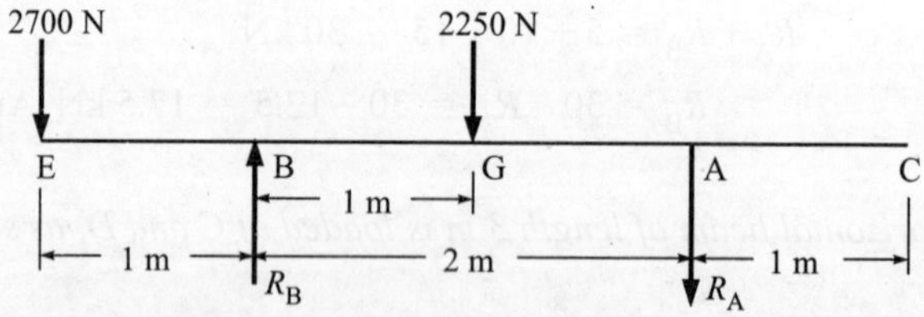


Fig. 3.10

Taking moments about $A$, we have

$$R_B \times 2 = 2700\,(1+2) + 2250\,(2-1) = 8100 + 2250 = 10\,350$$

$\therefore$ $$R_B = \frac{10\,350}{2} = 5175 \text{ N} \quad \textbf{Ans.}$$

We know that $$R_A + R_B = \text{Total load on the beam}$$

$$= 2700 + 2250 = 4950 \text{ N}$$

$\therefore$ $$R_A = 4950 - R_B = 4950 - 5175 = -225 \text{ N} \quad \textbf{Ans.}$$

The –ve sign shows that the reaction at $A$ ($R_A$) is downward, as shown in Fig. 3.10.

**Example 3.6** *A ladder is supported at A and B, as shown in Fig. 3.11. A vertical load W can have any position as defined by the distance AD = a, from the bottom. Neglecting friction and weight of the ladder, determine the magnitude of the reaction $R_B$ at B.*

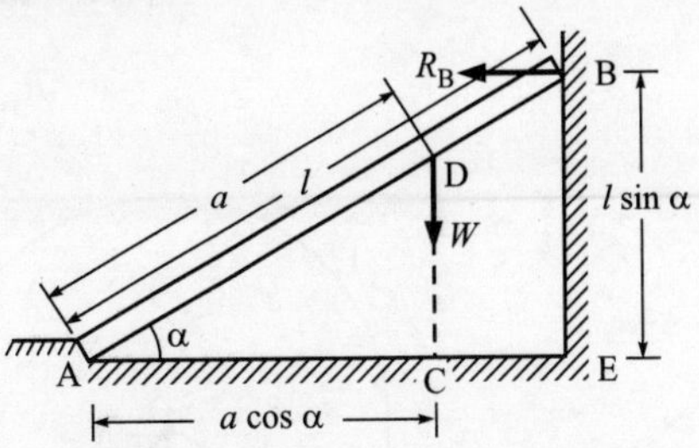


Fig. 3.11

***Solution*** Given : Load at $D = W$ ; $AB = l$ ; $AD = a$

Let $R_B$ be the magnitude of the reaction at $B$.

Taking moments about $A$, we have

$$W \times AC = R_B \times BE$$

or

$$W \times AD \cos \alpha = R_B \times AB \sin \alpha$$

$$W \times a \cos \alpha = R_B \times l \sin \alpha$$

$$\therefore \quad R_B = \frac{Wa \cos \alpha}{l \sin \alpha} = \frac{Wa}{l} \cot \alpha \quad \textbf{Ans.}$$

**Example 3.7** *Two identical cylinders, each weighing 500 N, are placed in a trough, as shown in Fig. 3.12. Determine the reaction developed at contact points A, B, C and D. Assume that all points of contact are smooth.*

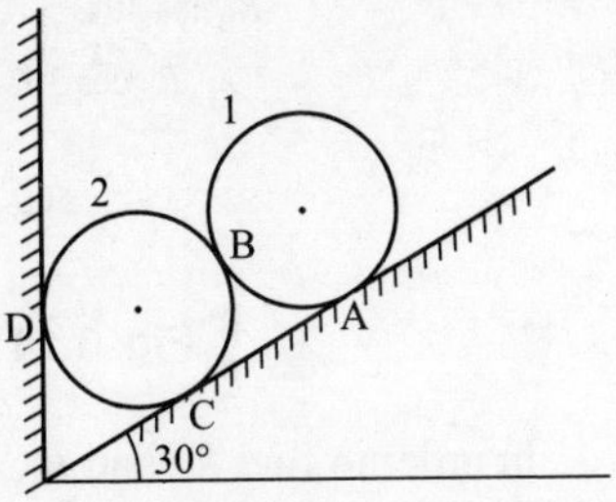


Fig. 3.12

***Solution*** Given : Weight of each cylinder = 500 N

Let $R_A$, $R_B$, $R_C$ and $R_D$ be the reactions at the contact points $A$, $B$, $C$ and $D$, respectively, as shown in Fig. 3.13.

Now consider both the identical cylinders as a single unit. This unit is in equilibrium under the action of their weights along with the reactions $R_A$, $R_B$, $R_C$ and $R_D$ at the contact points $A$, $B$, $C$ and $D$, respectively, as shown in Fig. 3.13.

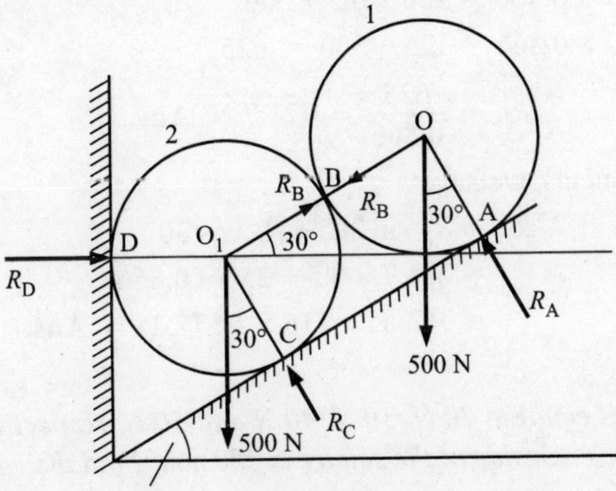


Fig. 3.13

In order to find $R_A$ and $R_B$, draw the free body diagram of cylinder 1, as shown in Fig. 3.14.

Taking moments about $O_1$, we have

$$R_A \times OO_1 = 500 \times O_1O_2$$

$$R_A \times OO_1 = 500 \times OO_1 \cos 30°$$

$$R_A = 500 \times \cos 30°$$

$$= 500 \times 0.866 = 433 \text{ N} \quad \textbf{Ans.}$$

Now equating horizontal components, we have

$$R_B \cos 30° = R_A \sin 30°$$
$$R_B \times 0.866 = 443 \times 0.5 = 221.5$$
$$\therefore \quad R_B = \frac{221.5}{0.866} = 250 \text{ N } \textbf{Ans.}$$

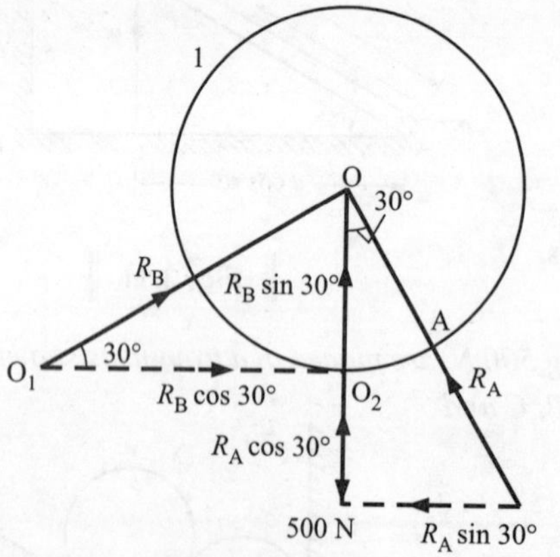


**Fig. 3.14**

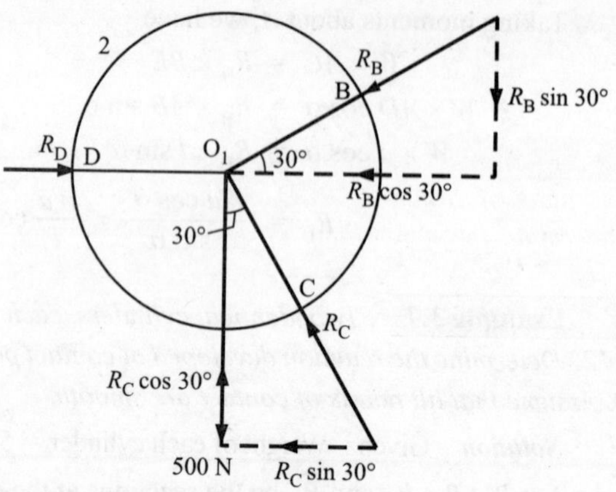


**Fig. 3.15**

In order to find $R_C$ and $R_D$, draw the free body diagram of cylinder 2, as shown in Fig. 3.15. Equating vertical components, we have

$$R_C \cos 30° = R_B \sin 30° + 500$$
$$R_C \times 0.866 = 250 \times 0.5 + 500$$
$$R_C \times 0.866 = 125 + 500 = 625$$
$$\therefore \quad R_C = \frac{625}{0.866} = 721.7 \text{ N } \textbf{Ans.}$$

Equating horizontal components, we have

$$R_D = R_C \sin 30° + R_B \cos 30°$$
$$= 721.7 \times 0.5 + 250 \times 0.866$$
$$= 360.85 + 216.5 = 577.35 \text{ N } \textbf{Ans.}$$

**Example 3.8** *Four forces equal to 20 N, 30 N, 40 N and 50 N, respectively, act along the side of a square, as shown in Fig. 3.16. The each side of the square is 200 mm. Find the value of the moments of these forces about*

*1. the centre of the square; and 2. the point of intersection of the forces 20 N and 50 N.*

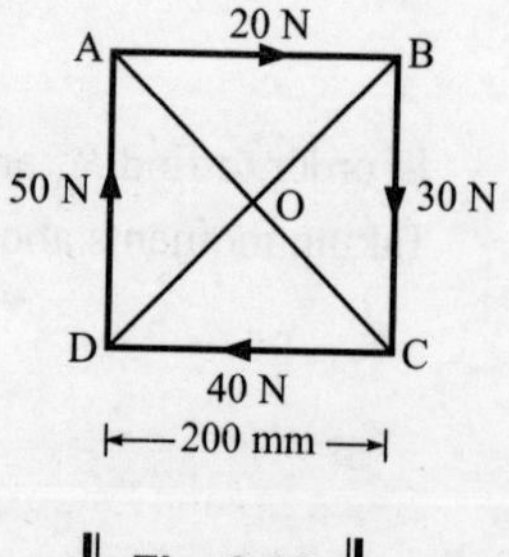


**Fig. 3.16**

***Solution*** Given : Force on $AB = 20$ N ; Force on $BC = 30$ N ; Force on $CD = 40$ N ; Force on $DA = 50$ N

**1. *Moment of the forces about the centre of the square***

We know that distance of each side of the square from the centre $O$

$$= \frac{200}{2} = 100 \text{ mm}$$

Now taking moments about the centre of the square,

$$M_O = 20 \times 10 + 30 \times 10 + 40 \times 10 + 50 \times 10 = 1400 \text{ N-mm} \quad \textbf{Ans.}$$

**2. *Moment of the forces at the point of intersection of the forces 20 N and 50 N***

We see that the forces 20 N and 50 N intersect at point *A*. Therefore, taking moment about point *A*,

$$M_A = 30 \times 20 + 40 \times 20 = 1400 \text{ N-mm} \quad \textbf{Ans.}$$

**Note:** Since the line of action of the forces 20 N and 50 N passes through point *A*, therefore the moment of these forces about point *A* is zero.

**Example 3.9** *Four forces equal to 10 N, 20 N, 30 N and 40 N are respectively acting along the four sides of a square ABCD, as shown in Fig. 3.17, taken in order. Determine the magnitude, direction and position of the resultant force.*

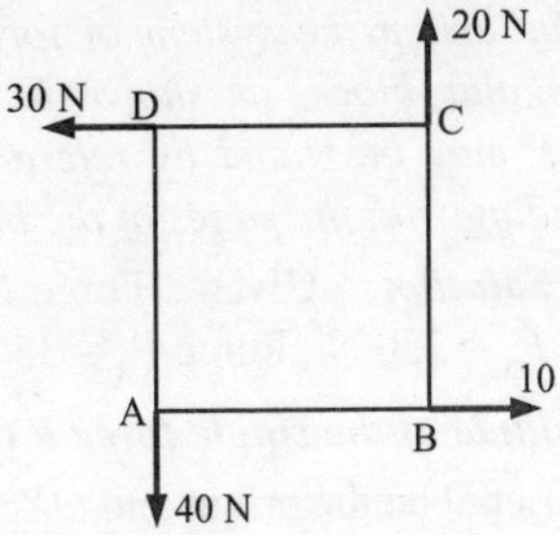


**Fig. 3.17**

***Solution*** Given : Force on $AB = 10$ N ; Force on $BC = 20$ N ; Force on $CD = 30$ N ; Force on $DA = 40$ N

***Magnitude of the resultant force***

Resolving all the forces horizontally,

$$\Sigma H = 10 - 30 = -20 \text{ N}$$

and resolving all the forces vertically,

$$\Sigma V = 20 - 40 = -20 \text{ N}$$

We know that the resultant force, $R = \sqrt{(\Sigma H)^2 + (\Sigma V)^2}$

$$= \sqrt{(-20)^2 + (-20)^2} = 2\sqrt{20} \text{ N} \quad \textbf{Ans.}$$

***Direction of the resultant force***

Let $\alpha$ = Angle of inclination of the resultant force with the horizontal.

We know that $\tan\alpha = \dfrac{\Sigma V}{\Sigma H} = \dfrac{-2}{-2} = 1$

$\therefore$ $\alpha = \tan^{-1} 1 = 45°$

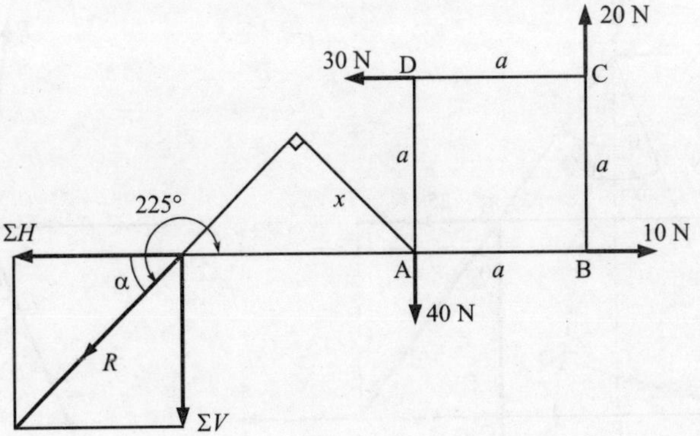


**Fig. 3.18**

Since $\Sigma H$ and $\Sigma V$ are both *negative*, therefore the resultant force ($R$) lies in the third quadrant (*i.e.* between 180° and 270°). Thus actual angle of the resultant force, as shown in Fig. 3.18, is 180° + 45° = 225°. **Ans.**

***Position of the resultant force***

Let

$a$ = Side of the square; and

$x$ = Perpendicular distance between $A$ and the line of application of the resultant force (see Fig. 3.18).

Now taking moments about $A$ and equating clockwise moments and anticlockwise moments, we have

$$2\sqrt{20} \times x = 20a + 30a = 50a$$

$$\therefore \quad x = \frac{50a}{2\sqrt{20}} = \frac{25a}{\sqrt{20}} \quad \textbf{Ans.}$$

**Example 3.10** *Determine the magnitude, direction and position of a single force P, which keeps in equilibrium the system of forces acting at the corners of a rectangular block, as shown in Fig. 3.19. The position of the force P may be stated by reference to axes with origin O and coinciding with the edges of the block.*

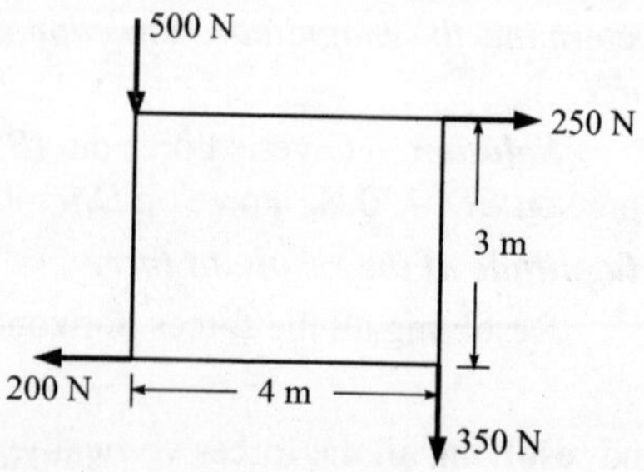


**Fig. 3.19**

***Solution*** Given : Force $P_1$ = 500 N ; Force $P_2$ = 200 N ; Force $P_3$ = 250 N ; Force $P_4$ = 350 N

***Magnitude of the single force P (or resultant force P)***

Let $O$ be the origin and $OX$ and $OY$ be the reference axis, as shown in Fig. 3.20.

The forces 500 N and 200 N form a system of concurrent forces and their line of action intersect at $O$. The resultant of these forces is

$$R_1 = \sqrt{P_1^2 + P_2^2} = \sqrt{(500)^2 + (200)^2} = 538.5 \text{ N}$$

and its inclination with $OA$ is given by

$$\tan \theta_1 = \frac{200}{500} = 0.4$$

$$\therefore \quad \theta_1 = \tan^{-1}(0.4) = 21.8°$$

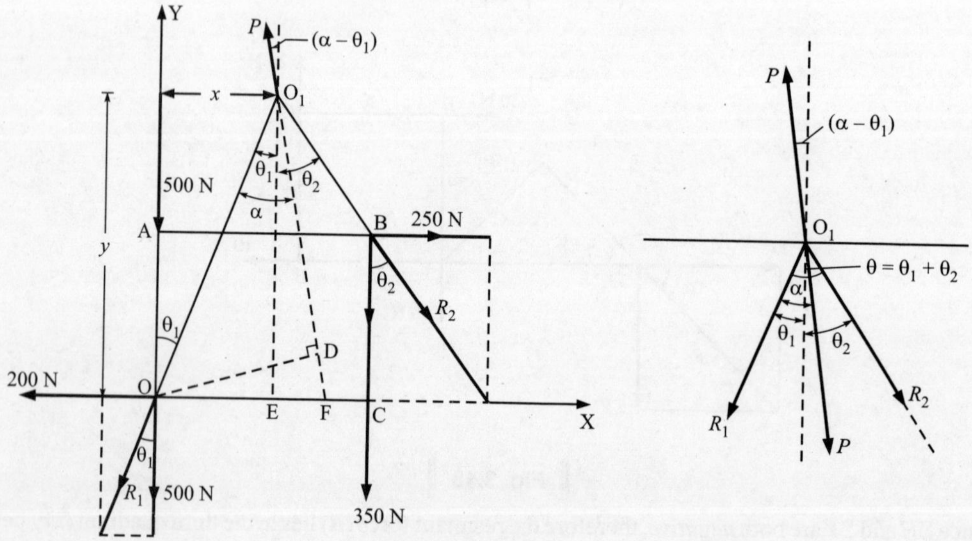


**Fig. 3.20**

Similarly, forces 250 N and 350 N form a system of concurrent forces and their line of action intersect at $B$. The resultant of these forces is given by

$$R_2 = \sqrt{P_3^2 + P_4^2} = \sqrt{(250)^2 + (350)^2} = 430.1 \text{ N}$$

and its inclination with $BC$ is given by

$$\tan \theta_2 = \frac{250}{350} = 0.714$$

$$\therefore \quad \theta_2 = \tan^{-1}(0.714) = 35.53°$$

These two forces *i.e.* $R_1$ and $R_2$ intersect at point $O_1$, as shown in Fig. 3.20, and make an angle $\theta = \theta_1 + \theta_2 = 21.8° + 35.53° = 57.33°$.

We know that, in such cases, the resultant force or single force,

$$P = \sqrt{R_1^2 + R_2^2 + 2R_1R_2 \cos \theta}$$

$$= \sqrt{(538.5)^2 + (430.1)^2 + 2 \times 538.5 \times 430.1 \cos 57.33°}$$

$$= \sqrt{289\,982 + 184\,986 + 250\,045} = 815.48 \text{ N} \quad \textbf{Ans.}$$

***Direction of the single force P (or resultant force P)***

Let $\quad \alpha$ = Angle of inclination which the single force $P$ makes with $R_1$

We know that

$$\tan \alpha = \frac{R_2 \sin \theta}{R_1 + R_2 \cos \theta} = \frac{430.1 \times \sin 57.33°}{538.5 + 430.1 \cos 57.33°}$$

$$= \frac{430.1 \times 0.8418}{538.5 + 430.1 \times 0.5398} = \frac{362}{538.5 + 232.17}$$

$$= \frac{362}{770.67} = 0.4697$$

$$\therefore \quad \alpha = \tan^{-1}(0.4697) = 25.16°$$

In other words, the single force $P$ makes an angle of 25.16° – 21.8° = 3.36° with the vertical in the anticlockwise direction to $O_1Y$.

***Position of the single force P***

The position of the single force ($P$) is obtained by equating the clockwise moments and anticlockwise moments about the origin $O$. The force $P$ is produced to meet the line $OX$ at $F$. Draw $OD$ perpendicular on line $O_1F$, as shown in Fig. 3.20.

Taking moment of all the forces about $O$ and equating clockwise moments equal to anticlockwise moment, we have

$$350 \times 4 + 250 \times 3 = P \times OD = 851.48 \times OD$$

$$\therefore \quad OD = \frac{350 \times 4 + 250 \times 3}{851.48} = 2.525 \text{ m}$$

From right angled triangle $ODO_1$,

$$\sin \alpha = \frac{OD}{OO_1}$$

$$\therefore \quad OO_1 = \frac{OD}{\sin \alpha} = \frac{2.525}{\sin 25.16°} = \frac{2.525}{0.4251} = 5.94 \text{ m}$$

Let $x$ and $y$ be the co-ordinates of the force $P$ with reference to the axes with origin $O$. From the right angled triangle $OEO_1$.

$$x = OE = OO_1 \sin \theta_1 = 5.94 \sin 21.8°$$
$$= 5.94 \times 0.371 = 2.2 \text{ m } \textbf{Ans.}$$

and
$$y = O_1E = OE \cos \theta_1 = 5.94 \cos 21.8°$$
$$= 5.94 \times 0.928 = 5.51 \text{ m } \textbf{Ans.}$$

**Example 3.11** *A rod AB is 1 m long and weighs 20 N. It is hinged at its lower end A to a fixed point. The rod rests in a position inclined at 60° to the horizontal by a string at B in a vertical position, as shown in Fig. 3.21. Find the tension in the string and reaction at point A.*

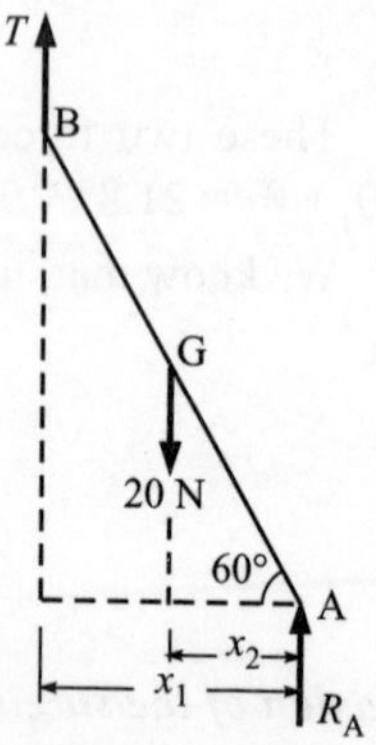


Fig. 3.21

***Solution*** Given : $AB = 1$ m ; $W = 20$ N ; $\theta = 60°$

***Tension in the string***

Let $T$ = Tension in the string at $B$.

Since the weight of rod $AB$ will act at its centre $G$, therefore

$$AG = BG = \frac{1}{2} = 0.5 \text{ m}$$

Now taking moments about point $A$ and equating clockwise moment equal to anticlockwise moment, we have

$$T \times x_1 = 20 \times x_2$$
$$T \times AB \cos 60° = 20 \times AG \cos 60°$$

or
$$T \times 1 \cos 60° = 20 \times 0.5 \cos 60°$$
$$T \times 0.5 = 20 \times 0.5 \times 0.5$$

$\therefore$
$$T = 10 \text{ N } \textbf{Ans.}$$

***Reaction at point A***

Let $R_A$ = Reaction at point $A$.

Equating upward forces equal to downward forces for equilibrium, we have

$$R_A + T = 20$$

or
$$R_A = 20 - T = 20 - 10 = 10 \text{ N } \textbf{Ans.}$$

**Example 3.12** *A man of weight 675 N stands on the middle rug of a ladder of weight 225 N, as shown in Fig. 3.22. The floor and the wall being perfectly smooth, the slipping is prevented by a string DE. Find the tension in the string and also the reactions $R_A$ and $R_B$ at A and B.*

Fig. 3.22

***Solution*** Given : $W_M = 675$ N ; $W_L = 225$ N

The arrangement of the system is shown in Fig. 3.23.

Let $T$ = Tension in the string $DE$, and

$R_A$ and $R_B$ = Reactions at $A$ and $B$, respectively.

Total weight acting at $C$,

$$W = W_1 + W_2 = 675 + 225 = 900 \text{ N}$$

Since the whole system is in equilibrium, therefore equating the sum of all the horizontal ($\Sigma H$) and vertical components ($\Sigma V$) equal to zero.

Equating $\Sigma H = 0$, we have

$$R_B = T \cos 30° = T \times \frac{\sqrt{3}}{2} = 0.866\, T \quad ...(i)$$

and equating $\Sigma V = 0$, we have

$$R_A = 900 + T \sin 30° = 900 + T \times 0.5 = 900 + 0.5T \quad ...(ii)$$

Draw perpendiculars *AF* and *CH* on *ED* and *AE*, respectively. We see that

$$AF = x = 1.8 \sin 30° = 1.8 \times 0.5 = 0.9 \text{ m}$$

and
$$AH = \frac{AE}{2} = \frac{1.8}{2} = 0.9 \text{ m}$$

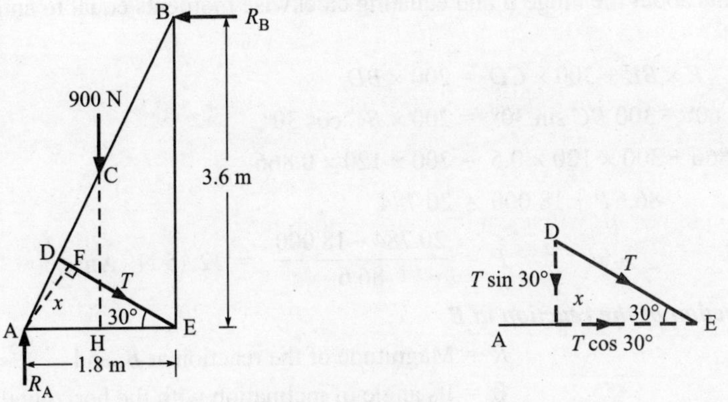


**Fig. 3.23**

Now taking moments about *A*,

$$R_B \times 3.6 = 900 \times AH + T \times x$$

$$0.866\,T \times 3.6 = 900 \times 0.9 + T \times 0.9$$

$$3.1176T - 0.9T = 810 \quad \text{or} \quad 2.2176\,T = 810$$

$$\therefore \quad T = \frac{810}{2.2176} = 365.26 \text{ N} \quad \textbf{Ans.}$$

Substituting the value of *T* in equations (*i*) and (*ii*), we have

$$R_B = 0.866\,T = 0.866 \times 365.26 = 316.3 \text{ N} \quad \textbf{Ans.}$$

and
$$R_A = 900 + 0.5\,T = 900 \times 0.5 \times 365.26 = 1082.6 \text{ N} \quad \textbf{Ans.}$$

**Example 3.13** *A lever ABC of a component of a machine is hinged at B and is subjected to a system of coplanar forces, as shown in Fig. 3.24.*

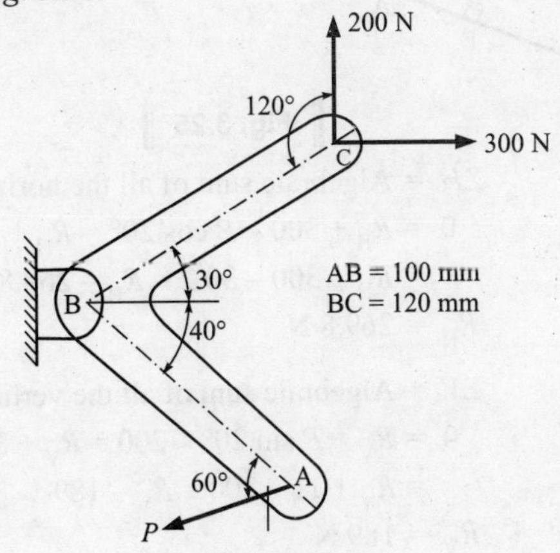


**Fig. 3.24**

*Neglecting friction, find the magnitude of the force P to keep the lever in equilibrium. Also determine the magnitude and direction of the reaction at B.*

***Solution*** Given : Force at $C$ = 200 N (vertically upwards) and 300 N (horizontally rightward) ; Force at $A = P$ ; $AB$ = 100 mm ; $BC$ = 120 mm

The line diagram of the lever is shown in Fig. 3.25.

***Magnitude of the force P to keep the lever in equilibrium***

Taking moments about the hinge $B$ and equating clockwise moments equal to anticlockwise moments, we have

$$P \times BE + 300 \times CD = 200 \times BD$$

$$P \times AB \sin 60° + 300\, BC \sin 30° = 200 \times BC \cos 30°$$

$$P \times 100 \times 0.866 + 300 \times 120 \times 0.5 = 200 \times 120 \times 0.866$$

$$86.6P + 18\,000 = 20\,784$$

$$\therefore \quad P = \frac{20\,784 - 18\,000}{86.6} = 32.15 \text{ N} \quad \textbf{Ans.}$$

***Magnitude and direction of the reaction at B***

Let $R$ = Magnitude of the reaction at $B$, and

$\theta$ = Its angle of inclination with the horizontal.

Resolve $R$ into horizontal and vertical components. Let $R_H$ be the horizontal component and $R_V$ be the vertical component. Also resolve the force $P$ into horizontal component (*i.e.* $P \cos 20°$) and vertical component ($P \sin 20°$). For the equilibrium of the lever $ABC$, we know that $\Sigma H = 0$ and $\Sigma V = 0$ (refer Fig. 3.25).

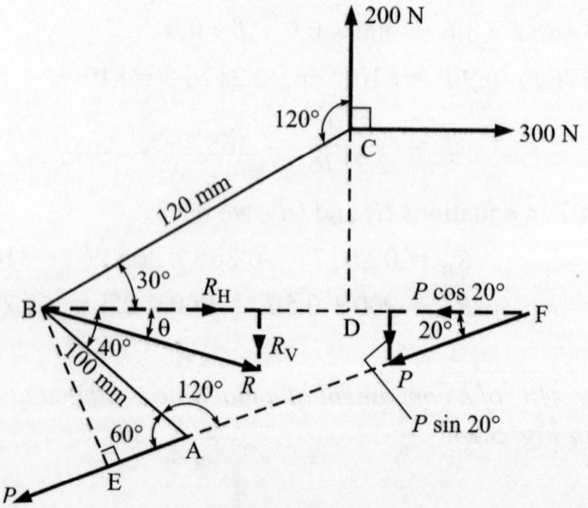


**Fig. 3.25**

$\therefore$ $\Sigma H$ = Algebraic sum of all the horizontal forces

or $0 = R_H + 300 - P \cos 20° = R_H + 300 - 32.15 \times 0.9397$

$= R_H + 300 - 30.2 = R_H - 269.8$

$\therefore$ $R_H = 269.8$ N

Similarly, $\Sigma V$ = Algebraic sum of all the vertical forces

$0 = R_V + P \sin 20° - 200 = R_V + 32.15 \times 0.342 - 200$

$= R_V + 11 - 200 = R_V - 189$

$\therefore$ $R_V = 189$ N

We know that $$R = \sqrt{(R_H)^2 + (R_V)^2} = \sqrt{(269.8)^2 + (189)^2}$$

$$= \sqrt{72\,792 \quad 35\,721} = 329.4 \text{ N} \textbf{ Ans.}$$

and $$\tan\theta = \frac{R_V}{R_H} = \frac{189}{269.8} = 0.7005$$

∴ $$\theta = \tan^{-1}(0.7005) = 35.01° \textbf{ Ans.}$$

**Example 3.14** *Two forces P and Q are applied to the corners of a 100 mm² square plate, as shown in Fig. 3.26.*

*Find forces P, Q and angle θ if the resultant of two forces has a magnitude of 140 N, passing through the centroid of the plate and making an angle of 30° with positive X-axis.*

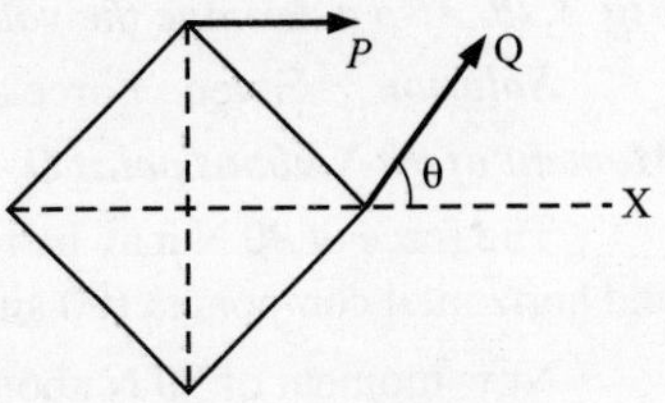


Fig. 3.26

***Solution*** Given : $R = 140$ N ; $\alpha = 30°$

Let angle of inclination of force $Q$ with $X$-axis = $\theta$

Refer Fig. 3.27.

Resolving $Q$ and $R$ into horizontal and vertical components and equating the same, we have

$$P + Q\cos\theta = R\cos 30° = 140 \times 0.866 = 121.24 \text{ N} \quad ...(i)$$

and $$Q\sin\theta = R\sin 30° = 140 \times 0.5 = 70 \text{ N} \quad ...(ii)$$

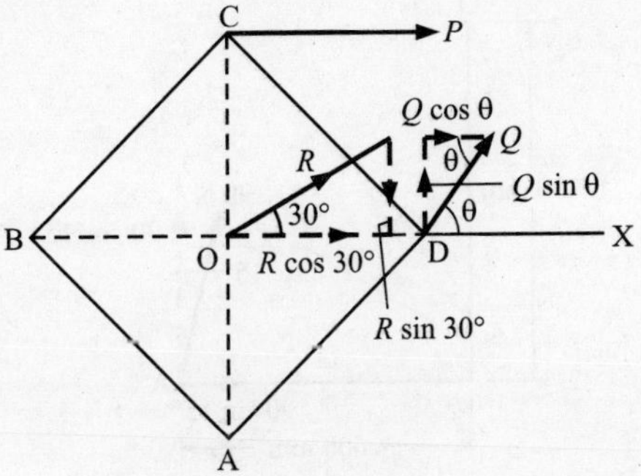


Fig. 3.27

Let $a$ be the side of the square. Therefore diagonal of the square,

$$AC = BD = \sqrt{a^2 + a^2} = \sqrt{2a^2} = \sqrt{2}a$$

Now taking moment about the centroid $O$ of the square, we have

$$P \times CO = Q\sin\theta \times DO$$

$$P \times \frac{\sqrt{2}a}{2} = Q\sin\theta \times \frac{\sqrt{2}a}{2}$$

or $$P = Q\sin\theta = 70 \text{ N} \textbf{ Ans.} \quad ...[\text{From equation } (ii)]$$

Now from equation ($i$), $P + Q\cos\theta = 121.24$

or $$Q\cos\theta = 121.24 - P = 121.24 - 70 = 51.24 \quad ...(iii)$$

Dividing equation ($ii$) by equation ($iii$), we have

$$\frac{Q\sin\theta}{Q\cos\theta} = \frac{70}{51.24}$$

or $\tan\theta = 1.3661$

∴ $\theta = \tan^{-1}(1.3661) = 53.79°$ **Ans.**

Substituting this value of θ in equation (*iii*),

$$Q\cos 53.79° = 51.24$$

$$Q \times 0.59 = 51.24$$

∴ $$Q = \frac{51.24}{0.59} = 86.85 \text{ N} \textbf{ Ans.}$$

**Example 3.15** *Calculate the moment of 90 N about point O for the condition θ = 15°, as shown in Fig. 3.28. Also determine the value of θ for which the moment about O is zero.*

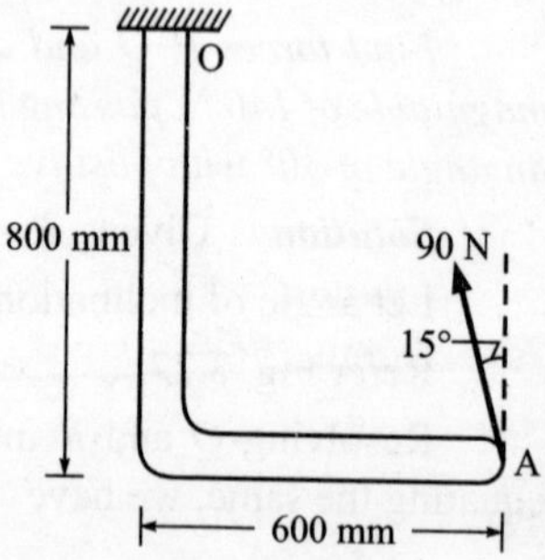

Fig. 3.28

**Solution** Given : Force at $A = 90$ N

***Moment of 90 N about point O***

The force of 90 N may be resolved into vertical component (90 cos 15°) and horizontal component (90 sin 15°), as shown in Fig. 3.29.

Now moment of 90 N about point *O*,

$$M_O = 90\cos 15° \times 600 - 90\sin 15° \times 800$$

$$= 90 \times 0.966 \times 600 - 90 \times 0.2588 \times 800$$

$$= 52\,164 - 18\,634$$

$$= 33\,530 \text{ N-mm} \textbf{ Ans.}$$

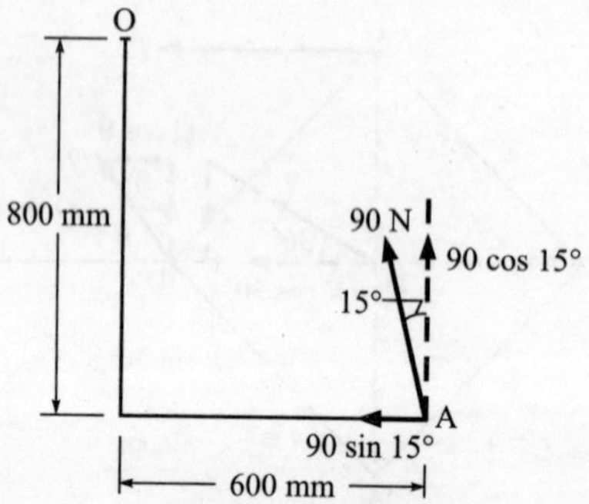

Fig. 3.29

***Value of θ for which the moment about O is zero***

We know that the moment of 90 N about *O* is

$$M_O = 90\cos\theta \times 600 - 90\sin\theta \times 800$$

For $M_O$ = zero, we have

$$0 = 90\cos\theta \times 600 - 90\sin\theta \times 800$$

or $$90\sin\theta \times 800 = 90\cos\theta \times 600$$

$$\frac{\sin}{\cos} = \frac{90 \times 600}{90 \times 800} = 0.75$$

or $\tan\theta = 0.75$

∴ $\theta = \tan^{-1}(0.75) = 36.87°$ **Ans.**

**Example 3.16** *Calculate the force P on the crank, as shown in Fig. 3.30, for equilibrium. Find the reaction at the hinge.*

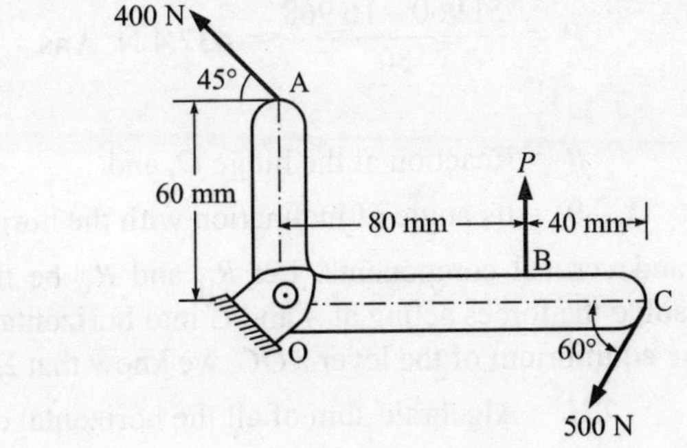


**Fig. 3.30**

***Solution*** Given : Force at $A$ = 400 N ; Force at $B = P$ ; Force at $C$ = 500 N ; $OA$ = 60 mm ; $OB$ = 80 mm ; $BC$ = 40 mm

***Force P on the crank***

The line diagram of the crank is shown in Fig. 3.31.

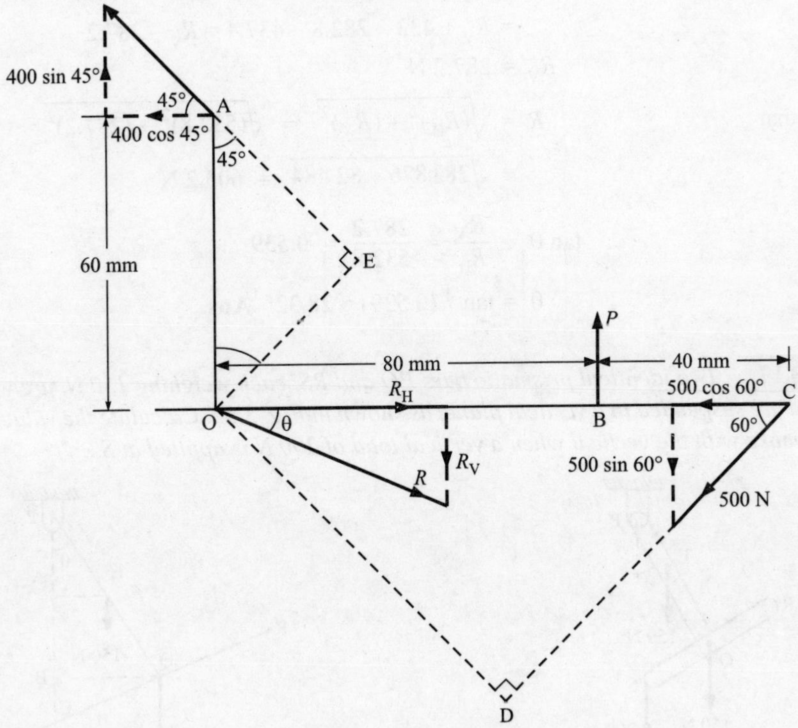


**Fig. 3.31**

Taking moments about the hinge $O$ and equating clockwise moments equal to anticlockwise moments, we have

$$500 \times OD = P \times OB + 400 \times OE$$

$$500 \times OC \sin 60° = P \times 80 + 400 \times OA \sin 45°$$

$$500\,(80 + 40)\,0.866 = P \times 80 + 400 \times 60 \times 0.707$$

$$51\,960 = P \times 80 + 16\,968$$

$$\therefore \quad P = \frac{51\,960 - 16\,968}{80} = 437.4 \text{ N} \quad \textbf{Ans.}$$

***Reaction at the hinge***

Let $R$ = Reaction at the hinge $O$, and

$\theta$ = Its angle of inclination with the horizontal.

Resolve $R$ into horizontal and vertical components. Let $R_H$ and $R_V$ be the horizontal and vertical components, respectively. Also resolve the forces acting at $A$ and $C$ into horizontal and vertical components, as shown in Fig. 3.31. Now, for the equilibrium of the lever $AOC$, we know that $\Sigma H = 0$ and $\Sigma V = 0$.

$\therefore$ $\Sigma H$ = Algebraic sum of all the horizontal components

$$0 = R_H - 400 \cos 45° - 500 \cos 60°$$

$$= R_H - 400 \times 0.707 - 500 \times 0.5 = R_H - 532.8$$

or $R_H = 532.8$ N

Similarly, $\Sigma V$ = Algebraic sum of all the vertical components

$$0 = R_V + 500 \sin 60° - 400 \sin 45° - P$$

$$= R_V + 500 \times 0.866 - 400 \times 0.707 - 437.4$$

$$= R_V + 433 - 282.8 - 437.4 = R_V - 287.2$$

or $R_V = 287.2$ N

We know that

$$R = \sqrt{(R_H)^2 + (R_V)^2} = \sqrt{(532.8)^2 + (287.2)^2}$$

$$= \sqrt{283\,876 + 82\,484} = 605.2 \text{ N}$$

and

$$\tan\theta = \frac{R_V}{R_H} = \frac{287.2}{532.8} = 0.539$$

$$\therefore \quad \theta = \tan^{-1}(0.539) = 28.32° \quad \textbf{Ans.}$$

**Example 3.17** *Two identical prismatic bars PQ and RS, each weighing 150 N, are welded together to form a Tee and are suspended in a vertical plane, as shown in Fig. 3.32. Calculate the value of angle θ that the bar PQ will make with the vertical when a vertical load of 200 N is applied at S.*

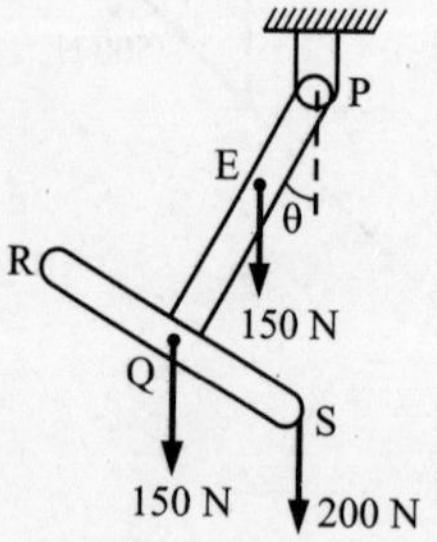


Fig. 3.32

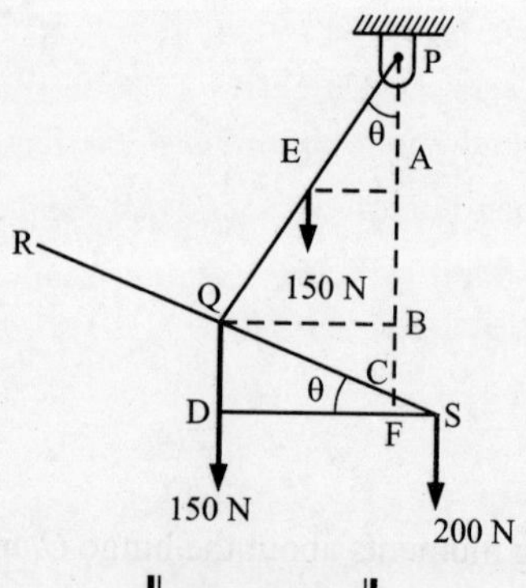


Fig. 3.33

***Solution*** Given : Force at the centre of gravity of bar $PQ$ (*i.e.* at $E$) = 150 N ; Force at the centre of gravity of bar $RS$ (*i.e.* at $Q$) = 150 N ; Force at $S$ = 200 N

The line diagram of the bar $PQ$ and $RS$ is shown in Fig. 3.33. Let $l$ be the length of each bar.

$$\therefore \quad PE = EQ = RQ = QS = \frac{l}{2}$$

***Angle θ that the bar PQ makes with the vertical***

Let $\angle QPC$ = Required angle θ.

From $E$ and $Q$, draw horizontal lines intersecting the vertical line $PC$ at $A$ and $B$, respectively. Similarly, draw horizontal line from $S$ intersecting the vertical line from $Q$ at $D$. In triangle $QPB$,

$$\angle QPB = \theta;\ \angle QBP = 90°$$

$$\therefore \quad \angle PQB = 180° - (90° + \theta) = 90° - \theta$$

Since $\angle PQB + \angle BQS = 90°$, therefore

$$\angle BQS = 90° - \angle PQB = 90° - [90° - \theta] = \theta$$

and $\angle QSD = \angle BQS = \theta$ ...(Alternate angles)

Now taking moments about $P$ and equating clockwise moments equal to anticlockwise moment for the equilibrium of the structure $PQRS$,

$$200 \times SF = 150 \times EA + 150 \times DF$$

$$200 \times (DS - DF) = 150 \times EP \sin\theta + 150 \times QB \quad ...(\because DF = QB)$$

$$200\left(\frac{l}{2}\cos\theta - l\sin\theta\right) = 150 \times \frac{l}{2} \times \sin\theta + 150 \times l\sin\theta \quad ...(\because DF = QB = l\sin\theta)$$

$$100\cos\theta - 200\sin\theta = 75\sin\theta + 150\sin\theta \quad ...(\text{Cancelling } l \text{ throughout})$$

or $$100\cos\theta = 425\sin\theta$$

$$\tan\theta = \frac{\sin\theta}{\cos\theta} = \frac{100}{425} = 0.2353$$

$$\therefore \quad \theta = \tan^{-1}(0.2353) = 13.24° \ \textbf{Ans.}$$

**Example 3.18** *A prismatic bar of length 2.1 m and weight 365 N rests within a hemispherical bowl of radius 0.7 m, as shown in Fig. 3.34. Determine angle α for position of equilibrium.*

***Solution*** Given : $AB(l) = 2.1$ m ; $W = 365$ N ; $r = 0.7$ m

The prismatic bar is in equilibrium under the action of the following three forces, as shown in Fig. 3.35.

1. Weight of the bar ($W = 365$ N) acting vertically downwards through the midpoint $G$ of the bar $AB$,
2. Reaction at $A$ ($R_A$) acting in the direction $AO$, and
3. Reaction at $C$ ($R_C$) acting in the direction $CO$, at right angles to the bar $AB$.

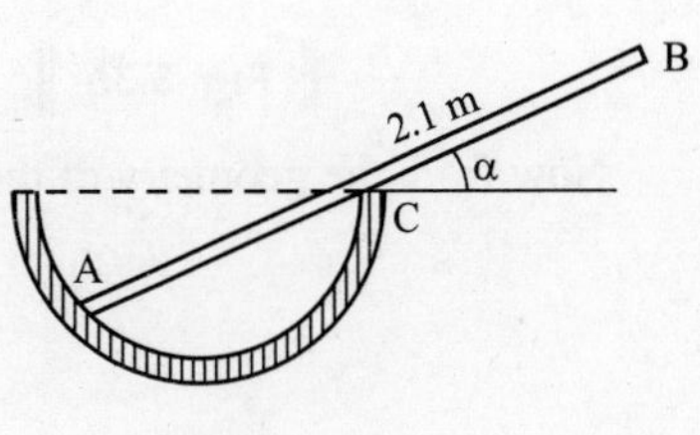


‖ Fig. 3.34 ‖

The line of action of these three forces intersect at point $O$, as shown in Fig. 3.35.

First of all, let us find the angles between these three forces as follows:

In a right angled triangle $OAH$,

$$\angle OAH = 2\alpha$$

$$\therefore \quad \angle AOH = 90° - 2\alpha$$

Now, in a right angled triangle $OAC$,

$$\angle OAC = \alpha$$

$$\therefore \quad \angle AOC = 90° - \alpha$$

From Fig. 3.36, we find that

Angle between $R_A$ and $R_C$ $= 90° - \alpha$

Angle between $R_A$ and $W = (180° + \alpha) - (90° - \alpha) = 90° + 2\alpha$

and angle between $R_C$ and $W$ $= 180° - \alpha$

Now applying Lami's theorem for the forces acting at $O$, we have

$$\frac{R_A}{\sin(180° - \alpha)} = \frac{R_C}{\sin(90° + 2\alpha)} = \frac{W}{\sin(90° - \alpha)}$$

$$\frac{R_A}{\sin \alpha} = \frac{R_C}{\cos 2\alpha} = \frac{365}{\cos \alpha}$$

$$\therefore \quad R_A = 365 \times \frac{\sin \alpha}{\cos \alpha} = 365 \tan \alpha \quad \text{...(i)}$$

and

$$R_C = 365 \times \frac{\cos 2\alpha}{\cos \alpha} \quad \text{...(ii)}$$

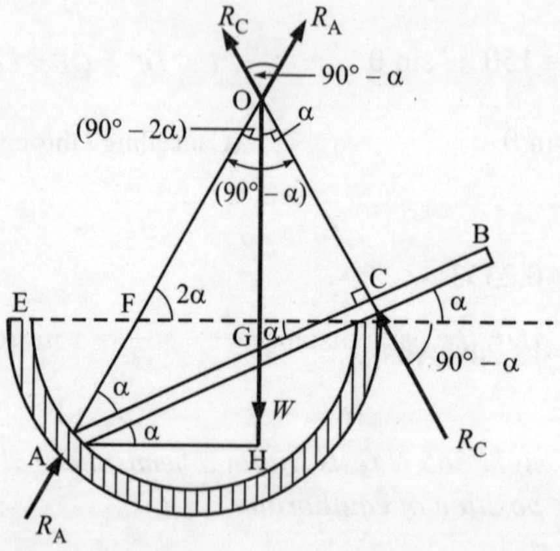


**Fig. 3.35**

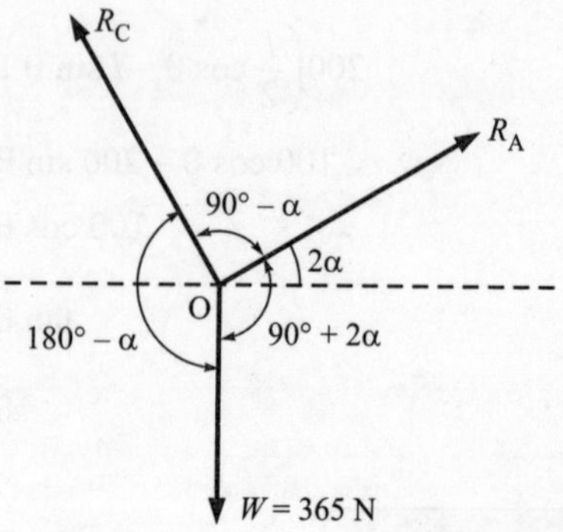


**Fig. 3.36**

Now from the geometry of the figure,

$$AC = OA \cos \alpha = 2r \cos \alpha$$

$$= 2 \times 0.7 \cos \alpha = 1.4 \cos \alpha \quad \text{...}(\because OA = EC = 2r)$$

and

$$AG = \frac{AB}{2} = \frac{2.1}{2} = 1.05 \text{ m}$$

Taking moment of the various forces about point $A$, we have

$$W \times AH = R_C \times A_C$$

$$W \times AG \cos \alpha = R_C \times 1.4 \cos \alpha$$

$$365 \times 1.05 \cos \alpha = 365 \times \frac{\cos 2\alpha}{\cos \alpha} \times 1.4 \cos \alpha \quad \text{...[From equation (ii)]}$$

$$383.25 \cos \alpha = 511 \cos 2\alpha = 511 (2 \cos^2 \alpha - 1) \quad \text{...}[\cos 2\alpha = 2 \cos^2 \alpha - 1]$$

$$\cos\alpha = \frac{511}{383.25}(2\cos^2\alpha - 1) = 2.66\cos^2\alpha - 1.33$$

or $$2.66\cos^2\alpha - \cos\alpha - 1.33 = 0$$

or $$\cos\alpha = \frac{1 \pm \sqrt{1^2 + 4 \times 2.66 \times 1.33}}{2 \times 2.66} = \frac{1 \pm \sqrt{1 + 14.15}}{5.32}$$

$$= \frac{1 \pm 3.89}{5.32} = \frac{4.89}{5.32} = 0.9196 \qquad \text{...(Taking + sign only)}$$

∴ $\alpha = 23.13°$ **Ans.**

**Example 3.19** *Determine the support reactions in the given cantilever beam, as shown in Fig. 3.37.*

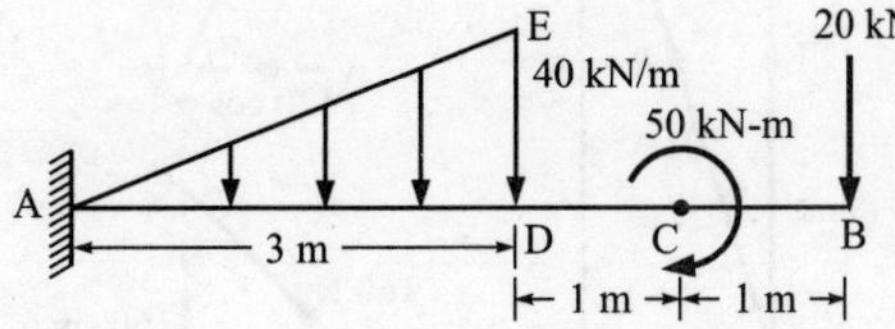


**Fig. 3.37**

***Solution*** Given : Load at $B = 20$ kN ; Moment at $C = 50$ kN-m (clockwise about $A$)

Since the load on beam length $AD$ is uniformly increasing from zero at $A$ to 40 kN/m at $D$, therefore the total load acting downwards.

$$= \frac{1}{2} \times AD \times DE = \frac{1}{2} \times 3 \times 40 = 60 \text{ kN}$$

This load will act at the centre of gravity of the triangle $ADE$, *i.e.*, at a distance $\frac{AD}{3} = \frac{3}{3} = 1$ m from $D$. The equivalent beam is now shown in Fig. 3.38.

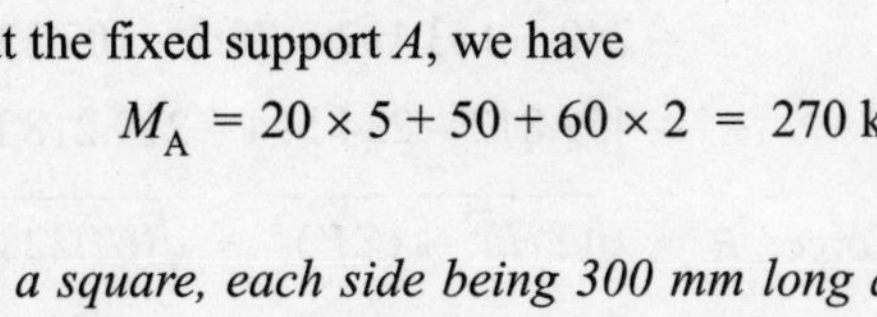


**Fig. 3.38**

Now taking moments about the fixed support $A$, we have

$$M_A = 20 \times 5 + 50 + 60 \times 2 = 270 \text{ kN-m (anticlockwise)} \quad \textbf{Ans.}$$

**Example 3.20** *ABCD is a square, each side being 300 mm long and E is the middle point of AB. Forces of 140 N ; 160 N ; 240 N ; 100 N ; 180 N and 120 N are acting on a body along the line AB, EC, BC, BD, CA and DE, respectively. Find the magnitude, direction and position of the resultant force.*

***Solution*** Given : $AB = BC = CD = DA = 300$ mm ; $AE = EB = 150$ mm ; Force on $AB = 140$ N ; Force on $EC = 160$ N ; Force on $BC = 240$ N ; Force on $BD = 100$ N ; Force on $CA = 180$ N ; Force on $DE = 120$ N

***Magnitude of the resultant force***

The various forces acting on a square are shown in Fig. 3.39. First of all, let us find the angle of inclination of $EC$ and $ED$ with the horizontal. Since $E$ is the mid-point of $AB$, therefore

$$\angle AED = \angle BEC = \alpha$$

We know that $\quad EC = ED = \sqrt{(BC)^2 + (EB)^2} = \sqrt{(300)^2 + (150)^2} = 335.4 \text{ mm}$

and $\quad \sin\alpha = \dfrac{BC}{EC} = \dfrac{300}{335.4} = 0.89\,445$

$\therefore \quad \alpha = \sin^{-1}(0.89\,445) = 63.44°$

and $\quad \cos\alpha = \cos 63.44° = 0.4471$

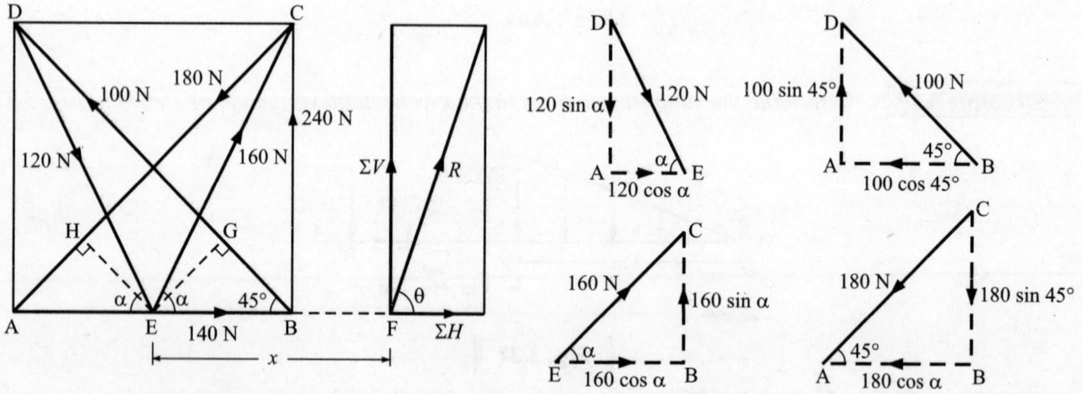


**Fig. 3.39**

Now resolving all the force horizontally and vertically, as shown separately in Fig. 3.39. We know that

$$\Sigma H = \text{Algebraic sum of all the horizontal forces}$$
$$= 140 + 160\cos\alpha - 100\cos 45° + 120\cos\alpha - 180\cos 45°$$
$$= 140 + 160 \times 0.4471 - 100 \times 0.707 + 120 \times 0.4471 - 180 \times 0.707$$
$$= 140 + 71.536 - 70.7 + 53.652 - 127.26$$
$$= 265.188 - 197.96 = 67.228 \text{ N} \rightarrow$$

Similarly, $\quad \Sigma V = \text{Algebraic sum of all the vertical forces}$

$$= 240 + 160\sin\alpha + 100\sin 45° - 120\sin\alpha - 180\sin 45°$$
$$= 240 + 160 \times 0.89\,445 + 100 \times 0.707 - 120 \times 0.89\,445 - 180 \times 0.707$$
$$= 240 + 143.112 + 70.7 - 107.334 - 127.26$$
$$= 453.812 - 234.594 = 219.218 \text{ N} \uparrow$$

We know that the resultant force, $R = \sqrt{(\Sigma H)^2 + (\Sigma V)^2} = \sqrt{(67.228)^2 + (219.218)^2}$

$= 229.3$ N **Ans.**

***Direction of the resultant force***

Let $\quad \theta$ = Angle of inclination of the resultant force with the horizontal $AB$

We know that $\quad \tan\theta = \dfrac{\Sigma V}{\Sigma H} = \dfrac{219.218}{67.228} = 3.2608$

$\therefore \quad \theta = \tan^{-1}(3.2608) = 72.95°$ **Ans.**

***Position of the resultant force***

Let $x = EF$ = Distance of the resultant force from $E$.

According to Varignon's theorem, the algebraic sum of the moments of all the forces about any point is equal to the moment of their resultant force about the same point.

Draw perpendiculars $EG$ and $EH$ on $BD$ and $AC$, respectively. From Fig. 3.39, we find that

$$EG = BE \sin 45°; \text{ and } EH = AE \sin 45°$$

Taking moments about $E$, we have

$$240 \times BE + 100 \times EG + 180 \times EH = \Sigma V \times EF$$

$$240 \times 150 + 100 \times BE \sin 45° + 180 \times AE \sin 45° = 219.218 \times x$$

$$32\,000 + 100 \times 150 \times 0.707 + 180 \times 150 \times 0.707 = 219.218x$$

$$32\,000 + 10\,605 + 19\,089 = 219.218x$$

$$61\,694 = 219.218x$$

$$\therefore \quad x = \frac{61\,694}{219.218} = 281.4 \text{ mm} \quad \textbf{Ans.}$$

**Note:** Since the line of action of the forces 140 N, 160 N and 120 N pass through point $E$, therefore their moments about $E$ is zero.

**Example 3.21** *A string AB is passed over two pulleys mounted on a L-shaped frame which is hinged to the ground at its foot E, as shown in Fig. 3.40. A weight W is hung at the one end of the string whose other end is fixed to the ground. The frame is kept in equilibrium by another string CD as shown. Neglecting the self weight of the frame, the friction in the pulleys and the diameters of the latter, find the tension in CD and the reaction at the frame base E, in terms of W.*

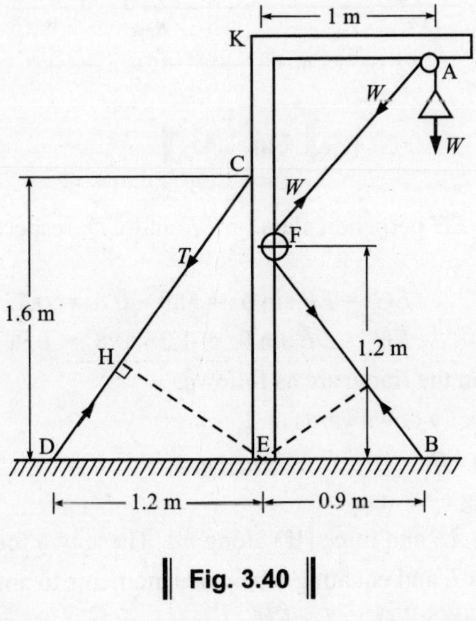


**Fig. 3.40**

***Solution*** Given : $DE = 1.2$ m ; $EB = 0.9$ m ; $CE = 1.6$ m ; $KA = 1$ m ; $EF = 1.2$ m

***Tension in CD***

Let $T$ = Tension in the string $CD$.

The line diagram of L-shaped frame is shown in Fig. 3.41. From the geometry of the figure, we find that

$$BF = \sqrt{(BE)^2 + (EF)^2} = \sqrt{(0.9)^2 + (1.2)^2} = 1.5 \text{ m}$$

$$CD = \sqrt{(DE)^2 + (CE)^2} = \sqrt{(1.2)^2 + (1.6)^2} = 2 \text{ m}$$

From triangle *BEF*, $\sin \alpha = \dfrac{EF}{BF} = \dfrac{1.2}{1.5} = 0.8$

$\therefore$ $\alpha = 53.13°$

and $\alpha_1 = 180° - (90° + \alpha) = 180° - (90 + 53.13°) = 36.47°$

Similarly, from triangle *CDE*, $\sin \theta = \dfrac{CE}{CD} \dfrac{1.6}{\quad} = 0.8$

$\therefore$ $\theta = 53.13°$

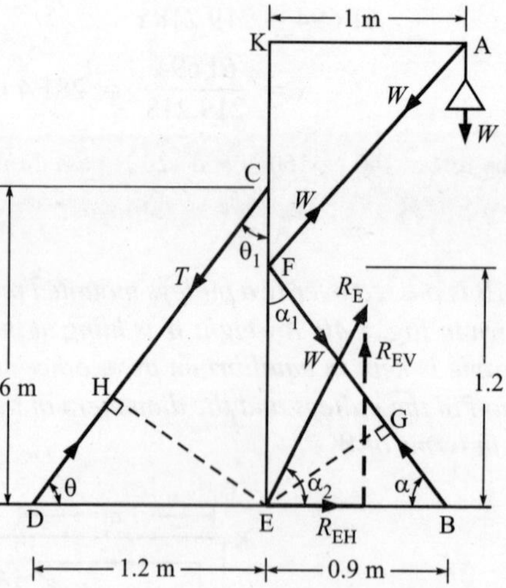


**Fig. 3.41**

Now from *E* draw *EG* and *EH* perpendiculars on *BF* and *CD*, respectively. From the geometry of the figure,

$$EG = EB \sin \alpha = 0.9 \times 0.8 = 0.72 \text{ m}$$

and

$$EH = DE \sin \theta = 1.2 \times 0.8 = 0.96 \text{ m}$$

The various forces acting on the frame are as follows:

(*a*) Force (*W*) acting vertically downwards at *A*;

(*b*) Force (*W*) acting along *FB*;

(*c*) Tension (*T*) in the string *CD*; and

(*d*) Force (*W*) acting along *AF* and force (*W*) along *FA*. These two forces neutralise each other.

Now taking moments about *E* and equating clockwise moments to anticlockwise moments, we have

$$W \times AK + W \times EG = T \times EH$$

$$W \times 1 + W \times 0.72 = T \times 0.96$$

$$W(1 + 0.72) = T \times 0.96 \quad \text{or} \quad 1.72W = 0.96T$$

$\therefore$ $T = \dfrac{1.72}{0.96} = 1.792W$ **Ans.**

***Reaction at the frame base E***

Let $R_E$ = Reaction at the frame base $E$,

$R_{EH}$ = Horizontal components of reaction $R_E$, and

$R_{EV}$ = Vertical component of reaction $R_E$.

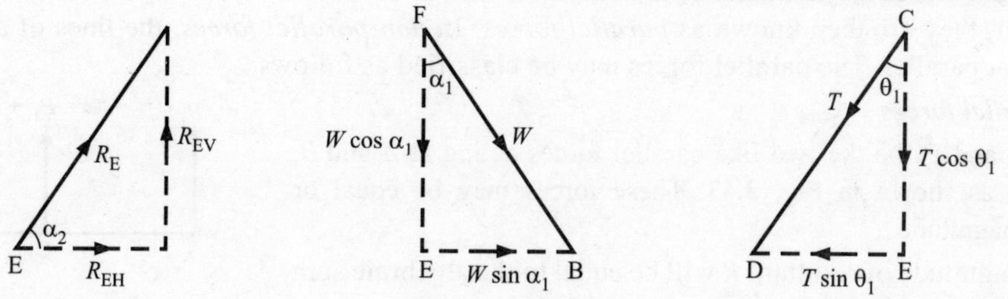


**Fig. 3.42**

Now resolving the various forces acting at $A$ (*i.e.* $W$), along $FB$ (*i.e.* $W$) and along $CD$ (*i.e.* $T$) in the horizontal and vertical components, as shown in Fig. 3.42. We know that

$$\Sigma H = \text{Algebraic sum of all the horizontal forces}$$

$$= R_{EH} + W \sin \alpha_1 - T \sin \theta_1$$

$$= R_{EH} + W \times \frac{BE}{BF} - 1.792\, W \times \frac{DE}{CD}$$

$$= R_{EH} + W \times \frac{0.9}{1.5} - 1.792\, W \times \frac{1.2}{2}$$

$$= R_{EH} + 0.6\,W - 1.0752\,W = R_{EH} - 0.4752\,W \qquad ...(i)$$

and

$$\Sigma V = \text{Algebraic sum of all the vertical forces}$$

$$= R_{EV} - W - W \cos \alpha_1 - T \cos \theta_1$$

$$= R_{EV} - W - W \times \frac{EF}{BF} - T \times \frac{CE}{CD}$$

$$= R_{EV} - W - W \times \frac{1.2}{1.5} - 1.792\, W \times \frac{1.6}{2}$$

$$= R_{EV} - W - 0.8\,W - 1.4336\,W = R_{EV} - 3.2336\,W \qquad ...(ii)$$

For equilibrium, $\Sigma H = 0$ and $\Sigma V = 0$. Therefore from equations (*i*) and (*ii*),

$$R_{EH} = 0.4752\,W$$

and

$$R_{EV} = 3.2336\,W$$

We know that resultant at the frame base $E$,

$$R_E = \sqrt{(R_{EH})^2 + (R_{EV})^2} = \sqrt{(0.4752\,W)^2 + (3.2336\,W)^2}$$

$$= \sqrt{0.2258\,W^2 + 10.4562\,W^2} = 3.268\,W \quad \textbf{Ans.}$$

Let $\alpha_2$ = Angle of inclination of the resultant $R_E$ with the horizontal $EB$.

We know that

$$\tan \alpha_2 = \frac{R_{EV}}{R_{EH}} = \frac{3.2336}{0.4752} = 6.8047$$

$$\therefore \quad \alpha_2 = \tan^{-1}(6.8047) = 81.64° \quad \textbf{Ans.}$$

## 3.7 METHOD OF DETERMINING THE POSITION OF RESULTANT OF COPLANAR NON-CONCURRENT FORCES

We have already discussed that when the lines of action of all the forces lie in the same plane but do not pass through a common point, then the forces are called coplanar non-concurrent forces. The non-concurrent forces may be parallel or non-parallel. When the lines of action of the forces are parallel to each other, they are then known as *parallel forces*. In *non-parallel forces*, the lines of action of the forces are not parallel. The parallel forces may be classified as follows:

**1. *Like parallel forces***

Let $P_1$ and $P_2$ be the two like parallel forces acting at $A$ and $B$, respectively, as shown in Fig. 3.43. These forces may be equal or unequal in magnitude.

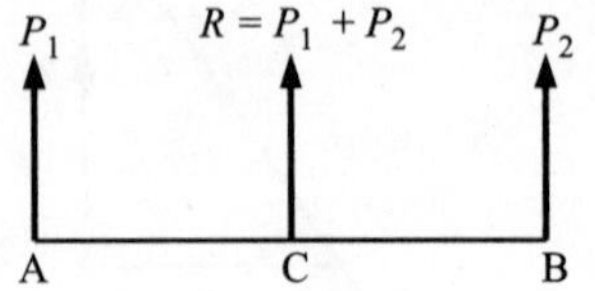


**Fig. 3.43**

The magnitude of resultant $R$ will be equal to the algebraic sum of all the like forces, *i.e.*

$$R = P_1 + P_2$$

The direction of the resultant is in the direction of all the like parallel forces.

The position of the resultant $R$ may be obtained by taking moments of all the forces about a point and equating the algebraic sum of the moments of all the forces to that of their resultant about the *same point. Let the resultant $R$ acts at point $C$. Now taking moments about point $C$,

$$P_1 \times AC = P_2 \times BC$$

or $$\frac{P_1}{P_2} = \frac{BC}{AC}$$

**Notes:** (*a*) The point of application of the resultant of two like parallel forces lies between the two forces.

(*b*) The resultant divides the distance between the two forces internally in the inverse ratio of the magnitude of the forces.

**2. *Unlike parallel forces***

The resultant of two unlike parallel forces $P_1$ and $P_2$, as shown in Fig. 3.44, may be determined in the similar way as discussed above. In this case,

$$R = P_1 - P_2$$

The point of application of $R$ will be outside of $A$ and in the direction of the larger force $P_1$. Let the resultant $R$ acts at point $C$. Now taking moments about point $C$,

$$P_1 \times AC = P_2 \times BC$$

or $$\frac{P_1}{P_2} = \frac{BC}{AC}$$

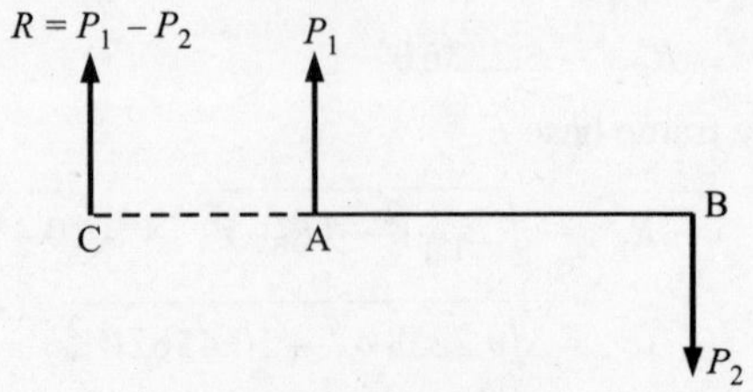


**Fig. 3.44**

* This is according to Varignon's theorem. Thus, Varignon's theorem may also be applied to parallel forces.

**Notes:** **(a)** The point of application of the resultant of two unlike parallel forces lies outside the two forces and nearer to the larger force.

**(b)** The resultant divides the distance between the two forces externally in the inverse ratio of the magnitude of the forces.

**Example 3.22** *A beam 200 mm long carries two like parallel forces of 60 N and 90 N at points A and B, as shown in Fig. 3.45. Find the resultant of these forces and the point at which it acts. Also determine the magnitude and position of the resultant when the forces are unlike.*

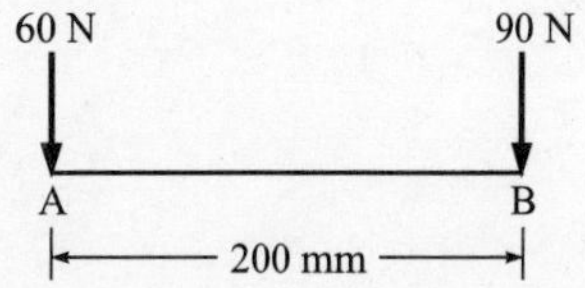

**Fig. 3.45**

***Solution*** Given : $AB = 200$ mm ; Force at $A$ ($P_1$) = 60 N ; Force at $B$ ($P_2$) = 90 N

***Resultant when the forces are like***

Since the forces are like, then their resultant,

$$R = P_1 + P_2 = 60 + 90 = 150 \text{ N} \quad \textbf{Ans.}$$

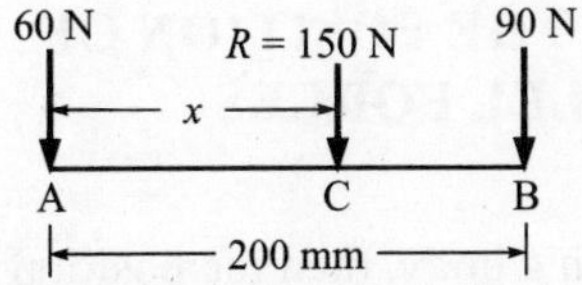

**Fig. 3.46**

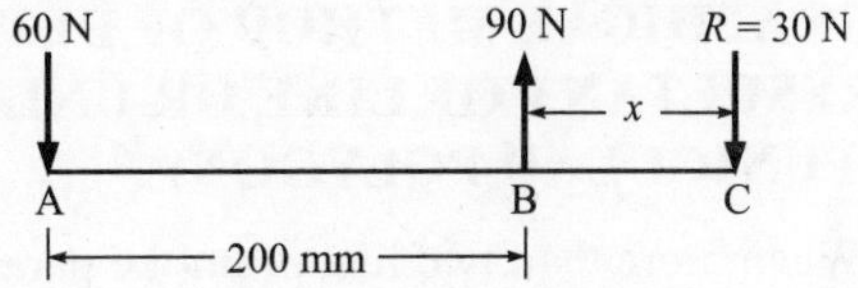

**Fig. 3.47**

Let $x$ be the position of this resultant from point $A$, as shown in Fig. 3.46. Now taking moments about the point $C$, we have

$$60x = 90(200 - x) = 90 \times 200 - 90x$$

or

$$60x + 90x = 18\,000$$

$$\therefore \quad x = \frac{18\,000}{150} = 120 \text{ mm} \quad \textbf{Ans.}$$

**Note:** The position of the resultant of two like parallel forces may also be obtained as follows:

We know that

$$\frac{P_1}{P_2} = \frac{BC}{AC}$$

$$\frac{60}{90} = \frac{200 - x}{x}$$

or

$$60x = 90 \times 200 - 90x$$

$$60x + 90x = 18\,000$$

$$\therefore \quad x = \frac{18\,000}{150} = 120 \text{ mm} \quad \textbf{Ans.}$$

***Resultant when the forces are unlike***

Since the forces are unlike, therefore their resultant,

$$R = P_2 - P_1 = 90 - 60 = 30 \text{ N}$$

We know that the resultant of unlike forces lies outside the two forces and nearer to the larger force (*i.e.* 90 N). Let $x$ be the position of the resultant $R$ from $B$, as shown in Fig. 3.47. Taking moments about $A$, we have

$$90 \times 200 = R \times AC = 30(200 + x) = 6000 + 30x$$

$$\therefore \quad x = \frac{90 \times 200 - 6000}{30} = \frac{12\,000}{30} = 400 \text{ mm } \textbf{Ans.}$$

**Note:** The position of the resultant of two unlike parallel forces may also be obtained as follows:

We know that $\dfrac{P_1}{P_2} = \dfrac{BC}{AC}$

$$\frac{60}{90} = \frac{x}{200 + x}$$

$$60\,(200 + x) = 90x$$

$$12\,000 + 60x = 90x$$

$$12\,000 = 90x - 60x = 30x$$

$$\therefore \quad x = \frac{12\,000}{30} = 400 \text{ mm } \textbf{Ans.}$$

## 3.8 GRAPHICAL METHOD OF DETERMINING THE POSITION OF RESULTANT OF LIKE OR UNLIKE PARALLEL FORCES (FUNICULAR POLYGON)

When more than two like or unlike parallel forces act on a body, then the position of the resultant of these forces may be determined graphically by drawing the funicular polygon, as discussed below:

Consider three parallel forces $P_1$, $P_2$ and $P_3$, as shown in Fig. 3.48 (*a*), whose resultant $R$ is required to be determined. First of all, draw the space diagram of the given forces $P_1$, $P_2$ and $P_3$, as shown in Fig. 3.48 (*a*). Name the forces according to Bow's notation. Now the force $P_1$ may be read as force $AB$, force $P_2$ as force $BC$ and force $P_3$ as force $CD$. In order to draw the vector diagram, as shown in Fig. 3.48 (*b*), choose a suitable scale. Mark $ab$ equal to force $AB$ (*i.e.* $P_1$), $bc$ equal to force $BC$ (*i.e.* $P_2$) and $cd$ equal to force $CD$ (*i.e.* $P_3$).

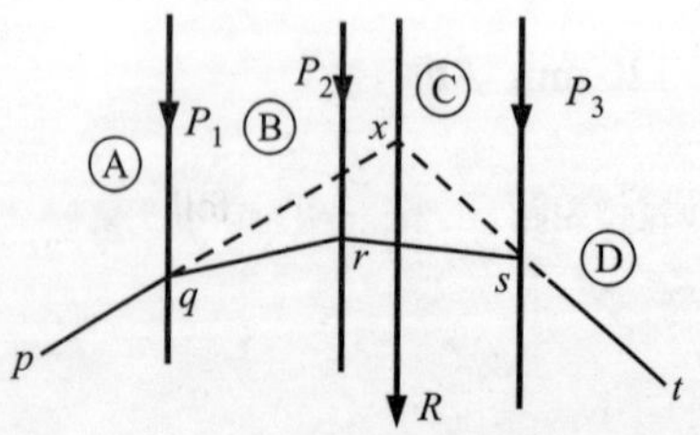


(*a*) Space diagram

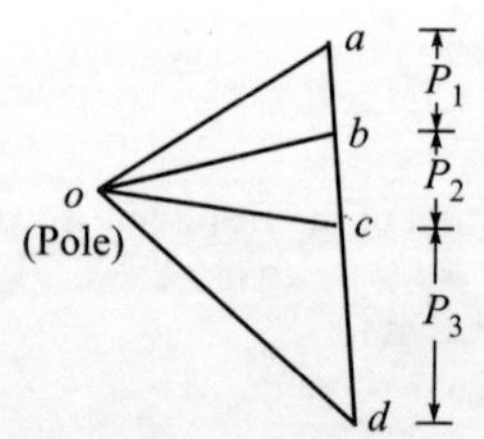


(*b*) Vector diagram

**Fig. 3.48**

Now take a point $o$ (known as pole) outside the vector diagram and join $oa$, $ob$, $oc$ and $od$. Choose any point $q$ on the force $P_1$, as shown in the space diagram. From point $q$, draw $pq$ parallel to $oa$, $qr$ parallel to $ob$, $rs$ parallel to $oc$ and $st$ parallel to $od$. Now produce $pq$ and $ts$ to meet at point $x$ which is the point through which the resultant passes. Through point $x$, draw a line parallel to $ad$.

The magnitude of the resultant force is given by $ad$ to the scale *i.e.*

$$R = ad = P_1 + P_2 + P_3$$

**Note:** This method may be used for parallel, like, unlike and inclined forces also.

**Example 3.23** *Determine the magnitude, direction and position of the resultant of the parallel system of forces, as shown in Fig. 3.49, by graphical method.*

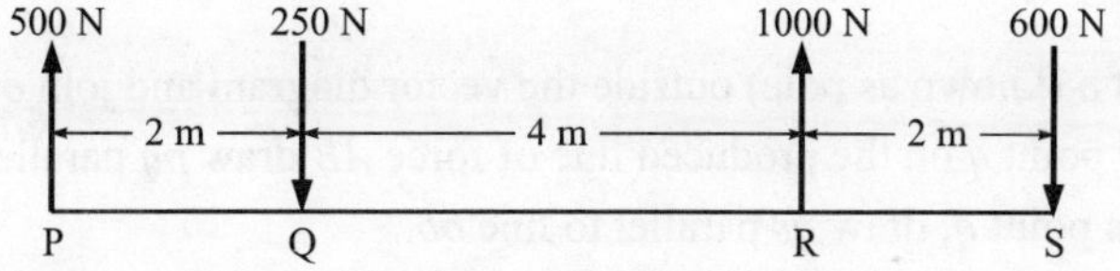


**Fig. 3.49**

***Solution*** Given : Force at $P$ = 500 N (upward) ; Force at $Q$ = 250 N (downward) ; Force at $R$ = 1000 N (upward) ; Force at $S$ = 600 N (downward) ; $PQ$ = 2 m; $QR$ = 4 m ; $RS$ = 2 m

***Magnitude and direction of the resultant***

First of all, draw the space diagram, as shown in Fig. 3.50 (*a*), to some suitable scale and name the forces according to Bow's notation. Now draw the vector diagram, as shown in Fig. 3.50 (*b*), to some suitable scale, as discussed below:

1. Take any point *a* and draw *ab* parallel to the force *AB* and equal to 500 N to the scale, in the upward direction.
2. From point *b*, draw *bc* parallel to the force *BC* and equal to 250 N, to the scale, in the downward direction.
3. Now from point *c*, draw *cd* parallel to the force *CD* and equal to 1000 N, to the scale, in the upward direction.
4. Lastly, from point *d*, draw *de* parallel to the force *DE* and equal to 600 N, to the scale, in the downward direction.
5. Now *ae* represents the required resultant *R*. By measurement, we find that

$$R = ae = 650 \text{ N (upwards)}$$ **Ans.**

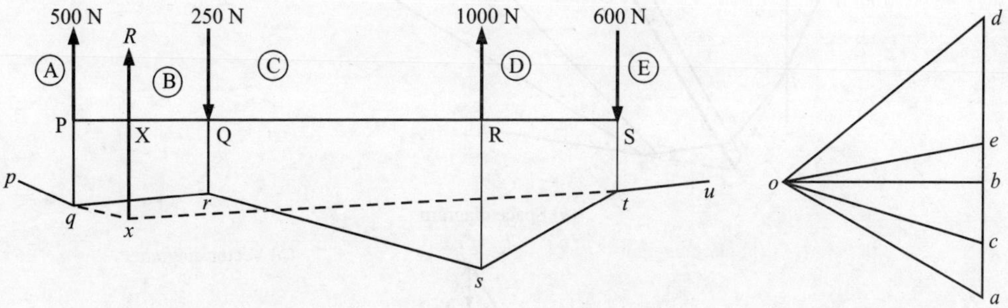


(*a*) Space diagram (*b*) Vector diagram

**Fig. 3.50**

**Note:** The resultant *R* is also given by

$$R = 500 - 250 + 1000 - 600 = 650 \text{ N}$$ **Ans.**

The upward forces are taken as *positive* and downward forces as *negative.* Since the magnitude of the upward forces is greater than the downward forces, therefore the resultant *R* will act in the upward direction.

***Position of the resultant***

In order to find the position of the resultant, draw the funicular polygon as discussed below:

1. First of all, produce the lines of action of the forces *AB*, *BC*, *CD* and *DE*, as shown in the space diagram.
2. Take any point *o* (known as pole) outside the vector diagram and join *oa, ob, oc, od* and *oe*.
3. Now from any point *q* on the produced line of force *AB* draw *pq* parallel to the line *oa*.
4. Similarly, from point *q*, draw *qr* parallel to line *ob*.
5. From point *r*, draw *rs* parallel to line *oc*, and from point *s*, draw *st* parallel to line *od*.
6. Lastly, draw *tu* parallel to line *oe*. Now produce *pq* and *tu* to intersect at point *x* as shown in Fig. 3.50 (*a*).
7. Now point *x* is the point of application of the resultant *R*. From point *x*, draw a line parallel to *ae* intersecting the line *PS* at *X*. On measurement, we find that

$$PX = 0.8 \text{ m} \quad \textbf{Ans.}$$

**Example 3.24** *On a horizontal line PQRS 120 mm long, where PQ = QR = RS = 40 mm, forces of 10 kN, 15 kN, 10 kN and 5 kN are acting at P, Q, R and S, respectively, all downwards. The lines of action of these forces make angles of 90°, 60°, 45° and 30°, respectively, with PS. Find the resultant of the system completely in magnitude, direction and position graphically and check the answer analytically.*

***Solution*** Given : $PQ = QR = RS = 40$ mm ; Force at $P = 10$ kN ; Force at $Q = 15$ kN ; Force at $R = 10$ kN ; Force at $S = 5$ kN

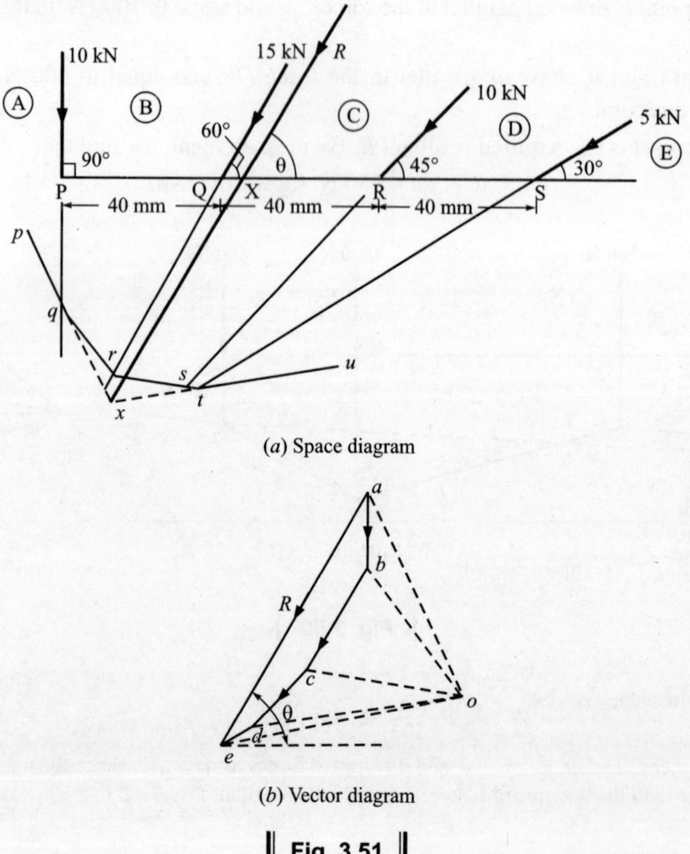


(*a*) Space diagram

(*b*) Vector diagram

**Fig. 3.51**

***Magnitude and direction of the resultant***

First of all, draw the space diagram, as shown in Fig. 3.51 (*a*), to some suitable scale and name the forces according to Bow's notation. Now draw the vector diagram, as shown in Fig. 3.51 (*b*), to some suitable scale, as discussed below:

1. Take any point *a* and draw *ab* parallel to the force *AB* and equal to 10 kN, to a suitable scale, in the downward direction.
2. From point *b*, draw *bc* parallel to the force *BC* (which is inclined at 60° to the horizontal) and equal to 15 kN, to the scale, in the downward direction.
3. Now from point *c*, draw *cd* parallel to the force *CD* (which is inclined at 45° to the horizontal) and equal to 10 kN, to the scale, in the downward direction.
4. Lastly, from point *d*, draw *de* parallel to the force *DE* (which is inclined at 30° to the horizontal) and equal to 5 kN, to the scale, in the downward direction.
5. Now *ae* represents the required resultant *R*. By measurement, we find that

$$R = ae = 37.5 \text{ kN} \quad \textbf{Ans.}$$

and angle of inclination with the horizontal,

$$\theta = 61.5° \quad \textbf{Ans.}$$

***Position of the resultant***

In order to find the position of the resultant, draw the funicular polygon as discussed below:

1. First of all, produce the lines of the forces *AB*, *BC*, *CD* and *DE*, as shown in the space diagram.
2. Take any point *o* (known as pole) outside the vector diagram and join *oa, ob, oc, od* and *oe*.
3. Now from any point *q* on the produced line of force *AB*, draw *pq* parallel to the line *oa*.
4. Similarly, from point *q*, draw *qr* parallel to the line *ob*.
5. From point *r*, draw *rs* parallel to the line *oc* and from point *s*, draw *st* parallel to the line *od*.
6. Lastly, draw *tu* parallel to the line *oe*. Now produce *pq* and *tu* to intersect at point *x*, as shown in Fig. 3.51 (*a*).
7. Now point *x* is the point of application of the resultant *R*. From point *x*, draw a line parallel to the resultant *ae* intersecting the line *PS* at *X*. On measurement, we find that

$$PX = 45 \text{ mm} \quad \textbf{Ans.}$$

***Analytical method***

The system of forces is shown in Fig. 3.52.

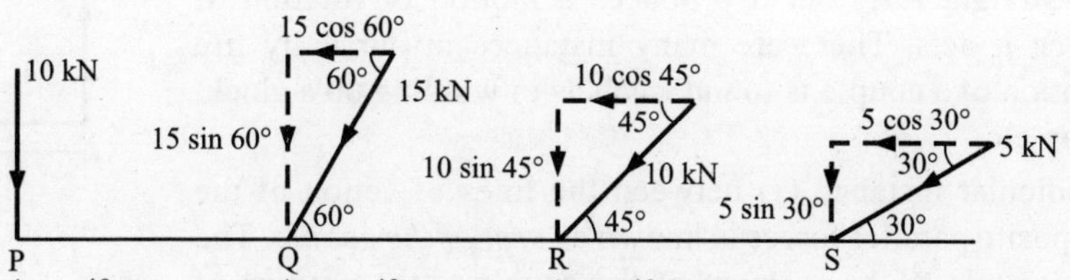


**Fig. 3.52**

***Magnitude and direction of the resultant***

Resolving the forces horizontally, we have

$$\Sigma H = \text{Algebraic sum of all the horizontal components}$$
$$= 15 \cos 60° + 10 \cos 45° + 5 \cos 30°$$
$$= 15 \times 0.5 + 10 \times 0.707 + 5 \times 0.866$$
$$= 7.5 + 7.07 + 4.33 = 18.9 \text{ kN} \leftarrow$$

Resolving the forces vertically, we have

$$\Sigma V = \text{Algebraic sum of all the vertical components}$$
$$= 10 + 15 \sin 60° + 10 \sin 45° + 5 \sin 30°$$
$$= 10 + 15 \times 0.866 + 10 \times 0.707 + 5 \times 0.5$$
$$= 10 + 12.99 + 7.07 + 2.5 = 32.56 \text{ kN} \downarrow$$

We know that the resultant,

$$R = \sqrt{(\Sigma H)^2 + (\Sigma V)^2} \quad \text{...(Refer Fig. 3.53)}$$
$$= \sqrt{(18.9)^2 + (32.56)^2}$$
$$= \sqrt{357.21 + 1060.15} = \sqrt{1417.36}$$
$$= 37.65 \text{ kN} \quad \textbf{Ans.}$$

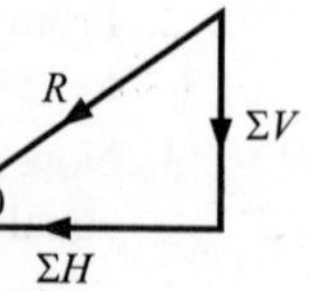


Fig. 3.53

Let $\theta$ = Angle which the resultant $R$ makes with the horizontal

We know that $\tan \theta = \dfrac{\Sigma V}{\Sigma H} = \dfrac{32.56}{18.9} = 1.723$

$\therefore$ $\theta = \tan^{-1} (1.723) = 59.89°$ **Ans.**

***Position of the resultant***

Let $x$ be the distance of the resultant $R$ from $P$ (*i.e.* 10 kN). Now taking moments of all the vertical components about $P$ and the vertical component of resultant force $R$, according to Varignon's theorem, we have

$$R \sin \theta \times x = 10 \times 0 + 15 \sin 60° \times 40 + 10 \sin 45° \times 80 + 5 \sin 30° \times 120$$
$$37.65 \text{ Sin } 59.89° \times x = 0 + 15 \times 0.866 \times 40 + 10 \times 0.707 \times 80 + 5 \times 0.5 \times 120$$
$$37.65 \times 0.865 = 519.6 + 565.6 + 300 = 1385.2$$

$\therefore$ $x = \dfrac{1385.2}{37.65 \times 0.865} = 42.53 \text{ mm}$ **Ans.**

## 3.9 COUPLE

When two equal and opposite parallel forces act on a body and their lines of action are different, as shown in Fig. 3.54, then they form a *couple*. A couple does not produce any translatory motion (*i.e.* motion in a straight line) but it produces a motion of rotation of the body on which it acts. There are many instances in our daily life where the application of a couple is found, such as in winding up a clock, turning a water tap etc.

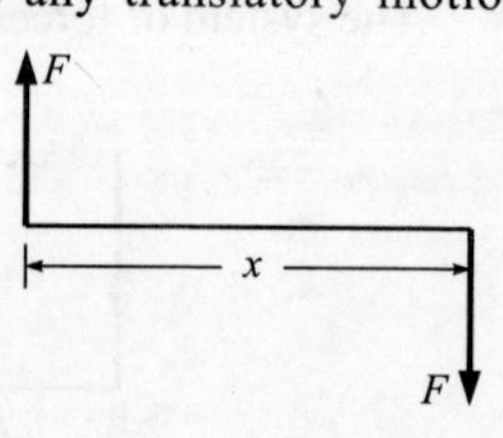


Fig. 3.54

The perpendicular distance ($x$) between the lines of action of the two equal and opposite parallel forces is known as *arm of the couple.* The magnitude of the couple or the moment of the couple is the product of either of the two forces and the arm of the couple. Mathematically,

$$\text{Moment of a couple} = F \times x$$

The unit of a couple depends upon the unit of force and the unit of arm of the couple. When the force is in newton (N) and arm of the couple in metres (m), then the unit of couple will be N-m.

## 3.10 CLASSIFICATION OF COUPLES

The couples may be *clockwise couple* or *anticlockwise couple* depending upon the direction in which they tend to rotate the body.

When a couple tends to rotate the body in a clockwise direction, it is then known as *clockwise couple*, as shown in Fig. 3.55 (*a*). The clockwise couples are taken as *positive*.

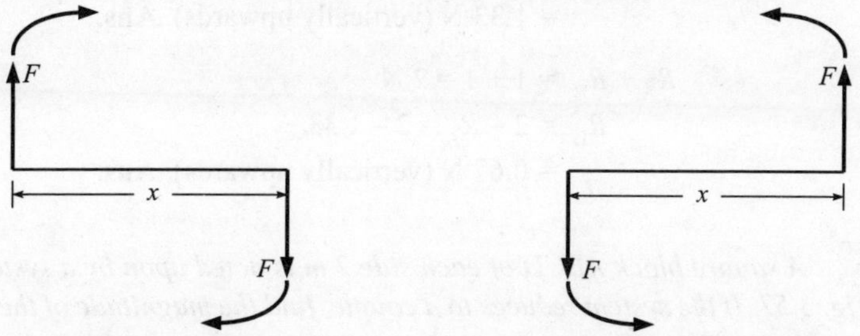


(*a*) Clockwise couple (*b*) Anticlockwise couple

**Fig. 3.55**

Similarly, when a couple tends to rotate the body in an anticlockwise direction, it is then known as *anticlockwise couple*, as shown in Fig. 3.55 (*b*). The anticlockwise couples are taken as *negative*.

## 3.11 PROPERTIES OF A COUPLE

A couple has the following properties:

1. The algebraic sum of the forces forming the couple is equal to zero.
2. The algebraic sum of the moments of the forces, forming a couple, about any point in their plane, is constant.
3. The two couples acting in one plane upon a rigid body, whose moments are equal but opposite in sense, balance each other.
4. A single force and a couple in the same plane acting on a rigid body cannot balance each other, but both are equivalent to a single force acting parallel to its original direction at such a distance that it gives moments equal to the moment of the couple.
5. A couple cannot be balanced by a single force, but can be balanced only by a couple, of opposite sense.
6. Any two couples of equal moments and sense, in the same plane are equivalent in their effect.

**Example 3.25** *Find the reactions at the supports A and B of a beam, as shown in Fig. 3.56. The length of the beam is 4 m and a couple of 3 kN-m is applied at a distance of 2 m from the support A.*

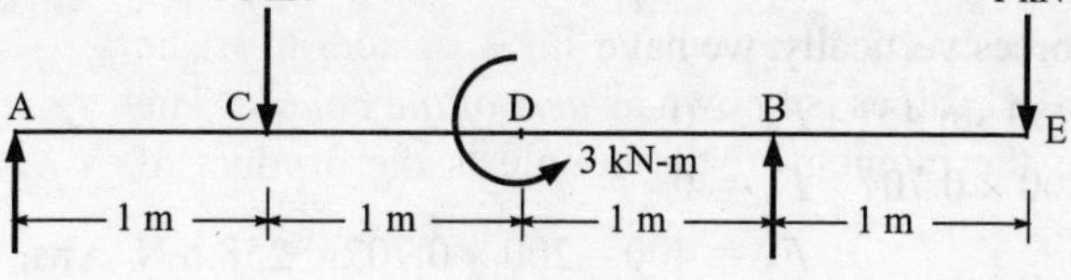


**Fig. 3.56**

***Solution*** Given : $L = AE = 4$ m ; Couple applied at $D = 3$ kN-m (anticlockwise) ; $AD = 2$ m

Let $R_A$ and $R_B$ = Reactions at the supports $A$ and $B$, respectively.

Taking moments about $B$, and equating clockwise moments equal to anticlockwise moments, we have

$$R_A \times 3 + 1 \times 1 = 1 \times 2 + 3 = 5$$

$$\therefore \qquad R_A = \frac{5-1}{3} = \frac{4}{3}$$

$$= 1.33 \text{ N (vertically upwards)} \textbf{ Ans.}$$

We know that $\qquad R_A + R_B = 1 + 1 = 2 \text{ N}$

$$\therefore \qquad R_B = 2 - R_A = 2 - 1.33$$

$$= 0.67 \text{ N (vertically upwards)} \textbf{ Ans.}$$

**Example 3.26** *A square block ABCD of each side 2 m is acted upon by a system of forces along its sides, as shown in Fig. 3.57. If the system reduces to a couple, find the magnitude of the forces $F_1$ and $F_2$ and the couple.*

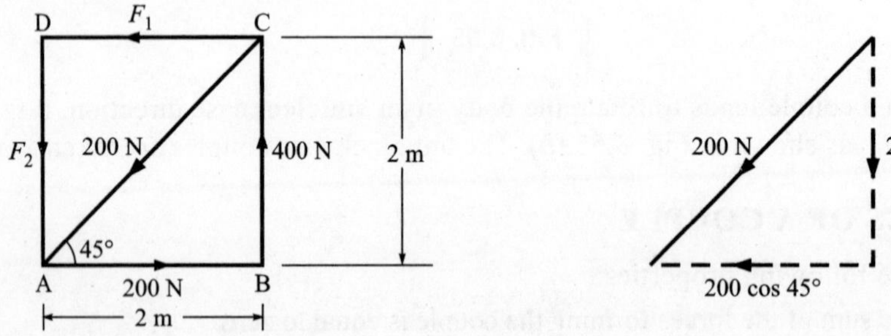


**Fig. 3.57**

***Solution*** Given : $AB = BC = CD = DA = 2$ m ; Force on $AB = 200$ N ; Force on $BC = 400$ N ; Force on $CA = 200$ N

***Magnitude of the forces $F_1$ and $F_2$***

Let $\qquad F_1$ = Force on *CD*, and

$\qquad F_2$ = Force on *DA*.

We know that when the system reduces to a couple, the algebraic sum of the horizontal and vertical components of the forces must be equal to zero.

Resolving the forces horizontally, we have

$$200 - 200 \cos 45° - F_1 = 0$$

$$200 - 200 \times 0.707 - F_1 = 0$$

$$\therefore \qquad F_1 = 200 - 200 \times 0.707 = 58.6 \text{ N} \textbf{ Ans.}$$

Now resolving the forces vertically, we have

$$400 - 200 \sin 45° - F_2 = 0$$

$$400 - 200 \times 0.707 - F_2 = 0$$

$$\therefore \qquad F_2 = 400 - 200 \times 0.707 = 258.6 \text{ N} \textbf{ Ans.}$$

***Magnitude of the couple***

We know that moment of the couple = Algebraic sum of the moments about any corner of the square (say *A*)

$$= -400 \times 2 - F_1 \times 2 = -400 \times 2 - 58.6 \times 2$$

$$= -800 - 117.2 = -917.2 \text{ N-m} \textbf{ Ans.}$$

The *negative* sign indicates that the moment of the couple is *anticlockwise*.

**Example 3.27** *A horizontal force 200 N is applied to the sloping bar BCD whose bottom rests on a horizontal plane, as shown in Fig. 3.58.*

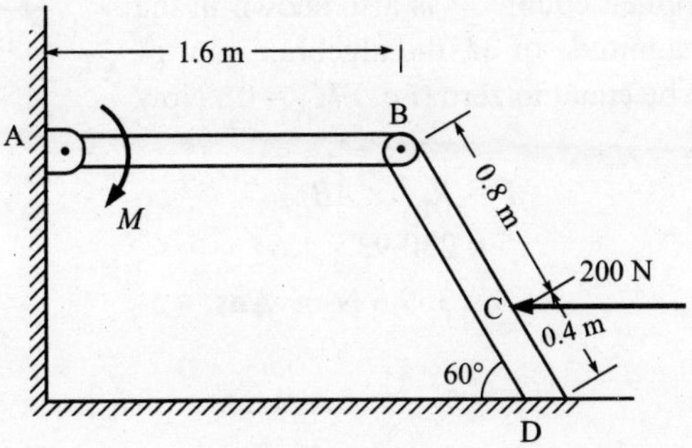


**Fig. 3.58**

*Its upper end is pinned at B to the horizontal bar AB which has a pinned support at A. What couple M must be applied to AB to hold the system in equilibrium? What is the magnitude of the pin reaction at B? Assume the bars to be weightless and pins at A and B to be smooth.*

***Solution*** Given : Horizontal force at $C = 200$ N ; $CD = 0.4$ m ; $CB = 0.8$ m ; $AB = 1.6$ m ; $BD = 0.4 + 0.8 = 1.2$ m

Let $M$ = Couple applied at $A$ to the bar $AB$, to hold the system in equilibrium, and

$R_B$ = Reaction at $B$.

The free body or line diagram of the sloping bar is shown in Fig. 3.59.

The reaction at $D$ ($R_D$) is vertical, $R_{BH}$ and $R_{BV}$ are the horizontal and vertical components of the reaction at $B$ ($R_B$).

Since the bar $BCD$ is in equilibrium, therefore equating horizontal and vertical forces, we have

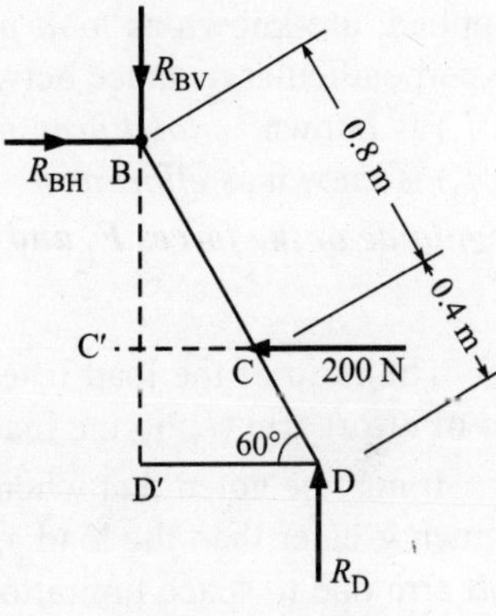


**Fig. 3.59**

$$R_{BH} = 200 \text{ N}$$

and
$$R_{BV} = R_D \qquad ...(i)$$

From the geometry of the figure, we find that

$$DD' = BD \cos 60° = 1.2 \times 0.5 = 0.6 \text{ m}$$

$$BC' = BC \sin 60° = 0.8 \times 0.866 = 0.6928 \text{ m}$$

Now in order to find the value of $R_{BV}$ and $R_D$, taking moments of all the forces about $B$ and equating clockwise moments equal to anticlockwise moments, we have

$$200 \times BC' = R_D \times DD'$$

$$200 \times 0.6928 = R_D \times 0.6$$

$$\therefore \quad R_D = \frac{200 \times 0.6928}{0.6} = 230.93 \text{ N}$$

From equation (*i*), $R_{BV} = R_D = 230.93$ N

We know that reaction at $B$, $R_B = \sqrt{(R_{BH})^2 + (R_{BV})^2}$

$$= \sqrt{(200)^2 + (230.93)^2} = 305.5 \text{ N} \quad \textbf{Ans.}$$

The line diagram of the horizontal bar $AB$ is shown in Fig. 3.60. Let the reactions $R_{BH}$ and $R_{BV}$ are at $B$ and the reactions $R_{AH}$ and $R_{AV}$ are at $A$. The applied couple $M$ is also shown in the figure. In order to find the magnitude of $M$, the algebraic sum of all the moments about $A$ must be equal to zero (*i.e.* $\Sigma M_A = 0$). Now taking moment about $A$

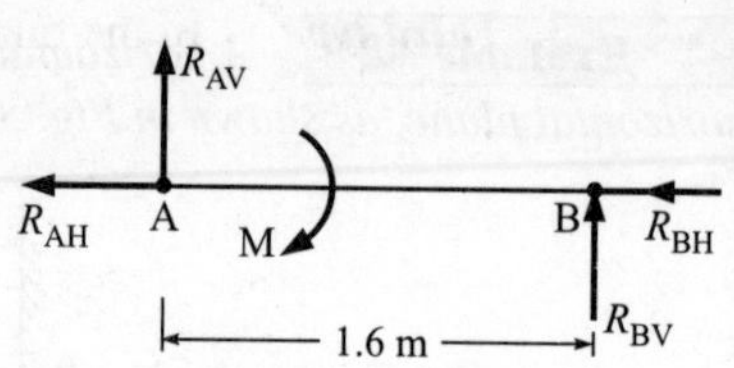


**Fig. 3.60**

$$M = R_{BV} \times AB$$
$$= 230.93 \times 1.6$$
$$= 369.5 \text{ N-m} \quad \textbf{Ans.}$$

## 3.12 LEVERS

A lever is a rigid bar capable of turning about a fixed point called *fulcrum*. It may be straight or curved and the forces applied on the lever (or by the lever) may be parallel or inclined to one another. Sometimes, a lever is merely used to facilitate the application of force in a desired direction. The principle on which the lever works is same as that of moments.

The lever is used as a machine to lift the *load* by the application of a small *effort.* The levers may be of the following two types:

1. Simple levers; and 2. Compound levers.

In a simple lever, there is one point $F$, known as *fulcrum*, about which the lever is capable of turning, as shown in Fig. 3.61. The points $A$ and $B$, through which the load ($W$) and the effort ($P$) is applied, are known as *load point* and *effort point*, respectively. The perpendicular distance between the fulcrum and the load point (*i.e.* $l_1$) is known as *load arm* and the perpendicular distance between the fulcrum and the effort point (*i.e.* $l_2$) is known as *effort arm.* According to the principle of moments

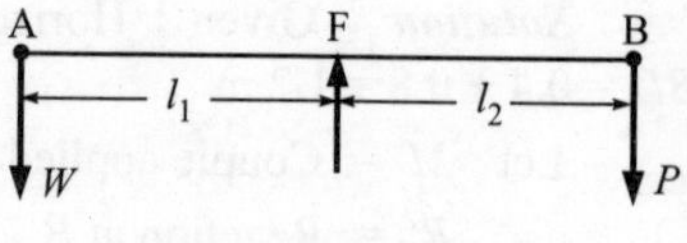


**Fig. 3.61**

$$W \times l_1 = P \times l_2 \quad \text{or} \quad \frac{W}{P} = \frac{l_2}{l_1}$$

The ratio of the load lifted ($W$) to the effort applied ($P$) is called *mechanical advantage* and the ratio of effort arm ($l_2$) to the load arm ($l_1$) is known as *leverage.*

It may be noted that when a large load is to be lifted by a small effort, then the effort arm should be much greater than the load arm. In some cases, it may not be possible to provide a lever with large effort arm due to space limitation. In such cases, a compound lever may be used.

A compound lever is made of straight pieces which may be attached to one another by pin joints. In other words, a compound lever is a combination of a number of simple levers. In a compound lever, the leverage is the produce of leverages of various levers.

## 3.13 TYPES OF SIMPLE LEVERS

Following are the three types of simple levers:

1. ***First type*** In the first type of levers, the fulcrum $F$ is in between the load ($W$) and effort ($P$) as shown in Fig. 3.62 (*a*). This type of lever is found in bell cranked levers used in railway signalling arrangement, handle of a hand pump, beam of a balance etc. Since the effort arm ($l_2$) is greater than the load arm ($l_1$), therefore, the mechanical advantage, in this case, is always more than one.
2. ***Second type*** In the second type of levers, the load ($W$) is in between the effort ($P$) and fulcrum $F$, as shown in Fig. 3.62 (*b*). This type of lever is found in levers of loaded safety valves. From the figure, we see that in this case also the effort arm ($l_2$) is greater than the load arm ($l_1$), therefore, the mechanical advantage of this type of lever will be more than one.

3. ***Third type*** In the third type of levers, the effort ($P$) is in between the load ($W$) and fulcrum $F$, as shown in Fig. 3.62 ($c$). This type of lever is usually found in a pair of tongs, treadle of a sewing machine. From the figure, we see that load arm ($l_1$) is greater than the effort arm ($l_2$), therefore, the mechanical advantage of this type of lever is always less than one.

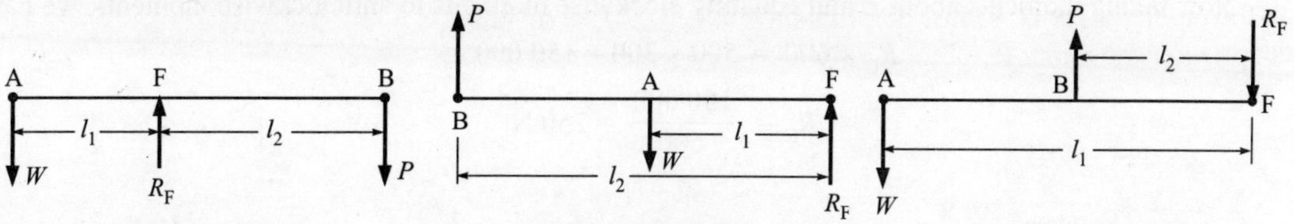


**Fig. 3.62**

**Example 3.28** *A crow bar made of iron has a weight of 150 N and is 6 m long. It is required to lift a load of 1000 N placed at one end, by applying a effort at the other end. If the fulcrum of the bar is at a distance of 1 m from the load end, find the effort required.*

***Solution*** Given : $W_1 = 150$ N ; $l = AB = l_1 + l_2 = 6$ m; $W = 1000$ N ; $AF = l_1 = 1$ m

The crow bar is shown in Fig. 3.63.

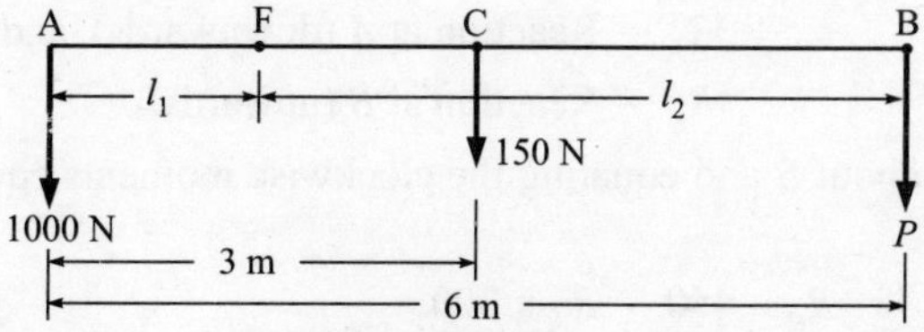


**Fig. 3.63**

Let $P$ = Effort required to lift a load of 1000 N.

The weight of the crow bar ($W_1$) will act at the centre *i.e.* 3 m from either end.

Taking moments about the fulcrum $F$ and equating clockwise moments equal to anticlockwise moments, we have

$$P \times BF + 150 \times FC = 1000 \times AF$$
$$P(AB - AF) + 150(AC - AF) = 1000 \times AF$$
$$P(6 - 1) + 150(3 - 1) = 1000 \times 1$$
$$5P + 300 = 1000$$

or $$5P = 1000 - 300 = 700$$

$\therefore$ $$P = \frac{700}{5} = 140 \text{ N} \quad \textbf{Ans.}$$

**Example 3.29** *Determine the reaction of the supports at A and B, for the system of levers, as shown in Fig. 3.64.*

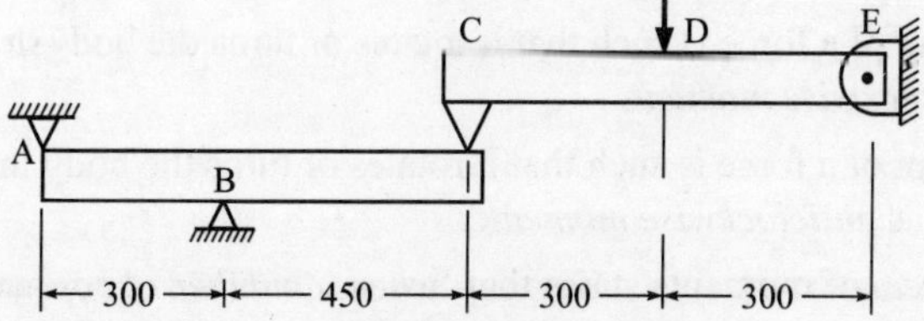


**Fig. 3.64**

***Solution*** Given : Load at $D = 500$ N ; $AB = 300$ mm ; $BC = 450$ mm ; $CD = 300$ mm ; $DE = 300$ mm

First of all, let us consider the top lever *CDE*, the line diagram of which is shown in Fig. 3.65.

Let $R_C$ = Reaction on the top lever at *C* in the upward direction.

Now taking moments about *E* and equating clockwise moments to anticlockwise moments, we have

$$R_C \times 600 = 500 \times 300 = 150\,000$$

$$\therefore \quad R_C = \frac{150\,000}{600} = 250 \text{ N}$$

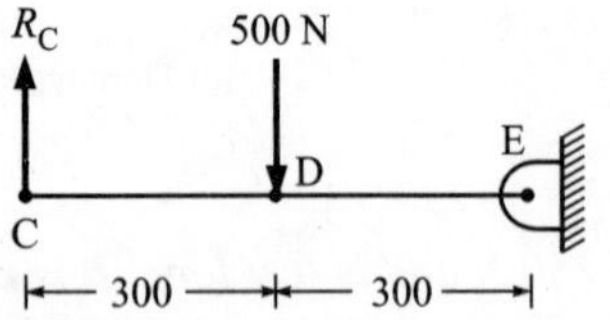


Fig. 3.65

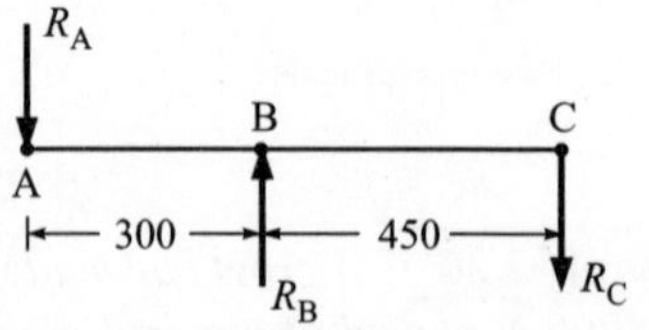


Fig. 3.66

Similarly, considering the bottom lever *ABC*, the line diagram of which is shown in Fig. 3.66.

Let $R_A$ = Reaction at *A* (downwards), and

$R_B$ = Reaction at *B* (upwards).

Now taking moments about *B* and equating the clockwise moments equal to anticlockwise moments, we have

$$R_C \times 450 = R_A \times 300$$

or

$$250 \times 450 = R_A \times 300$$

$$\therefore \quad R_A = \frac{250 \times 450}{300} = 375 \text{ N } \textbf{Ans.}$$

We know that

$$R_B = R_A + R_C$$

$$= 375 + 250 = 625 \text{ N } \textbf{Ans.}$$

## HIGHLIGHTS

1. The *moment* of a force is the turning effect produced by a force, on the body, on which it acts. It is equal to the product of the force and the perpendicular distance of the point about which moment is required and the line of action of the force.
2. The unit of moment depends upon the unit of force and the perpendicular distance. If the force is in newtons and the perpendicular distance in metres, then the unit of moment will be newton-metre (briefly written as N-m).
3. The moment of a force is a vector quantity, as it has both magnitude as well as direction.
4. When the moment of a force is such that it rotates or turns the body in the clockwise direction, it is then known as *clockwise moment.*
5. When the moment of a force is such that it rotates or turns the body in the anticlockwise direction, it is then known as *anticlockwise moment.*
6. Varignon's theorem of moments states that '*when a number of coplanar forces are acting about a point in their plane, then the algebraic sum of the moments of all the forces about a point is equal to the moment of their resultant force about the same point*'.

**7.** According to the principle of moments, '*when a body acted upon by a system of coplanar forces is in equilibrium, then the algebraic sum of the moments of all the forces, about any point in their plane is equal to zero, i.e.* $\Sigma M = 0$'.

**8.** A body is said to be in equilibrium when it is completely at rest or continues to be in steady linear motion. The conditions of equilibrium are $\Sigma H = 0$; $\Sigma V = 0$; and $\Sigma M = 0$.

**9.** The non-concurrent forces may be parallel or non-parallel. When the lines of action of the forces are parallel to each other, they are known as *parallel forces*. When the lines of action of the forces are not parallel, then they are *non-parallel forces*. The parallel forces may be like parallel forces and unlike parallel forces.

**10.** When two equal and opposite parallel forces act on a body and their lines of action are different, then they are said to form a *couple*.

**11.** A couple does not produce any *translatory motion* (*i.e.* motion in a straight line).

**12.** The couple may be *clockwise couple* or *anticlockwise couple*.

**13.** The couple has the following properties:

(*a*) The algebraic sum of the forces forming the couple is equal to zero.

(*b*) The algebraic sum of the moments of the forces, forming a couple, about any point in their plane, is constant.

(*c*) The two couples acting in one plane upon a rigid body, whose moments are equal but opposite in senses, balance each other.

(*d*) A couple cannot be balanced by a single force, but can be balanced only by a couple of opposite sense.

(*e*) A couple has no resultant force, but has resultant moment.

(*f*) Any two couples of equal moments and sense, in the same plane are equivalent in their effect.

**14.** A lever is a rigid bar capable of turning about a fixed point called *fulcrum*. It may be straight or curved.

**15.** The lever is used as a machine to lift the *load* ($W$) by the application of a small *effort* ($P$).

**16.** The perpendicular distance between the fulcrum and the load point is known as *load arm*.

**17.** The perpendicular distance between the fulcrum and the effort point is known as *effort arm*.

**18.** The ratio of the load lifted ($W$) to the effort applied ($P$) is called *mechanical advantage* (*M.A.*)

**19.** The ratio of the effort arm to the load arm is known as *leverage*.

**20.** The levers may be simple levers and compound levers. The compound lever is a combination of a number of simple levers. In a compound lever, the leverage is the product of leverages of various levers.

**21.** The simple levers are of the following three types:

(*a*) ***First type*** In the first type of levers, the fulcrum $F$ is in between the load ($W$) and the effort ($P$).

(*b*) ***Second type*** In the second type of levers, the load ($W$) is in between the effort ($P$) and the fulcrum $F$.

(c) ***Third type*** In the third type of levers, the effort ($P$) is in between the load ($W$) and the fulcrum $F$.

## NUMERICAL PROBLEMS

1. A force of 200 N is acting at a point *A*, as shown in Fig. 3.67. Find the moment of this force about *A*. [**Ans.** 282.8 N-m]

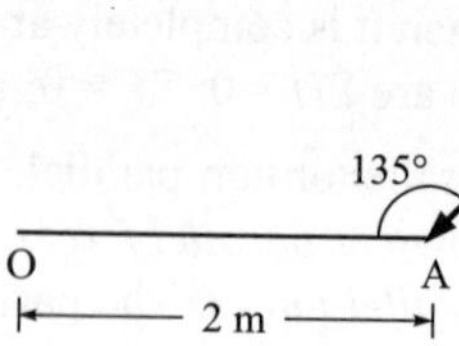


**Fig. 3.67**

2. Find the reactions at the supports *A* and *B* of a beam carrying a load as shown in Fig. 3.68. [**Ans.** 13 kN, 5 kN]

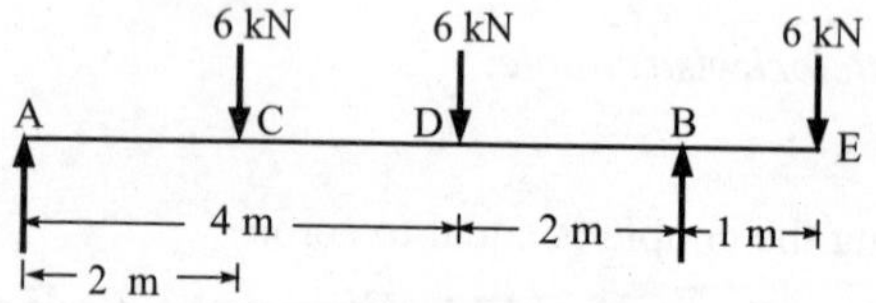


**Fig. 3.68**

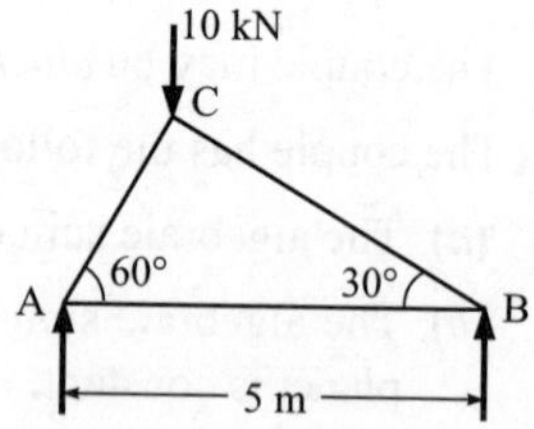


**Fig. 3.69**

3. A truss, as shown in Fig. 3.69, has a span of 5 m. It carries a load of 10 kN at the apex. Find the reactions at *A* and *B*. [**Ans.** 7.5 kN, 2.5 kN]
4. A rod weighing 400 N is 8 m long. If weights of 150 N and 200 N are suspended at its ends, find the position of the point at which the rod must be supported so that it remain horizontal. [**Ans.** 3.73 m from 200 N weight]
5. Four forces of magnitude 30 N, 40 N, 50 N and 60 N, respectively, act along the sides of a square, each side of which is 500 mm. Find the value of the moments of these forces about (*a*) the centre of the square; and (*b*) the point of intersection of the forces 30 N and 60 N. [**Ans.** (*a*) 45 N-m; (*b*) 45 N-m]
6. Determine the magnitude, direction and position of a single force *P*, which keeps in equilibrium the system of forces acting at the corners of a rectangular block, as shown in Fig. 3.70. The position of the force *P* may be stated by reference to axis with origin *O* and coinciding with the edges of the block. [**Ans.** 851.48 N; 3.36°; 2.2 m, 5.51 m]

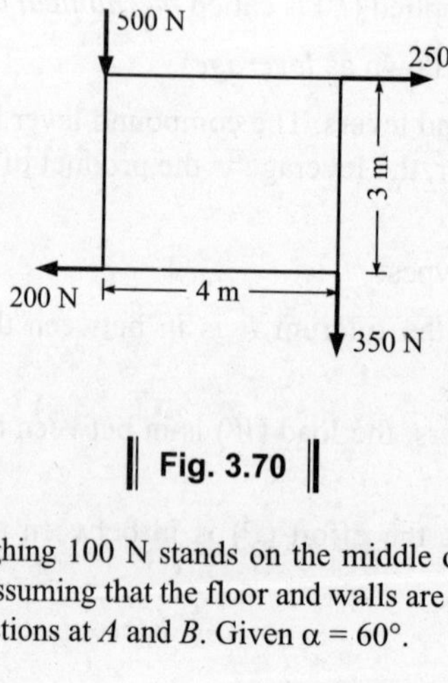


**Fig. 3.70**

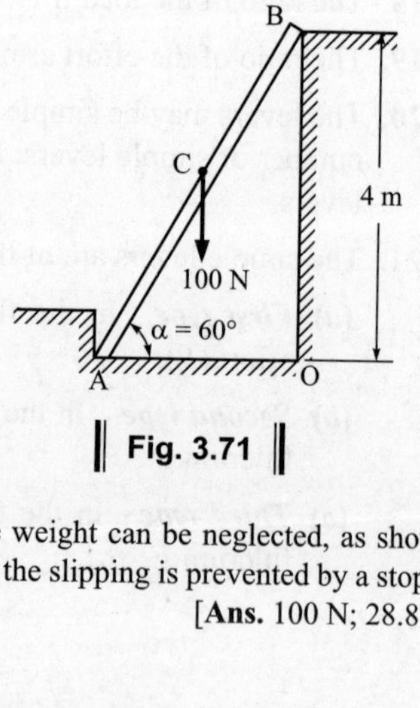


**Fig. 3.71**

7. A man weighing 100 N stands on the middle of a ladder *AB* whose weight can be neglected, as shown in Fig. 3.71. Assuming that the floor and walls are perfectly smooth and the slipping is prevented by a stop at *A*, find the reactions at *A* and *B*. Given α = 60°. [**Ans.** 100 N; 28.875 N]

**8.** A regular hexagon *ABCDEF* of each side 200 mm is subjected to forces of 10 N, 20 N, 30 N, 40 N, 50 N and 60 N, respectively, along the sides *AB*, *CB*, *DC*, *DE*, *EF* and *FA*. Determine the sum of moments of these forces about point *A*. [**Ans.** 1732 N-mm (anticlockwise)]

**9.** Three forces equal to 2 kN, 3 kN, and 4 kN act along the three sides of an equilateral triangle of side 100 mm taken in order. Find the magnitude, direction and position of the resultant. [**Ans.** 17.32 N; 90°; 200 mm]

**10.** An angle bracket, as shown in Fig. 3.72, is subjected to four forces. Determine the magnitude, direction and position of the resultant from the point *A*. [**Ans.** 55.42 kN; 32.77°; 400 mm]

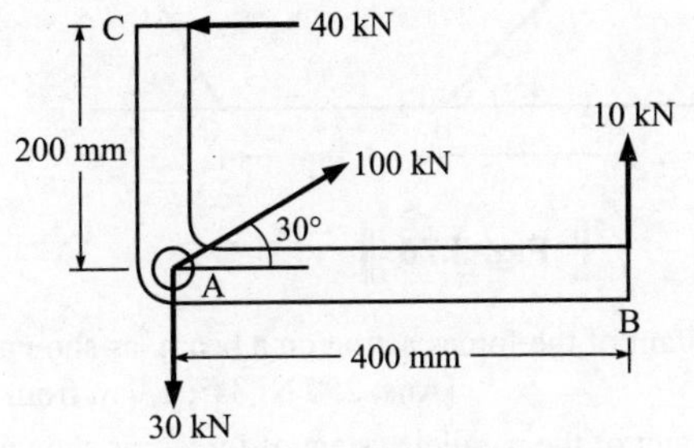


**Fig. 3.72**

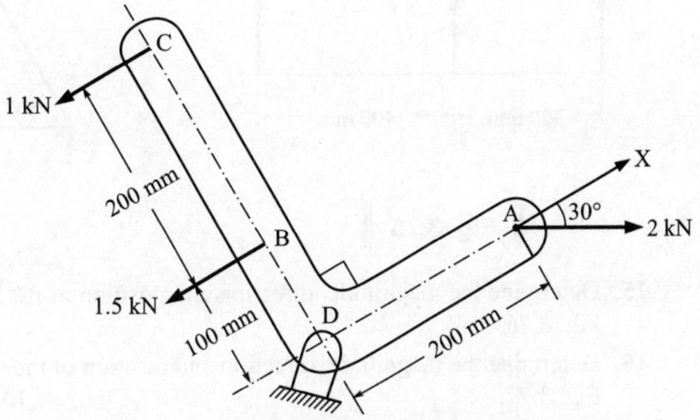


**Fig. 3.73**

**11.** A lever *ABC* of a component of a machine is hinged at *D* and is subjected to a system of forces, as shown in Fig. 3.73. Find the resultant of the forces and the moment through point *D*.

[**Ans.** 1.26 kN, 250 kN-mm (anticlockwise)]

**12.** A lever *ABC* of a component of a machine is hinged at *B* and is subjected to a system of forces, as shown in Fig. 3.74. Neglecting friction, find

1. Magnitude of the force *P* to keep the lever in equilibrium; and
2. Magnitude and direction of the reaction at *B*. [**Ans.** 895.7 N; 1444.74 N; 3.717°]

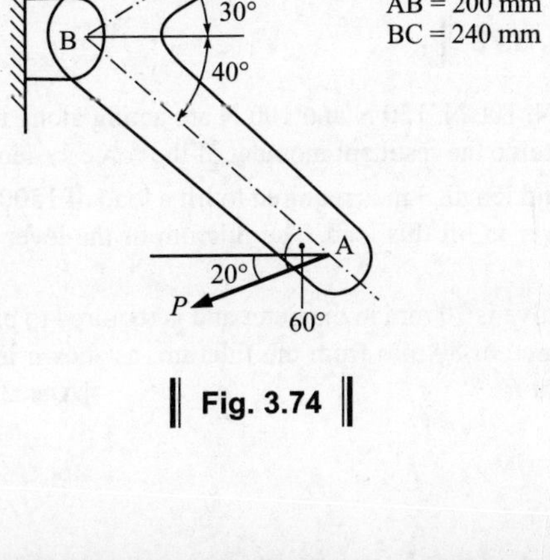


**Fig. 3.74**

**13.** The resultant of two like parallel forces $P_1$ and $P_2$ is 20 N acting at a distance of 500 mm from one of the forces. Determine the magnitude of these forces, if they are acting at a distance of 2.5 m from each other. [**Ans.** 16 N, 4 N]

**14.** The three like parallel forces of 100 N, 200 N and 300 N are acting on a beam, as shown in Fig. 3.75. Determine the magnitude and position of the resultant by graphical method. [**Ans.** 600 N at 450 mm from 100 N force]

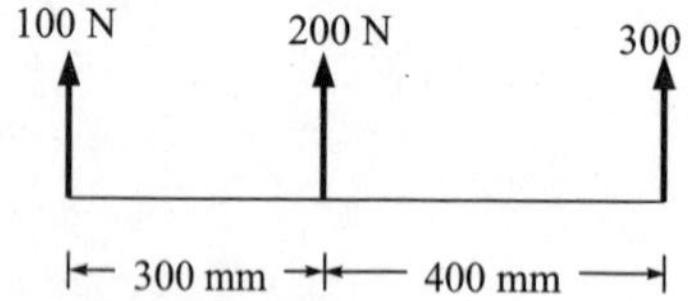


**Fig. 3.75**

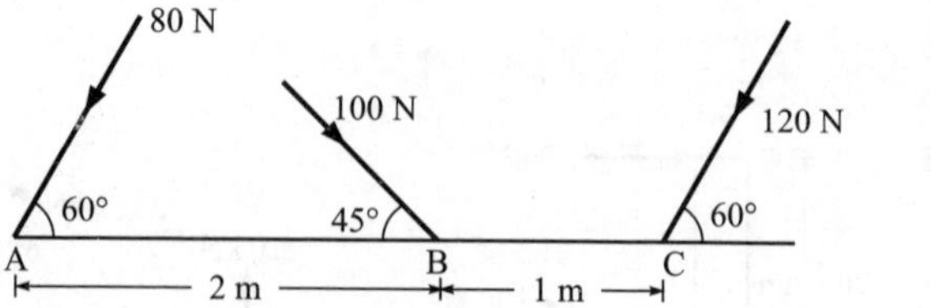


**Fig. 3.76**

**15.** Determine the magnitude, direction and position of the resultant of the forces acting on a beam, as shown in Fig. 3.76. [**Ans.** 250 N; 84°; 1.9 m from A]

**16.** Determine the magnitude, direction and position of the resultant of the parallel system of forces, as shown in Fig. 3.77. [**Ans.** 300 N acting downward at 6 m from 50 N force]

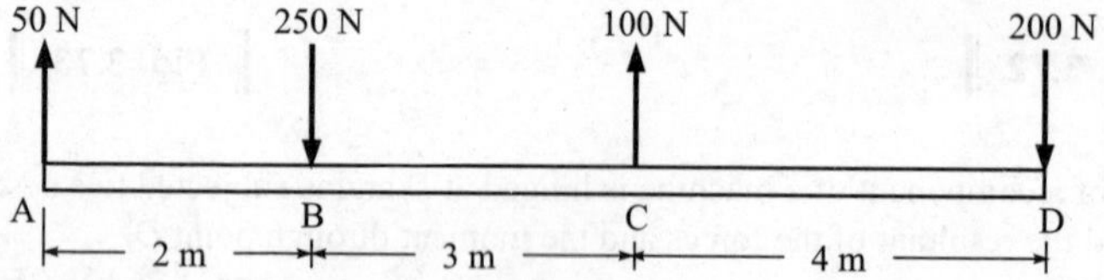


**Fig. 3.77**

**17.** A beam supported at *A* and *B* is loaded, as shown in Fig. 3.78. If a couple of 12 kN-m is applied at *C* in the anticlockwise direction, find the reactions at the supports *A* and *B*. [**Ans.** 3 kN, 9 kN]

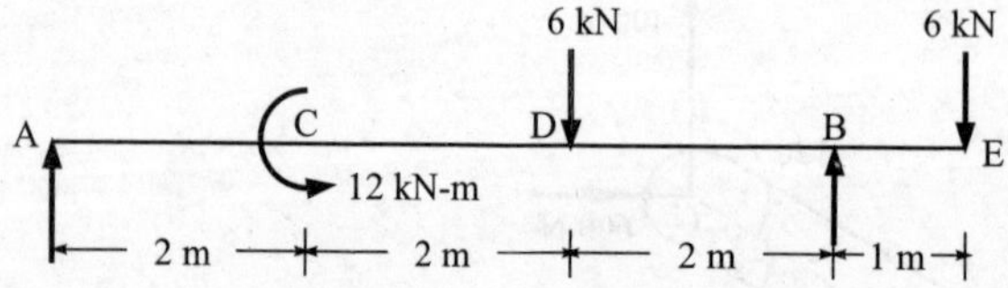


**Fig. 3.78**

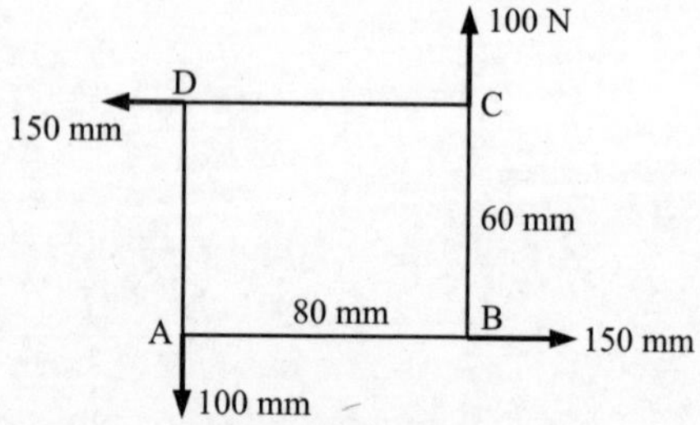


**Fig. 3.79**

**18.** Four forces equal to 150 N, 100 N, 150 N and 100 N are acting along the four sides of a rectangle *ABCD*, as shown in Fig. 3.79. Determine the resultant moment of the force system. [**Ans.** 17 kN-m]

**19.** A lever of weight 200 N and length 3 m is required to lift a load of 1500 N at one end. Find the effort required at the other end of the lever to lift this load. The fulcrum of the lever is at a distance of 1 m from the load point. [**Ans.** 700 N]

**20.** A lever of loaded safety valve is 70 mm in diameter and is required to blow-off at a pressure of 1 N/mm$^2$. The valve is located at a distance of 80 mm from the fulcrum, as shown in Fig. 3.80. Determine the value of *P* required and the reaction at *F*. [**Ans.** 350 N; 3500 N (vertically downward)]

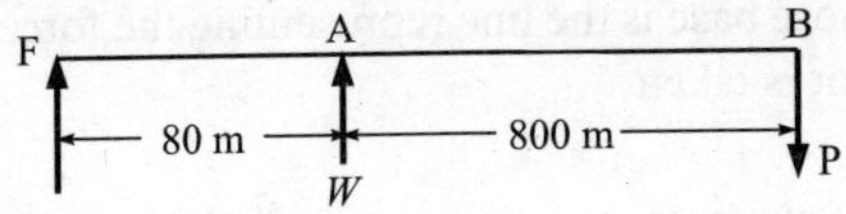


**Fig. 3.80**

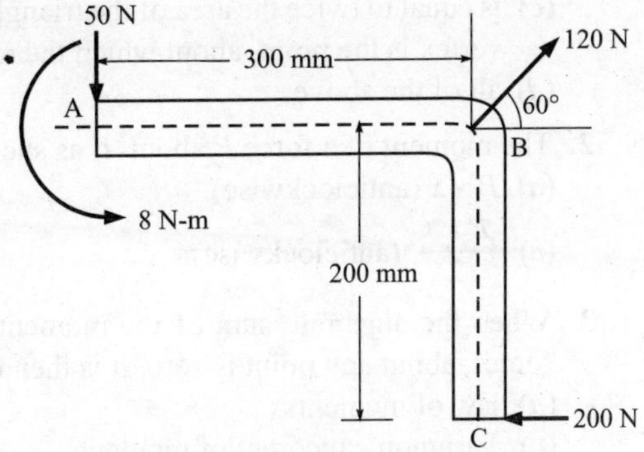


**Fig. 3.81**

**21.** An angle bracket is subjected to three forces and a couple, as shown in Fig. 3.81. Find the resultant of this system of forces. Also determine the position of this resultant in the line *AB* and *BC*.

[**Ans.** 150 N; 314.7 from *B* towards *A* and 121.4 mm from *B* towards *C*]

**22.** A lever *AB* rests horizontally on a fulcrum at *C* such that *AC* = *a* and *CB* = *b*, as shown in Fig. 3.82. Two loads, each of magnitude *Q*, are acting on the left of *C*, one at the free end *A* and the other mid-way between *A* and *C*.

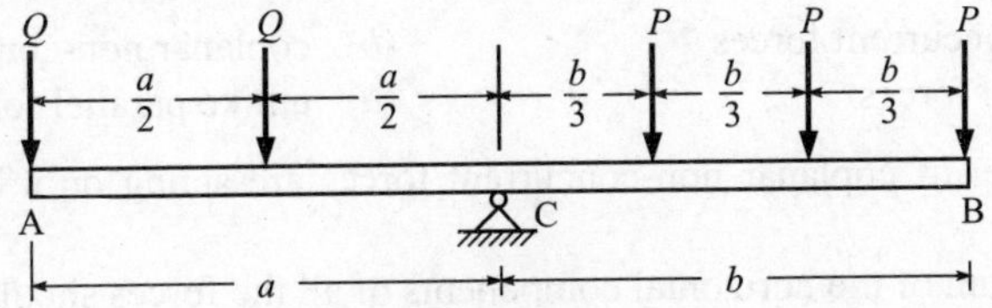


**Fig. 3.82**

Three loads, each of magnitude *P*, are acting on the right of *C*, one at the free end *B*, the second at a distance of *b*/3 on the left of *B* and the third at a distance of *b*/3 on the right of *C*. If *Q* = 2 *P* and the weight of lever is negligible, determine the ratio *a* : *b* of an arm of the lever, if it is in equilibrium. [**Ans.** 2 : 3]

## REVIEW QUESTIONS

1. What do you understand by moment of a force?
2. Explain clearly the difference between clockwise moment and anticlockwise moment.
3. State Varignon's theorem of moments and principle of moments.
4. State the conditions of equilibrium for a coplanar non-concurrent system of forces.
5. What do you understand by parallel forces and their classifications?
6. Describe the meaning and use of funicular polygon.
7. Differentiate between moment of a force and a couple.
8. What are the properties of a couple?
9. Distinguish between simple lever and compound lever.
10. Explain the various types of levers.

## OBJECTIVE TYPE QUESTIONS

1. The moment of a force
   (*a*) is the turning effect produced by a force, on the body, on which it acts
   (*b*) is equal to the product of the force on the body and the perpendicular distance of the point and the line of application of the force

(*c*) is equal to twice the area of the triangle, whose base is the line representing the force and whose vertex is the point, about which the moment is taken
(*d*) all of the above

**2.** The moment of a force $P$ about $A$, as shown in Fig. 3.83, is
(*a*) $P \times x$ (anticlockwise) (*b*) $P \times x$ (clockwise)
(*c*) $\frac{P \times x}{2}$ (anticlockwise) (*d*) $\frac{P \times x}{2}$ (clockwise)

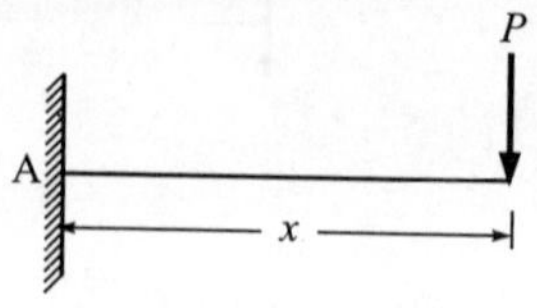

**Fig. 3.83**

**3.** When the algebraic sum of the moments of a number of coplanar forces, about any point is zero, it is then known as
(*a*) law of moments (*b*) Lami's theorem
(*c*) Varignon's theorem of moments (*d*) none of these

**4.** When a number of coplanar forces acting on a particle are in equilibrium, then the algebraic sum of their moments about any point is equal to the moment of their resultant force about the same point. This statement is known as
(*a*) law of moments (*b*) Lami's theorem
(*c*) Varignon's theorem of moments (*d*) none of these

**5.** The forces, whose lines of action are parallel to each other and act in the same directions, are known as
(*a*) coplanar concurrent forces (*b*) coplanar non-concurrent forces
(*c*) like parallel forces (*d*) unlike parallel forces

**6.** When a number of coplanar non-concurrent forces are acting on a body, then the condition of equilibrium is
(*a*) algebraic sum of the horizontal components of all the forces should be zero
(*b*) algebraic sum of the vertical components of all the forces should be zero
(*c*) algebraic sum of moments of all the forces about any point should be zero
(*d*) all of the above

**7.** A couple produces ............... motion.
(*a*) translatory (*b*) rotational (*c*) combined translatory and rotational

**8.** When the arm of the couple is doubled, then the moment of that couple will
(*a*) remain same (*b*) be halved (*c*) be doubled (*d*) none of these

**9.** Two couples will balance each other when they are in the same plane and
(*a*) have equal moments and their direction of rotation is opposite
(*b*) have equal moments and their direction of rotation is same
(*c*) have unequal moments and their direction of rotation is opposite
(*d*) none of the above

**10.** In the lever $AB$, as shown in Fig. 3.84, $F$ is the fulcrum, $AF$ is the effort arm and $BF$ is the load arm. The leverage of the lever is given by
(*a*) $AF \times BF$ (*b*) $AF + FB$
(*c*) $\frac{BF}{AF}$ (*d*) $\frac{AF}{BF}$

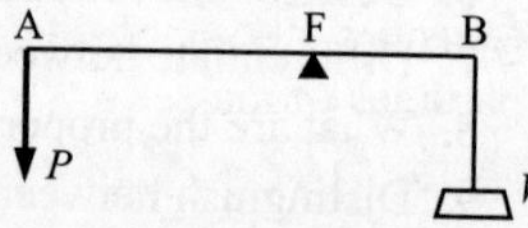

**Fig. 3.84**

***Answers to Objective Type Questions***

| | | | | |
|---|---|---|---|---|
| **1.** (*d*) | **2.** (*b*) | **3.** (*a*) | **4.** (*c*) | **5.** (*c*) |
| **6.** (*d*) | **7.** (*b*) | **8.** (*c*) | **9.** (*a*) | **10.** (*d*) |

# 4

# CENTRE OF GRAVITY



## 4.1 INTRODUCTION

A body is considered to be made up of infinite number of small particles and each particle is attracted towards the centre of the earth. The resultant of all these forces of attraction, acting vertically downwards, is known as weight of the body. Since the distance of the different particles of the body is same from the centre of the earth, therefore, all the forces of attraction are assumed to be parallel and their resultant constitutes the weight of the body. The point through which the whole weight of the body acts, irrespective of the position of the body, is termed as *centre of gravity* (briefly written as C.G.).

The centre of gravity of a plane figure (like triangle, rectangle, circle etc.) which has area only and no mass is referred to as *centroid* or *centre of area* and the whole area of the plane figure is assumed to be acting through this point.

**Note:** It is a common practice to use centre of gravity for centroid and *vice versa.*

## 4.2 CENTRE OF GRAVITY OF A RIGID BODY BY METHOD OF MOMENTS

Consider a body of total mass $M$. Suppose, it is composed of small particles of masses $m_1$, $m_2$, $m_3$, ..., etc. Let $x_1$, $x_2$, $x_3$, ..., etc. be the distances of these masses from the fixed line or *reference axis OY* and $y_1$, $y_2$, $y_3$, ..., etc. be the distances from the fixed line or *reference axis OX,* as shown in Fig. 4.1. Let the centre of gravity (or centre of mass) of the whole mass $M$ lies at $G$ which is at a distance of $\bar{x}$ and $\bar{y}$ from the axes $OY$ and $OX$, respectively.

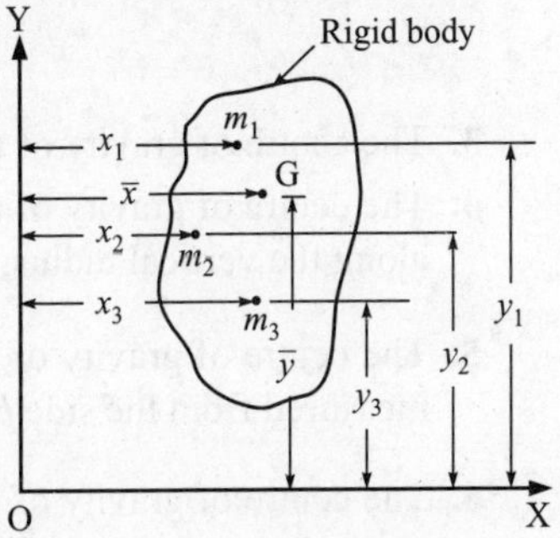


Fig. 4.1

We know that moment of all the small masses about $OY$,

$$= m_1 x_1 + m_2 x_2 + m_3 x_3 + ...$$

and moment of total mass $M$ about $OY = M\bar{x}$

Now, according to principle of moments,

$$M\bar{x} = m_1 x_1 + m_2 x_2 + m_3 x_3 + \ldots$$

$$\therefore \quad \bar{x} = \frac{m_1 x_1 + m_2 x_2 + m_3 x_3 + \ldots}{M} = \frac{\Sigma m x}{M}$$

Similarly, when the moments are taken about *OX*, then

$$\bar{y} = \frac{m_1 y_1 + m_2 y_2 + m_3 y_3 + \ldots}{M} = \frac{\Sigma m y}{M}$$

where $M = m_1 + m_2 + m_3 + \ldots$

**Note:** The centre of gravity is always determined about some fixed line. This fixed line is known as *reference axis.* It may be noted that for different axis of reference, the centre of gravity will be different.

## 4.3 CENTRE OF GRAVITY (CENTROID) OF COMMONLY USED PLANE SECTIONS

The centre of gravity of most commonly used plane sections are as follows:

1. The centre of gravity (*G*) of a rectangle (or parallelogram) lies at the intersection of its diagonals, *i.e.* at the midpoint of width (*b*) and depth (*d*), as shown in Fig. 4.2 (*a*).
2. The centre of gravity (*G*) of a triangle lies at the point of intersection of three medians. If *h* is the height of the triangle, then its C.G. lies at a distance of *h*/3 from its base along the vertical axis, as shown in Fig. 4.2 (*b*).

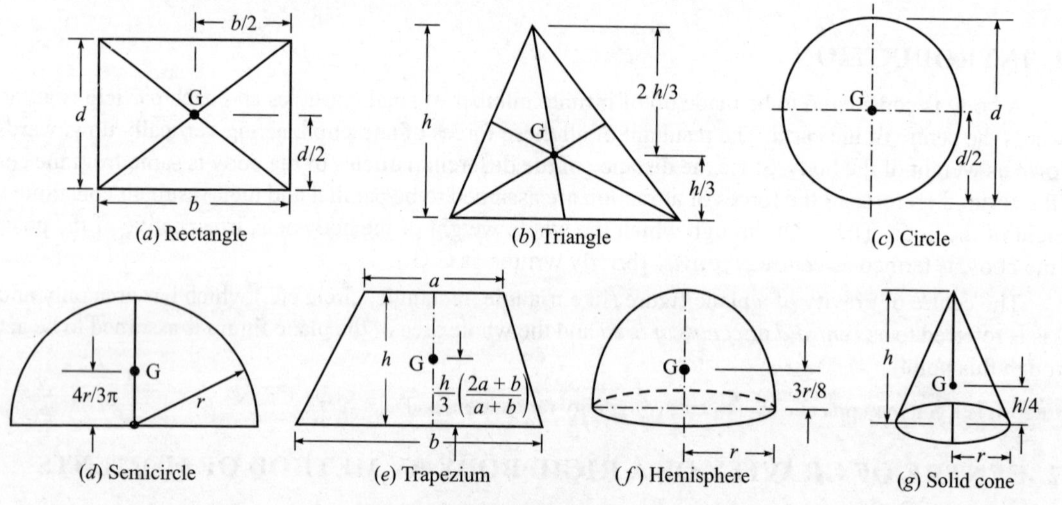


**Fig. 4.2**

3. The centre of gravity of a circle lies at its centre, as shown in Fig. 4.2 (*c*).
4. The centre of gravity of a semicircle of radius *r* lies at a distance of $4r/3\pi$ from its base measured along the vertical radius, as shown in Fig. 4.2 (*d*).
5. The centre of gravity of a trapezium with parallel sides *a* and *b*, lies at a distance of $\frac{h}{3}\left[\frac{2a+b}{a+b}\right]$ measured from the side *b*, as shown in Fig. 4.2 (*e*).
6. The centre of gravity of a hemisphere of radius *r* lies at a distance of $\frac{3r}{8}$ from its base measured along the vertical radius, as shown in Fig. 4.2 (*f*).

7. The centre of gravity of a right circular solid cone of height $h$, lies at a distance of $h/4$ from its base measured along the vertical axis, as shown in Fig. 4.2 (*g*).
8. The centre of gravity of a uniform rod lies at its midpoint.

## 4.4 METHOD OF FINDING CENTRE OF GRAVITY OF PLANE SECTIONS

The centre of gravity (or centroid) of plane geometrical figures (which have only areas but no mass) may be determined in the similar way as discussed in Sec. 4.2.

Let $a_1, a_2, a_3, ...$, etc. are the areas of the figure whose centre of gravity is to be determined. If $\bar{x}$ and $\bar{y}$ are the distances of centre of gravity of the whole section, from the reference axis *OY* and *OX*, respectively, then

$$\bar{x} = \frac{a_1x_1 + a_2x_2 + a_3x_3 + ...}{a_1 + a_2 + a_3 + ...}$$

and

$$\bar{y} = \frac{a_1y_1 + a_2y_2 + a_3y_3 + ...}{a_1 + a_2 + a_3 + ...}$$

where $x_1$, $x_2$ and $x_3$, ..., etc. are the distances of individual centre of gravity of areas, $a_1, a_2, a_3$, ..., etc. from the reference axis *OY*, and $y_1, y_2, y_3$, ..., etc. are the distances of individual centre of gravity of areas $a_1, a_2, a_3$, ..., etc. from the reference axis *OX*.

## 4.5 CENTRE OF GRAVITY OF SYMMETRICAL AND UNSYMMETRICAL SECTIONS

When the section is symmetrical about *X*-axis or *Y*-axis, then the centre of gravity will lie on that axis about which the section is symmetrical. In other words, the centre of gravity will lie on the axis of symmetry (*i.e.* either *X*-axis or *Y*-axis). Thus we have to determine only $\bar{x}$ or $\bar{y}$.

When the section is not symmetrical about *X*-axis or *Y*-axis, then we have to determine both $\bar{x}$ and $\bar{}$ (See Example 4.4).

**Note:** When the section is symmetrical about both the axes of symmetry (*i.e.* *X*-axis and *Y*-axis), then the centre of gravity lies at the intersection of both the axes. The areas such as squares, rectangles, and circles have centroid at the intersection of their centroidal axis.

**Example 4.1** *Find the centroid of an inverted T-section of flange 100 mm × 10 mm and web 120 mm × 12 mm.*

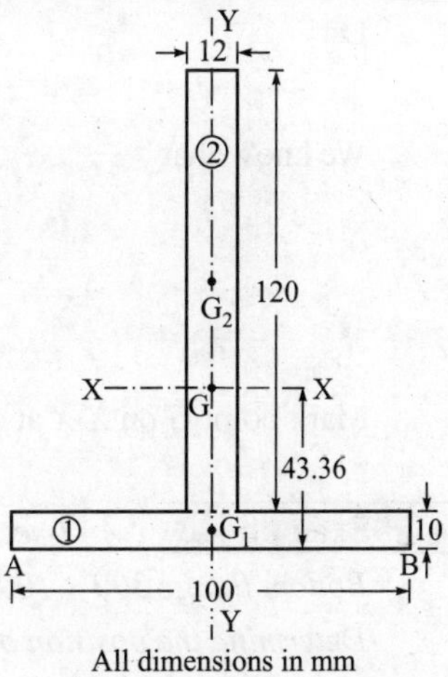

**Fig. 4.3**

***Solution*** Given : Inverted T-section, as shown in Fig. 4.3.

Since the section is symmetrical about *Y-Y* axis, therefore the centroid will lie on this axis. Let $\bar{y}$ be the distance of the centroid from the bottom face of the section *AB* which is taken as reference line.

Divide the section into two rectangles (1) and (2), as shown in the figure.

Area of rectangle (1), $a_1 = 100 \times 10 = 1000 \text{ mm}^2$

Distance of its C.G. ($G_1$) from *AB*,

$$y_1 = \frac{10}{2} = 5 \text{ mm}$$

Area of rectangle (2), $a_2 = 120 \times 12 = 1440 \text{ mm}^2$

Distance of its C.G. ($G_2$) from $AB$, $y_2 = 10 + \frac{120}{2} = 70$ mm

We know that $\bar{y} = \frac{a_1 y_1 + a_2 y_2}{a_1 + a_2} = \frac{1000 \times 5 + 1440 \times 70}{1000 + 1440}$

$$= \frac{5000 + 100\,800}{2440} = 43.36 \text{ mm } \textbf{Ans.}$$

Mark point $G$ on $Y$-$Y$ axis at a distance of 43.36 mm from $AB$, which is the required centroid of the given section.

**Example 4.2** *Find the centre of gravity of the channel section, as shown in Fig. 4.4.*

***Solution*** Given : A channel section, as shown in Fig. 4.4.

Since the section is symmetrical about $X$-$X$ axis, therefore C.G. of the section will lie on this axis.

Divide the section into three rectangles (1), (2) and (3), as shown in the figure. Let $AB$ be the axis of reference.

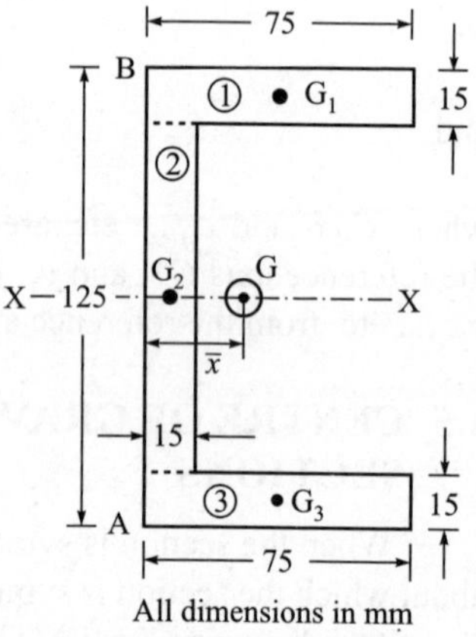


**Fig. 4.4**

Area of rectangle (1), $a_1 = 75 \times 15 = 1125 \text{ mm}^2$

Distance of its C.G. ($G_1$) from $AB$,

$$x_1 = \frac{75}{2} = 37.5 \text{ mm}$$

Area of rectangle (2), $a_2 = (125 - 2 \times 15) \times 15$

$= 95 \times 15 = 1425 \text{ mm}^2$

Distance of its C.G. ($G_2$) from $AB$,

$$x_2 = \frac{15}{2} = 7.5 \text{ mm}$$

Area of rectangle (3), $a_3 = 75 \times 15 = 1125 \text{ mm}^2$

Distance of its C.G. ($G_3$) from $AB$,

$$x_3 = \frac{75}{2} = 37.5 \text{ mm}$$

Let $\bar{x}$ = Distance of C.G. ($G$) of the whole section from $AB$.

We know that $\bar{x} = \frac{a_1 x_1 + a_2 x_2 + a_3 x_3}{a_1 + a_2 + a_3} = \frac{1125 \times 37.5 + 1425 \times 7.5 + 1125 \times 37.5}{1125 + 1425 + 1125}$

$$= \frac{42\,187.5 + 10\,687.5 + 42\,187.5}{3675} = \frac{95\,062.5}{3675}$$

$= 25.867$ mm **Ans.**

Mark point $G$ on $X$-$X$ at a distance of 25.867 mm from $AB$, which is the required C.G. of the section.

**Example 4.3** *An I-section has the following dimensions in mm :*

*Bottom flange 300 × 100 ; Top flange 150 × 50 and web 450 × 40.*

*Determine the position of centre of gravity of the section.*

***Solution*** Given : Bottom flange = 300 mm × 100 mm ; Top flange = 150 mm × 50 mm ; Web = 450 mm × 40 mm

The *I*-section is shown in Fig. 4.5.

Since the section is symmetrical about *Y-Y* axis, therefore the centre of gravity of the section will lie on this axis. Let $\bar{y}$ be the distance of centre of gravity from the base *AB* from the bottom flange. Split the whole section into three rectangles (1), (2) and (3).

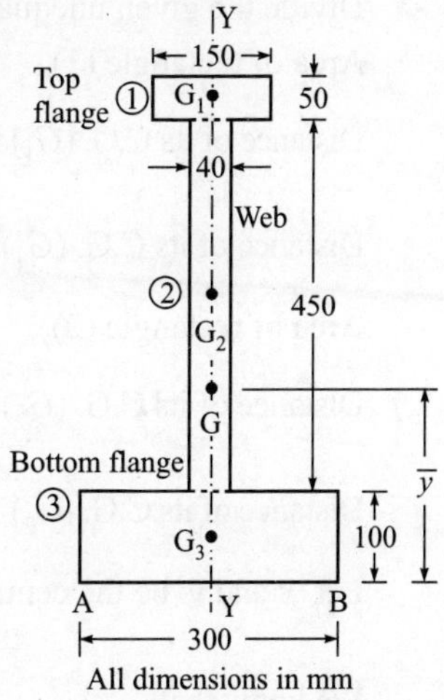


Fig. 4.5

Area of rectangle (1), $a_1 = 150 \times 50 = 7500 \text{ mm}^2$

Distance of its C.G. ($G_1$) from *AB*,

$$y_1 = 100 + 450 + \frac{50}{2} = 575 \text{ mm}$$

Area of rectangle (2), $a_2 = 450 \times 40 = 18\,000 \text{ mm}^2$

Distance of its C.G. ($G_2$) from *AB*,

$$y_2 = 100 + \frac{450}{2} = 325 \text{ mm}$$

Are of rectangle (3), $a_3 = 300 \times 100 = 30\,000 \text{ mm}^2$

Distance of its C.G. ($G_3$) from *AB*, $y_3 = \frac{100}{2} = 50 \text{ mm}$

We know that

$$\bar{y} = \frac{a_1 y_1 + a_2 y_2 + a_3 y_3}{a_1 + a_2 + a_3}$$

$$= \frac{7500 \times 575 + 18\,000 \times 325 + 30\,000 \times 50}{7500 + 18\,000 + 30\,000}$$

$$= \frac{4\,312\,500 + 5\,850\,000 + 1\,500\,000}{55\,500} = \frac{11\,662\,500}{55\,500}$$

$= 210$ mm **Ans.**

**Example 4.4** *Find the centroid of the unequal angle 200 mm × 150 mm × 12 mm, as shown in Fig. 4.6.*

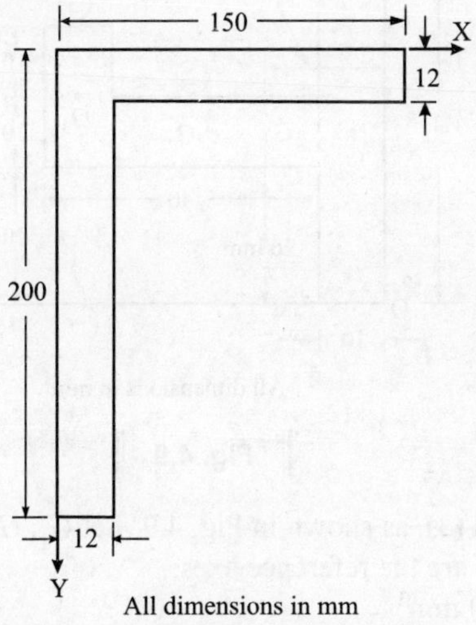


Fig. 4.6

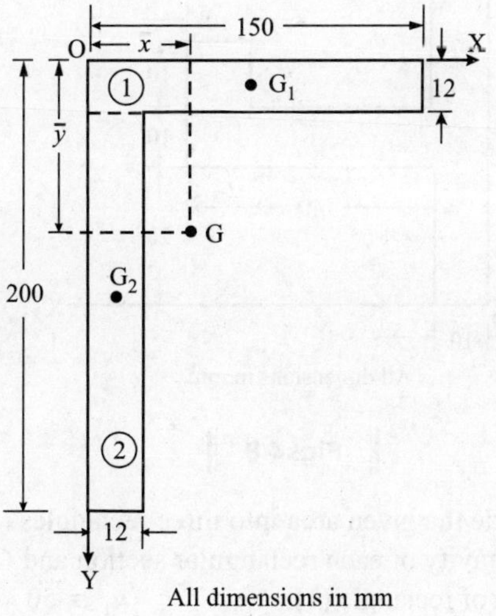


Fig. 4.7

***Solution*** Given : Unequal angle section, as shown in Fig. 4.6.

Divide the given unequal angle section into two rectangles (1) and (2), as shown in Fig. 4.7.

Area of rectangle (1), $a_1 = 150 \times 12 = 1800 \text{ mm}^2$

Distance of its C.G. ($G_1$) from $OY$, $x_1 = \dfrac{150}{2} = 75$ mm

Distance of its C.G. ($G_1$) from $OX$, $y_1 = \dfrac{12}{2} = 6$ mm

Area of rectangle (2), $a_2 = (200 - 12)\,12 = 188 \times 12 = 2256 \text{ mm}^2$

Distance of its C.G. ($G_2$) from $OY$, $x_2 = \dfrac{12}{2} = 6$ mm

Distance of its C.G. ($G_2$) from $OX$, $y_2 = 12 + \dfrac{(200-12)}{2} = 106$ mm

Let $\overline{x}$ and $\overline{y}$ be the centroid of the section from the reference axis $OY$ and $OX$, respectively.

We know that
$$\overline{x} = \frac{a_1 x_1 + a_2 x_2}{a_1 + a_2} = \frac{1800 \times 75 + 2256 \times 6}{1800 + 2256}$$
$$= \frac{135\,000 + 13\,536}{4056} = 36.62 \text{ mm } \textbf{Ans.}$$

and
$$\overline{y} = \frac{a_1 y_1 + a_2 y_2}{a_1 + a_2} = \frac{1800 \times 6 + 2256 \times 106}{1800 + 2256} = \frac{10\,800 + 239\,136}{4056}$$
$$= 61.62 \text{ mm } \textbf{Ans.}$$

**Example 4.5** *Locate the centre of gravity of the area, as shown in Fig. 4.8, with respect to coordinate axes. All dimensions are in mm.*

***Solution*** Given : Area as shown in Fig. 4.8.

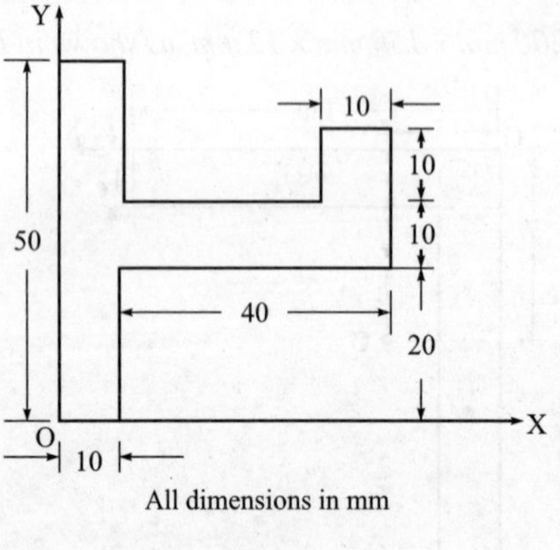


**Fig. 4.8**

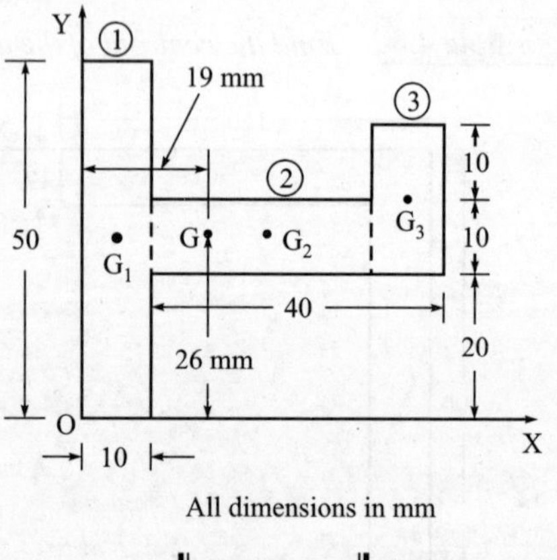


**Fig. 4.9**

Divide the given area into three rectangles (1), (2) and (3), as shown in Fig. 4.9. Let $G_1$, $G_2$, $G_3$ be the centre of gravity of each rectangular section and $OX$ and $OY$ are the reference axes.

Area of rectangle (1), $a_1 = 50 \times 10 = 500 \text{ mm}^2$

Distance of its C.G. ($G_1$) from $OY$, $x_1 = \dfrac{10}{2} = 5$ mm

Distance of its C.G. ($G_1$) from $OX$, $y_1 = \dfrac{50}{2} = 25$ mm

Area of rectangle (2), $a_2 = (40 - 10) \times 10 = 300 \text{ mm}^2$

Distance of its C.G. ($G_2$) from $OY$, $x_2 = 10 + \left(\dfrac{40 - 10}{2}\right) = 25 \text{ mm}$

Distance of its C.G. ($G_2$) from $OX$, $y_2 = 20 + \dfrac{10}{2} = 25 \text{ mm}$

Area of rectangle (3), $a_3 = (10 + 10) \times 10 = 200 \text{ mm}^2$

Distance of its C.G. ($G_3$) from $OY$, $x_3 = 10 + 30 + \dfrac{10}{2} = 45 \text{ mm}$

Distance of its C.G. ($G_3$) from $OX$, $y_3 = 20 + 10 = 30 \text{ mm}$

Let $\bar{x}$ and $\bar{y}$ be the distances of centre of gravity ($G$) from the reference axes $OY$ and $OX$, respectively.

We know that

$$\bar{x} = \frac{a_1x_1 + a_2x_2 + a_3x_3}{a_1 + a_2 + a_3} = \frac{500 \times 5 + 300 \times 25 + 200 \times 45}{500 + 300 + 200}$$

$$= \frac{2500 + 7500 + 9000}{1000} = \frac{19\,000}{1000} = 19 \text{ mm}$$

and

$$\bar{y} = \frac{a_1y_1 + a_2y_2 + a_3y_3}{a_1 + a_2 + a_3} = \frac{500 \times 25 + 300 \times 25 + 200 \times 30}{500 + 300 + 200}$$

$$= \frac{12\,500 + 7500 + 6000}{1000} = \frac{26\,000}{1000} = 26 \text{ mm}$$

Mark point $G$ on Fig. 4.9, such that $\bar{x} = 19$ mm and $\bar{y} = 26$ mm. **Ans.**

**Example 4.6** *Find the centre of gravity of the E-section, as shown in Fig. 4.10, with respect to the point O. All dimensions are in mm.*

***Solution*** Given : *E*-section, as shown in Fig. 4.10

Divide the section into four rectangles, as shown in Fig. 4.11.

Let $G_1$, $G_2$, $G_3$ and $G_4$ be the centroid of each rectangular section and $OX$ and $OY$ be the reference axes.

Area of rectangle (1), $a_1 = 100 \times 20 = 2000 \text{ mm}^2$

Distance of its C.G. ($G_1$) from $OY$, $x_1 = \dfrac{100}{2} = 50 \text{ mm}$

Distance of its C.G. ($G_1$) from $OX$, $y_1 = 300 - \dfrac{20}{2} = 290 \text{ mm}$

Area of rectangle (2), $a_2 = 50 \times 20 = 1000 \text{ mm}^2$

Distance of its C.G. ($G_2$) from $OY$, $x_2 = 20 + \dfrac{50}{2} = 45 \text{ mm}$

Distance of its C.G. ($G_2$) from $OX$, $y_2 = 150 - \dfrac{20}{2} = 140 \text{ mm}$

Area of rectangle (3), $a_3 = (300 - 40) \times 20 = 5200 \text{ mm}^2$

Distance of its C.G. ($G_3$) from $OY$, $x_3 = \dfrac{20}{2} = 10 \text{ mm}$

Distance of its C.G. ($G_3$) from $OX$, $y_3 = 150 - 20 = 130$ mm

Area of rectangle (4), $a_4 = 200 \times 20 = 4000$ mm²

Distance of its C.G. ($G_4$) from $OY$, $x_4 = \frac{200}{2} = 100$ mm

Distance of its C.G. ($G_4$) from $OX$, $y_4 = \frac{20}{2} = 10$ mm

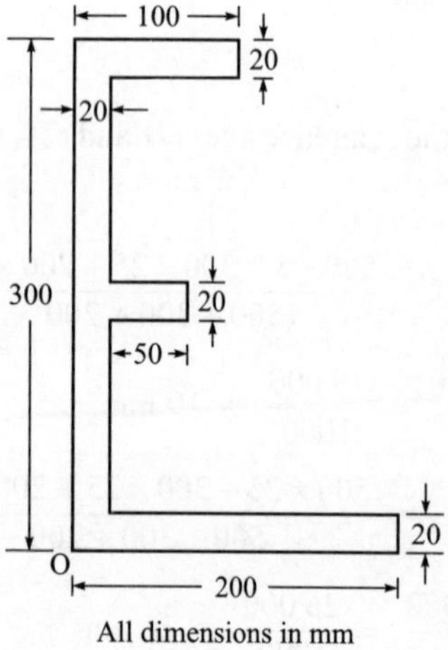


**Fig. 4.10**

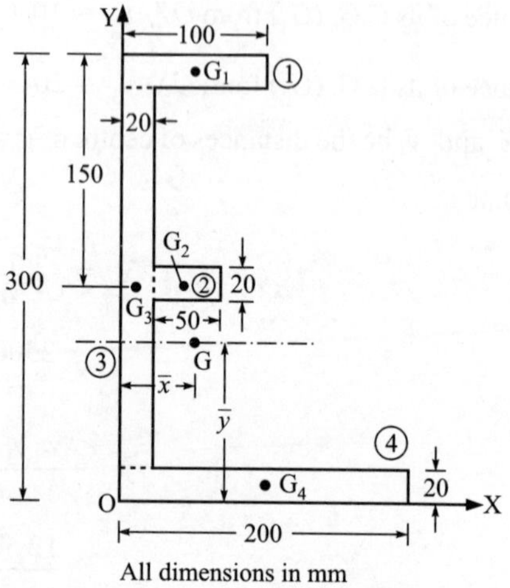


**Fig. 4.11**

Let $\overline{x}$ and $\overline{y}$ be the distances of centre of gravity $G$ from the reference axes $OY$ and $OX$, respectively.

We know that

$$\overline{x} = \frac{a_1x_1 + a_2x_2 + a_3x_3 + a_4x_4}{a_1 + a_2 + a_3 + a_4}$$

$$= \frac{2000 \times 50 + 1000 \times 45 + 5200 \times 10 + 4000 \times 100}{2000 + 1000 + 5200 + 4000}$$

$$= \frac{100\,000 + 45\,000 + 52\,000 + 400\,000}{12\,200} = \frac{597\,000}{12\,200}$$

$$= 48.93 \text{ mm}$$

and

$$\overline{y} = \frac{a_1y_1 + a_2y_2 + a_3y_3 \;\; a_4y_4}{a_1 + a_2 + a_3 + a_4}$$

$$= \frac{2000 \times 290 + 1000 \times 140 + 5200 \times 130 + 4000 \times 10}{2000 + 1000 + 5200 + 4000}$$

$$= \frac{580\,000 + 140\,000 + 676\,000 + 40\,000}{12\,200}$$

$$= \frac{1\,436\,000}{12\,200} = 117.7 \text{ mm}$$

Mark point $G$ in Fig. 4.11, such that $\overline{x} = 48.93$ mm and $\overline{y} = 117.7$ mm. **Ans.**

**Example 4.7** *Determine the centre of gravity of the plane uniform lamina consisting of a rectangle, a triangle and a semicircle, as shown in Fig. 4.12.*

***Solution*** Given : A plane uniform lamina, as shown in Fig. 4.12.

Let *OX* and *OY* be the reference axes, as shown in Fig. 4.13. Since the section in not symmetrical about any axis, therefore we have to determine the value of both $\bar{x}$ and $\bar{y}$. Let $\bar{x}$ is the distance of C.G. from axis *OY* and $\bar{y}$ be the distance of C.G. from axis *OX*.

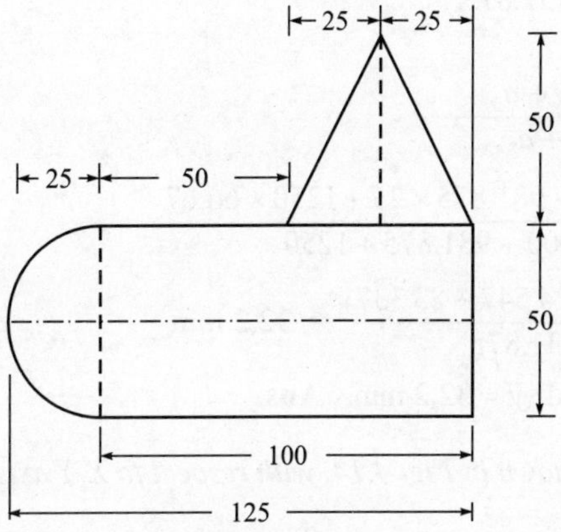


Fig. 4.12

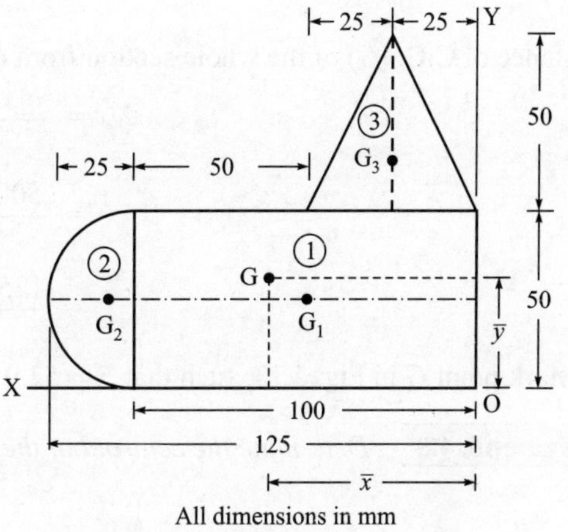


Fig. 4.13

***For rectangular section***

$$\text{Area, } a_1 = 100 \times 50 = 5000 \text{ mm}^2$$

$$\text{Distance of its C.G. } (G_1) \text{ from } OY,\ x_1 = \frac{100}{2} = 50 \text{ mm}$$

$$\text{Distance of its C.G. } (G_1) \text{ from } OX,\ y_1 = \frac{50}{2} = 25 \text{ mm}$$

***For semicircular section***

$$\text{Area, } a_2 = \frac{\pi r^2}{2} = \frac{\pi (25)^2}{2} = 981.875 \text{ mm}^2$$

$$\text{Distance of its C.G. } (G_2) \text{ from } OY,\ x_2 = 100 + \frac{4r}{3\pi} = 100 + \frac{4 \times 25}{3\pi} = 110.61 \text{ mm}$$

$$\text{Distance of its C.G. } (G_2) \text{ from } OX,\ y_2 = \frac{50}{2} = 25 \text{ mm}$$

***For triangular section***

$$\text{Area, } a_3 = \frac{1}{2} \times 50 \times 50 = 1250 \text{ mm}^2$$

$$\text{Distance of its C.G. } (G_3) \text{ from } OY,\ x_3 = \frac{50}{2} = 25 \text{ mm}$$

$$\text{Distance of its C.G. } (G_3) \text{ from } OX,\ y_3 = 50 + \frac{50}{3} = 66.67 \text{ mm}$$

We know that distance of C.G. ($G$) of the whole section from $OY$,

$$\bar{x} = \frac{a_1x_1 + a_2x_2 + a_3x_3}{a_1 + a_2 + a_3}$$

$$= \frac{5000 \times 50 + 981.875 \times 110.61 + 1250 \times 25}{5000 + 981.875 + 1250}$$

$$= \frac{250\,000 + 108\,605 + 31\,250}{7231.875} = 53.9 \text{ mm}$$

and distance of C.G. ($G$) of the whole section from $OX$,

$$\bar{y} = \frac{a_1y_1 + a_2y_2 + a_3y_3}{a_1 + a_2 + a_3}$$

$$= \frac{5000 \times 25 + 981.875 \times 25 + 1250 \times 66.67}{5000 + 981.875 + 1250}$$

$$= \frac{125\,000 + 24\,547 + 83\,337}{7231.875} = 32.2 \text{ mm}$$

Mark point $G$ in Fig. 4.13, such that $\bar{x} = 53.9$ mm and $\bar{y} = 32.2$ mm. **Ans.**

**Example 4.8** *Determine the centroid of the area, shown in Fig. 4.14, with respect to X-Y axis.*

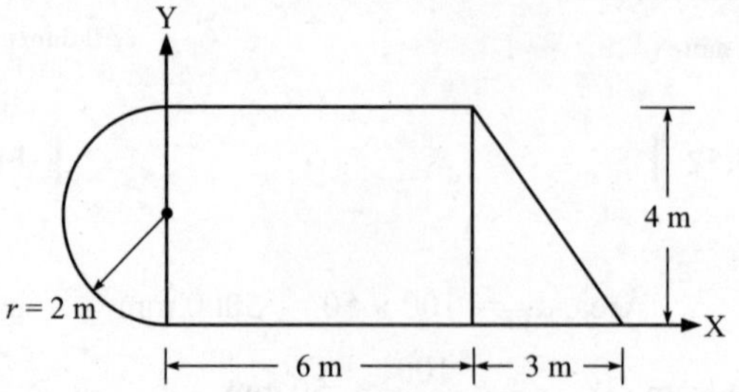


**Fig. 4.14**

***Solution*** Given : Area, as shown in Fig. 4.14.

The given area consists of a triangle $BCD$, rectangle $ABDE$ and a semicircle $AFE$, as shown in Fig. 4.15. The reference axes are $AX$ and $AY$ from which the centroid of the whole area is to be determined.

***For triangle BCD***

$$\text{Area, } a_1 = \frac{1}{2} \times 3 \times 4 = 6 \text{ m}^2$$

Distance of its C.G. ($G_1$) from $AY$, $x_1 = 6 + \frac{1}{3} \times 3 = 7$ m

Distance of its C.G. ($G_1$) from $AX$, $y_1 = \frac{1}{3} \times 4 = \frac{4}{3}$ m

***For rectangle ABDE***

$$\text{Area, } a_2 = 6 \times 4 = 24 \text{ m}^2$$

Distance of its C.G. ($G_2$) from $AY$, $x_2 = \frac{6}{2} = 3$ m

Distance of its C.G. ($G_2$) from $AX$, $y_2 = \frac{4}{2} = 2$ m

***For semicircle AFE***

$$\text{Area, } a_3 = \frac{\pi r^2}{2} = \frac{\pi(2)^2}{2} = 6.284 \text{ m}^2$$

$$\text{Distance of its C.G. } (G_3) \text{ from } AY,\ x_3 = -\frac{4r}{3\pi} = -\frac{4 \times 2}{3 \times 3.142} = -0.85 \text{ m}$$

...(Minus sign due to on the opposite side of $AY$)

$$\text{Distance of its C.G. } (G_3) \text{ from } AX,\ y_3 = \frac{AE}{2} = \frac{4}{2} = 2 \text{ m}$$

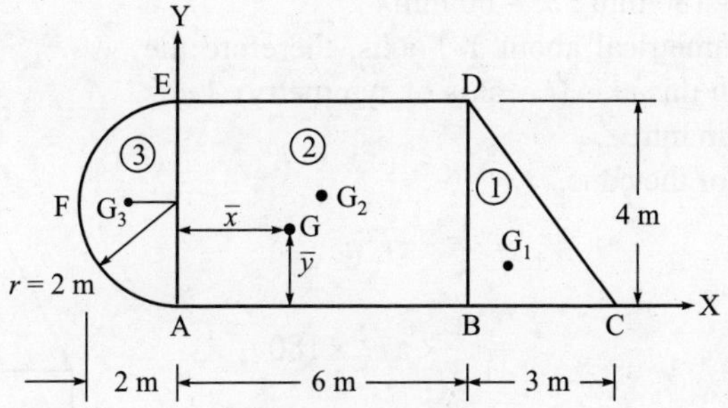


**Fig. 4.15**

Let $\bar{x}$ and $\bar{y}$ be the distances of centre of gravity ($G$) from the reference axes $AY$ and $AX$, respectively.

We know that

$$\bar{x} = \frac{a_1x_1 + a_2x_2 + a_3x_3}{a_1 + a_2 + a_3} = \frac{6 \times 7 + 24 \times 3 + 6.284(-0.85)}{6 + 24 + 6.284}$$

$$= \frac{42 + 72 - 5.341}{36.284} = \frac{108.659}{36.284} = 2.9947 \text{ m } \textbf{Ans.}$$

and

$$\bar{y} = \frac{a_1y_1 + a_2y_2 + a_3y_3}{a_1 + a_2 + a_3} = \frac{6 \times \frac{4}{3} + 24 \times 2 + 6.284 \times 2}{6 + 24 + 6.284}$$

$$= \frac{8 + 48 + 12.568}{36.284} = \frac{68.568}{36.284} = 1.8897 \text{ m } \textbf{Ans.}$$

Mark point $G$ on the given area such that $\bar{x} = 2.9947$ m and $\bar{y} = 1.8897$ m, as shown in Fig. 4.15.

## 4.6 CENTRE OF GRAVITY OF SOLID BODIES

The centre of gravity of solid bodies is determined in the similar way as already discussed in Sec. 4.4, *i.e.*

$$\bar{x} = \frac{V_1x_1 + V_2x_2 + V_3x_3 + ...}{V_1 + V_2 + V_3 + ...}$$

and

$$\bar{y} = \frac{V_1y_1 + V_2y_2 + V_3y_3 + ...}{V_1 + V_2 + V_3 + ...}$$

where $V_1$, $V_2$, $V_3$, ..., etc. are the volumes of the solid bodies.

The volumes of commonly used solids are as follows:

**1.** Volume of a right circular solid cone $= \frac{\pi}{3} \times r^2 h$

where $r$ = Radius of base of the cone, and
$h$ = Height of the cone.

**2.** Volume of a cylinder $= \pi r^2 h$

where $r$ = Radius of the cylinder, and
$h$ = Height of the cylinder.

3. Volume of a hemisphere $= \frac{2\pi}{3} \times r^3$

where $r$ = Radius of the hemisphere.

**Example 4.9** *A right circular cone and a cylinder are formed by their base to form a new solid body, as shown in Fig. 4.16. The height of the cone is 180 mm and the height of cylinder is 60 mm. Determine the distance of the centre of mass of the solid body from the plane face.*

***Solution*** Given : $H = 180$ mm ; $h = 60$ mm

Since the body is symmetrical about *Y-Y* axis, therefore the centre of mass (*G*) will lie on this axis (*i.e.* axis of symmetry). Let *r* be the radius of the cylinder in mm.

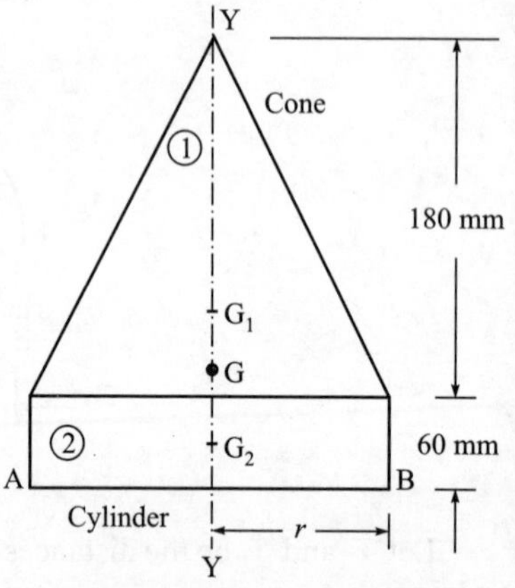


Fig. 4.16

We know that volume of the cone,

$$V_1 = \frac{1}{3} \times \pi r^2 h$$

$$= \frac{1}{3} \times \pi r^2 \times 180$$

$$= 60\,\pi r^2 \text{ mm}^3$$

Distance of its C.G. ($G_1$) from the base *AB* of the cylinder,

$$y_1 = 60 + \frac{180}{4} = 105 \text{ mm}$$

Volume of the cylinder, $V_2 = \pi r^2 h = \pi r^2 \times 60$

$$= 60\,\pi r^2 \text{ mm}^3$$

Distance of its C.G. ($G_2$) from *AB*, $y_2 = \frac{60}{2} = 30$ mm

Let $\bar{y}$ = Distance of centre of mass of the solid body (*G*) from the plane face *AB*.

We know that

$$\bar{y} = \frac{V_1 y_1 + V_2 y_2}{V_1 + V_2} = \frac{60\,\pi r^2 \times 105 + 60\,\pi r^2 \times 30}{60\,\pi r^2 + 60\,\pi r^2}$$

$$= \frac{135}{2} = 62.5 \text{ mm} \quad \textbf{Ans.}$$

**Example 4.10** *A right circular cone of height 120 mm is placed on a hemisphere of radius 100 mm. The cone and the hemisphere have a common base and are made of the same material. Find the position of the centre of gravity of the body.*

***Solution*** Given : $h = 120$ mm ; $r = 100$ mm

The cone and the hemisphere having a common base *AB* is shown in Fig. 4.17. Since the body is symmetrical about *Y-Y* axis, therefore, the centre of gravity (*G*) will lie on this axis (axis of symmetry).

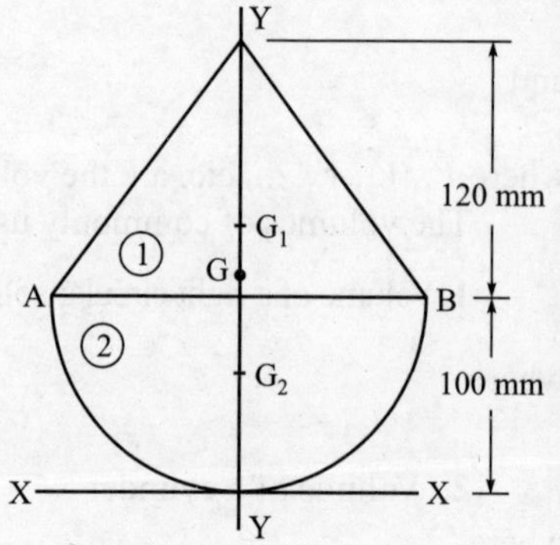


Fig. 4.17

We know that volume of the cone,

$$V_1 = \frac{1}{3} \times \pi r^2 h$$

$$= \frac{1}{3} \times \pi\,(100)^2 \times 120$$

$$= 1.2568 \times 10^6 \text{ mm}^3$$

Distance of its centre of gravity ($G_1$) from $X$-$X$ axis,

$$y_1 = 100 + \frac{120}{4} = 130 \text{ mm}$$

Volume of the hemisphere, $V_2 = \frac{2}{3} \times \pi r^3 = \frac{2}{3} \times \pi(100)^3 = 2.095 \times 10^6 \text{ mm}^3$

Distance of its centre of gravity ($G_2$) from $X$-$X$ axis,

$$y_2 = r - \frac{3r}{8} = \frac{5r}{8} = \frac{5 \times 100}{8} = 62.5 \text{ mm}$$

Let $\bar{y}$ = Distance of centre of gravity ($G$) of the body from $X$-$X$ axis.

We know that $$\bar{y} = \frac{V_1 y_1 + V_2 y_2}{V_1 + V_2} = \frac{1.2568 \times 10^6 \times 130 + 2.095 \times 10^6 \times 62.5}{1.2568 \times 10^6 + 2.095 \times 10^6}$$

$$= \frac{163.384 + 130.94}{3.3518} = 87.81 \text{ mm } \textbf{Ans.}$$

**Example 4.11** *A solid hemisphere of diameter 120 mm is placed on the top of a solid cylinder whose diameter is also 120 mm. The height of the cylinder is 60 mm. Find the centre of gravity of the combined solid from the base of the cylinder.*

***Solution*** Given : $d = 120$ mm or $r = 60$ mm ; $D = 120$ mm or $R = 60$ mm ; $h = 60$ mm

The solid is shown in Fig. 4.18.

We know that volume of the cylinder,

$$V_1 = \pi R^2 h = \pi(60)^2 \times 60$$

$$= 216\pi \times 10^3 \text{ mm}^3$$

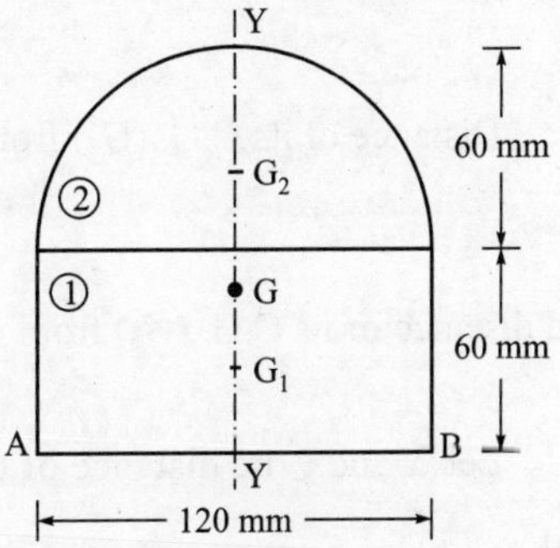


**Fig. 4.18**

Distance of its C.G. ($G_1$) from $AB$,

$$y_1 = \frac{60}{2} = 30 \text{ mm}$$

Volume of the hemisphere, $V_2 = \frac{2}{3} \times \pi r^3 = \frac{2}{3} \times \pi(60)^3$

$$= 144\pi \times 10^3 \text{ mm}^3$$

Distance of its centre of gravity ($G_2$) from $AB$,

$$y_2 = 60 + \frac{3 \times 60}{8} = 82.5 \text{ mm}$$

Let $\bar{y}$ = Distance of centre of gravity ($G$) of the combined solid from $AB$.

We know that $$\bar{y} = \frac{V_1 y_1 + V_2 y_2}{V_1 + V_2} = \frac{216\pi \times 10^3 \times 30 + 144\pi \times 10^3 \times 82.5}{216\pi \times 10^3 + 144\pi \times 10^3}$$

$$= \frac{6480 + 11\,880}{360} = 51 \text{ mm } \textbf{Ans.}$$

## 4.7 CENTRE OF GRAVITY OF BODIES AND AREAS WITH PORTIONS REMOVED

The centre of gravity of bodies and areas with portions removed (or cut out portions) may be obtained by taking the mass or area of the removed portions as negative. For example, if $a_1$ is the area of the main section and $a_2$ is the area of the removed portion, then

$$\bar{x} = \frac{a_1 x_1 - a_2 x_2}{a_1 - a_2}$$

and
$$\bar{y} = \frac{a_1 y_1 - a_2 y_2}{a_1 - a_2}$$

**Example 4.12** *A square hole is punched out of a circular lamina, the diagonal of the square being the radius of the circle. Find the centre of gravity of the remainder, if r is the radius of the circle. The diagonal of the square is on radius of the circle.*

***Solution*** Given : A square hole punched out of a circular lamina, as shown in Fig. 4.19.

The diagonal of the square is equal to radius ($r$) of the circle and the diagonal of the square is on radius of the circle. Mark *OX* and *OY* as the coordinate axes, as shown in Fig. 4.19.

We know that area of the circular lamina,

$$a_1 = \pi r^2$$

Distance of its C.G. ($G_1$) from *OY* and *OX* are

$$x_1 = r \text{ and } y_1 = r$$

Area of the square hole,

$$a_2 = \frac{r}{\sqrt{2}} \times \frac{r}{\sqrt{2}} = \frac{r^2}{2}$$

$$\ldots\left(\because \text{ Side of square} = \frac{r}{\sqrt{2}}\right)$$

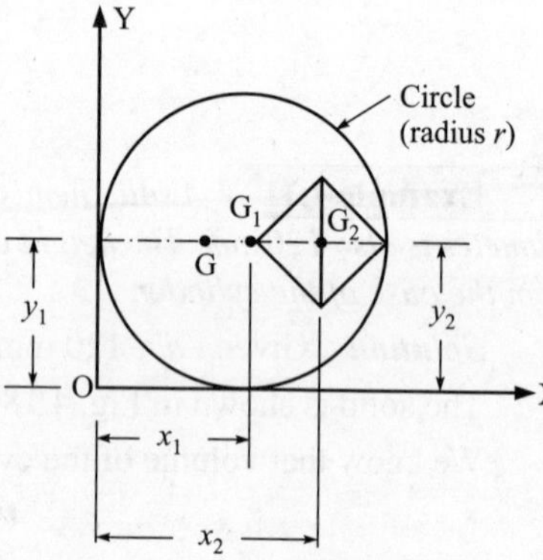


**Fig. 4.19**

Distance of its C.G. ($G_2$) from *OY*,

$$x_2 = x_1 + \frac{r}{2} = r + \frac{r}{2} = \frac{3r}{2}$$

and distance of its C.G. ($G_2$) from *OX*,

$$y_2 = y_1 = r$$

Let $\bar{x}$ and $\bar{y}$ be distance of the centre of gravity ($G$) from axes *OY* and *OX*, respectively. We know that

$$\bar{x} = \frac{a_1 x_1 - a_2 x_2}{a_1 - a_2} = \frac{\pi r^2 \times r - \frac{r^2}{2} \times \frac{3r}{2}}{\pi r^2 - \frac{r^2}{2}} = \frac{\pi r^3 - \frac{3r^3}{4}}{\pi r^2 - \frac{r^2}{2}}$$

$$= \frac{(3.142 - 0.75) r^3}{(3.142 - 0.5) r^2} = \frac{2.392 r^3}{2.642 r^2} = 0.905 r$$

Similarly,
$$\bar{y} = \frac{a_1 y_1 - a_2 y_2}{a_1 - a_2} = \frac{\pi r^2 \times r - \frac{r^2}{2} \times r}{\pi r^2 - \frac{r^2}{2}} = \frac{\pi r^3 - \frac{r^3}{2}}{\pi r^2 - \frac{r^2}{2}}$$

$$= \frac{(3.142 - 0.5)\, r^3}{(3.142 - 0.5)\, r^2} = r$$

Thus the centre of gravity ($G$) of the remainder from the coordinate axes is $\bar{x} = 0.905 r$ and $\bar{y} = r$. **Ans.**

**Example 4.13** *Locate the centroid of the shaded area, as shown in Fig. 4.20.*

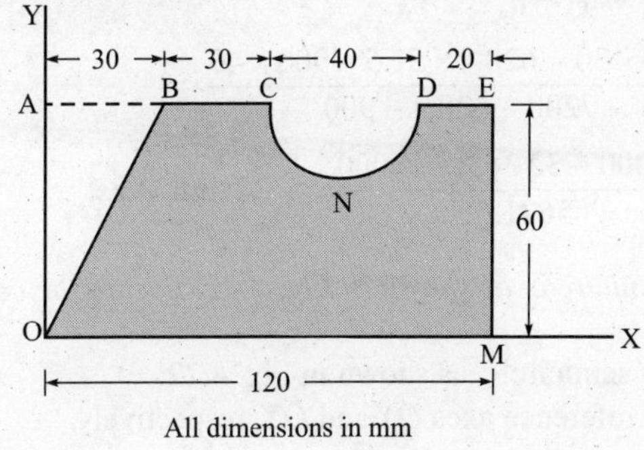


Fig. 4.20

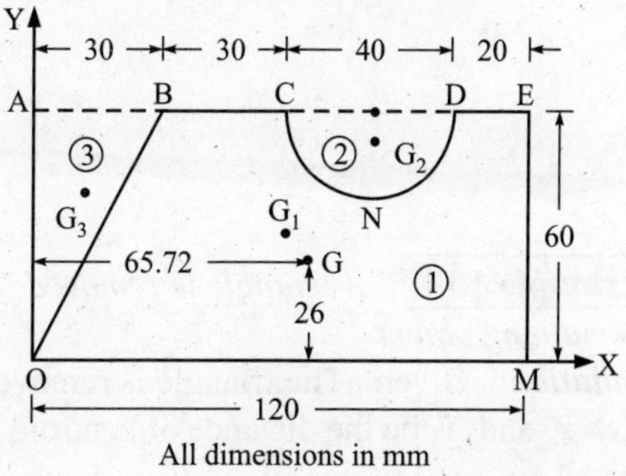


Fig. 4.21

***Solution*** Given : A section, as shown in Fig. 4.20.

Let $\bar{x}$ and $\bar{y}$ be the distance of centroid from the reference axes *OY* and *OX*, respectively. Divide the section into the following three parts, as shown in Fig. 4.21.

1. Rectangle *OAEM*; 2. Semicircle *CND*; and 3. Triangle *OAB*.

***For rectangle OAEM***

Area of rectangle *OAEM*, $a_1 = 120 \times 60 = 7200 \text{ mm}^2$

Distance of its C.G. ($G_1$) from *OY*, $x_1 = \dfrac{120}{2} = 60 \text{ mm}$

and distance of its C.G. ($G_1$) from *OX*, $y_1 = \dfrac{60}{2} = 30 \text{ mm}$

***For semicircle CND***

Area of semicircle *CND*, $a_2 = \dfrac{\pi(20)^2}{2} = 628.4 \text{ mm}^2$

Distance of its C.G. ($G_2$) from *OY*, $x_2 = 60 + 20 = 80 \text{ mm}$

and distance of its C.G. ($G_2$) from *OX*, $y_2 = 60 - \dfrac{4 \times 20}{3\pi} = 60 - 8.5 = 51.5 \text{ mm}$

... $\left(\because \text{C.G. of semicircle} = \dfrac{4r}{3\pi}\right)$

***For triangle OAB***

We know that area of triangle *OAB*, $a_3 = \dfrac{1}{2} \times 30 \times 60 = 900 \text{ mm}^2$

Distance of its C.G. ($G_3$) from *OY*, $x_3 = \dfrac{30}{3} = 10 \text{ mm}$

and distance of its C.G. ($G_3$) from *OX*, $y_3 = \dfrac{2 \times 60}{3} = 40 \text{ mm}$

We know that
$$\bar{x} = \frac{a_1x_1 - a_2x_2 - a_3x_3}{a_1 - a_2 - a_3} = \frac{7200 \times 60 - 628.4 \times 80 - 900 \times 10}{7200 - 628.4 - 900}$$

$$= \frac{432\,000 - 50\,272 - 9000}{5671.6} = 65.72 \text{ mm } \textbf{Ans.}$$

and
$$\bar{y} = \frac{a_1 y_1 - a_2 y_2 - a_3 y_3}{a_1 - a_2 - a_3}$$
$$= \frac{7200 \times 30 - 628.4 \times 51.5 - 900 \times 40}{7200 - 628.4 - 900}$$
$$= \frac{216\,000 - 32\,363 - 36\,000}{5671.6} = 26 \text{ mm} \quad \textbf{Ans.}$$

**Example 4.14** *A triangle is removed from a semicircle, as shown in Fig. 4.22. Locate the centroid of the remaining object.*

***Solution*** Given : The triangle is removed from a semicircle, as shown in Fig. 4.22.

Let $\bar{x}$ and $\bar{y}$ be the distance of centroid from the reference axes *OY* and *OX*, respectively.

Divide the section into the following three parts, as shown in Fig. 4.23.

1. Semicircle *OAB*; 2. Triangle *OAD*; and 3. Triangle *ABD*.

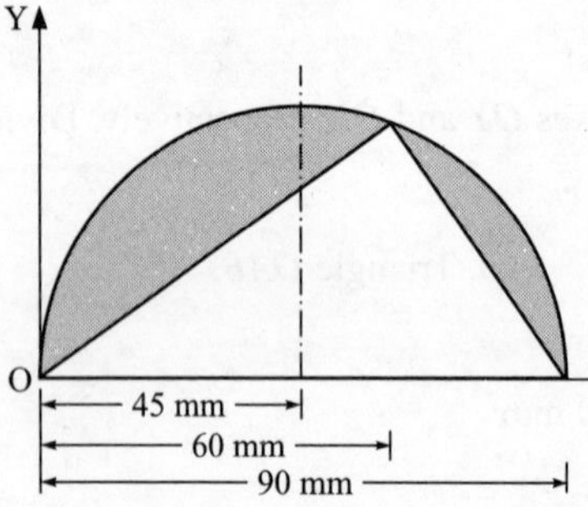


**Fig. 4.22**

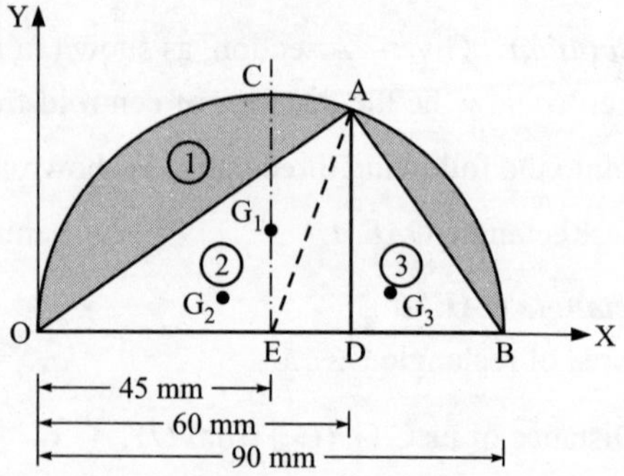


**Fig. 4.23**

***For semicircle OAB***

Area of semicircle *OAD*, $a_1 = \pi r^2 = \pi(45)^2 = 6362.55 \text{ mm}^2$ ...($\because r = EC = OE$)

Distance of its C.G. ($G_1$) from *OY*,
$$x_1 = \frac{90}{2} = 45 \text{ mm}$$

and distance of its C.G. ($G_1$) from *OX*, $y_1 = \frac{4r}{3\pi} = \frac{4 \times 45}{3\pi} = 19.1 \text{ mm}$

***For triangle OAD***

First of all, let us find the height *AD* of the triangle *OAD*. We know that
$$AD = \sqrt{(AE)^2 - (ED)^2} = \sqrt{(45)^2 - (60-45)^2}$$
$$= \sqrt{2025 - 225} = \sqrt{1800} = 30\sqrt{2} \text{ mm}$$

$\therefore$ Area of triangle *OAD* $a_2 = \frac{1}{2} \times OD \times AD = \frac{1}{2} \times 60 \times 30\sqrt{2} = 1272.6 \text{ mm}^2$

Distance of its C.G. ($G_2$) from *OY*,
$$x_2 = - \times 60 = 40 \text{ mm}$$

and distance of its C.G. ($G_2$) from *OX*,
$$y_2 = \frac{AD}{3} = \frac{30\sqrt{2}}{3} = 14.14 \text{ mm}$$

***For triangle ABD***

Area of triangle *ABD*, $a_3 = \frac{1}{2} \times BD \times AD = \frac{1}{2}(90 - 60)30\sqrt{2} = 636.3 \text{ mm}^2$

Distance of its C.G. ($G_3$) from *OY*, $x_3 = 60 + \frac{1}{3} \times DB = 60 + \frac{1}{3}(90 - 60) = 70 \text{ mm}$

and distance of its C.G. ($G_3$) from *OX*, $y_2 = \frac{1}{3} \times AD = \frac{30\sqrt{2}}{3} = 14.14 \text{ mm}$

We know that

$$\bar{x} = \frac{a_1x_1 - a_2x_2 - a_3x_3}{a_1 - a_2 - a_3}$$

$$= \frac{6362.55 \times 45 - 1272.6 \times 40 - 636.3 \times 70}{4453.65}$$

$$= \frac{286\,315 - 50\,904 - 44\,541}{4453.65} = \frac{190\,870}{4453.65} = 42.86 \text{ mm} \quad \textbf{Ans.}$$

and

$$\bar{y} = \frac{a_1y_1 - a_2y_2 - a_3y_3}{a_1 - a_2 - a_3}$$

$$= \frac{6362.55 \times 19.1 - 1272.6 \times 14.14 - 636.3 \times 14.14}{6362.55 - 1272.6 - 636.3}$$

$$= \frac{121\,525 - 17\,995 - 8997}{4453.65} = \frac{94\,533}{4453.65} = 21.22 \text{ mm} \quad \textbf{Ans.}$$

**Example 4.15** *Locate the centroid of the given composite area, as shown in Fig. 4.24.*

***Solution*** Given : An area, as shown in Fig. 4.24.

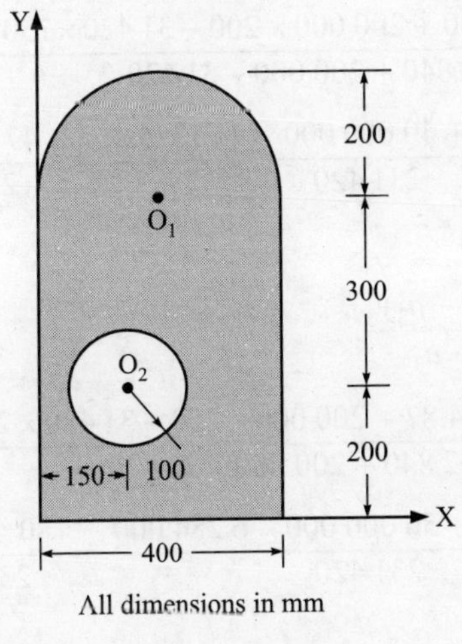


Fig. 4.24

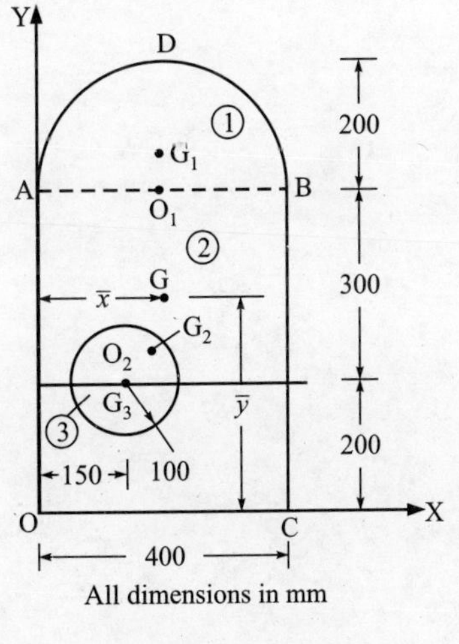


Fig. 4.25

Divide the given composite area into the following three parts, as shown in Fig. 4.25.

1. A semicircle of radius $r = 200$ mm;
2. A rectangle of area 400 mm × 500 mm; and
3. A circular hole of radius 100 mm.

***For semicircle ADB***

Area of semicircle *ADB*, $a_1 = \dfrac{\pi r^2}{2} = \dfrac{\pi (200)^2}{2} = 62\,840 \text{ mm}^2$

Distance of its C.G. ($G_1$) from *AB* $= \dfrac{4r}{3\pi} = \dfrac{4 \times 200}{3\pi} = 84.87 \text{ mm}$

Distance of its C.G. ($G_1$) from *OX*, $y_1 = 200 + 300 + 84.87 = 584.87 \text{ mm}$

Distance of its C.G. ($G_1$) from *OY*, $x_1 = \dfrac{400}{2} = 200 \text{ mm}$

***For rectangle OABC***

Area of rectangle *OABC*, $a_2 = (300 + 200)\,400 = 200\,000 \text{ mm}^2$

Distance of its C.G. ($G_2$) from *OX*, $y_2 = \dfrac{300 + 200}{2} = 250 \text{ mm}$

Distance of its C.G. ($G_2$) from *OY*, $x_2 = \dfrac{400}{2} = 200 \text{ mm}$

***For circular hole***

Area of circular hole, $a_3 = \pi r^2 = \pi (100)^2 = 31\,420 \text{ mm}^2$

Distance of its C.G. ($G_3$) from *OX*, $y_3 = 200 \text{ mm}$

Distance of its C.G. ($G_3$) from *OY*, $x_3 = 150 \text{ mm}$

Let $\bar{x}$ and $\bar{y}$ be the distance of centre of gravity from axes *OY* and *OX*, respectively. We know that

$$\bar{x} = \frac{a_1 x_1 + a_2 x_2 - a_3 x_3}{a_1 + a_2 - a_3}$$

$$= \frac{62\,840 \times 200 + 200\,000 \times 200 - 31\,420 \times 150}{62\,840 + 200\,000 - 31\,420}$$

$$= \frac{12\,568\,000 + 40\,000\,000 - 4\,713\,000}{231\,420} = \frac{47\,855\,000}{231\,420}$$

$$= 206.8 \text{ mm}$$

$$\bar{y} = \frac{a_1 y_1 + a_2 y_2 - a_3 y_3}{a_1 + a_2 - a_3}$$

$$= \frac{62\,840 \times 584.87 + 200\,000 \times 250 - 31\,420 \times 200}{62\,840 + 200\,000 - 31\,420}$$

$$= \frac{36\,753\,231 + 50\,000\,000 - 6\,284\,000}{231\,420} = \frac{80\,469\,231}{231\,420}$$

$$= 347.7 \text{ mm}$$

Mark point *G* on Fig. 4.25, such that $\bar{x} = 206.8$ mm and $\bar{y} = 347.7$ mm **Ans.**

**Example 4.16** *A semicircular area is removed from a trapezium, as shown in Fig. 4.26. Determine the centroid of the remaining area shown shaded.*

***Solution*** Given : A semicircular area removed from a trapezium, as shown in Fig. 4.26.

Let $\bar{x}$ and $\bar{y}$ be the distance of centroid from the reference axes *OY* and *OX*, respectively. Divide the section into the following three parts, as shown in Fig. 4.27.

1. Rectangle *OACD*; 2. Triangle *ABC*; and 3. Semicircle, *EFD*.

***For rectangle OACD***

Area of rectangle *OACD*, $a_1 = 80 \times 40 = 3200 \text{ mm}^2$

Distance of its C.G. ($G_1$) from *OY*, $x_1 = \dfrac{80}{2} = 40 \text{ mm}$

Distance of its C.G. ($G_1$) from *OX*, $y_1 = \dfrac{40}{2} = 20 \text{ mm}$

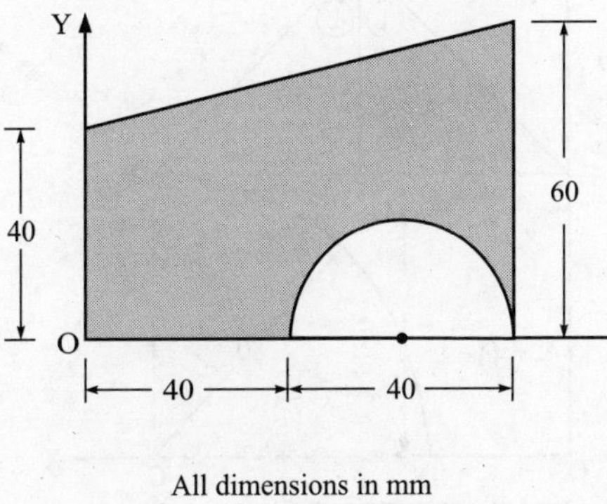


**Fig. 4.26**

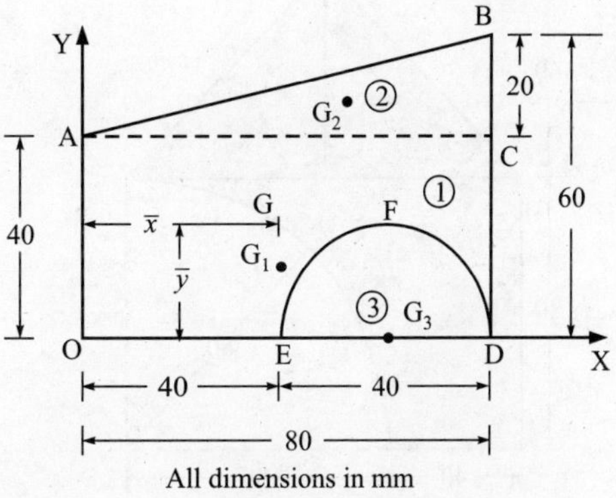


**Fig. 4.27**

***For triangle ABC***

Area of triangle *ABC*, $a_2 = \dfrac{1}{2} \times 80 \times 20 = 800 \text{ mm}^2$

Distance of its C.G. ($G_2$) from *OY*, $x_2 = 2 \times \dfrac{80}{3} = 53.3 \text{ mm}$

Distance of its C.G. ($G_2$) from *OX*, $y_2 = 40 + \dfrac{20}{3} = 46.67 \text{ mm}$

***For semicircle EFD***

Area of semicircle, $a_3 = \dfrac{\pi(20)^2}{2} = 628.3 \text{ mm}^2$

Distance of its C.G. ($G_3$) from *OY*, $x_3 = 40 + \dfrac{40}{2} = 60 \text{ mm}$

Distance of its C.G. ($G_3$) from *OX*, $y_3 = \dfrac{4(20)}{3\pi} = 8.5 \text{ mm}$

We know that $\bar{x} = \dfrac{a_1x_1 + a_2x_2 - a_3x_3}{a_1 + a_2 - a_3} = \dfrac{3200 \times 40 + 800 \times 53.3 - 628.3 \times 60}{3200 + 800 - 628.3}$

$$= \frac{128\,000 + 42\,640 - 37\,698}{3371.7} = \frac{132\,942}{3371.7} = 39.43 \text{ mm}$$ **Ans.**

and $$\bar{y} = \frac{a_1y_1 + a_2y_2 - a_3y_3}{a_1 + a_2 - a_3} = \frac{3200 \times 20 + 800 \times 46.67 - 628.3 \times 8.5}{3200 + 800 - 628.3}$$

$$= \frac{64\,000 + 37\,336 - 5340.55}{3371.7} = \frac{95\,995.45}{3371.7}$$

$$= 28.47 \text{ mm } \textbf{Ans.}$$

**Example 4.17** *Locate the centroid of the shaded area, as shown in Fig. 4.28.*

***Solution*** Given : Area, as shown in Fig. 4.28.

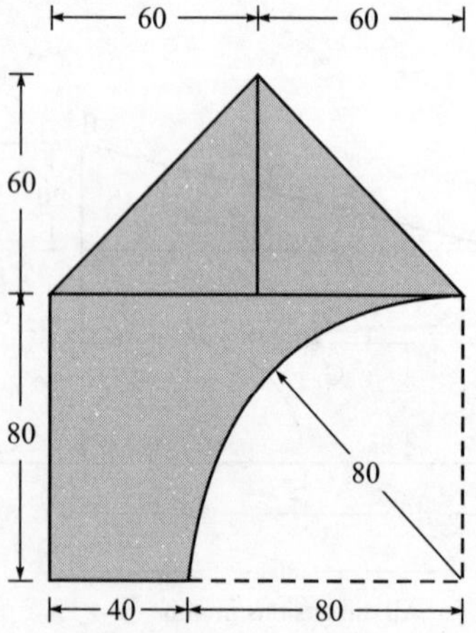


All dimensions in mm

**Fig. 4.28**

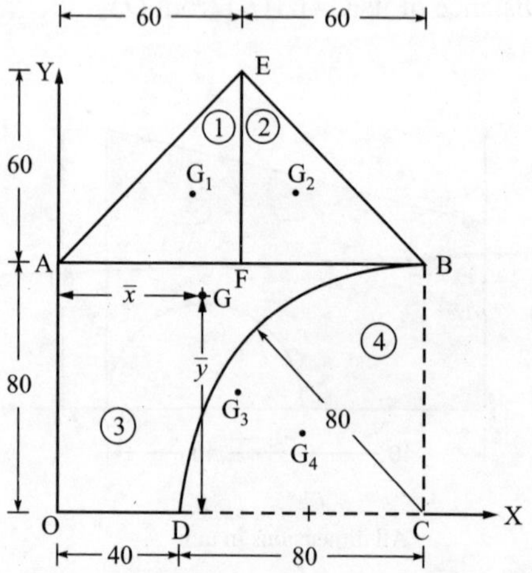


All dimensions in mm

**Fig. 4.29**

Divide the section into the following four parts, as shown in Fig. 4.29.

1. Triangle *AEF*; 2. Triangle *BEF*; 3. Rectangle *OABC*; and 4. Quadrant *CDB*

***For triangle AEF***

Area of triangle *AEF*, $a_1 = \frac{1}{2} \times AF \times EF = \frac{1}{2} \times 60 \times 60 = 1800 \text{ mm}^2$

Distance of its C.G. ($G_1$) from *OY*, $x_1 = \frac{2}{3} \times 60 = 40 \text{ mm}$

Distance of its C.G. ($G_1$) from *OX*, $y_1 = 80 + \frac{1}{3} \times 60 = 100 \text{ mm}$

***For triangle BEF***

Area of triangle *BEF* $a_2 = \frac{1}{2} \times FB \times EF = \frac{1}{2} \times 60 \times 60 = 1800 \text{ mm}^2$

Distance of its C.G. ($G_2$) from *OY*, $x_2 = 60 + \frac{1}{3} \times 60 = 80 \text{ mm}$

Distance of its C.G. ($G_2$) from *OX*, $y_2 = 80 + \frac{1}{3} \times 60 = 100 \text{ mm}$

***For rectangle OABC***

Area of rectangle *OABC*, $a_3 = (40 + 80) \times 80 = 9600 \text{ mm}^2$

Distance of its C.G. ($G_3$) from *OY*, $x_3 = \dfrac{40 + 80}{2} = 60 \text{ mm}$

Distance of its C.G. ($G_3$) from *OX*, $y_3 = \dfrac{80}{2} = 40 \text{ mm}$

***For quadrant CDB***

Area of quadrant *CDB*, $a_4 = \dfrac{1}{4} \times \pi(80)^2 = 5027.2 \text{ mm}^2$

Distance of its C.G. ($G_4$) from *OY*, $x_4 = 40 + \left(80 - \dfrac{4 \times 80}{3\pi}\right) = 86.05 \text{ mm}$

Distance of its C.G. ($G_4$) from *OX*, $y_4 = \dfrac{4 \times 80}{3\pi} = 33.95 \text{ mm}$

Let $\bar{x}$ and $\bar{y}$ be the distance of centroid (*G*) from *OY* and *OX*, respectively. We know that

$$\bar{x} = \frac{a_1 x_1 + a_2 x_2 + a_3 x_3 - a_4 x_4}{a_1 + a_2 + a_3 - a_4}$$

$$= \frac{1800 \times 40 + 1800 \times 80 + 9600 \times 60 - 5027.2 \times 86.05}{1800 + 1800 + 9600 - 5027.2}$$

$$= \frac{72\,000 + 144\,000 + 576\,000 - 432\,590}{8172.8} = \frac{359\,410}{8172.8}$$

$= 44$ mm **Ans.**

and

$$\bar{y} = \frac{a_1 y_1 + a_2 y_2 + a_3 y_3 - a_4 y_4}{a_1 + a_2 + a_3 - a_4}$$

$$= \frac{1800 \times 100 + 1800 \times 100 + 9600 \times 40 - 5027.2 \times 33.95}{1800 + 1800 + 9600 - 5027.2}$$

$$= \frac{180\,000 + 180\,000 + 384\,000 - 170\,673}{8172.8} = \frac{573\,327}{8172.8}$$

$= 70.15$ mm **Ans.**

**Example 4.18** *A semicircular area is shown in Fig. 4.30. Find the ratio of a to b so that* $\bar{y} = \dfrac{3}{4}b$.

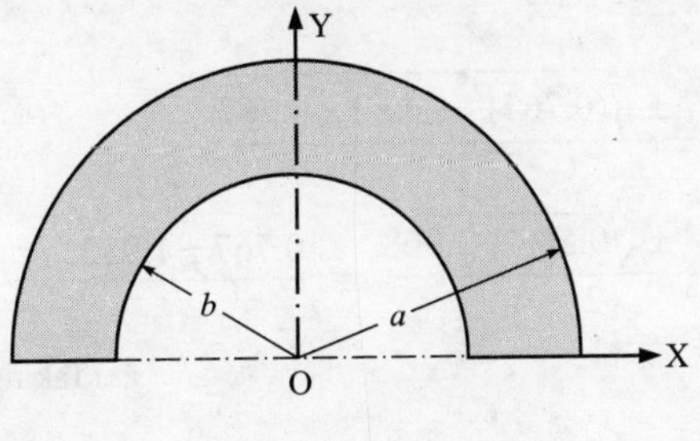


**Fig. 4.30**

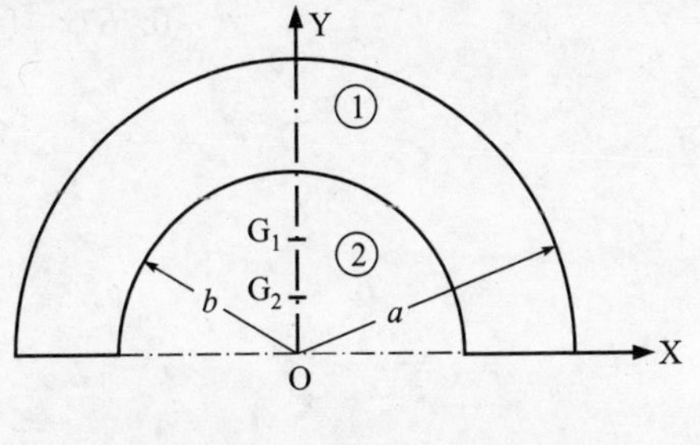


**Fig. 4.31**

***Solution*** (Refer Fig. 4.31) Given : $r_1 = a; r_2 = b\,;\; \bar{y} = \frac{3}{4}b$

We know that area of outer semicircle, $a_1 = \frac{\pi a^2}{2}$

and its centre of gravity ($G_1$) from $X$-axis, $y_1 = \frac{4a}{3\pi}$

Similarly, area of inner semicircle, $a_2 = \frac{\pi b^2}{2}$

and its centre of gravity ($G_2$) from $X$-axis, $y_2 = \frac{4b}{3\pi}$

We know that the centre of gravity of the whole section,

$$\bar{y} = \frac{a_1 y_1 - a_2 y_2}{a_1 - a_2}$$

$$\frac{3}{4}b = \frac{\frac{\pi a^2}{2} \times \frac{4a}{3\pi} - \frac{\pi b^2}{2} \times \frac{4b}{3\pi}}{\frac{\pi a^2}{2} - \frac{\pi b^2}{2}} = \frac{(a^3 - b^3)}{(a^2 - b^2)} \times \frac{4}{3\pi}$$

$$= \frac{(a-b)\,(a^2 + ab + b^2)}{(a+b)\,(a-b)} \times \frac{4}{3\pi} = \frac{(a^2 + ab + b^2)}{(a+b)} \times \frac{4}{3\pi}$$

or $$a^2 + ab + b^2 = \frac{3}{4}b \times \frac{3\pi}{4}(a+b)$$

$$a^2 + ab + b^2 = \frac{9\pi}{16}(ab + b^2) = 1.767(ab + b^2)$$

$$a^2 + ab - 1.767ab + b^2 - 1.767b^2 = 0$$

$$a^2 - 0.767ab - 0.767b^2 = 0$$

Dividing throughout by $b^2$

$$\left(\frac{a}{b}\right)^2 - 0.767\left(\frac{a}{b}\right) - 0.767 = 0$$

Putting $\frac{a}{b} = x$, we have

$$x^2 - 0.767x - 0.767 = 0$$

or $$x = \frac{+0.767 \pm \sqrt{(0.767)^2 + 4 \times 1 \times 0.767}}{2}$$

$$= \frac{+0.767 \pm \sqrt{0.588 + 3.068}}{2} = \frac{0.767 \pm 1.912}{2}$$

$$= 1.34 \qquad \text{...(Taking + sign)}$$

$$\frac{a}{b} = 1.34 \textbf{ Ans.}$$

**Example 4.19** *A body in the form of a frustrum of a right circular cone is shown in Fig. 4.32. Find the centre of gravity of the body.*

***Solution*** Given : The frustrum *ABCD* of a cone *AOD* is shown in Fig. 4.32.

First of all, let us find the height *OE* = *h* of the cone *BOC*.

From similar triangles, *OAD* and *OBC*,

$$\frac{AD}{BC} = \frac{OF}{OE}$$

or

$$\frac{1200}{600} = \frac{600+h}{h}$$

$$2h = 600 + h$$

$$\therefore \quad h = 600 \text{ mm}$$

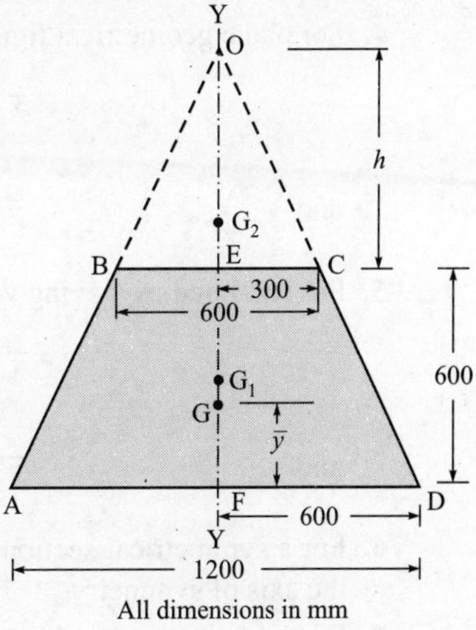


**Fig. 4.32**

Since the frustum *ABCD* is symmetrical about *Y-Y* axis, therefore, centre of gravity (*G*) will lie on this axis. We know that

Volume of the cone *AOD*, $V_1 = \frac{1}{3} \times \pi(600)^2 \times (600 + 600)$

$$= 452 \times 10^6 \text{ mm}^3$$

Distance of its C.G. ($G_1$) from *AD*,

$$y_1 = \frac{1200}{4} = 300 \text{ mm}$$

Volume of the cone *BOC*, $V_2 = \frac{1}{3} \times \pi(300)^2 \ (600) = 56.5 \times 10^6 \text{ mm}^3$

Distance of its C.G. ($G_2$) from *AD*,

$$y_2 = 600 + \frac{600}{4} = 750 \text{ mm}$$

Let $\bar{y}$ be the distance of centre of gravity (*G*) of the frustrum.

We know that $\bar{y} = \frac{V_1 y_1 - V_2 y_2}{V_1 - V_2} = \frac{452 \times 10^6 \times 300 - 56.5 \times 10^6 \times 750}{452 \times 10^6 - 56.5 \times 10^6}$

$$= \frac{135\,600 - 42\,375}{395.5} = 235.7 \text{ mm} \ \textbf{Ans.}$$

## HIGHLIGHTS

1. The centre of gravity of a body may be defined as the point through which the whole weight of the body acts, irrespective of the position of the body. It is denoted by C.G.
2. The centre of gravity of a plane figure (like triangle, rectangle, circle etc.) which has area only but no mass, is referred to as *centroid* or *centre of area* and the whole area of the plane figure is assumed to act through this point. It may be noted that it is a common practice to usc centre of gravity for centroid or vice versa.
3. If $\bar{x}$ and $\bar{y}$ are the co-ordinates of C.G. from the *Y*-axis and *X*-axis, respectively, for a rigid body having masses $m_1, m_2, m_3$, ..., etc., then

$$\bar{x} = \frac{m_1 x_1 + m_2 x_2 + m_3 x_3 + ...}{m_1 + m_2 + m_3 + ...} \quad \text{and} \quad \bar{y} = \frac{m_1 y_1 + m_2 y_2 + m_3 y_3 + ...}{m_1 + m_2 + m_3 + ...}$$

where $x_1$, $x_2$, $x_3$, ..., etc. are the distances of C.G. of masses $m_1$, $m_2$, $m_3$, .., etc. from $Y$-axis and $y_1$, $y_2$, $y_3$, ..., etc. are the distances of C.G. from $X$-axis.

4. For plane geometrical figures, having areas $a_1$, $a_2$, $a_3$, ..., etc.,

$$\bar{x} = \frac{a_1x_1 + a_2x_2 + a_3x_3 + ...}{a_1 + a_2 + a_3 + ...}$$

and
$$\bar{y} = \frac{a_1y_1 + a_2y_2 + a_3y_3 + ...}{a_1 + a_2 + a_3 + ...}$$

5. For solid bodies, having volumes $V_1$, $V_2$, $V_3$, ..., etc.,

$$\bar{x} = \frac{V_1x_1 + V_2x_2 + V_3x_3 + ...}{V_1 + V_2 + V_3 + ...}$$

and
$$\bar{y} = \frac{V_1y_1 + V_2y_2 + V_3y_3 + ...}{V_1 + V_2 + V_3 + ...}$$

6. For a symmetrical section about $X$-axis or $Y$-axis, the centre of gravity of the section will lie on the axis of symmetry.
7. For an unsymmetrical section, we should determine both $\bar{x}$ and $\bar{y}$. In other words, we should determine the centre of gravity about both the axes *i.e.* $X$-axis and $Y$-axis.

## NUMERICAL PROBLEMS

1. Find the centre of a gravity of a $T$-section having dimensions 200 mm × 300 mm × 50 mm as shown in Fig. 4.33. **[Ans.** $\bar{y}$ = 191.66 mm]

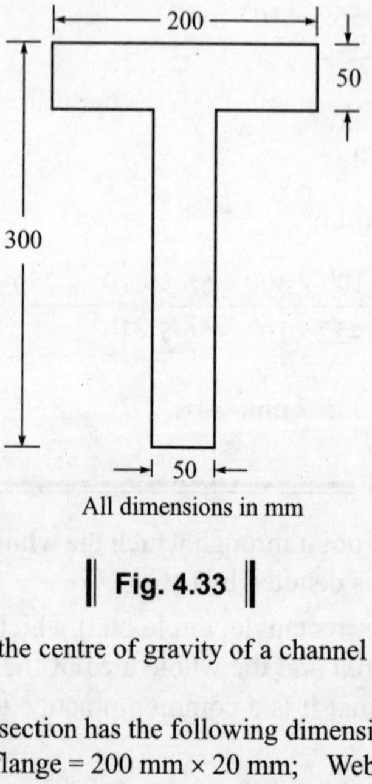


**Fig. 4.33**

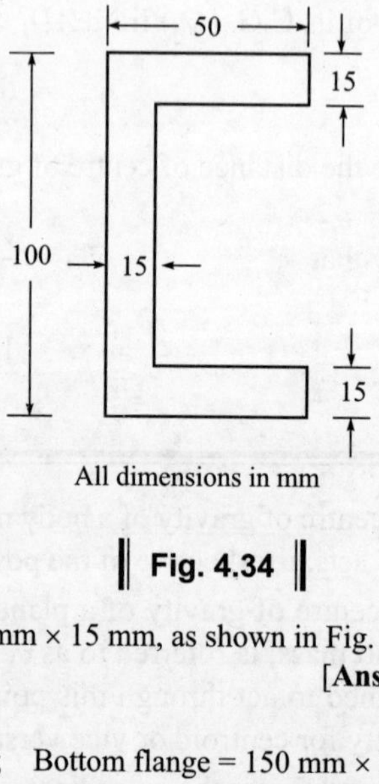


**Fig. 4.34**

2. Find the centre of gravity of a channel section 100 mm × 50 mm × 15 mm, as shown in Fig. 4.34. **[Ans.** $\bar{x}$ = 17.8 mm]
3. An $I$-section has the following dimensions:
   Top flange = 200 mm × 20 mm; Web = 300 mm × 100 mm; Bottom flange = 150 mm × 30 mm
   Determine the position of the centroid of the section. **[Ans.** $\bar{y}$ = 171.08 mm]
4. Find the centroid of an unequal angle section 100 mm × 60 mm × 20 mm. **[Ans.** $\bar{x}$ = 18.57 mm; $\bar{y}$ = 38.57 mm]

**5.** Find the centre of gravity of the Z-section, as shown in Fig. 4.35. [**Ans.** $\bar{x}$ = 137.5 mm; $\bar{y}$ = 68.5 mm]

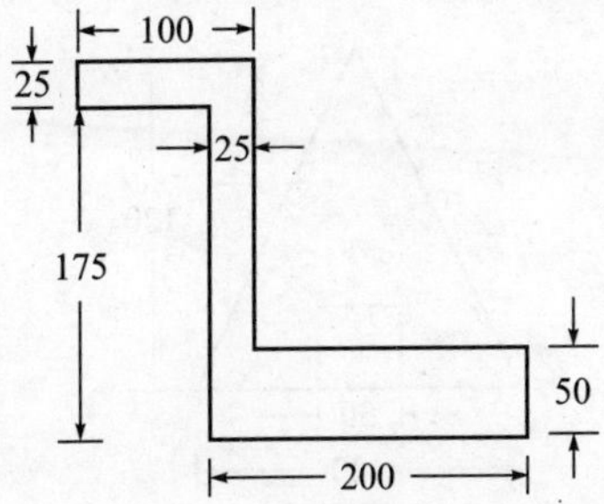


All dimensions in mm

**Fig. 4.35**

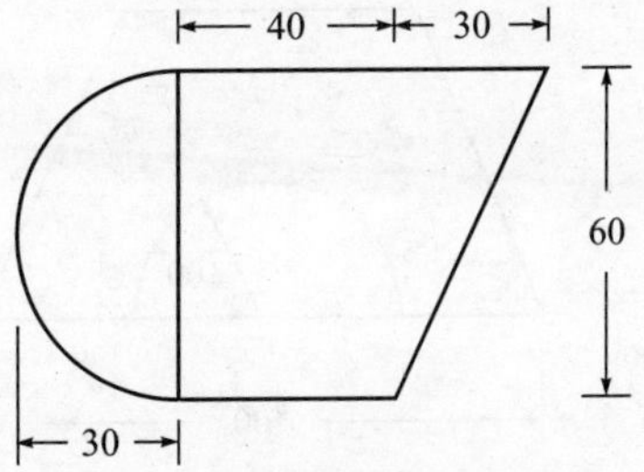


All dimensions in mm

**Fig. 4.36**

**6.** Locate the centroid of a piece of sheet metal, as shown in Fig. 4.36. [**Ans.** $\bar{x}$ = 15.91 mm; $\bar{y}$ = 31.9 mm]

**7.** Find the coordinates of the centroid of the thin wire, as shown in Fig. 4.37, with respect to reference axes. [**Ans.** $\bar{x}$ = 6.14 mm; $\bar{y}$ = 34.27 mm]

**8.** A solid hemisphere of diameter 160 mm is placed on the top of a solid cylinder whose diameter is also 160 mm. The height of the cylinder is 200 mm. Find the common centre of gravity of the combined solid from the base of a cylinder. [**Ans.** 141.8 mm]

**9.** A right circular cone of height 50 mm is placed over a hemisphere of radius 20 mm. The materials of both are same. Find the position of the centre of gravity of the body. [**Ans.** 23.6 mm]

**10.** A right circular cylinder of 120 mm diameter is joined with a hemisphere of the same diameter face to face, as shown in Fig. 4.38. Find the greatest height ($h$) of the cylinder so that the centre of gravity of the composite section coincides with the plane of joining the two sections. The density of the material of hemisphere is twice of the material of the cylinder. [**Ans.** 60 mm]

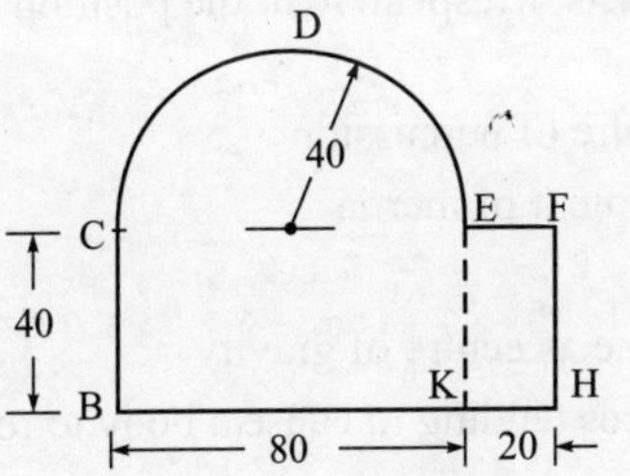


All dimensions in mm

**Fig. 4.37**

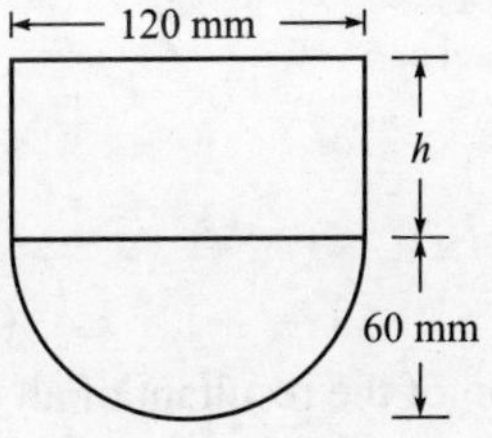


All dimensions in mm

**Fig. 4.38**

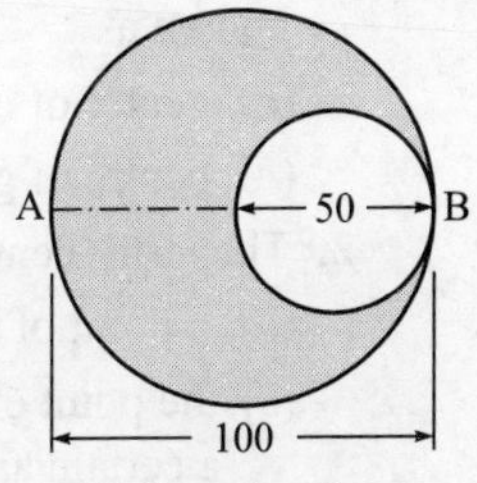


All dimensions in mm

**Fig. 4.39**

**11.** A solid consists of a cylinder and a hemisphere of equal radius fixed base to base. Find the ratio of the radius to the height of the cylinder so that the centre of gravity of the entire solid is lying at the common face of the hemisphere and cylinder. [**Ans.** $\sqrt{2}$ ]

**12.** A circular hole of 50 mm diameter is cut from a circular disc of 100 mm diameter, as shown in Fig. 4.39. Find the centre of gravity of the remaining section from $A$. [**Ans.** 41.7 mm]

**13.** A semicircle of 100 mm radius is cut from a trapezium, as shown in Fig. 4.40. Find the centroid of the shaded area. **[Ans.** 105.44 mm]

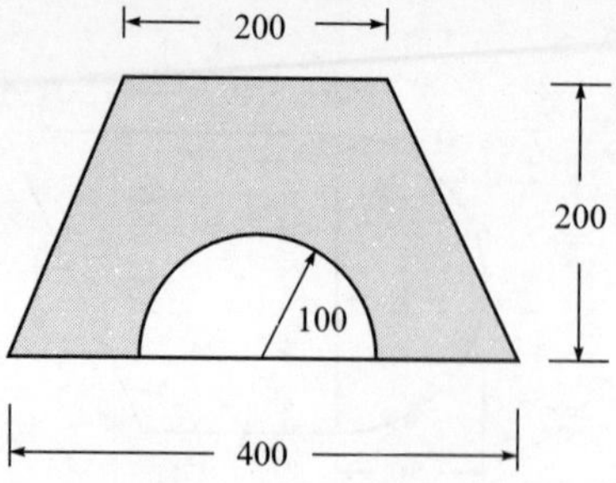


**Fig. 4.40**

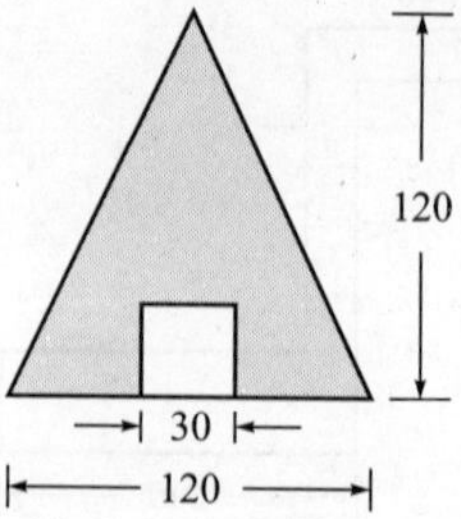


**Fig. 4.41**

**14.** A solid steel cone is made lighter by removing a part of the material from it in the form of a cylinder, as shown in Fig. 4.41. Determine the centre of gravity of the remaining metal. **[Ans.** 30.7 mm]

**15.** A solid right circular cone of height 120 mm and base 40 mm is cut by a plane perpendicular to its axis so that its upper half portion is removed. Find the position of the centre of gravity of the remaining portion from its base. **[Ans.** 23.6 mm]

## REVIEW QUESTIONS

1. What do you understand by centre of gravity?
2. What do you mean by centroid? How does it differ from centre of gravity?
3. Derive the expression for determining the centre of gravity of a rigid body by the method of moments.
4. What is axis of reference?
5. Write the centre of gravity for the following standard sections:
   (*a*) Rectangle; (*b*) Triangle; (*c*) Semicircle; and (*d*) Hemisphere

## OBJECTIVE TYPE QUESTIONS

1. The point, through which the whole weight of the body acts, irrespective of the position of body, is known as
   (*a*) centre of mass (*b*) centre of percussion
   (*c*) centre of gravity (*d*) moment of inertia
2. The term 'centroid' is
   (*a*) a point of suspension (*b*) same as centre of gravity
   (*c*) the point of application of the resultant of all the forces tending to cause a body to rotate about a certain axis
   (*d*) none of the above
3. The centre of gravity of a triangle of height *h*, lies at a distance of ................ from the base.
   (*a*) $\frac{h}{2}$ (*b*) $\frac{h}{3}$ (*c*) $\frac{h}{4}$ (*d*) $\frac{2h}{3}$
4. The centre of gravity of an equilateral triangle with each side *a*, from any of the three sides is
   (*a*) $\frac{\sqrt{3}a}{2}$ (*b*) $2\sqrt{3}a$ (*c*) $\frac{a}{2\sqrt{3}}$ (*d*) $3\sqrt{2}a$

**5.** The centre of gravity of a semicircle of radius $r$ lies at a distances of ................ from its base measured along the vertical radius.

(*a*) $\frac{3r}{8}$ (*b*) $\frac{4r}{3\pi}$ (*c*) $\frac{8r}{3}$ (*d*) $\frac{3r}{4\pi}$

**6.** The centre of gravity of a hemisphere of radius $r$ lies at a distance of $\frac{3r}{8}$ from its base measured along the vertical radius.

(*a*) True (*b*) False

**7.** The centre of gravity of a right circular solid cone of height $h$, is at a distance of ................ from its base, measured along the vertical axis.

(*a*) $\frac{h}{2}$ (*b*) $\frac{h}{3}$ (*c*) $\frac{h}{4}$ (*d*) $\frac{h}{6}$

**8.** The centre of gravity of a quadrant of a circle lies along its central radius ($r$) at a distance of

(*a*) $0.5r$ (*b*) $0.6r$ (*c*) $0.7r$ (*d*) $0.8r$

**9.** The centre of gravity of a T-section 100 mm × 150 mm × 50 mm from its bottom is

(*a*) 50 mm (*b*) 75 mm (*c*) 87.5 mm (*d*) 125 mm

**10.** The centroid of the shaded area between two circles of radii 20 mm and 10 mm, as shown in Fig. 4.42, lies

(*a*) in the shaded area (*b*) in the hole (*c*) at $O$

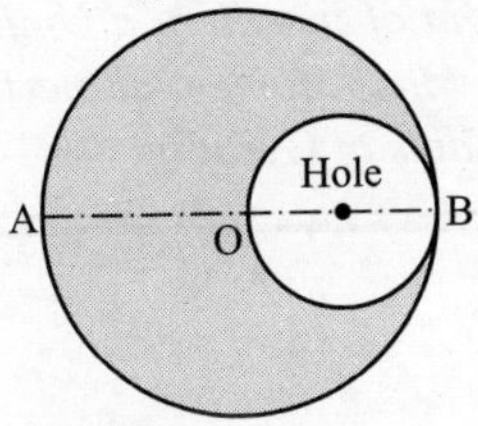


**Fig. 4.42**

***Answers to Objective Type Questions***

| | | | | |
|---|---|---|---|---|
| **1.** (*c*) | **2.** (*b*) | **3.** (*b*) | **4.** (*c*) | **5.** (*b*) |
| **6.** (*a*) | **7.** (*c*) | **8.** (*b*) | **9.** (*c*) | **10.** (*a*) |

# 5

# MOMENT OF INERTIA

**1.** *Introduction* **2.** *Theorem of Perpendicular Axis* **3.** *Theorem of Parallel Axis* **4.** *Moment of Inertia of a Rectangular Section* **5.** *Moment of Inertia of a Hollow Rectangular Section* **6.** *Moment of Inertia of a Circular Section* **7.** *Moment of Inertia of a Hollow Circular Section* **8.** *Moment of Inertia of a Semicircular Section* **9.** *Moment of Inertia of a Triangular Section* **10.** *Moment of Inertia of a Composite Section* **11.** *Mass Moment of Inertia of a Uniform Thin Rod* **12.** *Mass Moment of Inertia of a Rectangular Plate* **13.** *Mass Moment of Inertia of a Uniform Solid Cylinder* **14.** *Mass Moment of Inertia of a Uniform Circular Plate* **15.** *Mass Moment of Inertia of a Solid Sphere* **16.** *Mass Moment of Inertia of a Uniform Solid Cone* **17.** *Radius of Gyration*

## 5.1 INTRODUCTION

The moment of inertia may be defined as the moment of the moment or second moment of the mass or area of the body. It is usually denoted by $I$. The moment of inertia is always determined with reference to a certain axis of rotation and has different values about different axis. Thus, the moment of inertia about $X$-$X$ axis is written as $I_{XX}$ and about $Y$-$Y$ axis, it is written as $I_{YY}$. When the moment of inertia is determined about an axis passing through the centre of gravity, then it is denoted by $I_G$ or simply $I$.

Consider a body of total mass $m$. Let it is composed of small particles of masses $m_1$, $m_2$, $m_3$, ..., etc. If $x_1$, $x_2$, $x_3$, ..., etc. are the distances from the $Y$-$Y$ axis and $y_1$, $y_2$, $y_3$, ..., etc. are the distances from the $X$-$X$ axis, as shown in Fig. 5.1, then moment of inertia of the whole body about $X$-$X$ axis,

$$I_{XX} = m_1 y_1^2 + m_2 y_2^2 + m_3 y_3^2 + ... = \Sigma m y^2$$

Similarly, moment of inertia of the whole body about $Y$-$Y$ axis,

$$I_{YY} = m_1 x_1^2 + m_2 x_2^2 + m_3 x_3^2 + ... = \Sigma m x^2$$

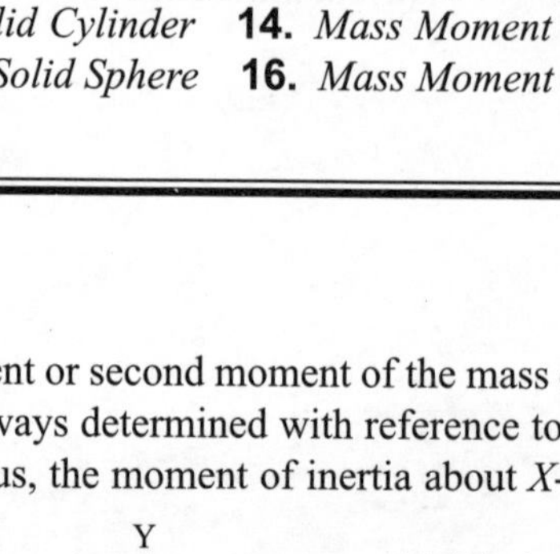


**Fig. 5.1**

If, instead of mass, the area of plane figure is taken into consideration, then moment of inertia of the whole area is given by

$$I = A.k^2$$

where $k$ is called the *radius of gyration*. It is defined as the distance from a given reference where the whole area (or mass) of the body is assumed to be concentrated to give the same value of $I$.

If $k_1$ and $k_2$ are the two radii of gyration of the area about *X-X* axis and *Y-Y* axis, respectively, then

$$I_{XX} = Ak_1^2; \text{ and } I_{YY} = Ak_2^2$$

In S.I. units, the unit of mass moment of inertia is kg-m$^2$ and the moment of inertia of the area is expressed in m$^4$ or mm$^4$.

## 5.2 THEOREM OF PERPENDICULAR AXIS

According to the theorem of mutually perpendicular axes, '*the moment of inertia of a plane lamina about an axis perpendicular to the plane of the lamina and passing through its centroid is equal to the sum of the moment of inertia of the lamina about the two mutually perpendicular axes passing through the centroid and in the plane of the lamina*'.

***Proof*** Consider a plane section of area *a* and lying in a plane *X-Y*, as shown in Fig. 5.2.

Let *OZ* be the perpendicular axis to the two mutually perpendicular axes *OX* and *OY*. Consider a small area *da*.

Let $x$ = Distance of *da* from axis *OY*,

$y$ = Distance of *da* from axis *OX*, and

$r$ = Distance of *da* from axis *OZ*.

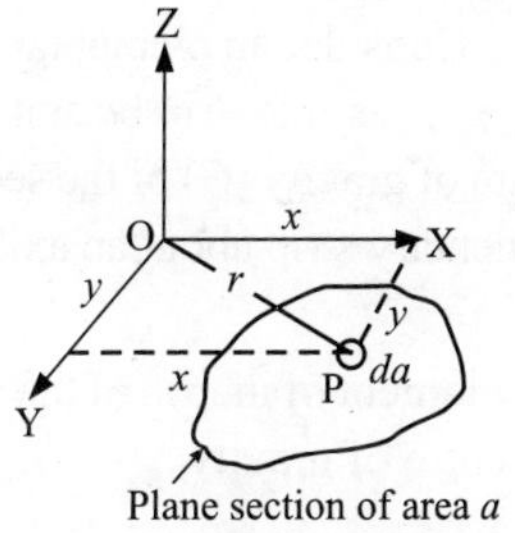


**Fig. 5.2**

From the geometry of the figure,

$$r^2 = x^2 + y^2 \quad ...(i)$$

Moment of inertia of *da* about *X*-axis,

$$I_{XX} = da \times (\text{Distance of } da \text{ from } X\text{-axis})^2 = da \cdot y^2$$

Similarly, moment of inertia of *da* about *Y*-axis,

$$I_{YY} = da \cdot x^2$$

and moment of inertia of *du* about *Z*-axis (or polar moment of inertia),

$$I_{ZZ} = da \cdot r^2$$

$$= da(x^2 + y^2) = da \cdot x^2 + da \cdot y^2 \quad \text{...[From equation } (i)]$$

$$= I_{YY} + I_{XX}$$

Thus, if $I_{XX}$ and $I_{YY}$ are the moment of inertia about *X-X* axis and *Y-Y* axis perpendicular to each other, then the moment of inertia about the third axis *Z-Z* perpendicular to the plane of lamina is

$$I_{ZZ} = I_{XX} + I_{YY}$$

**Notes:** **1.** All the three axes *X-X, Y-Y* and *Z-Z* are mutually perpendicular to each other.

**2.** The third axis *Z-Z* is known as *polar axis* and is always perpendicular to the plane figure passing through its centroid.

**3.** The moment of inertia about *Z-Z* axis (*i.e.* $I_{ZZ}$) is known as *polar moment of inertia.*

## 5.3 THEOREM OF PARALLEL AXIS

It states that '*the moment of inertia of a given plane lamina about a given axis is equal to the sum of moment of inertia of the lamina about a parallel axis passing through its centroid and the product of the area of the lamina and square of the perpendicular distance between the two axes*'.

Consider a plane figure, as shown in Fig. 5.3, whose moment of inertia is to be determined about *A-A*.

Let $I_G$ = Moment of inertia of the figure about an axis parallel to *A-A* and passing through the centroid,

$A$ = Area of the figure, and

$h$ = Distance between the two axes.

Now according to parallel axis theorem, moment of inertia of the whole area about *AA*,

$$I_{AA} = I_G + Ah^2$$

Similarly, moment of inertia of the whole area about *BB*,

$$I_{BB} = I_G + Ah_1^2$$

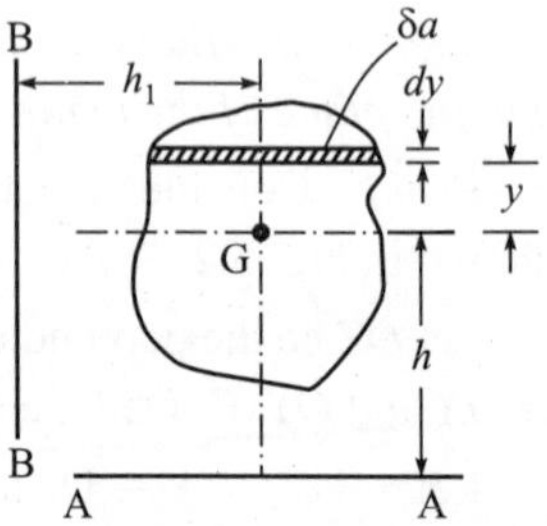


Fig. 5.3

***Proof***

Consider an elementary strip of area $\delta a$ of the section, as shown in Fig. 5.3. Let this strip be at a distance $y$ from the axis passing through the centre of gravity (*G*) of the section. We know that moment of inertia of the elementary strip about an axis passing through the C.G. of the section

$$= \delta a \cdot y^2$$

and moment of inertia of the whole section about an axis passing through its centre of gravity,

$$I_G = \Sigma \delta a \cdot y^2$$

∴ Moment of inertia of the section about *AA*,

$$I_{AA} = \Sigma \delta a\,(h + y)^2 = \Sigma \delta a\,(h^2 + y^2 + 2hy)$$

$$= (\Sigma \delta a . h^2) + (\Sigma \delta a . y^2) + (\Sigma \delta a . 2hy)$$

$$= Ah^2 + I_G$$ ...(Neglecting $\Sigma \delta a.2hy$, being very small)

## 5.4 MOMENT OF INERTIA OF A RECTANGULAR SECTION

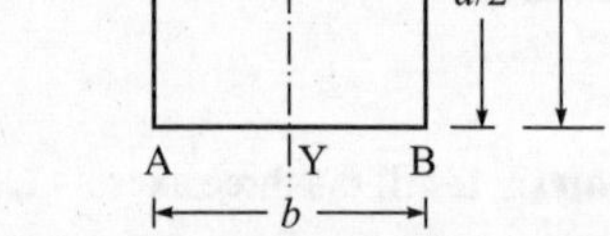


Fig. 5.4

A rectangular section *ABCD* having width $b$ and depth $d$ resting on a horizontal surface with its width parallel to the horizontal line, is shown in Fig. 5.4.

Let *X-X* be the horizontal axis passing through the centre of gravity *G* of the section.

Consider a rectangular elementary strip *PQ* of thickness $dy$ at a distance $y$ from *X-X* axis, as shown in Fig. 5.4.

Area of the strip $= b \cdot dy$

Moment of inertia of the area of the strip about *X-X* axis

$$= \text{Area of the strip} \times y^2$$

$$= (b \cdot dy)\, y^2 = b \cdot y^2\, dy$$

Integrating this expression between the limits $-\dfrac{d}{2}$ and $+\dfrac{d}{2}$, in order to determine the moment of inertia of the whole section.

$$\therefore \quad I_{XX} = \int_{-d/2}^{+d/2} b\,y^2\,dy = b\left[\frac{y^3}{3}\right]_{-d/2}^{+d/2} = \frac{b}{3}\left[y^3\right]_{-d/2}^{+d/2}$$

$$= \frac{b}{3}\left[\left(\frac{d}{2}\right)^3 - \left(\frac{-d}{2}\right)^3\right] = \frac{b}{3}\left[\frac{d^3}{8} + \frac{d^3}{8}\right] = \frac{b}{3} \times \frac{2d^3}{8} = \frac{bd^3}{12}$$

Similarly, the moment of the inertia of the rectangular section about *Y-Y* axis passing through the C.G. of the section is given by

$$I_{YY} = \frac{db^3}{12}$$

**Notes:** **1.** The moment of inertia of the rectangular section passing through the base *AB* may be determined by using parallel axis theorem, *i.e.*

$$I_{AB} = I_G + A.h^2 = \frac{bd^3}{12} + bd\left(\frac{d}{2}\right)^2 = \frac{bd^3}{12} + \frac{bd^3}{4} = \frac{bd^3}{3}$$

...(Here $I_G = I_{XX}$; and $h = d/2$)

**2.** $I_{AB}$ may also be obtained by integrating the equation (*i*) between the limits 0 and *d*.

$$\therefore \; I_{AB} = \int_0^d b\,y^2\,dy = b\left[\frac{y^3}{3}\right]_0^d = \frac{bd^3}{3}$$

## 5.5 MOMENT OF INERTIA OF A HOLLOW RECTANGULAR SECTION

A hollow rectangular section is shown in Fig. 5.5. Rectangle *PQRS* is the main section having width *B* and depth *D* while the rectangle *EFGH* is the cut out section having width *b* and depth *d*. We know that moment of inertia of the rectangular section *PQRS* about *X-X* axis,

$$I_1 = \frac{BD^3}{12}$$

and moment of inertia of the rectangular section *EFGH* about *X-X* axis,

$$I_2 = \frac{bd^3}{12}$$

∴ Moment of inertia of the hollow rectangular section (shaded area) about *X-X* axis,

$$I_{XX} = I_1 - I_2 = \frac{BD^3}{12} - \frac{bd^3}{12}$$

Similarly,

$$I_{YY} = \frac{DB^3}{12} - \frac{db^3}{12}$$

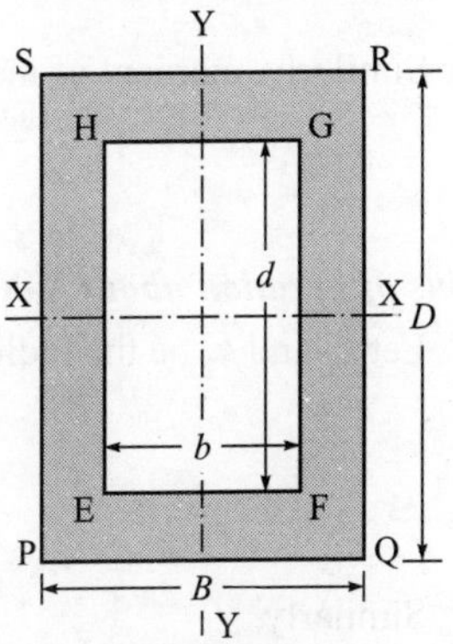


**Fig. 5.5**

**Example 5.1** *Find the moment of inertia of the rectangular section of width 40 mm and depth 60 mm about X-X axis and Y-Y axis. What will be the moment of inertia of the section about the base AB and the edge AD. Also determine the radius of gyration of the section about X-X axis and Y-Y axis.*

***Solution*** Given : $b = 40$ mm ; $d = 60$ mm

The rectangular section is shown in Fig. 5.6.

***Moment of inertia about X-X and Y-Y axis***

We know that moment of inertia of the section about *X-X* axis passing through its centroid is given by

$$I_{XX} = \frac{bd^3}{12}$$

$$= \frac{40(60)^3}{12}$$

$$= 720 \times 10^3 \text{ mm}^4 \; \textbf{Ans.}$$

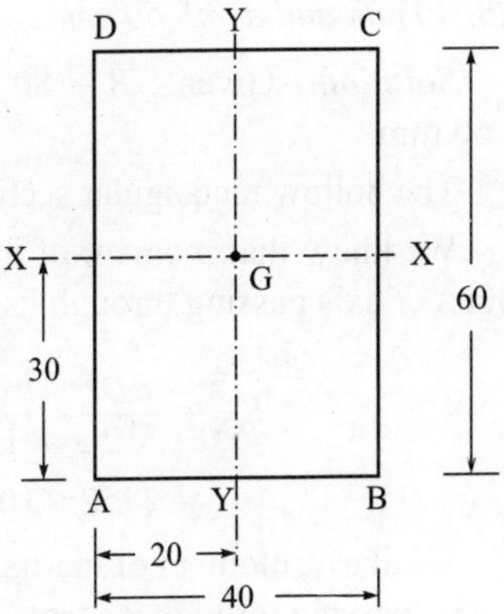


All dimensions in mm

**Fig. 5.6**

Similarly, moment of inertia of the section about $Y$-$Y$ axis passing through its centroid,

$$I_{YY} = \frac{d\,b^3}{12} = \frac{60\,(40)^3}{12} = 320 \times 10^3 \text{ mm}^4 \quad \textbf{Ans.}$$

***Moment of inertia about the base AB and edge AD***

We know that area of the rectangle,

$$A = b \times d = 40 \times 60 = 2400 \text{ mm}^2$$

Distance of the centroid ($G$) from the base $AB$,

$$h_1 = \frac{60}{2} = 30 \text{ mm}$$

and distance of the centroid ($G$) from the edge $AD$,

$$h_2 = \frac{40}{2} = 20 \text{ mm}$$

According to parallel axis theorem, moment of inertia of the section about base $AB$,

$$I_{AB} = I_G + A(h_1)^2 = I_{XX} + A(h_1)^2$$
$$= 720 \times 10^3 + 2400\,(30)^2 = 2880 \times 10^3 \text{ mm}^4 \quad \textbf{Ans.}$$

Similarly, moment of inertia of the section about edge $AD$,

$$I_{AD} = I_G + A(h_2)^2 = I_{YY} + A(h_2)^2$$
$$= 320 \times 10^3 + 2400\,(20)^2 = 1280 \times 10^3 \text{ mm}^4 \quad \textbf{Ans.}$$

***Radius of gyration about X-X axis and Y-Y axis***

Let $k_1$ and $k_2$ be the radius of gyration about $X$-$X$ axis and $Y$-$Y$ axis, respectively. We know that

$$I_{XX} = Ak_1^2 \quad \text{or} \quad k_1 = \sqrt{\frac{I_{XX}}{A}} = \sqrt{\frac{720 \times 10^3}{2400}} = 17.32 \text{ mm} \quad \textbf{Ans.}$$

Similarly,

$$I_{YY} = Ak_2^2 \quad \text{or} \quad k_2 = \sqrt{\frac{I_{YY}}{A}} = \sqrt{\frac{320 \times 10^3}{2400}} = 11.547 \text{ mm} \quad \textbf{Ans.}$$

**Example 5.2** *Find the moment of inertia of a hollow rectangular section about X-X axis and Y-Y axis passing through the centroid. The outer dimensions of the section are: width 80 mm and depth 120 mm while the inner dimensions are: width 40 mm and depth 60 mm.*

***Solution*** Given : $B = 80$ mm ; $D = 120$ mm ; $b = 40$ mm ; $d = 60$ mm

The hollow rectangular section is shown in Fig. 5.7.

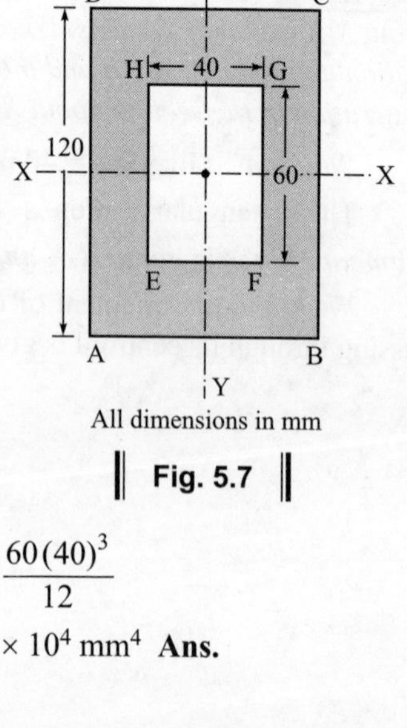


**Fig. 5.7**

We know that moment of inertia of a hollow rectangular section about $X$-$X$ axis passing through its centroid,

$$I_{XX} = \frac{BD^3}{12} - \frac{b\,d^3}{12} = \frac{80\,(120)^3}{12} - \frac{40\,(60)^3}{12}$$
$$= 1152 \times 10^4 - 72 \times 10^4 = 1080 \times 10^4 \text{ mm}^4 \quad \textbf{Ans.}$$

Similarly, moment of inertia of a hollow rectangular section about $Y$-$Y$ axis passing through its centroid,

$$I_{YY} = \frac{D\,B^3}{12} - \frac{d\,b^3}{12} = \frac{120\,(80)^3}{12} - \frac{60\,(40)^3}{12}$$
$$= 512 \times 10^4 - 32 \times 10^4 = 480 \times 10^4 \text{ mm}^4 \quad \textbf{Ans.}$$

## 5.6 MOMENT OF INERTIA OF A CIRCULAR SECTION

A circular section of radius $R$ with centre $O$ is shown in Fig. 5.8. Consider an elementary ring of radius $r$ and having a thickness $dr$.

$\therefore$ Area of this elementary ring, $da = 2\pi r\, dr$

and moment of inertia of this area about the *polar axis (Z-Z axis)

$$= \text{Area} \times (\text{Distance})^2$$

$$= (2\pi r . dr) r^2 = 2\pi r^3\, dr$$

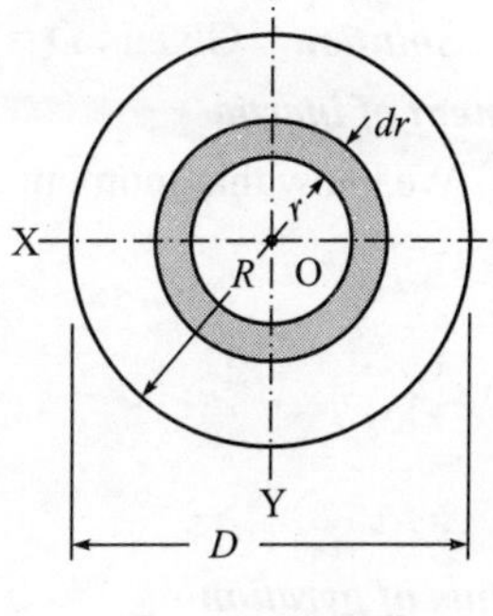

Fig. 5.8

The polar moment of inertia of the whole section may be obtained by integrating the above expression between the limits from 0 to $R$.

$$\therefore \quad I_{ZZ} = \int_0^R 2\pi r^3\, dr$$

$$= 2\pi \left[\frac{r^4}{4}\right]_0^R = 2\pi \times \frac{R^4}{4} = \frac{\pi R^4}{2}$$

$$= \frac{\pi}{2}\left(\frac{D}{2}\right)^4 = \frac{\pi}{2} \times \frac{D^4}{16} = \frac{\pi}{32} D^4$$

where $D$ = Diameter of the circular section.

From the theorem of perpendicular axis, we know that

$$I_{ZZ} = I_{XX} + I_{YY}$$

But due to symmetry, $I_{XX} = I_{YY}$

$$\therefore \quad I_{XX} = I_{YY} = \frac{I_{ZZ}}{2} = \frac{1}{2}\left(\frac{\pi}{32}\right) D^4 = \frac{\pi}{64} D^4$$

## 5.7 MOMENT OF INERTIA OF A HOLLOW CIRCULAR SECTION

A hollow circular section having outer diameter ($D$) and inner diameter ($d$), is shown in Fig. 5.9.

We know that the moment of inertia of the outer circle about both $X$-$X$ axis and $Y$-$Y$ axis,

$$I_1 = \frac{\pi}{64} D^4$$

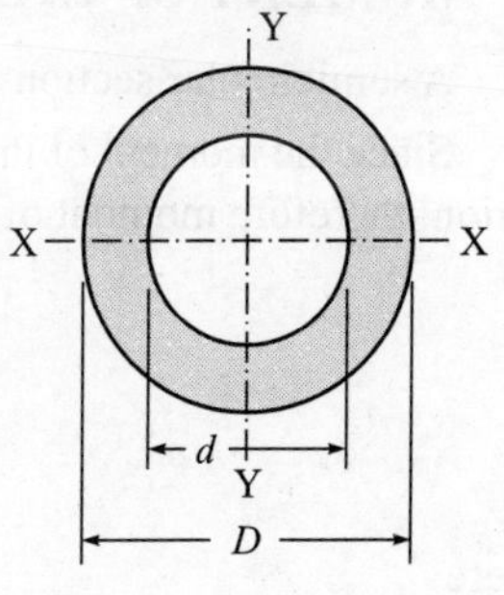

Fig. 5.9

and moment of inertia of the inner circle about both $X$-$X$ axis and $Y$-$Y$ axis

$$I_2 = \frac{\pi}{64} d^4$$

$\therefore$ Moment of inertia of the hollow circular section (shaded area) about both $X$-$X$ axis and $Y$-$Y$ axis,

$$I_{XX} = I_{YY} = I_1 - I_2$$

$$= \frac{\pi}{64} D^4 - \frac{\pi}{64} d^4 = \frac{\pi}{64}(D^4 - d^4)$$

*The polar axis ($Z$-$Z$ axis) is perpendicular to the plane of the circle and it passes through the point of intersection of $X$-$X$ and $Y$-$Y$ axis.

**Example 5.3** *The outer and inner diameters of the circle are 100 mm and 80 mm, respectively. Determine the moment of inertia of this hollow circular section about an axis passing through its centroid. Also determine the radius of gyration and polar moment of inertia.*

***Solution*** Given : $D = 100$ mm ; $d = 80$ mm

***Moment of inertia***

We know that moment of inertia of a hollow circular section about an axis passing through its centroid,

$$I_{XX} = I_{YY} = \frac{\pi}{64}(D^4 - d^4)$$

$$= \frac{\pi}{64}\left[(100)^4 - (80)^4\right] = \frac{\pi}{64} \times 59.04 \times 10^6 \text{ mm}^4$$

$$= 2.898 \times 10^6 \text{ mm}^4 \textbf{ Ans.}$$

***Radius of gyration***

Let $k$ = Radius of gyration.

We know that $I = Ak^2$

$$2.898 \times 10^6 = \frac{\pi}{4}(D^2 - d^2)\, k^2 \qquad ...(\because I = I_{XX} = I_{YY})$$

$$= \frac{\pi}{4}\left[(100)^2 - (80)^2\right]k^2 = 2827.8\, k^2$$

or $$k^2 = \frac{2.898 \times 10^6}{2827.8} = 1024.8$$

$\therefore$ $$k = \sqrt{1024.8} = 32 \text{ mm} \textbf{ Ans.}$$

***Polar moment of inertia***

We know that polar moment of inertia,

$$I_{ZZ} = I_{XX} + I_{YY} = 2.898 \times 10^6 + 2.898 \times 10^6$$

$$= 5.796 \times 10^6 \text{ mm}^4 \textbf{ Ans.}$$

## 5.8 MOMENT OF INERTIA OF A SEMICIRCULAR SECTION

A semicircular section *ACB* having radius *r*, is shown in Fig. 5.10.

Since the moment of inertia of the semicircular section is one-half the moment of inertia of the circular section, therefore moment of inertia of the semicircular section *ACB* about base *AB*,

$$I_{AB} = — \times \text{M.I. of circular section} = \frac{1}{2}\left[\frac{\pi}{64}d^4\right]$$

$$= \frac{1}{2} \times \frac{\pi}{64}(2r)^4 = \frac{\pi}{8}r^4 \qquad ...(d = 2r)$$

where $d$ = Diameter of the circle.

We know that area of the semicircle,

$$a = ——$$

and distance of centre of gravity (*G*) of the semicircular section from the base *AB*,

$$h = \frac{4r}{3\pi}$$

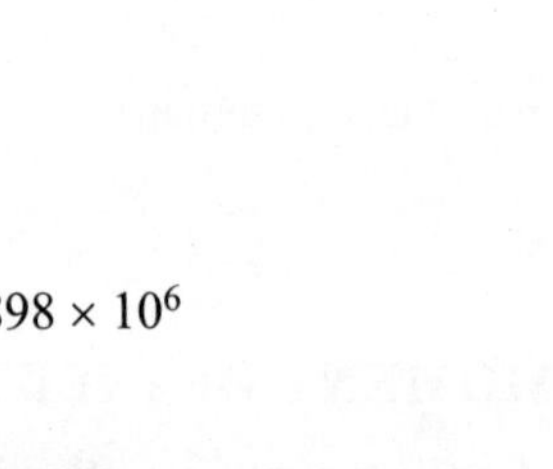


**Fig. 5.10**

∴ Moment of inertia of the semicircular section about an axis passing through its centroid and parallel to X-X axis,

$$I_G = I_{AB} - ah^2$$

$$= \frac{\pi}{8}(r^4) - \frac{\pi r^2}{2}\left(\frac{4r}{3\pi}\right)^2$$

$$= \frac{\pi}{8}r^4 - \frac{8}{9\pi}r^4 = 0.11r^4$$

and moment of inertia of the semicircular section about an axis passing through its centroid and parallel to Y-Y axis

$$= \frac{\pi}{8}r^4 = 0.393r^4 \quad \text{...(Same as about base } AB \text{ i.e. } I_{AB})$$

## 5.9 MOMENT OF INERTIA OF A TRIANGULAR SECTION

A triangular section *ABC* having base *b*, and height *h* is shown in Fig. 5.11.

Consider an elementary strip of width $b_1$ and thickness *dy* parallel to the base *AB*, at a distance *y* from the base *AB*.

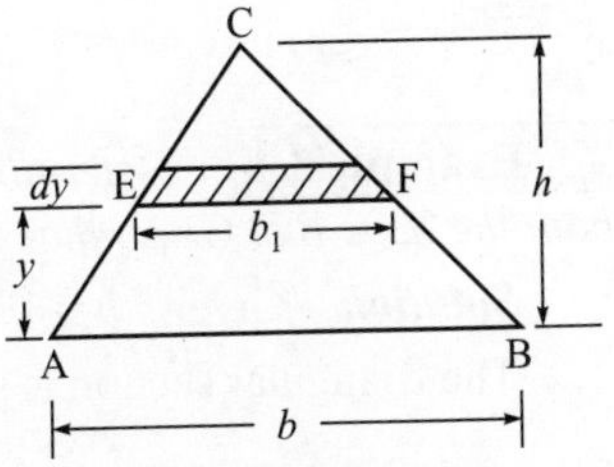


Fig. 5.11

Area of this elementary strip, $da = b_1 \times dy$

Moment of this strip about *AB* $= b_1\, dy \times y$

and second moment or moment of inertia of this strip about *AB*,

$$= b_1\, dy \times y \times y$$

$$= b_1 dy \times y^2 \quad \text{...}(i)$$

Now from similar triangles *ABC* and *CEF*,

$$\frac{b_1}{b} = \frac{h-y}{h}$$

or

$$b_1 = b\left(\frac{h-y}{h}\right)$$

Substituting the value of $b_1$ in equation (*i*), we find that the moment of inertia of the elementary strip about *AB*

$$= b\left(\frac{h-y}{h}\right)dy \times y^2$$

$$= \frac{b}{h}(h-y)\,dy\,y^2 = \frac{b}{h}(hy^2 - y^3)\,dy$$

The moment of inertia of the whole triangular section about *AB* ($I_{AB}$) is obtained by integrating the above expression within the limits from 0 to *h*.

$$\therefore \quad I_{AB} = \int_0^h \frac{b}{h}(hy^2 - y^3)\,dy = \frac{b}{h}\left[\frac{h \times y^3}{3} - \frac{y^4}{4}\right]_0^h$$

$$= \frac{b}{h}\left[\frac{h \times h^3}{3} - \frac{h^4}{4}\right] = \frac{b}{h}\left[\frac{h^4}{3} - \frac{h^4}{4}\right]$$

$$= \frac{b}{h}\left[\frac{4h^4 - 3h^4}{12}\right] = \frac{b}{h} \times \frac{h^4}{12} = \frac{bh^3}{12}$$

In order to find the moment of inertia of a triangular section about an axis passing through its centroid (*i.e.* about an axis passing through G, as shown in Fig. 5.12) may be obtained by applying the theorem of parallel axis. According to this theorem,

$$I_{AB} = \text{M.I. about C.G.} + \text{Area} \times (\text{Distance between } X\text{-}X \text{ and } AB)^2$$

$$= I_G + A\left(\frac{h}{3}\right)^2 = I_G + \frac{b \times h}{2}\left(\frac{h}{3}\right)^2$$

$$\therefore \quad I_G = I_{AB} - \frac{bh}{2}\left(\frac{h^2}{9}\right) = \frac{bh^3}{12} - \frac{bh^3}{18} = \frac{bh^3}{36}$$

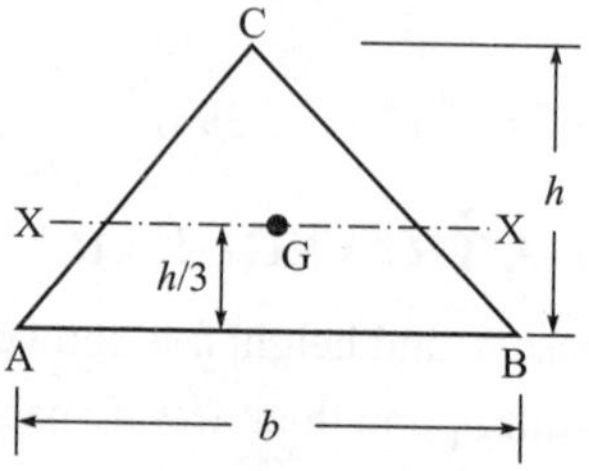


**Fig. 5.12**

**Example 5.4** *Determine the moment of inertia of the triangular section ABC about its centroid and about the base BC. Given that base BC = 100 mm and height 90 mm.*

***Solution*** Given : $b = 100$ mm ; $h = 90$ mm

The triangular section is shown in Fig. 5.13.

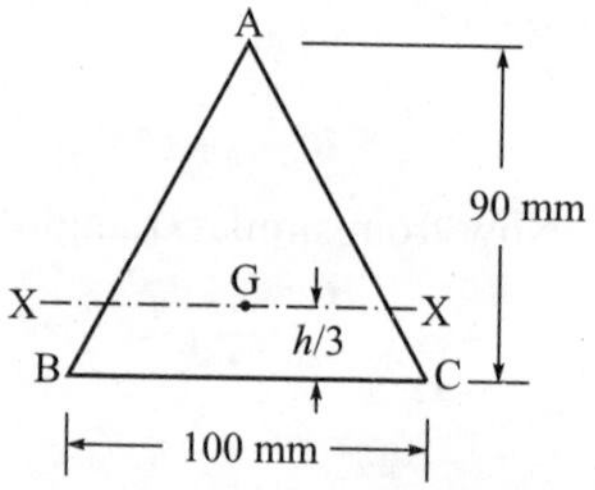


**Fig. 5.13**

***Moment of inertia of the triangular section about its centroid***

We know that moment of inertia of the triangular section about its centroid,

$$I_G = \frac{bh^3}{36} = \frac{100(90)^3}{36}$$

$$= 2025 \times 10^3 \text{ mm}^4 \quad \textbf{Ans.}$$

***Moment of inertia of the section about BC***

We know that the moment of inertia of the triangular section about the base *BC*,

$$I_{BC} = \frac{bh^3}{12} = \frac{100(90)^3}{12} = 6075 \times 10^3 \text{ mm}^4 \quad \textbf{Ans.}$$

## 5.10 MOMENT OF INERTIA OF A COMPOSITE SECTION

The moment of inertia of a composite section may be determined by using the following steps:

1. Divide the given section into regular sections (such as rectangles, triangles, circles etc.)
2. Determine the centre of gravity of the composite section.
3. Find the moment of inertia of the regular section about their centroidal axis.
4. Now transfer these moment of inertias about the required axis by using theorem of parallel axis.
5. The moment of inertia of the composite section may be obtained by the algebraic sum of the moment of inertia of the regular section about the required axis.

**Example 5.5** *Find the centroid of a 150 mm × 125 × 10 mm T-section, as shown in Fig. 5.14. Also determine the moment of inertia of the section about the centroidal X-X and Y-Y axis.*

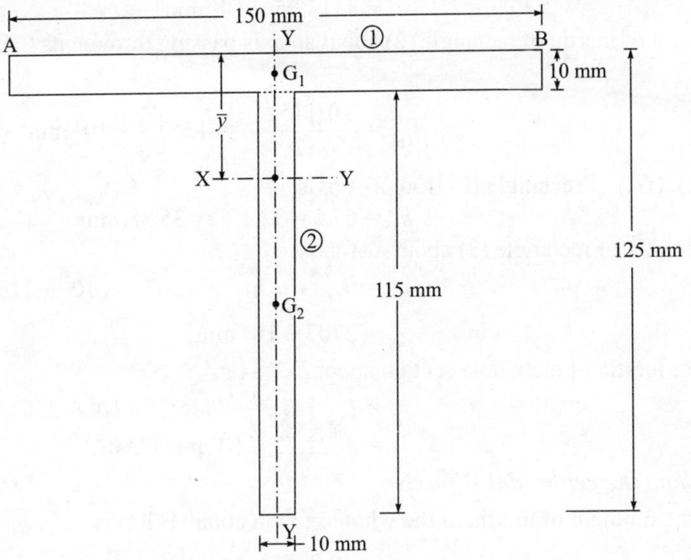


**Fig. 5.14**

***Solution*** Given : 150 mm × 125 mm × 10 mm *T*-section.

***Centroid of the section***

Since the section is symmetrical about *Y-Y* axis, therefore the centroid will lie on this axis. Let $\bar{y}$ be the distance of centroid *G* from the top face of the section *AB* which is taken as the reference axis.

Divide the section into two rectangles (1) and (2), as shown in the figure.

Area of rectangle (1), $a_1 = 150 \times 10 = 1500 \text{ mm}^2$

Distance of its C.G. ($G_1$) from *AB*, $y_1 = \frac{10}{2} = 5 \text{ mm}$

Area of rectangle (2), $a_2 = (125 - 10)\,10 = 115 \times 10 = 1150 \text{ mm}^2$

Distance of its C.G. ($G_2$) from *AB*,

$$y_2 = 10 + \frac{115}{2} = 67.5 \text{ mm}$$

We know that

$$\bar{y} = \frac{a_1 y_1 + a_2 y_2}{a_1 + a_2} = \frac{1500 \times 5 + 1150 \times 67.5}{1500 + 1150}$$

$$= \frac{7500 + 77\,625}{2650} = 32.12 \text{ mm} \quad \textbf{Ans.}$$

***Moment of inertia about the centroidal X-X axis***

We know that moment of inertia of rectangle (1) about an axis passing through its C.G. ($G_1$) and parallel to *X-X* axis,

$$I_{G1} = \frac{150(10)^3}{12} = 12\,500 \text{ mm}^4$$

Distance of C.G. ($G_1$) of rectangle (1) from *X-X* axis,

$$h_1 = 32.12 - \frac{10}{2} = 27.12 \text{ mm}$$

∴ Moment of inertia of rectangle (1) about $X$-$X$ axis,

$$I_{X1} = I_{G1} + a_1 h_1^2 = 12\,500 + 1500\,(27.12)^2$$
$$= 1115.75 \times 10^3 \text{ mm}^4$$

Similarly, moment of inertia of rectangle (2) about an axis passing through its C.G. ($G_2$) and parallel to $X$-$X$ axis,

$$I_{G2} = \frac{10(115)^3}{12} = 1267.4 \times 10^3 \text{ mm}^4$$

Distance of C.G. ($G_2$) of rectangle (2) from $X$-$X$ axis,

$$h_2 = 67.5 - 32.12 = 35.38 \text{ mm}$$

∴ Moment of inertia of rectangle (2) about $X$-$X$ axis,

$$I_{X2} = I_{G2} + a_2 h_2^2 = 1267.4 \times 10^3 + 1150\,(35.38)^2$$
$$= 2707 \times 10^3 \text{ mm}^4$$

Total moment of inertia of the whole section about $X$-$X$ axis,

$$I_{XX} = I_{X1} + I_{X2} = 1115.75 \times 10^3 + 2707 \times 10^3$$
$$= 3822.75 \times 10^3 \text{ mm}^4 \quad \textbf{Ans.}$$

***Moment of inertia about the centroidal Y-Y axis***

We know that the moment of inertia of the whole section about $Y$-$Y$ axis,

$$I_{YY} = \frac{10(150)^3}{12} + \frac{115(10)^3}{12}$$
$$= \frac{33\,750\,000}{12} + \frac{115\,000}{12} = 2822.08 \times 10^3 \text{ mm}^4 \quad \textbf{Ans.}$$

**Example 5.6** *Determine the moment of inertia of an inverted T-section of flange 100 mm × 10 mm and web 120 mm × 12 mm about both centroidal axes.*

***Solution*** Given : Inverted $T$-section, as shown in Fig. 5.15.

First of all, let us find the centroid of the given section.

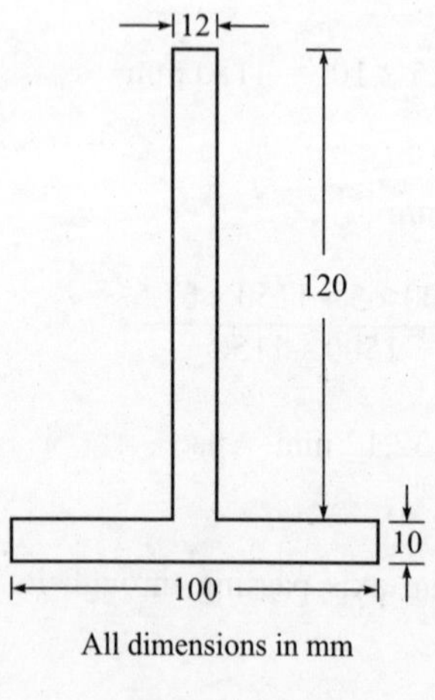


**Fig. 5.15**

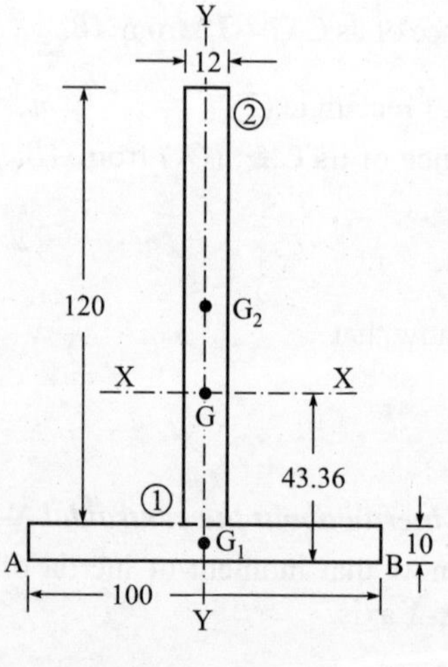


**Fig. 5.16**

Since the section is symmetrical about $Y$-$Y$ axis, therefore the centroid will lie on this axis. Let $\bar{y}$ be the distance of the centroid from the bottom face $AB$ of the section which is taken as reference line.

Divide the section into two rectangles (1) and (2), as shown in Fig. 5.16.

Area of rectangle (1), $a_1 = 100 \times 10 = 1000 \text{ mm}^2$

Distance of its C.G. ($G_1$) from $AB$,

$$y_1 = \frac{10}{2} = 5 \text{ mm}$$

Area of rectangle (2), $a_2 = 120 \times 12 = 1440 \text{ mm}^2$

Distance of its C.G. ($G_2$) from $AB$,

$$y_2 = 10 + \frac{120}{2} = 70 \text{ mm}$$

We know that

$$\bar{y} = \frac{a_1 y_1 + a_2 y_2}{a_1 + a_2} = \frac{1000 \times 5 + 1440 \times 70}{1000 + 1440}$$

$$= \frac{5000 + 100\,800}{2440} = 43.36 \text{ mm}$$

Let $I_{XX}$ and $I_{YY}$ be the moment of inertia about the centroidal axes $X$-$X$ and $Y$-$Y$, respectively.

We know that moment of inertia of rectangle (1) about an axis passing through its C.G. ($G_1$) and parallel to $X$-$X$ axis,

$$I_{G1} = \frac{100\,(10)^3}{12} = 8333.33 \text{ mm}^4$$

Distance of C.G. ($G_1$) of rectangle (1) from $X$-$X$ axis,

$$h_1 = 43.36 - \frac{10}{2} = 38.36 \text{ mm}$$

∴ Moment of inertia of rectangle (1) about $X$-$X$ axis,

$$I_{X1} = I_{G1} + a_1 h_1^2 = 8333.33 + 1000\,(38.36)^2 = 1\,479\,823 \text{ mm}^4$$

Similarly, moment of inertia of rectangle (2) about an axis passing through its C.G. ($G_2$) and parallel to $X$-$X$ axis,

$$I_{G2} = \frac{12\,(120)^3}{12} = 1\,728\,000 \text{ mm}^4$$

Distance of C.G. ($G_2$) of rectangle (2) from $X$-$X$ axis,

$$h_2 = \frac{120}{2} - (43.36 - 10) = 26.64 \text{ mm}$$

∴ Moment of inertia of rectangle (2) about $X$-$X$ axis,

$$I_{X2} = I_{G2} + a_2 h_2^2 = 1\,728\,000 + 1440\,(26.64)^2 = 2\,749\,953 \text{ mm}^4$$

Total moment of inertia of the whole section about $X$-$X$ axis,

$$I_{XX} = I_{X1} + I_{X2} = 1\,479\,823 + 2\,749\,953 = 4\,229\,776 \text{ mm}^4 \quad \textbf{Ans.}$$

Since the section is symmetrical about $Y$-$Y$ axis, therefore the moment of inertia of the whole section about $Y$-$Y$ axis,

$$I_{YY} = I_{Y1} + I_{Y2} = \frac{10\,(100)^3}{12} + \frac{120\,(12)^3}{12}$$

$$= 833\,333.33 + 17\,280 = 850\,613.33 \text{ mm}^4 \quad \textbf{Ans.}$$

**Example 5.7** *A channel section 150 mm × 75 mm × 10 mm is shown in Fig. 5.17. Find the centroid and moment of inertia about a vertical axis passing through the centroid of the section.*

**Solution** Given : 150 mm × 75 mm × 10 mm channel section.

***Centroid of the section***

Since the section is symmetrical about *X-X* axis, therefore the centroid will lie on this axis. Let $\bar{x}$ be the distance of centroid ($G$) from the face $AC$, the axis of reference.

Split the whole area into three rectangles (1), (2) and (3).

Area of rectangle (1), $a_1 = 75 \times 10 = 750 \text{ mm}^2$

Distance of its C.G. ($G_1$) from $AC$,

$$x_1 = \frac{75}{2} = 37.5 \text{ mm}$$

Area of rectangle (2), $a_2 = (150 - 2 \times 10) \times 10 = 1300 \text{ mm}^2$

Distance of its C.G. ($G_2$) from $AC$,

$$x_2 = \frac{10}{2} = 5 \text{ mm}$$

Area of rectangle (3), $a_3 = 75 \times 10 = 750 \text{ mm}^2$

Distance of its C.G. ($G_3$) from $AC$,

$$x_3 = \frac{75}{2} = 37.5 \text{ mm}$$

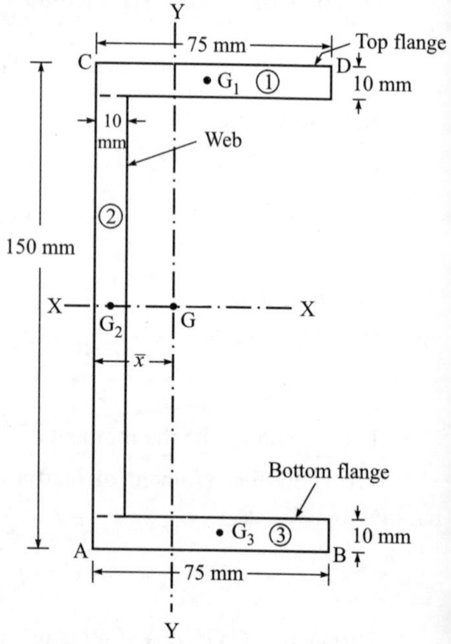


**Fig. 5.17**

We know that

$$\bar{x} = \frac{a_1 x_1 + a_2 x_2 + a_3 x_3}{a_1 + a_2 + a_3} = \frac{750 \times 37.5 + 1300 \times 5 + 750 \times 37.5}{750 + 1300 + 750}$$

$$= \frac{28\,125 + 6500 + 28\,125}{2800} = 22.41 \text{ mm } \textbf{Ans.}$$

***Moment of inertia about a vertical axis passing through the centroid of the section, i.e. Y-Y axis***

We know that moment of inertia of rectangle (1) about an axis passing through its C.G. ($G_1$) and parallel to *Y-Y* axis,

$$I_{G1} = \frac{10(75)^3}{12} = 0.35 \times 10^6 \text{ mm}^4$$

Distance of C.G. ($G_1$) of rectangle (1) from *Y-Y* axis,

$$h_1 = \frac{75}{2} - 22.41 = 15.09 \text{ mm}$$

∴ Moment of inertia of rectangle (1) about *Y-Y* axis,

$$I_{Y1} = I_{G1} + a_1 h_1^2 = 0.35 \times 10^6 + 750\,(15.09)^2 = 0.52 \times 10^6 \text{ mm}^4$$

Similarly, moment of inertia of rectangle (2) about an axis passing through its C.G. and parallel to *Y-Y* axis,

$$I_{G2} = \frac{10(130)^3}{12} = 1.83 \times 10^6 \text{ mm}^4$$

Distance of C.G. ($G_2$) of rectangle (2) from *Y-Y* axis,

$$h_2 = 22.41 - \frac{10}{2} = 17.41 \text{ mm}$$

∴ Moment of inertia of rectangle (2) about *Y-Y* axis,

$$I_{Y2} = I_{G2} + a_2 h_2^2 = 1.83 \times 10^6 + 1300\,(17.41)^2 = 2.224 \times 10^6 \text{ mm}^4$$

Since rectangle (3) is similar to rectangle (1), therefore

$$I_{Y3} = I_{Y1} = 0.52 \times 10^6 \text{ mm}^4$$

∴ Total moment of inertia of the whole section about *Y-Y* axis,

$$I_{YY} = I_{Y1} + I_{Y2} + I_{Y3} = 0.52 \times 10^6 + 2.224 \times 10^6 + 0.52 \times 10^6$$

$$= 3.264 \times 10^6 \text{ mm}^4 \quad \textbf{Ans.}$$

**Note:** The moment of inertia of the whole section about *X-X* axis may be obtained by using the relation.

$$I_{YY} = \frac{BD^3}{12} - \frac{bd^3}{12}$$

where $B = 75$ mm; $D = 150$ mm; $b = 65$ mm; and $d = 130$ mm.

$$\therefore \quad I_{XX} = \frac{75(150)^3}{12} - \frac{65(130)^3}{12} = 21.1 \times 10^6 - 11.9 \times 10^6 = 9.2 \times 10^6 \text{ mm}^4 \quad \textbf{Ans.}$$

**Example 5.8** *An I-section has the following dimensions:*

*Top flange : 150 mm × 25 mm ; Bottom flange : 300 mm × 50 mm ; and Web : 200 mm × 25 mm. Determine the centroid and moment of inertia of the section about the horizontal axis passing through the centroid of the section.*

***Solution*** Given : Top flange = 150 mm × 25 mm ; Bottom flange = 300 × 50 mm ; Web = 200 mm × 25 mm

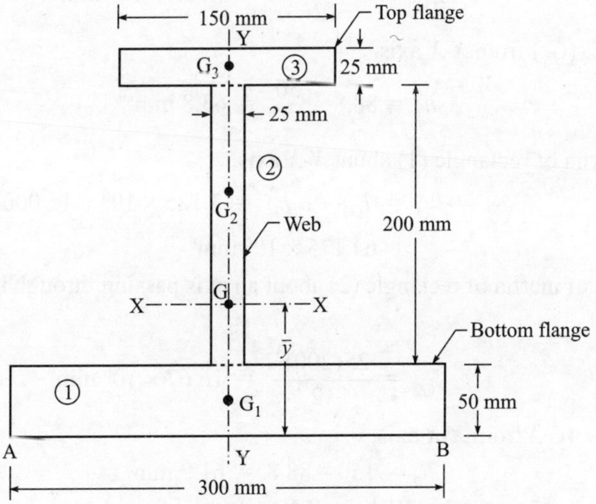


**Fig. 5.18**

***Centroid of the section***

The *I*-section is shown in Fig. 5.18. Since the section is symmetrical about *Y-Y* axis, therefore the centroid of the section will lie on this axis. Let $\bar{y}$ be the distance of centroid from the base *AB* of the bottom flange.

Split the whole section into three rectangles, (1), (2) and (3).

Area of rectangle (1), $a_1 = 300 \times 50 = 15\,000 \text{ mm}^2$

Distance of its C.G. ($G_1$) from $AB$,

$$y_1 = \frac{50}{2} = 25 \text{ mm}$$

Area of rectangle (2), $a_2 = 200 \times 25 = 5000 \text{ mm}^2$

Distance of its C.G. ($G_2$) from $AB$,

$$y_2 = 50 + \frac{200}{2} = 150 \text{ mm}$$

Area of rectangle (3), $a_3 = 150 \times 25 = 3750 \text{ mm}^2$

Distance of its C.G. ($G_3$) from $AB$,

$$y_3 = 50 + 200 + \frac{25}{2} = 262.5 \text{ mm}$$

We know that

$$\bar{y} = \frac{a_1 y_1 + a_2 y_2 + a_3 y_3}{a_1 + a_2 + a_3}$$

$$= \frac{15\,000 \times 25 + 5000 \times 150 + 3750 \times 262.5}{15\,000 + 5000 + 3750}$$

$$= \frac{375\,000 + 750\,000 + 984\,375}{23\,750} = 88.8 \text{ mm} \quad \textbf{Ans.}$$

***Moment of inertia about the centroidal horizontal axis, i.e. X-X axis***

We know that moment of inertia of rectangle (1) about an axis passing through its C.G. and parallel to $X$-$X$ axis,

$$I_{G1} = \frac{300(50)^3}{12} = 3.125 \times 10^6 \text{ mm}^4$$

Distance of its C.G. ($G_1$) from $X$-$X$ axis,

$$h_1 = 88.8 - \frac{50}{2} = 63.8 \text{ mm}$$

∴ Moment of inertia of rectangle (1) about $X$-$X$ axis,

$$I_{X1} = I_{G1} + a_1 h_1^2 = 3.125 \times 10^6 + 15\,000\,(63.8)^2$$

$$= 64.175 \times 10^6 \text{ mm}^4$$

Similarly, moment of inertia of rectangle (2) about an axis passing through its C.G. and parallel to $X$-$X$ axis,

$$I_{G2} = \frac{25(200)^3}{12} = 16.67 \times 10^6 \text{ mm}^4$$

Distance of its C.G. ($G_2$) from $X$-$X$ axis,

$$h_2 = 150 - 88.8 = 61.2 \text{ mm}$$

∴ Moment of inertia of rectangle (2) about $X$-$X$ axis,

$$I_{X2} = I_{G2} + a_2 h_2^2 = 16.67 \times 10^6 + 5000\,(61.2)^2 = 35.4 \times 10^6 \text{ mm}^4$$

Moment of inertia of rectangle (3) about an axis passing through its C.G. and parallel to $X$-$X$ axis,

$$I_{G3} = \frac{150(25)^3}{12} = 0.195 \times 10^6 \text{ mm}^4$$

Distance of its C.G. ($G_3$) from $X$-$X$ axis,

$$h_3 = 262.5 - 88.8 = 173.7 \text{ mm}$$

$\therefore$ Moment of inertia of rectangle (3) about $X$-$X$ axis,

$$I_{X3} = I_{G3} + a_3 h_3^2 = 0.195 \times 10^6 + 3750(173.7)^2 = 113.338 \times 10^6 \text{ mm}^4$$

Total moment of inertia of the whole section about $X$-$X$ axis,

$$I_{XX} = I_{X1} + I_{X2} + I_{X3} = 64.175 \times 10^6 + 35.4 \times 10^6 + 113.338 \times 10^6$$
$$= 212.913 \times 10^6 \text{ mm}^4 \quad \textbf{Ans.}$$

**Note:** The moment of inertia of the whole section about $Y$-$Y$ axis is given by

$$I_{YY} = I_{Y1} + I_{Y2} + I_{Y3} = \frac{50(300)^3}{12} + \frac{200(25)^3}{12} + \frac{25(150)^3}{12}$$
$$= 112.5 \times 10^6 + 0.26 \times 10^6 + 7.03 \times 10^6 = 119.79 \times 10^6 \text{ mm}^4 \quad \textbf{Ans.}$$

**Example 5.9** *Determine the centroid and moment of inertia about centroidal X-X axis and Y-Y axis of 150 mm × 100 mm × 20 mm L-section, as shown in Fig. 5.19.*

***Solution*** Given : $L$-section of 150 × 100 mm × 20 mm as shown in Fig. 5.19.

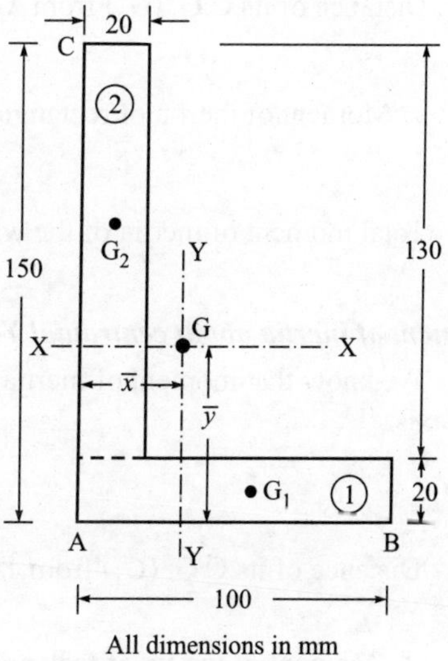


**Fig. 5.19**

***Centroid of the section***

Let $\overline{x}$ and $\overline{y}$ be the distance of centroid from the reference axes $AC$ and $AB$, respectively. Divide the section into two rectangles (1) and (2), as shown in the figure.

Area of rectangle (1), $a_1 = 100 \times 20$
$= 2000 \text{ mm}^2$

Distance of its C.G. ($G_1$) from $AC$, $x_1 = \dfrac{100}{2} = 50$ mm

and distance of its C.G. ($G_1$) from $AB$, $y_1 = \dfrac{20}{2} = 10$ mm

Area of rectangle (2), $a_2 = (150 - 20)\,20$
$= 2600 \text{ mm}^2$

Distance of its C.G. ($G_2$) from $AC$, $x_2 = \dfrac{20}{2} = 10$ mm

and distance of its C.G. ($G_2$) from $AB$, $y_2 = 20 + \left(\dfrac{150 - 20}{2}\right) = 85$ mm

We know that

$$\overline{x} = \frac{a_1 x_1 + a_2 x_2}{a_1 + a_2} = \frac{2000 \times 50 + 2600 \times 10}{2000 + 2600}$$
$$= \frac{100\,000 + 26\,000}{4600} = 27.4 \text{ mm} \quad \textbf{Ans.}$$

and

$$\overline{y} = \frac{a_1 y_1 + a_2 y_2}{a_1 + a_2} = \frac{2000 \times 10 + 2600 \times 85}{2000 + 2600}$$
$$= \frac{20\,000 + 221\,000}{4600} = 52.4 \text{ mm} \quad \textbf{Ans.}$$

***Moment of inertia about centroidal X-X axis***

We know that moment of inertia of rectangle (1) about an axis passing through its C.G. and parallel to *X-X* axis,

$$I_{G1} = \frac{100(20)^3}{12} = 0.067 \times 10^6 \text{ mm}^4$$

Distance of its C.G. ($G_1$) from *X-X* axis,

$$h_1 = 52.4 - \frac{20}{2} = 42.4 \text{ mm}$$

∴ Moment of inertia of rectangle (1) about *X-X* axis,

$$I_{X1} = I_{G1} + a_1 h_1^2 = 0.067 \times 10^6 + 2000\,(42.4)^2 = 3.667 \times 10^6 \text{ mm}^4$$

Similarly, moment of inertia of rectangle (2) about an axis passing through its C.G. and parallel to *X-X* axis,

$$I_{G2} = \frac{20(130)^3}{12} = 3.66 \times 10^6 \text{ mm}^4$$

Distance of its C.G. ($G_2$) from *X-X* axis,

$$h_2 = 85 - 52.4 = 32.6 \text{ mm}$$

∴ Moment of inertia of rectangle (2) about *X-X* axis,

$$I_{X2} = I_{G2} + a_2 h_2^2 = 3.66 \times 10^6 + 2600\,(32.6)^2 = 6.42 \times 10^6 \text{ mm}^4$$

Total moment of inertia of the whole section about *X-X* axis,

$$I_{XX} = I_{X1} + I_{X2} = 3.667 \times 10^6 + 6.42 \times 10^6 = 10.087 \times 10^6 \text{ mm}^4 \quad \textbf{Ans.}$$

***Moment of inertia about centroidal Y-Y axis***

We know that moment of inertia of rectangle (1) about an axis passing through its C.G. and parallel to *Y-Y* axis,

$$I_{G1} = \frac{20(100)^3}{12} = 1.67 \times 10^6 \text{ mm}^4$$

Distance of its C.G. ($G_1$) from *Y-Y* axis,

$$h_1 = 50 - 27.4 = 22.6 \text{ mm}$$

∴ Moment of inertia of rectangle (1) about *Y-Y* axis,

$$I_{Y1} = I_{G1} + a_1 h_1^2 = 1.67 \times 10^6 + 2000\,(22.6)^2 = 2.69 \times 10^6 \text{ mm}^4$$

Similarly, moment of inertia of rectangle (2) about an axis passing through its C.G. and parallel to *Y-Y* axis,

$$I_{G2} = \frac{130(20)^3}{12} = 0.087 \times 10^6 \text{ mm}^4$$

Distance of its C.G. ($G_2$) from *Y-Y* axis,

$$h_2 = 27.4 - \frac{20}{2} = 17.4 \text{ mm}$$

∴ Moment of inertia of rectangle (2) about *Y-Y* axis,

$$I_{Y2} = I_{G2} + a_2 h_2^2 = 0.087 \times 10^6 + 2600\,(17.4)^2 = 0.874 \times 10^6 \text{ mm}^4$$

Total moment of inertia of the whole section about *Y-Y* axis,

$$I_{YY} = I_{Y1} + I_{Y2} = 2.69 \times 10^6 + 0.874 \times 10^6 = 3.564 \times 10^6 \text{ mm}^4 \quad \textbf{Ans.}$$

**Example 5.10** *Determine the moment of inertia of beam's cross-sectional area, as shown in Fig. 5.20, about X-X and Y-Y axes.*

***Solution*** Given : Beam's cross-sectional area, as shown in Fig. 5.20.

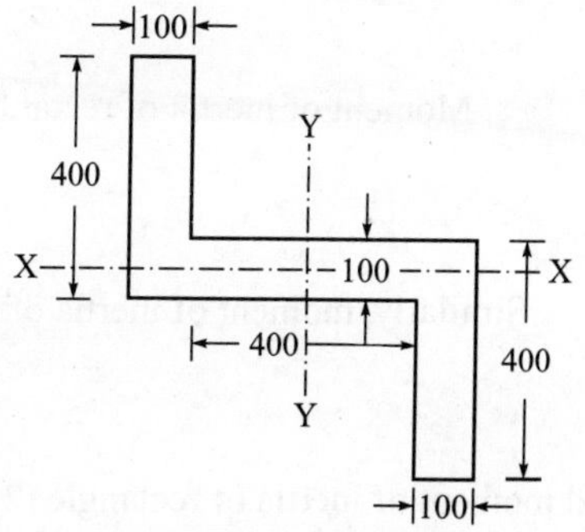


**Fig. 5.20**

Since the section is symmetrical about *X-X* and *Y-Y* axis, therefore, centre of gravity of the given section lies at the intersection of *X-X* and *Y-Y* (*i.e.* at point $G_2$), as shown in Fig. 5.21.

Split the whole section into three rectangles (1), (2) and (3).

Area of rectangle (1), $a_1 = 100 \times 300 = 30\,000 \text{ mm}^2$

Distance of its C.G. ($G_1$) from *X-X* axis,

$$h_1 = \frac{100}{2} + \frac{300}{2} = 200 \text{ mm}$$

and distance of its C.G. ($G_1$) from *Y-Y* axis,

$$h_{1'} = -\frac{100}{2} - 200$$

$$= -250 \text{ mm}$$ ...(minus sign due to left of *Y-Y*)

Area of rectangle (2), $a_2 = (100 + 400 + 100) \times 100 = 60\,000 \text{ mm}^2$

Distance of its C.G. ($G_2$) from *X-X* and *Y-Y* axes,

$$h_2 = h_2' = 0$$

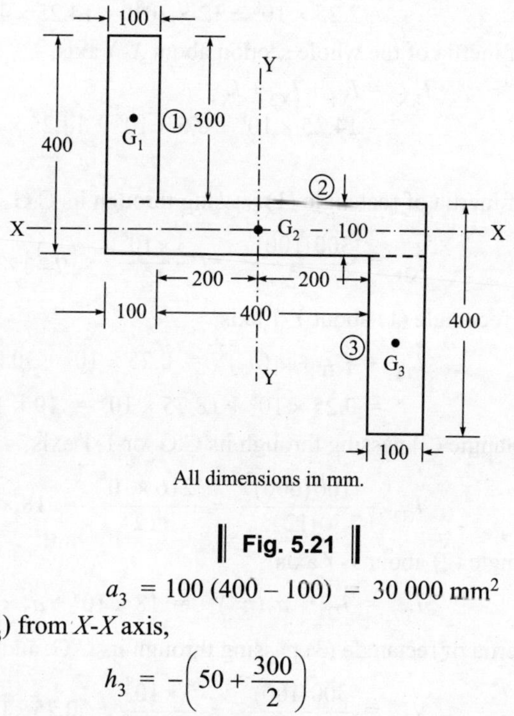


**Fig. 5.21**

Area of rectangle (3), $a_3 = 100\,(400 - 100) = 30\,000 \text{ mm}^2$

Distance of its C.G. ($G_3$) from *X-X* axis,

$$h_3 = -\left(50 + \frac{300}{2}\right)$$

$$= -200 \text{ mm}$$ ...(minus sign due to distance below *X-X*)

and distance of its C.G. ($G_3$) from *Y-Y* axis,

$$h_{3'} = 200 + \frac{100}{2} = 250 \text{ mm}$$

***Moment of inertia about X-X***

We know that moment of inertia of rectangle (1) passing through its C.G. and parallel to *X-X* axis,

$$I_{G1} = \frac{100(300)^3}{12} = \frac{27\times10^8}{12} = 2.25 \times 10^8 \text{ mm}^4$$

∴ Moment of inertia of rectangle (1) about *X-X* axis,

$$I_{X1} = I_{G1} + a_1h_1^2 = 2.25\times10^8 + 30\,000(200)^2$$
$$= 2.25 \times 10^8 + 12 \times 10^8 = 14.25 \times 10^8 \text{ mm}^4$$

Similarly, moment of inertia of rectangle (2) passing through its C.G. or *X-X* axis,

$$I_{G2} = \frac{(100 + 400 + 100)\,(100)^3}{12} = \frac{6\times10^8}{12} = 0.5 \times 10^8 \text{ mm}^4$$

and moment of inertia of rectangle (2) about *X-X* axis,

$$I_{X2} = I_{G2} + a_2h_2^2 = \frac{6\times10^8}{12} + a_2\times0 = 0.5 \times 10^8 \text{ mm}^4$$

Moment of inertia of rectangle (3) passing through its C.G. and parallel to *X-X* axis,

$$I_{G3} = \frac{100(300)^3}{12} = \frac{27\times10^8}{12} = 2.25 \times 10^8 \text{ mm}^4$$

and moment of inertia of rectangle (3) about *X-X*,

$$I_{X3} = I_{G3} + a_3h_3^2 = 2.25\times10^8 + 30\,000\,(-200)^2$$
$$= 2.25 \times 10^8 + 12 \times 10^8 = 14.25 \times 10^8 \text{ mm}^4$$

We know that moment of inertia of the whole section about *X-X* axis,

$$I_{XX} = I_{X1} + I_{X2} + I_{X3}$$
$$= 14.25 \times 10^8 + 0.5 \times 10^8 + 14.25 \times 10^8 = 29 \times 10^8 \text{ mm}^4$$ **Ans.**

***Moment of inertia about Y-Y***

We know that moment of inertia of rectangle (1) passing through its C.G. and parallel to *Y-Y* axis,

$$I_{G1} = \frac{300(100)^3}{12} = \frac{3\times10^8}{12} = 0.25 \times 10^8 \text{ mm}^4$$

∴ Moment of inertia of rectangle (1) about *Y-Y* axis,

$$I_{Y1} = I_{G1} + a_1(h_{1'})^2 = 0.25 \times 10^8 + 30\,000\,(-250)^2$$
$$= 0.25 \times 10^8 + 18.75 \times 10^8 = 19 \times 10^8 \text{ mm}^4$$

Moment of inertia of rectangie (2) passing through its C.G. or *Y-Y* axis,

$$I_{G2} = \frac{100(600)^3}{12} = \frac{216\times10^8}{12} = 18 \times 10^8 \text{ mm}^4$$

and moment of inertia of rectangle (2) about *Y-Y* axis,

$$I_{Y2} = I_{G2} + a_2(h_{2'})^2 = 18 \times 10^8 + a_2 \times 0 = 18 \times 10^8 \text{ mm}^4$$

Similarly, moment of inertia of rectangle (3) passing through its C.G. and parallel to *YY* axis,

$$I_{G3} = \frac{300(100)^3}{12} = \frac{3\times10^8}{12} = 0.25\times10^8 \text{ mm}^4$$

and moment of inertia of rectangle (3) about *Y-Y* axis,

$$I_{Y3} = I_{G3} + a_3(h_{3'})^2 = 0.25 \times 10^8 + 30\,000\,(250)^2$$
$$= 0.25 \times 10^8 + 18.75 \times 10^8 = 19 \times 10^8 \text{ mm}^4$$

We know that moment of inertia of the whole section about *Y-Y* axis,

$$I_{YY} = I_{Y1} + I_{Y2} + I_{Y3}$$
$$= 19 \times 10^8 + 18 \times 10^8 + 19 \times 10^8 = 56 \times 10^8 \text{ mm}^4 \textbf{ Ans.}$$

**Example 5.11** *Determine the moment of inertia of the built-up section, as shown in Fig. 5.22, about its centroidal axis X-X and Y-Y.*

***Solution*** Given : Built-up section, as shown in Fig. 5.22.

First of all, let us find the centroid of the section. Since the section is symmetrical about *Y-Y* axis, therefore the centroid of the section will lie on this axis. Let $\bar{y}$ be the distance of the centroid from the bottom of the section, *i.e. AB*.

Split the whole section into three rectangles (1), (2) and (5) and two triangles (3) and (4), as shown in Fig. 5.23.

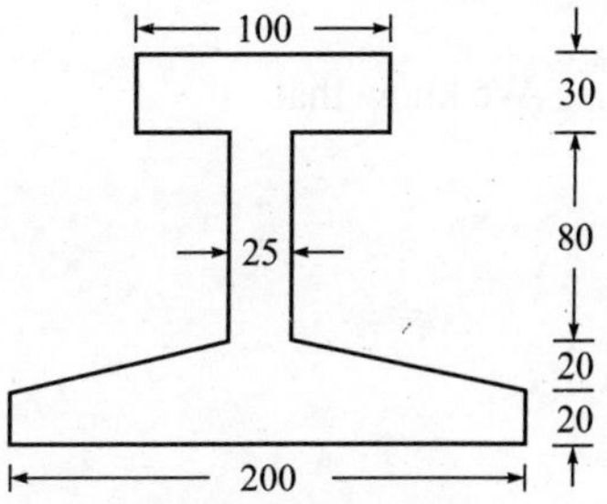

All dimensions in mm

**Fig. 5.22**

We know that area of rectangle (1),

$$a_1 = 100 \times 30 = 3000 \text{ mm}^2$$

Distance of its C.G. ($G_1$) from *AB*,

$$y_1 = 20 + 20 + 80 + \frac{30}{2}$$
$$= 135 \text{ mm}$$

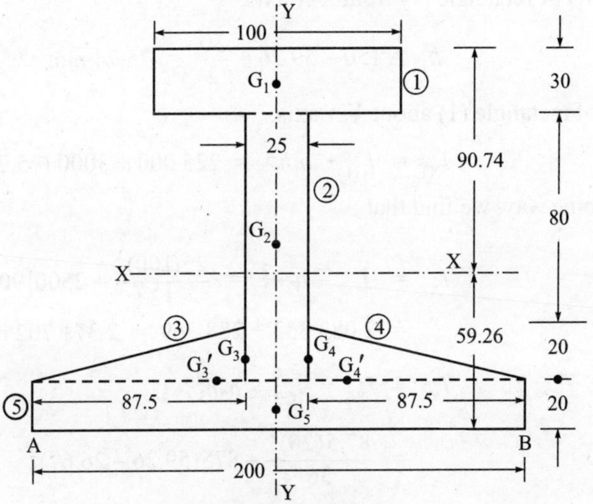

All dimensions in mm

**Fig. 5.23**

Area of rectangle (2), $a_2 = 25\,(80 + 20) = 2500 \text{ mm}^2$

Distance of its C.G. ($G_2$) from *AB*,

$$y_2 = 20 + \left(\frac{20 + 80}{2}\right) = 70 \text{ mm}$$

Area of triangles (3) and (4), $a_3 = a_4 = \frac{1}{2} \times 87.5 \times 20 = 875 \text{ mm}^2$

Distance of its C.G. ($G_3$ and $G_4$) from $AB$,

$$y_3 = y_4 = 20 + \frac{1}{3} \times 20 = 26.67 \text{ mm}$$

Area of rectangle (5), $a_5 = 200 \times 20 = 4000 \text{ mm}^2$

Distance of its C.G. ($G_5$) from $AB$, $y_5 = \dfrac{20}{2} = 10 \text{ mm}$

We know that

$$\bar{y} = \frac{a_1 y_1 + a_2 y_2 + a_3 y_3 + a_4 y_4 + a_5 y_5}{a_1 + a_2 + a_3 + a_4 + a_5}$$

$$= \frac{3000 \times 135 + 2500 \times 70 + 875 \times 26.67 \times 2 + 4000 \times 10}{3000 + 2500 + 875 \times 2 + 4000}$$

...($\because$ $a_3 = a_4$ and $y_3 = y_4$)

$$= \frac{666\,672.5}{11\,250} = 59.26 \text{ mm}$$

***Moment of inertia about the centroidal axis X-X***

We know that moment of inertia of rectangle (1) about an axis passing through its C.G. and parallel to *X-X* axis,

$$I_{G1} = \frac{100(30)^3}{12} = 225\,000 \text{ mm}^4$$

Distance of C.G. ($G_1$) of rectangle (1) from *X-X* axis,

$$h_1 = 150 - 59.26 - \frac{30}{2} = 75.74 \text{ mm}$$

Moment of inertia of rectangle (1) about *X-X* axis,

$$I_{X1} = I_{G1} + a_1 h_1^2 = 225\,000 + 3000\,(75.74)^2 = 17\,434\,643 \text{ mm}^4$$

Proceeding in the same way, we find that

$$I_{X2} = I_{G2} + a_2 h_2^2 = \frac{25(100)^3}{12} + 2500[90.74 - (30 + 50)]^2$$

$$= 2\,083\,333 + 288\,369 = 2\,371\,702 \text{ mm}^4$$

$$I_{X3} = I_{X4} = I_{G3} + a_3 h_3^2$$

$$= \frac{87.5(20)^3}{36} + 875(59.26 - 26.67)^2$$

$$= 19\,444 + 929\,345 = 948\,789 \text{ mm}^4$$

and

$$I_{X5} = I_{G5} + a_5 h_5^2 = \frac{200(20)^3}{12} + 4000\left(59.26 - \frac{20}{2}\right)^2$$

$$= 133\,333 + 9\,706\,190 = 9\,839\,523 \text{ mm}^4$$

Total moment of inertia of the whole section about *X-X*,

$$I_{XX} = I_{X1} + I_{X2} + 2 \times I_{X3} + I_{X5}$$

...($\because I_{X3} = I_{X4}$)

$$= 17\,434\,643 + 2\,371\,702 + 2 \times 948\,789 + 9\,839\,523$$

$$= 31\,543\,446 \text{ mm}^4$$ **Ans.**

***Moment of inertia about the centroidal axis Y-Y***

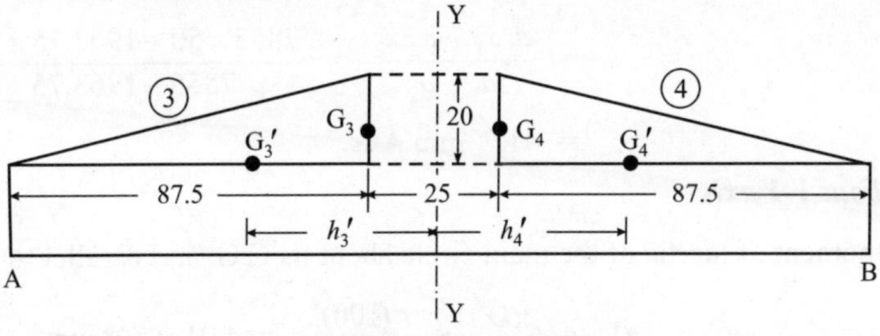


**Fig. 5.24**

We know that moment of inertia of the whole section about *Y-Y*,

$$I_{YY} = I_{Y1} + I_{Y2} + 2 \times {}^*I_{Y3} + I_{Y5}$$

$$= \frac{30(100)^3}{12} + \frac{100(25)^3}{12} + 2\left[\frac{20(87.5)^3}{36} + 875(41.67)^2\right] + \frac{20(200)^3}{12}$$

$$= 2\,500\,000 + 130\,208 + 2(372\,179 + 1\,519\,340) + 133\,333\,333$$

$$= 19\,746\,579 \text{ mm}^4 \quad \textbf{Ans.}$$

**Example 5.12** *A circular hole of 50 mm diameter is cut out from a circular disc of 100 mm diameter, as shown in Fig. 5.25. Find the centroid of the section from A. Also find the moment of inertia of the hollow section about Y-Y axis passing through the centroid.*

***Solution*** Given : $D = 100$ mm or $R = 50$ mm; $d = 50$ mm or $r = 25$ mm

***Centroid of the section***

Since the section is symmetrical about *X-X* axis, therefore the centroid of the section will lie on this axis. The point *A* is the point of reference.

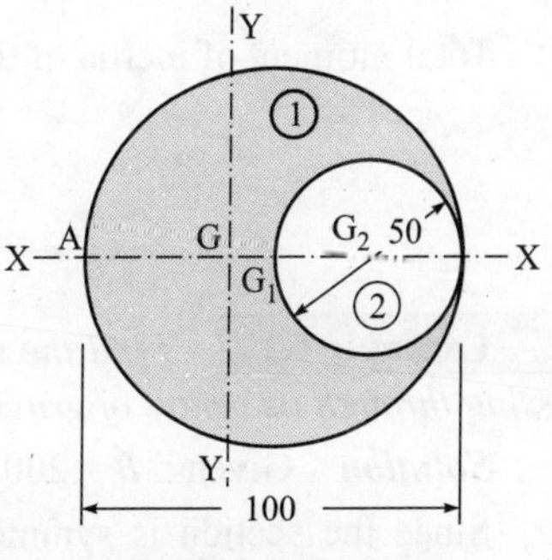


**Fig. 5.25**

Area of main circle, $a_1 = \pi R^2 = \pi(50)^2$

$= 7855 \text{ mm}^2$

Distance of its C.G. ($G_1$) from *A*, $x_1 = \dfrac{100}{2} = 50$ mm

Area of cut out circle, $a_2 = \pi r^2 = \pi(25)^2$

$= 1963.75 \text{ mm}^2$

Distance of its C.G. ($G_2$) from *A*, $x_2 = 50 + \dfrac{50}{2} = 75$ mm

---

* We know that moment of inertia of triangles (3) and (4) about its centroidal axis and parallel to *Y-Y* axis (see Fig. 5.24)

$$I_{G3'} = \frac{20(87.5)^3}{36}$$

Distance of $G_{3'}$ and $G_{4'}$ of the triangles (3) and (4) from *Y-Y* axis,

$$h_{3'} = h_{4'} = \frac{25}{2} + \frac{87.5}{3} = 41.67 \text{ mm} \quad \ldots\left(\text{In this case C.G. of triangle (3) and (4)} = \frac{87.5}{3} \text{ mm}\right)$$

$$\therefore \quad I_{Y3} = I_{Y4} = 2\left[I_{G3'} + a_3(h_{3'})^2\right]$$

Since $a_2$ is cut out from $a_1$, therefore centroid of the section from $A$,

$$\bar{x} = \frac{a_1 x_1 - a_2 x_2}{a_1 - a_2} = \frac{7855 \times 50 - 1963.75 \times 75}{7855 - 1963.75} = \frac{245\,468.75}{5891.25}$$

$$= 41.7 \text{ mm } \textbf{Ans.}$$

***Moment of inertia about Y-Y axis***

We know that moment of inertia of the main circle about its C.G. and parallel to $Y$-$Y$ axis,

$$I_{G1} = \frac{\pi D^4}{64} = \frac{\pi (100)^4}{64} = 4.91 \times 10^6 \text{ mm}^4$$

Distance of C.G. ($G_1$) of main circle from $Y$-$Y$ axis,

$$h_1 = 50 - 41.7 = 8.3 \text{ mm}$$

∴ Moment of inertia of the main circle about $Y$-$Y$ axis

$$I_{Y1} = I_{G1} + a_1 h_1^2 = 4.91 \times 10^6 + 7855\,(8.3)^2 = 5.45 \times 10^6 \text{ mm}^4$$

Similarly, moment of inertia of the cut out circle about its C.G. and parallel to $Y$-$Y$ axis,

$$I_{G2} = \frac{\pi (50)^4}{64} = 0.307 \times 10^6 \text{ mm}^4$$

Distance of C.G. ($G_2$) of the cut out circle from $Y$-$Y$ axis,

$$h_2 = 75 - 41.7 = 33.3 \text{ mm}$$

∴ Moment of inertia of the cut out circle about $Y$-$Y$ axis,

$$I_{Y2} = I_{G2} + a_2 h_2^2 = 0.307 \times 10^6 + 1963.75\,(33.3)^2 = 2.487 \times 10^6 \text{ mm}^4$$

Total moment of inertia of the hollow section about $Y$-$Y$ axis,

$$I_{YY} = I_{Y1} - I_{Y2} \quad \text{...(–ve sign due to cut out section)}$$

$$= 5.45 \times 10^6 - 2.487 \times 10^6 = 2.963 \times 10^6 \text{ mm}^4 \textbf{ Ans.}$$

**Example 5.13** *Find the moment of inertia of a hollow section, shown in Fig. 5.26, about an axis passing through its centre of gravity or parallel to X-axis.*

***Solution*** Given : $B = 200$ mm ; $D = 300$ mm ; $d = 150$ mm

Since the section is symmetrical about $Y$-$Y$ axis, therefore the centroid of the section will lie on this axis.

We know that moment of inertia of the rectangular section (1) about $Y$-$Y$ axis,

$$I_{G1} = \frac{DB^3}{12} = \frac{300\,(200)^3}{12}$$

$$= 200 \times 10^6 \text{ mm}^4$$

and moment of inertia of the circular hole (2) about $Y$-$Y$ axis,

$$I_{G2} = \frac{\pi d^4}{64} = \frac{\pi}{64}(150)^4$$

$$= 24.85 \times 10^6 \text{ mm}^4$$

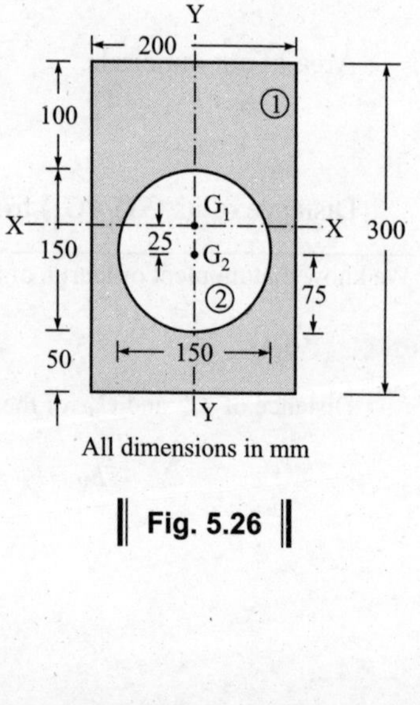


**Fig. 5.26**

∴ Moment of inertia of the hollow section about an axis passing through it, *i.e.* about *Y-Y* axis,

$$I_{YY} = I_{G1} - I_{G2} = 200 \times 10^6 - 24.85 \times 10^6 = 175.15 \times 10^6 \text{ mm}^4 \textbf{ Ans.}$$

Moment of inertia of the rectangular section (1) about *X-X* axis,

$$I_{X1} = \frac{BD^3}{12} = \frac{200(300)^3}{12} = 450 \times 10^6 \text{ mm}^4$$

We know that cross-sectional area of the circular hole,

$$a_2 = \frac{\pi d^2}{4} = \frac{\pi}{4}(150)^2 \text{ mm}^2$$

and distance of its C.G. ($G_2$) from *X-X* axis,

$$h_2 = G_1G_2 = 150 - (50 + 75) = 25 \text{ mm}$$

∴ Moment of inertia of the circular hole (2) about *X-X* axis,

$$I_{X2} = I_{G2} + a_2h_2^2 = \frac{\pi(150)^4}{64} + \frac{\pi}{4}(150)^2\,(25)^2$$

$$= 24.85 \times 10^6 + 11.05 \times 10^6 = 35.9 \times 10^6 \text{ mm}^4$$

∴ Moment of inertia of the hollow section about an axis passing through it, *i.e.* about *X-X* axis,

$$I_{XX} = I_{X1} - I_{X2} = 450 \times 10^6 - 35.9 \times 10^6 = 414.1 \times 10^6 \text{ mm}^4 \textbf{ Ans.}$$

**Example 5.14** *Determine the second moment of the shaded area, as shown in Fig. 5.27, about the centroidal axis parallel to the X-X axis. Take R = 1 mm.*

***Solution*** Given : Area, as shown in Fig. 5.27.

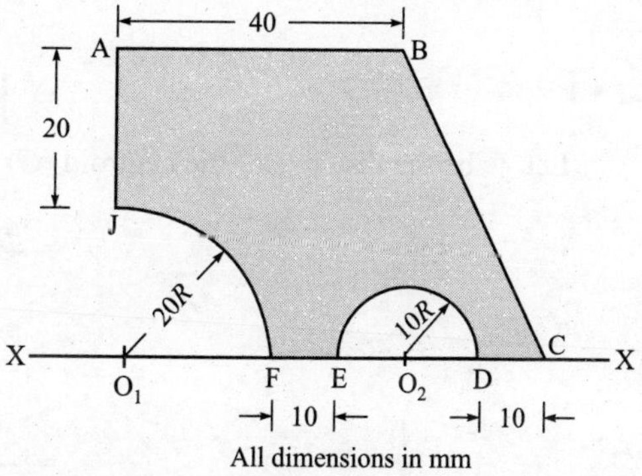


**Fig. 5.27**

First of all, let us find the position of centroidal axis (*i.e.* position of centre of gravity of the shaded area) from *X-X* axis.

Complete the rectangle $O_1AHC$, as shown in Fig. 5.28.

The whole area consists of the following four parts

1. Rectangle $O_1AHC$ having dimensions 60 mm × 40 mm;
2. Quadrant $O_1JF$ of radius 20 mm;
3. Semicircle of radius 10 mm; and
4. Triangle $BCH$ of base 20 mm and height 40 mm.

Area of rectangle $O_1AHC$ $\quad a_1 = 60 \times 40 = 2400 \text{ mm}^2$

Distance of its centre of gravity ($G_1$) from $O_1C$,

$$y_1 = \frac{40}{2} = 20 \text{ mm}$$

Area of quadrant $O_1JF$ $\quad a_2 = \dfrac{\pi r^2}{4} = \dfrac{\pi(20)^2}{4} = 314.2 \text{ mm}^2$

Distance of its centre of gravity ($G_2$) from $O_1F$,

$$y_2 = \frac{4r}{3\pi} = \frac{4 \times 20}{3\pi} = 8.487 \text{ mm}$$

Area of semicircle, $a_3 = \dfrac{\pi r^2}{2} = \dfrac{\pi(10)^2}{2} = 157.1 \text{ mm}^2$

Distance of its centre of gravity ($G_3$) from *ED*,

$$y_3 = \frac{4r}{3\pi} = \frac{4 \times 10}{3\pi} = 4.243 \text{ mm}$$

Area of triangle *BCH*, $a_4 = \dfrac{1}{2} \times BH \times HC = \dfrac{1}{2} \times 20 \times 40 = 400 \text{ mm}^2$

Distance of its centre of gravity ($G_4$) from $O_1X$,

$$y_4 = \frac{2 \times \text{height}}{3} = \frac{2 \times 40}{3} = 26.67 \text{ mm}$$

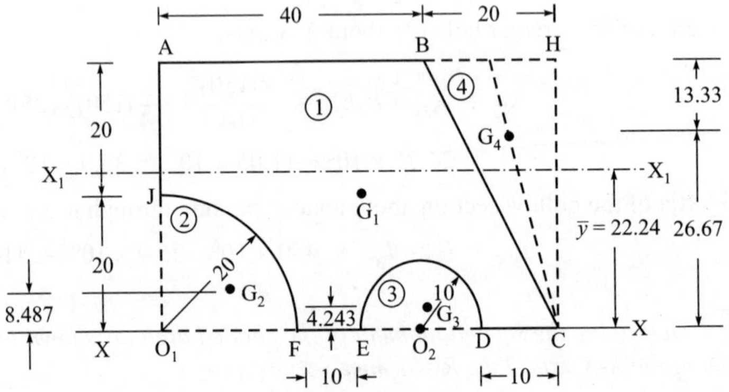


All dimensions in mm

**Fig. 5.28**

Let $\bar{y}$ be the distance of the centroid (*G*) of the shaded area from *X-X* axis. We know that

$$\bar{y} = \frac{a_1 y_1 - a_2 y_2 - a_3 y_3 - a_4 y_4}{a_1 - a_2 - a_3 - a_4}$$

$$= \frac{2400 \times 20 - 314.2 \times 8.487 - 157.1 \times 4.243 - 400 \times 26.67}{2400 - 314.2 - 157.1 - 400}$$

$$= \frac{48\,000 - 2666.6 - 666.6 - 10\,668}{1528.7} = 22.24 \text{ mm}$$

Now an axis $X_1$-$X_1$ at a distance $\bar{y} = 22.24$ mm gives the centroidal axis parallel to *X-X* axis.

We know that moment of inertia of rectangle $O_1AHC$ about an axis passing through its centre of gravity ($G_1$) and parallel to the centroidal axis $X_1$-$X_1$,

$$I_{G1} = \frac{60(40)^3}{12} = 320\,000 \text{ mm}^4$$

Distance of its centre of gravity ($G_1$) from axis $X_1$-$X_1$,

$$h_1 = 22.24 - 20 = 2.24 \text{ mm}$$

∴ Moment of inertia of rectangle $O_1AHC$ about centroidal axis $X_1$-$X_1$,

$$I_1 = I_{G1} + a_1 h_1^2 = 320\,000 + 2400\,(2.24)^2 = 332\,042 \text{ mm}^4$$

Similarly, moment of inertia of quadrant $O_1JF$ about an axis passing through its centre of gravity ($G_2$) and parallel to the centroidal axis $X_1$-$X_1$,

$$I_{G2} = \frac{1}{4} \times \frac{\pi(40)^4}{64} = 31\,420 \text{ mm}^4$$

Distance of its centre of gravity ($G_2$) from axis $X_1$-$X_1$,

$$h_2 = 22.24 - 8.487 = 13.753 \text{ mm}$$

∴ Moment of inertia of quadrant $O_1JF$ about centroidal axis $X_1$-$X_1$,

$$I_2 = I_{G2} + a_2 h_2^2 = 31\,420 + 314.2\,(13.753)^2 = 90\,849 \text{ mm}^4$$

Moment of inertia of semicircle about an axis passing through its centre of gravity ($G_3$) and parallel to the centroidal axis $X_1$-$X_1$,

$$I_{G3} = \frac{1}{2} \times \frac{\pi(20)^4}{64} = 3927.5 \text{ mm}^4 \quad ...(\because \text{ Diameter of semicircle} = 20 \text{ mm})$$

Distance of its centre of gravity ($G_3$) from axis $X_1$-$X_1$,

$$h_3 = 22.24 - 4.243 = 17.997 \text{ mm}$$

∴ Moment of inertia of semicircle about centroidal axis $X_1$-$X_1$

$$I_3 = I_{G3} + a_3 h_3^2 = 3927.5 + 157.1\,(17.997)^2 = 54\,811 \text{ mm}^4$$

Moment of inertia of triangle $BCH$ about an axis passing through its centre of gravity ($G_4$) and parallel to the centroidal axis $X_1$-$X_1$,

$$I_{G4} = \frac{20(40)^3}{36} = 35\,556 \text{ mm}^4$$

Distance of its centre of gravity ($G_4$) from axis $X_1$-$X_1$,

$$h_4 = 26.67 - 22.24 = 4.43 \text{ mm}$$

∴ Moment of inertia of triangle $BCH$ about centroidal axis $X_1$-$X_1$,

$$I_4 = I_{G4} + a_1 h_4^2 = 35\,556 + 400\,(4.43)^2 = 43\,406 \text{ mm}^4$$

We know that moment of inertia (second moment) of the shaded area about centroidal axis $X_1$-$X_1$ parallel to $X$-$X$ axis,

$$I = I_1 - I_2 - I_3 - I_4$$
$$= 332\,042 - 90\,849 - 54\,811 - 43\,406 = 142\,976 \text{ mm}^4 \quad \textbf{Ans.}$$

**Example 5.15** *The cross-section of a beam is shown in Fig. 5.29. Locate the centroid and calculate the moment of inertia of the section about the horizontal and vertical centroidal axes.*

***Solution*** Given : A section, as shown in Fig. 5.29.

***Centroid of the section***

Since the section is symmetrical about $Y$-$Y$ axis, therefore, centroid of the section will lie on this axis.

Divide the section into two parts, *i.e.* a semicircle $ABC$ of diameter 120 mm (or radius $r$ = 60 mm) and a rectangle $PQRS$ of dimensions 40 mm × 20 mm, as shown in Fig. 5.30.

We know that area of the semicircle $ABC$,

$$a_1 = \frac{\pi r^2}{2} = \frac{\pi(60)^2}{2} = 5655.6 \text{ mm}^2$$

and distance of its C.G. ($G_1$) from $AC$, $y_1 = \dfrac{4r}{3\pi} = \dfrac{4 \times 60}{3\pi} = 25.46$ mm

Area of rectangle $PQRS$, $a_2 = 40 \times 20 = 800 \text{ mm}^2$

and distance of its C.G. ($G_2$) from $AC$, $y_2 = \dfrac{20}{2} = 10$ mm

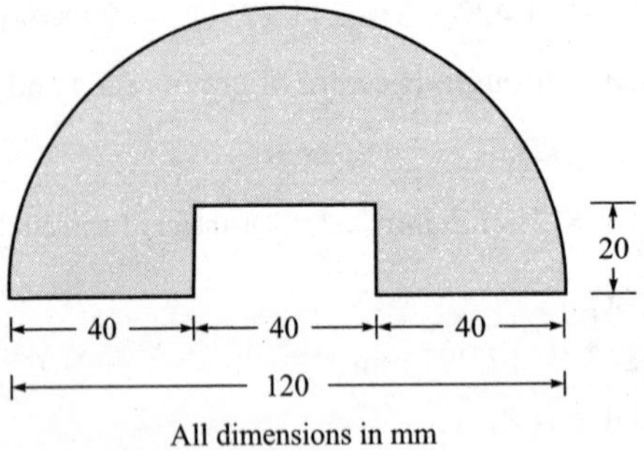


**Fig. 5.29**

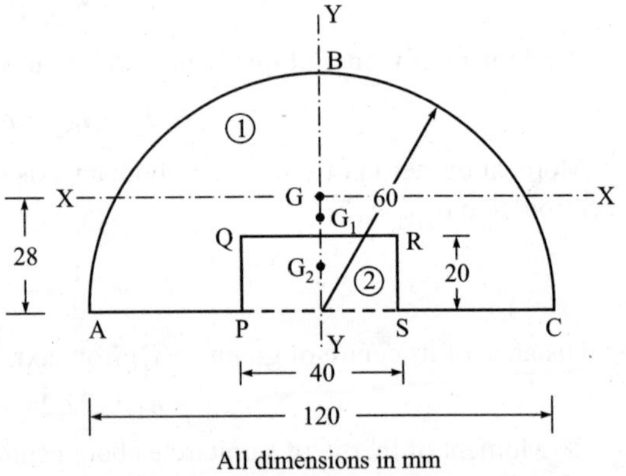


**Fig. 5.30**

Since a rectangular section is removed from the semicircular section, therefore, the distance of centroid ($G$) of the whole section from $AC$ along $Y$-$Y$ axis,

$$\bar{y} = \frac{a_1 y_1 - a_2 y_2}{a_1 - a_2} = \frac{5655.6 \times 25.46 - 800 \times 10}{5655.6 - 800}$$

$$= \frac{143\,991.6 - 8000}{4855.6} = 28 \text{ mm } \textbf{Ans.}$$

We know that moment of inertia of the semicircle about its centroid ($G_1$),

$$I_{G1} = 0.11\, r^4 = 0.11(60)^4 = 1425.6 \times 10^3 \text{ mm}^4$$

Distance of its C.G. ($G_1$) from $X$-$X$ axis,

$$h_1 = 28 - 25.46 = 2.54 \text{ mm}$$

∴ Moment of inertia of semicircle about $X$-$X$ axis,

$$I_{X1} = I_{G1} + a_1 h_1^2 = 1425.6 \times 10^3 + 5655.6\,(2.54)^2$$

$$= 1462.1 \times 10^3 \text{ mm}^4$$

Similarly, moment of inertia of the rectangle about its centroid ($G_2$),

$$I_{G2} = \frac{40\,(20)^3}{12} = 26.7 \times 10^3 \text{ mm}^4$$

Distance of its C.G. ($G_2$) from $X$-$X$ axis,

$$h_2 = 28 - \frac{20}{2} = 18 \text{ mm}$$

∴ Moment of inertia of the rectangle about $X$-$X$ axis,

$$I_{X2} = I_{G2} + a_2 h_2^2 = 26.7 \times 10^3 + 800\,(18)^2 = 285.9 \times 10^3 \text{ mm}^4$$

We know that moment of inertia of the whole section about *X-X* axis,

$$I_{XX} = I_{X1} - I_{X2} = 1462.1 \times 10^3 - 285.9 \times 10^3 = 1176.2 \times 10^3 \text{ mm}^4 \textbf{ Ans.}$$

**Example 5.16** *Find the moment of inertia of the lamina shown in Fig. 5.31, about a centroidal axis parallel to X-X axis. A circular hole 40 mm is cut in the lamina.*

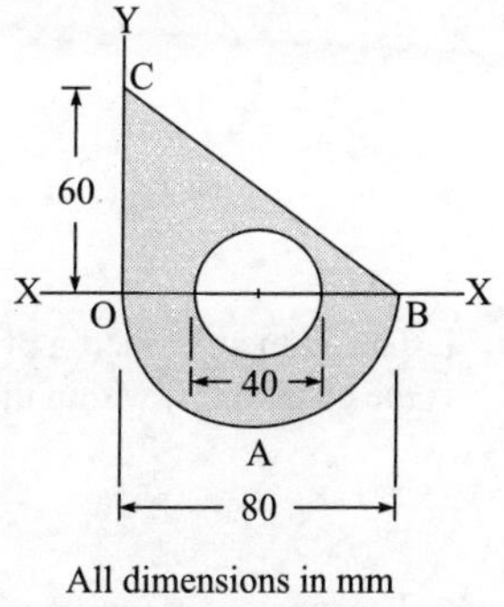


Fig. 5.31

***Solution*** Given : A lamina, as shown in Fig. 5.31.

Divide the section into the following three parts, as shown in Fig. 5.32.

1. A triangle *OBC* with base *OB* ($b$) = 80 mm and height *OC*($h$) = 60 mm;
2. A semicircle *OAB* with diameter *OB* ($d$) = 80 mm, or radius ($r$) = 40 mm; and
3. A circle of diameter ($d_1$) = 40 mm.

We know that moment of inertia of the triangle about *X-X* axis,

$$I_{X1} = \frac{bh^3}{12} = \frac{80(60)^3}{12} = 1440 \times 10^3 \text{ mm}^4$$

Moment of inertia of the semicircle about *X-X* axis,

$$I_{X2} = \frac{\pi}{8}(r^4) = \frac{\pi}{8}(40)^4 = 1005.4 \times 10^3 \text{ mm}^4$$

and moment of inertia of the circle about *X-X* axis,

$$I_{X3} = \frac{\pi(d_1)^4}{64} = \frac{\pi(40)^4}{64} = 125.7 \times 10^3 \text{ mm}^4$$

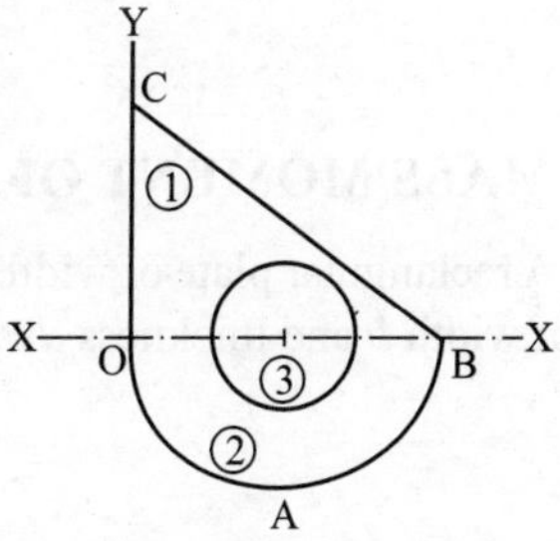


Fig. 5.32

Since the circular portion is removed from the whole section, therefore total moment of inertia of the section about *X-X* axis,

$$I_{XX} = I_{X1} + I_{X2} - I_{X3}$$
$$= 1440 \times 10^3 + 1005.4 \times 10^3 - 125.7 \times 10^3$$
$$= 2319.7 \times 10^3 \text{ mm}^4 \textbf{ Ans.}$$

## 5.11 MASS MOMENT OF INERTIA OF A UNIFORM THIN ROD

A uniform thin rod of length $l$ is shown in Fig. 5.33.

Let $m$ = Mass of the rod per unit length

$\therefore$ Total mass of the rod, $M = m \times l$

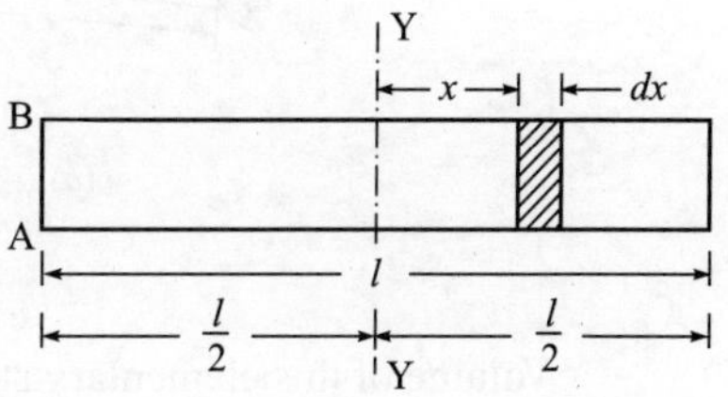


Fig. 5.33

Consider an elementary strip of thickness $dx$ at a distance $x$ from the centroidal axis *Y-Y* about which the moment of inertia is to be determined. We know that moment of inertia of the strip about the centroidal axis and normal to the rod (*i.e.* $I_{YY}$)

$$= \text{Mass of the strip} \times (\text{Distance from Y-Y axis})^2$$

$$= (m.dx)x^2 = mx^2\,dx \qquad ...(i)$$

The moment of inertia of the whole rod about the centroidal axis and normal to it (*i.e.* $I_{YY}$) may be obtained by integrating the equation (*i*) within the limits from $-\frac{l}{2}$ to $+\frac{l}{2}$.

$$\therefore \quad I_{YY} = \int_{-\frac{l}{2}}^{+\frac{l}{2}} m\,x^2\,dx = m\left[\frac{x^3}{3}\right]_{-\frac{l}{2}}^{+\frac{l}{2}}$$

$$= m\left[\frac{\left(\frac{l}{2}\right)^3}{3} - \frac{\left(-\frac{l}{2}\right)^3}{3}\right] = m\left[\frac{l^3}{24} + \frac{l^3}{24}\right] = \frac{ml^3}{12}$$

$$= \frac{M\,l^2}{12} \qquad ...(M = m\,.\,l)$$

**Notes:** **1.** The moment of inertia of the whole rod about the end of the rod *AB* (*i.e.* $I_{AB}$) may be obtained by integrating the equation (*i*) within the limits from 0 to *l*.

$$\therefore \quad I_{AB} = \int_0^l m\,x^2\,dx = m\left[\frac{x^3}{3}\right]_0^l = \frac{ml^3}{3} = \frac{Ml^2}{3} \qquad ...(\because M = m.l)$$

**2.** The moment of inertia of the whole rod about the end of the rod *AB* (*i.e.* $I_{AB}$) may be obtained by using the parallel axis theorem, *i.e.*

$$I_{AB} = I_{YY} + M\left(\frac{l}{2}\right)^2 + \frac{Ml^2}{12} + \frac{Ml^2}{4} = \frac{Ml^2}{3}$$

## 5.12 MASS MOMENT OF INERTIA OF A RECTANGULAR PLATE

A rectangular plate of width *b*, depth *d* and thickness *t* is shown in Fig. 5.34 (*a*). Consider an elementary strip of width *b* and thickness *dy* at a distance *y* from *X-X* axis.

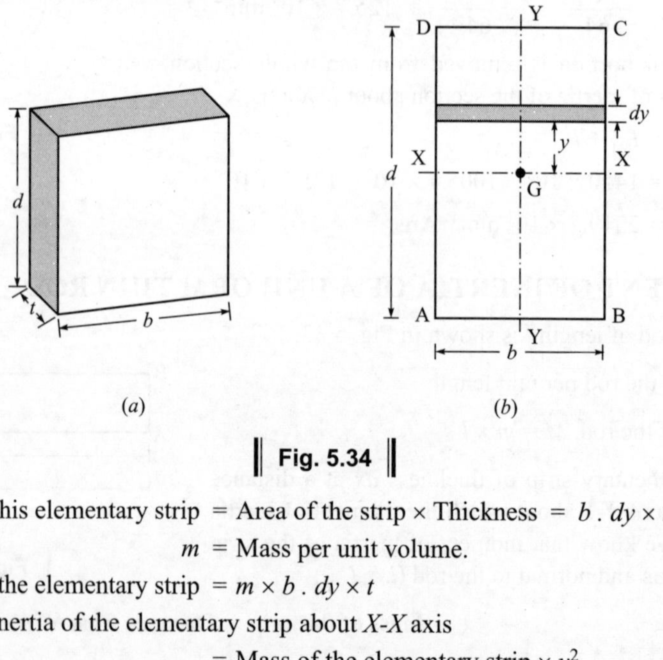


**Fig. 5.34**

Volume of this elementary strip $= \text{Area of the strip} \times \text{Thickness} = b\,.\,dy \times t$

Let $m$ = Mass per unit volume.

$\therefore$ Mass of the elementary strip $= m \times b\,.\,dy \times t$

and mass moment of inertia of the elementary strip about *X-X* axis

$$= \text{Mass of the elementary strip} \times y^2$$

$$= m \times b\,.\,dy \times t \times y^2 = mbt \times y^2\,dy \qquad ...(i)$$

The mass moment of inertia of the whole plate may be obtained by integrating the above equation (*i*) within the limits from $-\frac{d}{2}$ to $+\frac{d}{2}$.

$$\therefore \quad I_{XX} = \int_{-\frac{d}{2}}^{+\frac{d}{2}} mbt\, y^2\, dy = mbt\left[\frac{y^3}{3}\right]_{-\frac{d}{2}}^{+\frac{d}{2}}$$

$$= \frac{mbt}{3}\left[\left(\frac{d}{2}\right)^3 - \left(\frac{-d}{2}\right)^3\right] = \frac{mbt}{3}\left[\frac{d^3}{8} + \frac{d^3}{8}\right]$$

$$= \frac{mbt}{3} \times \frac{d^3}{4} = mbtd \times \frac{d^2}{12} = \frac{M.d^2}{12}$$

where $M$ = Total mass of the rectangular plate $= m \times b.d \times t$

Similarly, we can find the mass moment of inertia about *Y-Y* axis passing through the centre of gravity *G*.

$$\therefore \quad I_{YY} = \frac{M.b^2}{12}$$

According to the theorem of perpendicular axis, the mass moment of inertia about polar axis, *i.e.* *Z-Z* axis,

$$I_{ZZ} = I_{XX} + I_{YY} = \frac{Md^2}{12} + \frac{Mb^2}{12} = \frac{M}{12}(d^2 + b^2)$$

**Note:** The mass moment of inertia of the rectangular plate about its base *AB* may be obtained by using the parallel axis theorem.

$$\therefore \quad I_{AB} = I_{XX} + M\left(\frac{d}{2}\right)^2 = \frac{Md^2}{12} + \frac{M.d^2}{4} = \frac{Md^2}{3}$$

## 5.13 MASS MOMENT OF INERTIA OF A UNIFORM SOLID CYLINDER

A solid cylinder of radius *R*, length *l* and mass *M* is shown in Fig. 5.35. Consider an elementary circular cylindrical strip of radius *r*, thickness *dr* and length *l*, as shown in Fig. 5.35.

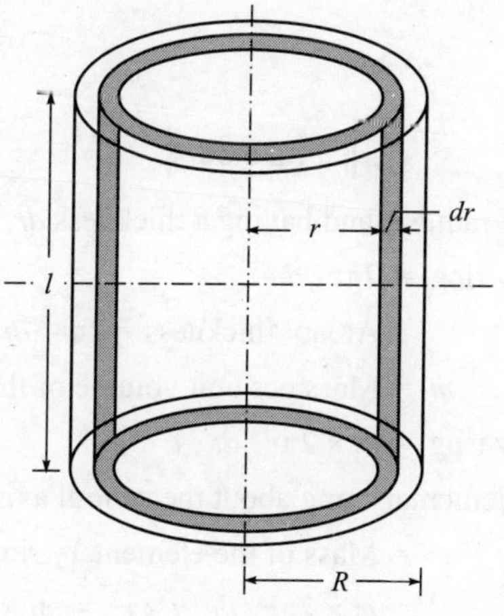


**Fig. 5.35**

Volume of this cylindrical strip = Area of the strip × Length of the strip $= 2\pi r \,.\, dr \times l$

Let $m$ = Mass per unit volume of the section.

$\therefore$ Mass of the elementary strip $= m \times 2\pi r \,.\, dr \times l$

and mass moment of inertia of the elementary strip about the polar axis, *i.e.* *Z-Z* axis

= Mass of the elementary strip $\times r^2$

$$= m \times 2\pi r\, dr \,.\, l \times r^2 = m \times 2\pi l \,.\, r^3\, dr \qquad ...(i)$$

The mass moment of inertia of the whole cylinder may be obtained by integrating the above equation (*i*) within the limits from 0 to *R*.

$$\therefore \qquad I_{ZZ} = \int_0^R m \times 2\pi l . r^3\, dr = m \times 2\pi l \left[\frac{r^4}{4}\right]_0^R$$

$$= m \times 2\pi l \times \frac{R^4}{4} = m \times \pi R^2 l \times \frac{R^2}{2} = \frac{MR^2}{2} = 0.5MR^2$$

where $M$ = Total mass of the cylinder

= Mass per unit volume × Volume of the cylinder = $m \times \pi R^2 l$

**Note:** A thin solid disc of radius *R* and mass *M*, is a special case of solid circular cylinder.

## 5.14 MASS MOMENT OF INERTIA OF A UNIFORM CIRCULAR PLATE

A circular section of radius *R* and thickness *t* with centre *O* is shown in Fig. 5.36 (*a*).

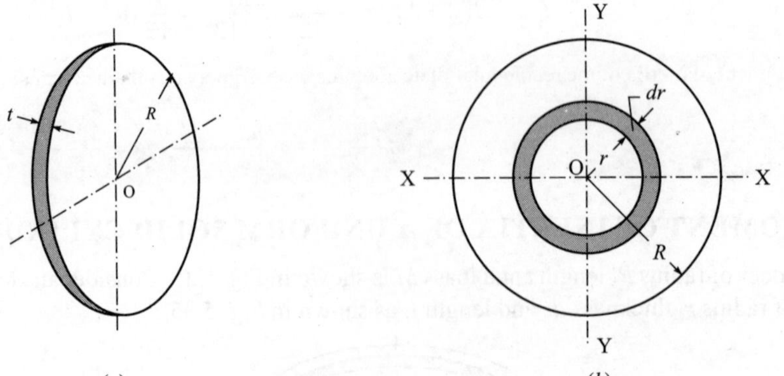


**Fig. 5.36**

Consider an elementary ring of radius *r* and having a thickness *dr*, as shown in Fig. 5.36 (*b*).

∴ Area of this elementary ring = $2\pi r . dr$

and volume of the ring = Area × thickness = $2\pi r . dr . t$

Let $m$ = Mass per unit volume of the section.

∴ Mass of this elementary ring = $m \times 2\pi r . dr . t$

and mass moment of inertia of this elementary ring about the central axis

= Mass of the elementary ring × $r^2$

= $m \times 2\pi r . dr . t \times r^2 = m \times 2\pi r^3 \cdot t\, dr$

The mass moment of inertia of the whole circular section about the central axis ($I_{ZZ}$) may be obtained by integrating the above equation between the limits from 0 to *R*.

$$\therefore \qquad I_{ZZ} = \int_0^R m \times 2\pi t r^3\, dr = m \times 2\pi t \int_0^R r^3\, dr$$

$$= m \times 2\pi t \left[\frac{r^4}{4}\right]_0^R = m \times 2\pi t \times \frac{R^4}{4} = mt \times \frac{\pi R^4}{2}$$

We know that total mass of the section,

$$M = \text{Mass per unit volume of the section} \times \text{volume of the section}$$

$$= m \times \pi R^2 t$$

Thus the above expression may be written as

$$I_{ZZ} = m \times \pi R^2 t \times \frac{R^2}{2} = M \times \frac{R^2}{2} = 0.5MR^2$$

According to theorem of perpendicular axes, mass moment of inertia about *X-X* and *Y-Y* axis,

$$I_{XX} = I_{YY} = \frac{I_{ZZ}}{2} = \frac{0.5MR^2}{2} = 0.25MR^2$$

## 5.15 MASS MOMENT OF INERTIA OF A SOLID SPHERE

A solid sphere of radius $R$ with centre $O$ is shown in Fig. 5.37. Consider an elementary strip $PQ$ of thickness $dx$ at a distance $x$ from $O$, as shown in Fig. 5.37. The strip under consideration is of circular section of radius $y$, such that

$$y = \sqrt{R^2 - x^2}$$

Let $m$ = Mass per unit volume of the sphere.

$\therefore$ Mass of the elementary strip

$$= m \times \text{Volume of the elementary strip}$$

$$= m \times \text{Area of the elementary strip} \times \text{Thickness}$$

$$= m \times \pi y^2 \times dx = m \times \pi (R^2 - x^2) dx$$

$$\ldots(\because y = \sqrt{R^2 - x^2}\,)$$

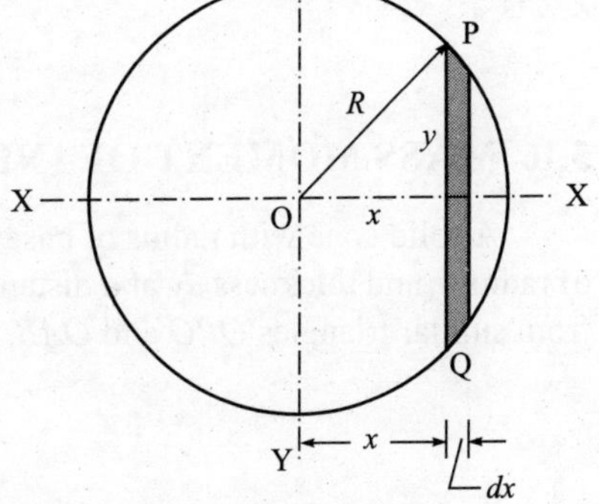


Fig. 5.37

and mass moment of inertia of this elementary strip about *X-X* axis

$$= \text{Mass} \times \frac{(\text{Radius})^2}{2} = m \times \pi (R^2 - x^2) dx \times \frac{y^2}{2}$$

$$= m \times \pi (R^2 - x^2) dx \times \frac{R^2 - x^2}{2}$$

$$= \frac{m \times \pi}{2} (R^2 - x^2)^2 \, dx = \frac{m\pi}{2} \left[ R^4 + x^4 - 2R^2 x^2 \right] dx$$

The mass moment of inertia of the whole sphere is obtained by integrating the above expression between the limits from $-R$ to $+R$.

$$\therefore \quad I_{XX} = \int_{-R}^{+R} \frac{m\pi}{2} \left[ R^4 + x^2 - 2R^2 x^2 \right] dx$$

$$= \frac{m \times \pi}{2} \int_{-R}^{+R} (R^4 + x^4 - 2R^2 x^2) dx = m \times \frac{\pi}{2} \left[ R^4 x + \frac{x^5}{5} - \frac{2R^2 x^3}{3} \right]_{-R}^{+R}$$

$$= m \times \frac{\pi}{2} \left[ \left( R^4 \times R + \frac{R^5}{5} - \frac{2R^2 \times R^3}{3} \right) - \left( R^4 \times -R + \frac{(-R)^5}{5} - \frac{2R^2 (-R)^3}{3} \right) \right]$$

$$= m \times \frac{\pi}{2}\left[R^5 + \frac{R^5}{5} - \frac{2R^5}{3} + R^5 + \frac{R^5}{5} - \frac{2R^5}{3}\right]$$

$$= m \times \frac{\pi}{2}\left[2R^5 + \frac{2R^5}{5} - \frac{4R^5}{3}\right] = m \times \frac{\pi}{2}\left[\frac{30R^5 + 6R^5 - 20R^5}{15}\right]$$

$$= m \times \frac{\pi}{2} \times \frac{16R^5}{15} = m \times \pi \times \frac{8R^5}{15}$$

We know that total mass of the whole sphere,

$$M = \text{Mass per unit volume} \times \text{Volume of the sphere}$$

$$= m \times \frac{4\pi R^3}{3}$$

$\therefore$ From the above expression, $I_{XX} = m \times \frac{4\pi R^3}{3} \times \frac{2R^2}{5} = M \times \frac{2R^2}{5} = 0.4MR^2$

It may be noted that the mass moment of inertia is same about any axis as the sphere is symmetrical.

According to the perpendicular axis theorem, moment of inertia of a solid sphere about the polar axis *Z-Z* is

$$I_{ZZ} = I_{XX} + I_{YY} = 0.4MR^2 + 0.4MR^2 = 0.8MR^2$$

## 5.16 MASS MOMENT OF INERTIA OF A UNIFORM SOLID CONE

A solid cone with radius of base *R* and height *h* is shown in Fig. 5.38. Consider an elementary strip *PQ* of radius *r* and thickness *dy* at a distance *y* from the base *AB*. The radius *r* of the elementary strip is obtained from similar triangles *OPQ* and *OAB*. We know that

$$\frac{r}{R} = \frac{h-y}{h}$$

or

$$r = R\left(\frac{h-y}{h}\right) \qquad ...(i)$$

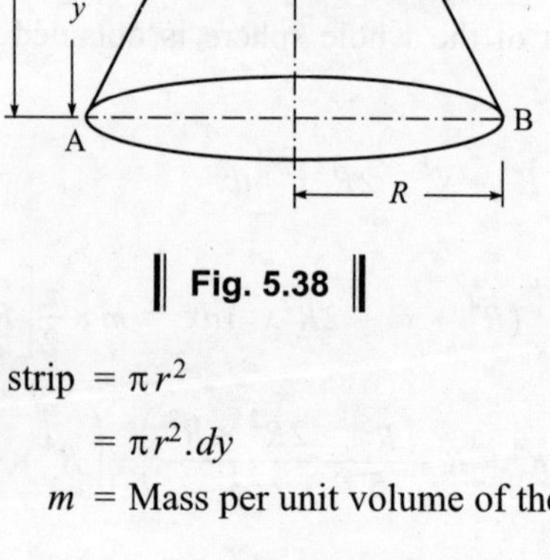


Fig. 5.38

$\therefore$ Area of the elementary strip $= \pi r^2$

and volume of the elementary strip $= \pi r^2 . dy$

Let $m$ = Mass per unit volume of the cone

$\therefore$ Mass of the elementary strip $= m \times$ Volume of the elementary strip

$$= m \times \pi r^2 . dy$$

$$= m \times \pi R^2 \left( \frac{h-y}{h} \right)^2 dy$$

$$= \frac{m \pi R^2}{h^2} (h-y)^2 \, dy$$

$$= \frac{m \pi R^2 h}{h^3} (h-y)^2 \, dy \qquad \text{...(Multiplying and dividing by } h)$$

and mass moment of inertia of the elementary strip about the axis of the cone,

$$dI = \text{Mass} \times \frac{(\text{Radius})^2}{2}$$

$$= \frac{m\pi R^2 h}{h^3} (h-y)^2 \, dy \times \frac{r^2}{2}$$

$$= \frac{m\pi R^2 h}{h^3} (h-y)^2 \, dy \times \frac{R^2}{2} \left( \frac{h-y}{h} \right)^2$$

...[Substituting the value of $r$ from equation (*i*)]

$$= \frac{m\pi R^2 h}{h^3} \times \frac{R^2}{2h^2} (h-y)^4 \, dy \qquad \text{...(ii)}$$

The mass moment of inertia of the whole section may be obtained by integrating the above equation (*ii*) within the limits from 0 to $h$.

$$I = \int_0^h \frac{m \pi R^2 h}{h^3} \times \frac{R^2}{2h^2} (h-y)^4 \, dy$$

$$\therefore \quad = \frac{m\pi R^2 h}{h^3} \times \frac{R^2}{2h^2} \left[ \frac{(h-y)^5}{5} \right]_0^h$$

$$= \frac{m \pi R^2 h}{h^3} \times \frac{R^2}{2h^2} \times \frac{h^5}{5} \qquad \text{...(iii)}$$

We know that volume of the cone, $V = \frac{1}{3} \pi R^2 h$

$$\therefore \quad \text{Total mass of the cone, } M = m \times \frac{1}{3} \pi R^2 h = \frac{m \times \pi R^2 h}{3}$$

or

$$3M = m \times \pi R^2 h$$

Substituting $m \times \pi R \; h = \; M$ in equation (*iii*), we have

$$I = \frac{3MR^2}{10}$$

## 5.17 RADIUS OF GYRATION

We have already discussed that the radius of gyration of a body is defined as the distance from a given reference where the whole mass of the body is assumed to be concentrated to give the same value of moment of inertia ($I$). Thus the mass moment of inertia in terms of radius of gyration ($k$) is given by

$$I = M.k^2$$

$$\therefore \qquad k = \sqrt{\frac{I}{M}}$$

The radius of gyration for some important sections are given as follows:

**1. *Radius of gyration for a circular section***

We know that mass moment of inertia of a circular section of radius $R$ and mass $M$, about the central axis is given by

$$I_{ZZ} = 0.5MR^2$$

or

$$Mk^2 = 0.5MR^2$$

$$k^2 = 0.5R^2 = \frac{R^2}{2}$$

$$\therefore \qquad k = \frac{R}{\sqrt{2}}$$

**2. *Radius of gyration for a solid sphere***

We know that the mass moment of inertia of a solid sphere of radius $R$ about the central axis is given by

$$I = 0.4MR^2$$

or

$$Mk^2 = 0.4MR^2$$

$$k^2 = 0.4R^2$$

$$\therefore \qquad k = \sqrt{0.4}\,R = 0.6324R$$

**3. *Radius of gyration for a solid cone***

We know that mass moment of inertia of a solid cone of base radius $R$ about the axis of cone is given by

$$I = \frac{3}{10}MR^2 \quad \text{or} \quad Mk^2 = \frac{3}{10}MR^2$$

$$\therefore \qquad k^2 = \frac{3}{10}R^2 \quad \text{or} \quad k = \sqrt{\frac{3}{10}}R$$

**Example 5.17** *Calculate the moment of inertia and radius of gyration with respect to the axis of rotation, for a solid sphere of diameter 500 mm and mass 1500 kg/m³.*

***Solution*** Given : $D = 500$ mm or $R = 250$ mm $= 0.25$ m; $m = 1500$ kg/m³

***Moment of inertia***

We know that the total mass of a solid sphere,

$$M = m \times \frac{4\pi R^3}{3} = 1500 \times \frac{4\pi(0.25)^3}{3} = 98.2 \text{ kg}$$

$\therefore$ Moment of inertia about the axis of rotation

$$I = 0.4MR^2$$

$$= 0.4 \times 98.2\,(0.25)^2 = 2.455 \text{ kg-m}^2$$

***Radius of gyration***

We know that radius of gyration, $k = 0.6324R$

$$= 0.6324 \times 0.25 = 0.1581 \text{ m} \quad \textbf{Ans.}$$

**Example 5.18** *Calculate the moment of inertia and radius of gyration of a grinding stone 900 mm in diameter and 100 mm thick with respect to its axis of rotation. The stone mass is 2600 kg/m³.*

***Solution*** Given : $D = 900$ mm or $R = 450$ mm $= 0.45$ m; $t = 100$ mm $= 0.1$ m ; $m = 2600$ kg/m$^3$

***Moment of inertia***

We know that total mass of a grinding stone (similar to a cylinder),

$$M = m \times \pi R^2 t = 2600 \times \pi (0.45)^2 \times 0.1 = 165.4 \text{ kg}$$

∴ Moment of inertia about the axis of rotation,

$$I = 0.5MR^2 = 0.5 \times 165.4\,(0.45)^2 = 16.747 \text{ kg-m}^2 \quad \textbf{Ans.}$$

***Radius of gyration***

We know that radius of gyration, $k = \dfrac{R}{\sqrt{2}} = \dfrac{0.45}{\sqrt{2}} = 0.3182$ m **Ans.**

## HIGHLIGHTS

1. The moment of inertia ($I$) of a body may be defined as the moment of moment or second moment of the mass or area of the body. Mathematically,

$$I = mk^2 \quad \text{or} \quad I = A.k^2$$

where $k$ is called the *radius of gyration.*

2. The radius of gyration ($k$) is defined as the distance from a given reference where the whole area (or mass) of the body is assumed to be concentrated to give the same value of $I$. Mathematically,

$$k = \sqrt{\frac{I}{A}}$$

3. According to theorem of perpendicular axis, the moment of inertia about $Z$-$Z$ axis (known as polar moment of inertia), *i.e.* perpendicular to the plane of lamina is equal to the sum of the moment of inertia of the lamina about the two mutually perpendicular axes (*i.e.* $I_{XX}$ and $I_{YY}$) passing through the centroid and in the plane of the lamina. Mathematically, polar moment of inertia,

$$I_{ZZ} = I_{XX} + I_{YY}$$

4. According to theorem of parallel axis, the moment of inertia about a given axis is equal to the sum of moment of inertia about the parallel axis passing through the C.G. ($I_G$) and the product of the area of the lamina ($A$) and square of the perpendicular distance ($h$) between the two axes.

5. The moment of inertia of a rectangular section of width $b$ and depth $d$, about an axis passing through its C.G. is

$$I_{XX} = \frac{bd^3}{12}; \text{ and } I_{YY} = \frac{db^3}{12}$$

and moment of inertia about an axis passing through the base of the rectangular section $= \dfrac{bd^3}{3}$.

6. The moment of inertia of a hollow rectangular section having outer width ($B$) and depth ($D$) and inner width ($b$) and depth ($d$) is given by

$$I_{XX} = \frac{BD^3}{12} - \frac{bd^3}{12}; \text{ and } I_{YY} = \frac{DB^3}{12} - \frac{db^3}{12}$$

7. The moment of inertia of a solid circular section of diameter $D$ is

$$I_{XX} = I_{YY} = \frac{\pi D^4}{64}$$

and moment of inertia of a hollow circular section of external diameter $D$ and internal diameter $d$ is

$$I_{XX} = I_{YY} = \frac{\pi}{64}(D^4 - d^4)$$

8. The moment of inertia of a triangular section of base $b$ and height $h$ about the base of the triangle is given by

$$I = \frac{bh^3}{12}$$

and about an axis passing through the centroid and parallel to the base,

$$I = \frac{bh^3}{36}$$

9. The moment of inertia of a semicircle about its base diameter is equal to one-half the moment of inertia of a circular section.

10. The moment of inertia of a semicircular section of radius $r$ about an axis passing through the centroid and parallel to $X$-$X$ axis

$$= 0.11\,r^4$$

and moment of inertia about an axis passing through the centroid and parallel to $Y$-$Y$ axis

$$= \frac{\pi}{8}(r^4) = 0.393\,r^4$$

11. The mass moment of inertia of a uniform thin rod about the centroidal axis and normal to the rod is given by

$$I_{YY} = \frac{Ml^2}{12}$$

and mass moment of inertia about the end of the rod and normal to it is given by

$$I_{YY} = \frac{M\,l^2}{3}$$

If $m$ is the mass of the rod per unit length, then total mass of the rod,

$$M = m \cdot l$$

12. The mass moment of inertia of a rectangular plate of width $b$, depth $d$ and thickness $t$, about $X$-$X$ axis passing through the centre of gravity,

$$I_{XX} = \frac{M \cdot d^2}{12}$$

and mass moment of inertia about $Y$-$Y$ axis passing through the centre of gravity,

$$I_{YY} = \frac{M \cdot b^2}{12}$$

where $M$ = Total mass of the rectangular plate $= m \times b\,.\,d\,.\,t$

13. The mass moment of inertia of a circular section or a solid cylinder about an axis of rotation is

$$I_{ZZ} = 0.5MR^2$$

and radius of gyration, $k = \frac{R}{\sqrt{2}}$ or $k^2 = \frac{R^2}{2} = 0.5R^2$ ...($\because\ I = Mk^2$)

where $M$ is the total mass of the circular section or a solid cylinder.

**14.** The mass moment of inertia of a solid sphere about the central axis is

$$I = 0.4MR^2$$

and radius of gyration, $k = 0.6324R$ or $k^2 = 0.4R^2$ $(\because\ I = Mk^2)$

where $M$ is the total mass of the sphere.

**15.** The mass moment of inertia of a uniform solid cone about the axis of the cone is

$$I = \frac{3}{10}MR^2$$

and radius of gyration, $k = \sqrt{\frac{3}{10}}R$ or $k^2 = \frac{3}{10}R^2$

## NUMERICAL PROBLEMS

**1.** Find the moment of inertia for a rectangular section of 80 mm wide and 160 mm deep about $X$-$X$ axis and $Y$-$Y$ axis passing through the centroid. Also determine the radius of gyration about the same axes.

[**Ans.** $27.3 \times 10^6$ mm$^4$; $6.827 \times 10^6$ mm$^4$; 46.18 mm; 22.16 mm]

**2.** Find the moment of inertia of a rectangular section $ABCD$ about base $AB$ and edge $AD$. It is given that $AB = 30$ mm, and $BC = 40$ mm. [**Ans.** $640 \times 10^3$ mm$^4$; $360 \times 10^3$ mm$^4$]

**3.** Find the moment of inertia of a hollow rectangular section about $X$-$X$ axis and $Y$-$Y$ axis passing through its centroid. It is given that the external dimensions of the section are: width 60 mm and depth 80 mm while the internal dimensions are: width 30 mm and depth 40 mm. [**Ans.** $240 \times 10^4$ mm$^4$; $135 \times 10^4$ mm$^4$]

**4.** Determine the moment of inertia of a hollow circular section about an axis passing through its centroid. The outer and inner diameters of the circle are 80 mm and 60 mm, respectively. [**Ans.** $1.374 \times 10^6$ mm$^4$]

**5.** Determine the moment of inertia of the triangular section $ABC$ about the base $BC$. It is given that the base $BC = 80$ mm and the height is 60 mm. [**Ans.** $1440 \times 10^3$ mm$^4$]

**6.** Calculate the centroid and moment of inertia about $X$-$X$ axis of a $T$-section having flange dimensions 60 mm × 20 mm and web 20 mm × 40 mm. [**Ans.** 38 mm; 578 666 mm$^4$]

**7.** Find the moment of inertia about the centroidal axes $X$-$X$ and $Y$-$Y$ of the $L$-section, as shown in Fig. 5.39.

[**Ans.** 387 757 mm$^4$]

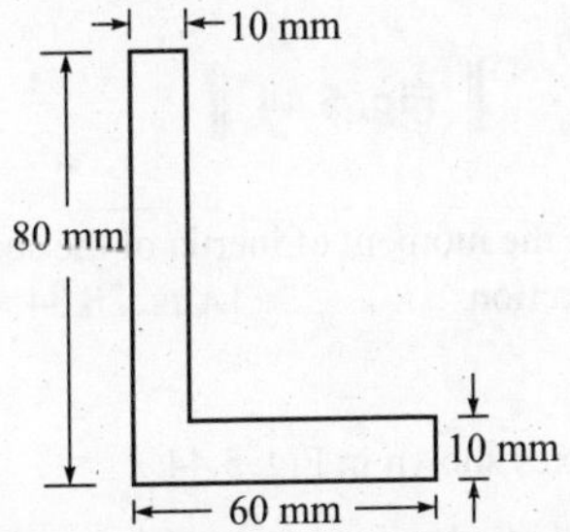


**Fig. 5.39**

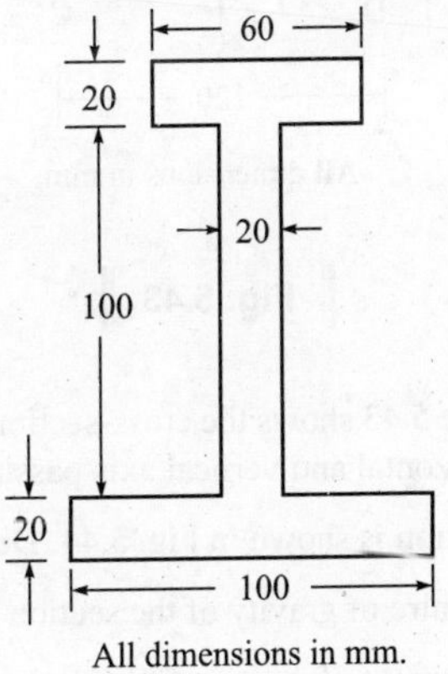


**Fig. 5.40**

**8.** An $I$-section, as shown in Fig. 5.40, is made up of three rectangles. Find the moment of inertia of the section about the horizontal axis passing through the C.G. of the section. [**Ans.** $1285 \times 10^4$ mm$^4$]

**9.** A channel section is shown in Fig. 5.41. Determine the moment of inertia about *X-X* and *Y-Y* axis passing through its centroid. Also find the least radius of gyration for the section.

[**Ans.** $9.84 \times 10^6$ mm$^4$; $11.4 \times 10^6$ mm$^4$; 48.4 mm]

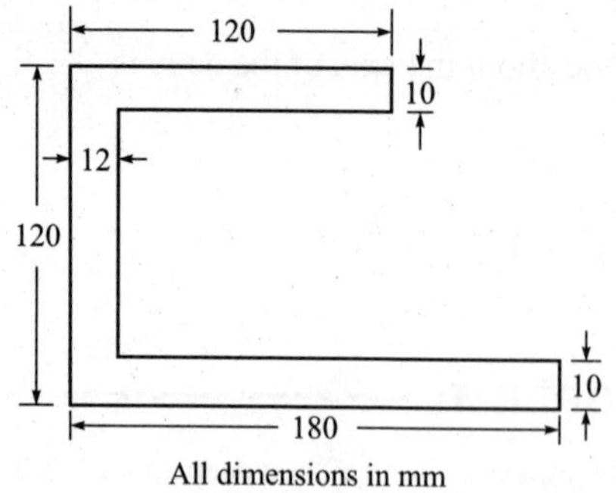


Fig. 5.41

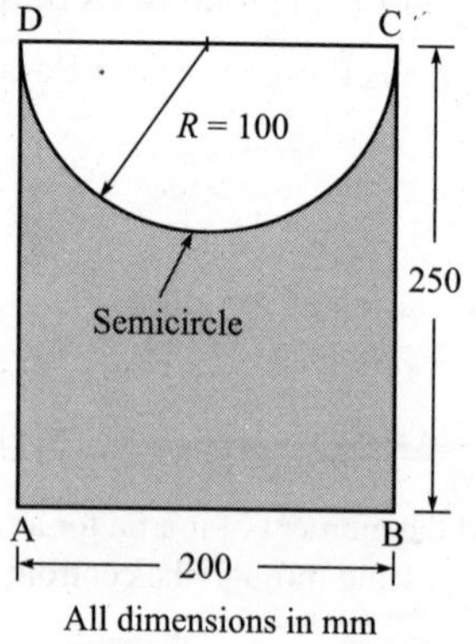


Fig. 5.42

**10.** Find the moment of inertia of the area shown shaded in Fig. 5.42, about the edge *AB*. [**Ans.** $353 \times 10^6$ mm$^4$]

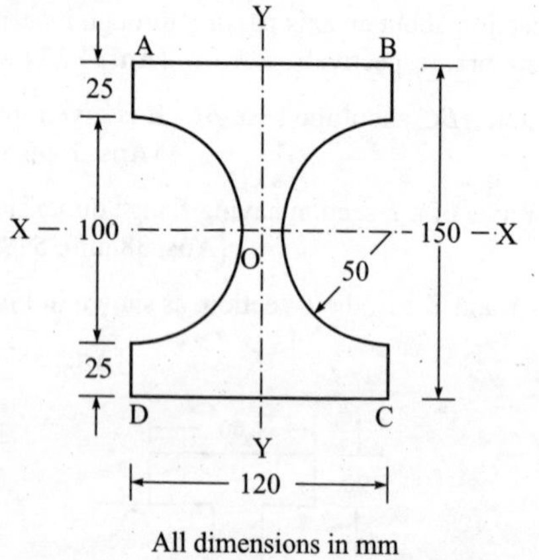


Fig. 5.43

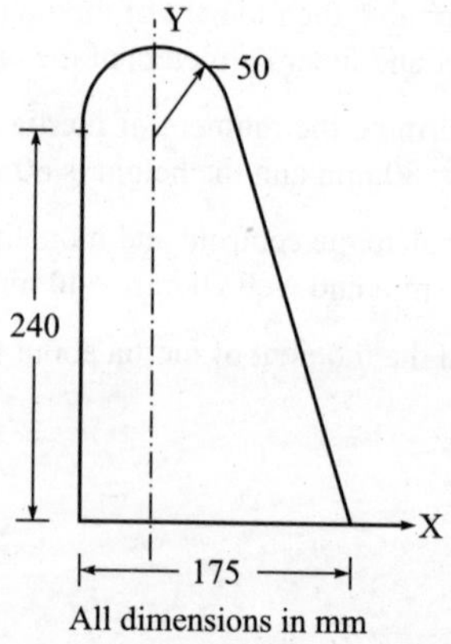


Fig. 5.44

**11.** Figure 5.43 shows the cross-section of a cast iron beam. Determine the moment of inertia of the section about a horizontal and vertical axis passing through the centroid of the section. [**Ans.** $28.84 \times 10^6$ mm$^4$]

**12.** A section is shown in Fig. 5.44. Determine:

1. Centre of gravity of the section with respect to the coordinate axes shown in Fig. 5.44.
2. Moment of inertia of the section about the centroidal axis parallel to the *Y*-axis.

[**Ans.** $\bar{x} = 18.28$ mm; $\bar{y} = 125.27$ mm; 61 786 468 mm$^4$]

**13.** The cross-section of a culvert is shown in Fig. 5.45. Determine the moment of inertia of the shaded area about *X-X* and *Y-Y* axis. [**Ans.** 68 643 mm$^4$; 233 252 mm$^4$]

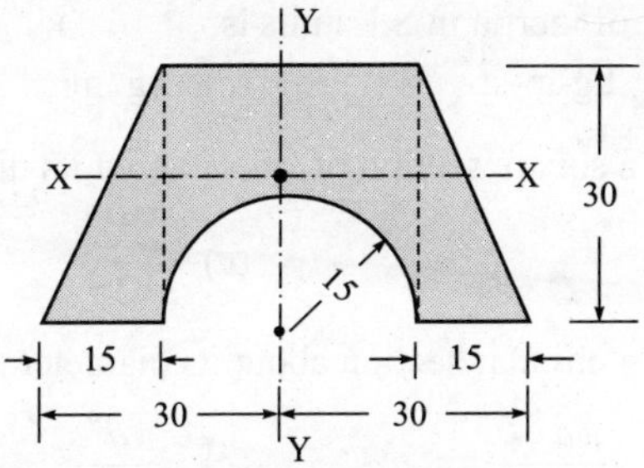


**Fig. 5.45**

**14.** Calculate the moment of inertia and radius of gyration with respect to the axis of rotation, for the following sections:

(*a*) a circular lamina of diameter 600 mm and of mass 1000 kg/m$^3$;

(*b*) a circular cylinder of diameter 800 mm, and height 150 mm and of mass 2000 kg/m$^3$; and

(*c*) a sphere of diameter 400 mm and of mass 1500 kg/m$^3$.

[**Ans.** (*a*) 12.725 kg-m$^2$, 212 mm; (*b*) 12.064 kg-m$^2$, 282.8 mm; (*c*) 0.804 kg-m$^2$, 126.48 mm]

## REVIEW QUESTIONS

**1.** Explain what do you understand by the terms 'moment of inertia' and 'radius of gyration'.

**2.** State and prove the theorem of perpendicular axis.

**3.** Define polar moment of inertia.

**4.** State and prove the parallel axis theorem related to the moment of inertia of plane figures.

**5.** Write the expressions used for finding the moment of inertia of the following sections about centroidal axis:

(*a*) Rectangular section; and (*b*) Hollow rectangular section.

**6.** Derive, from first principles, the moment of inertia of a circular lamina about its centroidal axis.

**7.** What is the moment of inertia of a hollow circular lamina of external diameter $D$ and internal diameter $d$, about a diametral axis?

**8.** From the fundamental principles, derive an expression for the moment of inertia of an area in the shape of a semicircle of radius $r$, about its centroidal axis taken parallel to the base diameter.

**9.** Determine the moment of inertia of a triangle of a base $b$ and height $h$ with respect to (*i*) a centroidal axis parallel to the base; and (*ii*) an axis coinciding with the base.

**10.** (*a*) Derive the expressions for mass moment of inertia of a solid circular lamina and a solid sphere, about the axis of rotation.

(*b*) What is the value of radius of gyration for these sections?

## OBJECTIVE TYPE QUESTIONS

**1.** The moment of inertia is the

(*a*) second moment of force

(*b*) second moment of area

(*c*) second moment of mass

(*d*) all of these

2. The unit of mass moment of inertia in S.I. units is

(*a*) kg-m$^2$ (*b*) kg-m-s$^2$ (*c*) kg/m$^2$ (*d*) m$^4$

3. The moment of inertia of a square section of side *a* about its diagonal is

(*a*) $\frac{a^2}{8}$ (*b*) $\frac{a^3}{12}$ (*c*) $\frac{a^4}{12}$ (*b*) $\frac{a^4}{16}$

4. The moment of inertia of a circular section about its diameter (*d*) is

(*a*) $\frac{\pi d^3}{16}$ (*b*) $\frac{\pi d^3}{32}$ (*c*) $\frac{\pi d^4}{32}$ (*d*) $\frac{\pi d^4}{64}$

5. The moment of inertia of a triangular section of base (*b*) and height (*h*) about an axis passing through its C.G. and parallel to the base is

(*a*) $\frac{bh^3}{4}$ (*b*) $\frac{bh^3}{8}$ (*c*) $\frac{bh^3}{12}$ (*d*) $\frac{bh^3}{36}$

6. The moment of inertia of a triangular section of base (*b*) and height (*h*) about an axis coinciding with the base is

(*a*) $\frac{bh^3}{4}$ (*b*) $\frac{bh^3}{8}$ (*c*) $\frac{bh^3}{12}$ (*d*) $\frac{bh^3}{36}$

7. The moment of inertia of a rectangular section of width (*b*) and depth (*d*) about an axis passing through its centroid and parallel to the width (*b*) is

(*a*) $\frac{db^3}{12}$ (*b*) $\frac{bd^3}{12}$ (*c*) $\frac{db^3}{36}$ (*d*) $\frac{bd^3}{36}$

8. The moment of inertia of a rectangular section of width (*b*) and depth (*d*) about an axis passing through its centroid and parallel to the depth (*d*) is

(*a*) $\frac{db^3}{12}$ (*b*) $\frac{bd^3}{12}$ (*c*) $\frac{db^3}{36}$ (*d*) $\frac{bd^3}{36}$

9. The moment of inertia of a rectangular section 30 mm wide and 40 mm deep about an axis passing through its centroid and parallel to the 30 mm width is

(*a*) $9 \times 10^3$ mm$^4$ (*b*) $12 \times 10^3$ mm$^4$ (*c*) $160 \times 10^3$ mm$^4$ (*d*) $200 \times 10^3$ mm$^4$

10. The mass moment of inertia of a solid sphere about the axis of rotation is

(*a*) $0.1MR^2$ (*b*) $0.2MR^2$ (*c*) $0.3MR^2$ (*d*) $0.4MR^2$

where $R$ = Radius of the sphere,

$M$ = Total mass of the sphere = $m \times \frac{4\pi R^3}{3}$, and

$m$ = Mass per unit volume of the sphere.

***Answers to Objective Type Questions***

| | | | | |
|---|---|---|---|---|
| **1.** (*d*) | **2.** (*a*) | **3.** (*c*) | **4.** (*d*) | **5.** (*d*) |
| **6.** (*c*) | **7.** (*b*) | **8.** (*a*) | **9.** (*c*) | **10.** (*d*) |

# 6

# FRICTION—I (PRINCIPLES)



## 6.1 INTRODUCTION

When a body slides or is made to slide over the surface of another body, some resistance is encountered, which is known as *friction,* and the force which resists the motion is known as *force of friction*. This force always acts in opposite direction to the relative motion of the body at the point of contact, and it is always tangential to the contacting surfaces. It may be noted that when the surfaces in contact are smooth, there will be less frictional forces and when surfaces in contact are rough, then there will be large frictional forces. Since it is impossible to produce perfectly smooth surfaces, therefore the frictional forces cannot be neglected.

In engineering applications, friction is both desirable and undesirable. In other words, we can also say that *friction is a friend and also a foe.* The friction is desirable (friend) in the movement of the wheels of a train on the railway track, friction between the wheels of a vehicle and the road, friction between the two mating gears, friction between the belt and the pulley, friction in the friction clutches and brakes etc. On the other hand, friction is undesirable (foe) in movement of the piston in the cylinder of an engine, friction in the bearings, friction at the crank-pin of a crank etc.

## 6.2 TYPES OF FRICTION

The friction, depending upon the condition of the surfaces, may be classified as follows:

**1.** ***Dry friction*** When there is a relative motion between the two completely dry (or unlubricated) surfaces in contact, it is then known as *dry friction*. The dry friction is further divided into the following two groups:

(*a*) ***Sliding friction*** When the two surfaces have a sliding motion relative to each other, it is called as *sliding friction* such as nuts and bolts.

(*b*) ***Rolling friction*** When the two surfaces have a rolling motion relative to each other, it is called *rolling friction* such as ball and roller bearings. It may be noted that rolling friction is always less than sliding friction.

2. ***Greasy friction*** When the two surfaces in contact have a fine layer of lubricant between them, the friction experienced between the surfaces is called *greasy friction*. It is also known as *skin* or *boundary friction*.
3. ***Film friction*** When the two surfaces in contact are completely separated by a film of lubricant, the friction occurs due to resistance to relative motion between the lubricant and the surfaces in contact. This type of friction is known as *film friction*. It is also known as *viscous friction* or *fluid friction*.

**Note:** The film friction is not due to the surfaces in contact, but it is due to the viscosity and oiliness of the lubricant.

## 6.3 STATIC FRICTION

It is defined as a self-adjusting opposing force of friction, experienced by a body, when at rest.

## 6.4 KINETIC OR DYNAMIC FRICTION

It is defined as the opposing force of friction, experienced by a body, when in motion. It may be noted that kinetic friction is somewhat less than the maximum static frictional force.

## 6.5 LIMITING FRICTION

It may be defined as the maximum value of the frictional force (or the opposing force) developed between the contact surfaces at which the body just begins to slide over the surface of the other body. It may be noted that when the force applied is less than the limiting friction, then the body will be at rest and in such a case, the friction is said to be static friction.

## 6.6 NORMAL REACTION

Consider a body *A* of weight *W* is lying on a horizontal and an inclined plane *B*, as shown in Fig. 6.1 (*a*) and (*b*), respectively.

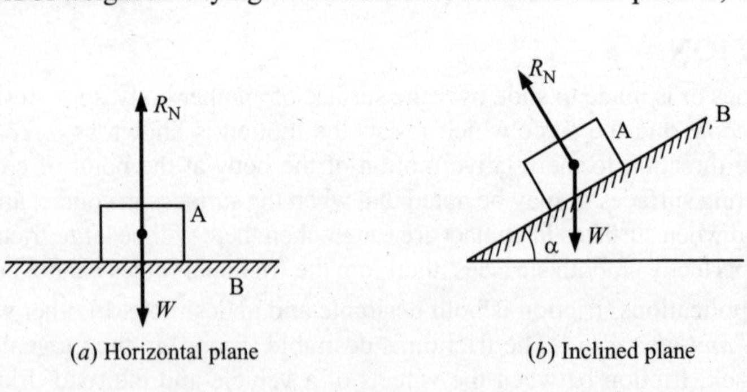


**Fig. 6.1**

It may be noted that weight of the body acts vertically downwards through its centre of gravity. In order to keep the body in equilibrium, an upward reaction perpendicular to the surface must act on the body. This reaction is known as *normal reaction* and is denoted by $R_N$, as shown in Fig. 6.1.

## 6.7 COEFFICIENT OF FRICTION

It is defined as a constant ratio which a limiting force of friction bears to the normal reaction. It is usually denoted by μ (Mu). It is also known as *coefficient of static friction.* Mathematically, coefficient of friction,

$$\mu = \frac{F}{R_N} \quad \text{or} \quad F = \mu.R_N$$

where $F$ = Limiting force of friction, and

$R_N$ = Normal reaction.

**Note:** The ratio of the kinetic friction to the normal reaction between the two surfaces when sliding motion is taking place, is called *coefficient of kinetic friction.*

## 6.8 ANGLE OF FRICTION

Consider a body $A$ of weight $W$ is lying on a horizontal plane $B$, as shown in Fig. 6.2. When the horizontal force $P$ is applied and the body moves or has the tendency to move relative to $B$, in the direction as shown, the force of friction ($F$) acts in the opposite direction of motion of the body. The normal reaction ($R_N$) acts perpendicular to the surface $B$. The angle between the normal reaction ($R_N$) and the resultant ($R$) of force of friction ($F$) and normal reaction ($R_N$), is known as *angle of friction.* It is usually denoted by ϕ (Phi). From the geometry of the figure, we find that

$$\tan\phi = \frac{F}{R_N} = \frac{\mu.R_N}{R_N} = \mu \qquad ...(\because F = \mu.R_N)$$

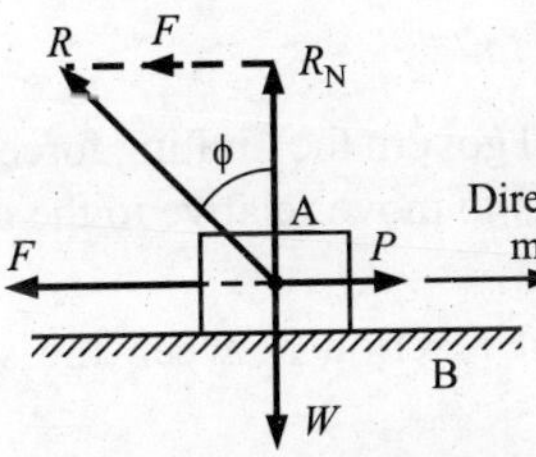


Fig. 6.2

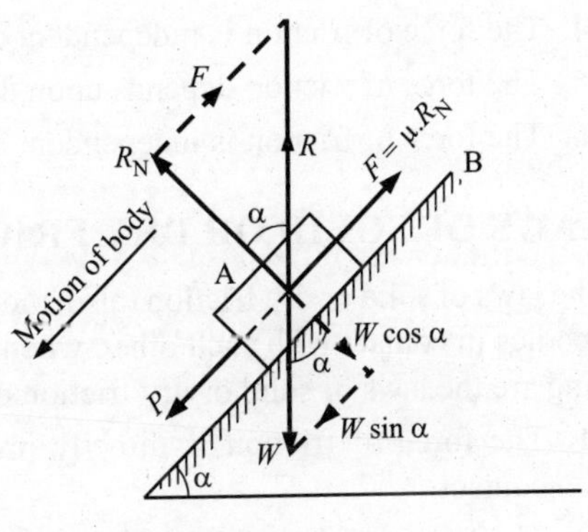


Fig. 6.3

In case of an inclined plane, as shown in Fig. 6.3, which makes an angle α with the horizontal, it can be proved that the angle of inclination of the plane (α) is equal to the angle of friction (ϕ), as follows:

The weight ($W$) can be resolved into the following two components

*(a)* $W \sin\alpha$, parallel to the plane; and

*(b)* $W \cos\alpha$, perpendicular to the plane.

From the geometry of the figure, $F = W\sin\alpha$

and $R_N = W\cos\alpha$

We know that $\tan\phi = \dfrac{F}{R_N} = \dfrac{W\sin\alpha}{W\cos\alpha} = \tan\alpha$

$\therefore \quad \alpha = \phi$

It may be noted that when the angle of inclination of the plane with the horizontal ($\alpha$) is such that body begins to move itself down the plane, then the angle $\alpha$ is known as *angle of repose.* In other words, the angle of repose is defined as the minimum angle of inclination of a plane with the horizontal such that the body placed on the plane just begins to slide down.

## 6.9 CONE OF FRICTION

The cone of friction may be defined as the right circular cone with vertex at the point of contact of the two bodies, axis in the direction of normal reaction ($R_N$) and semi-vertical angle equal to the angle of friction ($\phi$), as shown in Fig. 6.4.

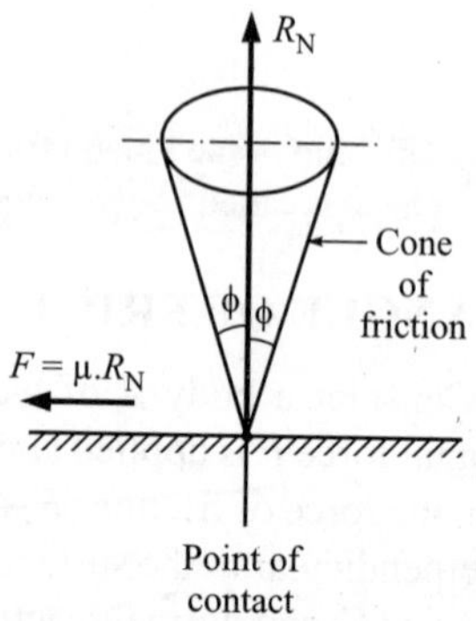

**Fig. 6.4** *Cone of friction*

## 6.10 LAWS OF STATIC FRICTION

The various laws of static friction are as follows:

1. The force of friction always acts in a direction, opposite to that in which the body is moving.
2. The magnitude of the force of friction is exactly equal to the force, which tends the body to move.
3. The magnitude of the limiting friction ($F$) bears a constant ratio to the normal reaction ($R_N$) between the two contacting surfaces. Mathematically,

$$\frac{F}{R_N} = \text{Constant}$$

4. The force of friction is independent of the area of contact, between the two surfaces.
5. The force of friction depends upon the material or roughness of the surfaces in contact.
6. The force of friction is independent of velocity of sliding.

## 6.11 LAWS OF SOLID OR DRY FRICTION

The laws of solid or dry friction (also known as *Coulomb friction) govern the limiting force of friction of two bodies in contact with each other, when one body moves or tends to move relative to the other body. Following are the laws of solid or dry friction determined experimentally.

1. The force of friction is directly proportional to the normal reaction between the surfaces in contact.
2. The force of friction depends upon the material of the surfaces in contact.
3. The force of friction is independent of the area of contact, between the two surfaces.
4. The force of friction is independent of the velocity of sliding of one body relative to the other body.

## 6.12 LAWS OF DYNAMIC OR KINETIC FRICTION

The various laws of dynamic or kinetic friction are as follows:

1. The force of friction always acts in the opposite direction of the motion of the body.
2. The magnitude of kinetic friction bears a constant ratio to the normal reaction between the two surfaces, but this ratio is slightly less than that in case of limiting friction.
3. The force of friction remains constant for moderate speeds, but it decreases slightly with the increase of speed.

* The force of friction between two solid surfaces is known as *Coulomb friction,* named after Charles-Augustin de Coulomb.

## 6.13 CONDITIONS OF EQUILIBRIUM OF A BODY ON A ROUGH INCLINED PLANE

Consider a body *A* resting on an inclined plane *OB* which makes an angle $\alpha$ with the horizontal. For the motion of the body down the plane, the force of friction (*F*) will act upwards, as shown in Fig. 6.5. The normal reaction ($R_N$) acts perpendicular to the inclined plane *OB*.

The conditions of equilibrium of a body on an inclined plane when no other external forces are acting on a body, are as follows:

1. The components of all the forces acting on the body parallel to the inclined plane must be equal to zero, *i.e.*

$$F = \mu \cdot R_N = W \sin \alpha \qquad ...(i)$$

2. The components of all the forces acting on the body perpendicular to the inclined plane must be equal to zero, *i.e.*

$$R_N = W \cos \alpha \qquad ...(ii)$$

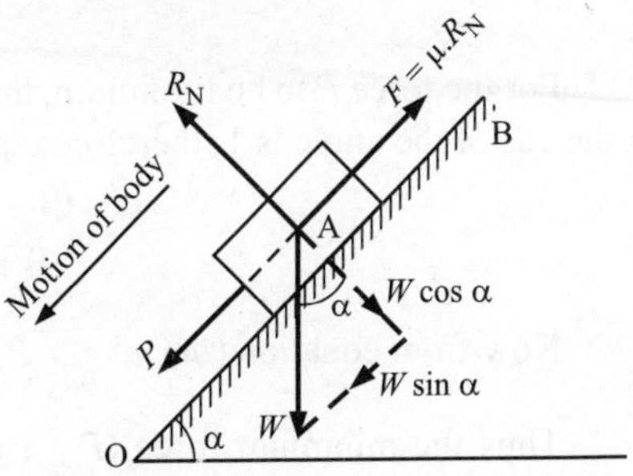


**Fig. 6.5** *Inclined plane*

The following points are worth noting:

(*a*) In case the angle of inclination of the plane with the horizontal (*i.e.* angle $\alpha$) is less than the angle of friction (*i.e.* angle $\phi$), then the body will remain in equilibrium without any external force. If the body is required to be moved upwards or downwards in this condition, then an external force is required.

(*b*) In case the angle of inclination of the plane with the horizontal (*i.e.* angle $\alpha$) is more than the angle of friction (*i.e.* angle $\phi$), then the body will not remain in equilibrium. The body will move downward and an upward external force is required to keep the body in equilibrium.

## 6.14 MINIMUM FORCE REQUIRED TO SLIDE A BODY

Let a force *P* making an angle $\theta$ with the horizontal is applied on a body *A* to slide it on a rough horizontal plane *OB*, as shown in Fig. 6.6. The weight of the body is *W* and the normal reaction is $R_N$. The force of friction ($F - \mu.R_N$) acts in the opposite direction of motion of the body. The force *P* is resolved into two components, *i.e.*

Horizontal component of $P = P \cos \theta$

and vertical component of $P = P \sin \theta$

Now for the equilibrium of the body *A*, equating the various forces acting on the body, we have

$$F = \mu \cdot R_N = P \cos \theta$$

or

$$R_N = \frac{P \cos \theta}{\mu} \qquad ...(i)$$

and

$$R_N + P \sin \theta = W$$

or

$$^*R_N = W - P \sin \theta \qquad ...(ii)$$

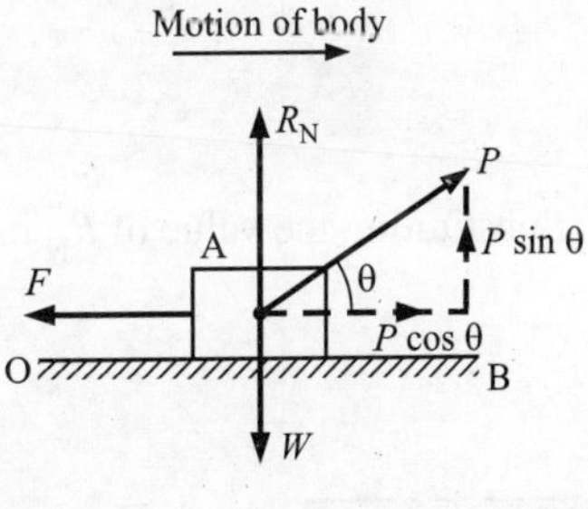


**Fig. 6.6** *Minimum force required*

Substituting the value of $R_N$ from equation (*i*), we have

$$\frac{P \cos \theta}{\mu} = W - P \sin \theta$$

or

$$P \cos \theta = \mu (W - P \sin \theta) = \tan \phi (W - P \sin \theta) \qquad ...(\because \mu = \tan \phi)$$

$$= \frac{\sin \phi}{\cos \phi} (W - P \sin \theta)$$

* This expression shows that normal reaction ($R_N$) is not always equal to weight of the body (*W*).

or $$P \cos\theta \cos\phi = W \sin\phi - P \sin\theta \sin\phi$$

$$P \cos\theta \cos\phi + P \sin\theta \sin\phi = W \sin\phi$$

$$P \cos(\theta - \phi) = W \sin\phi \quad \text{...}[\because \cos(\theta - \phi) = \cos\theta \cos\phi + \sin\theta \sin\phi]$$

$$\therefore \quad P = \frac{W \sin\phi}{\cos(\theta - \phi)} \quad \text{...(iii)}$$

For the force $P$ to be minimum, the value of $\cos(\theta - \phi)$ should be maximum. Since the maximum value of the cos of the angle is 1, therefore equating

$$\cos(\theta - \phi) = 1$$

or $$\theta - \phi = 0° \quad \text{or} \quad \theta = \phi$$

Now from equation (*iii*), $$P_{min} = \frac{W \sin\phi}{\cos(\phi - \phi)} = W \sin\phi$$

Thus the minimum force ($P_{min}$) required to slide the body on a rough horizontal plane is $W \sin\phi$, where $\phi$ is the angle of friction. In other words, $P$ will be minimum when $\theta = \phi$.

**Example 6.1** *A body of weight 300 N is resting on a rough horizontal plane. A pull of 100 N is applied at angle of 15° to the horizontal, which just causes to move the body. Find the value of the normal reaction and the coefficient of friction.*

***Solution*** Given : $W = 300$ N ; $P = 100$ N ; $\alpha = 15°$

A force of 100 N can be resolved into horizontal component (100 cos 15°) and vertical component (100 sin 15°), as shown in Fig. 6.7.

Equating horizontal components, $F = \mu.R_N$

$$= 100 \cos 15°$$

$$= 100 \times 0.966$$

$$= 96.6 \text{ N} \quad \text{...(i)}$$

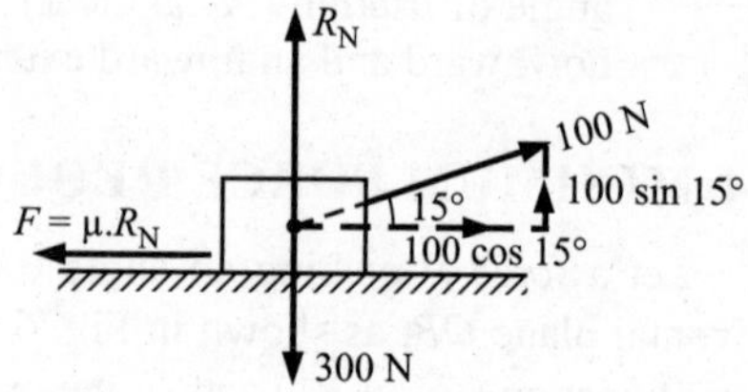


Fig. 6.7

and equating vertical components,

$$R_N + 100 \sin 15° = 300$$

$$\therefore \quad R_N = 300 - 100 \sin 15°$$

$$= 300 - 100 \times 0.259$$

$$= 274.1 \text{ N} \quad \textbf{Ans.}$$

Substituting the value of $R_N$ in equation (*i*), we have

$$\mu.R_N = 96.6$$

$$\therefore \quad \mu = \frac{96.6}{R_N} = \frac{96.6}{274.1} = 0.3524 \quad \textbf{Ans.}$$

**Example 6.2** *Determine the horizontal force (P) to be applied to the block weighing 1500 N to hold it in position on a smooth inclined plane AB which makes angle of 30° with the horizontal, as shown in Fig. 6.8.*

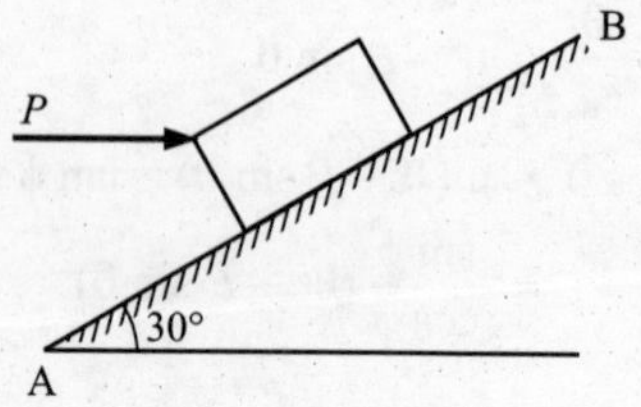


Fig. 6.8

***Solution*** Given : $W = 1500$ N ; $\alpha = 30°$

Resolving the force $P$ parallel to the plane (Fig. 6.9),

$$P \cos 30° = 1500 \sin 30°$$

$$P \times 0.866 = 1500 \times 0.5$$

$$= 750$$

$$\therefore \quad P = \frac{750}{0.866} = 866 \text{ N} \quad \textbf{Ans.}$$

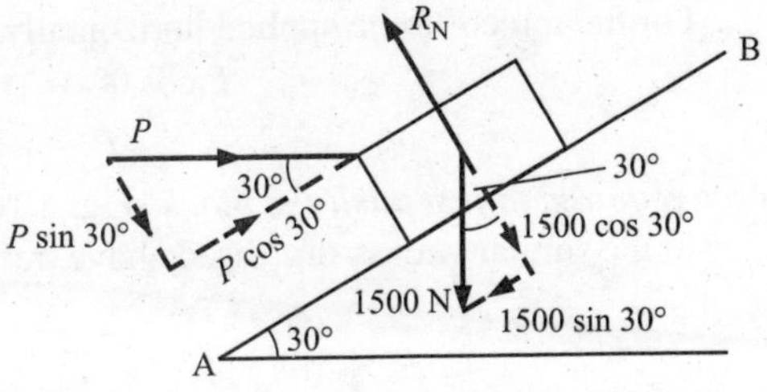


**Fig. 6.9**

**Note:** In order to find the normal reaction ($R_N$), resolving the forces perpendicular to the plane,

$$R_N = 1500 \cos 30° + P \sin 30° = 1500 \times 0.866 + 866 \times 0.5$$

$$= 1299 + 433 = 1732 \text{ N}$$

**Example 6.3** *A wooden block of weight 50 N rests on a horizontal plane. Determine the force required to just (a) pull it; and (b) push it.*

*Take coefficient of friction, $\mu = 0.4$ between the mating surfaces. Comment on the result.*

***Solution*** Given : $W = 50$ N ; $\mu = \tan \phi = 0.4$ or $\phi = \tan^{-1}(0.4) = 21.8°$

***Force required to just pull the block***

Let a force $P$ making an angle $\theta$ is applied on the block to pull it as shown in Fig. 6.10 (*a*). $R_N$ is the normal reaction and $F = \mu.R_N$ is the force of friction acting in the opposite direction of the motion of the block.

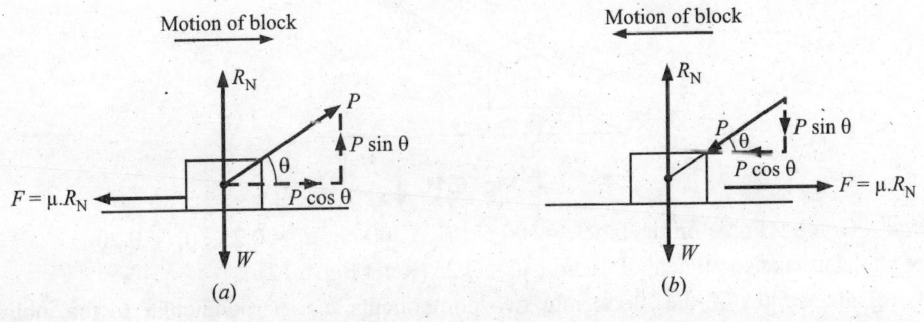


**Fig. 6.10**

Resolving the forces horizontally, we have

$$F = \mu.R_N = P \cos \theta \quad \text{or} \quad R_N = \frac{P \cos \theta}{\mu} \qquad ...(i)$$

and resolving the forces vertically, we have

$$R_N + P \sin \theta = W \quad \text{or} \quad R_N = W - P \sin \theta \qquad ...(ii)$$

From equations (*i*) and (*ii*), we have

$$\frac{P \cos \theta}{\mu} = W - P \sin \theta$$

or

$$P \cos \theta = \mu(W - P \sin \theta)$$

For the force $P$ to be applied horizontally, $\theta = 0°$

$$\therefore \quad P \cos 0° = \mu (W - P \sin 0°)$$

or
$$P = \mu W = 0.4 \times 50 = 20 \text{ N} \quad \textbf{Ans.}$$

***Force required to just push the block*** [Fig. 6.10 (*b*)]

In the similar way as discussed above, resolving the forces horizontally,

$$F = \mu . R_N = P \cos \theta \quad \text{or} \quad R_N = \frac{P \cos \theta}{\mu} \qquad ...(iii)$$

and resolving the forces vertically,

$$W + P \sin \theta = R_N \qquad ...(iv)$$

From equations (*iii*) and (*iv*), we have

$$W + P \sin \theta = \frac{P \cos \theta}{\mu}$$

or
$$P \cos \theta = \mu (W + P \sin \theta)$$

Substituting $\theta = 0°$, we have

$$P \cos 0° = \mu (W + P \sin 0°)$$

or
$$P = \mu W = 0.4 \times 50 = 20 \text{ N} \quad \textbf{Ans.}$$

We see from above that in both the cases, the force required is equal to 20 N. This shows that the block is in static condition.

**Example 6.4** *A 100 N force acts, as shown in Fig. 6.11, on a 200 N block placed on an inclined plane. The coefficients of friction between the block and the plane are $\mu_S = 0.25$ and $\mu_K = 0.20$. Determine whether the block is in equilibrium and find the value of the friction force.*

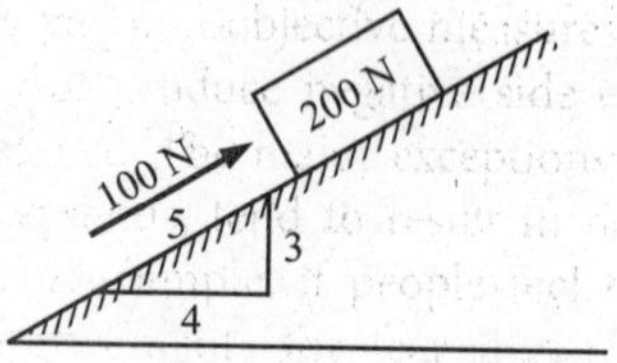


**Fig. 6.11**

***Solution*** Given : Force on the block = 100 N ; $W = 200$ N ; $\mu_S = 0.25$ ; $\mu_K = 0.20$

First of all, let us take coefficient of friction, $\mu_S = 0.25$ (Refer Fig. 6.12).

Resolving the weight of the block into two components, *i.e.* perpendicular to the inclined plane ($W \cos \alpha$), and parallel to the inclined plane ($W \sin \alpha$). We know that the frictional force,

$$F_S = \mu_S . R_N$$
$$= \mu_S W \cos \alpha \qquad ...(\because R_N = W \cos \alpha)$$
$$= 0.25 \times 200 \times \frac{4}{5} = 40 \text{ N}$$

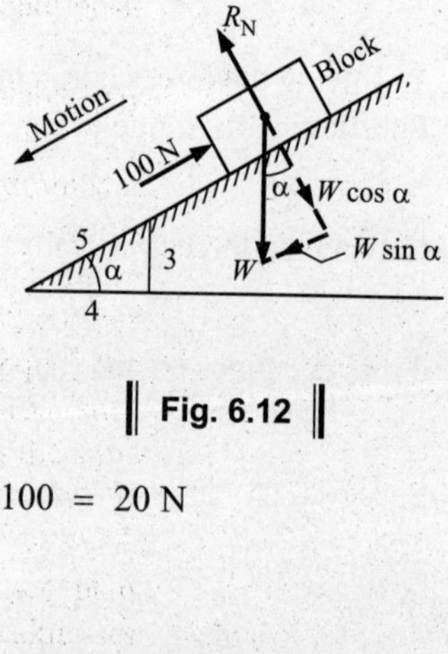


**Fig. 6.12**

Now let us consider the coefficient of friction, $\mu_K = 0.20$. We know that frictional force,

$$F_K = \mu_K . R_N = \mu_K W \cos \alpha$$
$$= 0.2 \times 200 \times \frac{4}{5} = 32 \text{ N}$$

Net force acting parallel to the plane $= W \sin \alpha - 100 = 200 \times \frac{3}{5} - 100 = 20$ N

It follows that the frictional force to maintain equilibrium, $F$ should be equal to 20 N.

Since $F_S$ and $F_K$ are greater than 20 N, therefore the block will remain in equilibrium. **Ans.**

**Example 6.5** *A body resting on a rough horizontal plane, as shown in Fig. 6.13, required a pull of 180 N inclined at 30° to the plane upwards just to move it. It was also found that a push of 220 N inclined at 30° to the plane downwards just moved the body. Find the weight of the body and the coefficient of friction.*

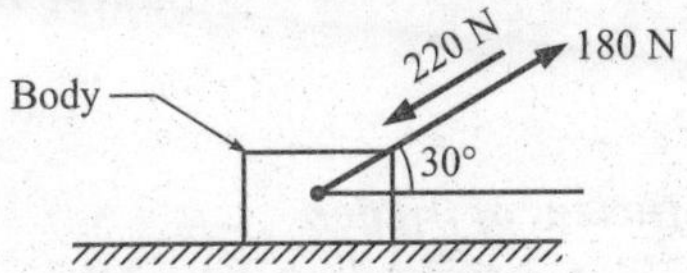


Fig. 6.13

***Solution*** Given : Pull = 180 N ; Angle of inclination = 30° ; Push = 220 N

***Weight of the body***

Let $W$ = Weight of the body,

$\mu$ = Coefficient of friction between the body and the rough horizontal plane, and

$R_N$ = Normal reaction.

First of all, let us consider the pull acting on the body, as shown in Fig. 6.14.

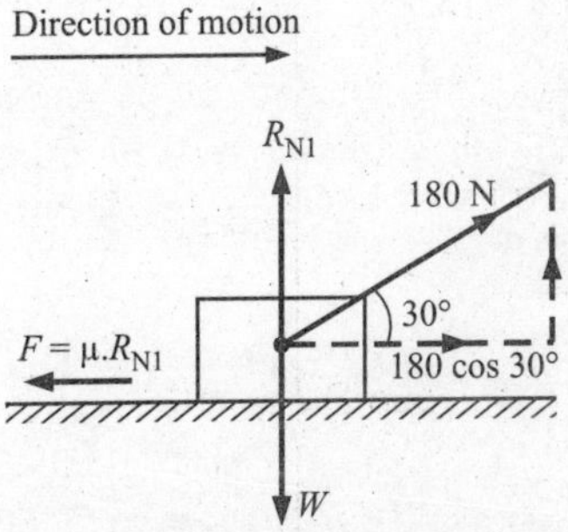


Fig. 6.14

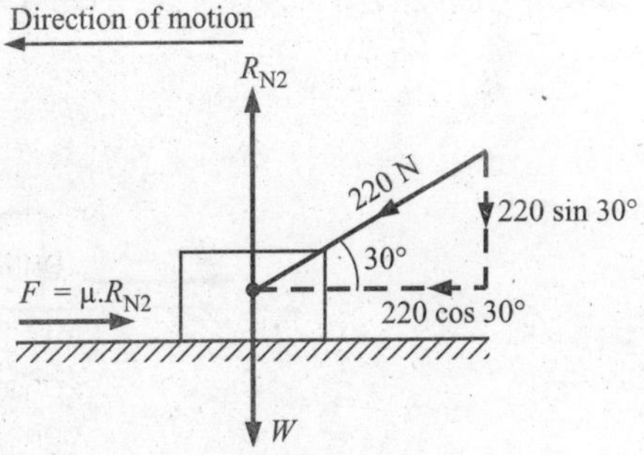


Fig. 6.15

Resolving the forces horizontally,

$$\mu.R_{N1} = 180 \cos 30° = 180 \times 0.866 = 156 \quad ...(i)$$

and resolving the forces vertically,

$$R_{N1} + 180 \sin 30° = W$$

or

$$R_{N1} + 180 \times 0.5 = W$$

$$R_{N1} = W - 180 \times 0.5 = W - 90$$

Substituting the value of $R_{N1}$ in equation (*i*), we have

$$\mu(W - 90) = 156 \quad ...(ii)$$

Now considering the push acting on the body, as shown in Fig. 6.15.

Resolving the forces horizontally,

$$\mu.R_{N2} = 220 \cos 30° = 220 \times 0.866 = 190.5 \quad ...(iii)$$

and resolving the forces vertically,

$$R_{N2} = W + 220 \sin 30°$$

$$= W + 220 \times 0.5 = W + 110$$

Substituting the value of $R_{N2}$ in equation (*iii*), we have

$$\mu(W + 110) = 190.5 \quad ...(iv)$$

Now dividing equation (*iv*) by equation (*ii*),

$$\frac{\mu(W+110)}{\mu(W-90)} = \frac{190.5}{156} = 1.22$$

or $$W + 110 = 1.22\,(W - 90) = 1.22\,W - 1.22 \times 90 = 1.22\,W - 109.8$$

or $$1.22\,W - W = 110 + 109.8 = 219.8$$

$$0.22\,W = 219.8 \text{ or } W = \frac{219.8}{0.22} = 999 \text{ N } \textbf{Ans.}$$

***Coefficient of friction***

Substituting the value of $W$ in equation (*iv*), we have

$$\mu(W + 110) = 190.5$$

$$\therefore \quad \mu = \frac{190.5}{W+110} = \frac{190.5}{999+110} = 0.172 \textbf{ Ans.}$$

**Example 6.6** *A block of stone weighing 50 kN rests on a horizontal floor. If the coefficient of friction between the floor and the block is 0.3 and if a man pulls the block through a string which makes an angle α with the horizontal, find for what value of the angle α, the force necessary to move the block will be minimum. Find the force also.*

***Solution*** Given : $W = 50$ kN ; $\mu = \tan\phi = 0.3$ or $\phi = \tan^{-1} 0.3 = 16.7°$

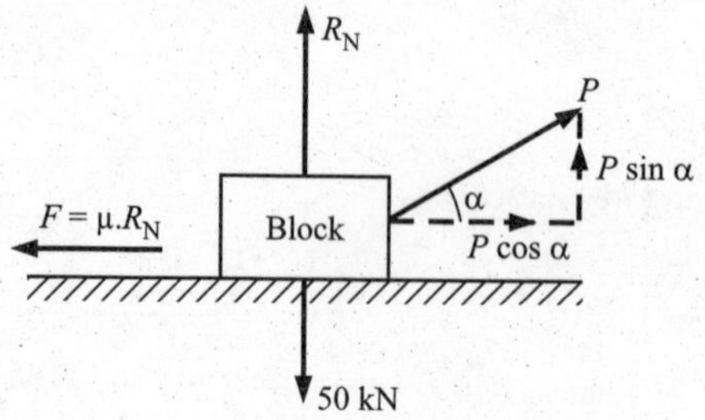


**Fig. 6.16**

Refer Fig. 6.16.

Let $P$ = Force required to move the block, and

α = Angle which the force makes with the horizontal.

The force $P$ is resolved into two components, *i.e.*

Horizontal component of $P = P\cos\alpha$

and vertical component of $P = P\sin\alpha$

Now for the equilibrium of the block, equating the horizontal and vertical forces acting on the block, we have

$$F = \mu.R_N = P\cos\alpha \quad \text{or} \quad R_N = \frac{P\cos\alpha}{\mu} \qquad ...(i)$$

and $$R_N + P\sin\alpha = W$$

or $$R_N = W - P\sin\alpha$$

or $$\frac{P\cos\alpha}{\mu} = 50 - P\sin\alpha \qquad ...[\text{From equation } (i)]$$

$$P\cos\alpha = \mu(50 - P\sin\alpha)$$

$$= \frac{\sin\phi}{\cos\phi}(50 - P\sin\alpha) \qquad ...(\because \mu = \tan\phi)$$

$$\therefore \quad P \cos \alpha \cos \phi = 50 \sin \phi - P \sin \alpha \sin \phi$$

$$P(\cos \alpha \cos \phi + \sin \alpha \sin \phi) = 50 \sin \phi$$

$$P \cos (\alpha - \phi) = 50 \sin \phi \qquad [\because \cos (\alpha - \phi) = \cos \alpha \cos \phi + \sin \alpha \sin \phi]$$

$$\therefore \quad P = \frac{50 \sin \phi}{\cos (\alpha - \phi)} \qquad ...(ii)$$

For the force to be minimum, the value of cos (α – ϕ) should be maximum. Since the maximum of the cos of the angle is 1, therefore equating

$$\cos (\alpha - \phi) = 1 \quad \text{or} \quad \alpha - \phi = 0$$

$$\therefore \quad \alpha = \phi = 16.7° \ \textbf{Ans.}$$

Substituting the value of $\alpha = \phi = 16.7°$ in equation (*ii*), we have minimum force required to move the block,

$$P_{min} = 50 \sin 16.7° = 50 \times 0.2874 \qquad ...[\because \cos (\alpha - \phi) = 1]$$

$$= 14.37 \text{ kN } \textbf{Ans.}$$

**Example 6.7** *Block A weighing 1000 N rests over block B which weighs 2000 N, as shown in Fig. 6.17. Block A is tied to a wall with a horizontal string. If the coefficient of friction between blocks A and B is 0.25 and between B and the floor is* $\frac{1}{3}$, *what should be the value of P to move the block B, if P is horizontal?*

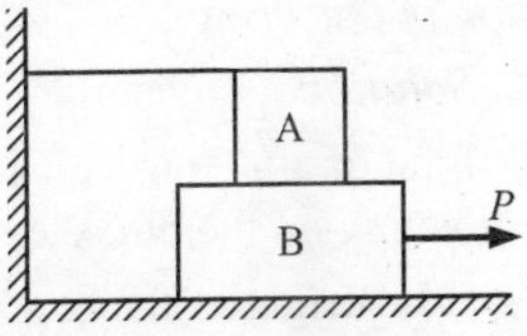


Fig. 6.17

***Solution*** Given : $W_A = 1000$ N ; $W_B = 2000$ N ; $\mu_1 = 0.25$; $\mu_2 = \frac{1}{3}$

First of all, let us consider the equilibrium of the block *A*. The various forces acting on the block *A*, as shown in Fig. 6.18, are as follows:

(*a*) Weight of the block *A* (1000 N) acting downwards,

(*b*) Normal reaction ($R_{N1}$) acting upwards,

(*c*) Force of friction ($F_1 = \mu_1 \cdot R_{N1}$) acting at the contact surface of the block *A* and *B*, and

(*d*) Tension (*T*) in the string.

The block *B* tends to move rightward, it will drag block *A* with it. Since the block *A* is tied to the wall with a string, it will not move and the tension (*T*) will be equal to $F_1 = \mu_1 \cdot R_{N1}$.

Equating vertical components, $R_{N1} = W_A = 1000$ N

and equating horizontal components, $T = F_1 = \mu_1 \cdot R_{N1} = 0.25 \times 1000 = 250$ N

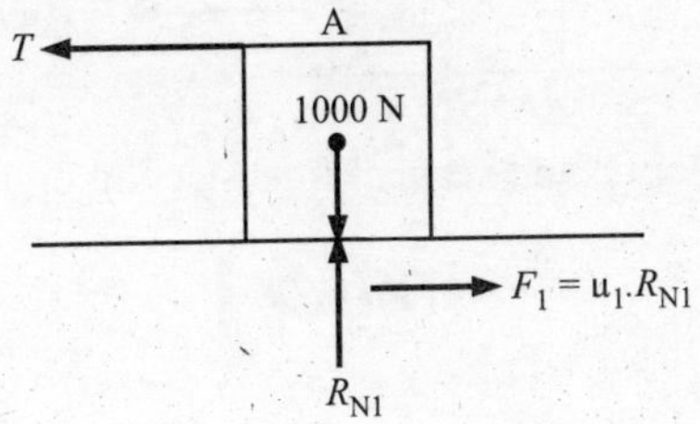


Fig. 6.18

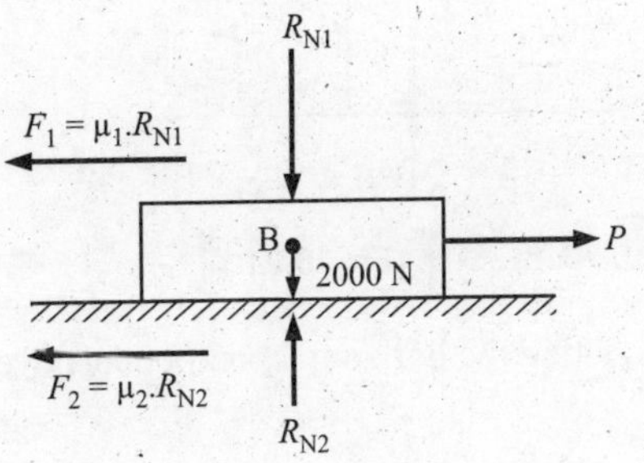


Fig. 6.19

Now consider the equilibrium of the block *B*. The various forces acting on the block *B*, as shown in Fig. 6.19, are as follows:

(*a*) Weight of the block *B* (2000 N) acting downwards,

(*b*) Normal reaction ($R_{N2}$) acting upwards,

(*c*) Force of friction ($F_2 = \mu_2 \cdot R_{N2}$) acting at the surface of the block *B* and the floor,

(*d*) Force (*P*) required to move the body.

The block *B* is also acted upon the normal reaction $R_{N1}$ and the force of friction ($F_1 = \mu_1 \cdot R_{N1}$), as shown in the figure.

Equating the vertical components,

$$R_{N2} = R_{N1} + 2000 = 1000 + 2000 = 3000 \text{ N}$$

and equating the horizontal components,

$$P = F_1 + F_2 = \mu_1 . R_{N1} + \mu_2 . R_{N2}$$

$$= 0.25 \times 1000 + \frac{1}{3} \times 3000 = 250 + 1000 = 1250 \text{ N} \quad \textbf{Ans.}$$

**Example 6.8** *The block A of weight 200 N is resting on the top of the block B of weight 300 N and is prevented from slipping by a string attached to the block at an angle of 30° with the horizontal, as shown in Fig. 6.20. If the coefficient of friction for all the sliding surfaces is 0.4, find the force P required to move the block B and tension in the string.*

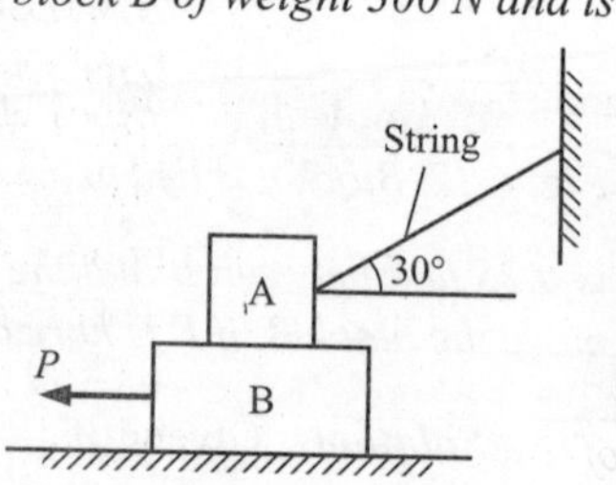


**Fig. 6.20**

***Solution*** Given : $W_A = 200$ N ; $W_B = 300$ N ; $\mu = \tan \phi = 0.4$; $\alpha = 30°$

First of all, let us consider the equilibrium of the block *A*. The various forces acting on the block *A*, as shown in Fig. 6.21, are as follows:

(*a*) Weight of the block *A* (200 N) acting downwards,

(*b*) Normal reaction ($R_{N1}$) acting upwards,

(*c*) Force of friction ($F_1 = \mu . R_{N1}$) acting at the contact surface of block *A* and *B*, and

(*d*) Tension (*T*) in the string acting at 30° to the horizontal. It is resolved into two components, *i.e.* horizontal component (*T* cos 30°) and vertical component (*T* sin 30°).

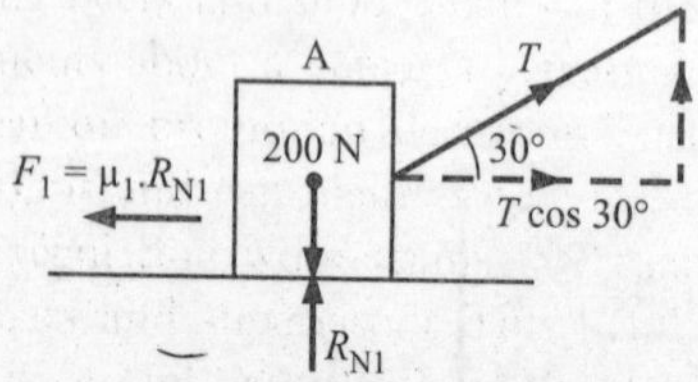


**Fig. 6.21**

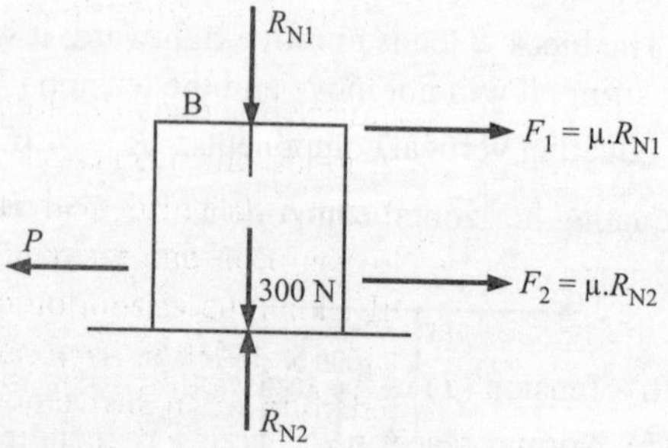


**Fig. 6.22**

Equating the horizontal components, we have

$$F_1 = \mu \cdot R_{N1} = T \cos 30°$$

$$\therefore \quad R_{N1} = \frac{T \cos 30°}{\mu} = \frac{T \times 0.866}{0.4} = 2.165\, T \quad ...(i)$$

Now equating vertical components, we have

$$R_{N1} + T \sin 30° = 200$$

$$2.165T + T \times 0.5 = 200 \qquad \text{...[From equation } (i)]$$

$$\therefore \qquad T = \frac{200}{2.665} = 75 \text{ N} \quad \textbf{Ans.}$$

and $$R_{N1} = 2.165\,T = 2.165 \times 75 = 162.4 \text{ N}$$

Now consider the equilibrium of the block $B$. The various forces acting on the block $B$, as shown in Fig. 6.22, are as follows:

(*a*) Weight of the block $B$ (300 N) acting downwards,

(*b*) Normal reaction ($R_{N2}$) acting upwards,

(*c*) Force of friction ($F_2 = \mu \cdot R_{N2}$) acting at the contact surface of block $B$ and the horizontal surface. The block $B$ is also acted upon the normal reaction ($R_{N1}$) and force of friction ($F_1 = \mu \cdot R_{N1}$), as shown in the figure.

(*d*) Force ($P$) required to move the body.

Equating the vertical components,

$$R_{N2} = R_{N1} + 300 = 162.4 + 300 = 462.4 \text{ N}$$

and equating the horizontal components,

$$P = F_1 + F_2 = \mu . R_{N1} + \mu \cdot R_{N2} = \mu\,(R_{N1} + R_{N2})$$
$$= 0.4\,(162.4 + 462.4) = 250 \text{ N} \quad \textbf{Ans.}$$

**Example 6.9** *Two planes AC and BC inclined at 60° and 30° to the horizontal meet at C. A load of 100 N rests on the inclined plane BC and is tied by a rope passing over a pulley to a block weighing W newtons and resting on the plane AC, as shown in Fig. 6.23. If the coefficient of friction between the load and the plane BC is 0.28 and between the block and the plane AC is 0.20, find the least and the greatest value of W for the equilibrium of the system.*

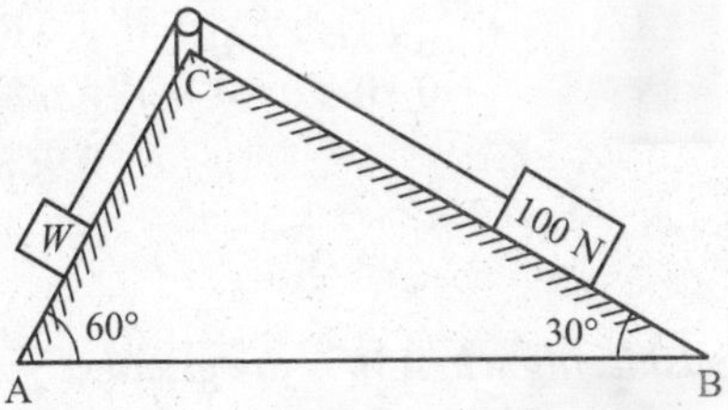


**Fig. 6.23**

***Solution*** Given : Load on plane $BC$ = 100 N ; Weight of block on plane $AC = W$; $\mu_{BC} = 0.28$; $\mu_{AC} = 0.20$

***Considering when W is the least***

When $W$ is the least, then the load of 100 N will be at the point of slipping downwards.

Let $T$ be the tension in the rope which provide the necessary minimum force to keep the load in equilibrium. The various forces acting on the inclined plane $BC$, as shown in Fig. 6.24, are as follows:

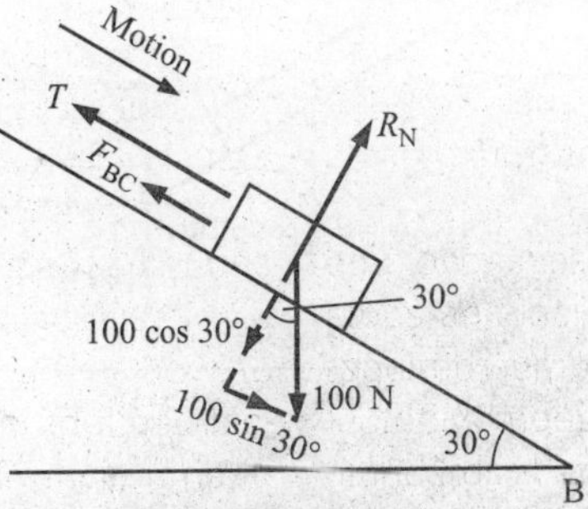


**Fig. 6.24**

(*a*) Tension ($T$) in the rope acting along the plane.

(*b*) Normal reaction $R_N$, acting perpendicular to the plane.

(*c*) Load 100 N acting vertically downwards which can be resolved into two components, one perpendicular to the plane (100 cos 30°) and the other parallel to the plane (100 sin 30°).

(*d*) Force of friction ($F_{BC} = \mu_{BC} \cdot R_N$) acting opposite to the motion.

Equating forces perpendicular to the plane,

$$R_N = 100 \cos 30° = 100 \times 0.866 = 86.6 \text{ N}$$

Equating forces parallel to the plane,

$$T + F_{BC} = 100 \sin 30° = 100 \times 0.5 = 50$$

$$T + \mu_{BC} \cdot R_N = 50$$

$$T + 0.28 \times 86.6 = 50$$

$$T + 24.248 = 50$$

$$\therefore \quad T = 50 - 24.248 = 25.752 \text{ N}$$

This tension ($T$) causes the load ($W$) to be in limiting equilibrium with a tendency to move up the plane, as shown in Fig. 6.25.

Since the pulley is smooth, the tension ($T$) in the rope is uniform throughout. In the similar way, as discussed above, the load ($W$) may be resolved into two components, one perpendicular to the plane ($W \cos 60°$) and the other parallel to the plane ($W \sin 60°$). The force of friction ($F_{AC} = \mu_{AC} . R_{N1}$) acting downwards (opposite to the motion up the plane), as shown in Fig. 6.25.

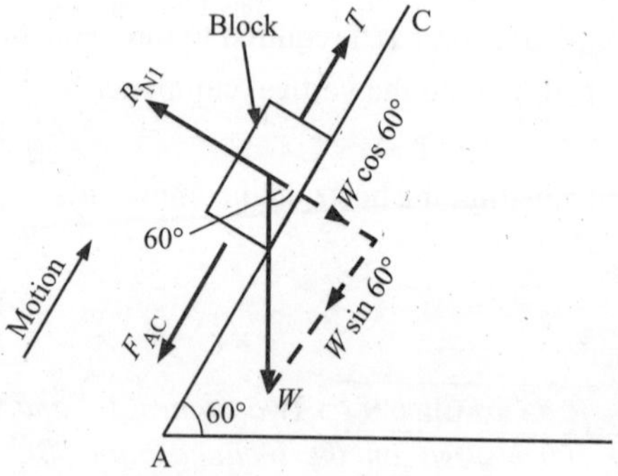


Fig. 6.25

Equating forces perpendicular to the plane,

$$R_{N1} = W \cos 60°$$

$$= W \times 0.5$$

$$= 0.5\, W$$

Equating forces parallel to the plane,

$$F_{AC} + W \sin 60° = T$$

$$\mu_{AC} \times R_{N1} + W \times 0.866 = 25.752$$

$$0.20 \times 0.5W + W \times 0.866 = 25.752$$

$$0.966W = 25.752$$

$$\therefore \quad W = \frac{25.752}{0.966} = 26.66 \text{ N } \textbf{Ans.}$$

***Considering when W is the greatest***

When $W$ is the greatest, then the load of 100 N will be at the point of moving up the plane. The various forces acting on the inclined plane $BC$, are shown in Fig. 6.26.

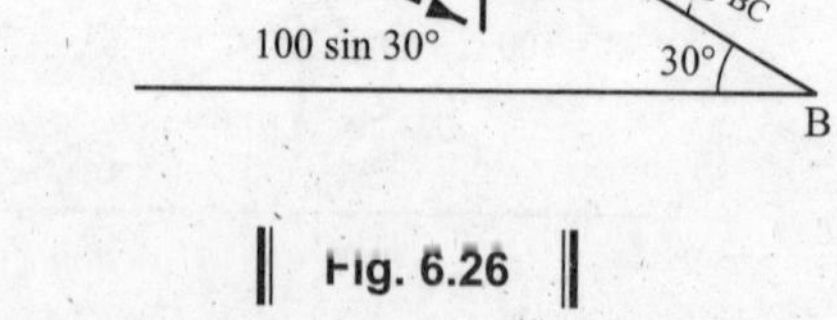


Fig. 6.26

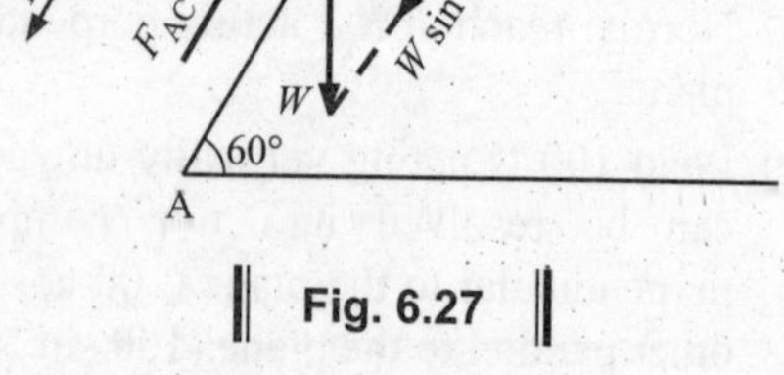


Fig. 6.27

Equating the forces perpendicular to the plane,

$$R_N = 100 \cos 30° = 100 \times 0.866 = 86.6 \text{ N}$$

Equating the forces parallel to the plane,

$$T = F_{BC} + 100 \sin 30° = \mu_{BC} \cdot R_N + 100 \sin 30°$$

$$= 0.28 \times 86.6 + 100 \times 0.5 = 74.248 \text{ N}$$

The block of weight $W$ is now at the point of slipping down the inclined plane. The various forces acting on the block are shown in Fig. 6.27.

Equating the forces perpendicular to the plane,

$$R_{N1} = W \cos 60° = W \times 0.5 = 0.5\,W$$

Equating the forces parallel to the plane,

$$F_{AC} + T = W \sin 60°$$

$$\mu_{AC} \cdot R_{N1} + T = W \sin 60°$$

$$0.20 \times 0.5W + 74.248 = W \times 0.866$$

$$74.248 = 0.866W - 0.1W = 0.766\,W$$

$$\therefore \quad W = \frac{74.248}{0.766} = 96.93 \text{ N}$$

Thus, the least value of $W$ is 26.66 N and the greatest value of $W$ is 96.93 N. **Ans.**

## 6.15 EQUILIBRIUM OF A BODY LYING ON A ROUGH INCLINED PLANE

We have already discussed in Sec. 6.13, that when the angle of inclination of the plane is more than the angle of friction, then the body will move downwards and an upward external force is required to keep the body in equilibrium. It may be noted that an inclined plane is frequently used in one form or the other. Thus it is very necessary to examine the various forces that act on a body when it moves up or down the inclined plane. The force may act in the following three ways:

1. The force may act at an angle with the inclined plane;
2. The force may act along the inclined plane; and
3. The force may act horizontally.

We shall now discuss the equilibrium of the body, in detail, in the above three cases.

## 6.16 MOTION OF A BODY UP THE PLANE WHEN THE FORCE IS ACTING AT AN ANGLE TO THE INCLINED PLANE

Consider a body $A$ of weight $W$ resting on an inclined plane which makes an angle α to the horizontal, as shown in Fig. 6.28 (*a*). Let us consider the following two cases:

1. When the friction between the body and the inclined plane is neglected, and
2. When the friction between the body and the inclined plane is taken into account.

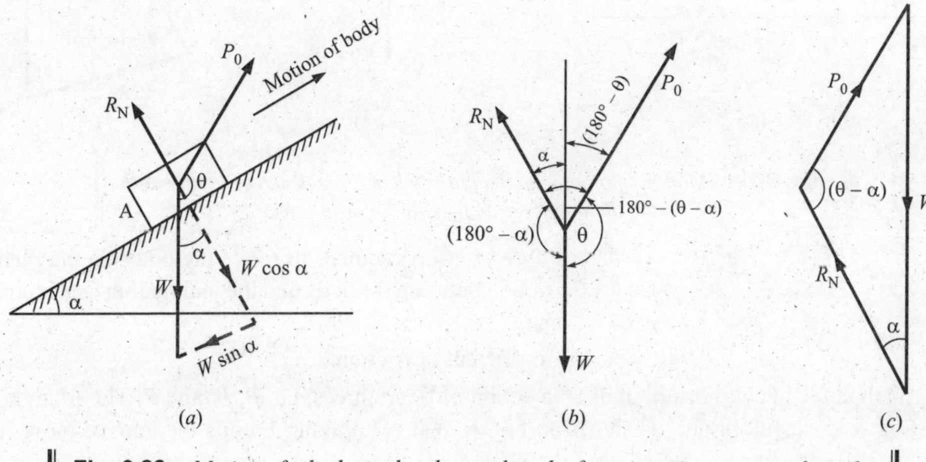


**Fig. 6.28** *Motion of a body up the plane, when the force is acting at an angle to the plane and when friction is neglected.*

These two cases are discussed as follows:

**1. *Neglecting friction***

Let $P_0$ = Force or effort required to move the body up the plane, when friction between the body and the inclined surface is neglected,

$\theta$ = Inclination of $P_0$ with the line of action of weight $W$, and

$R_N$ = Normal reaction exerted by the inclined plane on the body.

When friction is neglected, the body is in equilibrium under the action of three forces, *i.e.* $P_0$, $W$ and $R_N$, as shown in Fig. 6.28 (*b*). The triangle of these forces must close, for equilibrium, as shown in Fig. 6.28 (*c*). Now applying Lami's theorem to these forces, as shown in Fig. 6.28 (*b*), we have

$$\frac{P_0}{\sin(180° - \alpha)} = \frac{W}{\sin[180° - (\theta - \alpha)]}$$

or

$$\frac{P_0}{\sin \alpha} = \frac{W}{\sin(\theta - \alpha)}$$

$\therefore$

$$P_0 = \frac{W \sin \alpha}{\sin(\theta - \alpha)} \quad ...(i)$$

**2. *Considering friction***

When friction between the body and the inclined plane is considered, then the force of friction ($F = \mu.R_N$) will act in the opposite direction in which the body moves, as shown in Fig. 6.29 (*a*). The resultant of normal reaction ($R_N$) and the force of friction ($F$) is given by $R$ making an angle $\phi$ (known as angle of friction) with the normal reaction ($R_N$).

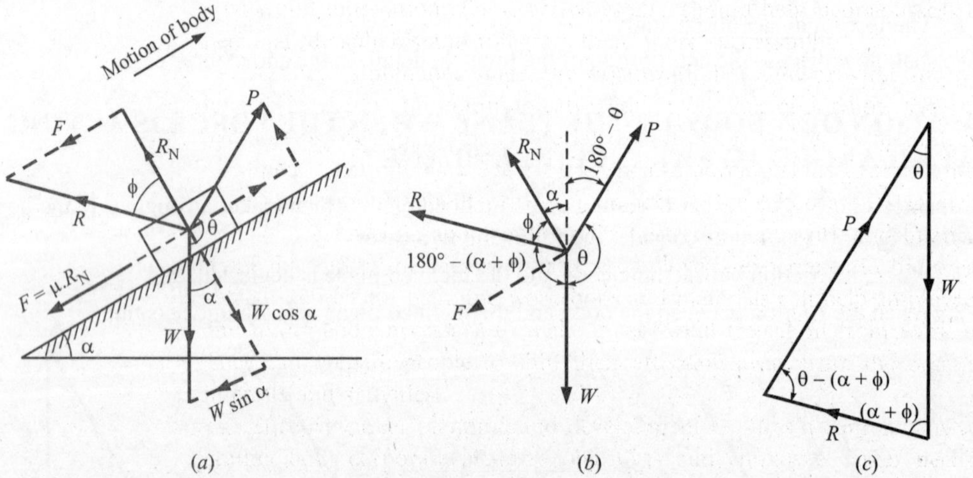


**Fig. 6.29** *Motion of a body up the plane when force is acting at an angle to the plane and when friction is considered.*

Let $P$ = Force or effort required to move the body up the plane, when friction between the body and the inclined surface is considered, and

$\mu$ = Coefficient of friction.

Now the body is in equilibrium under the action of three forces, *i.e.* $P$, $W$ and $R$. The triangle of these forces must close, for equilibrium, as shown in Fig. 6.29 (*c*). Applying Lami's theorem to these forces, as shown in Fig. 6.29 (*b*),

$$\frac{P}{\sin[180° - (\alpha + \phi)]} = \frac{W}{\sin(180° - \theta + \alpha + \phi)}$$

$$= \frac{W}{\sin[180° - (\theta - \alpha - \phi)]}$$

or

$$\frac{P}{\sin(\alpha + \phi)} = \frac{W}{\sin[\theta - (\alpha + \phi)]}$$

$$\therefore \quad P = \frac{W \sin(\alpha + \phi)}{\sin[\theta - (\alpha + \phi)]} \qquad ...(ii)$$

## 6.17 MOTION OF A BODY UP THE PLANE WHEN THE FORCE IS ACTING HORIZONTALLY

When the force or effort required to move the body up the plane, is acting horizontally, as shown in Fig. 6.30 (*a*), then the angle between the effort ($P_0$) and weight ($W$) is 90°, *i.e.* $\theta = 90°$.

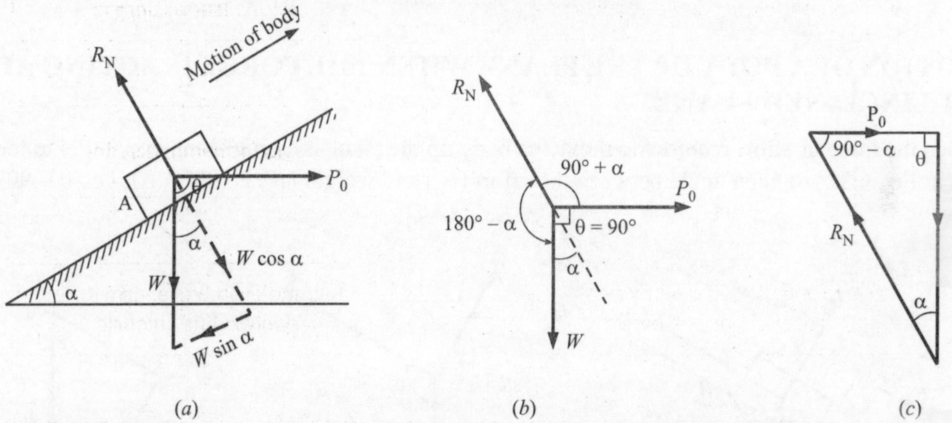


**Fig. 6.30** *Motion of the body up the plane when the force is acting horizontally and when friction is neglected.*

We have already discussed in the previous article, that when friction between the body and the inclined surface is neglected, then

$$P_0 = \frac{W \sin \alpha}{\sin(\theta - \alpha)}$$

and when friction is considered, as shown in Fig. 6.31, then the force or effort required to move the body,

$$P = \frac{W \sin(\alpha + \phi)}{\sin[\theta - (\alpha + \phi)]}$$

Now substituting $\theta = 90°$, we have

$$P_0 = \frac{W \sin \alpha}{\sin(90° - \alpha)} = \frac{W \sin \alpha}{\cos \alpha} = W \tan \alpha$$

Similarly, when friction is considered, then

$$P = \frac{W \sin(\alpha + \phi)}{\sin[90° - (\alpha + \phi)]}$$

$$= \frac{W \sin(\alpha + \phi)}{\cos(\alpha + \phi)} = W \tan(\alpha + \phi)$$

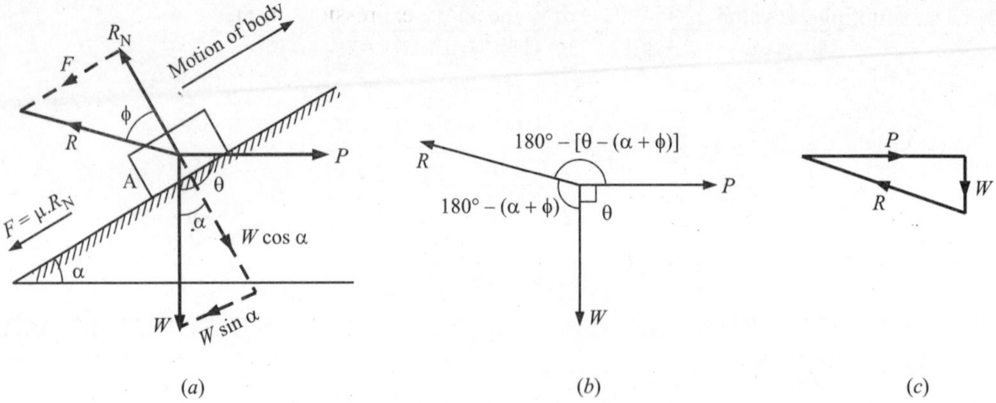


**Fig. 6.31** *Motion of the body up the plane when the force is acting horizontally and when friction is considered.*

## 6.18 MOTION OF A BODY UP THE PLANE WHEN THE FORCE IS ACTING ALONG THE INCLINED PLANE

When the force or effort required to move the body up the plane is acting along or parallel to the plane, as shown in Fig. 6.32 (*a*), then angle between the effort ($P_0$) and weight ($W$) is ($90° + \alpha$), *i.e.*, $\theta = 90° + \alpha$.

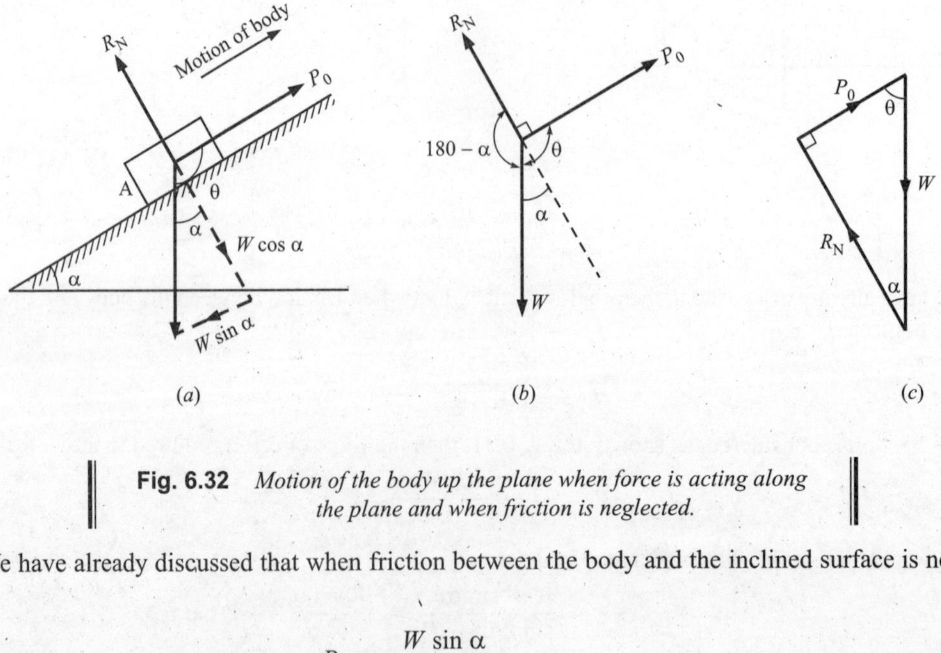


**Fig. 6.32** *Motion of the body up the plane when force is acting along the plane and when friction is neglected.*

We have already discussed that when friction between the body and the inclined surface is neglected, then

$$P_0 = \frac{W \sin \alpha}{\sin (\theta - \alpha)}$$

and when friction is considered, as shown in Fig. 6.33, then the force or effort required to move the body,

$$P = \frac{W \sin (\alpha + \phi)}{\sin [\theta - (\alpha + \phi)]}$$

Now substituting the value of $\theta = 90° + \alpha$, in the above expressions, we have

$$P_0 = \frac{W \sin \alpha}{\sin (90° + \alpha - \alpha)} = \frac{W \sin \alpha}{\sin 90°} = W \sin \alpha$$

and

$$P = \frac{W \sin (\alpha + \phi)}{\sin [(90° + \alpha) - (\alpha + \phi)]} = \frac{W \sin (\alpha + \phi)}{\sin (90° - \phi)}$$

$$= \frac{W \sin (\alpha + \phi)}{\cos \phi} = \frac{W (\sin \alpha \cos \phi + \cos \alpha \sin \phi)}{\cos \phi}$$

$$= W(\sin \alpha + \cos \alpha \tan \phi) = W (\sin \alpha + \mu \cos \alpha)$$

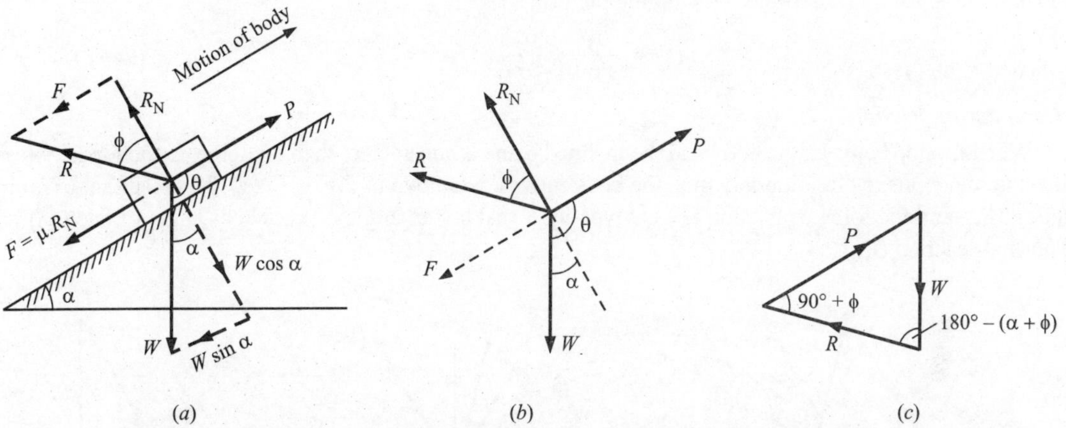


**Fig. 6.33** *Motion of the body up the plane when the force is acting along the plane and when friction is considered.*

## 6.19 MOTION OF A BODY DOWN THE PLANE WHEN THE FORCE IS ACTING AT AN ANGLE TO THE INCLINED PLANE

Consider a body $A$ of weight $W$ resting on an inclined plane which makes an angle $\alpha$ to the horizontal, as shown in Fig. 6.34. Let us consider the following two cases:

1. When the friction between the body and the inclined plane is neglected, and
2. When the friction between the body and the inclined plane is considered.

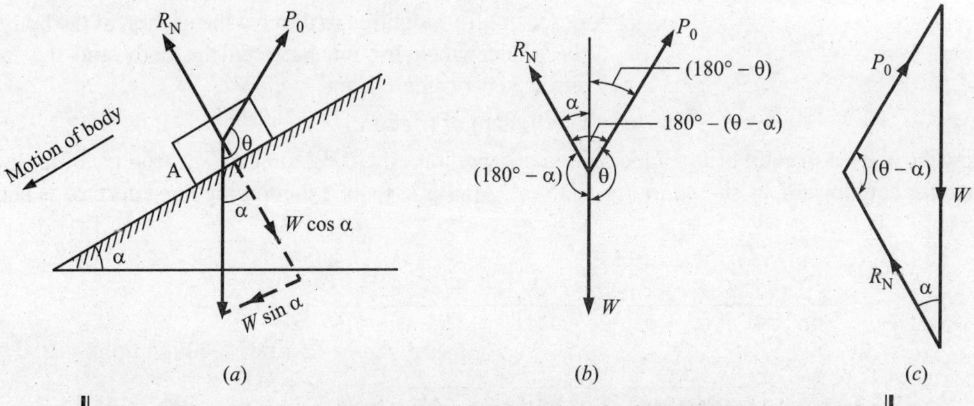


**Fig. 6.34** *Motion of a body down the plane when the force is acting at an angle to the plane and when friction is neglected.*

These two cases are discussed as follows:

**1. *Neglecting friction***

Let $P_0$ = Force or effort required to prevent the motion of the body down the plane when the friction between the body and the inclined plane is neglected,

$\theta$ = Inclination of $P_0$ with the line of action of the weight $W$, and

$R_N$ = Normal reaction exerted by the inclined surface on the body.

When friction is neglected, the force or effort required to the prevent the motion of the body down the plane is same as for motion of the body up the plane, *i.e.*

$$P_0 = \frac{W \sin \alpha}{\sin (\theta - \alpha)} \quad \text{...(Refer Sec. 16.16)}$$

**2. *Considering friction***

When friction between the body and the inclined plane is considered, then the force of friction ($F = \mu.R_N$) will act in the opposite direction in which the body moves, as shown in Fig. 6.35 (*a*). The resultant of normal reaction ($R_N$) and the force of friction ($F$) is given by $R$ making an angle $\phi$ (known as angle of friction) with the normal reaction ($R_N$).

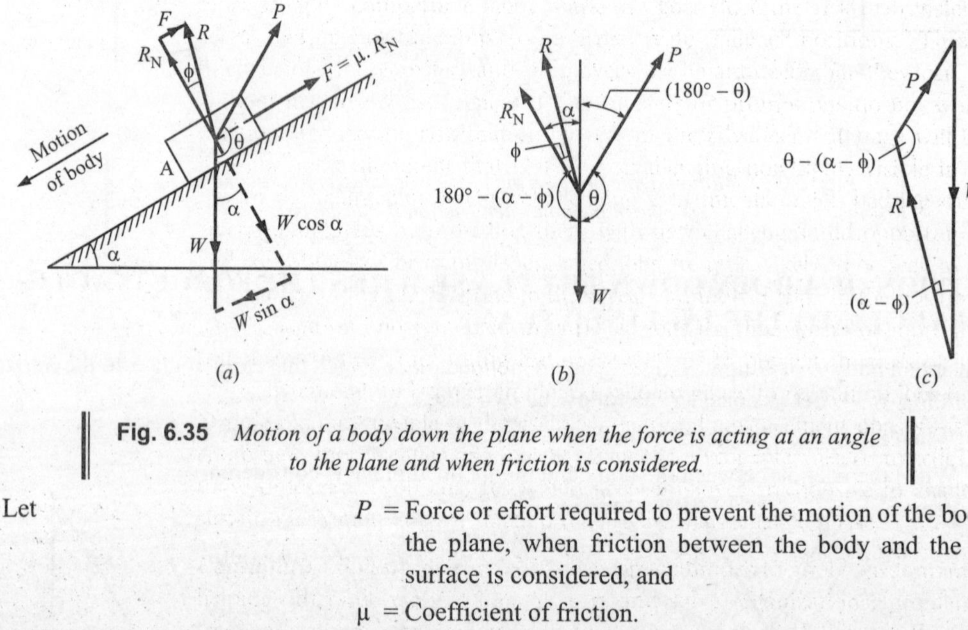


**Fig. 6.35** *Motion of a body down the plane when the force is acting at an angle to the plane and when friction is considered.*

Let $P$ = Force or effort required to prevent the motion of the body down the plane, when friction between the body and the inclined surface is considered, and

$\mu$ = Coefficient of friction.

Now the body is in equilibrium under action of three forces, *i.e.* $P$, $W$ and $R$. The triangle of these forces must close for equilibrium, as shown in Fig. 6.35 (*c*). Applying Lami's theorem to these forces, as shown in Fig. 6.35 (*b*),

$$\frac{P}{\sin [180° - (\alpha - \phi)]} = \frac{W}{\sin [180° - (\theta - \alpha + \phi)]}$$

$$\frac{P}{\sin (\alpha - \phi)} = \frac{W}{\sin [\theta - (\alpha - \phi)]}$$

$$\therefore \quad P = \frac{\sin (\alpha - \phi)}{\sin [\theta - (\alpha - \phi)]}$$

## 6.20 MOTION OF A BODY DOWN THE PLANE WHEN THE FORCE IS ACTING HORIZONTALLY

When the force or effort required to prevent the motion of the body down the plane, is acting horizontally, as shown in Fig. 6.36 (*a*), then the angle between the effort ($P_0$) and the weight ($W$) is 90°, *i.e.* $\theta = 90°$.

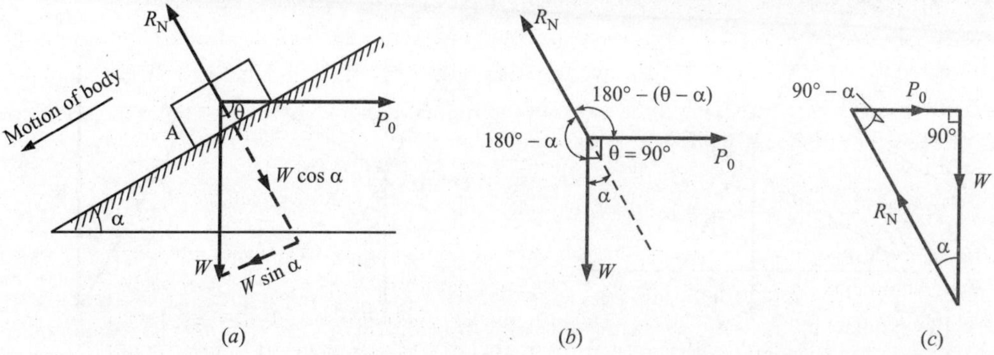


**Fig. 6.36** *Motion of the body down the plane when the force is acting horizontally and when friction is neglected.*

We have already discussed in the previous article that when the friction between the body and the inclined surface is neglected, then

$$P_0 = \frac{W \sin \alpha}{\sin (\theta - \alpha)}$$

and when friction is considered, as shown in Fig. 6.37, then the force or effort required to prevent the motion of the body down the plane,

$$P = \frac{W \sin (\alpha - \phi)}{\sin [\theta - (\alpha - \phi)]}$$

Now substituting $\theta = 90°$, we have $P_0 = \dfrac{W \sin \alpha}{\sin (90° - \alpha)} = \dfrac{W \sin \alpha}{\cos \alpha} = W \tan \alpha$

Similarly, when friction is considered,

$$P = \frac{W \sin (\alpha - \phi)}{\sin [90° - (\alpha - \phi)]} = \frac{W \sin (\alpha - \phi)}{\cos (\alpha - \phi)} = W \tan (\alpha - \phi)$$

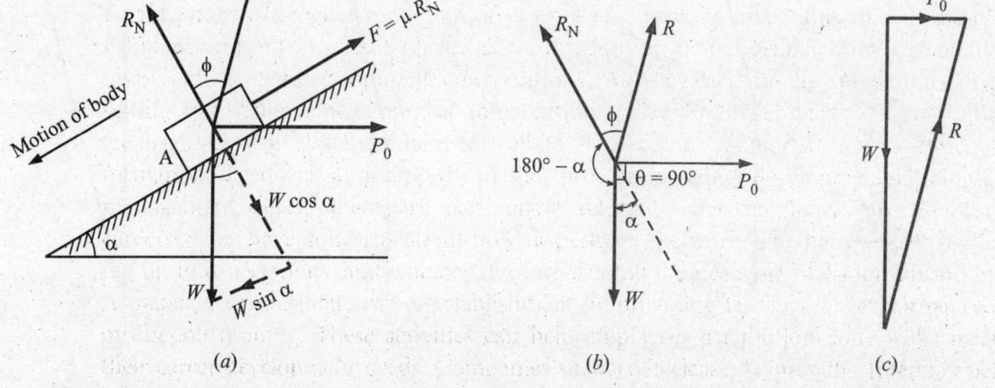


**Fig. 6.37** *Motion of the body down the plane when the force is acting horizontally and when friction is considered.*

## 6.21 MOTION OF A BODY DOWN THE PLANE WHEN THE FORCE IS ACTING ALONG THE INCLINED PLANE

When the force or effort required to prevent the motion of the body down the plane, is acting along or parallel to the plane, as shown in Fig. 6.38 (*a*), then the angle between the effort ($P_0$) and the weight ($W$) is $90° + \alpha$, *i.e.*, $\theta = 90° + \alpha$.

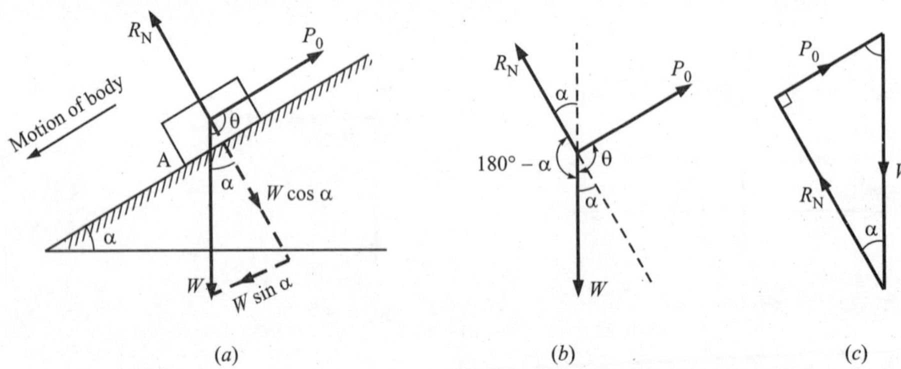


**Fig. 6.38** *Motion of the body down the plane when the force is acting along the plane and when friction is neglected*

We have already discussed that when friction between the body and the inclined surface is neglected, then

$$P_0 = \frac{W \sin \alpha}{\sin (\theta - \alpha)}$$

and when friction is considered, as shown in Fig. 6.39, then the force or effort required to prevent the motion of the body down the plane,

$$P = \frac{W \sin (\alpha - \phi)}{\sin [\theta - (\alpha - \phi)]}$$

Now substituting $\theta = 90° + \alpha$, we have

$$P_0 = \frac{W \sin \alpha}{\sin (90° + \alpha - \alpha)} = W \sin \alpha$$

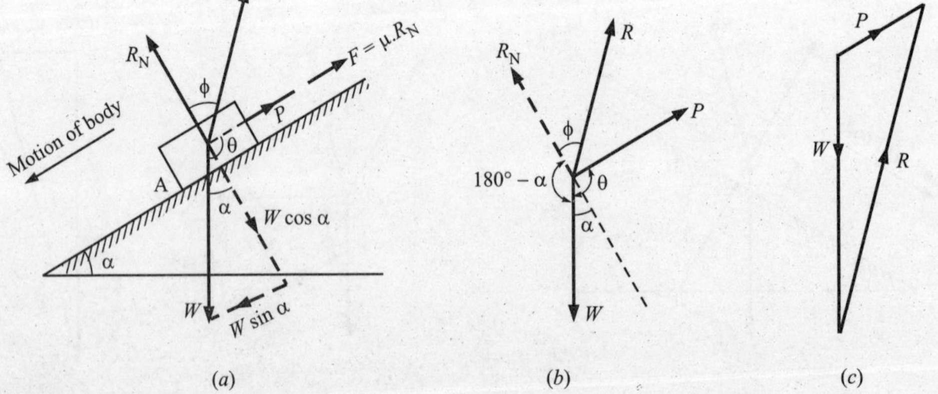


**Fig. 6.39** *Motion of the body down the plane when the force is acting along the plane and when friction is considered*

Similarly, when friction is considered,

$$P = \frac{W \sin (\alpha - \phi)}{\sin [90° + \alpha - (\alpha - \phi)]} = \frac{W \sin (\alpha - \phi)}{\cos \phi}$$

$$= \frac{W (\sin \alpha \cos \phi - \cos \alpha \sin \phi)}{\cos \phi}$$

$$= W (\sin \alpha - \cos \alpha \tan \phi) = W (\sin \alpha - \mu \cos \alpha) \quad ...(\because \tan \phi = \mu)$$

## 6.22 EFFICIENCY OF INCLINED PLANE

The inclined plane is commonly used in machines in one form or the other. For example, a screw jack used for lifting heavy loads by applying a smaller effort at its handle, works on the principle similar to that of an inclined plane. Let us consider the following two cases:

**1. *When the body moves up the plane***

A little consideration will show that for the motion of a body up the plane, the force or effort required to move the body considering friction (*i.e.* $P$) is greater than effort required neglecting friction (*i.e.* $P_0$). Thus the efficiency in this case, is defined as the ratio of the effort required neglecting friction to the effort required considering friction. Mathematically,

$$\text{Efficiency, } \eta_U = \frac{P_0}{P} = \frac{W \sin \alpha}{\sin (\theta - \alpha)} \times \frac{\sin [\theta - (\alpha + \phi)]}{W \sin (\alpha + \phi)}$$

$$= \frac{\sin \alpha [\sin \theta \cos (\alpha + \phi) - \cos \theta \sin (\alpha + \phi)]}{[\sin \theta \cos \alpha - \cos \theta \sin \alpha] \sin (\alpha + \phi)}$$

$$= \frac{\sin \alpha [\sin \theta \cot (\alpha + \phi) - \cos \theta]}{\sin \theta \cos \alpha - \cos \theta \sin \alpha}$$

Dividing the numerator and denominator by $\sin \theta \sin \alpha$, we get

$$\eta_U = \frac{\sin \alpha [\sin \theta \cot (\alpha + \phi) - \cos \theta]}{\sin \theta \cos \alpha - \cos \theta \sin \alpha} \times \frac{\sin \theta \sin \alpha}{\sin \theta \sin \alpha}$$

$$= \frac{\cot (\alpha + \phi) - \cot \theta}{\cot \alpha - \cot \theta} \quad ...(i)$$

This expression gives the value of efficiency of the inclined plane when the body moves up the plane and the force is acting at an angle $\theta$ with the line of action of weight $W$.

Let us now consider when the force or effort is acting horizontally and along the inclined plane as discussed below:

***(a)*** When the force or effort is acting horizontally, then $\theta = 90°$.

$$\therefore \quad \eta_U = \frac{\cot (\alpha + \phi) - \cot 90°}{\cot \alpha - \cot 90°} = \frac{\cot (\alpha + \phi)}{\cot \alpha} \quad ...[\text{From equation } (i)]$$

$$= \frac{\tan \alpha}{\tan (\alpha + \phi)}$$

***(b)*** When the force or effort is applied along (or parallel to) the plane, then $\theta = 90° + \alpha$.

$$\therefore \quad \eta_U = \frac{\cot (\alpha + \phi) - \cot (90° + \alpha)}{\cot \alpha - \cot (90° + \alpha)} \quad ...[\text{From equation } (i)]$$

$$= \frac{\cot (\alpha + \phi) + \tan \alpha}{\cot \alpha + \tan \alpha}$$

$$= \frac{\dfrac{\cos(\alpha+\phi)}{\sin(\alpha+\phi)} + \dfrac{\sin\alpha}{\cos\alpha}}{\dfrac{\cos\alpha}{\sin\alpha} + \dfrac{\sin\alpha}{\cos\alpha}}$$

$$= \frac{\dfrac{\cos(\alpha+\phi)\cos\alpha + \sin(\alpha+\phi)\sin\alpha}{\sin(\alpha+\phi)\cos\alpha}}{\dfrac{\cos^2\alpha + \sin^2\alpha}{\sin\alpha\cos\alpha}}$$

$$= \frac{\cos(\alpha+\phi-\alpha)}{\sin(\alpha+\phi)\cos\alpha} \times \frac{\sin\alpha\cos\alpha}{1}$$

...[$\because$ $\cos A \cos B + \sin A \sin B = \cos(A - B)$]

$$= \frac{\cos\phi}{\sin(\alpha+\phi)\cos\alpha} \times \sin\alpha\cos\alpha = \frac{\sin\alpha\cos\phi}{\sin(\alpha+\phi)}$$

**2. *When the body moves down the plane***

A little consideration will show that for the motion down the plane, the force or effort required to prevent the motion considering friction (*i.e.* $P$) is less than the effort required neglecting friction (*i.e.* $P_0$). Thus the efficiency in this case, is defined as the ratio of effort required considering friction to the effort required neglecting friction. Mathematically,

$$\text{Efficiency, } \eta_D = \frac{P}{P_0}$$

$$= \frac{W\sin(\alpha-\phi)}{\sin[\theta-(\alpha-\phi)]} \times \frac{\sin(\theta-\alpha)}{W\sin\alpha}$$

$$= \frac{\sin(\alpha-\phi)[\sin\theta\cos\alpha - \cos\theta\sin\alpha]}{[\sin\theta\cos(\alpha-\phi) - \cos\theta\sin(\alpha-\phi)]\sin\alpha}$$

Dividing the numerator and denominator by $\sin(\alpha - \phi)$, we get

$$\eta_D = \frac{\sin\theta\cos\alpha - \cos\theta\sin\alpha}{[\sin\theta\cot(\alpha-\phi) - \cos\theta]\sin\alpha}$$

Again dividing the numerator and denominator by $\sin\theta\sin\alpha$, we get

$$\eta_D = \frac{\sin\theta\cos\alpha - \cos\theta\sin\alpha}{[\sin\theta\cot(\alpha-\phi) - \cos\theta]\sin\alpha} \times \frac{\sin\theta\sin\alpha}{\sin\theta\sin\alpha}$$

$$= \frac{\cot\alpha - \cot\theta}{\cot(\alpha-\phi) - \cot\theta} \quad \text{...}(ii)$$

This expression gives the efficiency of the inclined plane when the body moves down the plane and the force is acting at an angle $\theta$ with the line of action of $W$.

Let us now consider when the force or effort is acting horizontally and along the inclined plane as discussed below:

***(a)*** When the force or effort is acting horizontally, then $\theta = 90°$.

$$\therefore \quad \eta_D = \frac{\cot\alpha - \cot 90°}{\cot(\alpha-\phi) - \cot 90°} = \frac{\cot\alpha}{\cot(\alpha-\phi)} \quad \text{...[From equation } (ii)]$$

$$= \frac{\tan(\alpha-\phi)}{\tan\alpha}$$

(*b*) When the force or effort is acting along (or parallel to) the plane, then $\theta = 90° + \alpha$.

$$\therefore \quad \eta_D = \frac{\cot\alpha - \cot(90° + \alpha)}{\cot(\alpha - \phi) - \cot(90° + \alpha)}$$

$$= \frac{\cot\alpha + \tan\alpha}{\cot(\alpha - \phi) + \tan\alpha} = \frac{\dfrac{\cos\alpha}{\sin\alpha} + \dfrac{\sin\alpha}{\cos\alpha}}{\dfrac{\cos(\alpha - \phi)}{\sin(\alpha - \phi)} + \dfrac{\sin\alpha}{\cos\alpha}}$$

$$= \frac{\dfrac{\cos^2\alpha + \sin^2\alpha}{\sin\alpha\cos\alpha}}{\dfrac{\cos(\alpha - \phi)\cos\alpha + \sin(\alpha - \phi)\sin\alpha}{\sin(\alpha - \phi)\cos\alpha}}$$

$$= \frac{1}{\sin\alpha\cos\alpha} \times \frac{\sin(\alpha - \phi)\cos\alpha}{\cos(\alpha - \phi - \alpha)} = \frac{\sin(\alpha - \phi)}{\sin\alpha\cos\phi}$$

## 6.23 MAXIMUM EFFICIENCY OF THE INCLINED PLANE FOR MOTION UP THE PLANE

We have already discussed that the efficiency of an inclined plane for the motion up the plane, when effort is applied at an angle $\theta$ with the line of action of weight $W$ is given by

$$\eta_U = \frac{P_0}{P} = \frac{W\sin\alpha}{\sin(\theta - \alpha)} \times \frac{\sin[\theta - (\alpha + \phi)]}{W\sin(\alpha + \phi)}$$

$$= \frac{2\sin\alpha\sin[\theta - (\alpha + \phi)]}{2\sin(\theta - \alpha)\sin(\alpha + \phi)}$$

...(Multiplying the numerator and denominator by 2)

$$= \frac{\cos[\alpha - \{\theta - (\alpha + \phi)\}] - \cos[\alpha + \{\theta - (\alpha + \phi)\}]}{\cos[(\theta - \alpha) - (\alpha + \phi)] - \cos[(\theta - \alpha) + (\alpha + \phi)]}$$

...[$\because$ $2\sin A\sin B = \cos(A - B) - \cos(A + B)$]

$$= \frac{{}^*\cos(2\alpha - \theta + \phi) - \cos(\theta - \phi)}{\cos(-2\alpha + \theta - \phi) - \cos(\theta + \phi)}$$

$$= \frac{\cos(\theta - \phi - 2\alpha) - \cos(\theta - \phi)}{\cos(\theta - \phi - 2\alpha) - \cos(\theta + \phi)} \qquad ...(i)$$

For the efficiency to be maximum, $\cos(\theta - \phi - 2\alpha)$ should be maximum, *i.e.*

$$\cos(\theta - \phi - 2\alpha) = 1$$

or $$\theta - \phi - 2\alpha = 0$$

$$\therefore \quad \alpha = \frac{\theta - \phi}{2}$$

Substituting this value of $\alpha$ in equation (*i*), we have maximum efficiency,

$$\eta_{U(max)} = \frac{\cos\left[\theta - \phi - \left(2 \times \dfrac{\theta - \phi}{2}\right)\right] - \cos(\theta - \phi)}{\cos\left[\theta - \phi - \left(2 \times \dfrac{\theta - \phi}{2}\right)\right] - \cos(\theta + \phi)}$$

---

* $\cos(2\alpha - \theta + \phi) = \cos[-(\theta - \phi - 2\alpha)] = \cos(\theta - \phi - 2\alpha)$ ...[$\because$ $\cos(-\theta) = \cos\theta$]

$$= \frac{\cos(\theta - \phi - \theta + \phi) - \cos(\theta - \phi)}{\cos[\theta - \phi - \theta + \phi] - \cos(\theta + \phi)}$$

$$= \frac{\cos 0 - \cos(\theta - \phi)}{\cos 0 - \cos(\theta + \phi)} = \frac{1 - \cos(\theta - \phi)}{1 - \cos(\theta + \phi)} \quad ...(ii)$$

It may be noted that

**1.** When effort is applied horizontally, *i.e.* when $\theta = 90°$, then

$$\eta_{U(max)} = \frac{1 - \cos(90° - \phi)}{1 - \cos(90° + \phi)} = \frac{1 - \sin\phi}{1 + \sin\phi} \quad ...[\text{From equation } (ii)]$$

**2.** When effort is applied along (or parallel to) the inclined plane, *i.e.* when $\theta = 90° + \alpha$, then

$$\eta_{U(max)} = \frac{1 - \cos[(90° + \alpha) - \phi]}{1 - \cos[(90° + \alpha) + \phi]}$$

$$= \frac{1 - \cos[90° + (\alpha - \phi)]}{1 - \cos[90° + (\alpha + \phi)]} = \frac{1 + \sin(\alpha - \phi)}{1 + \sin(\alpha + \phi)}$$

## 6.24 MAXIMUM EFFICIENCY OF THE INCLINED PLANE FOR MOTION DOWN THE PLANE

We have already discussed that efficiency of an inclined plane for the motion down the plane, when effort is applied at an angle $\theta$ with the line of action of weight ($W$) is given by

$$\eta_D = \frac{P}{P_0} = \frac{W\sin(\alpha - \phi)}{\sin[\theta - (\alpha - \phi)]} \times \frac{\sin(\theta - \alpha)}{W\sin\alpha}$$

$$= \frac{2\sin(\alpha - \phi)\sin(\theta - \alpha)}{2\sin[\theta - (\alpha - \phi)]\sin\alpha}$$

...(Multiplying the numerator and denominator by 2)

$$= \frac{\cos[(\alpha - \phi) - (\theta - \alpha)] - \cos[(\alpha - \phi) + (\theta - \alpha)]}{\cos[\theta - (\alpha - \phi) - \alpha] - \cos[\theta - (\alpha - \phi) + \alpha]}$$

$$= \frac{{}^{*}\cos(2\alpha - \theta - \phi) - \cos(\theta - \phi)}{\cos(-2\alpha + \theta + \phi) - \cos(\theta + \phi)}$$

$$= \frac{\cos(\theta + \phi - 2\alpha) - \cos(\theta - \phi)}{\cos(\theta + \phi - 2\alpha) - \cos(\theta + \phi)} \quad ...(i)$$

For the efficiency to be maximum, $\cos(\theta + \phi - 2\alpha)$ should be maximum,

*i.e.* $\cos(\theta + \phi - 2\alpha) = 1$

or $\theta + \phi - 2\alpha = 0$

$\therefore$ $\alpha = \frac{\theta + \phi}{2}$

* $\cos(2\alpha - \theta - \phi) = \cos[(-(\theta + \phi - 2\alpha)] = \cos(\theta + \phi - 2\alpha)$ ...[$\because \cos(-\theta) = \cos\theta$]

Substituting this value of $\alpha$ in equation (*i*), we have maximum efficiency,

$$\eta_{D(max)} = \frac{\cos\left[\theta + \phi - \left(2 \times \frac{\theta + \phi}{2}\right)\right] - \cos(\theta - \phi)}{\cos\left[\theta + \phi - \left(2 \times \frac{\theta + \phi}{2}\right)\right] - \cos(\theta + \phi)}$$

$$= \frac{\cos(\theta + \phi - \theta - \phi) - \cos(\theta - \phi)}{\cos(\theta + \phi - \theta - \phi) - \cos(\theta + \phi)}$$

$$= \frac{\cos 0 - \cos(\theta - \phi)}{\cos 0 - \cos(\theta + \phi)} = \frac{1 - \cos(\theta - \phi)}{1 - \cos(\theta + \phi)} \quad ...(ii)$$

From the above expression, we see that

$$\eta_{U(max)} = \eta_{D(max)}$$

This shows that the maximum efficiency is *same irrespective of the motion of the body up or down the inclined plane.

It may be noted that

**1.** When the effort is applied horizontally, *i.e.* when $\theta = 90°$, then

$$\eta_{D(max)} = \frac{1 - \cos(90° - \phi)}{1 - \cos(90° + \phi)} \quad ...[\text{From equation } (ii)]$$

$$= \frac{1 - \sin\phi}{1 + \sin\phi} \quad ...[\text{same as } \eta_{U(max)}]$$

**2.** When effort is applied along (or parallel to) the plane, then $\theta = 90° + \alpha$.

$$\therefore \quad \eta_{D(max)} = \frac{1 - \cos[(90° + \alpha) - \phi]}{1 - \cos[(90° + \alpha) + \phi]} = \frac{1 - \cos[90° + (\alpha - \phi)]}{1 - \cos[90° + (\alpha + \phi)]}$$

$$= \frac{1 + \sin(\alpha - \phi)}{1 + \sin(\alpha + \phi)} \quad ...[\text{same as } \eta_{U(max)}]$$

It is necessary to mention here that $\eta_D$ is not equal to $\eta_U$, in both cases.

**Example 6.10** *A body of weight 500 N is pulled up an inclined plane by a force of 50 N. This inclination of the plane is 20° to the horizontal and the force is applied in upward direction and parallel to the plane. Determine the coefficient of friction.*

***Solution*** Given : $W = 500$ N ; $P = 50$ N ; $\alpha = 20°$

Let $\mu$ = Coefficient of friction, and

$R_N$ = Normal reaction.

* The efficiency of the inclined plane for the motion down the plane ($\eta_D$) is *not equal to* the efficiency of the inclined plane for the motion up the plane ($\eta_U$).

Refer Fig. 6.40.

Resolving weight of the body ($W$) into two components, *i.e.* parallel to the plane ($W \sin \alpha$) and perpendicular to the plane ($W \cos \alpha$).

We know that component of $W$ parallel to the plane

$$= W \sin \alpha = 500 \sin 20° = 500 \times 0.342 = 171 \text{ N}$$

Since this component is more than the effort applied ($P = 50$ N), therefore the body will slide downwards. Thus the force of friction ($F = \mu \cdot R_N$) will act upwards, as shown in Fig. 6.40.

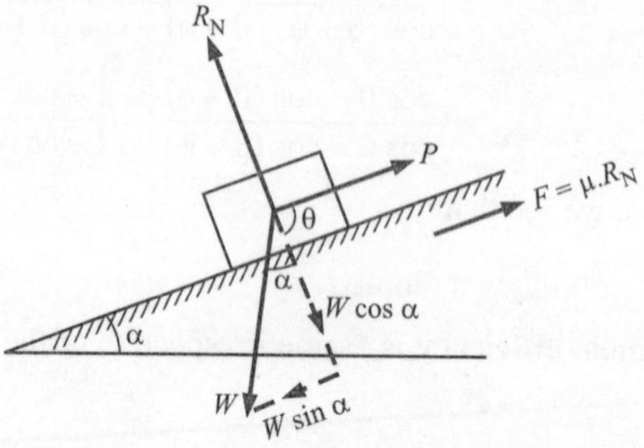


**Fig. 6.40**

Now, for the equilibrium of the body, equating components parallel to the plane, we have

$$P + \mu \cdot R_N = W \sin \alpha$$

$$50 + \mu \cdot R_N = 500 \sin 20° = 500 \times 0.342 = 171 \text{ N}$$

$$\therefore \quad \mu \cdot R_N = 171 - 50 = 121 \text{ N} \quad \quad ...(i)$$

Equating vertical components, we have

$$R_N = W \cos \alpha = 500 \cos 20° = 500 \times 0.9397 = 469.85 \text{ N}$$

Substituting the value of $R_N$ in equation ($i$),

$$\mu \times 469.85 = 121$$

$$\therefore \quad \mu = \frac{121}{469.85} = 0.2575 \quad \textbf{Ans.}$$

**Example 6.11** *An object of weight 100 N is kept in position on a plane inclined at 30° to the horizontal by horizontally applied force P. The coefficient of friction of the surface of the inclined plane is 0.25. Determine the minimum magnitude of the force P.*

***Solution*** Given : $W = 100$ N ; $\alpha = 30°$ ; $\mu = 0.25$

The object $A$ on an inclined plane is shown in Fig. 6.41.

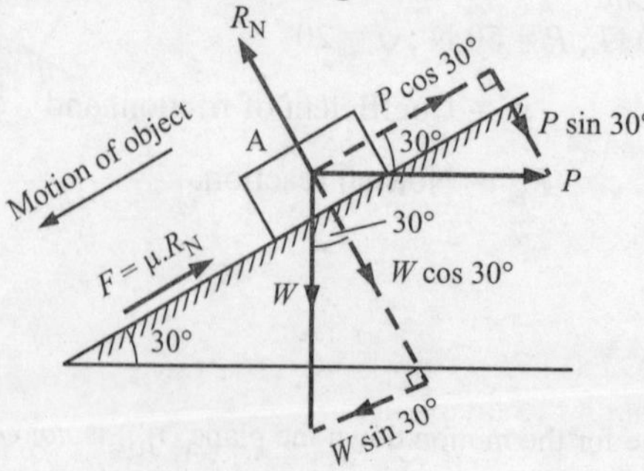


**Fig. 6.41**

The component of weight $W$ parallel to the inclined plane is $W \sin \alpha$ and perpendicular to the plane is $W \cos \alpha$. Similarly, component of force $P$ parallel to the inclined plane is $P \cos \alpha$ and perpendicular to the plane is $P \sin \alpha$. Let $R_N$ be the normal reaction exerted by the inclined plane on the body. It may be noted that for minimum value of the force $P$, the body must be at the point of sliding downwards. Thus, the force of friction, $F = \mu.R_N$ will act upwards, to keep the object in equilibrium, as shown in Fig. 6.41.

Now equating the components parallel to the plane,

$$F + P \cos 30° = W \sin 30°$$

$$\mu.R_N + P \times 0.866 = 100 \times 0.5 = 50 \quad ...(i)$$

Equating the components perpendicular to the plane,

$$R_N = W \cos 30° + P \sin 30°$$

$$= W \times 0.866 + P \times 0.5 = 100 \times 0.866 + 0.5P$$

$$= 86.6 + 0.5P \quad ...(ii)$$

Substituting the value of $R_N$ in equation (*i*), we have

$$0.25\,(86.6 + 0.5P) + 0.866P = 50$$

$$21.65 + 0.125P + 0.866P = 50 \quad \text{or} \quad 0.991P = 50 - 21.65 = 28.35$$

$$P = \frac{28.35}{0.991} = 28.6 \text{ N} \quad \textbf{Ans.}$$

**Example 6.12** *A block of stone, weighing 20 N, is resting on a plane which is inclined to the horizontal at an angle of 30°. A force (P) is acting on the block in direction parallel to the inclined plane. The coefficient of static friction between the block and the inclined plane is 0.2. Find the magnitude of P when the block is on the point of sliding: 1. up the plane; and 2. down the plane.*

***Solution*** Given : $W = 20$ N ; $\alpha = 30°$ ; $\mu = 0.2$

**1. *Force (P) when the block is on the point of sliding up the plane***

The block, as shown in Fig. 6.42 (*a*), is in equilibrium under the following forces:

(*a*) Weight of the block (20 N) acting downward,

(*b*) Normal reaction ($R_N$) acting perpendicular to the plane,

(*c*) Force ($P$) acting upwards parallel to the plane, and

(*d*) Force of friction ($F = \mu.R_N$) acting downwards parallel to the plane.

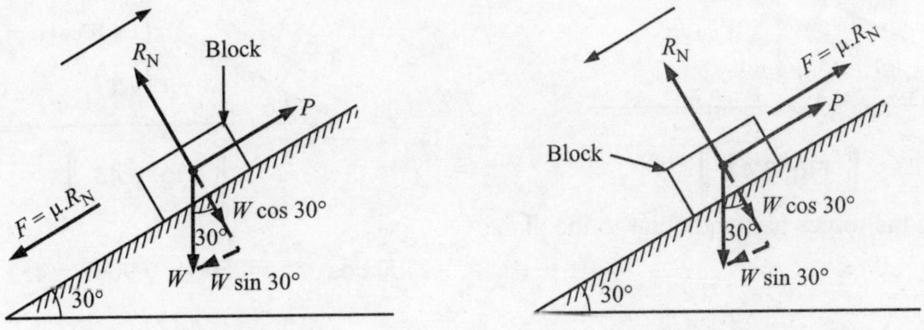


(*a*) Motion up the plane

(*b*) Motion down the plane

**Fig. 6.42**

Equating the forces perpendicular to the plane,

$$R_N = W \cos 30° = 20 \times 0.866 = 17.32 \text{ N}$$

Equating the forces parallel to the plane,

$$P = W \sin 30° + \mu \cdot R_N$$
$$= 20 \times 0.5 + 0.2 \times 17.32 = 10 + 3.464 = 13.464 \text{ N } \textbf{Ans.}$$

**2. *Force (P) when the block is on the point of sliding down the plane***

Since the motion of the block is down the plane, the force of friction ($F = \mu \cdot R_N$) will act upwards parallel to the plane, as shown in Fig. 6.42 (*b*).

Equating the forces perpendicular to the plane,

$$R_N = W \cos 30° = 20 \times 0.866 = 17.32 \text{ N}$$

and equating the forces parallel to the plane,

$$P + \mu \cdot R_N = W \sin 30°$$

or

$$P = W \sin 30° - \mu \cdot R_N$$
$$= 20 \times 0.5 - 0.2 \times 17.32 = 10 - 3.464 = 6.536 \text{ N } \textbf{Ans.}$$

**Example 6.13** *A body of weight 500 N is lying on the rough inclined plane making an angle of 25° with the horizontal. It is supported by an effort (P) parallel to the plane, as shown in Fig. 6.43. Determine the minimum and the maximum values of P for which the equilibrium can exist, if the angle of friction is 20°.*

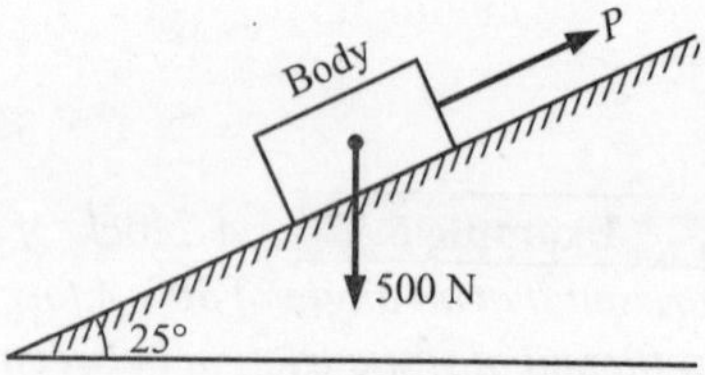


Fig. 6.43

***Solution*** Given : $W = 500$ N ; $\alpha = 25°$ ; $\phi = 20°$

We know that coefficient of friction,

$$\mu = \tan \phi = \tan 20° = 0.364$$

***Minimum value of P***

For minimum value of the effort (*P*), the body must be at the point of sliding downwards. Thus, the force of friction ($F = \mu \cdot R_N$) will act upwards, as shown in Fig. 6.44.

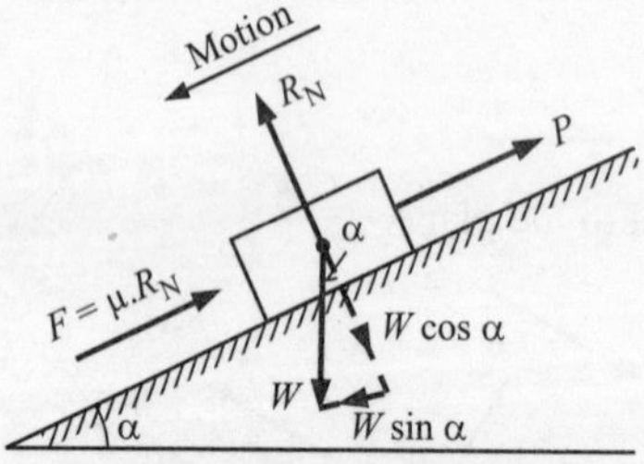


Fig. 6.44

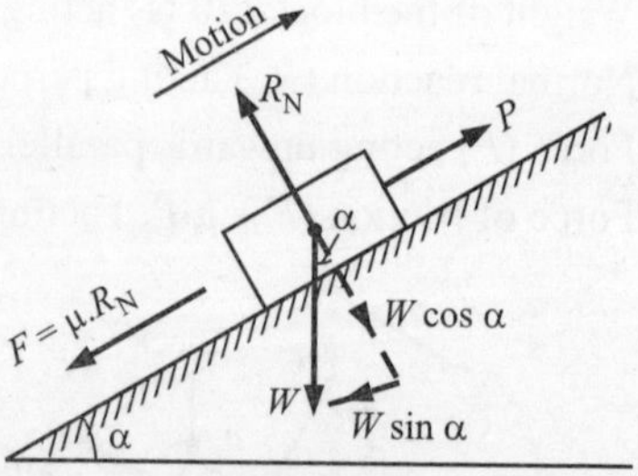


Fig. 6.45

Equating the forces perpendicular to the plane,

$$R_N = W \cos \alpha = 500 \cos 25° = 500 \times 0.9063 = 453.15 \text{ N}$$

and equating the forces parallel to the plane,

$$P + \mu \cdot R_N = W \sin \alpha$$

or

$$P = W \sin \alpha - \mu \cdot R_N$$
$$= 500 \sin 25° - 0.364 \times 453.15$$
$$= 500 \times 0.4226 - 165 = 211.3 - 165 = 46.3 \text{ N } \textbf{Ans.}$$

***Maximum value of P***

For the maximum value of $P$, the body must be at the point of sliding upwards. Thus, the force of friction ($F = \mu \cdot R_N$) will act downwards, as shown in Fig. 6.45.

Equating the forces perpendicular to the plane,

$$R_N = W \cos \alpha = 500 \cos 25° = 500 \times 0.9063 = 453.15 \text{ N}$$

and equating the forces parallel to the plane,

$$P = \mu \cdot R_N + W \sin \alpha = 0.364 \times 453.15 + 500 \sin 25°$$
$$= 165 + 500 \times 0.4226 = 165 + 211.3 = 376.3 \text{ N} \quad \textbf{Ans.}$$

It may be noted that the above example may be solved as discussed below:

1. For minimum value of $P$, the body must be at the point of sliding downwards. We know that for the motion down the plane, when $P$ acts parallel to the plane,

$$P = \frac{W \sin (\alpha - \phi)}{\cos \phi} = \frac{500 \sin (25° - 20°)}{\cos 20°} = \frac{500 \times 0.087}{0.9397}$$
$$= 46.3 \text{ N} \quad \textbf{Ans.}$$

2. For maximum value of $P$, the body must be at the point of sliding upwards. We know that for the motion up the plane, when $P$ acts parallel to the plane,

$$P = \frac{W \sin (\alpha + \phi)}{\cos \phi} = \frac{W \sin (25° + 20°)}{\cos 20°} = \frac{500 \times 0.707}{0.9397}$$
$$= 376.2 \text{ N} \quad \textbf{Ans.}$$

**Example 6.14** *A body of weight 500 N is required to just move the body up an inclined plane making an angle of 30° to the horizontal. If the coefficient of friction between the body and the plane is 0.25, find the effort required to move the body up the plane assuming that the effort is applied parallel to the plane.*

***Solution*** Given : $W = 500$ N ; $\alpha = 30°$ ; $\mu = 0.25$

Let $P$ = Effort required to move the body, parallel to the plane, and

$R_N$ = Normal reaction.

Refer Fig. 6.46.

Resolving weight of the body ($W$) into two components, *i.e.* parallel to plane ($W \sin \alpha$) and perpendicular to the plane ($W \cos \alpha$). Now equating the forces perpendicular to the plane, we have

$$R_N = W \cos \alpha = 500 \cos 30°$$
$$= 500 \times 0.866 = 433 \text{ N}$$

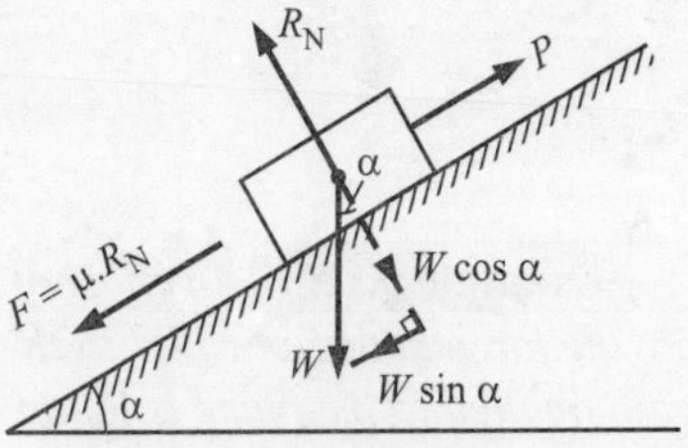


**Fig. 6.46**

and equating the forces parallel to the plane,

$$P = \mu \cdot R_N + W \sin \alpha$$
$$= 0.25 \times 433 + 500 \times \sin 30°$$
$$= 108 + 500 \times 0.5 = 358 \text{ N} \quad \textbf{Ans.}$$

**Example 6.15** *Find the force required to move a load of 30 kN on a rough inclined plane, the force being applied parallel to the plane. The inclination of the plane is such that when the same load is kept on a perfectly smooth plane inclined at the same angle, a force of 6 kN applied at an angle of 30° to the plane, keeps the same load in equilibrium. Assume that the coefficient of friction between the rough plane and the load is equal to 0.3.*

***Solution*** Given : $W = 30$ kN ; $P$ at 30° to the plane = 6 kN ; $\mu = 0.3$

Let $\alpha$ = Angle of inclination of the plane.

First of all, let us consider that the load is kept on a perfectly smooth plane, as shown in Fig. 6.47 (*a*). It may be noted that in this case, there will be no force of friction. Resolving $W$ and $P$ into two components, as shown in the figure.

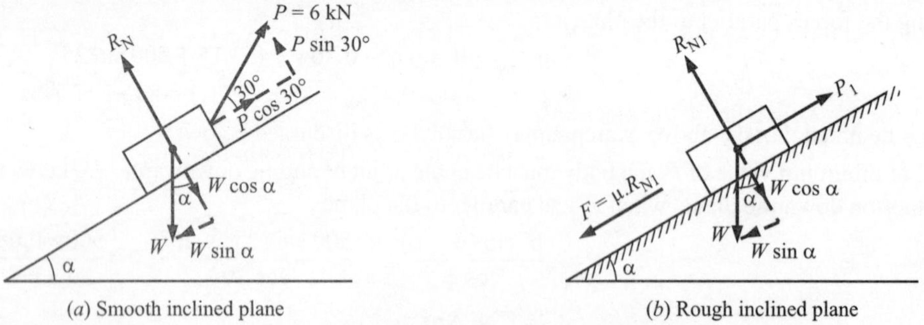


**Fig. 6.47**

Equating the forces parallel to the plane,

$$W \sin \alpha = P \cos 30°$$

$$30 \sin \alpha = 6 \cos 30° = 6 \times 0.866 = 5.196$$

$$\therefore \quad \sin \alpha = \frac{5.196}{30} = 0.1732 \text{ or } \alpha = 9.97°$$

Now considering the rough inclined plane, as shown in Fig. 6.47 (*b*). In this case, force of friction ($F = \mu \cdot R_{N1}$) will act downwards, as shown in the figure.

Equating the forces perpendicular to the plane,

$$R_{N1} = W \cos \alpha = 30 \cos 9.97° = 30 \times 0.985 = 29.55 \text{ kN}$$

and equating the forces parallel to the plane,

$$P_1 = F + W \sin \alpha = \mu . R_{N1} + W \sin \alpha$$

$$= 0.3 \times 29.55 + 30 \sin 9.97°$$

$$= 8.865 + 30 \times 0.1732 = 14.06 \text{ kN} \quad \textbf{Ans.}$$

**Example 6.16** *An effort of 200 N is required to move a certain body up an inclined plane of an angle 15°, the force acting parallel to the plane. If the angle of inclination of the plane is 20°, the effort required again applied to the plane is found to be 230 N. Find the weight of the body and the coefficient of friction.*

***Solution*** **First case:** $P_1 = 200$ N ; $\alpha_1 = 15°$

Let $W$ = Weight of the body,

$\mu$ = Coefficient of friction,

$R_{N1}$ = Normal reaction, and

$F_1$ = Force of friction = $\mu \cdot R_{N1}$

First equating the forces perpendicular to the plane [Refer Fig. 6.48 (*a*)],

$$R_{N1} = W \cos \alpha_1$$

Now equating the forces parallel to the plane,

$$P_1 = F_1 + W \sin \alpha_1 = \mu . R_{N1} + W \sin \alpha_1$$

$$200 = \mu W \cos \alpha_1 + W \sin 15° = \mu W \cos 15° + W \sin 15°$$

$$= W(\mu \cos 15° + \sin 15°) \quad ...(i)$$

**Second case:** Given: $P_2 = 230$ N; $\alpha_2 = 20°$

Let $R_{N2}$ = Normal reaction, and

$F_2$ = Force of friction = $\mu . R_{N2}$

Equating the forces perpendicular the plane [Refer Fig. 6.48 (*b*)],

$$R_{N2} = W \cos \alpha_2$$

and equating the forces parallel to the plane,

$$P_2 = F_2 + W \sin \alpha_2 = \mu \cdot R_{N2} + W \sin \alpha_2$$

$$= \mu W \cos \alpha_2 + W \sin \alpha_2$$

$$230 = \mu W \cos 20° + W \sin 20°$$

$$= W(\mu \cos 20° + \sin 20°) \quad ...(ii)$$

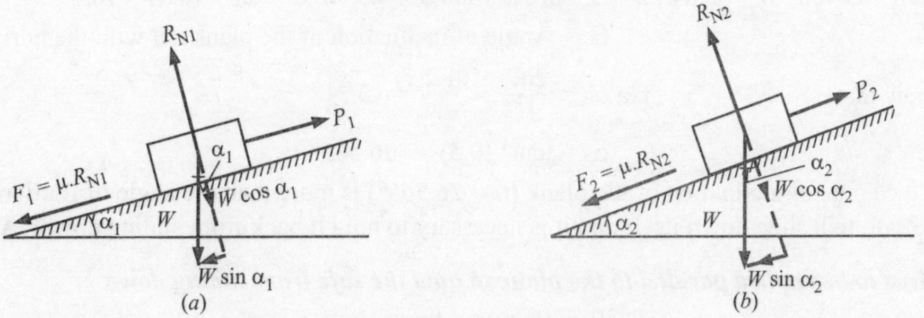


**Fig. 6.48**

***Coefficient of friction***

Dividing equation (*ii*) by equation (*i*),

$$\frac{230}{200} = \frac{W(\mu \cos 20° + \sin 20°)}{W(\mu \cos 15° + \sin 15°)}$$

$$1.15 = \frac{\mu \times 0.9397 + 0.342}{\mu \times 0.966 + 0.259}$$

$$1.15(0.966\,\mu + 0.259) = 0.9397\mu + 0.342$$

$$1.11\,\mu + 0.298 = 0.9397\mu + 0.342$$

or $$1.11\mu - 0.9397\,\mu = 0.342 - 0.298$$

$$0.1703\,\mu = 0.044$$

$$\therefore \quad \mu = \frac{0.044}{0.1703} = 0.258 \textbf{ Ans.}$$

***Weight of the body***

Substituting the value of μ in equation (*ii*),

$$230 = W(0.258 \cos 20° + \sin 20°)$$

$$= W(0.258 \times 0.9397 + 0.342) = 0.5844\,W$$

$$\therefore \quad W = \frac{230}{0.5844} = 393.6 \text{ N } \textbf{Ans.}$$

**Example 6.17** *A heavy safe of weight 4 kN is to be unloaded slowly from a truck 1.2 m high from the ground, by using an inclined plank, as shown in Fig. 6.49. The coefficient of friction between the contact surfaces of the safe and plank is 0.3. State whether it is necessary to push the safe down the plank or hold it back from sliding down. What force is required to be applied parallel to the plank for this purpose.*

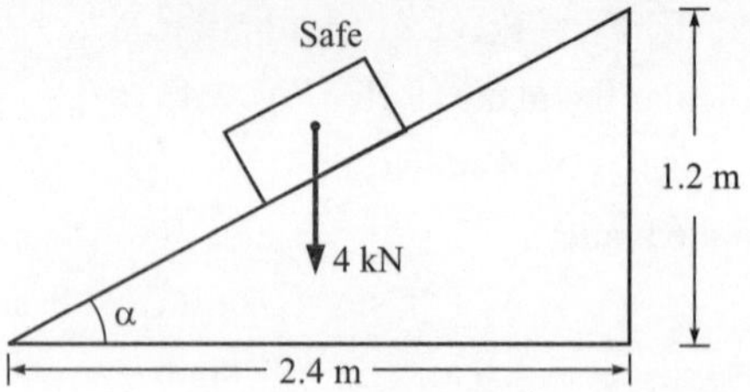


**Fig. 6.49**

***Solution*** Given : $W = 4$ kN ; $h = 1.2$ m ; $\mu = \tan\phi = 0.3$ or $\phi = \tan^{-1}(0.3) = 16.7°$

Let $\alpha$ = Angle of inclination of the plank $AB$ with the horizontal.

We know that $$\tan\alpha = \frac{BC}{AC} = \frac{1.2}{2.4} = 0.5$$

$\therefore$ $$\alpha = \tan^{-1}(0.5) = 26.565°$$

Since the angle of inclination of the plank ($\alpha = 26.565°$) is more than the angle of friction ($\phi = 16.7°$), therefore the safe will slide down itself and it is necessary to hold it back from sliding down. **Ans.**

***Force required to be applied parallel to the plane to hold the safe from sliding down***

Let $P$ = Required force.

Since the safe is to be held from sliding down the plank, therefore the force of friction ($F = \mu \cdot R_N$) will act upwards, as shown in Fig. 6.50, where $R_N$ is the normal reaction.

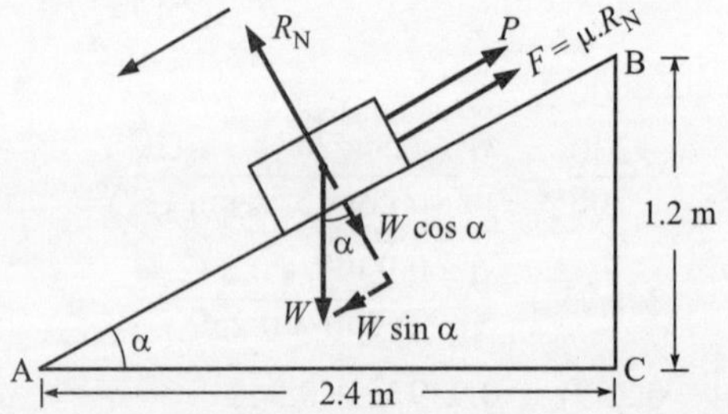


**Fig. 6.50**

Now equating the forces perpendicular to the plane,

$$R_N = W\cos\alpha = 4\cos 26.565° = 4 \times 0.8944 = 3.58 \text{ kN}$$

and equating the forces parallel to the plane,

$$P + \mu \cdot R_N = W\sin\alpha$$

$$P + 0.3 \times 3.58 = 4\sin 26.565°$$

$$P + 1.074 = 4 \times 0.4472 = 1.789$$

$\therefore$ $$P = 1.789 - 1.074 = 0.715 \text{ kN} = 715 \text{ N}$$ **Ans.**

**Example 6.18** *A block weighing 200 N is 1.8 m wide and 2.5 m high and rests on a rough horizontal floor, the coefficient of friction for which is 0.2. A pull P is applied at 25° with the horizontal floor at the middle point of the top edge, as shown in Fig. 6.51. If P is gradually increased till equilibrium is broken, state clearly whether the block will be at the point of sliding or at the point of toppling about A.*

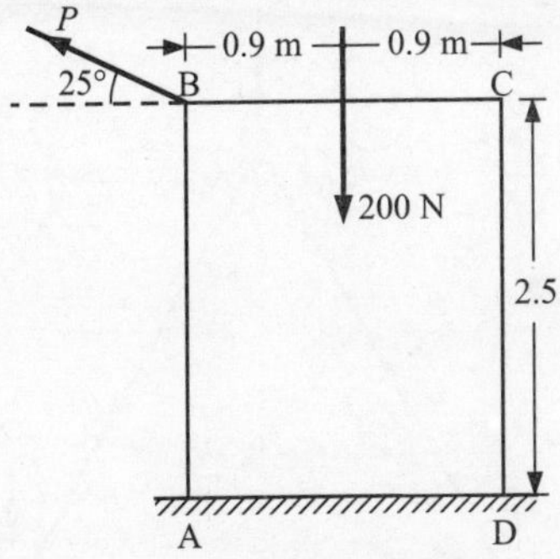


Fig. 6.51

***Solution*** Given : $W = 200$ N ; $b = 1.8$ m ; $h = 2.5$ m ; $\mu = 0.2$

Angle at which pull $P$ is applied with the horizontal floor,

$$\theta = 25°$$

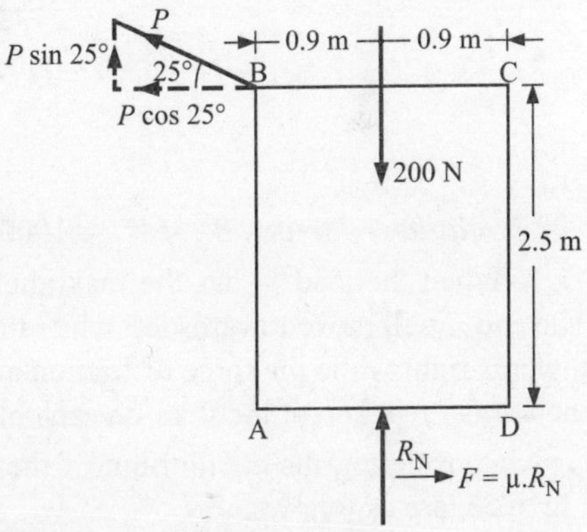


Fig. 6.52

First of all, let us consider the block at the point of sliding.

Let $R_N$ be the normal reaction exerted by the floor, as shown in Fig. 6.52. The force of friction is given by

$$F = \mu \cdot R_N = 0.2\, R_N$$

Resolving the forces horizontally, we have

$$P \cos 25° = F = 0.2 R_N$$

$$P \times 0.9063 = 0.2\, R_N$$

or

$$R_N = \frac{0.9063P}{0.2} = 4.5315P \quad ...(i)$$

Resolving the forces vertically,

$$R_N + P \sin 25° = 200$$

$$4.5315P + P \times 0.4226 = 200$$

$$4.9541P = 200$$

$$\therefore \quad P = \frac{200}{4.9541} = 40.37 \text{ N}$$

Now, let us consider the block to just topple about the edge $A$. Taking moments about the edge $A$, we have

$$P \cos 25° \times 2.5 = 200 \times 0.9$$

$$P \times 0.9063 \times 2.5 = 180 \quad \text{or} \quad 2.2657P = 180$$

$$\therefore \quad P = \frac{180}{2.2657} = 79.44 \text{ N}$$

From above, it is clear that the block will be at the point of sliding at $A$ before toppling. The maximum pull before equilibrium is just broken is $P = 40.38$ N **Ans.**

**Example 6.19** *A load W is supported at C by two symmetrically placed identical rods AC and BC hinged to each other at C, as shown in Fig. 6.53. The lower ends of the rods are hinged to two blocks each weighing 1000 N and resting on a rough horizontal floor, whose coefficient of friction is 0.25. Ignoring the weight of the rods, find the maximum value of W.*

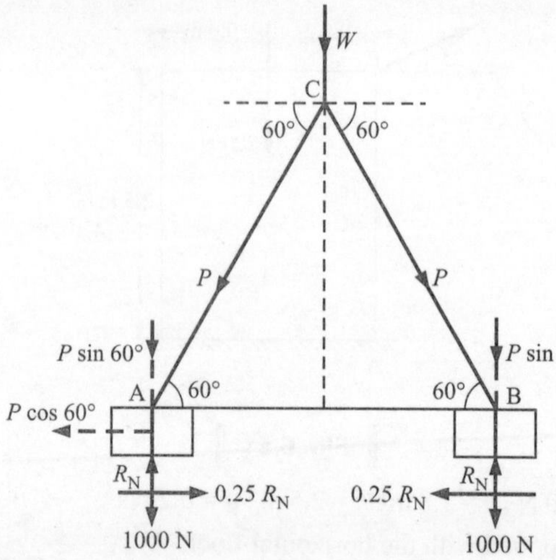


**Fig. 6.53**

***Solution*** Given : $W_A = W_B = 1000$ N ; $\mu = 0.25$

When the load $W$ has the maximum value, then each of the blocks will be at the point of slipping. The end $A$ will move towards left while the end $B$ will move towards right. The force of friction at $A$ will act towards right while the force of friction at $B$ will act towards left. Let $P$ be the thrust in each rod and $R_N$ be the normal reaction of the floor on each block.

Considering the equilibrium of the block $A$. The various forces acting on the block $A$, as shown in Fig. 6.53, are as follows:

(*a*) Weight of block $A$ (1000 N), acting downwards,

(*b*) Normal reaction $R_N$ acting upwards,

(*c*) Force of friction ($F = \mu \cdot R_N = 0.25\, R_N$) acting towards right,

(*d*) Horizontal component of thrust $P$ (*i.e.* $P \cos 60°$) acting towards left, and

(*e*) Vertical component of thrust $P$ (*i.e.* $P \sin 60°$) acting downwards.

Now equating horizontal forces, we have

$$0.25\, R_N = P \cos 60° = 0.5\, P$$

or $$R_N = \frac{0.5P}{0.25} = 2P \qquad ...(i)$$

Equating vertical forces, we have

$$P \sin 60° + 1000 = R_N$$

$$P \times 0.866 + 1000 = 2P \qquad ...(\because\ R_N = 2P)$$

or $$2P - 0.866P = 1000$$

$$1.134\, P = 1000$$

$\therefore$ $$P = \frac{1000}{1.134} = 881.83 \text{ N}$$

Considering the equilibrium at point $C$, we have

$$W = 2P \sin 60° = 2 \times 881.83 \times 0.866 = 1527.33 \text{ N} \quad \textbf{Ans.}$$

**Example 6.20** *Two blocks A and B, as shown in Fig. 6.54, weigh 75 N and 275 N, respectively. Assume the coefficient of friction between block A and the plane to be 0.25 and that between block B and the plane to be 0.10. Calculate the force between the two as they slide down the incline.*

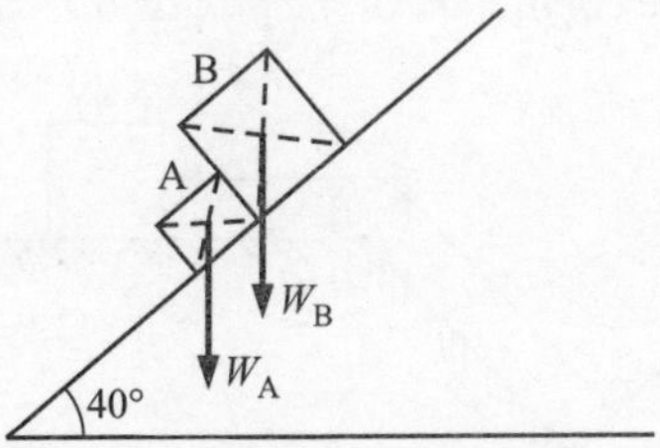


Fig. 6.54

***Solution*** Given : $W_A = 75$ N ; $W_B = 275$ N ; $\mu_A = 0.25$ ; $\mu_B = 0.10$

Refer Fig. 6.55.

Let $P$ = Force between the two as they slide down the incline, and

$R_A$ and $R_B$ = Normal reaction of the plane on the blocks $A$ and $B$, respectively.

We know that force of friction on block $A$,

$$F_A = \mu_A \cdot R_A = 0.25 \times W_A \cos 40° \quad ...(\because R_A = W_A \cos 40°)$$
$$= 0.25 \times 75 \times 0.766 = 14.36 \text{ N}$$

Similarly, force of friction on block $B$,

$$F_B = \mu_B \cdot R_B = 0.10 \times W_B \cos 40°$$
$$= 0.10 \times 275 \times 0.766 = 21.06 \text{ N}$$

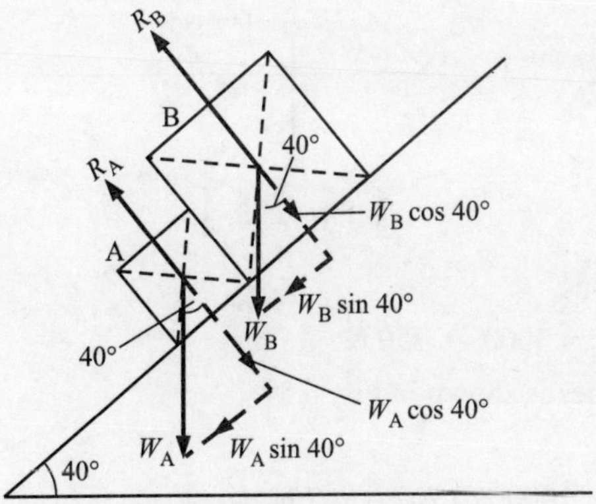


Fig. 6.55

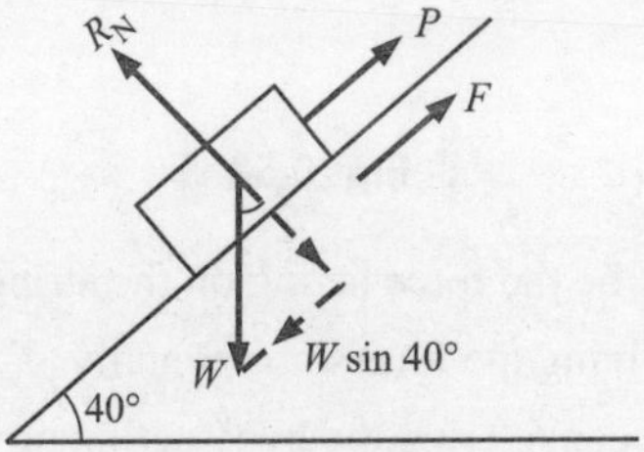


Fig. 6.56

Considering that the two blocks $A$ and $B$ form a single body, as shown in Fig. 6.56.

∴ Total weight of the blocks $A$ and $B$,

$$W = W_A + W_B = 75 + 275 = 350 \text{ N}$$

and total force of friction on the system,

$$F = F_A + F_B = 14.36 + 21.06 = 35.42 \text{ N}$$

∴ Net force on the blocks parallel to the plane

$$P = W \sin 40° - F$$
$$= 350 \times 0.643 - 35.42 = 225.05 - 35.42 = 189.63 \text{ N} \textbf{ Ans.}$$

**Example 6.21** *Two bodies of weight 1 kN and 3 kN are acting on a rough inclined plane and a horizontal plane, respectively, as shown in Fig. 6.57. Find the necessary force P acting parallel to the plane to cause the motion to impend. Assume coefficient of friction as 0.25 and the pulley to be smooth.*

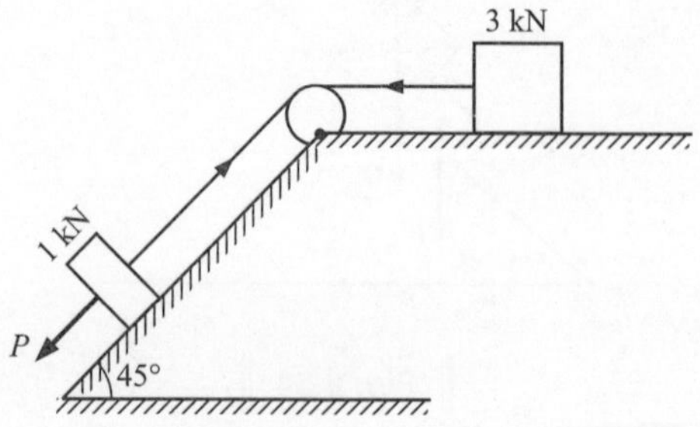


**Fig. 6.57**

***Solution*** Given : $W_1 = 1$ kN = 1000 N ; $W_2 = 3$ kN = 3000 N ; $\mu = 0.25$

First of all, let us consider the body acting on the horizontal plane, as shown in Fig. 6.58.

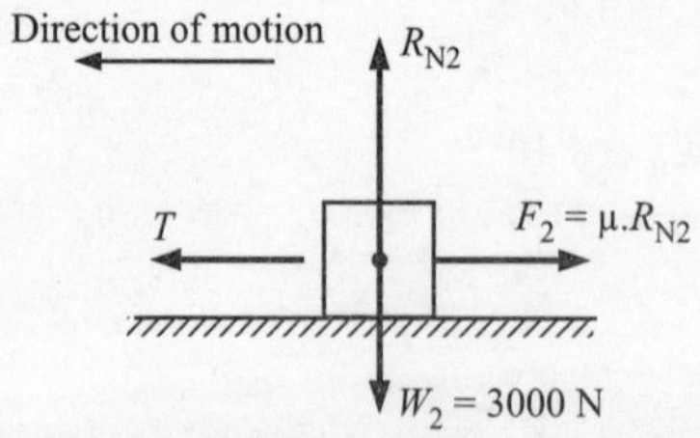


**Fig. 6.58**

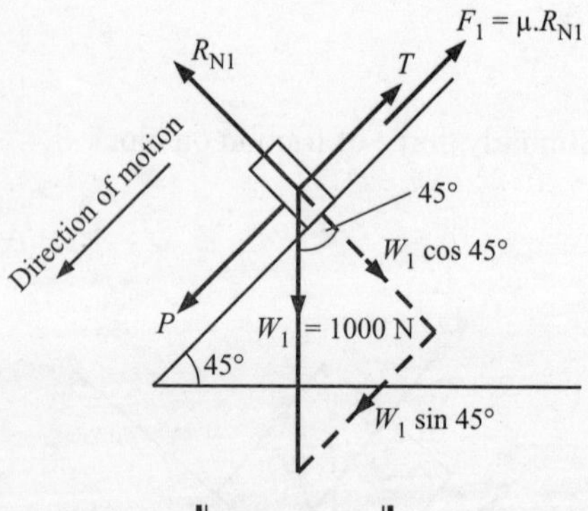


**Fig. 6.59**

Let $T$ be the force acting on the string to move the body.

Resolving the forces horizontally, $T = \mu.R_{N2} = 0.25 \times 3000 = 750$ N

Now considering the body acting on the inclined plane, as shown in Fig. 6.59.

Resolving the forces perpendicular to the plane,

$$R_{N1} = W_1 \cos 45° = 1000 \times 0.707 = 707 \text{ N}$$

Now resolving the forces parallel to the plane,

$$T + \mu.R_{N1} = P + W_1 \sin 45°$$
$$750 + 0.25 \times 707 = P + 1000 \times 0.707$$
$$750 + 176.75 = P + 707$$

∴ $$P = 750 + 176.75 - 707 = 219.75 \text{ N} \textbf{ Ans.}$$

**Example 6.22** *Two blocks A and B are connected to each other by a rope which passes over a fixed drum. Find the least value of the force (P) that will just start the system of the blocks, as shown in Fig. 6.60, moving to the right. The weights of the blocks A and B are 3 kN and 5 kN, respectively. The coefficient of friction under each block and between the rope and fixed drum is 0.20.*

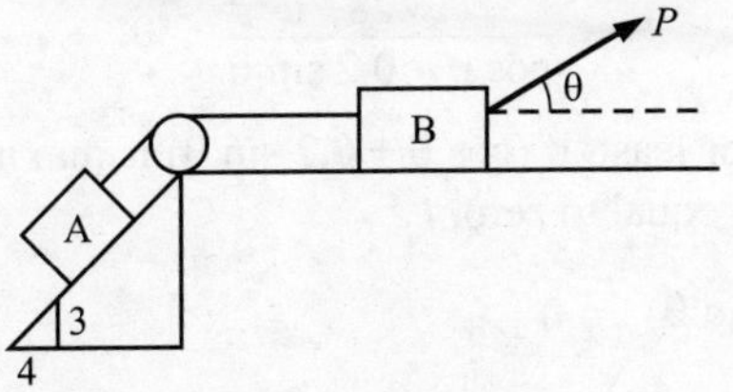


**Fig. 6.60**

***Solution*** Given : $W_1 = 3$ kN ; $W_2 = 5$ kN ; $\mu = 0.20$

First of all, let us consider the block acting on the inclined plane, as shown in Fig. 6.61. From the geometry of the figure,

$$\tan\alpha = \frac{3}{4};\ \sin\alpha = \frac{3}{5} \ \text{and}\ \cos\alpha = \frac{4}{5}$$

Let $T$ be the force in the rope to move the block upwards. The force of friction, $F = \mu.R_{N1}$ will act downwards, as shown in the figure.

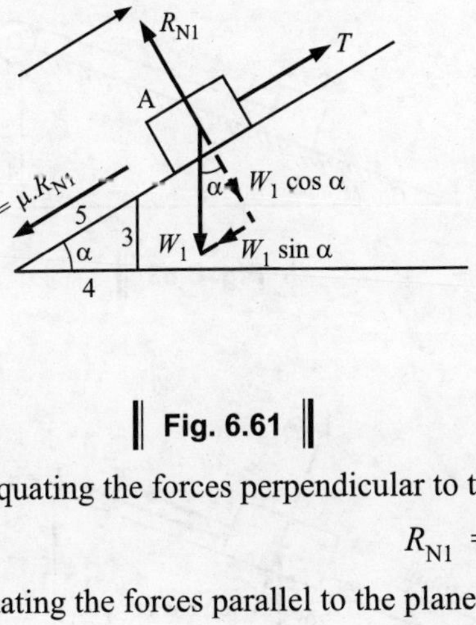


**Fig. 6.61**

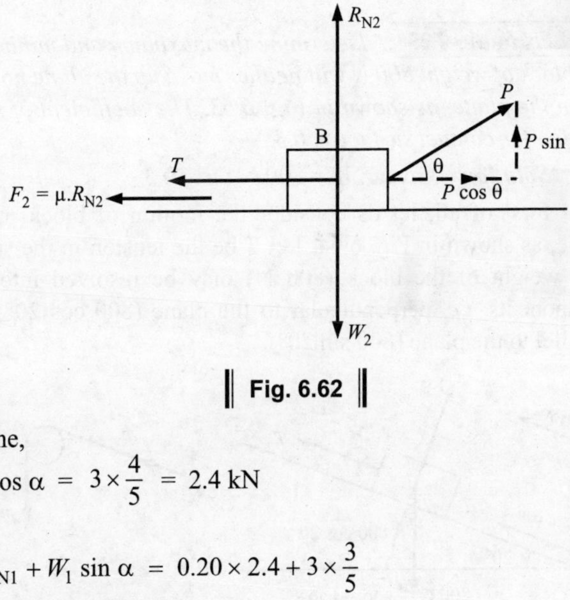


**Fig. 6.62**

Equating the forces perpendicular to the plane,

$$R_{N1} = W_1\cos\alpha = 3\times\frac{4}{5} = 2.4 \text{ kN}$$

and equating the forces parallel to the plane,

$$T = \mu.R_{N1} + W_1\sin\alpha = 0.20\times 2.4 + 3\times\frac{3}{5}$$

$$= 0.48 + 1.8 = 2.28 \text{ kN}$$

Now consider the block $B$ acting on the horizontal plane, as shown in Fig. 6.62. Since the block $B$ tends to move rightward, therefore force of friction, $F_2 = \mu\cdot R_{N2}$ will act leftward, as shown in the figure.

Equating the forces perpendicular to the plane,

$$R_{N2} + P\sin\theta = W_2$$

$$\therefore \quad R_{N2} = W_2 - P\sin\theta = 5 - P\sin\theta \qquad \text{...}(i)$$

and equating the forces parallel to the plane,

$$T + \mu\cdot R_{N2} = P\cos\theta$$

$$2.28 + 0.20\,(5 - P\sin\theta) = P\cos\theta$$
$$2.28 + 1 - 0.2\,P\sin\theta = P\cos\theta$$
or
$$P\cos\theta + 0.2\,P\sin\theta = 3.28$$
$$P\,(\cos\theta + 0.2\sin\theta) = 3.28$$
$$\therefore \quad P = \frac{3.28}{\cos\theta + 0.2\sin\theta} \qquad ...(ii)$$

The force $P$ will be minimum or least, if $(\cos\theta + 0.2\sin\theta)$ is maximum. For $(\cos\theta + 0.2\sin\theta)$ to be maximum, its differentiation must be equal to zero, *i.e.*

$$\frac{d}{d\theta}(\cos\theta + 0.2\sin\theta) = 0$$
or
$$-\sin\theta + 0.2\cos\theta = 0$$
$$0.2\cos\theta = \sin\theta$$
or
$$\frac{\sin\theta}{\cos\theta} = \tan\theta = 0.2$$
$$\therefore \quad \theta = \tan^{-1}(0.2) = 11.31°$$

Substituting the value of $\theta$ in equation (*ii*), the minimum value of $P$ is

$$P_{(min)} = \frac{3.28}{\cos 11.31° + 0.2\sin 11.31°}$$
$$= \frac{3.28}{0.98 + 0.2 \times 0.196} = \frac{3.28}{1.0192} = 3.218 \text{ kN } \textbf{Ans.}$$

**Example 6.23** *Determine the maximum and minimum values which the weight w may have so that the block of weight 800 N will neither move up the plane nor slip down the plane, as shown in Fig. 6.63. The coefficient of static friction for contact surface is 0.3.*

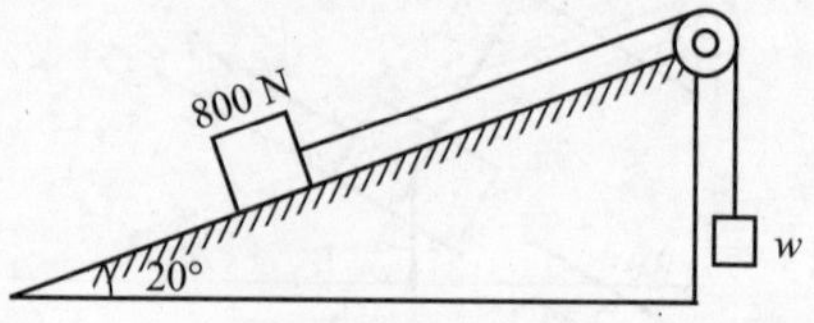

**Fig. 6.63**

***Solution*** Given : $W = 800$ N ; $\mu = 0.3$

First of all, let us consider the motion of block up the plane, as shown in Fig. 6.64. Let $T$ be the tension in the string. The weight of the block (800 N) may be resolved into two components, *i.e.* perpendicular to the plane (800 cos 20°) and parallel to the plane (800 sin 20°).

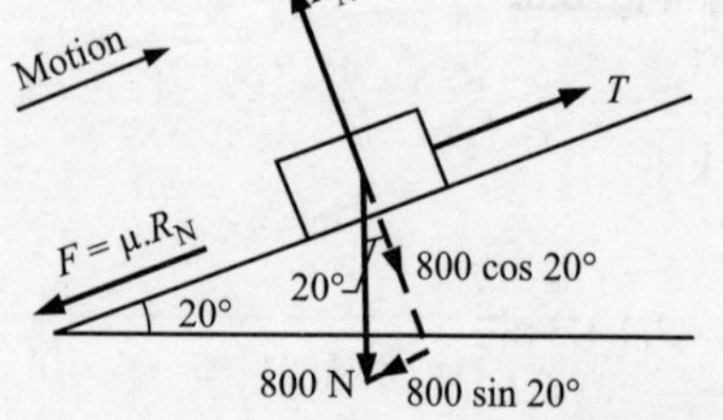

**Fig. 6.64**

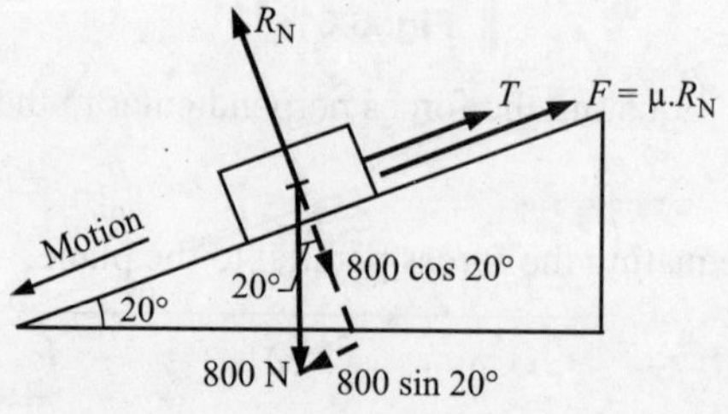

**Fig. 6.65**

Now equating the forces perpendicular to the plane, we have

$$R_N = 800\cos 20° = 800 \times 0.9397 = 751.76 \text{ N}$$

and equating the forces parallel to the plane,

$$T = \mu.R_N + 800\sin 20°$$
$$= 0.3 \times 751.76 + 800 \times 0.342 = 225.53 + 273.6 = 499.13 \text{ N}$$

Since the tension ($T$) in the string will be equal to the weight ($w$), therefore, maximum value of the weight,

$$w = T = 499.13 \text{ N} \quad \textbf{Ans.}$$

Now let us consider motion of the block down the plane, as shown in Fig. 6.65. First equating the forces perpendicular to the plane,

$$R_N = 800 \cos 20° = 800 \times 0.9397 = 751.76 \text{ N}$$

and equating the forces parallel to the plane,

$$\mu \cdot R_N + T = 800 \sin 20°$$

$$0.3 \times 751.76 + T = 800 \times 0.342$$

$$225.53 + T = 273.6$$

or $$T = 273.6 - 225.53 = 48.07 \text{ N}$$

Since $T = w$, therefore minimum value of the weight = 48.07 N **Ans.**

**Example 6.24** *A body of weight 1000 N is resting on a rough inclined plane which rises 10 mm for every 40 mm of its length. If the coefficient of friction is 0.2, find the effort required to move the body up the plane when it is applied : (a) horizontally; and (b) parallel to the plane.*

***Solution*** Given : $W = 1000$ N ; $\mu = \tan \phi = 0.2$ or $\phi = \tan^{-1} (0.2) = 11.31°$

Since the rise of inclined plane is 10 mm for every 40 mm of its length, as shown in Fig. 6.66, therefore inclination of the plane with the horizontal is given by

$$\sin \alpha = \frac{10}{40} = 0.25$$

$$\therefore \quad \alpha = \sin^{-1} (0.25) = 14.48°$$

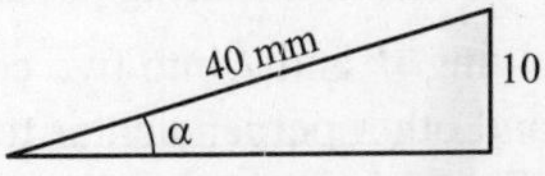


Fig. 6.66

***(a) Effort required when it is applied horizontally***

Let $P$ = Effort required.

We know that when the effort ($P$) is applied horizontally, as shown in Fig. 6.67, then

$$P = W \tan (\alpha + \phi) = 1000 \tan (14.48° + 11.31°) = 1000 \tan 25.79°$$

$$= 1000 \times 0.4832 = 483.2 \text{ N} \quad \textbf{Ans.}$$

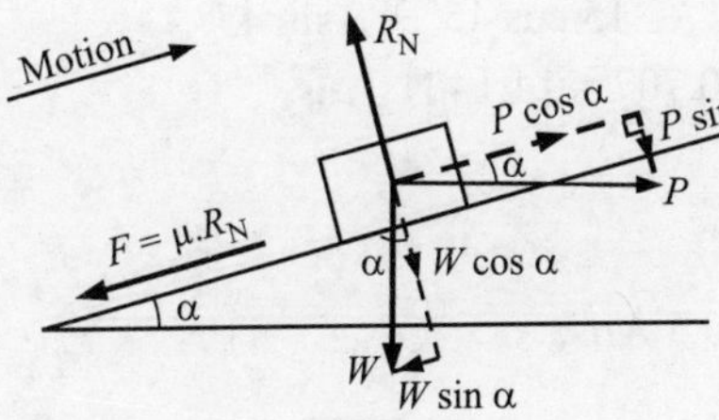


Fig. 6.67

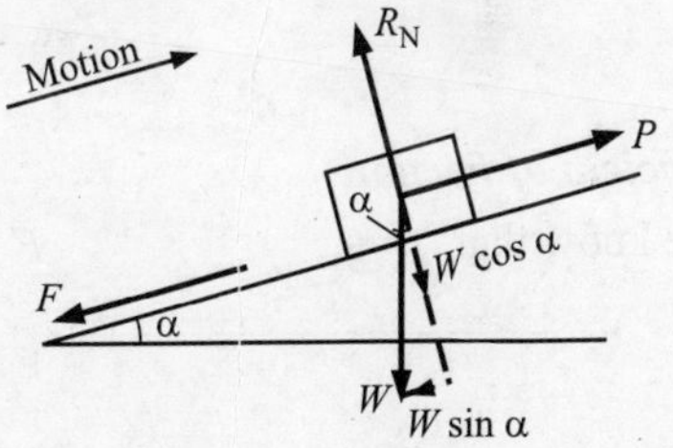


Fig. 6.68

***(b) Effort required when it is applied parallel to the plane***

Let $P$ = Effort required.

We know that when the effort ($P$) is applied parallel to the plane, as shown in Fig. 6.68, then

$$P = \frac{W \sin (\alpha + \phi)}{\cos \phi} = \frac{1000 \sin (14.48° + 11.31°)}{\cos 11.31°}$$

$$= \frac{1000 \times 0.4351}{0.98} = 444 \text{ N} \quad \textbf{Ans.}$$

**Example 6.25** *A rectangular prism block weighing 15 N is resting on a rough inclined plane, as shown in Fig. 6.69. The block is tied up by a horizontal string which has a tension of 5 N. Find :*

*1. Frictional force on the block ;*

*2. Normal reaction of the inclined plane ; and*

*3. Coefficient of friction between the surfaces of contact.*

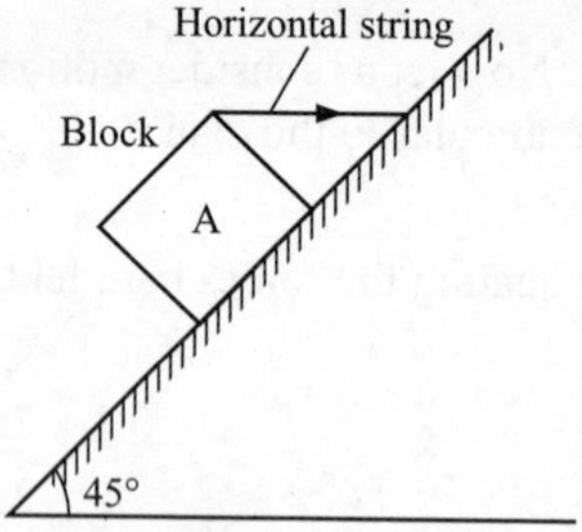


**Fig. 6.69**

***Solution*** Given : $W = 15$ N ; $T = 5$ N ; $\alpha = 45°$

**1. *Frictional force on the block***

Let $F$ = Frictional force on the block $= \mu \cdot R_N$,

$R_N$ = Normal reaction of the inclined plane, and

$\mu$ = Coefficient of friction between the contact surfaces.

It may be noted that if the string is removed, the block will slide down the plane. Therefore force of friction ($F = \mu \cdot R_N$) will act upward, as shown in Fig. 6.70.

The block is in equilibrium under the action of weight of the block ($W$), tension in the string ($T$) and the normal reaction ($R_N$).

Resolving $W$ and $T$ into two components, *i.e.* one parallel to the plane and other perpendicular to the plane. Now equating the forces parallel to the plane,

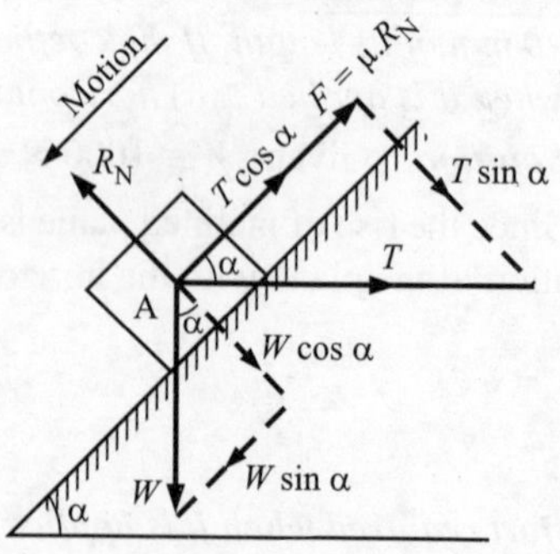


**Fig. 6.70**

$$F + T\cos\alpha = W\sin\alpha$$

or

$$F = W\sin\alpha - T\cos\alpha$$
$$= 15\sin 45° - 5\cos 45°$$
$$= 15 \times 0.707 - 5 \times 0.707 = 7.07 \text{ N} \quad \textbf{Ans.}$$

**2. *Normal reaction of the inclined plane***

Equating the forces perpendicular to the plane,

$$R_N = W\cos\alpha + T\sin\alpha = 15\cos 45° + 5\sin 45°$$
$$= 15 \times 0.707 + 5 \times 0.707 = 14.14 \text{ N} \quad \textbf{Ans.}$$

**3. *Coefficient of friction***

We know that

$$F = \mu \cdot R_N$$

$$\therefore \quad \mu = \frac{F}{R_N} = \frac{7.07}{14.14} = 0.5 \quad \textbf{Ans.}$$

**Example 6.26** *Two blocks, one of weight $W_A$ and the other of weight 2000 N, are supported on two rough planes and connected by a horizontal link AB, as shown in Fig. 6.71. The coefficient of friction for block of weight $W_A$ on the horizontal plane is 0.4 and the angle of friction for the block of 2000 N on the inclined plane is 15°. What is the smallest weight $W_A$ for the equilibrium of the system.*

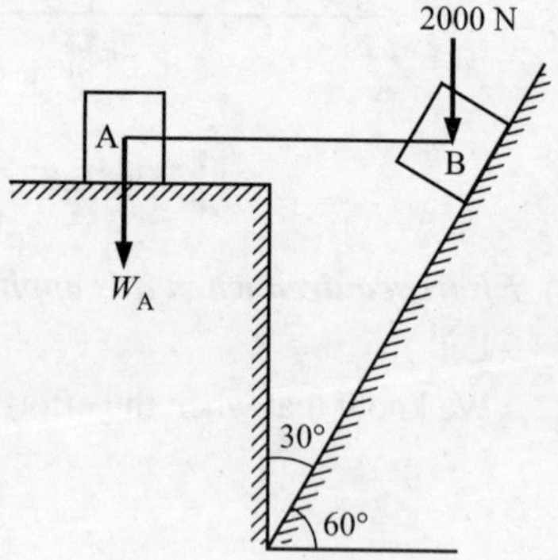


**Fig. 6.71**

***Solution*** Given : $W_B = 2000$ N ; $\mu_A = 0.4$ ; $\phi_B = 15°$

We know that coefficient of friction for the block $B$,

$$\mu_B = \tan\phi_B = \tan 15° = 0.268$$

For the smallest weight ($W_A$) of the block $A$ required for the equilibrium, the block $B$ will slide downwards and the block $A$ will move towards left. Thus the force of friction ($F_A = \mu \cdot R_{NA}$) on the block $A$ will be acting towards right and force of friction ($F_B = \mu.R_{NB}$) will act upwards, as shown in Fig. 6.72.

We know that force of friction acting on block $A$,

$$F_A = \mu_A \cdot R_{NA} = \mu_A \cdot W_A = 0.4W_A$$

This force is transmitted to the block $B$. Now the block $B$ is in equilibrium under the action of the following forces, as shown in Fig. 6.73.

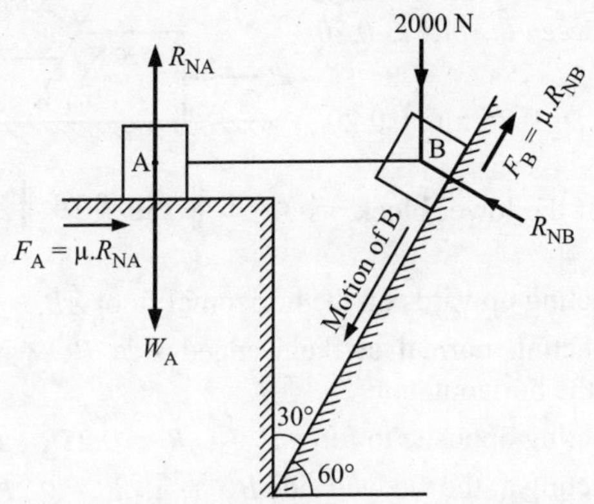


**Fig. 6.72**

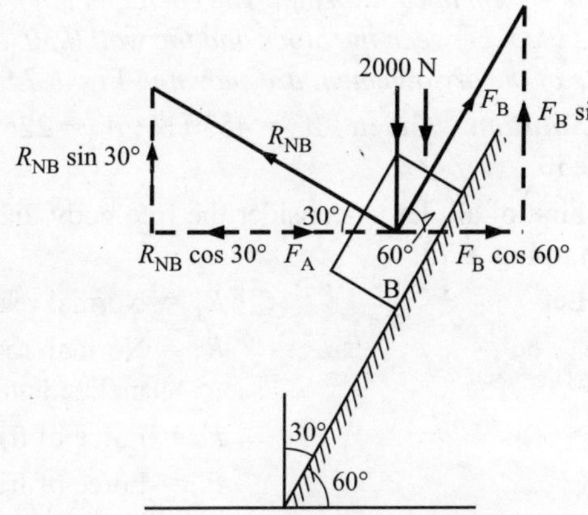


**Fig. 6.73**

(*a*) Weight of the block $B$ ($W_B$ = 2000 N),

(*b*) Normal reaction ($R_{NB}$) of the inclined plane, which can be resolved into two components, *i.e.* horizontal component ($R_{NB}$ cos 30°) and vertical component ($R_{NB}$ sin 30°),

(*c*) Force of friction ($F_B = \mu_B \cdot R_{NB}$) acting up the inclined plane, which can be resolved into two components, *i.e.* horizontal component ($F_B$ cos 60°) and vertical component ($F_B$ sin 60°), and

(*d*) Horizontal force ($F_A = 0.4W_A$).

Now equating the horizontal components,

$$F_A + F_B \cos 60° = R_{NB} \cos 30°$$

$$0.4W_A + \mu_B \cdot R_{NB} \cos 60° = R_{NB} \cos 30°$$

$$0.4W_A + 0.268R_{NB} \times 0.5 = R_{NB} \times 0.866$$

$$0.4W_A + 0.134R_{NB} = 0.866R_{NB}$$

or

$$W_A = \frac{0.866R_{NB} - 0.134R_{NB}}{0.4} = 1.83R_{NB}$$

∴

$$R_{NB} = \frac{W_A}{1.83} = 0.546W_A$$

Equating vertical components,

$$R_{NB} \sin 30° + F_B \sin 60° = 2000$$

$$R_{NB} \times 0.5 + \mu_B \cdot R_{NB} \times 0.866 = 2000$$

$$0.5R_{NB} + 0.268\, R_{NB} \times 0.866 = 2000$$

$$0.5 \times 0.546W_A + 0.268 \times 0.546W_A \times 0.866 = 2000$$

$$0.273W_A + 0.1267W_A = 2000$$

$$0.3997 W_A = 2000$$

$$\therefore \quad W_A = \frac{2000}{0.3997} = 5003.75 \text{ N} \quad \textbf{Ans.}$$

**Example 6.27** *A heavy block of weight 4530 N resting on a horizontal floor supports another block of weight 2265 N against the vertical wall and the wedged side. Find the minimum value of a horizontal force (P) applied to the lower block that will hold the system in equilibrium. The coefficient of friction between the block and floor is 0.25, between the block and the wall 0.30, and between the blocks 0.20. Details of the arrangement are shown in Fig. 6.74.*

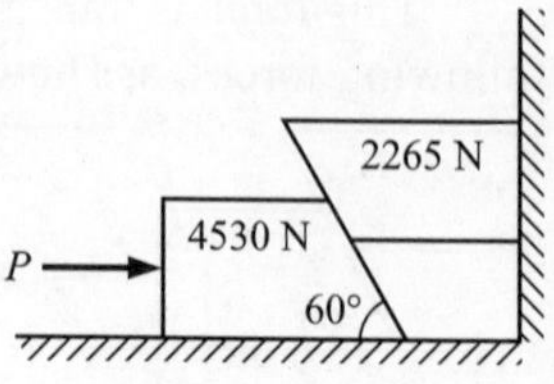


**Fig. 6.74**

***Solution*** Given : $W_1 = 4530$ N ; $W_2 = 2265$ N ; $\mu_1 = 0.25$ ; $\mu_2 = 0.20$ ; $\mu_3 = 0.30$

First of all, let us consider the free body diagram of the lower block, as shown in Fig. 6.75.

Let $R_1$ = Normal reaction acting upwards on the horizontal floor $AB$,

$R_2$ = Normal reaction acting normal at the wedged side $BC$ which is inclined at 30° to the horizontal,

$F_1$ = Force of friction acting opposite to force $P = \mu_1.R_1 = 0.25R_1$, and

$F_2$ = Force of friction acting at the wedged side $BC$ = $\mu_2.R_2 = 0.2R_2$.

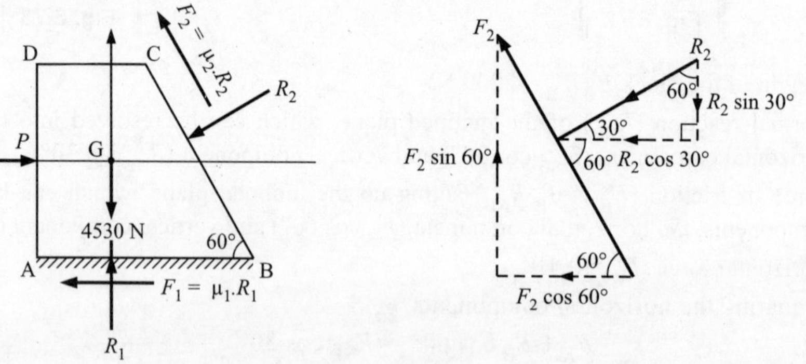


**Fig. 6.75** *Lower block*

Resolving the forces $R_2$ and $F_2$ into horizontal and vertical components, as shown in Fig. 6.75. Considering the equilibrium of the lower block and first of all equating the horizontal components, we have

$$P = F_1 + F_2 \cos 60° + R_2 \cos 30°$$

$$= \mu_1.R_1 + \mu_2.R_2 \cos 60° + R_2 \cos 30°$$

$$= 0.25R_1 + 0.2R_2 \times 0.5 + R_2 \times 0.866$$

$$= 0.25R_1 + 0.966R_2 \quad \text{...(i)}$$

Now equating vertical components,

$$R_1 + F_2 \sin 60° = 4530 + R_2 \sin 30°$$

$$R_1 + \mu_2.R_2 \sin 60° = 4530 + R_2 \sin 30°$$

$$R_1 + 0.2R_2 \times 0.866 = 4530 + R_2 \times 0.5$$

$$R_1 + 0.1732R_2 - 0.5R_2 = 4530$$

$$R_1 - 0.3268R_2 = 4530 \quad \text{...(ii)}$$

The free body diagram of the upper block is shown in Fig. 6.76.

Let $R_3$ = Normal reaction at the wall,

$F_3$ = Force of friction at the wall = $\mu_3 \cdot R_3 = 0.3\, R_3$

Resolving the forces $R_2$ and $F_2$ into horizontal and vertical components, as shown in Fig. 6.76. Considering the equilibrium of the upper block.

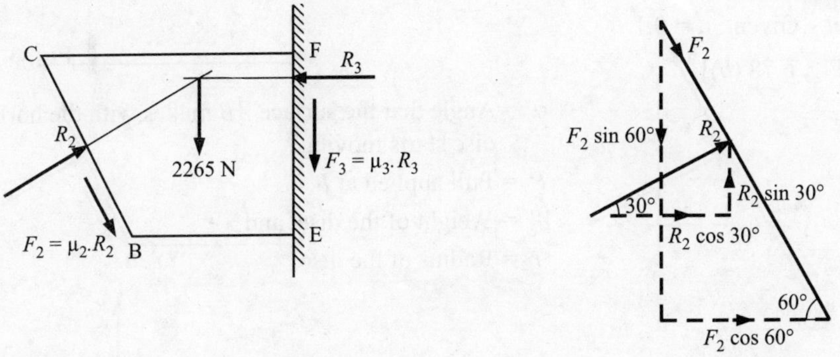


**Fig. 6.76** *Upper block*

Equating the horizontal components, we have

$$R_2 \cos 30° + F_2 \cos 60° = R_3$$

$$R_2 \cos 30° + \mu_2 \cdot R_2 \cos 60° = R_3$$

$$R_2 \times 0.866 + 0.2 R_2 \times 0.5 = R_3$$

$$0.966 R_2 = R_3 \qquad \ldots(iii)$$

Equating the vertical components,

$$R_2 \sin 30° = F_2 \sin 60° + F_3 + 2265$$

$$= \mu_2 \cdot R_2 \sin 60° + \mu_3 \cdot R_3 + 2265$$

$$R_2 \times 0.5 = 0.2 R_2 \times 0.866 + 0.3 R_3 + 2265$$

$$= 0.1732 R_2 + 0.3 \times 0.966 R_2 + 2265 \qquad \ldots\text{[From equation } (iii)\text{]}$$

$$= 0.1732 R_2 + 0.2898 R_2 + 2265$$

$$= 0.4630 R_2 + 2265$$

$$0.5 R_2 - 0.463 R_2 = 2265$$

$$0.037 R_2 = 2265$$

$$\therefore \quad R_2 = \frac{2265}{0.037} = 61\,216 \text{ N}$$

Substituting the value of $R_2$ in equation (*ii*),

$$R_1 - 0.3268 \times 61\,216 = 4530$$

$$R_1 - 20\,005 = 4530$$

$$\therefore \quad R_1 = 20\,005 + 4530 = 24\,535 \text{ N}$$

Now substituting $R_1$ and $R_2$ in equation (*i*),

$$P = 0.25 \times 24\,535 + 0.966 \times 61\,216$$

$$= 6133.75 + 59\,134.65 = 65\,268.4 \text{ N} \quad \textbf{Ans.}$$

**Example 6.28** *A homogeneous semi-circular disc of uniform thickness rests on a horizontal surface, as shown in Fig. 6.77. The coefficient of friction between the disc and the surface is 0.1. A pull whose direction always remains horizontal, is applied to one edge, in order to move the disc along the surface. Determine the angle that the surface AB will make with the horizontal just before the disc starts moving.*

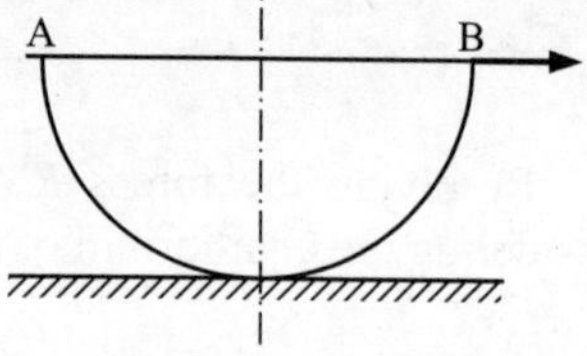


**Fig. 6.77**

***Solution*** Given : $\mu = 0.1$

[Refer Fig. 6.78 (*b*)]

Let $\alpha$ = Angle that the surface *AB* makes with the horizontal before the disc starts moving,

$P$ = Pull applied at *B*,

$W$ = Weight of the disc, and

$r$ = Radius of the disc.

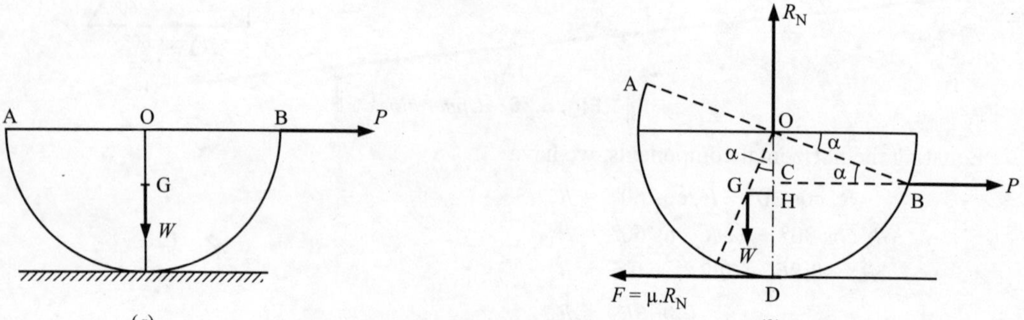


**Fig. 6.78**

The weight of the disc ($W$) will act at its centre of gravity ($G$) whose value is given by

$$OG = \frac{4r}{3\pi}$$

Since the pull ($P$) is applied at $B$, therefore force of friction ($F = \mu \cdot R_N$) will act at the contact surface of the disc and horizontal surface (*i.e.* at $D$). The normal reaction acts at $O$.

Now equating horizontal forces, we have

$$P = F = \mu \cdot R_N \qquad ...(i)$$

and equating vertical forces,

$$W = R_N \qquad ...(ii)$$

Taking moment of the forces about $O$,

$$F \times OD = P \times OC + W \times GH$$

$$F \times r = P \times OB \sin\alpha + W \times OG \sin\alpha \qquad ...(\because\ OD = OB = r)$$

$$\mu.R_N \times r = P \times r \sin\alpha + W \times \frac{4r}{3\pi} \times \sin\alpha$$

$$= \mu.R_N \times r \sin\alpha + R_N \times \frac{4r}{3\pi} \times \sin\alpha \qquad ...\text{[From equations } (i) \text{ and } (ii)]$$

$$\mu.r = \mu \cdot r \sin\alpha + \frac{4r}{3\pi} \times \sin\alpha = r \sin\alpha \left(\mu + \frac{4}{3\pi}\right)$$

$$\mu = \sin\alpha\left(\mu + \frac{4}{3\pi}\right) = \sin\alpha\,(0.1 + 0.424)$$

$$0.1 = 0.524 \sin\alpha$$

$$\sin\alpha = \frac{0.1}{0.524} = 0.1908$$

$$\therefore \quad \alpha = 11° \quad \textbf{Ans.}$$

## HIGHLIGHTS

1. When a body is made to slide over the surface of another body, then some resistance is encountered, which is known as *friction* and the force which resists the motion is known as *force of friction*.
2. The force of friction always acts in opposite direction to the relative motion of the body at the point of contact and it is always tangential to the contacting surfaces.
3. The *static friction* is the friction experienced by a body, when it is at rest.
4. The *kinetic* or *dynamic friction* is the friction experienced by a body, when it is in motion.
5. The *limiting friction* ($F$) is the maximum value of the frictional force (or the opposing force) which comes into play when the body just begins to slide over the surface of the other body.
6. An upward reaction perpendicular to the surface of the body is known as normal reaction ($R_N$).
7. The ratio of the limiting force of friction ($F$) to the normal reaction ($R_N$) is called *coefficient of friction.* It is denoted by μ (Mu). Mathematically,

$$\mu = \frac{F}{R_N}$$

8. The angle between the normal reaction ($R_N$) and the resultant ($R$) of the force of friction ($F$) and the normal reaction ($R_N$) is known as *angle of friction*. It is denoted by ϕ (Phi). Mathematically,

$$\tan\phi = \frac{F}{R_N} = \frac{\mu \cdot R_N}{R_N} = \mu$$

9. In case of an inclined plane, the angle of inclination of the plane (say α) is equal to the angle of friction (ϕ), *i.e.* $\alpha = \phi$.
10. When the angle of inclination of the plane with the horizontal (*i.e.* α) is such that the body begins to move itself down the plane, then the angle α is called *angle of repose.*
11. When the angle of inclination of the plane with the horizontal (*i.e.* α) is *less than* the angle of friction (*i.e.* ϕ), then the body will remain in equilibrium without any external force. If the body is required to move upwards or downwards in this condition, then the external force is required.
12. When the angle of inclination of the plane with the horizontal (*i.e.* α) is *more than* the angle of friction (*i.e.* ϕ), then the body will not remain in equilibrium. The body will move downwards and an upward external force is required to keep the body in equilibrium.
13. The minimum force ($P_{min}$) required to slide a body of weight $W$, on a rough horizontal plane is given by

$$P_{min} = W \sin\phi$$

where ϕ is the angle of friction.

**14.** The force required to move the body of weight $W$ on an inclined plane, when the force is acting at an angle to the inclined plane is given by

$$P = \frac{W \sin \alpha}{\sin (\theta - \alpha)}$$ ...(when friction between the body and the inclined plane is neglected)

$$= \frac{W \sin (\alpha + \phi)}{\sin [\theta - (\alpha + \phi)]}$$ ...(when friction between the body and the inclined plane is taken into account)

where $\alpha$ = Angle of inclination of the plane with the horizontal,

$\theta$ = Inclination of the force $P$ with the line of action of weight $W$ of the body, and

$\phi$ = Angle of friction.

The following points are worthnoting:

**(*a*)** When the force ($P$) is acting horizontally, *i.e.* when $\theta = 90°$. Substituting the value of $\theta$ in the above expression, we have

$$P = \frac{W \sin \alpha}{\sin (90° - \alpha)} = \frac{W \sin \alpha}{\cos \alpha} = \tan \alpha$$

...(when friction is neglected)

$$= \frac{W \sin (\alpha + \phi)}{\sin [90° - (\alpha + \phi)]} = \frac{W \sin (\alpha + \phi)}{\cos (\alpha + \phi)} = W \tan (\alpha + \phi)$$

...(when friction is taken into account)

**(*b*)** When the force ($P$) is acting along the inclined plane, then $\theta = 90° + \alpha$. Substituting the value of $\theta$ in the above expression, we have

$$P = \frac{W \sin \alpha}{\sin [90° + \alpha - \alpha]} = \frac{W \sin \alpha}{\sin 90°} = W \sin \alpha$$

...(when friction is neglected)

$$= \frac{W \sin (\alpha + \phi)}{\sin [(90° + \alpha) - (\alpha + \phi)]} = \frac{W \sin (\alpha + \phi)}{\sin (90° - \phi)} = \frac{W \sin (\alpha + \phi)}{\cos \phi}$$

...(when friction is taken into account)

**15.** In order to prevent the motion of the body down the plane, the force or effort required,

$$P = \frac{W \sin \alpha}{\sin (\theta - \alpha)}$$ ...(when friction is neglected)

$$= \frac{W \sin (\alpha - \phi)}{\sin [\theta - (\alpha - \phi)]}$$ ...(when friction is taken into account)

The following points are worth noting for motion of a body down the plane:

**(*a*)** When the force ($P$) is acting horizontally, then $\theta = 90°$. Substituting the value of $\theta$ in the above expression, we have

$$P = \frac{W \sin \alpha}{\sin [90° - \alpha]} = \frac{W \sin \alpha}{\cos \alpha} = W \tan \alpha$$

...(when friction is neglected)

$$= \frac{W \sin (\alpha - \phi)}{\sin [90° - (\alpha - \phi)]} = \frac{W \sin (\alpha - \phi)}{\cos (\alpha - \phi)} = W \tan (\alpha - \phi)$$

...(when friction is taken into account)

(*b*) When the force (*P*) is acting along the inclined plane, then $\theta = 90° + \alpha$. Substituting the value of $\theta$ in the above expression, we have

$$P = \frac{W \sin \alpha}{\sin (90° + \alpha - \alpha)} = \frac{W \sin \alpha}{\sin 90°} = W \sin \alpha$$

...(when friction is neglected)

$$= \frac{W \sin (\alpha - \phi)}{\sin [(90° + \alpha) - (\alpha - \phi)]} = \frac{W \sin (\alpha - \phi)}{\cos \phi}$$

...(when friction is taken into account)

**16.** The efficiency of inclined plane, when the body moves up the plane is given by

$$\eta_U = \frac{\cot (\alpha + \phi) - \cot \theta}{\cot \alpha - \cot \theta}$$

The following points are worthnoting:

(*a*) When the force (*P*) is acting horizontally, then $\theta = 90°$. In such a case,

$$\eta_U = \frac{\tan \alpha}{\tan (\alpha + \phi)}$$

(*b*) When force (*P*) is applied along (or parallel to) the plane, then $\theta = 90° + \alpha$. In such a case,

$$\eta_U = \frac{\sin \alpha \cos \phi}{\sin (\alpha + \phi)}$$

**17.** The efficiency of the inclined plane, when the body moves down the plane is given by

$$\eta_D = \frac{\cot \alpha - \cot \theta}{\cot (\alpha - \phi) - \cot \theta}$$

The following points are worthnoting:

(*a*) When the force (*P*) is acting horizontally, then $\theta = 90°$. In such a case,

$$\eta_D = \frac{\tan (\alpha - \phi)}{\tan \alpha}$$

(*b*) When the force (*P*) is acting along (or parallel to) the plane, then $\theta = 90° + \alpha$. In such a case,

$$\eta_D = \frac{\sin (\alpha - \phi)}{\sin \alpha \cos \phi}$$

**18.** The efficiency of the inclined plane for the motion up the plane, is maximum, when the angle of inclination of plane with the horizontal is

$$\alpha = \frac{\theta - \phi}{2}$$

and maximum efficiency is given by

$$\eta_{U(max)} = \frac{1 - \cos (\theta - \phi)}{1 - \cos (\theta + \phi)}$$

When force is applied horizontally, then $\theta = 90°$, and

$$\eta_{U(max)} = \frac{1 - \sin \phi}{1 + \sin \phi}$$

In the similarly way, when force is applied along the plane, then $\theta = 90° + \alpha$, and

$$\eta_{U(max)} = \frac{1 + \sin(\alpha - \phi)}{1 + \sin(\alpha + \phi)}$$

**19.** The efficiency of the inclined plane for the motion down the plane is maximum, when the angle of inclination of the plane with the horizontal is

$$\alpha = \frac{\theta + \phi}{2}$$

and maximum efficiency is given by

$$\eta_{D(max)} = \frac{1 - \cos(\theta - \phi)}{1 - \cos(\theta + \phi)} = \eta_{U(max)}$$

This shows that the maximum efficiency is same irrespective of the motion of the body up or down the inclined plane. It may be noted here that $\eta_D$ is not equal to $\eta_U$ (*i.e.* $\eta_D \neq \eta_U$).

## NUMERICAL PROBLEMS

1. A body of weight 200 N is lying on a rough horizontal surface. The coefficient of friction between the body and the horizontal rough surface is 0.3. Determine the force required to just move the body, while acting at an angle of 20° with the horizontal. [**Ans.** 57.6 N]
2. A body of weight 70 N is placed on a rough horizontal plane. To just move the body on the horizontal plane, a push of 20 N inclined at 20° to the horizontal plane is required. Find coefficient of friction. [**Ans.** 0.297]
3. A body lying on a rough horizontal plane is moved slowly along the plane when a horizontal force of 100 N is applied to the body. Also, a force of 90 N is applied to the body at an angle of 30° to the horizontal and moves the body slowly along the same plane. Find the weight of the body and the coefficient of friction. [**Ans.** 204.1 N; 0.49]
4. A body resting on a rough horizontal plane required a pull of 150 N inclined at 30° to the plane just to move it. It is found that a push of 200 N inclined at 30° to the plane, just moved the body. Determine the weight of the body and the coefficient of friction. [**Ans.** 601.3 N; 0.247]
5. A body of weight 100 N rests on a rough horizontal plane having a coefficient of friction as 0.3. It is acted upon by a force applied at an angle of 30° to the horizontal. What force is required to just cause the body to slide over the surface?

   Determine, also, the inclination and magnitude of the minimum force required to set the body into impending motion. [**Ans.** 29.52 N; 16.7°, 28.74 N]
6. Two blocks *A* and *B* of weight 1000 N and 2000 N, respectively, are positioned, as shown in Fig. 6.79. If the coefficient of friction for all the contact surfaces is 0.3, find the force *P* required to move the block *B* and tension in the string. [**Ans.** 1111 N; 851.7 N]

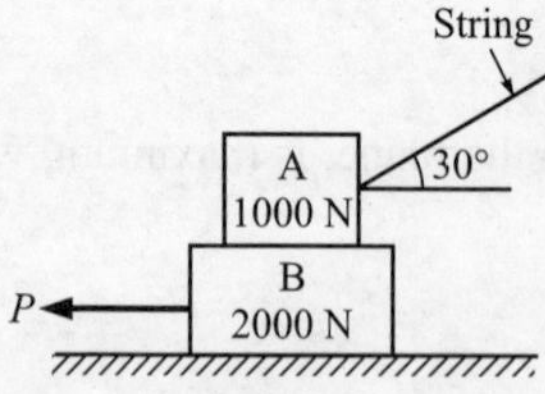


**Fig. 6.79**

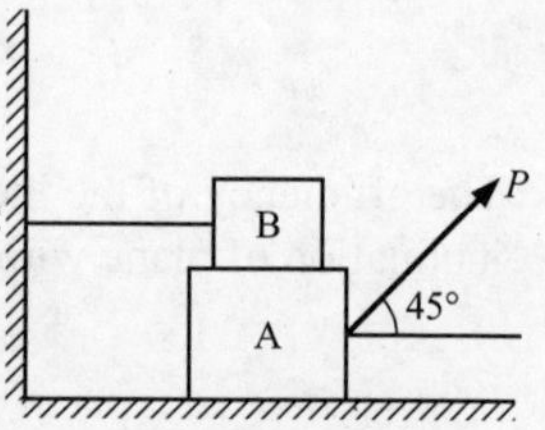


**Fig. 6.80**

7. A block *B* which is resting on block *A* is attached by a horizontal rope *BC* to the wall, as shown in Fig. 6.80. What force *P* is necessary to cause motion of *A* to impend. The coefficient of friction between *A* and *B* is 0.25 and that between *A* and the floor is 0.33. The block *A* weighs 300 N and the block *B* weighs 200 N. [**Ans.** 228.6 N]

**8.** A body of weight 500 N is at rest on a rough inclined plane which makes an angle of 30° from the horizontal. Determine the minimum horizontal force which makes the body to move upward on the plane. Take $\mu = 0.25$. **[Ans.** 163.7 N]

**9.** An effort of 350 N is required to just move a body of weight 500 N, up an inclined plane of an angle 30°. When the effort is acting parallel to the plane, determine the coefficient of friction between the body and the plane. **[Ans.** 0.231]

**10.** A body weighing 100 N is lying on a rough inclined plane, making an angle of 12° with the horizontal. If the coefficient of friction is 0.2, find the effort required to drag the body up the plane when the effort is applied (*a*) horizontally; and (*b*) parallel to the plane. **[Ans.** 43.1 N; 40.38 N]

**11.** A body of weight 40 kN rests in limiting equilibrium on a rough plane whose slope is 30°. The plane is raised to a slope of 60°. What force applied to the body parallel to the inclined plane will support the body on the plane? **[Ans.** 23.1 N]

[**Hint.** Angle 30° is the angle of repose = angle of friction ($\phi$). Therefore, coefficient of friction ($\mu$) = tan $\phi$ = tan 30°]

**12.** A body of weight 30 kN is on the point of motion down a rough inclined plane when supported by a force of 12 kN acting parallel to the plane. When a force of 18 kN is applied parallel to the plane, the body is at the point of motion up the inclined plane. Find the coefficient of friction between the body and the inclined plane. **[Ans.** 0.115]

**13.** Find the force acting parallel to the inclined plane required to move a body of weight of 500 N up the plane. The inclination of the plane is such that a force of 100 N applied to the body at an angle 30° could support it, if the plane were smooth. Assume coefficient of friction as 0.25. **[Ans.** 209.625 N]

**14.** A body of weight $W$ is supported by friction on an inclined plane which is inclined at an angle $\alpha$ with the horizontal. Prove that the body cannot be moved up the plane by any horizontal force whose value is less than $W \tan 2\alpha$.

**15.** Two blocks $A$ and $B$ weighing 45 N and 90 N, respectively, slide down together under gravity down an inclined plane that makes with the horizontal an angle of 30°. The block $A$ is behind the block $B$. If the values of coefficient of friction between the block $A$ and the plane, and between the block $B$ and the plane are 0.15 and 0.3, respectively, find the pressure existing between the blocks during the motion. **[Ans.** 38.273 N]

**16.** A block $A$ rests on a rough horizontal surface and is connected to another block $B$ standing on a rough inclined plane, by means of a rope passing over a frictionless pulley, as shown in Fig. 6.81. What minimum force $P$ applied to the block $B$, parallel to the plane will cause the blocks to move? The weight of block $A$ is 3 kN and weight of block $B$ is 1 kN. The coefficient of friction for all contacting surfaces is 0.25. **[Ans.** 219.75 N]

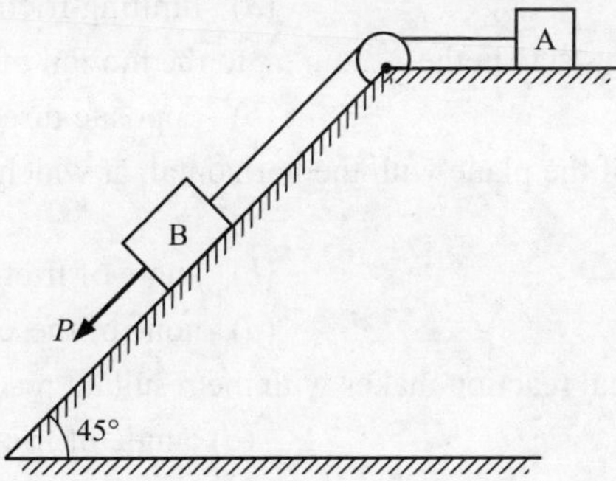


**Fig. 6.81**

## REVIEW QUESTIONS

**1.** 'Friction is a friend and also a foe'. Give three distinct engineering examples to justify each of the above two points of view.

**2.** What do you understand by the term force of friction?

**3.** State the various types of friction.

**4.** Define the following:

(*a*) Static friction; (*b*) Kinetic friction; and (*c*) Limiting friction.

**5.** Explain the following:

(*a*) Angle of friction; (*b*) Coefficient of static friction; (*c*) Coefficient of kinetic friction; (*d*) Cone of friction; and (*e*) Angle of repose.

**6.** Prove that when a body is about to slide down a rough inclined plane, the inclination of the plane to the horizontal is equal to the angle of friction.

**7.** State and explain the laws of static friction, kinetic friction and solid friction.

**8.** State Coulomb's law of dry friction.

**9.** State the conditions of equilibrium of a body on an inclined plane when no other external forces are acting on the body.

**10.** Derive an expression for the minimum force required to slide a body on a rough horizontal plane.

**11.** Show that the force or effort ($P$) required to move the body of weight ($W$) up an inclined plane, when force is acting horizontally is

$$P = W \tan(\alpha + \phi)$$

where $\alpha$ = Angle of inclination of the inclined plane with the horizontal,

$\phi$ = Angle of friction.

**12.** Prove that the efficiency of the inclined plane for the motion up the plane, is maximum, when angle of inclination of the plane with the horizontal is

$$\alpha = \frac{\theta - \phi}{2}$$

where $\theta$ is the angle between the effort applied ($P$) and the line of action of load ($W$).

## OBJECTIVE TYPE QUESTIONS

**1.** The maximum frictional force, which comes into play, when a body just begins to slide over the surface of another body, is called

(*a*) coefficient of friction (*b*) static friction

(*c*) dynamic friction (*d*) limiting friction

**2.** The force of friction always acts in the ............... to the motion of the body.

(*a*) same direction (*b*) opposite direction

**3.** The angle of inclination of the plane with the horizontal, at which the body begins to move down the plane, is called

(*a*) angle of projection (*b*) angle of friction

(*c*) angle of repose (*d*) none of these

**4.** The angle which the normal reaction makes with the resultant reaction is called

(*a*) angle of projection (*b*) angle of friction

(*c*) angle of repose (*d*) none of these

**5.** The dynamic or kinetic friction is always ............... static friction.

(*a*) less than (*b*) greater than

**6.** The minimum force required to slide a body of weight ($W$) on a rough horizontal plane is

(*a*) $W \sin\phi$ (*b*) $W \cos\phi$

(*c*) $W \tan\phi$ (*d*) none of these

where $\phi$ = Angle of friction.

**7.** The static friction

(*a*) always acts in the same direction in which the body tends to move

(*b*) depends upon the area of contact between the two surfaces

(*c*) bears a constant ratio to the normal reaction between the two surfaces

(*d*) all of the above

**8.** Which of the following statements regarding laws governing the friction between dry surfaces are correct?

(*a*) The friction force is dependent on materials of the contact surfaces.

(*b*) The friction force is directly proportional to the normal force.

(*c*) The friction force is independent of the area of contact.

(*d*) All of the above.

**9.** A body will begin to move down an inclined plane, if the angle of inclination of the plane is ............... the angle of friction.

(*a*) equal to (*b*) less than (*c*) greater than

**10.** A body of weight $W$ is required to move up the rough inclined plane whose angle of inclination with the horizontal is $\alpha$. The effort applied parallel to the plane is given by

(*a*) $P = W \tan \alpha$ (*b*) $P = W \tan (\alpha + \phi)$

(*c*) $P = W (\cos \alpha + \mu \sin \alpha)$ (*d*) $P = W (\sin \alpha + \mu \cos \alpha)$

where $\mu = \tan \phi$ = Coefficient of friction between the plane and the body.

***Answers to Objective Type Questions***

| | | | | |
|---|---|---|---|---|
| **1.** (*d*) | **2.** (*b*) | **3.** (*c*) | **4.** (*b*) | **5.** (*a*) |
| **6.** (*a*) | **7.** (*c*) | **8.** (*d*) | **9.** (*c*) | **10.** (*d*) |

7

# FRICTION—II (APPLICATIONS)



## 7.1 INTRODUCTION

We have discussed the principles of friction in the previous chapter. The applications of friction such as screw friction, ladder friction and wedge friction are discussed, in detail, in this chapter.

## 7.2 SCREW FRICTION

The screws, nuts, bolts, studs etc. are widely used in various machines and structures for *temporary fastenings. These fastening have screw threads, which are made by cutting a continuous helical groove on a cylindrical surface, as shown in Fig. 7.1. It may be noted that if one complete turn of a screw thread be imagined to be unwound from the body of the screw and developed, it will form an inclined plane whose base $AC$ is equal to the circumference of the cylinder (*i.e.* $\pi d$), where $d$ is the mean diameter of the screw and its altitude $AB$ is equal to the **pitch ($p$) of the screw in case of single-start threads and $\alpha$ is the angle of helix. Thus, a screw is an inclined plane when developed.

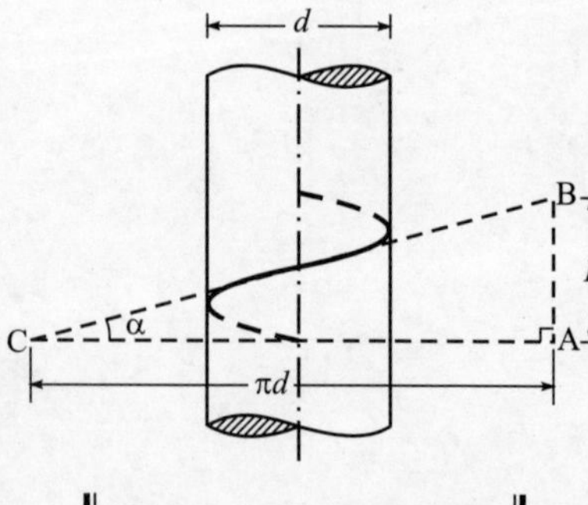


**Fig. 7.1** *Screw friction*

The screw threads when cut on the outer surface of a solid rod, then these are known as *external threads.* When the threads are cut on the internal surface of a hollow rod, then these are known as *internal threads.* The screw threads are mainly of two types, *i.e.* V-threads and square threads. The V-threads are stronger and offer more frictional resistance to motion than square threads. In general, the V-threads are used for the purpose of tightening pieces together, *e.g.* bolts and nuts etc. But the square-threads are used for the power transmission in the machine elements such as screw jack, vice screws etc.

The screws may be right handed or left handed according to its movement.

* The temporary fastenings are those fastenings which can be disassembled without destroying the connecting components.

** In case of multi-start threads, it is equal to lead ($l$) of the screw.

## 7.3 IMPORTANT TERMS OF SCREW

The following terms are important in the study of screw:

1. ***Helix*** It is the curve traced by a particle while moving along a screw thread.
2. ***Pitch*** It is the axial distance from a point on a screw thread to a corresponding point on the next thread. It is denoted by $p$.
3. ***Lead*** It is the distance moved by the screw when rotated in a fixed nut through one complete revolution. For the single start threads, lead and pitch are the same. For double start threads, lead is equal to twice the pitch and so on. Mathematically,

$$\text{Lead, } l = n \cdot p$$

where $n$ = Number of starts.

4. ***Lead or helix angle*** It is the slope or inclination of the thread with the horizontal. Mathematically,

$$\tan\alpha = \frac{\text{Lead of screw}}{\text{Circumference of screw}} = \frac{l}{\pi d}$$

$$= \frac{p}{\pi d} \qquad \text{...(For single start threads)}$$

We have already discussed that for single start threads,

$$l = p$$

and for multi-start threads, $l = n \cdot p$

where $n$ = Number of starts, and

$d$ = Mean diameter of the screw.

## 7.4 SCREW JACK

The screw jack, as shown in Fig. 7.2, is a device for lifting heavy loads, by applying a comparatively smaller effort at its handle. It works on the same principle as an inclined plane. A common type of a screw jack has square threads.

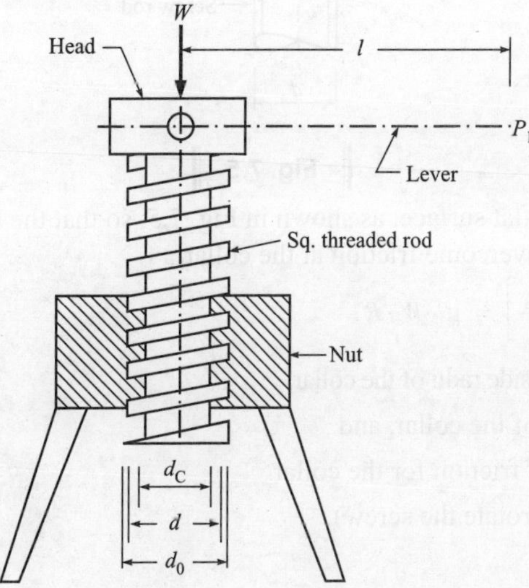


**Fig. 7.2** *Screw jack*

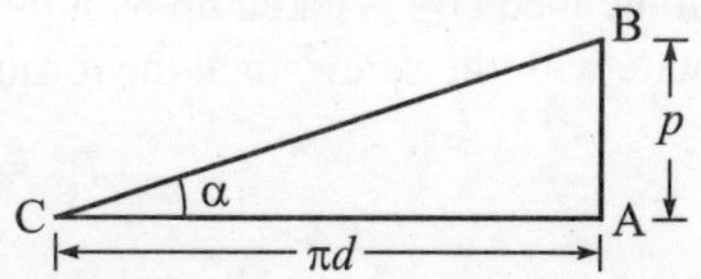


**Fig. 7.3** *Development of a screw*

If one complete turn of a screw thread is unwound and developed, it will form an inclined plane, as shown in Fig. 7.3, in which $AB$ is the pitch ($p$), $CA$ is the circumference ($\pi d$), where $d$ is the mean diameter of the screw. The angle $\alpha$ is the helix angle.

Since the screw jack works on the same principle as that of an inclined plane, therefore we shall discuss the following two cases:

**1. *When the load (W) is being lifted or raised***

The force or effort ($P$) applied on the lever may be considered as horizontal, as shown in Fig. 7.4. When the load is being lifted, then the condition is similar to the body moving up the inclined plane when the force is acting horizontally. Proceeding in the same way as discussed in Chapter 6 (Sec. 6.17), we find that the force required to overcome friction between the screw and nut, or the force required at the circumference of the screw to lift the load,

$$P = W \tan (\alpha + \phi)$$

where $\alpha$ = Helix angle, and

$\phi$ = Angle of friction.

The turning moment or torque ($T$) required to overcome friction between the screw and nut is given by

$$T = P \times \frac{d}{2} = W \tan (\alpha + \phi) \frac{d}{2}$$

If an effort $P_1$ is applied at the end of the lever at a distance $l$ from the axis of the screw, then

$$T = P \times \frac{d}{2} = P_1 \times l$$

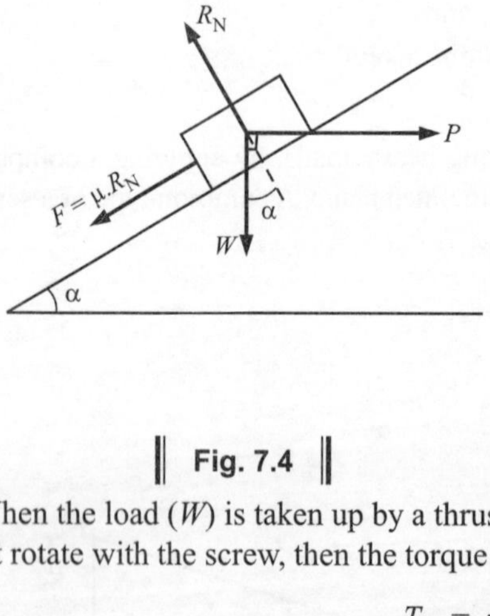


**Fig. 7.4**

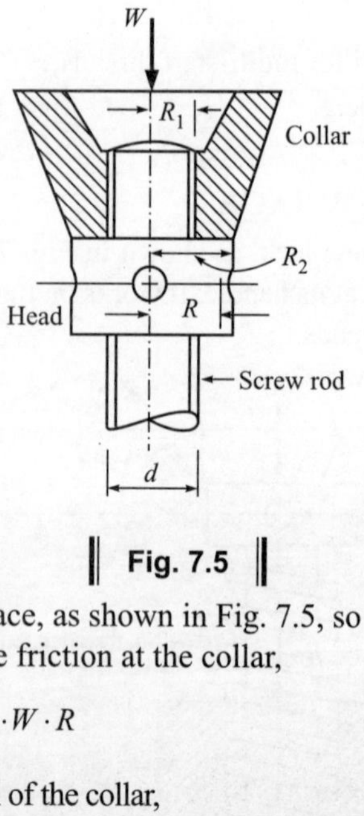


**Fig. 7.5**

When the load ($W$) is taken up by a thrust collar or a flat surface, as shown in Fig. 7.5, so that the load does not rotate with the screw, then the torque required to overcome friction at the collar,

$$T_C = \mu_1 \cdot W \left[ \frac{R_1 + R_2}{2} \right] = \mu_1 \cdot W \cdot R$$

where $R_1$ and $R_2$ = Outside and inside radii of the collar,

$R$ = Mean radius of the collar, and

$\mu_1$ = Coefficient of friction for the collar.

∴ Total torque required to overcome friction (*i.e.* to rotate the screw),

$$= T + T_C$$

**Note:** Sometimes outside diameter, $d_0$ (or nominal diameter or major diameter) and core diameter, $d_c$ (or root diameter or minor diameter) is given. In such a case, mean diameter of the screw,

$$d = \frac{d_0 + d_c}{2} = d_0 - \frac{p}{2} = d_c + \frac{p}{2}$$

**2. *When the load (W) is being lowered***

When the load ($W$) is being lowered, then the condition is similar to the body moving down the inclined plane when the force is acting horizontally. Proceeding in the same way as discussed in Chapter 6 (Sec. 6.20), we find that the force required to overcome friction between the screw and nut or the force required at the circumference of the screw to lower the load,

$$P = W \tan(\alpha - \phi)$$

and the turning moment or torque ($T$) required to overcome friction between the screw and nut,

$$T = P \times \frac{d}{2} = W \tan(\alpha - \phi)\frac{d}{2}$$

If an effort $P_1$ is applied at the end of the lever at a distance $l$ from the axis of the screw, then

$$T = P \times \frac{d}{2} = P_1 \times l$$

**Note:** If $\phi > \alpha$, then the above expression may be written as

$$P = W \tan(\phi - \alpha)$$

## 7.5 EFFICIENCY OF A SCREW JACK

The efficiency of a screw jack may be defined as the ratio of the effort required to move the load neglecting friction (*i.e.* ideal effort, $P_0$) to the effort required to move the load considering friction (*i.e.* actual effort, $P$). Mathematically, efficiency of a screw jack,

$$\eta_S = \frac{\text{Ideal effort } (P_0)}{\text{Actual effort } (P)}$$

We have already discussed that the effort required to move the load neglecting friction, *i.e.* ideal effort,

$$P_0 = W \tan \alpha$$

and effort required to move the load considering friction, *i.e.* actual effort,

$$P = W \tan(\alpha + \phi) \quad \text{...(For raising the load)}$$

$$= W \tan(\alpha - \phi) \quad \text{...(For lowering the load)}$$

$$\therefore \quad \eta_S = \frac{P_0}{P} = \frac{W \tan \alpha}{W \tan(\alpha + \phi)}$$

$$= \frac{\tan \alpha}{\tan(\alpha + \phi)} \quad \text{...(For raising the load)}$$

$$= \frac{\tan \alpha}{\tan(\alpha - \phi)} \quad \text{...(For lowering the load)}$$

This expression shows that the efficiency of a screw jack is independent of the load moved and the effort applied.

## 7.6 MAXIMUM EFFICIENCY OF A SCREW JACK

We have already discussed that efficiency of a screw jack is

$$\eta_S = \frac{\tan \alpha}{\tan(\alpha + \phi)} \quad \text{...}(i)$$

For maximum efficiency, differentiate the above equation with respect to $\alpha$ and equate to zero, *i.e.*

$$\frac{d\eta_S}{d\alpha} = 0$$

or

$$\frac{d}{d\alpha}\left[\frac{\tan \alpha}{\tan(\alpha + \phi)}\right] = 0$$

$$\frac{\tan(\alpha+\phi)\sec^2\alpha - \tan\alpha\,.\sec^2(\alpha+\phi)}{[\tan(\alpha+\phi)]^2} = 0 \qquad ...\left[\because \frac{d}{dx}\left(\frac{u}{v}\right) = \frac{v\times\frac{du}{dx} - u\times\frac{dv}{dx}}{v^2}\right]$$

$$\tan(\alpha+\phi)\sec^2\alpha - \tan\alpha\sec^2(\alpha+\phi) = 0$$

$$\frac{\sin(\alpha+\phi)}{\cos(\alpha+\phi)}\times\frac{1}{\cos^2\alpha} - \frac{\sin\alpha}{\cos\alpha}\times\frac{1}{\cos^2(\alpha+\phi)} = 0$$

$$\frac{\sin(\alpha+\phi)}{\cos(\alpha+\phi)\cos^2\alpha} = \frac{\sin\alpha}{\cos\alpha\cos^2(\alpha+\phi)}$$

$$\frac{\sin(\alpha+\phi)}{\cos\alpha} = \frac{\sin\alpha}{\cos(\alpha+\phi)}$$

$$\sin(\alpha+\phi)\cos(\alpha+\phi) = \sin\alpha\cos\alpha$$

Multiplying both sides by 2, we have

$$2\sin(\alpha+\phi)\cos(\alpha+\phi) = 2\sin\alpha\cos\alpha$$

$$\sin[2(\alpha+\phi)] = \sin 2\alpha \qquad ...[\because 2\sin A\cos A = \sin 2A]$$

or $$\sin 2(\alpha+\phi) = \sin(\pi-2\alpha) \qquad ...[\because \sin(\pi-2\alpha) = \sin 2\alpha]$$

$\therefore$ $$2(\alpha+\phi) = \pi - 2\alpha$$

$$4\alpha = \pi - 2\phi$$

and $$\alpha = \frac{\pi}{4} - \frac{\phi}{2}$$

Substituting this value of $\alpha$ in equation (*i*), we have

Maximum efficiency of the screw jack,

$$\eta_{S(max)} = \frac{\tan\left(\frac{\pi}{4}-\frac{\phi}{2}\right)}{\tan\left(\frac{\pi}{4}-\frac{\phi}{2}+\phi\right)} = \frac{\tan\left(\frac{\pi}{4}-\frac{\phi}{2}\right)}{\tan\left(\frac{\pi}{4}+\frac{\phi}{2}\right)}$$

$$= \frac{\tan\frac{\pi}{4} - \tan\frac{\phi}{2}}{1+\tan\frac{\pi}{4}\times\tan\frac{\phi}{2}} \times \frac{1-\tan\frac{\pi}{4}\times\tan\frac{\phi}{2}}{\tan\frac{\pi}{4}+\tan\frac{\phi}{2}}$$

$$= \frac{1-\tan\frac{\phi}{2}}{1+\tan\frac{\phi}{2}}\times\frac{1-\tan\frac{\phi}{2}}{1+\tan\frac{\phi}{2}} = \frac{\left[1-\tan\frac{\phi}{2}\right]^2}{\left[1+\tan\frac{\phi}{2}\right]^2} \qquad ...\left(\because \tan\frac{\pi}{4} = 1\right)$$

$$= \frac{\left[1-\frac{\sin\frac{\phi}{2}}{\cos\frac{\phi}{2}}\right]^2}{\left[1+\frac{\sin\frac{\phi}{2}}{\cos\frac{\phi}{2}}\right]^2} = \frac{\left(\cos\frac{\phi}{2}-\sin\frac{\phi}{2}\right)^2}{\left(\cos\frac{\phi}{2}+\sin\frac{\phi}{2}\right)^2}$$

$$= \frac{\cos^2 \frac{\phi}{2} + \sin^2 \frac{\phi}{2} - 2\cos\frac{\phi}{2}\sin\frac{\phi}{2}}{\cos^2 \frac{\phi}{2} + \sin^2 \frac{\phi}{2} + 2\cos\frac{\phi}{2}\sin\frac{\phi}{2}}$$

$$= \frac{1 - 2\sin\frac{\phi}{2}\cos\frac{\phi}{2}}{1 + 2\sin\frac{\phi}{2}\cos\frac{\phi}{2}}$$

$$= {}^*\frac{1-\sin\phi}{1+\sin\phi} \qquad \ldots[\because 2\sin A\cos B = \sin(A+B) + \sin(A-B)]$$

**It may be noted that the above relation of maximum efficiency may be obtained as discussed below:**

**We know that efficiency of a screw jack,**

$$\eta_S = \frac{\tan\alpha}{\tan(\alpha+\phi)} = \frac{\dfrac{\sin\alpha}{\cos\alpha}}{\dfrac{\sin(\alpha+\phi)}{\cos(\alpha+\phi)}}$$

$$= \frac{\sin\alpha \times \cos(\alpha+\phi)}{\cos\alpha\times\sin(\alpha+\phi)}$$

**Multiplying the numerator and denominator by 2, we have**

$$\eta_S = \frac{2\sin\alpha\times\cos(\alpha+\phi)}{2\cos\alpha\times\sin(\alpha+\phi)}$$

$$= \frac{\sin(\alpha+\alpha+\phi) + \sin(\alpha-\alpha-\phi)}{\sin(\alpha+\alpha+\phi) - \sin(\alpha-\alpha-\phi)}$$

...[$\because$ $2\sin A\cos B = \sin(A+B) + \sin(A-B)$ and $2\cos A\sin B = \sin(A+B) - \sin(A-B)$]

$$= \frac{\sin(2\alpha+\phi) + \sin(-\phi)}{\sin(2\alpha+\phi) - \sin(-\phi)}$$

$$= \frac{\sin(2\alpha+\phi) - \sin\phi}{\sin(2\alpha+\phi) + \sin\phi} \qquad \ldots[\because \sin(-\phi) = -\sin\phi] \quad \ldots(i)$$

**The efficiency of the screw jack will be maximum, if the value of sin (2α + φ), in the above equation (*i*), is maximum, *i.e.* when**

$$\sin(2\alpha+\phi) = 1$$

or $$2\alpha + \phi = 90° = \frac{\pi}{2}$$

$$\therefore \qquad \alpha = \frac{\pi}{4} - \frac{\phi}{2}$$

**Substituting this value of α in equation (*i*), we have maximum efficiency of the screw jack,**

$$\eta_{S\,(max)} = \frac{\sin\left[2\left(\frac{\pi}{4}-\frac{\phi}{2}\right)+\phi\right] - \sin\phi}{\sin\left[2\left(\frac{\pi}{4}-\frac{\phi}{2}\right)+\phi\right] + \sin\phi} = \frac{\sin\left(\frac{\pi}{2}\right) - \sin\phi}{\sin\left(\frac{\pi}{2}\right) + \sin\phi} = \frac{1-\sin\phi}{1+\sin\phi}$$

---

* We also know that $\sin A = 2\sin\left(\frac{A}{2}\right)\cos\left(\frac{A}{2}\right)$

## 7.7 OVERHAULING AND SELF-LOCKING SCREWS

Sometimes, the load starts moving downwards without the application of any effort or torque. Such a condition is known as *overhauling of screws.* On the other hand, when the effort is applied to lower the load, then the screw is known as *self-locking screw.*

It may also be noted that for overhauling screws, the efficiency is greater than 50% and for self locking screws, the efficiency is less than 50%.

**Note:** When the angle of inclination of the plane or helix angle ($\alpha$) is greater than angle of friction ($\phi$), then it is known as overhauling screws. On the other hand, when $\alpha < \phi$, then the screw is said to self locking screws.

**Example 7.1** *A screw jack has a double thread of mean diameter 50 mm. The pitch is 5 mm. If a force of 200 N is applied to the device, by a lever of length 200 mm, what load W can be raised? With this load on the device, what will happen if the applied force is released? Take coefficient of friction as 0.3 for all surfaces of contact.*

***Solution*** Given : $d = 50$ mm ; $p = 5$ mm ; $P_1 = 200$ N ; $l = 200$ mm ; $\mu = \tan\phi = 0.3$ or $\phi = \tan^{-1}(0.3) = 16.7°$

We know that for double start threads, lead of the screw

$$= 2p = 2 \times 5 = 10 \text{ mm}$$

and

$$\tan\alpha = \frac{\text{Lead of screw}}{\text{Circumference of screw}} = \frac{10}{\pi d} = \frac{10}{\pi \times 50} = 0.0636$$

$$\alpha = \tan^{-1}(0.0636) = 3.64°$$

Let $W$ be the load raised by the device. We know that effort required at the circumference of the screw,

$$P = W\tan(\alpha + \phi) = W\tan(3.64° + 16.7°) = W\tan 20.34°$$

$$= W \times 0.37 = 0.37\,W$$

We also know that

$$P_1 \times l = P \times \frac{d}{2}$$

$$200 \times 200 = 0.37W \times \frac{50}{2} = 9.25W$$

$$\therefore \quad W = \frac{200 \times 200}{9.25} = 4324.3 \text{ N} \quad \textbf{Ans.}$$

When the applied force is released, then the effort required at the circumference of the screw,

$$P = W\tan(\alpha + \phi) = 0.37\,W$$

$$= 0.37 \times 4324.3 = 1600 \text{ N} \quad \textbf{Ans.}$$

**Example 7.2** *A load of 2500 N is to be raised by a screw jack with mean diameter of 75 mm and pitch of 12 mm. Find the efficiency of the screw jack, if the coefficient of friction between the screw and nut is 0.075.*

***Solution*** Given : $W = 2500$ N ; $d = 75$ mm ; $p = 12$ mm ; $\mu = \tan\phi = 0.075$ or $\phi = \tan^{-1}(0.075) = 4.289°$

We know that

$$\tan\alpha = \frac{p}{\pi d} = \frac{12}{\pi \times 75} = 0.0509$$

$$\therefore \quad \alpha = \tan^{-1}(0.0509) = 2.915°$$

We know that efficiency of the screw jack

$$\eta_S = \frac{\tan\alpha}{\tan(\alpha + \phi)} = \frac{\tan 2.915°}{\tan(2.915° + 4.289°)} = \frac{\tan 2.915°}{\tan 7.204°}$$

$$= \frac{0.0509}{0.1264} = 0.4027 \text{ or } 40.27\% \quad \textbf{Ans.}$$

**Example 7.3** *The pitch of a screw jack is 15 mm, the mean diameter of the threads is 90 mm, the effective length of the operating lever is 850 mm and the coefficient of friction is 0.08. Calculate the necessary effort to be applied at the end of the lever when a load of 25 kN is to be lowered.*

***Solution*** Given : $p$ = 15 mm ; $d$ = 90 mm ; $l$ = 850 mm ; $\mu = \tan\phi = 0.08$ or $\phi = \tan^{-1}(0.08) = 4.574°$ ; $W$ = 25 kN

Let $P_1$ = Effort to be applied at the end of the lever.

We know that
$$\tan\alpha = \frac{p}{\pi d} = \frac{15}{\pi \times 90} = 0.053$$

$$\therefore \quad \alpha = \tan^{-1}(0.053) = 3.036°$$

and effort required at the circumference of the screw,

$$P = W\tan(\phi - \alpha) = 25\tan(4.574° - 3.036°) \quad ...(\because \phi > \alpha)$$
$$= 25\tan 1.538° = 25 \times 0.0268 = 0.67 \text{ kN} = 670 \text{ N}$$

We know that
$$P_1 \times l = P \times \frac{d}{2}$$

$$\therefore \quad P_1 = \frac{P \times d}{2l} = \frac{670 \times 90}{2 \times 850} = 35.47 \text{ N} \quad \textbf{Ans.}$$

**Example 7.4** *A simple screw jack has threads of pitch 5 mm. The effort is applied at the end of a lever 500 mm long. What effort will be required to lift a load of 10 kN, if the efficiency at this load is 40%?*

***Solution*** Given : $p$ = 5 mm ; $l$ = 500 mm ; $W$ = 10 kN ; $\eta$ = 40% = 0.4

Let $P_1$ = Effort required at the end of a lever,

$P$ = Effort required at the circumference of the screw,

$P_0$ = Ideal effort *i.e.* when friction is neglected = $W\tan\alpha$, and

$d$ = Mean diameter of the screw.

We know that efficiency of the screw,

$$\eta = \frac{P_0}{P}$$

or
$$P = \frac{P_0}{\eta} = \frac{W\tan\alpha}{0.4} = \frac{W}{0.4} \times \frac{p}{\pi d} = \frac{10 \times 5}{0.4\pi d} = \frac{39.8}{d} \quad ...(i)$$

We also know that
$$P_1 \times l = P \times \frac{d}{2}$$

$$\therefore \quad P_1 = \frac{P \times d}{2l} = \frac{39.8 \times d}{d \times 2 \times 500} = 0.0398 \text{ kN} = 39.8 \text{ N} \quad \textbf{Ans.}$$

**Example 7.5** *A screw jack has a square threaded screw of 50 mm mean diameter and a pitch of 10 mm. The coefficient of friction of the screw threads is 0.15. Find the force required at the end of 700 mm long handle to raise a load of 20 kN. What is the force required if the screw jack is considered to be an ideal machine?*

***Solution*** Given : $d$ = 50 mm ; $p$ = 10 mm ; $\mu = \tan\phi = 0.15$ or $\phi = \tan^{-1}(0.15) = 8.53°$ ; $l$ = 700 mm ; $W$ = 20 kN

***Force required at the end of the lever***

Let $P_1$ = Force required at the end of the lever.

We know that $\tan \alpha = \frac{p}{\pi d} = \frac{10}{\pi \times 50} = 0.0636$

$\therefore$ $\alpha = \tan^{-1}(0.0636) = 3.64°$

We also know that force required at the circumference of the screw,

$$P = W \tan(\alpha + \phi) = 20 \tan(3.64° + 8.53°) = 20 \tan 12.17°$$
$$= 20 \times 0.2156 = 4.312 \text{ kN}$$

and $P_1 \times l = P \times \frac{d}{2}$

$\therefore$ $P_1 = \frac{P \times d}{2} \quad \frac{4.312 \times 50}{2 \times 700} = 0.154 \text{ kN} = 154 \text{ N}$ **Ans.**

***Force required if the screw jack is considered to be an ideal machine***

Let $P_2$ = Required force.

We know that ideal effort, $P_0 = W \tan \alpha = 20 \times 0.0636 = 1.272 \text{ kN}$ ...($\because \phi = 0$)

and $P_2 \times l = P_0 \times \frac{d}{2}$

$\therefore$ $P_2 = \frac{P_0 \times d}{2l} = \frac{1.272 \times 50}{2 \times 700} = 0.0454 \text{ kN} = 45.4 \text{ N}$ **Ans.**

**Example 7.6** *A load of 30 kN is lifted through 120 mm by a screw jack. The mean diameter of the thread is 60 mm and the pitch is 12 mm. If the coefficient of friction is 0.08, find :*

1. *The efficiency of the jack at this load, and*
2. *Work done by the effort to lift the load.*

**Solution** Given : $W = 30 \text{ kN} = 30 \times 10^3 \text{ N}$ ; $h = 120$ mm ; $d = 60$ mm ; $p = 12$ mm ; $\mu = \tan \phi = 0.08$ or $\phi = \tan^{-1}(0.08) = 4.574°$

**1. *Efficiency of the jack***

We know that $\tan \alpha = \frac{p}{\pi d} = \frac{12}{\pi \times 60} = 0.0637$

$\therefore$ $\alpha = \tan^{-1}(0.0637) = 3.645°$

We also know that effort required at the circumference of the screw to lift the load,

$$P = W \tan(\alpha + \phi) = 30 \tan(3.645° + 4.574°)$$
$$= 30 \tan 8.219° = 30 \times 0.1444 = 4.332 \text{ kN}$$

and ideal effort (*i.e.* when there is no friction),

$$P_0 = W \tan \alpha = 30 \tan 3.645° \quad ...(\because \phi = 0)$$
$$= 30 \times 0.0637 = 1.911 \text{ kN}$$

$\therefore$ Efficiency of the jack, $\eta = \frac{P_0}{P} = \frac{1.911}{4.332} = 0.4411$ or 44.11% **Ans.**

**2. *Work done by the effort***

We know that efficiency of the jack,

$$\eta = \frac{\text{Work done by the load } (W)}{\text{Work done by the effort } (P)}$$

$\therefore$ Work done by the effort $(P) = \dfrac{\text{Work done by the load } (W)}{\eta} = \dfrac{30 \times 120}{0.4411} = 8161.4$ kN-mm

$= 8.1614$ kN-m **Ans.**

**Example 7.7** *A screw jack with square threads has a mean diameter 62.5 mm. The pitch of the threads is 12.5 mm. The load on the jack revolves with the screw. The coefficient of friction at the screw thread is 0.05.*

1. *Find the tangential force which is applied at it at 300 mm radius required to lift a load of 5440 N.*
2. *State whether the jack is self-locking. If it is, find the torque necessary to lower the load. If it is not, find the torque which must be applied to keep the load from descending.*

**Solution** Given : $d = 62.5$ mm ; $p = 12.5$ mm ; $\mu = \tan\phi = 0.05$ or $\phi = \tan^{-1}(0.05) = 2.86°$ ; $l = 300$ mm ; $W = 5440$ N

**1. *Tangential force to be applied***

Let $P_1$ = Tangential force applied at 300 mm radius to lift the load.

We know that $\tan\alpha = \dfrac{p}{\pi d} = \dfrac{12.5}{\pi \times 62.5} = 0.0636$

or $\alpha = \tan^{-1}(0.0636) = 3.64°$

$\therefore$ Tangential force required at the circumference of the screw,

$$P = W\tan(\alpha + \phi) = 5440\tan(3.64° + 2.86°) = 5440\tan 6.50°$$

$$= 5440 \times 0.114 = 620 \text{ N}$$

We know that $P_1 \times l = P \times \dfrac{d}{2}$

$\therefore$ $P_1 = \dfrac{P \times d}{2l} = \dfrac{620 \times 62.5}{2 \times 300} = 64.6$ N **Ans.**

**2. *Condition for self-locking of the jack***

We know that efficiency of the screw jack,

$$\eta = \frac{P_0}{P} = \frac{\tan\alpha}{\tan(\alpha + \phi)} = \frac{\tan 3.64°}{\tan(3.64° + 2.86°)}$$

$$= \frac{\tan 3.64°}{\tan 6.50°} = \frac{0.0636}{0.114} = 0.558 \text{ or } 55.8\%$$

Since the efficiency of the jack is more than 50%, therefore the screw jack is not self-locking.

We know that force required at the circumference of the screw to keep the load from descending,

$$P = W\tan(\alpha - \phi) = 5440\tan(3.64° - 2.86°)$$

$$= 5440\tan(0.78)° = 5440 \times 0.0136 = 74 \text{ N}$$

$\therefore$ Torque required to keep the load from descending,

$$T = P \times \frac{d}{2} = 74 \times \frac{62.5}{2} = 2312.5 \text{ N-mm } \textbf{Ans.}$$

**Example 7.8** *A rectangular sluice gate weighing 5 kN is subjected to a normal pressure of 1 MN/m². The gate is 2.5 m high and 2 m wide and is raised by means of vertical square threaded screw which rotates at 105 r.p.m. The core diameter of the thread is 90 mm and the pitch is 20 mm. The coefficient of friction is 0.08 for the screw and 0.1 for the guides. Find the power required at the screw.*

***Solution*** Given : $W_1 = 5$ kN ; $p_1 = 1$ MN/m$^2$ = $1 \times 10^6$ N/m$^2$ ; $h = 2.5$ m ; $w = 2$ m ; $N = 105$ r.p.m.; $d_c = 90$ mm ; $p = 20$ mm ; $\mu = \tan\phi = 0.08$ or $\phi = \tan^{-1}(0.08) = 4.57°$ ; $\mu_1 = 0.1$

We know that total pressure on the gate,

$$W_2 = \text{Pressure} \times \text{Cross-sectional area of gate}$$
$$= p_1 \times w \times h = 1 \times 10^6 \times 2 \times 2.5 = 5 \times 10^6 \text{ N} = 5000 \text{ kN}$$

and total vertical load,

$$W = W_1 + \mu_1 W_2$$
$$= 5 + 0.1 \times 5000 = 505 \text{ kN} = 505 \times 10^3 \text{ N}$$

We know that mean diameter of the screw

$$d = d_c + \frac{p}{2} = 90 + \frac{20}{2} = 100 \text{ mm} = 0.1 \text{ m}$$

and

$$\tan\alpha = \frac{p}{\pi d} = \frac{20}{\pi \times 100} = 0.0636$$

$\therefore$

$$\alpha = \tan^{-1}(0.0636) = 3.64°$$

We know that effort required at the circumference of the screw to raise the load,

$$P = W\tan(\alpha + \phi)$$
$$= 505 \times 10^3 \tan(3.64° + 4.57°)$$
$$= 505 \times 10^3 \tan(8.21°)$$
$$= 505 \times 10^3 \times 0.1443 = 72\,872 \text{ N}$$

and the torque required,

$$T = P \times \frac{d}{2} = 72\,872 \times \frac{0.1}{2} = 3643.6 \text{ N-m}$$

Angular speed of the screw,

$$\omega = \frac{2\pi N}{60} = \frac{2\pi \times 105}{60} = 11 \text{ rad/s}$$

$\therefore$ Power required $= T.\omega = 3643.6 \times 11 = 40\,080$ N-m/s $= 40\,080$ W

...($\because$ 1 N-m/s = 1 watt)

$= 400.8$ kW **Ans.**

## 7.8 LADDER FRICTION

A ladder consists of a set of horizontal bars of wood or metal fixed between two uprights. It is used for climbing or scaling on the roofs or walls.

Consider a ladder *AC* resting on the ground *AB* and leaning against a wall *BC*, as shown in Fig. 7.6.

Let $R_A$ = Reaction at *A*,

$F_A$ = Force of friction at $A = \mu_A \cdot R_A$,

$\mu_A$ = Coefficient of friction between the ladder and the wall,

$R_C$ = Reaction at *C*,

$F_C$ = Force of friction at $C = \mu_C . R_C$, and

$\mu_C$ = Coefficient of friction between the ladder and the ground.

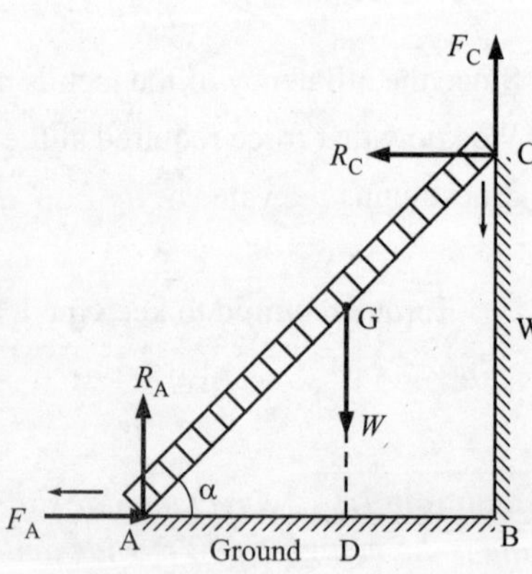


**Fig. 7.6** *Ladder friction*

Due to the self weight of the ladder or when a person stands on the ladder, the upper end *C* of the ladder tends to slip downwards, therefore the force of friction ($F_C = \mu_C . R_C$) between the wall and ladder will act upwards, as shown in Fig. 7.6.

Similarly, the lower end *A* of the ladder will tend to move towards left, therefore, the force of friction ($F_A = \mu_A . R_A$) between the ladder and ground will act towards right.

For the ladder to be in equilibrium, the following conditions must be satisfied:

(*a*) The algebraic sum of the horizontal components of the forces acting on the ladder must be equal to zero, *i.e.*

$$R_C = F_A = \mu_A \cdot R_A \qquad ...(i)$$

(*b*) The algebraic sum of the vertical components of the forces acting on the ladder must be equal to zero *i.e.*

$$W = R_A + F_C = R_A + \mu_C \cdot R_C$$

or

$$R_C = \frac{W - R_A}{\mu_C} = \frac{W - \dfrac{R_C}{\mu_A}}{\mu_C} = \frac{\mu_A \cdot W - R_C}{\mu_A \cdot \mu_C} \qquad ...[\text{From equation } (i)]$$

$$R_C(\mu_A \cdot \mu_C) = \mu_A \cdot W - R_C$$

or

$$R_C + R_C(\mu_A \cdot \mu_C) = \mu_A \cdot W$$

$$R_C(1 + \mu_A \cdot \mu_C) = \mu_A \cdot W$$

$$\therefore \quad R_C = \frac{\mu_A \cdot W}{1 + \mu_A \cdot \mu_C} \qquad ...(ii)$$

(*c*) The algebraic sum of the moments of the various forces acting on the ladder about any point must be equal to zero. Taking moments about *A*,

$$F_C \times AB + R_C \times BC - W \times AD = 0 \qquad ...(iii)$$

It may be noted that

**1.** If the vertical wall is smooth, then there will be no force of friction between the ladder and the vertical wall.

**2.** The angle of inclination (α) of the ladder with the ground, may be obtained as discussed below:

From equation (*iii*),

$$F_C \times AB + R_C \times BC - W \times AD = 0$$

Since $F_C = \mu_C \cdot R_C$; $AB = AC \cos\alpha$; $BC = AC \sin\alpha$; and $AD = \frac{AB}{2} = \frac{AC\cos\alpha}{2}$, therefore the above equation may be written as

$$\mu_C \cdot R_C \times AC \cos\alpha + R_C \times AC \sin\alpha - W \times \frac{AC\cos\alpha}{2} = 0$$

$$\mu_C \cdot R_C \times AC \cos\alpha + R_C \times AC \sin\alpha = \frac{W \times AC\cos\alpha}{2}$$

$$\mu_C \cdot R_C \cos\alpha + R_C \sin\alpha = \frac{W\cos\alpha}{2}$$

Substituting the value of $R_C$ from equation (*ii*),

$$\mu_C \times \frac{\mu_A \cdot W}{1 + \mu_A \cdot \mu_C} \times \cos\alpha + \frac{\mu_A \cdot W}{1 + \mu_A \cdot \mu_C} \times \sin\alpha = \frac{W\cos\alpha}{2}$$

$$\frac{\mu_A \cdot \mu_C \cos\alpha}{1 + \mu_A \cdot \mu_C} + \frac{\mu_A \sin\alpha}{1 + \mu_A \cdot \mu_C} = \frac{\cos\alpha}{2}$$

$$\frac{\mu_A \sin\alpha}{1 + \mu_A \cdot \mu_C} = \frac{\cos\alpha}{2} - \frac{\mu_A \cdot \mu_C \cos\alpha}{1 + \mu_A \cdot \mu_C}$$

$$= \cos\alpha\left[\frac{1}{2} - \frac{\mu_A \cdot \mu_C}{1 + \mu_A \cdot \mu_C}\right]$$

$$= \cos\alpha\left[\frac{1+\mu_A\cdot\mu_C-2\mu_A\cdot\mu_C}{2(1+\mu_A\cdot\mu_C)}\right]$$

$$= \cos\alpha\left[\frac{1-\mu_A\cdot\mu_C}{2(1+\mu_A\cdot\mu_C)}\right]$$

or
$$\tan\alpha = \frac{\sin\alpha}{\cos\alpha}$$

$$= \frac{1-\mu_A\cdot\mu_C}{2(1+\mu_A\cdot\mu_C)} \times \frac{1+\mu_A\cdot\mu_C}{\mu_A} = \frac{1-\mu_A\cdot\mu_C}{2\mu_A}$$

**Example 7.9** *A long ladder 15 m long resting on horizontal floor and leans against a vertical wall for which $\mu_F = 0.25$ and $\mu_W = 0.15$, respectively. Determine the angle of inclination of ladder with the floor when the ladder is just about to slip.*

***Solution*** Given : *Length of ladder, $AC = 15$ m ; $\mu_F = \mu_A = 0.25$ ; $\mu_W = \mu_C = 0.15$

Let $\alpha$ = Angle of inclination of ladder with the floor.

We know that
$$\tan\alpha = \frac{1-\mu_A\cdot\mu_C}{2\mu_A} = \frac{1-0.25\times0.15}{2\times0.25} = \frac{0.9625}{0.5} = 1.9250$$

$\therefore$ $\alpha = \tan^{-1}(1.9250) = 62.55°$ **Ans.**

**Example 7.10** *A uniform ladder of length 10 m and weighing 20 N is placed against a smooth vertical wall with its lower end 8 m from the wall. In this position, the ladder is just to slip. Determine*

1. *Coefficient of friction between the ladder and the floor ; and*
2. *Frictional force acting on the ladder at the point of contact between the ladder and the floor.*

***Solution*** Given : $AC = 10$ m ; $W = 20$ N ; $AB = 8$ m

**1. *Coefficient of friction between the ladder and the floor***

Let $\mu_A$ = Coefficient of friction between the ladder and the floor.

Since the ladder is placed against a smooth vertical wall, therefore there will be no force of friction between the ladder and the vertical wall, *i.e.*

$$\mu_C = 0$$

From Fig. 7.7, we find that
$$\cos\alpha = \frac{AB}{AC} = \frac{8}{10} = 0.8$$

$\therefore$ $\alpha = 36.87°$

We know that
$$\tan\alpha = \frac{1-\mu_A\cdot\mu_C}{2\mu_A}$$

$$\tan 36.87° = \frac{1-\mu_A\times 0}{2\mu_A} \qquad ...(\because \mu_C = 0)$$

$$0.75 = \frac{1}{2\mu_A}$$

$\therefore$ $\mu_A = \dfrac{1}{2\times0.75} = 0.67$ **Ans.**

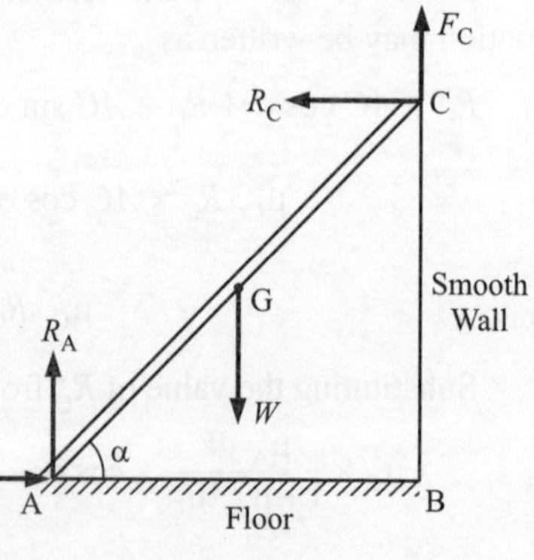

Fig. 7.7

* Superfluous data.

**2. *Frictional force acting on the ladder***

Let $F_A$ = Frictional force acting on the ladder at the point of contact between the ladder and the floor, *i.e.* at point *A*.

Resolving the forces horizontally, we have

$$F_A = R_C = \mu_A \cdot R_A = 0.67 R_A \qquad ...(i)$$

and resolving the forces vertically, we have

$$W = R_A + F_C = R_A + \mu_C \times R_C$$

$$20 = R_A + 0 \times R_C = R_A$$

Substituting the value of $R_A$ = 20 N in equation (*i*), we get

$$F_A = 0.67 R_A = 0.67 \times 20 = 13.4 \text{ N} \quad \textbf{Ans.}$$

**Example 7.11** *A ladder of 5 m length and 50 N weight rest on a horizontal ground and against a smooth vertical wall at an angle of 30° with the vertical. When a man of 100 N stands on a ring 2 m from the foot of the ladder, it is on the point of slipping. Determine the coefficient of friction between the ladder and the ground.*

***Solution*** Given : $AB$ = 5 m ; $W$ = 50 N ; $\angle ABC$ = 60° ; $W_M$ = 100 N ; $x$ = 2 m

Let *$\mu = \mu_A$ = Coefficient of friction between the ladder and the floor.

Considering the equilibrium of the ladder. The various forces acting on the ladder, as shown in Fig. 7.8, are as follows:

(*a*) Normal reaction ($R_A$) between the ladder and the floor at *A*,

(*b*) Normal reaction ($R_B$) between the ladder and the wall at *B*,

(*c*) Force of friction ($F_A$) between the floor and the ladder, such that

$$F_A = \mu \cdot R_A$$

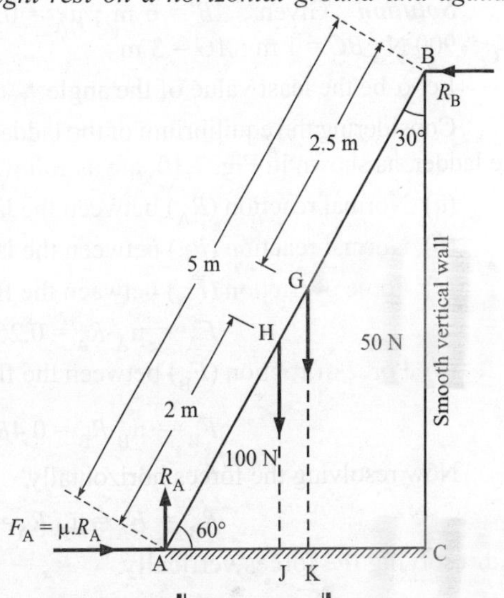


**Fig. 7.8**

Resolving the forces vertically, $R_A$ = 100 + 50 = 150 N

Resolving the forces horizontally, $R_B = F_A = \mu \cdot R_A = \mu \times 150$ N ...(*i*)

Since *G* is the middle point of *AB*, therefore $AG = \frac{5}{2} = 2.5$ m. From the geometry of the figure, we find that

$$AC = AB \cos 60° = 5 \times 0.5 = 2.5 \text{ m} \qquad ...(\because \angle BAC = 60°)$$

$$AK = AG \cos 60° = 2.5 \times 0.5 = 1.25 \text{ m}$$

$$AJ = AH \cos 60° = 2 \times 0.5 = 1 \text{ m}$$

$$BC = AB \sin 60° = 5 \times 0.866 = 4.33 \text{ m}$$

Now taking moments about A,

$$100 \times AJ + 50 \times AK = R_B \times BC$$

$$100 \times 1 + 50 \times 1.25 = R_B \times 4.33$$

$$100 + 62.5 = R_B \times 4.33$$

* Since the end *B* of the ladder rests against a smooth vertical wall *CB*, therefore there will be no force of friction acting at *B*. In other words, coefficient of friction between the ladder and the wall ($\mu_B$) is zero.

$$\therefore \qquad R_B = \frac{100 + 62.5}{4.33} = 37.53$$

Now from equation (*i*), $R_B = \mu \times 150$

$$\therefore \qquad \mu = \frac{R_B}{150} = \frac{37.53}{150} = 0.25 \textbf{ Ans.}$$

**Example 7.12** *The ladder, as shown in Fig. 7.9, is 6 m long and is supported by a horizontal floor and a vertical wall. The coefficient of friction between the floor and the ladder is 0.25 and between the wall and ladder is 0.4. The weight of the ladder is 200 N and acts at C.G. The ladder also supports a vertical load of 900 N at C which is at a distance of 1 m from B. Determine the least value of α at which the ladder may be placed without slipping.*

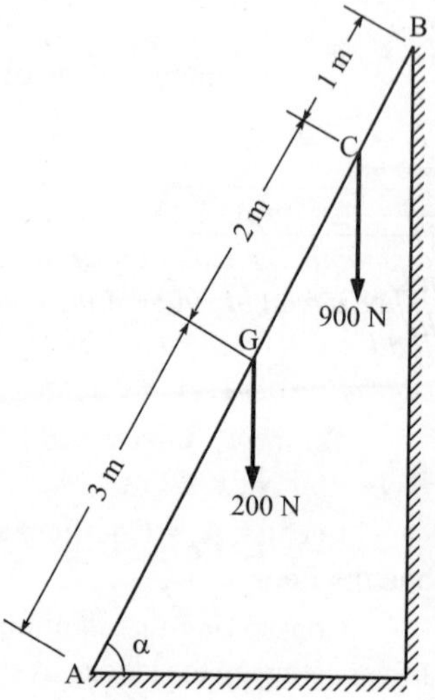


Fig. 7.9

***Solution*** Given : $AB = 6$ m ; $\mu_A = 0.25$ ; $\mu_B = 0.4$ ; $W = 200$ N ; $W_C = 900$ N ; $BC = 1$ m ; $AG = 3$ m

Let α be the least value of the angle.

Considering the equilibrium of the ladder. The various forces acting on the ladder, as shown in Fig. 7.10, are as follows:

(*a*) Normal reaction ($R_A$) between the ladder and the floor at *A*,

(*b*) Normal reaction ($R_B$) between the ladder and the wall at *B*,

(*c*) Force of friction ($F_A$) between the floor and the ladder, such that

$$F_A = \mu_A \cdot R_A = 0.25 R_A$$

(*d*) Force of friction ($F_B$) between the floor and the ladder, such that

$$F_B = \mu_B \cdot R_B = 0.4 R_B$$

Now resolving the forces horizontally,

$$R_B = F_A = \mu_A . R_A = 0.25 R_A \qquad \text{...}(i)$$

and resolving the forces vertically,

$$R_A + F_B = 200 + 900$$

$$R_A + \mu_B \cdot R_B = 1100$$

$$R_A + 0.4(0.25 R_A) = 1100$$

$$R_A + 0.1 R_A = 1100$$

$$\therefore \qquad R_A = \frac{1100}{1.1} = 1000 \text{ N}$$

and $$R_B = 0.25 R_A = 0.25 \times 1000 = 250 \text{ N}$$

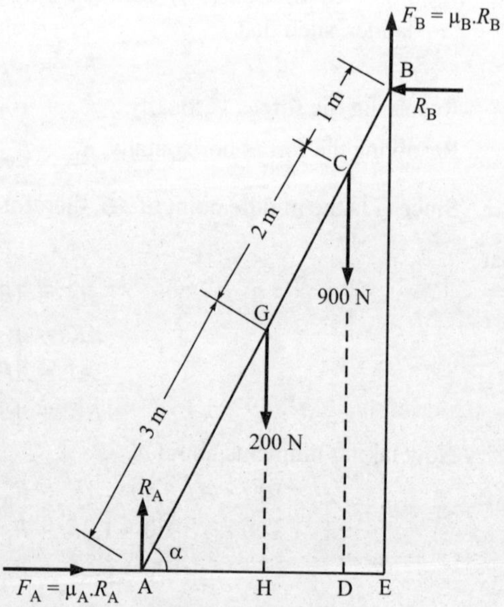


Fig. 7.10

Since *G* is the middle point of *AB*, therefore $AG = \frac{6}{2} = 3$ m. From the geometry of the figure, we find that

$$AE = AB \cos \alpha = 6 \cos \alpha$$

$$BE = AB \sin \alpha = 6 \sin \alpha$$

$$AD = AC \cos \alpha = 5 \cos \alpha$$

$$AH = AG \cos \alpha = 3 \cos \alpha$$

Now taking moments about $A$, we have

$$200 \times AH + 900 \times AD = F_B \times AE + R_B \times BE$$

$$200 \times 3 \cos\alpha + 900 \times 5 \cos\alpha = 0.4 R_B \times 6 \cos\alpha + R_B \times 6 \sin\alpha$$

$$600 \cos\alpha + 4500 \cos\alpha = 0.4 \times 250 \times 6 \cos\alpha + 250 \times 6 \sin\alpha$$

$$5100 \cos\alpha = 600 \cos\alpha + 1500 \sin\alpha$$

$$5100 \cos\alpha - 600 \cos\alpha = 1500 \sin\alpha$$

$$4500 \cos\alpha = 1500 \sin\alpha$$

or $$\frac{\sin\alpha}{\cos\alpha} = \tan\alpha = \frac{4500}{1500} = 3$$

$\therefore$ $$\alpha = \tan^{-1}(3) = 71.56° \quad \textbf{Ans.}$$

**Example 7.13** *A ladder 5 m long and 250 N weight is placed against a vertical wall in a position where its inclination to the vertical is 30°. A man weighing 800 N climbs the ladder. At what position will he induce slipping? The coefficient of friction for both the contact surfaces of the ladder, i.e. with the wall and the floor is 0.2.*

***Solution*** Given : $AB = 5$ m ; $W = 250$ N ; $\angle ABC = 30°$, $\angle BAC = 60°$ ; $\mu_A = \mu_B = 0.2$

Let $x = AH$ = Distance climbed by the man from $A$, when the ladder is at the point of slipping.

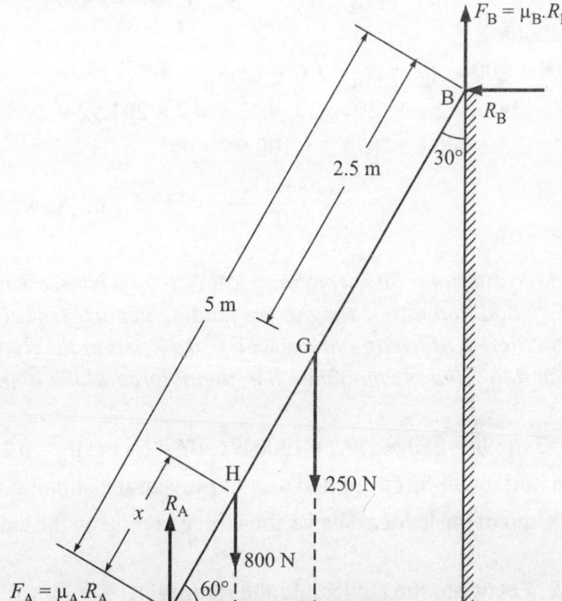


**Fig. 7.11**

Considering the equilibrium of the ladder. The various forces acting on the ladder, as shown in Fig. 7.11, are as follows:

(*a*) Normal reaction ($R_A$) between the ladder and the floor at $A$,

(*b*) Normal reaction ($R_B$) between the ladder and the wall at $B$,

(*c*) Force of friction ($F_A$) between the floor and the ladder, such that

$$F_A = \mu_A \cdot R_A = 0.2 R_A$$

(*d*) Force of friction ($F_B$) between the wall and the ladder, such that

$$F_B = \mu_B \cdot R_B = 0.2 R_B$$

Now resolving the forces horizontally,

$$R_B = F_A = \mu_A \cdot R_A = 0.2R_A \quad ...(i)$$

and resolving the forces vertically,

$$R_A + F_B = 250 + 800$$

$$R_A + \mu_B . R_B = 1050 \quad ...(\because R_B = 0.2R_A)$$

$$R_A + 0.2\,(0.2R_A) = 1050$$

$$1.04R_A = 1050$$

$$\therefore \quad R_A = \frac{1050}{1.04} = 1009.6 \text{ N}$$

and

$$R_B = 0.2R_A = 0.2 \times 1009.6 = 201.92 \text{ N}$$

Since $G$ is the middle point of $AB$, therefore $AG = \frac{5}{2} = 2.5$ m

From the geometry of the figure, we find that

$$AC = AB \cos 60° = 5 \times 0.5 = 2.5 \text{ m}$$

$$AK = AG \cos 60° = 2.5 \times 0.5 = 1.25 \text{ m}$$

$$AJ = AH \cos 60° = x \times 0.5 = 0.5x$$

$$BC = AB \sin 60° = 5 \times 0.866 = 4.33 \text{ m}$$

Now taking moments about $A$,

$$250 \times AK + 800 \times AJ = R_B \times BC + \mu_B . R_B \times AC$$

$$250 \times 1.25 + 800 \times 0.5x = 201.92 \times 4.33 + 0.2 \times 201.92 \times 2.5$$

$$312.5 + 400x = 874.3 + 100.96 = 975.26$$

$$\therefore \quad x = \frac{975.26 - 312.5}{400} = 1.657 \text{ m}$$ **Ans.**

**Example 7.14** *A ladder 3 m long and weighing 250 N is placed against a wall with end A at the floor level and B on the wall. In addition to self weight, the ladder supports a man weighing 1200 N at 2.5 m from A on the ladder. If the coefficient of friction at the wall is 0.25 and at the floor is 0.35 and if the ladder makes an angle of 60° with the floor, find the minimum horizontal force which if applied at A will prevent the slipping of the ladder.*

***Solution*** Given : $AB = 3$ m ; $W = 250$ N ; $W_M = 1200$ N ; $AH = 2.5$ m ; $\mu_B = 0.25$ ; $\mu_A = 0.35$ ; $\angle BAC = 60°$

Let $P$ = Minimum horizontal force applied at $A$ to prevent the slipping of the ladder.

Considering the equilibrium of the ladder. The various forces acting on the ladder, as shown in Fig. 7.12, are as follows:

(*a*) Normal reaction ($R_A$) between the ladder and the floor at $A$,

(*b*) Normal reaction ($R_B$) between the ladder and the wall at $B$,

(*c*) Horizontal force ($P$) and force of friction ($F_A$) between the floor and the ladder, such that

$$F_A = \mu_A \cdot R_A = 0.35R_A$$

(*d*) Force of friction ($F_B$) between the wall and the ladder, such that

$$F_B = \mu_B \cdot R_B = 0.25R_B$$

Since $G$ is the middle point of $AB$, therefore $AB = \frac{3}{2} = 1.5$ m. From the geometry of the figure,

$$AC = AB \cos 60° = 3 \times 0.5 = 1.5 \text{ m}$$

$$AJ = AH \cos 60° = 2.5 \times 0.5 = 1.25 \text{ m}$$

$$AK = AG\cos 60° = 1.5 \times 0.5 = 0.75 \text{ m}$$

and
$$BC = AB\sin 60° = 3 \times 0.866 = 2.598 \text{ m}$$

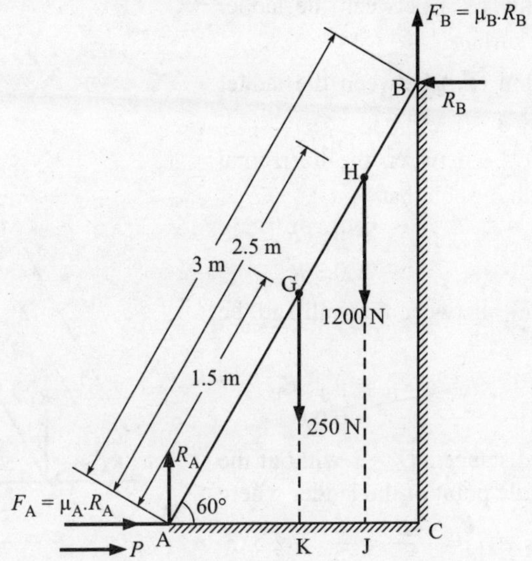


**Fig. 7.12**

Taking moments about *A*,

$$1200 \times AJ + 250 \times AK = R_B \times BC + \mu_B . R_B \times AC$$
$$1200 \times 1.25 + 250 \times 0.75 = R_B \times 2.598 + 0.25 R_B \times 1.5$$
$$1500 + 187.5 = 2.598 R_B + 0.375 R_B$$
$$1687.5 = 2.973 R_B$$
$$\therefore \quad R_B = \frac{1687.5}{2.973} = 567.6 \text{ N}$$

Now resolving the forces horizontally,

$$P + \mu_A \cdot R_A = R_B$$
$$P + 0.35 R_A = 567.6 \qquad ...(i)$$

and resolving the forces vertically,

$$R_A + \mu_B \cdot R_B = 1200 + 250$$
$$R_A + 0.25 \times 567.6 = 1450$$
$$\therefore \quad R_A = 1450 - 0.25 \times 567.6 = 1308 \text{ N}$$

Substituting the value of $R_A$ in equation (*i*),

$$P + 0.35 \times 1308 = 567.6$$
$$P = 567.6 - 0.35 \times 1308 = 576.6 - 457.8 = 109.8 \text{ N} \quad \textbf{Ans.}$$

**Example 7.15** *A man wishes to climb a 5 m long ladder placed at 60° on a horizontal surface ($\mu$ = 0.3) against a vertical wall ($\mu$ = 0.2). How far he can climb without the ladder slipping? The man and the ladder weigh 800 N and 150 N, respectively.*

***Solution*** Given : $AB = 5$ m ; $\angle BAC = 60°$ ; $\mu_A = 0.3$ ; $\mu_B = 0.2$ ; $W_M = 800$ N ; $W_L = 150$ N

Considering the equilibrium of the ladder. The various forces acting on the ladder, as shown in Fig. 7.13, are as follows:

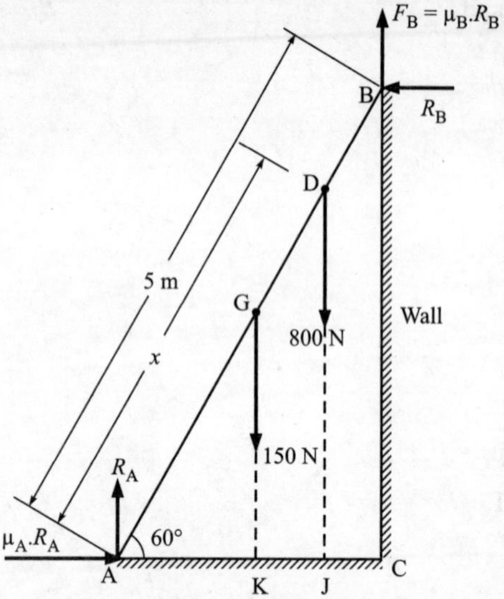


Fig. 7.13

(*a*) The normal reaction ($R_A$) between the ladder and the horizontal surface,

(*b*) The normal reaction ($R_B$) between the ladder and the wall,

(*c*) Force of friction ($F_A$) between the horizontal surface and the ladder, such that

$$F_A = \mu_A \cdot R_A = 0.3\, R_A$$

(*d*) Force of friction ($F_B$) between the wall and the ladder, such that

$$F_B = \mu_B \cdot R_B = 0.2 R_B$$

Let the man climbs a distance $AD = x$ without the ladder slipping. $G$ is the middle point of the ladder where weight of the ladder acts. Thus $AG = \frac{AB}{2} = \frac{5}{2} = 2.5$ m

From the geometry of the figure,

$$AC = AB \cos 60° = 5 \times 0.5 = 2.5 \text{ m}$$
$$AJ = AD \cos 60° = x \times 0.5 = 0.5x \text{ m}$$
$$AK = AG \cos 60° = 2.5 \times 0.5 = 1.25 \text{ m}$$
$$BC = AB \sin 60° = 5 \times 0.866 = 4.33 \text{ m}$$

Taking moments about $A$,

$$800 \times AJ + 150 \times AK = R_B \times BC + \mu_B \cdot R_B \times AC$$
$$800 \times 0.5x + 150 \times 1.25 = R_B \times 4.33 + 0.2 R_B \times 2.5$$
$$400x + 187.5 = 4.33 R_B + 0.5 R_B = 4.83 R_B \quad ...(i)$$

Now resolving the forces horizontally

$$F_A = R_B \text{ or } 0.3 R_A = R_B \text{ or } R_A = \frac{R_B}{0.3} \quad ...(ii)$$

and resolving the forces vertically, $R_A + F_B = 150 + 800$

$$R_A + 0.2 R_B = 950$$
$$\frac{R_B}{0.3} + 0.2 R_B = 950$$
$$R_B + 0.2 \times 0.3 R_B = 950 \times 0.3 = 285$$
$$1.06\, R_B = 285 \text{ or } R_B = \frac{285}{1.06} = 268.87 \text{ N}$$

Substituting the value of $R_B$ in equation (*i*),

$$400\,x + 187.5 = 4.83 R_B = 4.83 \times 268.87 = 1298.6$$

$$\therefore \quad x = \frac{1298.6 - 187.5}{400} = 2.778 \text{ m} \quad \textbf{Ans.}$$

**Example 7.16** *A uniform ladder 3 m long weighs 180 N. It is placed against a wall making an angle of 60° with the floor, as shown in Fig. 7.14. The coefficient of friction between the wall and the ladder is 0.25 and that between the floor and ladder is 0.35. The ladder, in addition to its own weight, has to support a man 900 N at its top end. Calculate*

1. *The horizontal force P to be applied to the ladder at the floor level to prevent slipping; and*
2. *If the force P is not applied, what should be the minimum inclination of the ladder with the horizontal so that there is no slipping of it, with the man at its top?*

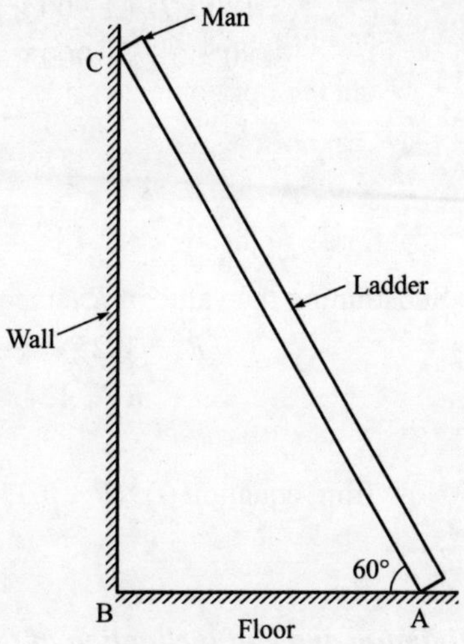


Fig. 7.14

***Solution*** Given : $AC = 3$ m ; $W = 180$ N ; $\alpha = 60°$ ; $\mu_C = 0.25$ ; $\mu_A = 0.35$ ; $W_M = 900$ N

**1. *Horizontal force (P) applied to the ladder at the floor level to prevent slipping***

The various forces acting on the ladder are shown in Fig. 7.15.

Let
$R_A$ = Normal reaction between the ladder and the floor at *A*,
$R_C$ = Normal reaction between the ladder and the wall at *C*,
$F_A$ = Force of friction between the floor and the ladder
$= \mu_A \cdot R_A = 0.35 R_A$, and
$F_C$ = Force of friction between the wall and the ladder
$= \mu_C \cdot R_C = 0.25 R_C$

Resolving the forces horizontally,

$$P + F_A = R_C$$

or
$$P + 0.35 R_A = R_C \quad \text{...(i)}$$

Now resolving the forces vertically,

$$R_A + F_C = W + W_M$$
$$= 180 + 900$$

or
$$R_A + 0.25\,R_C = 1080 \quad \text{...(ii)}$$

In Fig. 7.15, *G* is the middle point of the ladder where a weight of the ladder acts. Thus $AG = \dfrac{AB}{2} = \dfrac{3}{2} = 1.5$ m.

From the geometry of the figure,

$$BA = AC \cos 60°$$
$$= 3 \times 0.5$$
$$= 1.5 \text{ m}$$
$$DA = AG \cos 60°$$
$$= 1.5 \times 0.5$$
$$= 0.75 \text{ m}$$
$$CB = AC \sin 60°$$
$$= 3 \times 0.866$$
$$= 2.598 \text{ m}$$

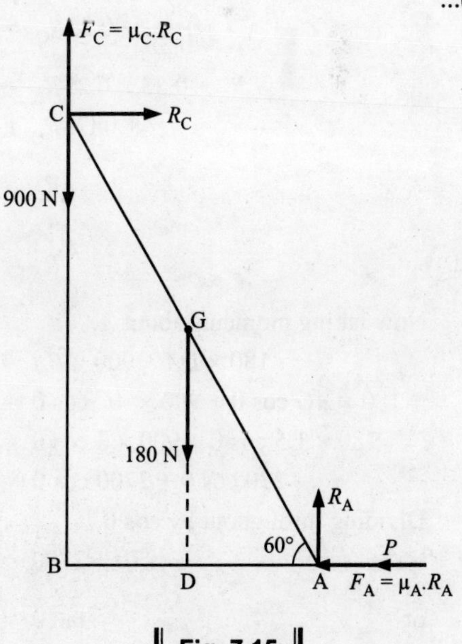


Fig. 7.15

Taking moments about $A$, we have

$$180 \times DA + 900 \times BA = F_C \times BA + R_C \times CB$$
$$180 \times 0.75 + 900 \times 1.5 = 0.25 R_C \times 1.5 + R_C \times 2.598$$
$$135 + 1350 = 0.375 R_C + 2.598 R_C$$
$$1485 = 2.973 R_C$$
$$\therefore \quad R_C = \frac{1485}{2.973} = 499.5 \text{ N}$$

Substituting the value of $R_C$ in equation (*ii*),

$$R_A + 0.25 \times 499.5 = 1080$$
$$R_A + 124.875 = 1080$$

or

$$R_A = 1080 - 124.875 = 955.125 \text{ N}$$

Now from equation (*i*), $P + 0.35 R_A = R_C$

$$\therefore \quad P = R_C - 0.35 R_A = 499.5 - 0.35 \times 955.125$$
$$= 499.5 - 334.3 = 165.2 \text{ N} \quad \textbf{Ans.}$$

**2. *Minimum angle of inclination of the ladder with the horizontal (Refer Fig. 7.16)***

Let $\theta$ = Minimum angle of inclination of the ladder with the floor to prevent slipping.

In this case, horizontal force $P$ is not applied.

Resolving the forces horizontally,

$$F_A = R_C$$
$$\mu_A . R_A = R_C$$
$$0.35 R_A = R_C \quad ...(i)$$

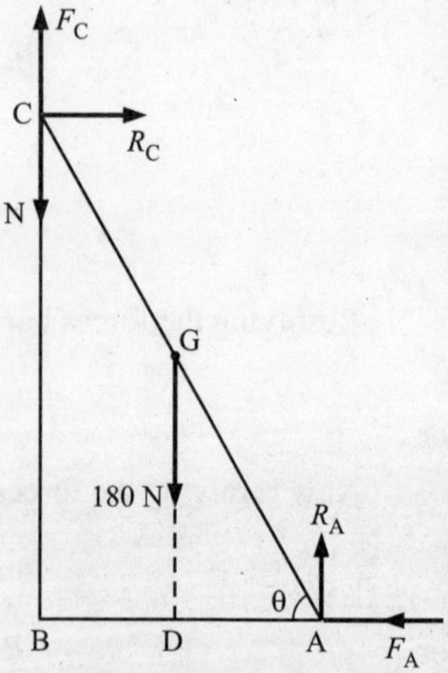


Fig. 7.16

Now resolving the forces vertically,

$$R_A + F_C = 900 + 180$$
$$R_A + \mu_C . R_C = 1080$$
$$\frac{R_C}{0.35} + 0.25 R_C = 1080 \quad \text{...[From equation } (i)]$$
$$R_C + 0.35 \times 0.25 R_C = 0.35 \times 1080$$
$$1.0875 R_C = 378$$
$$\therefore \quad R_C = \frac{378}{1.0875} = 347.6 \text{ N}$$

and

$$R_A = \frac{R_C}{0.35} = \frac{347.6}{0.35} = 933 \text{ N}$$

Now taking moments about $A$,

$$180 \times DA + 900 \times BA = F_C \times BA + R_C \times CB$$
$$180 \times AG \cos\theta + 900 \times AC \cos\theta = 0.25 R_C \times AC \cos\theta + R_C \times AC \sin\theta$$
$$180 \times 1.5 \cos\theta + 900 \times 3 \cos\theta = 0.25 \times 347.6 \times 3 \cos\theta + 347.6 \times 3 \sin\theta$$
$$270 \cos\theta + 2700 \cos\theta = 260.7 \cos\theta + 1042.8 \sin\theta$$

Dividing throughout by $\cos\theta$,

$$270 + 2700 = 260.7 + 1042.8 \tan\theta$$

or

$$\tan\theta = \frac{270 + 2700 - 260.7}{1042.8} = 2.598$$
$$\therefore \quad \theta = \tan^{-1}(2.598) = 69° \quad \textbf{Ans.}$$

**Example 7.17** *Two equal ladders OA and OB, whose length are l and weight W, are joined at point O. They rest on a rough horizontal plane for which the coefficient of friction is μ. If the ladders are in equilibrium condition, prove that the angle AOB is equal to* $\sin^{-1}\left[\dfrac{4\mu}{1+4\mu^2}\right]$.

***Solution*** Given : $OA = OB = l$

The ladders resting on a rough horizontal plane $AB$ is shown in Fig. 7.17.

Let $\angle AOB = \alpha$

Consider the equilibrium of the ladder $OA$. The various forces acting on this ladder are as follows:

(*a*) Weight $W$ acting downwards at $D$,

(*b*) Normal reaction at $A$ ($R_{NA} = W$),

(*c*) Force of friction ($F_A = \mu \cdot R_{NA}$) acting at $A$ towards right, as shown in the figure.

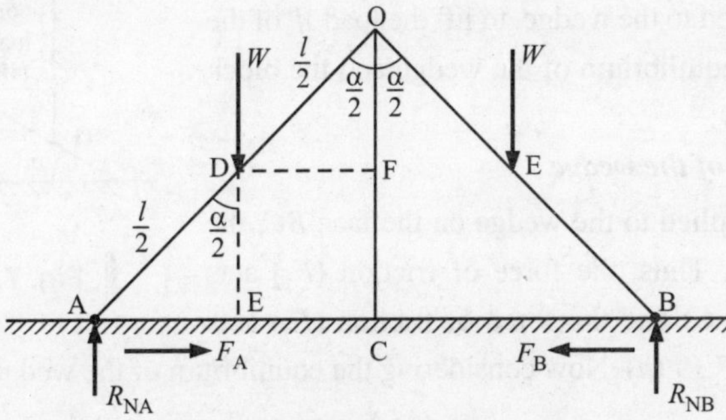


**Fig. 7.17**

Now taking moment of all the forces about $O$, and equating clockwise moment to anticlockwise moment,

$$R_{NA} \times AC = W \times EC + F_A \times OC$$

$$W \times OA \sin\frac{\alpha}{2} = W \times OD \sin\frac{\alpha}{2} + \mu.R_{NA} \times OA \cos\frac{\alpha}{2}$$

$$W \times l \sin\frac{\alpha}{2} = W \times \frac{l}{2} \sin\frac{\alpha}{2} + \mu W \times l \cos\frac{\alpha}{2}$$

$$\sin\frac{\alpha}{2} = \frac{1}{2}\sin\frac{\alpha}{2} + \mu\cos\frac{\alpha}{2}$$

$$\sin\frac{\alpha}{2} - \frac{1}{2}\sin\frac{\alpha}{2} = \mu\cos\frac{\alpha}{2}$$

or $$\frac{1}{2}\sin\frac{\alpha}{2} = \mu\cos\frac{\alpha}{2}$$

or $$\frac{\sin\frac{\alpha}{2}}{\cos\frac{\alpha}{2}} = 2\mu$$

∴ $$\tan\frac{\alpha}{2} = 2\mu$$

We know that $$^*\sin\alpha = \frac{2\tan\frac{\alpha}{2}}{1+\tan^2\frac{\alpha}{2}} = \frac{2\times 2\mu}{1+4\mu^2} = \frac{4\mu}{1+4\mu^2}$$

$$\therefore \qquad \alpha = \angle AOB = \sin^{-1}\left[\frac{4\mu}{1+4\mu^2}\right]$$

## 7.9 WEDGE FRICTION

A wedge is a piece of wood or metal and is, usually, of a triangular or trapezoidal in cross-section, as shown in Fig. 7.18. It is used either for lifting heavy loads or for slight adjustment in the position of a body, *i.e.* for tightening fits or keys for shafts.

Consider a wedge *ABC* lifting a block *DEFG* of weight ($W$) by an effort or force $P$, acting on the wedge, as shown in Fig. 7.19 (*a*). In order to find the force $P$ applied to the wedge, to lift the load $W$ of the block, we shall consider the equilibrium of the wedge and the block separately, as discussed below:

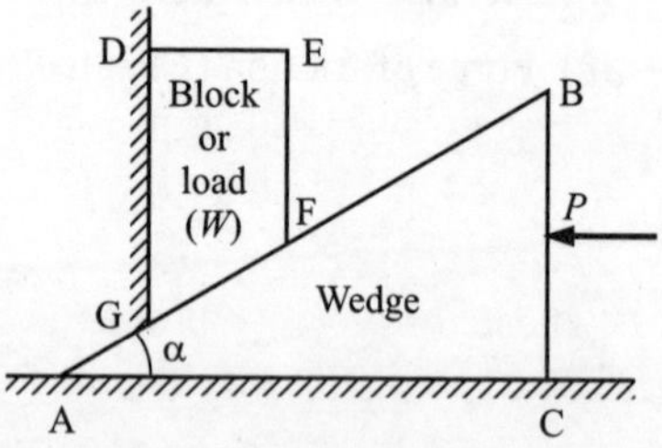


**Fig. 7.18.** *Wedge friction*

***Considering the equilibrium of the wedge***

When the force $P$ is applied to the wedge on the face *BC*, the face *CA* moves towards left. Thus, the force of friction ($F_1$) acts towards right (*i.e.* from *A* to *C*) and the force of friction ($F_2$) acts from *A* to *B* as shown in Fig. 7.19 (*a*). Now considering the equilibrium of the wedge, the various forces acting on this, are as follows:

**(*a*)** Force $P$ which acts horizontally on the face *BC*.

**(*b*)** Reaction ($R_1$), which is the resultant of normal reaction ($R_{N1}$) and force of friction ($F_1$) on the face *AC*. Let this reaction $R_1$ makes an angle $\phi_1$ with the normal reaction ($R_{N1}$).

**(*c*)** Reaction ($R_2$), which is the resultant of normal reaction ($R_{N2}$) and force of friction ($F_2$) on the face *AB*. Let this reaction $R_2$ makes an angle $\phi_2$ with the normal reaction ($R_{N2}$).

Now resolving the forces horizontally, we have

$$P = R_1\sin\phi_1 + R_2\sin(\phi_2+\alpha) \qquad \text{...[Refer Fig. 7.19 (b)]}$$

and resolving the forces vertically, we have

$$R_1\cos\phi_1 = R_2\cos(\phi_2+\alpha)$$

---

$$^* \quad \frac{2\tan\frac{\alpha}{2}}{1+\tan^2\frac{\alpha}{2}} = \frac{2\sin\frac{\alpha}{2}}{\cos\frac{\alpha}{2}} \times \frac{1}{\left(1+\dfrac{\sin^2\frac{\alpha}{2}}{\cos^2\frac{\alpha}{2}}\right)}$$

$$= \frac{2\sin\frac{\alpha}{2}}{\cos\frac{\alpha}{2}} \times \frac{\cos^2\frac{\alpha}{2}}{\left(\cos^2\frac{\alpha}{2}+\sin^2\frac{\alpha}{2}\right)} = 2\sin\frac{\alpha}{2}\times\cos\frac{\alpha}{2} = \sin\alpha$$

$$\text{...}\left(\because \sin A = 2\sin\frac{A}{2}\cos\frac{A}{2}\right)$$

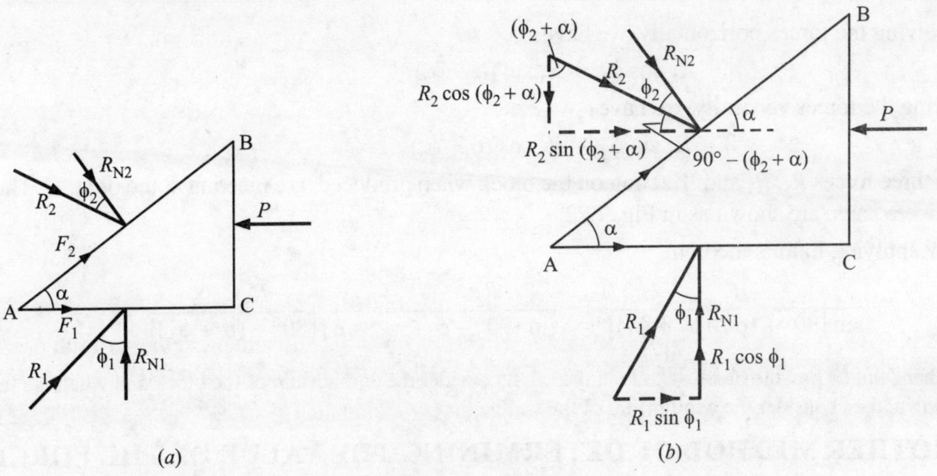


**Fig. 7.19**

The value of $P$ may also be obtained by applying Lami's theorem to the three forces, *i.e.* $R_1$, $R_2$ and $P$. These forces when produced meet at a point $O$ and the various angles between them are as shown in Fig. 7.20.

Now applying Lami's theorem, we have

$$\frac{P}{\sin[180° - (\phi_1 + \phi_2 + \alpha)]} = \frac{R_1}{\sin[90° + (\phi_2 + \alpha)]} = \frac{R_2}{\sin(90° + \phi_1)}$$

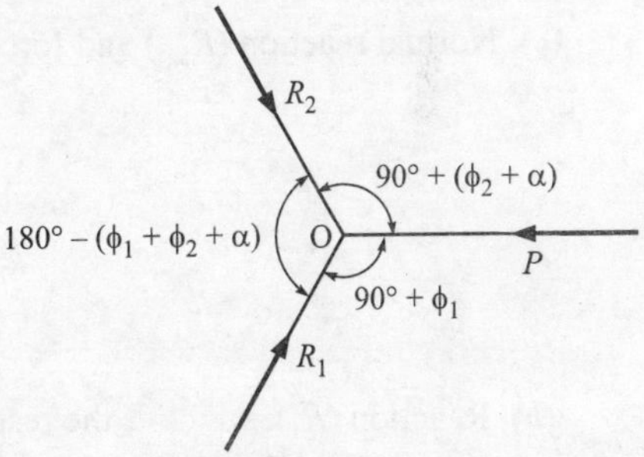


**Fig. 7.20**

***Considering the equilibrium of the block***

The various forces acting on the face $DG$ and $FG$ of the block $DEFG$, as shown in Fig. 7.21, are as follows:

**(*a*)** Normal reaction ($R_{N3}$) and force of friction ($F_3$) on the face $DG$. $R_3$ is the resultant reaction of $R_{N3}$ and $F_3$, as shown in the figure.

**(*b*)** Normal reaction ($R_{N2}$) and force of friction ($F_2$) on the face $FG$. $R_2$ is the resultant reaction of $R_{N2}$ and $F_2$.

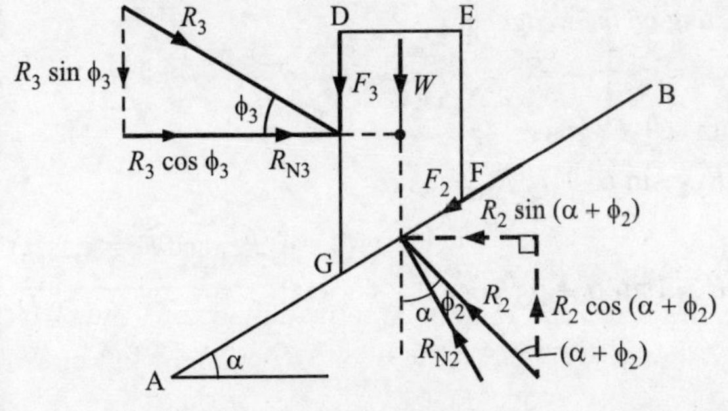


**Fig. 7.21**

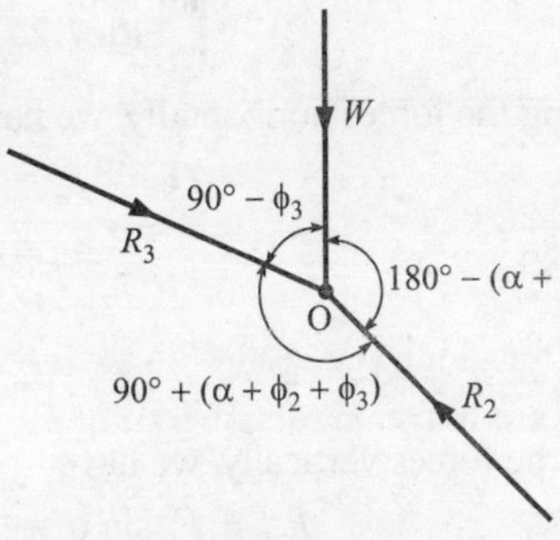


**Fig. 7.22**

Resolving the forces horizontally, we have

$$R_3 \cos \phi_3 = R_2 \sin (\alpha + \phi_2)$$

and resolving the forces vertically, we have

$$W + R_3 \sin \phi_3 = R_2 \cos (\alpha + \phi_2)$$

The three forces $R_2$, $R_3$ and $W$ acting on the block when produced, are meeting at the point $O$. The various angles between them are shown as in Fig. 7.22.

Now applying Lami's theorem,

$$\frac{W}{\sin [90° + (\alpha + \phi_2 + \phi_3)]} = \frac{R_2}{\sin (90° - \phi_3)} = \frac{R_3}{\sin [180° - (\alpha + \phi_2)]}$$

**Note:** When load ($W$) on the block is known, then first consider the equilibrium of the block and when the force $P$ is known, then consider the equilibrium of the wedge first.

## 7.10 ANOTHER METHOD OF DETERMINING THE VALUE OF THE FORCE *P*

First of all considering the equilibrium of the wedge. The various forces acting on the wedge $ABC$, as shown in Fig. 7.23, are as follows:

***(a)*** Force $P$ on face $BC$,

***(b)*** Normal reaction ($R_{N1}$) and force of friction ($F_1 = \mu_1 \cdot R_{N1}$) on the face $AC$, and

***(c)*** Normal reaction ($R_{N2}$) and force of friction ($F_2 = \mu_2 \cdot R_{N2}$) on the face $AB$.

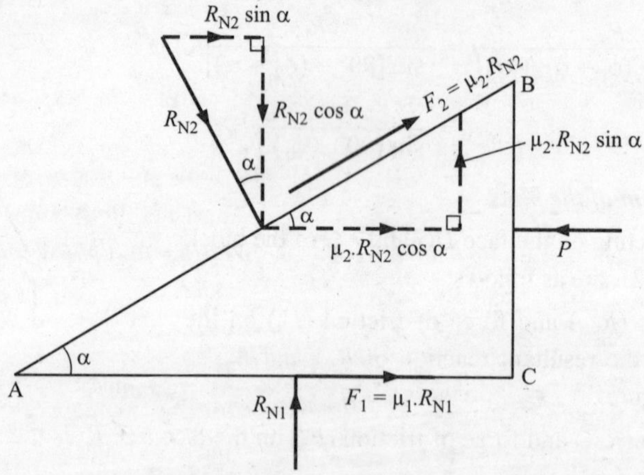


**Fig. 7.23** *Forces acting on the wedge*

Resolving the forces horizontally, we have

$$P = F_1 + R_{N2} \sin \alpha + F_2 \cos \alpha$$

$$= \mu_1.R_{N1} + R_{N2} \sin \alpha + \mu_2.R_{N2} \cos \alpha$$

...($\because$ $F_1 = \mu_1.R_{N1}$ and $F_2 = \mu_2.R_{N2}$)

$$= \mu_1.R_{N1} + R_{N2} (\sin \alpha + \mu_2 \cos \alpha) \quad ...(i)$$

and resolving the forces vertically, we have

$$R_{N1} + F_2 \sin \alpha = R_{N2} \cos \alpha$$

or

$$R_{N1} + \mu_2 \cdot R_{N2} \sin \alpha = R_{N2} \cos \alpha$$

$\therefore$

$$R_{N1} = R_{N2} \cos \alpha - \mu_2 \cdot R_{N2} \sin \alpha$$

Substituting the value of $R_{N1}$ in equation (*i*), we have

$$P = \mu_1 [R_{N2} \cos\alpha - \mu_2 . R_{N2} \sin\alpha] + R_{N2}(\sin\alpha + \mu_2 \cos\alpha)$$

$$= R_{N2}[\mu_1 \cos\alpha - \mu_1\mu_2 \sin\alpha + \sin\alpha + \mu_2 \cos\alpha]$$

$$= R_{N2}[\cos\alpha(\mu_1 + \mu_2) + \sin\alpha(1 - \mu_1\mu_2)] \quad ...(ii)$$

Now considering the equilibrium of the block *DEFG*. The various forces acting on the block, as shown in Fig. 7.24, are as follows:

(*a*) Weight of the block ($W$),

(*b*) Normal reaction ($R_{N2}$) and force of friction ($F_2 = \mu \cdot R_{N2}$) on the face *FG*, and

(*c*) Normal reaction ($R_{N3}$) and force of friction ($F_3 = \mu_3 \cdot R_{N3}$) on the face *DG*.

Resolving the forces horizontally, we have

$$R_{N3} = R_{N2} \sin\alpha + F_2 \cos\alpha$$

$$= R_{N2} \sin\alpha + \mu_2 \cdot R_{N2} \cos\alpha$$

$$= R_{N2} (\sin\alpha + \mu_2 \cos\alpha) \quad ...(iii)$$

and resolving the forces vertically, we have

$$R_{N2} \cos\alpha = W + F_3 + F_2 \sin\alpha$$

or

$$R_{N2} \cos\alpha = W + \mu_3 \cdot R_{N3} + \mu_2 \cdot R_{N2} \sin\alpha \quad ...(\because F_3 = \mu_3 \cdot R_{N3})$$

$$\therefore \quad W = R_{N2} \cos\alpha - \mu_3 \cdot R_{N3} - \mu_2 \cdot R_{N2} \sin\alpha$$

$$= R_{N2} (\cos\alpha - \mu_2 \sin\alpha) - \mu_3 \cdot R_{N3}$$

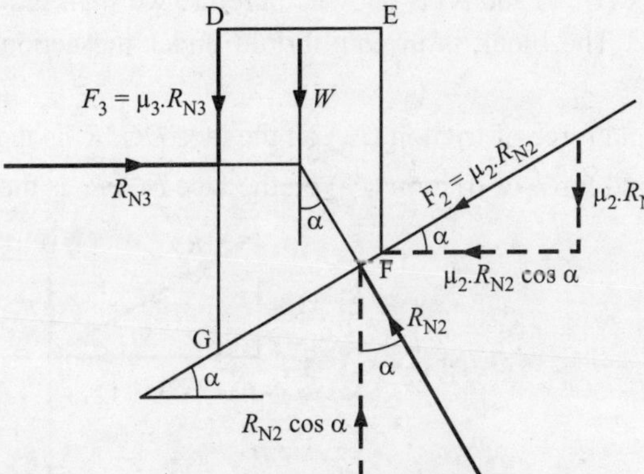


**Fig. 7.24** *Forces acting on the block*

Substituting the value of $R_{N3}$ from equation (*iii*), we have

$$W = R_{N2} (\cos\alpha - \mu_2 \sin\alpha) - \mu_3[R_{N2} (\sin\alpha + \mu_2 \cos\alpha)]$$

$$= R_{N2} \cos\alpha - \mu_2 \cdot R_{N2} \sin\alpha - \mu_3 \cdot R_{N2} \sin\alpha - \mu_3\mu_2 \cdot R_{N2} \cos\alpha$$

$$= R_{N2} (\cos\alpha - \mu_2 \sin\alpha - \mu_3 \sin\alpha - \mu_2\mu_3 \cos\alpha)$$

$$= R_{N2} [\cos\alpha(1 - \mu_2\mu_3) - \sin\alpha(\mu_2 + \mu_3)] \quad ...(iv)$$

Dividing equation (*ii*) by equation (*iv*),

$$\frac{P}{W} = \frac{R_{N2} [\cos\alpha(\mu_1 + \mu_2) + \sin\alpha(1 - \mu_1\mu_2)]}{R_{N2} [\cos\alpha(1 - \mu_2\mu_3) - \sin\alpha(\mu_2 + \mu_3)]}$$

$$= \frac{\cos\alpha\,(\mu_1+\mu_2)+\sin\alpha\,(1-\mu_1\mu_2)}{\cos\alpha\,(1-\mu_2\mu_3)-\sin\alpha\,(\mu_2+\mu_3)}$$

From this expression, the value of $P$ may be evaluated.

**Note:** In case the coefficient of friction between all the surfaces is same, then $\mu_1 = \mu_2 = \mu_3 = \mu$ (say).

$$\therefore \quad \frac{P}{W} = \frac{2\mu\cos\alpha + \sin\alpha\,(1-\mu^2)}{\cos\alpha\,(1-\mu^2) - 2\mu\sin\alpha}$$

**Example 7.18** *A block lying over a 10° wedge on a horizontal floor and leaning against a vertical wall and weighing 1500 N is to be raised by applying a horizontal force to the wedge. Assuming coefficient of friction between all the surfaces in contact to be 0.3, determine the minimum horizontal force to be applied to raise the block.*

***Solution*** Given : $\alpha = 10°$ ; $W = 1500$ N ; $\mu_1 = \mu_2 = \mu_3 = \mu = 0.3$

A block lying over the wedge is shown in Fig. 7.25.

Since coefficient of friction, $\mu = \tan\phi$, therefore

$$\tan\phi = \tan\phi_1 = \tan\phi_2 = \tan\phi_3 = 0.3$$

or $$\phi = \tan^{-1}(0.3) = 16.7°$$

This example may be solved by any one of the following three methods:

**First Method:** ***Considering equilibrium of forces***

Since the weight of the block ($W = 1500$ N) is known, therefore we shall first consider the equilibrium of the forces acting on the block. The block is in equilibrium under the action of the following forces (Refer Fig. 7.26).

(*a*) Normal reaction ($R_{N3}$) and force of friction ($F_3$) on the face *DG*. $R_3$ is the resultant of $R_{N3}$ and $F_3$.

(*b*) Normal reaction ($R_{N2}$) and force of friction ($F_2$) on the face *FG*. $R_2$ is the resultant of $R_{N2}$ and $F_2$.

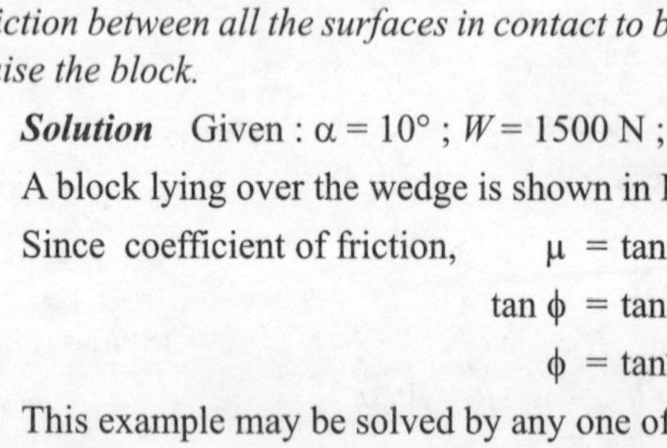


Fig. 7.25

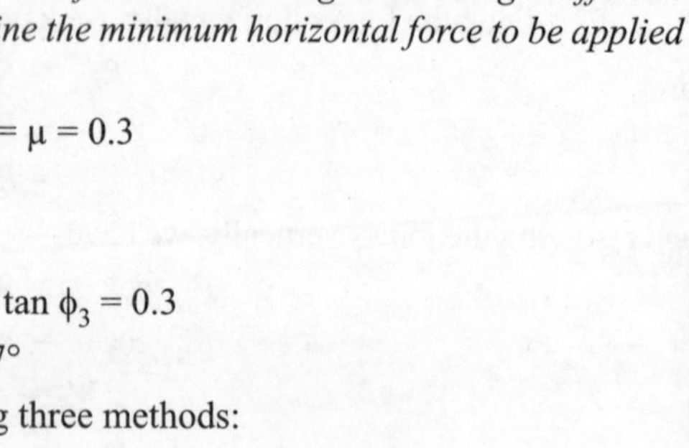


Fig. 7.26

Now resolving the forces horizontally,

$$R_3\cos\phi_3 = R_2\sin(\alpha+\phi_2)$$

or $$R_3\cos\phi = R_2\sin(\alpha+\phi) \qquad ...(\because\ \phi_3=\phi_2=\phi)$$

$$R_3\cos 16.7° = R_2\sin(10° + 16.7°)$$

$$\therefore \quad R_3 = \frac{R_2\sin 26.7°}{\cos 16.7°} = \frac{R_2\times 0.4493}{0.9578} = 0.469R_2 \qquad ...(i)$$

Again resolving the forces vertically,

$$W + R_3 \sin \phi_3 = R_2 \cos (\alpha + \phi_2)$$

or

$$W + R_3 \sin \phi = R_2 \cos (\alpha + \phi)$$

$$1500 + R_3 \sin 16.7° = R_2 \cos (10° + 16.7°)$$

$$1500 + R_3 \times 0.2874 = R_2 \times 0.8934$$

Substituting the value of $R_3$ from equation (*i*), we have

$$1500 + 0.469R_2 \times 0.2874 = R_2 \times 0.8934$$

$$1500 + 0.1348R_2 = 0.8934R_2$$

$$\therefore \quad R_2 = \frac{1500}{0.8934 - 0.1348} = 1977.3 \text{ N}$$

and

$$R_3 = 0.469R_2 = 0.469 \times 1977.3 = 927.4 \text{ N}$$

Let $P$ = Minimum horizontal force to be applied to raise the block.

Now considering the equilibrium of the wedge *ABC*. The wedge is in equilibrium under the action of the following forces (Refer Fig. 7.27):

(*a*) Force $P$ acting horizontally,

(*b*) Normal reaction ($R_{N1}$) and force of friction ($F_1$) on the face *AC*. $R_1$ is the resultant of $R_{N1}$ and $F_1$, and

(*c*) Normal reaction ($R_{N2}$) and force of friction ($F_2$) on the face *AB*. $R_2$ is the resultant of $R_{N2}$ and $F_2$.

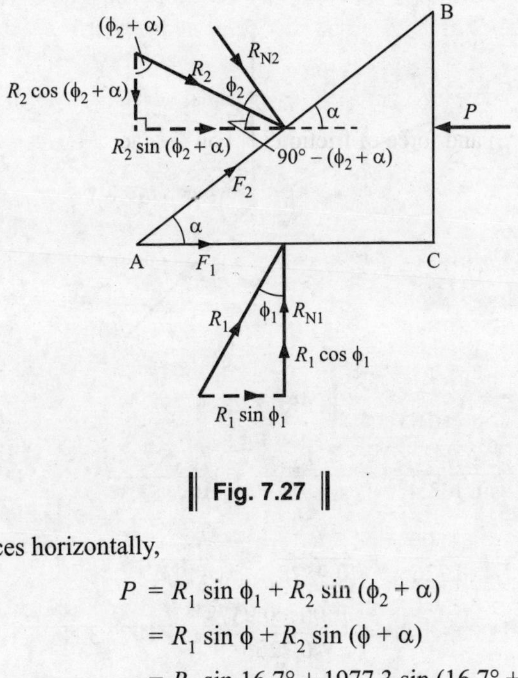


**Fig. 7.27**

Now resolving the forces horizontally,

$$P = R_1 \sin \phi_1 + R_2 \sin (\phi_2 + \alpha)$$

$$= R_1 \sin \phi + R_2 \sin (\phi + \alpha)$$

$$= R_1 \sin 16.7° + 1977.3 \sin (16.7° + 10°)$$

$$= R_1 \times 0.2874 + 1977.3 \times 0.4493$$

$$= 0.2874R_1 + 888.4 \qquad ...(i)$$

Again resolving the forces vertically,

$$R_1 \cos \phi_1 = R_2 \cos (\phi_2 + \alpha)$$

or

$$R_1 \cos \phi = R_2 \cos (\phi + \alpha)$$

$$R_1 \cos 16.7° = 1977.3 \cos (16.7° + 10°)$$

$$R_1 \times 0.9578 = 1977.3 \times 0.8934 = 1766.5$$

$$\therefore \quad R_1 = \frac{1766.5}{0.9578} \quad 1844.3 \text{ N}$$

Substituting the value of $R_1$ in the above equation (*i*), we have

$$P = 0.2874 \times 1844.3 + 888.4 = 1418.4 \text{ N} \textbf{ Ans.}$$

**Second Method: *Lami's Theorem***

First of all, consider the equilibrium of the block under the action of three forces, *i.e.* $R_2$, $R_3$ and $W$. These forces when produced meet at a point $O$. The various angles between them are shown in Fig. 7.28.

Now applying Lami's theorem,

$$\frac{W}{\sin [90° + (\alpha + \phi_2 + \phi_3)]} = \frac{R_2}{\sin (90° - \phi_3)} = \frac{R_3}{\sin [180° - (\alpha + \phi_2)]}$$

$$\frac{W}{\sin [90° + (\alpha + 2\phi)]} = \frac{R_2}{\sin (90° - \phi)} = \frac{R_3}{\sin [180° - (\alpha + \phi)]}$$

$$\frac{W}{\sin [90° + (10° + 2 \times 16.7°)]} = \frac{R_2}{\sin (90° - 16.7°)} = \frac{R_3}{\sin [180° - (10 + 16.7°)]}$$

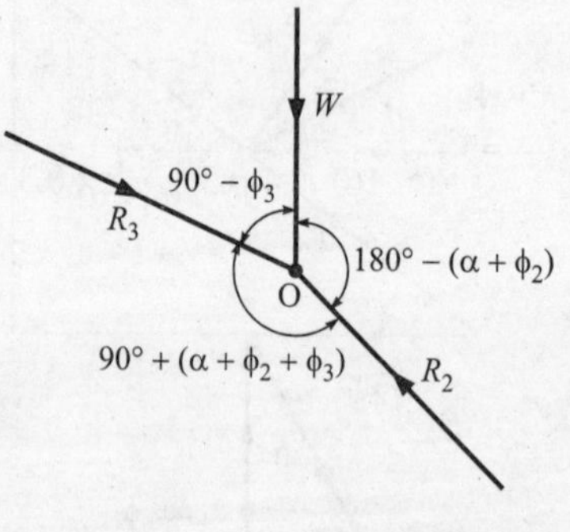


**Fig. 7.28**

$$\frac{W}{\sin 133.4°} = \frac{R_2}{\sin 73.3°} = \frac{R_3}{\sin 153.3°}$$

$$\frac{1500}{0.7266} = \frac{R_2}{0.9578} = \frac{R_3}{0.4493}$$

$$\therefore \quad R_2 = \frac{1500 \times 0.9578}{0.7266} = 1977.3 \text{ N}$$

and

$$R_3 = \frac{1500 \times 0.4493}{0.7266} = 927.5 \text{ N}$$

Now considering the equilibrium of the wedge under the action of three forces, *i.e.* $R_1$, $R_2$ and $P$. These forces when produced meet at a point $O$. The various angles between them are shown in Fig. 7.29.

Now applying Lami's theorem, we have

$$\frac{P}{\sin[180° - (\phi_1 + \phi_2 + \alpha)]} = \frac{R_1}{\sin[90° + (\phi_2 + \alpha)]}$$

$$= \frac{R_2}{\sin(90° + \phi_1)}$$

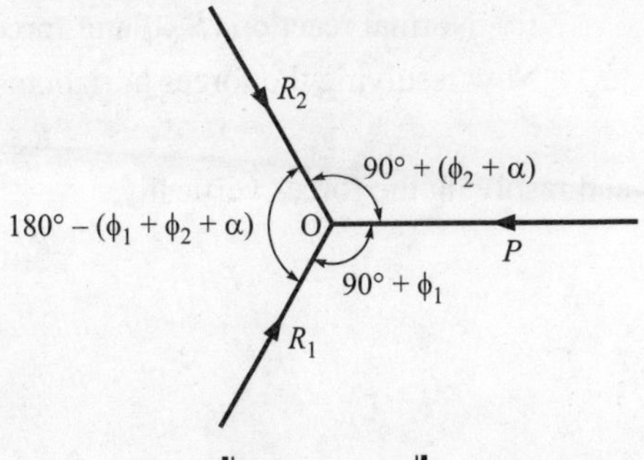


Fig. 7.29

Since the value of $R_2$ = 1977.3 N has been calculated above, therefore we shall take

$$\frac{P}{\sin[180° - (2\phi + \alpha)]} = \frac{R_2}{\sin(90° + \phi)}$$

$$\frac{P}{\sin[180° - (2 \times 16.7° + 10°)]} = \frac{1977.3}{\sin(90° + 16.7°)}$$

$$\frac{P}{\sin 136.6°} = \frac{1977.3}{\sin 106.7°}$$

$$\therefore \quad P = \frac{1977.3 \times \sin 136.6°}{\sin 106.7°} = \frac{1977.3 \times 0.687}{0.9578}$$

$$= 1418.3 \text{ N} \quad \textbf{Ans.}$$

**Third Method:** ***Applying the formula as derived in Sec. 7.10***

We know that

$$\frac{P}{W} = \frac{2\mu \cos\alpha + \sin\alpha(1 - \mu^2)}{\cos\alpha(1 - \mu^2) - 2\mu \sin\alpha}$$

$$= \frac{2 \times 0.3 \cos 10° + \sin 10° (1 - 0.3^2)}{\cos 10° (1 - 0.3^2) - 2 \times 0.3 \sin 10°}$$

$$= \frac{0.6 \times 0.9848 + 0.1736(1 - 0.09)}{0.9848(1 - 0.09) - 0.6 \times 0.1736} = \frac{0.591 + 0.158}{0.896 - 0.104} = 0.9457$$

$$\therefore \quad P = 0.9457\,W = 0.9457 \times 1500 = 1418.5 \text{ N} \quad \textbf{Ans.}$$

**Example 7.19** *Figure 7.30 shows a wedge and a block. Calculate the force P that will cause the wedge and block to skid. Take μ = 0.2 for all surfaces.*

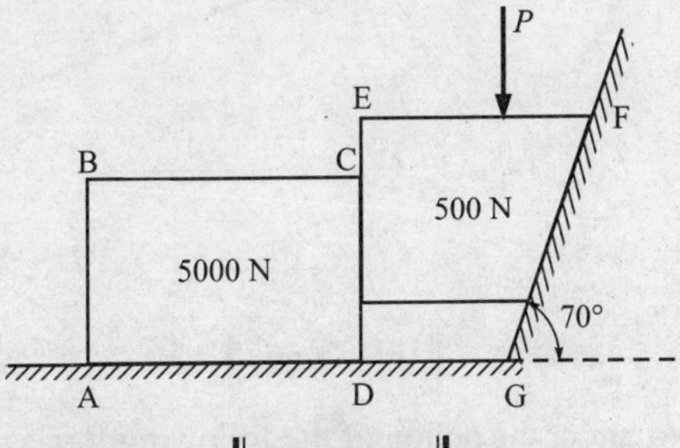


Fig. 7.30

***Solution*** Given : $W$ = 5000 N ; $W_1$ = 500 N ; μ for all surfaces = 0.2

First of all, let us consider the equilibrium of the block *ABCD*, as shown in Fig. 7.31. The block is in equilibrium under the action of the following forces:

(*a*) Weight of the block ($W = 5000$ N),
(*b*) Normal reaction ($R_{N1}$) and force of friction ($F_1 = \mu \cdot R_{N1}$) acting on the face *AD*, and
(*c*) Normal reaction ($R_{N2}$) and force of friction ($F_2 = \mu \cdot R_{N2}$) acting on the face *CD*.

Now resolving the forces horizontally,

$$R_{N2} = F_1 = \mu \cdot R_{N1} = 0.2R_{N1}$$

and resolving the forces vertically,

$$R_{N1} = 5000 + F_2$$
$$= 5000 + \mu \cdot R_{N2}$$
$$= 5000 + 0.2\,(0.2R_{N1})$$
$$= 5000 + 0.04R_{N1}$$
$$\dots(\because R_{N2} = 0.2R_{N1})$$
$$R_{N1} - 0.04R_{N1} = 5000$$

or
$$0.96R_{N1} = 5000$$

$\therefore$
$$R_{N1} = \frac{5000}{0.96} = 5208$$

and
$$R_{N2} = 0.2R_{N1}$$
$$= 0.2 \times 5208 = 1041.6 \text{ N}$$

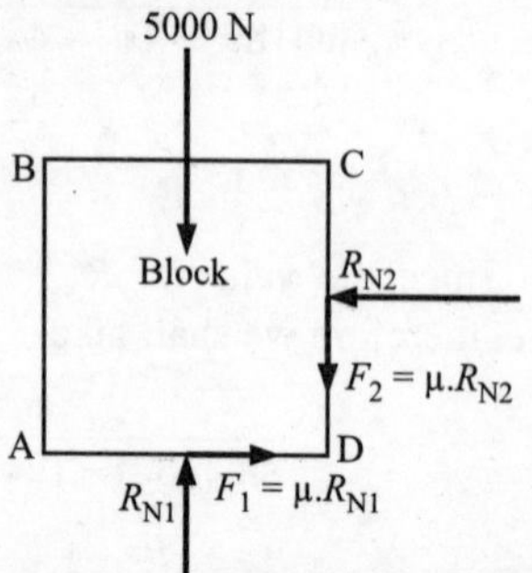


Fig. 7.31

Now considering the equilibrium of the wedge *DEFG*, as shown in Fig. 7.32 (*a*).

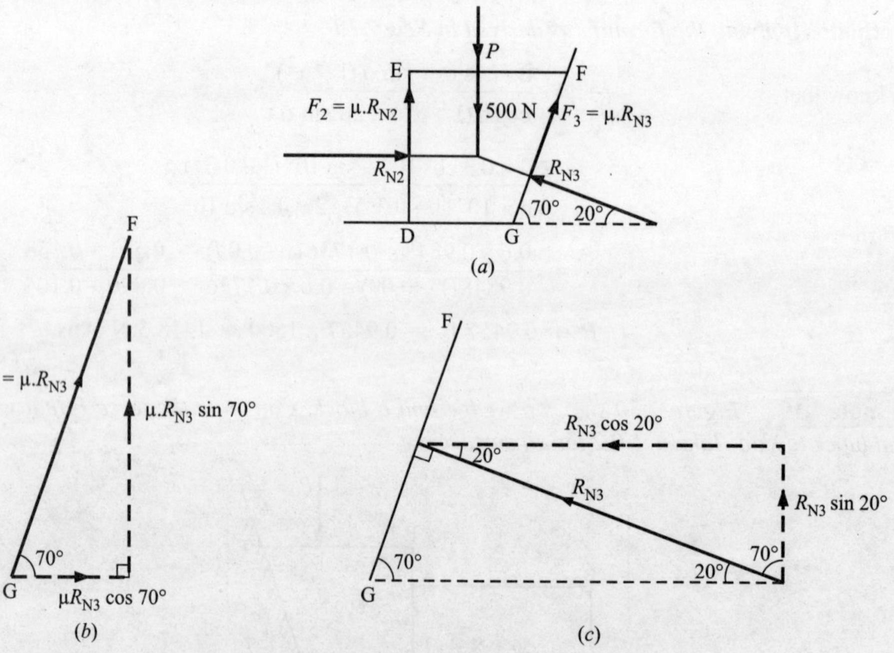


Fig. 7.32

The wedge is in equilibrium under the action of the following forces:
(*a*) Weight of the wedge ($W_1 = 500$ N) and the force *P*,
(*b*) Normal reaction ($R_{N3}$) and force of friction ($F_3 = \mu \cdot R_{N3}$) acting on the face *FG*, and
(*c*) Normal reaction ($R_{N2}$) and force of friction ($F_2 = \mu \cdot R_{N2}$) acting on the face *DE*.

Now resolving the forces horizontally,

$$R_{N2} + \mu \cdot R_{N3} \cos 70° = R_{N3} \cos 20°$$

$$1041.6 + 0.2R_{N3} \times 0.342 = R_{N3} \times 0.9397$$

$$1041.6 + 0.0684R_{N3} = 0.9397R_{N3}$$

$$\therefore \quad R_{N3} = \frac{1041.6}{0.9397 - 0.0684} = 1195.5 \text{ N}$$

Again resolving the forces vertically,

$$P + 500 = \mu \cdot R_{N2} + \mu \cdot R_{N3} \sin 70° + R_{N3} \sin 20°$$

$$= 0.2 \times 1041.6 + 0.2 \times 1195.5 \times 0.9397 + 1195.5 \times 0.342$$

$$= 208.32 + 224.68 + 408.86 = 841.86$$

$$\therefore \quad P = 841.86 - 500 = 341.86 \text{ N} \quad \textbf{Ans.}$$

**Example 7.20** *Two wedges A and B are employed to raise a weight of 2 kN acting on block C, by applying a force P, as shown in Fig. 7.33. The inclination of the inclined face of each wedge is 15° to the horizontal. Assuming coefficient of friction for all the surfaces as 0.25, find the value of force P required for the upward motion of the block.*

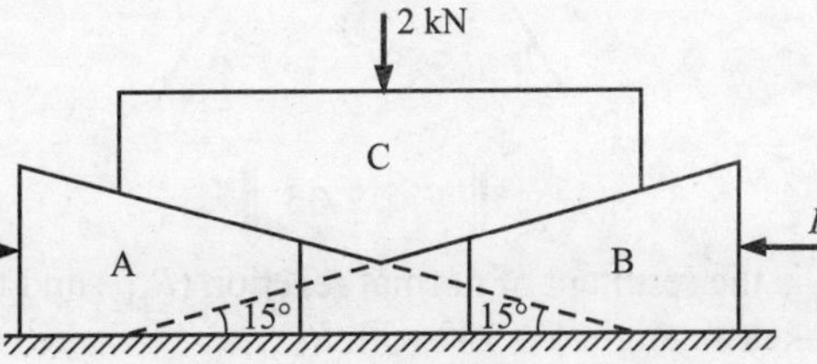


**Fig. 7.33**

***Solution*** Given : $W = 2$ kN ; $\alpha = 15°$ ; $\mu$ for all surfaces = 0.25

We know that Coefficient of friction,

$$\mu = \tan \phi = 0.25$$

$$\therefore \quad \phi = \tan^{-1} (0.25) = 14°$$

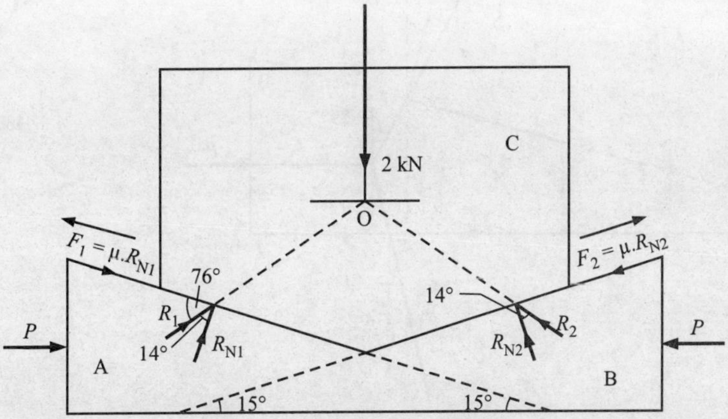


**Fig. 7.34**

Since the block $C$ tends to move upwards under the action of the forces $P$ on the wedges $A$ and $B$, therefore, the force of friction $F_1 = \mu \cdot R_{N1}$ and $F_2 = \mu \cdot R_{N2}$ will act downwards, as shown in Fig. 7.34. $R_{N1}$ and $R_{N2}$ are the normal reactions. Consider the equilibrium of the block $C$. The various forces acting on the block $C$ are as follows:

(*a*) Load on the block $C$, *i.e.* 2 kN.

(*b*) Reaction $R_1$ which is the resultant of normal reaction ($R_{N1}$) and the force of friction ($F_1 = \mu \cdot R_{N1}$). The reaction $R_1$ makes an angle $\phi = 14°$ with $R_{N1}$.

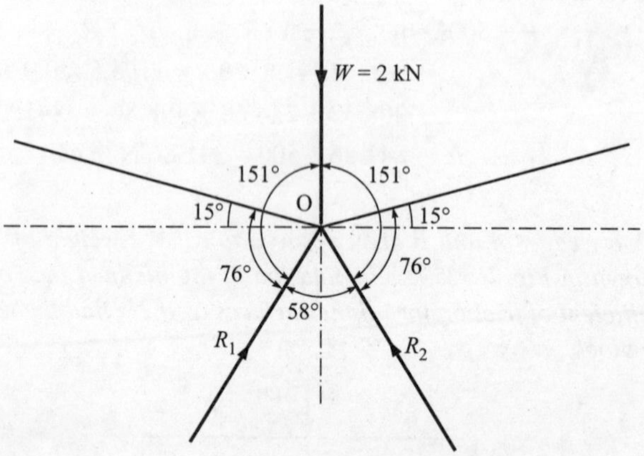


**Fig. 7.35**

(*c*) Reaction $R_2$ which is the resultant of normal reaction ($R_{N2}$) and the force of friction ($F_2 = \mu \cdot R_{N2}$). The reaction $R_2$ makes an angle $\phi = 14°$ with $R_{N2}$.

All the above three forces meet at a point $O$, as shown in Fig. 7.35. The angles between these forces are as under:

(*i*) Angle between $W = 2$ kN and $R_1$ and $R_2$ is [$90° + (76 - 15°) = 151°$], and

(*ii*) Angle between $R_1$ and $R_2$ is ($360° - 2 \times 151° = 58°$).

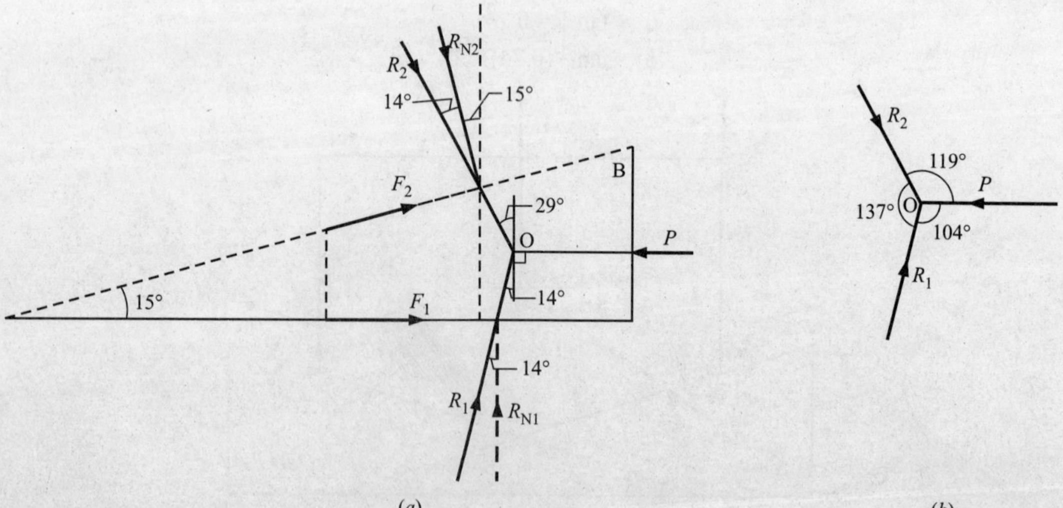


**Fig. 7.36**

Now applying Lami's theorem,

$$\frac{R_1}{\sin 151^\circ} = \frac{R_2}{\sin 151^\circ} = \frac{W}{\sin 58^\circ}$$

$$\therefore \quad R_1 = R_2 = \frac{W \times \sin 151^\circ}{\sin 58^\circ} = \frac{2 \times 0.4848}{0.848} = 1.143 \text{ kN}$$

Let us now consider the equilibrium of the wedge *B*. The various forces acting on the wedge *B*, as shown in Fig. 7.36, are as follows:

(*a*) Force *P* acting horizontally.

(*b*) Reaction $R_1$ which is the resultant of normal reaction ($R_{N1}$) and the force of friction ($F_1 = \mu \cdot R_{N1}$). The reaction $R_1$ makes an angle of $\phi = 14^\circ$ with $R_{N1}$.

(*c*) Reaction $R_2$ which is the resultant of normal reaction ($R_{N2}$) and the force of friction ($F_2 = \mu \cdot R_{N2}$). The reaction $R_2$ makes an angle of $\phi = 14^\circ$ with $R_{N2}$.

All the above three forces meet at a point *O*, as shown in Fig. 7.36 (*b*). The angles between these forces are as under:

(*i*) Angle between *P* and $R_1$ is ($90^\circ + 14^\circ = 104^\circ$); and

(*ii*) Angle between $R_1$ and $R_2$ is [$180^\circ - (29^\circ + 14^\circ) = 137^\circ$].

Now applying Lami's theorem,

$$\frac{P}{\sin 137^\circ} = \frac{R_2}{\sin 104^\circ}$$

$$\therefore \quad P = \frac{R_2 \times \sin 137^\circ}{\sin 104^\circ} = \frac{1.143 \times 0.682}{0.97} = 0.8036 \text{ kN} \quad \textbf{Ans.}$$

## HIGHLIGHTS

1. The screw jack is a device for lifting heavy loads, by applying a comparatively smaller effort at its handle.
2. Since the screw jack works on the same principle as that of an inclined plane (Chapter 6, Sec. 6.17), therefore proceeding in the same way, we find that the force required to overcome friction between the screw and nut or the force required at the circumference of the screw

$$P = W \tan (\alpha + \phi) \quad \text{...(to raise the load)}$$

$$= W \tan (\alpha - \phi) \quad \text{...(to lower the load)}$$

where $\alpha$ = Helix angle, and

$\phi$ = Angle of friction.

3. The torque required to overcome friction between the screw and nut is given by

$$T = P \times \frac{d}{2} = W \tan (\alpha + \phi) \frac{d}{2} \quad \text{...(to raise the load)}$$

$$= W \tan (\alpha - \phi) \frac{d}{2} \quad \text{...(to lower the load)}$$

where $d$ = Mean diameter of the screw.

4. If an effort $P_1$ is applied at the end of the lever at a distance *l* from the axis of the screw, then

$$T = P \times \frac{d}{2} = P_1 \times l$$

5. When the load (*W*) is taken up by a thrust collar or a flat surface, then torque required to overcome friction at the collar,

$$T_C = \mu_1 W \left( \frac{R_1 + R_2}{2} \right) = \mu W R$$

where $R_1$ and $R_2$ = Outside and inside radii of the collar,

$R$ = Mean radius of the collar, and

$\mu_1$ = Coefficient of friction for the collar.

$\therefore$ Total torque required to overcome friction (*i.e.* to rotate the screw)

$$= T + T_C$$

6. The efficiency of the screw jack may be defined as the ratio of the effort required to move the load neglecting friction (*i.e.* ideal effort, $P_0$) to the effort required to move the load considering friction (*i.e.* actual effort, $P$).
7. The effort required to move the load neglecting frictions, *i.e.* ideal effort,

$$P_0 = W \tan \alpha$$

8. The effort required to move the load considering friction, *i.e.* actual effort,

$$P = W \tan (\alpha + \phi) \quad \text{...(for raising the load)}$$

$$= W \tan (\alpha - \phi) \quad \text{...(for lowering the load)}$$

9. Efficiency of the screw jack is given by

$$\eta_S = \frac{P_0}{P} = \frac{W \tan \alpha}{W \tan (\alpha + \phi)} = \frac{\tan \alpha}{\tan (\alpha + \phi)} \quad \text{...(for raising the load)}$$

$$= \frac{\tan \alpha}{\tan (\alpha - \phi)} \quad \text{...(for lowering the load)}$$

This shows that the efficiency of the screw jack is independent of the load moved and the effort applied.

10. The efficiency of the screw jack is maximum, when

$$\alpha = \frac{\pi}{4} - \frac{\phi}{2}$$

and maximum efficiency of the screw jack is given by

$$\eta_{S\,(max)} = \frac{1 - \sin \phi}{1 + \sin \phi}$$

11. When load starts moving downwards without the application of any effort or torque, (or when $\alpha > \phi$), then it is known as *overhauling of screws*. On the other hand, when the effort is applied to lower the load (or when $\alpha < \phi$), then screw is known as *self-locking screw*.
12. For overhauling screws, the efficiency is greater than 50% and for self-locking screws, the efficiency is less than 50%.
13. A ladder consists of a set of horizontal bars of wood or metal fixed between two uprights.
14. For the ladder to be in equilibrium, the following conditions must be satisfied:
    - ***(a)*** The algebraic sum of the horizontal components of the forces, acting on the ladder, must be equal to zero.
    - ***(b)*** The algebraic sum of the vertical components of the forces, acting on the ladder, must be equal to zero.
    - ***(c)*** The algebraic sum of the moments of the various forces, acting on the ladder, must be equal to zero.
15. A wedge is a piece of wood or metal and is, usually, of a triangular or trapezoidal in cross-section. It is used either for lifting heavy loads or for slight adjustment in the position of a body, *i.e.* for tightening fits or keys for shafts.

## NUMERICAL PROBLEMS

1. It is required to lift a weight of 30 kN by a square threaded screw jack. The mean diameter of the screw is 40 mm whose pitch is 8 mm. If the coefficient of friction is 0.1, find the effort applied at the end of a lever of 500 mm length. [**Ans.** 197.6 N]
2. A screw jack has a square thread of 75 mm mean diameter and a pitch of 15 mm. If the coefficient of friction between the screw and thread is 0.05, find the force required at the end of a 500 mm long lever to lower a load of 25 kN. [**Ans.** 25.5 N]
3. A square threaded screw jack is required to lift a load of 4 kN. The outer diameter of the screw is 50 mm and pitch is 10 mm. If the coefficient of friction is 0.1, determine the force required to be applied at the end of a 700 mm long lever. What will be its magnitude for the load to be lowered? [**Ans.** 22.1 N; 3.75 N]
4. A screw jack is required to lift a load of 2000 N. The mean diameter and pitch of the screw are 70 mm and 10 mm. If the coefficient of friction is 0.03, find the efficiency of the screw jack. State whether the screw jack is self-locking or not. [**Ans.** 60%]
5. A ladder 6.75 m long rests with its lower end on rough horizontal floor and its upper end leaning against a smooth vertical wall, the lower end being at a distance of 2.25 m from the wall. A man weighing 680 N stands on a ring of the ladder at a distance of 2 m measured along the ladder from its foot. The weight of the ladder is 320 N and acts at its midpoint. Calculate the minimum frictional resistance that needs to be developed by the floor to ensure equilibrium. If actual resistance offered is somewhat less than this calculated value, how will you place the ladder to lessen its tendency to slip. Justify your answer.
   [**Ans.** 127.75 N; Foot of ladder should be shifted towards wall to increase the moment of frictional resistance]
6. The bar, as shown in Fig. 7.37, is non-homogeneous. Its weight of 900 N may be considered to be concentrated at the point 6 m from the bottom. It rests against a small wall at *C* and on a rough floor at *A*. The coefficient of static friction between the bar and floor is 1/3. Establish by calculations if the bar would stand in the 60° position shown. [**Ans.** The bar would stand in the 60° position]

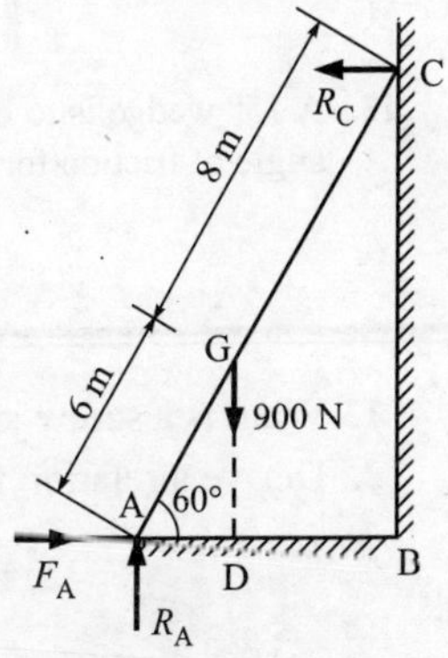

**Fig. 7.37**

7. A uniform ladder of 7 m length rests against a vertical wall with which it makes an angle of 45°. The coefficient of friction between the ladder and the wall is 0.3 and that between the ladder and the floor is 0.5. If a man, whose weight is half of that of the ladder, ascends it, how high will he be when the ladder just starts to slip? [**Ans.** 4.867 m]
8. A uniform ladder of length 13 m and weighing 25 N is placed against a smooth vertical wall with its lower end 5 m from the wall. The coefficient of friction between the ladder and the floor is 0.3. Show that the ladder will remain in equilibrium in this position. What is the frictional force acting on the ladder at the point of contact between the ladder and the floor? [**Ans.** 7.5 N]
9. A ladder of length 6 m and weight 500 N rests on the rough wall and smooth ground making an angle of 60° with the ground. A force of 200 N is applied to the end of the ladder on the ground opposite to the *X*-direction. If the ladder is in equilibrium condition, find the frictional force acting on the ladder resting on the wall. [**Ans.** 96.4 N]
10. A block weighing 3 kN is to be raised by means of a 10° wedge, as shown in Fig. 7.38. Assuming coefficient of friction as 0.3 for all surfaces in contact, find the minimum horizontal force *P* required to raise the block. Determine also the normal reactions. [**Ans.** 2.8365 kN; 3.5327 kN, 3.788 kN, 1.776 kN]

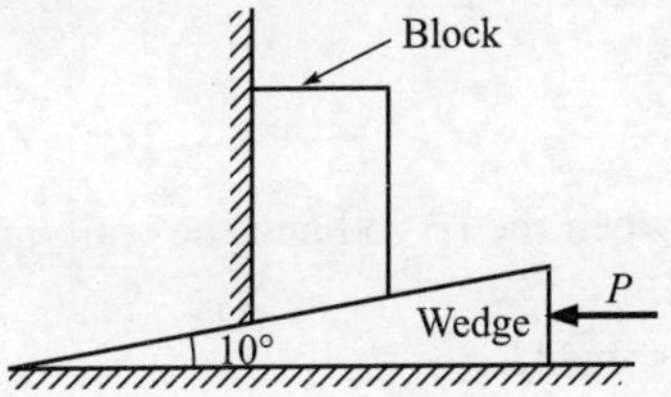

**Fig. 7.38**

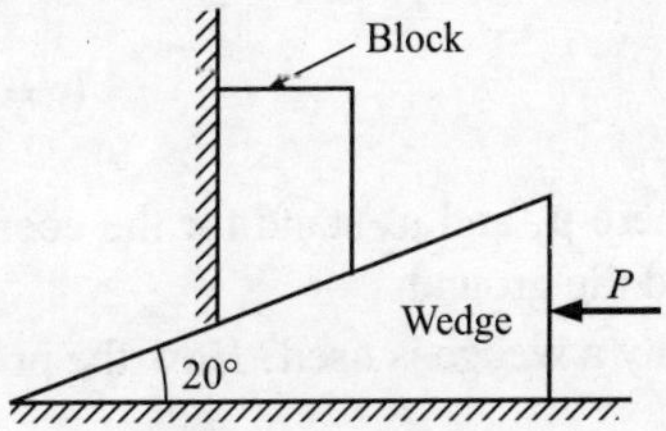

**Fig. 7.39**

**11.** A block weighing 10 kN is to be raised with the help of a wedge of 20°, as shown in Fig. 7.39. The coefficient of friction between the vertical wall and the surface of block is 0.2 while the coefficient of friction between the floor and the surface of the block is 0.3. Assuming the coefficient of friction between the block and the wedge surface as 0.25, determine the magnitude and direction of the minimum horizontal force required to lift the block. [**Ans.** 11.635 kN]

**12.** A block weighing 500 N is to be raised by means of a 15° wedge weighing 250 N, as shown in Fig. 7.40. Assuming coefficient of friction between all contact surfaces to be 0.2, determine the minimum horizontal force *P* to be applied to raise the block. [**Ans.** 434.8 N]

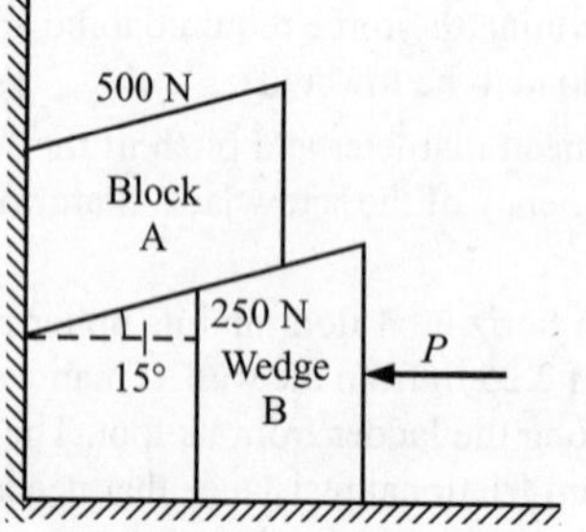


**Fig. 7.40**

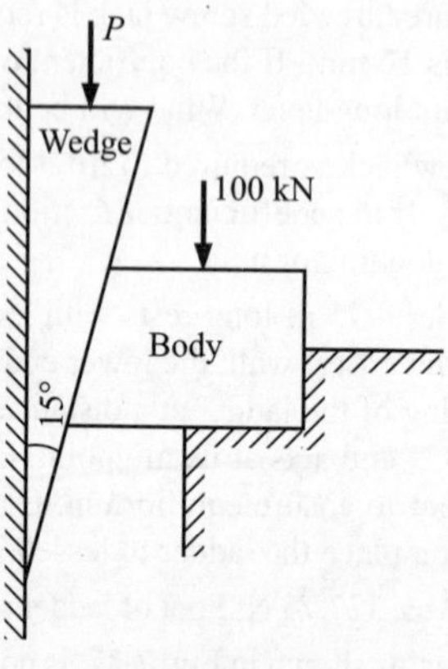


**Fig. 7.41**

**13.** A 15° wedge is to be driven for tightening a body loaded with 100 kN weight, as shown in Fig. 7.41. If the angle of friction for all the surfaces is 14°, find the force *P* which should be applied to the wedge. [**Ans.** 23.185 kN]

## REVIEW QUESTIONS

1. What is a screw jack? Explain the principle on which it works.
2. Derive a relation for the efficiency of a screw jack. Show that the efficiency is maximum, when

$$2\alpha = \frac{\pi}{2} - \phi$$

where $\alpha$ = Helix angle; and

$\phi$ = Angle of friction.

3. Show that the maximum efficiency of the screw jack is $\frac{1-\sin\phi}{1+\sin\phi}$, where $\phi$ is the angle of friction.
4. What is a ladder? Derive a relation to determine the frictional force between the ladder and the floor.
5. A uniform rod is resting against a rough vertical wall and a rough ground. Show that the least angle that the rod can make with the ground without slipping is given by

$$\theta = \tan^{-1}\left[\frac{1-\mu_1\mu_2}{2\mu_2}\right]$$

where $\mu_1$ and $\mu_2$ stand for the coefficient of friction between the (*i*) rod and the wall, and (*ii*) rod and the ground.

6. Why a wedge is used? How the problem on wedges are solved?

## OBJECTIVE TYPE QUESTIONS

**1.** In a screw jack, the effort required to lift the load $W$ is given by

(*a*) $P = W \tan(\alpha - \phi)$ (*b*) $P = W \tan(\alpha + \phi)$

(*c*) $P = W \tan(\phi - \alpha)$ (*d*) $P = W \cos(\alpha + \phi)$

**2.** In a screw jack, the effort required to lower the load $W$ is given by

(*a*) $P = W \tan(\alpha - \phi)$ (*b*) $P = W \tan(\alpha + \phi)$

(*c*) $P = W \tan(\phi - \alpha)$ (*d*) $P = W \cos(\alpha + \phi)$

**3.** The efficiency of a screw jack is given by

(*a*) $\dfrac{\tan(\alpha + \phi)}{\tan \alpha}$ (*b*) $\dfrac{\tan \alpha}{\tan(\alpha + \phi)}$

(*c*) $\dfrac{\tan(\alpha - \phi)}{\tan \alpha}$ (*d*) $\dfrac{\tan \alpha}{\tan(\alpha - \phi)}$

where $\alpha$ = Helix angle; and $\phi$ = Angle of friction.

**4.** The maximum efficiency of a screw jack is

(*a*) $\dfrac{1 - \sin \phi}{1 + \sin \phi}$ (*b*) $\dfrac{1 + \sin \phi}{1 - \sin \phi}$

(*c*) $\dfrac{1 - \tan \phi}{1 + \tan \phi}$ (*d*) $\dfrac{1 + \tan \phi}{1 - \tan \phi}$

**5.** A ladder is resting on a rough ground and leaning against a smooth vertical wall. The force of friction will act

(*a*) downward at its upper end (*b*) upwards at its upper end

(*c*) zero at its upper end (*d*) perpendicular to the wall at its upper end

***Answers to Objective Type Questions***

**1.** (*b*) **2.** (*a*) **3.** (*b*) **4.** (*a*) **5.** (*c*)

8

# SIMPLE MACHINES



## 8.1 INTRODUCTION

A machine is a device by means of which a heavy load is being lifted or dragged by applying a small effort at some other convenient point. The invention of this device gave a great relief to mankind whose much time and labour were saved to a great extent. It may be noted that when the effort is applied at one point to lift a load acting at some other point, then such a device is known as a *simple machine.* On the other hand, when the effort is applied at more than one point to lift the load, then the machine is said to be a *compound machine.*

## 8.2 IMPORTANT TERMS USED IN SIMPLE MACHINES

In order to understand clearly the working of simple machines, the following terms are widely used:

1. ***Mechanical advantage*** It is defined as the ratio of the load lifted to the effort applied. If $W$ is the load lifted by the application of effort $P$, then mechanical advantage,

$$M.A. = \frac{W}{P}$$

2. ***Velocity ratio*** It is defined as the ratio of distance moved by the effort to the distance moved by the load. If $y$ is the distance moved by the effort and $x$ is the distance moved by the load, then velocity ratio,

$$V.R. = \frac{y}{x}$$

3. ***Input of a machine*** It is defined as the total work done on the machine. This is the energy supplied to the machine. This is equal to the work done on the machine by the effort. Thus, the product of effort ($P$) and the distance moved by the effort ($y$) gives the input of the machine. Mathematically,

$$\text{Input of a machine} = P.y$$

4. ***Output of a machine*** It is defined as the actual work done by the machine. Since the machine is used to lift the load, therefore work done by the machine is equal to the product of load lifted ($W$) and distance ($x$) through which the load is lifted. Mathematically,

$$\text{Output of a machine} = W.x$$

5. ***Efficiency of a machine*** It is defined as the ratio of the output of the machine to the input of the machine. It is denoted by a Greek letter eta ($\eta$) and generally expressed as a percentage. Mathematically,

$$\text{Efficiency of a machine} = \frac{\text{Output}}{\text{Input}} \times 100\% = \frac{W \times x}{P \times y} \times 100\%$$

$$= \frac{M.A.}{V.R.} \times 100\% \qquad \ldots\left(\because M.A. = \frac{W}{P};\ \text{and}\ V.R. = \frac{y}{x}\right)$$

6. ***Ideal machine*** A machine whose efficiency is 100%, *i.e.* output is equal to input, then it is called a perfect or ideal machine. In actual practice, no machine is an ideal machine.

**Example 8.1** *In a certain weight lifting machine, a weight of 1000 N is lifted by an effort of 25 N. While the weight moves up by 100 mm, the point of application of effort moves by 8 m. Find mechanical advantage, velocity ratio and efficiency of the machine.*

***Solution*** Given : $W = 1000$ N ; $P = 25$ N ; $x = 100$ mm ; $y = 8$ m $= 8000$ mm

***Mechanical advantage***

We know that mechanical advantage,

$$M.A. = \frac{W}{P} = \frac{1000}{25} = 40 \ \textbf{Ans.}$$

***Velocity ratio***

We know that velocity ratio, $$V.R. = \frac{y}{x} = \frac{8000}{100} = 80 \ \textbf{Ans.}$$

***Efficiency of the machine***

We know that efficiency of the machine,

$$\eta = \frac{M.A.}{V.R.} = \frac{40}{80} = 0.5 \text{ or } 50\% \ \textbf{Ans.}$$

## 8.3 EFFECT OF FRICTION IN MACHINES

We have already discussed that no machine is an ideal machine. Thus, in all the practical machines, a part of the energy supplied (input) is wasted in overcoming the frictional resistances. Hence, the useful workdone (*i.e.* output) is always less than the energy supplied.

Let $P$ = Actual effort applied to lift a load $W$, considering friction,

$P_0$ = Ideal effort, *i.e.* effort required to lift the same load under ideal conditions (neglecting friction),

$W$ = Actual load lifted considering friction in the machine, and

$W_0$ = Ideal load, *i.e.* load which is lifted under ideal conditions (when there is no friction in the machine).

It may be noted that $P$ is greater than $P_0$ and $W_0$ is greater than $W$.

∴ Loss of effort due to friction in the machine,

$$P_1 = P - P_0 \qquad \ldots(i)$$

and loss of load due to friction in the machine,

$$W_1 = W_0 - W \qquad \text{...}(ii)$$

We know that for an ideal machine,

$$M.A. = V.R.$$

or

$$\frac{W}{P_0} = V.R. \text{ or } P_0 = \frac{W}{V.R.} \qquad \text{...}(iii)$$

...($\because$ To lift a load $W$ in case of an ideal machine, the effort required is $P_0$)

Substituting the value of $P_0$ in equation (*i*), we have

$$P_1 = P - \frac{W}{V.R.}$$

or

$$P_1 \times V.R. = P \times V.R. - W \qquad \text{...}(iv)$$

Also, the effort $P$ will lift a load $W_0$ when friction in the machine is neglected. Thus for an ideal machine,

$$M.A. = \frac{W_0}{P} = V.R.$$

or

$$W_0 = P \times V.R. \qquad \text{...}(v)$$

Subtracting $W$ from both sides, we have

$$W_0 - W = P \times V.R. - W$$

From equation (*ii*), loss of load due to friction,

$$W_1 = W_0 - W = P \times V.R. - W \qquad \text{...}(vi)$$

Now equating equations (*iv*) and (*vi*), we have

$$W_1 = P_1 \times V.R.$$

We know that efficiency of the machine,

$$\eta = \frac{M.A.}{V.R.} = \frac{W}{P \times V.R.} = \frac{W}{W_0} = \frac{\text{Actual load}}{\text{Ideal load}}$$

Also

$$\eta = \frac{W}{V.R.} \times \frac{1}{P} = \frac{P_0}{P} = \frac{\text{Ideal effort}}{\text{Actual effort}}$$

## 8.4 REVERSIBLE MACHINE

A machine capable of doing work in the reversed direction after the effort is removed, is known as *reversible machine.* The action of such a machine is called *reversibility of the machine.*

Let $W$ = Load lifted by the machine,

$P$ = Effort applied,

$x$ = Distance through which the load is lifted, and

$y$ = Distance through which the effort acts.

We know that input of the machine $= P.y$

and output of the machine $= W.x$

$\therefore$ Work lost due friction in machine = Input – Output $= P.y - W.x$

It may be noted that for a machine to be reversible, it should be capable of doing work on the removal of the effort, *i.e.* when $P = 0$. This is possible only when the output of the machine is more than the work lost due to friction in machine, *i.e.*

$$W \times x > P \times y - W \times x$$

or

$$2W \times x > P \times y$$

or $$\frac{W}{P} \times \frac{x}{y} > \frac{1}{2}$$

$$\frac{M.A.}{V.R.} > \frac{1}{2} \qquad \ldots \left( \because V.R. = \frac{y}{x} \right)$$

$\therefore$ $$\eta > \frac{1}{2} \text{ or } 50\%$$

Thus, the condition for a machine to be reversible, its efficiency should be more than 50%.

## 8.5 IRREVERSIBLE OR SELF-LOCKING MACHINE

A machine which is not capable of doing any work, in the reversed direction after the effort is removed, is known as *irreversible* or *self-locking machine*. In this case, the work lost due to friction in machine should be greater than the output of the machine, *i.e.*

$$P \times y - W \times x > W \times x$$

or $$P \times y > W \times x + W \times x$$

or $$P \times y > 2W \times x$$

or $$2W \times x < P \times y$$

$$\frac{W}{P} \times \frac{x}{y} < \frac{1}{2}$$

$$\frac{M.A.}{V.R.} < \frac{1}{2}$$

$\therefore$ $${}^{*}\eta < \frac{1}{2} \text{ or } 50\%$$

Thus, the condition for a machine to be irreversible, its efficiency should be less than 50%.

**Example 8.2** *The velocity ratio of a lifting machine is 20. With a load of 800 N; the frictional resistance in it has a value of 40 N. Find the effort needed to lift the load. Find also the efficiency of the machine at that load.*

***Solution*** Given : $V.R. = 20$ ; $W = 800$ N ; $W_1 = 40$ N

---

* It may be proved as follows:

We know that the loss of effort due to friction,

$$P_1 = P - P_0 = P - \frac{W}{V.R.} \qquad \ldots \left( \because P_0 = \frac{W}{V.R.} \right)$$

where $P$ is the actual effort and $P_0$ is the ideal effort.

The machine will be irreversible or self-locking when the load does not fall down when the actual effort ($P$) is removed. This will happen only when the loss of effort due to friction ($P_1$) is greater than the ideal effort ($P_0$), *i.e.*

$$P_1 > P_0$$

or $$P - P_0 > P_0$$

or $$P > 2P_0 > 2 \times \frac{W}{V.R.}$$

or $$\frac{2W}{V.R.} < P \text{ or } \frac{W}{P \times V.R.} < \frac{1}{2}$$

$\therefore$ $$\frac{M.A.}{V.R.} < \frac{1}{2} \text{ or } \eta < \frac{1}{2} \text{ or } 50\%$$

***Effort needed to lift the load***

We know that efficiency of the machine,

$$\eta = \frac{\text{Actual load } (W)}{\text{Ideal load } (W_0)} = \frac{W}{W + W_1} = \frac{800}{800 + 40} = 0.95 \text{ or } 95\%$$

Let $P$ = Actual effort needed to lift the load.

We know that efficiency of the machine,

$$\eta = \frac{\text{Ideal effort } (P_0)}{\text{Actual effort } (P)} = \frac{W}{V.R.} \times \frac{1}{P}$$

or $$0.95 = \frac{800}{20} \times \frac{1}{P} = \frac{40}{P}$$

$\therefore$ $$P = \frac{40}{0.95} = 42.1 \text{ N} \quad \textbf{Ans.}$$

***Efficiency of the machine***

We have calculated above that efficiency of the machine is 95%. **Ans.**

**Example 8.3** *In a simple machine, an effort of 500 N is applied to raise a load of 4 kN. If the velocity ratio of the machine is 20, determine whether this machine is reversible or irreversible. Also determine the ideal load and loss of load due to friction.*

***Solution*** Given : $P = 500$ N ; $W = 4$ kN $= 4000$ N ; $V.R. = 20$

We know that mechanical advantage of the machine,

$$M.A. = \frac{W}{P} = \frac{4000}{500} = 8$$

and efficiency of the machine, $$\eta = \frac{M.A.}{V.R.} = \frac{8}{20} = 0.4 \text{ or } 40\%$$

Since the efficiency of the machine is less than 50%, therefore the machine is irreversible (*i.e.* self-locking). **Ans.**

***Ideal load***

We know that ideal load, $$W_0 = P \times V.R. = 500 \times 20 = 10\,000 \text{ N} \quad \textbf{Ans.}$$

***Loss of load due to friction***

We know that loss of load due to friction,

$$W_1 = W_0 - W = 10\,000 - 4000 = 6000 \text{ N} \quad \textbf{Ans.}$$

## 8.6 LAW OF A MACHINE

The relationship between the load lifted and the effort applied is called *law of the machine.* This is obtained by plotting the various values of the efforts (along *Y*-axis) and corresponding loads along *X*-axis, as shown in Fig. 8.1. It is seen that the graph for an *actual machine is a straight line *AB*. The intercept *OA* represents the effort lost in overcoming frictional resistance. If the effort applied is less than *OA*, then the load will not be lifted. Mathematically, the equation of a straight line is given as

$$P = mW + C$$

where $P$ = Effort applied,

$W$ = Load lifted,

* For an ideal machine, the straight line passes through the origin *O*, as shown in Fig. 8.1.

$$m = \text{Slope of the line } AB = \tan\theta = \frac{P - C}{W}, \text{ and}$$

$C$ = Intercept $OA$ which represents the effort required to overcome friction in the machine.

The values of $m$ and $C$ may be determined as discussed below:

Consider two points 1 and 2 on the line $AB$, as shown in Fig. 8.1. Let at point 1, the effort is $P_1$ and the corresponding load is $W_1$. Similarly, at point 2, the effort is $P_2$ and the corresponding load is $W_2$. According to law of the machine,

$$P_1 = mW_1 + C \quad ...(i)$$

and

$$P_2 = mW_2 + C \quad ...(ii)$$

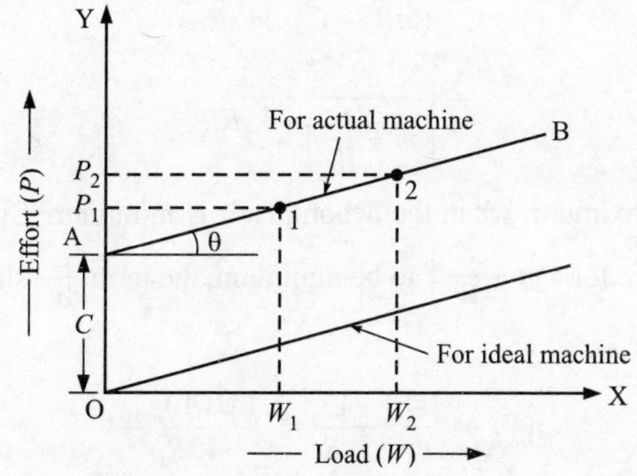


**Fig. 8.1** *Law of a machine*

Subtracting equation (*ii*) from equation (*i*), we have

$$P_1 - P_2 = m(W_1 - W_2)$$

$$\therefore \quad m = \frac{P_1 - P_2}{W_1 - W_2} \quad ...(iii)$$

Substituting this value of $m$ in equation (*i*), we get

$$P_1 = \frac{P_1 - P_2}{W_1 - W_2} \times W_1 + C$$

$$\therefore \quad C = P_1 - \frac{P_1 - P_2}{W_1 - W_2} \times W_1$$

## 8.7 MAXIMUM MECHANICAL ADVANTAGE OF THE MACHINE

We know that mechanical advantage,

$$M.A. = \frac{W}{P}$$

According to law of the machine, $P = mW + C$

Substituting the value of $P$ in the above expression, we have mechanical advantage,

$$M.A. = \frac{W}{mW + C} = \frac{1}{m + \frac{C}{W}}$$

When the load $W$ is very high, then the value of $\frac{C}{W}$ will be very small and may be neglected and the mechanical advantage will be maximum. Thus, maximum mechanical advantage,

$$(M.A.)_{max} = \frac{1}{m}$$

## 8.8 MAXIMUM EFFICIENCY OF THE MACHINE

We know that efficiency of the machine,

$$\eta = \frac{M.A.}{V.R.} = \frac{W}{P} \times \frac{1}{V.R.}$$

$$= \frac{W}{(mW + C)} \times \frac{1}{V.R.} \qquad ...(\because P = mW + C)$$

$$= \left[\frac{1}{m + \frac{C}{W}}\right] \times \frac{1}{V.R.}$$

The efficiency will be maximum, when the denominator is minimum. Since the velocity ratio ($V.R.$) is constant for a machine, therefore for $\left(m + \frac{C}{W}\right)$ to be minimum, the term $\frac{C}{W}$ should be zero. Thus, maximum efficiency of the machine,

$$\eta_{max} = \frac{1}{m} \times \frac{1}{V.R.} = \frac{(M.A.)_{max}}{V.R.}$$

**Example 8.4** *In a lifting machine, an effort of 30 N raised a load of 1 kN. If the efficiency of the machine is 75%, what is its velocity ratio? If on this machine, an effort of 60 N raised a load of 2 kN, what is now the efficiency. What will be the effort required to raise a load of 5 kN?*

***Solution*** Given : $P_1 = 30$ N ; $W_1 = 1$ kN $= 1000$ N ; $\eta_1 = 75\% = 0.75$ ; $P_2 = 60$ N ; $W_2 = 2$ kN $= 2000$ N ; $W_3 = 5$ kN $= 5000$ N

***Velocity ratio (1st case)***

We know that mechanical advantage,

$$M.A. = \frac{W_1}{P_1} = \frac{1000}{30} = 33.33$$

$$\therefore \quad \text{Velocity ratio, } V.R. = \frac{M.A.}{\eta_1} = \frac{33.33}{0.75} = 44.44 \textbf{ Ans.}$$

***Efficiency (2nd case)***

We know that mechanical advantage,

$$M.A. = \frac{W_2}{P_2} = \frac{2000}{60} = 33.33$$

Velocity ratio, $V.R. = 44.44$ ...(as calculated above)

$$\therefore \quad \text{Efficiency, } \eta = \frac{M.A.}{V.R.} = \frac{33.33}{44.44} = 0.75 \text{ or } 75\% \textbf{ Ans.}$$

***Effort required to raise a load of $W_3$ = 5 kN = 5000 N***

We know that law of the machine is

$$P = mW + C$$

∴ For 1st case, $30 = m \times 1000 + C$ ...(i)

and for 2nd case $60 = m \times 2000 + C$ ...(ii)

Subtracting equation (i) from equation (ii), we have

$$30 = 1000\,m$$

$$\therefore \quad m = \frac{30}{1000} = \frac{3}{100}$$

Substituting this value of $m$ in equation (i), we have

$$30 = \frac{3}{100} \times 1000 + C \quad \text{or} \quad C = 0$$

The law of machine now becomes

$$P_3 = mW_3$$

$$\therefore \quad \text{Effort required, } P_3 = \frac{3}{100} \times 5000 = 150 \text{ N} \textbf{ Ans.}$$

**Example 8.5** *Observations in an experiment with a machine are tabulated below:*

| *Load (W) in N* | *20* | *60* | *100* | *140* | *180* | *220* | *260* | *300* |
|---|---|---|---|---|---|---|---|---|
| *Effort (P) in N* | *13* | *18* | *21* | *24* | *29* | *33.5* | *37* | *40* |

*The velocity ratio is 20. With the help of a graph between load and effort, find the efficiency of the machine when the lifting load is 240 N. Derive the law of this machine.*

***Solution*** Given : $V.R. = 20$ ; $W = 240$ N

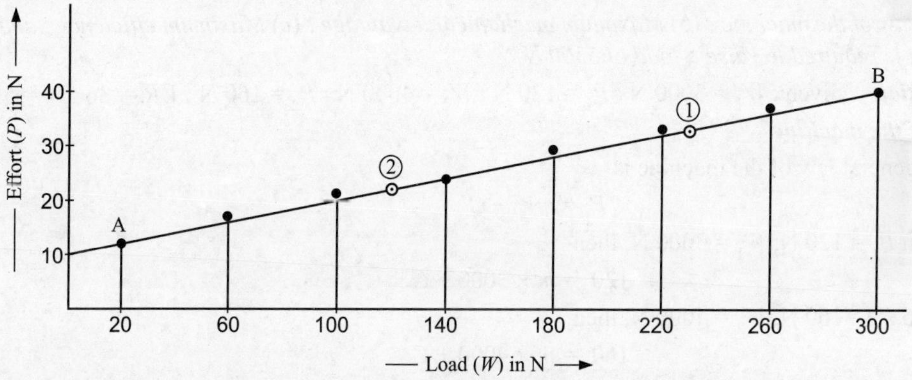


**Fig. 8.2**

Figure 8.2 shows the graph drawn through the load verses effort for the given values. Join the first point $A$ (20 N, 13 N) and last point $B$ (300 N, 40 N). It may be noted that line $AB$ will not pass through all the points. Consider any two points (1) and (2) on the line $AB$. From the graph, we see that

At point 1, $W_1 = 230$ N, and $P_1 = 33$ N

At point 2, $W_2 = 120$ N, and $P_2 = 22$ N

We know that the general equation for the law of machine is

$$P = mW + C$$

Thus for points 1 and 2, the law of machine becomes

$33 = m \times 230 + C$ ...(i)

and $22 = m \times 120 + C$ ...(ii)

Subtracting equation (*ii*) from equation (*i*), we have

$$11 = 110\,m$$

$$\therefore \quad m = \frac{11}{110} = 0.1$$

Substituting this value in equation (*i*), we have

$$33 = 0.1 \times 230 + C = 23 + C$$

$$\therefore \quad C = 33 - 23 = 10$$

Hence the law of the given machine becomes

$$P = 0.1W + 10 \quad \textbf{Ans.}$$

***Efficiency of the machine***

We know that effort required to lift a load of 240 N,

$$P = 0.1 \times 240 + 10 = 34 \text{ N}$$

$$\therefore \text{ Mechanical advantage,} \quad M.A. = \frac{W}{P} = \frac{240}{34} = 7.06$$

We know that efficiency of the machine,

$$\eta = \frac{M.A.}{V.R.} = \frac{7.06}{20} = 0.353 \text{ or } 35.3\% \quad \textbf{Ans.}$$

**Example 8.6** *It is found that in a machine, a load of 3000 N is raised by an effort of 120 N and a load of 4000 N is raised by the effort of 160 N. If the velocity ratio of the machine is 36, then determine the following :*

*(a) Law of the machine ; (b) Maximum mechanical advantage ; (c) Maximum efficiency ; and (d) How much effort is required to raise a load of 5500 N ?*

***Solution*** Given : $W_1 = 3000$ N ; $P_1 = 120$ N ; $W_2 = 4000$ N ; $P_2 = 160$ N ; $V.R. = 36$ ; $W = 5500$ N

***(a) Law of the machine***

The general law of the machine is

$$P = mW + C$$

When $P_1 = 120$ N; $W_1 = 3000$ N, then

$$120 = m \times 3000 + C \quad \text{...}(i)$$

When $P_2 = 160$ N; $W_2 = 4000$ N, then

$$160 = m \times 4000 + C \quad \text{...}(ii)$$

Subtracting equation (*i*) from equation (*ii*), we have

$$40 = 1000m$$

$$\therefore \quad m = \frac{40}{1000} = 0.04$$

Substituting this value of *m* in equation (*i*), we have

$$120 = 0.04 \times 3000 + C = 120 + C$$

$$\therefore \quad C = 120 - 120 = 0$$

Thus the law of machine is $P = 0.04\,W$ **Ans.**

***(b) Maximum mechanical advantage***

We know that maximum mechanical advantage,

$$(M.A.)_{max} = \frac{1}{m} = \frac{1}{0.04} = 25 \quad \textbf{Ans.}$$

**(c) *Maximum efficiency***

We know that maximum efficiency,

$$\eta_{max} = \frac{1}{m \times V.R.} = \frac{1}{0.04 \times 36} = 0.6944 \text{ or } 69.44\% \textbf{ Ans.}$$

**(d) *Effort required to raise a load W = 5500 N***

We know that law of the machine is

$$P = 0.04\,W$$

$\therefore$ Effort required, $P = 0.04 \times 5500 = 220$ N **Ans.**

**Example 8.7** *The values of the effort required to lift loads of 3600 N and 7200 N are 80 N and 120 N, respectively. Determine the law of the machine and the values of the efficiency at the above loads, when velocity ratio is 90. Find also the load created by friction when the load being lifted is 5400 N.*

***Solution*** Given : $W_1 = 3600$ N ; $W_2 = 7200$ N ; $P_1 = 80$ N ; $P_2 = 120$ N ; $V.R. = 90$ ; $W = 5400$ N

***Law of the machine***

The general law of the machine is $P = mW + C$

When $P_1 = 80$ N; $W_1 = 3600$ N, then

$$80 = m \times 3600 + C \quad ...(i)$$

When $P_2 = 120$ N; $W_2 = 7200$ N, then

$$120 = m \times 7200 + C \quad ...(ii)$$

Subtracting equation (*i*) from equation (*ii*), we have

$$40 = 3600m$$

$$\therefore \quad m = \frac{40}{3600} = \frac{1}{90}$$

Substituting this value of *m* in equation (*i*), we have

$$80 = \frac{1}{90} \times 3600 + C = 40 + C$$

$$\therefore \quad C = 80 - 40 = 40$$

Thus the law of machine becomes $P = \dfrac{W}{90} + 40$ **Ans.**

***Values of the efficiency at the above loads***

We know that mechanical advantage in the first case

$$(M.A.)_1 = \frac{W_1}{P_1} = \frac{3600}{80} = 45$$

Mechanical advantage in the second case,

$$(M.A.)_2 = \frac{W_2}{P_2} = \frac{7200}{120} = 60$$

$\therefore$ Efficiency in the first case, $\eta_1 = \dfrac{(M.A.)_1}{V.R.} = \dfrac{45}{90} = 0.5$ or 50% **Ans.**

and efficiency in the second case, $\eta_2 = \dfrac{(M.A.)_2}{V.R.} = \dfrac{60}{90} = 0.6667$ or 66.67% **Ans.**

***Load created by friction***

We know that law of the machine is

$$P = \frac{W}{90} + 40 = \frac{5400}{90} + 40 = 100 \text{ N} \qquad ...(\because \text{ Given } W = 5400 \text{ N})$$

$\therefore$ Ideal load, $W_0 = P \times V.R. = 100 \times 90 = 9000$ N

We know that

$$\text{Load created by friction} = \text{Ideal load} - \text{Actual load}$$
$$= 9000 - 5400 = 3600 \text{ N} \quad \textbf{Ans.}$$

**Example 8.8** *The law of a lifting machine is* $P = \frac{W}{50} + 15$. *The velocity ratio of the machine is 100. Find the maximum possible mechanical advantage and the maximum possible efficiency of the machine. Determine the effort required to overcome friction while lifting a load of 1500 N. Also calculate the efficiency at this load.*

***Solution*** Given : $P = \frac{W}{50} + 1.5$ ; $V.R. = 100$

***Maximum possible advantage***

Comparing the given law of the machine, $P = \frac{W}{50} + 15$, with the general law of machine $P = mW + C$, we find that

$$m = \frac{1}{50}; \text{ and } C = 15$$

We know that maximum possible mechanical advantage,

$$(M.A.)_{max} = \frac{1}{m} = 50 \quad \textbf{Ans.}$$

***Maximum possible efficiency***

We know that maximum possible efficiency,

$$\eta_{max} = \frac{1}{m \times V.R.} = \frac{50}{100} = 0.5 \text{ or } 50\% \quad \textbf{Ans.}$$

***Effort required to overcome friction***

Given: Load lifted, $W = 1500$ N

$\therefore$ Effort required to overcome friction,

$$P = \frac{W}{50} + 15 = \frac{1500}{50} + 15 = 45 \text{ N} \quad \textbf{Ans.}$$

and mechanical advantage,

$$M.A. = \frac{W}{P} = \frac{1500}{45} = 33.33$$

***Efficiency***

We know that

$$\text{Efficiency, } \eta = \frac{M.A.}{V.R.} = \frac{33.33}{100} = 0.3333 \text{ or } 33.33\% \quad \textbf{Ans.}$$

## 8.9 IMPORTANT LIFTING MACHINES

The following are some lifting machines which are important from the subject point of view:

1. Simple wheel and axle.
2. Differential wheel and axle.

3. **Pulleys:**
   (*a*) First system of pulleys,
   (*b*) Second system of pulleys, and
   (*c*) Third system of pulleys.
4. Weston's differential pulley block.
5. Worm and worm wheel.
6. Simple screw jack.
7. Differential screw jack.
8. Single purchase winch crab.
9. Double purchase winch crab.

These lifting machines are discussed, in detail, in the following pages.

## 8.10 SIMPLE WHEEL AND AXLE

A simple wheel and axle, as shown in Fig. 8.3, consists of a wheel *A* and an axle *B* mounted on the same spindle. The whole assembly is mounted on ball bearings in order to reduce frictional resistance. A string is wound round the wheel *A*. One end of the string is tied to the pin provided on the wheel while the other end is free for the application of effort *P*. A second string is wound on the axle *B* in the opposite direction to that of string on wheel *A*. One end of this string is tied to the pin provided on the axle while the other end carries the load *W* to be lifted.

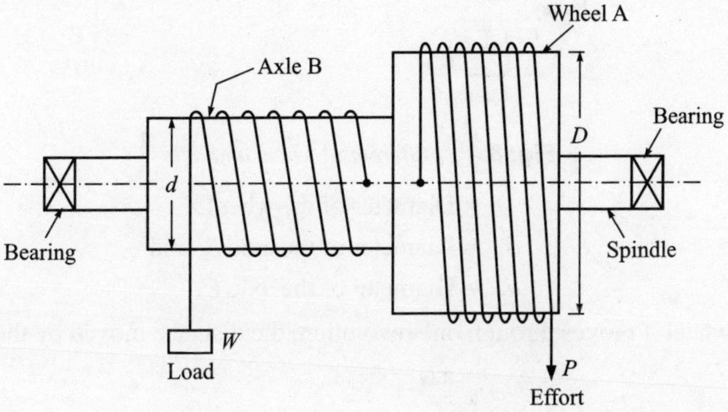


**Fig. 8.3** *Simple wheel and axle*

Let $D$ = Diameter of the wheel *A*, and

$d$ = Diameter of the axle *B*.

Since the wheel and axle are mounted on the same spindle, therefore they move together. Now, when the wheel moves through our revolution, the axle will also move through one revolution. Thus the distance moved by the effort in one revolution

$$= \pi D$$

and the distance moved by the load $= \pi d$

We know that velocity ratio, $$V.R. = \frac{\text{Distance moved by the effort}}{\text{Distance moved by the load}} = \frac{\pi D}{\pi d} = \frac{D}{d}$$

Mechanical advantage, $$M.A. = \frac{\text{Load lifted}}{\text{Effort applied}} = \frac{W}{P}$$

and efficiency, $$\eta = \frac{M.A.}{V.R.}$$

## 8.11 DIFFERENTIAL WHEEL AND AXLE

In a differential wheel and axle, as shown in Fig. 8.4, the axle is made of two different diameters. The axle *B* is of larger diameter ($d_1$) and the other axle *C* is of smaller diameter ($d_2$). The wheel and the axles are mounted on the same shaft which is mounted on the ball bearings. A string is wound round the wheel *A*. One end of the string is fixed to the wheel and on the other end, effort *P* is applied. A second string is tied to axle *C* and is wound a few turns in the same direction as that of wheel *A*. The remaining portion of the second string is wound on the axle *B* in the opposite direction and the other end of the string is fixed to the axle *B*, as shown in Fig. 8.4. The load *W* to be lifted is attached to the pulley provided on the second string. A little consideration will show that when the string unwinds from the wheel *A*, the other string also unwinds from the axle *C* but it winds on the axle *B*.

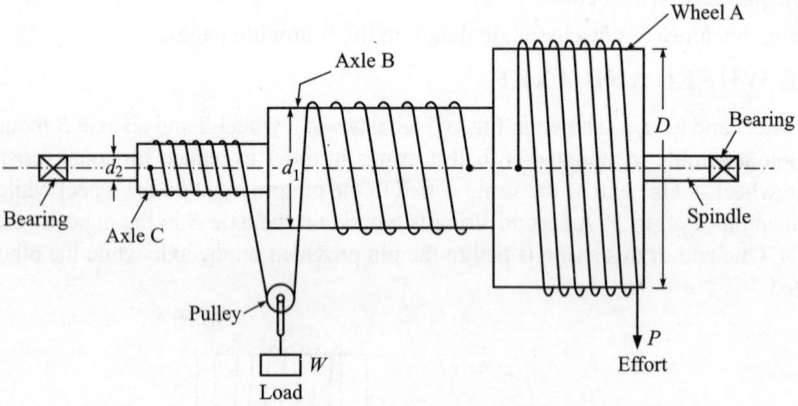


**Fig. 8.4** *Differential wheel and axle*

Let $D$ = Diameter of the wheel $A$,

$d_1$ = Diameter of the axle $B$, and

$d_2$ = Diameter of the axle $C$.

Now, when the wheel *A* moves through one revolution, the distance moved by the effort

$$= \pi D$$

Length of the string wound on the axle *B* in one revolution

$$= \pi d_1$$

and length of the string unwound from the axle *C* in one revolution

$$= \pi d_2$$

∴ Net length of the string which will wound on the axle

$$= \pi d_1 - \pi d_2$$

Thus, the load *W* is lifted up through a distance

$$= \frac{\pi d_1 - \pi d_2}{2} = \frac{\pi}{2}(d_1 - d_2)$$

$$\therefore \quad \text{Velocity ratio, } V.R. = \frac{\text{Distance moved by the effort}}{\text{Distance moved by the load}}$$

$$= \frac{\pi D}{\frac{\pi}{2}(d_1 - d_2)} = \frac{2D}{d_1 - d_2}$$

Mechanical advantage, $M.A. = \frac{W}{P}$

and efficiency, $\eta = \frac{M.A.}{V.R.}$

**Note:** In case the diameter of the string on the wheel $A$ is $D'$ and diameter of the string on the axles $B$ and $C$ is $d'$, then velocity ratio is given by

$$V.R. = \frac{\pi(D + D')}{\frac{\pi}{2}[(d_1 + d') - (d_2 + d')]} = \frac{2(D + D')}{d_1 - d_2}$$

**Example 8.9** *The diameter of a simple wheel and axle are 500 mm and 100 mm, respectively. If a load of 1800 N is lifted by an effort of 450 N ; find the efficiency of the machine.*

***Solution*** Given : $D = 500$ mm ; $d = 100$ mm ; $W = 1800$ N ; $P = 450$ N

We know that velocity ratio, $V.R. = \frac{D}{d} = \frac{500}{100} = 5$

and mechanical advantage, $M.A. = \frac{W}{P} = \frac{1800}{450} = 4$

$\therefore$ Efficiency, $\eta = \frac{M.A.}{V.R.} = \frac{4}{5} = 0.8$ or 80% **Ans.**

**Example 8.10** *In a differential wheel and axle, the diameter of the effort wheel is 400 mm. The radii of the axle are 150 mm and 100 mm, respectively. The diameter of the ropes is 10 mm. Find the load which can be lifted by an effort of 196 N, assuming the efficiency of the machine to be 75%.*

***Solution*** Given : $D = 400$ mm ; $r_1 = 150$ mm or $d_1 = 300$ mm ; $r_2 = 100$ mm or $d_2 = 200$ mm ; $D' = d' = 10$ mm ; $P = 196$ N ; $\eta = 75\% = 0.75$

Let $W$ = Load lifted.

We know that velocity ratio, $V.R. = \frac{2(D + D')}{d_1 - d_2} = \frac{2(400 + 10)}{300 - 200} = 8.2$

Mechanical advantage, $M.A. = \frac{W}{P} = \frac{W}{196}$

$\therefore$ Efficiency, $\eta = \frac{M.A.}{V.R.}$

$$0.75 = \frac{\frac{W}{196}}{8.2} = \frac{W}{196 \times 8.2}$$

$\therefore$ $W = 0.75 \times 196 \times 8.2 = 1205.4$ N **Ans.**

**Example 8.11** *In a differential wheel and axle, the difference between the axle diameters is 50 mm and the diameter of the effort wheel is 750 mm. If the laod of 2500 N is lifted by an effort of 160 N and a load of 3250 N is lifted by an effort of 200 N, calculate the load lifted by an effort of 225 N. Also calculate the mechanical advantage and efficiency.*

***Solution*** Given : $d_1 - d_2 = 50$ mm ; $D = 750$ mm ; $W_1 = 2500$ N ; $P_1 = 160$ N ; $W_2 = 3250$ N ; $P_2 = 200$ N ; $P = 225$ N

***Load lifted by an effort of P = 225 N***

We know that law of the machine is

$$P = mW + C$$

When $W_1 = 2500$ N and $P_1 = 160$ N; then

$$160 = m \times 2500 + C \quad ...(i)$$

When $W_2 = 3250$ N and $P_2 = 200$ N; then

$$200 = m \times 3250 + C \quad ...(ii)$$

Subtracting equation (*i*) from equation (*ii*), we have

$$40 = m \times 750$$

$$\therefore \quad m = \frac{40}{750} = 0.053$$

Substituting the value of *m* in equation (*i*), we have

$$160 = 0.053 \times 2500 + C = 132.5 + C$$

$$\therefore \quad C = 160 - 132.5 = 27.5$$

Now the law of machine becomes, $P = 0.053W + 27.5$

When $P = 225$ N, then $\quad 225 = 0.053W + 27.5$

$$\therefore \quad W = \frac{225 - 27.5}{0.053} = 3726.4 \text{ N} \quad \textbf{Ans.}$$

***Mechanical advantage***

We know that mechanical advantage,

$$M.A. = \frac{W}{P} = \frac{3726.4}{225} = 16.56 \quad \textbf{Ans.}$$

***Efficiency***

We know that velocity ratio,

$$V.R. = \frac{2D}{d_1 - d_2} = \frac{2 \times 750}{50} = 30$$

$$\therefore \quad \text{Efficiency, } \eta = \frac{M.A.}{V.R.} = \frac{16.56}{30} = 0.552 \text{ or } 55.2\% \quad \textbf{Ans.}$$

## 8.12 PULLEYS

A pulley is a wheel on an axle or shaft that is generally made up of wood or metal. The outer side of the wheel is grooved around its circumference to accommodate a rope or chain. The axle about which the pulley rotates freely, passes through the centre of the pulley and is perpendicular to its surface plane. It is supported by a frame of metal or wood called pulley block. A simple pulley with the block supporting the pulley is shown in Fig. 8.5. The following assumptions are made while dealing with the pulleys.

1. The weight of the pulley, pulley block, axle and rope are negligible as compared to the weight to be lifted.
2. The pulleys are smooth and frictionless. Thus, the tension in the two sides of the rope, passing round the pulley, may be taken equal.

It may be noted that a pulley whose pulley block is fixed, is known as *fixed pulley* and when the pulley block can move up and down, then it is called *movable pulley.*

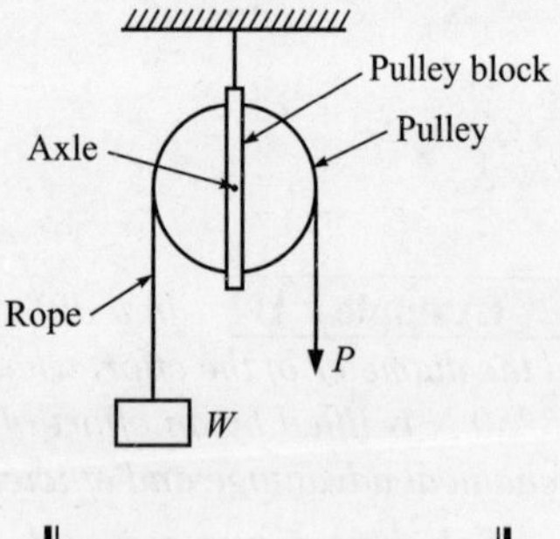


**Fig. 8.5** *Simple pulley*

In a single fixed pulley, as shown in Fig. 8.6, the mechanical advantage as well as velocity ratio is unity. In this case, the effort $P$ can be applied in any convenient direction.

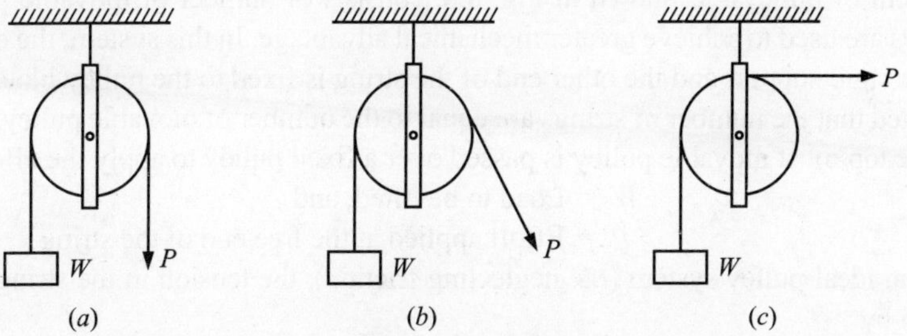


Fig. 8.6

In a single movable pulley, as shown in Fig. 8.7 (*a*), one end of the rope is attached to the fixed support and the effort $P$ is applied at the other end. The load $W$ will be shared equally by the two portions of the rope.

$$\therefore \qquad P = \frac{W}{2} \text{ or } M.A. = \frac{W}{P} = 2$$

Thus, the mechanical advantage is more than unity. But the application of force is not easy. In order to make the application of force easy, a fixed pulley as shown in Fig. 8.7 (*b*) is introduced which will not increase the mechanical advantage, but will help the operator to use the force conveniently. A little consideration will show that to raise the load $W$ through a certain distance, the effort $P$ will have to traverse double the distance. Thus the velocity ratio in this case will be equal to 2.

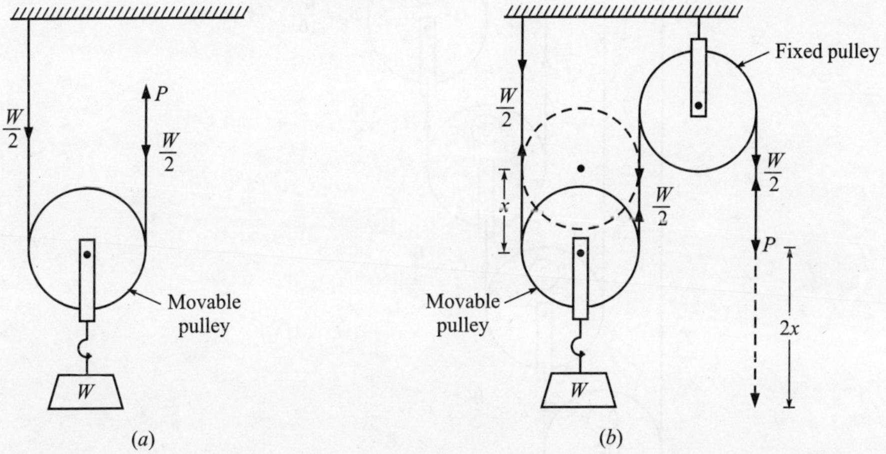


Fig. 8.7

## 8.13 SYSTEM OF PULLEYS

In order to obtain higher mechanical advantage and efficiency, a combination of fixed and movable pulley are used. These combination of pulleys are known as *system of pulleys.* In general, the following three system of pulleys are commonly used:

1. First system of pulleys,
2. Second system of pulleys, and
3. Third system of pulleys.

These systems are now discussed, in detail, in the following pages:

## 8.14 FIRST SYSTEM OF PULLEYS

The first system of pulleys, as shown in Fig. 8.8, consists of number of movable pulleys (say, 1, 2, 3 and 4). These pulleys are used to achieve greater mechanical advantage. In this system, the end of each string of pulley is fixed to the rigid support and the other end of the string is fixed to the pulley block of the next higher pulley. It may be noted that the number of strings are equal to the number of movable pulley. The free end of the string supporting the top most movable pulley is passed over a fixed pulley to apply the effort.

Let $W$ = Load to be lifted, and

$P$ = Effort applied at the free end of the string.

Considering an ideal pulley system (*i.e.* neglecting friction), the tension in the string which passes over the pulley 1 is given by

$$T_1 = \frac{W}{2}$$

and tension in the string which passes over pulley 2 is

$$T_2 = \frac{W}{4} = \frac{W}{2^2}$$

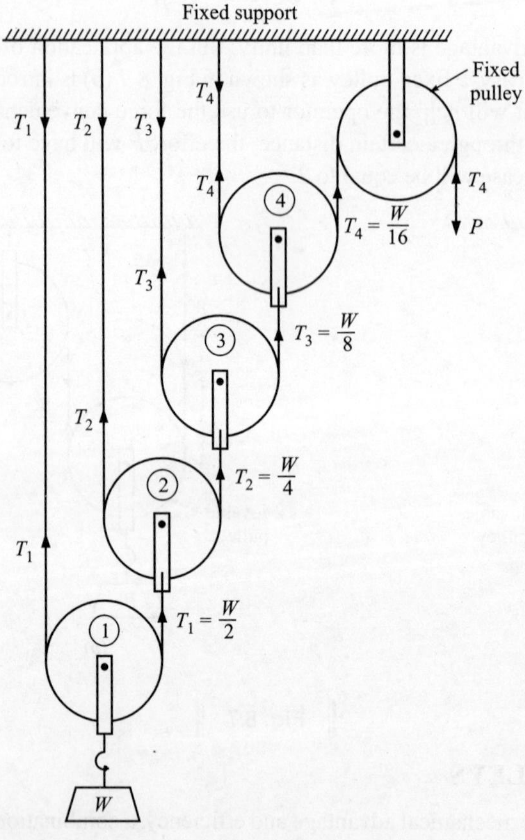


**Fig. 8.8** *First system of pulleys*

Similarly, tension in the string which passes over pulley 3 is

$$T_3 = \frac{W}{8} = \frac{W}{2^3}$$

and tension in the string which passes over pulley 4 is

$$T_4 = \frac{W}{16} = \frac{W}{2^4}$$

Thus, an effort of $\frac{W}{16}$ or $\frac{W}{2^4}$ is required to lift a load $W$

$$\therefore \quad P = \frac{W}{2^4}$$

In general, if there are $n$ number of movable pulleys, then

$$P = \frac{W}{2^n}$$

We know that mechanical advantage,

$$M.A. = \frac{W}{P} = \frac{W}{\frac{W}{2^n}} = 2^n$$

But in an ideal machine, $M.A. = V.R. = 2^n$

**Note:** If the weight of each movable pulley (say $w$) is given, then effort required to lift a load $W$ is given by

$$P = \frac{W + w(2^n - 1)}{2^n}$$

and mechanical advantage, $M.A. = \frac{W}{P} = W \times \frac{2n}{W + w(2^n - 1)} = \frac{W \times 2^n}{W + w(2^n - 1)}$

## 8.15 SECOND SYSTEM OF PULLEYS

The second system of pulleys, as shown in Fig. 8.9, consists of two blocks, *i.e.* upper pulley block and lower pulley block. The upper pulley block is fixed in position to the top support where as the lower pulley block can move vertically along with the load ($W$) which is attached to it. When both the blocks have equal number of pulley, as shown in Fig. 8.9 ($a$), the string which passes round all the pulleys is fixed to the upper block. On the other hand, when lower block has one pulley less than the upper block, then the string is attached to the lower block, as shown in Fig. 8.9 ($b$).

Considering an ideal pulley system (*i.e.* neglecting friction), the tension in each string ($T$) will remain equal to the applied effort. If $n$ is equal to the number of segments supporting the lower block with weight $W$, then for equilibrium of the lower block,

$$W = n \cdot T = n \cdot P \quad ...(\because T = P)$$

$\therefore$ Mechanical advantage, $M.A. = \frac{W}{P} = \frac{n \cdot P}{P} = n$

For ideal pulley, velocity ratio, $V.R. = M.A. = n$

It may be noted that number of segments supporting the lower block is always equal to the total number of pulleys in the two blocks

**Note:** If weight ($w$) of the lower block is taken into consideration, then for equilibrium

$$W + w = n \cdot P$$

or

$$W = n \cdot P - w$$

$\therefore$ Mechanical advantage, $M.A. = \frac{W}{P} = \frac{nP - w}{P} = n - \frac{W}{P}$

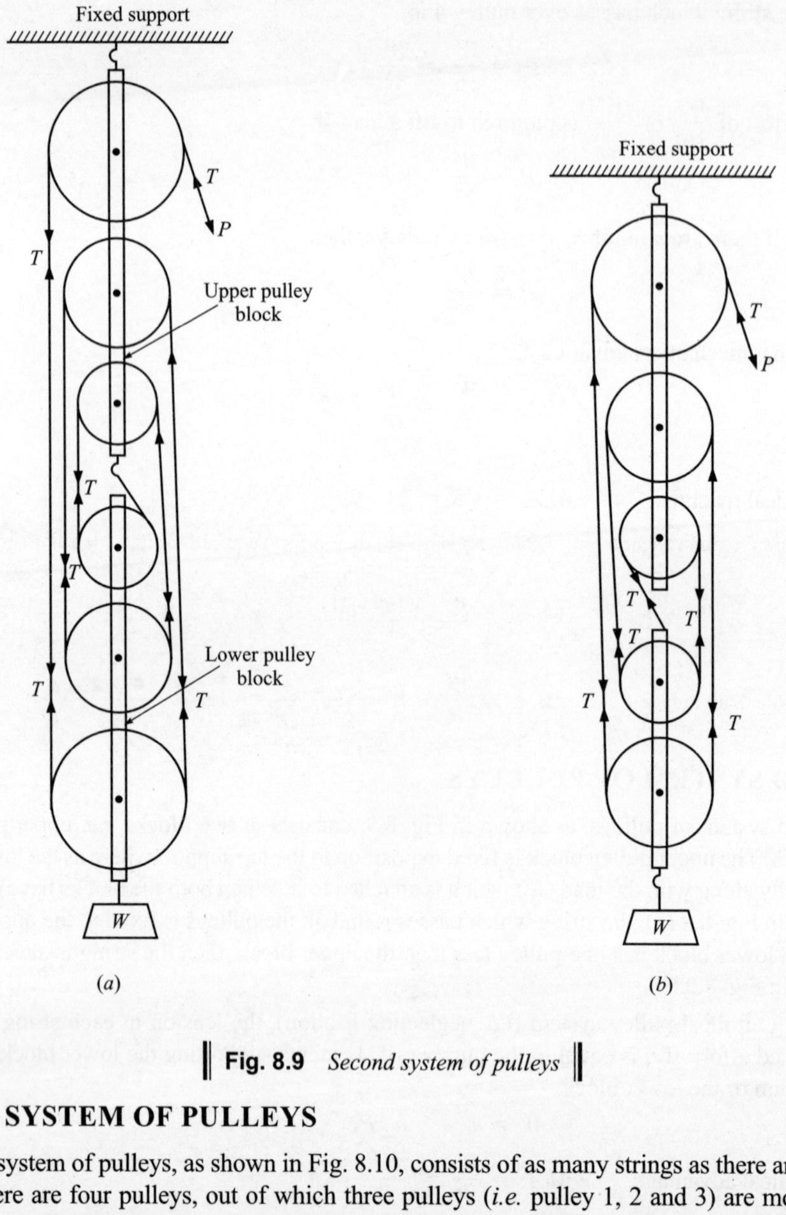


**Fig. 8.9** *Second system of pulleys*

## 8.16 THIRD SYSTEM OF PULLEYS

The third system of pulleys, as shown in Fig. 8.10, consists of as many strings as there are pulleys. In the present case, there are four pulleys, out of which three pulleys (*i.e.* pulley 1, 2 and 3) are movable while the pulley number 4 is a fixed pulley. One end of each string is fixed to a beam, to which the load $W$ is attached and the other end of each rope passing round the upper periphery of the pulley is fixed to the pulley block of the next lower pulley, as shown in Fig. 8.10.

Let $T_1$, $T_2$, $T_3$ and $T_4$ be the tensions in the strings.

Considering the equilibrium of the fixed beam,

$$T_1 + T_2 + T_3 + T_4 = W \quad ...(i)$$

Considering the equilibrium of the pulleys 1, 2, 3 and 4, we have

$$T_1 = P$$

$$T_2 = 2T_1 = 2P$$

$$T_3 = 2T_2 = 2 \times 2P = 4P = 2^2P$$

and $$T_4 = 2T_3 = 2 \times 4P = 8P = 2^3P$$

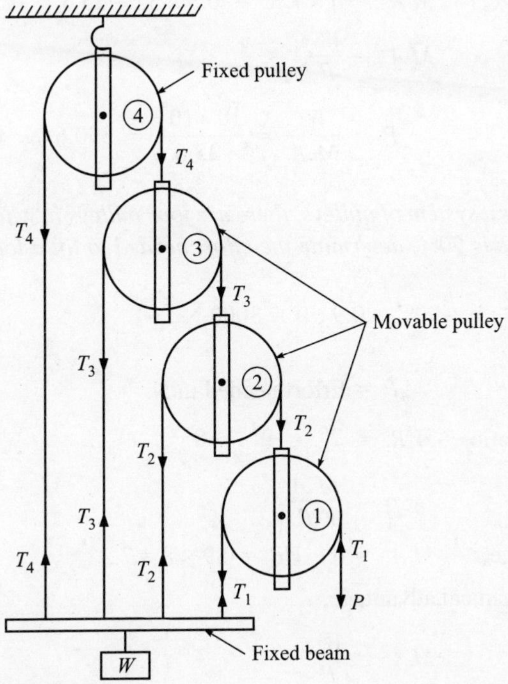


**Fig. 8.10** *Third system of pulleys*

Substituting the value of $T_1$, $T_2$, $T_3$ and $T_4$ in equation (*i*), we have

$$P + 2P + 2^2P + 2^3P = W$$

or $$P(1 + 2 + 2^2 + 2^3) = W$$

∴ Mechanical advantage, $$M.A. = \frac{W}{P} = 1 + 2 + 2^2 + 2^3$$

If there are *n* number of pulleys, then

$$M.A. = \frac{W}{P} = 1 + 2^1 + 2^2 + 2^3 + ... + 2^n - 1$$

Since the above expression is in geometric progression, therefore,

$$M.A. = \frac{2^n - 1}{2 - 1} = 2^n - 1$$

For ideal pulley, velocity ratio, $V.R. = M.A. = 2^n - 1$

**Example 8.12** *Five movable pulleys are arranged as in the first system of pulleys. If the efficiency of the system is 80%, find the effort required to raise a weight of 10 kN.*

***Solution*** Given : $n = 5$ ; $\eta = 80\% = 0.8$ ; $W = 10$ kN $= 10 \times 10^3$ N

Let $P$ = Effort required in N.

We know that velocity ratio for the first system of pulleys,

$$V.R. = 2^n = 2^5 = 32$$

and efficiency, $\eta = \dfrac{M.A.}{V.R.}$

∴ Mechanical advantage, $M.A. = \eta \times V.R. = 0.8 \times 32 = 25.6$

We also know that $M.A. = \dfrac{W}{P}$

∴ $P = \dfrac{W}{M.A.} = \dfrac{10 \times 10^3}{25.6} = 390.6$ N **Ans.**

**Example 8.13** *In a first system of pulleys, there are four pulleys in which three pulleys are movable. If the efficiency of the machine is 90%, determine the effort needed to lift a load of 3000 N. Also determine the friction of the machine.*

***Solution*** Given : $n = 3$ ; $\eta = 90\% = 0.9$ ; $W = 3000$ N

***Effort needed***

Let $P$ = Effort needed in N.

We know that velocity ratio, $V.R. = 2^n = 2^3 = 8$

and efficiency, $\eta = \dfrac{M.A.}{V.R.}$

∴ Mechanical advantage, $M.A. = \eta \times V.R. = 0.9 \times 8 = 7.2$

We also know that mechanical advantage,

$$M.A. = \frac{W}{P}$$

∴ $P = \dfrac{W}{M.A.} = \dfrac{3000}{7.2} = 416.67$ N **Ans.**

***Friction of the machine***

We know that effort lost in friction,

$$P_0 = P - \frac{W}{V.R.} = 416.67 - \frac{3000}{8}$$

$$= 416.67 - 375 = 41.67 \text{ N}$$ **Ans.**

**Example 8.14** *In a first system of pulleys, there are four movable pulleys. If the weight of each pulley is 0.5 N, determine the effort required to lift a load of 1000 N.*

***Solution*** Given : $n = 4$ ; $w = 0.5$ ; N ; $W = 1000$ N

We know that effort required, $P = \dfrac{W + w(2^n - 1)}{2^n}$

$$= \frac{1000 + 0.5(2^4 - 1)}{2^4} = \frac{1000 + 7.5}{16} = 62.97 \text{ N}$$ **Ans.**

**Example 8.15** *In a second system of pulleys, there are three pulleys in each block. A weight of 3000 N is to be lifted by an effort of 700 N. Find : 1. Efficiency of the system ; 2. Effort wasted in friction ; and 3. Amount of frictional load.*

***Solution*** Given : $n = 3 + 3 = 6$ ; $W = 3000$ N ; $P = 700$ N

**1. *Efficiency of the system***

We know that velocity ratio for a second system of pulleys,

$$V.R. = n = 6$$

and mechanical advantange,

$$M.A. = \frac{W}{P} = \frac{3000}{700} = 4.28$$

∴ Efficiency of the system,

$$\eta = \frac{M.A.}{V.R.} = \frac{4.28}{6} = 0.7133 \text{ or } 71.33\% \textbf{ Ans.}$$

**2. *Effort wasted in friction***

We know that effort wasted in friction,

$$P_0 = P - \frac{W}{V.R.} = 700 - \frac{3000}{6} = 700 - 500 = 200 \text{ N } \textbf{Ans.}$$

**3. *Amount of frictional load***

We know that amount of frictional load,

$$W_0 = P \times V.R. - W = 700 \times 6 - 3000 = 1200 \text{ N } \textbf{Ans.}$$

**Example 8.16** *There are 4 pulleys arranged in third system. Find the effort required to lift a load of 5000 N. Assume efficiency of the machine is 85%.*

***Solution*** Given : $n = 4$ ; $W = 5000$ N ; $\eta = 85\% = 0.85$

Let $P$ = Effort required in N.

We know that velocity ratio for the third system of pulley,

$$V.R. = 2^n - 1 = 2^4 - 1 = 15$$

and efficiency,

$$\eta = \frac{M.A.}{V.R.}$$

∴ Mechanical advantage, $M.A. = \eta \times V.R. = 0.85 \times 15 = 12.75$

We also know that

$$M.A. = \frac{W}{P}$$

∴

$$P = \frac{W}{M.A.} = \frac{5000}{12.75} = 392.157 \text{ N } \textbf{Ans.}$$

**Example 8.17** *In a third system of pulleys, there are four pulleys. Find the effort required to lift a load of 10 kN, if the efficiency of the machine is 80%. Also find the amount of effort wasted in friction.*

***Solution*** Given : $n = 4$ ; $W = 10 \text{ kN} = 10 \times 10^3$ N ; $\eta = 80\% = 0.8$

***Effort required***

Let $P$ = Effort required in N.

We know that velocity ratio for the third system of pulleys,

$$V.R. = 2^n - 1 = 2^4 - 1 = 15$$

and mechanical advantage,

$$M.A. = \frac{W}{P} = \frac{10 \times 10^3}{P} \text{ N}$$

We also know that efficiency,

$$\eta = \frac{M.A.}{V.R.}$$

∴

$$M.A. = \eta \times V.R.$$

or $$\frac{10 \times 10^3}{P} = 0.8 \times 15 = 12$$

$\therefore$ $$P = \frac{10 \times 10^3}{12} = 833.33 \text{ N } \textbf{Ans.}$$

***Amount of effort wasted in friction***

We know that amount of effort wasted in friction

$$P_0 = P - \frac{W}{V.R.} = 833.33 - \frac{10 \times 10^3}{15}$$

$$= 833.33 - 666.67 = 166.66 \text{ N } \textbf{Ans.}$$

## 8.17 WESTON'S DIFFERENTIAL PULLEY BLOCK

The Weston's differential pulley block is shown in Fig. 8.11. It consists of two pulley blocks, the upper fixed block *A* and a lower moving block *B*. The upper block *A* has two pulleys 1 and 2 of slightly different diameters and are casted integrally. The lower block *B* also carries one movable pulley to which load (*W*) to be lifted, is attached.

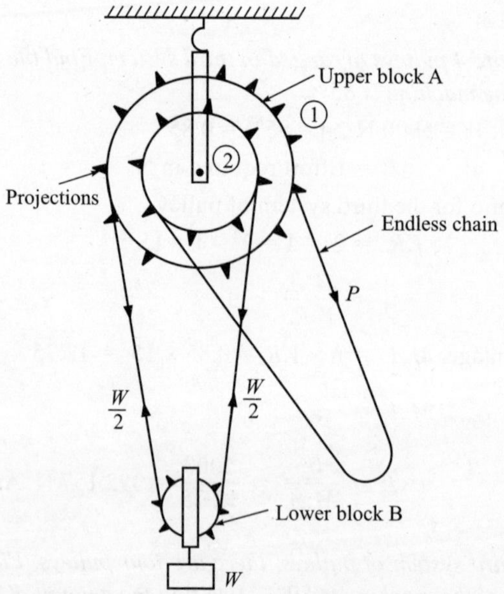


**Fig. 8.11** *Weston's differential pulley block*

An endless chain passes around the upper pulley 1, and then the lower movable pulley and finally around the pulley 2, as shown in Fig. 8.11. The remaining chain hangs slack. The effort *P* is applied to the slack portion of the chain passing over the pulley 1, as shown in the figure. In order to prevent slipping of chain, small projections are provided over the pulleys.

Let $D$ = Diameter of pulley 1, and

$d$ = Diameter of pulley 2.

For one complete revolution of the pulley 1, the distance moved by the effort ($P$) is $\pi D$ and the chain moves over the pulley 2 through a distance $\pi d$. Thus, the chain released by the pulley 2 is equal to $\pi d$.

$\therefore$ Net shortening of the chain $= \pi D - \pi d = \pi(D - d)$

This shortening of the chain is divided equally into two portions of the chain passing round the lower pulley which support the load ($W$). Thus, the distance moved by the load

$$= \frac{\pi}{2}(D - d)$$

$$\therefore \quad \text{Velocity ratio, } V.R. = \frac{\text{Distance moved by the effort } P}{\text{Distance moved by the load } W} = \frac{\pi D}{\frac{\pi}{2}(D-d)} = \frac{2D}{D-d}$$

and mechanical advantage,

$$M.A. = \frac{W}{P}$$

$$\therefore \quad \text{Efficiency, } \eta = \frac{M.A.}{V.R.} = \frac{W}{P} \times \frac{(D-d)}{2D}$$

**Note:** If $T_1$ and $T_2$ are the number of teeth on pulleys 1 and 2, respectively, then

$$\text{Velocity ratio, } V.R. = \frac{2T_1}{T_1 - T_2}$$

**Example 8.18** *In Weston's differential pulley block, the diameters of larger and smaller pulley are 300 mm and 225 mm, respectively. If the efficiency of the machine is 75%, determine the effort required to raise a load of 1500 N and effort wasted in friction.*

***Solution*** Given : $D = 300$ mm ; $d = 225$ mm ; $\eta = 75\% = 0.75$ ; $W = 1500$ N

***Effort required***

Let $P$ = Effort required in N.

We know that velocity ratio for Weston's differential pulley block,

$$V.R. = \frac{2D}{D-d} = \frac{2 \times 300}{300 - 225} = 8$$

and mechanical advantage,

$$M.A. = \frac{W}{P} = \frac{1500}{P}$$

We also know that efficiency,

$$\eta = \frac{M.A.}{V.R.}$$

$$\therefore \quad M.A. = \eta \times V.R.$$

$$\frac{1500}{P} = 0.75 \times 8 = 6$$

or

$$P = \frac{1500}{6} = 250 \text{ N } \textbf{Ans.}$$

***Effort wasted in friction***

We know that effort wasted in friction,

$$P_0 = P - \frac{W}{V.R.} = 250 - \frac{1500}{8} = 250 - 187.5 = 62.5 \text{ N } \textbf{Ans.}$$

**Example 8.19** *In a differential pulley block, there are 20 teeth in the larger pulley and 16 teeth in the smaller pulley. If a load of 1500 N is raised by an effort of 500 N, determine the velocity ratio, mechanical advantage and efficiency of the machine.*

***Solution*** Given : $T_1 = 20$ ; $T_2 = 16$ ; $W = 1500$ N ; $P = 500$ N

***Velocity ratio***

We know that velocity ratio,

$$V.R. = \frac{2T_1}{T_1 - T_2} = \frac{2 \times 20}{20 - 16} = 10 \textbf{ Ans.}$$

***Mechanical advantage***

We know that mechanical advantage,

$$M.A. = \frac{W}{P} = \frac{1500}{500} = 3 \textbf{ Ans.}$$

***Efficiency***

We know that efficiency,
$$\eta = \frac{M.A.}{V.R.} = \frac{3}{10} = 0.3 \text{ or } 30\% \textbf{ Ans.}$$

**Example 8.20** *In Weston's differential pulley block, the diameters of larger and smaller pulley are 320 mm and 280 mm. If the pulley runs smooth, calculate how much load will be lifted by applying an effort of 400 N. Also calculate the efficiency and the effort wasted in friction, if the actual load raised is 5600 N.*

**Solution** Given : $D = 320$ mm ; $d = 280$ mm ; $P = 400$ N

***Load lifted***

Let $W$ = Load lifted in N.

We know that velocity ratio,
$$V.R. = \frac{2D}{D-d} = \frac{2 \times 320}{320 - 280} = 16$$

and mechanical advantage,
$$M.A. = \frac{W}{P} = \frac{W}{400}$$

Since the pulley runs smooth, therefore

$$M.A. = V.R.$$

or
$$\frac{W}{400} = 16$$

$\therefore$
$$W = 400 \times 16 = 6400 \text{ N } \textbf{Ans.}$$

It may be noted that this is an ideal load.

***Efficiency***

Since the actual load raised is 5600 N, therefore

$$\text{Efficiency, } \eta = \frac{\text{Actual load}}{\text{Ideal load}} = \frac{5600}{6400} = 0.875 \text{ or } 87.5\% \textbf{ Ans.}$$

***Effort wasted in friction***

We know that effort wasted in friction,

$$P_0 = P - \frac{W}{V.R.} = 400 - \frac{5600}{16} = 400 - 350 = 50 \text{ N } \textbf{Ans.}$$

...[Here $W$ is the actual load raised which is 5600 N]

## 8.18 WORM AND WORM WHEEL

A worm and worm wheel, as shown in Fig. 8.12, consists of a square threaded screw known as *worm* and a toothed wheel known as *worm wheel.* Both the worm and worm wheel are geared with each other. The worm can be rotated by means of an effort wheel attached at one end of the worm over which passes a rope. Sometimes, a handle is fixed to the worm axle (instead of an effort wheel) to rotate the worm. A load drum is keyed to the worm wheel. The rope is taken round the drum and its free end carries the load.

Let $T$ = Number of teeth on the worm wheel,

$l$ = Radius of the effort wheel or length of the handle,

$r$ = Radius of the load drum,

$W$ = Load lifted, and

$P$ = Effort applied to lift the load.

If the worm is single threaded, then for one complete revolution of the effort wheel (or handle), the distance moved by the effort $= 2\pi l$

During this period, the worm wheel and the load drum will rotate by $\frac{1}{T}$ of a revolution. Thus, the distance moved by the load $W$ $= \frac{2\pi r}{T}$

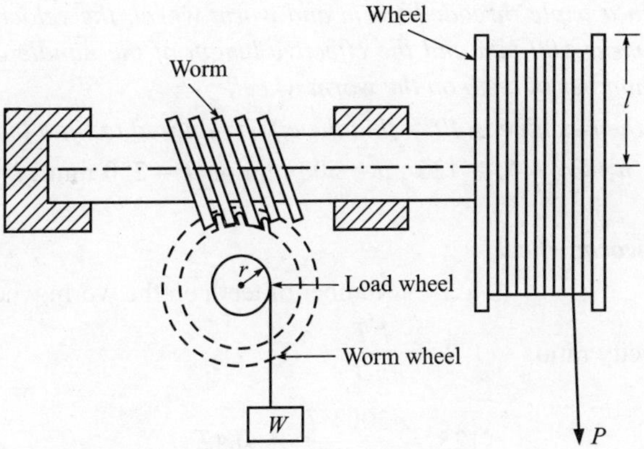


**Fig. 8.12** *Worm and worm wheel*

$$\therefore \quad \text{Velocity ratio, } V.R. = \frac{\text{Distance moved by the effort}}{\text{Distance moved by the load}} = \frac{2\pi l}{\frac{2\pi r}{T}} = \frac{l.T}{r}$$

$$\text{Mechanical advantage, } M.A. = \frac{W}{P}$$

$$\therefore \quad \text{Efficiency, } \eta = \frac{M.A.}{V.R.} = \frac{W}{P} \times \frac{r}{l.T}$$

**Note:** If the worm is double threaded, then for one revolution of the effort wheel (or handle), the worm wheel will move by two teeth. Thus, the load drum will rotate through $\frac{2}{T}$ revolutions.

$$\therefore \quad \text{Distance moved by the load } W = 2\pi r \times \frac{2}{T}$$

and velocity ratio,
$$V.R. = \frac{2\pi l}{2\pi r} \times \frac{T}{2} = \frac{l.T}{2r}$$

In general, if the worm is $n$ threaded, then velocity ratio,
$$V.R. = \frac{l.T}{n.r}$$

**Example 8.21** *In a single threaded worm and worm wheel, the effective diameter of the effort wheel is 280 mm and that of drum is 80 mm. If the worm wheel has 100 teeth, determine the efficiency of the machine, if a load of 7500 N is lifted by an effort of 50 N.*

***Solution*** Given : $2l = 280$ mm or $l = 140$ mm ; $d = 80$ mm or $r = 40$ mm ; $T = 100$ ; $W = 7500$ N ; $P = 50$ N

We know that velocity ratio,
$$V.R. = \frac{l.T}{r} = \frac{140 \times 100}{40} = 350$$

and mechanical advantage, $M.A. = \dfrac{W}{P} = \dfrac{7500}{50} = 150$

$\therefore$ Efficiency, $\eta = \dfrac{M.A.}{V.R.} = \dfrac{150}{350} = 0.4286$ or $42.86\%$ **Ans.**

**Example 8.22** *In a triple threaded worm and worm wheel, the velocity ratio is given as 125. The diameter of the load drum is 500 mm and the effective length of the handle attached to the worm axle is 300 mm. Determine the number of teeth on the worm wheel.*

*If the efficiency of the machine is 40%, find the effort required to lift a load of 20 kN.*

***Solution*** Given : $n = 3$ ; $V.R. = 125$ ; $d = 500$ mm or $r = 250$ mm ; $l = 300$ mm ; $\eta = 40\% = 0.4$ ; $W = 20$ kN $= 20 \times 10^3$ N

***Number of teeth on the worm wheel***

Let $T$ = Number of teeth on the worm wheel.

We know that velocity ratio, $V.R. = \dfrac{l.T}{n.r}$

$$125 = \frac{300 \times T}{3 \times 250} = 0.4T$$

$$\therefore \quad T = \frac{125}{0.4} = 312.5$$

Since the number of teeth cannot be fractional, therefore, we shall take

$T = 312$ **Ans.**

***Effort required***

Let $P$ = Effort required in N.

We know that efficiency, $\eta = \dfrac{M.A.}{V.R.} = \dfrac{W}{P \times V.R.}$

$$0.4 = \frac{20 \times 10^3}{P \times 125} = \frac{160}{P}$$

$$\therefore \quad P = \frac{160}{0.4} = 400 \text{ N}$$ **Ans.**

## 8.19 SIMPLE SCREW JACK

A *screw jack, as shown in Fig. 8.13, is a device for lifting heavy loads by applying comparatively a smaller effort at its handle (or lever). It works on the same principle as an inclined plane. It mainly consists of a nut which forms the body of the jack, and a square threaded screw is fitted into it. The load $W$ is placed on the head of the screw. The load is lifted or lowered by applying the effort ($P$) at the end of the handle.

Let $l$ = Length of the handle,

$W$ = Load to be lifted,

$P$ = Effort applied at the end of the handle or lever,

$p$ = Pitch of the screw.

Consider that a single threaded screw has taken one full revolution. Therefore the distance moved by the load $= p$

and the distance moved by the effort $= 2\pi l$

* Refer also Chapter 7 on Friction (Sec. 7.4)

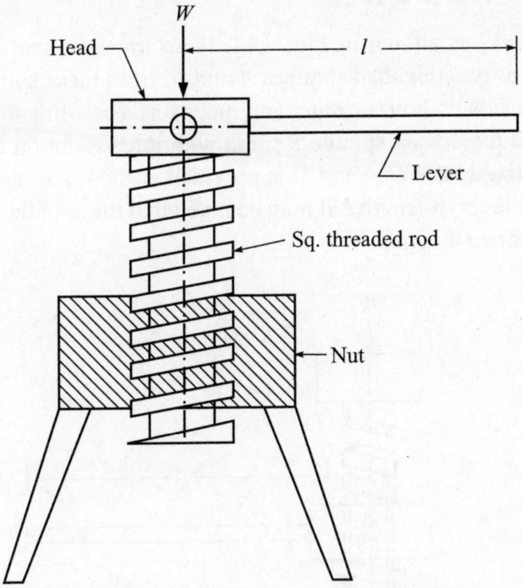


**Fig. 8.13** *Simple screw jack*

We know that velocity ratio, $V.R. = \dfrac{\text{Distance moved by the effort}}{\text{Distance moved by the load}} = \dfrac{2\pi l}{p}$

and mechanical advantage, $M.A. = \dfrac{W}{P}$

$\therefore$ Efficiency, $\eta = \dfrac{M.A.}{V.R.} = \dfrac{W}{P} \times \dfrac{p}{2\pi l}$

**Note:** When the screw is double threaded, then the distance moved by the load in one full revolution $= 2p$

and velocity ratio, $V.R. = \dfrac{2\pi l}{2p} = \dfrac{\pi l}{p}$

$\therefore$ Efficiency, $\eta = \dfrac{M.A.}{V.R.} = \dfrac{W}{P} \times \dfrac{p}{\pi l}$

**Example 8.23** *A simple screw jack has threads of pitch 5 mm. The effort is applied at the end of a lever 500 mm long. What effort will be required to lift a load of 10 kN, if the efficiency at this load is 40%.*

***Solution*** Given : $p = 5$ mm ; $l = 500$ mm ; $W = 10$ kN $= 10\,000$ N ; $\eta = 40\% = 0.4$

Let $P$ = Effort required in N.

We know that velocity ratio, $V.R. = \dfrac{2\pi l}{p} = \dfrac{2\pi \times 500}{5} = 628.4$

and efficiency, $\eta = \dfrac{M.A.}{V.R.} = \dfrac{W}{P \times V.R.}$

or $0.4 = \dfrac{10\,000}{P \times 628.4}$

$\therefore$ $P = \dfrac{10\,000}{0.4 \times 628.4} = 39.8$ N **Ans.**

## 8.20 DIFFERENTIAL SCREW JACK

The differential screw jack, as shown in Fig. 8.14, is an improvement of simple screw jack to give more velocity ratio. It consists of two threaded spindles *A* and *B*. Both these spindles have threads in the same direction. The spindle *A* has threads on both its outer and inner surfaces while the spindle *B* has threads on the outer surface only. The external threads on spindle *B* mesh with threads cut on the inner surface of spindle *A*. The spindle *A* is screwed to a fixed base *C*. A nut *D* is provided at the top of spindle *A*. The nut is rotated by applying effort *P* at the end of a lever of length *l*. It may be noted that the spindle *B* does not rotate, but it moves only in vertical direction and carries the load *W*.

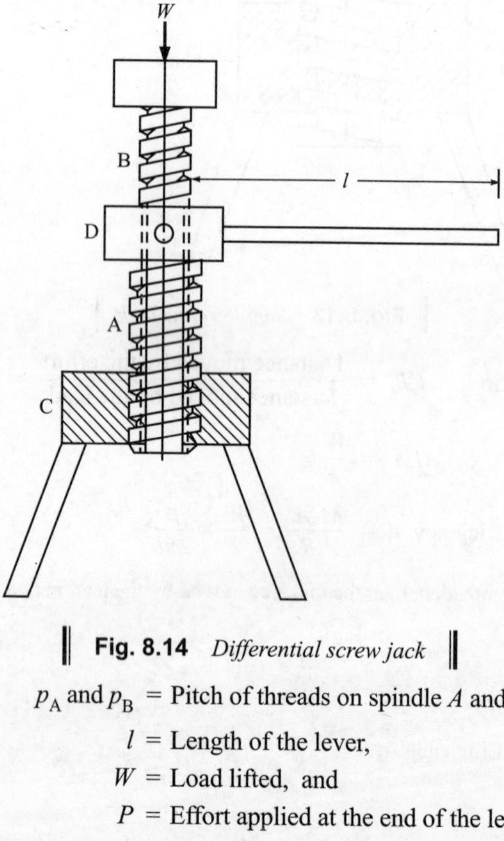


**Fig. 8.14** *Differential screw jack*

Let $p_A$ and $p_B$ = Pitch of threads on spindle *A* and *B*, respectively,

$l$ = Length of the lever,

$W$ = Load lifted, and

$P$ = Effort applied at the end of the lever.

Consider that the lever is rotated through one complete revolution. During this time, the spindle *A* moves up and the spindle *B* moves down. Therefore the net distance moved by the load

$$= p_A - p_B \quad ...(\because p_A > p_B)$$

and the distance moved by the effort *P* $= 2\pi l$

We know that velocity ratio, $$V.R. = \frac{\text{Distance moved by the effort}}{\text{Distance moved by the load}} = \frac{2\pi l}{p_A - p_B}$$

and mechanical advantage, $$M.A. = \frac{W}{P}$$

$$\therefore \quad \text{Efficiency, } \eta = \frac{M.A.}{V.R.} = \frac{W}{P} \times \frac{p_A - p_B}{2\pi l}$$

**Example 8.24** *In a differential screw jack, the smaller and larger screws have pitches equal to 5 mm and 10 mm, respectively. The effort arm has a radius equal to 500 mm. What effort will be required to lift a load of 30 kN at an efficiency of 50%?*

***Solution*** Given : $p_B = 5$ mm ; $p_A = 10$ mm ; $l = 500$ mm ; $W = 30$ kN $= 30 \times 10^3$ N ; $\eta = 50\% = 0.5$

Let $P$ = Effort required in N.

We know that velocity ratio, $$V.R. = \frac{2\pi l}{p_A - p_B} = \frac{2\pi \times 500}{10 - 5} = 628.4$$

and efficiency, $$\eta = \frac{M.A.}{V.R.} = \frac{W}{P \times V.R.}$$

or $$0.5 = \frac{30 \times 10^3}{P \times 628.4}$$

$$\therefore \quad P = \frac{30 \times 10^3}{0.5 \times 628.4} = 95.48 \text{ N} \quad \textbf{Ans.}$$

## 8.21 SINGLE PURCHASE WINCH CRAB

A single purchase winch crab, as shown in Fig. 8.15, has two axles $A$ and $B$. The axle $A$ is known as effort axle on which a handle is fitted. A toothed wheel mounted on the effort axle is known as pinion. The axle $B$ is called a load axle on which a load drum is mounted and a toothed wheel mounted on this load axle is called a spur wheel. It may be noted that the pinion is always smaller in size than the spur wheel.

Let $T_1$ = Number of teeth on the pinion,
$T_2$ = Number of teeth on the spur wheel,
$D$ = Diameter of the load drum, and
$l$ = Length of the handle.

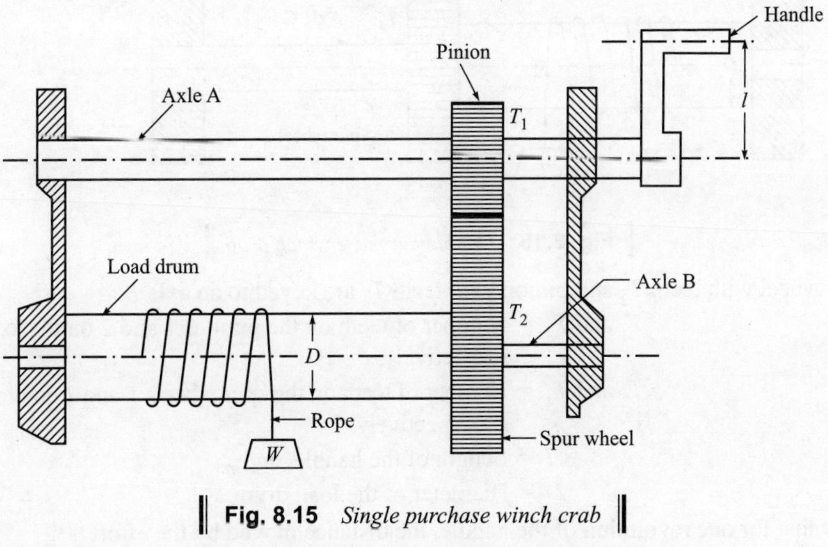


**Fig. 8.15** *Single purchase winch crab*

We know that for one revolution of the handle, the distance moved by the effort ($P$)

$$= 2\pi l$$

It may be noted that when the handle makes one revolution, the pinion also makes one revolution and the spur wheel as well as the load drum will make $\frac{T_1}{T_2}$ revolutions. Thus, the distance moved by the load

$$= \pi D \times \frac{T_1}{T_2}$$

$$\therefore \quad \text{Velocity ratio, } V.R. = \frac{\text{Distance moved by the effort}}{\text{Distance moved by the load}} = \frac{2\pi l}{\pi D \times \frac{T_1}{T_2}} = \frac{2l}{D} \times \frac{T_2}{T_1}$$

$$\text{Mechanical advantage, } M.A. = \frac{W}{P}$$

and efficiency,
$$\eta = \frac{M.A.}{V.R.}$$

## 8.22 DOUBLE PURCHASE WINCH CRAB

A double purchase winch crab, as shown in Fig. 8.16, has three axles $A$, $B$ and $C$. It consists of two pinions having teeth $T_1$ and $T_3$, which are geared to two spur wheels having teeth $T_2$ and $T_4$. The pinion with teeth $T_1$ is keyed to axle $A$ which is known as the effort axle and is rotated by the handle of length $l$ by the application of effort $P$. The spur wheel with teeth $T_4$ is keyed to the load drum carrying the load $W$.

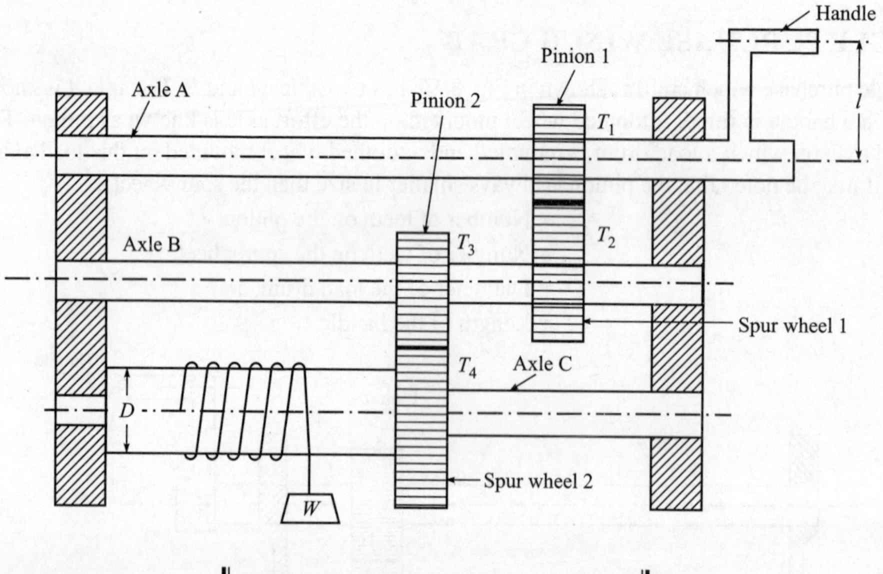


**Fig. 8.16** *Double purchase winch crab*

The spur wheel with teeth $T_2$ and pinion with teeth $T_3$ are keyed to an axle $B$.

Let
$T_1$ and $T_3$ = Number of teeth on the pinions 1 and 2 on the axles $A$ and $B$, respectively,
$T_2$ and $T_4$ = Number of teeth on the spur wheels 1 and 2 on the axles $B$ and $C$, respectively,
$l$ = Length of the handle, and
$D$ = Diameter of the load drum.

We know that for one revolution of the handle, the distance moved by the effort ($P$)

$$= 2\pi l$$

It may be noted that when the handle makes one revolution, the pinion 1 with teeth $T_1$, also makes one revolution and the spur wheel 1 on axle $B$ will make $\frac{T_1}{T_2}$ revolutions.

Now, if pinion with teeth $T_3$ makes one revolution, the spur wheel with teeth $T_4$ will make $\frac{T_3}{T_4}$ revolution.

Thus for one revolution of the handle, the number of revolutions made by the load drum (or load axle)

$$= \frac{T_3}{T_4} \times \frac{T_1}{T_2}$$

$$\therefore \quad \text{Distance moved by the load} = \pi D \times \frac{T_3}{T_4} \times \frac{T_1}{T_2}$$

We know that velocity ratio,

$$V.R. = \frac{\text{Distance moved by the effort}}{\text{Distance moved by the load}} = \frac{2\pi l}{\pi D \times \frac{T_3}{T_4} \times \frac{T_1}{T_2}}$$

$$= \frac{2l}{D} \times \frac{T_4 \times T_2}{T_3 \times T_1}$$

$$\text{Mechanical advantage, } M.A. = \frac{W}{P}$$

and efficiency,

$$\eta = \frac{M.A.}{V.R.}$$

**Example 8.25** *A single purchase crab-winch has the following details :*

*Length of lever = 700 mm ; Number of pinion teeth = 16 ; Number of spur gear teeth = 96 ; Diameter of load axle = 200 mm.*

*It is observed that an effort of 60 N lift a load of 1800 N and an effort of 120 N lifts a load of 3960 N.*

*1. Establish the law of the machine ;*

*2. Find the efficiencies in the two cases ; and*

*3. Determine the loss of load and loss of effort in friction when lifting a load of 3960 N.*

***Solution*** Given : $l$ = 700 mm = 0.7 m ; $T_1$ = 16 ; $T_2$ = 96 ; $D$ = 200 mm = 0.2 m ; $P_1$ = 60 N ; $W_1$ = 1800 N ; $P_2$ = 120 N ; $W_2$ = 3960 N

**1. *Law of the machine***

We know that law of the machine is

$$P = mW + C$$

When $P_1$ = 60 N; $W_1$ = 1800, then

$$60 = m \times 1800 + C \quad \text{...}(i)$$

When $P_2$ = 120 N; $W_2$ = 3960 , then

$$120 = m \times 3960 + C \quad \text{...}(ii)$$

Subtracting equation (*i*) from equation (*ii*),

$$60 = 2160\, m$$

$$\therefore \quad m = \frac{60}{2160} = 0.0278$$

Substituting this value of $m$ in equation (*i*), we have

$$60 = 0.0278 \times 1800 + C = 50 + C$$

$$\therefore \quad C = 60 - 50 = 10$$

Thus law of the machine is $P = 0.0278W + 10$ **Ans.**

**2. *Efficiency in the two cases***

We know that velocity ratio, $V.R. = \frac{2l}{D} \times \frac{T_2}{T_1} = \frac{2 \times 0.7}{0.2} \times \frac{96}{16} = 42$

Efficiency when load $W_1 = 1800$ N,

$$\eta_1 = \frac{M.A.}{V.R.} = \frac{W_1}{P_1 \times V.R.} = \frac{1800}{60 \times 42} = 0.7143 \text{ or } 71.43\% \textbf{ Ans.}$$

Similarly, efficiency when load $W_2 = 3960$ N,

$$\eta_2 = \frac{W_2}{P_2 \times V.R.} = \frac{3960}{120 \times 42} = 0.7857 \text{ or } 78.57\% \textbf{ Ans.}$$

**3. *Loss of load and loss of effort in friction when lifting a load of $W_2 = 3960$ N***

We know that loss of load in friction = Ideal load – Actual load

$$= P_2 \times V.R. - W_2 = 120 \times 42 - 3960$$

$$= 5040 - 3960 = 1080 \text{ N } \textbf{Ans.}$$

Loss of effort in friction = Actual effort – Ideal effort

$$= P_2 - \frac{W_2}{V.R.} = 120 - \frac{3960}{42} = 120 - 94.28 = 25.72 \text{ N } \textbf{Ans.}$$

**Example 8.26** *In a double purchase crab, the pinions have 15 and 20 teeth while the spur wheels have 45 and 40 teeth. The effort handle is 200 mm while the effective diameter of the drum is 150 mm. If the efficiency of the winch is 40%, what load will be lifted by an effort of 250 N applied at the end of the handle. Also, find the effort lost in friction.*

***Solution*** Given : $T_1 = 15$ ; $T_3 = 20$ ; $T_2 = 45$ ; $T_4 = 40$ ; $l = 200$ mm = 0.2 m ; $D = 150$ mm = 0.15 m ; $\eta = 40\% = 0.4$ ; $P = 250$ N

***Load lifted***

Let $W$ = Load lifted in N.

We know that velocity ratio, $V.R. = \frac{2l}{D} \times \frac{T_4 \times T_2}{T_3 \times T_1} = \frac{2 \times 0.2}{0.15} \times \frac{40 \times 45}{20 \times 15} = 16$

and efficiency, $\eta = \frac{M.A.}{V.R.} = \frac{W}{P \times V.R.}$

or $0.4 = \frac{W}{250 \times 16} = \frac{W}{4000}$

∴ $W = 0.4 \times 4000 = 1600$ N **Ans.**

***Effort lost in friction***

We know that effort lost in friction = Actual effort – Ideal effort

$$= P - \frac{W}{V.R.} = 250 - \frac{1600}{16} = 250 - 100 = 150 \text{ N } \textbf{Ans.}$$

## HIGHLIGHTS

1. A *machine* is a device by means of which a heavy load is being lifted or dragged by applying a small effort at some other convenient point.
2. When the effort is applied at one point to lift a load acting at some other point, then the machine is known as a *simple machine.*

3. When the effort is applied at more than one point to lift the load, then the machine is said to be a *compound machine.*
4. The ratio of load ($W$) lifted to the effort ($P$) applied is known *mechanical advantage* (briefly written as *M.A.*). Mathematically,

$$M.A. = \frac{W}{P}$$

5. The ratio of the distance moved by the effort ($y$) to the distance moved by the load ($x$) is known as *velocity ratio* (briefly written as *V.R.*). Mathematically,

$$V.R. = \frac{y}{x}$$

6. The total work done on the machine is called *input of the machine*. It is equal to the product of effort ($P$) and the distance moved by the effort ($y$). Mathematically,

$$\text{Input of a machine} = P \times y$$

7. The actual work done by the machine is called *output of the machine*. It is equal to the product of the load lifted ($W$) and the distance ($x$) through which the load is lifted. Mathematically,

$$\text{Output of a machine} = W \times x$$

8. The ratio of output to input of a machine is termed as *efficiency of a machine*. It is expressed by a Greek letter eta ($\eta$). Mathematically,

$$\eta = \frac{\text{Output of a machine}}{\text{Input of a machine}} = \frac{W \times x}{P \times y}$$

$$= \frac{M.A.}{V.R.} \qquad \ldots\left(\because M.A. = \frac{W}{P} \text{ and } V.R. = \frac{y}{x}\right)$$

9. A machine whose efficiency is 100%, is known as an *ideal machine.* Since no machine is an ideal machine, therefore a part of energy supplied (input) is wasted in overcoming frictional resistance. Thus useful work done (output) is always less than energy supplied (input).
10. If $P$ and $P_0$ are the actual effort (considering friction) and ideal effort (neglecting friction) and $W$ and $W_0$ are the actual load (considering friction) and ideal load (neglecting friction), then

$$\text{Ideal effort, } P_0 = \frac{W}{V.R.}$$

$$\text{and loss of effort due to friction, } P_1 = P - P_0 = P - \frac{W}{V.R.}$$

$$\text{Ideal load, } W_0 = P \times V.R.$$

$$\text{and loss of load due to friction, } W_1 = W_0 - W = P \times V.R. - W$$

11. The efficiency of a machine in terms of actual and ideal conditions is given by

$$\eta = \frac{W}{W_0} = \frac{\text{Actual load}}{\text{Ideal load}}$$

$$= \frac{P_0}{P} = \frac{\text{Ideal effort}}{\text{Actual effort}}$$

12. A machine capable of doing work in the reverse direction after the effort is removed, is known as *reversible machine.* The condition for a machine to be reversible, its efficiency should be more than 50%.
13. A machine which is not capable of doing any work in the reversed direction, after the effort is removed, is known as *irreversible* or *self-locking machine.* The condition for a machine to be irreversible or self-locking, its efficiency should be less than 50%.

**14.** The law of a machine is

$$P = mW + C$$

where $P$ = Effort applied,
$W$ = Load lifted,
$m$ = Slope of the straight line, and
$C$ = Constant.

**15.** Maximum mechanical advantage,

$$(M.A.)_{max} = \frac{1}{m}$$

and maximum efficiency, $\eta_{max} = \frac{1}{m} \times \frac{1}{V.R.} = \frac{(M.A.)_{max}}{V.R.}$

**16.** Velocity ratio for the simple wheel and axle is given by

$$V.R. = \frac{D}{d}$$

where $D$ = Diameter of the wheel, and
$d$ = Diameter of the axle.

**17.** Velocity ratio for the differential wheel and axle is given by

$$V.R. = \frac{2D}{d_1 - d_2}$$

where $D$ = Diameter of the wheel, and
$d_1$ and $d_2$ = Diameter of the larger and smaller axles, respectively.

**18.** The velocity ratio in case of various system of pulleys is as follows:

***(a)*** For first system of pulleys,

$$V.R. = 2n$$

where $n$ = Number of movable pulleys.

***(b)*** For second system of pulleys,

$$V.R. = n$$

where $n$ = Total number of pulleys in the upper and lower pulley blocks.

***(c)*** For third system of pulleys,

$$V.R. = 2^n - 1$$

where $n$ = Total number of pulleys including the fixed pulley.

**19.** In a Weston's differential pulley block, the velocity ratio is given by

$$V.R. = \frac{2D}{D - d}$$

where $D$ and $d$ are diameter of pulleys.

**20.** The velocity ratio in case of worm and worm wheel is

$$V.R. = \frac{l.T}{r}$$

where $l$ = Radius of effort wheel or length of the handle,
$T$ = Number of teeth on the worm wheel, and
$r$ = Radius of the load drum.

**21.** The velocity ratio for a simple screw jack is given by

$$V.R. = \frac{2\pi l}{p} \quad \text{...(For single threaded screw)}$$

$$= \frac{\pi l}{p} \quad \text{...(For double threaded screw)}$$

where $l$ = Length of the handle, and

$p$ = Pitch of the screw.

**22.** For a differential screw jack, velocity ratio is given by

$$V.R. = \frac{2\pi l}{p_A - p_B}$$

where $p_A$ and $p_B$ are the pitch of threads on the two spindles.

**23.** The velocity ratio for a single purchase winch crab is

$$V.R. = \frac{2l}{D} \times \frac{T_2}{T_1}$$

where $l$ = Length of the handle,

$D$ = Diameter of the load drum,

$T_1$ = Number of teeth on the pinion, and

$T_2$ = Number of teeth on the spur wheel.

**24.** For a double purchase winch crab, velocity ratio is given by

$$V.R. = \frac{2l}{D} \times \frac{T_4 \times T_2}{T_3 \times T_1}$$

where $T_1$ and $T_3$ = Number of teeth on the pinions, and

$T_2$ and $T_4$ = Number of teeth on the spur wheels.

## NUMERICAL PROBLEMS

**1.** In a simple lifting machine, a load of 1200 N is raised by an effort of 30 N. Find the mechanical advantage. If the machine has an efficiency of 60%, what is the velocity ratio? [**Ans.** 40; 66.67]

**2.** In a lifting machine, an effort of 500 N is to be moved by a distance of 20 m to raise a load of 10 kN by a distance of 0.8 m. Determine the mechanical advantage, velocity ratio and efficiency of the machine. [**Ans.** 20; 25; 80%]

**3.** In a certain machine, an effort of 100 N is required to lift a load of 840 N. If the velocity of the machine is 10, determine the ideal effort, effort lost in friction, ideal load and frictional resistance. [**Ans.** 84 N; 16 N; 1000 N; 160 N]

**4.** In a simple machine, whose velocity ratio is 30, a load of 3.4 kN is lifted by an effort of 200 N. State whether this machine is reversible or irreversible. Also determine the frictional resistance. [**Ans.** Reversible; 1600 N]

**5.** In a lifting machine, an effort of 98.2 N raised a load of 1000 N and an effort of 498.2 N raised a load of 6000 N. Find the law of the machine. What is the effort required to lift a load of 10 000 N ? Find the maximum mechanical advantage. [**Ans.** $P = 0.08W + 18.2$; 818.2 N; 12.5]

**6.** In a lifting machine, an effort of 150 N raised a load of 7700 N. What is the mechanical advantage? Find the velocity ratio if the efficiency at this load is 60%.

If on this machine, an effort of 250 N raised a load of 13.2 kN, what is the efficiency?

Calculate the maximum mechanical advantage and the maximum efficiency.

[**Ans.** 51.33; 85.55; 61.72%; 55; 64.3%]

7. The law of the machine is given by $P = \frac{W}{25} + 10$. The velocity ratio of the machine is 40. Determine the mechanical advantage and efficiency of the machine to lift a load 2000 N. Also determine the maximum efficiency and the effort lost in friction. [**Ans.** 22.2; 55.5%; 62.5%; 40 N]

8. In a certain experiment, the following values of load and effort are observed:

| Load lifted (N) | 100 | 200 | 300 | 400 | 500 | 600 |
|---|---|---|---|---|---|---|
| Effort applied (N) | 16 | 22.5 | 28 | 34 | 40.5 | 46.5 |

Determine the law of machine and draw the graph between effort and load. If the velocity ratio is 25, determine the effort required and efficiency when the load is 550 N. Also determine the maximum mechanical advantage and maximum efficiency. [**Ans.** $P = 0.06\,W + 10$; 43 N; 51.1%; 16.66; 66.67%]

9. In a simple wheel and axle, the diameter of wheel is 500 mm and the diameter of axle is 90 mm. If a weight of 140 N is raised by an effort of 40 N, find mechanical advantage, velocity ratio and efficiency of the machine. [**Ans.** 5.55; 3.5; 63.06%]

10. In a differential wheel and axle, the diameter of wheel is 250 mm. The diameter of the larger and smaller axles are 100 mm and 90 mm respectively. A load of 750 N is lifted by an effort of 30 N. Determine: 1. velocity ratio; 2 mechanical advantage; and 3. efficiency. [**Ans.** 50; 25; 50%]

11. In a differential wheel and axle, the larger and smaller diameters of axles are 85 mm and 75 mm, respectively. The effort is applied to the wheel at the end of a lever which is 100 mm long. If an effort of 30 N can lift a load of 900 N; determine the mechanical advantage, velocity ratio and efficiency of the machine. [**Ans.** 30; 40; 75%]

12. In a differential wheel and axle, the diameter of the effort wheel is 400 mm and the difference between the axle diameters is 50 mm. If a load of 2000 N is lifted by an effort of 500 N and a load of 5000 N is lifted by an effort of 1000 N, determine law of the machine and load lifted by an effort of 1500 N. Also calculate the mechanical advantage and efficiency of the machine. [**Ans.** $P = 0.1W + 300$; 12 kN; 8; 50%]

13. In a first system of pulleys, there are three movable pulleys. Determine the effort required to lift a load of 5 kN, if the efficiency of the system is 84%. Also determine the effort lost in friction. [**Ans.** 744 N; 119 N]

14. In a system of pulleys of the first type, there are total six number of pulleys and the weight of each pulley is 14 N. Determine the load that can be lifted up by an effort of 170 N. [**Ans.** 5006 N]

15. In a second system of pulley, there are three pulleys in the upper block and two pulleys in the lower block. If the efficiency of the system is 80%, find the effort required to lift a load of 1000 N and the effort lost in friction. [**Ans.** 250 N; 50 N]

16. In a third system of pulleys, there are four pulleys. If the efficiency of the machine is 75%, find the effort needed to lift a load of 9 kN. Also determine the amount of effort wasted in friction. [**Ans.** 800 N; 200 N]

17. A Weston's differential pulley block consists of a lower block and an upper block. The upper block has two cogged grooves, one of which has a diameter of 250 mm and the other has a diameter of 230 mm. If the efficiency of the machine is 80%, calculate the effort required to raise a load of 1500 N. [**Ans.** 75 N]

18. In a differential pulley block, the number of teeth on the larger and smaller pulley are 12 and 11, respectively. Determine the velocity ratio. If the mechanical efficiency of the machine is 75%, determine the amount of load lifted by an effort of 100 N. [**Ans.** 24; 1800 N]

19. The diameter of a larger pulley of a Weston's differential pulley is 1.5 times of the diameter of smaller pulley. Determine the effort required to lift a load of 6 kN. The efficiency of the machine is 80%. Also find the effort wasted in friction. [**Ans.** 1250 N; 250 N]

20. In a single threaded worm and worm wheel, the number of teeth on the worm wheel is 60. The diameter of the effort wheel is 250 mm and that of load drum is 125 mm. Calculate the velocity ratio. If the efficiency of the machine is 40%, determine the effort required to lift a load of 1200 N. [**Ans.** 120; 250 N]

21. In a double threaded worm and worm wheel, the number of teeth on the worm wheel is 80. The length of the effort handle is 200 mm and the diameter of the load drum is 100 mm. Determine the efficiency of the machine, when a load of 3500 N is lifted by an effort of 75 N. [**Ans.** 29.16%]

**22.** A screw jack has a thread of 12 mm pitch. What effort will be applied at the end of a handle of 450 mm to lift a load of 2.5 kN, if the corresponding efficiency is 50%? **[Ans.** 21.22 N]

**23.** In a differential screw jack, the screw threads have pitch of 12 mm and 8 mm and the length of the lever arm is 400 mm. If the efficiency of the machine is 40%, find the effort required to lift a load of 1250 N. **[Ans.** 4.973 N]

**24.** Following are the specifications of a single purchase crab:

Diameter of load drum, $D$ = 200 mm

Length of lever, $l$ = 1.2 m

Number of teeth on pinion, $T_1$ = 10

Number of teeth on spur wheel, $T_2$ = 100

Find the velocity ratio of this machine. On this machine, efforts of 100 N and 160 N are required to lift the loads of 3 kN and 9 kN, respectively. Find the law of the machine and the efficiencies at the above loads. Also find the maximum mechanical advantage and maximum efficiency.

[**Ans.** $P = 0.01\ W + 70$; 25%, 46.87%; 100; 83.3%]

**25.** In a double purchase crab, the pinions have 10 and 15 teeth while the spur wheels have 45 and 50 teeth. The effort handle is 600 mm long and the effective diameter of the drum is 200 mm. Find the velocity ratio of the machine. If the efficiency of the machine is 40%, what load will be lifted by an effort of 100 N at the end of handle. [**Ans.** 90; 3600 N]

## REVIEW QUESTIONS

1. What is the difference between a simple machine and a compound machine?
2. Define the following terms as applied to a simple machine:
   (*a*) Mechanical advantage; (*b*) Velocity ratio; and (*c*) Efficiency.
3. Derive the relationship between efficiency, mechanical advantage, and velocity ratio in case of a lifting machine.
4. What do you understand by an ideal machine?
5. Discuss the effect of friction in machines. What is an ideal effort and load?
6. Write the expressions for the efficiency of a simple machine in terms of ideal and actual values of load and effort.
7. What is a reversible machine? Show that a machine is reversible if its efficiency is more than 50%.
8. Prove that for a machine to be irreversible, its efficiency should be below 50%.
9. Distinguish between reversible and self-locking machine.
10. What is law of a machine? Derive an expression for the same.
11. Obtain an expression for the maximum mechanical advantage and maximum efficiency of the machine.
12. Describe briefly the working of differential wheel and axle. Derive the expression for the velocity ratio of the machine. How does it differ from the simple wheel and axle?
13. Explain with the help of a neat sketch, the commonly used system of pulleys.
14. Write a short note on Weston's differential pulley block.
15. Draw a diagram of a worm and worm wheel and explain its principle and working.
16. Explain with the help of a neat sketch, the working of a simple screw jack.
17. Explain the construction and working of a differential screw jack.
18. Describe the principle and working of the following:
    (*a*) Single purchase winch crab; (*b*) Double purchase winch crab.

## OBJECTIVE TYPE QUESTIONS

1. The mechanical advantage of a lifting machine is the ratio of
   (*a*) load lifted to the effort applied
   (*b*) output to the input
   (*c*) distance moved by the effort to the distance moved by the load
   (*d*) all of the above

2. The velocity ratio of a lifting machine is the ratio of
   (*a*) load lifted to the effort applied
   (*b*) output to the input
   (*c*) distance moved by the effort to the distance moved by the load
   (*d*) all of the above

3. The efficiency of a lifting machine is the ratio of
   (*a*) mechanical advantage to the velocity ratio
   (*b*) output to the input
   (*c*) work done by the machine to the work done on the machine
   (*d*) all of the above

4. In ideal machines, mechanical advantage is ............. velocity ratio.
   (*a*) equal to (*b*) less than (*c*) greater than

5. A machine having efficiency greater than 50% is known as
   (*a*) irreversible machine
   (*b*) reversible machine
   (*c*) neither reversible nor irreversible machine
   (*d*) ideal machine

6. A machine having efficiency .............. 50% is known as self-locking machine.
   (*a*) equal to (*b*) less than (*c*) greater than

7. A weight of 1000 N can be lifted by an effort of 80 N. If the velocity ratio is 20, the machine is
   (*a*) reversible (*b*) irreversible (*c*) ideal

8. The law of the machine is
   (*a*) $P = mW - C$ (*b*) $P = \frac{m}{W} + C$ (*c*) $P = mW + C$ (*d*) $P = C - mW$

   where $P$ = Effort applied,
   $W$ = Load lifted,
   $m$ = A constant which is equal slope of the line, and
   $C$ = Another constant which represents the machine friction.

9. The maximum mechanical advantage of a lifting machine is
   (*a*) $\frac{1}{m}$ (*b*) $1 + m$ (*c*) $1 - m$ (*d*) $m$

10. The maximum efficiency of a lifting machine is
    (*a*) $\frac{1}{m}$ (*b*) $\frac{V.R.}{m}$ (*c*) $\frac{m}{V.R.}$ (*d*) $\frac{1}{m \times V.R.}$

**11.** The velocity ratio of a simple wheel and axle is

(*a*) $\frac{D}{d}$ (*b*) $D \times d$ (*c*) $D + d$ (*d*) $D - d$

where $D$ = Diameter of effort wheel, and

$d$ = Diameter of load axle.

**12.** The velocity ratio of a differential wheel and axle is

(*a*) $\frac{D}{d_1 - d_2}$ (*b*) $\frac{D}{d_1 + d_2}$ (*c*) $\frac{2D}{d_1 - d_2}$ (*d*) $\frac{2D}{d_1 + d_2}$

where $D$ = Diameter of effort wheel, and

$d_1$ and $d_2$ = Diameters of larger and smaller axles, respectively.

**13.** The velocity ratio of a first system of pulleys with 3 pulleys is

(*a*) 4 (*b*) 8 (*c*) 16 (*d*) 20

**14.** The velocity ratio for the second system of pulley having $n$ number of pulleys is

(*a*) $n$ (*b*) $n^2$ (*c*) $2^n$ (*d*) $2^n - 1$

**15.** The velocity ratio for the third system of pulley is

(*a*) $n$ (*b*) $n^2$ (*c*) $2^n$ (*d*) $2^n - 1$

where $n$ = Number of pulleys.

**16.** The velocity ratio of a Weston's differential pulley block is

(*a*) $\frac{D}{D + d}$ (*b*) $\frac{D}{D - d}$ (*c*) $\frac{2D}{D - d}$ (*d*) $\frac{2D}{D + d}$

where $D$ and $d$ = Diameters of larger and smaller pulley, respectively.

**17.** In a double threaded worm and worm wheel, the number of teeth on the worm is $T$. The length of the handle is $l$ and radius of the load drum is $r$. The velocity ratio is

(*a*) $\frac{l.T}{r}$ (*b*) $\frac{l.T}{2r}$ (*c*) $\frac{l.T}{4r}$ (*d*) $\frac{l.T}{16r}$

**18.** The velocity ratio of a simple screw jack is

(*a*) $\frac{2\pi l}{p}$ (*b*) $\frac{\pi l}{p}$ (*c*) $\frac{2\pi p}{l}$ (*d*) $\frac{\pi l}{2p}$

**19.** The velocity ratio of a single purchase winch crab can be increased by ............. length of the handle.

(*a*) decreasing (*b*) increasing

**20.** In a double purchase winch crab, the number of teeth on pinions is $T_1$ and $T_3$, while $T_2$ and $T_4$ are the number of teeth on the spur wheels. The length of the handle is $l$ and diameter of the load drum is $D$. The velocity ratio is

(*a*) $\frac{l}{D} \times \frac{T_3 \times T_1}{T_4 \times T_2}$ (*b*) $\frac{l}{D} \times \frac{T_4 \times T_2}{T_3 \times T_1}$ (*c*) $\frac{2l}{D} \times \frac{T_3 \times T_1}{T_4 \times T_2}$ (*d*) $\frac{2l}{D} \times \frac{T_4 \times T_2}{T_3 \times T_1}$

***Answers to Objective Type Questions***

| | | | | |
|---|---|---|---|---|
| **1.** (*a*) | **2.** (*c*) | **3.** (*d*) | **4.** (*b*) | **5.** (*b*) |
| **6.** (*b*) | **7.** (*a*) | **8.** (*c*) | **9.** (*a*) | **10.** (*d*) |
| **11.** (*a*) | **12.** (*c*) | **13.** (*b*) | **14.** (*a*) | **15.** (*d*) |
| **16.** (*c*) | **17.** (*b*) | **18.** (*a*) | **19.** (*b*) | **20.** (*d*) |

# 9

# ANALYSIS OF PERFECT FRAMES

1. *Introduction* 2. *Types of Frames* 3. *Tensile and Compressive Forces* 4. *Members with Zero Force* 5. *Assumptions Made in Analysis of Perfect Frames* 6. *Determination of Force in Frames* 7. *Method of Joints* 8. *Method of Sections* (*or Method of Moments*) 9. *Graphical Method*

## 9.1 INTRODUCTION

A structure made of several members, riveted or welded together is known as a *framed structure* or simply a *frame* or *truss.* If the frame is composed of such members which keep the frame in equilibrium after subjecting to the external load, then the frame is called a *perfect frame*. Though the members are riveted or welded together at their joints, yet for calculation purposes, the joints are assumed to be hinged or pin joints.

Actually there is no difference between a truss and a beam. The trusses are used as frames for the roof of buildings while beams are frames used in bridges, building structures, cranes, etc. The principle of their design is same.

## 9.2 TYPES OF FRAMES

The following two types of the frame are important from the subject point of view:

1. ***Perfect frame*** The frame which is made up of such members which keep the frame in equilibrium for all position of the loads, is known as a perfect frame. For example, consider a triangle *ABC* consisting of three members *AB*, *BC* and *AC*, pin jointed to each other at *A*, *B* and *C*, as shown in Fig. 9.1 (*a*).

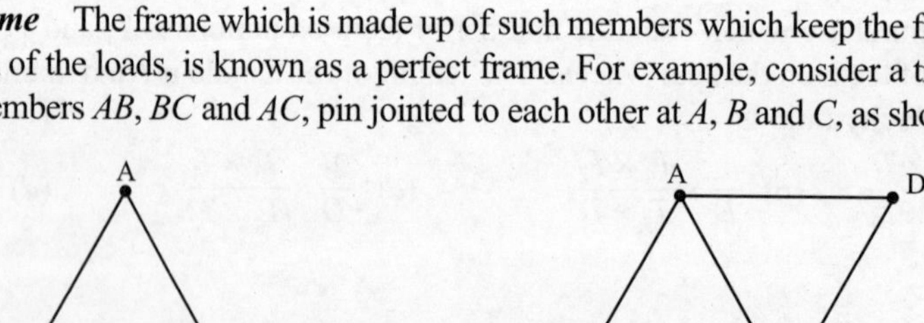


**Fig. 9.1** *Perfect frame*

This frame is called a *basic perfect frame.* If we add to this basic perfect frame, two members *AD* and *CD* and joint *D*, then we get a frame as shown in Fig. 9.1 (*b*). This frame is also a perfect frame.

Thus, we see that for every additional joint, two members are required. The relation between the number of members ($n$) and the numbers of joints ($j$) for a perfect frame is given by

$$n = 2j - 3 \quad \text{...}(i)$$

In Fig. 9.1 ($a$), $n = 3$, and $j = 3$

$$\therefore \quad n = 2j - 3 = 2 \times 3 - 3 = 3$$

Similarly, in Fig. 9.1 ($b$), $n = 5$, $j = 4$

$$\therefore \quad n = 2j - 3 = 2 \times 4 - 3 = 5$$

Thus both the frames, as shown in Fig. 9.1, are perfect frames.

2. ***Imperfect frame*** A frame in which the number of members and number of joints does not satisfy the equation ($i$), is called an imperfect frame. Consider a frame $ABCD$, as shown in Fig. 9.2 ($a$). It has 4 members (*i.e.* $n = 4$) and 4 joints (*i.e.* $j = 4$). For a perfect frame,

$$n = 2j - 3 = 2 \times 4 - 3 = 5$$

Thus, it is not a perfect frame.

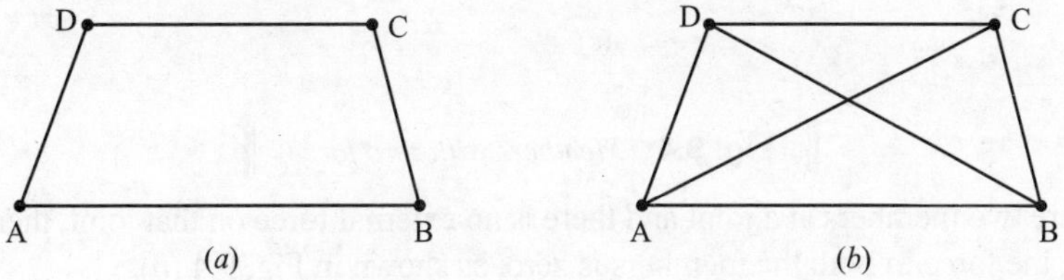


**Fig. 9.2** *Imperfect frame*

Now consider a frame as shown in Fig. 9.2 ($b$). It has 6 members (*i.e.* $n = 6$) and 4 joints (*i.e.* $j = 4$). For a perfect frame,

$$n = 2j - 3 = 2 \times 4 - 3 = 5$$

Thus it is also not a perfect frame.

The imperfect frame may be further classified as *deficient frame* and *redundant frame.*

The frame which has number of members less than $(2j - 3)$, is known as deficient frame. The frame shown in Fig. 9.2 ($a$) has one member less (*i.e.* $n = 4$) than required for the perfect frame. Thus it is a *deficient frame.*

The frame which has number of members more than $(2j - 3)$, is known as redundant frame. The frame as shown in Fig. 9.2 ($b$) has one member more (*i.e.* $n = 6$) than required for the perfect frame. Thus, it is a *redundant frame.*

## 9.3 TENSILE AND COMPRESSIVE FORCES

The force in the members of the frame may be tensile or compressive. If the force in the member pulls the joint to which it is connected, then it is said to be a *tensile force.* On the other hand, if the member pushes the joint to which it is connected, then it is said to be a *compressive force.* The tensile and compressive forces are marked by arrows on the member of a frame, as shown in Fig. 9.3 ($a$) and ($b$), respectively.



($a$) Tensile force ($b$) Compressive force

**Fig. 9.3**

**Note:** In solving problems, the tensile force in the member is shown by the arrow head away from the joint. On the other hand, the compressive force in the member is shown by the arrowhead towards the joint.

## 9.4 MEMBERS WITH ZERO FORCE

Sometimes in a truss, there are certain members in which the force is zero. Such members are identified by using the following principles:

1. If there are two members at a joint and the external force is acting only at one of the members, as shown in Fig. 9.4 (*a*), then for equilibrium of the joint, the force in the other member is zero.

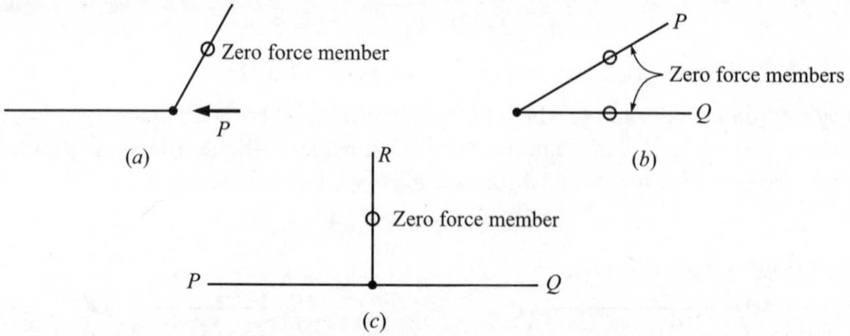


**Fig. 9.4** *Members with zero force*

2. If there are two members at a joint and there is no external force on that joint, then for equilibrium of the joint, the force in both the members is zero, as shown in Fig. 9.4 (*b*).
3. If there are three members at a joint and two of them are collinear (*i.e.* along the same straight line), then for equilibrium of the joint, the force in the third number is zero, as shown in Fig. 9.4 (*c*).

## 9.5 ASSUMPTIONS MADE IN ANALYSIS OF PERFECT FRAMES

The following assumptions are made in the analysis of perfect frames:

1. All the members are straight and have uniform cross-section.
2. All the members have negligible weight.
3. The framed structure is a perfect one.
4. All the members of the frame are pin jointed.
5. The external forces must be applied at the pin joints.

## 9.6 DETERMINATION OF FORCE IN FRAMES

The force in the members of a frame may be determined by the following three methods:

**1.** Method of joints; **2.** Method of sections; and **3.** Graphical method.

These methods are discussed in detail, one by one, in the following pages.

## 9.7 METHOD OF JOINTS

This method of analysing of force is generally known as *method of resolution.* In this method, the following procedure is adopted:

1. First of all, the reactions at the supports are determined.
2. Select a joint in such a way that at any time, there are only two members in which the forces are unknown, as shown in Fig. 9.5. These unknown forces are now determined by applying the conditions of equilibrium, *i.e.* by resolving the forces in the vertical and horizontal directions and equating it to zero, *i.e.* $\Sigma V = 0$ and $\Sigma H = 0$, where $\Sigma V$ is the algebraic sum of all the vertical forces and $\Sigma H$ is the algebraic sum of all the horizontal forces.

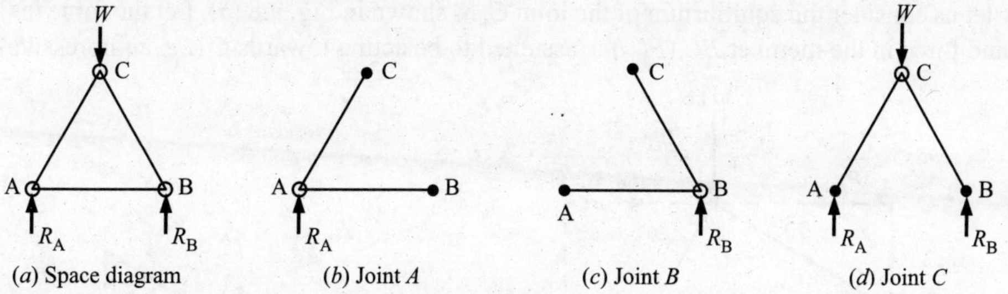


Fig. 9.5

3. Every joint is treated separately and the force in the members connected to the joints are determined along with its nature, *i.e.* tensile or compressive.
4. Mark arrows on both ends of every member at that joint to show the nature of the force as discussed in Sec. 9.3.

**Example 9.1** *A triangular shaped truss ABC, as shown in Fig. 9.6, is simply supported at A and B. Find the forces in the members AB, AC and BC by using the method of joints.*

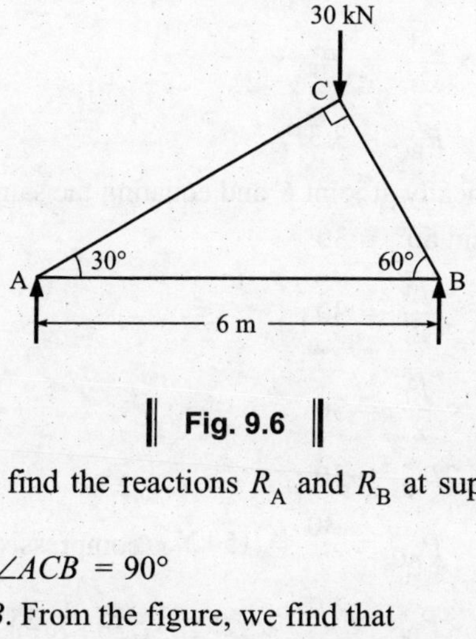


Fig. 9.6

***Solution*** First of all, let us find the reactions $R_A$ and $R_B$ at supports $A$ and $B$, respectively. From Fig. 9.7, we find that

$$\angle ACB = 90°$$

Draw $CD$ perpendicular to $AB$. From the figure, we find that

$$AC = AB \cos 30° = 6 \times \frac{\sqrt{3}}{2} = 3\sqrt{3} \text{ m}$$

and $$AD = AC \cos 30° = 3\sqrt{3} \times \frac{\sqrt{3}}{2} = 4.5 \text{ m}$$

Now taking moments about $A$, we have

$$R_B \times 6 = 30 \times AD = 30 \times 4.5$$

$$\therefore \quad R_B = \frac{30 \times 4.5}{6} = 22.5 \text{ kN}$$

We know that $$R_A + R_B = \text{Total load at } C = 30 \text{ kN}$$

$$\therefore \quad R_A = 30 - R_B$$

$$= 30 - 22.5 = 7.5 \text{ kN}$$

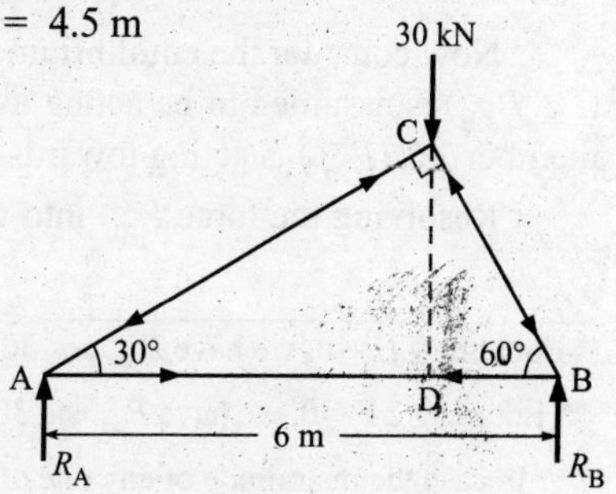


Fig. 9.7

Now let us consider the equilibrium of the joint $C$, as shown in Fig. 9.8 ($a$). Let the force in the member $AC$ ($P_{AC}$) and force in the member $BC$ ($P_{BC}$) is assumed to be acting towards $C$ (*i.e.* compressive).

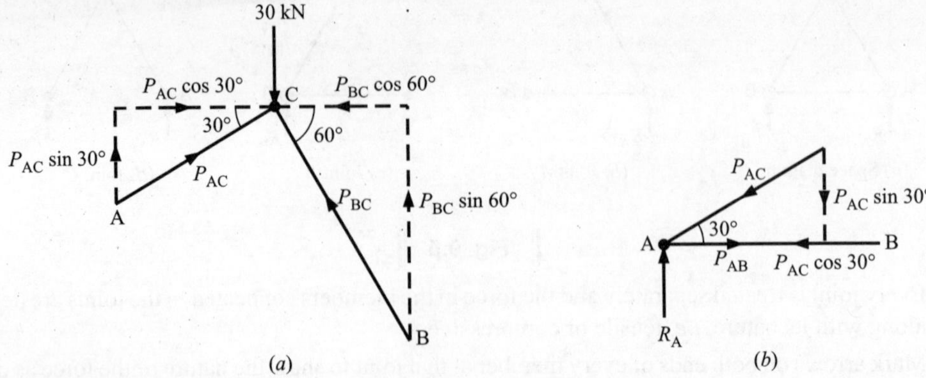


**Fig. 9.8**

Resolving the forces horizontally at joint $C$ and equating the same, we have

$$^{*}P_{AC}\cos 30° = P_{BC}\cos 60°$$

$$P_{AC}\times\frac{\sqrt{3}}{2} = P_{BC}\times\frac{1}{2}$$

$$\therefore \quad P_{BC} = \sqrt{3}P_{AC} \qquad ...(i)$$

Now resolving the forces vertically at joint $C$ and equating the same, we have

$$^{**}P_{AC}\sin 30° + P_{BC}\sin 60° = 30$$

$$P_{AC}\times\frac{1}{2} + P_{BC}\times\frac{\sqrt{3}}{2} = 30$$

$$P_{AC}\times\frac{1}{2} + \sqrt{3}P_{AC}\times\frac{\sqrt{3}}{2} = 30 \qquad ...[\text{From equation } (i)]$$

or $$2P_{AC} = 30$$

$$\therefore \quad P_{AC} = \frac{30}{2} = 15 \text{ kN (compressive) } \textbf{Ans.}$$

and $$P_{BC} = \sqrt{3}\,P_{AC} = \sqrt{3}\times 15 = 25.98 \text{ kN (compressive) } \textbf{Ans.}$$

Since the magnitude of $P_{AC}$ and $P_{BC}$ is ***positive, therefore the assumed directions of $P_{AC}$ and $P_{BC}$ is correct. Mark these directions near the joint $C$ considered and reverse direction at the other end, as shown in Fig. 9.7. Thus we see that the force in the members $AC$ and $BC$ is compressive.

Now consider the equilibrium of the joint $A$, as shown in Fig. 9.8 ($b$). Let the force in the member $AB$ (*i.e.* $P_{AB}$) is assumed to be acting away from the joint $A$ (*i.e.* tensile). We have seen above that force in the member $AC$ ($P_{AC}$) is acting towards $A$.

Resolving the force $P_{AC}$ into vertical component and equating the same, we have

$$R_A = P_{AC}\sin 30°$$

---

* Equating $\Sigma H = 0$, we have $P_{AC}\cos 30° - P_{BC}\cos 60° = 0$.

** Equating $\Sigma V = 0$, we have $P_{AC}\sin 30° + P_{BC}\sin 60° - 30° = 0$.

*** In case the magnitude of any one of the force is negative, then the direction of force in that member will be opposite to the assumed direction.

or $$P_{AC} = \frac{R_A}{\sin 30°} = \frac{7.5}{0.5} = 15 \text{ kN}$$ ...(Same as before)

Resolving the force $P_{AC}$ into horizontal component and equating the same, we have

$$P_{AB} = P_{AC} \cos 30° = 15 \times \frac{\sqrt{3}}{2} = 12.99 \text{ kN (tensile)} \textbf{ Ans.}$$

**Example 9.2** *Determine the forces and their nature in each member of the truss, as shown in Fig. 9.9.*

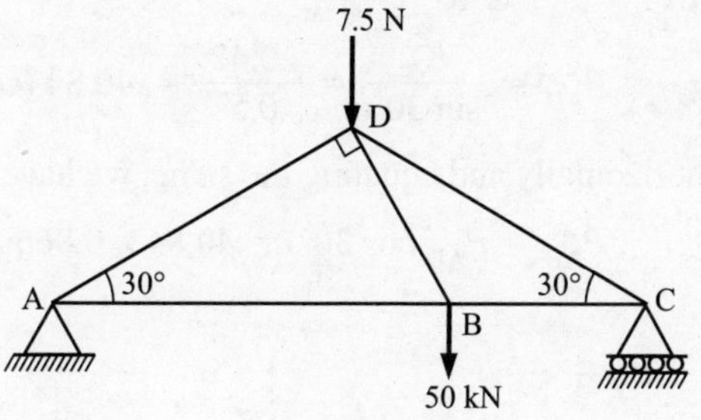


**Fig. 9.9**

***Solution*** First of all, let us find the reactions $R_A$ and $R_C$ at the supports $A$ and $C$, respectively. From the right angled triangle $ABD$, as shown in Fig. 9.10,

$$BD = AB \sin 30° = 0.5AB$$

or $$AB = 2BD = 2x$$ ...(Taking $BD = x$)

$$AD = AB \cos 30° = 0.866AB$$

and $${}^*BD = BC = x$$

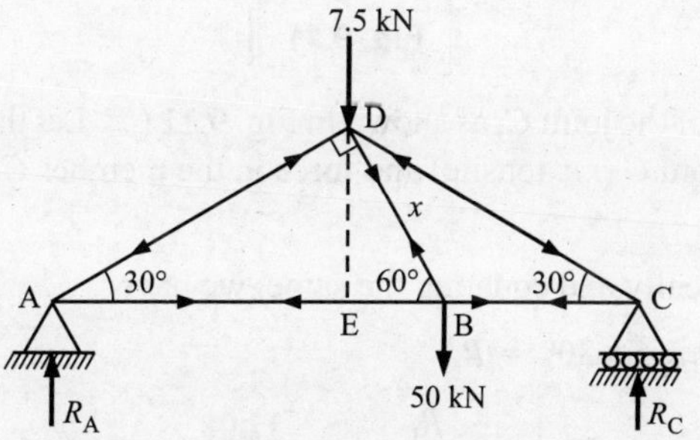


**Fig. 9.10**

Now taking moments about $A$, we have

$$R_C \times AC = 7.5 \times AE + 50 \times AB$$

$$R_C (AB + BC) = 7.5 \times AD \cos 30° + 50 \times AB$$

$$R_C (2x + x) = 7.5 \times 0.866 \times 2x \cos 30° + 50 \times 2x$$

$$= 7.5 \times 1.5x + 100x = 111.25x$$

$\therefore$ $$R_C = \frac{111.25x}{3x} = 37.08 \text{ kN}$$

* Since $\angle BCD = \angle BDC = 30°$, therefore triangle $BCD$ is an isosceles triangle.

We know that $R_A + R_C = 7.5 + 50 = 57.5$ kN

$\therefore \quad R_A = 57.5 - R_C = 57.5 - 37.08 = 20.42$ kN

Let us now consider the equilibrium of the joint *A*, as shown in Fig. 9.11 (*a*). Let the force in the member *AD* ($P_{AD}$) is assumed to be acting towards *A* (*i.e.* compressive) and force in the member *AB* ($P_{AB}$) is assumed to be acting away from *A* (*i.e.* tensile).

Resolving the forces vertically and equating the same, we have

$$P_{AD} \sin 30° = R_A$$

$$\therefore \quad P_{AD} = \frac{R_A}{\sin 30°} = \frac{20.42}{0.5} = 40.84 \text{ (compressive)} \quad \textbf{Ans.}$$

Now resolving the forces horizontally and equating the same, we have

$$P_{AB} = P_{AD} \cos 30° = 40.84 \times 0.866 = 35.36 \text{ kN (tensile)} \quad \textbf{Ans.}$$

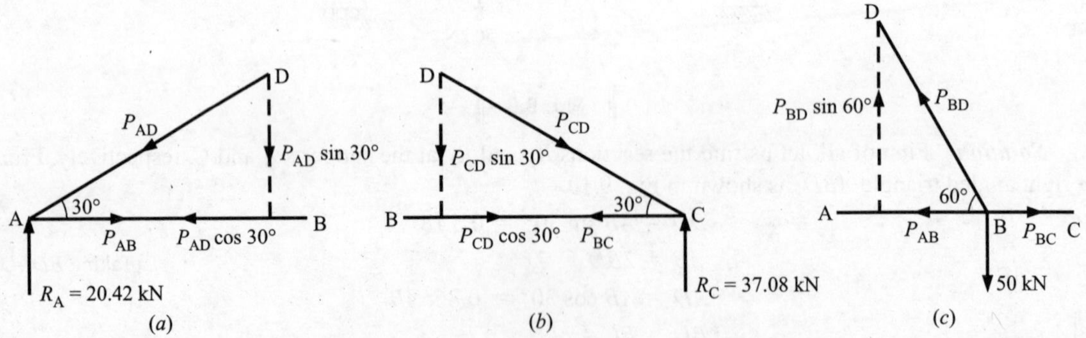


**Fig. 9.11**

Consider the equilibrium of the joint *C*, as shown in Fig. 9.11 (*b*). Let the force in the member *BC* ($P_{BC}$) is assumed to be acting away from *C* (*i.e.* tensile) and force in the member *CD* ($P_{CD}$) is assumed to be acting towards *C* (*i.e.* compressive).

Resolving the forces vertically and equating the same, we have

$$P_{CD} \sin 30° = R_C$$

$$\therefore \quad P_{CD} = \frac{R_C}{\sin 30°} = \frac{37.08}{0.5} = 74.16 \text{ kN (compressive)} \quad \textbf{Ans.}$$

Resolving the forces horizontally and equating the same, we have

$$P_{BC} = P_{CD} \cos 30° = 74.16 \times 0.866 = 64.22 \text{ kN (tensile)} \quad \textbf{Ans.}$$

Now consider the equilibrium of the joint *B*, as shown in Fig. 9.11 (*c*). Let the force in the member *BD* ($P_{BD}$) is assumed to be acting away from *B* (*i.e.* tensile).

Resolving the forces vertically and equating the same, we have

$$P_{BD} \sin 60° = 50$$

$$\therefore \quad P_{BD} = \frac{50}{\sin 60°} = \frac{50}{0.866} = 57.7 \text{ kN (tensile)}$$

Now the magnitude and nature of forces in the various members of the truss (Refer Fig. 9.10) are tabulated as below:

| *S. No.* | *Member of the truss* | *Magnitude of force in kN* | *Nature of force* |
|---|---|---|---|
| 1. | *AB* | 35.36 | Tensile |
| 2. | *AD* | 40.84 | Compressive |
| 3. | *BC* | 64.22 | Tensile |
| 4. | *CD* | 74.16 | Compressive |
| 5. | *BD* | 57.7 | Tensile |

**Example 9.3** *A truss is loaded as shown in Fig. 9.12. Find the reactions and forces in the members of the truss.*

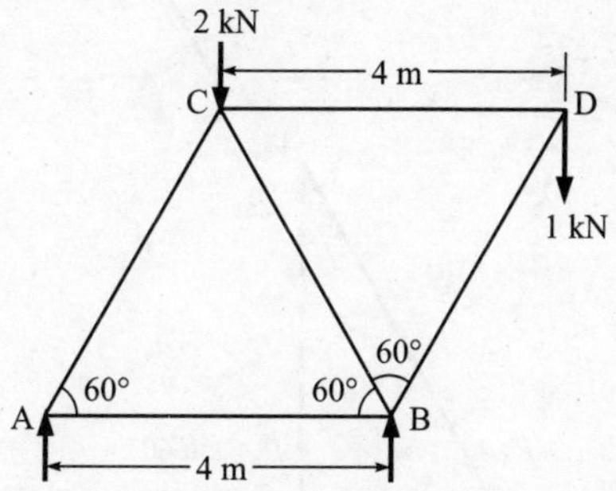


**Fig. 9.12**

***Solution*** First of all, let us find the reactions $R_A$ and $R_B$ at supports *A* and *B*, respectively (Refer Fig. 9.13).

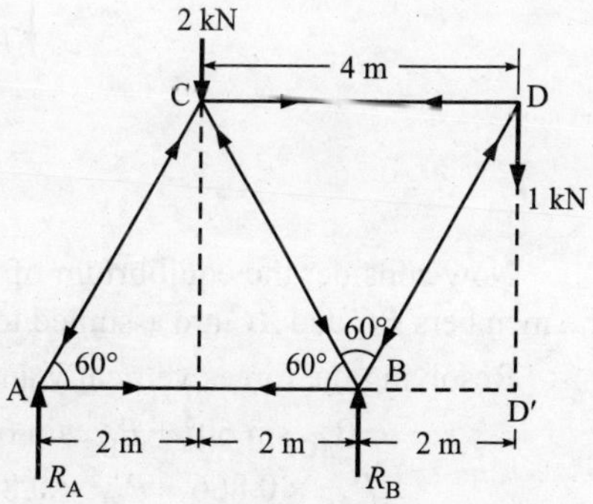


**Fig. 9.13**

Taking moments about *A*, we have

$$R_B \times 4 = 2 \times 2 + 1 \times 6 = 10$$

$$\therefore \quad R_B = \frac{10}{4} = 2.5 \text{ kN}$$

We know that $R_A + R_B = 2 + 1 = 3$ kN

$$\therefore \quad R_A = 3 - R_B$$

$$= 3 - 2.5 = 0.5 \text{ kN}$$

Now consider the equilibrium of the point *D*, as shown in Fig. 9.14 (*a*). Let the forces $P_{BD}$ and $P_{CD}$ in the members *BD* and *CD* are assumed to be acting away from *D* (*i.e.* tensile).

Resolving the forces vertically and equating the same, we have

$$P_{BD} \sin 60° = -1$$

$$\therefore \quad P_{BD} = -\frac{1}{\sin 60°} = -\frac{1}{0.866} = -1.1547 \text{ kN}$$

$$= 1.1547 \text{ kN (compressive)}$$ **Ans.**

...(–ve sign shows that the direction of $P_{BD}$ is opposite to that assumed)

Now resolving the forces horizontally and equating the same, we have

$$P_{CD} = -P_{BD}\cos 60° = -(-1.1547\cos 60°)$$
$$= 1.1547 \times 0.5 = 0.57735 \text{ say } 0.577 \text{ kN (tensile)} \textbf{ Ans.}$$

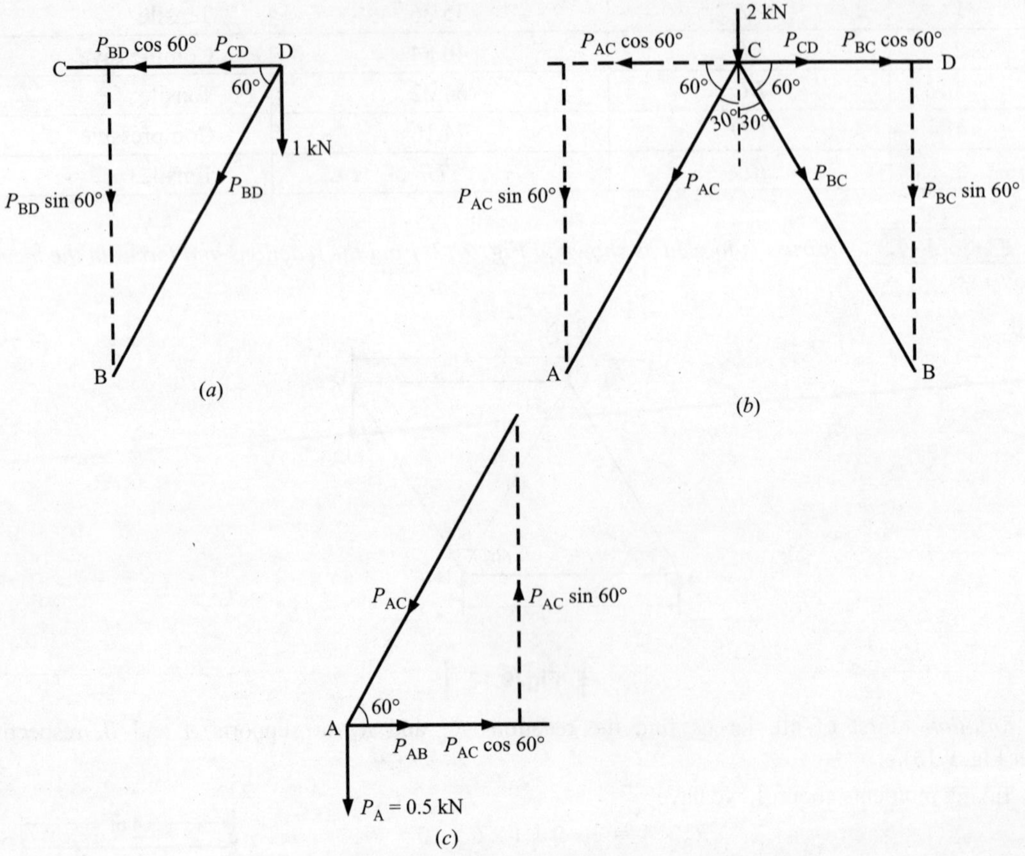


**Fig. 9.14**

Now consider the equilibrium of the point *C*, as shown in Fig. 9.14 (*b*). Let the force $P_{BC}$ and $P_{AC}$ in the members *BC* and *AC* are assumed to be acting away from *C* (*i.e.* tensile).

Resolving the forces vertically and equating the same, we have

$$P_{AC}\sin 60° + P_{BC}\sin 60° = -2$$
$$P_{AC} \times 0.866 + P_{BC} \times 0.866 = -2$$
$$P_{AC} + P_{BC} = \frac{-2}{0.866} = -2.3094 \qquad ...(i)$$

Now resolving the forces horizontally and equating the same, we have

$$P_{AC}\cos 60° - P_{BC}\cos 60° = P_{CD}$$
$$P_{AC} \times 0.5 - P_{BC} \times 0.5 = 0.577$$
$$\therefore \qquad P_{AC} - P_{BC} = \frac{0.577}{0.5} = 1.1547 \qquad ...(ii)$$

Adding equations (*i*) and (*ii*), we have

$$2P_{AC} = -2.3094 + 1.1547 = -1.1547$$

$\therefore \qquad P_{AC} = \dfrac{-1.1547}{2} = -0.57735 \text{ say } -0.577 \text{ kN}$

$= 0.577$ kN (compressive) **Ans.**

...(–ve sign shows that the direction of $P_{AC}$ is opposite to that assumed)

From equation (*i*), $\qquad P_{AC} + P_{BC} = -2.3094$

$\therefore \qquad P_{BC} = -2.3094 - P_{AC} = -2.3094 - (-0.577) = -1.7324$ kN

$= 1.7324$ kN (compressive) **Ans.**

...(–ve sign shows that direction of $P_{BC}$ is opposite to that assumed)

Now consider the equilibrium of the point *A*, as shown in Fig. 9.14 (*c*). Let the force in the member *AB* ($P_{AB}$) is assumed to be acting away from *A* (*i.e.* tensile).

Resolving the forces horizontally and equating the same, we have

$P_{AB} = -P_{AC} \cos 60°$

$= -(-0.577 \times 0.5) = 0.2885$ kN (tensile) **Ans.**

The magnitude and nature of forces in the various members of the truss (Refer Fig. 9.13) are now tabulated as below:

| *S. No.* | *Member of the truss* | *Magnitude of force in kN* | *Nature of force* |
|---|---|---|---|
| 1. | *AB* | 0.2885 | Tensile |
| 2. | *AC* | 0.577 | Compressive |
| 3. | *BC* | 1.7324 | Compressive |
| 4. | *BD* | 1.1547 | Compressive |
| 5. | *CD* | 0.577 | Tensile |

**Example 9.4** *A truss is loaded as shown in Fig. 9.15. Determine the magnitude and nature of forces in all the members of the truss.*

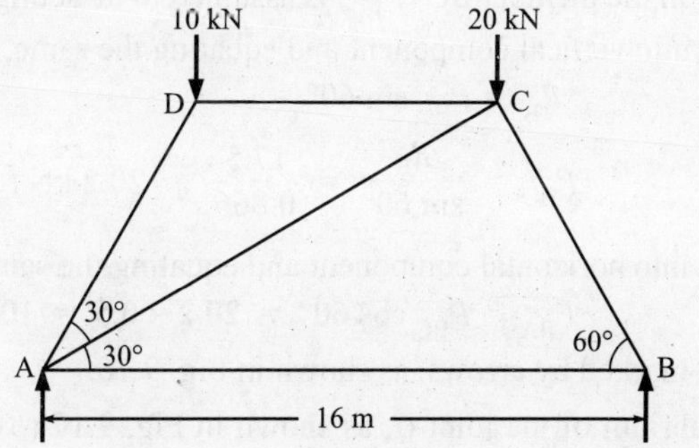


**Fig. 9.15**

***Solution*** First of all, let us find the reactions $R_A$ and $R_B$ at supports *A* and *B*, respectively (Refer Fig. 9.16).

From triangle *BCE*, $\qquad \tan 60° = \dfrac{CE}{EB}$ or $\sqrt{3} = \dfrac{CE}{EB}$

$\therefore \qquad CE = \sqrt{3}\, EB$

Now from triangle $ACE$, $\tan 30° = \frac{CE}{AE}$ or $\frac{1}{\sqrt{3}} = \frac{CE}{AE}$

$$\therefore \quad CE = \frac{AE}{\sqrt{3}} = \frac{AB - EB}{\sqrt{3}} = \frac{16 - EB}{\sqrt{3}} \quad ...(ii)$$

From equations ($i$) and ($ii$), we have

$$\sqrt{3}\, EB = \frac{16 - EB}{\sqrt{3}}$$

or $$3EB = 16 - EB$$

$$3EB + EB = 16$$

$$\therefore \quad EB = \frac{16}{4} = 4 \text{ m}$$

and $$AE = AB - EB$$
$$= 16 - 4$$
$$= 12 \text{ m}$$

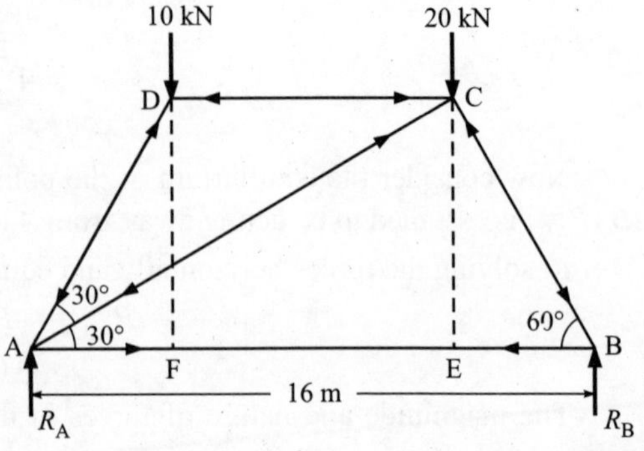


**Fig. 9.16**

Now taking moments about $A$,

$$R_B \times 16 = 20 \times AE + 10 \times AF$$
$$= 20 \times 12 + 10 \times 4 = 280 \quad ...(\because AF = EB = 4 \text{ m})$$

$$\therefore \quad R_B = \frac{280}{16} = 17.5 \text{ kN}$$

We know that $$R_A + R_B = 10 + 20 = 30 \text{ kN}$$

$$\therefore \quad R_A = 30 - R_B = 30 - 17.5 = 12.5 \text{ kN}$$

Since there are two unknown forces at joint $B$, therefore we shall first consider the equilibrium of the joint $B$ as shown in Fig. 9.17 ($a$). Let the force in the member $BA$ ($P_{BA}$) is assumed to be acting away from point $B$ ($i.e.$ tensile), and force in the member $BC$ ($P_{BC}$) is assumed to be acting towards $B$ ($i.e.$ compressive).

Resolving the force $P_{BC}$ into vertical component and equating the same, we have

$$\therefore \quad R_B = P_{BC} \sin 60°$$

or $$P_{BC} = \frac{R_B}{\sin 60°} = \frac{17.5}{0.866} = 20.2 \text{ kN (compressive)} \quad \textbf{Ans.}$$

Resolving the force $P_{BC}$ into horizontal component and equating the same, we have

$$P_{BA} = P_{BC} \cos 60° = 20.2 \times 0.5 = 10.1 \text{ (tensile)} \quad \textbf{Ans.}$$

The nature of forces are marked by arrows, as shown in Fig. 9.16.

Now consider the equilibrium of the joint $C$, as shown in Fig. 9.17 ($b$). Let the force in the member $AC$ ($P_{AC}$) and force in the member $CD$ ($P_{CD}$) is assumed to be acting towards $C$ ($i.e.$ compressive).

The force in the member $BC$ ($P_{BC}$) has already been determined.

Resolving the forces vertically and equating the same, we have

$$P_{AC} \sin 30° + P_{BC} \sin 60° = 20$$
$$P_{AC} \times 0.5 + 20.2 \times 0.866 = 20$$
$$P_{AC} \times 0.5 = 20 - 20.2 \times 0.866 = 20 - 17.5 = 2.5$$

$$\therefore \quad P_{AC} = \frac{2.5}{0.5} = 5 \text{ kN (compressive)} \quad \textbf{Ans.}$$

Now resolving the forces horizontally and equating the same, we have

$$P_{AC}\cos 30° + P_{CD} = P_{BC}\cos 60°$$

$$5 \times 0.866 + P_{CD} = 20.2 \times 0.5 = 10.1$$

$$\therefore \quad P_{CD} = 10.1 - 5 \times 0.866 = 10.1 - 4.33 = 5.77 \text{ kN (compressive)}$$ **Ans.**

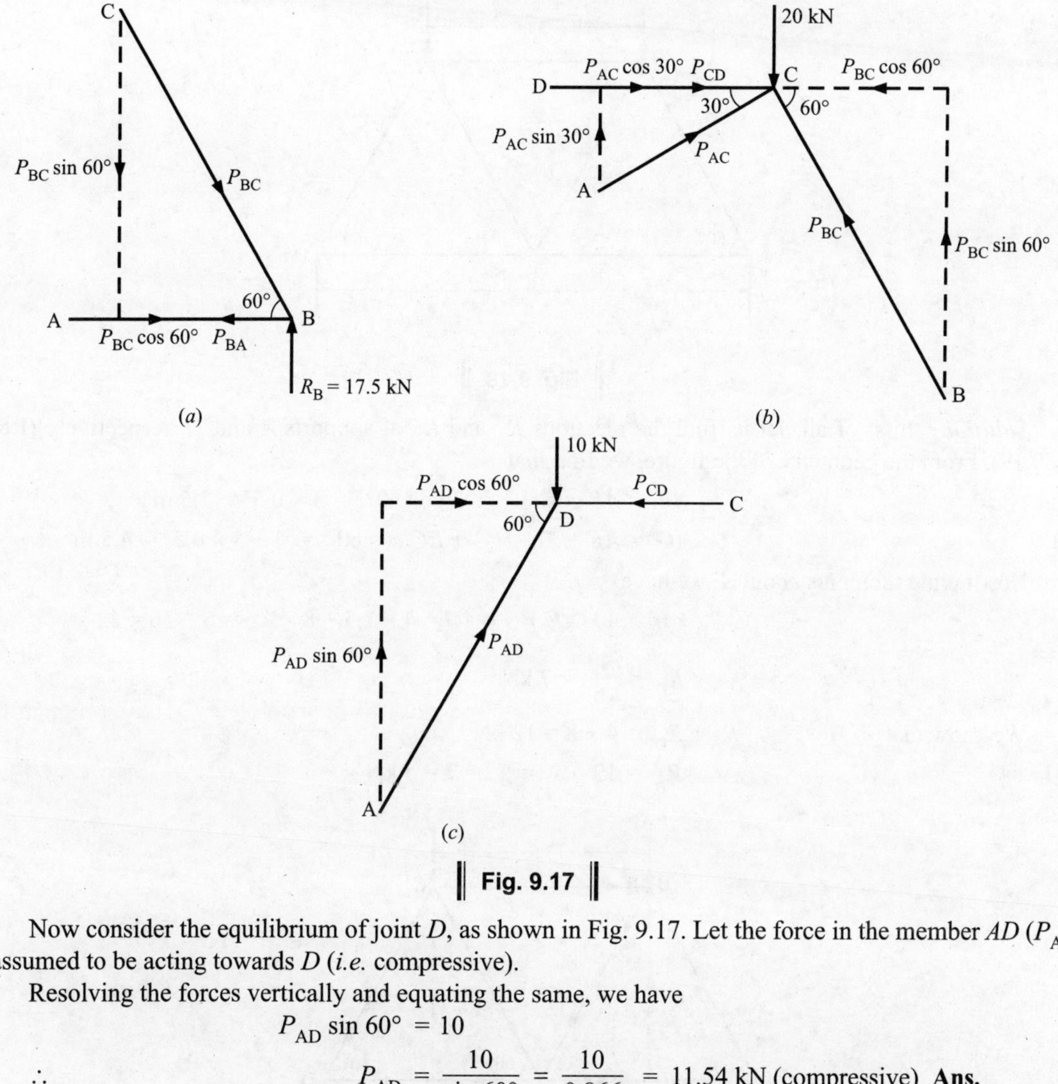


**Fig. 9.17**

Now consider the equilibrium of joint *D*, as shown in Fig. 9.17. Let the force in the member *AD* ($P_{AD}$) is assumed to be acting towards *D* (*i.e.* compressive).

Resolving the forces vertically and equating the same, we have

$$P_{AD}\sin 60° = 10$$

$$\therefore \quad P_{AD} = \frac{10}{\sin 60°} = \frac{10}{0.866} = 11.54 \text{ kN (compressive)}$$ **Ans.**

The magnitude and nature of forces in the various members of the truss (Refer Fig. 9.16) are now tabulated as below:

| *S. No.* | *Member of the truss* | *Magnitude of force in kN* | *Nature of force* |
|---|---|---|---|
| 1. | *BC* | 20.2 | Compressive |
| 2. | *BA* | 10.1 | Tensile |
| 3. | *AC* | 5 | Compressive |
| 4. | *CD* | 5.77 | Compressive |
| 5. | *AD* | 11.54 | Compressive |

**Example 9.5** *A truss of seven members each of 3 m length is freely supported at the ends, as shown in Fig. 9.18. The points B and C are loaded with 4 kN and 8 kN, respectively. Determine the magnitude of forces and nature in all the members of the truss. Use method of joints.*

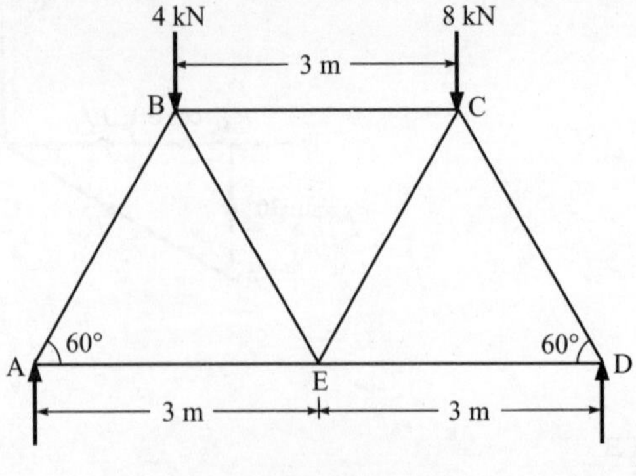


Fig. 9.18

***Solution*** First of all, let us find the reactions $R_A$ and $R_D$ at supports $A$ and $D$, respectively (Refer Fig. 9.19). From the geometry of the figure, we find that

$$AF = AB \cos 60° = 3 \cos 60° = 3 \times 0.5 = 1.5 \text{ m}$$

and

$$AG = AE + EG = 3 + EC \cos 60° = 3 + 3 \times 0.5 = 4.5 \text{ m}$$

Now taking moments about $A$, we have

$$R_D \times 6 = 4 \times AF + 8 \times AG = 4 \times 1.5 + 8 \times 4.5 = 6 + 36 = 42$$

$$\therefore \quad R_D = \frac{42}{6} = 7 \text{ kN}$$

We know that

$$R_A + R_D = 4 + 8 = 12 \text{ kN}$$

$$\therefore \quad R_A = 12 - R_D = 12 - 7 = 5 \text{ kN}$$

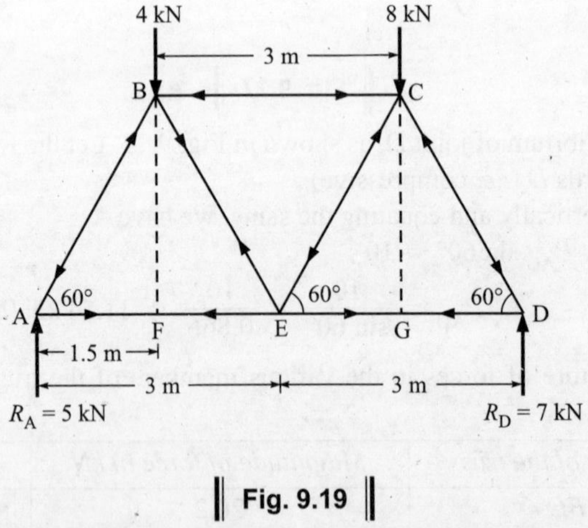


Fig. 9.19

Now let us consider the equilibrium of the joint $A$, as shown in Fig. 9.20 (*a*). Let the force in the member $AB$ ($P_{AB}$) is assumed to be acting towards point $A$ (*i.e.* compressive) and force in the member $AE$ ($P_{AE}$) to be acting away from point $A$ (*i.e.* tensile).

Resolving the force $P_{AB}$ into vertical component and equating the same, we have

$$\therefore \quad R_A = P_{AB} \sin 60°$$

or $$P_{AB} = \frac{R_A}{\sin 60°} = \frac{5}{0.866} = 5.774 \text{ kN (compressive)} \quad \textbf{Ans.}$$

Resolving the force $P_{AB}$ into horizontal component and equating the same, we have

$$P_{AE} = P_{AB} \cos 60° = 5.774 \times 0.5 = 2.887 \text{ kN (tensile)} \quad \textbf{Ans.}$$

Consider the equilibrium of the joint *B*, as shown in Fig. 9.20 (*b*). Let the force in the member *BC* ($P_{BC}$) is assumed to be acting towards *B* (*i.e.* compressive) and force in the member *BE* ($P_{BE}$) is assumed to be acting away from *B* (*i.e.* tensile).

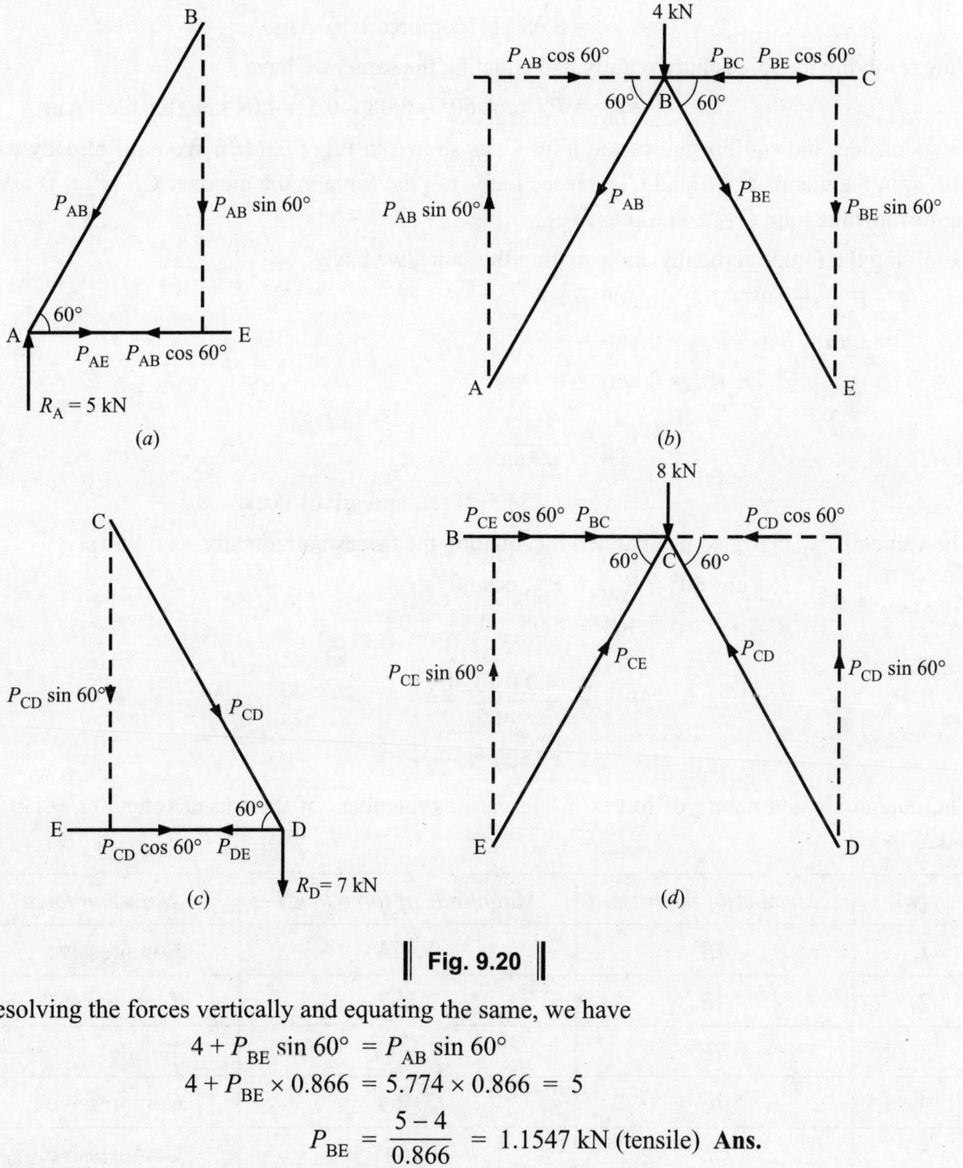


**Fig. 9.20**

Resolving the forces vertically and equating the same, we have

$$4 + P_{BE} \sin 60° = P_{AB} \sin 60°$$

$$4 + P_{BE} \times 0.866 = 5.774 \times 0.866 = 5$$

$$\therefore \quad P_{BE} = \frac{5-4}{0.866} = 1.1547 \text{ kN (tensile)} \quad \textbf{Ans.}$$

Now resolving the forces horizontally and equating the same, we have

$$P_{BC} = P_{BE} \cos 60° + P_{AB} \cos 60°$$

$$= 1.1547 \times 0.5 + 5.774 \times 0.5 = 3.464 \text{ kN (compressive)} \quad \textbf{Ans.}$$

Now consider the equilibrium of the joint *D*, as shown in Fig. 9.20 (*c*). Let the force in the member *CD* ($P_{CD}$) is assumed to be acting towards *D* (*i.e.* compressive) and force in the member *DE* ($P_{DE}$) is assumed to be acting away from *D* (*i.e.* tensile).

Resolving the forces vertically and equating the same, we have

$$P_{CD} \sin 60° = R_D = 7$$

$$\therefore \quad P_{CD} = \frac{7}{\sin 60°} = \frac{7}{0.866}$$

$$= 8.08 \text{ kN (compressive)} \quad \textbf{Ans.}$$

Now resolving the forces horizontally and equating the same, we have

$$P_{DE} = P_{CD} \cos 60° = 8.08 \times 0.5 = 4.04 \text{ kN (tensile)} \quad \textbf{Ans.}$$

Now consider the equilibrium of the joint *C*, as shown in Fig. 9.20 (*d*). We have already calculated $P_{BC}$ and $P_{CD}$ in the members *BC* and *CD*, respectively. Let the force in the member *CE* ($P_{CE}$) is assumed to be acting towards the joint *C* (*i.e.* compressive).

Resolving the forces vertically and equating the same, we have

$$P_{CD} \sin 60° + P_{CE} \sin 60° = 8$$

$$8.08 \times 0.866 + P_{CE} \times 0.866 = 8$$

$$7 + P_{CE} \times 0.866 = 8$$

$$\therefore \quad P_{CE} = \frac{8-7}{0.866}$$

$$= 1.1547 \text{ kN (compressive)} \quad \textbf{Ans.}$$

The value of $P_{CE}$ may also be obtained by equating the forces horizontally, as follows:

$$P_{CE} \cos 60° + P_{BC} = P_{CD} \cos 60°$$

$$P_{CE} \times 0.5 + 3.464 = 8.08 \times 0.5 = 4.04$$

$$\therefore \quad P_{CE} = \frac{4.04 - 3.464}{0.5}$$

$$= 1.152 \text{ kN}$$

The magnitude and nature of forces in the various members of the truss (Refer Fig. 9.19) are now tabulated as below:

| *S. No.* | *Member of the truss* | *Magnitude of force in kN* | *Nature of force* |
|---|---|---|---|
| 1. | *AB* | 5.774 | Compressive |
| 2. | *AE* | 2.887 | Tensile |
| 3. | *BE* | 1.1547 | Tensile |
| 4. | *BC* | 3.464 | Compressive |
| 5. | *CE* | 1.1547 | Compressive |
| 6. | *CD* | 8.08 | Compressive |
| 7. | *DE* | 4.04 | Tensile |

**Example 9.6** *Determine the forces in each member of the truss, as shown in Fig. 9.21, by using the method of joints.*

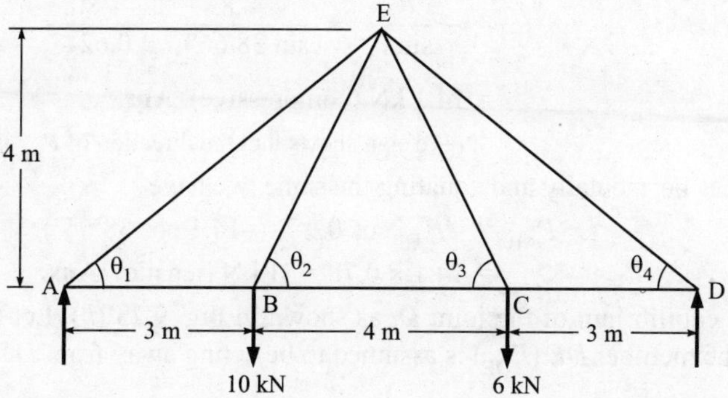


**Fig. 9.21**

***Solution*** Since the load at $B$ and $C$ is acting downward, therefore the reactions $R_A$ and $R_D$ at the supports $A$ and $D$, respectively, will be upward as shown in Fig. 9.22.

Taking moments about $A$, $R_D \times 10 = 6 \times 7 + 10 \times 3 = 42 + 30 = 72$ kN

$\therefore$ $R_D = \dfrac{72}{10} = 7.2$ kN

We know that $R_A + R_D = 10 + 6 = 16$ kN

$\therefore$ $R_A = 16 - R_D = 16 - 7.2 = 8.8$ kN

From triangle $AEF$,

$$\tan\theta_1 = \frac{EF}{AF} = \frac{EF}{AB + BF} = \frac{4}{3+2} = \frac{4}{5} = 0.8$$

$\therefore$ $\theta_1 = \tan^{-1}(0.8) = 38.66° = \theta_4$

Now from triangle $BEF$, $\tan\theta_2 = \dfrac{EF}{BF} = \dfrac{4}{2} = 2$

$\therefore$ $\theta_2 = \tan^{-1}(2) = 63.43° = \theta_3$

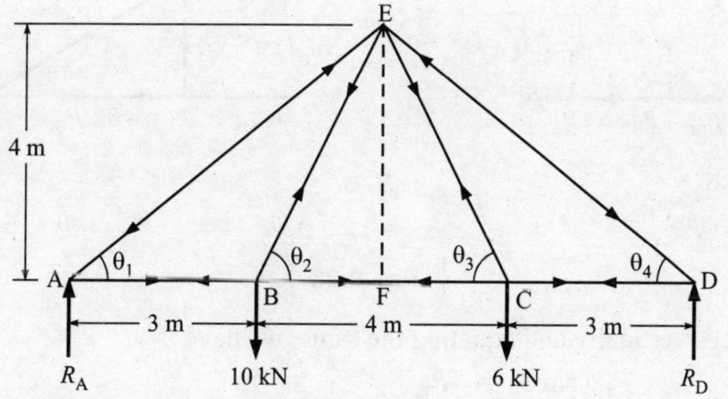


**Fig. 9.22**

Let us now consider the equilibrium of the joint $A$, as shown in Fig. 9.23 (*a*). Let the force in the member $AB$ ($P_{AB}$) and force in the member $AE$ ($P_{AE}$) is assumed to be acting away from $A$ (*i.e.* tensile).

Resolving the forces vertically and equating the same, we have

$$P_{AE} \sin \theta_1 = -R_A$$

$$\therefore \quad P_{AE} = -\frac{R_A}{\sin \theta_1} = \frac{-8.8}{\sin 38.66°} = \frac{-8.8}{0.6247} = -14.1 \text{ kN}$$

$$= 14.1 \text{ kN (compressive)} \quad \textbf{Ans.}$$

...(–ve sign shows that the direction of $P_{AE}$ is opposite to that assumed)

Resolving the forces horizontally and equating the same, we have

$$P_{AB} = -P_{AE} \cos \theta_1 = -(-14.1 \cos 38.66°)$$

$$= 14.1 \times 0.78 = 11 \text{ kN (tensile)} \quad \textbf{Ans.}$$

Now consider the equilibrium of the joint *D*, as shown in Fig. 9.23 (*b*). Let the force in the member *CD* ($P_{CD}$) and force in the member *DE* ($P_{DE}$) is assumed to be acting away from *D* (*i.e.* tensile).

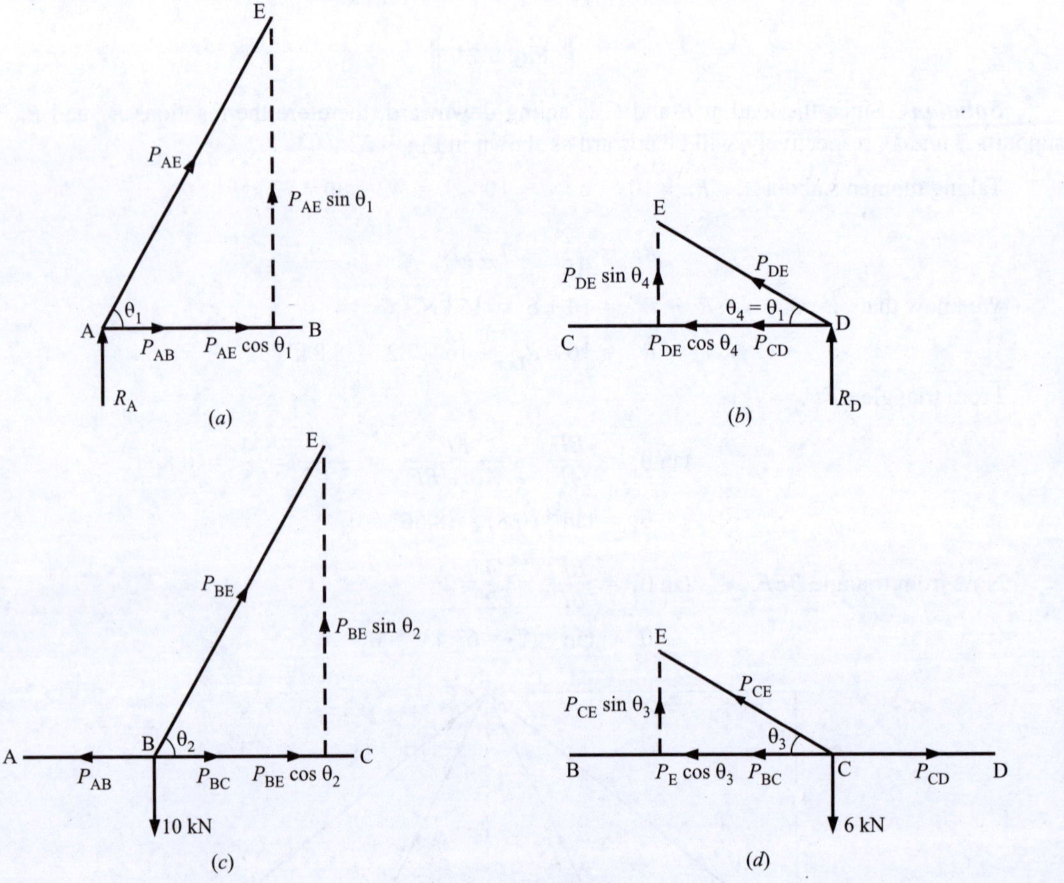


**Fig. 9.23**

Resolving the forces vertically and equating the same, we have

$$P_{DE} \sin \theta_4 = -R_D$$

$$\therefore \quad P_{DE} = \frac{-R_D}{\sin \theta_4} = \frac{-7.2}{\sin 38.66°} = \frac{-7.2}{0.6247} = -11.52 \text{ kN} \quad ...(\because \theta_4 = \theta_1)$$

$$= 11.52 \text{ kN (compressive)} \quad \textbf{Ans.}$$

...(–ve sign shows that the direction of $P_{DE}$ is opposite to that assumed)

Resolving the forces horizontally and equating the same, we have

$$P_{CD} = -P_{DE} \cos \theta_4 = -(-11.52 \cos 38.66°)$$

$$= 11.52 \times 0.78 = 9 \text{ kN (tensile)} \textbf{ Ans.}$$

Now consider the equilibrium of the joint *B*, as shown in Fig. 9.23 (*c*). Let the force in the member *BE* ($P_{BE}$) and force in the member *BC* ($P_{BC}$) are assumed to be acting away from *B* (*i.e.* tensile).

Resolving the forces vertically and equating the same, we have

$$P_{BE} \sin \theta_2 = 10 \text{ kN}$$

$$\therefore \quad P_{BE} = \frac{10}{\sin \theta_2} = \frac{10}{\sin 63.43°} = \frac{10}{0.894}$$

$$= 11.18 \text{ kN (tensile)} \textbf{ Ans.}$$

Resolving the forces horizontally and equating the same, we have

$$P_{BC} + P_{BE} \cos \theta_2 = P_{AB}$$

$$P_{BC} + 11.18 \cos 63.43° = 11$$

$$P_{BC} + 11.18 \times 0.4473 = 11$$

$$P_{BC} + 5 = 11$$

$$\therefore \quad P_{BC} = 11 - 5 = 6 \text{ kN (tensile)} \textbf{ Ans.}$$

Now consider the equilibrium of the joint *C*, as shown in Fig. 9.23 (*d*). Let the force in the member *CE* ($P_{CE}$) is assumed to be acting away from *C* (*i.e.* tensile).

Resolving the forces vertically and equating the same, we have

$$P_{CE} \sin \theta_3 = 6$$

$$\therefore \quad P_{CE} = \frac{6}{\sin \theta_3} = \frac{6}{\sin 63.43°} = \frac{6}{0.894} = 6.71 \text{ kN (tensile)} \textbf{ Ans.}$$

$$...(\because \theta_3 = \theta_2)$$

***Check*:** Resolving the forces horizontally and equating the same, we have

$$P_{CE} \cos \theta_3 + P_{BC} = P_{CD}$$

$$P_{CE} \cos \theta_3 = P_{CD} - P_{BC} = 9 - 6 = 3$$

$$\therefore \quad P_{CE} = \frac{3}{\cos \theta_3} = \frac{3}{\cos 63.43°} = \frac{3}{0.4473} = 6.7 \text{ (tensile)} \textbf{ Ans.}$$

The magnitude and nature of forces in the various members of the truss (Refer Fig. 9.22) are tabulated as below:

| *S. No.* | *Member of the truss* | *Magnitude of force in kN* | *Nature of force* |
|---|---|---|---|
| 1. | *AE* | 14.1 | Compressive |
| 2. | *AB* | 11 | Tensile |
| 3. | *DE* | 11.52 | Compressive |
| 4. | *CD* | 9 | Tensile |
| 5. | *BE* | 11.18 | Tensile |
| 6. | *BC* | 6 | Tensile |
| 7. | *CE* | 6.7 | Tensile |

**Example 9.7** *Find the forces in all the members of the pin-jointed truss supported and loaded, as shown in Fig. 9.24.*

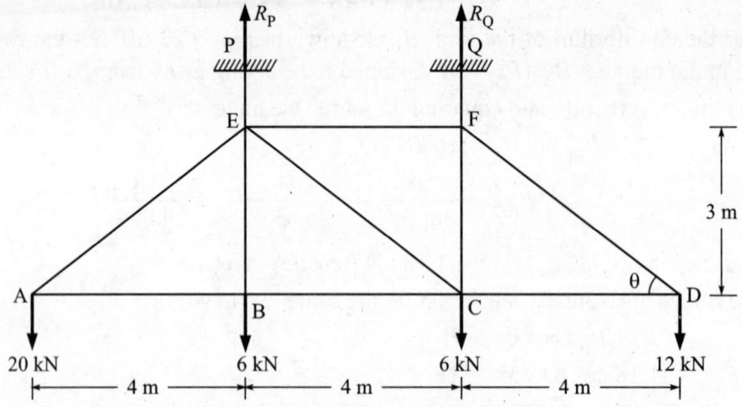


**Fig. 9.24**

***Solution*** Let θ be the inclination of the member $AE$ with the horizontal line $AB$, as shown in Fig. 9.25.

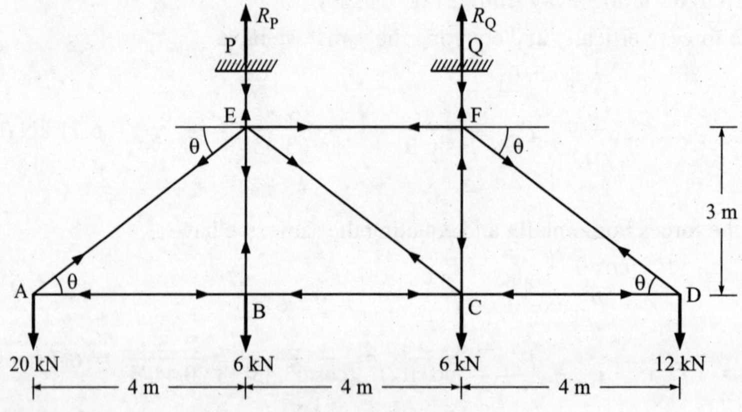


**Fig. 9.25**

From a right angled triangle $ABE$,

$$AE = \sqrt{(AB)^2 + (BE)^2} = \sqrt{4^2 + 3^2} = \sqrt{25} = 5 \text{ m}$$

$$\therefore \qquad \sin\theta = \frac{BE}{AE} = \frac{3}{5} = 0.6$$

and

$$\cos\theta = \frac{AB}{AE} = \frac{4}{5} = 0.8$$

Similarly, by symmetry, at the joint $D$, $\sin\theta = 0.6$ and $\cos\theta = 0.8$.

First of all, let us consider the equilibrium of the joint $A$, as shown in Fig. 9.26 (*a*). Let the forces $P_{AB}$ and $P_{AE}$ in the members $AB$ and $AE$ are assumed to be acting away from $A$ (*i.e.* tensile).

Resolving the force $P_{AE}$ into vertical component and equating the same, we have

$$P_{AE} \sin \theta = 20$$

$$\therefore \quad P_{AE} = \frac{20}{\sin \theta} = \frac{20}{0.6} = 33.33 \text{ kN (tensile) } \textbf{Ans.}$$

Now resolving the force $P_{AE}$ into horizontal component and equating the same, we have

$$P_{AB} = -P_{AE} \cos \theta$$

$$= -33.33 \times 0.8 = -26.67 \text{ kN}$$

$$= 26.67 \text{ kN (compressive) } \textbf{Ans.}$$

...(–ve sign shows that the direction of $P_{AB}$ is opposite to that assumed)

Consider the equilibrium of the joint *B*, as shown in Fig. 9.26 (*b*). Let the force in the member *BE* ($P_{BE}$) is assumed to be acting away from *B* (*i.e.* tensile) and force in the member *BC* ($P_{BC}$) is assumed to be acting towards *B* (*i.e.* compressive). Equating vertical components, we have

$$P_{BE} = 6 \text{ kN (tensile) } \textbf{Ans.}$$

Equating horizontal components, $P_{BC} = P_{AB} = 26.67$ kN (compressive) **Ans.**

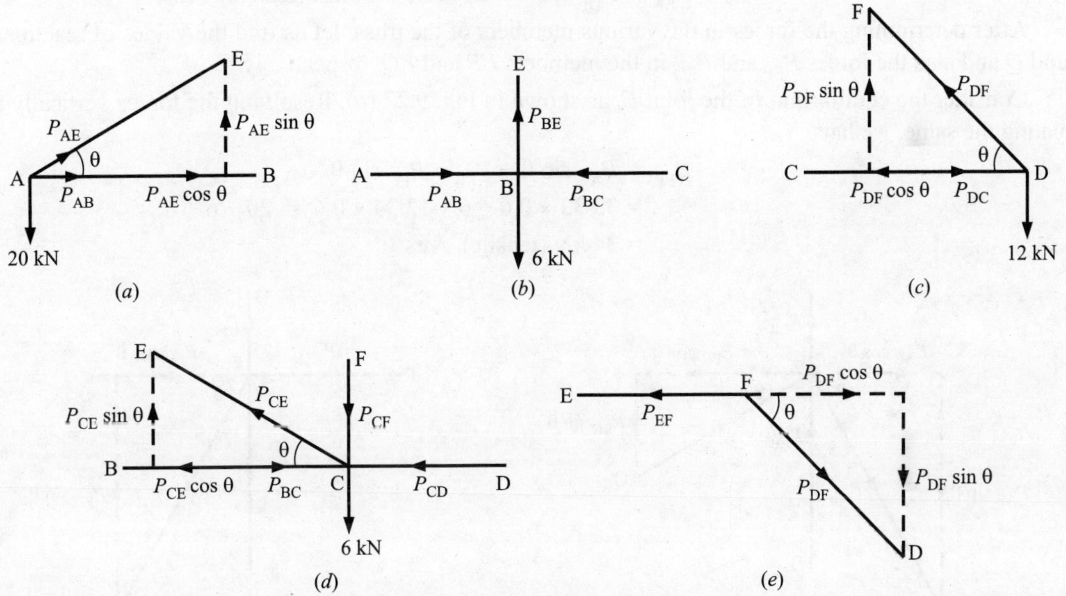


**Fig. 9.26**

Now consider the equilibrium of the joint *D*, as shown in Fig. 9.26 (*c*). Let the force in member *DC* ($P_{DC}$) is assumed to be acting towards *D* (*i.e.* compressive) and force in the member *DF* ($P_{DF}$) is assumed to be acting away from *D* (*i.e.* tensile).

Equating vertical components, we have

$$P_{DF} \sin \theta = 12$$

$$\therefore \quad P_{DF} = \frac{12}{\sin \theta} = \frac{12}{0.6} = 20 \text{ kN (tensile) } \textbf{Ans.}$$

Equating horizontal components, we have

$$P_{DC} = P_{DF} \cos \theta = 20 \times 0.8 = 16 \text{ kN (compressive) } \textbf{Ans.}$$

Consider the equilibrium of the joint *C*, as shown in Fig. 9.26 (*d*). Let the force in the member *CF* ($P_{CF}$) is assumed to be acting towards *C* (*i.e.* compressive) and force in the member *CE* ($P_{CE}$) is assumed to be acting away from *C* (*i.e.* tensile). Resolve $P_{CE}$ into vertical and horizontal component. Equating horizontal components, we have

$$P_{CD} + P_{CE} \cos\theta = P_{BC}$$

$$16 + P_{CE} \times 0.8 = 26.67$$

$$\therefore \quad P_{CE} = \frac{26.67 - 16}{0.8} = 13.34 \text{ (tensile)} \quad \textbf{Ans.}$$

Now equating vertical components, we have

$$\therefore \quad P_{CF} + 6 = P_{CE} \sin\theta = 13.34 \times 0.6 = 8$$

$$\therefore \quad P_{CF} = 8 - 6 = 2 \text{ kN (compressive)} \quad \textbf{Ans.}$$

Consider the equilibrium of the joint *F*, as shown in Fig. 9.26 (*e*). Let the force in the member *EF* ($P_{EF}$) is assumed to be acting away from *F* (*i.e.* tensile).

Equating horizontal components, we have

$$P_{EF} = P_{DF} \cos\theta = 20 \times 0.8 = 16 \text{ kN (tensile)} \quad \textbf{Ans.}$$

After determining the forces in the various members of the truss, let us find the values of reactions at *P* and *Q* and also the forces $P_{EP}$ and $P_{FQ}$ in the members *EP* and *FQ*, respectively.

Consider the equilibrium of the joint *E*, as shown in Fig. 9.27 (*a*). Resolving the forces vertically and equating the same, we have

$$P_{EP} = P_{AE} \sin\theta + P_{EB} + P_{EC} \sin\theta$$

$$= 33.33 \times 0.6 + 6 + 13.34 \times 0.6 = 20 + 6 + 8$$

$$= 34 \text{ kN (tensile)} \quad \textbf{Ans.}$$

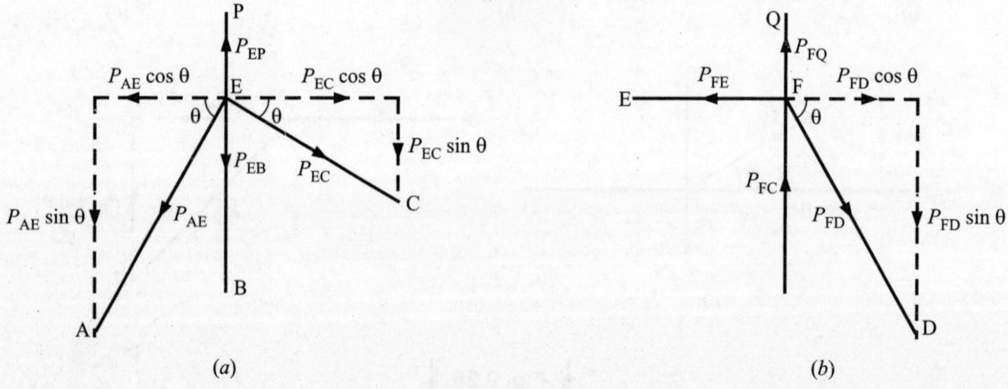


**Fig. 9.27**

The reaction at *P* ($R_P$) is equal to force in the member *EP*.

$$\therefore \quad R_P = 34 \text{ kN (upwards)} \quad \textbf{Ans.}$$

Now, consider the equilibrium of the joint *F*, as shown in Fig. 9.27 (*b*).

Resolving the forces vertically and equating the same, we have

$$P_{FQ} + P_{FC} = P_{FD} \sin\theta$$

$$P_{FQ} + 2 = 20 \times 0.6 = 12$$

$$\therefore \quad P_{FQ} = 12 - 2 = 10 \text{ kN (tensile)} \quad \textbf{Ans.}$$

The reaction at $Q$ ($R_Q$) is equal to force in member $FQ$.

$$\therefore \quad R_Q = 10 \text{ kN (upwards)} \quad \textbf{Ans.}$$

***Check:***

$$R_P + R_Q = 34 + 10 = 44 \text{ kN}$$

$$\text{Total external load} = \text{Sum of loads at } A, B, C \text{ and } D$$

$$= 20 + 6 + 6 + 12 = 44 \text{ kN}$$

The magnitude and nature of forces in the various members of the truss (Refer Fig. 9.25) are tabulated as below:

| *S. No.* | *Member of the truss* | *Magnitude of force in kN* | *Nature of force* |
|---|---|---|---|
| 1. | *AE* | 33.33 | Tensile |
| 2. | *AB* | 26.67 | Compressive |
| 3. | *BE* | 6 | Tensile |
| 4. | *BC* | 26.67 | Compressive |
| 5. | *DF* | 20 | Tensile |
| 6. | *DC* | 16 | Compressive |
| 7. | *CE* | 13.34 | Tensile |
| 8. | *CF* | 2 | Compressive |

**Example 9.8** *A cantilever truss of 4 m space is loaded, as shown in Fig. 9.28. Find the forces in the various members of the truss along with their nature.*

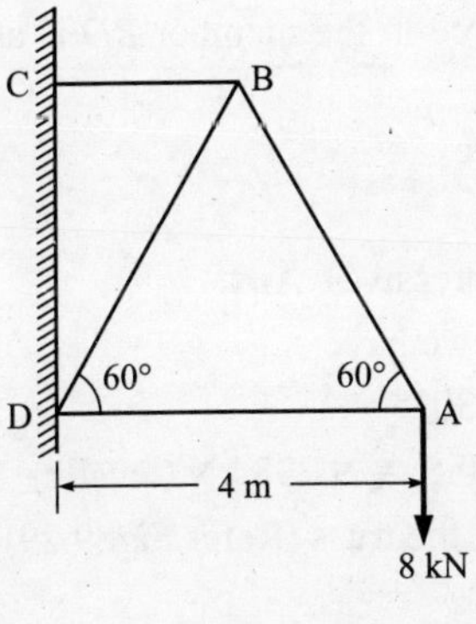


**Fig. 9.28**

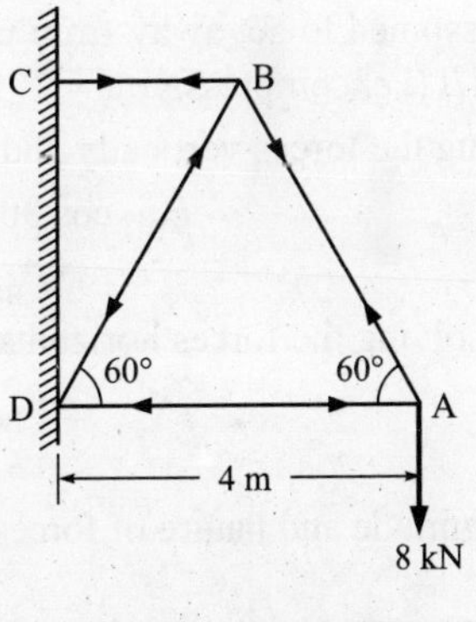


**Fig. 9.29**

***Solution*** The force in the different members of cantilever *truss are shown in Fig. 9.29. First of all, let us consider the equilibrium of the joint $A$, as shown in Fig. 9.30 (*a*). Let the force in the member $AB$ ($P_{AB}$)

* A truss which is connected to a wall or column at one end, and free at the other end is known as *cantilever truss.*

In case of cantilever trusses, it is not necessary to determine the support reactions. The forces in the members of cantilever truss can be obtained by starting the calculations from the free end of the cantilever.

is assumed to act away from point *A* (*i.e.* tensile) and force in the member *AD* ($P_{AD}$) is assumed to act towards point *A* (*i.e.* compressive).

Resolving the force $P_{AB}$ into vertical component and equating the same, we have

$$P_{AB} \sin 60° = 8$$

or
$$P_{AB} = \frac{8}{\sin 60°} = \frac{8}{0.866} = 9.238 \text{ kN (tensile)} \textbf{ Ans.}$$

Resolving the force $P_{AB}$ into horizontal component and equating the same, we have

or
$$P_{AD} = P_{AB} \cos 60° = 9.238 \times 0.5 = 4.619 \text{ kN (compressive)} \textbf{ Ans.}$$

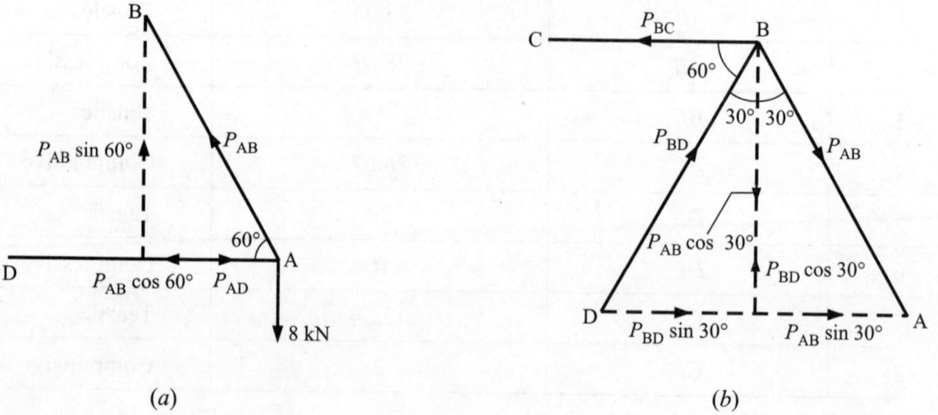


**Fig. 9.30**

Considering the equilibrium of the joint *B*, as shown in Fig. 9.30 (*b*). Let the force in the member *BC* ($P_{BC}$) is assumed to act away from point *B* (*i.e.* tensile) and force in the member *BD* is assumed to act towards point *B* (*i.e.* compressive).

Resolving the forces vertically and equating the same, we have

$$P_{BD} \cos 30° = P_{AB} \cos 30°$$

$$\therefore \quad P_{BD} = P_{AB} = 9.238 \text{ kN (compressive)} \textbf{ Ans.}$$

Now resolving the forces horizontally and equating the same, we have

$$P_{BC} = P_{BD} \sin 30° + P_{AB} \sin 30°$$
$$= 9.238 \times 0.5 + 9.238 \times 0.5 = 9.238 \text{ kN (tensile)} \textbf{ Ans.}$$

The magnitude and nature of forces in the various members of the truss (Refer Fig. 9.29) are tabulated as below:

| *S. No.* | *Member of the truss* | *Magnitude of force in kN* | *Nature of force* |
|---|---|---|---|
| 1. | *AB* | 9.238 | Tensile |
| 2. | *AD* | 4.619 | Compressive |
| 3. | *BC* | 9.238 | Tensile |
| 4. | *BD* | 9.238 | Compressive |

**Example 9.9** *Determine the magnitude and nature of forces in all the members of a cantilever truss, as shown in Fig. 9.31.*

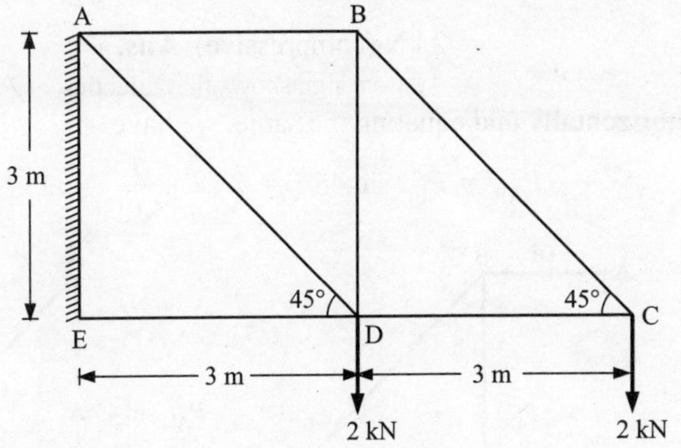


**Fig. 9.31**

***Solution*** We shall use the method of joints to determine the magnitude and nature of forces in all the members.

The forces in the different members of cantilever truss are shown in Fig. 9.32. At joint *C*, there are only two unknown forces. Therefore first of all, let us consider the equilibrium of joint *C*, as shown in Fig. 9.33 (*a*).

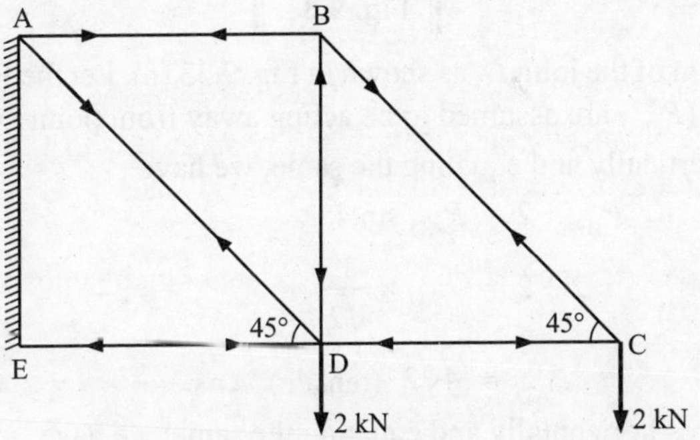


**Fig. 9.32**

Let the force in the member *BC* ($P_{BC}$) and force in the member *CD* ($P_{CD}$) are assumed to be acting away from the point *C* (*i.e.* tensile).

Resolving the force $P_{BC}$ into vertical component and equating the same, we have

$$P_{BC} \sin 45° = 2$$

or $$P_{BC} = \frac{2}{\sin 45°} = 2\sqrt{2} \text{ kN (tensile) } \textbf{Ans.}$$

Resolving the force $P_{BC}$ into horizontal component and equating the same, we have

or $$P_{CD} = -P_{BC} \cos 45° = -2\sqrt{2} \times \frac{1}{\sqrt{2}} = -2 \text{ kN}$$

$$= 2 \text{ kN (compressive) } \textbf{Ans.}$$

...(–ve sign shows that the direction of $P_{CD}$ is opposite to that assumed)

Considering the equilibrium of the joint *B*, as shown in Fig. 9.33 (*b*). Let the force in the member *AB* ($P_{AB}$) and force in the member *BD* ($P_{BD}$) are assumed to be acting away from point *B* (*i.e.* tensile).

Resolving the forces vertically and equating the same, we have

$$P_{BD} = -P_{BC}\cos 45° = -2\sqrt{2}\times\frac{1}{\sqrt{2}} = -2 \text{ kN}$$

$$= 2 \text{ kN (compressive)} \textbf{ Ans.}$$

...(–ve sign shows that direction of $P_{BD}$ is opposite to that assumed)

Resolving the forces horizontally and equating the same, we have

$$P_{AB} = P_{BC}\sin 45° = 2\sqrt{2}\times\frac{1}{\sqrt{2}} = 2 \text{ kN (tensile)} \textbf{ Ans.}$$

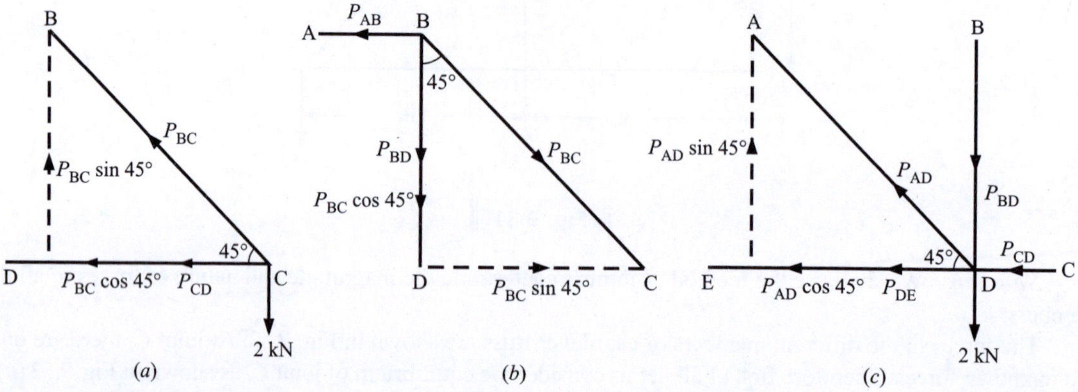


**Fig. 9.33**

Consider the equilibrium of the joint *D*, as shown in Fig. 9.33 (*c*). Let the force in the member *AD* ($P_{AD}$) and force in the member *DE* ($P_{DE}$) are assumed to be acting away from point *D* (*i.e.* tensile).

Resolving the forces vertically and equating the same, we have

$$P_{BD} + 2 = P_{AD}\sin 45°$$

$$2 + 2 = P_{AD}\times\frac{1}{\sqrt{2}}$$

$$\therefore \quad P_{AD} = 4\sqrt{2} \text{ (tensile)} \textbf{ Ans.}$$

Now resolving the forces horizontally and equating the same, we have

$$P_{DE} = -P_{AD}\cos 45° - P_{CD}$$

$$= -4\sqrt{2}\times\frac{1}{\sqrt{2}} - 2 = -6 \text{ kN} = 6 \text{ kN (compressive)} \textbf{ Ans.}$$

...(–ve sign shows that the direction of *DE* is opposite to that assumed)

The magnitude and nature of forces in the various members of the truss (Refer Fig. 9.32) are tabulated as below:

| *S. No.* | *Member of the truss* | *Magnitude of force in kN* | *Nature of force* |
|---|---|---|---|
| 1. | *BC* | $2\sqrt{2}$ | Tensile |
| 2. | *CD* | 2 | Compressive |
| 3. | *AB* | 2 | Tensile |
| 4. | *BD* | 2 | Compressive |
| 5. | *AD* | $4\sqrt{2}$ | Tensile |
| 6. | *DE* | 6 | Compressive |

**Example 9.10** *Find the forces in all the members of the pin-jointed truss supported and loaded, as shown in Fig. 9.34, and present your results on a sketch of the truss distinguishing clearly between those under tension and those under compression.*

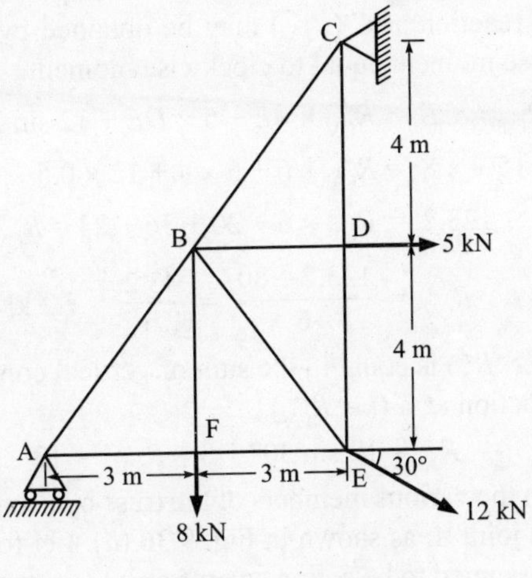


Fig. 9.34

***Solution*** Since there is a roller support at $A$, therefore reaction at $A$ will be a vertical reaction $R_A$ only. It will not take any horizontal reaction. The truss at $C$ is hinged, therefore there will be a vertical reaction $R_{CV}$ and a horizontal reaction $R_{CH}$, as shown in Fig. 9.35. The reaction at $C$ will be the resultant of $R_{CV}$ and $R_{CH}$ which will be inclined at an angle with the horizontal.

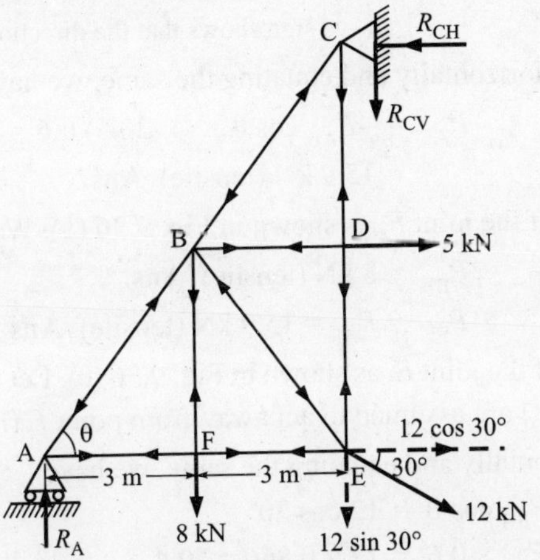


Fig. 9.35

At joint $E$, the load of 12 kN is inclined at an angle of 30° to the horizontal. Resolving this load horizontally and vertically, we have

Horizontal component $= 12 \cos 30°$

and vertical component $= 12 \sin 30°$

In order to maintain equilibrium of the truss in the horizontal direction, horizontal component of reaction at $C$,

$$R_{CH} = \text{Sum of horizontal loads at } D \text{ and } E = 5 + 12 \cos 30°$$

$$= 5 + 12 \times \frac{\sqrt{3}}{2} = 5 + 10.4 = 15.4 \text{ kN}$$

The vertical component of reaction at *C* ($R_{CV}$) may be obtained by taking moment of all the forces about *A* and equating anticlockwise moment equal to clockwise moment.

$$\therefore \quad R_{CH} \times CE = R_{CV} \times AE + 5 \times DE + 12 \sin 30° \times AE + 8 \times AF$$

$$15.4 \times 8 = R_{CV} \times 6 + 5 \times 4 + 12 \times 0.5 \times 6 + 8 \times 3$$

$$123.2 = R_{CV} \times 6 + 20 + 36 + 24 = R_{CV} \times 6 + 80$$

$$\therefore \quad R_{CV} = \frac{123.2 - 80}{6} = \frac{43.2}{6} = 7.2 \text{ kN}$$

The vertical reaction at *A* (*i.e.* $R_A$) is equal to the sum of vertical component of 12 kN at *E*, 8 kN load at *F* and vertical component of reaction at *C* (*i.e.* $R_{CV}$).

$$\therefore \quad R_A = 12 \sin 30° + 8 + R_{CV} = 12 \times 0.5 + 8 + 7.2 = 21.2 \text{ kN}$$

Now, let us find the forces in the various members of the truss by using method of joints. First of all, let us consider the equilibrium of the joint *A*, as shown in Fig. 9.36 (*a*). Let force in the member *AB* ($P_{AB}$) and force in the member *AF* ($P_{AF}$) is assumed to be acting away from *A* (*i.e.* tensile).

Resolving the forces vertically and equating the same, we have

$$P_{AB} \times {}^{*}\sin\theta = -R_A$$

$$P_{AB} \times 0.8 = -21.2$$

$$\therefore \quad P_{AB} = \frac{-21.2}{0.8} = -26.5 \text{ kN} = 26.5 \text{ kN (compressive)} \quad \textbf{Ans.}$$

...(–ve sign shows that the direction of $P_{AB}$ is opposite to that assumed)

Now resolving the forces horizontally and equating the same, we have

$$P_{AF} = -P_{AB} \cos\theta = -(-26.5)\, 0.6$$

$$= 15.9 \text{ kN (tensile)} \quad \textbf{Ans.}$$

Consider the equilibrium of the joint *F*, as shown in Fig. 9.36 (*b*). We see that for equilibrium,

Force in the member *BF*, $P_{BF} = 8$ kN (tensile) **Ans.**

and force in the member *EF*, $P_{EF} = P_{AF} = 15.9$ kN (tensile) **Ans.**

Consider the equilibrium of the joint *E*, as shown in Fig. 9.36 (*c*). Let the force in the member *EB* ($P_{EB}$) and force in the member *ED* ($P_{ED}$) are assumed to act away from point *E* (*i.e.* tensile).

Resolving the forces horizontally and equating the same, we have

$$P_{EF} + P_{EB} \cos\theta = 12 \cos 30°$$

$$15.9 + P_{EB} \times 0.6 = 12 \times 0.866 = 10.4$$

$$\therefore \quad P_{EB} = \frac{10.4 - 15.9}{0.6} = -9.17 \text{ kN} = 9.17 \text{ kN (compressive)} \quad \textbf{Ans.}$$

...(–ve sign shows that the direction of $P_{EB}$ is opposite to that assumed)

---

*From right angled triangle *ABF*, $AB = \sqrt{(AF)^2 + (BF)^2} = \sqrt{3^2 + 4^2} = 5$

$$\therefore \quad \sin\theta = \frac{BF}{AB} = \frac{4}{5} = 0.8$$

and $\cos\theta = \frac{AF}{AB} = \frac{3}{5} = 0.6$

Now resolving the forces vertically and equating the same, we have

$$P_{ED} + P_{EB} \sin \theta = 12 \sin 30°$$

$$P_{ED} + (-9.17) \times 0.8 = 12 \times 0.5 = 6$$

$$\therefore \quad P_{ED} = 6 + 9.17 \times 0.8 = 13.34 \text{ kN (tensile)} \quad \textbf{Ans.}$$

Considering the equilibrium of the joint *D*, as shown in Fig. 9.36 (*d*). We see that

Force in the member *DC*, $P_{DC} = P_{ED} = 13.34$ kN (tensile) **Ans.**

and force in the member *DB*, $P_{DB} = 5$ kN (tensile) **Ans.**

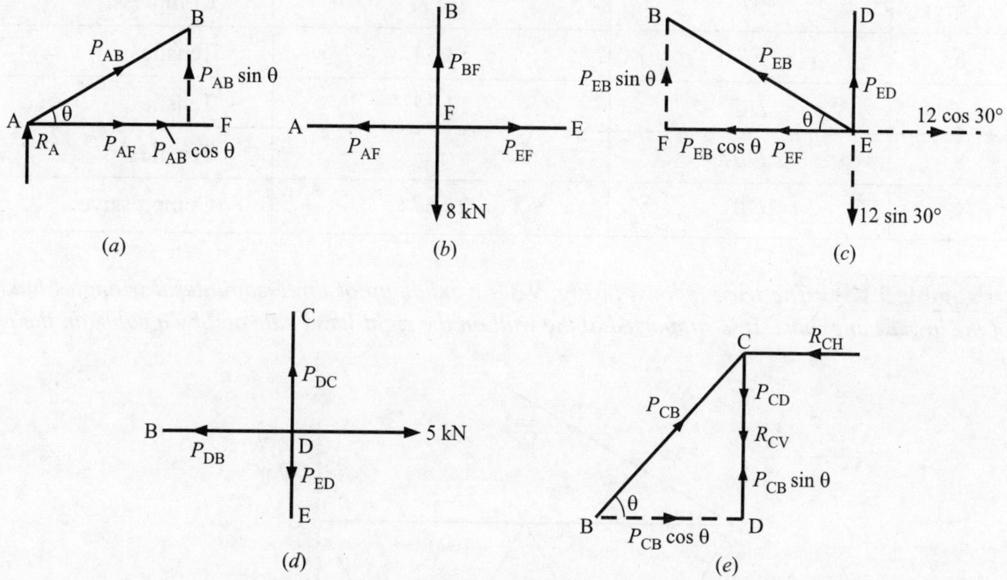


**Fig. 9.36**

Now consider the equilibrium of the joint *C*, as shown in Fig. 9.36 (*e*). Let the force in the member *CB* ($P_{CB}$) is assumed to be acting towards *C* (*i.e.* compressive).

Resolving the forces horizontally and equating the same, we have

$$P_{CB} \cos \theta = R_{CH} = 15.4$$

$$\therefore \quad P_{CB} = \frac{15.4}{\cos \theta} = \frac{15.4}{0.6}$$

$$= 25.67 \text{ kN (compressive)} \quad \textbf{Ans.}$$

***Check*:** Resolving the forces vertically and equating the same, we have

$$P_{CD} + R_{CV} = P_{CB} \sin \theta$$

$$13.34 + 7.2 = P_{CB} \sin \theta$$

$$\therefore \quad P_{CB} = \frac{13.34 + 7.2}{\sin \theta} = \frac{20.54}{0.8} = 25.67 \text{ kN}$$

The magnitude and nature of forces in the various members of the truss (Refer Fig. 9.35) are tabulated as below:

| *S. No.* | *Member of the truss* | *Magnitude of force in kN* | *Nature of force* |
|---|---|---|---|
| 1. | *AB* | 26.5 | Compressive |
| 2. | *AF* | 15.9 | Tensile |
| 3. | *BF* | 8 | Tensile |
| 4. | *EF* | 15.9 | Tensile |
| 5. | *EB* | 9.17 | Compressive |
| 6. | *ED* | 13.34 | Tensile |
| 7. | *DC* | 13.34 | Tensile |
| 8. | *DB* | 5 | Tensile |
| 9. | *CB* | 25.67 | Compressive |

**Example 9.11** *The truss, shown in Fig. 9.37, is made up of three equilateral triangles loaded at each of the lower panel pins. It is supported at the wall on the right-hand side and by a cable on the left, as shown.*

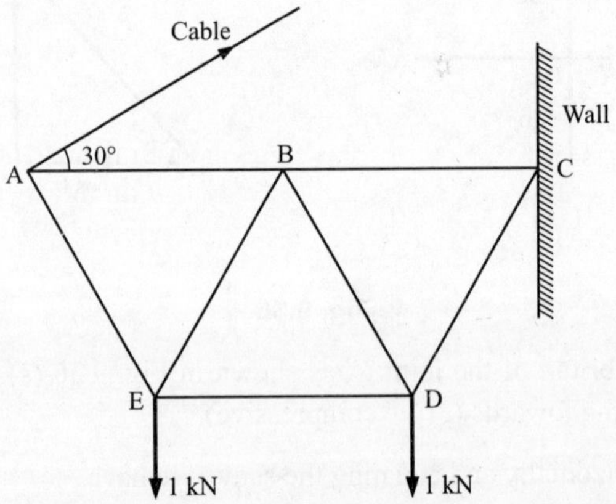


**Fig. 9.37**

*Determine : (a) the tension in the cable ; (b) the reaction at the wall ; and (c) the nature and magnitude of force in each bar.*

***Solution***

***(a) The tension in the cable***

Let $T$ be the tension in the cable and $a$ is the side of each equilateral triangle.

From right angled triangle $ACC'$ (Fig. 9.38),

$$\frac{CC'}{AC} = \sin 30°$$

or $$CC' = AC \sin 30° = 2a \times 0.5 = a$$

Taking moments about the joint $C$ and equating clockwise moment equal to anticlockwise moments, we have

$$T \times CC' = 1 \times CD' + 1 \times CE'$$

$$T \times a = 1 \times 0.5a + 1 \times 1.5a = 2a$$

$$\therefore \quad T = \frac{2a}{a} = 2 \text{ kN} \quad \textbf{Ans.}$$

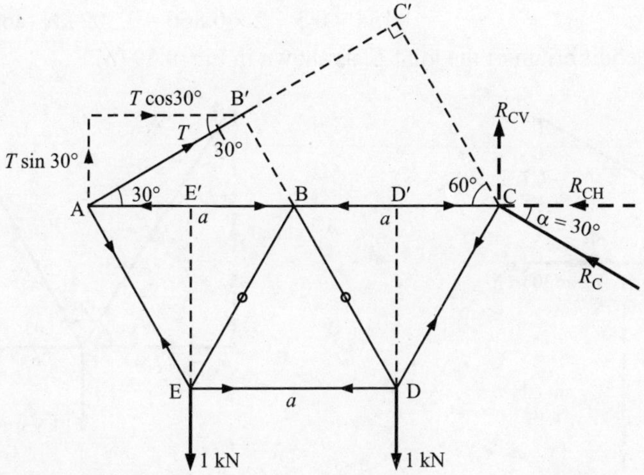


**Fig. 9.38**

***(b) Reaction at the wall***

Let $R_{CV}$ and $R_{CH}$ be the vertical and horizontal component of the reaction of the wall at $C$ ($R_C$) which will be the resultant of $R_{CV}$ and $R_{CH}$ and inclined at an angle α with the horizontal, as shown in Fig. 9.38. Resolve tension in the cable ($T$) into vertical and horizontal components. We know that

$$\text{Vertical component of } T = T \sin 30°$$

$$\text{and horizontal component of } T = T \cos 30°$$

Now equating vertical components, we have

$$T \sin 30° + R_{CV} = 1 + 1 = 2$$

$$\therefore \quad R_{CV} = 2 - T \sin 30° = 2 - 2 \times 0.5 = 1 \text{ kN}$$

Equating horizontal components, we have

$$R_{CH} = T \cos 30° = 2 \times 0.866 = 1.732 \text{ kN}$$

$$\therefore \text{ Reaction at the wall at } C, \quad R_C = \sqrt{(R_{CV})^2 + (R_{CH})^2} = \sqrt{1^2 + (1.732)^2} = 2 \text{ kN} \quad \textbf{Ans.}$$

and its inclination with the horizontal is given by

$$\tan \alpha = \frac{R_{CV}}{R_{CH}} = \frac{1}{1.732} = 0.577$$

$$\therefore \quad \alpha = \tan^{-1}(0.577) = 30° \quad \textbf{Ans.}$$

***(c) Nature and magnitude of force in each bar***

First of all, let us consider the equilibrium of the joint $A$, as shown in Fig. 9.39 (*a*). Let the force in the member $AB$ ($P_{AB}$) is assumed to be acting towards $A$ (*i.e.* compressive) and force in the member $AE$ ($P_{AE}$) is assumed to be acting away from $A$ (*i.e.* tensile).

Resolving the forces vertically and equating the same, we have

$$P_{AE} \sin 60° = T \sin 30°$$

$$\therefore \qquad P_{AE} = \frac{T \sin 30°}{\sin 60°} = \frac{2 \times 0.5}{0.866} = 1.154 \text{ kN (tensile)} \quad \textbf{Ans.}$$

Resolving the forces horizontally and equating the same, we have

$$P_{AB} = P_{AE} \cos 60° + T \cos 30°$$

$$= 1.154 \times 0.5 + 2 \times 0.866 = 2.309 \text{ kN (compressive)} \quad \textbf{Ans.}$$

Now consider the equilibrium of the joint *E*, as shown in Fig. 9.39 (*b*).

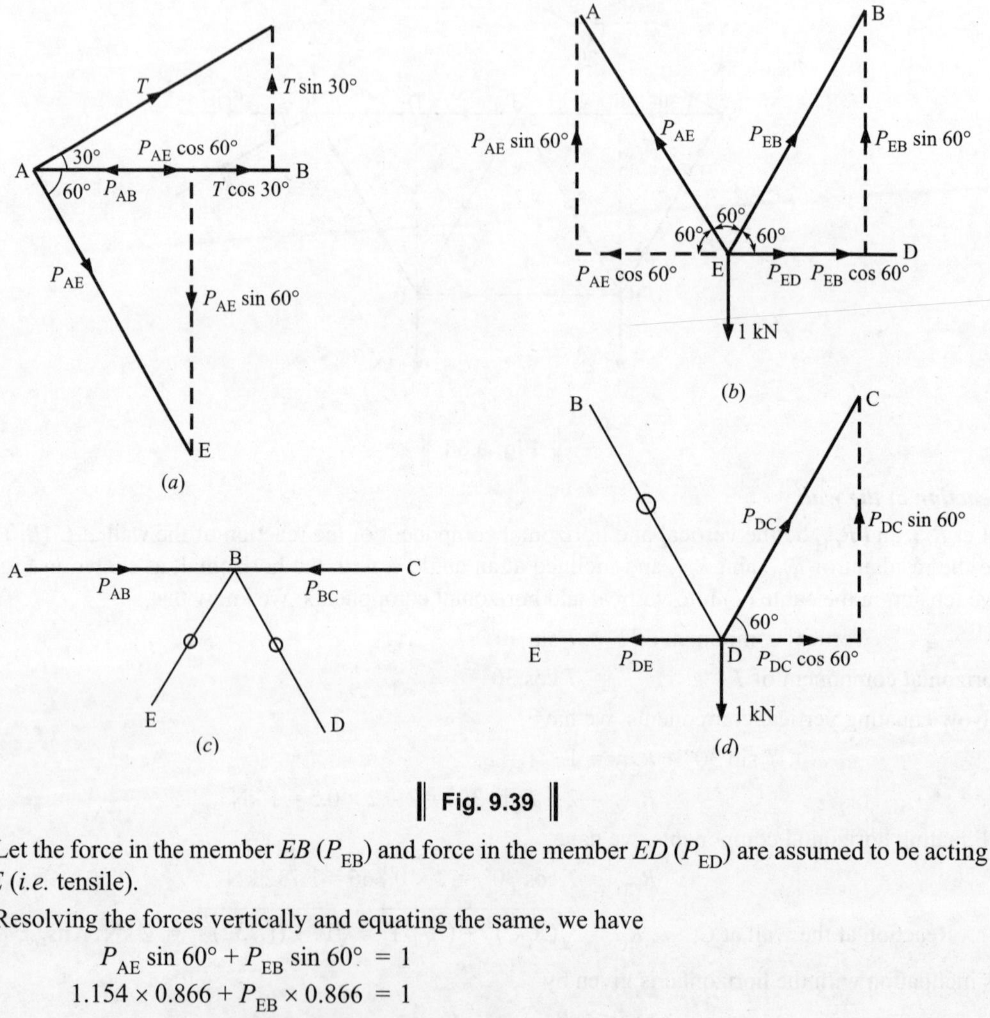


**Fig. 9.39**

Let the force in the member *EB* ($P_{EB}$) and force in the member *ED* ($P_{ED}$) are assumed to be acting away from *E* (*i.e.* tensile).

Resolving the forces vertically and equating the same, we have

$$P_{AE} \sin 60° + P_{EB} \sin 60° = 1$$

$$1.154 \times 0.866 + P_{EB} \times 0.866 = 1$$

$$1 + P_{EB} \times 0.866 = 1$$

$$\therefore \qquad {}^*P_{EB} = \frac{1-1}{0.866} = 0 \quad \textbf{Ans.}$$

Now resolving the forces horizontally and equating the same,

$$P_{ED} + P_{EB} \cos 60° = P_{AE} \cos 60°$$

$$\therefore \qquad P_{ED} = 1.154 \times 0.5 = 0.577 \text{ kN (tensile)} \quad \textbf{Ans.} \qquad ...(\because P_{EB} = 0)$$

* Refer Sec. 9.4

Consider the equilibrium of the joint *B*, as shown in Fig. 9.39 (*c*). Let the force in the member *BC* is assumed to be acting towards the joint *B* (*i.e.* compressive).

A little consideration will show that force in the member *BD* ($P_{BD}$) is *zero. Now equating the horizontal forces, we have,

$$P_{BC} = P_{AB} = 2.309 \text{ kN (compressive)} \quad \textbf{Ans.}$$

Now consider the equilibrium of the joint *D*, as shown in Fig. 9.39 (*d*).

Let the force in the member *DC* ($P_{DC}$) is assumed to be acting away from the joint *D* (*i.e.* tensile). The force in the member *BD* ($P_{BD}$) is zero. Equating the vertical components, we have

$$P_{DC} \sin 60° = 1$$

$$\therefore \quad P_{DC} = \frac{1}{\sin 60°} = \frac{1}{0.866} = 1.154 \text{ kN (tensile)} \quad \textbf{Ans.}$$

***Check*:** Resolving the horizontal forces, we have

$$P_{DC} \cos 60° = P_{DE}$$

$$\therefore \quad P_{DC} = \frac{P_{DE}}{\cos 60°} = \frac{0.577}{0.5} = 1.154 \text{ kN} \qquad \text{...(Same as before)}$$

The magnitude and nature of forces in the various members of the truss (Refer Fig. 9.38) are tabulated as below:

| *S. No.* | *Member of the truss* | *Magnitude of force in kN* | *Nature of force* |
|---|---|---|---|
| 1. | *AE* | 1.154 | Tensile |
| 2. | *AB* | 2.309 | Compressive |
| 3. | *EB* | 0 | — |
| 4. | *ED* | 0.577 | Tensile |
| 5. | *BC* | 2.309 | Compressive |
| 6. | *BD* | 0 | — |
| 7. | *DC* | 1.154 | Tensile |

**Example 9.12** *Find the forces in the members of the truss, as shown in Fig. 9.40, by analytical or graphical method.*

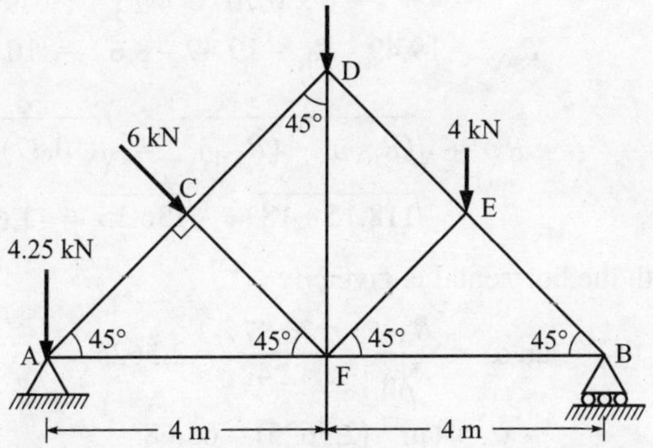


**Fig. 9.40**

* Refer Sec. 9.4

***Solution*** Since there is a roller support at *B*, therefore reaction at *B* ($R_B$) will be vertical. The truss at *A* is hinged, therefore there will be a vertical reaction ($R_{AV}$) and a horizontal reaction ($R_{AH}$), as shown in Fig. 9.41. The reaction at *A* ($R_A$) will be the resultant of $R_{AV}$ and $R_{AH}$ which will be inclined at an angle α with the horizontal.

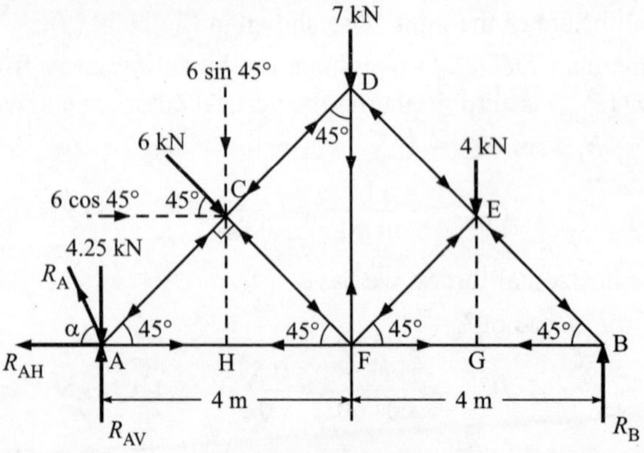


**Fig. 9.41**

First of all, let us find the reaction at *B* ($R_B$). Draw perpendiculars *CH* and *EG* on line *AB*, as shown in Fig. 9.41.

Taking moments about *A*, we have

$$R_B \times 8 = 4 \times AG + 7 \times AF + 6 \times AC$$
$$= 4 \times 6 + 7 \times 4 + 6 \times AF \sin 45°$$
$$= 24 + 28 + 6 \times 4 \times 0.707 = 68.97$$

$$\therefore \quad R_B = \frac{68.97}{8} = 8.62 \text{ kN}$$

From Fig. 9.41, we see that horizontal component of $R_A$,

$$R_{AH} = 6 \cos 45° = 6 \times 0.707 = 4.242 \text{ kN}$$

Now equating vertical components, we have

$$R_{AV} + R_B = 4 + 7 + 6 \sin 45° + 4.25$$
$$= 4 + 7 + 6 \times 0.707 + 4.25 = 19.49$$

$$\therefore \quad R_{AV} = 19.49 - R_B = 19.49 - 8.62 = 10.87 \text{ kN}$$

We know that

$$R_A = \sqrt{(R_{AV})^2 + (R_{AH})^2} = \sqrt{(10.87)^2 + (4.242)^2}$$
$$= \sqrt{118.15 + 18} = \sqrt{136.15} = 11.67 \text{ kN}$$

and its angle of inclination with the horizontal is given by

$$\tan \alpha = \frac{R_{AV}}{R_{AH}} = \frac{10.87}{4.242} = 2.5625$$

$$\therefore \quad \alpha = \tan^{-1} (2.5625) = 68.68°$$

Now let us find the forces in the various members of the truss. First of all, let us consider the equilibrium of joint *B*, as shown in Fig. 9.42 (*a*).

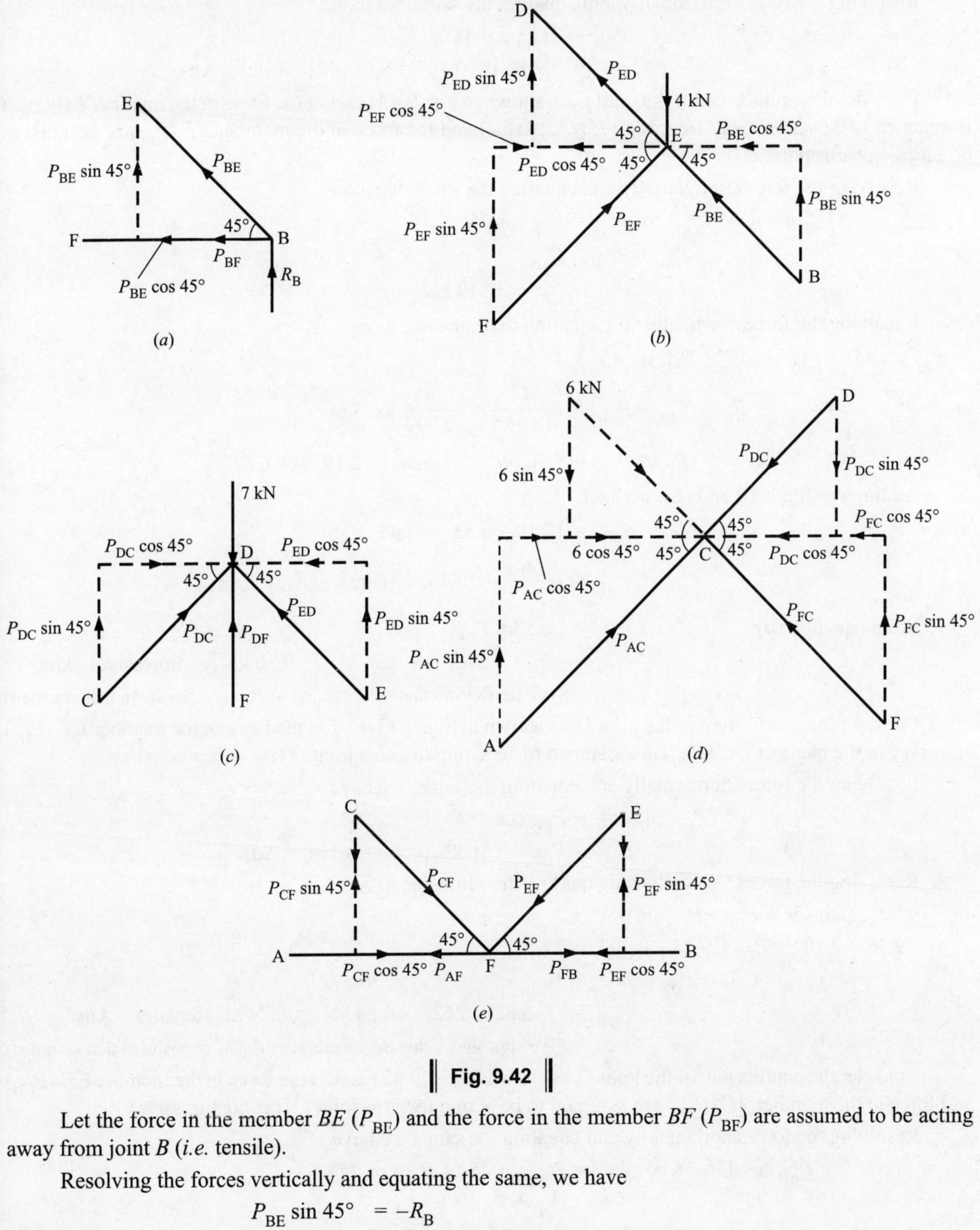


**Fig. 9.42**

Let the force in the member $BE$ ($P_{BE}$) and the force in the member $BF$ ($P_{BF}$) are assumed to be acting away from joint $B$ (*i.e.* tensile).

Resolving the forces vertically and equating the same, we have

$$P_{BE} \sin 45° = -R_B$$

$$\therefore \quad P_{BE} = \frac{-R_B}{\sin 45°} = \frac{-8.62}{0.707} = -12.19 \text{ kN} = 12.19 \text{ kN (compressive)} \textbf{ Ans.}$$

...(–ve sign shows that the direction of $P_{BE}$ is opposite to that assumed)

Resolving the forces horizontally and equating the same, we have

$$P_{BF} = -P_{BE} \cos 45°$$
$$= -(-12.19)\ 0.707 = 8.62 \text{ kN (tensile)} \quad \textbf{Ans.}$$

Consider the equilibrium of the joint *E*, as shown in Fig. 9.42 (*b*). Let the force in the member *ED* ($P_{ED}$) is assumed to be acting away from joint *E* (*i.e.* tensile) and the force in the member *EF* ($P_{EF}$) is assumed to be acting towards point *E* (*i.e.* compressive).

Resolving the forces horizontally and equating the same, we have

$$P_{BE} \cos 45° + P_{ED} \cos 45° = P_{EF} \cos 45°$$

or $$P_{BE} + P_{ED} = P_{EF}$$

∴ $$P_{EF} - P_{ED} = P_{BE} = 12.19 \text{ kN} \quad ...(i)$$

Resolving the forces vertically and equating the same, we have

$$P_{EF} \sin 45° + P_{BE} \sin 45° + P_{ED} \sin 45° = 4$$

or $$P_{EF} + P_{BE} + P_{ED} = \frac{4}{\sin 45°} = \frac{4}{0.707} = 5.66$$

∴ $$P_{EF} + P_{ED} = 5.66 - P_{BE} = 5.66 - 12.19 = -6.53 \quad ...(ii)$$

Adding equations (*i*) and (*ii*), we have

$$2P_{EF} = 12.19 - 6.53 = 5.66$$

∴ $$P_{EF} = \frac{5.66}{2} = 2.83 \text{ kN (compressive)} \quad \textbf{Ans.}$$

From equation (*ii*), $$P_{ED} = -6.53 - P_{EF}$$
$$= -6.53 - 2.83 = -9.36 \text{ kN} = 9.36 \text{ kN (compressive)} \quad \textbf{Ans.}$$

...(–ve sign shows that the direction of $P_{ED}$ is opposite to that assumed)

Consider the equilibrium of the joint *D*, as shown in Fig. 9.42 (*c*). Let the force in the member *DF* ($P_{DF}$) and force in the member *DC* ($P_{DC}$) are assumed to be acting towards joint *D* (*i.e.* compressive).

Resolving the forces horizontally and equating the same, we have

$$P_{DC} \cos 45° = P_{ED} \cos 45°$$

∴ $$P_{DC} = P_{ED} = 9.36 \text{ kN (compressive)} \quad \textbf{Ans.}$$

Resolving the forces vertically and equating the same, we have

$$P_{DC} \sin 45° + P_{DF} + P_{ED} \sin 45° = 7$$
$$9.36 \times 0.707 + P_{DF} + 9.36 \times 0.707 = 7$$
$$6.62 + P_{DF} + 6.62 = 7$$

∴ $$P_{DF} = 7 - 6.62 - 6.62 = -6.24 \text{ kN} = 6.24 \text{ kN (tensile)} \quad \textbf{Ans.}$$

...(–ve sign shows that the direction of $P_{DF}$ is opposite to that assumed)

Consider the equilibrium of the joint *C*, as shown in Fig. 9.42 (*d*). Let the force in the member *CD* ($P_{CD}$) and force in the member *AC* ($P_{AC}$) are assumed to be acting towards joint *C* (*i.e.* compressive).

Resolving the forces horizontally and equating the same, we have

$$P_{AC} \cos 45° + 6 \cos 45° = P_{DC} \cos 45° + P_{FC} \cos 45°$$

or $$P_{AC} + 6 = P_{DC} + P_{FC}$$

∴ $$P_{AC} - P_{FC} = P_{DC} - 6 = 9.36 - 6 = 3.36 \quad ...(iii)$$

Resolving the forces vertically and equating the same, we have

$$P_{AC} \sin 45° + P_{FC} \sin 45° = 6 \sin 45° + P_{DC} \sin 45°$$

or $$P_{AC} + P_{FC} = 6 + P_{DC} = 6 + 9.36 = 15.36 \quad ...(iv)$$

Adding equations (*iii*) and (*iv*), we get

$$2P_{AC} = 3.36 + 15.36 = 18.72$$

$$\therefore \quad P_{AC} = \frac{18.72}{2} = 9.36 \text{ kN (compressive)} \quad \textbf{Ans.}$$

From equation (*iv*), we have $P_{FC} = 15.36 - P_{AC} = 15.36 - 9.36 = 6$ kN (compressive) **Ans.**

Let us now consider the equilibrium of the joint *F* as shown in Fig. 9.42 (*e*). Let the force in the member *AF* ($P_{AF}$) is assumed to be acting away from the joint *F* (*i.e.* tensile). Resolving the forces horizontally and equating the same, we have

$$P_{CF} \cos 45° + P_{FB} = P_{AF} + P_{EF} \cos 45°$$

$$6 \times 0.707 + 8.62 = P_{AF} + 2.83 \times 0.707$$

$$4.242 + 8.62 = P_{AF} + 2$$

$$\therefore \quad P_{AF} = 4.242 + 8.62 - 2 = 10.862 \text{ kN (tensile)} \quad \textbf{Ans.}$$

The magnitude and nature of forces in various members of the truss (Refer Fig. 9.41) are tabulated as below:

| *S. No.* | *Member of the truss* | *Magnitude of force in kN* | *Nature of force* |
|---|---|---|---|
| 1. | *BE* | 12.19 | Compressive |
| 2. | *BF* | 8.62 | Tensile |
| 3. | *EF* | 2.83 | Compressive |
| 4. | *ED* | 9.36 | Compressive |
| 5. | *DC* | 9.36 | Compressive |
| 6. | *DF* | 6.24 | Tensile |
| 7. | *AC* | 9.36 | Compressive |
| 8. | *FC* | 6 | Compressive |
| 9. | *AF* | 10.862 | Tensile |

**Note:** The vertical and horizontal reactions at *A* (*i.e.* $R_{AV}$ and $R_{AH}$) may be obtained by considering the equilibrium of the joint *A*, as shown in Fig. 9.43. Equating the vertical components, we have

$$R_{AV} = 4.25 + P_{AC} \sin 45°$$

$$= 4.25 + 9.36 \times 0.707$$

$$= 4.25 + 6.62$$

$$= 10.87 \text{ kN}$$

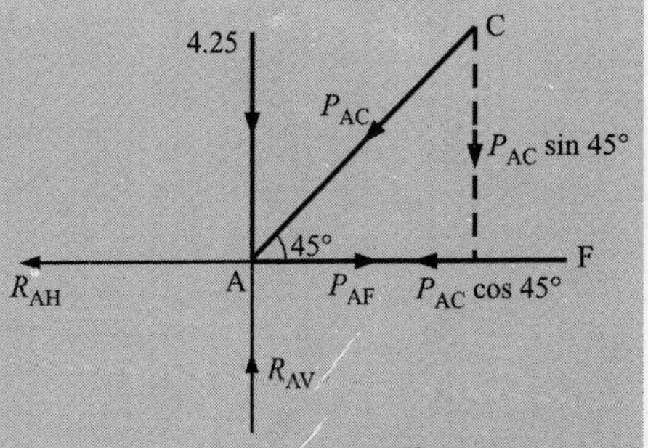


**Fig. 9.43**

Now equating horizontal components, we have

$$R_{AH} + P_{AC} \cos 45° = P_{AF}$$

or

$$R_{AH} = P_{AF} - P_{AC} \cos 45° = 10.862 - 9.36 \times 0.707$$

$$= 10.862 - 6.62 = 4.242 \text{ kN}$$

**Example 9.13** *The signboard truss, as shown in Fig. 9.44, is required to carry a total horizontal wind load of 8 kN. It may be assumed that* $\frac{5}{8}$ *of this load is transmitted to the joint C and the rest is equally divided between B and D. Calculate the forces in the members and give the result in a tabular form indicating the nature of each.*

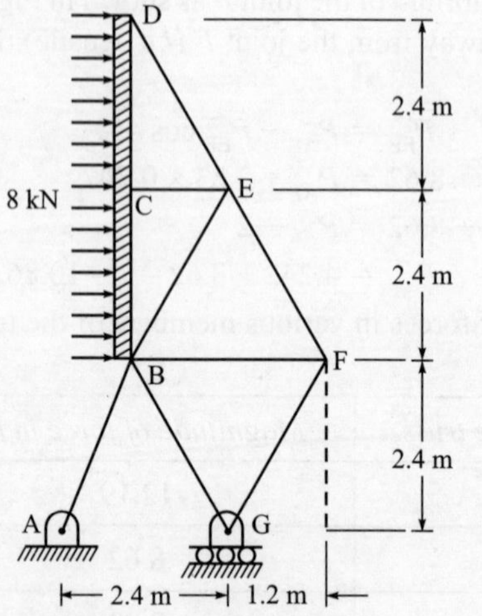


**Fig. 9.44**

***Solution*** Since the total horizontal load of 8 kN is distributed over *BCD*, therefore load transmitted on joint *C*

$$= \frac{5}{8} \times 8 = 5 \text{ kN}$$

and load transmitted on joint *B* and *D* $= \dfrac{8-5}{2} = \dfrac{3}{2} = 1.5 \text{ kN}$

This is shown in Fig. 9.45.

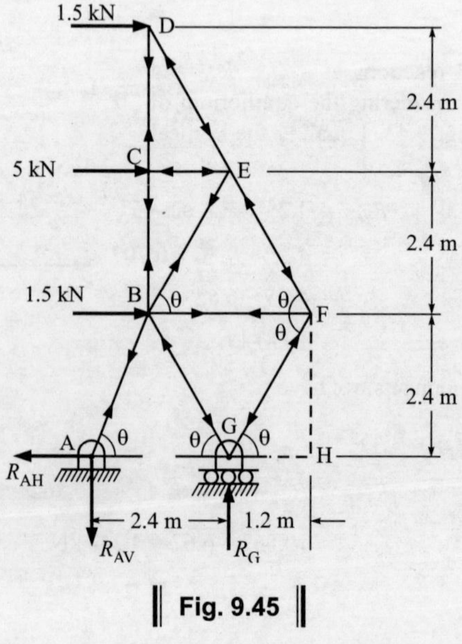


**Fig. 9.45**

The support at $A$ is a hinged support and the support at $G$ is a roller support. Let $R_{AV}$ and $R_{AH}$ be the vertical component and horizontal component of the reaction $R_A$ at the hinged support $A$. Let $R_G$ be the vertical reaction at the roller support $G$. From Fig. 9.45, we see that

$$R_{AH} = 8 \text{ kN}$$

Now taking moments about $A$, we have

$$R_G \times 2.4 = 1.5 \times 7.2 + 5 \times 4.8 + 1.5 \times 2.4 = 10.8 + 24 + 3.6 = 38.4$$

$$\therefore \quad R_G = \frac{38.4}{2.4} = 16 \text{ kN}$$

The vertical component of reaction at $A$ ($R_{AV}$) will also be equal to 16 kN downwards *i.e.*

$$R_{AV} = 16 \text{ kN}$$

From the right angled triangle $FGH$,

$$FG = \sqrt{(2.4)^2 + (1.2)^2} = \sqrt{5.76 + 1.44} = \sqrt{7.2} = 2.683 \text{ m}$$

$$\therefore \quad \sin\theta = \frac{FH}{FG} = \frac{2.4}{2.683} = 0.894$$

and $$\cos\theta = \frac{GH}{FG} = \frac{1.2}{2.683} = 0.447$$

Now consider the equilibrium of the joint $A$, as shown in Fig. 9.46 ($a$).

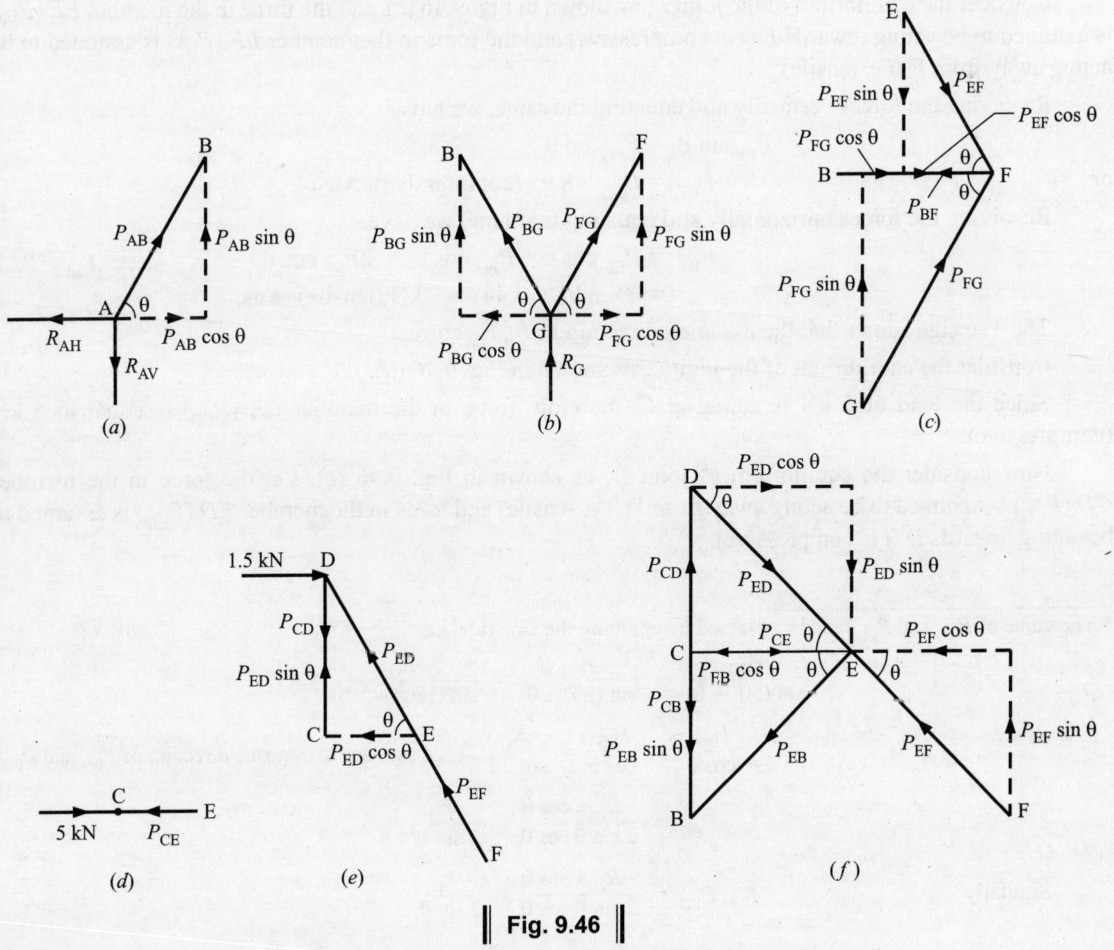


**Fig. 9.46**

Let the force in member *AB* ($P_{AB}$) is assumed to be acting away from *A* (*i.e.* tensile).

Resolving the force vertically, we have

$$P_{AB} \sin \theta = R_{AV}$$

$$\therefore \quad P_{AB} = \frac{R_{AV}}{\sin \theta} = \frac{16}{0.894} = 17.9 \text{ (tensile) } \textbf{Ans.}$$

Consider the equilibrium of the joint *G*, as shown in Fig. 9.46. Let the force in the member *BG* ($P_{BG}$) and force in the member *FG* ($P_{FG}$) are assumed to be acting away from *G* (*i.e.* tensile).

Resolving the forces horizontally and equating the same, we have

$$P_{BG} \cos \theta = P_{FG} \cos \theta$$

or $$P_{BG} = P_{FG} \qquad ...(\because P_{FG} = P_{BG})$$

Resolving the forces vertically and equating the same, we have

$$P_{BG} \sin \theta + P_{FG} \sin \theta = -R_G$$

$$2P_{BG} \sin \theta = -16$$

$$\therefore \quad {}^*P_{BG} = \frac{-16}{2 \sin \theta} = \frac{-8}{0.894} = -8.95 \text{ kN} = 8.95 \text{ kN (compressive) } \textbf{Ans.}$$

and $$P_{FG} = 8.95 \text{ kN (compressive) } \textbf{Ans.}$$

...(–ve sign shows that the direction of $P_{BG}$ and $P_{FG}$ is opposite to that assumed)

Consider the equilibrium of the joint *F*, as shown in Fig. 9.46 (*c*). Let the force in the member *EF* ($P_{EF}$) is assumed to be acting towards *F* (*i.e.* compressive) and the force in the member *BF* ($P_{BF}$) is assumed to be acting away from *F* (*i.e.* tensile).

Resolving the forces vertically and equating the same, we have

$$P_{EF} \sin \theta = P_{FG} \sin \theta$$

or $$P_{EF} = P_{FG} = 8.95 \text{ (compressive) } \textbf{Ans.}$$

Resolving the forces horizontally and equating the same, we have

$$P_{BF} = P_{EF} \cos \theta + P_{FG} \cos \theta = 2P_{EF} \cos \theta \qquad ...(\because P_{EF} = P_{FG})$$

$$= 2 \times 8.95 \times 0.447 = 8 \text{ kN (tensile) } \textbf{Ans.}$$

The +ve sign shows that the assumed direction of $P_{BF}$ is correct.

Consider the equilibrium of the joint *C*, as shown in Fig. 9.46 (*d*).

Since the load of 5 kN is acting at *C*, therefore force in the member *CE* ($P_{CE}$) is equal to 5 kN (compressive).

Now consider the equilibrium of joint *D*, as shown in Fig. 9.46 (*e*). Let the force in the member *CD* ($P_{CD}$) is assumed to be acting away from *D* (*i.e.* tensile) and force in the member *ED* ($P_{ED}$) is assumed to be acting towards *D* (*i.e.* compressive).

---

* The value of $P_{BG}$ and $P_{FG}$ may be obtained by applying the sine rule, *i.e.*

$$\frac{P_{BG}}{\sin (90° + \theta)} = \frac{P_{FG}}{\sin (90° + \theta)} = \frac{-R_G}{\sin (180° - 2\theta)}$$

$$\frac{P_{BG}}{\cos \theta} = \frac{P_{FG}}{\cos \theta} = \frac{-R_G}{\sin 2\theta} \quad \text{...(–ve sign due to opposite direction of } P_{BG} \text{ and } P_{FG})$$

$$\therefore \quad P_{BG} = \frac{-R_G \times \cos \theta}{2 \sin \theta \cos \theta} = \frac{-R_G}{2 \sin \theta}$$

Similarly $$P_{FG} = \frac{-R_G \times \cos \theta}{2 \sin \theta \cos \theta} = \frac{-R_G}{2 \sin \theta}$$

Resolving the forces horizontally and equating the same, we have

$$P_{ED}\cos\theta = 1.5$$

$$\therefore \quad P_{ED} = \frac{1.5}{\cos\theta} = \frac{1.5}{0.447} = 3.355 \text{ kN (compressive)} \quad \textbf{Ans.}$$

Resolving the forces vertically and equating the same, we have

$$P_{CD} = P_{ED}\sin\theta = 3.355 \times 0.894 = 3 \text{ kN (tensile)} \quad \textbf{Ans.}$$

The force in the member *CB* ($P_{CB}$) will be equal to force in the member *CD* ($P_{CD}$).

$$\therefore \quad P_{CB} = P_{CD} = 3 \text{ kN (tensile)} \quad \textbf{Ans.}$$

Now consider the equilibrium of the joint *E*, as shown in Fig. 9.46 (*f*). Let the force in the member *EB* ($P_{EB}$) is assumed to be acting away from *E* (*i.e.* tensile). Resolving the forces vertically and equating the same, we have

$$P_{EB}\sin\theta + P_{CB} + P_{ED}\sin\theta = P_{EF}\sin\theta + P_{CD}$$

or $$P_{EB} + P_{ED} = P_{EF} \qquad ...(\because P_{CB} = P_{CD})$$

$$\therefore \quad P_{EB} = P_{EF} - P_{ED} = 8.95 - 3.355 = 5.595 \text{ kN (tensile)} \quad \textbf{Ans.}$$

The +ve sign shows that the assumed direction of $P_{EB}$ is correct.

***Check*:** Resolving the forces horizontally and equating the same, we have

$$P_{ED}\cos\theta + P_{CE} = P_{EB}\cos\theta + P_{EF}\cos\theta$$

or $$P_{ED} + \frac{P_{CE}}{\cos\theta} = P_{EB} + P_{EF}$$

$$\therefore \quad P_{EB} = P_{ED} + \frac{P_{CE}}{\cos\theta} - P_{EF} = 3.355 + \frac{5}{0.447} - 8.95$$

$$= 3.355 + 11.185 - 8.95 = 5.59 \text{ kN}$$

The magnitude and nature of forces in the various members of the truss (Refer Fig. 9.45) are tabulated as below:

| *S. No.* | *Member of the truss* | *Magnitude of force in kN* | *Nature of force* |
|---|---|---|---|
| 1. | *AB* | 17.9 | Tensile |
| 2. | *BG* | 8.95 | Compressive |
| 3. | *FG* | 8.95 | Compressive |
| 4. | *EF* | 8.95 | Compressive |
| 5. | *BF* | 8 | Tensile |
| 6. | *CE* | 5 | Compressive |
| 7. | *ED* | 3.355 | Compressive |
| 8. | *CD* | 3 | Tensile |
| 9. | *CB* | 3 | Tensile |
| 10. | *EB* | 5.595 | Tensile |

**Example 9.14** *Determine the forces in the members AB, AC, EG and FG of the truss, shown in Fig. 9.47.*

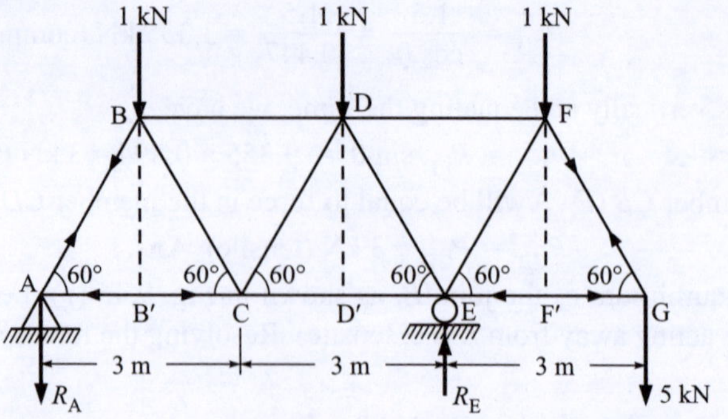


**Fig. 9.47**

***Solution*** First of all, let us find the reactions $R_A$ and $R_E$ at supports $A$ and $E$, respectively. The support at $A$ is hinged and the support at $E$ is on rollers. Therefore the reaction at $E$ ($R_E$) will be vertical. Further, all the loads are vertical, thus the reaction at $A$ ($R_A$) will also be vertical.

Taking moments about $A$, we have

$$R_E \times AE = 5 \times AG + 1 \times AF' + 1 \times AD' + 1 \times AB'$$

$$R_E \times 6 = 5 \times 9 + 1 \times 7.5 + 1 \times 4.5 + 1 \times 1.5 = 58.5$$

$$\therefore \quad R_E = \frac{58.5}{6} = 9.75 \text{ kN}$$

We know that $\quad R_A + R_E = 1 + 1 + 1 + 5 = 8$ kN

$$\therefore \quad R_A = 8 - R_E = 8 - 9.75 = -1.75 \text{ kN}$$

The –ve sign shows that $R_A$ will act downwards, as shown in Fig. 9.47.

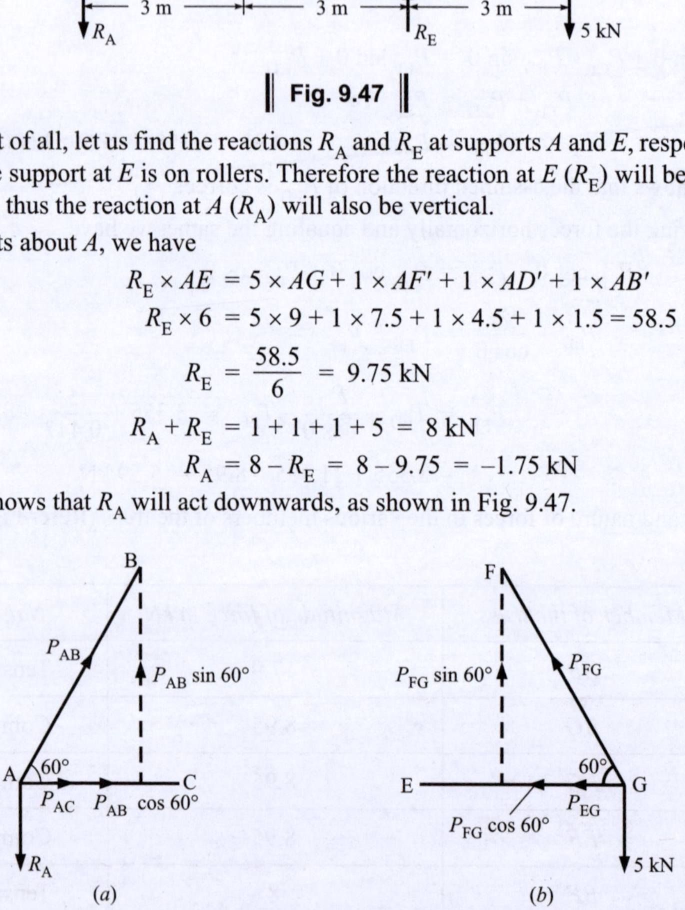


**Fig. 9.48**

Now consider the equilibrium of the joint $A$, as shown in Fig. 9.48 (*a*). Let the force in the member $AB$ ($P_{AB}$) and force in the member $AC$ ($P_{AC}$) are assumed to be acting away from $A$ (*i.e.* tensile).

Resolving the force $P_{AB}$ vertically, we have

$$P_{AB} \sin 60° = R_A$$

$$\therefore \quad P_{AB} = \frac{R_A}{\sin 60°} = \frac{1.75}{0.866} = 2.02 \text{ kN (tensile)} \quad \textbf{Ans.}$$

Resolving the force $P_{AB}$ horizontally, we have

$$P_{AC} = -P_{AB} \cos 60° = -2.02 \times 0.5 = -1.01 \text{ kN}$$
$$= 1.01 \text{ kN (compressive)} \quad \textbf{Ans.}$$

...(–ve sign shows that the direction of $P_{AC}$ in opposite to that assumed)

Now consider the equilibrium of the joint $G$, as shown in Fig. 9.48 (*b*). Let the force in the member $EG$ ($P_{EG}$) and force in the member $FG$ ($P_{FG}$) are assumed to be acting away from $G$ (*i.e.* tensile).

Resolving the force $P_{FG}$ vertically, we have

$$P_{FG} \sin 60° = 5$$

$$\therefore \quad P_{FG} = \frac{5}{\sin 60°} = \frac{5}{0.866} = 5.773 \text{ kN (tensile)} \quad \textbf{Ans.}$$

Resolving the force $P_{FG}$ horizontally, we have

$$P_{EG} = -P_{FG} \cos 60° = -5.773 \times 0.5 = -2.886 \text{ kN}$$
$$= 2.886 \text{ kN (compressive)} \quad \textbf{Ans.}$$

...(–ve sign shows that the direction of $P_{EG}$ is opposite to that assumed)

**Example 9.15** *Find the reactions at A and B for the truss, shown in Fig. 9.49. Determine also the nature and magnitude of the forces in all the members. Assume the pulley to be weightless and frictionless and having a diameter of 0.3 m. The load of 3 kN is held by a rope inclined at an angle of 30° with the horizontal.*

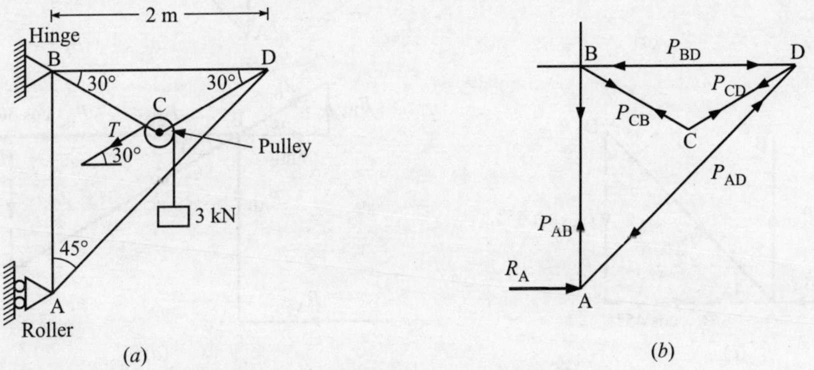


**Fig. 9.49**

***Solution*** First of all, let us find the nature and magnitude of the forces in all the members. Since the pulley is frictionless, therefore tension in the rope ($T$) will be equal to the load 3 kN.

Consider the equilibrium of the point $C$, as shown in Fig. 9.50 (*a*). Let the forces $P_{CB}$ and $P_{CD}$ in the members $CB$ and $CD$, respectively, are assumed to be acting away from $C$ (*i.e.* tensile).

Resolving the forces vertically and equating the same, we have

$$P_{CB} \sin 30° + P_{CD} \sin 30° = 3 + T \sin 30°$$
$$P_{CB} \times 0.5 + P_{CD} \times 0.5 = 3 + 3 \times 0.5 = 4.5$$

$$\therefore \quad P_{CB} + P_{CD} = \frac{4.5}{0.5} = 9$$

or $\quad P_{CD} = 9 - P_{CB} \quad$ ...(*i*)

Resolving the forces horizontally and equating the same, we have

$$P_{CB}\cos 30° + T\cos 30° = P_{CD}\cos 30°$$

or

$$P_{CB} + T = 9 - P_{CB} \qquad ...[\because P_{CD} = 9 - P_{CB}] \quad ...(ii)$$

$$\therefore \quad 2P_{CB} = 9 - 3 = 6$$

$$\therefore \quad P_{CB} = \frac{6}{2} = 3 \text{ kN (tensile)} \quad \textbf{Ans.}$$

From equation (*i*), $P_{CD} = 9 - 3 = 6$ kN (tensile) **Ans.**

Since the magnitude of $P_{CB}$ and $P_{CD}$ are both +ve, therefore the assumed direction is correct *i.e.* both the forces are tensile.

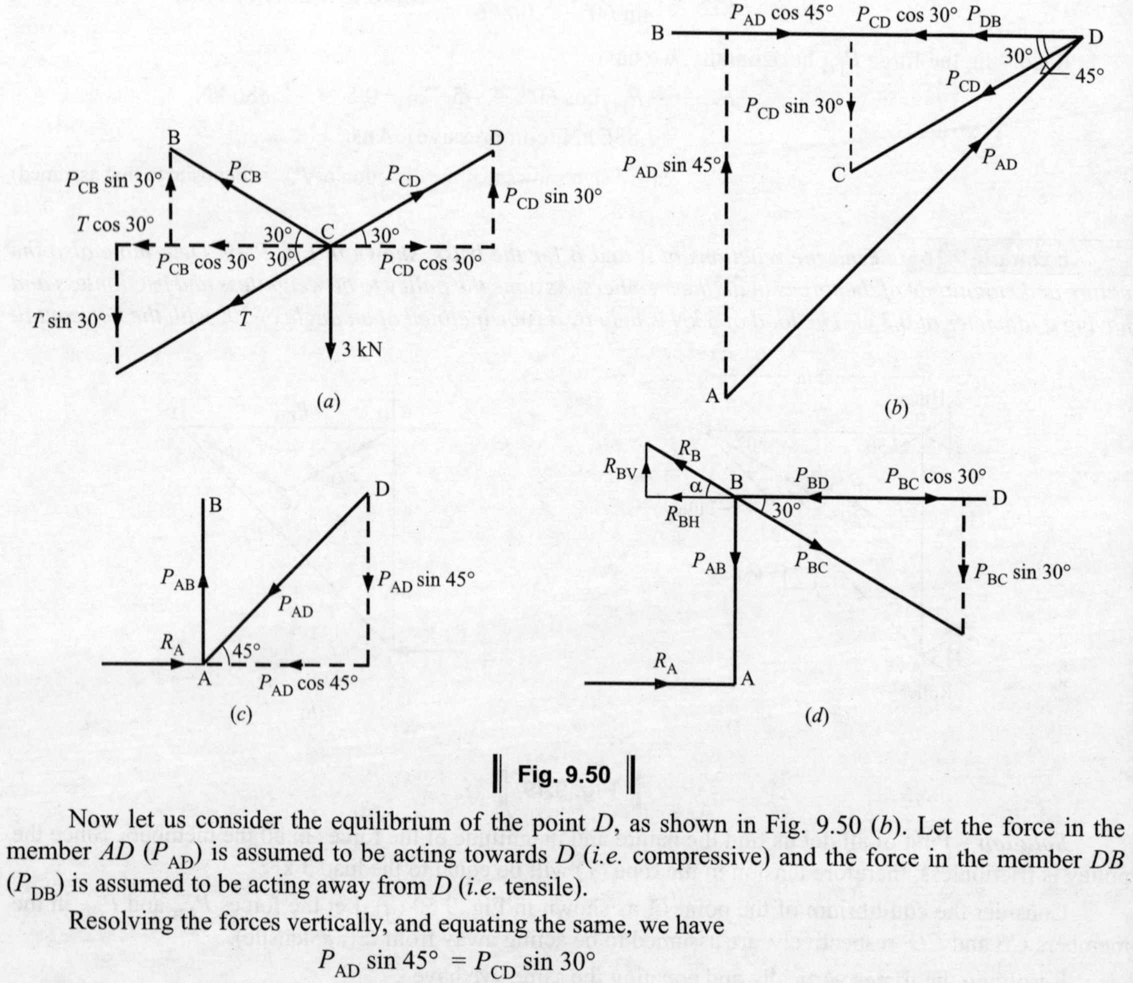


**Fig. 9.50**

Now let us consider the equilibrium of the point *D*, as shown in Fig. 9.50 (*b*). Let the force in the member *AD* ($P_{AD}$) is assumed to be acting towards *D* (*i.e.* compressive) and the force in the member *DB* ($P_{DB}$) is assumed to be acting away from *D* (*i.e.* tensile).

Resolving the forces vertically, and equating the same, we have

$$P_{AD}\sin 45° = P_{CD}\sin 30°$$

$$P_{AD} \times \frac{1}{\sqrt{2}} = P_{CD} \times \frac{1}{2}$$

$$\therefore \quad P_{AD} = P_{CD} \times \frac{\sqrt{2}}{2} = \frac{P_{CD}}{\sqrt{2}} = \frac{6}{\sqrt{2}} = 4.24 \text{ kN (compressive)} \quad \textbf{Ans.}$$

Resolving the forces horizontally and equating the same, we have

$$P_{AD}\cos 45° = P_{DB} + P_{CD}\cos 30°$$

$$4.24 \times \frac{1}{\sqrt{2}} = P_{DB} + 6 \times 0.866$$

$$3 = P_{DB} + 5.196$$

$$\therefore \quad P_{DB} = 3 - 5.196 = -2.196 \text{ kN} = 2.196 \text{ kN (compressive)} \quad \textbf{Ans.}$$

...(The –ve sign shows that the direction of $P_{DB}$ is opposite to that assumed)

Now let us consider the equilibrium of the joint *A*, as shown in Fig. 9.50 (*c*). Let the force in the member *AB* ($P_{AB}$) is assumed to be acting away from *A* (*i.e.* tensile).

Resolving the force $P_{AD}$ vertically and equating the same, we have

$$P_{AB} = P_{AD} \sin 45° = 4.24 \times \frac{1}{\sqrt{2}} = 3 \text{ kN (tensile)} \quad \textbf{Ans.}$$

Resolving the force $P_{AD}$ horizontally and equating the same, we have

$$R_A = \text{Reaction at } A = P_{AD} \cos 45° = 4.24 \times \frac{1}{\sqrt{2}} = 3 \text{ kN} \quad \textbf{Ans.}$$

Since the point *A* is supported on rollers, therefore reaction at *A* will be horizontal. The point *B* is supported on hinge, therefore the reaction at *B* ($R_B$) will be the resultant of vertical and horizontal components.

Consider the equilibrium of the point *B*, as shown in Fig. 9.50 (*d*).

We know that vertical component of reaction at *B*,

$$R_{BV} = P_{AB} + P_{BC} \sin 30°$$

$$= 3 + 3 \times 0.5 = 4.5 \text{ kN}$$

and horizontal component of the reaction at *B*,

$$R_{BH} = P_{BC} \cos 30° - P_{BD} = 3 \times 0.866 - 2.196 = 0.402 \text{ kN}$$

$$\therefore \text{ Reaction at } B, \quad R = \sqrt{(R_{BV})^2 + (R_{BH})^2} - \sqrt{(4.5)^2 + (0.402)^2}$$

$$= \sqrt{20.25 + 0.116} = \sqrt{20.41} = 4.518 \text{ kN} \quad \textbf{Ans.}$$

and its angle of inclination with the horizontal is given by

$$\tan \alpha = \frac{R_{BV}}{R_{BH}} = \frac{4.5}{0.402} = 11.19$$

$$\therefore \quad \alpha = \tan^{-1}(11.19) = 84.9° \quad \textbf{Ans.}$$

The magnitude and nature of forces in various members of the truss [Refer Fig. 9.49 (*b*)] are tabulated as below:

| *S. No.* | *Member of the truss* | *Magnitude of force in kN* | *Nature of force* |
|---|---|---|---|
| 1. | *CB* | 3 | Tensile |
| 2. | *CD* | 6 | Tensile |
| 3. | *AD* | 4.24 | Compressive |
| 4. | *BD* | 2.196 | Compressive |
| 5. | *AB* | 3 | Tensile |

**Example 9.16** *Find the forces in all members of a truss, as shown in Fig. 9.51 (a), which carries a horizontal load of 12 kN at point D and a vertical load of 18 kN at point C.*

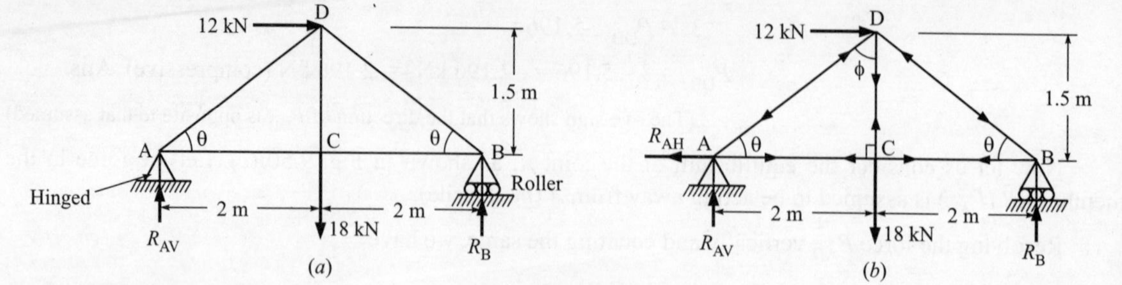


**Fig. 9.51**

***Solution*** Since the truss at *A* is hinged, therefore there will be horizontal reaction $R_{AH}$ and a vertical reaction $R_{AV}$, as shown in Fig. 9.51 (*b*). The reaction at this end *A* (*i.e.* $R_A$) will be the resultant of $R_{AV}$ and $R_{AH}$ which will be inclined at an angle with the horizontal. The end *B* is supported on rollers, therefore there will be a vertical reaction $R_B$.

From the right angled triangle *ACD*, we find that

$$\tan\theta = \frac{CD}{AC} = \frac{1.5}{2} = 0.75$$

$\therefore$ $$\theta = \tan^{-1}(0.75) = 36.87°$$

and $$\angle ADC = \phi = 90° - 36.87° = 53.13°$$

Now taking moments about *A*, we have

$$R_B \times 4 = 18 \times 2 + 12 \times 1.5 = 36 + 18 = 54$$

$\therefore$ $$R_B = \frac{54}{4} = 13.5 \text{ kN}$$

We know that $$R_{AV} + R_B = \text{Total vertical load} = 18 \text{ kN}$$

$\therefore$ $$R_{AV} = 18 - R_B = 18 \; 13.5 = 4.5 \text{ kN}$$

and $$R_{AH} = \text{Horizontal load at } D = 2 \text{ kN}$$

**Fig. 9.52**

Let us now consider the equilibrium of the joint *A*, as shown in Fig. 9.52 (*a*). Let the force in the member *AD* ($P_{AD}$) is assumed to be acting towards *A* (*i.e.* compressive) and force in the member *AC* ($P_{AC}$) is assumed to be acting away from *A* (*i.e.* tensile).

Resolving the forces vertically and equating the same, we have

$$P_{AD} \sin\theta = R_{AV}$$

$$P_{AD} \sin 36.87° = 4.5$$

$$\therefore \quad P_{AD} = \frac{4.5}{\sin 36.87°} = \frac{4.5}{0.6} = 7.5 \text{ kN (compressive)} \quad \textbf{Ans.}$$

Now resolving the forces horizontally and equating the same, we have

$$R_{AH} + P_{AD} \cos \theta = P_{AC}$$

$$12 + 7.5 \cos 36.87° = P_{AC}$$

$$12 + 7.5 \times 0.8 = P_{AC}$$

$$\therefore \quad P_{AC} = 12 + 7.5 \times 0.8 = 12 + 6 = 18 \text{ kN (tensile)} \quad \textbf{Ans.}$$

Now consider the equilibrium of the joint *C*, as shown in Fig. 9.52 (*b*). Let the force in the member *BC* ($P_{BC}$) and force in the member *CD* ($P_{CD}$) are assumed to be acting away from *C* (*i.e.* tensile).

For equilibrium, $P_{CD} = 18$ kN (tensile) **Ans.**

and $P_{BC} = P_{AC} = 18$ kN (tensile) **Ans.**

Now consider the equilibrium of the joint *B*, as shown in Fig. 9.52 (*c*). Let the force in the member *BD* ($P_{BD}$) is assumed to be acting towards *B* (*i.e.* compressive).

Resolving the force vertically and equating the same, we have

$$P_{BD} \sin \theta = R_B$$

$$P_{BD} = \frac{R_B}{\sin \theta} = \frac{13.5}{\sin 36.87°} = \frac{13.5}{0.6} = 22.5 \text{ kN (compressive)} \quad \textbf{Ans.}$$

Now the magnitude and nature of forces in the various members of the truss [Refer Fig. 9.51 (*b*)] are tabulated as below:

| *S. No.* | *Member of the truss* | *Magnitude of force in kN* | *Nature of force* |
|---|---|---|---|
| 1. | *AD* | 7.5 | Compressive |
| 2. | *AC* | 18 | Tensile |
| 3. | *CD* | 18 | Tensile |
| 4. | *BC* | 18 | Tensile |
| 5. | *BD* | 22.5 | Compressive |

**Example 9.17** *A roof truss with wind loads acting on it, is shown in Fig. 9.53 (a). Determine the reaction at the support and forces in all the members.*

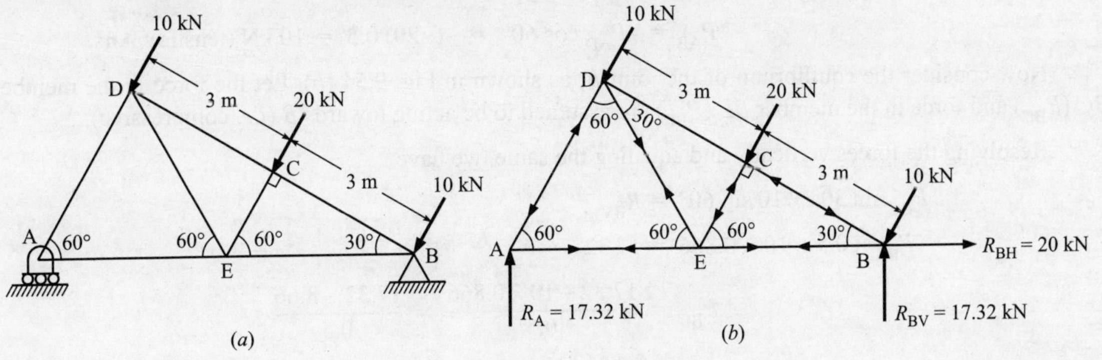


**Fig. 9.53**

***Solution*** Since the truss at $A$ is on rollers, therefore there will a vertical reaction $R_A$. The end $B$ of the truss is hinged, therefore there will be a vertical reaction $R_{BV}$ and a horizontal reaction $R_{BH}$, as shown in Fig. 9.53 (*b*). The reaction at this end (*i.e.* $R_B$) will be the resultant of $R_{BV}$ and $R_{BH}$, which will be inclined at an angle with the horizontal.

From right angled triangle $ADB$,

$$\frac{BD}{AB} = \cos 30° \quad \text{or} \quad AB = \frac{BD}{\cos 30°} = \frac{6}{0.866} = 6.928 \text{ m}$$

and

$$\frac{BC}{BE} = \cos 30° \quad \text{or} \quad BC = \frac{BE}{\cos 30°} = \frac{3}{0.866} = 3.464 \text{ m}$$

Taking moments about $A$, we have

$$R_{BV} \times AB = 10 \times BD + 20 \times CD$$

$$R_{BV} \times 6.928 = 10 \times 6 + 20 \times 3 = 120$$

$$\therefore \quad R_{BV} = \frac{120}{6.928} = 17.32 \text{ kN}$$

We know that horizontal component of reaction at $B$,

$$R_{BH} = \text{Sum of horizontal components of loads at } B, C \text{ and } D$$

$$= (10 + 20 + 10) \cos 60° = 40 \times 0.5 = 20 \text{ kN}$$

The reaction at $A$ (*i.e.* $R_A$) is equal to the algebraic sum of vertical components of loads at $B$, $C$ and $D$.

$$\therefore \quad R_A = (10 + 20 + 10) \sin 60° - R_{BV}$$

$$= 40 \times 0.866 - 17.32 = 17.32 \text{ kN}$$

Now let us consider the equilibrium of the joint $A$, as shown in Fig. 9.54 (*a*). Let the force in the member $AD$ ($P_{AB}$) and force in the member $AE$ ($P_{AE}$) are assumed to be acting away from $A$ (*i.e.* tensile).

Resolving the force $P_{AD}$ vertically and equating the same, we have

$$P_{AD} \sin 60° = -R_A$$

$$\therefore \quad P_{AD} = \frac{-R_A}{\sin 60°} = -\frac{17.32}{0.866} = -20 \text{ kN} = 20 \text{ kN (compressive)} \quad \textbf{Ans.}$$

...(–ve sign shows that the direction of $P_{AD}$ is opposite to that assumed)

Resolving the force $P_{AD}$ horizontally and equating the same, we have

$$P_{AE} = -P_{AD} \cos 60° = -(-20)\, 0.5 = 10 \text{ kN (tensile)} \quad \textbf{Ans.}$$

Now consider the equilibrium of the joint $B$, as shown in Fig. 9.54 (*b*). Let the force in the member $BC$ ($P_{BC}$) and force in the member $BE$ ($P_{BE}$) are assumed to be acting towards $B$ (*i.e.* compressive).

Resolving the forces vertically and equating the same, we have

$$P_{BC} \sin 30° + 10 \sin 60° = R_{BV}$$

$$P_{BC} \times 0.5 + 10 \times 0.866 = 17.32$$

$$\therefore \quad P_{BC} = \frac{17.32 - 10 \times 0.866}{0.5} = \frac{17.32 - 8.66}{0.5}$$

$$= 17.32 \text{ kN (compressive)} \quad \textbf{Ans.}$$

Now resolving the forces horizontally and equating the same, we have

$$P_{BC}\cos 30° + P_{BE} + R_{BH} = 10\cos 60°$$

$$17.32 \times 0.866 + P_{BE} + 20 = 10 \times 0.5$$

$$15 + P_{BE} + 20 = 5$$

$$\therefore \quad P_{BE} = 5 - 15 - 20 = -30 \text{ kN} = 30 \text{ kN (tensile)} \quad \textbf{Ans.}$$

...(–ve sign shows that the direction of $P_{BE}$ is opposite to that assumed)

Now consider the equilibrium of the joint *C*, as shown in Fig. 9.54 (*c*). Let the force in the member *CE* ($P_{CE}$) and force in the member *CD* ($P_{CD}$) are assumed to be acting towards *C* (*i.e.* compressive).

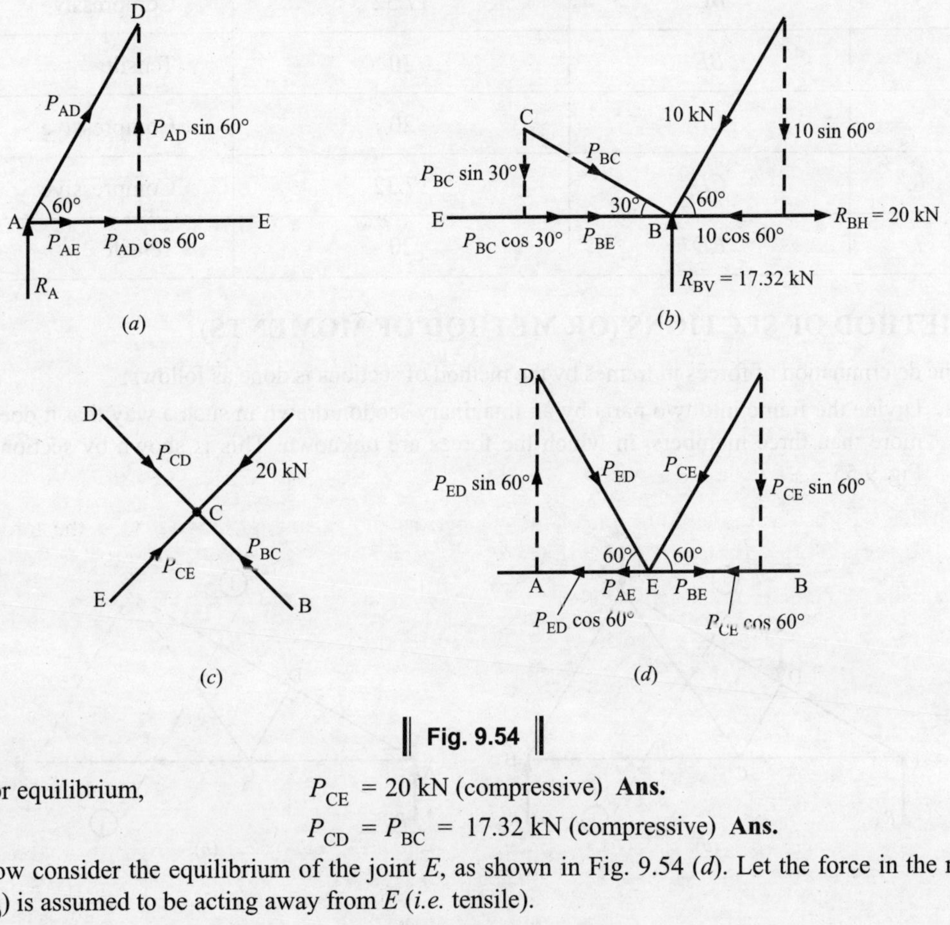


**Fig. 9.54**

For equilibrium, $P_{CE} = 20$ kN (compressive) **Ans.**

and $P_{CD} = P_{BC} = 17.32$ kN (compressive) **Ans.**

Now consider the equilibrium of the joint *E*, as shown in Fig. 9.54 (*d*). Let the force in the member *ED* ($P_{ED}$) is assumed to be acting away from *E* (*i.e.* tensile).

Resolving the forces vertically and equating the same, we have

$$P_{ED}\sin 60° = P_{CE}\sin 60°$$

or $$P_{ED} = P_{CE} = 20 \text{ kN (tensile)} \quad \textbf{Ans.}$$

***Check***: Resolving the forces horizontally and equating the same, we have

$$P_{ED}\cos 60° + P_{AE} + P_{CE}\cos 60° = P_{BE}$$

$$P_{ED} \times 0.5 + 10 + 20 \times 0.5 = 30$$

or $$P_{ED} \times 0.5 = 30 - 10 - 20 \times 0.5 = 10$$

$$\therefore \quad P_{ED} = \frac{10}{0.5} = 20 \text{ kN}$$

The magnitude and nature of forces in the various members of the truss [Refer Fig. 9.53 (*b*)] are tabulated as below:

| *S. No.* | *Member of the truss* | *Magnitude of force in kN* | *Nature of force* |
|---|---|---|---|
| 1. | *AD* | 20 | Compressive |
| 2. | *AE* | 10 | Tensile |
| 3. | *BC* | 17.32 | Compressive |
| 4. | *BE* | 30 | Tensile |
| 5. | *CE* | 20 | Compressive |
| 6. | *CD* | 17.32 | Compressive |
| 7. | *ED* | 20 | Tensile |

## 9.8 METHOD OF SECTIONS (OR METHOD OF MOMENTS)

The determination of forces in frames by the method of sections is done as follows:

1. Divide the frame into two parts by an imaginary section drawn in such a way that it does not cut more than three members, in which the forces are unknown. This is shown by section (1-1) in Fig. 9.55.

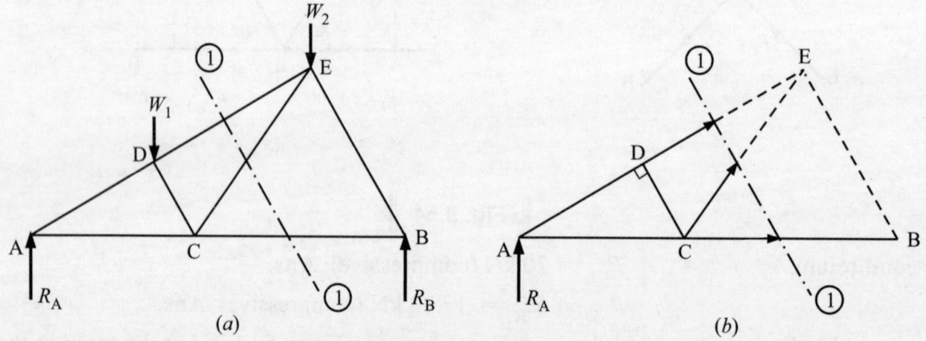


**Fig. 9.55**

2. Now apply the conditions of equilibrium (*i.e.* $\Sigma V = 0$; $\Sigma H = 0$ and $\Sigma M = 0$) to one of the parts which is in equilibrium under the action of external loads and the forces exerted on it by the members cut by the section.

The main difference in the method of joints and method of section is that former involves the analysis of concurrent forces system and the latter deals with analysis of non-concurrent forces system.

It may be noted that the method of section is mostly used when the forces in a few members of a large truss are required to be determined.

**Example 9.18** *A triangular shaped truss ABC, as shown in Fig. 9.56, is simply supported at A and B. Find the forces in the members AB, AC and BC by using the method of sections.*

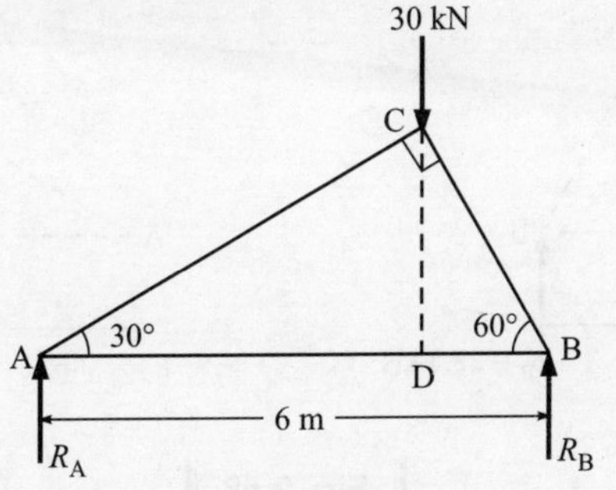


**Fig. 9.56**

***Solution*** First of all let us find the reactions $R_A$ and $R_B$ at supports $A$ and $B$, respectively, as shown in Fig. 9.57. From the geometry of the figure, we find that

$$\angle ACB = 90°$$

Draw $CD$ perpendicular on $AB$. We know that

$$AC = AB \cos 30° = 6 \times \frac{\sqrt{3}}{2} = 3\sqrt{3} \text{ m}$$

$$AD = AC \cos 30° = 3\sqrt{3} \times \frac{\sqrt{3}}{2} = 4.5 \text{ m}$$

$$BC = AB \sin 30° = 6 \times 0.5 = 3 \text{ m}$$

$$CD = AC \sin 30° = 3\sqrt{3} \times 0.5 = 1.5\sqrt{3} \text{ m}$$

$$BD = AB - AD = 6 - 4.5 = 1.5 \text{ m}$$

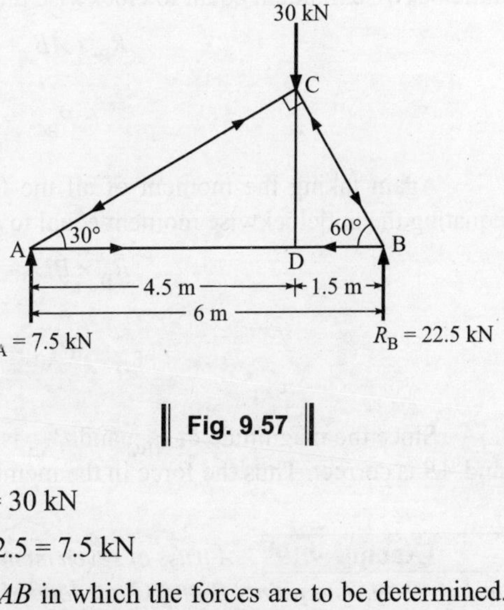


**Fig. 9.57**

Now taking moments about $A$, we have

$$R_B \times 6 = 30 \times AD = 30 \times 4.5$$

$$\therefore \quad R_B = \frac{30 \times 4.5}{6} = 22.5 \text{ kN}$$

We know that $\quad R_A + R_B = \text{Total load at } C = 30 \text{ kN}$

$$\therefore \quad R_A = 30 - R_B = 30 - 22.5 = 7.5 \text{ kN}$$

Now draw a section (1-1) cutting the members $AC$ and $AB$ in which the forces are to be determined. Consider the equilibrium of the left part of the truss, as shown in Fig. 9.58 (*a*). Let the forces $P_{AC}$ and $P_{AB}$ in the members $AC$ and $AB$, respectively, have directions as shown in the figure.

Now taking moment of all the forces acting on the left part of the truss about point $B$, and equating anticlockwise moment equal to clockwise moment, we have

$$P_{AC} \times BC = R_A \times AB$$

$$\therefore \quad P_{AC} = \frac{R_A \times AB}{BC} = \frac{7.5 \times 6}{3} = 15 \text{ kN (compressive)}$$ **Ans.**

Again taking moment of all the forces acting on the left part of the truss about point $C$ and equating anticlockwise moment equal to clockwise moment, we have

$$P_{AB} \times CD = R_A \times AD$$

$$P_{AB} = \frac{R_A \times AD}{CD} = \frac{7.5 \times 4.5}{1.5\sqrt{3}} = 12.99 \text{ kN (tensile)}$$ **Ans.**

Since the magnitude of $P_{AC}$ and $P_{AB}$ is positive, therefore the directions assumed for the members *AE* and *AB* is correct. Thus the force in the member *AC* is compressive and force in the member *AB* is tensile.

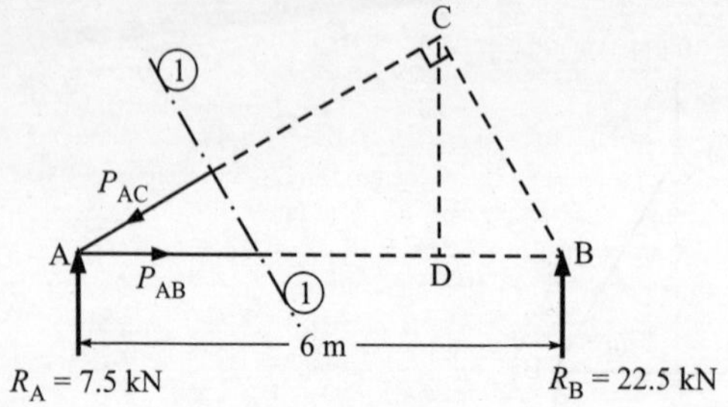


(*a*)

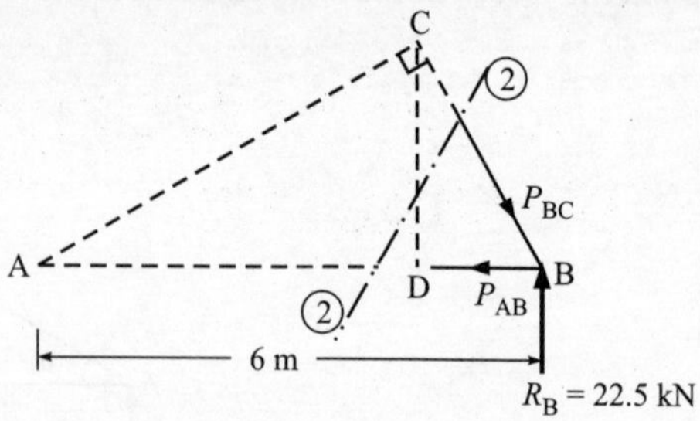


(*b*)

**Fig. 9.58**

Now draw a section (2-2) cutting the members *BC* and *AB* in which the forces are to be determined. Consider the equilibrium of the right part of the truss, as shown in Fig. 9.58 (*b*).

Let the forces $P_{BC}$ and $P_{AB}$ in the members *BC* and *AB*, respectively, have directions as shown in the figure.

Now taking moment of all the forces acting on the right part of the truss about point *A* and equating anticlockwise moment equal to clockwise moment, we have

$$R_B \times AB = P_{BC} \times AC$$

$$\therefore \quad P_{BC} = \frac{R_B \times AB}{AC} = \frac{22.5 \times 6}{3\sqrt{3}} = 25.98 \text{ kN (compressive)} \quad \textbf{Ans.}$$

Again taking the moment of all the forces acting on the right part of the truss about point *C* and equating the anticlockwise moment equal to clockwise moment, we have

$$R_B \times BD = P_{AB} \times CD$$

$$P_{AB} = \frac{R_B \times BD}{CD} = \frac{22.5 \times 1.5}{1.5\sqrt{3}} = 12.99 \text{ kN (tensile)} \quad \textbf{Ans.}$$

...(same as already calculated)

Since the magnitude of $P_{BC}$ and $P_{AB}$ is positive, therefore the directions assumed for the members *BC* and *AB* is correct. Thus the force in the member *BC* is compressive and force in the member *AB* is tensile.

**Example 9.19** *A truss of seven members each of 3 m length is freely supported at the ends, as shown in Fig. 9.59. The points B and C are loaded with 4 kN and 8 kN, respectively. Determine the magnitude of forces and nature in all the members of the truss. Use method of sections.*

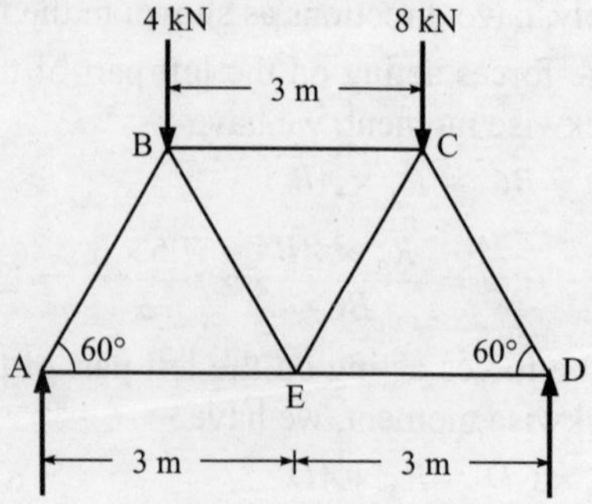


**Fig. 9.59**

***Solution*** First of all, let us find the reactions $R_A$ and $R_D$ at supports $A$ and $D$, respectively, as shown in Fig. 9.60. From the geometry of the figure, we find that

$$AF = AB \cos 60° = 3 \cos 60° = 3 \times 0.5 = 1.5 \text{ m}$$

and
$$AG = AE + EG = 3 + EC \cos 60° = 3 + 3 \times 0.5 = 4.5 \text{ m}$$

Now taking moments about $A$, we have

$$R_D \times 6 = 4 \times AF + 8 \times AG = 4 \times 1.5 + 8 \times 4.5 = 6 + 36 = 42$$

$$\therefore \quad R_D = \frac{42}{6} = 7 \text{ kN}$$

We know that
$$R_A + R_D = 4 + 8 = 12 \text{ kN}$$

$$\therefore \quad R_A = 12 - R_D = 12 - 7 = 5 \text{ kN}$$

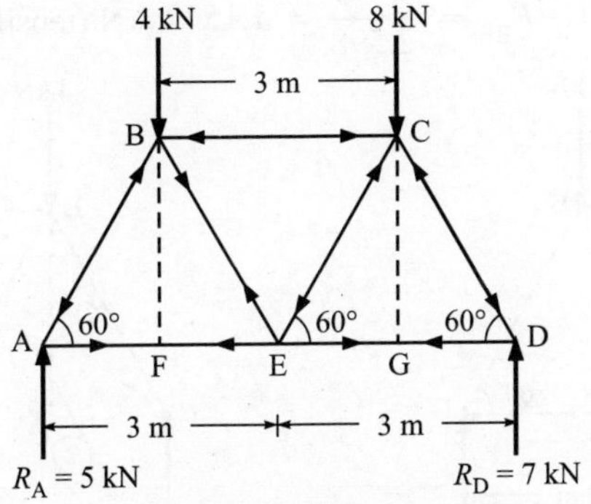


**Fig. 9.60**

Now draw a section (1-1) cutting the members $AB$ and $AE$, as shown in Fig. 9.61 (*a*). Consider the equilibrium of the left part of the truss. Let the forces $P_{AB}$ and $P_{AE}$ in the members $AB$ and $AE$, respectively, have directions as shown in the figure.

Taking moment of the forces acting on the left part of the truss about point $E$ and equating anticlockwise moment equal to clockwise moment, we have

$$P_{AB} \times EF = R_A \times AE$$

$$P_{AB} \times AE \sin 60° = 5 \times 3 = 15$$

$$P_{AB} = \frac{15}{AE \sin 60°} = \frac{15}{3 \times 0.866} = 5.774 \text{ (compressive)} \textbf{ Ans.}$$

Now taking moment of the forces acting on the left part of the truss about point $B$ and equating anticlockwise moment equal to clockwise moment, we have

$$P_{AE} \times BG = R_A \times AG$$

$$P_{AE} \times 3 \sin 60° = 5 \times 1.5$$

$$P_{AE} \times 3 \times 0.866 = 7.5$$

$$\therefore \quad P_{AE} = \frac{7.5}{3 \times 0.866} = 2.887 \text{ kN (tensile)} \textbf{ Ans.}$$

Now draw a section (2-2) cutting the members $BC$, $BE$ and $AE$, as shown in Fig. 9.61 (*b*). Consider the left part of the truss. Let the forces $P_{BC}$ and $P_{BE}$ in the members $BC$ and $BE$, respectively, have directions as shown in the figure. Taking moment of the forces acting on the left part of the truss about point $E$ and equating anticlockwise moment equal to clockwise moment, we have

$$P_{BC} \times BG + 4 \times GE = R_A \times AE$$

$$P_{BC} \times AB \sin 60° + 4 \times 1.5 = 5 \times 3$$
$$P_{BC} \times 3 \times 0.866 + 6 = 15$$
$$\therefore \quad P_{BC} = \frac{15-6}{3 \times 0.866} = 3.464 \text{ kN (compressive)} \textbf{ Ans.}$$

Now taking moment of the forces acting on the left part of the truss about point $A$ and equating anticlockwise moment equal to clockwise moment, we have

$$P_{BC} \times BG = 4 \times AG + P_{BE} \times AH$$

or
$$P_{BC} \times AB \sin 60° = 4 \times 1.5 + P_{BE} \times AB \sin 60° \quad ...(\because BG = AH = AB \sin 60°)$$
$$3.464 \times 3 \times 0.866 = 6 + P_{BE} \times 3 \times 0.866$$
$$9 = 6 + 2.598\, P_{BE}$$
$$\therefore \quad P_{BE} = \frac{9-6}{2.598} = 1.1547 \text{ kN (tensile)} \textbf{ Ans.}$$

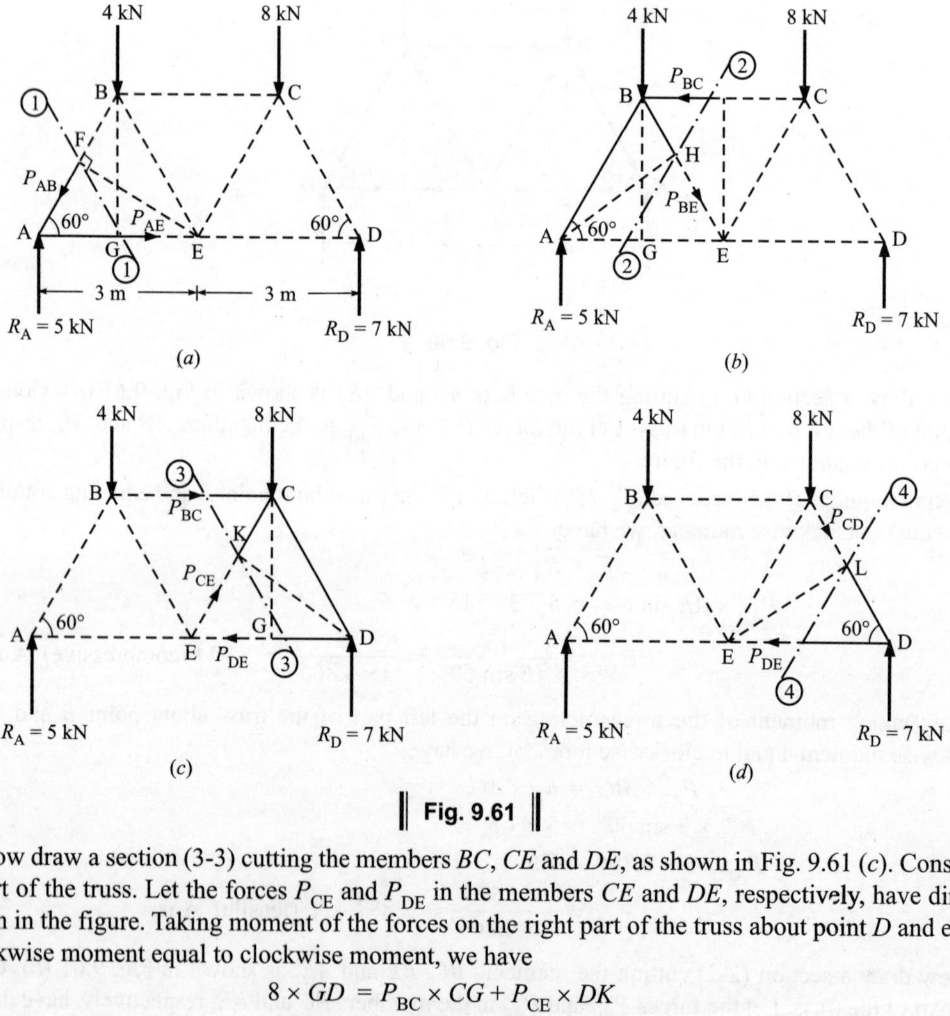


**Fig. 9.61**

Now draw a section (3-3) cutting the members $BC$, $CE$ and $DE$, as shown in Fig. 9.61 (c). Consider the right part of the truss. Let the forces $P_{CE}$ and $P_{DE}$ in the members $CE$ and $DE$, respectively, have directions as shown in the figure. Taking moment of the forces on the right part of the truss about point $D$ and equating anticlockwise moment equal to clockwise moment, we have

$$8 \times GD = P_{BC} \times CG + P_{CE} \times DK$$
$$8 \times 1.5 = 3.464 \times 3 \sin 60° + P_{CE} \times 3 \sin 60°$$
$$12 = 3.464 \times 3 \times 0.866 + P_{CE} \times 3 \times 0.866 = 9 + 2.598\, P_{CE}$$
$$\therefore \quad P_{CE} = \frac{12-9}{2.598} = 1.1547 \text{ kN (compressive)} \textbf{ Ans.}$$

Now taking the moment of the forces on the right part of the truss about point *C* and equating anticlockwise moment equal to clockwise moment, we have

$$R_D \times DG = P_{DE} \times GC$$

$$7 \times 1.5 = P_{DE} \times 3 \sin 60° = P_{DE} \times 3 \times 0.866 \qquad ...(\because \; GC = EC \sin 60°)$$

$$10.5 = 2.598 P_{DE}$$

$$\therefore \qquad P_{DE} = \frac{10.5}{2.598} = 4.04 \text{ kN (tensile)} \quad \textbf{Ans.}$$

Now draw a section (4-4) cutting the members *CD* and *DE*, as shown in Fig. 9.61 (*d*). Consider the right part of the truss. Let the forces $P_{CD}$ and $P_{DE}$ in the members *CD* and *DE*, respectively, have directions as shown in the figure. Taking moment of the forces acting on the right part of the truss about the point *E* and equating anticlockwise moment equal to clockwise moment, we have

$$P_{CD} \times EL = R_D \times DE$$

$$P_{CD} \times DE \sin 60° = 7 \times 3$$

$$P_{CD} \times 3 \times 0.866 = 21$$

$$\therefore \qquad P_{CD} = \frac{21}{3 \times 0.866} = 8.08 \text{ kN (compressive)} \quad \textbf{Ans.}$$

The magnitude and nature of forces in the various members of the truss (Refer Fig. 9.60) are tabulated as below:

| *S. No.* | *Member of the truss* | *Magnitude of force in kN* | *Nature of force* |
|---|---|---|---|
| 1. | *AB* | 5.774 | Compressive |
| 2. | *AE* | 2.887 | Tensile |
| 3. | *BE* | 1.1547 | Tensile |
| 4. | *BC* | 3.464 | Compressive |
| 5. | *CE* | 1.1547 | Compressive |
| 6. | *CD* | 8.08 | Compressive |
| 7. | *DE* | 4.04 | Tensile |

**Example 9.20** *Determine the forces in members DC, EC and EB of the truss, as shown in Fig. 9.62, using method of sections.*

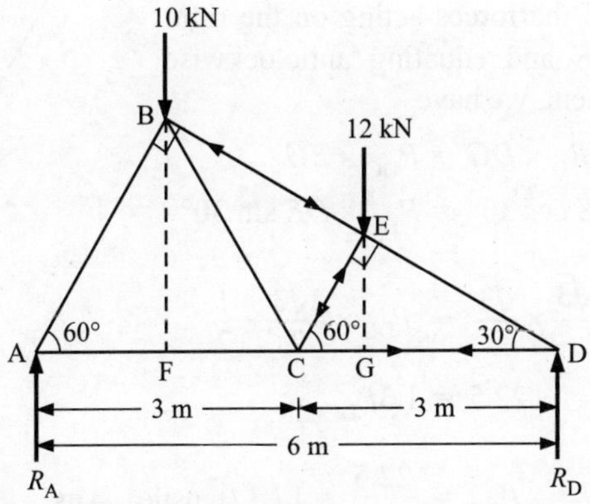


**Fig. 9.62**

***Solution*** First of all, let us find the reactions $R_A$ and $R_D$ at the supports $A$ and $D$, respectively. From a right angled triangle $ABD$,

$$\frac{BD}{AD} = \cos 30°$$

or $$BD = AD \cos 30° = 6 \times \frac{\sqrt{3}}{2} = 5.196 \text{ m}$$

and $$AB = AD \sin 30°$$
$$= 6 \times 0.5 = 3 \text{ m}$$

Now from triangle $ABF$, $$AF = AB \cos 60°$$
$$= 3 \times 0.5 = 1.5 \text{ m}$$

and from right angled triangle $CDE$, $$ED = CD \cos 30° = 3 \times \frac{\sqrt{3}}{2} \text{ m}$$

$$GD = ED \cos 30° = 3 \times \frac{\sqrt{3}}{2} \times \frac{\sqrt{3}}{2} = 2.25 \text{ m}$$

$\therefore$ $$AG = AD - GD = 6 - 2.25 = 3.75 \text{ m}$$

Now taking moments about $A$,

$$R_D \times AD = 12 \times AG + 10 \times AF$$
$$R_D \times 6 = 12 \times 3.75 + 10 \times 1.5 = 60$$

$\therefore$ $$R_D = \frac{60}{6} = 10 \text{ kN}$$

We know that $$R_A + R_D = 10 + 12 = 22 \text{ kN}$$

$\therefore$ $$R_A = 22 - R_D = 22 - 10 = 12 \text{ kN}$$

Now draw a section (1-1) cutting the members $DC$, $EC$ and $EB$ in which the forces are to be determined, as shown in Fig. 9.63.

Consider the equilibrium of the right part of the truss because it is smaller than the left part. Let the forces $P_{DC}$, $P_{EC}$ and $P_{EB}$ in the members $DC$, $EC$ and $EB$, respectively, have directions as shown in Fig. 9.63.

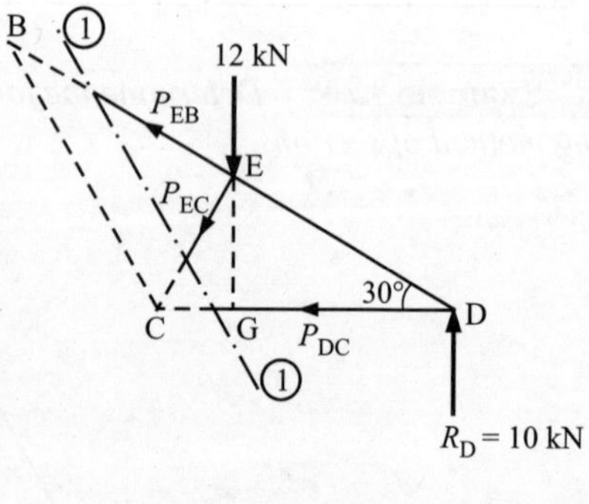


**Fig. 9.63**

Now taking moment of all the forces acting on the right part of the truss about point $E$ and equating anticlockwise moment equal to clockwise moment, we have

$$R_D \times DG = P_{DC} \times EG$$
$$R_D \times DE \cos 30° = P_{DC} \times DE \sin 30°$$

$$10 \times \frac{3 \times \sqrt{3}}{2} \times \frac{\sqrt{3}}{2} = P_{DC} \times \frac{3\sqrt{3}}{2} \times \frac{1}{2}$$

$$22.5 = 1.3 P_{DC}$$

$\therefore$ $$P_{DC} = \frac{22.5}{1.3} = 17.3 \text{ (tensile)} \textbf{ Ans.}$$

Now taking moment of all the forces acting on the right part of the truss about point $D$ and equating anticlockwise moment equal to clockwise moment, we have

$$12 \times DG + P_{EC} \times DE = 0$$

$$12 \times DE \cos 30° + P_{EC} \times DE = 0$$

$$12 \times \frac{3 \times \sqrt{3}}{2} \times \frac{\sqrt{3}}{2} + P_{EC} \times \frac{3\sqrt{3}}{2} = 0$$

$$\therefore \quad P_{EC} = -\frac{12 \times \sqrt{3}}{2} = -10.392 \text{ kN} = 10.392 \text{ kN (compressive)} \textbf{ Ans.}$$

...(–ve sign shows that the direction of $P_{EC}$ is opposite to that assumed)

Now taking moments of all the forces acting on the right part of the truss about point $C$ and equating anticlockwise moment equal to clockwise moment, we have

$$R_D \times CD + P_{EB} \times CE = 12 \times CG$$

$$10 \times 3 + P_{EB} \times CD \sin 30° = 12\,(CD - DG) = 12\,(3 - DE \cos 30°)$$

$$30 + P_{EB} \times 3 \times 0.5 = 12\left(3 - \frac{3\sqrt{3}}{2} \times \frac{\sqrt{3}}{2}\right) = 9$$

or $$P_{EB} = \frac{9 - 30}{3 \times 0.5} = -14 \text{ kN} = 14 \text{ kN (compressive)} \textbf{ Ans.}$$

...(–ve sign shows that the direction of $P_{ED}$ is opposite to that assumed)

**Example 9.21** *A truss of 4.5 m span is loaded, as shown in Fig. 9.64. Determine the force in the members CD, GD and GH of the truss. Use method of sections.*

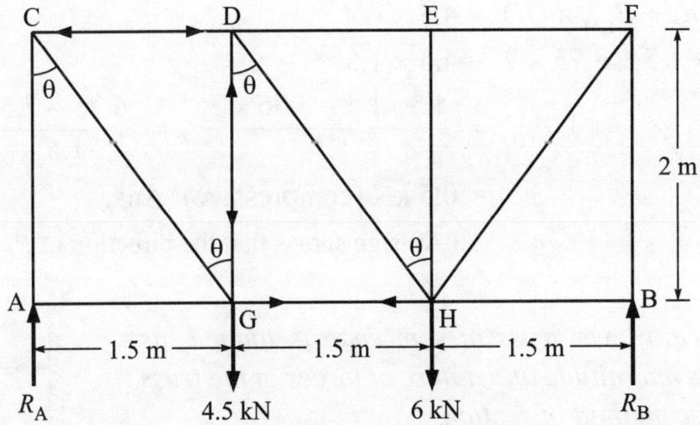


**Fig. 9.64**

***Solution*** First of all, let us find the reactions $R_A$ and $R_B$ at $A$ and $B$, respectively. Taking moments about $A$ (Refer Fig. 9.64),

$$R_B \times 4.5 = 6 \times 3 + 4.5 \times 1.5 = 18 + 6.75 = 24.75$$

$$\therefore \quad R_B = \frac{24.75}{4.5} = 5.5 \text{ kN}$$

We know that $$R_A + R_B = 4.5 + 6 = 10.5 \text{ kN}$$

$$\therefore \quad R_A = 10.5 - R_B = 10.5 - 5.5 = 5 \text{ kN}$$

Now draw a section (1-1) cutting the members *CD*, *DG* and *GH* in which the forces are to be determined. Consider the equilibrium of the left part of the truss, as shown in Fig. 9.65. Let the forces $P_{CD}$, $P_{GD}$ and $P_{GH}$ in the members *CD*, *DG* and *GH*, respectively, have directions as shown in the figure.

Taking moment of all the forces acting on the left part of the truss about point *D* and equating anticlockwise moment equal to clockwise moment, we have

$$R_A \times CD = P_{GH} \times 2$$
$$5 \times 1.5 = P_{GH} \times 2$$
$$\therefore \quad P_{GH} = \frac{5 \times 1.5}{2}$$
$$= 3.75 \text{ kN (tensile)} \quad \textbf{Ans.}$$

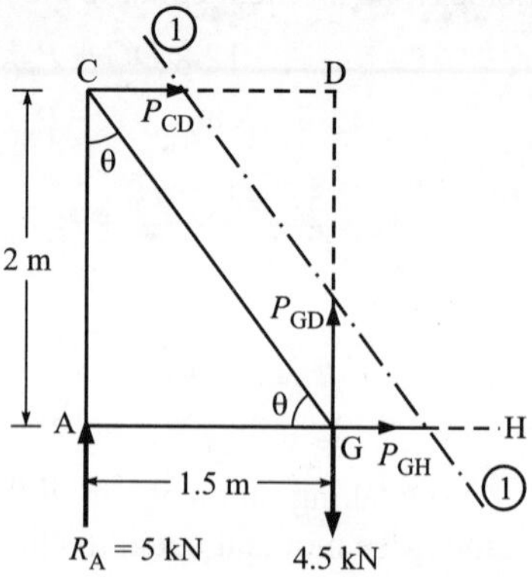


**Fig. 9.65**

Again taking moment of all the forces acting on the left part of the truss about point *G* and equating anticlockwise moment equal to clockwise moment, we have

$$R_A \times AG = -P_{CD} \times DG$$
$$5 \times 1.5 = -P_{CD} \times 2$$
$$\therefore \quad P_{CD} = -\frac{5 \times 1.5}{2} = -3.75 \text{ kN}$$
$$= 3.75 \text{ kN (compressive)} \quad \textbf{Ans.}$$

...(–ve sign shows that the direction of $P_{CD}$ is opposite to that assumed)

Now taking moment of all the forces acting on the left part of the truss about point *C* and equating anticlockwise moment equal to clockwise moment, we have

$$P_{GD} \times DC + P_{GH} \times GD = 4.5 \times GA$$
$$P_{GD} \times 1.5 + 3.75 \times 2 = 4.5 \times 1.5$$
$$\therefore \quad P_{GD} = \frac{4.5 \times 1.5 - 3.75 \times 2}{1.5} = \frac{6.75 - 7.5}{1.5} = -0.5 \text{ kN}$$
$$= 0.5 \text{ kN (compressive)} \quad \textbf{Ans.}$$

...(–ve sign shows that the direction of $P_{GD}$ is opposite to that assumed)

**Example 9.22** *A cantilever truss of 4 m span is loaded, as shown in Fig. 9.66. Find the magnitude and nature of forces of the truss and tabulate the results. Use method of sections.*

***Solution*** First of all, draw a section (1-1) cutting the member *AB* and *AD* in which the forces are to be determined. Consider the equilibrium of the right part of the truss, as shown in Fig. 9.67 (*a*). Let the force $P_{AB}$ and $P_{AD}$ in the members *AB* and *AD*, respectively, have directions as shown in the figure.

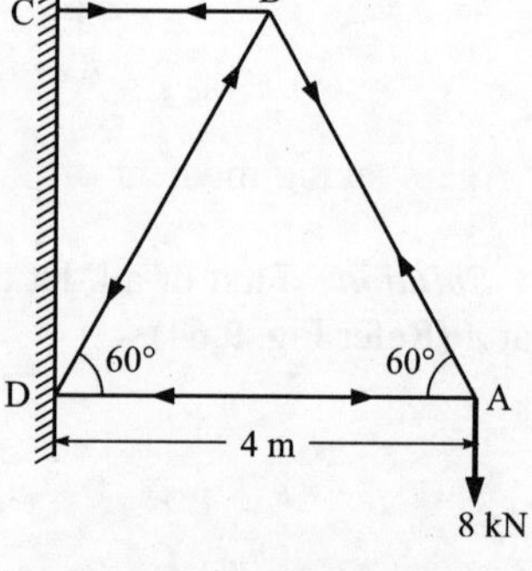


**Fig. 9.66**

Now taking moment of the forces acting on the right part of the truss about point *D* and equating anticlockwise moment equal to clockwise moment, we have

$$P_{AB} \times DE = 8 \times 4 = 32$$

$$\therefore \quad P_{AB} = \frac{32}{DE} = \frac{32}{AD \sin 60°} = \frac{32}{4 \times 0.866} = 9.238 \text{ kN (tensile) } \textbf{Ans.}$$

Again taking moment of the forces on the right part of the truss about point $B$ and equating anticlockwise moment equal to clockwise moment, we have

$$P_{AD} \times BF = 8 \times AF$$

$$P_{AD} \times 4 \sin 60° = 8 \times 2 = 16 \qquad ...(\because BF = CD = BD \sin 60°)$$

$$\therefore \quad P_{AD} = \frac{16}{4 \sin 60°} = \frac{4}{0.866} = 4.619 \text{ kN (compressive) } \textbf{Ans.}$$

Since the magnitude of $P_{AB}$ and $P_{AD}$ is positive, therefore the directions assumed for the members $AB$ and $AD$ is correct. Thus, the force in the member $AB$ is tensile and force in the member $AD$ is compressive.

Now draw a section (2-2) cutting the members $BC$, $BD$ and $AD$ in which the forces are to be determined. Consider the equilibrium of the right part of the truss, as shown in Fig. 9.67 (*b*). Let the forces $P_{BC}$ and $P_{BD}$ in the members $BC$ and $BD$, respectively, have directions as shown in the figure.

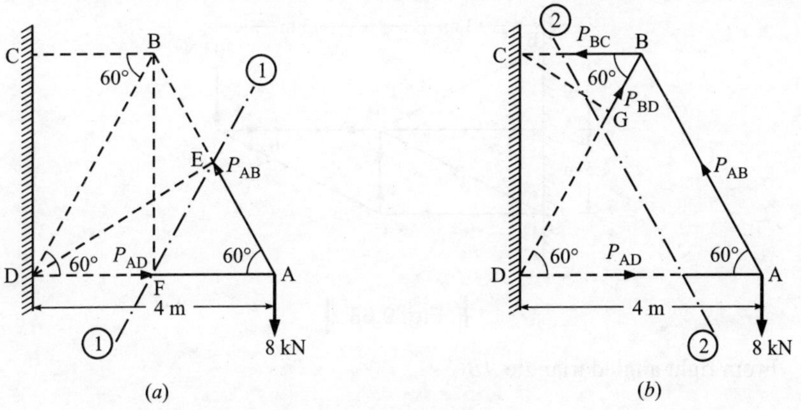


**Fig. 9.67**

Taking moment of the forces acting on the right part of the truss about point $D$ and equating anticlockwise moment equal to clockwise moment, we have

$$P_{BC} \times CD = 8 \times 4$$

$$P_{BC} \times 4 \sin 60° = 32 \qquad ...(\because CD = BD \sin 60°)$$

$$\therefore \quad P_{BC} = \frac{32}{4 \sin 60°} = \frac{8}{0.866} = 9.238 \text{ kN (tensile) } \textbf{Ans.}$$

Again taking moment of the forces acting on the right part of the truss about point $C$ and equating anticlockwise moment equal to clockwise moment, we have

$$P_{BD} \times CG + P_{AD} \times CD = 8 \times 4$$

$$P_{BD} \times BC \sin 60° + P_{AD} \times BD \sin 60° = 32 \qquad ...[\because BC = DF = 2 \text{ m}]$$

$$P_{BD} \times 2 \times 0.866 + 4.619 \times 4 \times 0.866 = 32$$

$$P_{BD} \times 1.732 + 16 = 32$$

$$\therefore \quad P_{BD} = \frac{32 - 16}{1.732} = 9.238 \text{ kN (compressive) } \textbf{Ans.}$$

Since the magnitude of $P_{BC}$ and $P_{BD}$ is positive, therefore the directions assumed for the members *BC* and *BD* are correct. Thus, the force in the member *BC* in tensile and force in the member *BD* is compressive.

The magnitude and nature of forces in the various member of the truss (Refer Fig. 9.66) are tabulated as below:

| *S. No.* | *Member of the truss* | *Magnitude of force in kN* | *Nature of force* |
|---|---|---|---|
| 1. | *AB* | 9.238 | Tensile |
| 2. | *AD* | 4.619 | Compressive |
| 3. | *BC* | 9.238 | Tensile |
| 4. | *BD* | 9.238 | Compressive |

**Example 9.23** *Find the magnitude and nature of force in each of the members of the truss, as shown in Fig. 9.68. Tabulate all the results. Assume all joints to be perfectly smooth hinges.*

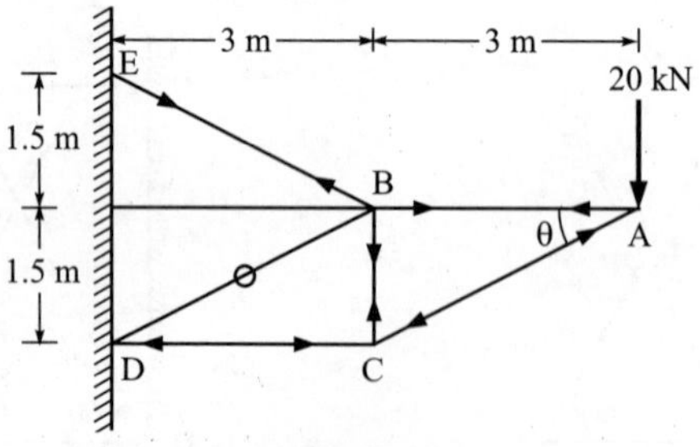


**Fig. 9.68**

***Solution*** From right angled triangle *ABC*,

$$\tan\theta = \frac{BC}{AB} = \frac{1.5}{3} = 0.5$$

$$\therefore \quad \theta = \tan^{-1}(0.5) = 26.56°$$

Now let us consider the equilibrium of the joint *A*, as shown in Fig. 9.69 (*a*). Let the force in the member *AB* ($P_{AB}$) and force in the member *AC* ($P_{AC}$) is assumed to be acting away from *A*.

Resolving the forces vertically and equating the same, we have

$$P_{AC}\sin\theta = -20$$

$$\therefore \quad P_{AC} = \frac{-20}{\sin\theta} = \frac{-20}{\sin 26.56°} = \frac{-20}{0.4471} = -44.73 \text{ kN}$$

$$= 44.73 \text{ kN (compressive)} \textbf{ Ans.}$$

...(–ve sign shows that the direction of $P_{AC}$ is opposite to that assumed)

Resolving the forces horizontally and equating the same, we have

$$P_{AB} = -P_{AC}\cos\theta = -(-44.73\cos\theta) = 4.73\cos 26.56°$$

$$= 44.73 \times 0.8944 = 40 \text{ kN (tensile)} \textbf{ Ans.}$$

In order to find the force in members *BC* and *CD*, we shall use the method of sections.

Draw a section (1-1) cutting the members *AB*, *BC* and *CD*, as shown in Fig. 9.69 (*b*).

Consider the equilibrium of right part of the truss. Let the forces $P_{BC}$ and $P_{CD}$ in the members *BC* and *CD*, respectively, have directions as shown in the figure.

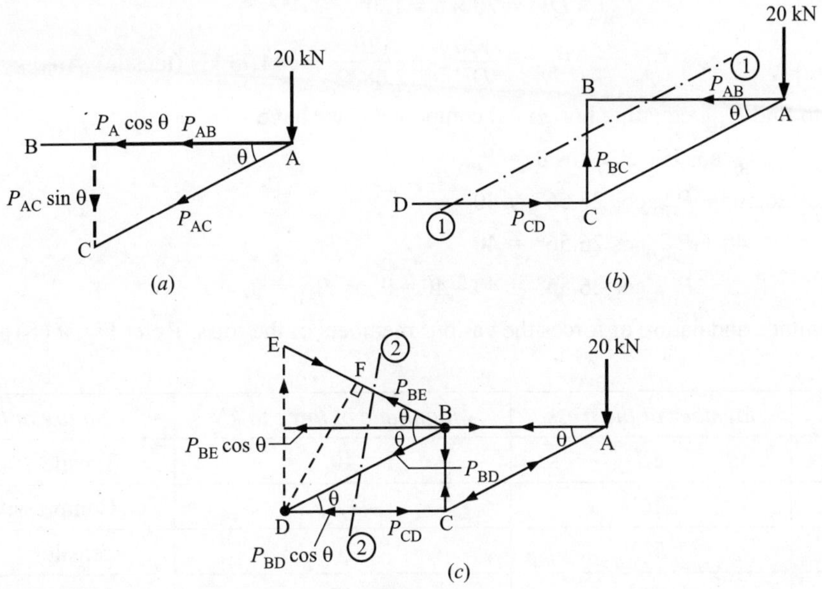


**Fig. 9.69**

Taking moment of all the forces acting on the right part of the truss about point *D* and equating anticlockwise moment equal to clockwise moment, we have

$$P_{BC} \times CD + P_{AB} \times BC = 20 \times 6$$

$$P_{BC} \times 3 + 40 \times 1.5 = 120$$

$$\therefore \quad P_{BC} = \frac{120 - 40 \times 1.5}{3} = 20 \text{ kN (tensile)} \ \textbf{Ans.}$$

Again taking moment of all the forces acting on the right part of the truss about point *B* and equating anticlockwise moment equal to clockwise moment, we have

$$P_{CD} \times BC = 20 \times AB$$

$$\therefore \quad P_{CD} = 20 \times \frac{AB}{BC} = 20 \times \frac{3}{1.5} = 40 \text{ kN (compressive)} \ \textbf{Ans.}$$

Since the magnitude of $P_{BC}$ and $P_{CD}$ is positive, therefore the directions assumed for the members *BC* and *CD* is correct. Thus the force in the member *BC* is tensile and force in the member *CD* is compressive.

Now draw a section (2-2) cutting the members *BE*, *BD* and *CD*, as shown in Fig. 9.69 (*c*). Consider the equilibrium of the right part of the truss. Let the forces $P_{BE}$ and $P_{BD}$ in the members *BE* and *BD*, respectively, have directions as shown in the figure.

First of all, let us find the perpendicular distance from point *D* on *BE* (*i.e. DF*). From the right angled triangle *DBF*,

$$DF = BD \sin 2\theta = BD \sin(2 \times 26.56°) = BD \sin 53.12°$$

$$= BD \times 0.8 = 0.8BD$$

and from right angled triangle *BCD*, $BD = \sqrt{(BC)^2 + (CD)^2} = \sqrt{(1.5)^2 + 3^2} = \sqrt{11.25} = 3.354$ m

$$\therefore \quad DF = 0.8BD = 0.8 \times 3.354 = 2.6832 \text{ m}$$

Now taking moment about point $D$ and equating anticlockwise moment equal to clockwise moment, we have

$$P_{BE} \times DF = 20 \times 6 = 120$$

$$\therefore \quad P_{BE} = \frac{120}{DF} = \frac{120}{2.6832} = 44.73 \text{ kN (tensile)} \quad \textbf{Ans.}$$

In order to find $P_{BD}$, equating horizontal components, we have

$$P_{BE} \cos\theta + P_{BD} \cos\theta = P_{CD}$$

$$44.73 \cos 26.56° + P_{BD} \cos 26.56° = 40$$

$$40 + P_{BD} \cos 26.56° = 40$$

$$\therefore \quad P_{BD} \cos 26.56° = 40 - 40 = 0 \text{ or } P_{BD} = 0$$

The magnitude and nature of forces the various members of the truss (Refer Fig. 9.68) are tabulated as below:

| *S. No.* | *Member of the truss* | *Magnitude of force in kN* | *Nature of force* |
|---|---|---|---|
| 1. | *AB* | 40 | Tensile |
| 2. | *AC* | 44.73 | Compressive |
| 3. | *BC* | 20 | Tensile |
| 4. | *CD* | 40 | Compressive |
| 5. | *BE* | 44.73 | Tensile |
| 6. | *BD* | 0 | — |

**Example 9.24** *Find the forces in all the members of the pin-joined truss supported and loaded, as shown in Fig. 9.70. AB = 0.9 m ; BC = 0.6 m and BD = 1.2 m.*

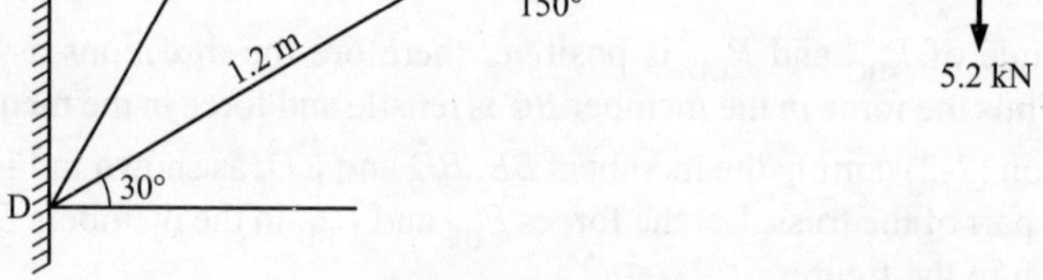


**Fig. 9.70**

***Solution*** First of all, let us find the value of $AC$ and angle $\theta$. From triangle $ABC$,

$$(AC)^2 = (AB)^2 + (BC)^2 - 2(AB)(BC)\cos 120°$$

$$= (0.9)^2 + (0.6)^2 - 2 \times 0.9 \times 0.6\,(-0.5)$$

$$= 0.81 + 0.36 + 0.54 = 1.71$$

$$\therefore \quad AC = \sqrt{1.71} = 1.31 \text{ m}$$

According to sine formula,

$$\frac{AC}{\sin 120°} = \frac{BC}{\sin \theta}$$

$$\therefore \quad \sin \theta = \frac{BC \times \sin 120°}{AC} = \frac{0.6 \times 0.866}{1.31} = 0.3966$$

or $$\theta = \sin^{-1}(0.3966) = 23.37°$$

and $$\cos \theta = \cos 23.37° = 0.918$$

Now consider the equilibrium of the joint $A$, as shown in Fig. 9.71 (*a*). Let the force in the member $AB$ ($P_{AB}$) and force in the member $AC$ ($P_{AC}$) are assumed to be acting away from $A$ (*i.e.* tensile).

Resolving the forces vertically and equating the same, we have

$$P_{AC} \sin \theta = 5.2$$

$$\therefore \quad P_{AC} = \frac{5.2}{\sin \theta} = \frac{5.2}{0.3966} = 13.1 \text{ kN (tensile)} \quad \textbf{Ans.}$$

Resolving the forces horizontally and equating the same, we have

$$P_{AB} = -P_{AC} \cos \theta = -13.1 \times 0.918 = -12 \text{ kN}$$
$$= 12 \text{ kN (compressive)} \quad \textbf{Ans.}$$

...(–ve sign shows that the direction of $P_{AB}$ is opposite to that assumed)

Consider the equilibrium of the joint $B$, as shown in the Fig. 9.71 (*b*). Let the forces in the member $BC$ ($P_{BC}$) and force in the member $BD$ ($P_{BD}$) are assumed to be acting towards point $B$ (*i.e.* compressive). Applying Lami's theorem, we have

$$\frac{P_{AB}}{\sin 90°} = \frac{P_{BC}}{\sin 150°} = \frac{P_{BD}}{\sin 120°}$$

or $$\frac{P_{AB}}{1} = \frac{P_{BC}}{0.5} = \frac{P_{BD}}{0.866}$$

$$\therefore \quad P_{BC} = P_{AB} \times 0.5 = 12 \times 0.5 = 6 \text{ kN (compressive)} \quad \textbf{Ans.}$$

and $$P_{BD} = P_{AB} \times 0.866 = 12 \times 0.866 = 10.4 \text{ kN (compressive)} \quad \textbf{Ans.}$$

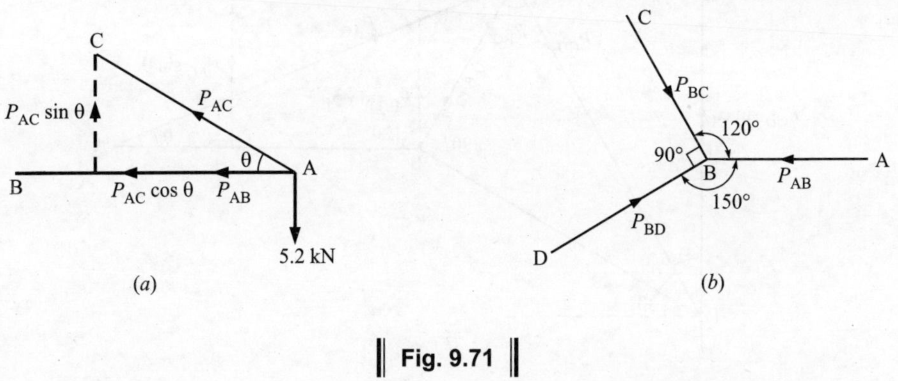


**Fig. 9.71**

Now let us find the lengths $CD$, $CE$ and $ED$. Let angle $BDC = \phi$ and angle $CDE = \alpha$. From right angled triangle $BCD$ [Refer Fig. 9.72 (*a*)],

$$CD = \sqrt{(BC)^2 + (BD)^2} = \sqrt{(0.6)^2 + (1.2)^2}$$
$$= \sqrt{0.36 + 1.44} = \sqrt{1.8} = 1.34 \text{ m}$$

and $$\tan\phi = \frac{BC}{BD} = \frac{0.6}{1.2} = 0.5$$

or $$\phi = \tan^{-1}(0.5) = 26.56°$$

∴ $$\alpha = 90° - (30° + \phi) = 90° - (30° + 26.56°) = 33.44°$$

From right angled triangle *CDE*,

$$\frac{CE}{CD} = \sin\alpha$$

∴ $$CE = CD\sin\alpha = 1.34 \times \sin 33.44° = 1.34 \times 0.551 = 0.738 \text{ m}$$

and $$\frac{ED}{CD} = \cos\alpha$$

∴ $$ED = CD\cos\alpha = 1.34\cos 33.44° = 1.34 \times 0.834 = 1.12 \text{ m}$$

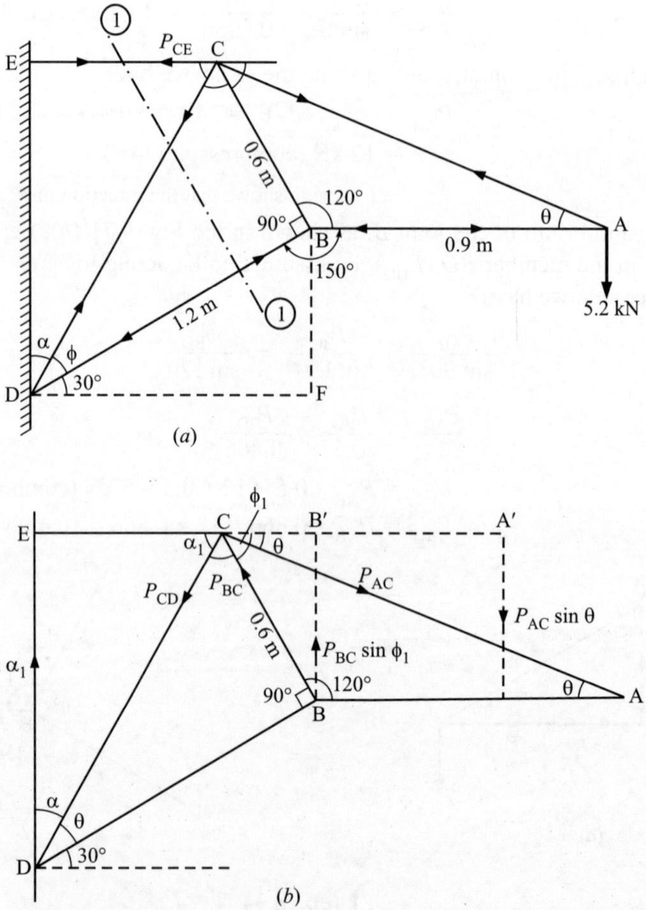


**Fig. 9.72**

In order to find the force in the member *CE*, we shall apply the method of sections. Draw a section (1-1) to cut the three members *CE*, *CD* and *BD* as shown in Fig. 9.72 (*a*). Now consider the equilibrium of the right side of the section. Let the force in the member *CE* ($P_{CE}$) is assumed to act away from point *C* (*i.e.* tensile).

Taking moment of all the forces about point $D$ and equating anticlockwise moment equal to clockwise moment, we have

$$P_{CE} \times ED = 5.2\,(AB + FD)$$

$$P_{CE} \times 1.12 = 5.2\,(0.9 + BD \cos 30°) = 5.2\,(0.9 + 1.2 \times 0.866) = 10.1$$

$$\therefore \quad P_{CE} = \frac{10.1}{1.12} = 9 \text{ kN (tensile)} \quad \textbf{Ans.}$$

Now consider the equilibrium of the joint $C$, as shown in Fig. 9.72 (*b*). Let the force in the member $CD$ ($P_{CD}$) is assumed to act away from point $C$ (*i.e.* tensile). Resolving vertically all the forces acting at $C$ and equating the same, we have

$$P_{CD} \sin \alpha_1 + P_{BC} \sin \phi_1 = P_{AC} \sin \theta$$

From right angled triangle $CED$, $\alpha_1 = \angle ECD = 180° - (90° + \alpha)$

$$= 180° - (90° + 33.44°) = 56.56°$$

and $\phi_1 = 180° - \angle BCE = 180° - 120° = 60°$

...[$\because \angle BCE = \angle ABC = 120°$]

Substituting the values of $\theta$, $\alpha_1$ and $\phi_1$ in the above expression, we have

$$P_{CD} \sin 56.56° + P_{BC} \sin 60° = P_{AC} \sin 23.37°$$

$$P_{CD} \times 0.834 + 6 \times 0.866 = 13.1 \times 0.397$$

$$P_{CD} \times 0.834 + 5.196 = 5.201$$

$$P_{CD} = \frac{5.201 - 5.196}{0.834} = 0.006 \text{ kN (tensile)} \quad \textbf{Ans.}$$

The magnitude and nature of forces in the various members of the truss [Refer Fig. 9.72 (*a*)] are tabulated as below:

| *S. No.* | *Member of the truss* | *Magnitude of force in kN* | *Nature of force* |
|---|---|---|---|
| 1. | *AC* | 13.1 | Tensile |
| 2. | *AB* | 12 | Compressive |
| 3. | *BC* | 6 | Compressive |
| 4. | *BD* | 10.4 | Compressive |
| 5. | *CE* | 9 | Tensile |
| 6. | *CD* | 0.006 | Tensile |

**Example 9.25** *Find the magnitude and nature of forces in the members BE, BF, CD, EF and CF of the roof truss, shown in Fig. 9.73. Tabulate the results. Assume all the joints to be perfectly smooth hinges.*

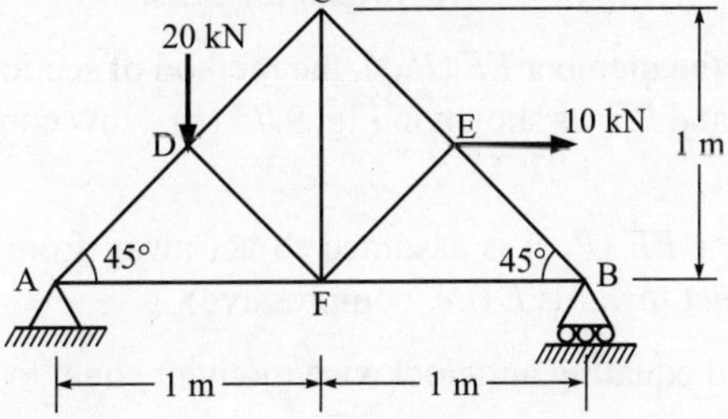


**Fig. 9.73**

***Solution*** Since a horizontal force of 10 kN acts at *E*, therefore in order to maintain equilibrium of the truss in the horizontal direction, an equal and opposite force of 10 kN is applied at the hinged support *A*, as shown in Fig. 9.74, as $R_{AH} = 10$ kN.

In order to find the vertical component of the reaction at *A* (*i.e.* $R_{AV}$), taking moment of all the external forces about *B*, we have

$$R_{AV} \times 2 = 20 \times HB - 10 \times EG = 20 \times 1.5 - 10 \times 0.5 = 30 - 5 = 25$$

$$\therefore \quad R_{AV} = \frac{25}{2} = 12.5 \text{ kN}$$

Let $R_B$ be the reaction at the support *B*.

We know that $\quad R_{AV} + R_B = 20$ kN

$$\therefore \quad R_B = 20 - R_{AV} = 20 - 12.5 = 7.5 \text{ kN}$$

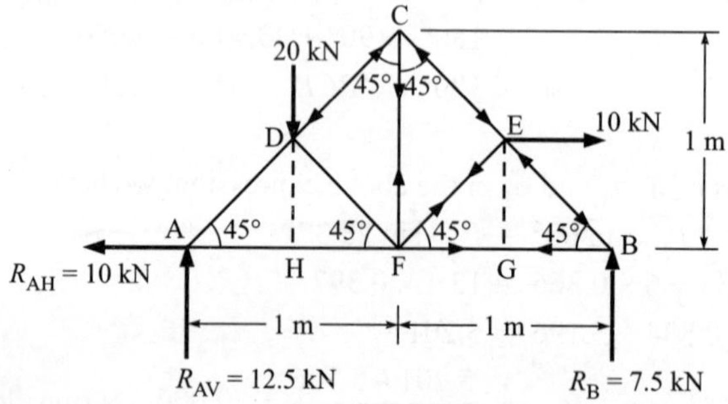


**Fig. 9.74**

First of all, let us consider the equilibrium of the joint *B*, as shown in Fig. 9.75 (*a*). Let the forces $P_{BE}$ and $P_{BF}$ in the members *BE* and *BF* are assumed to be acting away from *B* (*i.e.* tensile).

Resolving the force $P_{BE}$ into vertical component and equating the same, we have

$$P_{BE} \sin 45° = -R_B = -7.5$$

$$\therefore \quad P_{BE} = \frac{-7.5}{\sin 45°} = \frac{-7.5}{0.707} = -10.61 \text{ kN}$$

$$= 10.61 \text{ kN (compressive)} \quad \textbf{Ans.}$$

...(–ve sign shows that the direction of $P_{BE}$ is opposite to that assumed)

Resolving the force $P_{BE}$ into horizontal components and equating the same we have

$$P_{BF} = -P_{BE} \cos 45° = -(-10.61)\, 0.707$$

$$= 7.5 \text{ kN (tensile)} \quad \textbf{Ans.}$$

In order to find the force in the member *EF* ($P_{EF}$), the method of sections is applied. Draw a section 1-1 to cut the three members *EC*, *EF* and *BF*, as shown in Fig. 9.75 (*b*). Now consider the equilibrium of the right side of the section.

Let the force in the member *EF* ($P_{EF}$) is assumed to act away from *E* (*i.e.* tensile) and force in the member *EC* ($P_{EC}$) is assumed to act towards *E* (*i.e.* compressive).

Taking moment about *B* and equating anticlockwise moment equal to clockwise moment, we have

$$P_{EF} \times BE = 10 \times EG$$

$$P_{EF} \times \frac{EG}{\sin 45°} = 10 \times EG \qquad \ldots\left(\because \frac{EG}{BE} = \sin 45°\right)$$

$$\therefore \qquad P_{EF} = 10 \sin 45° = 10 \times 0.707 = 7.07 \text{ kN (tensile)} \ \textbf{Ans.}$$

Now taking moments about $F$ and equating the anticlockwise moment equal to clockwise movement, we have

$$R_B \times BF = 10 \times EG + P_{EC} \times {}^*EF$$

$$7.5 \times 1 = 10 \times 0.5 + P_{EC} \times \frac{1}{\sqrt{2}} = 5 + \frac{P_{EC}}{\sqrt{2}}$$

$$\therefore \qquad P_{EC} = (7.5 - 5)\sqrt{2} = 3.53 \text{ kN (compressive)} \ \textbf{Ans.}$$

Now draw a section (2-2) to cut the members $CD$, $FD$ and $AF$, as shown in Fig. 9.75 (*c*). Considering the equilibrium of the left side of the section.

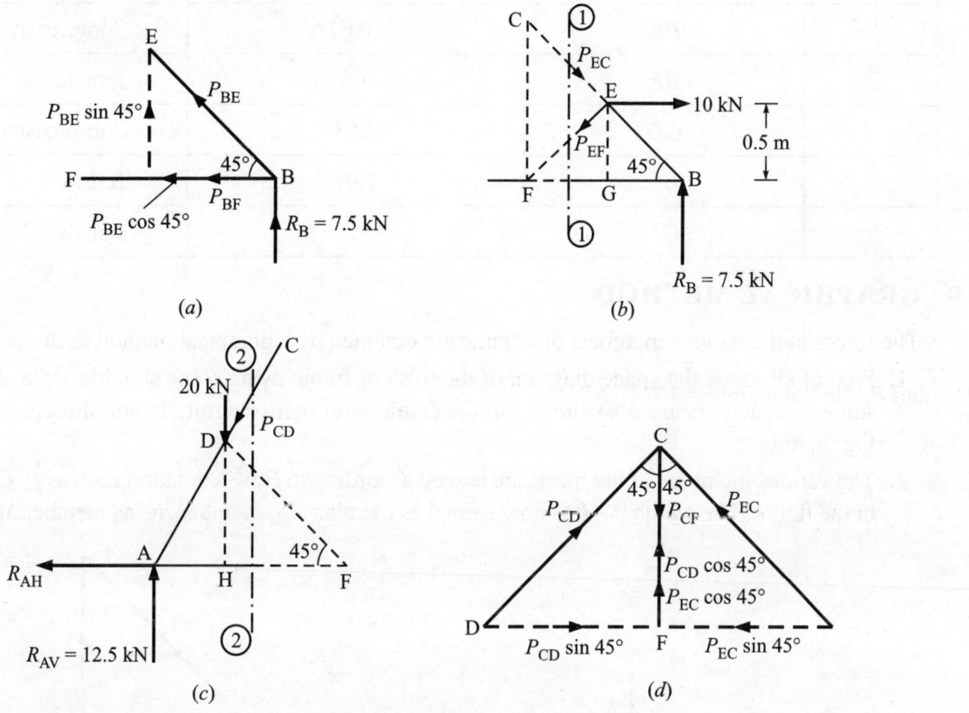


**Fig. 9.75**

Let the force in the member $CD$ ($P_{CD}$) is assumed to be acting towards $D$ (*i.e.* compressive). Taking moment about $F$ and equating anticlockwise moment equal to clockwise moment, we have

$$P_{CD} \times {}^*DF + 20 \times HF = R_{AV} \times AF$$

---

$$^* \ EF = \sqrt{(EG)^2 + (FG)^2} = \sqrt{\left(\frac{1}{2}\right)^2 + \left(\frac{1}{2}\right)^2} = \sqrt{\frac{1}{4} + \frac{1}{4}} = \sqrt{\frac{2}{4}} = \frac{1}{\sqrt{2}}$$

$$DF = \sqrt{(DH)^2 + (HF)^2} = \sqrt{\left(\frac{1}{2}\right)^2 + \left(\frac{1}{2}\right)^2} = \sqrt{\frac{1}{4} + \frac{1}{4}} = \sqrt{\frac{2}{4}} = \frac{1}{\sqrt{2}}$$

$$P_{CD} \times \frac{1}{\sqrt{2}} + 20 \times \frac{1}{2} = 12.5 \times 1$$

$$\therefore \quad P_{CD} = (12.5 - 10)\sqrt{2} = 3.53 \text{ kN (compressive)} \textbf{ Ans.}$$

Now consider the equilibrium of joint *C*, as shown in Fig. 9.75 (*d*). Let the force in the member *CF* ($P_{CF}$) is assumed to be acting away from *C* (*i.e.* tensile). Resolving the forces vertically and equating the same, we have

$$P_{CF} = P_{CD} \cos 45° + P_{EC} \cos 45°$$

$$= 3.53 \times \frac{1}{\sqrt{2}} + 3.53 \times \frac{1}{\sqrt{2}} = 2.5 + 2.5 = 5 \text{ kN (tensile)} \textbf{ Ans.}$$

The magnitude and nature of forces in the members of the truss (Refer Fig. 9.74) are tabulated as below:

| *S. No.* | *Member of the truss* | *Magnitude of force in kN* | *Nature of force* |
|---|---|---|---|
| 1. | *BE* | 10.61 | Compressive |
| 2. | *BF* | 7.5 | Tensile |
| 3. | *CD* | 3.53 | Compressive |
| 4. | *EF* | 7.07 | Tensile |
| 5. | *CF* | 5 | Tensile |

## 9.9 GRAPHICAL METHOD

The forces in the various members of a frame are obtained by a graphical method as discussed below:

1. First of all, draw the space diagram of the truss or frame using some suitable scale. The loads and supports reactions are also shown on the frame with their magnitude and direction, as shown in Fig. 9.76 (*a*).
2. The various members of the frame are named according to Bow's notation such as *P*, *Q*, *R*, as shown in the figure. The member *AB* is now named as member *PS*, member *BC* as member *SQ* and so on.

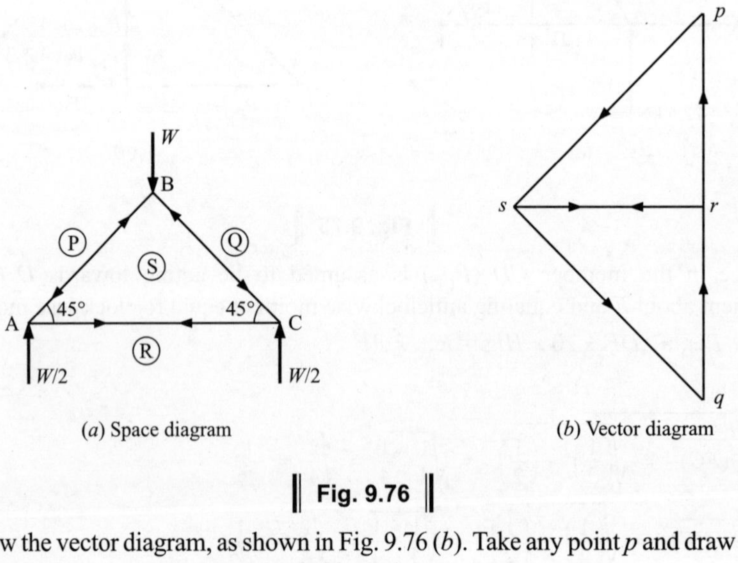


(*a*) Space diagram (*b*) Vector diagram

**Fig. 9.76**

3. Now draw the vector diagram, as shown in Fig. 9.76 (*b*). Take any point *p* and draw a line *pq* vertically downward and cut *pq* = *W*, to some suitable scale.

4. From point $q$, draw $qr$ parallel to $QR$ vertically upward and cut $qr = \frac{W}{2}$, to the same scale.
5. From point $r$, draw $rp$ parallel to $RP$ vertically upward and cut $rp = \frac{W}{2}$, to the same scale. Now vector $qp = qr + rp = W$.
6. Now from point $p$, draw $ps$ parallel to $PS$ and from point $r$ draw $rs$ parallel to $RS$ intersecting both the lines at $s$. The triangle $rps$ is the vector diagram for the joint $A$. For equilibrium of joint $A$, the directions of the vectors $ps$ and $sr$ will be in order, as shown in Fig. 9.76 (*b*). These directions are marked near the joint $A$ (the joint under consideration) on the space diagram and opposite direction near the other end of the joint. Thus, member $PS$ (or $AB$) is in compression and member $RS$ (or $AC$) is in tension.
7. Similarly, we can draw the vector diagram $qrs$ by considering the joint $C$ and mark the direction of force in the member $QS$ (or $CB$), as shown in the space diagram. We see that the force in member $CB$ is compressive.
8. The above steps are used to determine the forces in the truss or frame having any number of joints.

**Example 9.26** *A triangular shaped truss ABC, as shown in Fig. 9.77, is simply supported at A and B. Find the forces in the members AB, AC and BC by using graphical method.*

***Solution*** First of all, let us find the reactions $R_A$ and $R_B$ at supports $A$ and $B$, respectively. Refer Fig. 9.78 (*a*). From the geometry of the figure, we find that

$$\angle ACB = 90°$$

Draw $CD$ perpendicular on $AB$. We know that

$$AC = AB \cos 30°$$

$$= 6 \times \frac{\sqrt{3}}{2} = 3\sqrt{3} \text{ m}$$

and

$$AD = AC \cos 30° = 3\sqrt{3} \times \frac{\sqrt{3}}{2}$$

$$= 4.5 \text{ m}$$

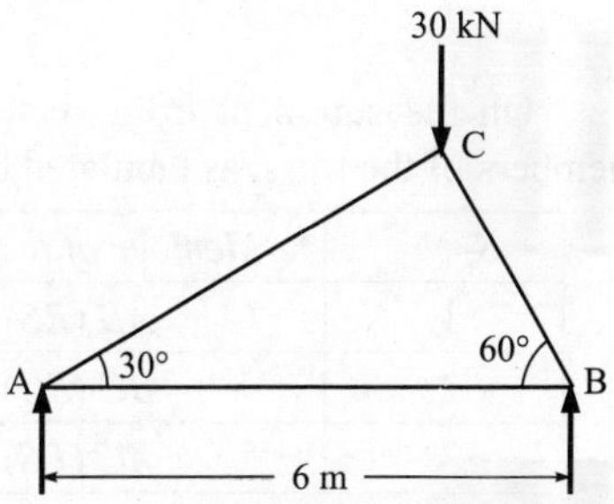


**Fig. 9.77**

Now taking moments about $A$, we have

$$R_B \times 6 = 30 \times AD = 30 \times 4.5$$

$$\therefore \quad R_B = \frac{30 \times 4.5}{6} = 22.5 \text{ kN}$$

We know that

$$R_A + R_B = 30 \text{ kN}$$

$$\therefore \quad R_A = 30 - R_B = 30 - 22.5 = 7.5 \text{ kN}$$

Now, draw the space diagram of the truss, as shown in Fig. 9.78 (*a*), using some suitable scale and name the members $AB$, $BC$ and $AC$ according to Bow's notation $P$, $Q$, $R$ and $S$, as shown in the space diagram. The vector diagram is now drawn as discussed below:

1. Take any point $q$ and draw a vertical line $qr$ upward equal to the load at $B$ (*i.e.* 22.5 kN) to some suitable scale.
2. From $q$, draw a line $qs$ parallel to the member $QS$ and from $r$ draw a line $rs$ parallel to the member $RS$, both line intersecting at $s$. Now $qrs$ is the vector diagram for the joint $B$. For the equilibrium of joint $B$, the direction of $rs$ and $sq$ will be in order, as shown in Fig. 9.78. These directions are marked near the joint $B$, on the space diagram. Thus, member $RS(AB)$ is in tension and member $SQ(BC)$ is in compression.

3. Similarly, draw the vector diagram for the joint $A$. Draw a line $rp$ upward parallel to load at $A$ (*i.e.* 7.5 kN) by using the same scale. Now from $p$ draw a line $ps$ parallel to the member $PS$ and from point $r$ draw a line $rs$ parallel to the member $RS$, both lines intersecting at $s$. Now $rps$ is the vector diagram for the joint $A$. For equilibrium of the joint $A$, directions of $ps$ and $sr$ will be in order, as shown in Fig. 9.78. These directions are marked near the joint $A$, on the space diagram. Thus member $PS(AC)$ is in compression and member $RS(AB)$ is in tension.

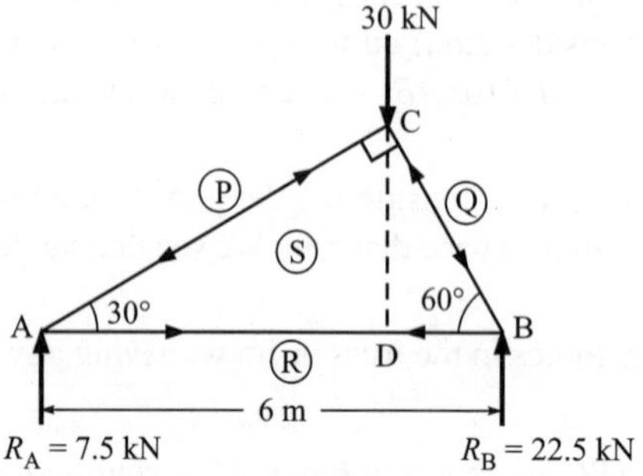

(*a*) Space diagram

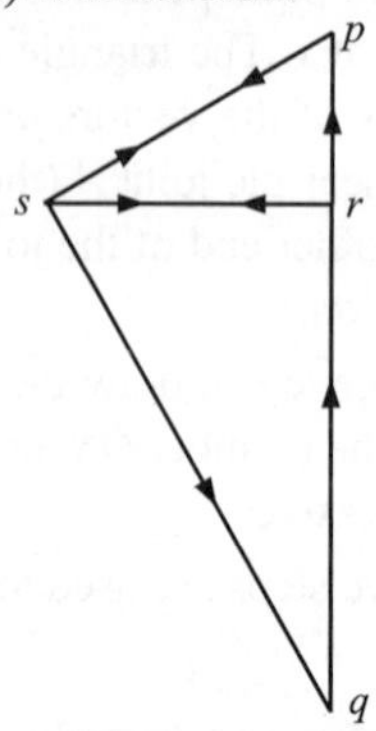

(*b*) Vector diagram

**Fig. 9.78**

On measurement from vector diagram, we find the magnitude and nature of forces in the various members of the truss, as tabulated below:

| *S. No.* | *Member of the truss* | *Magnitude of force in kN* | *Nature of force* |
|---|---|---|---|
| 1. | *AB* (*RS*) | 13 kN | Tensile |
| 2. | *BC* (*SQ*) | 26 kN | Compressive |
| 3. | *AC* (*PS*) | 15 kN | Compressive |

**Example 9.27** *A truss of seven members each of 3 m length is freely supported at the ends, as shown in Fig. 9.79. The points B and C are loaded with 4 kN and 8 kN, respectively. Determine the magnitude and nature of forces in all the members of the truss. Use graphical method.*

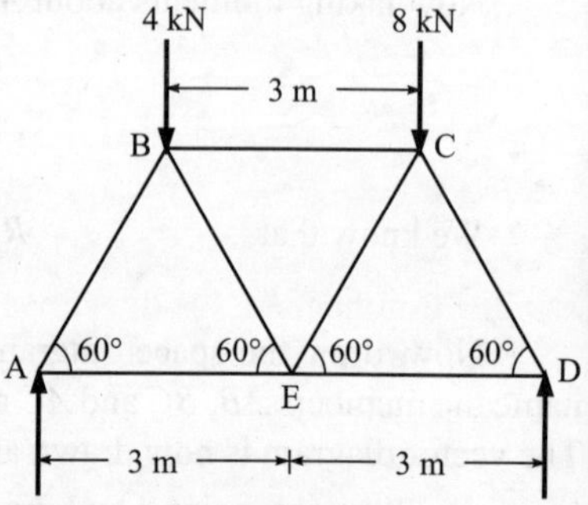

**Fig. 9.79**

***Solution*** First of all, let us find the reactions $R_A$ and $R_B$ at supports $A$ and $B$, respectively [Refer Fig. 9.80 (*a*)]. From the geometry of the figure, we find that

$$AF = AB \cos 60°$$
$$= 3 \times 0.5 = 1.5 \text{ m}$$

and
$$AG = AE + EG$$
$$= 3 + EC \cos 60°$$
$$= 3 + 3 \times 0.5 = 4.5 \text{ m}$$

Now taking moments about $A$, we have

$$R_D \times 6 = 4 \times AF + 8 \times AG$$
$$= 4 \times 1.5 + 8 \times 4.5 = 6 + 36 = 42$$

$\therefore$
$$R_D = \frac{42}{6} = 7 \text{ kN}$$

We know that
$$R_A + R_D = 4 + 8 = 12 \text{ kN}$$

$\therefore$
$$R_A = 12 - R_D = 12 - 7 = 5 \text{ kN}$$

Now draw the space diagram of the truss using some suitable scale and name the various members, according to Bow's notation, as shown in Fig. 9.80 (*a*).

The vector diagram is now drawn as discussed below:

1. Take any point *r* and draw a vertical line *rs* upward equal to the load at *D* (*i.e.* 7 kN) to some suitable scale.
2. From point *r*, draw *rv* parallel to the member *RV* and from point *s*, draw *sv* parallel to the member *SV*, intersecting both the lines at *v*. Now *rsv* is the vector diagram for the joint *D*. For equilibrium of joint *D*, the direction of *sv* and *vr* will be in order, as shown in Fig. 9.80 (*b*). These directions are marked near the joint *D* on the space diagram. Thus, member *RV* (*CD*) is in compression and member *SV* (*DE*) is in tension.

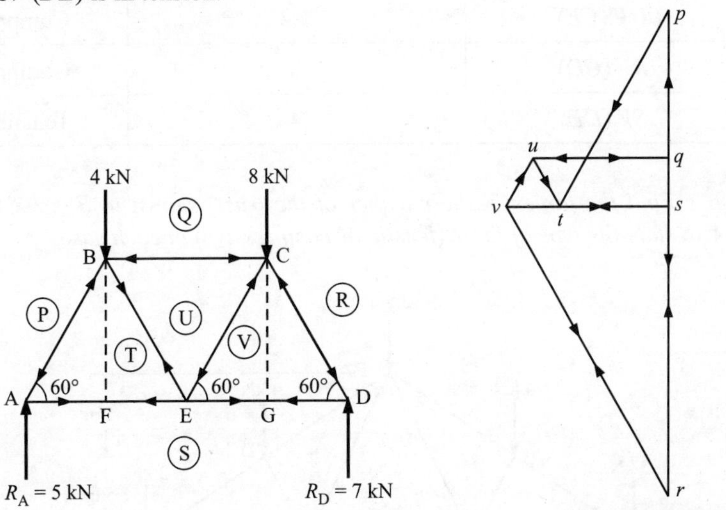


(*a*) Space diagram (*b*) Vector diagram

**Fig. 9.80**

3. Now draw a vertical line *sp* upward equal to the load at *A* (*i.e.* 5 kN). From point *p* draw *pt* parallel to the member *PT* and from point *s* draw *st* parallel to the member *ST*, intersecting both at *t*. Now *spt* is the vector diagram for the joint *A*. For equilibrium of the joint *A*, the directions of *pt* and *ts* will be in order, as shown in Fig. 9.80 (*b*). These directions are marked near the joint *A*. Thus the member *PT* (*AB*) is in compression and member *TS* (*AE*) is in tension.
4. On a line *pr*, mark point *q* such that *qr* is equal to the load at *C* (*i.e.* 8 kN) in the downward direction. Now from point *q*, draw *qu* parallel to the member *QU* and from point *v* draw *vu* parallel to the member *VU*, intersecting both at *u*. Now *qrvu* is the vector diagram for the joint *C*. For equilibrium of the joint, the directions of *uv* and *uq* will be in order, as shown in Fig. 9.80 (*b*). These directions are marked near the joint *C*. Thus the member *UV* (*CE*) is in compression and member *QU* (*BC*) is in compression.
5. Now from point *s*, draw *st* parallel to member *ST* and from point *p* draw *pt* parallel to member *PT*, intersecting both at *t*. For equilibrium of the joint *A*, the direction of *pt* and *ts* will be in order, as shown in Fig. 9.80 (*b*). These directions are marked near the joint *A*. Thus the member *PT* (*i.e. AB*) is in compression and member *TS* (*i.e. AE*) is in tension.
6. Finally from point *q*, draw *qu* parallel to member *QU* and from point *t* draw *tu* parallel to member *TU*, intersecting both at *u*. Now *squt* is the vector diagram for the joint *B*. For equilibrium of the joint *B*, the direction of the vector *qu* and *ut* will be in order, as shown in Fig. 9.80 (*b*). These directions are marked near the joint *B*. Thus, the member *QU* (*BC*) is in compression and member *TU* (*BE*) is in tension.

On measurement, from the vector diagram, we find the magnitude and nature of forces as tabulated below:

| *S. No.* | *Member of the truss* | *Magnitude of force in kN* | *Nature of force* |
|---|---|---|---|
| 1. | *PT* (*AB*) | 5.7 | Compressive |
| 2. | *ST* (*AE*) | 2.93 | Tensile |
| 3. | *TU* (*BE*) | 1.2 | Tensile |
| 4. | *QU* (*BC*) | 3.467 | Compressive |
| 5. | *UV* (*CE*) | 1.2 | Compressive |
| 6. | *RV* (*CD*) | 8 | Compressive |
| 7. | *SV* (*DE*) | 4 | Tensile |

**Example 9.28** *Find the forces in the member of the truss, shown in Fig. 9.81, due to the loads indicated there on. Also state the type of force (tensile or compressive) in each case.*

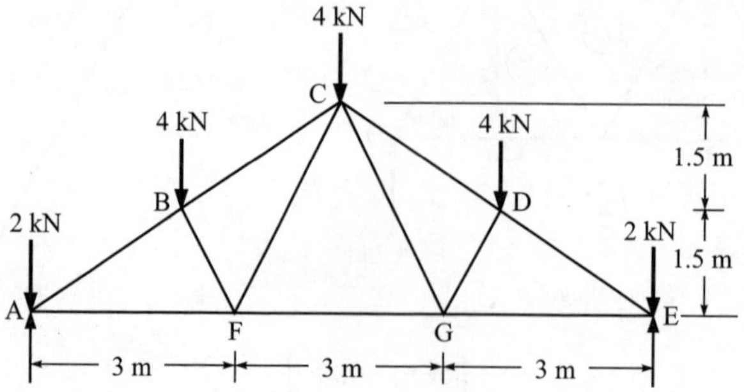


**Fig. 9.81**

***Solution*** First of all, let us find the reactions $R_A$ and $R_E$ at supports *A* and *E*, respectively [Refer Fig. 9.82 (*a*)]. Since the truss is symmetrical, therefore reaction at each support is given by

$$R_A = R_E = \frac{2+4+4+4+2}{2} = 8 \text{ kN}$$

Since the supports carries a downward vertical load at *A* and *E*, therefore net reaction at supports *A* and *E*,

$$R_A = R_E = 8 - 2 = 6 \text{ kN}$$

Draw the space diagram of the truss using some suitable scale and name the various members according to Bow's notation, as shown in Fig. 9.82 (*a*). Now draw a vector diagram as discussed below:

1. Take any point *p* and draw a vertical line *pt* upward and equal to the reaction at *A* (*i.e.* 6 kN) to some suitable scale.
2. From point *p*, draw a line *pu* parallel to the member *PU* and from point *t*, draw *tu* parallel to the member *TU*, intersecting both the lines at *u*. Now *ptu* is the vector diagram for the joint *A*. For the

equilibrium of the joint *A*, the directions of *tu* and *up* will be in order. These directions are marked near the joint *A* on the space diagram. Thus, the member *PU* (*AB*) is compression and the member *UT* (*AF*) is in tension.

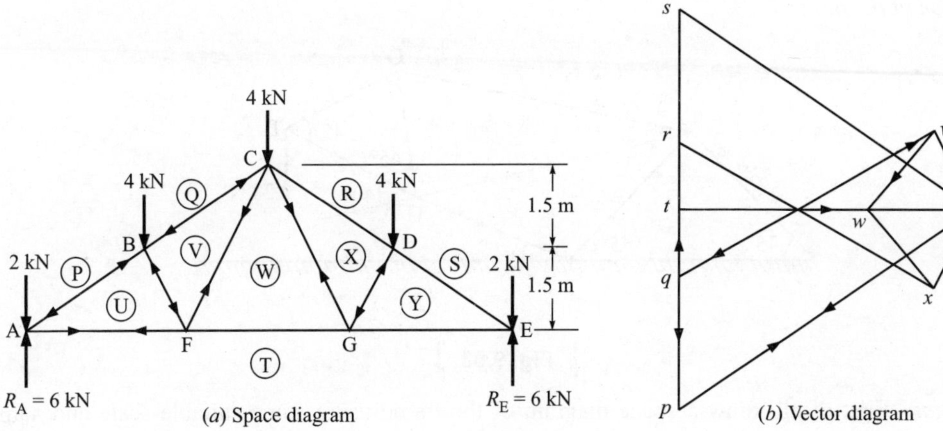


**Fig. 9.82**

3. From point *p*, draw *pq* parallel to the member *PQ* and equal to the downward load at *B* (*i.e.* 4 kN). Now from point *q*, draw *qv* parallel to the member *QV* and from point *u*, draw *uv* parallel to the member *UV*, intersecting both the lines at *v*. Now *uvqp* is the vector diagram for the joint *B*. For equilibrium of the joint *B*, the directions of *uv* and *vq* will be in order. These directions are marked near the joint *B* on the space diagram. Thus the members *QV* (*BC*) and *UV* (*BF*) are in compression.
4. Similarly, draw the vector diagrams for the equilibrium of the other joints, as shown in Fig. 9.82 (*b*).

On measurement from the vector diagram, the magnitude and nature of forces are tabulated below:

| *S. No.* | *Member of the truss* | *Magnitude of force in kN* | *Nature of force* |
|---|---|---|---|
| 1. | *PU* (*AB*) | 10.8 | Compressive |
| 2. | *TU* (*AF*) | 8.6 | Tensile |
| 3. | *QV* (*BC*) | 8.4 | Compressive |
| 4. | *UV* (*BF*) | 3.2 | Compressive |
| 5. | *VW* (*CF*) | 3.2 | Tensile |
| 6. | *WT* (*FG*) | 5.6 | Tensile |
| 7. | *RX* (*CD*) | 8.4 | Compressive |
| 8. | *WX* (*CG*) | 3.2 | Tensile |
| 9. | *XY* (*DG*) | 3.2 | Compressive |
| 10. | *SY* (*DE*) | 10.2 | Compressive |
| 11. | *TY* (*EG*) | 8.6 | Tensile |

**Example 9.29** *The loaded framework, as shown in Fig. 9.83, is hinged at the joint T and held by a cable RS. The members PU, UT and TR are all of equal length. Determine by graphical construction, the magnitude and nature of the forces in each member, tension in the cable RS and the magnitude and direction of the reaction at the hinge T.*

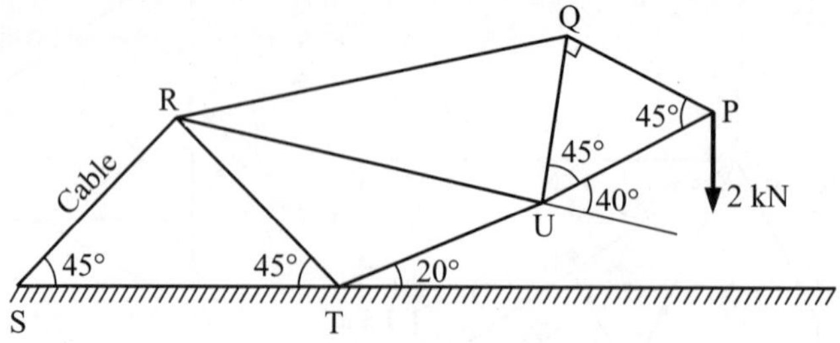


**Fig. 9.83**

***Solution*** First of all, draw a space diagram of the frame using some suitable scale and name the various members according to Bow's notation, as shown in Fig. 9.84 (*a*). Now draw a vector diagram as discussed below:

1. Take any point *a* and draw a vertical line *ab* downward and equal to the load at *P* (*i.e.* 2 kN) to some suitable scale.
2. From point *a*, draw *ac* parallel to the member *AC* from point *b* and draw *bc* parallel to the member *BC*, intersecting both the lines at *c*. Now *abc* is the vector diagram for the joint *P*. For equilibrium of the joint *P*, the direction of *bc* and *ca* will be in order, as shown in Fig. 9.84 (*b*). These directions are marked near the joint *P* on the space diagram. Thus, the member *AC* (*PQ*) is in tension and the member *BC* (*PU*) is in compression.

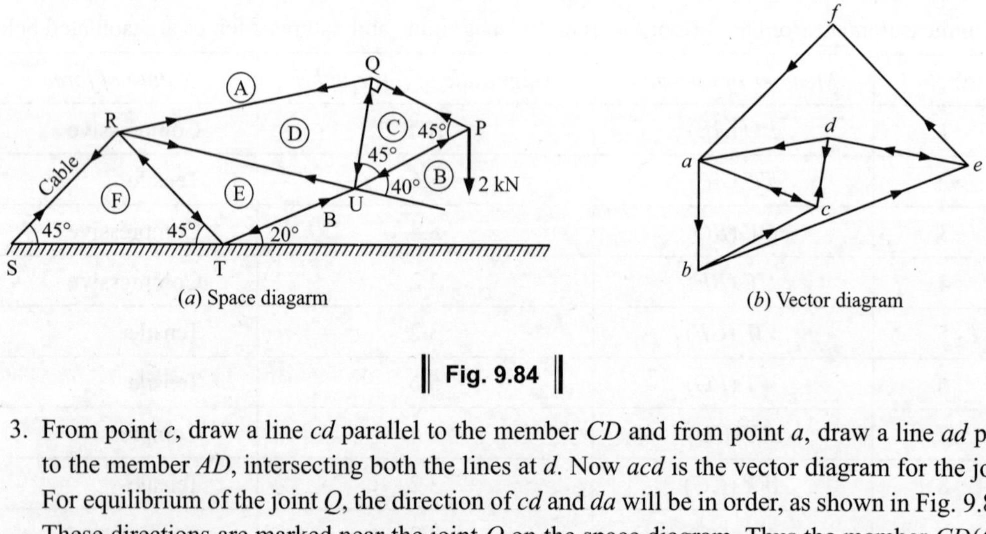


(*a*) Space diagarm (*b*) Vector diagram

**Fig. 9.84**

3. From point *c*, draw a line *cd* parallel to the member *CD* and from point *a*, draw a line *ad* parallel to the member *AD*, intersecting both the lines at *d*. Now *acd* is the vector diagram for the joint *Q*. For equilibrium of the joint *Q*, the direction of *cd* and *da* will be in order, as shown in Fig. 9.84 (*b*). These directions are marked near the joint *Q* on the space diagram. Thus the member *CD*(*QU*) is in compression and member *AD* (*QR*) is in tension.
4. Similarly, draw vector diagram *bed* for the joint *U* and vector diagram *efa* for the joint *R* and mark the directions.

On measurement from the vector diagram, the magnitude and nature of forces are tabulated as below:

| *S. No.* | *Member of the truss* | *Magnitude of force in kN* | *Nature of force* |
|---|---|---|---|
| 1. | *PU* (*BC*) | 2.6 | Compressive |
| 2. | *PQ* (*AC*) | 2.2 | Tensile |
| 3. | *QR* (*AD*) | 2.3 | Tensile |
| 4. | *QU* (*CD*) | 1.2 | Compressive |
| 5. | *UT* (*BE*) | 5.3 | Compressive |
| 6. | *RU* (*DE*) | 2.7 | Tensile |
| 7. | *RT* (*EF*) | 3.6 | Compressive |
| 8. | *RS* (*AF*) | 3.4 | Tensile |

**Example 9.30** *Determine by graphical construction, the stresses in the different members of the frame when loaded, as shown in Fig. 9.85.*

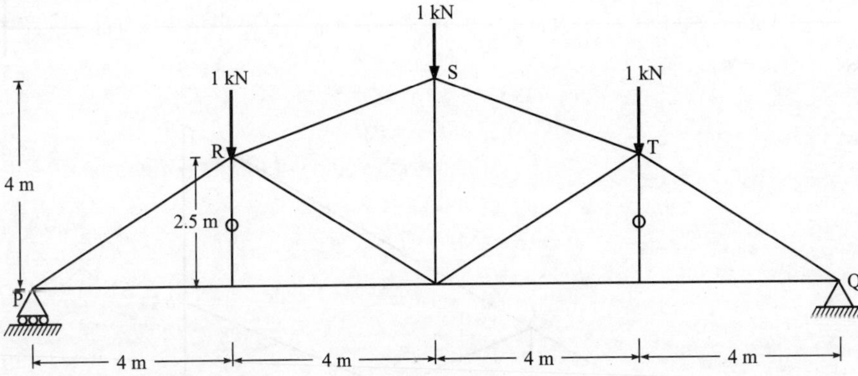


**Fig. 9.85**

***Solution*** First of all, let us find the reactions $R_A$ and $R_Q$ at supports *P* and *Q*, respectively. Since the frame is symmetrical, therefore reaction at each support is given by

$$R_P = R_Q = \frac{1+1+1}{2} = 1.5 \text{ kN}$$

Draw a space diagram of the frame using some suitable scale and name the various members according to Bow's notation, as shown in Fig. 9.86 (*a*).

Now draw a vector diagram as discussed below:

1. Take any point *d* and draw a vertical line *de* upward equal to the reaction at *Q* (*i.e.* 1.5 kN) to some suitable scale.
2. From point *d*, draw *dk* parallel to the member *DK* and from point *e*, draw *ek* parallel to the member *EK*, intersecting both the lines at *k*. Now *dek* is the vector diagram for the joint *Q*. For the equilibrium of the joint *Q*, the direction of *ek* and *kd* will be in order, as shown in Fig. 9.86 (*b*).

These directions are marked near the joint $Q$ on the space diagram. Thus, the member $DK(QT)$ is in compression and member $EK(NQ)$ is in tension.

3. Now draw a vertical line $ea$ upward equal to the reaction at $P$ (*i.e.* 1.5 kN) to the same scale. From point $a$, draw a line $af$ parallel to the member $AF$ and from point $e$ draw a line $ef$ parallel to the member $EF$, intersecting both lines at $f$. Now $eaf$ is the vector diagram for the joint $P$. For the equilibrium of the joint $P$, the directions of $af$ and $fe$ will be in order, as shown in Fig. 9.86 ($b$). These directions are marked near the joint $P$ on the space diagram. Thus, the member $AF(PR)$ is in compression and member $EF(PL)$ is in tension.

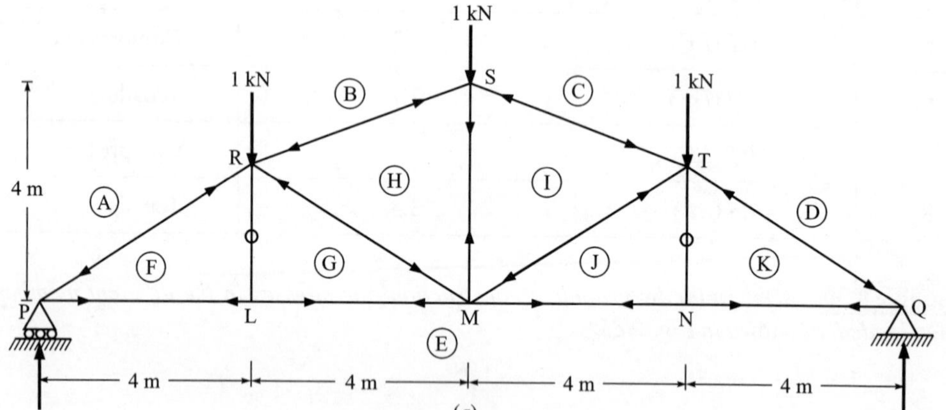


(*a*)

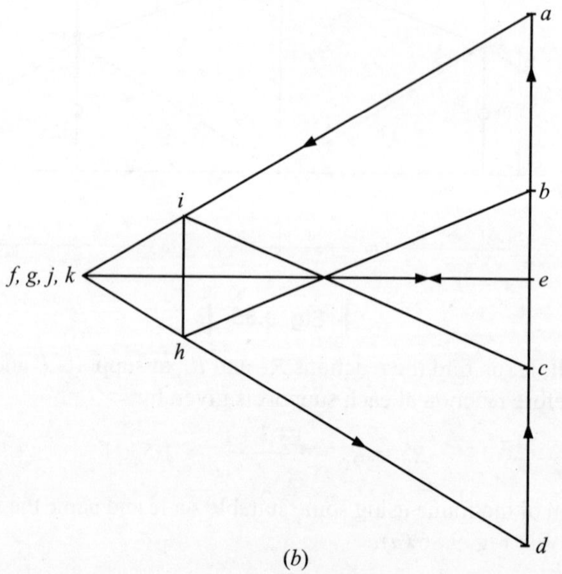


(*b*)

**Fig. 9.86**

4. Similarly, draw the vector diagram $abhg$, $bcih$ and $cdke$ for the joints $R$, $S$ and $T$ and mark the directions. Since the points $f$, $g$, $j$ and $k$ are at one point, therefore the force in the members $FG\ (RL)$ and $JK(TN)$ is zero.

On measurement from the vector diagram, the magnitude and nature of forces are tabulated below:

| *S. No.* | *Member of the truss* | *Magnitude of force in kN* | *Nature of force* |
|---|---|---|---|
| 1. | *PR* (*AF*) and *QT*(*KD*) | 2.9 kN | Compressive |
| 2. | *RS*(*BH*) and *ST*(*CI*) | 2.2 kN | Compressive |
| 3. | *RM* (*GH*) and *TM*(*IJ*) | 0.57 kN | Compressive |
| 4. | *PL*(*EF*), *LM*(*EG*), *MN*(*EJ*) *and NQ*(*EK*) | 2.5 kN | Tensile |
| 5. | *RL*(*FG*) and *TN*(*JK*) | 0 | — |
| 6. | *SM*(*IH*) | 0.57 kN | Tensile |

**Example 9.31** *Find the forces in the members of the truss, shown in Fig. 9.87.*

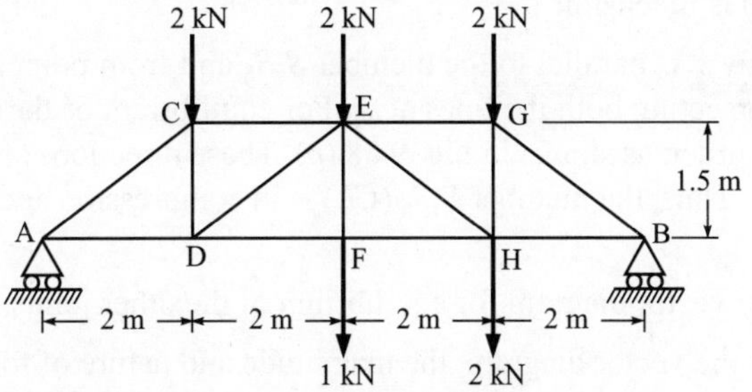


**Fig. 9.87**

***Solution*** First of all, let us find the reactions $R_A$ and $R_B$ at supports *A* and *B*, respectively [Refer Fig. 9.88 (*a*)].

Taking moments about *A*, we have

$$R_B \times AB = (2 + 2)AH + (2 + 1)AF + 2 \times AD$$

$$R_B \times 8 = 4 \times 6 + 3 \times 4 + 2 \times 2 = 24 + 12 + 4 = 40$$

$$\therefore \quad R_B = \frac{40}{8} = 5 \text{ kN}$$

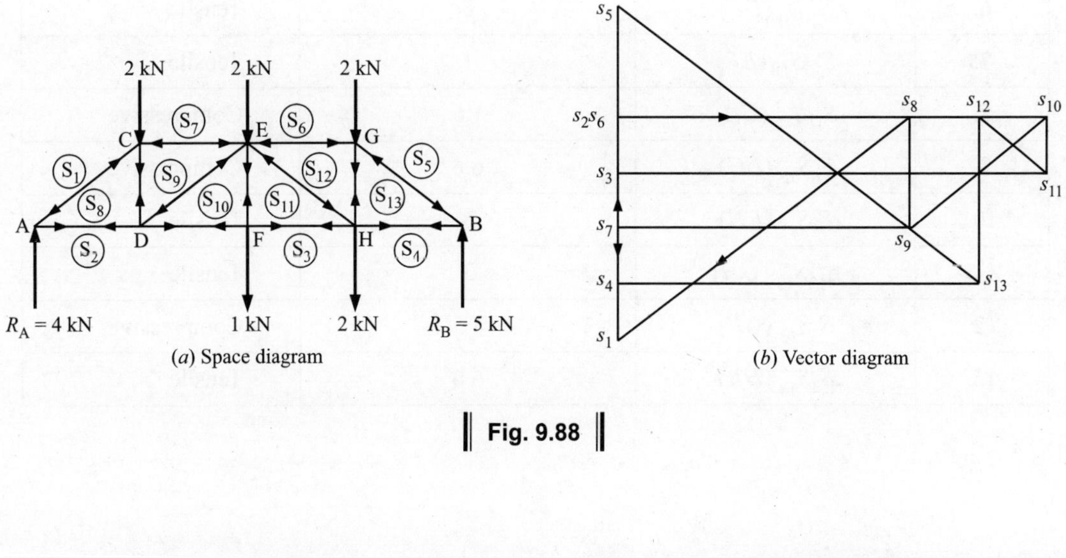


(*a*) Space diagram (*b*) Vector diagram

**Fig. 9.88**

We know that $R_A + R_B$ = Total load

$$= (2 + 2) + (2 + 1) + 2 = 9 \text{ kN}$$

$$\therefore \quad R_A = 9 - R_B = 9 - 5 = 4 \text{ kN}$$

Draw the space diagram of the truss to some suitable scale and name the various members according to Bow's notation, as shown in Fig. 9.88 (*a*). Now draw the vector diagram as discussed below:

1. Take any point $s_1$ and draw a vertical line $s_1s_5$ such that line $s_1s_2$ represents reaction $R_A$ = 4 kN to some suitable scale and line $s_4s_5$ represents reaction $R_B$ = 5 kN to the same scale. The lines $s_1s_7$, line $s_7s_6$ and line $s_6s_5$ represents loads at *C*, *E* and *G* each equal to 2 kN.
2. From point $s_1$, draw $s_1s_8$ parallel to the member $S_1S_8$ and $s_2s_8$ parallel to the member $S_2S_8$, intersecting both the lines at $s_8$. Now $s_1s_2s_8$ is the vector diagram for the joint at *A*. For equilibrium of the joint *A*, the direction of $s_2s_8$ and $s_8s_1$ will be in order, as shown in Fig. 9.88 (*b*). These directions are marked near the joint *A* on the space diagram. Thus, the member $S_1S_8$ (*AC*) is in compression and member $S_2S_8$ (*AD*) is in tension.
3. From point $s_7$, draw $s_7s_9$ parallel to the member $S_7S_9$ and from point $s_8$ draw a line parallel to the member $S_8S_9$, intersecting both the lines at $s_9$. For equilibrium of the joint *C*, the direction of $s_8s_9$ and $s_9s_7$ will be in order, as shown in Fig. 9.88 (*b*). These directions are marked near the joint *C* on the space diagram. Thus, the member $S_7S_9$ (*CE*) is in compression and the member $S_8S_9$ (*CD*) is in tension.
4. Similarly, draw the vector diagram for equilibrium of the other joints, as shown in Fig. 9.88 (*b*).

On measurement from the vector diagram, the magnitude and nature of forces are tabulated below:

| *S. No.* | *Member of the truss* | *Magnitude of force in kN* | *Nature of force* |
|---|---|---|---|
| 1. | $S_1S_8$ (*AC*) | 6.7 | Compressive |
| 2. | $S_2S_8$ (*AD*) | 5.4 | Tensile |
| 3. | $S_8S_9$ (*CD*) | 2 | Tensile |
| 4. | $S_7S_9$ (*CE*) | 5.4 | Compressive |
| 5. | $S_9S_{10}$ (*DE*) | 3.3 | Compressive |
| 6. | $S_2S_{10}$ (*DF*) | 8 | Tensile |
| 7. | $S_{10}S_{11}$ (*EF*) | 1 | Tensile |
| 8. | $S_{11}S_{12}$ (*EH*) | 1.6 | Compressive |
| 9. | $S_6S_{12}$ (*EG*) | 6.6 | Compressive |
| 10. | $S_3S_{11}$ (*FH*) | 8 | Tensile |
| 11. | $S_{12}S_{13}$ (*GH*) | 3 | Tensile |
| 12. | $S_5S_{13}$ (*GB*) | 8.5 | Compressive |
| 13. | $S_4S_{13}$ (*HB*) | 6.6 | Tensile |

**Example 9.32** *Determine the forces in all the members of the frame, as shown in Fig. 9.89.*

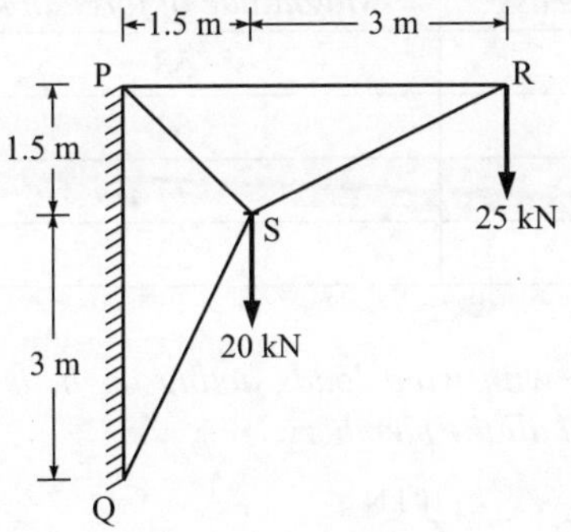


**Fig. 9.89**

***Solution*** First of all, draw the space diagram of the frame to some suitable scale and name the various members according to Bow's notation, as shown in Fig. 9.90 (*a*). Now draw the vector diagram as discussed below:

1. Take any point *a* and draw a vertical line downward such that *ab* = 25 kN and *bc* = 20 kN, to some suitable scale.
2. From point *a*, draw *ae* parallel to the member *AE* and from point *b*, draw *be* parallel to the member *BE*, intersecting both lines at *e*. Now *abe* is the vector diagram for the joint *R*. For equilibrium of the joint *R*, the directions of *be* and *ea* will be in order, as shown in Fig. 9.90 (*b*). These directions are marked near the joint *R* on the space diagram. Thus the member *AE* (*RP*) is in tension and member *BE* (*RS*) is in compression.
3. Now from point *c*, draw *cd* parallel to the member *CD* and from point *e*, draw *ed* parallel to the member *ED*, intersecting both the lines at *d*. Now *bc de* is the vector diagram for the joint *S*. For equilibrium of the joint *S*, the directions of *cd* and *de* will be in order. These directions are marked near the joint *S* on the space diagram. Thus the member *CD* (*SQ*) is in compression and member *DE* (*SP*) is in compression.

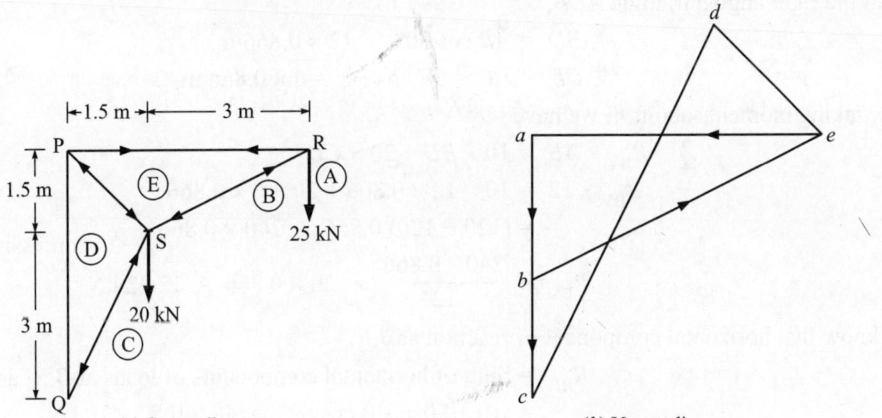


(*a*) Space diagram (*b*) Vector diagram

**Fig. 9.90**

On measurement from the vector diagram, the magnitude and nature of forces are tabulated below:

| *S. No.* | *Member of the truss* | *Magnitude of force in kN* | *Nature of force* |
|---|---|---|---|
| 1. | *RP* (*AE*) | 58 | Tensile |
| 2. | *RS* (*BE*) | 63 | Compressive |
| 3. | *SQ* (*CD*) | 77 | Compressive |
| 4. | *SP* (*DE*) | 33 | Compressive |

**Example 9.33** *A roof truss with wind loads acting on it, is shown in Fig. 9.91. Determine the reactions at the support and forces in all the members.*

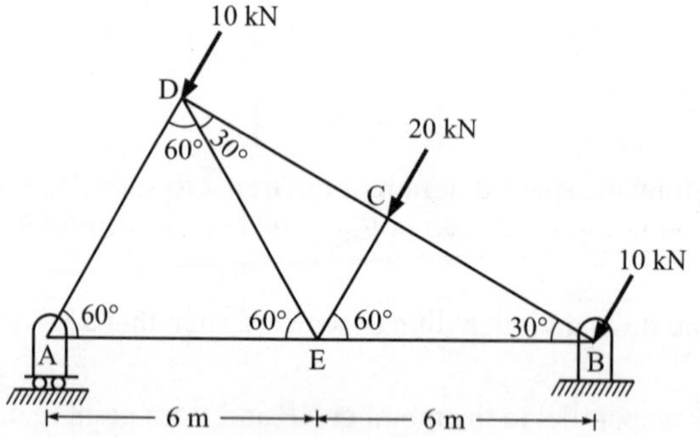


**Fig. 9.91**

***Solution***

***Reaction at the supports***

Since the truss at *A* is on rollers, therefore there will be a vertical reaction $R_A$. The end *B* of the truss is hinged, therefore there will be a vertical reaction $R_{BV}$ and a horizontal reaction $R_{BH}$, as shown in Fig. 9.92 (*a*). The reaction at this end *B* will be the resultant of $R_{BV}$ and $R_{BH}$, which will be inclined at an angle α with the horizontal.

From the right angled triangle *ADB*,

$$BD = AB \cos 30° = 12 \times 0.866 \text{ m}$$

$$CD = BC = BE \cos 30° = 6 \times 0.866 \text{ m}$$

Now taking moments about *A*, we have

$$R_{BV} \times AB = 10 \times BD + 20 \times CD$$

$$R_{BV} \times 12 = 10 \times 12 \times 0.866 + 20 \times 6 \times 0.866$$

$$= (120 + 120)\, 0.866 = 240 \times 0.866$$

$$\therefore \quad R_{BV} = \frac{240 \times 0.866}{12} = 20 \times 0.866 = 17.32 \text{ kN}$$

We know that horizontal component of reaction at *B*,

$$R_{BH} = \text{Sum of horizontal components of loads at } B, C \text{ and } D$$

$$= (10 + 20 + 10) \cos 60° = 40 \times 0.5 = 20 \text{ kN}$$

$$\therefore \quad \text{Reaction at } B,\ R_B = \sqrt{(R_{BV})^2 + (R_{BH})^2} = \sqrt{(17.32)^2 + (20)^2}$$

$$= \sqrt{300 + 400} = \sqrt{700} = 26.45 \text{ kN} \quad \textbf{Ans.}$$

and its inclination with the horizontal is given by

$$\tan \alpha = \frac{R_{BV}}{R_{BH}} = \frac{17.32}{20} = 0.866$$

$$\therefore \qquad \alpha = \tan^{-1}(0.866) = 41.9°$$

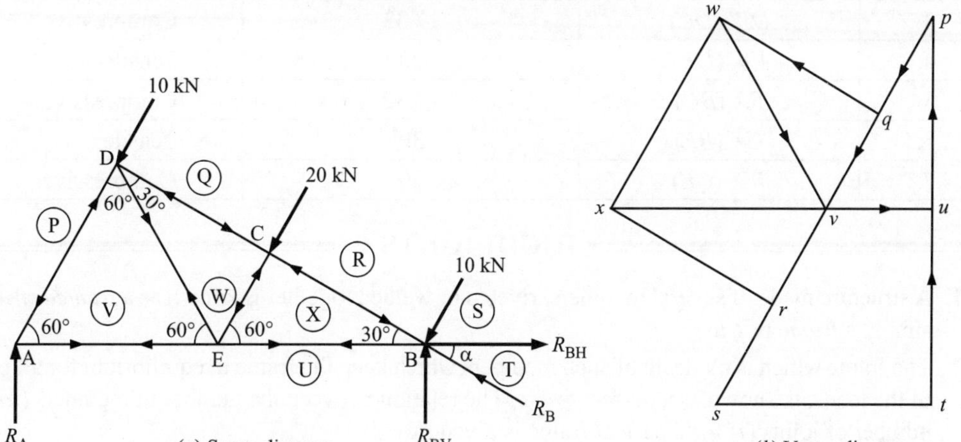


(*a*) Space diagram (*b*) Vector diagram

**Fig. 9.92**

The reaction at $A$ (*i.e.* $R_A$) is equal to the algebraic sum of vertical component of 10 kN at $D$; 20 kN at $C$ and 10 kN at $B$.

$$\therefore \qquad R_A = (10 \sin 60° + 20 \sin 60° + 10 \sin 60°) - R_{BV}$$
$$= (10 \times 0.866 + 20 \times 0.866 + 10 \times 0.866) - 17.32$$
$$= 40 \times 0.866 - 17.32 = 34.64 - 17.32 = 17.32 \text{ kN} \quad \textbf{Ans.}$$

***Forces in all the members***

First of all, draw the space diagram of the truss using some suitable scale and name the various members according to Bow's notation, as shown in Fig. 9.92 (*a*). Now draw a vector diagram as discussed below:

1. Take any point $t$ and draw a vertical line $tp$ upwards such that $tu$ and $up$ represents vertical reactions $R_{BV}$ and $R_A$ each equal to 17.32 kN, to some suitable scale.
2. From point $p$, draw $pv$ parallel to the member $PV$ and from point $u$ draw $uv$ parallel to the member $UV$, intersecting both the lines at $v$. Now $pvu$ is the vector diagram for the joint $A$. For equilibrium of the joint $A$, the direction of $pv$ and $uv$ will be in order, as shown in Fig. 9.92 (*b*). These directions are marked near the joint $A$ on the space diagram. Thus, the member $PV(AD)$ is in compression and the member $VU(AE)$ is in tension.
3. From point $p$, draw $pq$ parallel to the load at $D$ (*i.e.* 10 kN) to the same scale. Now from point $q$, draw $qw$ parallel to the member $QW$ and from point $v$ draw $vw$ parallel to the member $VW$, intersecting both the lines at $w$. Now $qwv$ is the vector diagram for the joint $D$. For equilibrium of the joint $D$, the direction of $qw$ and $wv$ will be in order, as shown in Fig. 9.92 (*b*). These directions are marked near the joint $D$ on the space diagram. Thus the member $QW(DC)$ is in compression and member $WV(DE)$ is in tension.
4. Draw $ts$ parallel to horizontal reaction $R_{BH}$ and equal to 20 kN and $sr$ parallel to load at $B$ and equal to 10 kN. It may be noted that $ps$ is one line such that $pq$ = 10 kN; $qr$ = 20 kN and $rs$ = 10 kN. Now complete the vector diagram as shown in Fig. 9.92 (*b*).

On measurement from the vector diagram, the magnitude and nature of forces are tabulated as below:

| *S. No.* | *Member of the truss* | *Magnitude of force in kN* | *Nature of force* |
|---|---|---|---|
| 1. | *PV* (*AD*) | 20 | Compressive |
| 2. | *UV* (*AE*) | 10 | Tensile |
| 3. | *QW* (*DC*) | 17.32 | Compressive |
| 4. | *VW* (*DE*) | 20 | Tensile |
| 5. | *RX* (*BC*) | 17.32 | Compressive |
| 6. | *UX* (*BE*) | 30 | Tensile |
| 7. | *WX* (*CE*) | 20 | Compressive |

## HIGHLIGHTS

1. A structure made of several members, riveted or welded together is known as a *framed structure* or simply a *frame* or *truss*.
2. The frame which is made up of such members which keep the frame in equilibrium for all positions of the loads, is known as a *perfect frame.* The relation between the number of members ($n$) and the number of joints ($j$) for a perfect frame is given by
$$n = 2j - 3$$
3. A frame in which the number of members ($n$) and the number of joints ($j$) does not satisfy the equation ($n = 2j - 3$), is called *imperfect frame.* The imperfect frame may be deficient or redundant frame.
4. The frame which has number of members ($n$) less than ($2j - 3$), is known as *deficient frame.*
5. The frame which has number of members ($n$) more than ($2j - 3$), is known as *redundant frame.*
6. If the force in the member pulls the joint to which it is connected, then it is said to be a *tensile force.* On the other hand, if the force in the member pushes the joint to which it is connected, then it is said to be a *compressive force.*
7. If three forces act at a joint and two of them are collinear, then for equilibrium of the joint, the third force should be equal to zero.
8. The force in the members of a truss are determined by the following three methods:
(*a*) Method of joints; (*b*) Method of sections; and (*c*) Graphical method.
The method of joints is mostly used when forces in a few members of a truss are to be determined.

## NUMERICAL PROBLEMS

1. Find the forces in the members *AB*, *AC* and *BC* of a triangular truss, as shown in Fig. 9.93.
[**Ans.** $P_{AB}$ = 8.66 kN (tensile); $P_{AC}$ = 17.32 kN (compressive); $P_{BC}$ = 10 kN (compressive)]

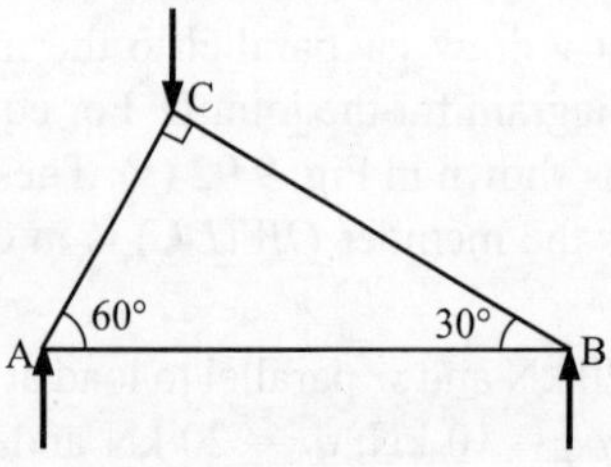


**Fig. 9.93**

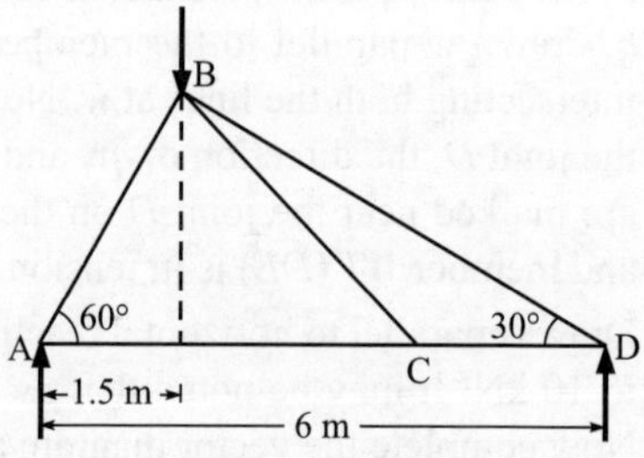


**Fig. 9.94**

**2.** Determine the forces in all members of the truss, as shown in Fig. 9.94.

[**Ans.** $P_{AB}$ = 17.32 kN (compressive); $P_{AC}$ = 8.66 kN (tensile); $P_{DB}$ = 10 kN (compressive); $P_{DC}$ = 8.66 kN (tensile); $P_{CB}$ = 18.03 kN (tensile)]

**3.** A frame, as shown in Fig. 9.95, carries a load of 20 kN at the joint *D* and 24 kN at joint *C*. It is simply supported at *A* and *E*. Find the load carried by each member.

[**Ans.** $P_{AC}$ = 40 kN (compressive); $P_{AB}$ = 34.64 kN (tensile); $P_{CD}$ = 28 kN (tensile); $P_{CB}$ = 20.784 kN (compressive); $P_{DB}$ = 20.784 kN (tensile); $P_{DE}$ = 27.712 kN (compressive); $P_{BE}$ = 13.856 kN (tensile)]

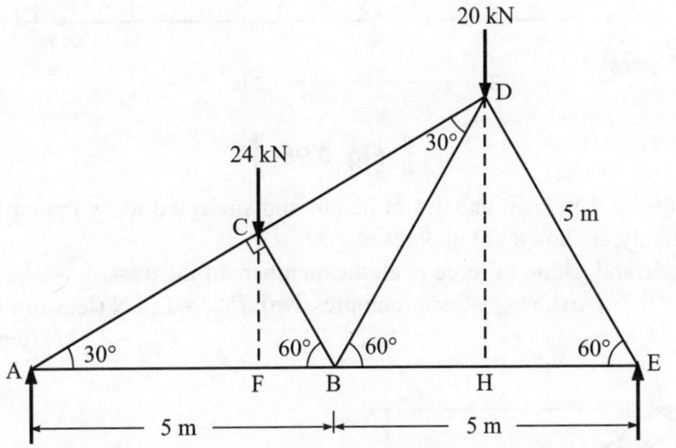


**Fig. 9.95**

**4.** Find the forces in all the members of the truss, as shown in Fig. 9.96.

[**Ans.** $P_{AB}$ = 2.887 kN (compressive); $P_{AE}$ = 1.44 kN (tensile); $P_{BE}$ = 0.577 kN (tensile); $P_{BC}$ = 1.732 kN (compressive); $P_{CE}$ = 0.577 kN (compressive); $P_{CD}$ = 4.04 kN (compressive); $P_{DE}$ = 2.02 kN (tensile)]

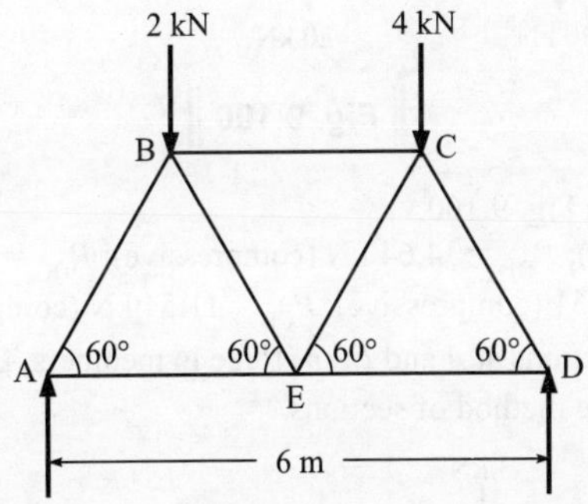


**Fig. 9.96**

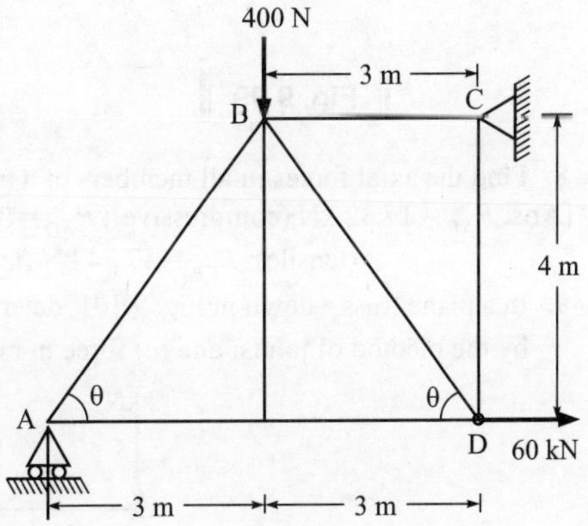


**Fig. 9.97**

**5.** A truss is loaded as shown in Fig. 9.97. Determine the force in each member of the truss.

[**Ans.** $P_{AB}$ = 750 N (compressive); $P_{AD}$ = 450 N (tensile); $P_{DB}$ = 250 N (tensile); $P_{DC}$ = 200 N (compressive); $P_{CB}$ = 600 N (compressive)]

**6.** Determine the forces in each member of the truss, shown in Fig. 9.98, and present your results in the form of a table.

[**Ans.** $P_{AB} = P_{BE} = P_{BC} = P_{EC} = P_{CD} = 0$; $P_{AE} = P_{ED} = 5$ kN (tensile)]

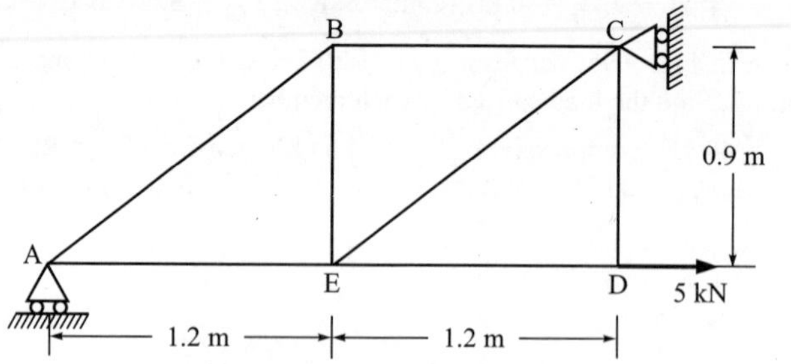


**Fig. 9.98**

**7.** A framed structure of 4 m span and 1.5 m height and subjected to two point loads of 8 kN and 12 kN at *B* and *D*, respectively, is shown in Fig. 9.99.

Find the magnitude and nature of force in all the members of the truss.

[**Ans.** $P_{AB} = 5$ kN (compressive); $P_{AD} = 12$ kN (tensile); $P_{BC} = 15$ kN (compressive) ; $P_{CD} = 12$ kN (tensile); $P_{BD} = 12$ kN (tensile)]

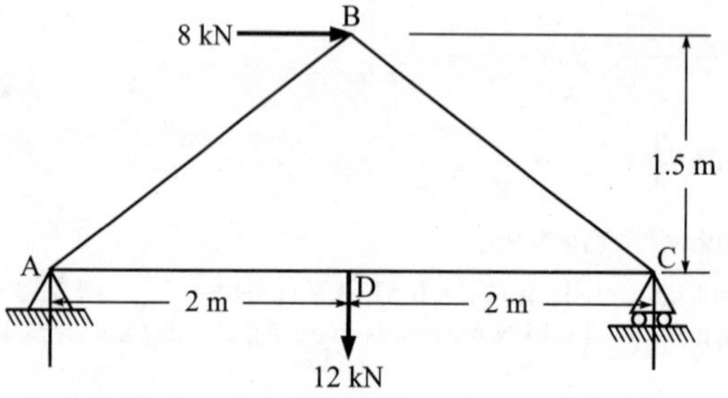


**Fig. 9.99**

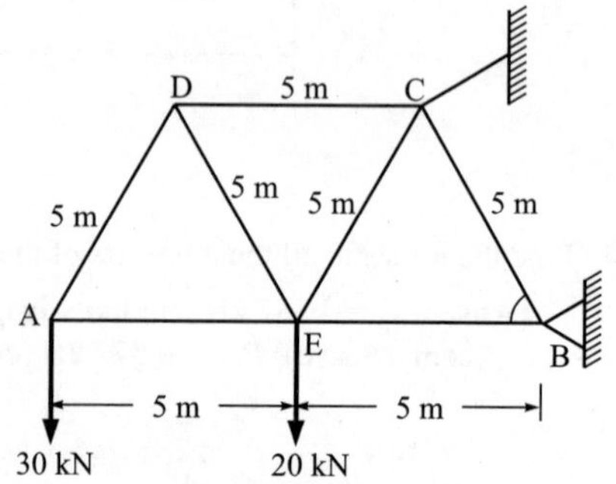


**Fig. 9.100**

**8.** Find the axial forces in all members of a truss, as shown in Fig. 9.100.

[**Ans.** $P_{AE} = 17.32$ kN (compressive); $P_{AD} = 34.64$ kN (tensile); $P_{DE} = 34.64$ kN (compressive); $P_{DC} = 34.64$ kN (tensile); $P_{CE} = 57.74$ kN (tensile); $P_{EB} = 63.51$ (compressive); $P_{CB} = 11.54$ kN (compressive)]

**9.** In a plane truss, shown in Fig. 9.101, determine (*a*) the reactions at *A* and *E*; (*b*) force in members *AB* and *AG* by the method of joints; and (*c*) force in member *CD* by the method of sections.

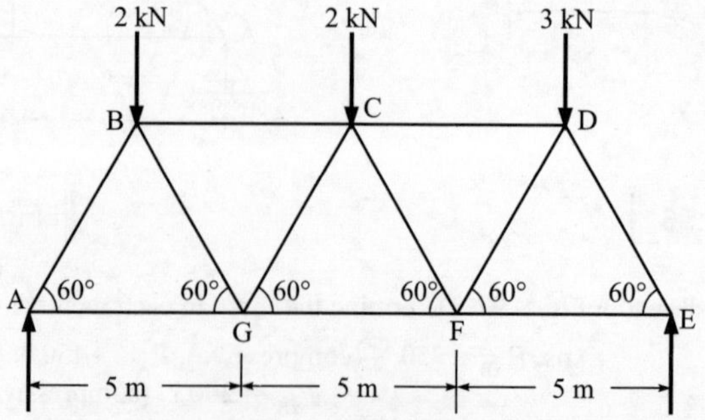


**Fig. 9.101**

[**Ans.** $R_A = 3.167$ kN; $R_E = 3.833$ kN; $P_{AG} = 3.6$ kN (tensile); $P_{AB} = 3.6$ kN (compressive); $P_{DC} = 4.42$ kN (compressive)]

10. Determine the magnitude and nature of forces in all member of the cantilever truss, as shown in Fig. 9.102.

[**Ans.** $P_{CD}$ = 12 kN (compressive); $P_{DE}$ = 6 kN (tensile); $P_{CB}$ = 12 kN (compressive); $P_{CE}$ = 12 kN (tensile); $P_{BE}$ = 24 kN (compressive); $P_{AE}$ = 24 kN (tensile)]

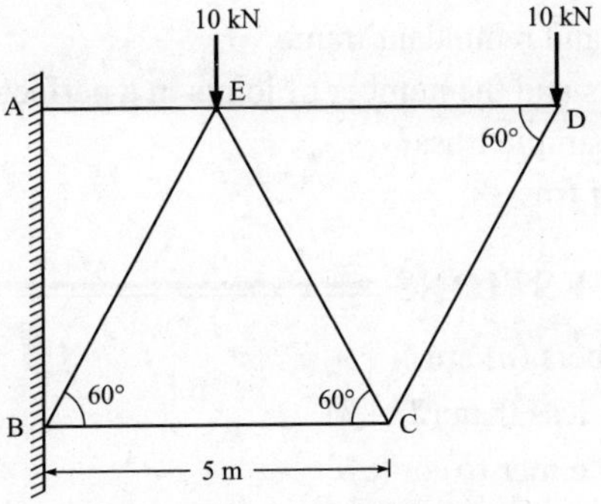


Fig. 9.102

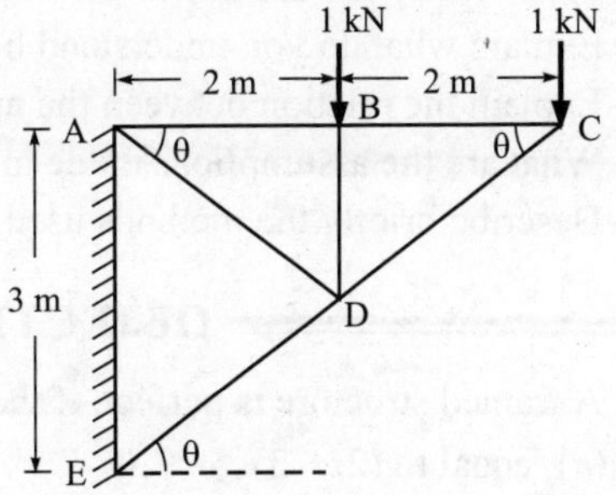


Fig. 9.103

11. Determine the forces in all the members of a cantilever truss, shown in Fig. 9.103.

[**Ans.** $P_{AB}$ = 1.33 kN (tensile); $P_{BC}$ = 1.33 kN (tensile); $P_{CD}$ = 1.67 kN (compressive); $P_{DE}$ = 2.5 kN (compressive); $P_{AD}$ =∴ 0.833 kN (tensile); $P_{BD}$ = 1 kN (compressive)]

12. Find the magnitude and nature of forces in all the members of the truss, shown in Fig. 9.104.

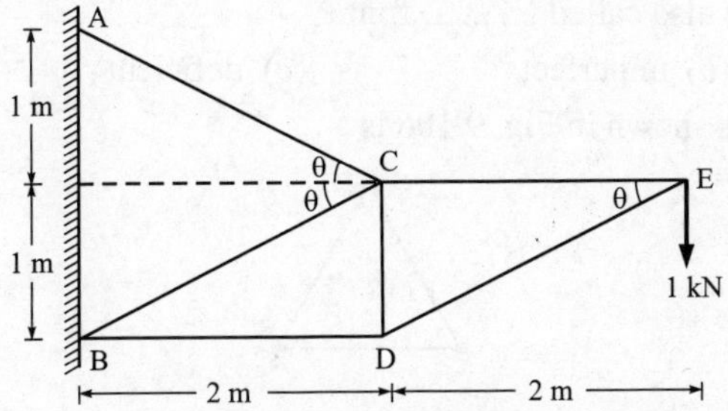


Fig. 9.104

[**Ans.** $P_{AB}$ and $P_{BC}$ = 0; $P_{AC}$ = 2.236 kN (tensile); $P_{BD}$ = 2 kN (compressive); $P_{CD}$ = 1 kN (tensile); $P_{ED}$ = 2.236 kN (compressive); $P_{EC}$ = 2 kN (tensile)]

13. Find the reactions at the supports and forces in all members of the truss, shown in Fig. 9.105.

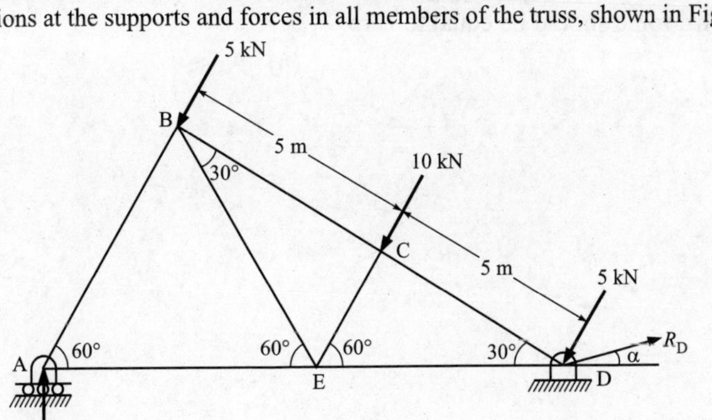


Fig. 9.105

[**Ans.** $R_A$ = 8.66 kN; $R_{DV}$ = 8.66 kN; $R_{DH}$ = 10 kN; $R_D$ = 13.23 kN; α = 40.9°; $P_{AB}$ = 10 kN (compressive); $P_{AE}$ = 5 kN (tensile); $P_{BC}$ = 8.66 kN (compressive); $P_{BE}$ = 10 kN (tensile); $P_{CD}$ = 8.66 kN (compressive); $P_{EC}$ = 10 kN (compressive); $P_{ED}$ = 15 kN (tensile)]

## REVIEW QUESTIONS

1. What do you mean by a frame or truss?
2. State clearly the difference between a perfect frame and an imperfect frame.
3. Explain what do you understand by deficient frame and redundant frame.
4. Explain the relation between the number of members and the number of joints in a perfect frame.
5. What are the assumptions made in the analysis of a simple truss?
6. Describe briefly the methods used for the analysis of trusses.

## OBJECTIVE TYPE QUESTIONS

1. A framed structure is perfect, if the number of members ($n$) are
   (*a*) equal to $(2j - 3)$ (*b*) less than $(2j - 3)$
   (*c*) greater than $(2j - 3)$ (*d*) either (*b*) or (*c*)
   where $j$ is the number of joints.
2. A framed structure is imperfect, if the number of members ($n$) are
   (*a*) equal to $(2j - 3)$ (*b*) less than $(2j - 3)$
   (*c*) greater than $(2j - 3)$ (*d*) either (*b*) or (*c*)
   where $j$ is the number of joints.
3. A redundant frame is also called ........... frame.
   (*a*) perfect (*b*) imperfect (*c*) deficient
4. A framed structure as shown in Fig. 9.106 is a

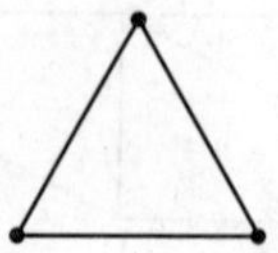

Fig. 9.106

   (*a*) perfect frame (*b*) deficient frame (*c*) redundant frame (*d*) none of these
5. When three forces in a truss act at a joint and two of them are collinear, then for equilibrium of the joint, the third force should be equal to zero.
   (*a*) True (*b*) False

***Answers to Objective Type Questions***

| | | | | |
|---|---|---|---|---|
| **1.** (*a*) | **2.** (*d*) | **3.** (*b*) | **4.** (*a*) | **5.** (*a*) |

# 10

# WORK, POWER, AND ENERGY



## 10.1 INTRODUCTION

When a force acting on a body is able to give it some displacement, then *work* is said to be done. For example, let a force $P$ acting on a body causes the body to move through a displacement $x$ in the direction of the force, as shown in Fig. 10.1 (*a*), then

$$\text{Work done} = \text{Force} \times \text{Distance moved in the direction of force} = P \times x$$

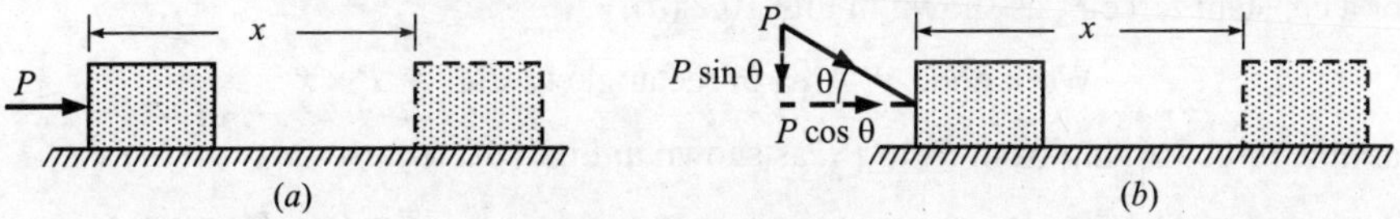


**Fig. 10.1**

In case the force acts at a certain angle (say θ) to the direction of displacement, as shown in Fig. 10.1 (*b*), then

$$\text{Work done} = \text{Component of the force in the direction of displacement} \times \text{Distance moved} = P\cos\theta \times x$$

It may be noted that

1. Only the component of the force in the direction of motion (*i.e.* $P\cos\theta$) does work.
2. No work is done by the normal component of the force ($P\sin\theta$) because, there is no displacement normal to the incline of its point of application.
3. When $\theta = 0$, the force $P$ acts in the direction of displacement.

$\therefore$ Work done $= P\cos\theta \times x = P\cos 0 \times x = P \times x$ ...(same as before)

4. When the point of application of the force moves in the direction of the force, then the work done by the force is said to be *positive.* On the other hand, if the point of application of the force moves in the opposite direction of the force, then the work done by the force is *negative.*

## 10.2 UNITS OF WORK

The unit of work depends upon the unit of force and displacement.

In S.I. system of units, the practical unit of work is N-m. It is the work done by a force of 1 newton, when it displaces a body through 1 metre. The work of 1 N-m is known as joule (briefly written as J) such that 1 N-m = 1 J.

It may be noted that while writing the unit of work, it is a general practice to put the unit of force first followed by the unit of displacement (*e.g.* N-m).

## 10.3 GRAPHICAL REPRESENTATION OF WORK DONE BY A CONSTANT AND VARIABLE FORCE

If we plot a graph between the force during any displacement along the *Y*-axis and the corresponding displacement along the *X*-axis, then area under the graph represents the work done.

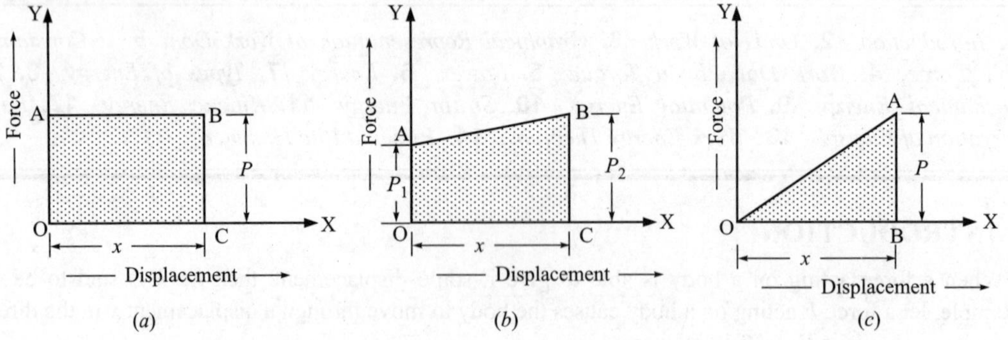


**Fig. 10.2**

Thus, for a constant force *P*, as shown in Fig. 10.2 (*a*),

$$\text{Work done} = \text{Area of rectangle } OABC = P \times x$$

For a variable force varying from $P_1$ to $P_2$, as shown in Fig. 10.2 (*b*),

$$\text{Work done} = \text{Area of trapezium } OABC = \frac{P_1 + P_2}{2} \times x$$

$$= \frac{\text{Initial force} + \text{Final force}}{2} \times \text{Displacement}$$

$$= \frac{\text{Average force}}{2} \times \text{Displacement}$$

and for a variable force varying from zero to *P*, as shown in Fig. 10.2 (*c*),

$$\text{Work done} = \text{Area of triangle } OAB = \frac{1}{2} P \times x = \text{Average force} \times \text{Displacement}$$

...(In this case, initial force is zero and final force is *P*)

**Note:** The classic example of work done by a variable force is a coil spring.

## 10.4 WORK DONE BY A TORQUE

Consider a circular body of radius $r$, which can rotate about the point $O$, as shown in Fig. 10.3. Let a tangential force $P$ is applied at $A$ and body moves through a small angle $\theta$. The distance moved by the force in rotating the body through an angle $\theta$ is equal to the length of the arc $AB$.

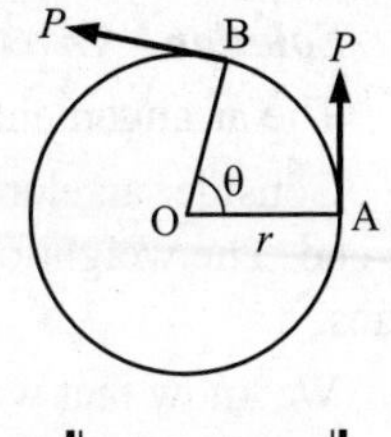


Fig. 10.3

We know that length of the arc $AB$ = Radius × Angle turned = $r \cdot \theta$

∴ Work done = Force × Distance moved = $P(r \cdot \theta)$

But $P.r$ = Torque ($T$)

∴ Work done = $P(r.\theta) = T.\theta$

where $\theta$ is the angular displacement in radians.

**Example 10.1** *Calculate the work done by the force on a body weighing 500 N, when it moves through a distance of 5 m along a horizontal surface*

*(a) by a horizontal force of 250 N ; and*

*(b) by a force of 200 N whose line of action makes an angle of 30° with the horizontal.*

***Solution*** Given : *$W$ = 500 N ; $x$ = 5 m

***(a) Work done by a horizontal force P = 250 N***

We know that Work done = $P \times x = 250 \times 5 = 1250$ N-m **Ans.**

***(b) Work done by a force P = 200 N making an angle (θ = 30°) with the horizontal***

We know that Work done = $P \cos \theta \times x = 200 \cos 30° \times 5 = 866$ N-m **Ans.**

**Example 10.2** *A body of weight of 2000 N is placed on a horizontal surface. An initial force of 750 N applied horizontally is sufficient just to move the body. The applied force is not constant but varies uniformly with the distance moved by the body. If the body moves a distance of 25 m and the force applied at any distance is given by the relation $P = P_0 + 10x$, find the work done by the applied force. $P_0$ is the initial force and x is the distance moved in metres.*

***Solution*** Given : *$W$ = 2000 N ; $P_0$ = 750 N ; $x$ = 25 m

The relation between the force and distance is

$$P = P_0 + 10x = 750 + 10 \times 25 = 1000 \text{ N}$$

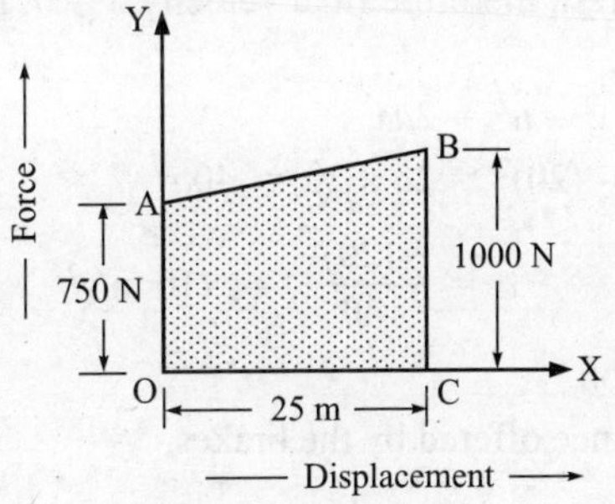


Fig. 10.4

The force-displacement diagram is shown in Fig. 10.4. The area of this diagram, which is a trapezium, gives the work done.

∴ Work done = Area of trapezium $OABC$ = $\frac{1}{2}[750 + 1000]\,25$

= 21 875 N-m = 21.875 kN-m **Ans.**

*Superfluous data.

**Example 10.3** *Determine the work done by an electric motor in winding up a uniform cable, which hangs from a hoisting drum if its free length is 10 m and weighs 500 N. The drum is rotated by the motor.*

***Solution*** Given : $l = 10$ m ; $W = 500$ N

The arrangement is shown in Fig. 10.5.

Consider an elementary length $dx$ of the cable at a distance $x$ from the free end. The weight of this element is to be raised to a distance of $(10 - x)$ metres.

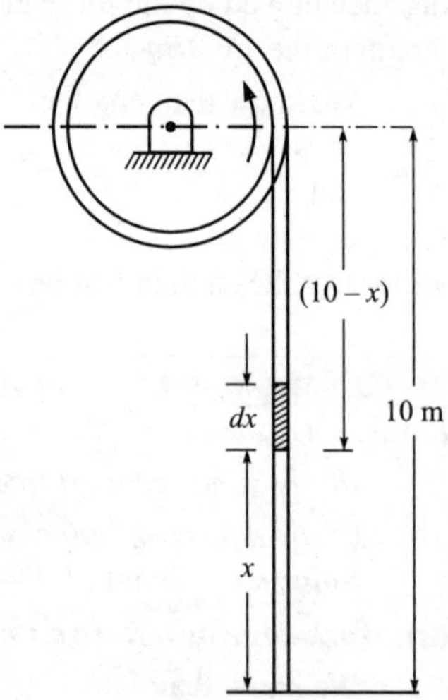


**Fig. 10.5**

We know that weight of this element = Weight of cable per metre length × Length of element

$$= \frac{500}{10} \times dx = 50\,dx$$

∴ Work done on the element, $dW$ = Weight × Distance raised

$$= 50\,dx\,(10 - x)$$
$$= 500\,dx - 50x\,.\,dx$$

The total work done by the electric motor is obtained by integrating the above expression within the limits from 0 to 10 m.

$$\therefore \quad \int dW = \int_0^{10} 500\,dx - 50x\,dx$$

or

$$W = \left[500\,x - \frac{50\,x^2}{2}\right]_0^{10}$$

$$= 500 \times 10 - \frac{50\,(10)^2}{2} = 5000 - 2500 = 2500 \text{ N-m} \quad \textbf{Ans.}$$

**Example 10.4** *A truck weighing 15 kN and moving at a speed of 72 km/h is brought to rest in 20 metres by the application of the brakes. Find the work done by the resistance offered by the brakes.*

***Solution*** Given : $W = 15$ kN $= 15 \times 10^3$ N ; $u = 72$ km/h $= \dfrac{72 \times 1000}{60 \times 60} = 20$ m/s ; $x = 20$ m

Since the truck is brought to rest, therefore final velocity, $v = 0$. Let $a$ be the retardation in m/s². Using the relation,

$$v^2 - u^2 = 2ax$$
$$0 - (20)^2 = 2a \times 20 = 40a$$
$$\therefore \quad a = \frac{-(20)^2}{40} = -10 \text{ m/s}^2$$

The –ve sign shows retardation.

We know that force or resistance offered by the brakes,

$$P = \text{Mass} \times \text{Retardation} = \frac{W}{g} \times a$$

$$= \frac{15 \times 10^3}{g} \times 10 = \frac{15 \times 10^3}{10} \times 10 = 15 \times 10^3 \text{ N}$$

...(Taking $g = 10$ m/s²)

$$\therefore \quad \text{Work done} = P \times x = 15 \times 10^3 \times 20 = 300 \times 10^3 \text{ N-m} = 300 \text{ kN-m} \quad \textbf{Ans.}$$

**Example 10.5** *A log of wood weighing 5 kN is pulled for 5 metres up an inclined plane having an angle of inclination of 30° with the horizontal. Calculate the work done.*

***Solution*** Given : $W = 5$ kN $= 5000$ N ; $x = 5$ m ; $\theta = 30°$

The weight of the wooden log has two components; one parallel to the inclined plane and the other perpendicular to the inclined plane, as shown in Fig. 10.6. The component parallel to the plane is $W \sin\theta$, against which the work is to be done, *i.e.*

$$P = W \sin\theta = 5000 \sin 30° = 2500 \text{ N}$$

$$\therefore \quad \text{Work done} = P \times x = 2500 \times 5 = 12\,500 \text{ N-m} = 12.5 \text{ kN-m} \quad \textbf{Ans.}$$

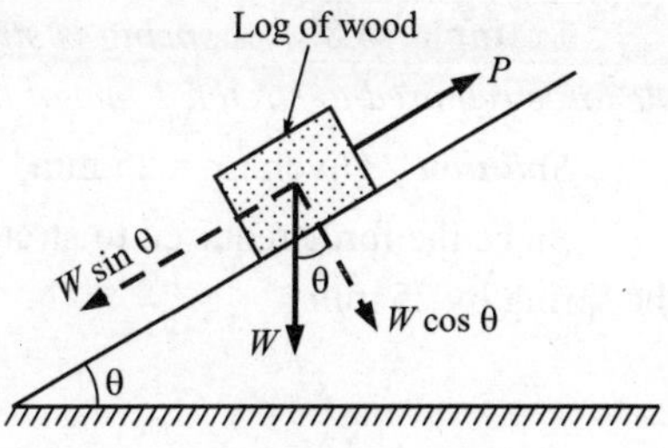


**Fig. 10.6**

**Example 10.6** *A wooden block weighing 2000 N is pulled horizontally on a floor by a rope through a distance of 10 m. If the coefficient of friction is 0.3, find the work done.*

***Solution*** Given : $W = 2000$ N ; $x = 10$ m ; $\mu = 0.3$

We know that force of friction (Refer Fig. 10.7),

$$F = \mu.R_N = \mu.W \quad ...(\because R_N = W) = 0.3 \times 2000 = 600 \text{ N}$$

Thus, the force required to pull the block horizontally,

$$P = F = 600 \text{ N}$$

$$\therefore \quad \text{Work done} = P \times x = 600 \times 10 = 6000 \text{ N-m} \quad \textbf{Ans.}$$

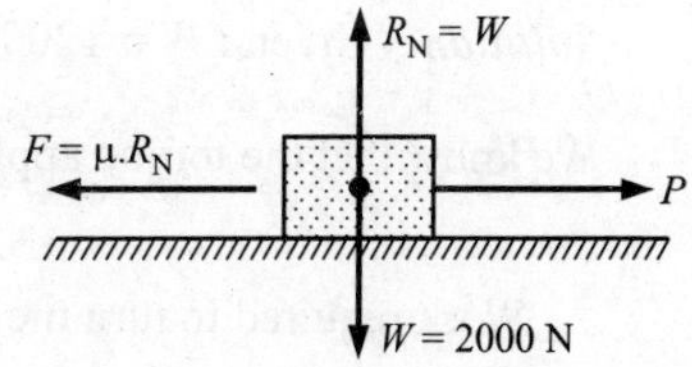


**Fig. 10.7**

**Example 10.7** *A block of wood of weight 800 N is pulled up on an inclined plane having an inclination of 30°. The surface of the inclined plane is rough having coefficient of friction 0.3. Determine the work done in pulling the block up for a distance of 5 metres.*

***Solution*** Given : $W = 800$ N ; $\theta = 30°$ ; $\mu = 0.3$ ; $x = 5$ m

The wooden block is subjected to the following forces, as shown in Fig. 10.8.

1. The weight of the block ($W$) acting vertically downwards;
2. Component of weight ($W$) perpendicular to the inclined plane and is equal to $W \cos\theta$;
3. Component of weight ($W$) parallel to the inclined plane and is equal to $W \sin\theta$; and
4. Friction force, $F = \mu.R_N = \mu W \cos\theta$.

where $R_N$ = Normal reaction = $W \cos\theta$

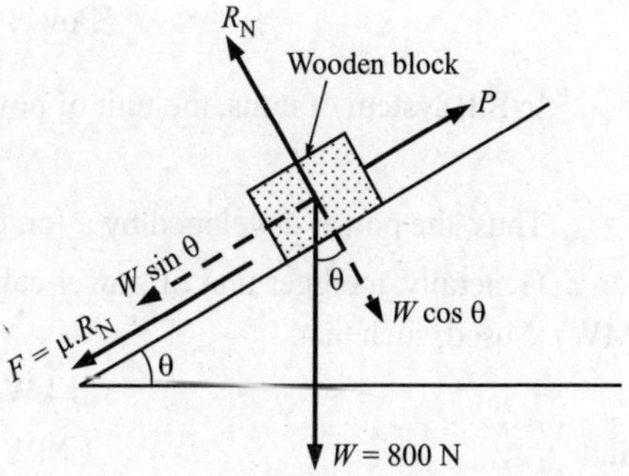


**Fig. 10.8**

We know that force required to pull the wooden block,

$$P = W \sin\theta + F = W \sin\theta + \mu W \cos\theta$$

$$= 800 \sin 30° + 0.3 \times 800 \cos 30°$$

$$= 800 \times 0.5 + 0.3 \times 800 \times 0.866 = 400 + 207.84 = 607.84 \text{ N}$$

$\therefore$ Work done in pulling the block up $= P \times x = 607.84 \times 5 = 3039.2$ N-m **Ans.**

**Example 10.8** *A spring is stretched by 25 mm by the application of a force. Find the work done, if the force required to stretch 1 mm of the spring is 20 N.*

***Solution*** Given : $x = 25$ mm

Since the force required to stretch 1 mm of the spring is 20 N, therefore total force required to stretch the spring by 25 mm

$$= 20 \times 25 = 500 \text{ N}$$

$$\text{Average force, } P = \frac{0 + 500}{2} = 250 \text{ N} \qquad \text{...}(\because \text{ Initial force} = 0)$$

$$\therefore \qquad \text{Work done} = P \times x = 250 \times 25 = 6250 \text{ N-mm} = 6.25 \text{ N-m} \textbf{ Ans.}$$

**Example 10.9** *A solid hemisphere of weight 120 N and radius 1 m rests with its flat face on a table. How many N-m of work is required to turn it over so that it may rest with its curved surface in tangential contact with the table?*

***Solution*** Given : $W = 120$ N ; $r = 1$ m ; $\theta = 90° = \dfrac{\pi}{2}$ rad

We know that the torque applied to the solid hemisphere,

$$T = W \times r = 120 \times 1 = 120 \text{ N-m}$$

$\therefore$ Work required to turn the solid hemisphere,

$$= \text{Torque} \times \text{Angular displacement} = T \cdot \theta$$

$$= 120 \times \frac{\pi}{2} = 188.52 \text{ N-m} \textbf{ Ans.}$$

## 10.5 POWER

The power is defined as the rate of doing work per unit time. Mathematically,

$$\text{Power} = \frac{\text{Work done}}{\text{Time taken}}$$

In S.I. system of units, the unit of power is watt (briefly written as W), such that

$$1 \text{ W} = 1 \text{ J/s or } 1 \text{ N-m/s} \qquad \text{...}(\because 1 \text{ J} = 1 \text{ N-m})$$

Thus, the power developed by a force of $P$ (in newtons) moving with a velocity $v$ (in m/s) in $P.v$ watt.

Generally, a bigger unit of power called kilowatt (briefly written as kW) or megawatt (briefly written as MW) is used, such that

$$1 \text{ kW} = 1000 \text{ W}$$

and
$$1 \text{ MW} = 1000 \text{ kW} = 10^6 \text{ W}$$

**Note:** If $T$ is the torque transmitted in N-m or J and $\omega$ is the angular speed in rad/s, then

$$\text{Power} = T.\omega = T \times \frac{2\pi N}{60} \text{ watt} \qquad \text{...}\left(\because \omega = \frac{2\pi N}{60}\right)$$

where $N$ is the speed in r.p.m.

**Example 10.10** *A train weighing 3000 kN is to be pulled by an engine on a horizontal track at constant speed of 60 km/h. If the resistance to motion is 6 N/kN of the weight of the train, calculate the power of the train.*

***Solution*** Given : $W = 3000$ kN ; $v = 60$ km/h $= \dfrac{60 \times 1000}{60 \times 60} = 16.67$ m/s

Since the resistance to motion is 6 N/kN, therefore force required to pull the train,

$$P = 6 \times 3000 = 18\,000 \text{ N} = 18 \text{ kN}$$

$\therefore$ Power of the engine $= P \,.\, v = 18 \times 16.67 = 300$ kW **Ans.**

**Example 10.11** *Determine the input power in kW of an electrical motor of a planer, if the length of the working stroke is 2 m and the time period is 10 seconds. The cutting force is equal to 12 kN and the efficiency of the machine is 0.8. Assume uniform motion.*

***Solution*** Given : $x = 2$ m ; $t = 10$ s ; $P = 12$ kN ; $\eta = 0.8$

We know that cutting speed of the motor,

$$v = \frac{x}{t} = \frac{2}{10} = 0.2 \text{ m/s}$$

and power of the motor $= P \,.\, v = 12 \times 0.2 = 2.4$ kN-m/s $= 2.4$ kW

Since the efficiency of the motor is 0.8, therefore input power

$$= \frac{2.4}{0.8} = 3 \text{ kW} \quad \textbf{Ans.}$$

**Example 10.12** *A train weighing 2500 kN moves on a level track with a velocity of 30 m/s. The tractive resistance is 5 N/kN. If the train is to move with an acceleration of 0.2 m/s², find the power of engine. Take g = 10 m/s².*

***Solution*** Given : $W = 2500$ kN $= 2500 \times 10^3$ N; $v = 30$ m/s ; $a = 0.2$ m/s$^2$

Since the tractive resistance is 5 N/kN, therefore force required to overcome resistance,

$$P_1 - 5 \times 2500 = 12\,500 \text{ N}$$

and force required to move the train with an acceleration ($a$),

$$P_2 = \text{Mass} \times \text{Acceleration} = \frac{W}{g} \times a$$

$$= \frac{2500 \times 10^3}{10} \times 0.2 = 50\,000 \text{ N}$$

$\therefore$ Total force required by the engine,

$$P = P_1 + P_2 = 12\,500 + 50\,000 = 62\,500 \text{ N} = 62.5 \text{ kN}$$

We know that power required $= P \,.\, v = 62.5 \times 30 = 1875$ kW **Ans.**

**Example 10.13** *A truck of weight 200 kN moves with a constant velocity of 60 km/h up an incline of 1 in 50. Find the power exerted by the engine, if the track resistance is 2.5 N/kN weight of the train.*

***Solution*** Given : $W = 200$ kN ; $v = 60$ km/h $= \dfrac{60 \times 1000}{60 \times 60} = 16.67$ m/s ; Slope of track, $\theta = \dfrac{1}{50}$

Since the track resistance is 2.5 N/kN, therefore total track resistance,

$$P_1 = 2.5 \times 200 = 500 \text{ N} = 0.5 \text{ kN}$$

The truck is moving up an incline of 1 in 50 or $\theta = \frac{1}{50}$. Therefore force required to move the truck up the plane,

$$P_2 = W \sin\theta = 200 \times \frac{1}{50} = 4 \text{ kN}$$

Total force required, $P = P_1 + P_2 = 0.5 + 4 = 4.5$ kN

$\therefore$ Power exerted by the engine $= P.v = 4.5 \times 16.67 = 75$ kW **Ans.**

**Example 10.14** *A tangential force of 180 kN is acting on a solid circular shaft of diameter 80 mm. Determine the power transmitted by the shaft when it runs at 160 r.p.m.*

***Solution*** Given : $P = 180$ kN $= 180 \times 10^3$ N ; $d = 80$ mm or $r = 40$ mm $= 0.04$ mm ; $N = 160$ r.p.m.

We know that the torque transmitted by the shaft,

$$T = P \times r = 180 \times 10^3 \times 0.04 = 7.2 \times 10^3 \text{ N-m}$$

$$\therefore \quad \text{Power transmitted} = \frac{T \times 2\pi N}{60} = \frac{7.2 \times 10^3 \times 2\pi \times 160}{60} = 120.65 \times 10^3 \text{ W}$$

$$= 120.65 \text{ kW} \quad \textbf{Ans.}$$

**Example 10.15** *A train of weight 2000 kN moves down a slope of 1 in 150 at 18 km/h and engine develops a power of 35 kW. If the train is pulled up at the same speed, what power will be required to pull the train?*

***Solution*** Given : $W = 2000$ kN $= 2000 \times 10^3$ N ; Slope $(\sin\theta) = \frac{1}{150}$ ; $v = 18$ km/h $= \frac{18 \times 1000}{60 \times 60}$ $= 5$ m/s ; Power $= 35$ kW $= 35 \times 10^3$ N-m/s

First of all, let us consider the motion of the train down the slope of 1 in 150, as shown in Fig. 10.9 (*a*). The forces acting on the train are as follows:

(*a*) Component of weight ($W$) in the direction of motion of the train (*i.e.* downwards) which is equal to $W \sin\theta$.

(*b*) Force exerted by the engine ($P$) in the direction of motion of the train, and

(*c*) Force of friction ($F$) in the opposite direction of motion of the train.

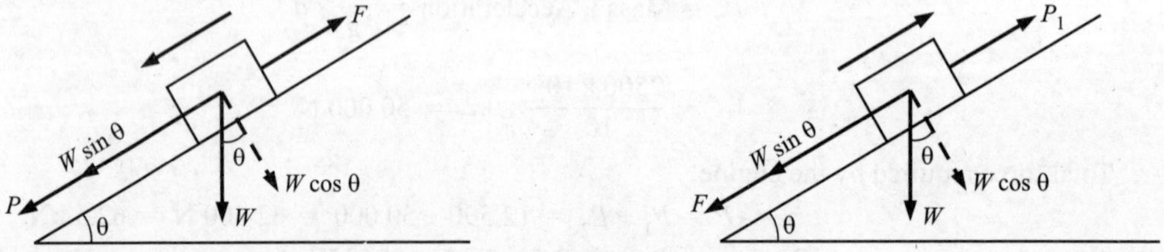


**Fig. 10.9**

$\therefore$ Net force acting on the engine in the direction of motion

$$= P + W\sin\theta - F = P + 2000 \times 10^3 \times \frac{1}{150} - F$$

$$= P + 13.33 \times 10^3 - F \qquad \text{...}(i)$$

We know that power exerted by the engine,

$$35 \times 10^3 = \text{Force exerted by the engine} \times \text{Velocity} = P \cdot v = P \times 5$$

$$\therefore \quad P = \frac{35 \times 10^3}{5} = 7 \times 10^3 \text{ N} \quad \text{...}(ii)$$

From equation (*i*),

$$F = P + 13.33 \times 10^3 = 7 \times 10^3 + 13.33 \times 10^3$$

$$= 20.33 \times 10^3 \text{ N} \quad \text{...}(iii)$$

Now consider the motion of the train up the inclined plane at the same speed, as shown in Fig. 10.9 (*b*). Let $P_1$ be the force exerted by the engine. In this case,

$$P_1 = F + W \sin\theta = 20.33 \times 10^3 + 2000 \times 10^3 \times \frac{1}{150}$$

$$= 20.33 \times 10^3 + 13.33 \times 10^3 = 33.66 \times 10^3 \text{ N} = 33.66 \text{ kN}$$

$\therefore$ Power developed by the engine $= P_1 \cdot v = 33.66 \times 5 = 168.3$ kN-m/s $= 168.3$ kW **Ans.**

**Example 10.16** *A locomotive draws a train of mass 400 tonnes, including its own mass, on a level ground with a uniform acceleration, until it acquires a velocity of 54 km/h in 5 minutes. If the frictional resistance is 40 N/tonne of mass and the air resistance varies with the square of the velocity, find the power of the engine. Take air resistance as 500 N at 18 km/h.*

***Solution*** Given : $m = 400$ t $= 400 \times 10^3$ kg ; $v = 54 \text{ km/h} = \frac{54 \times 1000}{60 \times 60} = 15$ m/s ; $t = 5$ min $= 300$ s ; Frictional resistance $= 40$ N/tonne $= 40 \times 400 = 16\,000$ N $= 16$ kN

Let $a$ be the uniform acceleration of the train. Since the initial velocity of the train is zero, therefore using the relation,

$$v = u + a \cdot t$$

$$15 = 0 + a \times 300$$

or

$$a = \frac{15}{300} = 0.05 \text{ m/s}^2$$

and force required for this acceleration, $F = m.a = 400 \times 10^3 \times 0.05 = 20 \times 10^3 \text{ N} = 20 \text{ kN}$

Since the air resistance varies with the square of the velocity, therefore air resistance at 54 km/h

$$= 500\left(\frac{54}{18}\right)^2 = 4500 \text{ N} = 4.5 \text{ kN}$$

Total resistance to overcome by the engine,

$$P = \text{Frictional resistance} + \text{Force required for accelerating} + \text{Air resistance}$$

$$= 16 + 20 + 4.5 = 40.5 \text{ kN}$$

$\therefore$ Power of the engine $= P \cdot v = 40.5 \times 15 = 607.5$ kW **Ans.**

## 10.6 ENERGY

The energy of a body is defined as its capacity of doing work. Any body which is capable of doing work is said to possess energy. In fact, work and energy are equivalent. Thus, the unit of energy is same as that of work.

## 10.7 TYPES OF ENERGY

The following are the various types of energy:

**1.** Mechanical energy;
**2.** Electrical energy;
**3.** Chemical energy;
**4.** Heat energy;
**5.** Pressure energy;
**6.** Sound energy;
**7.** Nuclear energy; and
**8.** Atomic energy.

It may be noted that energy can neither be created nor destroyed, but it can only be transformed from one form to another.

Though there are many types of energy, yet we shall only discuss mechanical energy, which is important from the subject point of view.

## 10.8 TYPES OF MECHANICAL ENERGY

The mechanical energy is equal to the work done on a body in altering either its position or its velocity. The following three types of mechanical energies are important:

1. Potential energy; 2. Strain energy; and 3. Kinetic energy.

These energies are discussed in the following articles.

## 10.9 POTENTIAL ENERGY

It is the energy possessed by a body doing work, by virtue of its position. It is also known as *datum energy* or *position energy.*

Consider a body of weight $W$ is raised to some height $h$ above the ground level. The work spent in rising the body is $W.h$. This work is stored in the body. When this body is allowed to fall back to the ground level, it will do the same amount of work, *i.e.* $W.h$.

$\therefore$ Potential energy, P.E. $= W.h = m.g.h$ ...($\because W = m.g$)

where $m$ = Mass of the body, and

$g$ = Acceleration due to gravity.

It may be noted that

***(a)*** When $W$ is in newtons and $h$ in metres, then potential energy will be in N-m.

***(b)*** When $m$ is in kg and $h$ in metres, then potential energy will also be in N-m as discussed below:

We know that potential energy, P.E. $= m.g.h = \text{kg} \times \frac{\text{m}}{\text{s}^2} \times \text{m} = \text{N-m}$ ...$\left(\because 1\text{ N} = \frac{1\text{ kg-m}}{\text{s}^2}\right)$

## 10.10 STRAIN ENERGY

It is the potential energy stored by an elastic body when deformed. A compressed spring possesses this type of energy, because it can do some work in recovering its original shape. Thus, if a compressed spring of stiffness $s$ newton per unit deformation (*i.e.* extension or compression) is deformed through a distance $x$ by a load $W$, then

$$\text{Strain energy} = \text{Work done} = \frac{1}{2}W \cdot x = \frac{1}{2}s \cdot x^2 \quad ...(\because W = s.x)$$

## 10.11 KINETIC ENERGY

It is the energy possessed by a body, for doing work, by virtue of its mass and velocity of motion. If a body of mass $m$ attains a velocity $v$ from rest in time $t$, under the influence of a force $F$ and moves a distance $x$, then

$$\text{Work done} = F.x = m.a.x \quad ...(\because F = m.a)$$

$\therefore$ Kinetic energy of the body or the kinetic energy of translation,

$$\text{K.E.} = m.a.x = m \times a \times \frac{{}^*v^2}{2a} = \frac{1}{2}m.v^2$$

---

* We know that $v^2 - u^2 = 2a.x$

Since the body starts from rest, therefore $u = 0$. Thus

$$v^2 = 2ax \text{ or } x = \frac{v^2}{2a}$$

It may be noted that when $m$ is in kg and $v$ in m/s, then the kinetic energy will be in N-m as discussed below:

We know that kinetic energy, $\text{K.E.} = \frac{1}{2}m \cdot v^2 = \text{kg} \times \frac{\text{m}^2}{\text{s}^2} = \frac{\text{kg-m}}{\text{s}^2} \times \text{m} = \text{N-m} \quad ...\left(\because \frac{1\text{ kg-m}}{\text{s}^2} = 1\text{ N}\right)$

## 10.12 LAW OF CONSERVATION OF ENERGY

It states that '*the energy can neither be created nor destroyed, but it can be transformed from one form to another form of energy*'.

Some examples of energy transformation are as follows:

**(*a*)** In heat engine (*i.e.* petrol engine, diesel engine, steam engine, etc.), the heat energy is converted into mechanical energy.

**(*b*)** In electric heater, the electrical energy is converted into heat energy.

**(*c*)** In electric lamps, the electrical energy is converted into light energy.

**(*d*)** When a fuel burns, the chemical energy is converted into heat energy.

The loss of energy in any one form is always accompanied by an equivalent increase in another form. When work is done on a rigid body, the work is converted into kinetic or potential energy or is used in overcoming friction. If the body is elastic, some of the work will be stored as strain energy. Thus the law of conservation of energy may also be expressed as '*the total energy possessed by a system of moving bodies remains constant at every instant provided no energy is added to or taken out from the system.*'

This can be proved as follows:

Consider a body of mass $m$ at position $A$ which is at height $h$ from the ground level or datum, as shown in Fig. 10.10. Since the velocity of body at $A$ is zero, therefore kinetic energy at $A$ is zero and the energy possessed by it, is wholly potential whose magnitude is $mgh$.

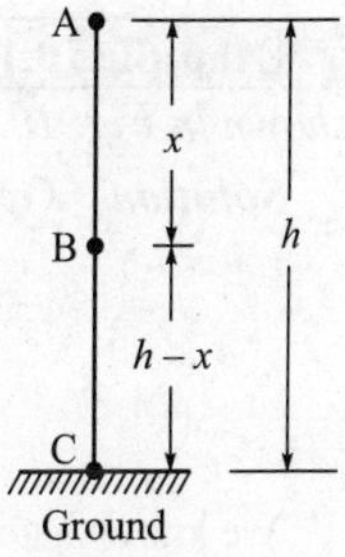


**Fig. 10.10**

$$\therefore \quad \text{Total energy at } A = \text{K.E.} + \text{P.E.}$$

$$= 0 + mgh = mgh \quad ...(i)$$

Let the body now falls from $A$ to $B$ through a distance $x$. Therefore the velocity of the body at $B$,

$$v = \sqrt{2gx}$$

and kinetic energy at $B$

$$= \frac{1}{2}mv^2 = \frac{1}{2} \times m \times 2gx = mgx$$

$$\text{Potential energy at } B = mg(h - x)$$

$$\therefore \quad \text{Total energy at } B = mgx + mg(h - x) = mgh \quad ...(ii)$$

Now the body reaches the ground at point $C$.

Velocity of the body at $C$, $\quad v_1 = \sqrt{2gh}$

and kinetic energy at $C$

$$= \frac{1}{2}mv_1^2 = \frac{1}{2}m \times 2gh = mgh$$

$$\text{Potential energy at } C = mg(h - h) = 0$$

$$\therefore \quad \text{Total energy at } C = mgh + 0 = mgh \quad ...(iii)$$

From equations (*i*), (*ii*) and (*iii*), we see that when the body falls, the potential energy decreases and the kinetic energy increases. The sum of both the energies at any instant remains constant.

## 10.13 WORK–ENERGY THEOREM

According to work-energy theorem, '*the work done by a body is equal to the change in kinetic energy*'.

We know that resultant force acting on a body,

$$F = \text{Mass} \times \text{Acceleration} = m.a \quad ...(i)$$

...(Newton's second law of motion)

We also know that acceleration, $a = \dfrac{dv}{dt} = \dfrac{dv}{dx} \times \dfrac{dx}{dt} = \dfrac{dv}{dx} \times v = v \times \dfrac{dv}{dx} \quad ...\left(\because \dfrac{dx}{dt} = v\right)$

Substituting the value of $a$ in equation ($i$), we have

$$F = m\,v \times \frac{dv}{dx}$$

or $$F \times dx = m.v\,dv$$

Integrating this expression, we have

$$\int_0^x F.dx = \int_u^v v\,dv = m\int_u^v v.dv$$

$$\therefore \quad F.x = m\left[\frac{v^2}{2}\right]_u^v = \frac{mv^2}{2} - \frac{mu^2}{2}$$

or Work done by the resultant force = Change in kinetic energy

**Example 10.17** *A particle is projected downward from a height of 100 m with a velocity of 30 m/s, as shown in Fig. 10.11. Find the velocity with which it strikes the ground.*

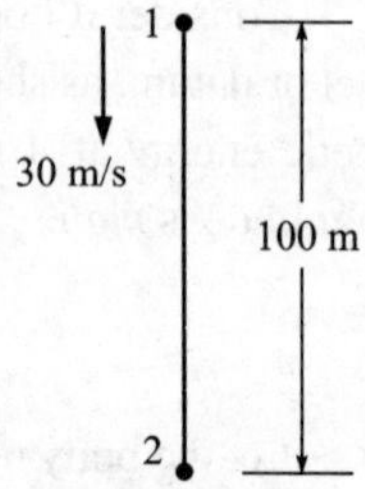


Fig. 10.11

***Solution*** Given : $h = 100$ m ; $u = 30$ m/s

Let $v$ = Velocity with which the particle strikes the ground, and

$m$ = Mass of the particle.

We know that initial kinetic energy, $E_1 = \dfrac{1}{2}mu^2$

and final kinetic energy, $E_2 = \dfrac{1}{2}mv^2$

Work done by the particle = $mgh$

Since the change in kinetic energy is equal to the work done, therefore

$$E_2 - E_1 = mgh$$

$$\frac{1}{2}mv^2 - \frac{1}{2}mu^2 = mgh$$

or $$v^2 - u^2 = 2gh$$

$$v^2 = u^2 + 2gh = (30)^2 + 2 \times 9.81 \times 100 = 900 + 1962 = 2862$$

$$\therefore \quad v = \sqrt{2862} = 53.5 \text{ m/s} \quad \textbf{Ans.}$$

**Example 10.18** *A car weighing 8000 N accelerates from rest to a speed of 45 km/h in a distance of 50 m against a resistance of 100 N. Find the average driving force acting on the car. Using this average force, find the greatest power developed by the engine.*

***Solution*** Given : $W = 8000$ N ; $u = 0$ ; $v = 45$ km/h = $\dfrac{45 \times 1000}{60 \times 60} = 12.5$ m/s; $x = 50$ m ; $R = 100$ N

***Average driving force acting on the car***

Let $P$ = Average driving force acting on the car.

$\therefore$ Net force acting on the car $= P - R = (P - 100)$ N

We know that initial kinetic energy $= \frac{1}{2} \times \frac{W}{g} u^2 = 0$ ...($\because$ $u = 0$)

and final kinetic energy $= \frac{1}{2} \times \frac{W}{g} v^2 = \frac{1}{2} \times \frac{8000}{9.81} (12.5)^2 = 63\ 710$ N-m ...(*i*)

Work done by the net force acting on the car

$$= (P - R)x = (P - 100)\ 50 \quad ...(ii)$$

Equating equations (*i*) and (*ii*), we have

$$(P - 100)\ 50 = 63\ 710$$

$$\therefore \quad P = \frac{63\ 710}{50} + 100 = 1274.2 + 100 = 1374.2 \text{ N } \textbf{Ans.}$$

***Greatest power developed by the engine***

We know that power developed by the engine

$$= P.v = 1374.2 \times 12.5 = 17\ 177.5 \text{ W} = 17.1775 \text{ kW } \textbf{Ans.}$$

**Example 10.19** *A pump lifts 40 m³ of water to a height of 50 m and delivers it with a velocity of 5 m/s. What is the amount of energy spent during this process? If the job is done in half an hour, what is the input power of the pump which has an overall efficiency of 70% ?*

***Solution*** Given : $V = 40$ m³ ; $h = 50$ m ; $v = 5$ m/s ; $\eta = 70\% = 0.7$

***Amount of energy spent***

Since 1 m³ of water weighs 9810 N, therefore, weight of water lifted,

$$W = 40 \times 9810 = 392.4 \times 10^3 \text{ N}$$

and work done in lifting the water or potential energy,

$$\text{P.E.} = W.h = 392.4 \times 10^3 \times 50 = 19.62 \times 10^6 \text{ N-m}$$

Kinetic energy at delivery, $\text{K.E.} = \frac{1}{2} mv^2 = \frac{1}{2} \times \frac{W}{g} v^2$

$$= \frac{1}{2} \times \frac{392.4 \times 10^3}{9.81} \times 5^2 = 0.5 \times 10^6 \text{ N-m}$$

Total energy spent = P.E. + K.E.

$$= 19.62 \times 10^6 + 0.5 \times 10^6 = 20.12 \times 10^6 \text{ N-m}$$

$$= 20.12 \text{ MN-m } \textbf{Ans.}$$

***Input power of the engine***

Since the total energy is spent in half an hour or $30 \times 60 = 1800$ seconds, therefore output power of the pump

= Output energy spent per second

$$= \frac{20.12 \times 10^6}{1800} = 11.2 \times 10^3 \text{ N-m/s or W} \quad ...(\because 1 \text{ N-m/s} = 1 \text{ W})$$

We know that input power of the engine

$$= \frac{\text{Output power}}{\text{Efficiency}} = \frac{11.2 \times 10^3}{0.7} = 16 \times 10^3 \text{ W} = 16 \text{ kW } \textbf{Ans.}$$

**Example 10.20** *A machine gun bullet of mass 30 g is fired with a velocity of 500 m/s. What is the kinetic energy of the bullet? If the bullet can penetrate 300 mm in a block of wood, what is the average force of resistance of wood?*

*What will be the exit velocity of the bullet if fired with the same initial velocity into a similar block of wood 150 mm thick?*

**Solution** Given : $m = 30$ g $= 0.03$ kg ; $u = 500$ m/s ; $x = 300$ mm $= 0.3$ m ; $x_1 = 150$ mm $= 0.15$ m

***Kinetic energy of the bullet***

We know that kinetic energy, $\text{K.E.} = \frac{1}{2}mu^2 = \frac{1}{2} \times 0.03(500)^2 = 3750$ N-m **Ans.**

***Average resistance of wood***

Let $R$ = Average force of resistance of wood.

We know that initial kinetic energy of bullet,

$$E_1 = \frac{1}{2}mu^2 = \frac{1}{2} \times 0.03(500)^2 = 3750 \text{ N-m}$$

and final kinetic energy of bullet, $E_2 = 0$ ...($\because$ Velocity after penetration is zero)

Work done against resistance of wood $= R \times x = R \times 0.3 = 0.3R$

Since the change in kinetic energy of bullet is equal to the work done, therefore

$$E_1 - E_2 = 0.3R$$

or

$$E_1 - 0 = 0.3R$$

$$\therefore \quad R = \frac{E_1}{0.3} = \frac{3750}{0.3} = 12\,500 \text{ N} = 12.5 \text{ kN} \quad \textbf{Ans.}$$

***Exit velocity of the bullet***

Let $V$ = Exit velocity of the bullet.

We know that final kinetic energy of the bullet,

$$E_3 = \frac{1}{2} m V^2 = \frac{1}{2} \times 0.03 V^2 = 0.015 V^2$$

and change in kinetic energy $= E_1 - E_3 = 3750 - 0.015 V^2$

We know that change in kinetic energy

$$= \text{Work done by the force of resistance } (R \times x_1)$$

or

$$3750 - 0.015 V^2 = 12\,500 \times 0.15 = 1875$$

$$0.015 V^2 = 3750 - 1875 = 1875$$

$$V^2 = \frac{1875}{0.015} = 125\,000$$

$$\therefore \quad V = \sqrt{125\,000} = 353.55 \text{ m/s} \quad \textbf{Ans.}$$

The exit velocity of the bullet may also be obtained as dicsused below:

Let $V$ = Exit velocity of the bullet, and

$a$ = Acceleration of the bullet.

In the above case, the final velocity of the bullet after penetration is zero (*i.e.* $v = 0$), and the initial velocity is given as 500 m/s (*i.e.* $u = 500$ m/s).

Using the relation, $v^2 - u^2 = 2ax$

$$0 - (500)^2 = 2a \times 0.3 = 0.6a$$

$$\therefore \quad a = -\frac{(500)^2}{0.6} \text{ m/s}^2$$

The –ve sign indicates retardation.

Again using the relation, $V^2 - u^2 = 2ax_1$

$$V^2 - (500)^2 = -2 \times \frac{(500)^2}{0.6} \times 0.15 = -125\,000$$

$$V^2 = (500)^2 - 125\,000 = 125\,000$$

$$\therefore \quad V = \sqrt{125\,000} = 353.55 \text{ m/s } \textbf{Ans.}$$

**Example 10.21** *A bullet of mass 1 g has a velocity of 1000 m/s as it enters a fixed block of wood. It comes to rest 0.002 s after entering the block. Find the average force that acted upon the bullet and the distance penetrated by it.*

***Solution*** Given : $m = 1$ g $= 0.001$ kg ; $u = 1000$ m/s ; $t = 0.002$ s

***Average force that acted upon the bullet***

First of all, let us find the acceleration ($a$) of the bullet after 0.002 seconds. Since the bullet comes to rest after entering the block, therefore its final velocity ($v$) is zero.

Using the relation, $v = u + a.t$

$$0 = 1000 + a \times 0.002$$

$$\therefore \quad a = \frac{-1000}{0.002} = -500 \times 10^3 \text{ m/s}^2$$

The –ve sign shows retardation.

We know that average force, $R$ = Mass × Acceleration

$$= 0.001 \times 500 \times 10^3 = 500 \text{ N } \textbf{Ans.}$$

***Distance penetrated***

Let $x$ = Distance penetrated in metres.

We know that the total kinetic energy of the bullet

= Initial kinetic energy of the bullet

$$= \frac{1}{2}mu^2 = \frac{1}{2} \times 0.001(1000)^2 = 500 \text{ N-m} \quad ...(i)$$

...(because final kinetic energy of the bullet is zero)

This energy is used in doing the work against resistance of wood ($R$).

∴ Work done against resistance of wood

$$= R \times x = 500 \times x \quad ...(ii)$$

Equating equations (*i*) and (*ii*), we have

$$500 = 500\,x$$

$$\therefore \quad x = \frac{500}{500} = 1 \text{ m } \textbf{Ans.}$$

## 10.14 PILE AND PILE HAMMER

Consider a pile having a conical shape, is driven into the ground by the impact of a pile hammer, as shown in Fig. 10.12.

Let $M$ = Mass of the pile hammer in kg,

$m$ = Mass of the pile in kg,
$h$ = Height through which the pile hammer falls before striking the pile in metres,
$v$ = Velocity of the pile hammer at which it strikes the pile in m/s,
$F$ = Force with which it strikes the pile in newtons,
$x$ = Distance through which the pile is driven in metres,
$V$ = Common velocity of the pile and pile hammer after impact in m/s, and
$R$ = Average resistance of the ground against the pile in newtons.

Since the initial velocity ($u$) of the pile hammer is zero, therefore from the relation,

$$v^2 - u^2 = 2gh$$

or

$$v^2 - 0 = 2gh$$

$$\therefore \quad v = \sqrt{2gh}$$

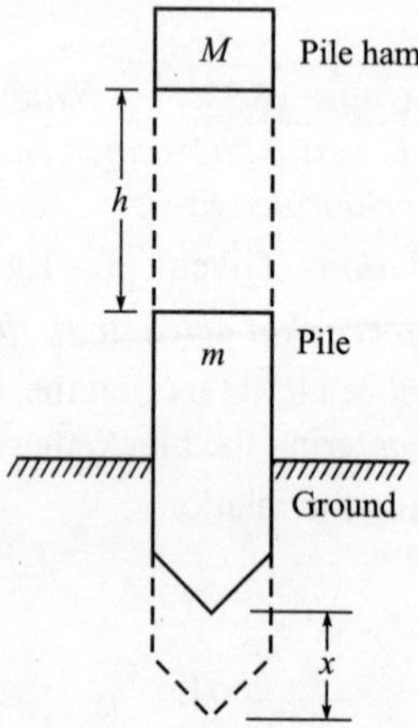


**Fig. 10.12** *Pile and pile hammer*

We know that linear momentum of the pile hammer before impact

$$= M.v$$

and linear momentum of the pile and pile hammer after impact

$$= (M + m)\,V$$

According to the principle of conservation of momentum,

$$\text{Momentum before impact} = \text{Momentum after impact}$$

$$M.v = (M + m)\,V$$

$$\therefore \quad V = \frac{M.v}{M + m} \quad ...(i)$$

The pile hammer has potential energy before it is released. After release, the potential energy is converted into kinetic energy with which the pile hammer strikes the pile head.

We know that total potential energy of the pile and pile hammer just after the impact,

$$\text{P.E.} = (M + m)\,g.x \text{ ...(in N-m)}$$

and kinetic energy of the pile and pile hammer just after the impact,

$$\text{K.E.} = \frac{1}{2}(M + m)V^2 \text{ ...(in N-m)}$$

$$\therefore \quad \text{Total energy, } E = \text{P.E.} + \text{K.E.} = (M + m)\,g.x + \frac{1}{2}(M + m)V^2 \quad ...(ii)$$

This energy is completely lost in penetrating the mass $m$ into the ground.

$\therefore$ Work done during penetration = Average resistance of the ground × Depth of penetration

$$= R \times x \quad ...(iii)$$

We know that total energy lost is equal to the work done during penetration, therefore equating equations (*ii*) and (*iii*), we get

$$(M + m)\,g.x + \frac{1}{2}(M + m)V^2 = R.x$$

Substituting the value of $V$ from equation (*i*), we have

$$(M + m)\,g \cdot x + \frac{1}{2}(M + m)\left(\frac{M \cdot v}{M + m}\right)^2 = R.x$$

$$(M + m)\,g \cdot x + \frac{M^2 v^2}{2(M + m)} = R.x$$

$$(M+m)\,g\cdot x+\frac{M^2}{2(M+m)}(\sqrt{2gh})^2 = R.x \qquad \ldots(\because\ v=\sqrt{2gh})$$

$$(M+m)\,g\cdot x+\frac{M^2}{M+m}\times gh = R.x$$

$$\therefore \qquad R = (M+m)\,g+\frac{M^2}{M+m}\times\frac{gh}{x} \qquad \ldots(iv)$$

We know that loss of kinetic energy during impact

$$= \text{*Kinetic energy before impact} - \text{Kinetic energy after impact}$$

$$= \frac{1}{2}M\,v^2 - \frac{1}{2}(M+m)V^2$$

$$= \frac{1}{2}M\,(\sqrt{2gh})^2 - \frac{1}{2}(M+m)\left(\frac{M\cdot v}{M+m}\right)^2$$

$$= \frac{1}{2}M\times 2gh - \frac{1}{2}(M+m)\left(\frac{M}{M+m}\right)^2\times 2gh \qquad \ldots(\because\ v^2=2gh)$$

$$= M.g.h - \left(\frac{M^2}{M+m}\right)gh = \frac{Mgh\,(M+m)-M^2gh}{M+m}$$

$$= \frac{Mgh}{M+m}\left[M+m-M\right] = \frac{M\cdot mgh}{M+m} \qquad \ldots(v)$$

In order to find the time ($t$) during which the pile is in motion, apply impulse-momentum equation.

We know that net force bringing the system in equilibrium

$$= R-(M+m)g$$

and

$$\text{Impulse} = \text{Net force}\times\text{time} = [R-(M+m)g]\,t \qquad \ldots(vi)$$

$$\text{Change of momentum} = (M+m)\,(0-V) = (M+m)\,V \qquad \ldots(vii)$$

Equating equations ($vi$) and ($vii$), we have

$$\text{Impulse} = \text{Change of momentum}$$

$$[R-(M+m)\,g]\,t = (M+m)\,V$$

Substituting the value of $R$ and $V$ from equation ($iv$) and equation ($i$), respectively, we have

$$(M+m)\,g\,t+\frac{M^2}{M+m}\times\frac{ght}{x}-(M+m)\,gt = (M+m)\left(\frac{M.v}{M+m}\right) = M.v$$

or

$$\frac{M^2}{M+m}\times\frac{ght}{x} = M\times\sqrt{2gh}$$

$$\therefore \qquad t = M\times\sqrt{2gh}\times\frac{(M+m)\,x}{M^2\times gh} = \frac{(M+m)\,x}{M}\times\sqrt{\frac{2}{gh}}$$

**Example 10.22** *A pile hammer of mass 1500 kg drops from a height of 600 mm on a pile of 750 kg. Find the depth of penetration into the ground if the average resistance of the ground is 140 kN. Assume that the ground resistance is constant.*

***Solution*** Given : $M = 1500$ kg ; $h = 600$ mm $= 0.6$ m ; $m = 750$ kg ; $R = 140$ kN $= 140\times10^3$ N

Let $x$ = Depth of penetration in metres.

* It is equal to potential energy because $\frac{1}{2}M\,v^2 = M.g.h$

We know that average resistance of the ground,

$$R = (M + m)\,g + \frac{M^2}{M + m} \times \frac{gh}{x}$$

$$140 \times 10^3 = (1500 + 750)\ 9.81 + \frac{(1500)^2}{1500 + 750} \times \frac{9.81 \times 0.6}{x}$$

$$140\ 000 = 22\ 072.5 + \frac{5886}{x}$$

$$\therefore \quad x = \frac{5886}{140\ 000 - 22\ 072.5} = \frac{5886}{117\ 927.5} = 0.05 \text{ m} = 50 \text{ mm} \quad \textbf{Ans.}$$

**Example 10.23** *A hammer of mass 1500 kg drops from a height of 600 mm on a pile of mass 750 kg. Find:*

1. *The common velocity after impact assuming plastic impact ; and*
2. *The average resistance of the ground if the pile comes to rest after penetration 50 mm into the ground.*

***Solution*** Given : $M = 1500$ kg ; $h = 600$ mm $= 0.6$ m ; $m = 750$ kg ; $x = 50$ mm $= 0.05$ m

**1. *Common velocity after impact***

Let $V$ = Common velocity of the pile and pile hammer.

We know that velocity of the pile hammer at which it strikes the pile,

$$v = \sqrt{2gh} = \sqrt{2 \times 9.81 \times 0.6} = \sqrt{11.772} = 3.43 \text{ m/s}$$

We also know that linear momentum of the pile hammer before impact

= Linear momentum of the pile and pile hammer after impact

or

$$M.v = (M + m)\,V$$

$$\therefore \quad V = \frac{M.v}{M + m} = \frac{1500 \times 3.43}{1500 + 750} = \frac{5145}{2250} = 2.287 \text{ m/s} \quad \textbf{Ans.}$$

**2. *Average resistance of the ground***

Let $R$ = Average resistance of the ground in newtons.

We know that total potential energy of the pile and pile hammer just after impact,

$$\text{P.E.} = (M + m)g.x = (1500 + 750)\ 9.81 \times 0.05 = 1103.6 \text{ N-m}$$

and kinetic energy of the pile and pile hammer just after impact,

$$\text{K.E.} = \frac{1}{2}(M + m)V^2 = \frac{1}{2}(1500 + 750)\ (2.287)^2 = 5884.2 \text{ N-m}$$

$$\therefore \quad \text{Total energy, } E = \text{P.E.} + \text{K.E.} = 1103.6 + 5884.2 = 6987.8 \text{ N-m} \qquad ...(i)$$

This energy is lost in penetrating the mass into the ground.

$$\therefore \quad \text{Work done during penetration} = R \times x = R \times 0.05 = 0.05\,R \qquad ...(ii)$$

Since the total energy lost is equal to the work done in penetration, therefore equating equations (*i*) and (*ii*), we have

$$6987.8 = 0.05\,R$$

$$\therefore \quad R = \frac{6987.8}{0.05} = 139\ 756 \text{ N} = 139.756 \text{ kN} \quad \textbf{Ans.}$$

**Note:** The average resistance of the ground may also be obtained by using the relation,

$$R = (M+m)g + \frac{M^2}{M+m} \times \frac{gh}{x}$$

$$= (1500+750)9.81 + \frac{(1500)^2}{1500+750} \times \frac{9.81 \times 0.6}{0.05}$$

$$= 22\,073 + 117\,720 = 139\,793 \text{ N} = 139.793 \text{ kN}$$

**Example 10.24** *A pile hammer of mass of 300 kg is allowed to fall vertically through 1 metre on to the top of a pile of mass 500 kg. If the pile is driven into the ground by 150 mm at each blow, find : 1. The energy lost in the blow ; and 2. The average resistance against the pile.*

***Solution*** Given : $M = 300$ kg ; $h = 1$ m ; $m = 500$ kg ; $x = 150$ mm $= 0.15$ m

We know that velocity of the pile hammer at which it strikes the pile,

$$v = \sqrt{2gh} = \sqrt{2 \times 9.81 \times 1} = 4.43 \text{ m/s}$$

Let $V$ = Common velocity of the pile and pile hammer after impact.

We know that linear momentum of the pile hammer before impact

= Linear momentum of the pile and pile hammer after impact

or $$M.v = (M+m)V$$

$\therefore$ $$V = \frac{M.v}{M+m} = \frac{300 \times 4.43}{300+500} = 1.661 \text{ m/s}$$

**1. *Energy lost in the blow***

We know kinetic energy before impact,

$$^*E_1 = \frac{1}{2} M v^2 = \frac{1}{2} M (\sqrt{2gh})^2 = M.g.h$$

$$= 300 \times 9.81 \times 1 = 2943 \text{ N-m}$$

and kinetic energy after impact, $$E_2 = \frac{1}{2}(M+m)V^2 = \frac{1}{2}(300+500)(1.661)^2 = 1103.6 \text{ N-m}$$

$\therefore$ Energy lost in the blow $= 2943 - 1103.6 = 1839.4$ N-m **Ans.**

**2. *Average resistance against the pile***

Let $R$ = Average resistance against the pile.

We know that total potential energy of the pile and pile hammer just after impact,

$$\text{P.E.} = (M+m)g \,.\, x = (300+500)\, 9.81 \times 0.15 = 1177.2 \text{ N-m}$$

and kinetic energy of the pile and pile hammer just after impact,

$$\text{K.E.} = \frac{1}{2}(M+m)V^2 = \frac{1}{2}(300+500)\,(1.661)^2 = 1103.6 \text{ N-m}$$

$\therefore$ Total energy, $E$ = P.E. + K.E. $= 1177.2 + 1103.6 = 2280.8$ N-m ...(*i*)

This energy is completely lost in penetrating the mass into the ground.

$\therefore$ Work done during penetration $= R.x = R \times 0.15 = 0.15R$ ...(*ii*)

Since the total energy lost in equal to the work done in penetration, therefore equating equations (*i*) and (*ii*), we have

$$2280.8 = 0.15R$$

$\therefore$ $$R = \frac{2280.8}{0.15} = 15\,205.3 \text{ N}$$ **Ans.**

*It is equal to the potential energy (*Mgh*).

It may be noted that the energy lost in the blow and the average resistance may also be obtained by applying the following relations:

(*i*) Energy lost in blow $= \dfrac{M.mgh}{M+m} = \dfrac{300 \times 500 \times 9.81 \times 1}{300+500} = 1839.4$ N-m **Ans.**

(*ii*) Average resistance against the pile,

$$R = (M+m)\,g + \frac{M^2}{M+m} \times \frac{gh}{x}$$

$$= (300+500)\,9.81 + \frac{(300)^2}{300+500} \times \frac{9.81 \times 1}{0.15}$$

$$= 7848 + 7357.5 = 15\ 205.5 \text{ N}$$ **Ans.**

## HIGHLIGHTS

1. When a force ($P$) acting on a body is able to give it some displacement ($x$), then *work* is said to be done. Mathematically,

   Work done $= P \times x$ ...(When $P$ and $x$ are in the same direction)

   $= P \cos \theta \times x$ ...(When $P$ acts at an angle $\theta$ with the direction of $x$)

2. The unit of work depends upon the unit of force and the displacement. If the force is in newton, and displacement in metres, then the unit of work is newton-metre (briefly written as N-m) or joule (briefly written as J), such that 1 N-m = 1 J.
3. If we plot a graph between the force during any displacement along the $Y$-axis, and the corresponding displacement along the $X$-axis, then area under the graph represents the work done.
4. The work done on a rotating body, by a torque ($T$), is given by

   Work done $= T.\theta$

   where $\theta$ is the angular displacement in radians.
5. The *power* is defined as the rate of doing work per unit time. Mathematically,

   $$\text{Power} = \frac{\text{Work done}}{\text{Time taken}}$$

   In S.I. system of units, the unit of power is watt (briefly written as W), such that

   1 W = 1 J/s or 1 N-m/s
6. The power developed by a force $P$ (in newtons) moving with a velocity $v$ (in m/s) is $P.v$ watt. Generally, a bigger unit of power called kilowatt (kW) or megawatt (MW) is used, such that

   1 kW = 1000 W

   and 1 MW = 1000 kW = $10^6$ W
7. If $T$ is the torque transmitted in N-m or J and $\omega$ is the angular speed in rad/s, then

   $$\text{Power} = T.\omega = T \times \frac{2\pi N}{60} \text{ watt}$$

   where $N$ is the speed in r.p.m.
8. The *energy* of a body is defined as its capacity of doing work. The unit of energy is same as that of work.

   Though there are many types of energies, yet the potential energy and kinetic energy are important from the subject of view.
9. The energy possessed by a body for doing work by virtue of its position, is called *potential energy.*

Mathematically,

Potential energy, P.E. $= W.h = m.gh$

where $W$ = Weight of the body $= m \cdot g$,

$m$ = Mass of the body, and

$h$ = Height of the body above the reference line or datumn line.

**10.** The potential energy stored by an elastic body when deformed, is called strain energy. A compressed spring possesses this type of energy. Mathematically,

$$\text{Strain energy} = \text{Work done} = \frac{1}{2}W \cdot x = \frac{1}{2}s \cdot x^2 \quad ...(\because W = s \cdot x)$$

where $s$ is the stiffness of the spring per unit deformation.

**11.** The energy possessed by a body, for doing work, by virtue of its mass and velocity of motion, is called *kinetic energy*. Mathematically,

$$\text{Kinetic energy, K.E.} = \frac{1}{2}m \cdot v^2 \text{ (in N-m)}$$

where $m$ = Mass of the body in kg, and

$v$ = Velocity of the body in m/s.

**12.** The law of conservation of energy states that '*the energy can neither be created nor destroyed, but it can be transformed from one form to another form of energy*'.

It can also be expressed as '*the total energy possessed by a system of moving bodies remains constant at every instant provided no energy is added to or taken out from the system*'.

**13.** According to the work-energy theorem,

Work done by the resultant force = Change in kinetic energy

**14.** When a pile is driven into the ground by the impact of a pile hammer, then the resistance offered by the ground against penetration into the ground is obtained by equating the work done during penetration to the total energy lost in penetrating the mass into the ground (*i.e.* sum of potential energy and kinetic energy of the pile and pile hammer just after the impact). Mathematically,

$$R \times x = (M + m)\, g \cdot x + \frac{1}{2}(M + m)V^2$$

where $M$ = Mass of the pile hammer,

$m$ = Mass of the pile,

$x$ = Depth of penetration,

$V$ = Common velocity of the pile and pile hammer after impact $= \dfrac{M \cdot v}{M + m}$,

$v$ = Velocity of the pile hammer at which it strikes the pile $= \sqrt{2gh}$,

$h$ = Height through which the pile hammer falls before striking the pile,

$R$ = Average resistance of the ground against the pile.

**15.** The average resistance of the ground ($R$) may be obtained by using the expression,

$$R = (M + m)g + \frac{M^2}{M + m} \times \frac{gh}{x}$$

and loss of kinetic energy during impact

$$= \frac{Mmgh}{M + m}$$

**16.** The time ($t$) during which the pile is in motion, is given by

$$t = \frac{(M+m)x}{M} \times \sqrt{\frac{2}{gh}}$$

## NUMERICAL PROBLEMS

**1.** A body of weight 1.5 kN moves on a level horizontal surface for a distance of 400 m. The resistance of the road is 10 N per kN weight of the body. Find the work done by the body. [**Ans.** 6 kN-m]

**2.** A wooden block is pulled on a smooth ground with the help of a string inclined at 30° to the horizontal. If the tension in the string is 40 N, find the work done to move the block through a distance of 15 metres.

[**Ans.** 519.6 N-m]

**3.** A train of weight 200 kN is pulled by an engine on a level track at a constant speed of 90 km/h. The resistance due to friction is 150 N/kN. Calculate the work done per minute by the engine. [**Ans.** 12.5 kN-m/min]

**4.** A man is drawing water from a well, with a light bucket which leaks uniformly. The bucket when full weighs 200 N and when it arrives at the top, it weighs 100 N. Find the work done, if the depth of the well is 30 metres.

[**Ans.** 4500 N-m]

**5.** A wooden box is pulled on the ground with the help of a rope through a distance of 15 m. The force required to pull the box is 400 N. Find the weight of the box and the work done, if the coefficient of friction is 0.25.

[**Ans.** 1600 N; 6000 N-m]

**6.** The force required to stretch 1 mm of a spring is 10 N. Find the work done required to stretch the spring by 50 mm. [**Ans.** 12.5 N-m]

**7.** A tangential force of 2000 N is acting on a shaft of diameter 10 mm. Calculate the work done by the force for one revolution of the shaft. [**Ans.** 62.84 N-m]

**8.** A train of weight 1000 kN is pulled by an engine on a level track at a constant speed of 45 km/h. The resistance due to friction is 10 N/kN of the weight of the train. Find the power of the engine. [**Ans.** 125 kW]

**9.** A pump lifts water from a tank of capacity 18 000 litres, to a height of 12 m in 20 minutes. If the efficiency of the pump is 65%, find the input power of the pump. (1 litre of water = 1 kg) [**Ans.** 2.7166 kW]

**10.** A train of weight 2000 kN is ascending a slope of 1 in 200 with a uniform velocity of 45 km/h. If the tractive resistance is 10 N/kN weight of the train, determine the power exerted by the engine. [**Ans.** 375 kW]

**11.** An engine of weight 500 kN pulls a train of weight 2500 kN up a gradient of 1 in 125 with a uniform speed of 36 km/h. Find the power transmitted by the engine, if the tractive resistance is 6 N/kN weight of the train. Also find the power transmitted by the engine, if the acceleration of the engine is 0.2 m/s$^2$ up the gradient.

[**Ans.** 420 kW; 1031.6 kW]

**12.** A body of mass 8 kg is thrown up vertically with a velocity of 12 m/s. Find its kinetic energy at the moment of propulsion; and after 1 second. Take $g$ = 9.8 m/s$^2$. [**Ans.** 576 N-m; 19.36 N-m]

**13.** A man of mass 60 kg dives vertically downwards into a swimming pool from a tower of height 20 m. He was found to go down in water by 2 m and started rising. Find the average resistance of water. Neglect air resistance. [**Ans.** 5886 N]

**14.** A bullet of mass 75 g moving with a velocity of 200 m/s is fired into a block of wood and gets embedded in it to a depth of 150 mm. What will be the resistance offered by the wood? [**Ans.** 10 kN]

**15.** A bullet of mass 80 g moving with a velocity of 300 m/s is fired into a block of wood and penetrates to a depth of 100 mm. Find the force of resistance, assuming it to be uniform.

If the bullet moving with the same velocity, is fired into a similar piece of wood 50 mm thick, with what velocity would it emerge? [**Ans.** 36 kN; 212.13 m/s]

**16.** A truck of weight 20 kN is travelling at 45 km/h on a level ground. It is brought to rest in 20 m. What is the average force of resistance acting on the road. [**Ans.** 7963.8 N]

**17.** A hammer of mass 200 kg falls through a height of 4 metres on a pile of negligible mass. If it drives the pile 500 mm into the ground, find the average resistance of the ground for penetration. [**Ans.** 17.658 kN]

**18.** A pile hammer of mass 2000 kg drops from a height of 750 mm on a pile of 1000 kg. Find
(*a*) the common velocity after impact, and
(*b*) the average resistance of the ground, if the pile comes to rest after penetrating 100 mm.
[**Ans.** 2.557 m/s; 127.53 kN]

**19.** A hammer of mass 750 kg drops from a height of 1.2 m on a pile of mass 200 kg. Find the depth of penetration of the pile into the ground, if the average resistance of the ground is 80 kN. The impact between the hammer and pile is assumed to be plastic. [**Ans.** 98.5 mm]

## REVIEW QUESTIONS

1. Explain the term 'work'. What is its unit?
2. Define power. Write the relation of power, in terms of torque transmitted ($T$) and angular speed ($\omega$).
3. What is the difference between potential energy and kinetic energy? Is strain energy a potential energy?
4. State the law of conservation of energy.
5. Show that the total energy possessed by a system of moving bodies remains constant at every instant.
6. Prove that loss of kinetic energy during impact in case of pile and pile hammer is $\frac{M.mgh}{M+m}$, where $M$ = Mass of the pile hammer; $m$ = Mass of the pile; and $h$ = Height through which the pile hammer falls before striking the pile.

## OBJECTIVE TYPE QUESTIONS

1. Whenever a force acts on a body and the body undergoes a displacement, then
   (*a*) body has kinetic energy (*b*) power is being transmitted
   (*c*) work is said to be done (*d*) none of these
2. The unit of work in S.I. units is
   (*a*) erg (*b*) kg-m (*c*) newton (*b*) joule
3. One joule is equal to
   (*a*) 0.1 N-m (*b*) 1 N-m (*c*) 10 N-m (*d*) 100 N-m
4. One joule means that work is done by a force of 1 N when it displaces a body through 1 m.
   (*a*) True (*b*) False
5. The rate of doing work is known as
   (*a*) potential energy (*b*) kinetic energy
   (*c*) power (*d*) none of these
6. The unit of power is S.I. units is
   (*a*) horse power (*b*) watt (*c*) joule (*d*) kg-m
7. One watt is equal to
   (*a*) 1 Joule/s (*b*) 10 Joule/s (*c*) 100 Joulc/s (*d*) 1000 joule/s
8. The power developed by a body acted upon by a torque $T$ newton-metre and revolving at $\omega$ rad/s is given by
   (*a*) $T.\omega$ (in kilowatt) (*b*) $T.\omega$ (in watt)
   (*c*) $\frac{T.\omega}{60}$ (in kilowatt) (*d*) $\frac{T.\omega}{1000}$ (in kilowatt)

9. The energy possessed by a body, for doing work, by virtue of its position, is called
   (*a*) potential energy  (*b*) kinetic energy
   (*c*) electrical energy  (*d*) chemical energy
10. The potential energy (in newtons) of a body of mass *m* raised through a height *h* is
   (*a*) $g.h$  (*b*) $m.gh$  (*c*) $m.h$  (*d*) none of these
11. The potential energy stored by a spring in compression is called strain energy.
   (*a*) True  (*b*) False
12. The kinetic energy of a body .................... upon its mass and velocity.
   (*a*) does not depend  (*b*) depends
13. When a body of mass *m* attains a velocity *v* from rest in time *t*, then kinetic energy of translation is
   (*a*) $mv^2$  (*b*) $mgv^2$  (*c*) $0.5mv^2$  (*d*) $0.5\ mgv^2$
14. The total energy possessed by a system of moving bodies
   (*a*) is constant at every instant
   (*b*) varies from point to point
   (*c*) is maximum in the start and minimum at the end
   (*d*) is minimum in the start and maximum at the end
15. When a pile of mass *m* is driven into the ground by the impact of a pile hammer of mass *M*, then the average resistance of the ground against the pile, is obtained by equating the work done during penetration to the
   (*a*) potential energy just after impact  (*b*) kinetic energy just after impact
   (*c*) sum of (*a*) and (*b*)  (*d*) none of these

***Answers to Objective Type Questions***

| | | | | |
|---|---|---|---|---|
| 1. (*c*) | 2. (*d*) | 3. (*b*) | 4. (*a*) | 5. (*c*) |
| 6. (*b*) | 7. (*a*) | 8. (*b*), (*d*) | 9. (*a*) | 10. (*b*) |
| 11. (*a*) | 12. (*b*) | 13. (*c*) | 14. (*a*) | 15. (*c*) |

# 11

# VIRTUAL WORK



## 11.1 INTRODUCTION

We have already discussed that work done by a force is equal to the product of the force and the distance moved in the direction of the force. But, if a number of forces are acting on a body and the body is in equilibrium, then the work done is zero. In case the body, in equilibrium, is given any arbitrary very small change in the position of the body (known as *virtual displacement*), then some work is imagined to be done. Such an imaginary work is known as *virtual work.* Thus, virtual work is defined as the work done by a force (or a couple) on a body due to a small imaginary or virtual displacement of the body.

## 11.2 VIRTUAL WORK DONE BY A FORCE

The virtual work done by a force is equal to the sum of virtual work done by its resolved components.

Consider a body at $A$ is subjected to a force $P$ inclined at an angle $\theta$ with the horizontal (*i.e.* $X$-axis), as shown in Fig. 11.1.

We know that component of the force along $X$-axis,

$$P_X = P\cos\theta$$

and component of the force along $Y$-axis,

$$P_Y = P\sin\theta$$

Now consider that the virtual displacement of the body from position $A$ to position $A'$. Let the line $AA'$ makes an angle $\alpha$ with the direction of force $P$. Draw $A'D$ perpendicular to the line of action of force $P$. Also draw $A'B$ and $A'C$ perpendicular to $X$-axis and $Y$-axis, respectively. From the geometry of the figure, we find that

$$AB = AA'\cos(\theta+\alpha)$$
$$= r\cos(\theta+\alpha)$$

...(Taking $AA' = r$)

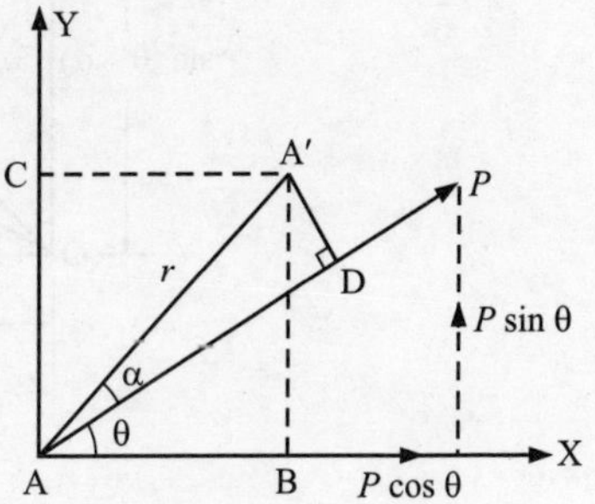


**Fig. 11.1**

and $$AC = A'B = AA' \sin(\theta + \alpha) = r \sin(\theta + \alpha)$$

We know that virtual work done by the component $P_X$,

$$= P_X \,.\, AB = P \cos\theta \,.\, r \cos(\theta + \alpha)$$

and virtual work done by the component $P_Y = P_Y \,.\, AC = P \sin\theta \,.\, r \sin(\theta + \alpha)$

∴ Sum of the virtual work done by the components $P_X$ and $P_Y$

$$= P_X \,.\, AB + P_Y \,.\, AC$$

$$= P \cos\theta \,.\, r \cos(\theta + \alpha) + P \sin\theta \,.\, r \sin(\theta + \alpha)$$

$$= P \,.\, r\,[\cos(\theta + \alpha)\cos\theta + \sin(\theta + \alpha)\sin\theta]$$

$$= P.r\,[\cos(\theta + \alpha - \theta)]$$

...[$\because \cos(A - B) = \cos A \cos B + \sin A \sin B$]

$$= P.r \cos\alpha = P \times AD$$ ...($\because AD = AA' \cos\alpha = r\cos\alpha$)

$$= \text{Virtual work done by the force } P$$

Thus, the virtual work done by a force is equal to the sum of virtual work done by its resolved components.

**Note:** For the sake of simplicity, we have discussed above the virtual work done by one force only, but it can be extended to any number of forces.

## 11.3 PRINCIPLE OF VIRTUAL WORK DONE

The principle of virtual work states that '*when a system of forces acting on a body or a system of bodies is in equilibrium and the system is imagined to undergo a small virtual displacement consistent with geometrical conditions of the system, then the algebraic sum of the virtual work done by all the forces of the system is equal to zero*'.

Conversely, *if the algebraic sum of the virtual works done for every displacement is zero, then the system is in equilibrium.*

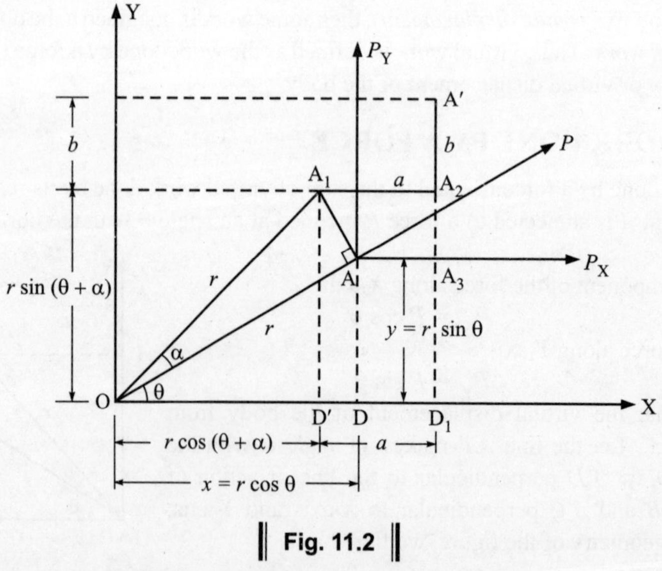


**Fig. 11.2**

***Proof*** Consider a number of forces $P, Q, R, S, ...,$ etc. keeps the body in equilibrium.

Let $A$ be the point of application of force $P$, as shown in Fig. 11.2. Let $\angle AOX = \theta$; $OA = OA_1 = r$; $OD = x = r\cos\theta$; and $AD = y = r\sin\theta$.

Let due to a virtual displacement of the body, the point $A$ is displaced to point $A'$. This displacement is assumed to have taken place as discussed below:

The point $A$ first moves to point $A_1$ such that $\angle AOA_1 = \alpha$. Now $A_1$ moves to $A_2$ parallel to $X$-axis and point $A_2$ moves to $A'$ parallel to $Y$-axis.

Let $A_1A_2 = a$; and $A_2A' = b$

From the geometry of the figure,

$$OD' = OA_1 \cos(\theta + \alpha) = r\cos(\theta + \alpha)$$

$$A_1D' = OA_1 \sin(\theta + \alpha) = r\sin(\theta + \alpha)$$

Now $$OD_1 = OD' + D'D_1 = OD' + A_1A_2 \qquad ...(\because D'D_1 = A_1A_2)$$

$$= r\cos(\theta + \alpha) + a = r(\cos\theta\cos\alpha - \sin\theta\sin\alpha) + a \qquad ...(i)$$

$$...[\because \cos(A+B) = \cos A\cos B - \sin A\sin B]$$

and $$A'D_1 = A_1D' + A'A_2 = r\sin(\theta + \alpha) + b$$

$$= r(\sin\theta\cos\alpha + \cos\theta\sin\alpha) + b \qquad ...(ii)$$

$$...[\because \sin(A+B) = \sin A\cos B + \cos A\sin B]$$

Since $\alpha$ is a small angle, therefore $\sin\alpha = \alpha$ in radians and $\cos\alpha = 1$. The equations (*i*) and (*ii*) may now be written as

$$OD_1 = r(\cos\theta - \alpha\,.\,\sin\theta) + a$$

$$= r\cos\theta - \alpha\,.\,r\sin\theta + a \qquad ...(iii)$$

and $$A'D_1 = r(\sin\theta + \alpha\,.\,\cos\theta) + b$$

$$= r\sin\theta + \alpha\,.\,r\cos\theta + b \qquad ...(iv)$$

Now after virtual displacement, $OD$ become $OD_1$

$\therefore$ Change in displacement, $$DD_1 = OD_1 - OD$$

$$= r\cos\theta - \alpha\,.\,r\sin\theta + a - r\cos\theta$$

$$= a - \alpha\,.\,r\sin\theta = a - \alpha.y \qquad ...(v)$$

Similarly, after virtual displacement, $AD$ becomes $A'D_1$

$\therefore$ Change in displacement, $$A_3A' = A'D_1 - A_3D_1$$

$$= r\sin\theta + \alpha\,.\,r\cos\theta + b - r\sin\theta \qquad ...(\because A_3D_1 = AD = r\sin\theta)$$

$$= b + \alpha.r\cos\theta = b + \alpha.x \qquad ...(vi)$$

Let $P_X$ and $P_Y$ be the components of $P$ along $X$-axis and $Y$-axis, respectively.

We know that virtual work done by the force $P$

= Virtual work done by the component $P_X$

+ Virtual work done by the component $P_Y$

$$= P_X \times DD_1 + P_Y \times A_3A'$$

$$= P_X(a - \alpha.y) + P_Y(b + \alpha.x)$$

$$= a.P_X - P_X\,.\,\alpha.y + bP_Y + P_Y\,.\,\alpha.x$$

$$= aP_X + bP_Y + \alpha(P_Y\,.\,x - P_X\,.\,y)$$

Similarly, virtual work done by the force $Q$

$$= a.Q_X + b.Q_Y + \alpha(Q_Y\,.\,x - Q_X\,.y)$$

$\therefore$ Total work done by all the forces $= a\Sigma P_X + b\Sigma P_Y + \alpha.\Sigma(P_Y.x - P_X.y)$

where $\Sigma P_X$ = Sum of the resolved components of the forces along $X$-axis

= 0, because the forces are in equilibrium,

$\Sigma P_Y$ = Sum of the resolved components of the forces along $Y$-axis

= 0, because the forces are in equilibrium, and

$$\Sigma(P_Y \cdot x - P_X \cdot y) = \text{Sum of the moments of all the forces about } O$$
$$= 0, \text{ because forces are in equilibrium.}$$

Thus, the total virtual work done $= 0$

## 11.4 APPLICATION OF THE PRINCIPLE OF VIRTUAL WORK

The principle of virtual work is applied for solving problems in beams, framed structures, ladders and lifting machines. It simplifies the solution of many problems, because the following forces can be omitted in forming the equation of virtual work:

1. The tension of an inextensible string.
2. The reaction between the body and any smooth surface with which it is in contact.
3. The reaction between the body and any surface on which it rolls without slipping.
4. The mutual action and reaction between the two bodies where equilibrium is being considered together.

## 11.5 CONDITIONS TO SATISFY BY VIRTUAL DISPLACEMENT

For solving a problem by the method of virtual work, a virtual displacement should satisfy the following conditions:

**(*a*)** Imagine that a very small displacement is given to the body.

**(*b*)** The displacement given to the body should be consistent with its geometrical conditions.

**(*c*)** The algebraic sum of the product of the force and the virtual displacement is equal to zero.

## 11.6 SIGN CONVENTIONS FOR VIRTUAL WORK DONE

When the force and virtual displacement are in the same direction, then the virtual work done is *positive*. On the other hand, when the force and virtual displacement are in opposite direction, then the virtual work done is *negative*.

## 11.7 APPLICATION OF THE PRINCIPLE OF VIRTUAL WORK ON SIMPLY SUPPORTED BEAM

We shall consider the following two cases:

**1. *When the beam carries a point load***

Consider a simply supported beam *AB* of length $l$ carrying a point load $W$ at point *C*, as shown in Fig. 11.3 (*a*).

Let $R_A$ be the reaction at the hinged support at *A* and $R_B$ be the reaction at the roller support *B*. Let the beam is given a vertically upward virtual displacement ($y$) at *B*, keeping the point *A* fixed, as shown in Fig. 11.3 (*b*). The beam at *C* also undergoes a virtual displacement ($x$) due to the point load ($W$), but the virtual displacement at *A* is zero. From the similar triangles *ABD* and *ACE*, we find that

$$\frac{BD}{CE} = \frac{AB}{AC} \quad \text{or} \quad \frac{y}{x} = \frac{l}{a} \quad \text{or} \quad x = \frac{a \cdot y}{l}$$

We know that virtual work done by the vertical reaction at *A*

$$= \text{Force} \times \text{Displacement} = R_A \times 0 = 0$$

and virtual work done by the vertical reaction at *B*

$$= R_B \times y$$

∴ Total virtual work done by the upward reactions at *A* and *B*

$$= 0 + R_B \times y = R_B \times y$$

...(+ve sign due to upward reaction $R_B$ and upward virtual displacement $y$)

Virtual work done by the point load at $C$

$$= -W \times x$$

...($\because$ –ve sign due to downward load $W$ and upward virtual displacement $x$)

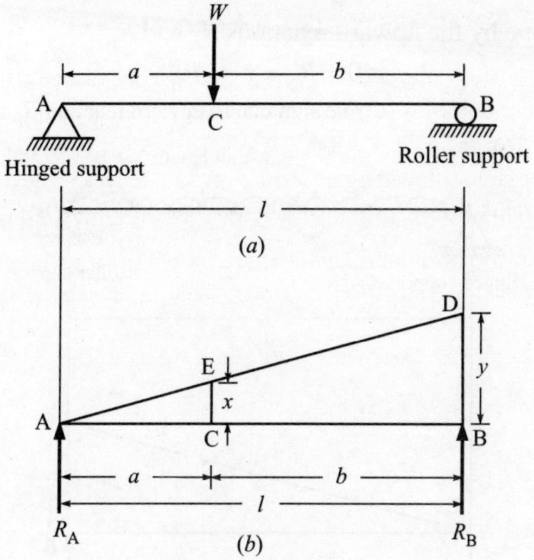


**Fig. 11.3**

We know that according to the principle of virtual work done, the algebraic sum of the virtual works done is equal to zero.

$$\therefore \quad R_B \times y - W \times x = 0$$

or $$R_B = \frac{W \times x}{y} = \frac{W \times a}{l} \qquad \ldots\left(\because x = \frac{a.y}{l}\right)$$

Similarly, it may be determined that

$$R_A = \frac{W \times b}{l}$$

**Notes: 1.** The value of $R_A$ may also be determined as follows:

We know that $$R_A + R_B = W$$

$$\therefore \quad R_A = W - R_B = W - \frac{W.a}{l}$$

$$= W\left[1 - \frac{a}{l}\right] = W\left[\frac{l-a}{l}\right] = \frac{W \times b}{l} \qquad \ldots(\because l - a = b)$$

**2.** In the above discussion, we have taken only one point load at $C$. But, the same principle may be applied, if there are number of point loads on the beam.

**2. *When the beam carries a uniformly distributed load***

Consider a simply supported beam $AB$ of length $l$ carrying a uniformly distributed $w$/unit length for the whole length of the beam, as shown in Fig. 11.4 ($a$).

Let $R_A$ and $R_B$ be the reactions at $A$ and $B$, respectively. Consider an upward virtual displacement ($y$) at $B$ due to the reaction at $B$, keeping the point $A$ fixed, as shown in Fig. 11.4 ($b$). Therefore the virtual displacement at $A$ is zero.

We know that virtual work done by the vertical reaction at $A$

$$= R_A \times 0 = 0$$

and virtual work done by the vertical reaction at $B$

$$= R_B \times y$$

∴ Total virtual work done by the upward reactions $R_A$ and $R_B$

$$= 0 + R_B \times y = R_B \times y$$

...(+ve sign due to upward reaction $R_B$ and upward virtual displacement $y$)

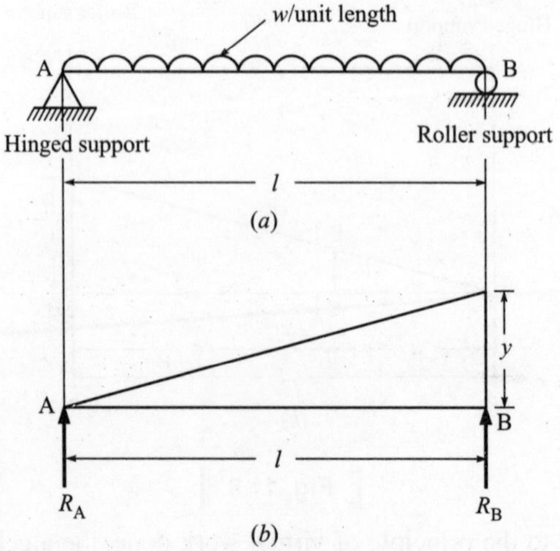


Fig. 11.4

Virtual work done by the uniformly distributed load

$$= -w\left(\frac{0+y}{2}\right)l = \frac{-w.y.l}{2}$$

...(–ve sign due to downward load and upward virtual displacement $y$)

According to the principle of virtual work, the algebraic sum of the virtual works done is equal to zero.

$$\therefore \quad R_B \times y - \frac{w.y.l}{2} = 0$$

or

$$R_B = \frac{w.y.l}{2 \times y} = \frac{w.l}{2}$$

We know that

$$R_A + R_B = w.l$$

$$R_A = w.l - R_B = w.l - \frac{wl}{2} = \frac{wl}{2}$$

**Example 11.1** *A simply supported beam AB of span 8 m is carrying a point load of 6 kN at a distance of 3 m from A. Using the principal of virtual work, find the reactions.*

***Solution*** Given : $AB = l = 8$ m ; $W = 6$ kN ; $AC = a = 3$ m ; $BC = b = 8 - 3 = 5$ m

The beam with a point load is shown in Fig. 11.5 (*a*). Let $R_A$ and $R_B$ be the reactions at the supports $A$ and $B$, respectively. Let $y$ and $x$ be the virtual vertical displacements of the beam at the end $B$ and $C$, respectively, as shown in Fig. 11.5 (*b*). Keeping the end $A$ of beam as fixed, we have from similar triangles $ABD$ and $AEC$,

$$\frac{BD}{EC} = \frac{AB}{AC} \quad \text{or} \quad \frac{y}{x} = \frac{8}{3}$$

$$\therefore \qquad x = \frac{3}{8}y$$

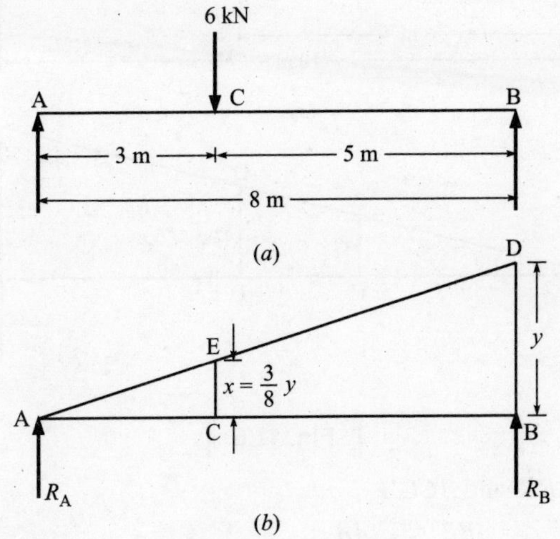


Fig. 11.5

We know that total virtual work done by the upwards reactions $R_A$ and $R_B$ at $A$ and $B$

$$= R_A \times 0 + R_B \times y = R_B \times y$$

...(+ve sign due to upward reaction $R_B$ and upward virtual displacement $y$)

and virtual work done by the downward load $W$ at $C$

$$= -W \times x = -6 \times \frac{3}{8}y = -\frac{9}{4}y$$

...(–ve sign due to downward load $W$ and upward virtual displacement $x$)

According to the principle of virtual work, the algebraic sum of the virtual works done is equal to zero.

$$\therefore \qquad R_B \times y - \frac{9}{4}y = 0$$

or $$R_B = \frac{9}{4} = 2.25 \text{ kN} \quad \textbf{Ans.}$$

We know that $$R_A + R_B = W$$

$$\therefore \qquad R_A = W - R_B = 6 - 2.25 = 3.75 \text{ kN} \quad \textbf{Ans.}$$

**Example 11.2** *A beam AB of span 10 m carries two points loads of 15 kN and 20 kN at 4 m and 6 m from the end A, respectively. Determine the beam reactions by the principle of virtual work.*

***Solution*** Given : $AB = l = 10$ m ; $W_1 = 15$ kN ; $W_2 = 20$ kN ; $AC = 4$ m ; $AD = 6$ m

The beam with point loads is shown in Fig. 11.6 (*a*).

Let $R_A$ and $R_B$ be the reactions at the supports $A$ and $B$, respectively. Let $y$ be the virtual vertical displacement of the beam at the end $B$, keeping the end $A$ of the beam as fixed. Let $x_1$ and $x_2$ be the virtual vertical displacement at the point loads at $C$ and $D$, respectively, as shown in Fig. 11.6 (*b*).

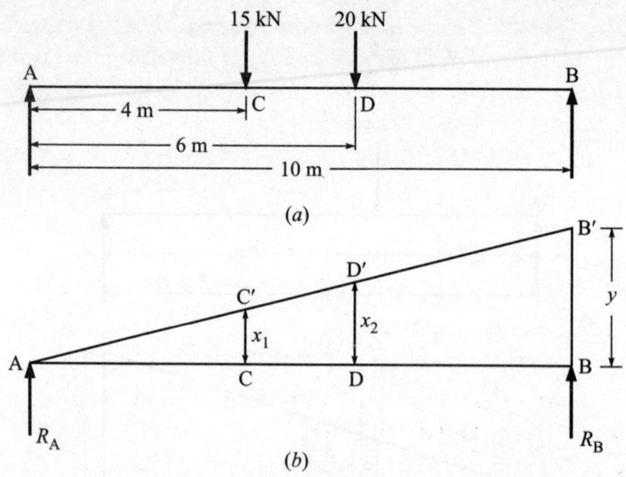


**Fig. 11.6**

From similar triangles $ABB'$ and $ACC'$,

$$\frac{BB'}{CC'} = \frac{AB}{AC}$$

$$\therefore \quad \frac{y}{x_1} = \frac{10}{4} \quad \text{or} \quad x_1 = \frac{4}{10}y$$

Similarly, from similar triangles $ABB'$ and $ADD'$,

$$\frac{BB'}{DD'} = \frac{AB}{AD}$$

$$\frac{y}{x_2} = \frac{10}{6} \quad \text{or} \quad x_2 = \frac{6}{10}y$$

We know that total virtual work done by the vertical reactions $R_A$ and $R_B$ at $A$ and $B$

$$= R_A \times 0 + R_B \times y = R_B \times y$$

...(+ve sign due to upward reaction $R_B$ and upward virtual displacement $y$)

Virtual work done by the downward load at $C$

$$= -W_1 \times x_1 = -15 \times \frac{4y}{10} = -6\,y$$

...(–ve sign due to downward load $W_1$ and upward virtual displacement $x_1$)

and virtual work done by the downward load at $D$

$$= -W_2 \times x_2 = -20 \times \frac{6}{10}y = -12y$$

...(–ve sign due to downward load $W_2$ and upward virtual displacement $x_2$)

According to the principle of virtual work, the algebraic sum of the virtual works done is equal to zero.

$$\therefore \quad R_B \times y - 6y - 12y = 0$$

or $\quad R_B \times y - 18y = 0$

and $\quad R_B = \frac{18y}{y} = 18 \text{ kN}$ **Ans.**

We know that $\quad R_A + R_B = 15 + 20 = 35 \text{ kN}$

$$\therefore \quad R_A = 35 - R_B = 35 - 18 = 17 \text{ kN}$$ **Ans.**

**Example 11.3** *Two beams AC and CD of lengths 9 m and 10 m, respectively, are hinged at C. These are supported on rollers at the left and right ends (A and D). A hinged support is provided at B, 7 m from A. Using the principle of virtual work, determine the reactions at the hinge C and the support B when a load of 700 N acts at a point 6 m from D.*

***Solution*** Given : $AC = 9$ m ; $CD = 10$ m ; $AB = 7$ m ; $ED = 6$ m ; $W_E = 700$ N

The beam is shown in Fig. 11.7 (*a*).

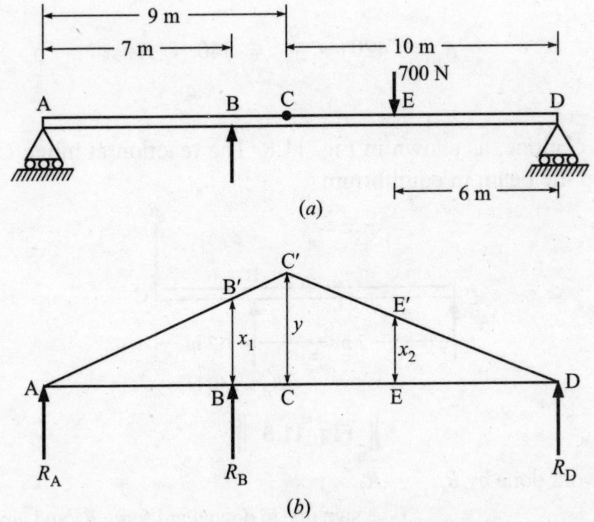


**Fig. 11.7**

Let $R_A$, $R_B$ and $R_D$ be the reactions at the points *A*, *B* and *D*, respectively. Let *y* be the virtual vertical displacement at the hinge *C*. Let $x_1$ and $x_2$ be the virtual vertical displacements at the hinge support at *B* and at the load point at *E*, respectively, as shown in Fig. 11.7 (*b*). Since the beams *AC* and *CD* are supported on rollers, therefore the virtual displacements at *A* and *D* will be zero.

***Reaction at support B***

From similar triangles *ACC'* and *ABB'*,

$$\frac{CC'}{BB'} = \frac{AC}{AB} \quad \text{or} \quad \frac{y}{x_1} = \frac{9}{7}$$

$$\therefore \qquad x_1 = \frac{7}{9}y$$

Similarly, from similar triangles *DCC'* and *DEE'*

$$\frac{CC'}{EE'} = \frac{CD}{ED} \quad \text{or} \quad \frac{y}{x_2} = \frac{10}{6}$$

$$\therefore \qquad x_2 = \frac{6}{10}y$$

We know that virtual work done by the vertical reactions at *A* and *D*

$= 0$ ...($\because$ Virtual displacement at *A* and *D* is zero)

Virtual work done by the upward reaction at *B*

$$= R_B \times x_1 = R_B \times \frac{7}{9}y$$

...(+ve sign due to upward reaction $R_B$ and upward virtual displacement $x_1$)

and virtual work done by the downward point load at $E$

$$= -W_E \times x_2 = -700 \times \frac{6}{10}y = -420y$$

...(–ve sign due to downward load $W_E$ and upward virtual displacement $x_2$)

According to the principle of virtual work, the algebraic sum of the virtual works done is equal to zero.

$$\therefore \quad R_B \times \frac{7}{9}y - 420y = 0$$

or

$$R_B = 420y \times \frac{9}{7y} = 540 \text{ N} \quad \textbf{Ans.}$$

***Reaction at hinge C***

Consider the beam $AC$ alone, as shown in Fig. 11.8. The reaction at hinge $C$ ($R_C$) must act vertically downwards in order to keep the beam in equilibrium.

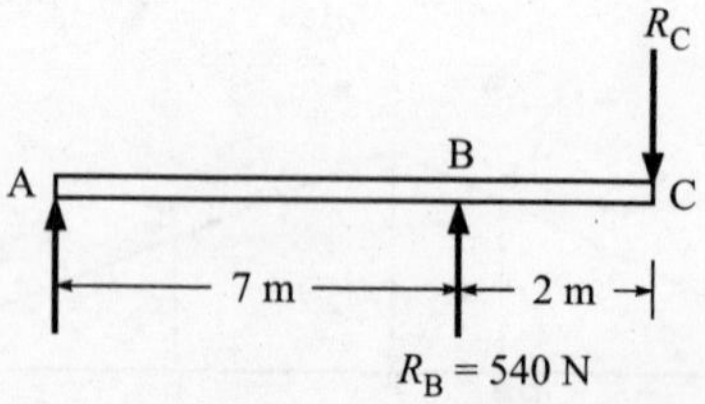


**Fig. 11.8**

We know that virtual work done by $R_C = -R_C \times y$

...(–ve sign due to downward force $R_C$ and upward virtual displacement $y$)

Virtual work done by $R_B = R_B \times x_1 = 540 \times \frac{7}{9}y$

and virtual work done by $R_A = 0$

According to the principle of virtual work, the algebraic sum of the virtual works done is equal to zero.

$$\therefore \quad -R_C \times y + 540 \times \frac{7}{9}y = 0$$

or

$$R_C \times y = 540 \times \frac{7}{9}y$$

$$\therefore \quad R_C = 540 \times \frac{7}{9} = 420 \text{ N} \quad \textbf{Ans.}$$

**Example 11.4** *Two beams AB and CD are supported on rollers at B and C, as shown in Fig. 11.9. Using the method of virtual work, determine the reactions at the rollers B and C.*

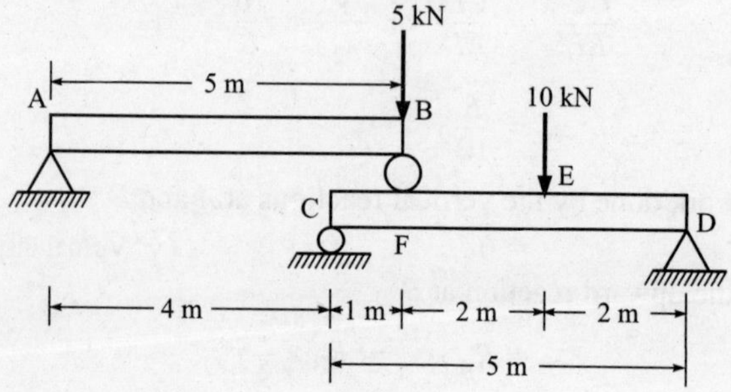


**Fig. 11.9**

***Solution*** Given : $AB = CD = 5$ m ; $AF = AC + CF = 4 + 1 = 5$ m ; $W_B = 5$ kN ; $W_E = 10$ kN ; $FE = 2$ m ; $ED = 2$ m

Let $R_B$ and $R_C$ be the reactions at the rollers $B$ and $C$, respectively.

First of all, let us consider the beam $AB$, as shown in Fig. 11.10 (*a*).

We know that virtual work done by the vertical reaction at $A$

$$= R_A \times 0 = 0 \quad \text{...(}\because \text{ Virtual displacement at } A \text{ is zero)}$$

Virtual work done by the vertical reaction at $B$

$$= +R_B \times y_B$$

...($\because$ +ve sign due to upward reaction $R_B$ and upward virtual displacement $y_B$)

Virtual work done by the downward load at $B$

$$= -5 \times y_B$$

...(–ve sign due to downward load and upward virtual displacement $y_B$)

According to the principle of virtual work, the algebraic sum of the virtual works done is equal to zero.

$$\therefore \quad R_A \times 0 + R_B \times y_B - 5 \times y_B = 0$$

or

$$R_B = 5 \text{ kN} \quad \textbf{Ans.}$$

Now consider the beam $CD$ with loads at $F$ and $E$, as shown in Fig. 11.10 (*b*).

From similar triangles $CDD'$ and $CFF'$,

$$\frac{DD'}{FF'} = \frac{CD}{CF} \quad \text{or} \quad \frac{y_D}{x_1} = \frac{5}{1}$$

$$\therefore \quad x_1 = \frac{y_D}{5} = 0.2 y_D$$

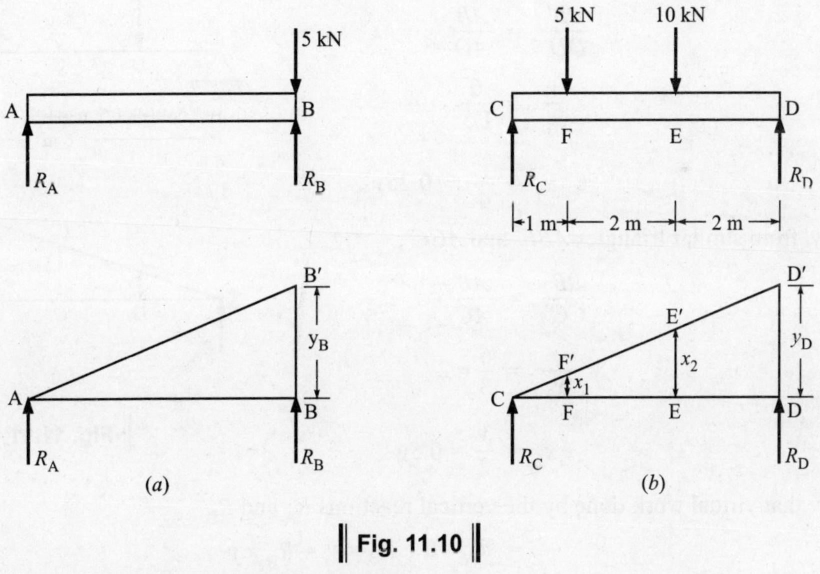


**Fig. 11.10**

Similarly, from similar triangles $CDD'$ and $CEE'$,

$$\frac{DD'}{EE'} = \frac{CD}{CE} \quad \text{or} \quad \frac{y_D}{x_2} = \frac{5}{3}$$

$$\therefore \quad x_2 = y_D \times \frac{3}{5} = 0.6 y_D$$

We know that virtual work done by the vertical reaction at $C$

$$= R_C \times 0 = 0 \quad \text{...(}\because \text{ Virtual displacement at } C \text{ is zero)}$$

Virtual work done by the vertical reaction at $D$

$$= +R_D \times y_D$$

...(+ve sign due to upward reaction $R_D$ and upward virtual displacement $y_D$)

Virtual work done by the downward load at $F$

$$= -5 \times x_1 = -5 \times 0.2y_D = -y_D$$

...(–ve sign due to downward load and upward virtual displacement $x_1$)

and virtual work done by the downward load at $E$

$$= -10 \times x_2 = -10 \times 0.6y_D = -6y_D$$

According to the principle of virtual work, the algebraic sum of the virtual works done is equal to zero.

$$\therefore \quad R_C \times 0 + R_D \times y_D - y_D - 6y_D = 0$$

or $\quad R_D = +1 + 6 = 7 \text{ kN}$ **Ans.**

**Example 11.5** *A simply supported beam AB is loaded, as shown in Fig. 11.11 (a). Determine the reactions at A and B, using the principle of virtual work.*

***Solution*** Given : $AB = 6$ m ; $AD = DC = 1.5$ m ; $CB = 3$ m ; $w = 2$ kN/m ; $W_D = 6$ kN

Let $R_A$ and $R_B$ be the reactions at the supports $A$ and $B$, respectively. Let $y$ be the virtual vertical displacement of beam at the end $B$, keeping the end $A$ of the beam as fixed. Let $x_1$ and $x_2$ be the virtual vertical displacement at the points $D$ and $C$, respectively, as shown in Fig. 11.11 (*b*).

From similar triangles $ABB'$ and $ADD'$,

$$\frac{BB'}{DD'} = \frac{AB}{AD}$$

or $$\frac{y}{x_1} = \frac{6}{1.5} = 4$$

$$\therefore \quad x_1 = \frac{y}{4} = 0.25y$$

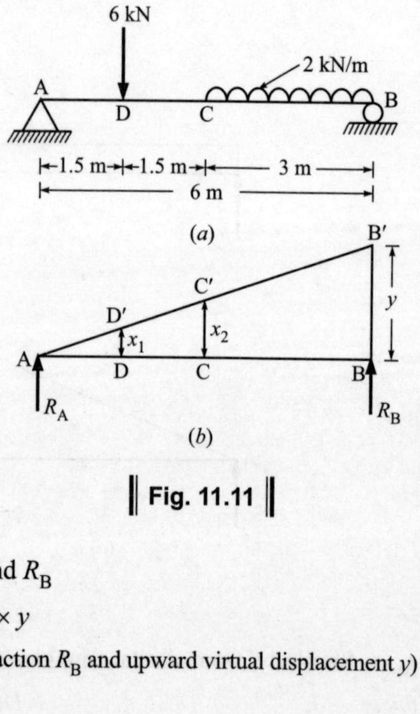


Fig. 11.11

Similarly, from similar triangles $ABB'$ and $ACC'$,

$$\frac{BB'}{CC'} = \frac{AB}{AC}$$

or $$\frac{y}{x_2} = \frac{6}{3} = 2$$

$$\therefore \quad x_2 = \frac{y}{2} = 0.5y$$

We know that virtual work done by the vertical reactions $R_A$ and $R_B$

$$= R_A \times 0 + R_B \times y = R_B \times y$$

...(+ve sign due to upward reaction $R_B$ and upward virtual displacement $y$)

Virtual work done by the point load at $D$

$$= -W_D \times x_1 = -6 \times 0.25y = -1.5\,y$$

...(–ve sign due to downward load $W_D$ and upward virtual displacement $x_1$)

and virtual work done by the uniformly distributed load between $B$ and $C$

$$= -2\left(\frac{y + x_2}{2}\right)3 = -2\left(\frac{y + 0.5y}{2}\right)3 = -4.5y$$

...(–ve sign due to downward load and upward virtual displacement $x_2$)

According to the principle of virtual work, the algebraic sum of the virtual works done is equal to zero.

$\therefore$ $R_B \times y - 1.5y - 4.5y = 0$

$R_B \times y - 6y = 0$

$\therefore$ $R_B = \dfrac{6y}{y} = 6$ kN **Ans.**

We know that $R_A + R_B = 6 + 2 \times 3 = 12$ kN

$\therefore$ $R_A = 12 - 6 = 6$ kN **Ans.**

## 11.8 APPLICATION OF THE PRINCIPLE OF VIRTUAL WORK ON LADDERS

Consider a ladder $AC$ resting on the ground $AB$ and leaning against a wall $BC$, as shown in Fig. 11.12 (*a*). In case of a ladder, no work is done by the normal reaction ($R_A$) at the foot $A$ of the ladder, because its foot moves on the floor towards or away from the wall. However, some work is done by the frictional force ($F_A$) at the foot of the ladder. Similarly, no work is done by the normal reaction ($R_C$) at the top of the ladder, but some work is done by the frictional force at the top of the ladder, if the surface of the wall is *rough (*i.e.* if the surface of wall has some coefficient of friction). By using the principle of virtual work, we can find virtual work done by the frictional force at the foot and top of the ladder.

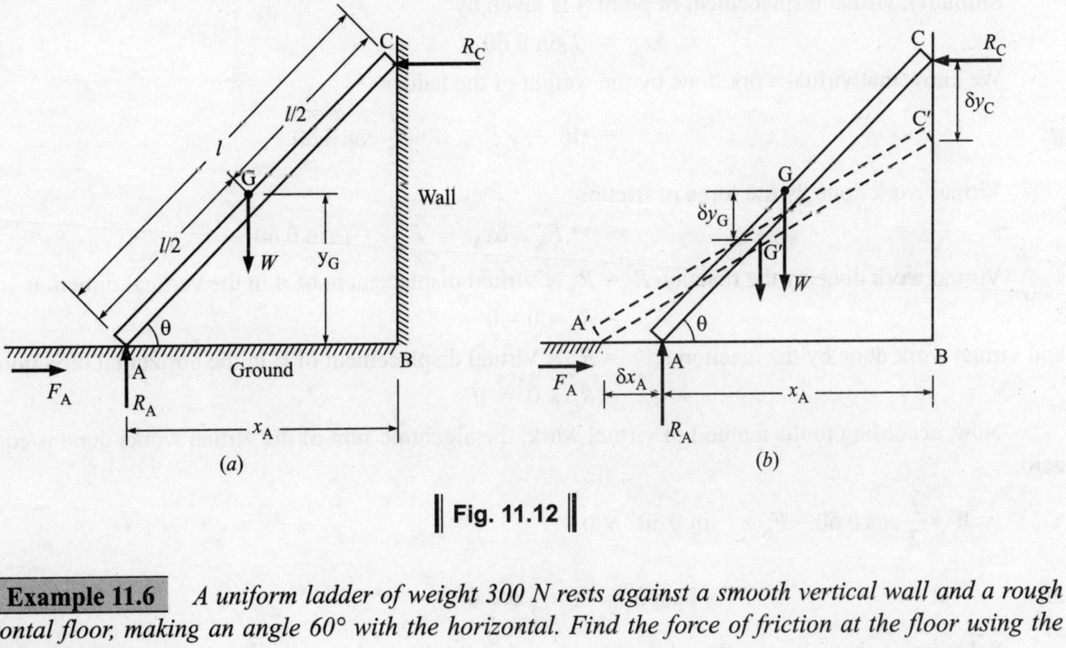


**Fig. 11.12**

**Example 11.6** *A uniform ladder of weight 300 N rests against a smooth vertical wall and a rough horizontal floor, making an angle 60° with the horizontal. Find the force of friction at the floor using the method of virtual work.*

***Solution*** Given : $W = 300$ N ; $\theta = 60°$ [Refer Fig. 11.12 (*a*)]

---

* In case the surface of the wall is smooth, then no frictional force acts at the top of the ladder. Thus no work is done at the top of the ladder.

Let $R_A$ and $R_C$ be the reactions at $A$ and $C$, respectively, $l$ be the length of the ladder and $F_A$ be the force of friction at $A$. Since the surface of the wall is smooth, therefore there will be no force of friction at $C$.

The weight of the ladder ($W$) is at $G$, the midpoint of $AC$.

$$\therefore \quad AG = \frac{AC}{2} = \frac{l}{2}$$

Let $x_A$ = Horizontal distance of $A$ from $B$, and

$y_G$ = Vertical distance of $G$ above $AB$.

From the geometry of the figure, $x_A = AB = AC \cos\theta = l \cos\theta$

and $$y_G = AG \sin\theta = \frac{l}{2}\sin\theta$$

The differential of $x_A$ and $y_G$ are obtained as follows:

$$dx_A = d(l\cos\theta) = -l\sin\theta\, d\theta$$

and $$dy_G = d\left(\frac{l}{2}\sin\theta\right) = \frac{l}{2}\cos\theta\, d\theta$$

Let the ladder is given a small virtual displacement in such a way that the point $A$ moves towards left by $\delta x_A$, keeping the length of the ladder fixed. Then the point $C$ and $G$ moves downwards by an amount $\delta y_C$ and $\delta y_G$, respectively, as shown in Fig. 11.12 (*b*).

Now, the virtual displacement of point $G$ is given by

$$\delta y_G = \frac{l}{2}\cos\theta\,\delta\theta$$ ...(Changing $d\theta$ to $\delta\theta$ for virtual displacement)

Similarly, virtual displacement of point $A$ is given by

$$\delta x_A = -l\sin\theta\,\delta\theta$$

We know that virtual work done by the weight of the ladder

$$= {}^{*}W \times \delta y_G = W \times \frac{l}{2}\cos\theta\,\delta\theta$$

Virtual work done by the force of friction

$$= {}^{**}F_A \times \delta x_A = F_A \times -l\sin\theta\,\delta\theta$$

Virtual work done by the reaction $R_A = R_A \times$ Virtual displacement of $A$ in the vertical direction

$= R_A \times 0 = 0$

and virtual work done by the reaction $R_C = R_C \times$ Virtual displacement of $C$ in the horizontal direction

$= R_C \times 0 = 0$

Now, according to the method of virtual work, the algebraic sum of the virtual works done is equal to zero.

$$\therefore \quad W \times \frac{l}{2}\cos\theta\,\delta\theta - F_A \times l\sin\theta\,\delta\theta = 0$$

or $$F_A \times l\sin\theta = W \times \frac{l}{2}\cos\theta$$

Substituting the values of $W$ and $\theta$ in the above expression, we have

---

* Since the weight ($W$) and virtual displacement ($\delta y_G$) is in the same direction, therefore virtual work done is +ve.

** Since the force $F_A$ and virtual displacement ($\delta x_A$) is in opposite direction, therefore virtual work done is –ve.

$$F_A \times \sin 60° = 300 \times \frac{1}{2} \cos 60°$$

$$F_A \times 0.866 = 300 \times \frac{1}{2} \times 0.5 = 75$$

$$\therefore \quad F_A = \frac{75}{0.866} = 86.6 \text{ N } \textbf{Ans.}$$

**Example 11.7** *A uniform ladder of 5 m long and weighing 200 N rests on a smooth floor at A and against a smooth wall at C, as shown in Fig. 11.13 (a). A horizontal rope PQ prevents the ladder from slipping. Using the method of virtual work, determine the tension in the rope.*

***Solution*** Given : $AC = l = 5$ m ; $W = 200$ N ; $AB = 3$ m ; $BC = 4$ m ; $AQ = CQ = 2.5$ m

Let $R_A$ and $R_C$ be the reactions at $A$ and $C$, respectively, and $T$ is the tension in the rope $PQ$, as shown in Fig. 11.13 (*b*). From the geometry of the figure,

$$x_Q = CQ \sin\theta = 2.5 \sin\theta$$

$$y_Q = AQ \sin(90° - \theta) = AQ \cos\theta = 2.5 \cos\theta$$

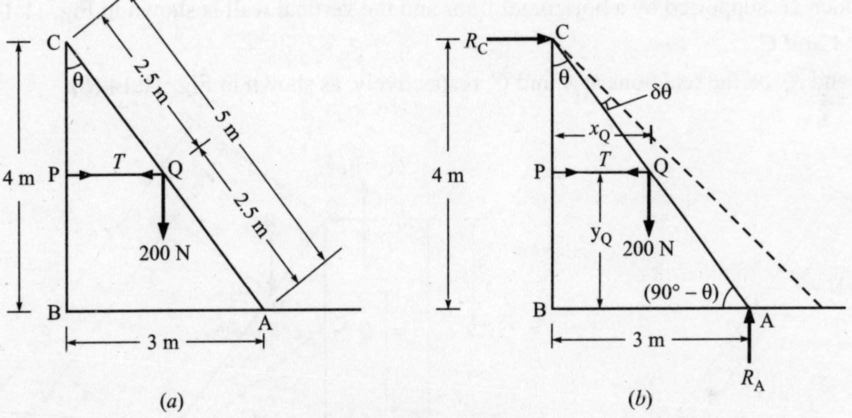


**Fig. 11.13**

The differentials of $x_Q$ and $y_Q$ are obtained as

$$dx_Q = d(2.5 \sin\theta) = 2.5 \cos\theta \, d\theta$$

and

$$dy_Q = d(2.5 \cos\theta) = -2.5 \sin\theta \, d\theta$$

Let the ladder is given a small displacement $\delta\theta$ by keeping the end $C$ fixed, as shown in Fig. 11.13 (*b*), such that $\theta$ becomes $(\theta + \delta\theta)$.

Now virtual displacement of $Q$ in the horizontal direction,

$$\delta x_Q = 2.5 \cos\theta \, \delta\theta \qquad \text{...(Changing } d\theta \text{ to } \delta\theta \text{ for virtual displacement)}$$

and virtual displacement of $Q$ in the vertical direction,

$$\delta y_Q = -2.5 \sin\theta \, \delta\theta$$

We know that virtual work done by the weight of ladder

$$= W \times \delta y_Q = 200 \times (-2.5 \sin\theta \, \delta\theta)$$

$$= -500 \sin\theta \, \delta\theta$$

Virtual work done by the tension $T$ $= T \times \delta x_Q = T \times 2.5 \cos\theta \, \delta\theta$

$$= 2.5\, T \cos\theta \, \delta\theta$$

Virtual work done by the reactions $R_A$ and $R_C$ at $A$ and $C$, respectively, is zero, because the virtual displacement in the vertical and horizontal directions is zero.

Now according to the method of virtual work, the algebraic sum of the virtual work done is equal to zero.

$$\therefore \quad -500 \sin\theta\, \delta\theta + 2.5\, T \cos\theta\, \delta\theta = 0$$

or
$$T = \frac{500 \sin\theta}{2.5 \cos\theta} = 200 \tan\theta = 200 \times \frac{AB}{BC}$$

$$= 200 \times \frac{3}{4} = 150 \text{ N } \textbf{Ans.} \quad \ldots\left[\text{From Fig. 11.13}(a),\ \tan\theta = \frac{AB}{BC}\right]$$

**Example 11.8** *A ladder AC is 6 m long and is supported by a horizontal floor and a vertical wall. The coefficient of friction between the floor and the ladder is 0.4 and between the wall and ladder is 0.25. The weight of the ladder is 200 N. The ladder also supports a load of 900 N at D which is at a distance of 1 m from C. Find the reactions at A and C and the value of angle θ for which the ladder may be placed without slipping.*

***Solution*** Given : $AC = l = 6$ m ; $\mu = 0.4$ ; $\mu_1 = 0.25$ ; $W = 200$ N ; $W_1 = 900$ N ; $CD = 1$ m ; $AD = AC - CD = 6 - 1 = 5$ m

The ladder $AC$ supported by a horizontal floor and the vertical wall is shown in Fig. 11.14 (*a*).

***Reactions at A and C***

Let $R_A$ and $R_C$ be the reactions at $A$ and $C$, respectively, as shown in Fig. 11.14 (*b*).

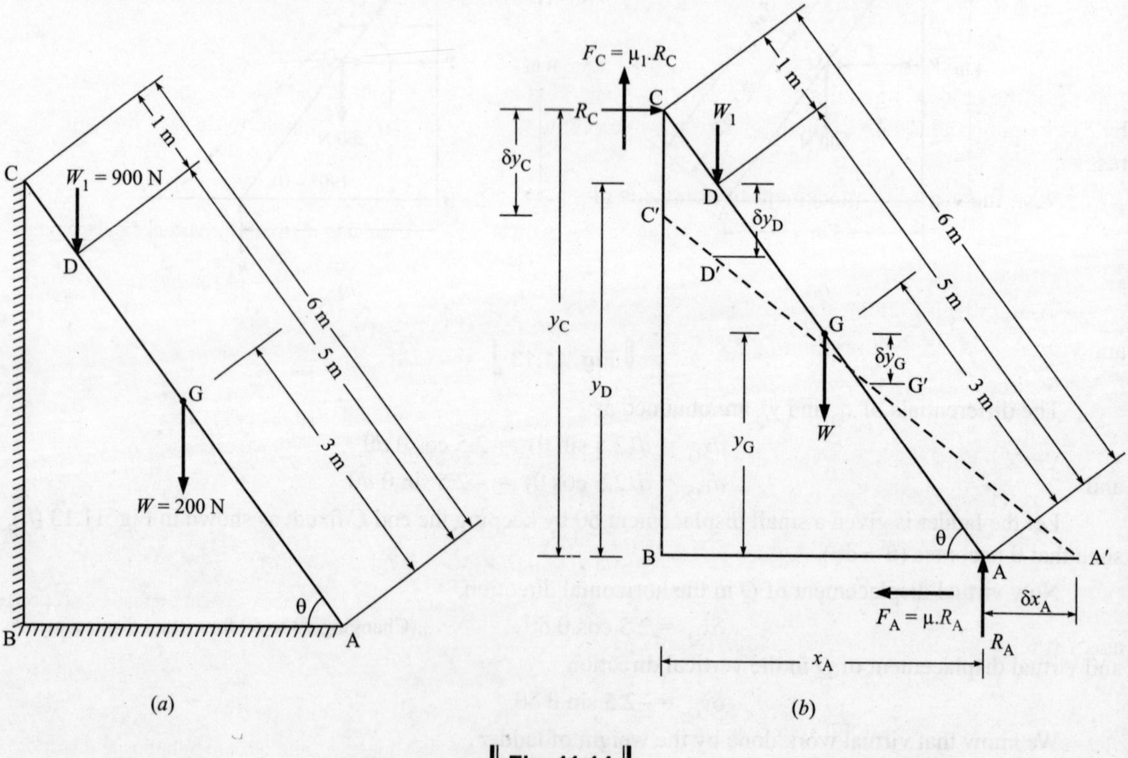


**Fig. 11.14**

We know that force of friction at $A$,

$$F_A = \mu.R_A = 0.4 R_A$$

and force of friction at $C$,

$$F_C = \mu_1.R_C = 0.25 R_C$$

For equilibrium of the ladder, $\Sigma H = 0$ and $\Sigma V = 0$.

For $\Sigma H = 0$, $\quad F_A = R_C$ or $0.4R_A = R_C$ ...(*i*)

For $\Sigma V = 0$,

$$R_A + F_C = 200 + 900$$
$$R_A + 0.25R_C = 1100$$
$$R_A + 0.25 \times 0.4R_A = 1100$$
$$1.1R_A = 1100$$
$$\therefore \quad R_A = \frac{1100}{1.1} = 1000 \text{ N } \textbf{Ans.}$$

and from equation (*i*), $\quad R_C = 0.4R_A = 0.4 \times 1000 = 400$ N **Ans.**

***Value of angle θ***

From the geometry of the Fig. 11.14 (*b*), we find that

Horizontal distance of *A* from *B*, $x_A = AB = AC \cos\theta = 6\cos\theta$

Vertical distance of *C* from *B*, $y_C = BC = AC \sin\theta = 6\sin\theta$

Vertical distance of *G* from *AB*, $y_G = AG \sin\theta = 3\sin\theta$

Vertical distance of *D* from *AB*, $y_D = AD \sin\theta = 5\sin\theta$

The differentials of $x_A$, $y_C$, $y_G$ and $y_D$ are obtained as follows:

$$dx_A = d\,(6\cos\theta) = -6\sin\theta\, d\theta$$
$$dy_C = d\,(6\sin\theta) = 6\cos\theta\, d\theta$$
$$dy_G = d\,(3\sin\theta) = 3\cos\theta\, d\theta$$
$$dy_D = d\,(5\sin\theta) = 5\cos\theta\, d\theta$$

Let the ladder is given a small virtual displacement in such a way that the point *A* moves to the right by $\delta x_A$ keeping the length of the ladder fixed. The points *C*, *G* and *D* moves by an amount $\delta y_C$, $\delta y_G$ and $\delta y_D$, respectively, as shown in Fig. 11.14 (*b*).

Now, the virtual displacement of point *A* is given by

$$\delta x_A = dx_A = -6\sin\theta\,\delta\theta \quad \text{...(Changing } d\theta \text{ to } \delta\theta \text{ for virtual displacement)}$$

Virtual displacement of point *C*, $\delta y_C = dy_C = 6\cos\theta\,\delta\theta$

Virtual displacement of point *G*, $\delta y_G = dy_G = 3\cos\theta\,\delta\theta$

and virtual displacement of point *D*, $\delta y_D = dy_D = 5\cos\theta\,\delta\theta$

We know that virtual work done by the weight of the ladder

$$= {}^*W \times \delta y_G = 200 \times 3\cos\theta\,\delta\theta = 600\cos\theta\,\delta\theta$$

Virtual work done by a load at *D* $\quad = {}^*W_1 \times \delta y_D = 900 \times 5\cos\theta\,\delta\theta = 4500\cos\theta\,\delta\theta$

Virtual work done by the force of friction ($F_A$) at *A*,

$$= F_A \times \delta x_A = R_C \times \delta x_A \quad \text{...}(\because F_A = R_C)$$
$$= 400 \times -6\sin\theta\,\delta\theta = -2400\sin\theta\,\delta\theta$$

...(–ve sign due to $F_A$ and $\delta x_A$ are in opposite direction)

and virtual work done by the force of friction ($F_C$) at *C*

$$= -F_C \times \delta y_C = -0.25R_C \times \delta y_C \quad \text{...}(\because F_C = 0.25R_C)$$
$$= -0.25 \times 400 \times 6\cos\theta\,\delta\theta = -600\cos\theta\,\delta\theta$$

...(–ve sign due to $F_C$ and $\delta y_C$ are in opposite direction)

According to the method of virtual work, the algebraic sum of the virtual works done is equal to zero.

---

* Since, the load *W* and $W_1$ are downward and the virtual displacement $\delta y_G$ and $\delta y_D$ are also downward, therefore virtual work done is +ve.

$\therefore \quad 600 \cos \theta \, \delta\theta + 4500 \cos \theta \, \delta\theta - 2400 \sin \theta \, \delta\theta - 600 \cos \theta \, \delta\theta = 0$

or $$4500 \cos \theta = 2400 \sin \theta$$

$$\tan \theta = \frac{4500}{2400} = 1.875$$

$\therefore$ $$\theta = \tan^{-1} (1.875) = 61.93° \textbf{ Ans.}$$

## 11.9 APPLICATION OF THE PRINCIPLE OF VIRTUAL WORK ON FRAMED STRUCTURES

In case of framed structures, the member in which the force is required to be determined, is assumed to be removed from the structure and the two forces each equal to *T* are applied at the two ends of the member (*See Example 11.9). Now the virtual work done by all the remaining members of the structure are determined. Apply the principle of virtual work to find the value of *T*.

**Example 11.9** *Five rods, AB, BC, CD, DA and BD each of equal length and cross-section are pin jointed together so as to form a plane ABCD. The frame ABCD has a rhombus shape with one horizontal diagonal BD. The frame is suspended from the top most point A.*

*A weight W is attached at the lower most joint C. Neglecting the self weight of the frame and using the method of virtual work, find the magnitude of the thrust in the member BD.*

***Solution*** Given : $AB = BC = CD = DA = BD = a$ ; $\angle ABD = \theta$ [Refer Fig. 11.15 (*a*)]

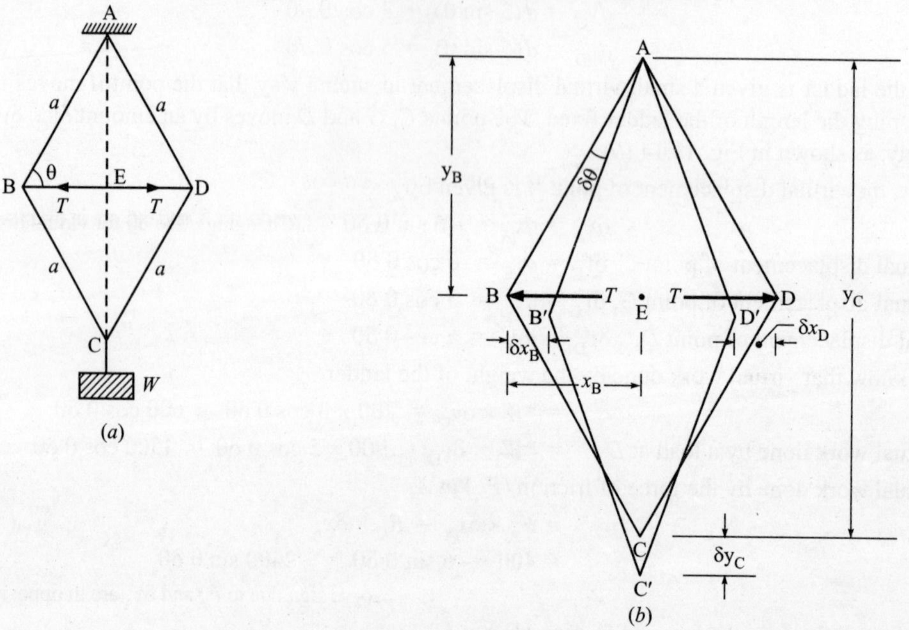


**Fig. 11.15**

From Fig. 11.15 (*b*), we find that

$$x_B = BE = a \cos \theta$$
$$y_B = AE = a \sin \theta$$

* In Example 11.9, the thrust in the member *BD* is to be determined. This member is assumed to be removed and the two forces each equal to *T* are applied at *B* and *D*.

and $$y_C = AC = AE + EC = a \sin\theta + a \sin\theta = 2a \sin\theta$$

The differentials of $x_B$, $y_B$ and $y_C$ are obtained as follows:

$$dx_B = d(a\cos\theta) = -a \sin\theta\, d\theta$$

$$dy_B = d(a\sin\theta) = a\cos\theta\, d\theta$$

and $$dy_C = d(2a\sin\theta) = 2a\cos\theta\, d\theta$$

Let $T$ be the thrust in the rod $BD$. Assume that the rod $BD$ is removed and the two forces, each equal to $T$ are applied at $B$ and $D$. Let the system be given a small virtual displacement $\delta\theta$ keeping the end $A$ fixed and the point $C$ exactly below $A$, as shown in Fig. 11.15 (*b*). The points $B$ and $D$ will move inward and the point $C$ will move downward.

Now the virtual horizontal displacement of point $B$ is given by

$$\delta x_B = -a\sin\theta\, \delta\theta \quad \text{...(Changing } d\theta \text{ to } \delta\theta \text{ for virtual displacement)}$$

Virtual horizontal displacement of point $D$

$$\delta x_D = \text{Virtual horizontal displacement of point } B$$

$$= \delta x_B = -a\sin\theta\,\delta\theta$$

Virtual vertical displacement of point $B$,

$$\delta y_B = a\cos\theta\,\delta\theta$$

Virtual vertical displacement of point $C$

$$\delta y_C = 2a\cos\theta\,\delta\theta$$

We know that virtual work done by the thrust $T$ at $B$

$$= T \times \delta x_B = T \times -a\sin\theta\,\delta\theta$$

$$= -T.\,a\sin\theta\,\delta\theta \quad \text{...(–ve sign due to } T \text{ and } \delta x_B \text{ are in opposite direction)}$$

Virtual work done by the thrust $T$ at $D$ $= T \times \delta x_D = T \times -a\sin\theta\,\delta\theta$

$$= -T.a\sin\theta\,\delta\theta \quad \text{...(–ve sign due to } T \text{ and } \delta x_D \text{ are in opposite direction)}$$

and virtual work done by weight $W$ at $C$ $= W \times \delta y_C = W \times 2a\cos\theta\,\delta\theta$

$$= 2W.\,a\cos\theta\,\delta\theta \quad \text{...(+ve sign due to } W \text{ and } \delta y_C \text{ are in the same direction)}$$

According to the method of virtual work, the algebraic sum of the virtual works done is equal to zero.

$$\therefore\ -T.a\sin\theta\,.\,\delta\theta - T.a\sin\delta\theta + 2W.a\cos\theta\,\delta\theta = 0$$

$$-2T.a\sin\theta + 2W.a\cos\theta = 0$$

or $$T = W \times \frac{\cos\theta}{\sin\theta} = W\cot\theta = \frac{W}{\tan\theta}$$

Since all the rods are equal, therefore $\theta = 60°$.

$$\therefore \quad T = \frac{W}{\tan 60°} = \frac{W}{1.732} \quad \textbf{Ans.}$$

**Example 11.10** *Six identical uniform rods AB, BC, CD, DE, EF and FA, each weighing W, are freely jointed at their extremities so as to form a hexagon. The rod AB is fixed in a horizontal position and the middle points of AB and DE are connected by a vertical string. Using the principle of virtual work, find the tension in the string.*

***Solution*** Given : $AB = BC = CD = DE = EF = FA = a$ (say) ; Weight of each rod = $W$

The weight $W$ of each rod is acting at the middle point of the rod. $GH$ is the string connecting the middle points $G$ and $H$ of the rods $AB$ and $DE$, respectively. Let $T$ be the tension acting at points $G$ and $H$, as shown in Fig. 11.16. Let $y_N$, $y_F$, $y_M$ and $y_E$ be the vertical distance of points $N$, $F$, $M$ and $E$, respectively, from the fixed rod $AB$.

From the geometry of the figure, $y_N = y_K = AN \sin\theta = \frac{a}{2}\sin\theta = 0.5a\sin\theta$

$$y_F = y_C = AF\sin\theta = a\sin\theta$$

$$y_M = y_L = y_N + y_F = 0.5a\sin\theta + a\sin\theta = 1.5a\sin\theta$$

and $$y_E = y_D = 2y_F = 2a\sin\theta$$

The differentials of $y_N$, $y_F$, $y_M$ and $y_E$ are obtained as follows:

$$dy_N = dy_K = d(0.5a\sin\theta)\,d\theta = 0.5a\cos\theta\,d\theta$$

$$dy_F = dy_C = d(a\sin\theta)\,d\theta = a\cos\theta\,d\theta$$

$$dy_M = dy_L = d(1.5a\sin\theta)\,d\theta = 1.5a\cos\theta\,d\theta$$

$$dy_E = dy_D = d(2a\sin\theta)\,d\theta = 2a\cos\theta\,d\theta$$

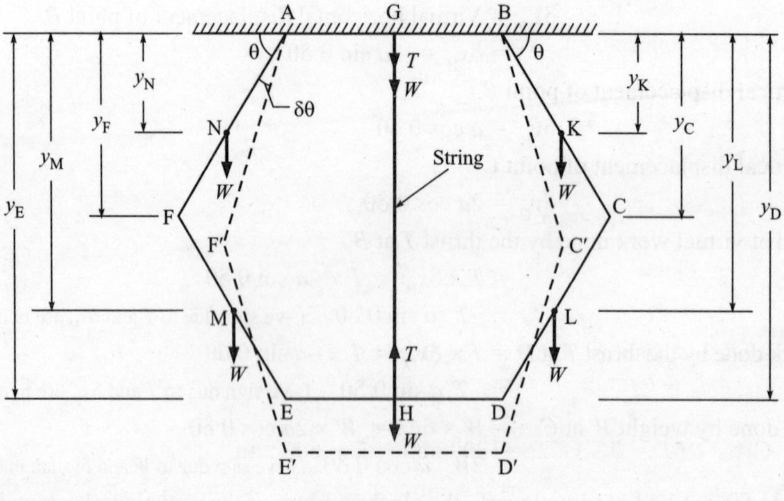


**Fig. 11.16**

Let the system is given a small virtual displacement $\delta\theta$ keeping the rod $AB$ fixed, as shown in Fig. 11.16. The points $F$ and $C$ move inward and the points $E$ and $D$ will move downward.

Now, the virtual vertical downward displacement of point $N$ or $K$ is given by

$$\delta y_N = \delta y_K = 0.5a\cos\theta\,\delta\theta$$ ...(Changing $d\theta$ to $\delta\theta$, for virtual displacement)

Virtual vertical downward displacement of point $F$ or $C$,

$$\delta y_F = \delta y_C = a\cos\theta\,\delta\theta$$

Virtual vertical downward displacement of point $M$ or $L$,

$$\delta y_M = \delta y_L = 1.5a\cos\theta\,\delta\theta$$

Virtual vertical downward displacement of point $E$ or $D$,

$$\delta y_E = \delta y_D = 2a\cos\theta\,\delta\theta$$

We know that virtual work done by the load $W$ at $N$ and $K$

$$= +W \times \delta y_N + W \times \delta y_K$$

$$= +W \times 0.5a\cos\theta\,\delta\theta + W \times 0.5a\cos\theta\,\delta\theta = W \times a\cos\theta\,\delta\theta$$

...(+ve sign due to $W$, $\delta y_N$ and $\delta y_K$ are in the same direction)

Virtual work done by the load $W$ at $M$ and $L$

$$= W \times \delta y_M + W \times \delta y_L$$

$$= W \times 1.5a \cos\theta\, \delta\theta + W \times 1.5a \cos\theta\, \delta\theta = 3W \times a \cos\theta\, \delta\theta$$

(+ve sign due to $W$, $\delta y_M$ and $\delta y_L$ are in the same direction)

Virtual work done by the load $W$ at point $H$

$$= +W \times \delta y_H = +W \times \delta y_E \qquad ...(\delta y_H = \delta y_E)$$

$$= +W \times 2a \cos\theta\, \delta\theta = +2W \times a \cos\theta\, \delta\theta$$

...(+ve sign due to $W$, $\delta y_H$ and $\delta y_E$ are in the same direction)

Virtual work done by the tension $T$ at point $H$

$$= -T \times \delta y_H = -T \times \delta y_E$$

$$= -T \times 2a \cos\theta\, \delta\theta = -2T \times a \cos\theta\, \delta\theta$$

(–ve sign due to $T$ and $\delta y_H$ are in the opposite direction)

Since the virtual displacement of point $G$ is zero, therefore virtual works done by the load $W$ and tension $T$ at $G$ is zero.

According to the method of virtual work, the algebraic sum of the virtual works done is zero.

$$\therefore \quad W \times a \cos\theta\, \delta\theta + 3W \times a \cos\theta\, \delta\theta + 2W \times a \cos\theta\, \delta\theta - 2T \times a \cos\theta\, \delta\theta = 0$$

$$6W - 2T = 0$$

or $$2T = 6\,W$$

$\therefore$ $$T = 3\,W \textbf{ Ans.}$$

**Example 11.11** *A structure with pin connected members is shown in Fig. 11.17(a). Using the principle of virtual work, determine the value of angle θ for equilibrium. Neglect friction and the weight of the members.*

***Solution*** Given : $BD = BC + CD = 1200$ mm ; $AE = 600$ mm

Draw perpendiculars $DD'$ and $EE'$ on the horizontal line $AB$. From the geometry of the Fig. 11.17 (*b*), we find that

$$x_D = BD' = BD \cos\theta = 1200 \cos\theta$$

and $$y_E = EE' = AE \sin\theta = 600 \sin\theta$$

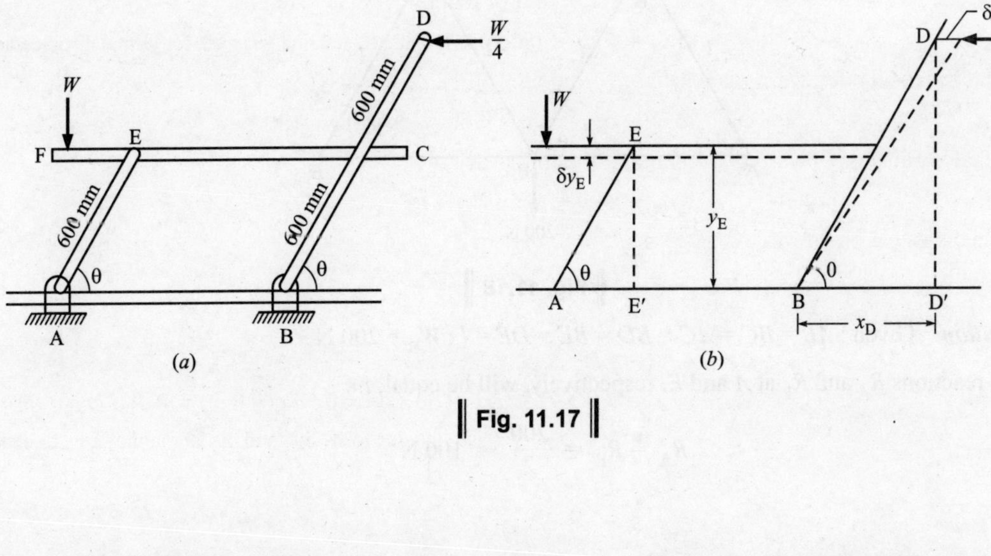


**Fig. 11.17**

The differentials of $x_D$ and $y_E$ are obtained as follows:

$$dx_D = d(1200 \cos\theta)\, d\theta = -1200 \sin\theta\, d\theta$$

and

$$dy_E = d(600 \sin\theta) d\theta = 600 \cos\theta\, d\theta$$

Let a small horizontal displacement $dx$ is given to $D$ towards right at the point of application of the load $\frac{W}{4}$. The point of application of load $W$ moves down by $dy$ in the direction of load, as shown in Fig. 11.17 (*b*).

∴ Virtual horizontal displacement of point $D$,

$$\delta x_D = -1200 \sin\theta\, \delta\theta \quad \text{...(Changing } d\theta \text{ to } \delta\theta)$$

and virtual vertical displacement of point $E$,

$$\delta y_E = 600 \cos\theta\, \delta\theta$$

We know that virtual work done by the force at $D$

$$= \frac{W}{4} \times \delta x_D = \frac{W}{4} \times -1200 \sin\delta\theta = -300W \sin\delta\theta$$

$$\text{...}\left(-\text{ve sign due to } \frac{W}{4} \text{ and } \delta x_D \text{ are in the opposite direction}\right)$$

and virtual work done by the load $W$ $= W \times \delta y_E = W \times 600 \cos\theta\, \delta\theta$

(+ve sign due to $W$ and $\delta y_E$ are in the same direction)

According to the method of virtual work, the algebraic sum of the virtual works done is equal to zero.

$$\therefore -300\, W \sin\theta\, \delta\theta + W \times 600 \cos\theta\, \delta\theta = 0$$

or

$$300\, W \sin\theta = 600\, W \cos\theta$$

$$\frac{\sin\theta}{\cos\theta} = \frac{600W}{300\,W} = 2$$

or

$$\tan\theta = 2$$

$$\therefore \quad \theta = \tan^{-1}(2) = 63.435° \ \textbf{Ans.}$$

**Example 11.12** *Determine the force developed in the member CD of the truss, as shown in Fig. 11.18, by using the method of virtual work.*

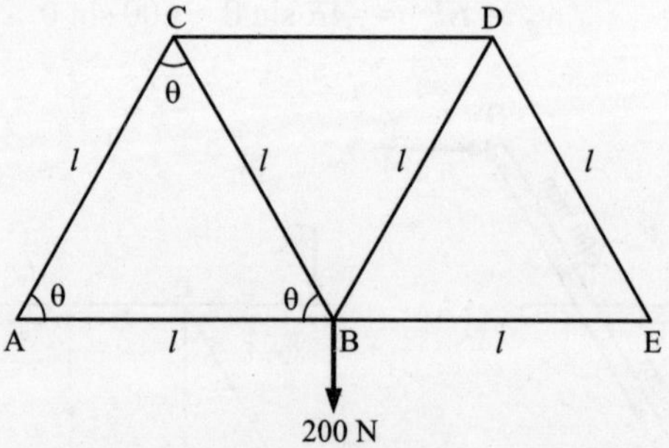


**Fig. 11.18**

***Solution*** Given : $AB = BC = AC = BD = BE = DE = l$ ; $W_B = 200$ N

The reactions $R_A$ and $R_E$ at $A$ and $E$, respectively, will be equal, *i.e.*

$$R_A = R_E = \frac{200}{2} = 100 \text{ N}$$

Let $P$ be the force developed in the member $CD$. Assume that the member $CD$ is removed and the two forces, each equal to $P$, are applied at $C$ and $D$, as shown in Fig. 11.19. Let the portion $BDE$ of the truss is given a small virtual displacement $\delta\theta$ at point $B$, such that the new position becomes $BD_1E_1$, as shown in Fig. 11.19.

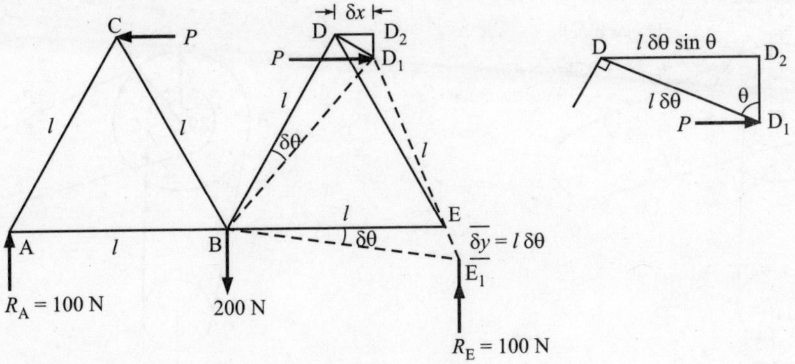


**Fig. 11.19**

Now, the virtual displacement of $P$ is given by

$$\delta x = DD_2 = DD_1 \sin\theta = l\,\delta\theta \sin\theta$$

and virtual displacement of reaction at $E$,

$$\delta y = EE_1 = l\,.\,\delta\theta$$

We know that virtual work done by the force $P$

$$= P \times \delta x = P \times DD_2 = P \times l\,\delta\theta \sin\theta$$

...(+ve sign due to $P$ and $\delta x$ are in the same direction)

Virtual work done by the reaction at $E = -R_E \times \delta y = -R_E \times EE_1 = -100 \times l\delta\theta$

...(–ve sign due to $R_E$ and $\delta y$ are in the opposite direction)

and virtual work done by the load at $B$ $= 0$ ...($\because$ Virtual displacement at $B$ is zero)

According to the method of virtual work, the algebraic sum of the virtual works done is zero.

$$\therefore \quad P \times l\delta\theta \sin\theta - 100 \times l\delta\theta = 0$$

or

$$P \sin\theta = 100$$

$$\therefore \quad P = \frac{100}{\sin\theta} = \frac{100}{\sin 60°} = \frac{100 \times 2}{\sqrt{3}} = 115.47 \text{ N} \quad ...(\because \theta = 60°)$$

## 11.10 APPLICATION OF THE PRINCIPLE OF VIRTUAL WORK ON LIFTING MACHINES

We know that in case of lifting machines, load moves upwards while the effort moves downward. In such cases, virtual work done by the effort and the load is determined. Now by applying the principle of virtual work, the unknown value of effort or load may be found out.

**Example 11.13** *The diameters of the two steps of the pulley of Weston's differential pulley block are 600 mm and 500 mm, respectively. Using the principle of virtual work and neglecting friction, determine the value of the effort (P) required to lift a load of 4 kN.*

***Solution*** Given : $D = 600$ mm or $R = 300$ mm ; $d = 500$ mm or $r = 250$ mm ; $W = 4$ kN

The differential pulley block is shown in Fig. 11.20 (*a*). Let the pulley block undergo a small virtual displacement $\delta\theta$ in the clockwise direction, as shown in Fig. 11.20 (*b*).

Virtual displacement of the effort $P = R\,\delta\theta = 300\,\delta\theta$

A little consideration will show that when effort $P$ is applied, the chain $AB$ will move upwards while the chain $CA$ will move downwards.

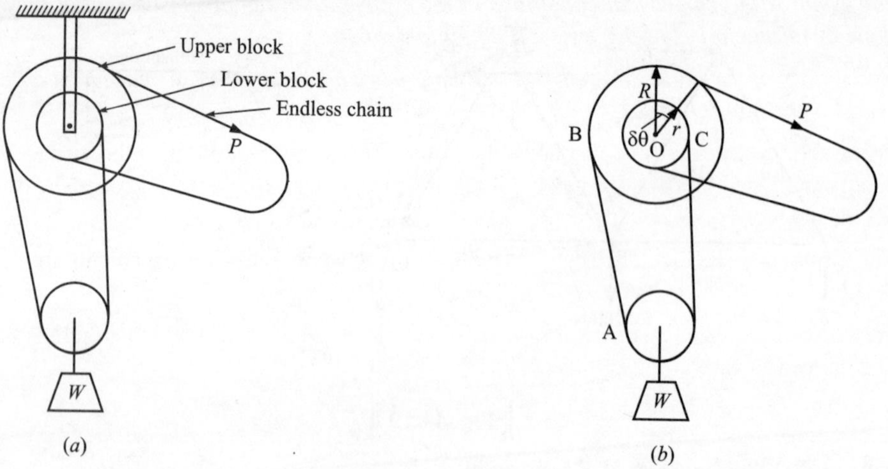


**Fig. 11.20**

Now distance moved in the upward direction by $AB$

$$= R\,\delta\theta$$

and distance moved in the downward direction by $CA$

$$= r\,\delta\theta$$

∴ Virtual displacement moved by the load ($W$) in the upward direction

$$= \frac{1}{2}[R\,\delta\theta - r\,\delta\theta] \quad \text{...}[\because R\,\delta\theta > r\,\delta\theta]$$

$$= \frac{1}{2}[300\,\delta\theta - 250\,\delta\theta] = 25\,\delta\theta$$

We know that virtual work done by the effort $P$

$$= P \times \text{Virtual displacement of } P$$

$$= P \times 300\,\delta\theta$$

...(+ve sign due to effort $P$ and virtual displacement are in the same direction)

and virtual work done by the load $W$ $= -W \times$ Virtual displacement of $W$

$$= -W \times 25\,\delta\theta$$

...(∵ –ve sign due to load $W$ and virtual displacement are in the opposite direction)

According to the method of virtual work, the algebraic sum of the virtual works done is zero.

$$\therefore \quad P \times 300\,\delta\theta - W \times 25\,\delta\theta = 0$$

or

$$P \times 300 = W \times 25 = 4 \times 25 = 100$$

$$\therefore \quad P = \frac{100}{300} = \frac{1}{3} \text{ kN} \quad \textbf{Ans.}$$

## HIGHLIGHTS

1. The virtual displacement is an imaginary displacement.
2. The virtual work is defined as the work done by a force on a body due to a small imaginary or virtual displacement of the body.

3. The principle of virtual work states that '*when a system of forces acting on a body or system of bodies is in equilibrium and the system is imagined to undergo a small virtual displacement consistent with geometrical conditions of the system, then the algebraic sum of the virtual work done by all the forces of the system is equal to zero*'.

   Conversely, *if the algebraic sum of the virtual work done for every displacement is zero, then the system is in equilibrium.*

4. When the force and virtual displacement are in the same direction, then the virtual work done is *positive.* On the other hand, when the force and virtual displacement are in opposite direction, then the virtual work done is *negative*.

5. The virtual displacements are denoted by $\delta x$ or $\delta y$, whereas actual displacements are denoted by $dx$ or $dy$.

6. For most of the problems solved by virtual work method, it is assumed that there is no friction in hinged, bearings or along sliding surfaces.

## NUMERICAL PROBLEMS

1. A simply supported beam *AB* of span 6 m carries a point load of 12 kN at 2 m from the end *A*. Determine the beam reactions by the principle of virtual work. [**Ans.** 8 kN; 4 kN]

2. A beam *AB* of span 12 m caries two point loads of 10 kN and 12 kN at 4 m and 8 m from the end *A*, respectively. Determine the beam reactions by the principle of virtual work. [**Ans.** 10.67 kN; 11.33 kN]

3. Two beams *AC* and *CD* of lengths 4.5 m and 5 m, respectively, are hinged at *C* and are supported as shown in Fig. 11.21. Using the principle of virtual work, determine the reactions at the hinge *C* and the support *B*.

   [**Ans.** 810 N; 630 N]

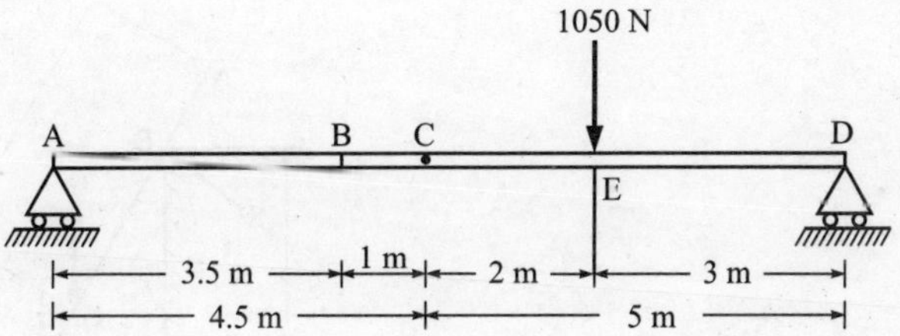


**Fig. 11.21**

4. Two beams *AB* and *CD* are supported as shown in Fig. 11.22. Find the reaction at the support *E* due to the vertical load *P*. $\left[\textbf{Ans. } P \times \frac{a}{b}\right]$

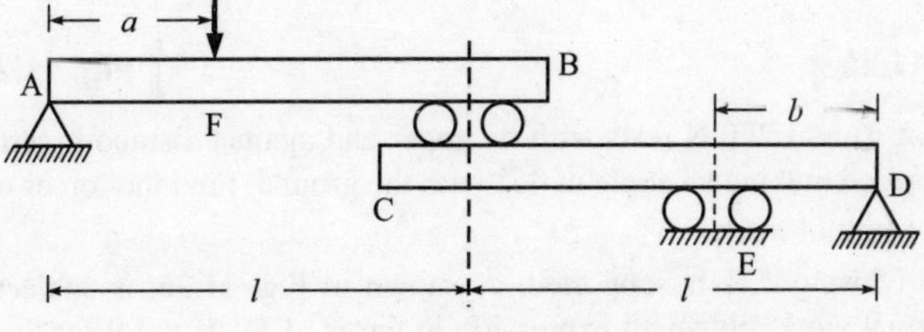


**Fig. 11.22**

5. Two beams *AB* and *DE* are supported on rollers at *C* and *D*, as shown in Fig. 11.23. Using the principle of virtual work, determine the reaction at the rollers *C* and *D*. **[Ans.** 600 N; 750 N]

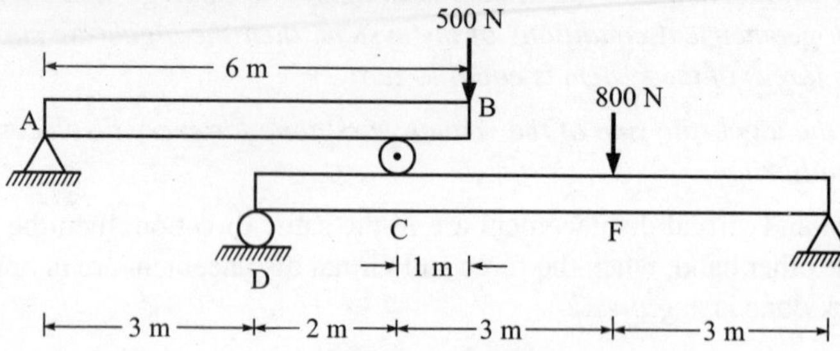


**Fig. 11.23**

6. A simply supported beam *AB* of span 5 m is loaded, as shown in Fig. 11.24. Determine the reactions at *A* and *B*, using the principle of virtual work. **[Ans.** 520 N; 580 N]

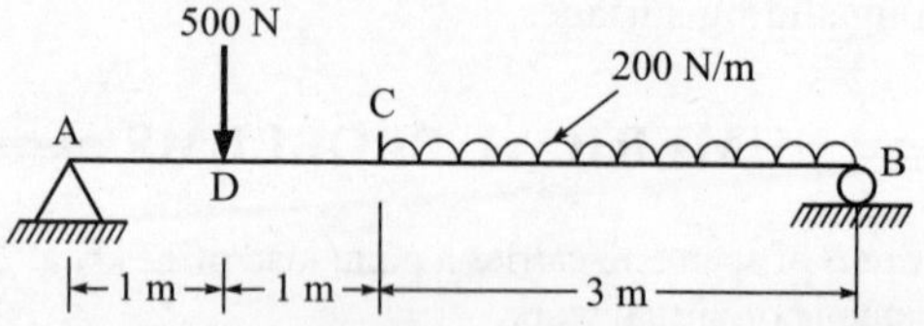


**Fig. 11.24**

7. A slender prismatic bar *AB* of length *l* and weight *W* stands in a vertical plane and is supported by smooth surfaces *A* and *B*, as shown in Fig. 11.25. Using the principle of virtual work, find the magnitude of the horizontal force applied at *A* to keep the bar in equilibrium. $\left[\textbf{Ans. } \dfrac{W}{2}\cot\theta\right]$

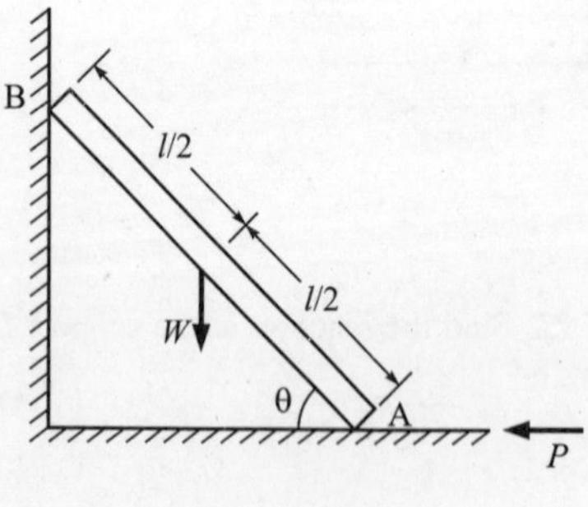


**Fig. 11.25**

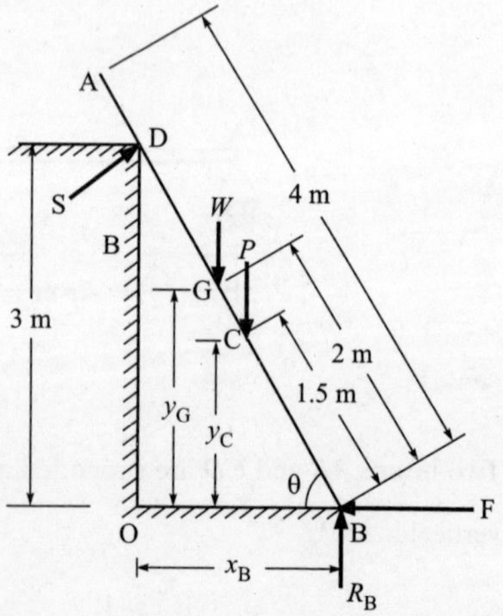


**Fig. 11.26**

8. A uniform ladder of weight 250 N rests with its upper end against a smooth vertical wall and its foot on a rough horizontal ground making an angle of 45° with the ground. Find the forces of friction of the ground by using the method of virtual work. **[Ans.** 125 N]

9. A uniform bar *AB* of weight *W* is supported, as shown in Fig. 11.26, is subjected to a load *P*. Applying the principle of virtual work, obtain an expression in terms of *P*, *W* and θ for the effort *F* required at *B* for equilibrium. Hence, calculate the value of *F* if *P* = 800 N, *W* = 600 N and θ = 60°. Neglect friction.

$$\left[\textbf{Ans. } \frac{(1.5P + 2W)\cos\theta\,.\,\sin^2\theta}{3};\ 300\text{ N}\right]$$

**10.** Four bars *AB*, *BC*, *CD* and *DA*, each of length *l*, are hinged together at their ends so as to form a rhombus, as shown in Fig. 11.27. Using the method of virtual work, find the relation between the forces *P* and *Q* for equilibrium in any configuration defined by the angle θ. Neglect the weight of the bars and friction.

$$\left[\textbf{Ans. } \frac{P}{Q} = \tan\theta\right]$$

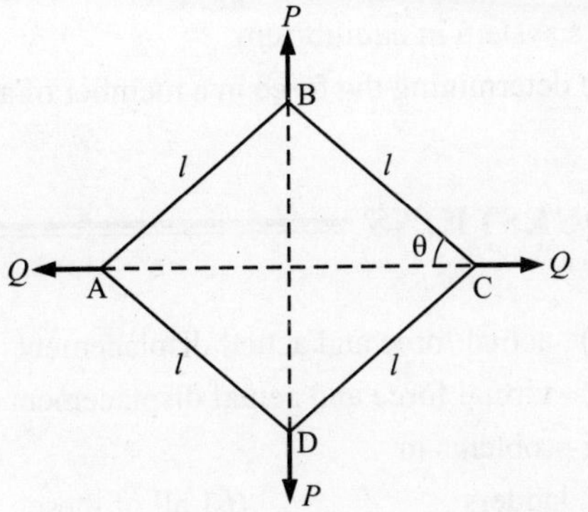


**Fig. 11.27**

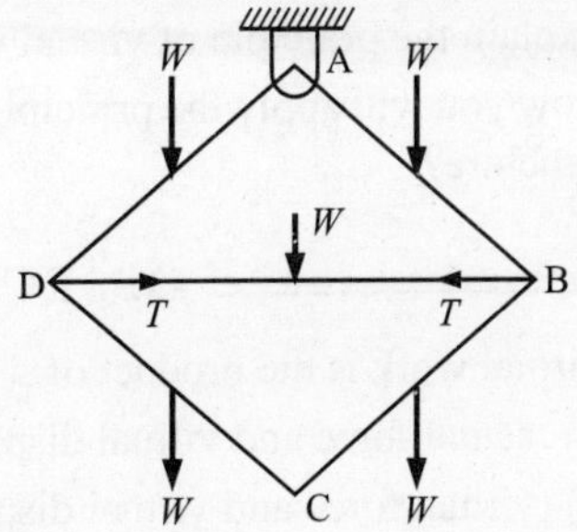


**Fig. 11.28**

**11.** Four identical uniform rods, each of weight *W*, are joined together to form a square *ABCD* and stiffened by a fifth rod *BD* also of weight *W*, as shown in Fig. 11.28. Using the method of virtual work, find the thrust in the rod *BD* when the frame is suspended from *A*. [**Ans.** $T = 2.5\ W$]

**12.** Using the principle of virtual work, determine the magnitude of forces $P_1$ to hold the frame in equilibrium under the action of a force *P* applied on the frame, as shown in Fig. 11.29. The length of each link is 2*a*. What is $P_1$ for $\theta = 45°$? [**Ans.** $P_1 = 1.5\ P \tan\theta$; $1.5\ P$]

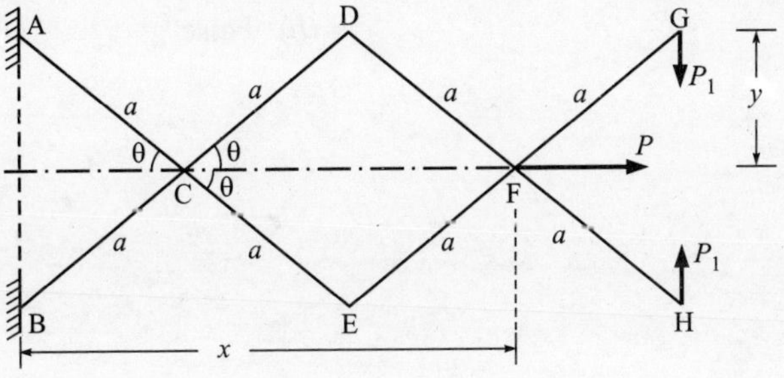


**Fig. 11.29**

**13.** A weight of 10 kN is raised by a system of two frictionless pulleys of the same diameter, as shown in Fig. 11.30. Using the principle of virtual work, find the effort *P* required to hold the weight in equilibrium. [**Ans.** 5 kN]

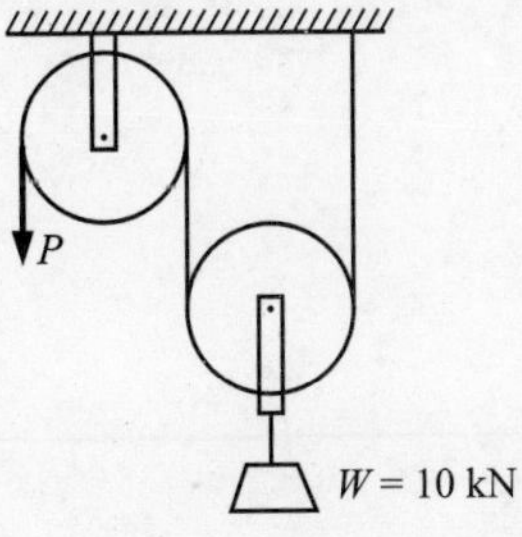


**Fig. 11.30**

## REVIEW QUESTIONS

1. Define virtual displacement and virtual work.
2. State and explain clearly the principle of virtual work. Also explain how it can be used in solving problems in statics.
3. What are the applications of the principle of virtual work?
4. Explain the principle of virtual work as applied to a system in equilibrium.
5. How you will apply the principle of virtual work in determining the force in a member of a framed structure?

## OBJECTIVE TYPE QUESTIONS

1. Virtual work is the product of
   (*a*) actual force and virtual displacement (*b*) actual force and actual displacement
   (*c*) virtual force and virtual displacement (*d*) virtual force and actual displacement
2. The principle of virtual work is applied for solving problems in
   (*a*) beams (*b*) framed structures (*c*) ladders (*b*) all of these
3. When the force and virtual displacement are in the same direction, then the virtual work done is
   (*a*) positive (*b*) negative (*c*) zero
4. When the force and virtual displacement are in the opposite direction, then the virtual work done is
   (*a*) positive (*b*) negative (*c*) zero
5. The principle of virtual work can be applied for finding out the reactions of a beam carrying simultaneously point loads and uniformly distributed load.
   (*a*) True (*b*) False

***Answers to Objective Type Questions***

**1.** (*a*) **2.** (*d*) **3.** (*a*) **4.** (*b*) **5.** (*a*)

# 12

# KINEMATICS OF LINEAR MOTION



## 12.1 INTRODUCTION

We have already discussed that a rigid body consists of a system of innumerable particles. It does not change its shape and size under the effect of forces acting on it. *Kinematics* deals with the bodies in motion without any reference to the forces responsible for the motion. In other words, kinematics of rigid bodies deals with the study of displacement, velocity and acceleration of rigid bodies without taking into consideration the forces causing the motion. The motion of a rigid body, when confined to one plane, is said to be *plane motion* which may be either rectilinear or curvilinear motion. When the motion of a body is along a straight line (such as a car moving on a road), it is said to be a *rectilinear motion* (also called *translatory* or *straight line motion*). On the other hand, when the motion takes place along a curved path, it is then known as *curvilinear motion.*

When all the particles of a body travel in concentric circular path of constant radii (about the axis of rotation perpendicular to the plane of motion) such as a pulley rotating about a fixed shaft, the motion is now said to be *plane rotational* or *rotary* motion.

The motion of a body, confined to one plane, may not be either completely rectilinear nor completely rotational. Such a type of motion is called *combined rectilinear and rotational motion.*

## 12.2 IMPORTANT TERMS USED IN KINEMATICS

The following terms are widely used while dealing with the kinematics of rigid body:

1. ***Displacement*** The displacement is the change of position of a particle in a definite direction. It may be along a straight path or curved path. It is measured by a straight distance between the initial and final positions of the particle. Since the displacement of a body has both magnitude and direction, therefore, it is a vector quantity. The displacement may be represented graphically by a straight line.

**Note:** The distance travelled should not be confused with the displacement of a body, because the distance travelled is a scalar quantity and represents the total distance moved by the body while the displacement is a vector quantity.

2. ***Speed*** The speed is the rate of change of displacement of a body with respect to the time. Since it does not take into account the direction of motion, therefore it is a scalar quantity.
3. ***Average speed*** The average speed is the ratio of the total distance to the total time taken by the body in going from the initial position to the final position.
4. ***Instantaneous speed*** The instantaneous speed is the speed at any instant of time.
5. ***Velocity*** The velocity is the rate of change of displacement of a body with respect to time in a particular direction. It is measured in metre per second (briefly written as m/s). Since it has both magnitude and direction, therefore it is a vector quantity.
6. ***Uniform velocity*** When the body covers equal distances in equal interval of time, then the velocity is called uniform velocity.
7. ***Non-uniform or variable velocity*** When the body covers unequal distances in equal intervals of time, then it is called non-uniform or variable velocity.
8. ***Acceleration*** The acceleration is the rate of change of velocity of a body with respect to time. It is measured in metre per second square (briefly written as m/s²). It is also a vector quantity.

   The acceleration may be positive or negative. When the velocity of the body increases with time, then acceleration is said to be *positive* and when the velocity decreases with time, then acceleration is said to be *negative.* The negative acceleration is known as *retardation* or *deceleration.*

**Notes:** 1. Consider that a body moves at a uniform rate in a definite direction and traversed a distance $s$ in time $t$. The velocity is given by

$$v = \frac{s}{t}$$

In differential form,

$$v = \frac{ds}{dt}$$

and acceleration is given by

$$a = \frac{dv}{dt} = \frac{d}{dt}\left(\frac{ds}{dt}\right) = \frac{d^2s}{dt^2} \qquad ...\left(\because v = \frac{ds}{dt}\right)$$

2. The acceleration may also be expressed as follows:

$$a = \frac{dv}{dt} = \frac{ds}{dt} \times \frac{dv}{ds} = v \times \frac{dv}{ds}$$

## 12.3 EQUATIONS OF LINEAR MOTION

Consider a body moving in a straight line from $A$ to $B$ with uniform acceleration, as shown in Fig. 12.1.

Let
$u$ = Initial velocity of the body in m/s,
$v$ = Final velocity of the body in m/s,
$t$ = Time (in seconds) during which the velocity changes from $u$ to $v$,
$s$ = Distance travelled by the body in $t$ seconds, and
$a$ = Uniform acceleration of the body in m/s².

**1. *Equation for final velocity* ($v$)**

We know that change of velocity = Final velocity – Initial velocity $= v - u$

and rate of change of velocity $= \dfrac{\text{Change of velocity}}{\text{Time taken}} = \dfrac{v-u}{t}$

Since the rate of change of velocity is the acceleration, therefore

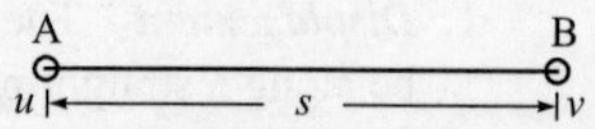


**Fig. 12.1**

$$\text{Acceleration, } a = \frac{v-u}{t}$$

or

$$at = v - u$$

$\therefore$

$$v = u + at \qquad ...(i)$$

**2. *Equation for distance travelled (s) in t seconds***

We know that average velocity $= \dfrac{\text{Initial velocity} + \text{Final velocity}}{2} = \dfrac{u+v}{2}$

and distance travelled, $s = \text{Average velocity} \times \text{Time} = \left(\dfrac{u+v}{2}\right) \times t$

Substituting the value of $v$ from equation ($i$), we have

$$s = \left(\frac{u+u+at}{2}\right) \times t = \frac{2ut + at^2}{2} = ut + \frac{1}{2}at^2 \quad ...(ii)$$

**3. *Derivation of $v^2 - u^2 = 2as$***

From equation ($i$), the value of $t$ is given by

$$t = \frac{v-u}{a}$$

Substituting this value of $t$ in equation ($ii$), we have

$$s = u\left(\frac{v-u}{a}\right) + \frac{1}{2}a\left(\frac{v-u}{a}\right)^2 = \frac{v-u}{a}\left[u + \frac{1}{2}a\left(\frac{v-u}{a}\right)\right]$$

$$= \frac{v-u}{a}\left[\frac{2u+v-u}{2}\right] = \frac{v-u}{a}\left(\frac{v+u}{2}\right) = \frac{v^2-u^2}{2a}$$

or $$v^2 - u^2 = 2as \quad ...(iii)$$

The above equations hold good for uniform linear acceleration. In case the negative acceleration (retardation or deceleration) is given, then the value of $a$ in the above equations will be *negative.*

It may be noted that the equations of linear motion [equations ($i$), ($ii$) and ($iii$)] may also be derived as discussed below:

(***a***) We have already discussed that acceleration

$$a = \frac{d^2s}{dt^2} = \frac{d}{dt}\left(\frac{ds}{dt}\right)$$

or $$d\left(\frac{ds}{dt}\right) = a.dt$$

Integrating this expression, we have

$$\int d\left(\frac{ds}{dt}\right) = \int a\, dt$$

or $$\frac{ds}{dt} = at + C_1 \quad ...(i)$$

where $C_1$ is the constant of integration.

Since $\dfrac{ds}{dt}$ is the velocity at any instant when $t = 0$, therefore this velocity is known as initial velocity ($u$).

Now, the equation ($i$) becomes

$$\frac{ds}{dt} = a \times 0 + C_1 = C_1$$

or $$C_1 = \frac{ds}{dt} = u$$

The equation (*i*) may now be written as

$$\frac{ds}{dt} = at + u$$

or $$ds = (at + u)\, dt$$

Integrating again this expression, $s = \frac{at^2}{2} + ut + C_2$ ...(*ii*)

where $C_2$ is another constant of integration.

For $t = 0$, $s = 0$. Therefore, $0 = 0 + 0 + C_2$ or $C_2 = 0$

Now the equation (*ii*) may be written as

$$s = \frac{at^2}{2} + ut = ut + \frac{1}{2}at^2$$ ...(Same as before) ...(*iii*)

**(*b*)** We have already discussed that

$$a = v \times \frac{dv}{ds}$$

or $$v\, dv = a\, ds$$

Integrating this expression, $\int v\, dv = \int a\, ds$

$$\frac{v^2}{2} = as + C_3$$ ...(*iv*)

where $C_3$ is the constant of integration.

When $s = 0$, the velocity is known as initial velocity ($u$). Therefore for $s = 0$, $v = u$. The equation (*iv*) is now written as

$$\frac{u^2}{2} = a \times 0 + C_3$$

or $$C_3 = \frac{u^2}{2}$$

Substituting the value of $C_3$ in equation (*iv*), we have

$$\frac{v^2}{2} = as + \frac{u^2}{2}$$

or $$\frac{v^2 - u^2}{2} = as$$

$\therefore$ $$v^2 - u^2 = 2as$$ ...(Same as before)

**(*c*)** From equation (*i*), we have $\frac{ds}{dt} = at + C_1$, and $C_1 = u$ when $t = 0$

$\therefore$ $$\frac{ds}{dt} = at + u$$

After time $t$ seconds, the velocity is called final velocity ($v$). Therefore, the above expression becomes

$$v = at + u$$

or $$v = u + at$$

**Example 12.1** *A particle starts from rest and moves along a straight line with constant acceleration. If it acquires a velocity of 3 m/s after having travelled a distance of 7.5 m, find the magnitude of the acceleration.*

***Solution*** Given : $u = 0$ ; $v = 3$ m/s ; $s = 7.5$ m

Let $a$ = Acceleration in m/s$^2$.

We know that $v^2 = u^2 + 2as$

$$3^2 = 0 + 2a \times 7.5 = 15a$$

$$\therefore \quad a = \frac{3^2}{15} = 0.6 \text{ m/s}^2 \quad \textbf{Ans.}$$

**Example 12.2** *A burglar's car had a start with an acceleration of 2 m/s². A police vigilant party came after 5 seconds and continued to chase the burglar's car with uniform velocity of 20 m/s. Find the time taken in which the police will overtake the car.*

***Solution*** Given : $a = 2$ m/s$^2$

Let $s_1$ be the distance covered by the burglar's car in $(t + 5)$ seconds. The initial velocity of the burglar's car, $u_1 = 0$ and acceleration, $a = 2$ m/s$^2$.

$$\therefore \quad s_1 = u_1 t_1 + \frac{1}{2} a t_1^2 = 0 + \frac{1}{2} \times 2(t+5)^2 \qquad ...(\because\ t_1 = t+5)$$

$$= (t + 5)^2 = t^2 + 25 + 10t \qquad ...(i)$$

Let $s_2$ be the distance covered by the police party and $t$ be the time taken by the police party to overtake burglar's car. The initial velocity for the police party is $u_2 = 20$ m/s and acceleration, $a = 0$ (because the velocity is uniform).

$$\therefore \quad s_2 = u_2 \cdot t_2 + \frac{1}{2} a t_2^2 = 20 \times t + 0 = 20t \qquad ...(\because\ t_2 = t) \ ...(ii)$$

From equations (*i*) and (*ii*),

$$t^2 + 25 + 10t = 20t$$

or

$$t^2 + 25 + 10t - 20t = 0$$

$$t^2 + 25 - 10t = 0$$

$$(t - 5)^2 = 0$$

$$t - 5 = 0$$

$$\therefore \quad t = 5 \text{ s} \quad \textbf{Ans.}$$

**Example 12.3** *An automobile is accelerated at the rate of 0.8 m/s² as it travels from station A to station B. If the speed of the automobile is 36 km/h as it passes station A, determine the time required for the automobile to reach station B and its speed as it passes station B. The distance between A and B is 250 m.*

***Solution*** Given : $a = 0.8$ m/s$^2$ ; $u = 36$ km/h $= \dfrac{36 \times 1000}{3600} = 10$ m/s ; $s = 250$ m

***Time required for the automobile to reach station B***

Let $t$ = Required time in seconds.

We know that $s = ut + \frac{1}{2} a t^2$

$$250 = 10t + \frac{1}{2} \times 0.8t^2 = 10t + 0.4t^2$$

or

$$0.4t^2 + 10t - 250 = 0$$

$$\therefore \qquad t = \frac{-10 \pm \sqrt{(10)^2 + 4 \times 0.4 \times 250}}{2 \times 0.4} = \frac{-10 \pm \sqrt{100 + 400}}{0.8}$$

$$= \frac{-10 \pm 22.36}{0.8} = \frac{12.36}{0.8} = 15.45 \textbf{ Ans.} \quad \text{...(Taking +ve sign only)}$$

***Speed as it passes station B***

Let $\qquad v$ = Speed as it passes station $B$.

We know that $\qquad v^2 - u^2 = 2as$

$$v^2 - (10)^2 = 2 \times 0.8 \times 250 = 400$$

or $\qquad v^2 = (10)^2 + 400 = 500$

$\therefore \qquad v = \sqrt{500} = 22.36$ m/s **Ans.**

**Example 12.4** *Two cars are travelling towards each other on a single lane road at velocities 12 m/s and 9 m/s, respectively. When 100 m apart, both drivers realise the situation and apply their brakes. They succeed in stopping simultaneously and just short of colliding. Assume constant retardation for each car, determine :*

1. *Time required for cars to stop ;*
2. *Retardation of each car ; and*
3. *Distance travelled by each car while slowing down.*

***Solution*** Given : $u_1 = 12$ m/s ; $u_2 = 9$ m/s ; $s_1 + s_2 = 100$ m

**1. *Time required for cars to stop***

Since both the cars stop simultaneously, therefore time taken by both the cars will be same. Let it be $t$ seconds. Also, the final velocity of both the cars will be zero, *i.e.* $v_1 = v_2 = 0$. Let $a_1$ and $a_2$ be the retardation of the first and second car, respectively. Using the relation for the first car,

$$v_1 = u_1 + a_1 t$$

$$0 = 12 - a_1 t \qquad \text{...(–ve sign due to retardation)}$$

or $\qquad a_1 t = 12 \qquad$ ...(*i*)

Similarly, using the relation for the second car,

$$v_2 = u_2 + a_2 t$$

$$0 = 9 - a_2 t$$

or $\qquad a_2 t = 9 \qquad$ ...(*ii*)

The distance travelled by the first car is given by the relation,

$$s_1 = u_1 t + \frac{1}{2} a_1 t_1^2 = 12t - \frac{1}{2}(a_1 t)t$$

$$= 12t - \frac{1}{2} \times 12t = 6t \qquad \text{...[From equation (i)] ...(iii)}$$

Similarly, the distance travelled by the second car is given by

$$s_2 = a_2 t + \frac{1}{2} a_2 t^2$$

$$= 9t - \frac{1}{2}(a_2 t)t$$

$$= 9t - \frac{1}{2} \times 9t = 4.5t \qquad \text{...[From equation (ii)] ...(iv)}$$

Adding equations (*iii*) and (*iv*)

$$s_1 + s_2 = 6t + 4.5t$$

$$100 = 10.5t$$

$$\therefore \quad t = \frac{100}{10.5} = 9.524 \text{ s} \quad \textbf{Ans.}$$

**2. *Retardation of each car***

From equation (*i*), retardation of the first car,

$$a_1 = \frac{12}{t} = \frac{12}{9.524} = 1.26 \text{ m/s}^2 \quad \textbf{Ans.}$$

From equation (*ii*), retardation of the second car,

$$a_2 = \frac{9}{t} = \frac{9}{9.524} = 0.945 \text{ m/s}^2 \quad \textbf{Ans.}$$

**3. *Distance travelled by each car***

From equation (*iii*), distance travelled by first car,

$$s_1 = 6t = 6 \times 9.524 = 57.144 \text{ m} \quad \textbf{Ans.}$$

From equation (*iv*), distance travelled by second car,

$$s_2 = 4.5t = 4.5 \times 9.524 = 42.858 \text{ m} \quad \textbf{Ans.}$$

**Example 12.5** *A car starts from rest and accelerates uniformly to a speed of 72 km/h over a distance of 500 m. Calculate the acceleration and the time taken to attain the speed.*

*If a further acceleration raises the speed to 90 km/h in 10 seconds, find this acceleration and the further distance moved. The brakes are now applied to bring the car to rest under uniform retardation in 5 seconds. Find the distance travelled during braking.*

***Solution*** Given : $u = 0$ ; $v = 72 \text{ km/h} = \dfrac{72 \times 1000}{3600} = 20 \text{ m/s}$ ; $s = 500$ m

**1.** First of all, let us consider the motion of the car from rest.

***Acceleration of the car***

Let $a$ = Acceleration of the car in m/s$^2$.

We know that

$$v^2 = u^2 + 2as$$

$$(20)^2 = 0 + 2a \times 500 = 1000a$$

$$\therefore \quad a = \frac{(20)^2}{1000} = 0.4 \text{ m/s}^2 \quad \textbf{Ans.}$$

***Time taken by the car to attain the speed***

Let $t$ = Time taken by the car to attain the speed.

We know that

$$v = u + at$$

$$20 = 0 + 0.4t = 0.4t$$

$$\therefore \quad t = \frac{20}{0.4} = 50\text{s} \quad \textbf{Ans.}$$

**2.** Now consider the motion of the car from 72 km/h to 90 km/h in 10 seconds. Here it is given that

$$^*u = 72 \text{ km/h} = 20 \text{ m/s}; \; v = 90 \text{ km/h} = \frac{90 \times 1000}{3600} = 25 \text{ m/s}; \; t = 10 \text{ s}$$

* It is the final velocity in the first case.

***Acceleration of the car***

Let $a$ = Acceleration of the car in m/s².

We know that

$$v = u + at$$

$$25 = 20 + a \times 10$$

or

$$a = \frac{25 - 20}{10} = 0.5 \text{ m/s}^2 \quad \textbf{Ans.}$$

***Distance moved by the car***

We know that distance moved by the car,

$$s = ut + \frac{1}{2}at^2$$

$$= 20 \times 10 + \frac{1}{2} \times 0.5(10)^2 = 200 + 25 = 225 \text{ m} \quad \textbf{Ans.}$$

**3.** Finally, consider the motion of the car during the application of brakes for bringing the car to rest. In this case, the final velocity in the second case (*i.e.* 25 m/s) will be the initial velocity (*i.e.* $u$ = 25 m/s) and the final velocity ($v$) is zero. The time taken ($t$) is 5 seconds.

We know that the distance travelled by the car during braking,

$$s = \text{Average velocity} \times \text{Time}$$

$$= \frac{u + v}{2} \times t = \frac{25 + 0}{2} \times 5 = 62.5 \text{ m} \quad \textbf{Ans.}$$

**Example 12.6** *Three marks A, B and C, spaced at a distance of 100 m are made along a straight road. A car starting from rest and accelerates uniformly passes the mark A and takes 10 seconds to reach the mark B and further 8 seconds to reach the mark C. Calculate :*

1. *The magnitude of the acceleration of the car ;*
2. *The velocity of the car at A ;*
3. *The velocity of the car at B ; and*
4. *The distance of the mark A from the starting point.*

***Solution*** Given : $s_2$ = 100 m ; $s_3$ = 100 m ; $t_2$ = 10 s ; $t_3$ = 8 s

**1. *Magnitude of the acceleration of the car***

Let the car start from point $O$ with initial velocity ($u_1$) zero, as shown in Fig. 12.2. After covering a distance $s_1$ metres, it reaches at point $A$ in time $t_1$ seconds. The three marks $A$, $B$ and $C$ spaced at a distance of 100 m are shown in Fig. 12.2. Considering the motion between $O$ and $A$. Let $v_1$ be the velocity of the car and the acceleration ($a$) is uniform.

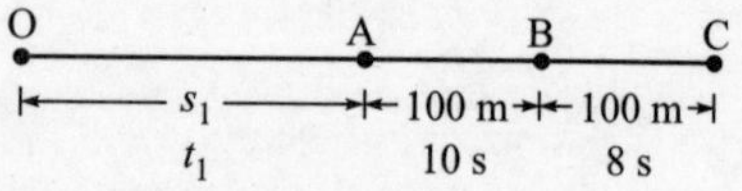


**Fig. 12.2**

We know that

$$s_1 = u_1 t_1 + \frac{1}{2}at_1^2$$

$$= 0 + \frac{1}{2}at_1^2 = \frac{1}{2}at_1^2 \qquad ...(\because u = 0) \quad ...(i)$$

Similarly for the motion between $O$ and $B$,

$$s_1 + s_2 = \frac{1}{2}a(t_1 + 10)^2$$

$$s_1 + 100 = \frac{1}{2}a(t_1^2 + 20t_1 + 100) \qquad ...(ii)$$

Substituting the value of $s_1$ from equation (*i*), we have

$$\frac{1}{2}at_1^2 + 100 = \frac{1}{2}at_1^2 + 10at_1 + 50a$$

or $$100 = 10at_1 + 50a$$

or $$10 = at_1 + 5a = a(t_1 + 5) \qquad \text{...(Dividing by 10 throughout) ...}(iii)$$

Now for the motion between *O* and *C*,

$$s_1 + s_2 + s_3 = \frac{1}{2}a(t_1 + 10 + 8)^2 = \frac{1}{2}a(t_1 + 18)^2$$

$$s_1 + 100 + 100 = \frac{1}{2}a(t_1^2 + 36t_1 + 324)$$

or $$s_1 + 200 = \frac{1}{2}a(t_1^2 + 36t_1 + 324)$$

Substituting the value of $s_1$ from equation (*i*), we have

$$\frac{1}{2}at_1^2 + 200 = \frac{1}{2}at_1^2 + 18at_1 + 162a$$

or $$200 = 18at_1 + 162a$$

$$11.11 = at_1 + 9a = a(t_1 + 9) \qquad \text{...(Dividing throughout by 18) ...}(iv)$$

Now dividing equation (*iv*) by equation (*iii*),

$$\frac{11.11}{10} = \frac{a(t_1 + 9)}{a(t_1 + 5)}$$

$$11.11(t_1 + 5) = 10(t_1 + 9)$$

$$11.11t_1 + 55.55 = 10t_1 + 90$$

$$11.11t_1 - 10t_1 = 90 - 55.55 = 34.45$$

$$\therefore \quad t_1 = \frac{34.45}{1.11} = 31 \text{ s}$$

Substituting the value of $t_1$ in equation (*iii*)

$$10 = a(t_1 + 5) = a(31 + 5) = 36a$$

$$\therefore \quad a = \frac{10}{36} = 0.278 \text{ m/s}^2 \quad \textbf{Ans.}$$

**2. *Velocity of the car at A***

We know that velocity of the car at *A*,

$$v_A = u_1 + at_1 = 0 + 0.278 \times 31 = 8.618 \text{ m/s} \quad \textbf{Ans.}$$

**3. *Velocity of the car at B***

We know that velocity of the car at *B*,

$$v_B = u_1 + a(t_1 + t_2)$$
$$= 0 + 0.278\,(31 + 10) = 11.4 \text{ m/s} \quad \textbf{Ans.}$$

**4. *Distance of mark A from the starting point***

The distance of mark *A* from the starting point *O* is given by equation (*i*), *i.e.*

$$s_1 = \frac{1}{2}at_1^2 = \frac{1}{2} \times 0.278(31)^2 = 133.58 \text{ m} \quad \textbf{Ans.}$$

**Example 12.7** *The acceleration of a body starting from rest moving along a straight line follows:*

$$a = \frac{t}{30} + \frac{2}{3}$$

*where a is in m/s² and t in seconds. Obtain the velocity and displacement at t = 10 s.*

***Solution*** Given : $a = \frac{t}{30} + \frac{2}{3}$; $t = 10$ s

We know that acceleration, $$a = \frac{dv}{dt} = \frac{t}{30} + \frac{2}{3}$$

or $$dv = \left(\frac{t}{30} + \frac{2}{3}\right) dt$$

Integrating this expression, we have

$$\int dv = \int \left(\frac{t}{30} + \frac{2}{3}\right) dt$$

or $$v = \frac{t^2}{2 \times 30} + \frac{2}{3}t + C_1 = \frac{t^2}{60} + \frac{2}{3}t + C_1 \quad \text{...}(i)$$

where $C_1$ is the constant of integration.

Since the body starts from rest, therefore $v = 0$, and $t = 0$.

Substituting these values in equation (*i*), we have

$$C_1 = 0$$

The equation (*i*) now becomes $$v = \frac{t^2}{60} + \frac{2}{3}t \quad \text{...}(ii)$$

***Velocity at t = 10 s***

Substituting $t = 10$ s in equation (*ii*), we have

$$\text{Velocity, } v = \frac{(10)^2}{60} + \frac{2}{3} \times 10 = \frac{100}{60} + \frac{20}{3} = 1.67 + 6.67 = 8.34 \text{ m/s } \textbf{Ans.}$$

***Displacement at t = 10 s***

From equation (*ii*), we have $$v = \frac{ds}{dt} = \frac{t^2}{60} + \frac{2}{3}t \quad \text{...}\left(\because v = \frac{ds}{dt}\right)$$

or $$ds = \left(\frac{t^2}{60} + \frac{2}{3}t\right) dt$$

Integrating this expression, we have

$$s = \frac{t^3}{3 \times 60} + \frac{2}{3} \times \frac{t^2}{2} + C_2 = \frac{t^3}{180} + \frac{t^2}{3} + C_2 \quad \text{...}(iii)$$

Since the body starts from rest, therefore $s = 0$, $t = 0$. Substituting these values in equation (*iii*), we have

$$C_2 = 0$$

Now the equation (*iii*) becomes, $$s = \frac{t^3}{180} + \frac{t^2}{3}$$

When $t = 10$ s, then $$s = \frac{(10)^3}{180} + \frac{(10)^2}{3} = \frac{1000}{180} + \frac{100}{3} = 5.56 + 33.33 = 38.89 \text{ m } \textbf{Ans.}$$

**Example 12.8** *A particle initially at rest is submitted to the action of a force $F = k \cdot t$, where t is the time and k is a constant. Prove that the ratio of displacement x at any time to the velocity of the particle at that time is a linear function of time.*

***Solution*** Given : $F = k \cdot t$

Since the particle is initially at rest, therefore initial velocity, $u = 0$. Let $a$ is the uniform acceleration. The displacement ($s = x$) is given by

$$s = ut + \frac{1}{2}at^2$$

or $$x = 0 + \frac{1}{2}at^2$$

∴ $$a = \frac{2x}{t^2} \qquad ...(i)$$

We also know that $$v^2 = u^2 + 2as = 0 + 2ax \qquad ...(\because s = x)$$

or $$a = \frac{v^2}{2x}$$

or $$v^2 = a \times 2x = \frac{2x}{t^2} \times 2x = \frac{4x^2}{t^2} \qquad ...\text{[From equation (i)]}$$

∴ $$v = \frac{2x}{t}$$

Now the ratio of displacement ($x$) at any time to the velocity of the particle ($v$) at that time is given by

$$\frac{x}{v} = \frac{x}{\frac{2x}{t}} = \frac{x.t}{2x} = \frac{t}{2}$$

which is a linear function of time ($t$). **Ans.**

**Example 12.9** *The velocity of a train travelling at 100 km/h decreases by 10% in the first 40 seconds after application of the brakes. Calculate the velocity at the end of a further 80 seconds, assuming that, during the whole period of 120 seconds, the retardation is proportional to the velocity.*

***Solution*** Given : Velocity (when $t = 0$), $v = 100$ km/h

Since the velocity decreases by 10% in the first 40 seconds (*i.e.* $t_1 = 40$ s) after the application of brakes, therefore velocity at the end of 40 seconds,

$$v_1 = 100 \times \frac{90}{100} = 90 \text{ km/h}$$

Let $v_2$ = Velocity at the end of 120 seconds (or further 80 seconds).

Since the retardation ($a$) is proportional to the velocity ($v$), therefore

$$a = -\frac{dv}{dt} = k.v$$

or $$\frac{dv}{v} = -k.dt \qquad ...(i)$$

where $k$ is a constant of proportionality whose value may be determined from the given conditions.

Integrating equation ($i$), $$\int \frac{dv}{v} = \int -k.dt$$

$$\log_e v = -k.t + C \qquad ...(ii)$$

where $C$ is a constant of integration. We know that when $t = 0$, $v = 100$ km/h. Substituting these values in equation (*ii*), we have

$$\log_e (100) = -k \times 0 + C$$

or
$$C = \log_e 100 = 2.3 \log 100 = 2.3 \times 2 = 4.6$$

We also know that when $t = t_1 = 40$ s, $v = v_1 = 90$ km/h. Substituting these values in equation (*ii*),

$$\log_e 90 = -k \times 40 + 4.6$$

or
$$k \times 40 = 4.6 - \log_e 90 = 4.6 - 2.3 \log 90 = 4.6 - 2.3 \times 1.9542$$
$$= 4.6 - 4.495 = 0.105$$

$\therefore$
$$k = \frac{0.105}{40} = 0.0026$$

Substituting the value of $k$ and $C$ in equation (*ii*),

$$\log_e v = -0.0026 \times t + 4.6$$

or
$$2.3 \log v = -0.0026t + 4.6 \quad ...(iii)$$

Now substituting the value of $t$ equal to 120 seconds in equation (*iii*), we have

$$2.3 \log v = -0.0026 \times 120 + 4.6 = -0.312 + 4.6 = 4.288$$

or
$$\log v = \frac{4.288}{2.3} = 1.8643$$

Taking antilog of 1.8643, we have $v = 73.17$ m/s **Ans.**

**Example 12.10** *The equation of motion of a particle moving in a straight line is given as $s = 15t + 3t^2 - t^3$, where s is the distance moved and t is the time in seconds. Find :*

1. *Velocity and acceleration at the start ;*
2. *Time at which the particle reaches maximum velocity ; and*
3. *Maximum velocity of the particle.*

***Solution*** Given : $s = 15t + 3t^2 - t^3$

**1. *Velocity and acceleration at the start***

Let $v$ and $a$ be the velocity and acceleration, respectively, at the start [*i.e.* when time ($t$)] is zero. Differentiating the given equation of motion with respect to time $t$, we have

$$\frac{ds}{dt} = 15 + 3 \times 2t - 3t^2$$

or
$$v = 15 + 6t - 3t^2 \quad ...(i)$$

Again differentiating this equation with respect to time $t$, we have

$$\frac{dv}{dt} = 6 - 3 \times 2t$$

or
$$a = 6 - 6t \quad ...(ii)$$

Now substituting $t = 0$ in equation (*i*) and (*ii*), we have

$$v = 15 + 6 \times 0 - 3 \times 0 = 15 \text{ m/s} \quad \textbf{Ans.}$$

and
$$a = 6 - 6 \times 0 = 6 \text{ m/s}^2 \quad \textbf{Ans.}$$

**2. *Time at which the particle reaches maximum velocity***

For velocity to be maximum, differentiate equation (*i*) and equate to zero, *i.e.*

$$\frac{dv}{dt} = 6 - 3 \times 2t = 0$$

or
$$6 - 6t = 0$$

$\therefore$
$$t = 1 \text{ s} \quad \textbf{Ans.}$$

**3. *Maximum velocity of the particle***

Substituting the value of $t = 1$ s in equation (*i*), we have maximum velocity of the particle,

$$v_{max} = 15 + 6 \times 1 - 3 \times 1^2 = 18 \text{ m/s} \quad \textbf{Ans.}$$

**Example 12.11** *The motion of a particle is given by $a = t^3 - 3t^2 + 5$, where a is the acceleration in m/s² and t is the time in seconds. The velocity of the particle at t = 1 s is 6.25 m/s, and the displacement is 8.3 m. Calculate the displacement and the velocity at t = 2 s.*

***Solution*** Given : $a = t^3 - 3t^2 + 5$

Since the acceleration, $a = \frac{dv}{dt}$, therefore the above equation may be written as

$$\frac{dv}{dt} = t^3 - 3t^2 + 5$$

or $$dv = (t^3 - 3t^2 + 5)\,dt$$

Integrating both sides, $$\int dv = \int (t^3 - 3t^2 + 5)\,dt$$

or $$v = \frac{t^4}{4} - \frac{3t^3}{3} + 5t + C_1 = \frac{t^4}{4} - t^3 + 5t + C_1 \quad ...(i)$$

where $C_1$ is the constant of integration. It is given that when $t = 1$ s, $v = 6.25$ m/s. Substituting these values in equation (*i*), we have

$$6.25 = \frac{(1)^4}{4} - (1)^3 + 5 \times 1 + C_1 = 0.25 - 1 + 5 + C_1 = 4.25 + C_1$$

$$\therefore \quad C_1 = 6.25 - 4.25 = 2$$

Now the equation (*i*) becomes $$v = \frac{t^4}{4} - t^3 + 5t + 2 \quad ...(ii)$$

***Velocity at t = 2 s***

Substituting the value of $t = 2$ s in equation (*ii*), we have

$$v = \frac{(2)^4}{4} - 2^3 + 5 \times 2 + 2 = 4 - 8 + 10 + 2 = 8 \text{ m/s} \quad \textbf{Ans.}$$

***Displacement at t = 2 s***

Since the velocity, $v = \frac{ds}{dt}$, therefore equation (*ii*) may be written as

$$\frac{ds}{dt} = \frac{t^4}{4} - t^3 + 5t + 2$$

or $$ds = \left(\frac{t^4}{4} - t^3 + 5t + 2\right) dt$$

Integrating both sides, $$\int ds = \int \left(\frac{t^4}{4} - t^3 + 5t + 2\right) dt$$

or $$s = \frac{t^5}{5 \times 4} - \frac{t^4}{4} + \frac{5t^2}{2} + 2t + C_2 \quad ...(iii)$$

where $C_2$ is another constant of integration. It is given that when $t = 1$ s, $s = 8.3$ m. Substituting these values in equation (*iii*), we have

$$8.3 = \frac{(1)^5}{20} - \frac{(1)^4}{4} + \frac{5(1)^2}{2} + 2 \times 1 + C_2$$

$$= \frac{1}{20} - \frac{1}{4} + \frac{5}{2} + 2 + C_2 = 0.05 - 0.25 + 2.5 + 2 + C_2 = 4.3 + C_2$$

or $$C_2 = 8.3 - 4.3 = 4$$

Now the equation (*iii*) may be written as

$$s = \frac{t^5}{20} - \frac{t^4}{4} + \frac{5t^2}{2} + 2t + 4 \qquad ...(iv)$$

Substituting the value of $t = 2$ s in equation (*iv*), we have

$$s = \frac{(2)^5}{20} - \frac{(2)^4}{4} + \frac{5(2)^2}{2} + 2 \times 2 + 4$$

$$= \frac{32}{20} - \frac{16}{4} + \frac{20}{2} + 4 + 4 = 1.6 - 4 + 10 + 8 = 15.6 \text{ m} \quad \textbf{Ans.}$$

**Example 12.12** *A body moves along a straight line and its acceleration (a) which varies with time (t) is given by a = (2 – 3t). After 5 seconds from start of observation, its velocity is observed to be 20 m/s. After 10 seconds from start of observation, the body was 85 m from the origin.*

*(a) Determine its acceleration, velocity and distance from the origin at the start of observation ;*

*(b) Determine the time after start of observation in which the velocity becomes zero and its distance from the origin ; and*

*(c) Describe the motion diagrammatically.*

***Solution*** Given : $a = 2 - 3t$

***(a) Acceleration, velocity and distance from the origin at the start of observation***

At the start of observation, time $(t) = 0$. Therefore, acceleration,

$$a = 2 - 3t = 2 - 3 \times 0 = 2 \text{ m/s}^2 \quad \textbf{Ans.}$$

We know that acceleration, $$a = \frac{dv}{dt}$$

Thus, the above equation may be written as

$$\frac{dv}{dt} = 2 - 3t$$

or $$dv = (2 - 3t)\, dt \qquad ...(i)$$

Integrating this equation, $$\int dv = \int (2 - 3t)\, dt$$

or $$v = 2t - \frac{3t^2}{2} + C_1 \qquad ...(ii)$$

where $C_1$ is the constant of integration. Since the initial conditions are given as $v = 20$ m/s when $t = 5$ seconds, therefore substituting these values in equation (*ii*), we have

$$\therefore \quad 20 = 2 \times 5 - \frac{3 \times 5^2}{2} + C_1 = -27.5 + C_1$$

$$\therefore \quad C_1 = 20 + 27.5 = 47.5$$

Substituting this value of $C_1$ in equation (*ii*), we have

$$v = 2t - \frac{3t^2}{2} + 47.5 \qquad ...(iii)$$

Now velocity at the start of observation (*i.e.* when $t = 0$),

$$v = 47.5 \text{ m/s} \quad \textbf{Ans.} \qquad [\text{Substituting } t = 0 \text{ in equation } (iii)]$$

The equation (*iii*) may now be written as,

$$\frac{ds}{dt} = 2t - \frac{3t^2}{2} + 47.5 \qquad \ldots\left(\because v = \frac{ds}{dt}\right)$$

or

$$ds = \left(2t - \frac{3t^2}{2} + 47.5\right) dt \qquad \ldots(iv)$$

Integrating this expression, we have

$$s = \frac{2t^2}{2} - \frac{3t^3}{2 \times 3} + 47.5t + C_2 = t^2 - \frac{t^3}{2} + 47.5t + C_2 \qquad \ldots(v)$$

where $C_2$ is another constant of integration. Since the body is 85 metres from the origin (*i.e.* $s = 85$ m), after 10 seconds from the start of observation (*i.e.* $t = 10$ s), therefore, substituting these values in equation ($v$), we have

$$85 = (10)^2 - \frac{(10)^3}{2} + 47.5 \times 10 + C_2$$

$$= 100 - 500 + 475 + C_2 = 75 + C_2$$

$$\therefore \quad C_2 = 85 - 75 = 10$$

Now equation ($v$) becomes,

$$s = t^2 - \frac{t^3}{2} + 47.5t + 10 \qquad \ldots(vi)$$

At the start of the observation, (*i.e.* when $t = 0$),

$$s = 10 \text{ m} \quad \textbf{Ans.} \qquad \ldots[\text{Substituting } t = 0 \text{ in equation } (vi)]$$

**(b) *Time after the start of observation in which the velocity become zero and its distance from the origin***

Substituting $v = 0$ in equation (*iii*), we have

$$0 = 2t - \frac{3t^2}{2} + 47.5$$

Multiplying both sides by 2 and rearranging the above equation,

$$3t^2 - 4t - 95 = 0$$

$$\therefore \quad t = \frac{4 \pm \sqrt{4^2 + 4 \times 3 \times 95}}{2 \times 3} = \frac{4 \pm 34}{6} = 6.33 \text{ s} \quad \textbf{Ans.}$$

...(Taking +ve sign only)

Substituting this value of $t$ in equation (*vi*),

$$s = t^2 - \frac{t^3}{2} + 47.5t + 10$$

$$= (6.33)^2 - \frac{(6.33)^3}{2} + 47.5 \times 6.33 + 10$$

$$= 40 - 126.8 + 300.7 + 10 = 223.9 \text{ m} \quad \textbf{Ans.}$$

**(c)** The motion is represented diagrammatically, as shown in Fig. 12.3. We know that $a = 2$ m/s² when $t = 0$ and when $a = 0$, $t = \frac{2}{3}$ s. The acceleration verses time graph is shown in Fig. 12.3 (*a*). We know that the maximum velocity occurs at zero acceleration. From the given equation of motion,

$$a = 2 - 3t \quad \text{or} \quad 0 = 2 - 3t$$

$$\therefore \qquad t = \frac{2}{3}\ \text{s}$$

Substituting this value in equation (*iii*),

$$v_{max} = 2\times\frac{2}{3} - \frac{3}{2}\left(\frac{2}{3}\right)^2 + 47.5 = \frac{4}{3} - \frac{2}{3} + 47.5 = 48.16\ \text{m/s}$$

The velocity verses time graph is shown in Fig. 12.3 (*b*).

We also know that the maximum displacement occurs at zero velocity (*i.e.* when $t = 6.33$ s). Substituting this value of $t$ in equation (*vi*),

$$s_{max} = (6.33)^2 - \frac{(6.33)^3}{2} + 47.5\times 6.33 + 10$$

$$= 40.07 - 126.81 + 300.67 + 10 = 223.93\ \text{m}$$

The displacement versus time graph is shown in Fig. 12.3 (*c*).

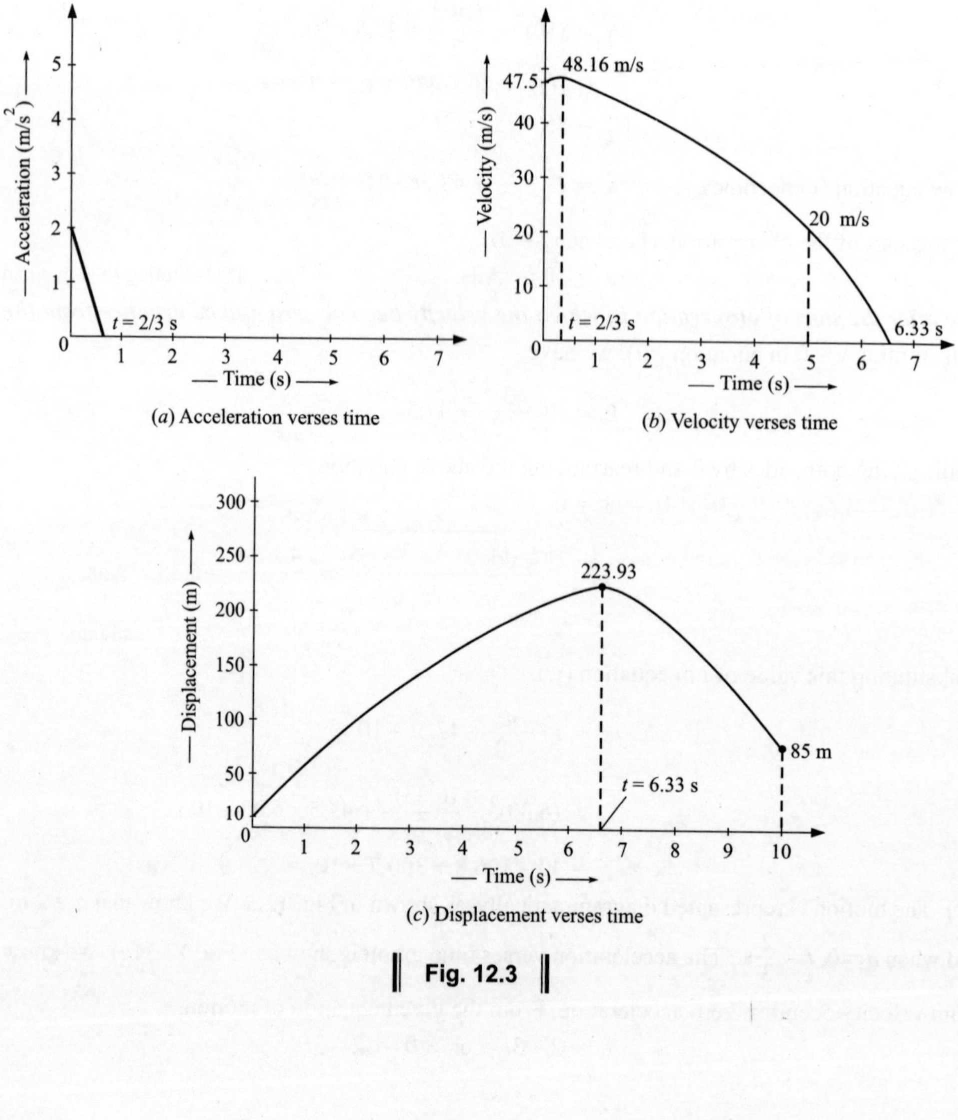


(*a*) Acceleration verses time

(*b*) Velocity verses time

(*c*) Displacement verses time

**Fig. 12.3**

## 12.4 DISTANCE COVERED BY A BODY IN *n*th SECOND

Let

$$S_n = \text{Distance covered by the body in } n \text{ seconds,}$$
$$S_{n-1} = \text{Distance covered by the body in } (n-1) \text{ seconds,}$$
$$u = \text{Initial velocity of the body, and}$$
$$a = \text{Acceleration of the body.}$$

We know that

$$s = ut + \frac{1}{2}at^2$$

∴ Distance covered by the body in *n* seconds,

$$S_n = un + \frac{1}{2}an^2 \qquad ...(\because t = n)$$

and distance covered by the body in (*n* – 1) seconds,

$$S_{n-1} = u(n-1) + \frac{1}{2}a(n-1)^2$$

∴ Distance covered by the body in the *n*th second

$$= S_n - S_{n-1}$$
$$= un + \frac{1}{2}an^2 - \left[u(n-1) + \frac{1}{2}a(n-1)^2\right]$$
$$= un + \frac{1}{2}an^2 - \left[un - u + \frac{1}{2}a(n^2 + 1 - 2n)\right]$$
$$= un + \frac{1}{2}an^2 - \left[un - u + \frac{1}{2}an^2 + \frac{a}{2} - an\right]$$
$$= un + \frac{1}{2}an^2 - un + u - \frac{1}{2}an^2 - \frac{a}{2} + an$$
$$= u - \frac{a}{2} + an = u + a\left(n - \frac{1}{2}\right) = u + \frac{a}{2}(2n-1)$$

**Example 12.13** *A body is moving from rest with a constant acceleration of 4 m/s². Find the distance travelled in the 6th second.*

***Solution*** Given : $u = 0$ ; $a = 4$ m/s²

We know that distance travelled in 6 seconds,

$$s_6 = un + \frac{1}{2}an^2 = 0 + \frac{1}{2} \times 4 \times 6^2 = 72 \text{ m}$$

and distance travelled in 5 seconds, $s_5 = 0 + \frac{1}{2} \times 4 \times 5^2 = 50$ m

∴ Distance travelled in 6th second $= s_6 - s_5 = 72 - 50 = 22$ m **Ans.**

**Example 12.14** *A particle travels 40 m during its 4 seconds and 42 m during its 6 seconds. Find its initial velocity and uniform acceleration. Also find the total distance travelled by the particle in 6 seconds.*

***Solution*** Given : $s_1 = 40$ m ; $t_1 = 4$ s ; $s_2 = 42$ m ; $t_2 = 6$ s

***Initial velocity and acceleration***

Let

$$u = \text{Initial velocity, and}$$
$$a = \text{Uniform acceleration.}$$

We know that distance travelled in $n$ second,

$$s_n = u + \frac{a}{2}(2n - 1)$$

$$\therefore \quad 40 = u + \frac{a}{2}(2 \times 4 - 1) = u + 3.5a \quad \text{...(Here } n = t_1\text{) ...(i)}$$

Similarly,

$$42 = u + \frac{a}{2}(2 \times 6 - 1) = u + 5.5a \quad \text{...(Here } n = t_2\text{) ...(ii)}$$

Subtracting equation (*i*) from equation (*ii*), we have

$$42 - 40 = u + 5.5a - (u + 3.5a)$$

or

$$2 = 2a$$

$$\therefore \quad a = \frac{2}{2} = 1 \text{ m/s}^2 \quad \textbf{Ans.}$$

Substituting this value of $a$ in equation (*i*),

$$40 = u + 3.5a = u + 3.5 \times 1$$

$$\therefore \quad u = 40 - 3.5 = 36.5 \text{ m/s} \quad \textbf{Ans.}$$

***Total distance travelled by the particle in 6 seconds***

Let $s$ = Total distance travelled by the particle.

Using the relation

$$s = ut + \frac{1}{2}at^2 = 36.5 \times 6 + \frac{1}{2} \times 1 \times 6^2 = 219 + 18 = 237 \text{ m} \quad \textbf{Ans.}$$

## 12.5 MOTION OF A BODY UNDER FORCE OF GRAVITY (ACCELERATION DUE TO GRAVITY)

It is a special case of motion under uniform acceleration. This acceleration is known as acceleration due to gravity. It is usually denoted by '$g$' whose value is generally taken as 9.81 m/s². It may be noted that when the body falls down, it has an acceleration of $g$. On the other hand, when the body is projected upwards, it is subjected to deceleration or retardation of $g$. Thus the equations of motion for acceleration due to gravity are written as follows:

**1. *When the body falls vertically down, i.e. when body moves vertically downwards***

$$v = u + gt$$

$$s = ut + \frac{1}{2}gt^2$$

and

$$v^2 - u^2 = 2gs$$

**2. *When the body is projected vertically upwards***

$$v = u - gt$$

$$s = ut - \frac{1}{2}gt^2$$

and

$$v^2 - u^2 = -2gs$$

**Notes: 1.** When a body falls freely from a height $h$, then its velocity $v$, with which it will hit the ground, is determined as follows:

We know that $v^2 - u^2 = 2gs$

Since the initial velocity $u = 0$, therefore

$$v^2 = 2gs \text{ or } 2gh \quad \text{...}(\because s = h)$$

$$\therefore \quad v = \sqrt{2gh}$$

**2.** The value of $g$ is taken as +9.81 m/s² for downward motion and –9.81 m/s² for upward motion of a body.

**Example 12.15** *A stone after falling from rest for 4 seconds breaks a glass pane and in breaking, it looses 25% of its velocity. How far will it fall in the next second?*

***Solution*** Given : $u = 0$ ; $t = 4$ s

Let $v$ = Velocity with which the stone strikes the pane.

We know that $v = u + gt = 0 + 9.81 \times 4 = 39.24$ m/s

Since the loss in velocity in breaking is 25% of the velocity with which the stone strikes, therefore

$$\text{Loss in velocity} = \frac{25}{100} \times 39.24 = 9.81 \text{ m/s}$$

$\therefore$ Initial velocity of the stone after breaking the glass pane (*i.e.* after 4 seconds),

$$u_1 = 39.24 - 9.81 = 29.43 \text{ m/s}$$

and the distance travelled by the stone or the height of fall in the next second *i.e.* in 1 second is given by

$$s = u_1 t + \frac{1}{2} g t^2 = 29.43 \times 1 + \frac{1}{2} \times 9.81 \times 1^2 \qquad \text{...(Here } t = 1 \text{ s)}$$

$$= 29.43 + 4.905 = 34.335 \text{ m} \quad \textbf{Ans.}$$

**Example 12.16** *A stone dropped into a well is heard to strike the water after 4 seconds. Find the depth of the well, if the velocity of sound is 350 m/s.*

***Solution*** Given : Velocity of sound = 350 m/s

Let $s$ = Depth of the well in metres,

$t$ = Time taken by the stone to reach the bottom of the well, in seconds,

$t_1$ = Time taken by the sound to reach the top of the well, in seconds.

Using the relation for downward motion,

$$s = ut + \frac{1}{2} g t^2 = 0 \times t + \frac{1}{2} \times 9.81 t^2 = 4.905 t^2$$

$$\text{...}(\because \text{ Initial velocity, } u = 0)$$

and

$$t_1 = \frac{\text{Depth of well}}{\text{Speed of sound}} = \frac{s}{350} = \frac{4.905 t^2}{350} = 0.014 t^2$$

$\therefore$ Total time $= t + t_1$

But this time is equal to 4 seconds, therefore

$$t + t_1 = 4$$

$$t + 0.014 t^2 = 4$$

or

$$0.014 t^2 + t - 4 = 0$$

Multiplying this equation by 100 throughout, we have

$$1.4 t^2 + 100\, t - 400 = 0$$

Solving this equation as a quadratic equation, we have

$$t = \frac{-100 \pm \sqrt{(100)^2 + 4 \times 1.4 \times 400}}{2 \times 1.4} = \frac{-100 \pm 110.6}{2.8} = 3.786 \text{ s}$$

$$\text{...(Taking +sign only)}$$

$\therefore$ Depth of the well, $s = 4.905 t^2 = 4.905\,(3.786)^2 = 70.3$ m **Ans.**

**Example 12.17** *A stone is thrown vertically upwards with a velocity of 19.6 m/s, from the top of a tower 24.5 m high. Calculate the following :*

1. *The maximum height at which the stone will rise in its flight.*
2. *Velocity of the stone at its downward travel at the point in the same level as the point of projection.*
3. *Time required for the stone to reach the ground.*

*Take* $g = 9.8 \text{ m/s}^2$

***Solution*** Given : $u_1 = 19.6$ m/s ; $h = s = 24.5$ m

**1. *Maximum height of flight***

When the stone reaches at the maximum height ($s_1$), its velocity will be zero (*i.e.* $v_1 = 0$).

Using the relation,

$$v_1^2 - u_1^2 = -2gs_1 \quad \text{...(–ve sign due to upward motion)}$$

$$0 - (19.6)^2 = -2 \times 9.8 s_1$$

$$\therefore \quad s_1 = \frac{(19.6)^2}{2 \times 9.8} = 19.6 \text{ m} \quad \textbf{Ans.}$$

**2. *Velocity of the stone at its downward travel***

The velocity of the stone during its downward travel at a point in the same level as the point of projection will be equal to the upward velocity of the stone at the top of the tower, *i.e.* 19.6 m/s. **Ans.**

**3. *Time required for the stone to reach the ground***

Let $t$ = Time required for the stone to reach the ground in seconds.

The total time ($t$) required to reach the ground will be equal to the sum of the time taken by the stone during upward travel ($t_1$) and the time taken by the stone during downward travel ($t_2$).

Using the relation, for upward motion,

$$v_1 = u_1 - gt_1$$

$$\text{or} \quad t_1 = \frac{u_1 - v_1}{g} = \frac{19.6 - 0}{9.8} = 2 \text{ s}$$

Now for the downward motion of the stone,

$$s_2 = ut_2 + \frac{1}{2}gt_2^2$$

$$s + s_1 = 0 \times t_2 + \frac{1}{2} \times 9.8 t_2^2$$

$$24.5 + 19.6 = 4.9 t_2^2$$

$$t_2^2 = \frac{24.5 + 19.6}{4.9} = 9$$

$$\text{or} \quad t_2 = 3 \text{ s}$$

∴ Total time required for the stone to reach the ground

$$t = t_1 + t_2 = 2 + 3 = 5 \text{ s} \quad \textbf{Ans.}$$

**Example 12.18** *A motor-cyclist want to jump over a ditch, as shown in Fig. 12.4. Find the necessary minimum velocity u at A in km/h of the motor-cycle.*

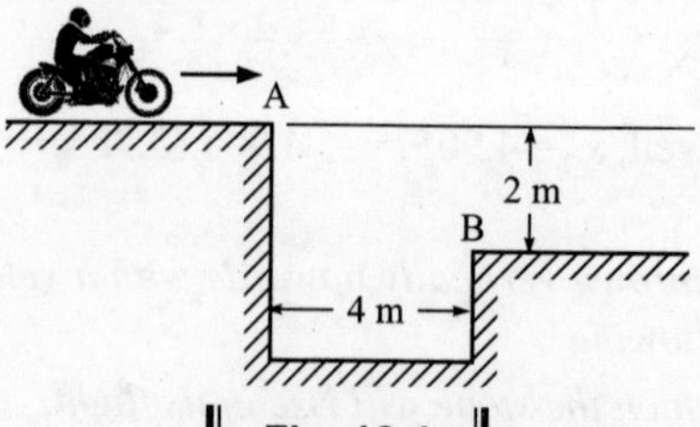


**Fig. 12.4**

*Also find the inclination and magnitude of the velocity of the motor-cycle just after clearing the ditch. Take g = 9.81 m/s². Assume any further data, if necessary.*

***Solution*** Given : $s = 2$ m. Width of ditch *i.e.* horizontal distance travelled = 4 m ; $a = g = 9.81$ m/s²

***Necessary minimum velocity (u) at A***

Let $t$ be the time taken to travel a vertical distance of 2 m. Since the vertical velocity (say $u$) at $A$ is zero, therefore using the relation

$$s = ut + \frac{1}{2}gt^2$$

$$2 = 0 + \frac{1}{2} \times 9.81t^2$$

$$\therefore \quad t = \sqrt{\frac{2 \times 2}{9.81}} = 0.6385 \text{ s}$$

We know that horizontal distance travelled

$$= \text{Width of ditch} = u \,.\, t$$

or

$$4 = u \times 0.6385$$

$$\therefore \quad u = \frac{4}{0.6385} = 6.26 \text{ m/s}$$

$$= \frac{6.26 \times 3600}{1000} = 22.54 \text{ km/h} \quad \textbf{Ans.}$$

This is the horizontal velocity of the motor-cycle at point $A$.

***Inclination and magnitude of the velocity of the motor-cycle just after clearing the ditch***

The final velocity of the motor-cycle in the vertical direction is given by

$$v = u + gt = 0 + 9.81 \times 0.6385 = 6.26 \text{ m/s}$$

∴ Magnitude of the velocity of the motor-cycle just after clearing the ditch

$$= \sqrt{u^2 + v^2} = \sqrt{(6.26)^2 + (6.26)^2} = 8.85 \text{ m/s}$$

$$= \frac{8.85 \times 3600}{1000} = 31.86 \text{ km/h} \quad \textbf{Ans.}$$

and its inclination with the horizontal is given by

$$\tan\theta = \frac{u}{v} = \frac{6.26}{6.26} = 1$$

$$\therefore \quad \theta = \tan^{-1}(1) = 45° \quad \textbf{Ans.}$$

**Example 12.19** *A large balloon is rising up with a velocity of 9.81 m/s at an altitude of 39.2 m from the ground. At that instant, a stone of mass 5 kg is dropped from it. After how many seconds will the stone reach the ground.*

***Solution*** Given : $u = 9.81$ m/s ; $AB = 39.2$ m ; $m = 5$ kg

(Refer Fig. 12.5)

First of all, let us consider the motion of stone from $B$ to $C$. Let $s_1$ is the distance covered by the stone from $B$ to $C$. Using the relation

$$v^2 = u^2 - 2gs_1$$

$$0 = (9.81)^2 - 2 \times 9.81 s_1 \qquad ...(\because v = 0)$$

$$\therefore \quad s_1 = \frac{(9.81)^2}{2 \times 9.81} = 4.905 \text{ m}$$

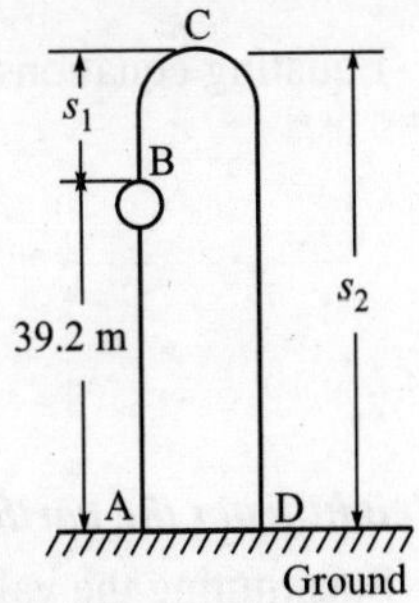


Fig. 12.5

Let $t_1$ is the time taken by the stone to reach the point *C* from point *B*. We know that

$$v = u - gt_1$$

$$0 = 9.81 - 9.81t_1$$

$$\therefore \quad t_1 = \frac{9.81}{9.81} = 1 \text{ s}$$

Let $t_2$ be the time taken by the stone to reach the ground at point *D* from the point *C*. Using the relation,

$$s_2 = u_2 t_2 + \frac{1}{2} g t_2^2$$

$$4.905 + 39.2 = 0 \times t_2 + \frac{1}{2} \times 9.81 t_2^2$$

or

$$44.1 = 4.905 t_2^2$$

or

$$t_2^2 = \frac{44.1}{4.905} = 8.99 \text{ say } 9$$

$$\therefore \quad t_2 = 3 \text{ s}$$

Thus, the total time taken by the stone to reach the ground from *B*

$$= t_1 + t_2 = 1 + 3 = 4 \text{ s} \quad \textbf{Ans.}$$

**Example 12.20** *An object is thrown vertically upwards with a velocity of 30 m/s. Four seconds later, a second object is thrown upwards with a velocity of 40 m/s. Determine :*

1. *The time (after the first object is thrown) when the two objects will meet each other in air.*
2. *The height from the earth at which the two objects will meet.*

***Solution*** Given : $u_1 = 30$ m/s ; $u_2$ after 4 seconds [*i.e.* after $(t + 4)$ seconds] = 40 m/s

**1. *Time when the two objects will meet each other in air***

Let the two objects meet after *t* seconds from their start at a height of *s* metres above the ground. Considering the upward motion of the first object,

$$s = u_1 t - \frac{1}{2} g t^2 = 30 \times t - \frac{1}{2} \times 9.81 t^2 = 30t - 4.905t^2 \quad \text{...(i)}$$

Now considering the upward motion of the second object after $(t + 4)$ seconds,

$$s = u_2 (t + 4) - \frac{1}{2} g (t + 4)^2$$

$$= 40(t + 4) - \frac{1}{2} \times 9.81 (t^2 + 16 + 8t)$$

$$= 40t + 160 - 4.905t^2 - 78.48 - 39.24t$$

$$= 0.76t - 4.905t^2 + 81.52 \quad \text{...(ii)}$$

Equating equations (*i*) and (*ii*), we have

$$30t - 4.905t^2 = 0.76t - 4.905t^2 + 81.52$$

or

$$30t - 0.76t = 81.52$$

$$29.24t = 81.52$$

$$\therefore \quad t = \frac{81.52}{29.24} = 2.788 \text{ s} \quad \textbf{Ans.}$$

**2. *Height from the earth at which the two objects will meet***

Substituting the value of *t* in equation (*i*), we have

$$s = 30t - 4.905t^2 = 30 \times 2.788 - 4.905 (2.788)^2$$

$$= 83.64 - 38.12 = 45.52 \text{ m} \quad \textbf{Ans.}$$

**Example 12.21** *A body falling freely under the action of gravity passes two points 9 m apart vertically in 0.2 second. Find what height, above the higher point, did it start to fall.*

***Solution*** Given : $u = 0$ ; $BC = 9$ m ; Time taken to pass distance $BC = 0.2$ s

Refer Fig. 12.6.

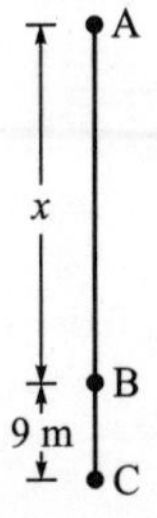

**Fig. 12.6**

Let the body start from $A$ and passes two points $B$ and $C$ in 0.2 second. Let

$$\text{Distance } AB = x$$

$$\therefore \quad \text{Distance } AC = (x + 9) \text{ m}$$

If the time taken by the body to travel from $A$ to $B$ is $t$ seconds, then the distance $x$ metres travelled by the body in time $t$ seconds is given by

$$x = ut + \frac{1}{2}gt^2 = 0 + \frac{1}{2} \times 9.81t^2$$

$$= 4.905t^2 \qquad ...(\because u = 0) \quad ...(i)$$

The distance $(x + 9)$ metres (*i.e.* from $A$ to $C$) travelled by the body in time $(t + 0.2)$ seconds is given by

$$(x + 9) = u(t + 0.2) + \frac{1}{2}g(t + 0.2)^2$$

$$= 0(t + 0.2) + \frac{1}{2} \times 9.81(t^2 + 0.04 + 0.4t)$$

$$= 4.905t^2 + 0.1962 + 1.962t \qquad ...(ii)$$

Subtracting equation (*i*) from equation (*ii*),

$$9 = 0.1962 + 1.962t$$

$$\therefore \quad t = \frac{9 - 0.1962}{1.962} = 4.487 \text{ s}$$

Substituting the value of $t$ in equation (*i*), we have

$$x = 4.905t^2 = 4.905\,(4.487)^2 = 98.76 \text{ m}$$

Thus, the body will start to fall when it is 98.76 m above the point $B$. **Ans.**

***Second method***

First of all, consider the motion of the body from $B$ to $C$. In this case, $s = 9$ m and $t = 0.2$ s. Let the initial velocity at $B = u$ m/s.

Using the relation,
$$s = ut + \frac{1}{2}gt^2$$

$$9 = u \times 0.2 + \frac{1}{2} \times 9.81(0.2)^2 = 0.2u + 0.1962$$

$$\therefore \quad u = \frac{9 - 0.1962}{0.2} = 44.02 \text{ m/s}$$

Now consider the motion from $A$ to $B$. In this case initial velocity at $A$, $u = 0$ and final velocity at $B$, $v = 44.02$ m/s and distance $AB = x$.

Using the relation,
$$v^2 = u^2 + 2gs$$

$$(44.02)^2 = 0 + 2 \times 9.81 \times x$$

$$\therefore \quad x = \frac{(44.02)^2}{2 \times 9.81} = 98.76 \text{ m} \quad \textbf{Ans.}$$

**Example 12.22** *The pilot of an aeroplane, flying horizontally with constant speed, v = 480 km/h at an elevation h = 610 m above a level plane, wishes to bomb a target B on the ground, as shown in Fig. 12.7. At what angle θ, below the horizontal should he see the target at the instant of releasing the bomb in order to score a hit? Neglect air resistance.*

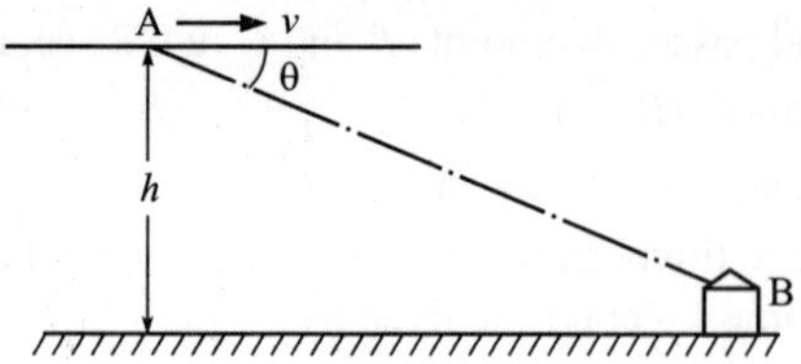


Fig. 12.7

***Solution*** Given : $v = 480 \text{ km/h} = \dfrac{480 \times 1000}{3600} = 133.33 \text{ m/s}$ ; $h = 610$ m

Let the time taken to hit the target $B$ on the ground be $t$ seconds. Since the aeroplane is moving horizontally with a constant speed $v$, therefore horizontal distance travelled,

$$s = \text{Horizontal velocity} \times \text{Time} = v.t \qquad ...(i)$$

The bomb is attracted downwards due to gravitational effect. The initial velocity of the bomb in the downward direction is zero.

Using the relation, $$h = ut + \frac{1}{2}gt^2 = 0 \times t + \frac{1}{2}gt^2$$

or $$610 = \frac{1}{2} \times 9.81t^2 = 4.905t^2$$

∴ $$t = \sqrt{\frac{610}{4.905}} = 11.15 \text{ s}$$

and $$s = v\,.\,t = 133.33 \times 11.15 = 1486.6 \text{ m}$$

Let θ be the required angle. We know that

$$\tan\theta = \frac{\text{Vertical distance}}{\text{Horizontal distance}} = \frac{610}{1486.6} = 0.41$$

∴ $$\theta = \tan^{-1}(0.41) = 22.3° \quad \textbf{Ans.}$$

## 12.6 GRAPHICAL REPRESENTATION OF DISPLACEMENT VERSUS TIME

The displacement verses time is represented graphically by taking displacement as ordinate and corresponding time as abscissa. We shall consider the following two cases:

**1. *When the body moves with uniform velocity***

When the body moves with uniform velocity, equal distances are covered in equal intervals of time. By plotting the distances ($s_1$, $s_2$, $s_3$, ..., etc.) on $Y$-axis and time ($t_1$, $t_2$, $t_3$, ..., etc.) on $X$-axis, a displacement-time ($s$-$t$) curve is drawn which is a straight line, as shown in Fig. 12.8 ($a$). The displacement of the body ($s$) is given by the equation

$$s = \text{Velocity} \times \text{Time}$$

∴ $$\text{Velocity at instant 1} = \frac{s_1}{t_1}$$

$$\text{Velocity at instant 2} = \frac{s_2}{t_2}$$

Since the velocity is uniform, therefore

$$\frac{s_1}{t_1} = \frac{s_2}{t_2} = \frac{s_3}{t_3} \dots = \tan\theta$$

where tan θ is the slope of the *s-t* curve. Thus, it is concluded that slope of the *s-t* curve at any instant gives the velocity.

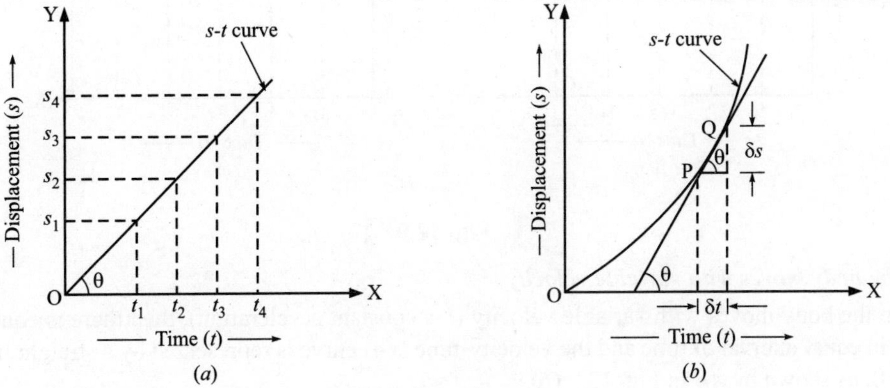


**Fig. 12.8**

**2. *When the body moves with variable velocity***

When the body moves with variable velocity, unequal distances are covered in equal intervals of time or equal distances are covered in unequal intervals of time. Thus, the displacement-time (*s-t*) curve, in this case, will be as shown in Fig. 12.8 (*b*).

Consider two points *P* and *Q* on the *s-t* curve, which are at a small distance δ*s* apart and the time interval between them is δ*t*. Let the chord *PQ* joining the points *P* and *Q* makes an angle θ with the horizontal. The average velocity during the interval *PQ* is given by

$$\tan\theta = \frac{\delta s}{\delta t} \qquad \text{...(From triangle } PQR\text{)}$$

In the limit, when δ*t* approaches to zero, *Q* will tend to approach *P* and the chord *PQ* becomes tangent to the curve at point *P*. Thus the velocity at *P*,

$$v_P = \tan\theta = \frac{ds}{dt}$$

where tan θ is the slope of the tangent at *P*. Thus, it is concluded that slope of the tangent at any instant on the *s-t* curve gives the velocity at that instant.

## 12.7 GRAPHICAL REPRESENTATION OF VELOCITY VERSUS TIME

The velocity verses time is represented graphically by taking velocity as ordinate and corresponding time as abscissa. We shall consider the following two cases:

**1. *When the body moves with uniform velocity***

When the body moves with uniform velocity (*i.e.* with zero acceleration), the velocity-time (*v-t*) curve is represented by a straight line, as shown by *AB* in Fig. 12.9 (*a*). We know that distance covered by a body in time *t* seconds

$$= \text{Area under the } v\text{-}t \text{ curve} = \text{Area of rectangle } OABC = v.t$$

Thus, it is concluded that the distance covered by the body during any interval of time is given by the area under the *v-t* curve.

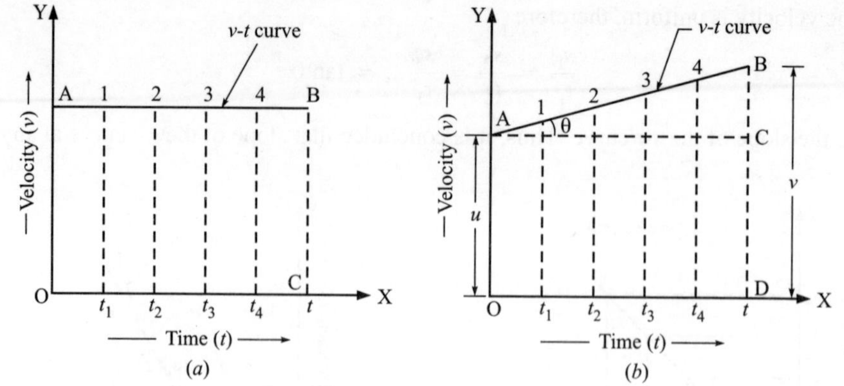


**Fig. 12.9**

**2. *When the body moves with variable velocity***

When the body moves with variable velocity (*i.e.* constant acceleration), then there is equal variation of velocity in equal interval of time and the velocity-time ($v$-$t$) curve is represented by a straight line inclined at an angle $\theta$, as shown by $AB$ in Fig. 12.9 ($b$).

Let $u$ = Initial velocity of the body, and

$v$ = Final velocity of the body after time $t$.

Then
$$\tan\theta = \frac{BC}{AC} = \frac{\text{Change in velocity}}{\text{Time taken}} = \text{Acceleration } (a)$$

Thus, the slope of the $v$-$t$ curve represents the acceleration of the body.

Now
$$a = \tan\theta = \frac{BC}{AC} = \frac{BD - CD}{AC} = \frac{v-u}{t}$$

or
$$v - u = at$$

$\therefore$
$$v = u + at$$

We know that the distance covered by a body is given by the area under the $v$-$t$ curve.

$\therefore$ Distance covered, $s$ = Area $OABD$ = Area of rectangle $OACD$ + Area of triangle $ABC$

$$= OA \times OD + \frac{1}{2} AC \times BC = u.t + \frac{1}{2} t(v-u)$$

$$= u.t + \frac{1}{2} a.t^2 \qquad ...(\because v - u = a.t)$$

## 12.8 GRAPHICAL REPRESENTATION OF ACCELERATION VERSUS TIME

The acceleration verses time is represented graphically by taking acceleration as ordinate and corresponding time as abscissa. We shall consider the following two cases:

**1. *When the body moves with uniform acceleration***

When the body moves with uniform acceleration, the acceleration-time ($a$-$t$) curve will be a straight horizontal line, as shown by $AB$ in Fig. 12.10 ($a$). We know that the change in velocity is the product of acceleration and time interval, therefore area under the $a$-$t$ curve (*i.e.* $OABC$) represents the change in velocity.

**2. *When the body moves with variable acceleration***

When the body moves with variable acceleration, the acceleration-time ($a$-$t$) curve may have any slope depending upon the values of acceleration at various instances, as shown in Fig. 12.10 ($b$). Consider two instances 1 and 2 having interval of time $\delta t$. Let $a$ is the mean value of acceleration during this interval.

∴ Change in velocity from instant 1 to instant 2

$$= a \times \delta t = \text{Area of rectangle } 12t_2t_1$$

Let $u$ is the velocity at $O$ and $v$ is the velocity at $C$, then

$$v - u = \text{Area under the } a\text{-}t \text{ curve } AB = \text{Area } OABC$$

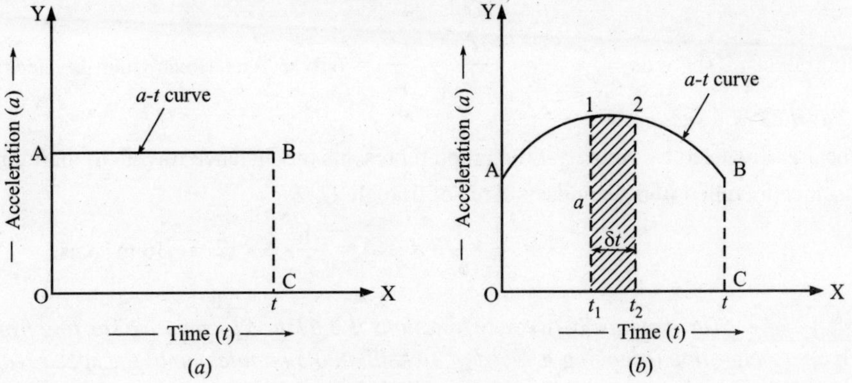


**Fig. 12.10**

**Example 12.23** *A particle moves with rectilinear motion and its speed rises from 0 to 12 m/s in 4 seconds, and then drops to 0 in 2 seconds. Plot the velocity-time curve and find :*

1. *Acceleration in first 4 seconds and that in last 2 seconds ; and*
2. *Distance travelled in 6 seconds.*

***Solution*** Given : Speed rises from 0 to 12 m/s in 4 seconds and then drops to 0 in 2 seconds.

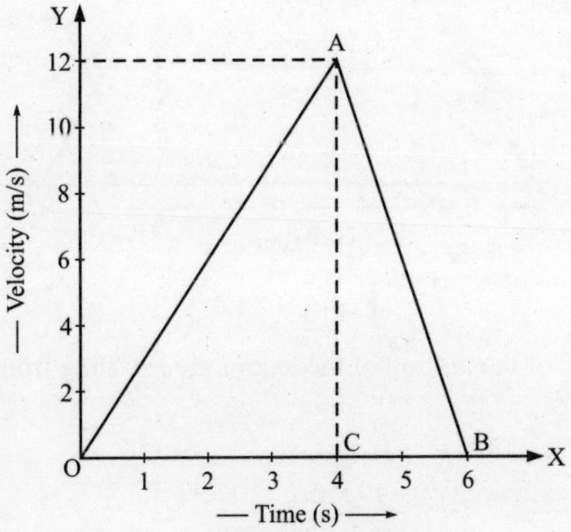


Velocity-time graph

**Fig. 12.11**

***Velocity-time curve***

The velocity-time curve is drawn, as shown in Fig. 12.11, by taking *X*-axis as time in seconds and *Y*-axis as velocity in m/s. From the graph, we see that when speed rises from 0 to 12 m/s in first 4 seconds, it is a straight line *OA*. Then the speed drops to 0 in 2 seconds, as shown by a straight line *AB*.

**1. *Acceleration in first 4 seconds and that in last 2 seconds***

We know that acceleration is the rate of change of velocity $\left(i.e.\ \frac{v}{t}\right)$.

$$\therefore \quad \text{Acceleration in first 4 seconds} = \frac{AC}{OC} = \frac{12}{4} = 3 \text{ m/s}^2 \textbf{ Ans.}$$

$$\text{and acceleration in last 2 seconds} = \frac{AC}{CB} = \frac{12}{2} = 6 \text{ m/s}^2 \textbf{ Ans.}$$ (Retardation as velocity drops to zero)

**2. *Distance travelled in 6 seconds***

Since the area enclosed by velocity-time graph represents the distance travelled, therefore

$$\text{Distance travelled in 6 seconds} = \text{Area of triangle } OAB$$

$$= \frac{1}{2} \times OB \times AC = \frac{1}{2} \times 6 \times 12 = 36 \text{ m } \textbf{Ans.}$$

**Example 12.24** *The distance between two stations is 2.6 km. A locomotive starting from one station, gives the train an acceleration (reaching a speed of 40 km/h in 0.5 minutes) until the speed reaches 48 km/h. The speed is maintained until brakes are applied and the train is brought to rest at the second station under a negative acceleration of 0.9 m/s². Find the time taken to perform this journey.*

***Solution*** Given : $s = s_1 + s_2 + s_3 = 2.6$ km $= 2600$ m ; $v = 40$ km/h $= \frac{40 \times 1000}{60 \times 60} = 11.11$ m/s ; $t = 0.5$ min $= 30$ s ; $v_1 = 48$ km/h $= \frac{48 \times 1000}{60 \times 60} = 13.33$ m/s

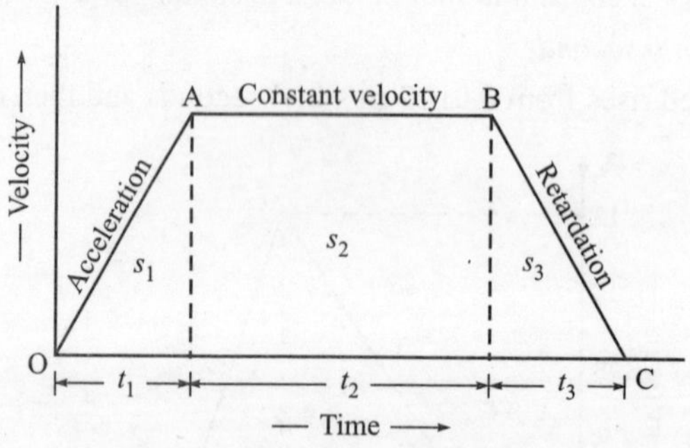


**Fig. 12.12**

First of all, let us consider of the motion of the locomotive starting from the first station (portion *OA* in Fig. 12.12). In this case,

$$\text{Initial velocity, } u = 0$$

$$\text{Final velocity, } v = 40 \text{ km/h} = 11.11 \text{ m/s}$$

$$\text{and time taken, } t = 0.5 \text{ min} = 30 \text{ s}$$

Let $a$ = Resulting acceleration.

Using the relation, $v = u + a \, . \, t$

or $11.11 = 0 + a \times 30$

$$\therefore \quad a = \frac{11.11}{30} = 0.37 \text{ m/s}^2$$

Let $t_1$ = Time taken to attain the speed of $v_1$ .

Again, using the relation, $v_1 = u + at_1$

or $13.33 = 0 + 0.37t_1$

$\therefore \quad t_1 = \dfrac{13.33}{0.37} = 36 \text{ s} \qquad \ldots(i)$

and distance covered during this time, $*s_1$ = Average velocity × time taken

$$= \frac{0 + 13.33}{2} \times 36 = 240 \text{ m}$$

Now consider the motion of the retarding period (*i.e.* negative acceleration) before the locomotive comes to rest at the second station (portion *BC* in Fig. 12.12). In this case,

Initial velocity, $u = 48 \text{ km/h} = 13.33 \text{ m/s}$

Final velocity, $v = 0$

and acceleration, $a = -0.9 \text{ m/s}^2$ ...(Given)

If $t_3$ is the time taken during this period, then from the relation,

$$v = u + at_3$$

$$0 = 13.33 - 0.9\, t_3$$

$$\therefore \quad t_3 = \frac{13.33}{0.9} = 14.81 \text{ s}$$

and distance covered during this time, $s_3$ = Average velocity × time

$$= \left(\frac{13.33 + 0}{2}\right) \times 14.81 = 98.7 \text{ m}$$

We know that distance covered ($s_2$) from *A* to *B* with a constant velocity of 13.33 m/s is given by

$$s_2 = s - (s_1 + s_3)$$

$$= 2600 - (240 + 98.7) = 2261.3 \text{ m}$$

and time taken to cover this distance, $t_2 = \dfrac{s_2}{v_2} = \dfrac{2261.3}{13.33} = 169.64 \text{ s} \qquad \ldots(\because v_2 = v_1)$

$\therefore$ Total time taken, $t = t_1 + t_2 + t_3 = 36 + 169.64 + 14.81 = 220.45$ s **Ans.**

**Example 12.25** *A pit cage is being raised in a mine shaft by a cable wound round a drum installed at the top of the shaft. The motion of the cage is as follows :*

*Velocity increases uniformly from rest to 3.3 m/s in the first 6.6 m ; constant speed for the next 26.4 m and velocity decreases uniformly to rest in further 13.4 m. Determine the time of the journey in each stage.*

***Solution*** The velocity-distance graph is shown in Fig. 12.13, by *OABC*, in which *OA* represents increase in velocity from zero to 3.3 m/s in 6.6 m ; *AB* represents constant speed for 26.4 m and *BC* represents decrease in velocity in 13.4 m.

First of all, let us consider the motion of the cage along *OA*. In this case,

Initial velocity at *O*, $u_1 = 0$

Final velocity at *A*, $v_1 = 3.3 \text{ m/s}$

and distance covered, $s_1 = 6.6 \text{ m}$

Let $a_1$ be the acceleration of the cage. We know that

$$v_1^2 = u_1^2 + 2a_1s_1$$

* The distance covered $s_1$ may also be obtained by using the relation,

$$s_1 = ut_1 + \frac{1}{2}at_1^2 = 0 + \frac{1}{2}(0.37)(36)^2 = 239.76 \text{ or } 240 \text{ m}$$

$$(3.3)^2 = 0 + 2a_1 \times 6.6$$

$$\therefore \quad a_1 = \frac{(3.3)^2}{2 \times 6.6} = 0.825 \text{ m/s}^2$$

and time taken during journey of the cage from $O$ to $A$,

$$t_1 = \frac{\text{Velocity } (u_1)}{\text{Acceleration } (a_1)} = \frac{3.3}{0.825} = 4 \text{ s} \quad \textbf{Ans.}$$

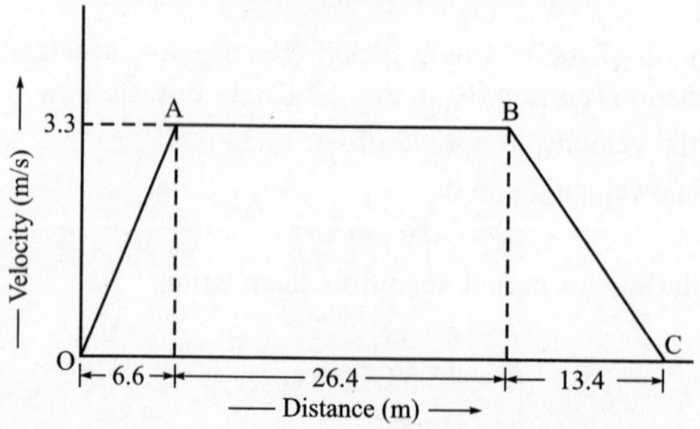


**Fig. 12.13**

Now consider the motion of the cage along $AB$. It is given that velocity is constant ($v_2 = v_1 = 3.3$ m/s) and the distance covered ($s_2$) is 26.4 m. Let $t_2$ be the time taken during journey of the cage from $A$ to $B$. We know that

$$s_2 = v_2 t_2$$

$$26.4 = 3.3\, t_2$$

$$\therefore \quad t_2 = \frac{26.4}{3.3} = 8 \text{ s} \quad \textbf{Ans.}$$

Lastly, consider the motion of the cage from $B$ to $C$. In this case,

Initial velocity at $B$, $\quad u_3 = v_2 = 3.3$ m/s

Final velocity at $C$, $\quad v_3 = 0$

and distance covered, $\quad s_3 = 13.4$ m

We know that $\quad v_3^2 = u_3^2 + 2a_3 s_3$

$$0 = (3.3)^2 + 2a_3 \times 13.4$$

$$\therefore \quad a_3 = \frac{-(3.3)^2}{2 \times 13.4} = -0.406 \text{ m/s}^2$$

The minus sign indicates retardation from $B$ to $C$.

Now using the relation, $\quad v_3 = u_3 - a_3 t_3$

$$0 = 3.3 - 0.406 t_3$$

$$\therefore \quad t_3 = \frac{3.3}{0.406} = 8.128 \text{ s} \quad \textbf{Ans.}$$

This example may be solved as follows:

Considering motion from $O$ to $A$. We know that

Distance covered, $\quad s_1$ = Average velocity × Time

$$6.6 = \frac{0 + 3.3}{2} \times t_1$$

$$\therefore \quad t_1 = \frac{2 \times 6.6}{3.3} = 4 \text{ s} \quad \textbf{Ans.}$$

Similarly, considering motion from $A$ to $B$. We know that distance covered,

$$s_2 = \text{Average velocity} \times \text{Time}$$

$$26.4 = \frac{3.3 + 3.3}{2} \times t_2$$

$$\therefore \quad t_2 = \frac{2 \times 26.4}{6.6} = 8 \text{ s} \quad \textbf{Ans.}$$

Again, considering motion from $B$ to $C$. We know that distance covered,

$$s_3 = \text{Average velocity} \times \text{Time}$$

$$13.4 = \frac{3.3 + 0}{2} \times t_3$$

$$\therefore \quad t_3 = \frac{2 \times 13.4}{3.3} = 8.12 \text{ s} \quad \textbf{Ans.}$$

**Example 12.26** *A train starts from station A and accelerates uniformly at 4 m/s² for the first 3 seconds and at 1.25 m/s² for the next 8 seconds. Then it travels with a uniform velocity for 5 minutes and comes to rest at station B in 10 seconds with uniform retardation. Determine the retardation and the distance travelled between the two stations A and B.*

***Solution*** The velocity-time graph is shown in Fig. 12.14 by *ACDEB* in which *AC* represents uniform acceleration of 4 m/s² for 3 seconds; *CD* represents acceleration of 1.25 m/s² for 8 seconds. *DE* represents uniform velocity for 5 minutes and *EB* represents retardation.

First of all, consider the motion of the train from $A$ to $C$. In this case,

Initial velocity at $A$, $u_1 = 0$

Acceleration, $a_1 = 4 \text{ m/s}^2$

Time, $t_1 = 3 \text{ s}$

Let $v_1$ be the velocity of the train during uniform acceleration of $a_1 = 4 \text{ m/s}^2$. We know that

$$v_1 = u_1 + a_1 t_1 = 0 + 4 \times 3 = 12 \text{ m/s}$$

and distance travelled from $A$ to $P$, $s_1 = u_1 t_1 + \frac{1}{2} a_1 t_1^2 = 0 + \frac{1}{2} \times 4 \times 3^2 = 18 \text{ m}$

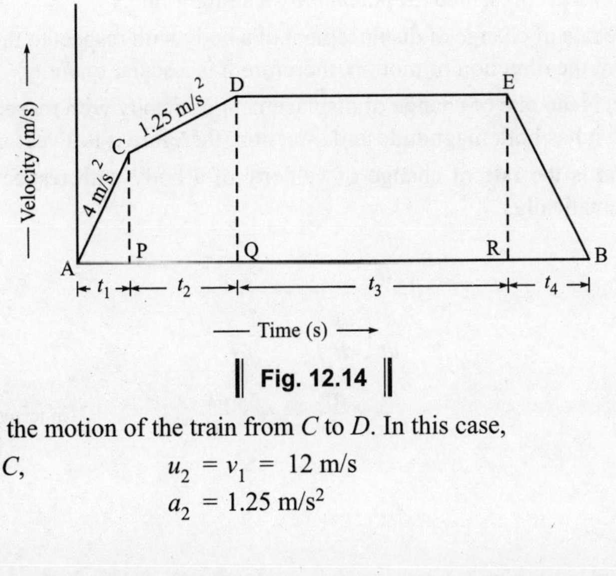


**Fig. 12.14**

Now, considering the motion of the train from $C$ to $D$. In this case,

Initial velocity at $C$, $u_2 = v_1 = 12 \text{ m/s}$

Acceleration, $a_2 = 1.25 \text{ m/s}^2$

Time, $t_2 = 8$ s

Let $v_2$ be the velocity of the train during uniform acceleration of $a_2 = 1.25$ m/s². We know that

$$v_2 = u_2 + a_2 t_2 = 12 + 1.25 \times 8 = 22 \text{ m/s}$$

and distance travelled from *P* to *Q*,
$$s_2 = u_2 t_2 + \frac{1}{2} a_2 t_2^2 = 12 \times 8 + \frac{1}{2} \times 1.25 \times 8^2 = 96 + 40 = 136 \text{ m}$$

Consider the motion of the train from *D* to *E*. In this case, velocity is uniform ($u_3 = v_2$) for 5 minutes (*i.e.* $t_3 = 300$ s) and acceleration ($a_3$) is zero.

We know that distance travelled from *Q* to *R*,

$$s_3 = u_3 t_3 + \frac{1}{2} a_3 t_3^2 = u_3 t_3 = 22 \times 300 = 6600 \text{ m} \qquad ...(\because a_3 = 0)$$

Lastly, consider the motion of the train from *E* to *B*. In this case,

Initial velocity at *E*, $u_4 = u_3 = 22$ m/s and the final velocity $v_4 = 0$. We know that

$$v_4 = u_4 + a_4 t_4$$

$$0 = 22 + a_4 \times 10 \qquad ...(\because t_4 = 10 \text{ s})$$

$$\therefore \quad a_4 = \frac{-22}{10} = -2.2 \text{ m/s}^2$$

The minus sign indicates retardation from *E* to *B*. Using the relation,

$$s_4 = u_4 t_4 + \frac{1}{2} a_4 t_4^2 = 22 \times 10 + \frac{1}{2} \times (-2.2)(10)^2$$

$$= 220 - 110 = 110 \text{ m}$$

∴ Total distance travelled from *A* to *B*,

$$s = s_1 + s_2 + s_3 + s_4 = 18 + 136 + 6600 + 110 = 6864 \text{ m} \quad \textbf{Ans.}$$

## HIGHLIGHTS

1. The motion of a rigid body, when confined to one plane, is said to be a *plane motion,* which may be either rectilinear or curvilinear motion.
2. When the motion of a body is along a straight line, it is said to be a *rectilinear* motion.
3. When the motion takes place along a curved path, it is known as a *curvilinear* motion.
4. The displacement ($s$) is the change of position of a particle in a definite direction. It is a vector quantity and may be represented graphically by a straight line.
5. The speed is the rate of change of displacement of a body with respect to the time. Since it does not take into account the direction of motion, therefore it is a scalar quantity.
6. The velocity ($v$) is the rate of change of displacement of a body with respect to time in a particular direction. Since it has both magnitude and direction, therefore it is a vector quantity.
7. The acceleration is the rate of change of velocity of a body with respect to time. It is a vector quantity. Mathematically,

$$a = \frac{dv}{dt} \qquad \text{...(in terms of velocity)}$$

$$= \frac{d}{dt}\left(\frac{ds}{dt}\right) = \frac{d^2 s}{dt^2} \qquad \text{...(in terms of displacement)}$$

$$= \frac{ds}{dt} \times \frac{dv}{ds} = v \times \frac{dv}{ds} \qquad \text{...(in terms of velocity and distance)}$$

**8.** When a body is moving in a straight line with uniform acceleration, then the equations of motion are as follows:

(*a*) $$v = u + at$$

(*b*) $$s = ut + \frac{1}{2}at^2$$

(*c*) $$v^2 - u^2 = 2as$$

where
$u$ = Initial velocity,
$v$ = Final velocity,
$a$ = Uniform acceleration,
$s$ = Distance travelled, and
$t$ = Time in seconds.

**9.** The distance covered by a body in *n*th second,

$$= u + \frac{a}{2}(2n - 1)$$

**10.** The equations of motion for acceleration due to gravity are written as follows:

(*a*) ***When the motion of the body is downward***

$$v = u + gt$$

$$s = ut + \frac{1}{2}gt^2$$

$$v^2 - u^2 = 2gs$$

(*b*) ***When the motion of the body is upward***

$$v = u - gt$$

$$s = ut - \frac{1}{2}gt^2$$

$$v^2 - u^2 = -2gs$$

**11.** When a body starts from rest, its initial velocity is zero (*i.e.* $u = 0$) and when it comes to rest, the final velocity is zero (*i.e.* $v = 0$).

**12.** When a body is projected vertically upwards, the final velocity of the body at the highest point is zero.

## NUMERICAL PROBLEMS

**1.** A particle starts from rest and moves along a straight line with constant acceleration. If the particle attains a velocity of 10 m/s after a distance of 25 m, find the acceleration. [**Ans.** 2 m/s$^2$]

**2.** A particle starts from rest and moves along a straight line with constant acceleration. If the particle covers a initial distance of 16 m in 10 seconds, find the time taken to cover the entire distance of 400 m. What will be its final velocity? [**Ans.** 50 s; 16 m/s]

**3.** A particle starts with initial velocity of 2 m/s and moves with a uniform acceleration of 0.1 m/s$^2$. Determine the distance described in a time of 30 seconds. [**Ans.** 105 m]

**4.** Car *A* at a gasoline station stays there for 10 minutes after a car *B* passes at an average speed of 17.6 m/s. How long will it take car *A* moving at an average speed of 22 m/s to over take car ? [**Ans.** 40 m]

**5.** A car accelerates from a velocity of 10 m/s to a velocity of 30 m/s, in a distance of 240 m. Calculate the average acceleration and the time required. [**Ans.** 1.67 m/s$^2$; 11.97 s]

**6.** A motorist is travelling at 80 km/h, where he observes a traffic light 200 m ahead of him turns red. The traffic light is timed to stay red for 10 seconds. If the motorist wishes to pass the light without stopping, just as it turns green, determine: 1. required uniform deceleration; 2. speed as it passes the light.

[**Ans.** –0.444 m/s$^2$; 17.78 m/s]

7. A bullet, moving at the rate of 300 m/s, is fired into a thick block of target and penetrates up to 0.5 m. If it is fired into a 0.25 m thick target, find the velocity of emergence. The two blocks are of the same material and take the resistance to be uniform in both the cases. **[Ans.** 212.13 m/s]

8. The relation $a = 2t$ defines the motion of a particle. It is given that $s = 1.2$ m and $v = 0.6$ m/s when $t = 1$ s. Find $s$ and $v$ at $t = 4$ s. **[Ans.** 21 m; 15.6 m/s]

9. A particle moving along a straight line has an acceleration $a = \sqrt{v}$. At time $t = 2$ s, its displacement is 42.67 m and velocity is 16 m/s. Find the displacement, velocity and acceleration at $t = 3$ s.

**[Ans.** 56 m; 10.67 m/s; –5.335 m/s²]

10. The acceleration of a particle along a straight line is given by the equation $a = 4 - \frac{t^2}{g}$. If the particle starts with velocity from a position $s = 0$; find: 1. its velocity after 6 seconds; and 2. distance travelled in 6 seconds. **[Ans.** 16.66 m/s²; 61 m]

11. A particle, moving in a straight line with an initial velocity of $u$ m/s is subjected at any instant to a retardation of $k.v^3$, where $k$ is a constant and $v$ is the velocity at that instant in m/s. Calculate the velocity when the distance described in $s$ metres and the time for doing so. $\left[\textbf{Ans. } \frac{u}{k.s.u+1}; \frac{1}{2u}(k.u.s^2 + 2)\right]$

12. The rectilinear motion of a particle is governed by $a = -8\,s^{-2}$, where $a$ is in m/s² and $s$ is in metres. When $t = 1$ s, $s = 4$ m and $v = 2$ m/s, determine the acceleration of the particle at $t = 4$ s. **[Ans.** –0.125 m/s²]

13. A car moves along a straight line whose equation of motion is given by $s = 12t + 3t^2 - 2t^3$, where $s$ is in metres and $t$ is in seconds. Calculate: 1. velocity and acceleration at the start; and 2. acceleration when the velocity is zero. **[Ans.** 12 m/s, 6 m/s²; –18 m/s²]

14. The motion of a particle is given by the acceleration, $a = t^3 - 2t^2 + 5$, where $a$ is in m/s² and $t$ is in seconds. The velocity is 3.8 m/s when $t = 1$ s and displacement, $s = 10$ m when $t = 1$ s. Calculate the velocity, acceleration and displacement at $t = 3$ s. **[Ans.** 16.47 m/s; 14 m/s²; 27.2 m]

15. A train moving with constant acceleration travels 7.2 m during the 10th second of its motion and 5.4 m during the 12th second of its motion. Find its initial velocity. **[Ans.** 15.75 m/s]

16. A stone after falling 5 seconds from rest breaks a glass pane and in breaking it loses 40% of its velocity. How far will it fall in the next second? **[Ans.** 34.335 m]

17. A stone is dropped down a well and 5 seconds later the sound of splash is heard. If the velocity of sound is 336 m/s, find the depth of well. **[Ans.** 107.43 m]

18. A particle is dropped from the top of a tower 60 m high and another particle is projected at the same time vertically upwards from the foot of the tower to meet the first particle at a height of 18 m. Find the velocity of projection of the second particle. **[Ans.** 20.5 m/s]

19. A stone is dropped into a well with no initial velocity and 4.5 seconds later, the splash is heard. Then, a second stone is thrown downwards into the well with an initial velocity $u$ and the splash is heard in 4 seconds. If the velocity of sound is constant at 336 m/s, determine the initial velocity of the second stone. **[Ans.** 5.215 m/s]

20. A ball is thrown vertically into the air at 36 m/s. After 3 seconds, another ball is thrown vertically. What initial velocity must the second ball have to pass the first ball at 30 m from the ground ? **[Ans.** 25.5 m/s]

21. A stone is thrown vertically upwards from the ground with a velocity of 50 m/s. After 2 seconds, another stone is thrown vertically upwards from the same place. If both the stones strike the ground at the same time, calculate the velocity with which the second stone was thrown upwards. Take $g = 10$ m/s². **[Ans.** 40 m/s]

22. A stone is dropped from the top of a tower 30 m high. At the same instant, a second stone is thrown upwards from the ground with an initial velocity of 15 m/s. When and where the two stones cross each other.

**[Ans.** 10.38 m from the ground]

**23.** A train travels between two stations 15 km apart in 18 minutes. If the train accelerates for a part of journey uniformly followed by uniform retardation, find graphically the maximum speed attained during journey.

[**Ans.** 100 km/h]

**24.** A local train starts from a station and reaches the next station 3.2 km away after 6 minutes. It accelerates during start and retards during stop at the uniform rate and attains a uniform speed of 48 km/h in between. Determine:

1. The time interval and distance for each motion; and

2. The acceleration of the train. [**Ans.** 120 s, 800 m, 1600 m, 800 m ; $\frac{1}{9}$ m/s$^2$]

**25.** A cage goes down a mine shaft 750 m deep in 45 seconds. For the first quarter of the distance only, the speed is being uniformly accelerated and during the last quarter uniformly retarded. The acceleration and retardation being equal. Find the uniform speed of the cage while traversing the central portion of the cage. [**Ans.** 25 m/s]

**26.** A train travels from station *A* to station *B* from rest to rest. At a point *C*, somewhere between *A* and *B*, it attains the highest speed of 45 km/h. If it travels with uniform acceleration from *A* to *C* and with uniform retardation from *C* to *B*, find the distance between *A* and *B*, if the total journey takes 45 minutes. [**Ans.** 16.875 km]

## REVIEW QUESTIONS

**1.** Distinguish between rectilinear motion and rotary motion.

**2.** Define the following terms:

(*a*) Displacement; (*b*) Speed; (*c*) Velocity; and (*d*) Acceleration.

**3.** What do you mean by positive acceleration and negative acceleration?

**4.** Derive the following equations of motion of a body moving in a straight line with uniform acceleration:

(*a*) $s = ut + \frac{1}{2}at^2$; and (*b*) $v^2 = u^2 - 2as$

**5.** What is meant by deceleration ?

**6.** Derive the relation for the distance covered by a body in *n*th second.

**7.** Show that the time taken by a body, projected vertically upwards with a velocity *u*, to reach a height *h* is $\frac{u}{g}$.

**8.** Discuss the graphical representation of displacement versus time

(*a*) When the body moves with uniform velocity, and

(*b*) When the body moves with variable velocity.

## OBJECTIVE TYPE QUESTIONS

**1.** The rate of change of displacement of a body is called

(*a*) velocity (*b*) acceleration (*c*) momentum (*d*) none of these

**2.** Which of the following is a vector quantity?

(*a*) Displacement (*b*) Velocity (*c*) Acceleration (*d*) all of these

**3.** The rate of change of velocity of a body with respect to time is called

(*a*) velocity (*b*) acceleration (*c*) momentum (*d*) none of these

**4. When the motion of a body is along a straight line, then it is called**

(*a*) straight line motion (*b*) rectilinear motion
(*c*) translatory motion (*d*) any one of these

**5. The ............... acceleration is called retardation.**

(*a*) positive (*b*) negative

**6. The acceleration of a body is given by**

(*a*) $\dfrac{ds}{dt}$ (*b*) $\dfrac{dv}{ds}$ (*c*) $\dfrac{dv}{dt}$ (*d*) $v \times \dfrac{dv}{ds}$

**7. The distance travelled by a body moving with uniform acceleration in the nth second is**

(*a*) $un + \dfrac{1}{2}an^2$ (*b*) $u + 2a\,(n-1)$ (*c*) $2u + a\,(2n-1)$ (*d*) $u + \dfrac{a}{2}(2n-1)$

**8. When a body is projected vertically upwards, the final velocity at the highest point is**

(*a*) zero (*b*) maximum (*c*) minimum (*d*) none of these

**9. When a body falls freely under gravity, then the gravitational acceleration is taken as**

(*a*) 8.9 m/s$^2$ (*b*) –8.9 m/s$^2$ (*c*) +9.81 m/s$^2$ (*d*) –9.81 m/s$^2$

**10. The velocity of a body on reaching the ground from a height *h* is**

(*a*) $2\sqrt{gh}$ (*b*) $\sqrt{2gh}$ (*c*) $\sqrt{gh}$ (*d*) $2g\sqrt{h}$

***Answers to Objective Type Questions***

| | | | | |
|---|---|---|---|---|
| **1.** (*a*) | **2.** (*d*) | **3.** (*b*) | **4.** (*d*) | **5.** (*b*) |
| **6.** (*c*) | **7.** (*d*) | **8.** (*a*) | **9.** (*c*) | **10.** (*b*) |

13

# PROJECTILES



## 13.1 INTRODUCTION

Whenever a particle is thrown upwards at a certain angle (other than vertical), it traces a curved path in the air and finally falls on the ground at some other point. A little consideration will show that the velocity with which the particle was projected, has two components, one is the vertical component and the other is horizontal component. The vertical component moves the particle vertically upwards and it is subjected to a acceleration due to gravity, while the horizontal component moves the particle horizontally in its direction and remains constant throughout the motion if there is no air resistance. The combined effect of the vertical and horizontal component of the velocity is to move the particle in a parabolic path. Thus, a particle which moves under the combined effect of vertical and horizontal component is known as a *projectile.* The study of motion of such a projectile is called *projectile motion.*

## 13.2 TECHNICAL TERMS USED IN PROJECTILES

The following technical terms, as shown in Fig. 13.1, are widely used in the study of a projectile motion.

1. ***Trajectory*** The curved path followed by the projectile when projected in space is known as *trajectory* or *path of the projectile*.
2. ***Velocity of projection*** The initial velocity with which the projectile is projected in space is called *velocity of projection*. It is denoted by $u$.
3. ***Angle of projection*** The angle between the direction of projection and the horizontal direction is called the *angle of projection*. In other words, it is the angle with the horizontal which the velocity of projection makes. It is denoted by $\alpha$.

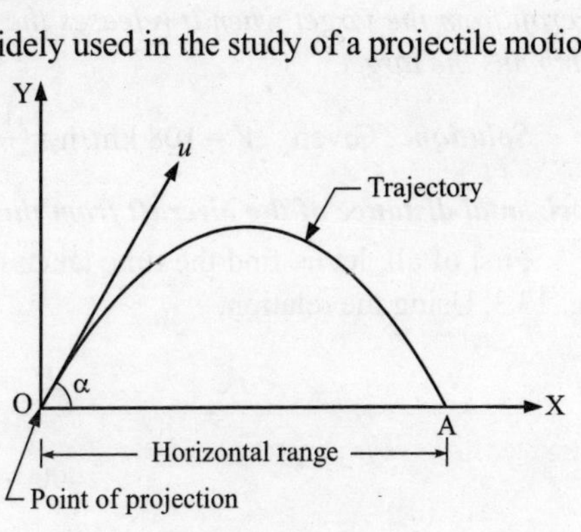


**Fig. 13.1**

4. ***Horizontal range*** The horizontal distance travelled by the projectile, from the point of projection ($O$) up to the point where it strikes the ground ($A$), is known as *range of the projectile*.
5. ***Time of flight*** The total time taken by the projectile for which it remains in space is known as *time of flight*.

## 13.3 MOTION OF A BODY PROJECTED HORIZONTALLY

Consider a body thrown horizontally into the air from the point $B$ at a height $H$ from the ground, with a velocity $u$ m/s, as shown in Fig. 13.2.

A little consideration will show that when the body moves in air towards ground, it is subjected to the following two components of velocity, *i.e.*

1. Horizontal component ($u$), and
2. Vertical component in the downward direction.

It may be noted that the horizontal component of velocity from $B$ to $A$ remains constant (*i.e.* $u$) while the vertical component (say $v$) in the downward direction is always subjected to acceleration due to gravity ($g$) and hence it is not constant. The velocity, with which the body strikes the ground at $A$ is the resultant of horizontal component ($u$) and the vertical component ($v$). Thus

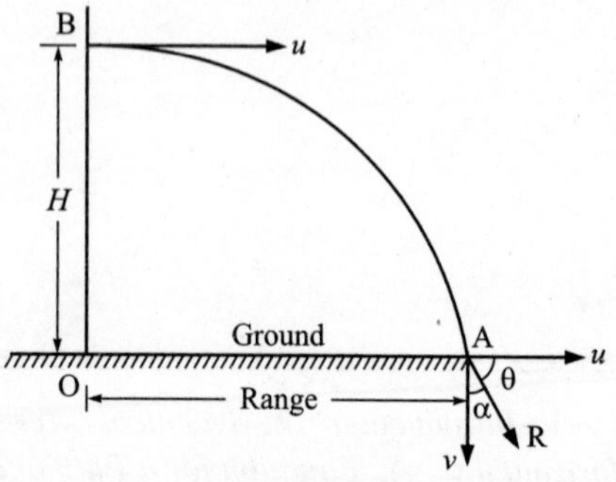


**Fig. 13.2**

$$\text{Resultant velocity at } A,\ R = \sqrt{u^2 + v^2}$$

and the angle made by the resultant velocity with the *horizontal is given by

$$\tan\theta = \frac{v}{u}$$

The height of point $B$ from the ground (*i.e.* $H$) may be obtained by considering the vertical motion, *i.e.*

$$H = 0 \times t + \frac{1}{2}gt^2 = \frac{1}{2}gt^2 \quad ...(\because \text{ Vertical component of velocity at } B = 0)$$

where $t$ is the time of flight, *i.e.* the time taken by the body to reach from $B$ to $A$. During this period, the body moves horizontally with uniform velocity $u$ m/s.

$$\therefore \quad \text{Range} = u \times t$$

**Example 13.1** *An aircraft moving horizontally at a speed of 108 km/h at an altitude of 1000 m towards a target on the ground releases a bomb which hits the target. Estimate the horizontal distance of the aircraft from the target when it releases the bomb. Calculate also the velocity and direction with which the bomb hits the target.*

***Solution*** Given : $V = 108 \text{ km/h} = \dfrac{108 \times 1000}{3600} = 30$ m/s ; $H = 1000$ m

***Horizontal distance of the aircraft from the target when it releases the bomb***

First of all, let us find the time taken ($t$) by the bomb to hit the target on the ground at $A$ as shown in Fig. 13.3. Using the relation,

$$H = \frac{1}{2}gt^2$$

$$1000 = \frac{1}{2} \times 9.81t^2 = 4.905t^2$$

* The angle made by the resultant velocity with the vertical will be $\tan\alpha = \dfrac{u}{v}$

$$\therefore \quad t = \sqrt{\frac{1000}{4.905}} = \sqrt{203.87} = 14.278 \text{ s}$$

We know that the horizontal distance *OA* (*i.e.* range)

$$= u \times t = 30 \times 14.278 = 428.34 \text{ m} \quad \textbf{Ans.}$$

***Velocity and direction with which the bomb hits the target***

We know that the final velocity of the bomb in the vertical direction when it hits target (*i.e.* after 14.278 s),

$$v = u_1 + g.t = 0 + 9.81 \times 14.278 = 140 \text{ m/s}$$

...(∵ Initial velocity of the bomb in vertical direction = 0)

The horizontal velocity remains constant throughout the motion and is equal to 30 m/s.

∴ Resultant velocity with which the bomb hits the target,

$$R = \sqrt{v^2 + u^2} = \sqrt{(140)^2 + (30)^2} = \sqrt{19\,600 + 900}$$

$$= 143.18 \text{ m/s}$$

Let $\theta$ = Angle which the resultant velocity makes with the horizontal when it hits the target, as shown in Fig. 13.3.

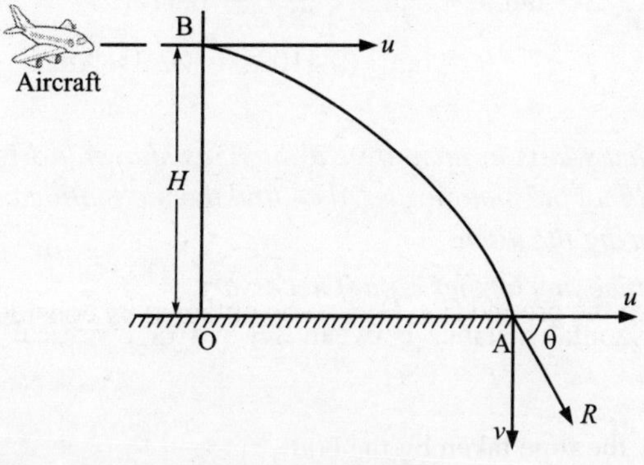


**Fig. 13.3**

We know that $\tan\theta = \dfrac{v}{u} = \dfrac{140}{30} = 4.667$

$$\therefore \quad \theta = \tan^{-1}(4.667) = 77.9° \quad \textbf{Ans.}$$

**Example 13.2** *The pilot of an aeroplane, flying horizontally with constant speed u = 480 km/h at an elevation H = 610 m above a level plane, wishes to bomb a target B on the ground, as shown in Fig. 13.4.*

*At what angle θ, below the horizontal should he see the target at the instant of releasing the bomb in order to score a hit? Neglect air resistance.*

***Solution*** Given : $u = 480 \text{ km/h} = \dfrac{480 \times 1000}{60 \times 60} = 133.33 \text{ m/s}$ ; $H = 610$ m

Let *t* seconds is the time taken by the bomb to hit the target *B* on the ground, as shown in Fig. 13.4. Considering vertical motion, we have

$$H = 0 \times t + \frac{1}{2} g t^2 \quad \text{...(∵ Initial velocity of the bomb in vertical direction = 0)}$$

$$610 = \frac{1}{2} \times 9.81 t^2 = 4.905 t^2$$

$$\therefore \qquad t^2 = \frac{610}{4.905} = 124.3 \text{ or } t = \sqrt{124.3} = 11.152 \text{ s}$$

During this period, horizontal distance travelled by the bomb,

$$x = \text{*Horizontal velocity of the bomb} \times t = u \cdot t$$
$$= 133.33 \times 11.152 = 1487 \text{ m}$$

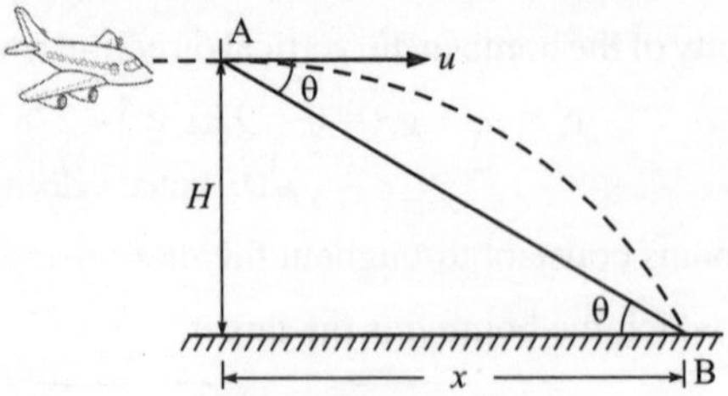


**Fig. 13.4**

Thus, the bomb should be released when the aeroplane is at a horizontal distance of 1487 m from the target and the angle θ is given by

$$\therefore \qquad \tan\theta = \frac{H}{x} = \frac{610}{1487} = 0.4102$$

or $\qquad \theta = \tan^{-1}(0.4102) = 22.3°$ **Ans.**

**Example 13.3** *A cyclist wants to jump over a ditch, as shown in Fig. 13.5 (a). Find the necessary minimum velocity u at A in km/h of the motorcycle. Also, find the inclination and magnitude of the velocity of the motor-cycle just after clearing the ditch.*

*Take g = 9.81 m/s². Assume any further data if necessary.*

***Solution*** Given : Horizontal distance between $AB$ = 4 m ; Vertical distance between $AB$ = 2 m ; $g = 9.81$ m/s²

***Necessary minimum velocity at A***

Let $\qquad u$ = Minimum velocity of the motorcycle at $A$ in the horizontal direction, and

$t$ = Time taken by the motorcycle to jump from $A$ to the other side $B$ of ditch.

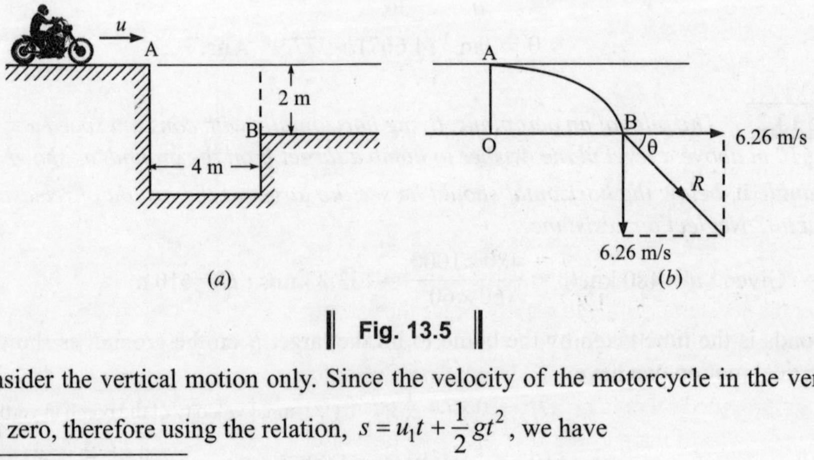


**Fig. 13.5**

First consider the vertical motion only. Since the velocity of the motorcycle in the vertical direction (say $u_1$) at $A$ is zero, therefore using the relation, $s = u_1 t + \frac{1}{2} g t^2$, we have

*The horizontal velocity of the bomb is same as that of aeroplane, *i.e.* 133.33 m/s.

$$2 = 0 \times t + \frac{1}{2} \times 9.81 t^2 = 4.905 t^2$$

or $$t^2 = \frac{2}{4.905} = 0.4077$$

∴ $$t = \sqrt{0.4077} = 0.6385 \text{ s} \quad ...(i)$$

Now considering the horizontal motion, we have

Horizontal distance between $AB$ = Horizontal velocity × Time

or $$4 = u.t = 4 \times 0.6385$$

∴ $$u = \frac{4}{0.6385} = 6.26 \text{ m/s}$$

$$= \frac{6.26 \times 3600}{1000} = 22.54 \text{ km/h} \quad \textbf{Ans.}$$

***Inclination and magnitude of the velocity of the motorcycle just after clearing the ditch***

We have calculated above that the horizontal velocity of the motorcycle from $A$ to $B$ is constant and is given by

$$u = 6.26 \text{ m/s}$$

The vertical velocity at $B$ ($v$) is obtained by considering the motion of the motorcycle from $A$ to $B$ in the vertical direction. Since the vertical velocity at $A$ ($u_1$) is zero, therefore

$$v = u_1 + gt = 0 + 9.81 \times 0.6385 = 6.26 \text{ m/s}$$

Resultant velocity at $B$, $$R = \sqrt{u^2 + v^2} = \sqrt{(6.26)^2 + (6.26)^2} = \sqrt{39.2 + 39.2}$$

$$= \sqrt{78.4} = 8.85 \text{ m/s}$$

$$= \frac{8.85 \times 3600}{1000} = 31.86 \text{ km/h} \quad \textbf{Ans.}$$

Let $\theta$ = Angle of inclination of the resultant $R$ with the horizontal.

From Fig. 13.5 ($b$), we have

$$\tan\theta = \frac{v}{u} = \frac{6.26}{6.26} = 1$$

∴ $$\theta = \tan^{-1}(1) = 45°$$

## 13.4 EQUATION FOR A PATH OF A PROJECTILE OR TRAJECTORY

Consider a particle is projected upwards from point $O$ at an angle $\alpha$ (angle of projection) with the horizontal with an initial velocity $u$ (velocity of projection), as shown in Fig. 13.6.

The velocity $u$ has two components, *i.e.* vertical and horizontal components. We know that

Vertical component = $u \sin\alpha$

and horizontal component = $u \cos\alpha$

We have already discussed that the vertical motion of the particle is under the acceleration due to gravity. Since the particle moves against the gravity, therefore it will decelerate with constant acceleration ($g$). The horizontal component of velocity remains constant provided there is no air resistance. Consider a point $P$ on the path $OBA$ of the particle after time $t$ seconds. Let $x$ and $y$ are the co-ordinates of the point $P$ with respect to $O$.

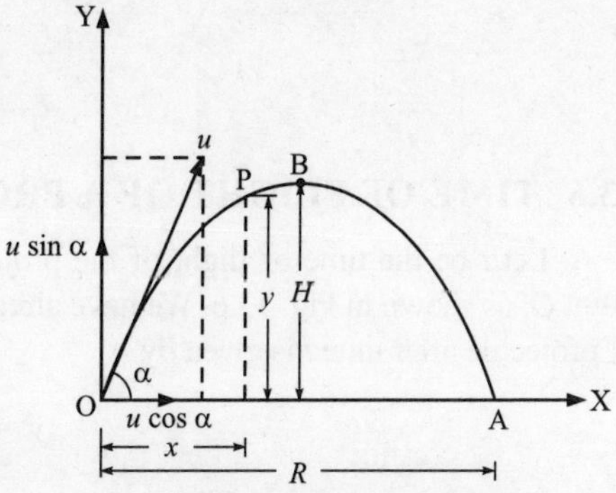


**Fig. 13.6**

∴ Horizontal distance travelled by the particle in time $t$ seconds,

$$x = (u \cos \alpha) \times t$$

or
$$t = \frac{x}{u \cos \alpha} \quad \text{...}(i)$$

and vertical distance travelled by the particle in time seconds,

$$y = (u \sin \alpha)t - \frac{1}{2}gt^2 \quad \text{...}(ii)$$

$$= u \sin \alpha \left( \frac{x}{u \cos \alpha} \right) - \frac{1}{2} g \left( \frac{x}{u \cos \alpha} \right)^2 \quad \text{...[From equation (}i\text{)]}$$

$$= \frac{x \sin \alpha}{\cos \alpha} - \frac{gx^2}{2u^2 \cos^2 \alpha}$$

$$= x \tan \alpha - \frac{gx^2}{2u^2 \cos^2 \alpha}$$

Since this is the equation of a parabola, therefore the path of a projectile or trajectory is parabolic.

## 13.5 MAXIMUM HEIGHT OF A PROJECTILE ON A HORIZONTAL PLANE

Let $B$ be the highest point of projectile from the horizontal plane $OA$ passing through the point of projection $O$. Let $H$ is the required maximum height attained by the projectile. We know that the vertical component of the initial velocity ($u$) of the projectile

$$= u \sin \alpha$$

and vertical component of final velocity $= 0$

We know that
$$v^2 = u^2 - 2gh$$

∴
$$0 = (u \sin \alpha)^2 - 2gH$$

or
$$2gH = u^2 \sin^2 \alpha$$

∴
$$H = \frac{u^2 \sin^2 \alpha}{2g}$$

Let $t_1$ be the time taken by the projectile to reach the maximum height ($H$), *i.e.* to attain zero velocity from $u \sin \alpha$. From the relation, $v = u - gt$, we have

$$0 = u \sin \alpha - gt_1$$

or
$$gt_1 = u \sin \alpha$$

∴
$$t_1 = \frac{u \sin \alpha}{g}$$

## 13.6 TIME OF FLIGHT OF A PROJECTILE

Let $t$ be the time of flight of the projectile, *i.e.* time taken by the projectile in reaching point $A$ from point $O$, as shown in Fig. 13.6. We have already discussed that the $y$-co-ordinate of any point lying on the path of projectile after time $t$ is given by

$$y = (u \sin \alpha)t - \frac{1}{2}gt^2$$

A little consideration will show that when the projectile reaches at point $A$, then $y = 0$. Substituting this value in the above equation,

$$0 = (u \sin \alpha)t - \frac{1}{2}gt^2$$

or $$\frac{1}{2}gt^2 = (u \sin \alpha)t$$

$\therefore$ $$^*t = \frac{2u \sin \alpha}{g}$$

Due to symmetry, time of ascent $(t_1)$ = time of descent $(t_2)$

and total time of flight from $O$ to $A$, $t = t_1 + t_2$

$$= \frac{u \sin \alpha}{g} + \frac{u \sin \alpha}{g} = \frac{2u \sin \alpha}{g}$$

## 13.7 HORIZONTAL RANGE OF THE PROJECTILE

Let $R$ be the horizontal range of the projectile. We have already discussed that it is the distance between the point of projection ($O$) and the point where it strikes back to the earth (*i.e.* point $A$). We know that horizontal component of velocity of projection

$$= u \cos \alpha$$

and time taken by the projectile to reach from $O$ to $A$, *i.e.* time of flight

$$t = \frac{2u \sin \alpha}{g}$$

$\therefore$ Horizontal range, $R$ = Horizontal component of velocity × Time of flight

$$= u \cos \alpha \times \frac{2u \sin \alpha}{g} = \frac{u^2 \times 2 \sin \alpha \cos \alpha}{g} = \frac{u^2 \sin 2\alpha}{g}$$

...($\because$ $2 \sin \alpha \cos \alpha = \sin 2\alpha$)

**Note:** We see that the horizontal range will be maximum when $\sin 2\alpha$ is maximum, *i.e.* when

$$\sin 2\alpha = 1 \text{ or } 2\alpha = 90° \text{ or } \alpha = 45°$$

Thus the horizontal range will be maximum when the angle of projection ($\alpha$) is 45°.

$\therefore$ Maximum horizontal range, $$R_{max} = \frac{u^2 \sin 2\alpha}{g} = \frac{u^2 \sin 90°}{g} = \frac{u^2 \times 1}{g} = \frac{u^2}{g}$$

**Example 13.4** *A ball is projected from the ground at an angle of 60° with the horizontal with a velocity of 40 m/s. Find the distance covered by the ball vertically and horizontally after 2 seconds.*

***Solution*** Given : $\alpha = 60°$ ; $u = 40$ m/s ; $t = 2$ s

***Distance covered by the ball vertically***

Let the ball $P$ is projected from the ground at $O$, at an angle of 60° with a velocity $u = 40$ m/s. Let it reaches a point $P$ after 2 seconds, as shown in Fig. 13.7. We know that distance covered by the ball vertically,

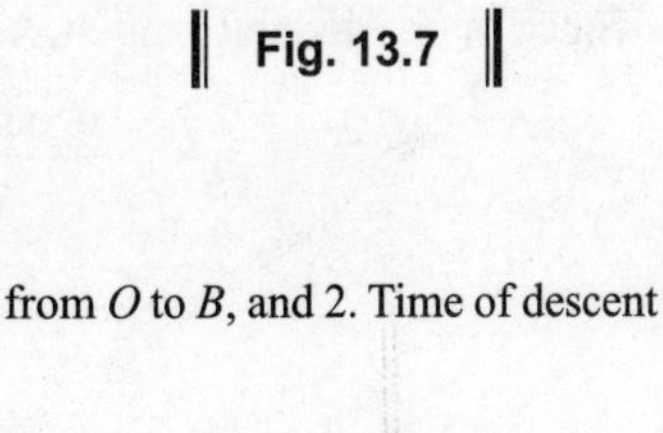


**Fig. 13.7**

$$y = (u \sin \alpha)t - \frac{1}{2}gt^2$$

$$= (40 \sin 60°)2 - \frac{1}{2} \times 9.81 \times (2)^2$$

$$= 69.28 - 19.62 = 49.66 \text{ m} \quad \textbf{Ans.}$$

* It may be noted that the total time of flight ($t$) consists of: 1. Time of ascent $(t_1)$ from $O$ to $B$, and 2. Time of descent $(t_2)$ from $B$ to $A$.

***Distance covered by the ball horizontally***

We know that distance covered by the ball horizontally,

$$x = u \cos \alpha \times t = 40 \cos 60° \times 2 = 40 \text{ m} \quad \textbf{Ans.}$$

**Note:** Distance $OP = \sqrt{x^2 + y^2} = \sqrt{(40)^2 + (49.66)^2} = 63.77$ m

**Example 13.5** *A bullet is fired from a rifle with an initial velocity of 50 m/s to clear a vertical wall 20 m high located at a distance of 30 m measured horizontally from the point of projection. Find the angle of projection.*

***Solution*** Given : $u = 50$ m/s ; $y = 20$ m ; $x = 30$ m

Let $\alpha$ be the angle of projection, as shown in Fig. 13.8.

We know that distance covered by the bullet vertically,

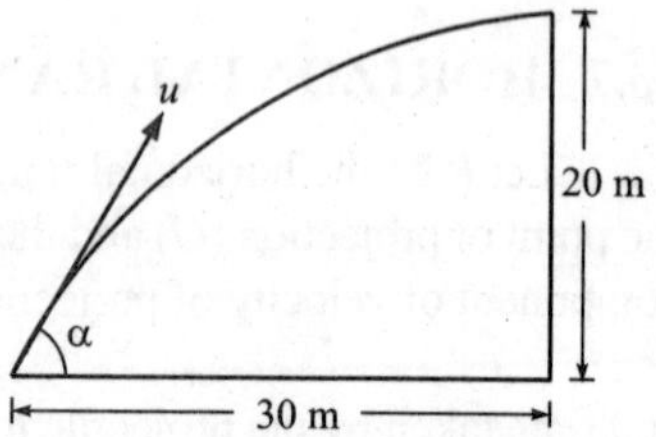


**Fig. 13.8**

$$y = x \tan \alpha - \frac{g x^2}{2u^2 \cos^2 \alpha}$$

$$20 = 30 \tan \alpha - \frac{9.81(30)^2}{2(50)^2} \sec^2 \alpha$$

$$= 30 \tan \alpha - \frac{8829}{5000}(1 + \tan^2 \alpha)$$

$$= 30 \tan \alpha - 1.7658 - 1.7658 \tan^2 \alpha$$

$$1.7658 \tan^2 \alpha - 30 \tan \alpha + 21.7658 = 0$$

or

$$\tan^2 \alpha - 17 \tan \alpha + 12.326 = 0$$

$$\therefore \quad \tan \alpha = \frac{17 \pm \sqrt{(17)^2 - 4 \times 1 \times 12.326}}{2} = \frac{17 \pm 15.48}{2} = 0.76$$

...(Taking –ve sign)

$$\alpha = \tan^{-1}(0.76) = 37.23° \quad \textbf{Ans.}$$

**Example 13.6** *The maximum horizontal range of a gun is R. Compute the firing angle to hit the enemy located at half distance of the maximum horizontal range.*

***Solution***

Let $\alpha$ = Firing angle to hit the enemy, and

$u$ = Velocity of projection.

We know that the horizontal range, $R = \dfrac{u^2 \sin 2\alpha}{g}$

and maximum horizontal range, $R_{max} = \dfrac{u^2}{g}$

Since the horizontal range ($R$) is equal to half the maximum horizontal range ($R_{max}$), therefore

$$\frac{u^2 \sin 2\alpha}{g} = \frac{u^2}{2g}$$

or $$\sin 2\alpha = \frac{1}{2} = 0.5$$

$\therefore$ $$2\alpha = \sin^{-1}(0.5) = 30° \text{ or } 150° \text{ or } \alpha = 15° \text{ or } 75° \quad \textbf{Ans.}$$

**Example 13.7** *A person throws a stone so as to clear a wall of height 4 m located at a distance of 5 m from the origin. The stone touches the ground at a distance of 3.5 m from the wall, away from the origin. Find the least initial velocity at which the stone to be thrown along with its direction.*

***Solution*** Given : $y = AB = 4$ m ; $x = OB = 5$ m ; $BC = 3.5$ m

Let $u$ = Least initial velocity, and

$\alpha$ = Angle of projection.

Consider that a stone is thrown from the origin $O$ to clear a wall $AB$ and it touches the ground at $C$, as shown in Fig. 13.9. We see from the figure that range of the stone,

$$R = OB + BC$$
$$= 5 + 3.5 = 8.5 \text{ m}$$

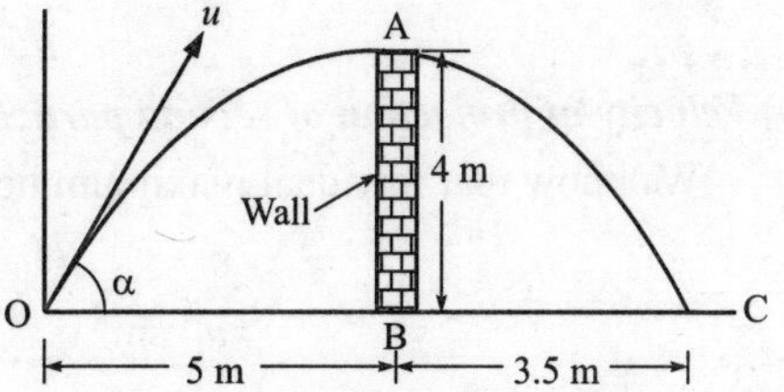


**Fig. 13.9**

We know that $$R = \frac{u^2 \sin 2\alpha}{g}$$

or $$8.5 = \frac{u^2 \sin 2\alpha}{9.81}$$

$\therefore$ $$u^2 \sin 2\alpha = 8.5 \times 9.81 = 83.4$$

or $$u^2 = \frac{83.4}{\sin 2\alpha} = \frac{83.4}{2 \sin \alpha \cos \alpha} \quad \text{...}(\because \sin 2\alpha = 2 \sin \alpha \cos \alpha)$$

Now using the equation of trajectory,

$$y = x \tan \alpha - \frac{g x^2}{2u^2 \cos^2 \alpha}$$

$$4 = 5 \tan \alpha - \frac{9.81 \times 5^2}{2\left(\dfrac{83.4}{2 \sin \alpha \cos \alpha}\right) \cos^2 \alpha} \quad \text{...[From equation } (i)]$$

$$= 5 \tan \alpha - \frac{245.25 \tan \alpha}{83.4}$$

$$= 5 \tan \alpha - 2.94 \tan \alpha = 2.06 \tan \alpha$$

$\therefore$ $$\tan \alpha = \frac{4}{2.06} = 1.942$$

or $$\alpha = \tan^{-1}(1.942) = 62.75° \quad \textbf{Ans.}$$

and $$u^2 = \frac{83.4}{\sin 2\alpha} = \frac{83.4}{\sin(2 \times 62.75°)} = \frac{83.4}{0.814} = 102.5$$

$\therefore$ $$u = \sqrt{102.5} = 10.12 \text{ m/s} \quad \textbf{Ans.}$$

**Example 13.8** *A particle is projected with a velocity of 5 m/s at an elevation of 60° to the horizontal. Find the velocity of projection of another particle thrown at an angle of 45° which will have equal (a) horizontal range ; (b) maximum height ; and (c) time of flight.*

***Solution*** Given : $u_1 = 5$ m/s ; $\alpha_1 = 60°$ ; $\alpha_2 = 45°$

Let $u_2$ = Velocity of projection of the second particle.

***(a) Velocity of projection of second particle for equal horizontal range***

We know that for equal horizontal range ($R$),

$$R_1 = R_2$$

or

$$\frac{u_1^2 \sin 2\alpha_1}{g} = \frac{u_2^2 \sin 2\alpha_2}{g}$$

$$u_2^2 = \frac{u_1^2 \sin 2\alpha_1}{\sin 2\alpha_2} = \frac{5^2 \sin(2 \times 60°)}{\sin(2 \times 45°)}$$

$$= \frac{25 \sin 120°}{\sin 90°} = \frac{25 \times 0.866}{1} = 21.65$$

$\therefore$ $$u_2 = \sqrt{21.65} = 4.653 \text{ m/s} \quad \textbf{Ans.}$$

***(b) Velocity of projection of second particle for equal maximum height***

We know that for equal maximum height ($H$),

$$H_1 = H_2$$

or

$$\frac{u_1^2 \sin^2 \alpha_1}{2g} = \frac{u_2^2 \sin^2 \alpha_2}{2g}$$

$$u_2^2 = \frac{u_1^2 \sin^2 \alpha_1}{\sin^2 \alpha_2} = \frac{5^2 \sin^2 60°}{\sin^2 45°}$$

$$= \frac{25(0.866)^2}{(0.707)^2} = \frac{25 \times 0.75}{0.5} = 37.5$$

$\therefore$ $$u_2 = \sqrt{37.5} = 6.124 \text{ m/s}$$

***(c) Velocity of projection of second particle for equal time of flight***

We know that for equal time of flight ($t$),

$$t_1 = t_2$$

or

$$\frac{2u_1 \sin \alpha_1}{g} = \frac{2u_2 \sin \alpha_2}{g}$$

$\therefore$ $$u_2 = \frac{u_1 \sin \alpha_1}{\sin \alpha_2} = \frac{5 \times \sin 60°}{\sin 45°} = \frac{5 \times 0.866}{0.707} = 6.124 \text{ m/s} \quad \textbf{Ans.}$$

**Example 13.9** *A ball is projected from the base of a hill whose shape is that of a right circular cone. The ball just touches the top of the hill and falls back to the ground on the other side of the hill. If the angle of projection of the ball is 30° with the horizontal and the vertical height of the hill is 40 m, find the semi-cone angle of the hill and the initial velocity with which the ball is projected.*

***Solution*** Given : $\alpha = 30°$ ; $H = 40$ m

Let $\theta$ = Semi-cone angle of the hill, and

$u$ = Initial velocity of projection.

In Fig. 13.10, $OAB$ is a right circular cone. The ball is projected from $O$ and falls at point $B$.

We know that maximum height through which the ball rises is given by

$$H = \frac{u^2 \sin^2 \alpha}{2g}$$

$$40 = \frac{u^2 \sin^2 30°}{2 \times 9.81} = \frac{u^2 (0.5)^2}{2 \times 9.81} = \frac{0.25u^2}{2 \times 9.81}$$

or $$u^2 = \frac{2 \times 40 \times 9.81}{0.25} = 3139.2$$

$\therefore$ $$u = \sqrt{3139.2} = 56.03 \text{ m/s } \textbf{Ans.}$$

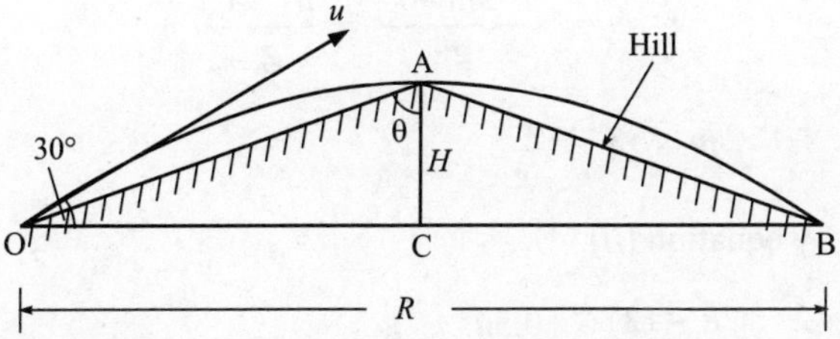


**Fig. 13.10**

We know that horizontal range, $R = \frac{u^2 \sin 2\alpha}{g} = \frac{(56.03)^2 \sin (2 \times 30°)}{9.81} = 277.12$ m

Since $OC = CB = \frac{R}{2} = \frac{277.12}{2} = 138.56$ m, therefore from triangle $OAC$,

$$\tan \theta = \frac{OC}{CA} = \frac{138.56}{40} = 3.464$$

$\therefore$ $$\theta = \tan^{-1}(3.464) = 73.9° \textbf{ Ans.}$$

**Example 13.10** *A projectile is aimed at a mark on the horizontal plane through the point of projection and falls 12 m short when the angle of projection is 15°, while it overshoots the mark by 24 m when the same angle is 45°. Find the angle of projection to hit the mark. Assume no air resistance.*

***Solution*** Given : When $\alpha_1 = 15°$, $R_1 = (R - 12)$ m ; when $\alpha_2 = 45°$, $R_2 = (R + 24)$ m, where $R$ is the horizontal range in metres.

Refer Fig. 13.11.

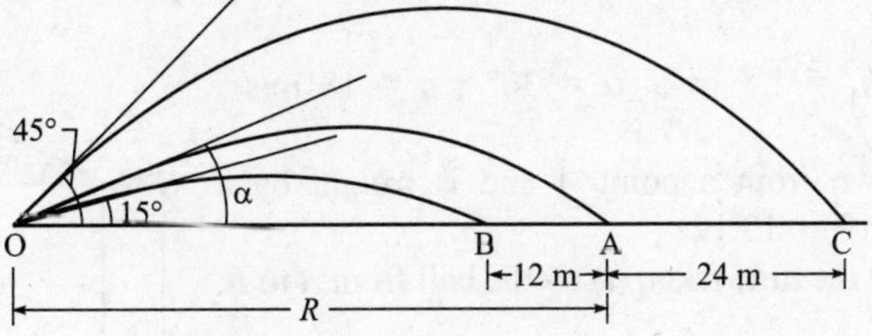


**Fig. 13.11**

We know that when the angle of projection is $\alpha_1 = 15°$, then horizontal range,

$$R_1 = OB = \frac{u^2 \sin 2\alpha_1}{g}$$

or $$R - 12 = \frac{u^2 \sin(2 \times 15°)}{g} = \frac{u^2 \sin 30°}{g} = \frac{0.5u^2}{g} \quad ...(i)$$

When the angle of projection is $\alpha_2 = 45°$, then horizontal range,

$$R_2 = OC = \frac{u^2 \sin 2\alpha_2}{g} = \frac{u^2 \sin(2 \times 45°)}{g}$$

$$= \frac{u^2 \sin 90°}{g} = \frac{u^2 \times 1}{g}$$

or $$R + 24 = \frac{u^2}{g} \quad ...(ii)$$

Dividing equation (*i*) by equation (*ii*),

$$\frac{R-12}{R+24} = \frac{0.5u^2}{g} \times \frac{g}{u^2} = 0.5$$

or $$R - 12 = 0.5(R + 24) = 0.5R + 12$$

$$R - 0.5R = 12 + 12 = 24 \text{ or } R = \frac{24}{0.5} = 48 \text{ m}$$

From equation (*ii*), $$u^2 = g(R + 24) = 9.81\ (48 + 24) = 706.32$$

$\therefore$ $$u = \sqrt{706.32} = 26.6 \text{ m/s}$$

Let $\alpha$ is the angle of projection to hit the mark at *A*. We know that

$$R = \frac{u^2 \sin 2\alpha}{g}$$

or $$\sin 2\alpha = \frac{R.g}{u^2} = \frac{48 \times 9.81}{(26.6)^2} = 0.667$$

$\therefore$ $$2\alpha = \sin^{-1}(0.667) = 41.8°$$

or $$\alpha = 20.9° \textbf{ Ans.}$$

**Example 13.11** *A cricket ball thrown from a height of 1.8 m at an angle of 30° from the horizontal with a velocity of 18 m/s is caught by a fieldsman at a height of 600 mm from the ground. How far apart were the two men?*

***Solution*** Given : $h_1 = 1.8$ m ; $\alpha = 30°$ ; $u = 18$ m/s ; $h_2 = 600$ mm $= 0.6$ m

Let the ball is thrown from a point *A* and is caught by a fieldsman at *C*, as shown in Fig. 13.12.

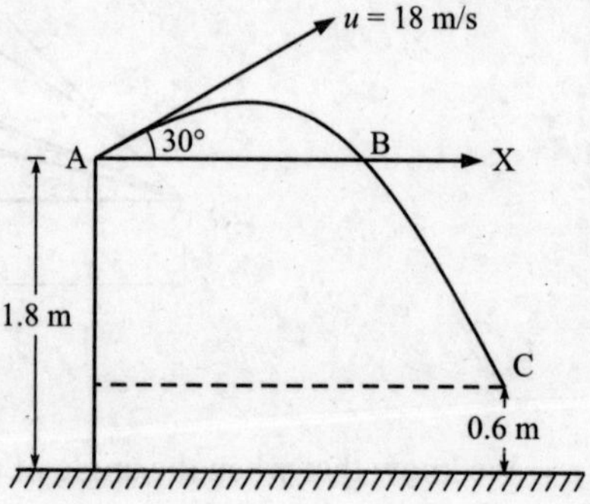


Fig. 13.12

First of all, let us find the time taken ($t$) by the ball from *A* to *B*. We know that

$$t_1 = \frac{2u \sin \alpha}{g}$$

$$= \frac{2 \times 18 \sin 30°}{9.81}$$

$$= 1.835 \text{ s}$$

Now considering downward motion of the ball from $B$ to $C$. Let $t_2$ is the time taken during this period. Using the relation,

$$h = (u \sin \alpha)t_2 + \frac{1}{2} g t_2^2$$

$$(1.8 - 0.6) = (18 \sin 30°)t_2 + \frac{1}{2} \times 9.81 t_2^2 \quad ...(\because h = h_1 - h_2)$$

$$1.2 = 9t_2 + 4.905t_2^2$$

or $$4.905t_2^2 + 9t_2 - 1.2 = 0$$

$$\therefore \quad t_2 = \frac{-9 \pm \sqrt{9^2 + 4 \times 4.905 \times 1.2}}{2 \times 4.905} = \frac{-9 \pm \sqrt{81 + 23.54}}{9.81}$$

$$= \frac{-9 \pm 10.22}{9.81} = 0.125 \text{ s} \quad ...(\text{Taking +ve sign})$$

Thus, total time taken by the ball from $A$ to $C$

$$^*t = t_1 + t_2 = 1.835 + 0.125 = 1.96 \text{ s}$$

Let $R$ = Horizontal distance between the two men, *i.e.* between the points $A$ and $C$

We know that $R = u \cos \alpha \times t = 18 \cos 30° \times 1.96 = 18 \times 0.866 \times 1.96$

$= 15.588 \times 1.96 = 30.55$ m **Ans.**

**Example 13.12** *From the top of a tower 60 m high, a bullet is fired at angle of 60° with the horizontal, with an initial velocity of 120 m/s, as shown in Fig. 13.13. Determine the maximum height above the ground that would be attained by the bullet and the time of travel up to this point.*

*Also determine the velocity of bullet 12 seconds after it is fired. Neglect air resistance.*

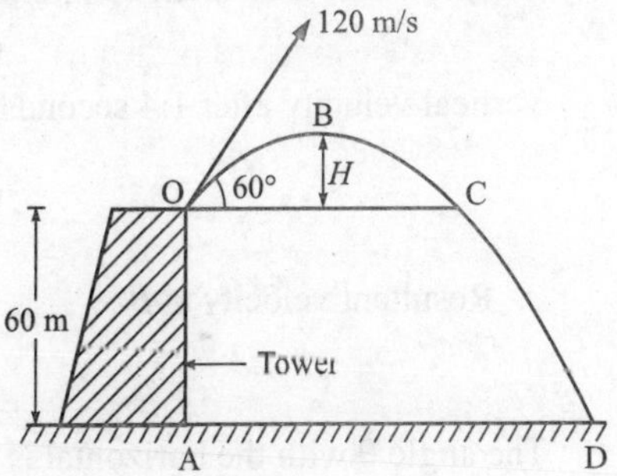


Fig. 13.13

***Solution*** Given : $h = OA = 60$ m ; $\alpha = 60°$ ; $u = 120$ m/s

***Maximum height above the ground***

We know that maximum height of the bullet above the horizontal plane $OC$,

$$H = \frac{u^2 \sin^2 \alpha}{2g}$$

* The total time taken by the ball from $A$ to $C$ may also be obtained as discussed below:
We know that the vertical distance travelled by the ball in time $t$ seconds is given by

$$h = (u \sin \alpha)t - \frac{1}{2}gt^2 \quad ... \left(\text{Using the relation } s = ut - \frac{1}{2}gt^2\right)$$

$$-(1.8 - 0.6) = (18 \sin 30°)t - \frac{1}{2} \times 9.81t^2$$

...($h$ is taken negative because it is below $AX$)

$$-1.2 = (18 \times 0.5)\, t - 4.905t^2 = 9t - 4.905t^2$$

or $$4.905t^2 - 9t - 1.2 = 0$$

$$\therefore \quad t = \frac{9 \pm \sqrt{9^2 + 4 \times 4.905 \times 1.2}}{2 \times 4.905} = \frac{9 \pm \sqrt{81 + 23.544}}{9.81}$$

$$= \frac{9 \pm 10.22}{9.81} = \frac{9 + 10.22}{9.81} = 1.96 \text{ s} \quad ...(\text{Taking +ve sign})$$

...(same as calculated above)

$$= \frac{(120)^2 \sin^2 60°}{9.81} = \frac{14\,400\,(0.866)^2}{2 \times 9.81} = 551 \text{ m}$$

∴ Maximum height above the ground $AD$

$$= H + h = 551 + 60 = 611 \text{ m} \quad \textbf{Ans.}$$

***Time of travel***

We know that time of travel up to the highest point $B$,

$$t = \frac{u \sin \alpha}{g} = \frac{120 \sin 60°}{9.81} = \frac{120 \times 0.866}{9.81} = 10.6 \text{ s} \quad \textbf{Ans.}$$

***Velocity of bullet after 12 seconds***

Let $B'$ is the point of the bullet after 12 seconds from point $O$, as shown in Fig. 13.14. Since the time taken by the bullet up to point $B$ is 10.6 s, therefore time taken by the bullet from $B$ to $B'$

$$= 12 - 10.6 = 1.4 \text{ s}$$

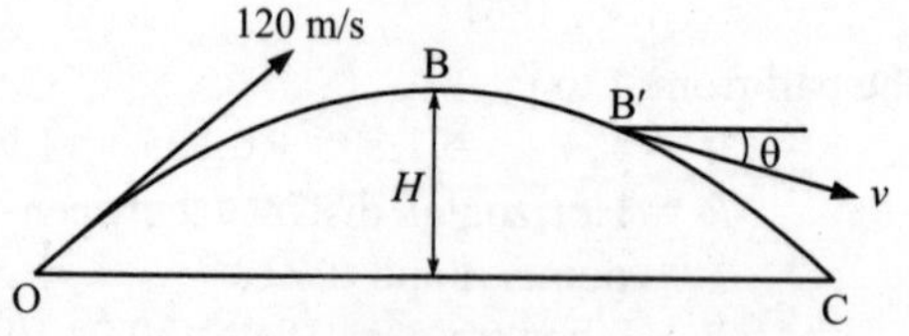


**Fig. 13.14**

We know that horizontal component of velocity at $B$,

$$v_H = u \cos \alpha = 120 \cos 60° = 60 \text{ m/s}$$

Vertical velocity after 1.4 seconds from $B$ (*i.e.* at $B'$),

$$v_V = 0 + \frac{1}{2} g t^2 = \frac{1}{2} \times 9.81\,(1.4)^2 = 9.6 \text{ m/s}$$

∴ Resultant velocity at $B'$,

$$v = \sqrt{(v_H)^2 + (v_V)^2} = \sqrt{(60)^2 + (9.6)^2}$$

$$= \sqrt{3600 + 92.16} = \sqrt{3692.16} = 60.76 \text{ m/s} \quad \textbf{Ans.}$$

The angle $\theta$ with the horizontal is given by

$$\tan \theta = \frac{v_V}{v_H} = \frac{9.6}{60} = 0.16$$

$$\theta = \tan^{-1}(0.16) = 9.1° \quad \textbf{Ans.}$$

**Example 13.13** *A bullet is fired from a height of 120 m at a velocity of 360 km/h at an angle of 30° upward. Neglecting air resistance, find :*

*1. Total time of flight ; 2. Horizontal range of the bullet ; 3. Maximum height reached by the bullet ; and 4. Final velocity of the bullet just before touching the ground.*

***Solution*** Given : $h = OA = 120$ m ; $u = 360$ km/h $= \frac{360 \times 1000}{60 \times 60} = 100$ m/s ; $\alpha = 30°$

**1. *Total time of flight* (*Refer Fig. 13.15*)**

Let $t$ = Total time of flight in seconds

= Time taken by the bullet from $O$ to $C$ ($t_1$) + Time taken by the bullet from $C$ to $D$ ($t_2$).

We know that time taken by the bullet from $O$ to $C$,

$$t_1 = \frac{2u \sin \alpha}{g} = \frac{2 \times 100 \sin 30°}{9.81} = \frac{100}{9.81} = 10.2 \text{ s}$$

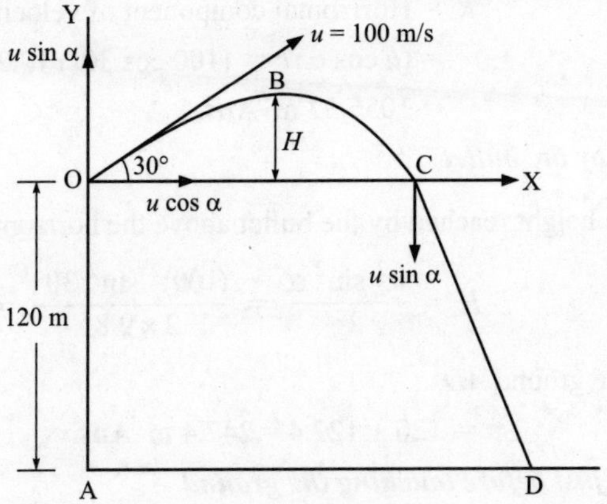


**Fig. 13.15**

Now consider the downward motion of the bullet from $C$ to $D$. The time taken during this period is $t_2$. Using the relation,

$$h = {}^*(u \sin \alpha)\, t_2 + \frac{1}{2} g t_2^2$$

$$120 = (100 \sin 30°)\, t_2 + \frac{1}{2} \times 9.81 t_2^2 = 50 t_2 + 4.905 t_2^2$$

or $$4.905 t_2^2 + 50 t_2 - 120 = 0$$

$\therefore$ $$t_2 = \frac{-50 \pm \sqrt{(50)^2 + 4 \times 4.905 \times 120}}{2 \times 4.905}$$

$$= \frac{-50 \pm \sqrt{2500 + 2354.4}}{9.81} = \frac{-50 \pm 69.7}{9.81} = \frac{19.7}{9.81} = 2 \text{ s}$$

...(Taking +ve sign)

$\therefore$ **Total time of flight, $t = t_1 + t_2 = 10.2 + 2 = 12.2$ s **Ans.**

---

* The initial velocity will be ($u \sin \alpha$).

** The total time of flight may also be obtained as follows:

Considering vertical motion of the bullet, the vertical distance of the bullet from $O$ at any time $t$ is given by

$$h = (u \sin \alpha) t - \frac{1}{2} g t^2$$

$$-120 = (100 \sin 30°) t - \frac{1}{2} \times 9.81 t^2$$

...($h$ is taken – ve because it is below horizontal plane $OX$)

$$-120 = (100 \times 0.5) t - 4.905 t^2 = 50t - 4.905 t^2$$

or $$4.905 t^2 - 50t - 120 = 0$$

$\therefore$ $$t = \frac{50 \pm \sqrt{(50)^2 + 4 \times 4.905 \times 120}}{2 \times 4.905}$$

$$= \frac{50 \pm \sqrt{2500 + 2354.4}}{9.81} = \frac{50 \pm 69.7}{9.81} = 12.2 \text{ s}$$ ...(Taking +ve sign)

***2. Horizontal range of the bullet***

We know that horizontal range of the bullet (Distance *AD*),

$$*R = \text{Horizontal component of velocity} \times \text{Time of flight}$$

$$= (u \cos \alpha) t = (100 \cos 30°)\, 12.2 = 100 \times 0.866 \times 12.2$$

$$= 1056.52 \text{ m}$$ **Ans.**

***3. Maximum height reached by the bullet***

We know that maximum height reached by the bullet above the horizontal plane *OX*,

$$H = \frac{u^2 \sin^2 \alpha}{2g} = \frac{(100)^2 \sin^2 30°}{2 \times 9.81} = \frac{(100)^2 (0.5)^2}{19.62} = 127.4 \text{ m}$$

and maximum height above the ground *AD*

$$= 120 + 127.4 = 247.4 \text{ m}$$ **Ans.**

***4. Final velocity of the bullet just before touching the ground***

We know that horizontal component of the velocity,

$$v_H = u \cos \alpha = 100 \cos 30° = 100 \times 0.866 = 86.6 \text{ m/s}$$

and vertical component of the velocity, $v_V = u \sin \alpha - gt$

$$= 100 \sin 30° - 9.81 \times 12.2$$

$$= -69.682 \text{ m/s} = 69.682 \text{ m/s downward}$$

∴ Final velocity of the bullet just before touching the ground at *D* (Refer Fig. 13.16),

$$v = \sqrt{(v_H)^2 + (v_V)^2}$$

$$= \sqrt{(86.6)^2 + (69.682)^2}$$

$$= \sqrt{7500 + 4856}$$

$$= 111.15 \text{ m/s}$$ **Ans.**

and $$\tan \theta = \frac{v_V}{v_H} = \frac{69.682}{86.6} = 0.8046$$

∴ $$\theta = \tan^{-1}(0.8046)$$

$$= 38.82°$$ **Ans.**

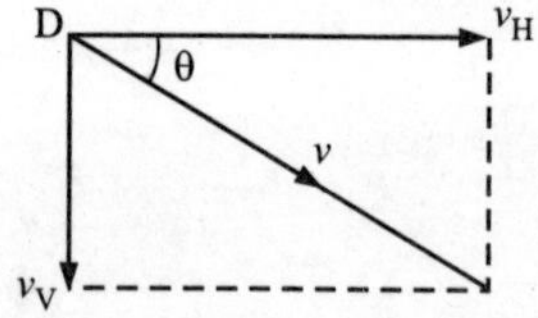


**Fig. 13.16**

**Example 13.14** *A missile projected from a rocket 20 km vertically and 10 km horizontally, from the launching point, when it was found that fuel of the rocket was exhausted. The missile had then acquired a velocity of 1600 m/s at an angle of 35° with the horizontal. Assuming that the rest of the flight is under the influence of gravity and neglecting air resistance and curvature of the earth, calculate :*

1. *The total horizontal range from the launching point ; and*
2. *The time of the flight after the fuel has completely burnt.*

***Solution*** Given : $h = 20$ km $= 20\,000$ m ; $s = 10$ km ; $u = 1600$ m/s ; $\alpha = 35°$

---

* The relation $R = \dfrac{u^2 \sin 2\alpha}{g}$ gives the distance *OC* only.

Let $O$ be the launching point of missile and $A$ is the point at which the missile was released. The missile moves up to point $B$ and then starts moving downwards and hits the earth at $D$ (Refer Fig. 13.17).

Considering the vertical motion, the vertical distance of the missile from $A$ at any time $t$ is given by

$$h = (u \sin \alpha)\, t - \frac{1}{2} g t^2$$

$$-20\,000 = (1600 \sin 35°)t - \frac{1}{2} \times 9.81 t^2$$

...($h$ is taken –ve as it is below horizontal axis $AC$)

$$= (1600 \times 0.5736)t - 4.905 t^2 = 917.76 t - 4.905 t^2$$

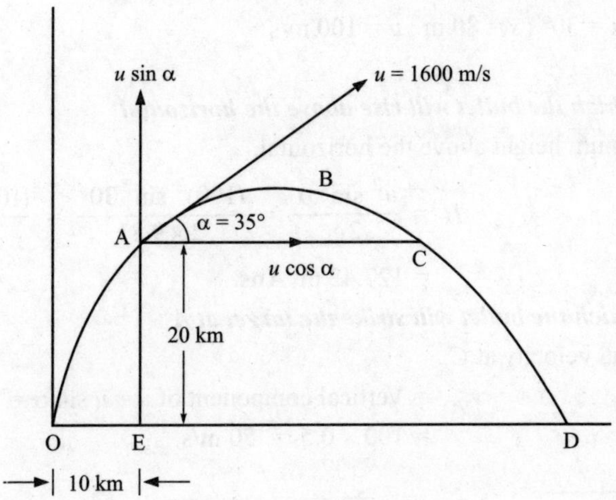


Fig. 13.17

Dividing throughout by 4.905 and rearranging,

$$t^2 - 187.1 t - 4077.5 = 0$$

$$\therefore \quad t = \frac{187.1 \pm \sqrt{(187.1)^2 + 4 \times 1 \times 4077.5}}{2}$$

$$= \frac{187.1 \pm \sqrt{35\,006.4 + 16\,310}}{2} = \frac{187.1 \pm 226.5}{2} = 206.8 \text{ s}$$

...(Taking +ve sign)

**1. *Total horizontal range from the launching point***

We know that the horizontal distance from point $E$ to $D$ (*i.e.* Distance $ED$)

$$= \text{Horizontal component of velocity} \times \text{Time} = (u \cos \alpha)\, t$$

$$= (1600 \cos 35°)\, 206.8 = 1600 \times 0.82 \times 206.8$$

$$= 271\,322 \text{ m} = 271.322 \text{ km}$$

$\therefore$ Total horizontal distance from the launching point $O$

$$= OE + ED = 10 + 271.322 = 281.322 \text{ km}$$ **Ans.**

**2. *Time of flight after the fuel has completely burnt***

We have already calculated above that the time of flight,

$$^{*}t = 206.8 \text{ s} \quad \textbf{Ans.}$$

**Example 13.15** *A bullet is fired upwards at an angle of 30° to the horizontal from a point P on a hill and it strikes the target which is 80 m lower than P. The initial velocity of the bullet is 100 m/s. Calculate:*

1. *The maximum height to which the bullet will rise above the horizontal.*
2. *The actual velocity with which it will strike the target.*
3. *The total time required for the flight of the bullet.*
4. *The horizontal distance between the hill position and the target.*

***Solution*** Given : $\alpha = 30°$ ; $s = 80$ m ; $u = 100$ m/s

Refer Fig. 13.18.

**1. *Maximum height to which the bullet will rise above the horizontal***

We know that maximum height above the horizontal,

$$H = \frac{u^2 \sin^2 \alpha}{2g} = \frac{(100)^2 \sin^2 30°}{2 \times 9.81} = \frac{(100)^2\ (0.5)^2}{19.62}$$

$$= 127.42 \text{ m} \quad \textbf{Ans.}$$

**2. *Actual velocity with which the bullet will strike the target at A***

We know that vertical velocity at $C$,

$$v_{VC} = \text{Vertical component of } u = u \sin \alpha = 100 \sin 30°$$

$$= 100 \times 0.5 = 50 \text{ m/s}$$

---

* The time of flight ($t$) may also be obtained as discussed below (See Example 13.11):

We know that time taken by the missile from $A$ to $C$,

$$t_1 = \frac{2u \sin \alpha}{g} = \frac{2 \times 1600 \sin 35°}{9.81} = \frac{2 \times 1600 \times 0.5736}{9.81}$$

$$= \frac{1835.5}{9.81} = 187.1 \text{ s}$$

Now considering the downward motion of the missile from $C$ to $D$. The time taken ($t_2$) during this period is given by the relation,

$$h = (u \sin \alpha) t_2 + \frac{1}{2} g t_2^2$$

$$20\,000 = (1600 \sin 35°) t_2 + \frac{1}{2} \times 9.81\, t_2^2$$

$$= (1600 \times 0.5736) t_2 + 4.905\, t_2^2 = 917.76\, t_2 + 4.905\, t_2^2$$

Dividing throughout by 4.905 and rearranging, we have

$$t_2^2 + 187.1 t_2 - 4077.5 = 0$$

$$t_2 = \frac{-187.1 \pm \sqrt{(187.1)^2 + 4 \times 1 \times 4077.5}}{2}$$

$$= \frac{-187.1 \pm \sqrt{35\,006.4 + 16\,310}}{2}$$

$$= \frac{-187.1 \pm 226.5}{2} = 19.7 \text{ s} \quad \text{...(Taking +ve sign)}$$

$$\therefore \quad \text{Total time, } t = t_1 + t_2 = 187.1 + 19.7 = 206.8 \text{ s}$$

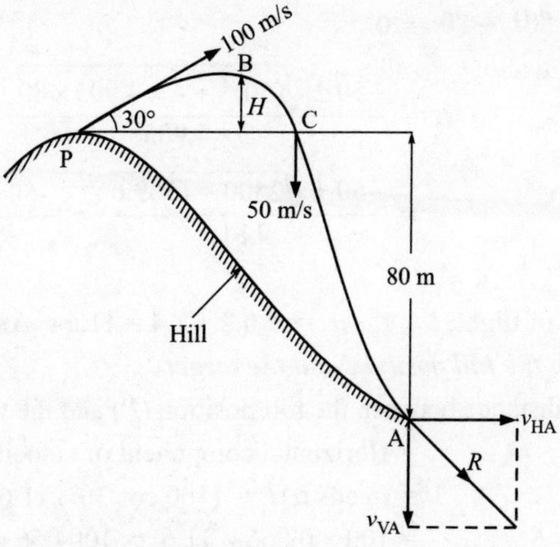


**Fig. 13.18**

The vertical velocity at $A$ ($v_{VA}$) may be determined by considering the downward motion from $C$ upto $A$.

Using the relation, $v_{VA}^2 - v_{VC}^2 = 2gs$

or $v_{VA}^2 = v_{VC}^2 + 2gs$

$$= (50)^2 + 2 \times 9.81 \times 80 = 2500 + 1569.6 = 4069.6$$

$\therefore$ $v_{VA} = \sqrt{4069.6} = 63.8$ m/s

Horizontal component of velocity,

$$v_{HA} = u \cos \alpha = 100 \cos 30° = 100 \times 0.866 = 86.6 \text{ m/s}$$

$\therefore$ Resultant velocity at $A$, $R = \sqrt{v_{VA}^2 + v_{HA}^2} = \sqrt{(63.8)^2 + (86.6)^2}$

$$= \sqrt{4069.6 + 7500} = 107.6 \text{ m/s} \quad \textbf{Ans.}$$

This is the actual velocity with which the bullet will hit the target at $A$.

**3. *Total time required for the flight of the bullet***

Let $t$ = Total time of flight in seconds

= Time taken by the bullet from $P$ to $C$ ($t_1$) + Time taken by the bullet from $C$ to $A$ ($t_2$).

We know that time taken by the bullet from $P$ to $C$,

$$t_1 = \frac{2u \sin \alpha}{g} = \frac{2 \times 100 \sin 30°}{9.81} = \frac{100}{9.81} = 10.2 \text{ s}$$

Now considering the downward motion of the bullet from $C$ to $A$. The time taken during this period ($t_2$) is obtained by using the relation,

$$s = (u \sin \alpha) t_2 + \frac{1}{2} g t_2^2$$

$$80 = (100 \sin 30°) t_2 + \frac{1}{2} \times 9.81 t_2^2 = 50 t_2 + 4.905 t_2^2$$

or $$4.905 t_2^2 + 50 t_2 - 80 = 0$$

$\therefore$ $$t_2 = \frac{-50 \pm \sqrt{(50)^2 + 4 \times 4.905 \times 80}}{2 \times 4.905}$$

$$= \frac{-50 \pm \sqrt{2500 + 1569.6}}{9.81} = \frac{-50 \pm 63.8}{9.81} = \frac{13.8}{9.81} = 1.4 \text{ s}$$

...(Taking +ve sign)

$\therefore$ Total time of flight, $t = t_1 + t_2 = 10.2 + 1.4 = 11.6$ s **Ans.**

**4. *Horizontal distance between the hill position and the target***

We know that horizontal distance between the hill position ($P$) and the target ($A$), *i.e.* $PA$

$$= \text{Horizontal component of velocity} \times \text{Time of flight}$$

$$= (u \cos \alpha) t = (100 \cos 30°)\, 11.6$$

$$= 100 \times 0.866 \times 11.6 = 1004.56 \text{ m}$$ **Ans.**

**Example 13.16** *Two guns are pointed at each other, one upward at an angle of 30° and the other at the same angle of depression, the muzzle being 30 m apart. If the guns are shot with velocities of 350 m/s upwards and 280 m/s downward, respectively, find when and where they will meet?*

***Solution*** Given: $\alpha_1 = \alpha_2 = 30°$ ; $AB = 30$ m ; $u_1 = 350$ m/s ; $u_2 = 280$ m/s

Let the two guns are shot from point $A$ and $B$ pointing each other and meet at point $C$ after time $t$ seconds, as shown in Fig. 13.19. From triangle $ABD$, horizontal distance,

$$AD = AB \cos \alpha_1 = 30 \cos 30° = 30 \times 0.866 = 25.98 \text{ m}$$

$\therefore$ $$ED = AD - AE = 25.98 - x$$

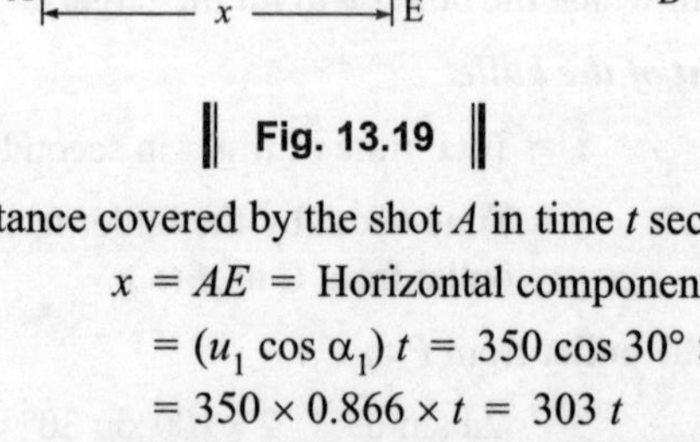


**Fig. 13.19**

We know that horizontal distance covered by the shot $A$ in time $t$ seconds,

$$x = AE = \text{Horizontal component of velocity at } A \times \text{time}$$

$$= (u_1 \cos \alpha_1) t = 350 \cos 30° \times t$$

$$= 350 \times 0.866 \times t = 303\, t \quad \text{...(i)}$$

Similarly, horizontal distance covered by the shot $B$ in time $t$ seconds,

$$ED = \text{Horizontal component of velocity at } B \times \text{time} = (u_2 \cos \alpha_2) t$$

or $$25.98 - x = 280 \cos 30° \times t = 280 \times 0.866 \times t = 242.48 t \quad \text{...(ii)}$$

Adding equations (*i*) and (*ii*),

$$25.98 = 303 t + 242.48 t = 545.48 t$$

$\therefore$ $$t = \frac{25.98}{545.48} = 0.0476 \text{ s}$$ **Ans.**

Thus the two shots will meet after time 0.0476 seconds. Now, substituting the value of $t$ in equation (*i*), we have

$$x = AE = 303t = 303 \times 0.0476 = 14.42 \text{ m} \quad \textbf{Ans.}$$

We know that vertical component of velocity at $A$,

$$u_V = u_1 \sin \alpha_1 = 350 \sin 30° = 350 \times 0.5 = 175 \text{ m/s}$$

$\therefore$ Vertical distance between $C$ and $E$,

$$y = u_V \times t - \frac{1}{2} g t^2 = 175 \times 0.0476 - \frac{1}{2} \times 9.81 (0.0476)^2$$

$$= 8.33 - 0.011 = 8.319 \text{ m} \quad \textbf{Ans.}$$

Thus, the two shots will meet at a point $C$ whose co-ordinates are $x = 14.42$ m, and $y = 8.319$ m. **Ans.**

## 13.8 MOTION OF A PARTICLE AFTER A GIVEN INTERVAL OF TIME FROM THE POINT OF PROJECTION

Consider a particle is projected from a point $O$ at an angle of projection $\alpha$, with an initial velocity $u$. Let after time $t$ seconds, it reaches a point $P$ when its velocity is $v$ which makes an angle $\theta$ with the horizontal at $P$, as shown in Fig. 13.20.

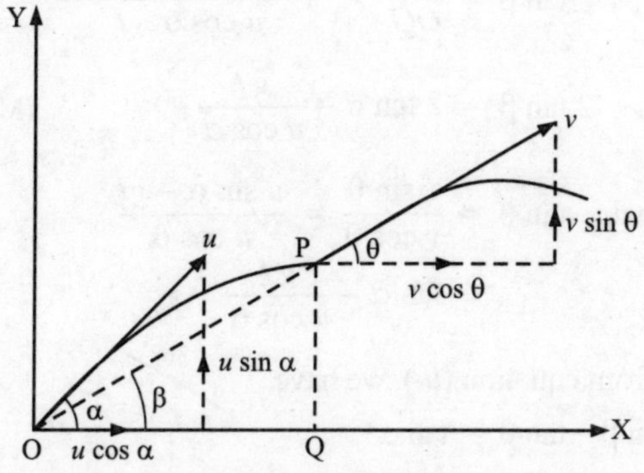


**Fig. 13.20**

From the geometry of the figure, we see that the horizontal component of final velocity $v$

$$= v \cos \theta$$

and horizontal component of initial velocity $u$

$$= u \cos \alpha$$

We have already discussed that the horizontal component of the velocity remains constant, therefore

$$v \cos \theta = u \cos \alpha \quad \text{or} \quad v = \frac{u \cos \alpha}{\cos \theta} \qquad ...(i)$$

We know that vertical component of final velocity $v$

$$= v \sin \theta$$

and vertical component of initial velocity $u$ $= u \sin \alpha$

This change in velocity from $u \sin \alpha$ to $v \sin \theta$ is due to retardation ($g$) due to gravity. Now from the general equation for upward motion ($v = u - gt$), we have

$$v \sin \theta = u \sin \alpha - gt \qquad ...(ii)$$

or

$$gt = u \sin \alpha - v \sin \theta$$

Substituting the value of $v$ from equation ($i$), we have

$$gt = u \sin \alpha - \frac{u \cos \alpha}{\cos \theta} \times \sin \theta = u \sin \alpha - u \cos \alpha \tan \theta$$

∴ Time required to reach point $P$ from point $O$,

$$t = \frac{u \sin \alpha - u \cos \alpha \tan \theta}{g}$$

We know that vertical distance of $P$ from the horizontal axis $OX$,

$$PQ = u \sin \alpha \times t - \frac{1}{2} g t^2$$

...(Using the relation $s = ut - \frac{1}{2}gt^2$, for upward motion)

and horizontal distance, $OQ = u \cos \alpha \times t$

∴ Distance of $P$ from $O$ *i.e.* $OP = \sqrt{(PQ)^2 + (OQ)^2}$

Let $\beta$ be the angle made by $OP$ with the horizontal. From the geometry of the figure, we find that

$$\tan \beta = \frac{PQ}{OQ} = \frac{u \sin \alpha \times t - \frac{1}{2} g t^2}{u \cos \alpha \times t} = \tan \alpha - \frac{g t}{2u \cos \alpha}$$

or $$2 \tan \beta = 2 \tan \alpha - \frac{g t}{u \cos \alpha}$$ ...(Multiplying both sides by 2) ...($iii$)

From Fig. 13.20, we see that $$\tan \theta = \frac{v \sin \theta}{v \cos \theta} = \frac{u \sin \alpha - gt}{u \cos \alpha}$$ ...[From equations ($i$) and ($ii$)]

$$= \tan \alpha - \frac{g t}{u \cos \alpha}$$ ...($iv$)

Subtracting equation ($iv$) from equation ($iii$), we have

$$2 \tan \beta - \tan \theta = \tan \alpha$$

or $$2 \tan \beta = \tan \alpha + \tan \theta$$

∴ $$\tan \beta = \frac{\tan \alpha + \tan \theta}{2}$$

**Example 13.17** *From a point A, a particle is projected upwards with a velocity of 100 m/s at an angle of 45° to the horizontal. When it reaches a certain point P, it is found to be moving at an angle of 30° to the horizontal. Find the time to reach P and the distance AP.*

***Solution*** Given : $u = 100$ m/s ; $\alpha = 45°$ ; $\theta = 30°$

Refer Fig. 13.21.

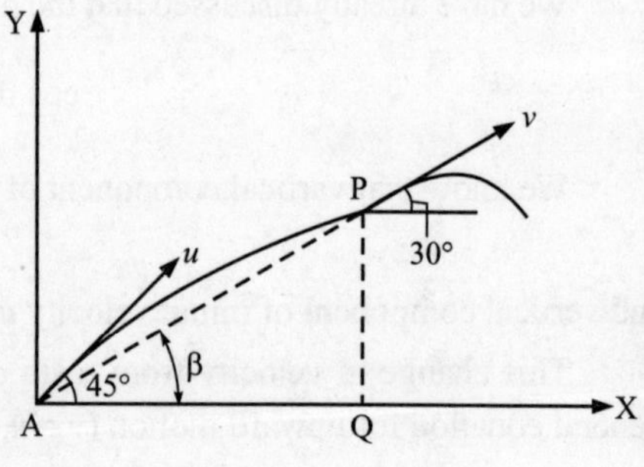


Fig. 13.21

***Time to reach P***

Time to reach point $P$ from point $A$,

$$t = \frac{u \sin \alpha - u \cos \alpha \tan \theta}{g}$$

$$= \frac{100 \sin 45° - 100 \cos 45° \tan 30°}{9.81}$$

$$= \frac{100 \times 0.707 - 100 \times 0.707 \times 0.577}{9.81}$$

$$= \frac{70.7 - 40.8}{9.81} = 3.05 \text{ s}$$ **Ans.**

***Distance AP***

We know that vertical distance $PQ = u \sin \alpha \times t - \frac{1}{2} g t^2$

$$= 100 \sin 45° \times 3.05 - \frac{1}{2} \times 9.81 (3.05)^2$$

$$= 215.6 - 45.6 = 170 \text{ m}$$

and horizontal distance, $AQ = u \cos \alpha \times t = 100 \cos 45° \times 3.05 = 215.6$ m

$\therefore$ Distance $AP = \sqrt{(PQ)^2 + (AQ)^2} = \sqrt{(170)^2 + (215.6)^2}$

$$= \sqrt{28\,900 + 46\,483} = \sqrt{75\,383} = 274.56 \text{ m} \quad \textbf{Ans.}$$

**Notes 1.** Let $\beta$ is the angle which the distance $AP$ makes with the horizontal. It may be determined as follows:

We know that $\tan \beta = \frac{PQ}{AQ} = \frac{170}{215.6} = 0.7885$

$\therefore$ $\beta = \tan^{-1} (0.7885) = 38.26°$

**2.** It is given that $\tan \beta = \frac{\tan \alpha + \tan \theta}{2} = \frac{\tan 45° + \tan 30°}{2} = \frac{1 + 0.577}{2} = 0.7885$

$\therefore$ $\beta = \tan^{-1} (0.7886) = 38.26°$

## 13.9 PATH OF A PROJECTILE ON AN INCLINED PLANE

Consider a particle is projected from point $O$ with an initial velocity $u$ (known as velocity of projection) on an inclined plane $OB$ which is inclined at an angle $\beta$ with the horizontal $OX$, as shown in Fig. 13.22. Let $\alpha$ be the angle of projection and $t$ is the time of flight from $O$ to $P$. Draw perpendicular $PA$ to the horizontal line $OX$.

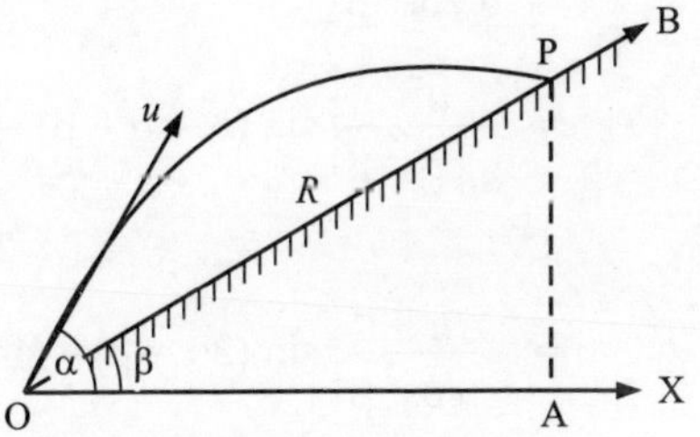


**Fig. 13.22**

From the geometry of the figure, horizontal distance travelled by the particle in time $t$ seconds,

$OA$ = Horizontal component of velocity × Time

$= u \cos \alpha \times t$ ...(*i*)

and the vertical distance travelled in time $t$ seconds,

$*PA = R \sin \beta = (u \sin \alpha) t - \frac{1}{2} g t^2$ ...(*ii*)

where $R = OP$ = Range of flight.

We know that $\tan \beta = \frac{PA}{OA} = \frac{(u \sin \alpha) t - \frac{1}{2} g t^2}{u \cos \alpha \times t} = \tan \alpha - \frac{g t^2}{2u \cos \alpha \times t}$

* It is same as $PQ$ in Fig. 13.20.

or $$\frac{gt^2}{2u \cos\alpha \times t} = \tan\alpha - \tan\beta = \frac{\sin\alpha}{\cos\alpha} - \frac{\sin\beta}{\cos\beta}$$

$$= \frac{\sin\alpha \cos\beta - \cos\alpha \sin\beta}{\cos\alpha \cos\beta} = \frac{\sin(\alpha - \beta)}{\cos\alpha \cos\beta}$$

...[$\because$ $\sin(A - B) = \sin A \cos B - \cos A \sin B$]

$\therefore$ $$\frac{gt}{2u} = \frac{\sin(\alpha - \beta)}{\cos\beta}$$

or $$t = \frac{2u \sin(\alpha - \beta)}{g \cos\beta} \quad ...(iii)$$

We know that horizontal distance travelled,

$$OA = u \cos\alpha \times t$$ ...[From equation (*i*)]

Substituting the value of *t* from equation (*iii*), we have

$$OA = u \cos\alpha \times \frac{2u \sin(\alpha - \beta)}{g \cos\beta}$$

and range of flight, $$R = OP = \frac{OA}{\cos\beta}$$

$$= u \cos\alpha \times \frac{2u \sin(\alpha - \beta)}{g \cos\beta} \times \frac{1}{\cos\beta}$$

$$= \frac{2u^2 \cos\alpha \sin(\alpha - \beta)}{g \cos^2\beta}$$

$$= \frac{u^2}{g \cos^2\beta}\left[2 \cos\alpha \sin(\alpha - \beta)\right]$$

$$= \frac{u^2}{g \cos^2\beta}\left[\sin(\alpha + \alpha - \beta) - \sin(\alpha - \alpha + \beta)\right]$$

...[$\because$ $2 \cos A \sin B = \sin(A + B) - \sin(A - B)$]

$$= \frac{u^2}{g \cos^2\beta}\left[\sin(2\alpha - \beta) - \sin\beta\right] \quad ...(iv)$$

The range ($R$) will be maximum when $\sin(2\alpha - \beta)$ is maximum. In other words,

$$\sin(2\alpha - \beta) = 1 \quad \text{or} \quad 2\alpha - \beta = 90°$$

$\therefore$ $$R_{max} = \frac{u^2}{g \cos^2\beta}\left[\sin 90° - \sin\beta\right]$$

$$= \frac{u^2}{g \cos^2\beta}\left[1 - \sin\beta\right]$$

$$= \frac{u^2(1 - \sin\beta)}{g(1 - \sin^2\beta)} = \frac{u^2(1 - \sin\beta)}{g(1 + \sin\beta)(1 - \sin\beta)}$$

$$= \frac{u^2}{g(1 + \sin\beta)}$$ ...($\because$ $\cos^2\theta = 1 - \sin^2\theta$)

It may be noted that when the particle is projected on a downward inclined plane, as shown in Fig. 13.23, then the angle β should be taken negative *i.e.* –β, in the expressions derived above. Thus

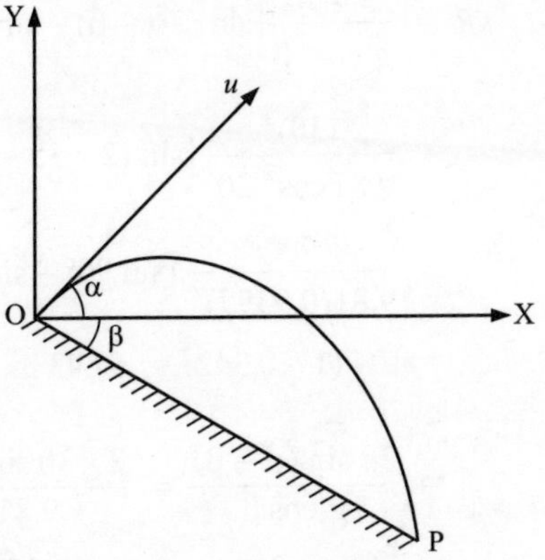


Fig. 13.23

1. Time of flight, $t = \dfrac{2u \sin [\alpha - (-\beta)]}{g \cos (-\beta)}$

$$= \frac{2u \sin (\alpha + \beta)}{g \cos \beta} \qquad ...[\because \cos(-\theta) = \cos\theta]$$

2. Range of flight, $R = \dfrac{u^2}{g \cos^2 (-\beta)}[\sin (2\alpha + \beta) - \sin (-\beta)]$

$$= \frac{u^2}{g \cos^2 \beta}[\sin (2\alpha + \beta) + \sin \beta] \qquad ...[\because \sin(-\theta) = -\sin\theta]$$

3. The condition for maximum range is

$$2\alpha - (-\beta) = 90°$$

or

$$2\alpha + \beta = 90°$$

and maximum range,

$$R_{max} = \frac{u^2}{g[1 + \sin(-\beta)]} = \frac{u^2}{g(1 - \sin\beta)}$$

**Example 13.18** *A particle is projected from a point on an inclined plane with a velocity of 30 m/s. The angle of projection and the angle of plane are 55° and 20° to the horizontal, respectively. Show that the range up the plane is the maximum one for the given plane. Find the range and the time of flight of the particle.*

***Solution*** Given : $u$ = 30 m/s ; α = 55° ; β = 20°

***Maximum range***

We know that for maximum range,

$$2\alpha - \beta = 90° \quad \text{or} \quad 2\alpha = 90° + \beta = 90° + 20° = 110°$$

$$\therefore \quad \alpha = \frac{110}{2} = 55°$$

Since the given angle of projection (α) is 55°, therefore the range up the plane is the maximum one for the given plane. **Ans.**

***Range***

We know that range of the projectile,

$$R = \frac{u^2}{g\cos^2\beta}[\sin(2\alpha - \beta) - \sin\beta]$$

$$= \frac{(30)^2}{9.81\cos^2 20°}[\sin(2 \times 55 - 20) - \sin 20°]$$

$$= \frac{900}{9.81(0.9397)^2}[\sin 90° - \sin 20°]$$

$$= 104\,(1 - 0.342) = 68.43 \text{ m} \quad \textbf{Ans.}$$

***Time of flight***

We know that time of flight, $t = \dfrac{2u\sin(\alpha - \beta)}{g\cos\beta} = \dfrac{2 \times 30\sin(55° - 20°)}{9.81\cos 20°}$

$$= \frac{60\sin 35°}{9.81\cos 20°} = \frac{60 \times 0.5736}{9.81 \times 0.9397} = 3.73 \text{ s} \quad \textbf{Ans.}$$

**Note:** In the present case, the range is maximum.

We know that maximum range, $R_{max} = \dfrac{u^2}{g(1 + \sin\beta)} = \dfrac{(30)^2}{9.81(1 + \sin 20°)} = \dfrac{900}{9.81(1 + 0.342)}$

$$= 68.36 \text{ m} \quad \textbf{Ans.}$$

**Example 13.19** *A ball is projected from A located on a slope with a velocity of 3 m/s at an angle of 25° with the horizontal, as shown in Fig. 13.24. Determine the co-ordinates of the point B at which it will hit the slope whose inclination is 25° to the horizontal, as shown.*

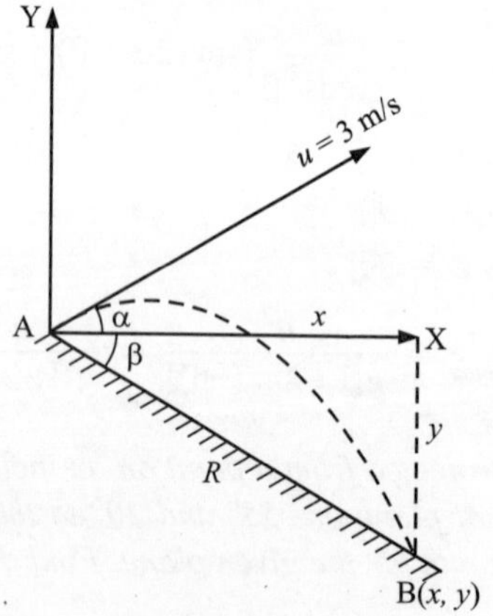


**Fig. 13.24**

***Solution*** Given : $u = 3$ m/s ; $\alpha = 25°$ ; $\beta = 25°$

Let $x$ and $y$ are the co-ordinates of the point $B$. The range of this slope $AB$ is given by

$$R = \frac{u^2}{g\cos^2\beta}[\sin(2\alpha + \beta) + \sin\beta]$$

$$= \frac{3^2}{9.81\cos^2 25°}[\sin(2 \times 25° + 25°) + \sin 25°]$$

$$= \frac{9}{9.81(0.9063)^2}[\sin 75° + \sin 25°]$$

$$= \frac{9}{9.81 \times 0.8214}[0.966 + 0.4226]$$

$$= 1.117 \times 1.3886 \; = \; 1.55 \text{ m}$$

From Fig. 13.24, we find that $\tan \beta = \dfrac{y}{x}$ or $\tan 25° = \dfrac{y}{x}$

$\therefore$ $$y = x \tan 25° = 0.4663x \qquad ...(i)$$

and $$R = \sqrt{x^2 + y^2}$$

or $$R^2 = x^2 + y^2$$

$$(1.55)^2 = x^2 + (0.4663x)^2 = x^2 + 0.2174x^2 = 1.2174x^2$$

$\therefore$ $$x^2 = \frac{(1.55)^2}{1.2174} = 1.973$$

or $$x = \sqrt{1.973} = 1.4 \text{ m} \quad \textbf{Ans.}$$

and $$y = 0.4663x = 0.4663 \times 1.4 = 0.653 \text{ m} \quad \textbf{Ans.}$$

## HIGHLIGHTS

1. A particle which moves under the combined effect of vertical and horizontal component of the velocity, is known as a *projectile.* The study of motion of such a projectile is called *projectile motion.*
2. The curved path followed by the projectile when projected in space is known as *trajectory* or *path of projectile.*
3. The initial velocity with which the projectile is projected in space is called *velocity of projection.*
4. The angle between the direction of projection and the horizontal direction is called *angle of projection.*
5. The horizontal distance travelled by the projectile from the point of projection up to the point where it strikes the ground, is known as *range* of the projectile.
6. The total time taken by the projectile for which it remains in space is known as *time of flight.*
7. The equation for the path of projectile or trajectory is given by

$$y = x \tan \alpha - \frac{g x^2}{2u^2 \cos^2 \alpha}$$

where $x$ and $y$ = Co-ordinates of a point on the path of a projectile with respect to the point from which the projectile is projected,

$\alpha$ = Angle of projection, and

$u$ = Velocity of projection.

8. The velocity of projection ($u$) has vertical component ($u \sin \alpha$) and horizontal component ($u \cos \alpha$). It may be noted that horizontal component of velocity ($u \cos \alpha$) remains constant during the flight of a projectile.
9. The maximum height ($H$) attained by the projectile from the horizontal plane is given by

$$H = \frac{u^2 \sin^2 \alpha}{2g}$$

**10.** The time of flight ($t$) of a projectile is given by

$$t = \frac{2u \sin \alpha}{g}$$

**11.** The time to reach the highest point (also called time of ascent, $t_1$) is equal to half the time of flight ($t$).

**12.** The horizontal range ($R$) of a projectile is given by

$$R = \frac{u^2 \sin 2\alpha}{g}$$

**13.** The horizontal range will be maximum when

$$\sin 2\alpha = 1 \quad \text{or} \quad 2\alpha = 90° \text{ or } \alpha = 45° \text{ or } \frac{\pi}{4}$$

∴ Maximum horizontal range,

$$R_{max} = \frac{u^2}{g}$$

**14.** When a particle is projected on an upward inclined plane, then

(*a*) Time of flight, $t = \dfrac{2u \sin(\alpha - \beta)}{g \cos \beta}$

(*b*) Range, $R = \dfrac{u}{g \cos^2 \beta}[\sin(2\alpha - \beta) - \sin \beta]$

(*c*) For maximum range, $2\alpha - \beta = 90°$

(*d*) Maximum range, $R_{max} = \dfrac{u^2}{g(1 + \sin \beta)}$

where $\alpha$ = Angle of projection with the horizontal,

$\beta$ = Angle of inclined plane with the horizontal, and

$u$ = Velocity of projection.

**15.** When the particle is projected on a downward inclined plane, then angle $\beta$ is negative. In such cases,

(*a*) Time of flight, $t = \dfrac{2u \sin(\alpha + \beta)}{g \cos \beta}$

(*b*) Range, $R = \dfrac{u^2}{g \cos^2 \beta}[\sin(2\alpha + \beta) + \sin \beta]$

(*c*) For maximum range,

$$2\alpha + \beta = 90°$$

(*d*) Maximum range, $R_{max} = \dfrac{u^2}{g(1 - \sin \beta)}$

## NUMERICAL PROBLEMS

**1.** An aircraft moving horizontally at a speed of 360 km/h at an altitude of 360 m towards a target on the ground releases a bomb which hits the target. Determine: 1. the time required for the bomb to reach the target on the ground; and 2. the horizontal distance of the aircraft from the target when it released the bomb. Take $g = 9.81$ m/s$^2$. **[Ans.** 8.56 s; 856 m**]**

**2.** A bomb is released from an aeroplane flying at a speed of 1000 km/h on a straight level course 2000 m above the ground. Find the time required for the bomb to reach the ground and the horizontal distance travelled by the bomb after its release. Assume $g = 9.8$ m/s$^2$ and ground to be horizontal. Neglect air resistance.

**[Ans.** 20.2 s; 5612.4 m**]**

3. A shot is fired horizontally from the top of a cliff 20 m high with a velocity of 30 m/s. Find: 1. the time of flight; 2. the distance from the foot of the cliff of the point at which it hits the ground; and 3. the velocity when it hits the ground. [**Ans.** 2.02 s; 35.95 m/s; 33.4 m/s]
4. A person wants to jump over a ditch as shown in Fig. 13.25. Find the minimum velocity with which he should jump. [**Ans.** 4.6985 m/s]

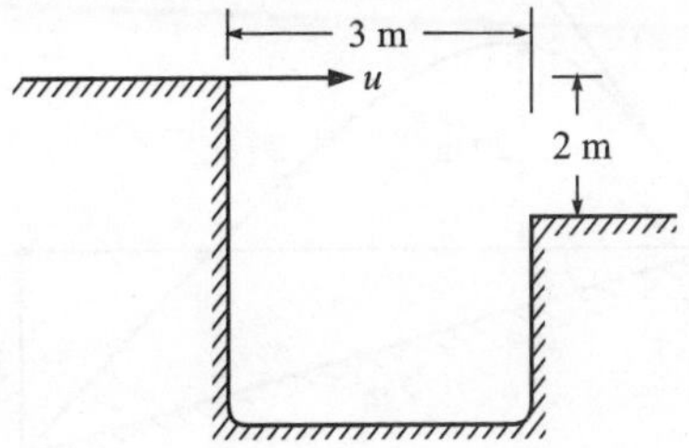


**Fig. 13.25**

5. A ball is projected from the ground at an angle of 60° to the horizontal with a velocity of 30 m/s. Find the distance of the ball from the point of projection after 2 seconds. [**Ans.** 44.1 m]
6. A body projected vertically upwards attains a maximum height of 475 m. Calculate the velocity of projection and determine the time of flight in air. [**Ans.** 96.538 m/s; 19.68 s]
7. A fireman holding a nozzle at a horizontal distance of 4.8 m from a vertical wall, wishes to send a jet of water through a small window in the wall located at 3.6 m vertically above the nozzle. If the inclination of the jet with the horizontal is 60° at the nozzle, calculate the required velocity of the jet at the exit nozzle. [**Ans.** 9.8 m/s]
8. A motor-cycle stunt driver tries to clear the ditch, as shown in Fig. 13.26. To do so, he drives his motor-cycle up an inclined slope at an angle of 20°. Find the minimum speed of the motor-cycle necessary to clear the ditch. [**Ans.** 5.07 m/s]

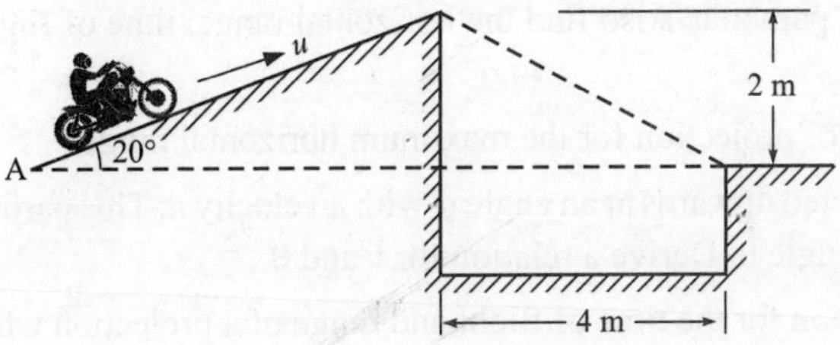


**Fig. 13.26**

9. A body is projected at such an angle that the horizontal range is three times the greatest height attained by the body. Find: 1. the angle of projection; 2. velocity of projection and time of flight, if the range, with this angle, is 360 m. [**Ans.** 53.13°; 60.65 m/s, 9.9 s]
10. A bomber was flying at an angle of 30° with the horizontal with a speed of 98 m/s at an altitude of 490 m when a bomb was dropped from it. After what time and at what horizontal distance would the bomb strike its object on the level surface of the ground? Also what would be the velocity of the bomb just before striking? Take $g$ = 9.8 m/s$^2$ and neglect the resistance of air. [**Ans.** 16.18 s; 1373.16 m; 179.8 m/s]
11. A projectile is fired from a point on a cliff to hit a mark 200 m horizontally from the point and 200 m vertically below it. The velocity of the projectile is 80 m/s. Determine the value of the firing angle. Neglect air resistance. [**Ans.** 82.2°]
12. A ball is thrown from the top of a tower 80 m high with a velocity of 25 m/s at an elevation of 30° above the horizon. Find the horizontal distance from the foot of the tower to the point where it hits the ground. [**Ans.** 119.3 m]
13. A particle is thrown with an initial velocity of 400 m/s at an angle of 60° with the horizontal. Find the velocity and direction of the particle after 20 seconds from the start. [**Ans.** 250.12 m/s; 37°]
14. A shot is fired with a velocity of 100 m/s at an angle of 45° on a plane inclined at an angle of 30° with the horizontal. Find the length of the inclined plane. [**Ans.** 497.8 m]

**15.** A ball rebounds at *A* and strikes the inclined plane at *B* at a distance of 75 m, as shown in Fig. 13.27. If the ball rises to a maximum height of 20 m above the point of projection, find the initial velocity and the angle of projection. [**Ans.** 24.3 m/s, 54.5°]

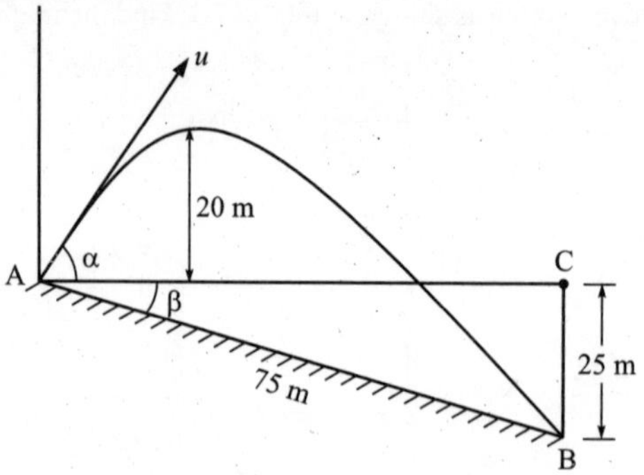


Fig. 13.27

## REVIEW QUESTIONS

1. What do you understand by projectile and trajectory?
2. For the projectile motion, define the following:
   (*a*) Velocity of projection; (*b*) Angle of projection; and (*c*) Horizontal range
3. Show that the path travelled by a projectile in air is parabolic. Neglect air resistance.
4. Prove that the path of a projectile thrown from the ground with a velocity *u* at an inclination α with the horizontal is a parabola. Also find the horizontal range, time of flight and maximum height of rise.
5. What is the angle of projection for the maximum horizontal range?
6. A particle is projected upwards at an angle α with a velocity *u*. This particle attains a velocity *v* after sometime *t* at an angle θ. Derive a relation for *v* and θ.
7. Derive an expression for the time of flight and range of a projection when the particle is projected upwards on an inclined plane. What will be the change in the above expressions when the particle is projected on the same plane but in the downward direction?
8. Show that the maximum range on an inclined plane when the particle is projected up the inclined plane, is given by

$$R_{max} = \frac{u^2}{g(1 + \sin \beta)}$$

where $u$ = Velocity of projection, and

$\beta$ = Angle of inclination of the plane with the horizontal.

## OBJECTIVE TYPE QUESTIONS

1. A path, traced by a projectile in space, is
   (*a*) hyperbola (*b*) parabola (*c*) linear (*d*) none of these

**2.** The time of flight ($t$) of a projectile on a horizontal plane is

(*a*) $\frac{2u \sin \alpha}{g}$ (*b*) $\frac{u^2 \sin 2\alpha}{g}$ (*c*) $\frac{u^2 \sin^2 \alpha}{2g}$ (*d*) $\frac{u^2 \cos \alpha}{g}$

where $u$ = Velocity of projection; and
$\alpha$ = Angle of projection.

**3.** The horizontal range ($R$) of a projectile is

(*a*) $\frac{2u \sin \alpha}{g}$ (*b*) $\frac{u^2 \sin 2\alpha}{g}$ (*c*) $\frac{u^2 \sin^2 \alpha}{2g}$ (*d*) $\frac{u^2 \cos \alpha}{g}$

**4.** The maximum height of a projectile on a horizontal plane is

(*a*) $\frac{2u \sin \alpha}{g}$ (*b*) $\frac{u^2 \sin 2\alpha}{g}$ (*c*) $\frac{u^2 \sin^2 \alpha}{2g}$ (*d*) $\frac{u^2 \cos \alpha}{g}$

**5.** The horizontal range of a projectile is maximum, when the angle of projection is

(*a*) 30° (*b*) 45° (*c*) 60° (*d*) 90°

**6.** The maximum horizontal range of a projectile is

(*a*) $\frac{u}{2g}$ (*b*) $\frac{u}{g}$ (*c*) $\frac{u^2}{2g}$ (*d*) $\frac{u^2}{g}$

**7.** The time of flight ($t$) of a projectile on an upward inclined plane is

(*a*) $\frac{g \cos \beta}{2u \sin (\alpha - \beta)}$ (*b*) $\frac{2u \sin (\alpha - \beta)}{g \cos \beta}$ (*c*) $\frac{g \cos \beta}{2u \sin (\alpha + \beta)}$ (*d*) $\frac{2u \sin (\alpha + \beta)}{g \cos \beta}$

where $\beta$ = Angle of inclination of the plane with the horizontal.

**8.** The time of flight of a projectile on a downward inclined plane depends upon

(*a*) angle of projection (*b*) angle of inclination of plane
(*c*) both (*a*) and (*b*) (*d*) none of these

**9.** The range of projectile ($R$) on an upward inclined plane is

(*a*) $\frac{g \cos^2 \beta}{2u^2 \sin (\alpha + \beta) \cos \alpha}$ (*b*) $\frac{2u^2 \sin (\alpha + \beta) \cos \alpha}{g \cos^2 \beta}$

(*c*) $\frac{g \cos^2 \beta}{2u^2 \sin (\alpha - \beta) \cos \alpha}$ (*d*) $\frac{2u^2 \sin (\alpha - \beta) \cos \alpha}{g \cos^2 \beta}$

**10.** The range of projectile will be maximum for a given velocity of projectile, when angle of projection ($\alpha$) is

(*a*) $\frac{\beta}{2}$ (*b*) $30° + \frac{\beta}{2}$ (*c*) $45° + \frac{\beta}{2}$ (*d*) $60° + \frac{\beta}{2}$

***Answers to Objective Type Questions***

| | | | | |
|---|---|---|---|---|
| **1.** (*b*) | **2.** (*a*) | **3.** (*b*) | **4.** (*c*) | **5.** (*b*) |
| **6.** (*d*) | **7.** (*b*) | **8.** (*c*) | **9.** (*d*) | **10.** (*c*) |

14

# CIRCULAR MOTION



## 14.1 INTRODUCTION

When a particle or a body moves along a circular path, then the motion of the body is said to have a *circular motion* or *curvilinear motion*. Such a motion takes place about a fixed axis of the body. The shafts, pulleys, flywheels etc. rotating about their geometric axis are the examples of the bodies moving in a circular path.

## 14.2 ANGULAR DISPLACEMENT

It is defined as the angle described by the particle from one point to another, with respect to the time. For example, let a line *OA* rotates about the centre *O* and has an inclination θ radians to the fixed line *OX*, as shown in Fig. 14.1.

If this line moves from *OA* to *OB* through an angle $\delta\theta$ during a short interval of time $\delta t$, then $\delta\theta$ is termed as the angular displacement of the line *OA*, measured in radians. Since the angular displacement has both magnitude and direction, therefore it is a vector quantity.

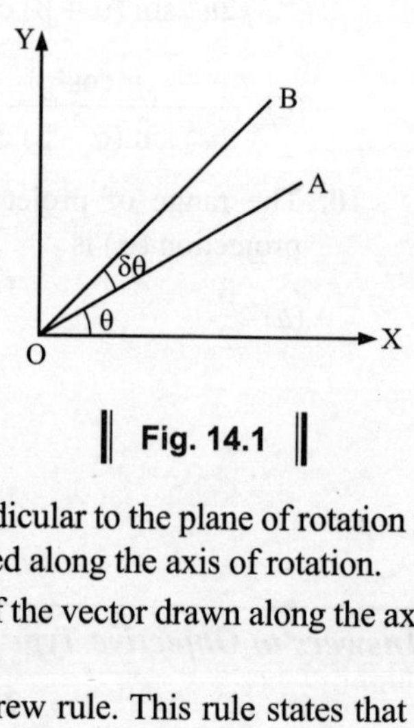


**Fig. 14.1**

In order to represent completely an angular displacement, by a vector, the following three conditions must be fixed:

1. ***Direction of the axis of rotation*** It is fixed by a line drawn perpendicular to the plane of rotation in which the angular displacement takes place. In other words, it is fixed along the axis of rotation.
2. ***Magnitude of the angular displacement*** It is fixed by the length of the vector drawn along the axis of rotation, to some suitable scale.
3. ***Sense of the angular displacement*** It is fixed by a right-hand screw rule. This rule states that if a screw rotates in a fixed nut in a clockwise direction, *i.e.* if the angular displacement is clockwise

and an observer is looking along the axis of rotation, then the arrowhead will point away from the observer. Similarly, if the angular displacement is anticlockwise, then the arrowhead will point towards the observer.

## 14.3 ANGULAR VELOCITY

It is defined as the rate of change of angular displacement with respect to time. It is usually expressed by a Greek letter ω (omega). It is measured in radians per second (rad/s). If θ is the angular displacement in radians and $t$ is the time taken in seconds, then angular velocity,

$$\omega = \frac{\theta}{t} \text{ or } \frac{d\theta}{dt} \text{ rad/s}$$

**Note:** If a body is rotating at the rate of $N$ r.p.m. (revolutions per minute), then the corresponding angular velocity is obtained as discussed below:

We know that 1 revolution = 2π radians

and 1 r.p.m. = 2π rad/min

or $N$ r.p.m. = $2\pi \times N$ rad/min

$$\therefore \quad \omega = 2\pi N \text{ rad/min} = \frac{2\pi N}{60} \text{ rad/s}$$

## 14.4 ANGULAR ACCELERATION

It is defined as the rate of change of angular velocity with respect to time. It is usually expressed by a Greek letter α (alpha). It is measured in radian per second per second (rad/s$^2$). Mathematically, angular acceleration,

$$\alpha = \frac{d\omega}{dt} = \frac{d}{dt}\left(\frac{d\theta}{dt}\right) = \frac{d^2\theta}{dt^2} \qquad ...\left(\because \omega = \frac{d\theta}{dt}\right)$$

## 14.5 RELATION BETWEEN LINEAR AND ANGULAR QUANTITIES OF MOTION

Consider a body moving along a circular path from $A$ to $B$, as shown in Fig. 14.2.

Let $r$ = Radius of the circular path,

θ = Angular displacement or angle traversed in radians,

ω = Angular velocity in rad/s,

$s$ = Linear displacement, in metres,

$v$ = Linear velocity in m/s, and

$a$ = Linear acceleration in m/s$^2$.

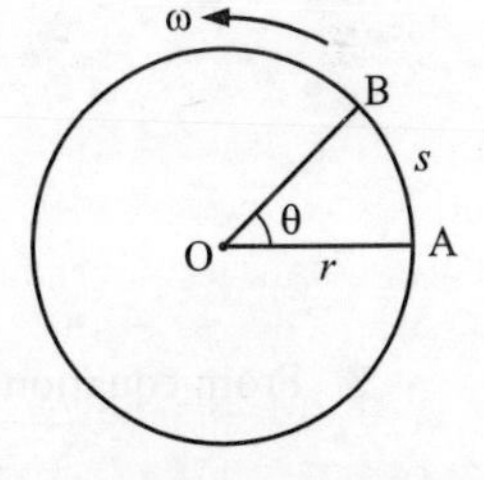


Fig. 14.2

From the geometry of the figure, we know that length of the arc $AB$ or linear displacement,

$$s = r\,.\,\theta$$

We also know that the linear velocity,

$$v = \frac{ds}{dt} = \frac{d(r.\theta)}{dt} = r \times \frac{d\theta}{dt} = r.\omega \qquad ...\left(\because \omega = \frac{d\theta}{dt}\right) \quad ...(i)$$

and linear acceleration,

$$a = \frac{dv}{dt} = \frac{d(r.\omega)}{dt} = r \times \frac{d\omega}{dt} = r.\alpha \qquad ...\left(\because \alpha = \frac{d\omega}{dt}\right) \quad ...(ii)$$

## 14.6 EQUATIONS OF ANGULAR MOTION

Consider a body moving along a circular path.

Let

$\omega_0$ = Initial angular velocity of the body in rad/s,

$\omega$ = Final angular velocity of the body in rad/s,

$t$ = Time (in seconds) taken by the body during which angular velocity changes from $\omega_0$ to $\omega$,

$\theta$ = Total angular displacement or angle traversed in radians, and

$\alpha$ = Constant angular acceleration in rad/s.

The various equations of angular motion are as follows:

**1.** We know that the angular acceleration ($\alpha$) is the rate of change of velocity. Therefore, angular acceleration,

$$\alpha = \frac{\text{Change of angular velocity}}{\text{Time}}$$

$$= \frac{\omega - \omega_0}{t}$$

or

$$\alpha.\,t = \omega - \omega_0$$

$\therefore$

$$\omega = \omega_0 + \alpha.t \qquad ...(i)$$

**2.** We know that average angular velocity,

$$= \frac{\omega_0 + \omega}{2}$$

$\therefore$ Angular displacement, $\theta$ = Average angular velocity × Time

$$= \frac{\omega_0 + \omega}{2} \times t \qquad ...(ii)$$

Substituting the value of $\omega$ from equation (*i*), we have

$$\theta = \frac{\omega_0 + \omega_0 + \alpha.t}{2} \times t$$

$$= \frac{2\omega_0 + \alpha.t}{2} \times t$$

$\therefore$

$$\theta = \omega_0.t + \frac{1}{2}\alpha.t^2 \qquad ...(iii)$$

**3.** From equation (*i*), we find that

$$t = \frac{\omega - \omega_0}{\alpha}$$

Substituting this value of $t$ in equation (*ii*), we have

$$\theta = \frac{\omega_0 + \omega}{2} \times \frac{\omega - \omega_0}{\alpha}$$

$$= \frac{\omega^2 - \omega_0^2}{2\alpha}$$

or

$$\theta \times 2\alpha = \omega^2 - \omega_0^2$$

$\therefore$

$$\omega^2 = \omega_0^2 + 2\alpha.\theta \qquad ...(iv)$$

## 14.7 RELATION BETWEEN LINEAR MOTION AND ANGULAR MOTION

The following table shows the relations between linear motion and corresponding angular motion of a body.

| *S. No.* | *Particulars* | *Linear motion* | *Angular motion* |
|---|---|---|---|
| **1.** | Initial velocity | $u$ | $\omega_0$ |
| **2.** | Final velocity | $v$ | $\omega$ |
| **3.** | Constant acceleration | $a$ | $\alpha$ |
| **4.** | Total distance traversed | $s$ | $\theta$ |
| **5.** | Equation for final velocity | $v = u + a.t$ | $\omega = \omega_0 + \alpha.t$ |
| **6.** | Equation for distance traversed | $s = ut + \frac{1}{2}a.t^2$ | $\theta = \omega_0.t + \frac{1}{2}\alpha.t^2$ |
| **7.** | Equation for final velocity | $v^2 = u^2 + 2a.s$ | $\omega^2 = \omega_0^2 + 2\alpha.\theta$ |

**Example 14.1** *A wheel accelerates uniformly from rest to 2000 r.p.m. in 20 seconds. What is its angular acceleration? How many revolutions does the wheel make in attaining the speed of 2000 r.p.m.?*

***Solution*** Given : $N_0 = 0$ or $\omega_0 = 0$ ; $N = 2000$ r.p.m. or $\omega = \dfrac{2\pi \times 2000}{60} = 209.5$ rad/s ; $t = 20$ s

***Angular acceleration***

Let $\alpha$ = Angular acceleration in rad/s².

We know that $\omega = \omega_0 + \alpha.t$

or $209.5 = 0 + \alpha \times 20$

$\therefore$ $\alpha = \dfrac{209.5}{20} = 10.475$ rad/s² **Ans.**

***Number of revolutions made by the wheel***

We know that angular distance moved by the wheel during 2000 r.p.m. (*i.e.* when $\omega = 209.5$ rad/s),

$$\theta = \frac{\omega + \omega_0}{2} \times t = \frac{209.5 + 0}{2} \times 20 = 2095 \text{ rad}$$

Since the angular distance moved by the wheel during one revolution is $2\pi$ radians, therefore number of revolutions made by the wheel,

$$n = \frac{\theta}{2\pi} = \frac{2095}{2\pi} = 333.4 \textbf{ Ans.}$$

**Example 14.2** *A horizontal bar 2 m long and of small cross-section rotates about a vertical axis through its one end. It accelerates uniformly from 1600 r.p.m. to 2000 r.p.m. in an interval of 10 seconds. What is the linear velocity of its other end at the beginning and at the end of the interval? Also find the angular acceleration of the bar during this time.*

***Solution*** Given : $r = 2$ m ; $N_0 = 1600$ r.p.m. or $\omega_0 = \dfrac{2\pi \times 1600}{60} = 167.6$ rad/s ; $N = 2000$ r.p.m. or $\omega = \dfrac{2\pi \times 2000}{60} = 209.5$ rad/s ; $t = 10$ s

***Linear velocity at the beginning and at the end of the interval***

We know that linear velocity at the beginning,

$$v_0 = \omega_0 . r = 167.6 \times 2 = 335.2 \text{ m/s} \quad \textbf{Ans.}$$

and linear velocity at the end of the interval,

$$v = \omega . r = 209.5 \times 2 = 419 \text{ m/s} \quad \textbf{Ans.}$$

***Angular acceleration of the bar during 10 s***

Let $\alpha$ = Angular acceleration in rad/s².

We know that $\omega = \omega_0 + \alpha.t$

$$209.5 = 167.6 + \alpha \times 10$$

$$\therefore \quad \alpha = \frac{209.5 - 167.6}{10} = 4.19 \text{ rad/s}^2 \quad \textbf{Ans.}$$

**Example 14.3** *A flywheel 460 mm in diameter is started from rest with constant angular acceleration to a speed of 280 r.p.m. in 15 seconds. Find the linear velocity and acceleration of a point on the rim 1 second after starting from rest.*

***Solution*** Given : $d = 460$ mm or $r = 230$ mm $= 0.23$ m ; $N_0 = 0$ or $\omega_0 = 0$ ; $N = 280$ r.p.m. or $\omega = \dfrac{2\pi \times 280}{60} = 29.32$ rad/s ; $t = 15$ s

Let $\alpha$ = Angular acceleration of the flywheel.

We know that $\omega = \omega_0 + \alpha.t$

$$29.32 = 0 + \alpha \times 15$$

$$\therefore \quad \alpha = \frac{29.32}{15} = 1.955 \text{ rad/s}^2$$

***Linear velocity***

We know that angular velocity of the flywheel 1 second after starting from rest,

$$\omega_1 = \alpha.t = 1.955 \times 1 = 1.955 \text{ rad/s} \quad \text{...(Here } t = 1 \text{ s)}$$

$$\therefore \quad \text{Linear velocity, } v = r.\omega_1 = 0.23 \times 1.955 = 0.45 \text{ m/s} \quad \textbf{Ans.}$$

***Linear acceleration***

We know that linear acceleration, $a = r.\alpha = 0.23 \times 1.955 = 0.45 \text{ m/s}^2$ **Ans.**

**Example 14.4** *A wheel is rotating about its axis with a constant angular acceleration of 1.05 rad/s². If the initial and final angular velocities are 5.23 rad/s and 10.47 rad/s, determine the total angle turned during the time interval this change of angular velocity took place.*

***Solution*** Given : $\alpha = 1.05$ rad/s² ; $\omega_0 = 5.23$ rad/s ; $\omega = 10.47$ rad/s

Let $\theta$ = Total angle turned in radians.

First of all, let us find the time ($t$) taken during the change of angular velocity from $\omega_0 = 5.23$ rad/s to $\omega = 10.47$ rad/s.

Using the relation, $\omega = \omega_0 + \alpha.t$

$$10.47 = 5.23 + 1.05t$$

$$\therefore \quad t = \frac{10.47 - 5.23}{1.05} = 5 \text{ s}$$

We know that $\theta$ = Average angular velocity in rad/s × Time

$$= \frac{\omega_0 + \omega}{2} \times t = \frac{5.23 + 10.47}{2} \times 5 = 39.25 \text{ rad } \textbf{Ans.}$$

**Note:** The total angle turned may also be obtained by using the relation:

$$\theta = \omega_0 . t + \frac{1}{2} \alpha . t^2 = 5.23 \times 5 + \frac{1}{2} \times 1.05 \times 5^2$$

$$= 26.15 + 13.125 = 39.275 \text{ rad } \textbf{Ans.}$$

**Example 14.5** *A wheel is making 180 r.p.m. and after 20 seconds it is running at 140 r.p.m. How many revolutions will it make, and what time will elapse before it stops, if retardation is uniform?*

***Solution*** Given : $N_0 = 180$ r.p.m. or $\omega_0 = \frac{2\pi \times 180}{60} = 18.85$ rad/s ; $t = 20$ s ; $N = 140$ r.p.m. or

$$\omega = \frac{2\pi \times 140}{60} = 14.66 \text{ rad/s}$$

***Number of revolutions made***

Let $N$ = Number of revolution made, and

$\alpha$ = Angular acceleration in rad/s².

We know that $\omega = \omega_0 + \alpha . t$

$$14.66 = 18.85 + \alpha \times 20$$

$$\therefore \quad \alpha = \frac{14.66 - 18.85}{20} = -0.21 \text{ rad/s}^2 \text{ (Retardation)}$$

If $\theta$ is the angular displacement of the flywheel before it stops (*i.e.* when final angular velocity, $\omega = 0$), then using the relation,

$$\omega^2 = \omega_0^2 + 2\alpha . \theta$$

$$0 = (18.85)^2 + 2 \times -0.21 \times \theta = 355.3 - 0.42\,\theta$$

$$\therefore \quad \theta = \frac{355.3}{0.42} = 846 \text{ rad}$$

We know that number of revolutions

$$N = \frac{\theta}{2\pi} = \frac{846}{2\pi} = 135 \text{ r.p.m. } \textbf{Ans.} \quad ...(\because 1 \text{ revolution} = 2\pi \text{ rad})$$

***Time taken before it stops (i.e. when $\omega = 0$)***

Let $t$ = Time taken in seconds.

We know that $\omega = \omega_0 + \alpha t$

$$0 = 18.85 - 0.21\,t$$

$$\therefore \quad t = \frac{18.85}{0.21} = 89.76 \text{ s } \textbf{Ans.}$$

**Example 14.6** *A wheel rotating about a fixed axis at 20 r.p.m. is uniformly accelerated for 70 seconds during which time it makes 50 revolutions. Find :*

1. *Angular velocity at the end of this interval ; and*
2. *Time required for the speed to reach 100 r.p.m.*

***Solution*** Given : $N_0 = 20$ r.p.m. or $\omega_0 = \frac{2\pi \times 20}{60} = 2.09$ rad/s ; $t = 70$ s

Since the number of revolutions made are 50, therefore angular displacement,

$$\theta = \text{Angular displacement in one revolution} \times \text{No. of revolutions}$$
$$= 2\pi \times 50 = 314.2 \text{ rad}$$

Let $\alpha$ = Angular acceleration in rad/s², and

$\omega$ = Final angular velocity in rad/s.

Using the relation,

$$\theta = \omega_0 .t + \frac{1}{2}\alpha .t^2$$

$$314.2 = 2.09 \times 70 + \frac{1}{2} \times \alpha (70)^2 = 146.3 + 2450\,\alpha$$

$$\therefore \quad \alpha = \frac{314.2 - 146.3}{2450} = 0.068 \text{ rad/s}^2$$

**1. *Angular velocity at the end of 70 seconds***

Let $\omega$ = Required angular velocity in rad/s.

We know that

$$\omega = \omega_0 + \alpha .t$$
$$= 2.09 + 0.068 \times 70 = 2.09 + 4.76 = 6.85 \text{ rad/s} \quad \textbf{Ans.}$$

**2. *Time required for the speed to reach 100 r.p.m.***

Let $t_1$ = Time required in seconds.

We know that angular speed corresponding to 100 r.p.m.,

$$\omega_1 = \frac{2\pi \times 100}{60} = 10.47 \text{ rad/s}$$

Using the relation,

$$\omega_1 = \omega_0 + \alpha .t_1$$
$$10.47 = 2.09 + 0.068\, t_1$$
$$\therefore \quad t_1 = \frac{10.47 - 2.09}{0.068} = 123.235 \text{ s} \quad \textbf{Ans.}$$

**Example 14.7** *A wheel rotates for 5 seconds with constant angular acceleration and describes 100 radians during this time. It then rotates with a constant angular velocity and during the next 5 seconds describes 80 radians. Find the initial angular velocity and the angular acceleration.*

***Solution*** Given : $t = 5$ s ; $\theta = 100$ rad

***Initial angular velocity***

Let $\omega_0$ = Initial angular velocity in rad/s,

$\omega$ = Angular velocity after 5 seconds in rad/s, and

$\alpha$ = Angular acceleration in rad/s².

First of all, let us consider that the wheel rotates with constant angular acceleration ($\alpha$). We know that angular displacement,

$$\theta = \omega_0 .t + \frac{1}{2}\alpha .t^2$$

$$100 = \omega_0 \times 5 + \frac{1}{2}\alpha \times 5^2 = 5\omega_0 + 12.5\alpha$$

$$20 = \omega_0 + 2.5\alpha \quad \text{...(Dividing by 5)}$$

$$\therefore \quad \alpha = \frac{20 - \omega_0}{2.5} \quad \text{...}(i)$$

and final angular velocity, $\omega = \omega_0 + \alpha.t = \omega_0 + \frac{20 - \omega_0}{2.5} \times 5$

$$= \omega_0 + 40 - 2\omega_0 = 40 - \omega_0 \qquad ...(ii)$$

Now consider that the wheel rotates with constant angular velocity ($\omega$) and describes $\theta = 80$ radians in time $t = 5$ seconds. We know that angular displacement,

$$\theta = \omega.t$$

$$80 = (40 - \omega_0)\,5 = 200 - 5\omega_0 \qquad ...[\text{From equation } (i)]$$

$$\therefore \quad \omega_0 = \frac{200 - 80}{5} = 24 \text{ rad/s} \quad \textbf{Ans.}$$

***Initial angular acceleration***

Substituting the value of $\omega_0$ in equation (*i*), we have initial angular acceleration,

$$\alpha = \frac{20 - \omega_0}{2.5} = \frac{20 - 24}{2.5} = -1.6 \text{ rad/s}^2 \quad \textbf{Ans.}$$

The –ve sign indicates retardation.

**Example 14.8** *The angular displacement of a body is a function of time and is given by the equation:*

$$\theta = 10 + 3t + 6t^2, \text{ where } t \text{ is in seconds.}$$

*Determine the angular velocity, displacement and acceleration when t = 5 seconds.*

***Solution*** Given : $\theta = 10 + 3t + 6t^2$

***Angular displacement when t = 5 s***

Substituting $t = 5$ s in the given relation, we have

$$\theta = 10 + 3 \times 5 + 6 \times 5^2 = 10 + 15 + 150 = 175 \text{ rad} \quad \textbf{Ans.}$$

***Angular velocity when t = 5 s***

Differentiating the given equation with respect to time ($t$), we have

$$\omega = \frac{d\theta}{dt} = 3 + 6 \times 2t = 3 + 12t$$

Substituting $t = 5$ s, in the above equation, we have angular velocity,

$$\omega = \frac{d\theta}{dt} = 3 + 12 \times 5 = 63 \text{ rad/s} \quad \textbf{Ans.}$$

***Angular acceleration when t = 5 s***

We know that angular acceleration, $\alpha = \frac{d\omega}{dt} = \frac{d}{dt}(3 + 12t) = 12 \text{ rad/s}^2$ **Ans.**

**Example 14.9** *A flywheel rotates with constant retardation due to braking. From t = 0 to t = 10 seconds, it made 300 revolutions. At time t = 7.5 seconds, its angular velocity was 40π rad/s. Determine:*

1. *The value of the constant retardation ;*
2. *The total time taken to come to rest ; and*
3. *The total revolutions made till it comes to rest.*

***Solution*** Given : $t = 10 - 0 = 10$ s ; $\theta = 300$ rev $= 300 \times 2\pi = 600\pi$ rad ; At $t = 7.5$ s, $\omega = 40\pi$ rad/s

**1.** ***Value of constant retardation***

Let $\alpha$ = Constant retardation in rad/s$^2$, and

$\omega_0$ = Initial angular velocity in rad/s.

Using the relation, $\theta = \omega_0 .t - \frac{1}{2}\alpha.t^2$ ...(–ve sign due to retardation)

$$600\pi = \omega_0 \times 10 - \frac{1}{2}\alpha\,(10)^2 = 10\,\omega_0 - 50\,\alpha$$

Dividing by 10 throughout, we have

$$60\pi = \omega_0 - 5\alpha \qquad ...(i)$$

It is given that when time $t = 7.5$ s, the angular velocity $\omega = 40\pi$ rad/s. We know that

$$\omega = \omega_0 - \alpha.t \qquad \text{...(–ve sign due to retardation)}$$

$$40\pi = \omega_0 - \alpha \times 7.5 = \omega_0 - 7.5\alpha \qquad ...(ii)$$

Subtracting equation (*ii*) from equation (*i*), we have

$$20\pi = 2.5\alpha$$

$$\therefore \quad \alpha = \frac{20\pi}{2.5} = 8\pi = 25.136 \text{ rad/s}^2 \quad \textbf{Ans.}$$

**2. *Total time taken to come to rest***

Let $T$ = Total time taken to come to rest in seconds.

First of all, let us find the initial angular velocity ($\omega_0$). Substituting the value of $\alpha = 25.136$ rad/s$^2$ in equation (*i*), we have

$$60\pi = \omega_0 - 5 \times 25.136 = \omega_0 - 125.68$$

$$\therefore \quad \omega_0 = 60\pi + 125.68 = 188.5 + 125.68 = 314.2 \text{ rad/s}$$

Since the flywheel comes to rest, therefore final angular velocity ($\omega$) = 0.

Now using the relation, $\omega = \omega_0 - \alpha.T$

$$0 = 314.2 - 25.136\,T$$

$$\therefore \quad T = \frac{314.2}{25.136} = 12.5 \text{ s} \quad \textbf{Ans.}$$

**3. *Total revolutions made till it comes to rest***

Let $\theta'$ be the angular displacement made by the flywheel till it comes to rest, *i.e.* in time $T = 12.5$ s. Again using the relation,

$$\theta' = \omega_0 .T - \frac{1}{2}\alpha.T^2$$

$$= 314.2 \times 12.5 - \frac{1}{2} \times 25.136(12.5)^2 = 3927.5 - 1963.75$$

$$= 1963.75 \text{ rad}$$

$\therefore$ Number of revolutions made $= \frac{\theta'}{2\pi} = \frac{1963.75}{2\pi} = 312.5$ rev **Ans.** ...($\because$ 1 rev = $2\pi$ rad)

**Example 14.10** *A grinding wheel is attached to a shaft of an electrified motor of rated speed of 1500 r.p.m. When the power is switched on, the unit attains the rated speed in 4 seconds and when the power is switched off, the unit comes to rest in 75 seconds. Assuming uniformly accelerated motion, determine the number of revolutions the unit turns to attain the rated speed and to come to rest.*

***Solution*** Given : $N = 1500$ r.p.m. or $\omega = \frac{2\pi \times 1500}{60} = 50\pi$ rad/s ; $t = 4$ s ; $t_1 = 75$ s

***Number of revolutions made by the grinding wheel at the rated speed of 1500 r.p.m.***

When the power is switched on, the initial angular velocity, $\omega_0 = 0$.

Let $\alpha$ be the angular acceleration in rad/s$^2$ and $\theta$ be the angular displacement in radians. We know that

$$\omega = \omega_0 + \alpha . t$$

or $$50\pi = 0 + \alpha \times 4$$

$\therefore$ $$\alpha = \frac{50\pi}{4} = 39.275 \text{ rad/s}^2$$

Now using the relation, $$\omega^2 = \omega_0^2 + 2\alpha.\theta$$

or $$(50\pi)^2 = 0 + 2 \times 39.275\theta$$

$\therefore$ $$\theta = \frac{(50\pi)^2}{2 \times 39.275} = 314.2 \text{ rad}$$

and number of revolutions made by the grinding wheel,

$$N_w = \frac{\theta}{2\pi} = \frac{314.2}{2\pi} = 50 \textbf{ Ans.}$$

***Number of revolutions made by the grinding wheel to come to rest***

When the power is switched off, the final angular velocity,

$$\omega = 0$$

Initial angular velocity, $\omega_0 = 50\pi$ rad/s

and time taken by the grinding wheel to come to rest,

$$t = 75 \text{ s}$$ ...(Given)

Let $\alpha$ be the angular acceleration in rad/s$^2$ and $\theta$ be the angular displacement in radians. We know that

$$\omega = \omega_0 + \alpha . t$$

or $$0 = 50\pi + \alpha \times 75$$

$\therefore$ $$\alpha = \frac{-50\pi}{75} = -\frac{2\pi}{3} \text{ rad/s}^2$$

The –ve sign indicates retardation.

Now using the relation, $$\omega^2 = \omega_0^2 + 2\alpha.\theta$$

$$0 = (50\pi)^2 - 2 \times \frac{2\pi}{3} \times \theta = (50\pi)^2 - \frac{4\pi}{3}\theta$$

$\therefore$ $$\theta = (50\pi)^2 \times \frac{3}{4\pi} = 5891.25 \text{ rad}$$

and number of revolution made by the grinding wheel

$$N_w = \frac{\theta}{2\pi} = \frac{5891.25}{2\pi} = 937.5 \text{ rev } \textbf{Ans.}$$

## 14.8 ACCELERATION OF A PARTICLE MOVING ALONG A CIRCULAR PATH

Consider a particle $A$ is moving along a circular path, and it takes position $B$ after some time say $\delta t$, as shown in Fig. 14.3 ($a$).

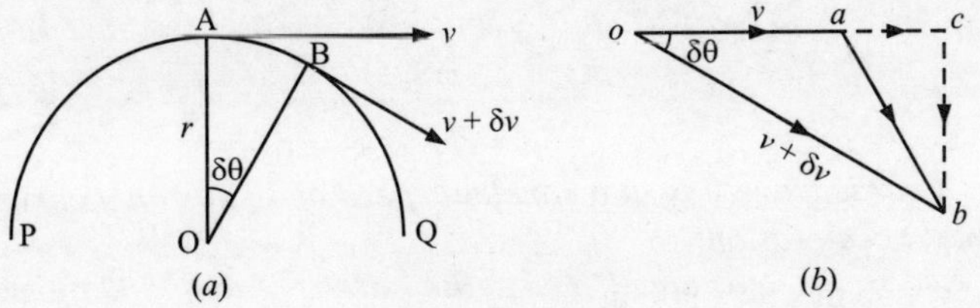


**Fig. 14.3**

Let $r$ = Radius of the circular path,

$v$ = Velocity of particle at $A$, and

$v+\delta v$ = Velocity of the particle at $B$.

The change of velocity as the particle moves from $A$ to $B$ may be obtained by drawing the vector triangle *oab*, as shown in Fig. 14.3 (*b*). In this triangle, *oa* represents the velocity $v$ and *ob* represents the velocity $v+\delta v$. The change of velocity in time $\delta t$ is represented by *ab*. Since this change of velocity has two mutually perpendicular components, therefore acceleration of a particle moving along a circular path has the following two accelerations which are perpendicular to each other.

1. ***Tangential (or linear) acceleration*** The acceleration of a particle at any instant moving along a circular path in a direction tangential to that instant, is known as *tangential* or *linear acceleration*. We know that tangential (or linear) acceleration,

$$a_t = \frac{dv}{dt} = \alpha . r$$

2. ***Normal acceleration*** The acceleration of a particle at any instant moving along a circular path in a direction normal to the tangent at that instant and directed towards the centre of the circular path (*i.e.* in the direction from $A$ to $O$) is known as *normal acceleration*. It is also called *radial* or *centripetal acceleration*.

∴ Normal acceleration of the particle at $A$,

$$a_n = v \times \frac{d\theta}{dt} = v.\omega = v \times \frac{v}{r} = \frac{v^2}{r} = \omega^2.r$$

$$\ldots\left(\because \frac{d\theta}{dt} = \omega \text{ and } \omega = \frac{v}{r}\right)$$

Since the tangential acceleration ($a_t$) and normal acceleration ($a_n$) of the particle at any instant $A$ are perpendicular to each other, as shown in Fig. 14.4, therefore total acceleration of the particle ($a$) is equal to the resultant acceleration of $a_t$ and $a_n$.

∴ Total acceleration or resultant acceleration,

$$a = \sqrt{(a_t)^2 + (a_n)^2}$$

and its inclination with the tangential acceleration is given by

$$\tan\theta = \frac{a_n}{a_t}$$

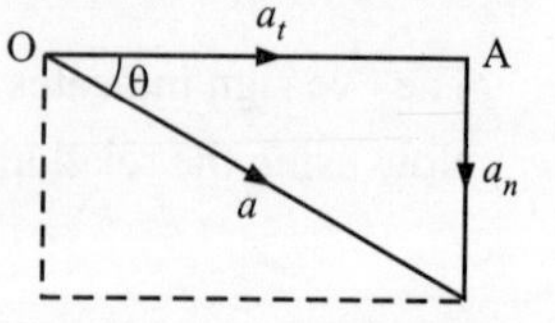


**Fig. 14.4**

**Notes: 1.** When a particle moves with a uniform velocity, then $\frac{dv}{dt}$ will be zero *i.e.* there will be no tangential acceleration and the particle will have only normal acceleration whose value is given by

$$a_n = \frac{v^2}{r} = v.\omega = \omega^2.r$$

**2.** When the particle moves along a straight path, then there will be no normal acceleration and the particle will have only tangential acceleration whose value is given by

$$a_t = \frac{dv}{dt} = \alpha.r$$

**Example 14.11** *A car is travelling at a constant speed of 30 m/s on a curve of 900 m. Determine the tangential and normal acceleration.*

***Solution*** Given : $v = 30$ m/s ; $r = 900$ m

***Tangential acceleration***

Since the car is travelling at a constant speed, therefore there will be no tangential acceleration. **Ans.**

***Normal acceleration***

We know that normal acceleration,

$$a_n = \frac{v^2}{r} = \frac{(30)^2}{900} = 1 \text{ m/s}^2 \text{ Ans.}$$

**Example 14.12** *A train starts from rest and moves along a curved track of radius 600 m with uniform acceleration until it attains a velocity of 70 km/h at the end of third minute. Determine the tangential, normal and total acceleration of the train at the end of second minute.*

***Solution*** Given : $\omega_0 = 0$ ; $r = 600$ m ; $v = 70$ km/h $= \frac{70 \times 1000}{3600} = 19.44$ m/s ; $t = 3$ min $= 180$ s

We know that angular velocity at the end of third minute,

$$\omega = \frac{v}{r} = \frac{19.44}{600} = 0.0324 \text{ rad/s}$$

Let $\alpha$ = Angular acceleration of the train in rad/s².

Using the relation, $\omega = \omega_0 + \alpha . t$

$$0.0324 = 0 + \alpha \times 180$$

$$\therefore \quad \alpha = \frac{0.0324}{180} = 0.000\,18 \text{ rad/s}^2$$

Now let us find the angular speed of the train at the end of second minute, *i.e.* when $t_1 = 2$ min $= 120$ s.

Again using the relation, $\omega_1 = \omega_0 + \alpha . t_1 = 0 + 0.000\,18 \times 120 = 0.0216$ rad/s

***Tangential acceleration***

We know that tangential acceleration,

$$a_t = \alpha . r = 0.000\,18 \times 600 = 0.108 \text{ m/s}^2 \text{ Ans.}$$

***Normal acceleration***

We know that normal acceleration.

$$a_n = \omega_1^2 . r - (0.0216)^2 600 = 0.28 \text{ m/s}^2 \text{ Ans.}$$

***Total acceleration***

We know that total acceleration, $a = \sqrt{(a_t)^2 + (a_n)^2} = \sqrt{(0.108)^2 + (0.28)^2}$

$$= \sqrt{0.0116 + 0.0784} = \sqrt{0.09} = 0.3 \text{ m/s}^2 \text{ Ans.}$$

**Example 14.13** *A train is travelling on a curved track of radius 1000 m at the speed of 180 km/h. The brakes are suddenly applied, causing the train to slow down at uniform rate. After 10 seconds, the speed has been reduced to 100 km/h. Determine the deceleration immediately after the brakes have been applied.*

***Solution*** Given : $r = 1000$ m ; $v_1 = 180$ km/h $= \frac{180 \times 1000}{3600} = 50$ m/s ; $v_2 = 100$ km/h $= \frac{100 \times 1000}{3600}$ $= 27.8$ m/s ; $t = 10$ s

Since the train slows down at uniform rate, therefore tangential deceleration,

$$a_t = \frac{\text{Change of velocity } (dv)}{\text{Time } (dt)} = \frac{v_1 - v_2}{t}$$

$$= \frac{50 - 27.8}{10} = \frac{22.2}{10} = 2.22 \text{ m/s}^2$$

Immediately after braking, normal deceleration,

$$a_n = \frac{v_1^2}{r} = \frac{(50)^2}{1000} = 2.5 \text{ m/s}^2$$

$\therefore$ Total deceleration,

$$a = \sqrt{(a_t)^2 + (a_n)^2} = \sqrt{(2.22)^2 + (2.5)^2}$$

$$= \sqrt{4.9284 + 6.25} = \sqrt{11.1784} = 3.343 \text{ m/s}^2 \textbf{ Ans.}$$

Let $\theta$ = Angle made by the total deceleration with the tangential deceleration.

We know that

$$\tan\theta = \frac{a_n}{a_t} = \frac{2.5}{2.22} = 1.126$$

$$\therefore \quad \theta = \tan^{-1}(1.126) = 48.4° \textbf{ Ans.}$$

## 14.9 CENTRIPETAL FORCE

Consider a body of mass $m$ moving with a linear velocity $v$ in a circular path of radius $r$, as shown in Fig. 14.5.

We have already discussed in Sec. 14.8 that the centripetal acceleration,

$$a_c = \frac{v^2}{r} = \omega^2 . r$$

$\therefore$

$$\text{Centripetal force} = \text{Mass} \times \text{Centripetal acceleration}$$

$$= m \times a_c$$

$$= \frac{m v^2}{r} = m \omega^2 r$$

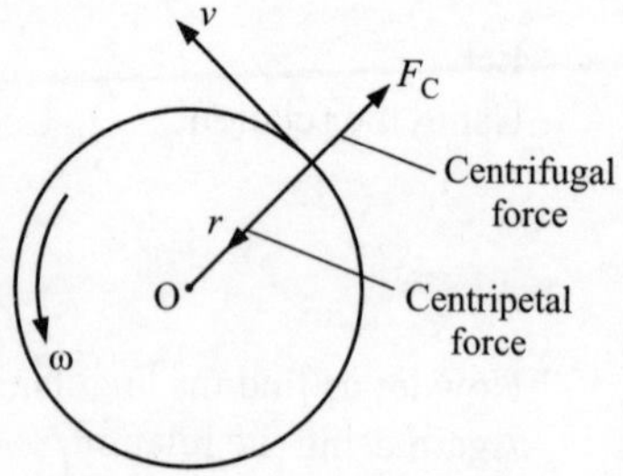


Fig. 14.5

where $\omega$ is the angular speed of the body in rad/s.

This force acts radially inwards and is essential for the circular motion.

It is important to note that centripetal force is not a fundamental force, but it is just a label given to the net force which causes the body to move in a circular path.

**Notes: 1.** The force of gravity acting on a satellite in orbit is an example of centripetal force.

**2.** When a cyclist goes around a curve, he leans inward to provide centripetal force required to push him sidewise towards the centre of the circle.

**3.** The friction of the tyres of a car making a turn similarly provides centripetal force on the car.

## 14.10 CENTRIFUGAL FORCE

We have discussed in the previous article that the centripetal force acts radially inwards. According to Newton's third law of motion, action and reaction are equal and opposite. Therefore, the body must exert a force radially outwards of equal magnitude. This force is known as *centrifugal force* whose magnitude is given by

$$F_C = \frac{m v^2}{r} = m \omega^2 r$$

## 14.11 MOTION OF A CYCLIST ALONG A CIRCULAR PATH

Consider a cyclist going around a circular path, as shown in Fig. 14.6.

Let $m$ = Mass of the cyclist,

$W$ = Weight of the cyclist = $mg$,

$r$ = Radius of the circular path,

$$v = \text{Linear velocity of the cyclist, and}$$

$$\omega = \text{Angular velocity of the cyclist} = \frac{v}{r}.$$

A little consideration will show that when a cyclist moves round a circular path, he has to bend himself a little from his vertical position in order to avoid overturning.

When he bends himself inward, he is acted upon by the centripetal force to keep him in the circular path. This centripetal force is provided by the frictional force exerted between the cycle wheel and the ground surface.

The centre of gravity of the cyclist lies at $G$ at a height $h$ from the ground surface. When the cyclist moves round a circular path, the following forces act on it:

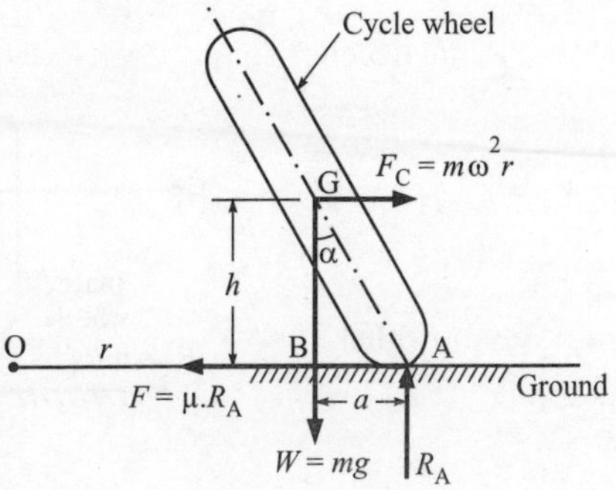


**Fig. 14.6**

1. Weight of the cyclist ($W = mg$), acting vertically downwards through the centre of gravity $G$,
2. The normal reaction of the ground surface at $A$ (*i.e.* $R_A$),
3. Frictional force $F = \mu.R_A$ exerted between the ground surface and the tyre of the cycle, and
4. Centrifugal force, $F_C = m\omega^2 r = m \times \frac{v^2}{r}$ acting horizontally outwards at the centre of gravity ($G$).

For equilibrium, taking moment of the forces about $A$, we have

$$W.a = F_C \,.\, h$$

$$mg \,.\, a = m\omega^2 r \times h = m \times \frac{v^2}{r^2} \times rh = m \times \frac{v^2}{r} \times h \quad \ldots\left(\because \omega = \frac{v}{r}\right)$$

$$\therefore \quad a = \frac{v^2 h}{g.r} \quad \ldots(i)$$

The angle of inclination ($\alpha$) of the cyclist with the vertical at $G$ is given by

$$\tan\alpha = \frac{a}{h}$$

From equation (*i*),

$$\frac{a}{h} = \frac{v^2}{g.r}$$

$$\therefore \quad \tan\alpha = \frac{v^2}{g.r}$$

or

$$v = \sqrt{g.r.\tan\alpha}$$

This is the maximum speed at which the cyclist can move.

**Example 14.14** *A cyclist is travelling around a circular path of radius 5 m with a speed of 10 km/h. Find the angle which the cyclist must make with the vertical in travelling around the given curved path.*

***Solution*** Given : $r = 5$ m ; $v = 10$ km/h $= \frac{10 \times 1000}{60 \times 60} = 2.78$ m/s

Let $\alpha$ = Angle which the cyclist make with the vertical.

We know that

$$\tan\alpha = \frac{v^2}{g.r} = \frac{(2.78)^2}{9.81 \times 5} = 0.1576$$

$$\therefore \quad \alpha = \tan^{-1}(0.1576) = 8.956° \quad \textbf{Ans.}$$

## 14.12 MOTION OF A VEHICLE ON A LEVEL CIRCULAR PATH

Consider a vehicle of mass $m$ moving with a velocity $v$ in a level circular path with $G$ as centre, as shown in Fig. 14.7.

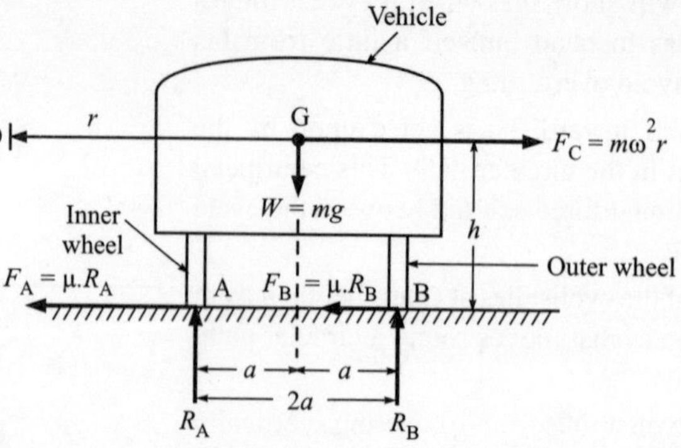


**Fig. 14.7**

The various forces acting on the vehicle are as follows:

***(a)*** Weight of the vehicle ($W = mg$),

***(b)*** Normal reaction $R_A$ and $R_B$ at the wheels $A$ and $B$, respectively,

***(c)*** Centrifugal force of magnitude $F_C = m\omega^2 r = m \times \dfrac{v^2}{r}$, acting at the centre of gravity ($G$) of the vehicle, and

***(d)*** Force of friction $F_A$ and $F_B$ between the wheels $A$ and $B$ and the road surface.

Let
$r$ = Radius of the curved path,
$\mu$ = Coefficient of friction between the wheels and the road surface,
$2a$ = Distance between the wheels $A$ and $B$, and
$h$ = Height of centre of gravity of the vehicle from the road surface.

When the vehicle moves with a velocity more than the permissible velocity, the vehicle may skid or overturn causing accident. Therefore, let us find the maximum velocity of the vehicle in order to avoid skidding or overturning as discussed below:

**1. *Maximum velocity of the vehicle in order to avoid skidding***

We know that for vertical equilibrium of the vehicle, $\Sigma V = 0$.

$$\therefore \quad R_A + R_B = W = mg \quad \text{...}(i)$$

and for horizontal equilibrium of the vehicle, $\Sigma H = 0$, *i.e.*

$$F_A + F_B = F_C$$

or

$$\mu.R_A + \mu.R_B = m\omega^2 r = m \times \frac{v^2}{r^2} \times r = m \times \frac{v^2}{r}$$

$$\mu(R_A + R_B) = m \times \frac{v^2}{r} \quad \text{...}(ii)$$

Substituting $R_A + R_B = mg$, from equation ($i$), we have

$$\mu \,.\, mg = m \times \frac{v^2}{r}$$

or $$\mu g = \frac{v^2}{r}$$

∴ $$v = \sqrt{\mu g r}$$

This gives the maximum velocity of the vehicle in order to avoid skidding. Thus maximum velocity,

$$v_{max} = \sqrt{\mu g r}$$

**2. *Maximum velocity of the vehicle to avoid overturning***

We know that centrifugal force acting at the C.G. of the vehicle,

$$F_C = m\omega^2 r = m \times \frac{v^2}{r^2} \times r = m \times \frac{v^2}{r}$$

Taking moments about the inner wheel *A*, we have

$$R_B \times 2a = F_C \times h + W.a$$

$$= \frac{mv^2 h}{r} + mga = m\left(\frac{v^2 h}{r} + g.a\right)$$

∴ $$R_B = \frac{m}{2a}\left(\frac{v^2 h}{r} + g.a\right) = \frac{m}{2}\left(\frac{v^2 h}{r.a} + g\right)$$

$$= \frac{mg}{2}\left(\frac{v^2 h}{gra} + 1\right)$$ ...(Multiplying and dividing by *g*)

We know that $$R_A + R_B = W = mg$$

∴ $$R_A = mg - R_B = mg - \frac{mg}{2}\left(\frac{v^2 h}{gra} + 1\right)$$

$$= mg\left[1 - \frac{1}{2}\left(\frac{v^2 h}{gra} + 1\right)\right] = \frac{mg}{2}\left[2 - \left(\frac{v^2 h}{gra} + 1\right)\right] = \frac{mg}{2}\left[1 - \frac{v^2 h}{gra}\right]$$

From this expression, we see that if the velocity of the vehicle ($v$) goes on increasing, then at one stage, $\frac{v^2 h}{gra}$ will be more than unity and the expression within the bracket will be negative and thus the vehicle will overturn. Thus the maximum velocity of the vehicle in order to avoid overturning may be obtained by equating $\frac{v^2 h}{gra} = 1$, which will make $R_A$ equal to zero.

∴ $$v = \sqrt{\frac{gra}{h}}$$

**Example 14.15** *The coefficient of friction between the mastic asphalt pavement on the tyres of a truck is found to be 0.29. At what maximum constant speed, the truck can move around a curve on the level path without skidding. Consider the radius of curvature as 350 m.*

***Solution*** Given : $\mu = 0.29$ ; $r = 350$ m

We know that maximum constant speed at which the truck can move around a curve on a level path without skidding,

$$v_{max} = \sqrt{\mu g r} = \sqrt{0.29 \times 9.81 \times 350} = \sqrt{995.7} = 31.6 \text{ m/s}$$ **Ans.**

**Example 14.16** *A vehicle of mass 1000 kg has its centre of gravity 0.6 m above the ground and the distance between the wheels is 1.8 m. If the coefficient of friction between the wheels and the road is 0.8, find the maximum velocity at which the vehicle can move around a curve of 120 m radius without skidding and overturning.*

***Solution*** Given : $m = 1000$ kg ; $h = 0.6$ m ; $2a = 1.8$ m or $a = 0.9$ m ; $\mu = 0.8$ ; $r = 120$ m

***Maximum velocity of the vehicle without skidding***

We know that maximum velocity of the vehicle without skidding,

$$v_{max} = \sqrt{\mu g r} = \sqrt{0.8 \times 9.81 \times 120} = \sqrt{941.76} = 30.7 \text{ m/s} \textbf{ Ans.}$$

***Maximum velocity of the vehicle without overturning***

We know that maximum velocity of the vehicle without overturning,

$$v_{max} = \sqrt{\frac{gra}{h}} = \sqrt{\frac{9.81 \times 120 \times 0.9}{0.6}} = \sqrt{1765.8} = 42.02 \text{ m/s} \textbf{ Ans.}$$

**Example 14.17** *A vehicle of mass 1000 kg is travelling along a circular path of radius 100 m on a level road with a speed of 10 m/s. The height of its centre of gravity above the road is 1 m and the distance between the wheels is 1.5 m. Find the reaction at the wheels. If the coefficient of friction between the wheels and the road is 0.4, find the maximum speed of vehicle in order to avoid skidding and overturning.*

***Solution*** Given : $m = 1000$ kg ; $r = 100$ m ; $v = 10$ m/s ; $h = 1$ m ; $2a = 1.5$ m or $a = 0.75$ m ; $\mu = 0.4$

***Reaction at the wheels***

Let $R_A$ and $R_B$ be the reactions at the wheels $A$ and $B$, respectively (Refer Fig. 14.7).

We know that

$$R_A = \frac{mg}{2}\left[1 - \frac{v^2 h}{gra}\right]$$

$$= \frac{1000 \times 9.81}{2}\left[1 - \frac{(10)^2 1}{9.81 \times 100 \times 0.75}\right]$$

$$= 4.905[1 - 0.136] = 4238 \text{ N} \textbf{ Ans.}$$

$$R_A + R_B = mg = 1000 \times 9.81 = 9810 \text{ N}$$

$$\therefore \quad R_B = 9810 - R_A = 9810 - 4238 = 5572 \text{ N} \textbf{ Ans.}$$

***Maximum speed of the vehicle in order to avoid skidding***

We know that maximum speed of the vehicle in order to avoid skidding,

$$v_{max} = \sqrt{\mu g r} = \sqrt{0.4 \times 9.81 \times 100} = \sqrt{392.4} = 19.8 \text{ m/s} \textbf{ Ans.}$$

***Maximum speed of the vehicle in order to avoid overturning***

We know that maximum speed of the vehicle in order to avoid overturning,

$$v_{max} = \sqrt{\frac{gra}{h}} = \sqrt{\frac{9.81 \times 100 \times 0.75}{1}} = \sqrt{735.75} = 27.12 \text{ m/s} \textbf{ Ans.}$$

## 14.13 BANKING OF ROADS—SUPERELEVATION

A little consideration will show that the friction between the wheels of the vehicle and the road surface causes the wheels to wear out soon on a level circular path. As the curvature is increased, the frictional force may not be sufficient to hold the wheels on the road surface and the vehicle may skid or overturn causing accident. Thus, in order to avoid accidents, the outer edge of the road is raised by some amount as compared to the inner edge of the road so that the equilibrium of the vehicle is maintained. This is known as *banking of roads.* It is done to distribute the load of the vehicle equally on both the wheels.

The angle at which the outer edge of the road is raised with respect to the inner edge of the road, is known as *angle of banking.*

In case of railway tracks negotiating a circular turn, a term *superelevation* is used. It is defined as the amount by which the outer edge of the rail is raised with respect to the inner rail.

The banking of roads or superelevation provides

1. Reduction in wear and tear of the wheels of the vehicle,
2. Safer turning of the fast moving vehicles,
3. Necessary inward centripetal force to counter balance the effect of centrifugal force, and
4. Equal distribution of load on the wheels of the vehicle.

## 14.14 MOTION OF A VEHICLE ON A BANKED ROAD

Consider a vehicle moving round a curve of radius $r$ on a banked road $OP$, as shown in Fig. 14.8. The road in cross-section becomes an inclined plane at angle $\alpha$ with the horizontal $OQ$ and it is known as *angle of banking*. The various forces acting on the vehicle are as follows:

1. Weight of the vehicle ($W$) such that $W = mg$, where $m$ is the mass of the vehicle,
2. Normal reactions $R_A$ and $R_B$ at the wheels $A$ and $B$, respectively,
3. Centrifugal force of magnitude $m \times \frac{v^2}{r}$ acting outwardly at the centre of gravity $G$ of the vehicle.

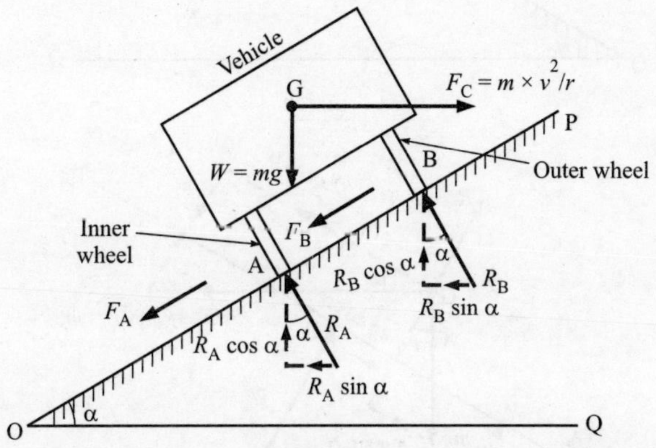


Fig. 14.8

Now resolving the reactions $R_A$ and $R_B$ into horizontal and vertical components and equating the same.

Equating the horizontal components, we have

$$R_A \sin\alpha + R_B \sin\alpha = F_C = m \times \frac{v^2}{r}$$

or

$$(R_A + R_B) \sin\alpha = m \times \frac{v^2}{r} \quad \text{...(i)}$$

Equating the vertical components, we have

$$R_A \cos\alpha + R_B \cos\alpha = W = mg$$

or

$$(R_A + R_B) \cos\alpha = mg \quad \text{...(ii)}$$

Dividing equation (*i*) by equation (*ii*), we have

$$\frac{(R_A + R_B)\sin\alpha}{(R_A + R_B)\cos\alpha} = \frac{m \times \frac{v^2}{r}}{mg}$$

or
$$\tan\alpha = \frac{v^2}{g.r}$$

$$\therefore \qquad v = \sqrt{g.r\tan\alpha}$$

This is the maximum speed of the vehicle.

## 14.15 MOTION OF A VEHICLE ON A BANKED ROAD TO AVOID SKIDDING

Consider a vehicle moving with a velocity *v*, round a curve of radius *r* on a banked road *OP*, as shown in Fig. 14.9 (*a*).

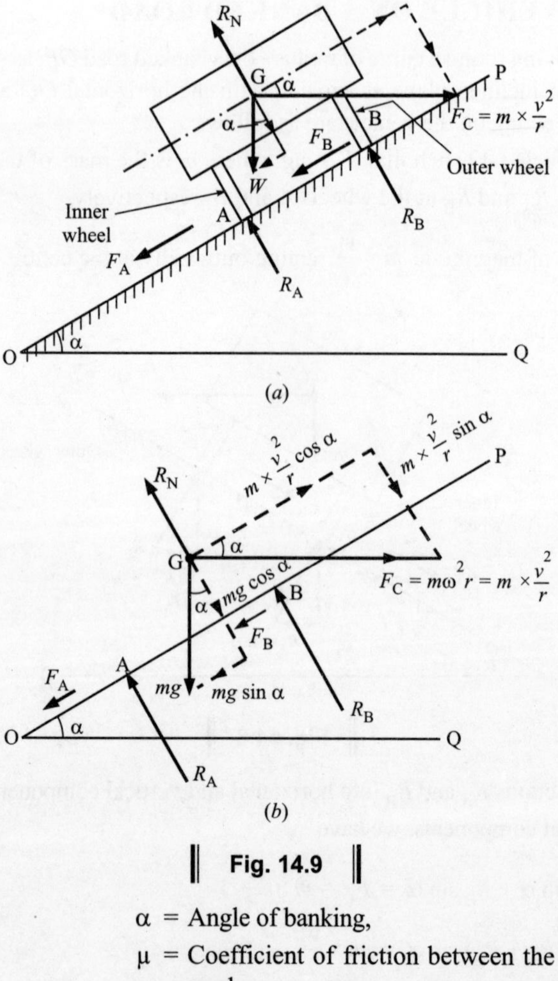


**Fig. 14.9**

Let
$\alpha$ = Angle of banking,

$\mu$ = Coefficient of friction between the wheels and surface of the road,

$F_A$ = Force of friction between the inner wheel and the surface of road = $\mu.R_A$, and

$F_B$ = Force of friction between the outer wheel and the surface of the road = $\mu.R_B$.

The various forces acting on the vehicle are shown separately in Fig. 14.9 (*b*).

The total normal reaction on the vehicle,

$$R_N = R_A + R_B \quad ...(i)$$

Resolving the forces parallel to the banked road *OP*, we have

$$F_A + F_B + mg\sin\alpha = m \times \frac{v^2}{r}\cos\alpha$$

or

$$F_A + F_B = m \times \frac{v^2}{r}\cos\alpha - mg\sin\alpha \quad ...(ii)$$

Resolving the forces perpendicular to the banked road *OP*, we have

$$R_N = mg\cos\alpha + m \times \frac{v^2}{r}\sin\alpha \quad ...(iii)$$

Since $F_A = \mu \,.\, R_A$ and $F_B = \mu . R_B$, therefore

$$F_A + F_B = \mu . R_A + \mu . R_B$$

$$= \mu(R_A + R_B) = \mu . R_N \quad ...\text{[From equation (i)]} \quad ...(iv)$$

Substituting the value of $R_N$ from equation (*iii*) in equation (*iv*), we have

$$F_A + F_B = \mu\left(mg\cos\alpha + m \times \frac{v^2}{r}\sin\alpha\right) \quad ...(v)$$

or

$$m \times \frac{v^2}{r}\cos\alpha - mg\sin\alpha = \mu\left(mg\cos\alpha + m \times \frac{v^2}{r}\sin\alpha\right) \quad ...\text{[From equation (ii)]}$$

$$= \tan\phi\left(mg\cos\alpha + m \times \frac{v^2}{r}\sin\alpha\right)$$

...($\because$ $\mu = \tan\phi$, where $\phi$ is angle of friction)

Dividing throughout by *mg* cos α, we have

$$\frac{v^2}{r.g} - \tan\alpha = \tan\phi\left(1 + \frac{v^2}{r.g}\tan\alpha\right) = \tan\phi + \frac{v^2}{r.g}\tan\alpha \,.\, \tan\phi$$

or

$$\frac{v^2}{r.g} - \frac{v^2}{r.g}\tan\alpha\tan\phi = \tan\alpha + \tan\phi$$

$$\frac{v^2}{r.g}(1 - \tan\alpha \,.\, \tan\phi) = \tan\alpha + \tan\phi$$

$$\frac{v^2}{r.g} = \frac{\tan\alpha + \tan\phi}{1 - \tan\alpha\tan\phi} = \tan(\alpha + \phi)$$

Thus, in order to avoid skidding, the speed is given by

$$v = \sqrt{r.g\tan(\alpha + \phi)}$$

**Example 14.18** *A racing car travels at a speed of 210 km/h on a track banked at 20° to the horizontal. If the coefficient of friction between the tyres and track surface is 0.4, find the minimum radius of curvature of the track, if the car does not skid outwards.*

***Solution*** Given : $v = 210 \text{ km/h} = \frac{210 \times 1000}{60 \times 60} = 58.3 \text{ m/s}$ ; $\alpha = 20°$ ; $\mu = \tan\phi = 0.4$ or $\phi = \tan^{-1}(0.4)$

$= 21.8°$

Let $r$ = Minimum radius of curvature of the track in metres.

We know that in order to avoid skidding,

$$\frac{v^2}{r.g} = \tan(\alpha + \phi)$$

or

$$r = \frac{v^2}{g \tan(\alpha + \phi)} = \frac{(58.3)^2}{9.81 \tan(20° + 21.8°)}$$

$$= \frac{3399}{9.81 \times 0.894} = 387.6 \text{ m} \quad \textbf{Ans.}$$

## 14.16 SUPERELEVATION FOR RAILWAYS

We have already discussed that the term superelevation is used in railway tracks and it is the amount by which the outer edge of the rails is raised as compared to inner rail.

Let $r$ = Radius of the circular path, and

$l$ = Distance between the inner and outer rails.

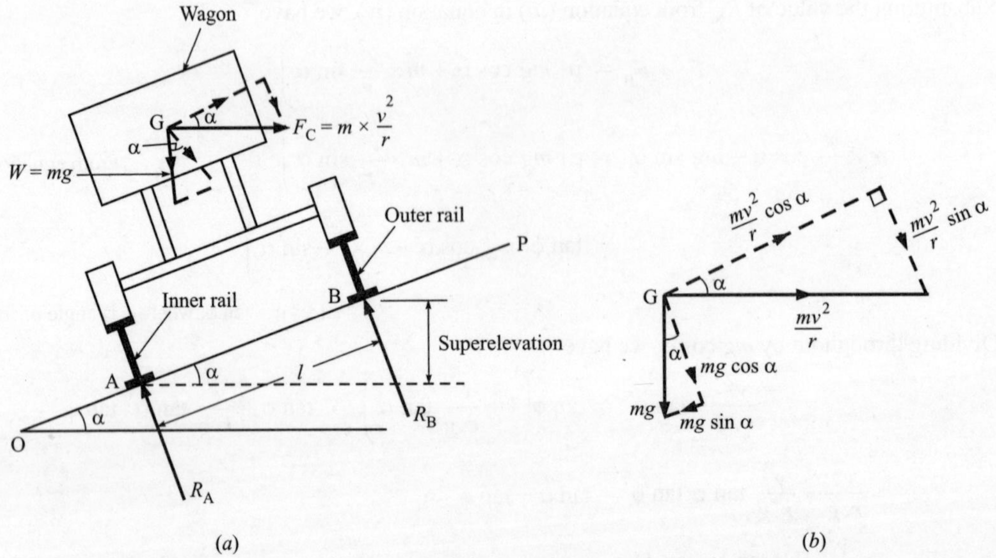


**Fig. 14.10**

The wagon is in equilibrium under the action of the following forces:

1. Weight of the wagon ($W$), such that $W = mg$, where $m$ is the mass of the wagon,
2. Normal reactions $R_A$ and $R_B$ at the inner and outer rails, respectively, and
3. Centrifugal force, $F_C = \frac{mv^2}{r}$

The weight of the wagon ($W = mg$) and the centrifugal force $\left(F_C = \frac{mv^2}{r}\right)$ acting at the centre of gravity ($G$) of wagon is shown in Fig. 14.10 ($a$). Resolve $W$ and $F_C$ into two components, one along the inclined plane $OP$ and the other perpendicular to the plane, as shown separately in Fig. 14.10 ($b$).

Considering components along the plane,

$$\frac{mv^2}{r} \cos \alpha = mg \sin \alpha$$

or
$$\frac{v^2}{r} = \frac{g \sin \alpha}{\cos \alpha} = g \tan \alpha$$

$$\tan \alpha = \frac{v^2}{r.g}$$

∴ Angle of superelevation,
$$\alpha = \tan^{-1}\left(\frac{v^2}{r.g}\right)$$

From Fig. 14.10 (*a*), we find that,

$$\text{Superelevation} = l \sin \alpha$$

Since α is very small, therefore sin α may be taken as tan α.

∴
$$\text{Superelevation} = l \tan \alpha = l \times \frac{v^2}{r.g}$$

**Example 14.19** *A locomotive of mass 50 tonnes travels round a curve of 250 m radius at a speed of 60 km/h. If the distance between the two rails is 1.2 m, find the amount by which the outer rail is raised above the level of the inner rail so that there is no thrust on the wheel flanges.*

***Solution*** Given : *$m = 50$ t ; $r = 250$ m ; $v = 60$ km/h $= \frac{60 \times 1000}{60 \times 60} = 16.67$ m/s ; $l = 1.2$ m

Let
$$\alpha = \text{Angle of superelevation.}$$

We know that
$$\tan \alpha = \frac{v^2}{r.g} = \frac{(16.67)^2}{250 \times 9.81} = 0.1133$$

or
$$\alpha = \tan^{-1}(0.1133) = 6.46°$$

∴ Height of the outer rail above the inner rail or superelevation

$$= l \tan \alpha = 1.2 \times 0.1133$$
$$= 0.136 \text{ m} = 136 \text{ mm} \quad \textbf{Ans.}$$

## HIGHLIGHTS

1. The *angular displacement* (θ) is defined as the angle described by the particle from one point to another, with respect to the time.
2. The *angular velocity* (ω) is defined as the rate of change of angular displacement with respect to time. Mathematically, angular velocity,

$$\omega = \frac{\theta}{t} \text{ or } \frac{d\theta}{dt} \text{ (in rad/s)}$$

If a body is rotating at the rate of $N$ r.p.m., then angular velocity,

$$\omega = \frac{2\pi N}{60} \text{ rad/s}$$

* Superfluous data.

3. The *angular acceleration* (α) is defined as the rate of change of angular velocity with respect to time. Mathematically, angular acceleration,

$$\alpha = \frac{d\omega}{dt} = \frac{d}{dt}\left(\frac{d\theta}{dt}\right) = \frac{d^2\theta}{dt^2}$$

4. The relation between linear and angular quantities of motion are as follows:
   (*a*) Linear displacement, $s = r \,.\, \theta$
   (*b*) Linear velocity, $v = \frac{ds}{dt} = r \times \frac{d\theta}{dt} = r.\omega$
   (*c*) Linear acceleration, $a = \frac{dv}{dt} = r \times \frac{d\omega}{dt} = r.\alpha$

   where *r* is the radius of the circular path.
5. The equations of angular motion are given as follows:
   (*a*) $\omega = \omega_0 + \alpha\, t$
   (*b*) $\theta = \omega_0 .t + \frac{1}{2}\alpha t^2 = \frac{\omega_0 + \omega}{2} \times t$
   (*c*) $\omega^2 = \omega_0^2 + 2\alpha.\theta$

   where
   $\omega_0$ = Initial angular velocity in rad/s,
   $\omega$ = Final angular velocity in rad/s,
   $\alpha$ = Angular acceleration in rad/s$^2$,
   $\theta$ = Angular displacement in radians, and
   $t$ = Time taken in seconds.
6. The acceleration of a particle moving along a circular path has the following two accelerations which are perpendicular to each other:
   (*a*) ***Tangential (or linear) acceleration*** The acceleration of a particle at any instant moving along a circular path in a direction tangential to that instant is known as *tangential* or *linear acceleration*. Mathematically,

   Tangential (or linear acceleration),

   $$a_t = \frac{dv}{dt} = \alpha.r$$

   (*b*) ***Normal acceleration*** The acceleration of a particle at any instant moving along a circular path in a direction normal to the tangent at that instant and directed towards the centre of the circular path is known as *normal acceleration*. Mathematically,

   $$\text{Normal acceleration, } a_n = v \times \frac{d\theta}{dt} = v.\omega = \frac{v^2}{r} = \omega^2.r$$
7. The total acceleration (*a*) of a body moving along a circular path is given by

   $$a = \sqrt{(a_t)^2 + (a_n)^2}$$

   and its inclination with the tangential acceleration is given by

   $$\tan\alpha = \frac{a_n}{a_t}$$
8. When a particle moves with a uniform velocity, then $\frac{dv}{dt}$ will be zero, *i.e.* there will be no tangential acceleration and the particle will have only normal acceleration whose value is given by

   $$a_n = \frac{v^2}{r} = v.\omega = \omega^2.r$$

**9.** When a particle moves along a straight path, then there will be no normal acceleration and the particle will have only tangential acceleration whose value is given by

$$a_t = \frac{dv}{dt} = \alpha \cdot r$$

**10.** The maximum speed ($v$) at which the cyclist can move round a circular path is given by

$$v = \sqrt{g \cdot r \tan \alpha}$$

and the angle of inclination ($\alpha$) of the cyclist with the vertical at the centre of gravity ($G$) is given by

$$\tan \alpha = \frac{v^2}{gr}$$

or

$$\alpha = \tan^{-1}\left(\frac{v^2}{g.r}\right)$$

**11.** The maximum velocity ($v$) of the vehicle moving in a level circular path is given by

$$v = \sqrt{\mu g r} \quad \text{...(in order to avoid skidding)}$$

$$= \sqrt{\frac{g r a}{h}} \quad \text{...(in order to avoid overturning)}$$

where $\mu$ = Coefficient of friction between the wheels and the road surface,

$r$ = Radius of the curved path,

$2a$ = Distance between the inner and outer wheels, and

$h$ = Height of centre of gravity of the vehicle from the road surface.

**12.** The maximum velocity ($v$) of a vehicle moving on a banked circular road, in order to avoid skidding, is given by

$$v = \sqrt{r \cdot g \tan(\alpha + \phi)}$$

where $\alpha$ = Angle of banking, and

$\phi$ = Angle of friction.

**13.** The term *superelevation* is used in railway tracks and it is defined as the amount by which the outer edge of the rails is raised as compared to the inner rail.

**14.** The angle of superelevation ($\alpha$) is given by

$$\tan \alpha = \frac{v^2}{r \cdot g}$$

or

$$\alpha = \tan^{-1}\left(\frac{v^2}{r \cdot g}\right)$$

and superelevation

$$= l \tan \alpha = l \times \frac{v^2}{r \cdot g}$$

where $l$ = Distance between the inner and outer rails, and

$r$ = Radius of the circular path.

## NUMERICAL PROBLEMS

1. A wheel accelerating uniformly at 5 rad/s$^2$ is found to rotate at 900 r.p.m. at the end of 10 seconds. Find the initial angular velocity and the angular displacement during this interval. [**Ans.** 44.26 rad/s; 781.12 rad]

2. A horizontal bar 1.5 m long and of small cross-section rotates about vertical axis through one end. It accelerates uniformly from 1200 r.p.m. to 1500 r.p.m. in an interval of 5 seconds. What is the linear velocity at the beginning and at the end of the interval? Also find the angular acceleration of the bar during this time.

   [**Ans.** 188.6 m/s; 235.5 m/s; 6.26 rad/s$^2$]

3. A wheel is rotating about its axis with a constant angular acceleration of 1 rad/s$^2$. If the initial and final angular velocities are 5.25 rad/s and 10.5 rad/s, determine the total angle turned through during the time interval when this change of angular velocity took place. [**Ans.** 41.35 rad]

4. A wheel is rotating at 240 r.p.m. and after 10 seconds, it is rotating at 180 r.p.m. Determine the number of revolutions made and time taken by the flywheel before it comes to rest, if the retardation is uniform.

   [**Ans.** 80 r.p.m.; 40 s]

5. A flywheel starts rotating from rest and is given an acceleration of 1.2 rad/s$^2$. Find the angular velocity and speed in r.p.m. after 1.5 minutes. If the flywheel is brought to rest with a uniform angular retardation of 0.6 rad/s$^2$, find the time taken by the flywheel to come to rest. [**Ans.** 108 rad/s, 1031.2 r.p.m.; 3 min]

6. A wheel that is rotating at 300 r.p.m. attains a speed of 180 r.p.m. after 20 seconds. Determine the acceleration of the flywheel assuming it to come to rest from a speed of 300 r.p.m. If the acceleration remains the same, find the number of revolutions made during this time. [**Ans.** 0.6284 (retardation); 125]

7. The angular velocity of a flywheel moving in a circular path is given by

   $$\omega = 3t^2 - 2t + 2$$

   where $\omega$ is in radians per second and $t$ is in seconds. After 1 second from the start, the angular displacement was 4 radians. Determine the angular displacement, angular velocity and angular acceleration of the flywheel when $t = 3$ seconds. [**Ans.** 26 rad; 23 rad/s; 16 rad/s$^2$]

8. The angular displacement of a particle moving in a circular path is given by

   $$\theta = 2t^3 - 5t^2 + 8t + 6$$

   where $\theta$ is the angular displacement in radians and $t$ is the time in seconds. Determine the angular velocity and acceleration when $t = 0$ and $t = 4$ seconds. [**Ans.** 8 rad/s, –10 rad/s$^2$; 64 rad/s, 38 rad/s$^2$]

9. A locomotive is running at a constant speed of 100 km/h. The diameter of the driving wheels is 1.8 m. The stroke of the piston of the steam engine cylinder is 600 mm. Find the centripetal acceleration of the crank pin relative to the engine frame. [**Ans.** 288.3 m/s$^2$]

10. A car enters a curve of 300 m radius at a speed of 50 km/h. If the speed of the car is increased at a rate of 2 m/s$^2$, what will be the magnitude and direction of its total acceleration when it has travelled 250 m along the curve? [**Ans.** 4.45 m/s$^2$; 63.3°]

11. A person, riding a motor-cycle at a speed of 72 km/h, negotiates a turn whose radius is 100 m. Find the centripetal acceleration and the inclination from the vertical while turning. [**Ans.** 4 m/s$^2$; 22.18°]

12. A vehicle of mass 1600 kg is travelling along an unbanked curve of 60 m radius. The wheel base is 1.6 m and the centre of gravity is 0.64 m above the ground and mid-way between the two wheels. The coefficient of friction between the wheels and the road is 0.4. Find: 1. reactions at the wheels at a speed of 17.5 m/s; and 2. the maximum speed at which the vehicle runs without skidding and overturning.

    [**Ans.** 5.398 kN, 10.298 kN; 17.7 m/s, 31.32 m/s]

13. A vehicle is moving round a curve of 80 m radius without any slip. At what speed the vehicle should move in the following cases:

    (*a*) on a level road; and (*b*) on a road banked an angle of 6° with the horizontal.

    The coefficient of friction between the wheels of the vehicle and the road surface may be taken as 0.24.

    [**Ans.** 13.7 m/s; 16.7 m/s]

**14.** A locomotive of mass 150 tonnes is rounding a curve of 800 m radius at a speed of 72 km/h. Calculate the radial pressure between the wheels and the rails. Find the angle at which the track should be banked in order to reduce the radial pressure to zero at the above speed. Determine also the height at which the outer rail should be raised above the inner rail. The distance between the two rails being 1.676 m.

**[Ans.** 75 kN; 2.92°; 85.4 mm]

## REVIEW QUESTIONS

1. Define the following terms:
   (*a*) Angular displacement; (*b*) Angular velocity; and (*c*) Angular acceleration.
2. What is the relation between angular velocity (ω) in rad/s and speed of rotation in revolutions per minute?
3. Derive a relation between linear velocity and angular velocity of a rotating body.
4. Compare the equations of linear motion and angular motion.
5. Explain the tangential acceleration and normal acceleration of a particle moving along a circular path.
6. Distinguish between centripetal force and centrifugal force.
7. Explain, why a cyclist has to lean inwards while moving along a circular path.
8. Derive the relation for maximum velocity of the vehicle while moving on a level circular path, in order to avoid skidding.
9. What do you understand by banking of roads? Define angle of banking.
10. Derive the expression for maximum velocity of a vehicle to avoid skidding on a banked road.
11. What is superelevation? Derive the relation for superelevation for railways.

## OBJECTIVE TYPE QUESTIONS

1. The angle described by a particle from one point to another, with respect to time is called
   (*a*) angular acceleration (*b*) angular displacement
   (*c*) angular velocity (*d*) none of these
2. The rate of change of angular velocity with respect to time is known as
   (*a*) angular acceleration (*b*) angular displacement
   (*c*) angular velocity (*d*) none of these
3. The angular acceleration of a rotating body is measured in
   (*a*) m/s (*b*) $m/s^2$ (*c*) rad/s (*d*) $rad/s^2$
4. The angular velocity (ω) of a rotating body is given by
   (*a*) $\omega = v \cdot r$ (*b*) $\omega = \dfrac{v}{r}$ (*c*) $\omega = \alpha \cdot r$ (*d*) $\omega = \dfrac{\alpha}{r}$

   where $v$ = Linear velocity of a body,
   $\alpha$ = Angular acceleration, and
   $r$ = Radius of a rotating body.
5. The linear acceleration ($a$) of a rotating body is given by
   (*a*) $a = v \cdot r$ (*b*) $a = \dfrac{v}{r}$ (*c*) $a = \alpha \cdot r$ (*d*) $a = \dfrac{\alpha}{r}$

6. When a body moving along a circular path has $\omega_0$ and $\omega$ as its initial and final velocities in rad/s and $t$ is the time taken (in seconds) during which angular velocity changes from $\omega_0$ to $\omega$, then which of the following equation is correct?

(a) $\omega = \omega_0 + \alpha . t$ (b) $\omega^2 = \omega_0^2 + 2\alpha \cdot \theta$ (c) $\theta = \omega_0 \cdot t + \frac{1}{2}\alpha \cdot t^2$ (d) all of these

where $\theta$ = Angular displacement in radians, and
$\alpha$ = Angular acceleration in rad/s$^2$.

7. The tangential acceleration ($a_t$) of a body moving along a circular path is given by

(a) $a_t = v \cdot r$ (b) $a_t = \frac{v}{r}$ (c) $a_t = \alpha \cdot r$ (d) $a = \frac{\alpha}{r}$

8. The normal acceleration ($a_n$) of a body moving along a circular path is given by

(a) $a_n = v \cdot \omega$ (b) $a_n = \omega^2 \cdot r$ (c) $a_n = \frac{v^2}{r}$ (d) all of these

9. The normal acceleration is also known as

(a) radial acceleration (b) centripetal acceleration
(c) either (a) or (b) (d) none of these

10. The total acceleration ($a$) of a body moving along a circular path is given by

(a) $\sqrt{(a_t)^2 + (a_n)^2}$ (b) $\sqrt{(a_t)^2 - (a_n)^2}$ (c) $\sqrt{a_t + a_n}$ (d) $\sqrt{a_t - a_n}$

11. The force which acts along the radius of a circle and directed ............. the centre of the circle is known as centripetal force.

(a) away from (b) towards

12. When a body of mass $m$ is moving with a linear velocity $v$ in a circular path of radius $r$, then centrifugal force acting on the body is

(a) $mv^2r$ (b) $\frac{mr}{v^2}$ (c) $\frac{mv^2}{r}$ (d) none of these

13. The angle of inclination ($\alpha$) of the cyclist moving round a circular path is

(a) $\tan\alpha = \frac{v^2}{g \cdot r}$ (b) $\tan\alpha = \frac{v^2 \cdot r}{g}$ (c) $\tan\alpha = \frac{g \cdot r}{v^2}$ (d) $\tan\alpha = v^2 \cdot g \cdot r$

14. The maximum velocity of a vehicle moving on a level circular path is

(a) $\frac{\mu}{g \cdot r}$ (b) $\frac{\mu \cdot g}{r}$ (c) $\mu \cdot g \cdot r$ (d) $\sqrt{\mu \cdot g \cdot r}$

15. The superelevation is given by the relation

(a) $\frac{r \cdot g}{l \cdot v^2}$ (b) $l \times \frac{v^2}{r \cdot g}$ (c) $\frac{l \cdot r}{v^2 \cdot g}$ (d) $\frac{v^2 \cdot g \cdot r}{l}$

***Answers to Objective Type Questions***

| | | | | |
|---|---|---|---|---|
| 1. (b) | 2. (a) | 3. (d) | 4. (b) | 5. (c) |
| 6. (d) | 7. (c) | 8. (b) | 9. (c) | 10. (a) |
| 11. (b) | 12. (c) | 13. (a) | 14. (d) | 15. (b) |

15

# SIMPLE HARMONIC MOTION

*1. Introduction 2. Velocity and Acceleration of a Particle Moving with Simple Harmonic Motion 3. Technical Terms Used in Simple Harmonic Motion 4. Oscillation of a Vertical Elastic Spring 5. Equivalent Spring Stiffness 6. Simple Pendulum 7. Compound Pendulum*

## 15.1 INTRODUCTION

When a particle moves in such a way that its acceleration is always proportional to its distance from a fixed point and is directed to the fixed point, then the particle is said to have *simple harmonic motion.*

Consider a particle moving round the circumference of a circle of radius $r$, in the anticlockwise direction at a constant angular velocity of $\omega$ radians per second, as shown in Fig. 15.1. Let $P$ be the position of the particle at any instant and $Q$ is its projection on the diameter $AC$. A little consideration will show that when the point $P$ moves round the circumference from $A$ to $B$, its projection $Q$ moves from $A$ to $O$ and when the point $P$ moves from $B$ to $C$, the point $Q$ moves from $O$ to $C$. Similarly, when $P$ moves from $C$ to $D$, then $Q$ moves from $C$ to $O$ and finally when $P$ moves from $D$ to $A$, the point $Q$ moves from $O$ to $A$. This to and fro motion of the point $Q$ along the diameter $AC$ is known as *simple harmonic motion.*

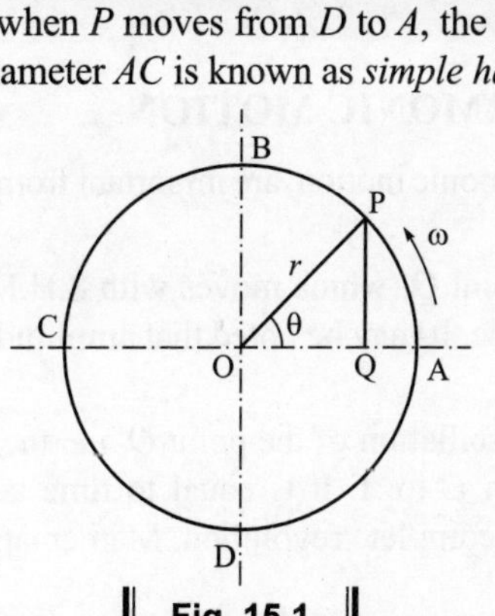


**Fig. 15.1**

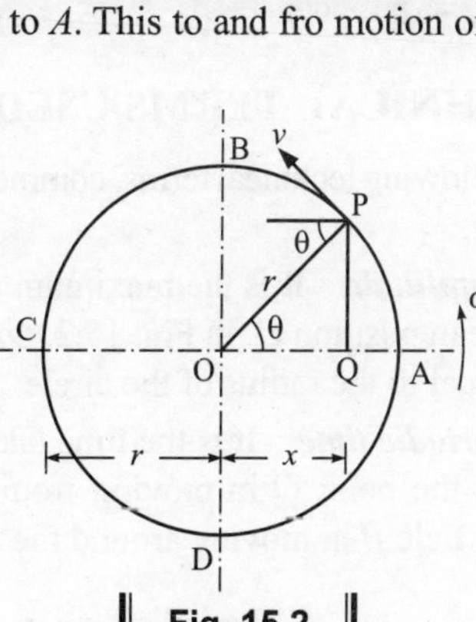


**Fig. 15.2**

## 15.2 VELOCITY AND ACCELERATION OF A PARTICLE MOVING WITH SIMPLE HARMONIC MOTION

Let $P$ be any position of the particle after time $t$ seconds and $\theta$ be the angle turned during this period. We know that
$$\theta = \omega.t$$

and displacement of point $Q$ from the centre of the circle (*i.e.* mean position), as shown in Fig. 15.2, is given by

$$x = r\cos\theta = r\cos\omega t \qquad ...(i)$$

Differentiating equation (*i*) with respect to *t*, we have velocity of $Q$,

$$v = \frac{dx}{dt} = -r.\omega\sin\omega t = -\omega.r\sin\theta \qquad ...(ii)$$

In a right angled triangle $OPQ$, $\sin\theta = \frac{PQ}{OP} = \frac{(OP)^2 - (OQ)^2}{OP} = \frac{\sqrt{r^2 - x^2}}{r}$

Substituting this value of $\sin\theta$ in equation (*ii*), we have

$$v = -\omega r \times \frac{\sqrt{r^2 - x^2}}{r} = -\omega\sqrt{r^2 - x^2} \qquad ...(iii)$$

The velocity is maximum, when $x = 0$, *i.e.* when the point $Q$ is at $O$. In other words, the velocity is maximum at the mean position. Thus, maximum velocity,

$$v_{max} = -\omega.r$$

In order to find the acceleration of point $Q$, differentiate equation (*ii*) with respect to *t*, we have

$$a = \frac{d^2x}{dt^2} = \frac{dv}{dt} = -r.\omega.\omega\cos\omega t$$

$$= -\omega^2 r\cos\omega t = -\omega^2 r\cos\theta = -\omega^2 x \qquad ...(iv)$$

$$...(\because x = r\cos\omega t = r\cos\theta)$$

The –ve sign shows that the direction of acceleration is opposite to the direction in which $x$ increases, *i.e.* the acceleration is always directed towards the centre of the circle (point $O$). In other words we can say that the point $Q$ is under retardation.

From equation (*iv*), we see that when $x = 0$, the acceleration is zero, when point $Q$ is at $O$ ( *i.e.* mean position) and the acceleration is maximum, when $x = r$.

∴ Maximum acceleration, $a_{max} = \omega^2 r$

**Note:** From equation (*iv*), acceleration, $a = -\omega^2 x$

Since $\omega$ is constant, therefore acceleration is directly proportional to the distance of $Q$ from the centre of the circle (*i.e.* mean position).

## 15.3 TECHNICAL TERMS USED IN SIMPLE HARMONIC MOTION

The following technical terms, commonly used in simple harmonic motion, are important from the subject point of view:

1. ***Amplitude*** It is the maximum displacement of the point $Q$ (which moves with S.H.M.) from its mean position $O$. In Fig. 15.2, $OA$ or $OC$ is the amplitude. It may be noted that amplitude is always equal to the radius of the circle.
2. ***Periodic time*** It is the time taken for one complete oscillation of the point $Q$, *i.e.* the time taken by the point $Q$ in moving from $A$ to $C$ and then from $C$ to $A$. It is equal to time taken by the particle $P$ in moving around the circumference for one complete revolution. Mathematically,

   Periodic time, $t_p = \frac{2\pi}{\omega}$ seconds

   where $\omega$ = Constant angular velocity of the particle $P$.

   We know that the acceleration, $a = \omega^2 . x$ or $\omega^2 = \frac{a}{x}$ or $\omega = \sqrt{\frac{a}{x}}$

$$\therefore \quad t_p = \frac{2\pi}{\omega} = 2\pi\sqrt{\frac{x}{a}} = 2\pi\sqrt{\frac{\text{Displacement}}{\text{Acceleration}}}$$

3. ***Frequency*** It is equal to the number of cycles per second and is reciprocal of the time period ($t_p$). Mathematically,

$$\text{Frequency}, f = \frac{\omega}{2\pi} = \frac{1}{t_p} = \frac{1}{2\pi}\sqrt{\frac{a}{x}}$$

**Note:** In S.I. units, the unit of frequency is hertz (briefly written as (Hz), which is equal to one cycle per second.

**Example 15.1** *Find the velocity and acceleration after 0.3 seconds from the extreme position of a body, moving with simple harmonic motion, with an amplitude of 0.8 m and period of complete vibration of 1.6 seconds.*

***Solution*** Given : $t = 0.3$ s ; $r = 0.8$ m ; $t_p = 1.6$ s

Let $\omega$ = Angular velocity of the body in rad/s.

We know that time period,

$$t_p = \frac{2\pi}{\omega}$$

$$\therefore \quad \omega = \frac{2\pi}{t_p} = \frac{2\pi}{1.6} = 3.93 \text{ rad/s}$$

***Velocity of the body***

Since the time is measured from the extreme position, the velocity after 0.3 seconds is given by

$$v_Q = -\omega \cdot r \sin\theta$$

$$= {}^{*}-\omega \cdot r \sin\omega t = -3.93 \times 0.8 \sin\left(3.93 \times 0.3 \times \frac{180}{\pi}\right)$$

$$= -3.144 \sin 67.54° = -3.144 \times 0.924$$

$$= -2.9 \text{ m/s} \quad \textbf{Ans.}$$

***Acceleration of the body***

We know that acceleration of the body after 0.3 seconds from the extreme position,

$$a_Q = -\omega^2 \cdot r \cos\theta = -\omega^2 \cdot r \cos\omega t$$

$$= -(3.93)^2\, 0.8 \cos 67.54°$$

$$= -12.356 \times 0.382 = -4.72 \text{ m/s}^2 \quad \textbf{Ans.}$$

**Example 15.2** *The piston of a reciprocating engine has a motion which is simple harmonic in character. The stroke is 225 mm and it makes 200 complete oscillations in a minute. If it weighs 4.54 kg, find :*

1. *The accelerating force upon it in newtons and its velocity in m/s when it is at a point 75 mm from mid-stroke position ;*
2. *The maximum accelerating force ; and*
3. *The maximum velocity.*

***Solution*** Given : $l = 2r = 225$ mm or $r = 112.5$ mm $= 0.1125$ m ; $N = 200$ r.p.m. or $\omega = \frac{2\pi \times 200}{60} = \frac{20\pi}{3}$ rad/s ; $m = 4.54$ kg

---

$* \; \theta = \omega t = 3.93 \times 0.3 = 1.179 \text{ rad} = 1.179 \times \frac{180}{\pi} = 67.54°$

**1. *Accelerating force and velocity when x = 75 mm = 0.075 m***

We know that acceleration of the piston,

$$a = \omega^2 x = \left(\frac{20\pi}{3}\right)^2 0.075 = 32.9 \text{ m/s}^2$$

$\therefore$ Accelerating force $= m.a = 4.54 \times 32.9 = 149.4$ N **Ans.**

We also know that velocity of the piston,

$$v = \omega\sqrt{r^2 - x^2} = \frac{20\pi}{3}\sqrt{(0.1125)^2 - (0.075)^2}$$

$$= \frac{20\pi}{3}\sqrt{0.0126 - 0.0056} = \frac{20\pi}{3}\sqrt{0.0070}$$

$$= \frac{20\pi}{3} \times 0.0837 = 1.753 \text{ m/s} \quad \textbf{Ans.}$$

**2. *Maximum accelerating force***

We know that maximum acceleration,

$$a_{max} = \omega^2 r = \left(\frac{20\pi}{3}\right)^2 0.1125 = 438.76 \times 0.1125 = 49.36 \text{ m/s}^2$$

$\therefore$ Maximum accelerating force $= m \,.\, a_{max} = 4.54 \times 49.36 = 224$ N **Ans.**

**3. *Maximum velocity***

We know that maximum velocity,

$$v_{max} = \omega.r = \frac{20\pi}{3} \times 0.1125 = 2.357 \text{ m/s} \quad \textbf{Ans.}$$

**Example 15.3** *In a mechanism, a crosshead moves in a straight guide with S.H.M. At a distance of 125 mm and 200 mm from its mean position, it has velocities of 6 m/s and 3 m/s, respectively. Find the amplitude, maximum velocity and period of vibration. If the crosshead weighs 0.2 kg, calculate the maximum force on it, in the direction of motion.*

***Solution*** Given : $x_1 = 125$ mm $= 0.125$ m ; $x_2 = 200$ mm $= 0.2$ m ; $v_1 = 6$ m/s ; $v_2 = 3$ m/s ; $m = 0.2$ kg

***Amplitude***

Let $\omega$ = Angular velocity of the crosshead in rad/s, and

$r$ = Amplitude in m.

We know that velocity of the crosshead when it is 0.125 m from its mean position,

$$v_1 = \omega\sqrt{r^2 - x_1^2} = \omega\sqrt{r^2 - (0.125)^2} \quad ...(i)$$

Similarly, velocity of the crosshead when it is at 0.2 m from the mean position,

$$v_2 = \omega\sqrt{r^2 - x_2^2} = \omega\sqrt{r^2 - (0.2)^2} \quad ...(ii)$$

Dividing equation (*i*) by equation (*ii*),

$$\frac{v_1}{v_2} = \frac{\omega\sqrt{r^2 - (0.125)^2}}{\omega\sqrt{r^2 - (0.2)^2}}$$

or

$$\frac{6}{3} = \frac{\sqrt{r^2 - 0.0156}}{\sqrt{r^2 - 0.04}}$$

Squaring both sides, we have $\frac{36}{9} = \frac{r^2 - 0.0156}{r^2 - 0.04}$

or $4(r^2 - 0.04) = r^2 - 0.0156$

$$4r^2 - 0.16 = r^2 - 0.0156$$

$$4r^2 - r^2 = 0.16 - 0.0156 = 0.1444$$

$$r^2 = \frac{0.1444}{3} = 0.048\,13$$

$$\therefore \quad r = \sqrt{0.048\,13} = 0.219 \text{ m} \quad \textbf{Ans.}$$

***Maximum velocity***

First of all, let us find the angular velocity (ω). Substituting the value of $r$ in equation (*i*), we have

$$v_1 = \omega\sqrt{r^2 - (0.125)^2}$$

$$6 = \omega\sqrt{(0.219)^2 - (0.125)^2} = \omega\sqrt{0.048\,13 - 0.0156}$$

$$= \omega\sqrt{0.03253}$$

Squaring both sides $6^2 = \omega^2 \times 0.03253$

$$\therefore \quad \omega = \sqrt{\frac{6^2}{0.03253}} = 33.27 \text{ rad/s}$$

We know that maximum velocity,

$$v_{max} = \omega \,.\, r = 33.27 \times 0.219 = 7.286 \text{ m/s} \quad \textbf{Ans.}$$

***Period of vibration***

We know that period of vibration, $t_p = \frac{2\pi}{\omega} = \frac{2\pi}{33.27} = 0.19 \text{ s}$ **Ans.**

***Maximum force***

We know that maximum acceleration,

$$a_{max} = \omega^2 r = (33.27)^2\, 0.219 = 242.4 \text{ m/s}^2$$

$\therefore$ Maximum force $= m \,.\, a = 0.2 \times 242.4 = 48.48$ N **Ans.**

## 15.4 OSCILLATION OF A VERTICAL ELASTIC SPRING

Consider a vertical elastic spring with upper end fixed, as shown in Fig. 15.3. If a weight $W$ is attached at the lower end of the spring, it will be extended (say by an amount δ) in the downward direction and finally the load will be in equilibrium position. Let *AA* be the equilibrium position of the spring. If the spring is further stretched up to *BB* and then released, the weight will move up and down with simple harmonic motion.

Let $m$ = Mass attached to the spring in kg,

$W$ = Weight attached in newtons = $m.g$,

$\delta$ = Extension or deflection of the spring when weight is attached to the spring, in metres and is also called static extension,

$x$ = Displacement of the load below the equilibrium position in metres,

$a$ = Acceleration of the weight $W$ in m/s², and

$s$ = Stiffness of the spring in N/m, *i.e.* *restoring force per unit displacement from the equilibrium position.

* It is the force which will bring back the load $W$ to the equilibrium position.

We know that extension or deflection of the spring,

$$\delta = \frac{W}{s} = \frac{m \cdot g}{s}$$

$$\text{Disturbing force} = m \cdot a \qquad ...(i)$$

and restoring force $= s \cdot x \qquad ...(ii)$

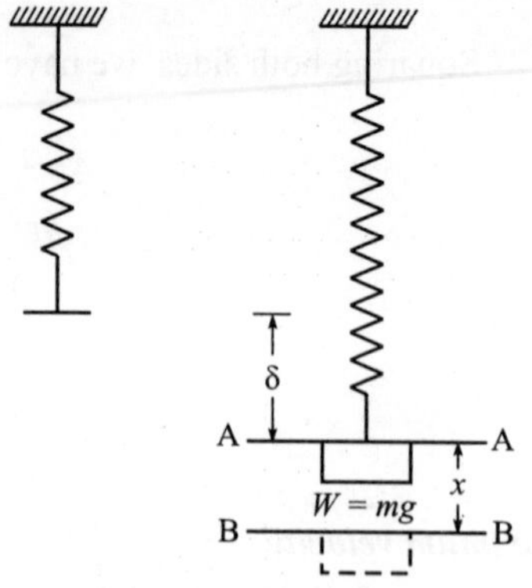


**Fig. 15.3**

Since the disturbing force must be equal to the restoring force, therefore equating equations (*i*) and (*ii*),

$$m.a = s \cdot x$$

or

$$\frac{x}{a} = \frac{m}{s}$$

We know that periodic time, $t_p = 2\pi\sqrt{\frac{\text{Displacement}}{\text{Acceleration}}}$

$$= 2\pi\sqrt{\frac{x}{a}}$$

$$= 2\pi\sqrt{\frac{m}{s}} = 2\pi\sqrt{\frac{\delta}{g}} \qquad ...\left(\because \delta = \frac{m.g}{s}\right)$$

and frequency, $f = \frac{1}{t_p} = \frac{1}{2\pi}\sqrt{\frac{s}{m}} = \frac{1}{2\pi}\sqrt{\frac{g}{\delta}}$

**Note:** If the mass of the spring ($m_1$) is also taken into consideration, then periodic time,

$$t_p = 2\pi\sqrt{\frac{m + \frac{m_1}{3}}{s}} \text{ seconds}$$

and frequency, $f = \frac{1}{t_p} = \frac{1}{2\pi}\sqrt{\frac{s}{m + \frac{m_1}{3}}}$ Hz

## 15.5 EQUIVALENT SPRING STIFFNESS

Sometimes, loads are supported with one or more springs simultaneously. It may be noted that in such cases, it is required to find out the equivalent spring stiffness ($s$) of the system. We shall now discuss the following two cases for finding out the equivalent spring stiffness.

### 1. *Springs in series*

When two or more springs are joined end to end, they are said to be in series. Let the two springs have stiffness $s_1$ and $s_2$, respectively, are connected in series, as shown in Fig. 15.4 (*a*). It may be noted that each spring will be subjected to the same load $W$ and the total extension or deflection produced by the spring is equal to the sum of the deflection of the individual springs.

∴ Total deflection of springs, $\delta = \delta_1 + \delta_2$

or

$$\frac{W}{s} = \frac{W}{s_1} + \frac{W}{s_2}$$

$$\frac{1}{s} = \frac{1}{s_1} + \frac{1}{s_2} = \frac{s_1 + s_2}{s_1 s_2}$$

or $$s = \frac{s_1 s_2}{s_1 + s_2}$$

where $s$ = Equivalent (combined) stiffness of the springs.

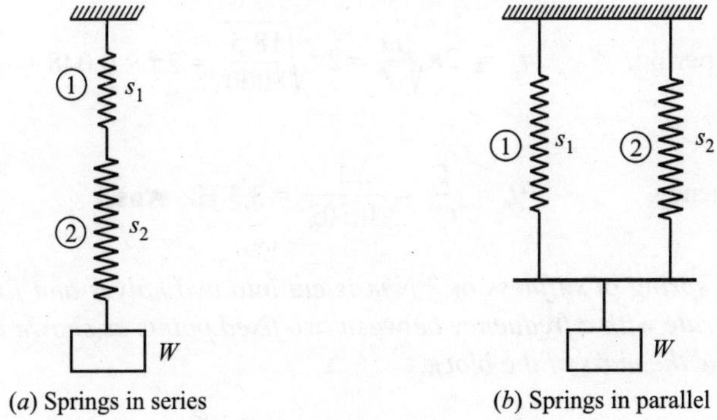


**Fig. 15.4**

### 2. *Springs in parallel*

When two or more springs are joined in such a way that they share the loads and deflect equally, then they are said to be in parallel. Let the two springs having stiffness $s_1$ and $s_2$, respectively, are connected in parallel, as shown in Fig. 15.4 (*b*). In this case, the load $W$ is shared by both the springs and the deflection produced by each spring will be same.

We know that $W = W_1 + W_2$

or $\delta.s = \delta.s_1 + \delta.s_2$

$\therefore$ $s = s_1 + s_2$

where $s$ – Equivalent (combined) stiffness of the springs,

$\delta$ = Deflection produced,

$W_1$ = Load shared by spring 1, and

$W_2$ = Load shared by spring 2.

**Example 15.4** *A weight of 31.5 N produces a static deflection of 30 mm in a given spring. Determine the period of vibration of a weight of 45 N suspended from the same spring.*

***Solution*** Given : $W = 31.5$ N ; $\delta = 30$ mm = 0.03 m ; $W_1 = 45$ N or $m_1 = \frac{45}{9.81} = 4.587$ kg

We know that stiffness of the spring,

$$s = \frac{W}{\delta} = \frac{31.5}{0.03} = 1050 \text{ N/m}$$

$\therefore$ Period of vibration of a weight of 45 N or mass ($m_1$),

$$t_p = 2\pi\sqrt{\frac{m_1}{s}} = 2\pi\sqrt{\frac{4.587}{1050}} = 0.415 \text{ s} \quad \textbf{Ans.}$$

**Example 15.5** *A 2 kW motor of mass 18.5 kg is mounted symmetrically on four identical springs, each of stiffness 2000 N/m. Determine the time period and the frequency of vibration of the motor.*

***Solution*** Given : *Power = 2 kW ; $m = 18.5$ kg ; Stiffness of each spring = 2000 N/m ; No. of spring = 4

Since there are four identical springs connected in parallel, therefore equivalent stiffness of the springs,

$$s = 4 \times 2000 = 8000 \text{ N/m}$$

***Time period of vibration***

We know that time period, $$t_p = 2\pi\sqrt{\frac{m}{s}} = 2\pi\sqrt{\frac{18.5}{8000}} = 2\pi \times 0.048 = 0.302 \text{ s} \quad \textbf{Ans.}$$

***Frequency of vibration***

We know that frequency, $$f = \frac{1}{t_p} = \frac{1}{0.302} = 3.3 \text{ Hz} \quad \textbf{Ans.}$$

**Example 15.6** *A spring of stiffness of 7 N/m is cut into two halves and fixed to a block of mass m so that the system can vibrate with a frequency between two fixed points as shown in Fig. 15.5. If the system vibrates at 3 Hz, determine the mass of the block.*

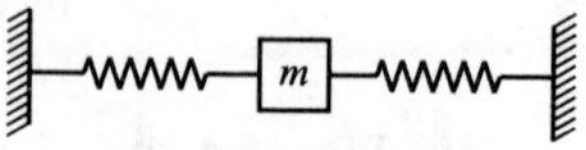


**Fig. 15.5**

***Solution*** Given : Stiffness of uncut spring = 7 N/m ; $f = 3$ Hz

When the spring is cut into two halves, the stiffness of each part is doubled, *i.e.* $2 \times 7 = 14$ N/m.

Now, the mass is attached to two springs (in parallel), each having a spring stiffness of 14 N/m. Hence equivalent spring stiffness of the springs,

$$s = 2 \times 14 = 28 \text{ N/m}$$

We know that frequency,

$$f = \frac{1}{2\pi}\sqrt{\frac{s}{m}}$$

$$3 = \frac{1}{2\pi}\sqrt{\frac{28}{m}}$$

Squaring both sides, we have $$(3)^2 = \left(\frac{1}{2\pi}\right)^2 \times \frac{28}{m}$$

$$\therefore \quad m = \frac{28}{2\pi \times (3)^2} = 0.079 \text{ kg} \quad \textbf{Ans.}$$

**Example 15.7** *A weight of 4.54 N is suspended vertically by a spring and vibrates with an amplitude of 75 mm and a frequency of 60 oscillations per minute. Find: 1. Spring constant; 2. Maximum tension induced in the spring; and 3. Maximum velocity of the weight.*

***Solution*** Given : $W = 4.54$ N or $m = \frac{W}{g} = \frac{4.54}{9.81} = 0.4628$ kg ; $r = 75$ mm $= 0.075$ m ; $f = 60$ osc/min $= \frac{60}{60} = 1$ osc/s or Hz

**1. *Spring constant***

Let $s$ = Spring constant in N/m.

* Superfluous data.

We know that frequency, $$f = \frac{1}{2\pi}\sqrt{\frac{s}{m}}$$

$$1 = \frac{1}{2\pi}\sqrt{\frac{s}{0.4628}}$$

Squaring both sides, we have $$(1)^2 = \left(\frac{1}{2\pi}\right)^2 \times \frac{s}{0.4628}$$

$$\therefore \quad s = (2\pi)^2\, 0.4628 = 39.49 \times 0.4628 = 18.276 \text{ N/m} \quad \textbf{Ans.}$$

**2. *Maximum tension induced in the spring***

Since the weight suspended vertically by the spring is 4.54 N, therefore maximum tension induced in the spring will be

$$T = 4.54 \text{ N} \quad \textbf{Ans.}$$

**3. *Maximum velocity of the weight***

We know that maximum velocity,

$$v_{max} = \omega.r = 2\pi.f.r \qquad ...(\because \omega = 2\pi f)$$

$$= 2\pi \times 1 \times 0.075 = 0.4713 \text{ m/s} \quad \textbf{Ans.}$$

**Example 15.8** *A vibration system consists of a mass m and a spring of stiffness s. It has natural frequency of 12 Hz. The natural frequency decreases by 2 Hz when an extra mass of 2 kg is attached to m. Find the mass of the body and stiffness of the spring.*

***Solution*** Given : $f_1 = 12$ Hz when mass, $m_1 = m$ kg ; $f_2 = 12 - 2 = 10$ Hz when mass, $m_2 = (m + 2)$ kg

***Mass of the body***

**First case:** We know that natural frequency ($f_1$),

$$12 = \frac{1}{2\pi}\sqrt{\frac{s}{m}} \qquad ...(i)$$

**Second case:** The frequency decreases by 2 Hz when an extra mass of 2 kg is attached.

$$\therefore \quad 10 = \frac{1}{2\pi}\sqrt{\frac{s}{m+2}} \qquad ...(ii)$$

Dividing equation (*i*) by equation (*ii*),

$$\frac{12}{10} = \frac{\frac{1}{2\pi}\sqrt{\frac{s}{m}}}{\frac{1}{2\pi}\sqrt{\frac{s}{m+2}}} = \sqrt{\frac{m+2}{m}}$$

Squaring both sides, we have $$\frac{144}{100} = \frac{m+2}{m}$$

$$144\,m = 100\,m + 200$$

or $$144\,m - 100\,m = 200 \quad \text{or} \quad 44\,m = 200$$

$$\therefore \quad m = \frac{200}{44} = 4.545 \text{ kg}$$

***Stiffness of the spring***

Substituting the value of $m$ in equation (*i*), we have

$$12 = \frac{1}{2\pi}\sqrt{\frac{s}{4.545}}$$

Squaring both sides,
$$144 = \frac{1}{4\pi^2} \times \frac{s}{4.545}$$

$$\therefore \quad s = 144 \times 4\pi^2 \times 4.545 = 25\,845 \text{ N/m} \quad \textbf{Ans.}$$

**Example 15.9** *A vibrating system is shown in Fig. 15.6. The block is pulled 40 mm down from its equilibrium position and released. For springs, $s_1$ = 4 kN/m and $s_2$ = 6 kN/m.*

*Determine the period of vibration, the minimum velocity of block, and the maximum acceleration of the block. The mass of the block is 50 kg.*

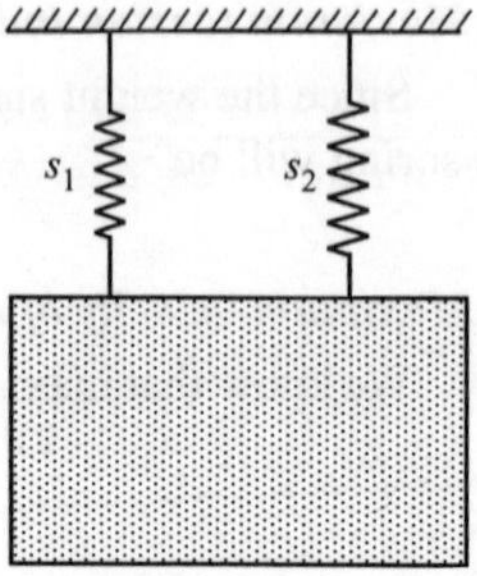

Fig. 15.6

***Solution*** Given : $\delta$ = 40 mm = 0.04 m ; $s_1$ = 4 kN/m = 4000 N/m ; $s_2$ = 6 kN/m = 6000 N/m ; $m$ = 50 kg or $W = m.g$ = 50 × 9.81 = 490.5 N

Let $W_1$ = Load supported by the first spring,

and $W_2$ = Load supported by the second spring.

We know that extension of the first spring,
$$\delta_1 = \frac{W_1}{s_1}$$

Similarly, extension of the second spring,
$$\delta_2 = \frac{W_2}{s_2}$$

Since the extension of the first spring is same as extension of the second spring, therefore
$$\delta_1 = \delta_2$$

or
$$\frac{W_1}{s_1} = \frac{W_2}{s_2}$$

$$\therefore \quad W_1 = \frac{s_1}{s_2} \times W_2 = \frac{4000}{6000} \times W_2 = \frac{2}{3} W_2 \qquad ...(i)$$

But
$$W = W_1 + W_2$$

$$490.5 = \frac{2}{3} W_2 + W_2 = \frac{5W_2}{3}$$

$$\therefore \quad W_2 = \frac{490.5 \times 3}{5} = 294.3 \text{ N}$$

Substituting the value of $W_2$ in equation (*i*), we have
$$W_1 = \frac{2}{3} W_2 = \frac{2}{3} \times 294.3 = 196.2 \text{ N}$$

$\therefore$ Static deflection,
$$\delta = \frac{W_1}{s_1} \text{ or } \frac{W_2}{s_2} = \frac{196.2}{4000} = 0.049\,05 \text{ m}$$

***Period of vibration***

We know that period of vibration,
$$t_p = 2\pi\sqrt{\frac{\delta}{g}} = 2\pi\sqrt{\frac{0.049\,05}{9.81}} = 2\pi \times 0.0707 = 0.444 \text{ s} \quad \textbf{Ans.}$$

***Minimum velocity of the block***

We know that velocity of the block, $v = -\omega \,.\, r \sin\theta$

The velocity of the block will be minimum, when $\theta = 0°$.

∴ Minimum velocity of the block,

$$v_{min} = -\omega \cdot r \sin 0° = 0 \textbf{ Ans.}$$

***Maximum acceleration of the block***

We know that angular velocity of the block,

$$\omega = \frac{2\pi}{t_p} = \frac{2\pi}{0.444} = 14.15 \text{ rad/s}$$

∴ Maximum acceleration of the block,

$$a_{max} = \pm\omega^2 r = \pm(14.15)^2\, 0.04 = 8 \text{ m/s}^2 \textbf{ Ans.}$$

**Example 15.10** *Determine the natural frequency of vibration of the system, as shown in Fig. 15.7. Take the following data for the system:*

*$s_1$ = 10 N/mm ; $s_2$ = 5 N/mm ; and W = 109 N*

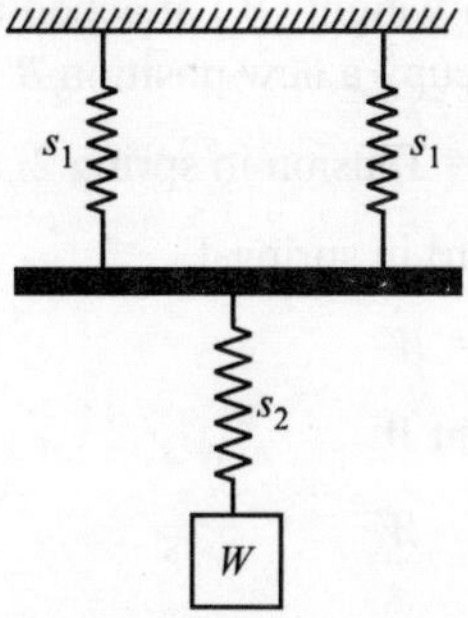


**Fig. 15.7**

***Solution*** Given : $s_1$ = 10 N/mm ; $s_2$ = 5 N/mm; $W$ = 109 N

Let $s$ = Equivalent stiffness of the system.

From Fig. 15.7, we see that the upper two springs are in parallel.

∴ Equivalent stiffness for upper two springs

$$= s_1 + s_1 = 2s_1$$

Now, two springs one of stiffness $2s_1$ and the other of stiffness $s_2$ are in series, as shown in Fig. 15.8.

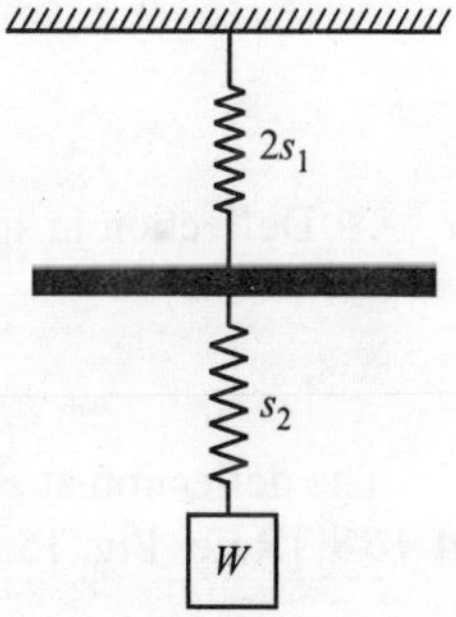


**Fig. 15.8**

We know that equivalent stiffness of the spring system is given as

$$\frac{1}{s} = \frac{1}{2s_1} + \frac{1}{s_2}$$

$$\frac{1}{s} = \frac{s_2 + 2s_1}{2s_1 . s_2}$$

$$\therefore \quad s = \frac{2s_1 . s_2}{2s_1 + s_2} = \frac{(2 \times 10) \times 5}{(2 \times 10) + 5} = \frac{100}{25} = 4 \text{ N/mm} = 4000 \text{ N/m}$$

We know that natural frequency of vibration of the system,

$$f = \frac{1}{2\pi}\sqrt{\frac{s}{m}} = \frac{1}{2\pi}\sqrt{\frac{g \cdot s}{W}} \quad \ldots\left(\because W = mg \text{ or } m = \frac{W}{g}\right)$$

$$= \frac{1}{2\pi}\sqrt{\frac{9.81 \times 4000}{109}} = 3.02 \text{ Hz } \textbf{Ans.}$$

**Example 15.11** *Determine the natural frequency of a vibrating system, as shown in Fig. 15.9.*

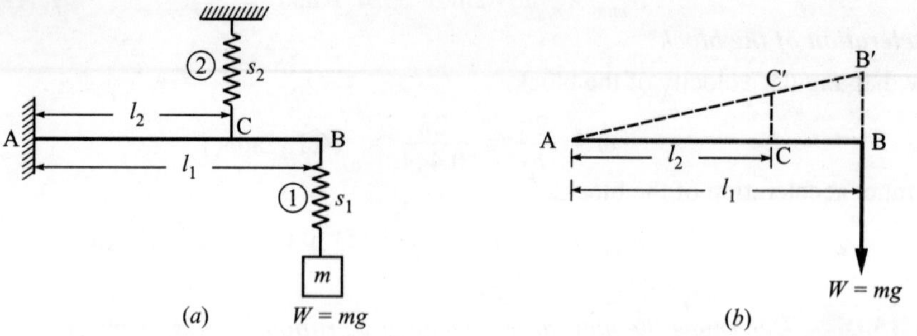


**Fig. 15.9**

***Solution*** It may be noted that when weight ($W$) attached to spring 1 is removed, then the beam $AB$ will spring upwards and the point $B$ will occupy a new position $B'$, as shown in Fig. 15.9 ($b$).

Let $T$ = Tension in spring 2.

We know that the spring force (tension) in spring 1,

$$= W$$

∴ Deflection in spring 1 due to weight $W$,

$$\delta_1 = \frac{W}{s_1}$$

The spring force (tension) in spring 2 may be calculated by taking moments about $A$,

$$W \times l_1 = T \times l_2$$

or
$$T = W \times \frac{l_1}{l_2} = \frac{W \cdot l_1}{l_2}$$

∴ Deflection in spring 2 due to tension $T$,

$$\delta_2 = \frac{T}{s_2} = \frac{W . l_1}{s_2 . l_2}$$

The deflection at $B$ due to deflection in spring 2 is obtained by considering the similar triangles $ACC'$ and $ABB'$ [Refer Fig. 15.9 ($b$)].

∴
$$\frac{CC'}{BB'} = \frac{l_2}{l_1} \quad \text{or} \quad BB' = \frac{l_1}{l_2} \times CC'$$

*i.e.* Deflection at $B$ due to deflection at $C$ (*i.e.* $BB'$)

$$= \frac{l_1}{l_2} \times \text{Deflection at } C = \frac{l_1}{l_2} \times \delta_2$$

Now, the total deflection of the system at $B$,

$$= \delta_1 + \frac{l_1}{l_2} \times \delta_2 = \frac{W}{s_1} + \frac{l_1}{l_2} \times \frac{W \cdot l_1}{s_2 \cdot l_2}$$

$$= W\left(\frac{1}{s_1} + \frac{1}{s_2} \times \frac{l_1^2}{l_2^2}\right) = W\left[\frac{s_2 + s_1\left(\frac{l_1}{l_2}\right)^2}{s_1 . s_2}\right]$$

We know that natural frequency of system,

$$f = \frac{1}{2\pi}\sqrt{\frac{g}{\delta}} = \frac{1}{2\pi}\sqrt{\frac{g}{W\left[\dfrac{s_2 + s_1\left(\dfrac{l_1}{l_2}\right)^2}{s_1 . s_2}\right]}}$$

$$= \frac{1}{2\pi}\sqrt{\frac{g \cdot s_1 \cdot s_2}{mg\left[s_2 + s_1\left(\dfrac{l_1}{l_2}\right)^2\right]}} \qquad ...(\because\ W = mg)$$

$$= \frac{1}{2\pi}\sqrt{\frac{s_1 \cdot s_2}{m\left[s_2 + s_1\left(\dfrac{l_1}{l_2}\right)^2\right]}} \quad \textbf{Ans.}$$

## 15.6 SIMPLE PENDULUM

A simple pendulum consists of a heavy bob suspended from one end of light inextensible and flexible string of negligible weight. The other end of the string is fixed in a rigid support at $O$, as shown in Fig. 15.10. The point $O$ is being referred to as the point of suspension of the pendulum.

Let
$m$ = Mass of the bob in kg,
$W$ = Weight of the bob in newtons = $mg$,
$l$ = Length of the string, and
$\theta$ = Angle through which the string is displaced.

When the bob is at $A$, the pendulum is in equilibrium position. But when the bob is displaced to $B$ or $C$ and released, it starts oscillating between $B$ and $C$ with $A$ as the mean position. It has been observed that if the angle $\theta$ is very small (less than 4°), then the bob will have simple harmonic motion.

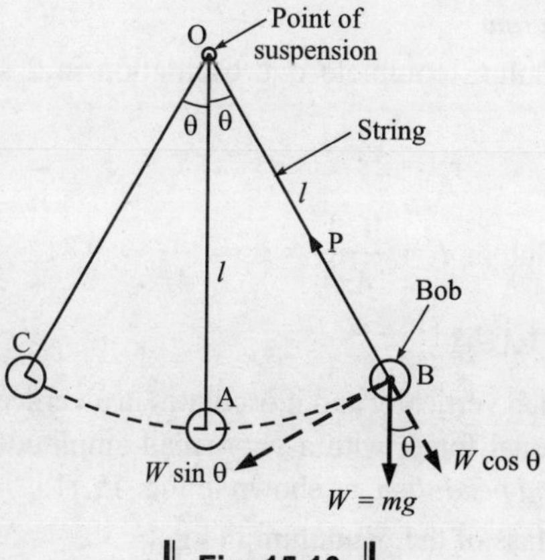


**Fig. 15.10**

Consider the bob at $B$. The forces acting on the bob are as follows:

1. Weight of the bob, $W = mg$, acting vertically downwards, and
2. Tension $P$ in the string.

The weight $W$ can be resolved into two components, *i.e.* $W \cos\theta$ along the string $OB$ and $W \sin\theta$ at right angles to $OB$. For the equilibrium of the bob,

$$P = W\cos\theta$$

The component $W \sin\theta$ acting towards $A$ will be unbalanced. The couple tending to restore the bob in equilibrium position or restoring torque,

$$T = W\sin\theta \times l = mgl\sin\theta$$

Since $\theta$ is a very small angle, therefore $\sin\theta = \theta$ radians.

$$\therefore \quad T = mgl\theta$$

We know that mass moment of inertia of the bob about an axis passing through the point of suspension,

$$I = \text{Mass} \times (\text{Length})^2 = m \, . \, l^2$$

$\therefore$ Angular acceleration of the string,

$$\alpha = \frac{T}{I} = \frac{mgl\theta}{ml^2} = \frac{g\theta}{l}$$

or

$$\frac{\theta}{\alpha} = \frac{l}{g}$$

*i.e.*

$$\frac{\text{Angular displacement}}{\text{Angular acceleration}} = \frac{l}{g}$$

We know that periodic time,

$$t_p = 2\pi\sqrt{\frac{\text{Displacement}}{\text{Acceleration}}} = 2\pi\sqrt{\frac{l}{g}} \qquad \text{...}(i)$$

and frequency of oscillation,

$$f = \frac{1}{t_p} = \frac{1}{2\pi}\sqrt{\frac{g}{l}} \qquad \text{...}(ii)$$

It may be noted that

1. The motion of the bob from one extremity to the other (*i.e.* from $B$ to $C$ or from $C$ to $B$) is known as *beat* or *swing*. Thus one beat $= \frac{1}{2}$ oscillation.

   $\therefore$ Periodic time for one beat $= \pi\sqrt{\frac{l}{g}}$

2. A pendulum, which executes one beat per second (*i.e.* one complete oscillation in two seconds) is known as *second's pendulum*.
3. Since the second's pendulum complete one oscillation in 2 seconds, therefore time period for second's pendulum,

$$t_p = 2\text{ s}$$

$$\therefore \text{ Length of second's pendulum, } l = \frac{(t_p)^2}{4\pi^2} \times g = \frac{2^2}{4\pi^2} \times 9.81 = 0.994\text{ m} \qquad \text{...}\left(\because t_p = 2\pi\sqrt{\frac{l}{g}}\right)$$

## 15.7 COMPOUND PENDULUM

When a rigid body is suspended vertically and it oscillates in a vertical plane under the action of gravitational force, with a very small amplitude, then the body is known as *compound pendulum*, as shown in Fig. 15.11.

Let $m$ = Mass of the pendulum in kg,

$W$ = Weight of the pendulum in newtons $= mg$,

$k_G$ = Radius of gyration of the pendulum about an axis passing through the centre of gravity $G$ and perpendicular to the plane of oscillation, and

$h$ = Distance between the point of suspension $O$ and centre of gravity $G$ of the pendulum.

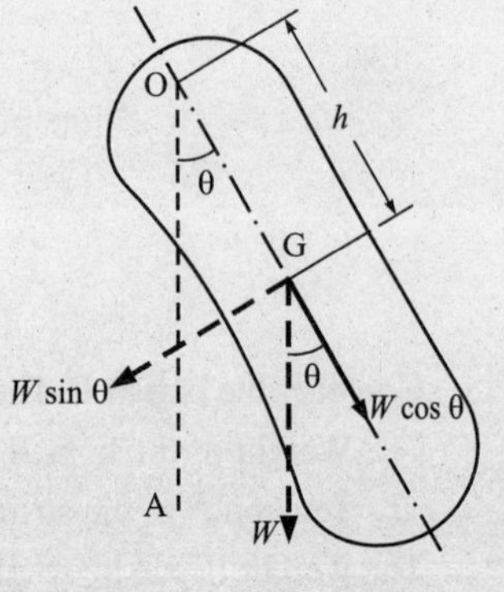


Fig. 15.11

We know that mass moment of inertia of the pendulum about the axis of suspension $O$ (by parallel axis theorem),

$$I = I_G + mh^2 = m \cdot k_G^2 + mh^2 = m(k_G^2 + h^2) \quad ...(i)$$

$$...(\because I_G = m \,.\, k_G^2)$$

Now consider that the pendulum is given a small angular displacement $\theta$ from the vertical equilibrium position $OA$. The pendulum will tend to restore itself to the equilibrium position $OA$ by the restoring couple,

$$T = W \sin\theta \times h = mg \sin\theta \times h$$

Since $\theta$ is a very small angle, therefore $\sin\theta = \theta$ radians.

$$\therefore \quad T = mgh\,\theta$$

We know that angular acceleration of the pendulum,

$$\alpha = \frac{T}{I} = \frac{mgh\,\theta}{m(k_G^2 + h^2)} = \frac{gh\,\theta}{k_G^2 + h^2} \quad ...(ii)$$

$$= \text{Constant} \times \theta$$

We see that the angular acceleration of the pendulum is directly proportional to angular displacement, therefore the pendulum executes simple harmonic motion.

From equation (*ii*),

$$\frac{\theta}{\alpha} = \frac{k_G^2 + h^2}{gh}$$

We know that periodic time,

$$t_p = 2\pi\sqrt{\frac{\text{Displacement}}{\text{Acceleration}}} = 2\pi\sqrt{\frac{\theta}{\alpha}} = 2\pi\sqrt{\frac{k_G^2 + h^2}{gh}}$$

and frequency,

$$f = \frac{1}{t_p} = \frac{1}{2\pi}\sqrt{\frac{gh}{k_G^2 + h^2}}$$

It may be noted that

1. In order to determine the equivalent length of a simple pendulum which gives the same frequency as that of a compound pendulum, we have

$$\frac{g}{l} = \frac{gh}{k_G^2 + h^2}$$

$\therefore$ Equivalent length of a simple pendulum,

$$l_e = \frac{k_G^2 + h^2}{h}$$

2. We know that time period of a compound pendulum,

$$t_p = 2\pi\sqrt{\frac{k_G^2 + h^2}{gh}}$$

The time period of the compound pendulum will be minimum, when $\dfrac{k_G^2 + h^2}{gh}$ is minimum. For $\dfrac{k_G^2 + h^2}{gh}$ to be minimum, differentiate this expression with respect to $h$ and equate to zero, *i.e.*

$$\frac{d}{dh}\left[\frac{k_G^2 + h^2}{h}\right] = 0$$

or $$\frac{d}{dh}\left[\frac{k_G^2}{h}+h\right] = 0$$

$\therefore$ $$k_G^2 \times -h^{-2} + 1 = 0$$

$$\frac{-k_G^2}{h^2}+1 = 0$$

$$k_G^2 = h^2$$

or $$k_G = h$$

Thus the periodic time of the compound pendulum will be minimum when the distance between the point of suspension and the centre of gravity is equal to the radius of gyration of the pendulum about its centre of gravity.

$\therefore$ Minimum periodic time of a compound pendulum

$$t_{p(min)} = 2\pi\sqrt{\frac{k_G^2+k_G^2}{g\,k_G}} = 2\pi\sqrt{\frac{2k_G}{g}}$$

**Example 15.12** *A flywheel of mass 90 kg is suspended in a vertical plane as a compound pendulum. The distance of centre of gravity from the point of suspension is 120 mm and the flywheel makes 100 oscillations in 135 seconds. Find the moment of inertia of the flywheel through the centre of gravity.*

***Solution*** Given : $m = 90$ kg ; $h = 120$ mm $= 0.12$ m

Since the flywheel makes 100 oscillations in 135 seconds, therefore frequency of oscillation,

$$f = \frac{100}{135} = 0.74 \text{ Hz}$$

Let $l_e$ = Equivalent length of the simple pendulum, and

$k_G$ = Radius of gyration through the centre of gravity.

We know that frequency, $$f = \frac{1}{2\pi}\sqrt{\frac{g}{l_e}}$$

$$0.74 = \frac{1}{2\pi}\sqrt{\frac{9.81}{l_e}}$$

Squaring both sides, $$(0.74)^2 = \left(\frac{1}{2\pi}\right)^2 \times \frac{9.81}{l_e} = \frac{0.25}{l_e}$$

$\therefore$ $$l_e = \frac{0.25}{(0.74)^2} = 0.456 \text{ m}$$

We also know that equivalent length of simple pendulum,

$$l_e = \frac{k_G^2+h^2}{h}$$

$$0.456 = \frac{k_G^2+(0.12)^2}{0.12}$$

or $$k_G^2 = 0.456 \times 0.12 - (0.12)^2 = 0.0547 - 0.0144 = 0.0403$$

$$k_G = \sqrt{0.0403} = 0.2007 \text{ m}$$

$\therefore$ Moment of inertia of the flywheel through the centre of gravity,

$$I = m.k_G^2 = 90(0.0403) = 3.627 \text{ kg-m}^2$$ **Ans.**

**Example 15.13** *A uniform slender rod 1.2 m long is fitted with a transverse pair of knife edges, so that it can swing in a vertical plane as a compound pendulum. The position of the knife edges is variable. Find the time of swing of the rod, if*

1. *the knife edges are 50 mm from one end of the rod, and*
2. *the knife edges are so placed that the time of swing is minimum.*

***Solution*** Given : $l = 1.2$ m

**1. *Time of swing of the rod when the knife edges are 50 mm from one end of the rod***

Since the distance between the knife edges from one end of the rod is 50 mm = 0.05 m, therefore the distance between the knife edge and centre of gravity of the rod,

$$h = \frac{1.2}{2} - 0.05 = 0.55 \text{ m}$$

We know that mass moment of inertia of the rod about an axis passing through its centre of gravity,

$$I = \frac{m.l^2}{12}$$

We also know that $\quad I = m.k_G^2$

$$\therefore \quad k_G^2 = \frac{l^2}{12} \text{ or } k_G = \frac{l}{\sqrt{12}}$$

Thus the radius of gyration of the rod about the centre of gravity,

$$k_G = \frac{l}{\sqrt{12}} = \frac{1.2}{\sqrt{12}} = \frac{1.2}{3.464} = 0.346 \text{ m}$$

$$\therefore \quad \text{Time of swing, } t_p = 2\pi\sqrt{\frac{k_G^2 + h^2}{g.h}} = 2\pi\sqrt{\frac{(0.346)^2 + (0.55)^2}{9.81 \times 0.55}}$$

$$= 6.284\sqrt{\frac{0.12 + 0.3025}{5.4}} = 6.284 \times 0.28 = 1.76 \text{ s } \textbf{Ans.}$$

**2. *Minimum time of swing***

We know that minimum time of swing,

$$t_{p(min)} = 2\pi\sqrt{\frac{2k_G}{g}} = 6.284\sqrt{\frac{2 \times 0.346}{9.81}}$$

$$= 6.284 \times 0.2656 = 1.67 \text{ s } \textbf{Ans.}$$

## HIGHLIGHTS

1. A body is said to have simple harmonic motion (briefly written as S.H.M.), if it moves or vibrates about a mean position such that
   (*a*) its acceleration is always proportional to its distance from the mean position; and
   (*b*) its acceleration is always directed towards the mean position.
2. The maximum displacement of a body from its mean position is known as *amplitude.* It is always equal to the radius of the circle.
3. The time taken for one complete revolution of the particle is known as *periodic time.* Mathematically,

$$\text{Periodic time, } t_p = \frac{2\pi}{\omega} \text{ seconds}$$

where $\quad \omega$ = Angular velocity of the particle in rad/s.

4. The number of oscillations (or cycles) per second is known as frequency and it is reciprocal of time period ($t_p$). Mathematically,

$$\text{Frequency, } f = \frac{1}{t_p} = \frac{\omega}{2\pi} \text{ cycles/s or Hz}$$

**Note:** In S.I. units, the unit of frequency is hertz (briefly written as Hz) which is equal to one cycle per second.

5. Consider a particle is moving round the circumference of a circle of radius $r$, with a uniform angular velocity $\omega$ rad/s, as shown in Fig. 15.12. Let $P$ be any position of the particle after $t$ seconds and $\theta$ be the angle turned by the particle during this time $t$ seconds. $Q$ is the projection of point $P$ on the diameter $AC$ and $x$ is its distance from the mean position $O$. We know that

$$\theta = \omega \cdot t \quad \text{and} \quad x = r\cos\theta = r\cos\omega t$$

Velocity of $Q$,

$$v_Q = \omega r \sin\theta = \omega r\sqrt{r^2 - x^2}$$

$$\ldots\left(\because r\sin\theta = PQ = \sqrt{r^2 - x^2}\right)$$

The velocity is maximum when $x = 0$, *i.e.* when $Q$ passes through $O$ (*i.e.* mean position).

$\therefore$ Maximum velocity, $v_{max} = \omega \cdot r$

Acceleration of $Q$, $a_Q = \omega^2 \cdot r\cos\theta = \omega^2 \cdot x$

$$\ldots(\because r\cos\theta = x)$$

The acceleration is maximum, when $x = r$.

$\therefore$ Maximum acceleration, $a_{max} = \omega^2 \cdot r$

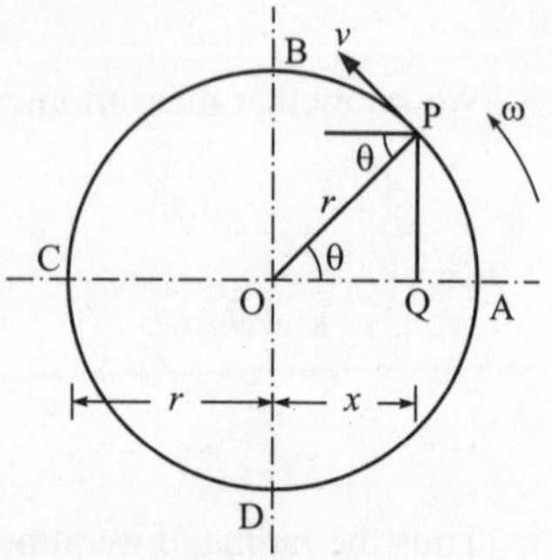


Fig. 15.12

6. When a vertical elastic spring of negligible mass with upper end fixed and carries a mass $m$ at the lower end, then

$$\text{Periodic time, } t_p = 2\pi\sqrt{\frac{\text{Displacement}}{\text{Acceleration}}} = 2\pi\sqrt{\frac{x}{a}}$$

$$= 2\pi\sqrt{\frac{m}{s}} = 2\pi\sqrt{\frac{\delta}{g}} \qquad \ldots\left(\because \delta = \frac{mg}{s}\right)$$

and frequency,

$$f = \frac{1}{t_p} = \frac{1}{2\pi}\sqrt{\frac{s}{m}} = \frac{1}{2\pi}\sqrt{\frac{g}{\delta}}$$

where
$m$ = Mass attached to the spring in kg,
$W$ = Weight attached in newtons = $mg$,
$\delta$ = Extension or deflection of the spring in metres,
$x$ = Displacement of the load below the equilibrium position in meters, and
$s$ = Stiffness of the spring in N/m = $\dfrac{W}{\delta} = \dfrac{m \cdot g}{\delta}$

In case, the mass of the spring ($m_1$) is taken into consideration, then

$$t_p = 2\pi\sqrt{\frac{m + \frac{m_1}{3}}{s}} \text{ seconds}$$

and

$$f = \frac{1}{t_p} = \frac{1}{2\pi}\sqrt{\frac{s}{m + \frac{m_1}{3}}} \text{ Hz}$$

**7.** For a simple pendulum, periodic time,

$$t_p = 2\pi\sqrt{\frac{l}{g}}$$

and frequency of oscillation, $f = \frac{1}{t_p} = \frac{1}{2\pi}\sqrt{\frac{g}{l}}$

where $l$ is the length of a light inextensible and flexible string.

**8.** The pendulum, which executes one beat per second (*i.e.* one complete oscillation in two seconds) is known as *second's pendulum*.

**Note:** The motion of the bob of the pendulum from one extremity to the other is known as beat or swing. Thus one beat $= \frac{1}{2}$ oscillation and periodic time for one beat, $t_p = \pi\sqrt{\frac{l}{g}}$.

**9.** When a rigid body is suspended vertically and it oscillates with small amplitude under the action of force of gravity, then the body is known as *compound pendulum*.

**10.** For a compound pendulum, the periodic time,

$$t_p = 2\pi\sqrt{\frac{k_G^2 + h^2}{g \cdot h}}$$

and frequency of oscillation, $f = \frac{1}{t_p} = \frac{1}{2\pi}\sqrt{\frac{gh}{k_G^2 + h^2}}$

where $k_G$ = Radius of gyration about an axis through the centre of gravity and perpendicular to the plane of oscillation, and

$h$ = Distance of point of suspension $O$ from the centre of gravity.

**11.** The equivalent length of a simple pendulum ($l_e$) which gives the same frequency as that of compound pendulum is given by

$$l_e = \frac{k_G^2 + h^2}{h}$$

**12.** The periodic time of compound pendulum is minimum, when $h = k_G$.

∴ Minimum periodic time of a compound pendulum,

$$t_{p(min)} = 2\pi\sqrt{\frac{2k_G}{g}}$$

## NUMERICAL PROBLEMS

1. A body is moving with simple harmonic motion. If it rotates at 120 r.p.m. and has an amplitude of 1 m, find the velocity and acceleration of the body, when it is at a distance of 0.75 m from the mean position. **[Ans.** 8.31 m/s; 118.46 m/s$^2$**]**
2. A point moves with simple harmonic motion. When this point is 0.75 m from the mid path, its velocity is 11 m/s and when 2 m from the centre of its path, its velocity is 3 m/s. Find its angular velocity, periodic time and its maximum acceleration. **[Ans.** 5.7 rad/s; 1.1 s; 67.25 m/s$^2$**]**
3. The amplitude of a body moving in a simple harmonic motion is 500 mm and its period of oscillation is 2 seconds. What would be its speed and acceleration 0.4 second after it has passed the middle position? Also, determine the frequency of oscillation. **[Ans.** 0.485 m/s; 4.7 m/s$^2$; 0.5 Hz**]**
4. The piston of the engine moves with simple harmonic motion. The crank rotates at 100 r.p.m. and the stroke is 1.8 m. Find the velocity and acceleration of the piston, when it is at a distance of 600 mm from the centre. **[Ans.** 7.02 m/s; 65.7 m/s$^2$**]**

5. A small motor of mass 20 kg is symmetrically mounted on four equal springs, each with a spring constant of 2500 N/m. Estimate the period of vibration and frequency of the motor. [**Ans.** 0.281 s; 3.56 Hz]
6. A helical spring of negligible mass is required to support a mass of 50 kg. The stiffness of the spring is 60 kN/m. The spring and the mass system is displaced vertically by 20 mm below the equilibrium position and released. Find: 1. the frequency of natural vibration of the system; and 2. the velocity and acceleration of the mass when it is 10 mm below the rest position. [**Ans.** 5.5 Hz; 0.6 m/s, 11.95 m/s$^2$]
7. A spring-pan system consists of a spring of stiffness *s* and pan of mass *m*, as shown in Fig. 15.13. If a mass of 1.5 kg is placed in the pan, it vibrates with a time period of 2.4 s and if a mass of 2 kg is placed, it vibrates with a time period of 2.6 s. Determine the values of *s* and *m*. [**Ans.** 19.7 N/m; $m$ = 1.38 kg]

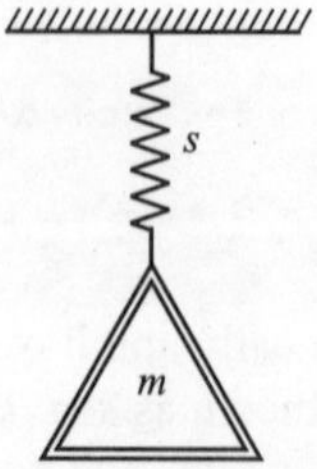


**Fig. 15.13**

8. A small flywheel of mass 85 kg is suspended in a vertical plane as a compound pendulum. The distance of centre of gravity from the knife edge support is 100 mm and the flywheel makes 100 oscillations in 145 seconds. Find the moment of inertia of the flywheel through the centre of gravity. [**Ans.** 3.6 kg-m$^2$]
9. The mass of a connecting rod of an engine is 20 kg and its length between centres is 1 m. The connecting rod oscillates in a vertical plane about the centre of the small end. The distance of its centre of gravity is 600 mm from the point of oscillation. From the experiment, it is observed that the rod completes 50 oscillations is 92 seconds. Find: 1. moment of inertia of the rod about an axis through the centre of gravity; and 2. length of the equivalent simple pendulum. [**Ans.** 2.904 kg-m$^2$; 0.842 m]

## REVIEW QUESTIONS

1. Explain 'simple harmonic motion' and derive the relation existing therein between acceleration and displacement at any time.
2. Define the following terms used in simple harmonic motion:
(*a*) Amplitude; (*b*) Periodic time; and (*c*) Frequency.
3. Show that when a particle moves with simple harmonic motion, its time for a complete oscillation is independent of the amplitude of its motion.
4. Derive an expression for the period of oscillation when a mass is attached to the vertical elastic spring.
5. Under what conditions can a simple pendulum be assumed to have simple harmonic motion? Derive an expression for the time period of a simple pendulum of length *l* performing simple harmonic motion. Find the length of a second's pendulum.
6. Derive an expression for the periodic time of a compound pendulum.

## OBJECTIVE TYPE QUESTIONS

1. The velocity of a particle (*v*) moving with simple harmonic motion, at any instant, is given by
(*a*) $\omega\sqrt{x^2 - r^2}$ (*b*) $\omega\sqrt{r^2 - x^2}$ (*c*) $\omega^2\sqrt{x^2 - r^2}$ (*d*) $\omega^2\sqrt{r^2 - x^2}$
where $r$ = Amplitude of the motion, and
$x$ = Displacement of the particle from the mean position.

**2.** The acceleration of a particle moving with simple harmonic motion, at any instant, is given by

(*a*) $\frac{\omega}{x}$ (*b*) $\omega \cdot x$ (*c*) $\frac{\omega^2}{x}$ (*d*) $\omega^2 x$

**3.** The maximum velocity of a particle moving with simple harmonic motion is

(*a*) $\omega \cdot r$ (*b*) $\omega^2 \cdot r$ (*c*) $\frac{\omega}{r}$ (*d*) $\frac{\omega^2}{r}$

**4.** The maximum acceleration of a particle moving with simple harmonic motion is

(*a*) $\omega \cdot r$ (*b*) $\omega^2 \cdot r$ (*c*) $\frac{\omega}{r}$ (*d*) $\frac{\omega^2}{r}$

**5.** The periodic time ($t_p$) is given by

(*a*) $\frac{\omega}{2\pi}$ (*b*) $2\pi \times \omega$ (*c*) $\frac{2\pi}{\omega}$ (*d*) $\frac{\pi}{\omega}$

where $\omega$ = Angular velocity of the particle in rad/s.

**6.** The velocity of a particle moving with simple harmonic motion is ......... at the mean position.

(*a*) zero (*b*) minimum (*c*) maximum

**7.** When one end of a vertical elastic spring is fixed and the other end carries a load $W$ which moves with simple harmonic motion, then the frequency of motion is given by

(*a*) $2\pi\sqrt{\frac{g}{\delta}}$ (*b*) $2\pi\sqrt{\frac{\delta}{g}}$ (*c*) $\frac{1}{2\pi}\sqrt{\frac{\delta}{g}}$ (*d*) $\frac{1}{2\pi}\sqrt{\frac{g}{\delta}}$

where $\delta$ = Extension or deflection of the spring.

**8.** The periodic time of one oscillation for a simple pendulum is

(*a*) $\frac{1}{2\pi}\sqrt{\frac{l}{g}}$ (*b*) $\frac{1}{2\pi}\sqrt{\frac{g}{l}}$ (*c*) $2\pi\sqrt{\frac{l}{g}}$ (*d*) $2\pi\sqrt{\frac{g}{l}}$

where $l$ = Length of the pendulum.

**9.** The frequency of oscillation of a compound pendulum is

(*a*) $\frac{1}{2\pi}\sqrt{\frac{gh}{k_G^2 + h^2}}$ (*b*) $\frac{1}{2\pi}\sqrt{\frac{k_G^2 + h^2}{gh}}$ (*c*) $2\pi\sqrt{\frac{gh}{k_G^2 + h^2}}$ (*d*) $2\pi\sqrt{\frac{k_G^2 + h^2}{gh}}$

where $k_G$ = Radius of gyration about the centre of gravity, and
$h$ = Distance between the point of suspension and the centre of gravity.

**10.** Which of the following statement is correct?

(*a*) The maximum displacement of the body, from its mean position is called amplitude.

(*b*) The length of a second's pendulum is 0.994 m.

(*c*) A body is said to move with simple harmonic motion if its acceleration is directed towards the mean position.

(*d*) All of the above.

***Answers to Objective Type Questions***

| | | | | |
|---|---|---|---|---|
| **1.** (*b*) | **2.** (*d*) | **3.** (*a*) | **4.** (*b*) | **5.** (*c*) |
| **6.** (*c*) | **7.** (*d*) | **8.** (*c*) | **9.** (*a*) | **10.** (*d*) |

# 16

# KINETICS OF LINEAR MOTION

**1.** *Introduction* **2.** *Newton's Laws of Motion* **3.** *Important Terms Used in Kinetics* **4.** *Law of Conservation of Momentum* **5.** *Force* **6.** *Impulse and Impulsive Force* **7.** *Pile and Pile Hammer* **8.** *D'Alembert's Principle* **9.** *Motion of a Lift* **10.** *Motion of a Body along a Rough Inclined Plane*

## 16.1 INTRODUCTION

We have discussed in Chapter 12, the kinematics of linear motion, *i.e.* the motion without considering the forces that causes the motion. In this chapter, we shall discuss the kinetics of motion, *i.e.* the motion which takes into consideration the forces or other factors such as mass or weight of the bodies. The force and motion are governed by the three Newton's laws of motion, as discussed below.

## 16.2 NEWTON'S LAWS OF MOTION

Sir Issac Newton, in 1647, formulated three laws of motion which are the basic postulates or assumptions on which the whole system of kinetics is based. These three laws of motion are as follows:

1. ***Newton's First Law of Motion*** It states that '*every body continues in its state of rest or of uniform motion in a straight line, unless it is acted upon by some external force*'.

   This law is also sometimes called as *Law of Inertia.*
2. ***Newton's Second Law of Motion*** It states that '*the rate of change of momentum is directly proportional to the impressed force and takes place in the same direction in which the force acts*'.
3. ***Newton's Third Law of Motion*** It states that '*to every action, there is always an equal and opposite reaction*'.

## 16.3 IMPORTANT TERMS USED IN KINETICS

The following terms are widely used in kinetics:

1. ***Mass*** It is the amount of matter present in the given body and it always remain same in all positions on the earth's surface. The mass of a body is measured by direct comparison with a standard mass by using a lever balance.

2. ***Weight*** It is the amount of pull, which the earth exerts upon a given body. Since the pull varies with the distance of the body from the centre of the earth, therefore, the weight of the body vary with its position on the earth's surface. It is thus obvious, that the weight is a force. The weight of a body is measured by the use of a spring balance which indicates the varying tension in the spring as the body is moved from place to place.

**Note:** In S.I. system of units, mass ($m$) is taken in kg and the weight ($W$) in newtons. The relation between mass ($m$) and weight ($W$) of a body is given by

$$W = mg \text{ or } m = \frac{W}{g}$$

where $W$ is in newtons, $m$ is in kg and $g$ is the acceleration due to gravity in m/s$^2$.

3. ***Momentum*** It is the total motion possessed by a body. Mathematically,

$$\text{Momentum} = \text{Mass} \times \text{Velocity}$$

The unit of momentum depends upon the units of mass and velocity.

4. ***Inertia*** It is that property of a matter, by virtue of which, a body cannot move of itself, nor change the motion imparted to it.

## 16.4 LAW OF CONSERVATION OF MOMENTUM

The law of conservation of momentum states that '*the total momentum of a system of masses* (*i.e. moving bodies*) *in any one direction remains constant, if no external force acts on them*'.

This law is applied to problems on impact, *i.e.* collision of two bodies. For example, if two bodies of masses $m_1$ and $m_2$ with linear velocities $v_1$ and $v_2$ are moving in the same straight line, and they collide and begin to move together with a common velocity $v$, then

$$\text{Momentum before impact} = \text{Momentum after impact}$$

*i.e.*

$$m_1 v_1 \pm m_2 v_2 = (m_1 + m_2) v$$

The *positive* sign is used when the two bodies moves in the same direction after collision. The *negative* sign is used when they move in the opposite direction after collision.

A familiar example on the application of law of conservation of momentum is the firing of a shot from a gun. Since initially the shot and the gun are at rest, the total momentum of the system is zero. Hence when shot is fired, the momentum of the gun must be equal and opposite to that of the shot so that the sum may remain zero.

## 16.5 FORCE

The force is an important factor in the field of Engineering-science, which may be defined as an agent, which produces or tends to produce, destroy or tends to destroy motion.

According to Newton's Second Law of Motion, the applied force or impressed force is directly proportional to the rate of change of momentum. Mathematically,

$$\text{Rate of change of momentum} = \frac{mv - mu}{t} = \frac{m(v-u)}{t} = m \cdot a \quad \ldots\left(\because \frac{v-u}{t} = a\right)$$

where

$m$ = Mass of the body,

$u$ and $v$ = Initial and final velocity of the body,

$t$ = Time taken (in seconds) by the body to change the velocity from $u$ to $v$, and

$a$ = Constant acceleration.

$$\therefore \quad \text{Force, } F \propto m \cdot a \quad \text{or} \quad F = k \cdot m \cdot a$$

where $k$ is a constant of proportionality.

For the sake of convenience, the unit of force adopted is such that it produces a unit acceleration to a body of unit mass.

$$\therefore \quad \text{Force} = m \cdot a = \text{Mass} \times \text{Acceleration}$$

In S.I. system of units, the unit of force is newton (briefly written as N). *A newton may be defined as the force while acting upon a mass of 1 kg produces an acceleration of 1 m/s² in the direction of which it acts.* Thus

$$1 \text{ N} = 1 \text{ kg} \times 1 \text{ m/s}^2 = 1 \text{ kg-m/s}^2$$

It may be noted that a force equal in magnitude but opposite in direction and collinear with the impressed force producing the acceleration is known as *inertia force.* Mathematically,

$$\text{Inertia force} = -m \cdot a$$

## 16.6 IMPULSE AND IMPULSIVE FORCE

The impulse is the product of force and time. Mathematically,

$$\text{Impulse} = F \times t$$

where $F$ = Force, and $t$ = Time.

Now consider a body of mass $m$. Let a force $F$ changes its velocity from $v_1$ to a final velocity $v_2$.

We know that force is equal to the rate of change of linear momentum.

$$\therefore \quad F = \frac{m(v_2 - v_1)}{t}$$

or

$$F \times t = m(v_2 - v_1)$$

*i.e.* Impulse = Change of linear momentum

If a force acts for a very short time, it is then known as *impulsive force* or blow. The impulsive force occurs in collisions, explosions, in the striking of a nail or a pile by a hammer.

**Example 16.1** *A body of mass 250 kg is pulled with an acceleration of 3 m/s² in the direction of the force. Find the force exerted to pull the body.*

***Solution*** Given : $m = 250$ kg ; $a = 3$ m/s²

We know that force exerted to pull the body,

$$F = m \cdot a = 250 \times 3 = 750 \text{ N} \quad \textbf{Ans.}$$

**Example 16.2** *A truck of weight 28 kN is travelling at 50 km/h on a level road. It is brought to rest in 15 m. What is the average force of resistance acting on the truck?*

***Solution*** Given : $W = 28$ kN $= 28 \times 10^3$ N ; $u = 50$ km/h $= \dfrac{50 \times 1000}{60 \times 60} = 13.9$ m/s ; $s = 15$ m

First of all, let us find the acceleration ($a$) at which the truck is travelling. Since the truck is brought to rest, therefore the final velocity of the truck is zero. Now, using the relation,

$$v^2 - u^2 = 2as$$

$$0 - (13.9)^2 = 2a \times 15 = 30a$$

$$\therefore \quad a = -\frac{(13.9)^2}{30} = -6.44 \text{ m/s}^2$$

The –ve sign indicates that there is a *retardation.*

We know that force of resistance acting on the truck,

$$F = m.a = \frac{W}{g} \times a = \frac{28 \times 10^3}{9.81} \times 6.44 = 18\,381 \text{ N}$$

$$= 18.381 \text{ kN} \quad \textbf{Ans.}$$

**Example 16.3** *A body of mass 200 kg has its velocity changed from 8 m/s to 12 m/s in the same direction in 20 seconds. Find :*

*1. The change in its momentum ; and 2. The force responsible for this change.*

***Solution*** Given : $m = 200$ kg ; $u = 8$ m/s ; $v = 12$ m/s ; $t = 20$ s

**1. *Change in momentum***

We know that the change in momentum

$$= m(v - u) = 200(12 - 8) = 800 \text{ kg-m/s}$$

**2. *Force responsible for this change***

First of all, let us find the acceleration ($a$).

Using the relation, $v = u + a \cdot t$

or $12 = 8 + a \times 20$

$\therefore$ $a = \dfrac{12 - 8}{20} = 0.2 \text{ m/s}^2$

and force responsible for this change, $F = m \cdot a = 200 \times 0.2 = 40$ N **Ans.**

**Example 16.4** *A body of mass 80 kg is moving with a velocity of 0.5 m/s. Find the force required to stop it in 8 seconds. Also, find the distance it will travel before coming to rest.*

***Solution*** Given : $m = 80$ kg ; $u = 0.5$ m/s ; $t = 8$ s

***Force required to stop***

Since the body is required to stop in 8 seconds, therefore the final velocity of the body ($v$) = 0, and there is a retardation ($a$).

Using the relation, $v = u + a \cdot t$

$$0 = 0.5 + a \times 8$$

$\therefore$ $a = -\dfrac{0.5}{8} = -\dfrac{1}{16} \text{ m/s}^2$

We know that force required to stop $= m.a = 80 \times \dfrac{1}{16} = 5$ N **Ans.**

***Distance travelled before coming to rest***

Let $s$ = Distance travelled before coming to rest.

Using the relation, $v^2 - u^2 = 2as$

$$0 - (0.5)^2 = 2 \times -\frac{1}{16} s$$

$$-0.25 = -\frac{1}{8} s$$

$\therefore$ $s = 0.25 \times 8 = 2$ m **Ans.**

**Example 16.5** *A hammer of mass 250 kg falls freely on a pile from a height of 6 m. If the hammer comes to rest in 0.015 second, find the change in linear momentum, impulse and average force.*

***Solution*** Given : $m = 250$ kg ; $h = 6$ m ; $t = 0.015$ s

***Change in linear momentum***

Let $v$ = Velocity with which the hammer falls on the pile.

Using the relation, $v^2 - u^2 = 2gh$

$$v^2 - 0 = 2 \times 9.81 \times 6 = 117.72 \quad \text{...(}\because \text{ Initial velocity, } u = 0\text{)}$$

$$\therefore \quad v = \sqrt{117.72} = 10.85 \text{ m/s}$$

We know that change in linear momentum

$$= \text{Mass} \times \text{Change of velocity} = m(v - u)$$

$$= 250\,(10.85 - 0) = 2712.5 \text{ N-s} \quad \textbf{Ans.}$$

***Impulse***

We know that Impulse = Change in linear momentum = 2712.5 N-s **Ans.**

***Average force***

Let $F$ = Average force.

We know that Impulse = Force × Time = $F \times t$

or $2712.5 = F \times 0.015$

$$\therefore \quad F = \frac{2712.5}{0.015} = 180\,833 \text{ N} = 180.833 \text{ kN} \quad \textbf{Ans.}$$

**Example 16.6** *The driver of an automobile, travelling straight level highway, suddenly applies the brakes so that the car slides for 2 seconds covering a distance of 9.81 m before coming to a stop. Assuming that during this time, the car moved with constant deceleration, find the coefficient of friction between the tyres and the pavement.*

***Solution*** Given : $t = 2$ s ; $s = 9.81$ m

Let $a$ = Deceleration in m/s²; and

$u$ = Initial velocity of the car in m/s.

Since the car covers a distance of 9.81 m before coming to a stop, therefore final velocity of the car is zero, *i.e.* $v = 0$

Using the relation, $v = u - at$ ...(–ve sign for deceleration)

$$0 = u - a \times 2$$

or $u = 2a$

Again, using the relation, $s = ut - \frac{1}{2}at^2$

$$9.81 = 2a \times 2 - \frac{1}{2}a \times 2^2 = 4a - 2a = 2a$$

$$\therefore \quad a = \frac{9.81}{2} = 4.905 \text{ m/s}^2$$

Let $W$ = Weight of the car in newtons, and

$\mu$ = Coefficient of friction between the tyres and the pavement.

We know that force required to produce this deceleration

$$= \text{Mass} \times \text{Deceleration} = \frac{W}{g} \times 4.905 \quad \text{...(}i\text{)}$$

Also force required to move the car $= \mu \cdot W$ ...(*ii*)

Equating equations (*i*) and (*ii*),

$$\frac{W}{g} \times 4.905 = \mu \cdot W$$

$$\therefore \quad \mu = \frac{4.905}{g} = \frac{4.905}{9.81} = 0.5 \quad \textbf{Ans.}$$

**Example 16.7** *A bullet of mass 10 g is fired horizontally with a velocity of 1000 m/s, from a gun of mass 50 kg. Find:*

1. *Velocity with which the gun will recoil;*
2. *Force necessary to bring the gun to rest in a distance of 250 mm; and*
3. *The time taken by the gun in coming to rest.*

***Solution*** Given : $m = 10$ g $= 0.01$ kg ; $v = 1000$ m/s ; $M = 50$ kg ; $s = 250$ mm $= 0.25$ m

**1. *Velocity with which the gun will recoil***

Let $V$ = Velocity with which the gun will recoil.

We know that according to law of conservation of momentum,

$$\text{Linear momentum of gun} = \text{Linear momentum of bullet}$$

$$M \cdot V = m \cdot v$$

$$\therefore \quad V = \frac{m \cdot v}{M} = \frac{0.01 \times 1000}{50} = 0.2 \text{ m/s} \quad \textbf{Ans.}$$

**2. *Force necessary to bring the gun to rest***

First of all, let us find the acceleration ($a$) of the gun. Considering the motion of the gun. The initial velocity of the gun ($u$) is 0.2 m/s and the final velocity of the gun ($v$) is zero because the gun comes to rest. Using the relation,

$$v^2 - u^2 = 2as$$

$$0 - (0.2)^2 = 2a \times 0.25$$

$$\therefore \quad a = \frac{-(0.2)^2}{2 \times 0.25} = -0.08 \text{ m/s}^2$$

The –ve sign indicates retardation.

We know that force necessary to bring the gun to rest,

$$F = \text{Mass} \times \text{Retardation} = 50 \times 0.08 = 4 \text{ N} \quad \textbf{Ans.}$$

**3. *Time taken by the gun in coming to rest***

Let $t$ = Time taken by the gun in coming to rest.

Using the relation,

$$v = u + at$$

$$0 = 0.2 - 0.08t$$

$$\therefore \quad t = \frac{0.2}{0.08} = 2.5 \text{ s} \quad \textbf{Ans.}$$

## 16.7 PILE AND PILE HAMMER

Consider a *pile having a conical shape, is driven into the ground by the impact of a pile hammer as shown in Fig. 16.1.

Let
$M$ = Mass of the pile hammer in kg,
$m$ = Mass of the pile in kg,
$h$ = Height through which the pile hammer falls before striking the pile in metres,
$v$ = Velocity of the pile hammer at which it strikes the pile in m/s,
$F$ = Force with which it strikes the pile in newtons,
$s$ = Distance through which the pile is driven in metres,

* See also Chapter 10 on Work, Power and Energy (Sec. 10.14).

$V$ = Common velocity of the pile and pile hammer after impact in m/s, and

$R$ = Average resistance of the ground against the pile in newtons.

Since the initial velocity ($u$) of the pile hammer is zero, therefore from the relation,

$$v^2 - u^2 = 2gh$$

or $$v^2 - 0 = 2gh$$

∴ $$v = \sqrt{2gh}$$

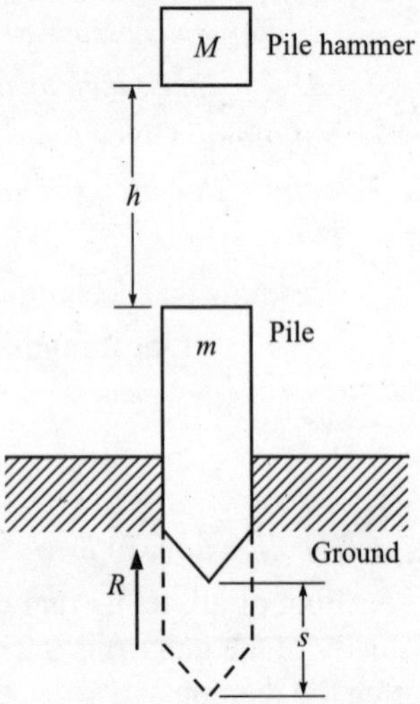


**Fig. 16.1**

We know that linear momentum of the pile hammer before impact

$$= M \cdot v$$

and linear momentum of the pile and pile hammer after impact

$$= (M + m)\, V$$

According to the principle of conservation of momentum,

$$\text{Momentum before impact} = \text{Momentum after impact}$$

or $$M \cdot v = (M + m) V$$

∴ $$V = \frac{M \cdot v}{M + m}$$

The average resistance of the ground to penetration will be in the opposite direction of the retarding force. In order to find the retardation ($a$), the following relation may be used.

$$v_1^2 - V^2 = -2a \cdot s$$

After the pile is driven into the ground up to a distance $s$, the velocity of the pile and pile hammer ($v_1$) becomes zero. Thus, the above relation becomes

$$0 - V^2 = -2a \cdot s$$

or $$a = \frac{V^2}{2s}$$

∴ Retarding force, $F = (M + m)\, a$

and average resistance of the ground to penetration,

$$R = \text{Retarding force} + \text{Weight of the pile and pile hammer}$$

$$= (M + m) a + (M + m)\, g$$

**Example 16.8** *A hammer of mass 200 kg falls from rest through a height of 5 m on a pile of mass 50 kg. If it drives the pile 0.5 m into the ground, find the average resistance of the ground for penetration.*

***Solution*** Given : $M = 200$ kg ; $m = 50$ kg ; $h = 5$ m ; $s = 0.5$ m

The initial velocity ($u$) of the hammer is zero. Let $v$ be the velocity of the pile hammer at which it strikes the pile. Using the relation,

$$v^2 - u^2 = 2gh$$

$$v^2 - 0 = 2 \times 9.81 \times 5 = 98.1$$

∴ $$v = \sqrt{98.1} = 9.9 \text{ m/s}$$

Let $V$ = Common velocity of the pile and hammer after impact.

We know that momentum before impact

$$= M \cdot v = 200 \times 9.9 = 1980 \text{ kg-m/s}$$

and momentum after impact $= (M + m)V = (200 + 50)V = 250V$ kg-m/s

According to law of conservation of momentum,

$$\text{Momentum before impact} = \text{Momentum after impact}$$

or $1980 = 250V$

$\therefore$ $V = \dfrac{1980}{250} = 7.92$ m/s

Let $R$ = Average resistance of the ground for penetration in newtons.

We know that the resistance of the ground for penetration is in the opposite direction of retarding force. Let $a$ be the retardation.

Using the relation, $u^2 - v^2 = -2a \cdot s$ ...(–ve sign due to retardation)

or $V^2 = 2a \cdot s$

$(7.92)^2 = 2a \times 0.5 = a$

$\therefore$ $a = (7.92)^2 = 62.7$ m/s$^2$

and retarding force, $F$ = Mass × Acceleration $= (M + m)\, a = (200 + 50)\, 62.7 = 15\,675$ N

$\therefore$ Average resistance of the ground for penetration,

$R$ = Retarding force + Weight of the hammer and pile

$= F + (M + m)g = 15\,675 + (200 + 50)\, 9.81$

$= 15\,675 + 2452.5 = 18\,127.5$ N **Ans.**

## 16.8 D'ALEMBERT'S PRINCIPLE

Consider a rigid body acted upon by a system of forces. The system may be reduced to a single resultant force acting on the body whose magnitude is given by the product of the mass of the body and the linear acceleration of the centre of mass of the body. We have already discussed that according to Newton's second law of motion,

$$P = m \cdot a \qquad ...(i)$$

where $P$ = Resultant force acting on the body,

$m$ = Mass of the body, and

$a$ = Linear acceleration of the centre of mass of the body.

The above equation (*i*) may also written as

$$P - m \cdot a = 0 \qquad ...(ii)$$

This equation is also known as *equation of dynamic equilibrium*.

A little consideration will show, that the quantity $-m \cdot a$ be treated as a force, equal, opposite and with the same line of action as the resultant force $P$, and include this force with the system of forces of which $P$ is the resultant, then the complete system of forces will be in equilibrium. This principle is known as **D'Alembert's principle*. The equal and opposite force $-m \cdot a$ is known as *reversed force* or the *inertia force*.

The equation (*ii*) may be written as

$$P + (-m \cdot a) = 0$$

Thus, D'Alembert's principle states that '*the resultant force acting on a body together with the reversed effective force (or inertia force) are in equilibrium.*'

This principle is used to reduce a dynamic problem into an equivalent static problem.

---

*A French mathematician, Jean Le Rouad D'Alembert (1717-1783), first published in 1743, a corollary of Newton's second law of motion, known as D'Alembert's principle.

**Example 16.9** *A train of weight 2850 kN starts from rest and reaches a velocity of 55 km/h after 1 minute. If the frictional resistance of the track is 5 N per kN of the train's weight, find the tractive force developed by the engine.*

***Solution*** Given : $W = 2850 \text{ kN} = 2850 \times 10^3 \text{ N}$ ; $v = 55 \text{ km/h} = \dfrac{55 \times 1000}{60 \times 60} = 15.28 \text{ m/s}$ ; Track resistance, $F = 5 \text{ N/kN} = 5 \times 2850 = 14\,250 \text{ N}$

Since the initial velocity of the train is zero and the velocity after 1 minute (*i.e.* after $t = 1 \text{ min} = 60 \text{ s}$) is 15.28 m/s, therefore acceleration of the train,

$$a = \frac{\text{Change in velocity}}{\text{Time}} = \frac{15.28 - 0}{60} = 0.2547 \text{ m/s}^2$$

Let $P$ = Tractive force developed by the engine.

We know that Net force $= P - F = (P - 14\,250) \text{ N}$

We also know that Force = Mass × Acceleration $= \dfrac{W}{g} \times a$

or
$$P - 14250 = \frac{2850 \times 10^3}{9.81} \times 0.2547 = 73\,995.4$$

$$\therefore \quad P = 14\,250 + 73\,995.4 = 88\,245.4 \text{ N} \quad \textbf{Ans.}$$

**Example 16.10** *A body of weight 500 N rests on a rough horizontal plane, as shown in Fig. 16.2. Determine the magnitude of the force P if it causes the body to move with an acceleration of 2 m/s². The coefficient of friction between the body and the plane is 0.4.*

***Solution*** Given : $W = 500 \text{ N}$ ; $a = 2 \text{ m/s}^2$ ; $\mu = 0.4$

The various forces acting on the body are shown in Fig. 16.2. The force $P$ has two components, one parallel to the plane (*i.e.* $P \cos 45°$) and the other perpendicular to the plane (*i.e.* $P \sin 45°$). The frictional force ($F = \mu \cdot R_N$) acts opposite to the motion of the body.

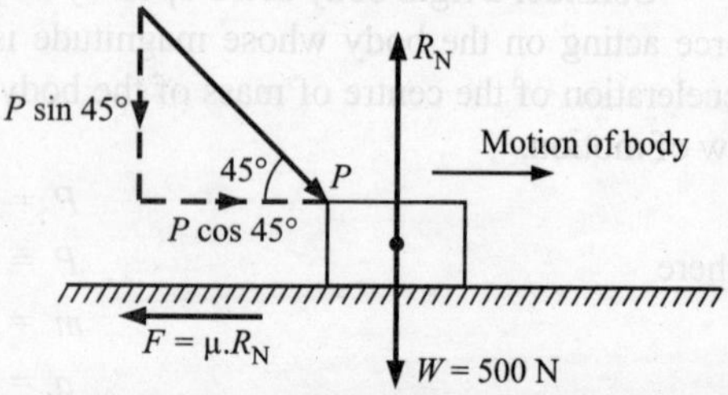


**Fig. 16.2**

Equating vertical forces, we have

$$R_N = W + P \sin 45°$$
$$= 500 + P \times 0.707$$

and frictional force, $F = \mu . R_N = 0.4(500 + P \times 0.707) = 200 + 0.2828\,P$

$\therefore$ Force required to move the body $= P \cos 45° - F = P \times 0.707 - (200 + 0.2828P)$

$$= 0.4242\,P - 200 \qquad ...(i)$$

We also know that Force = Mass × Acceleration

$$= \frac{W}{g} \times a = \frac{500}{9.81} \times 2 = 102 \text{ N} \qquad ...(ii)$$

Equating equations (*i*) and (*ii*), we have

$$0.4242\,P - 200 = 102$$

$$\therefore \quad P = \frac{102 + 200}{0.4242} = 712 \text{ N} \quad \textbf{Ans.}$$

**Example 16.11** *A block weighing 2 kN, rests on a horizontal surface for which coefficient of friction is 0.2. This block is pulled by a force of 1 kN acting at an angle of 30° to the horizontal, as shown in Fig. 16.3. Find the velocity of the block after it moves 20 m starting from rest. If the force of 1 kN is removed, how far it will move?*

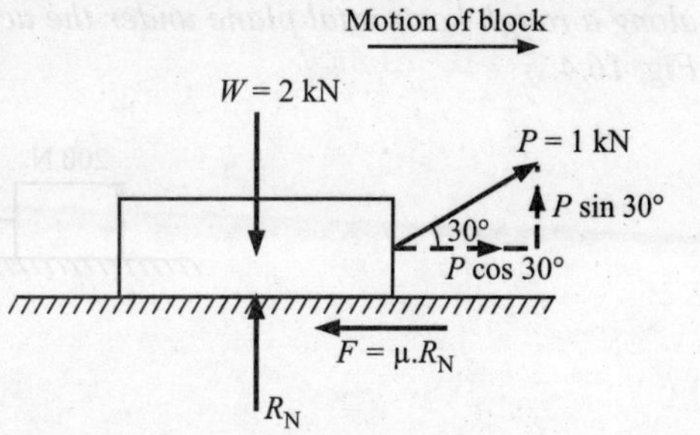


**Fig. 16.3**

***Solution*** Given : $W = 2$ kN $= 2000$ N ; $P = 1$ kN $= 1000$ N ; $\mu = 0.2$ ; $s = 20$ m

The various forces acting on the block are shown in Fig. 16.3. The force $P$ has two components, one parallel to the surface (*i.e.* $P \cos 30°$) and the other perpendicular to the surface (*i.e.* $P \sin 30°$). The frictional force ($F = \mu \cdot R_N$) acts opposite to the motion of the body.

***Velocity of the block***

Let $v$ = Velocity of the block after it moves 20 m starting from rest,

$a$ = Acceleration of the block, and

$R_N$ = Normal reaction on the block.

Equating vertical forces, we have $R_N = W - P \sin 30° = 2000 - 1000 \times 0.5 = 1500$ N

and frictional force, $F = \mu.R_N = 0.2 \times 1500 = 300$ N

$\therefore$ Force required to move the body $= P \cos 30° - F = 1000 \times 0.866 - 300 = 566$ N ...(*i*)

We know that Force = Mass × Acceleration $= \dfrac{W}{g} \times a = \dfrac{2000}{9.81} \times a$ ...(*ii*)

Equating equations (*i*) and (*ii*), we have

$$566 = \frac{2000}{9.81} \times a$$

$$\therefore \quad a = \frac{566 \times 9.81}{2000} = 2.776 \text{ m/s}$$

Now using the relation, $v^2 - u^2 = 2a \cdot s$

$$v^2 - 0 = 2 \times 2.776 \times 20 = 111.04 \qquad ...(\because u = 0)$$

$$\therefore \quad v = \sqrt{111.04} = 10.54 \text{ m/s } \textbf{Ans.}$$

***Distance moved after force of 1 kN is removed***

Let $s$ = Required distance.

If the force of 1 kN is removed, the block will move under the retarding force of $F = \mu \cdot R_N = 0.2 \times 2000 = 400$ N (Here, $R_N = W = 2000$ N).

We know that force, $F = \dfrac{W}{g} \times a$

$$\therefore \quad \text{Retardation, } a = -\frac{F \times g}{W} = -\frac{400 \times 9.81}{2000} = -1.962 \text{ m/s}^2$$

Now using the relation, $v^2 - u^2 = -2as$

$$(10.54)^2 - 0 = -2(-1.962)s = 3.924s$$

$$\therefore \quad s = \frac{(10.54)^2}{3.924} = 28.31 \text{ m } \textbf{Ans.}$$

**Example 16.12** *Two bodies of weights 200 N and 800 N are connected by a thread and they move along a rough horizontal plane under the action of a force 400 N applied to the weight 800 N, as shown in Fig. 16.4.*

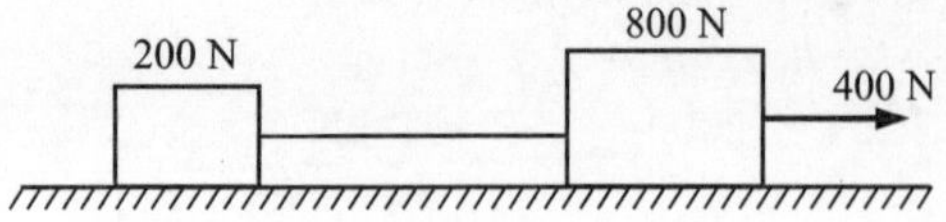


**Fig. 16.4**

*The coefficient of friction between the sliding surfaces of the bodies and the plane is 0.3. Using D'Alembert's principle, determine the acceleration of the two bodies and the tension in the string.*

***Solution*** Given : $W_1 = 200$ N ; $W_2 = 800$ N ; $P = 400$ N ; $\mu = 0.3$

***Acceleration of the two bodies***

Let $a$ = Acceleration of two bodies.

First of all, consider a body 1 of weight 200 N, as shown in Fig. 16.5 (*a*). The various forces acting on this body are

(*a*) Weight of the body = 200 N acting downwards,

(*b*) Normal reaction, $R_{N1} = 200$ N, acting upwards,

(*c*) Force of friction, $F_1 = \mu \cdot R_{N1} = 0.3 \times 200 = 60$ N acting towards left, and

(*d*) Tension ($T$) in the thread acting towards right.

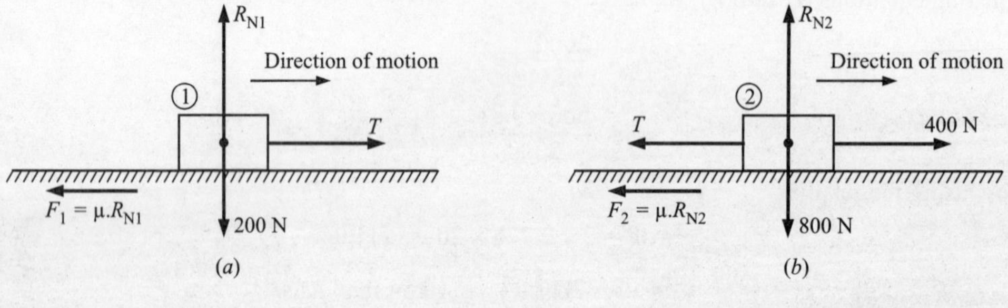


**Fig. 16.5**

We know that force causing acceleration to body 1,

$$= m_1 \cdot a = \frac{W_1}{g} \times a = \frac{200}{9.81} \times a$$

and resultant horizontal force, $P_1 = T - F_1 = T - 60$

According to D'Alembert's principle,

$$P_1 - m_1 \cdot a = 0$$

or $$T - 60 - \frac{200}{9.81} \times a = 0$$

$\therefore$ $$T = 60 + \frac{200}{9.81} \times a \qquad ...(i)$$

Now consider the body 2 of weight 800 N, as shown in Fig. 16.5 (*b*). The various forces acting on the body are

(*a*) Weight of the body = 800 N, acting downwards,

(*b*) Normal reaction, $R_{N2}$ = 800 N acting upwards,

(*c*) Force of friction, $F_2 = \mu \cdot R_{N2} = 0.3 \times 800 = 240$ N acting towards left,

(*d*) Force of 400 N acting towards right, and

(*e*) Tension (*T*) in the thread acting towards left.

We know that force causing acceleration to body 2

$$= m_2 \cdot a = \frac{W_2}{g} \times a = \frac{800}{9.81} \times a$$

and resultant horizontal force,
$$P_2 = 400 - T - F_2$$
$$= 400 - T - 240 = 160 - T$$

According to D'Alembert's principle,

$$P_2 - m_2 \cdot a = 0$$

or
$$160 - T - \frac{800}{9.81} \times a = 0$$

$$\therefore \quad T = 160 - \frac{800}{9.81} \times a \quad \text{...(ii)}$$

Equating equations (*i*) and (*ii*), we have

$$60 + \frac{200}{9.81} \times a = 160 - \frac{800}{9.81} \times a$$

or
$$\frac{200}{9.81} \times a + \frac{800}{9.81} \times a = 160 - 60$$

$$\frac{1000}{9.81} \times a = 100$$

$$\therefore \quad a = \frac{100 \times 9.81}{1000} = 0.981 \text{ m/s}^2 \quad \textbf{Ans.}$$

***Tension in the string***

Substituting the value of *a* in equation (*i*), we have

$$T = 60 + \frac{200}{9.81} \times 0.981 = 80 \text{ N} \quad \textbf{Ans.}$$

## 16.9 MOTION OF A LIFT

Consider a lift (or elevator or cage) is carrying some mass and is in motion with uniform acceleration.

Let $m$ = Mass carried by the lift,

$W$ = Load carried by the lift = $mg$

$R$ = Reaction of the lift or tension ($T$) in the cable supporting the lift, and

$a$ = Uniform acceleration of the lift.

We shall consider the following two cases:

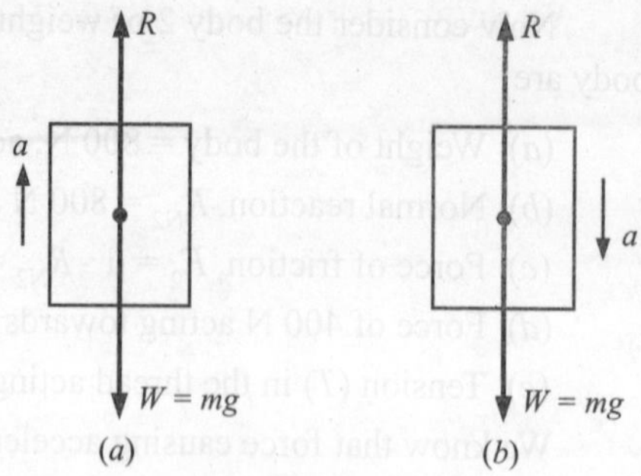


Fig. 16.6

**1. *When the lift is moving upwards***

When the lift is moving upwards, as shown in Fig. 16.6 (*a*), the downward force due to the load carried by the lift,

$$W = mg$$

Since the lift is moving upwards, therefore the reaction of the lift (*R*) must be greater than the weight carried by the lift.

$\therefore$ Net force acting upward on the lift $= R - W$

$$= R - mg \quad ...(i)$$

But the net force acting on the lift $=$ Mass $\times$ Acceleration

$$= m \cdot a \quad ...(ii)$$

Equating equations (*i*) and (*ii*), we have

$$R - m \cdot g = m.a$$

or $$R = ma + mg = m(a + g)$$

**2. *When the lift is moving downwards***

When the lift is moving downwards, the weight of the lift will be more than the reaction *R*, as shown in Fig. 16.6 (*b*).

$\therefore$ Net downward force acting on the lift

$$= W - R = mg - R \quad ...(iii)$$

Now equating equations (*ii*) and (*iii*), we have

$$mg - R = ma$$

or $$R = mg - ma = m(g - a)$$

**Example 16.13** *A lift of mass 600 kg is moving with an acceleration of 2 m/s². Find the tension in the string, when (a) lift is moving upwards ; and (b) when the lift is moving downwards.*

***Solution*** Given : $m = 600$ kg ; $a = 2$ m/s²

**(*a*) *When the lift is moving upwards***

We know that tension in the string, $T = m(g + a) = 600(9.81 + 2) = 7086$ N **Ans.**

**(*b*) *When the lift is moving downwards***

We know that tension in the string, $T = m(g - a) = 600(9.81 - 2) = 4686$ N **Ans.**

**Example 16.14** *A cage is allowed to go inside a mine shaft from rest which attains a speed of 12 m/s in covering a distance of 20 m. Find the pressure which a weight of 600 N exerts on the cage.*

***Solution*** Given : $u = 0$ ; $v = 12$ m/s ; $s = 20$ m ; $W = 600$ N or $m = \frac{600}{9.81}$ kg

Let $P$ = Pressure exerted on the cage.

First of all, let us find the acceleration (*a*) of the cage.

Using the relation, $$v^2 - u^2 = 2as$$

or $$(12)^2 - 0 = 2a \times 20 = 40a$$

$$\therefore \quad a = \frac{(12)^2}{40} = 3.6 \text{ m/s}^2$$

We know that pressure exerted on the cage when the cage is moving downwards,

$$P = m(g - a) = \frac{600}{9.81}(9.81 - 3.6) = 379.8 \text{ N} \textbf{ Ans.}$$

**Example 16.15** *An elevator weighing 6 kN is ascending with an acceleration of 4 m/s². The operator weighs 700 N. Assuming that he is standing on a weighing scale placed on the floor of the elevator during the ascent, what will be the scale reading? What is the tension in the cables at that time?*

***Solution*** Given : $W_1 = 6 \text{ kN} = 6000 \text{ N}$ or $m_1 = \frac{6000}{9.81}$ kg ; $a = 4 \text{ m/s}^2$ ; $W_2 = 700$ N or $m_2 = \frac{700}{9.81}$ kg

***Scale reading***

Let $R$ = Scale reading. This is equal to the reaction offered by the floor on the operator, as shown in Fig. 16.7.

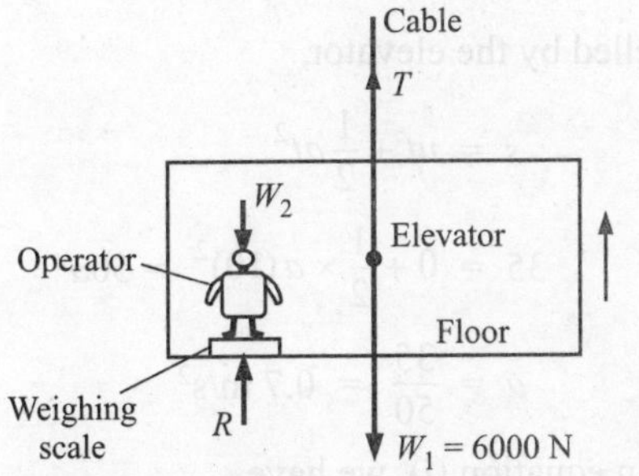


**Fig. 16.7**

When the elevator is ascending, the scale reading,

$$R = m_2(g + a) = \frac{700}{9.81}(9.81 + 4) = 985.4 \text{ N} \textbf{ Ans.}$$

***Total tension in the cables of the elevator***

Let $T$ = Total tension in the cables of the elevator.

When the elevator (mass $m_1$) with the operator (mass $m_2$) is moving upwards, the total tension in the cable of the elevator,

$$T = (m_1 + m_2)(g + a) = \left[\frac{6000 + 700}{9.81}\right](9.81 + 4) = 9432 \text{ N} \textbf{ Ans.}$$

**Example 16.16** *An elevator weighs 2500 N and is moving vertically downwards with a constant acceleration. Write the equation for the elevator cable tension. Starting from rest, it travels a distance of 35 m during an interval of 10 seconds. Find the cable tension during this time. Neglect all other resistance to motion. What are the limits of cable tension?*

***Solution*** Given : $W = 2500$ N ; $u = 0$ ; $s = 35$ m ; $t = 10$ s

***Equation for the elevator cable tension***

Let $T$ = Tension in the cable,

$a$ = Constant acceleration, and

$m$ = Mass of the elevator $= \frac{W}{g} = \frac{2500}{9.81}$ kg

Since the elevator is moving downwards, as shown in Fig. 16.8, therefore

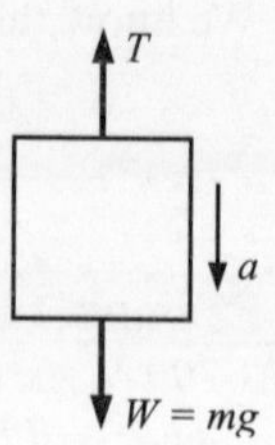


Fig. 16.8

$$W - T = m \cdot a$$

or
$$mg - T = m \cdot a$$

$$\therefore \quad T = mg - ma = m(g-a) = \frac{2500}{9.81}(9.81 - a)$$

$$= 2500\left(1 - \frac{a}{9.81}\right) \text{N} \quad \quad ...(i)$$

This is the required equation for the elevator cable tension when the elevator is moving downwards.

***Cable tension***

First of all, let us find the constant acceleration ($a$).

We know that distance travelled by the elevator,

$$s = ut + \frac{1}{2}at^2$$

or
$$35 = 0 + \frac{1}{2} \times a(10)^2 = 50a$$

$$\therefore \quad a = \frac{35}{50} = 0.7 \text{ m/s}^2$$

Substituting this value of $a$ in equation ($i$), we have

$$T = 2500\left(1 - \frac{0.7}{9.81}\right) = 2500 \times 0.93 = 2325 \text{ N} \quad \textbf{Ans.}$$

***Limits of cable tension***

When the elevator is moving with zero acceleration ($i.e.$ when $a = 0$), then tension in the cable,

$$T = 2500\left(1 - \frac{0}{9.81}\right) = 2500 \text{ N} \quad \textbf{Ans.} \quad \quad ...[\text{From equation } (i)]$$

and tension in the cable when acceleration is maximum ($i.e.$ when $a = 9.81$ m/s$^2$),

$$T = 2500\left(1 - \frac{9.81}{9.81}\right) = 0 \quad \textbf{Ans.}$$

## 16.10 MOTION OF A BODY ALONG A ROUGH INCLINED PLANE

Consider a body of weight $W$ sliding down a rough inclined plane, as shown in Fig. 16.9 ($a$).

Let $m$ = Mass of the body = $\frac{W}{g}$,

$\alpha$ = Angle of inclination of the plane with horizontal,

$a$ = Acceleration of the body, and

$\mu$ = Coefficient of friction between the body and the plane.

Since the body is moving down the plane, the force of friction ($F = \mu \cdot R_N$) will act up the plane.

Resolve the weight of the body ($W$) into two components: one parallel to plane and the other perpendicular to the plane.

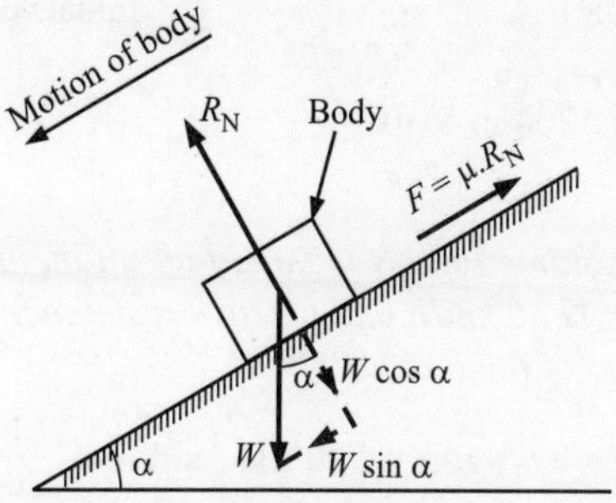


(a) Motion down the plane

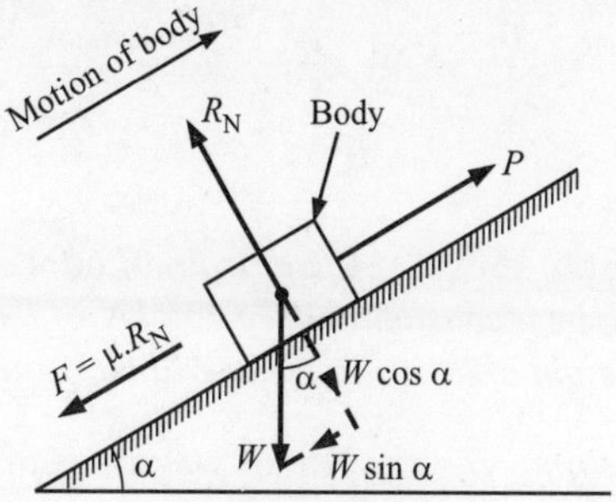


(b) Motion up the plane

**Fig. 16.9**

From Fig. 16.9 (a), we see that component of $W$ parallel to the plane

$$= W \sin \alpha$$

and component perpendicular to the plane $= W \cos \alpha = R_N$, the normal reaction

For the motion down the plane, the net force acting on the body

$$= W \sin \alpha - F = W \sin \alpha - \mu . R_N$$

$$= W \sin \alpha - \mu \cdot W \cos \alpha \quad ...(\because R_N = W \cos \alpha)$$

$$= W (\sin \alpha - \mu \cdot \cos \alpha) \quad ...(i)$$

We know that Force = Mass × Acceleration $= m \cdot a = \dfrac{W}{g} \times a$ ...(ii)

Equating equations (i) and (ii), we have

$$W (\sin \alpha - \mu \cos \alpha) = \frac{W}{g} \times a$$

$$\therefore \quad a = g (\sin \alpha - \mu \cos \alpha) \quad ...(iii)$$

In case the body is to move up the inclined plane under the action of pull $P$, as shown in Fig. 16.9 (b), the force of friction ($F = \mu \cdot R_N$) will act downwards. In this case,

Net force acting on the body for the motion up the plane

$$= P - (W \sin \alpha + \mu \cdot R_N) = P - (W \sin \alpha + \mu \cdot W \cos \alpha) \quad ...(iv)$$

Equating equations (ii) and (iv), we can find the acceleration of the body for the motion up the plane.

**Example 16.17** *A body of mass 20 kg is initially at rest on a 45° rough inclined plane. Find the distance the body travels along the inclined plane before attaining a speed of 2 m/s. The coefficient of friction between the body and the plane is 0.1.*

***Solution*** Given : $m = 20$ kg ; $\alpha = 45°$ ; $v = 2$ m/s ; $\mu = 0.1$

Let $s$ = Distance travelled by the body, and

$a$ = Acceleration of the body.

Since no external force is applied to the body, therefore the acceleration of the body for the motion down the plane is given by

$$a = g(\sin \alpha - \mu \cos \alpha) = 9.81 (\sin 45° - 0.1 \cos 45°)$$

$$= 9.81 (0.707 - 0.1 \times 0.707) = 6.242 \text{ m/s}^2$$

Now, using the relation, $v^2 = u^2 + 2a \cdot s$

$$2^2 = (0)^2 + 2 \times 6.242\,s \qquad ...(\because \text{ Initial velocity, } u = 0)$$

$$\therefore \quad s = \frac{2^2}{2 \times 6.242} = 0.3204 \text{ m} \quad \textbf{Ans.}$$

**Example 16.18** *A coal train of 60 wagons, each weighing 500 kN is pulled up an incline 1 in 200, 4.5 km long by a constant pull of 250 kN. If its initial velocity is 50 km/h and tractive resistance is 40 N per 10 kN of weight of the train, what will be its velocity at the top of the incline?*

***Solution*** Given : No. of wagons = 60 ; Weight of each wagon = 500 kN ; $\sin\alpha = \frac{1}{200} = 0.005$ ; $s = 4.5$ km = 4500 m ; $P = 250$ kN = $250 \times 10^3$ N ; $v = 50$ km/h $= \frac{50 \times 1000}{60 \times 60} = 13.9$ m/s ; $F = 40$ N/10 kN or 4 N/kN

Let $a$ = Acceleration of the train in m/s², and

$V$ = Velocity at the top of the incline in m/s.

We know that total weight of the wagons,

$$W = 60 \times 500 = 30\,000 \text{ kN}$$

$$= 30 \times 10^6 \text{ N}$$

∴ Force required for accelerating the train

$$= \text{Mass} \times \text{Acceleration}$$

$$= m \cdot a = \frac{W}{g} \times a$$

$$= \frac{30 \times 10^6}{9.81} \times a$$

$$= 3058 \times 10^3\, a \qquad ...(i)$$

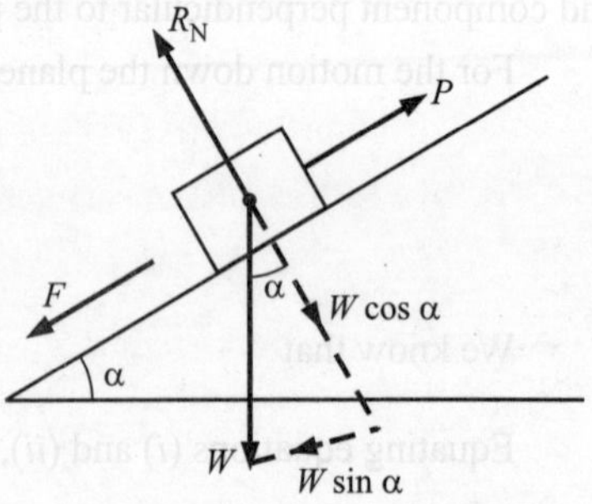


**Fig. 16.10**

Tractive or frictional resistance, $F = 4 \text{ N/kN} = 4 \times 30\,000 = 120 \times 10^3$ N

We also know that force required for accelerating the train (Refer Fig. 16.10)

$$= P - W\sin\alpha - F$$

$$= 250 \times 10^3 - 30 \times 10^6 \times 0.005 - 120 \times 10^3$$

$$= 250 \times 10^3 - 150 \times 10^3 - 120 \times 10^3 = -20 \times 10^3 \text{ N} \qquad ...(ii)$$

Equating equations (*i*) and (*ii*), we have

$$3058 \times 10^3\, a = -20 \times 10^3$$

$$\therefore \quad a = \frac{-20 \times 10^3}{3058 \times 10^3} = -0.006\,54 \text{ m/s}^2$$

The –ve sign indicates retardation.

Now using the relation, $V^2 - v^2 = -2a \cdot s$

or $$V^2 - (13.9)^2 = -2 \times 0.006\,54 \times 4500 = -58.86$$

$$V^2 = (13.9)^2 - 58.86 = 193 - 58.86 = 134$$

$$\therefore \quad V = \sqrt{134} = 11.58 \text{ m/s} = \frac{11.58 \times 60 \times 60}{1000} = 41.688 \text{ km/h} \quad \textbf{Ans.}$$

**Example 16.19** *An engine of mass 50 tonnes pulls a train of mass 250 tonnes up a gradient of 1 in 120 with a uniform speed of 36 km/h. Find the power exerted by the engine, if the tractive resistance is 60 N/per tonne.*

***Solution*** Given : $m_1 = 50$ t ; $m_2 = 250$ t ; $\sin\alpha = \frac{1}{120}$ ; $v = 36$ km/h $= \frac{36 \times 1000}{60 \times 60} = 10$ m/s ; Tractive resistance = 60 N/tonne

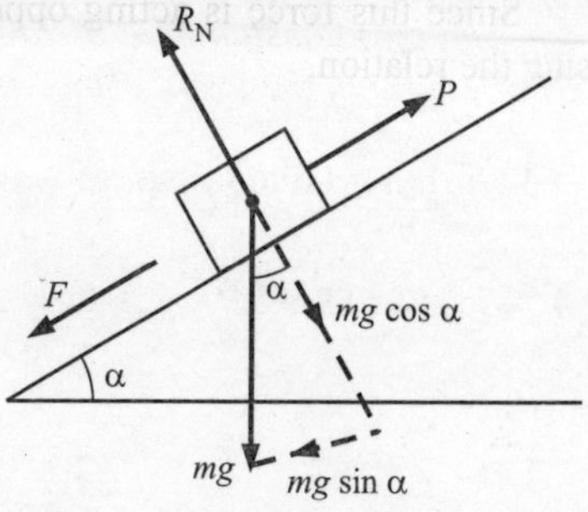


Fig. 16.11

We know that total mass of the engine and train,

$$m = m_1 + m_2 = 50 + 250$$
$$= 300 \text{ t} = 300 \times 10^3 \text{ kg}$$

Since the train is moving with a uniform speed, therefore the acceleration of the train is zero.

We know that force required for accelerating the train

$$= \text{Mass} \times \text{Acceleration}$$
$$= m \times 0 = 0 \qquad ...(i)$$

Let $P$ (in newtons) be the force exerted by the engine while moving up the gradient.

Tractive or frictional resistance acting in the opposite direction of $P$ is given by

$$F = 60 \text{ N/tonne} = 60 \times 300 = 18\,000 \text{ N}$$

We know that force required for accelerating the train (Refer Fig. 16.11)

$$= P - mg \sin\alpha - F$$
$$= P - 300 \times 10^3 \times 9.81 \times \frac{1}{120} - 18\,000$$
$$= P - 24\,525 - 18\,000 = P - 42\,525 \qquad ...(i)$$

From equations (*i*) and (*ii*), $\quad 0 = P - 42\,525$

or $\quad P = 42\,525$ N

∴ Power exerted by the engine $= P \cdot v = 42\,525 \times 10 = 425\,250 \text{ W} = 425.25 \text{ kW}$ **Ans.**

**Example 16.20** *A truck of weight 15 kN is moving on a level track and then climbs a hill having an inclination of 30° with the horizontal. At 50 m up the inclined plane, when it is moving at 30 km/h, the engine stops working and the brakes are not applied. If the tractive resistance is 10 N/kN, determine*

1. *Distance travelled by the truck up the hill after the engine stops ;*
2. *Velocity and acceleration of the truck when it is coming down ;*
3. *Retardation of the truck on the level track ; and*
4. *Distance travelled by the truck on the level track before coming to rest.*

***Solution*** Given : $W = 15$ kN $= 15 \times 10^3$ N ; $\alpha = 30°$ ; $u = 30$ km/h $= \frac{30 \times 1000}{60 \times 60} = 8.33$ m/s ; Track resistance = 10 N/kN = 10 × 15 = 150 N

**1. *Distance travelled by the truck up the hill after the engine stops***

We know that component of the weight ($W$) or resistance of the truck along the hill (Refer Fig. 16.12),

$$= W \sin\alpha = 15 \times 10^3 \sin 30° = 15 \times 10^3 \times 0.5 = 7500 \text{ N}$$

When the truck is moving up the hill, the resistance of the truck along the hill and the track resistance acts in the opposite direction of motion of the truck.

∴ Total resistance force acting against the upward motion of the truck,

$$F = 7500 + 150 = 7650 \text{ N}$$

Since this force is acting opposite to the motion of the truck, therefore it will be taken as negative. Using the relation,

$$F = \text{Mass} \times \text{Acceleration} = \frac{W}{g} \times a$$

$$-7650 = \frac{15 \times 10^3}{9.81} \times a$$

$$\therefore \quad a = \frac{-7650 \times 9.81}{15 \times 10^3} = -5 \text{ m/s}^2$$

The –ve sign indicates retardation.

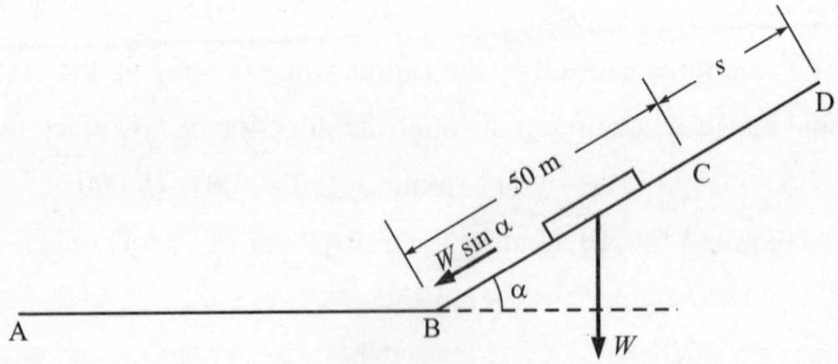


**Fig. 16.12**

Now for displacement ($s$) from $C$ to $D$, the initial velocity $u = 8.33$ m/s and final velocity, $v = 0$. Using the relation,

$$v^2 - u^2 = 2as$$

$$0 - (8.33)^2 = 2 \times -5\, s$$

or

$$s = \frac{(8.33)^2}{10} = 6.94 \text{ m}$$

∴ Distance travelled up the hill after the engine stops

$$= 6.94 \text{ m} \quad \textbf{Ans.}$$

**2. *Velocity and acceleration of the truck when it is coming down***

Let $v_1$ and $a_1$ = Velocity and acceleration of the truck when it coming down.

After stopping on the hill, the truck will start moving downwards due to its own weight. In this case, the track resistance will act in the upward direction, *i.e.* in the opposite direction of the motion of the truck. In this case, the total resistance force acting downwards,

$$F_1 = 7500 - 150 = 7350 \text{ N}$$

We know that

$$F_1 = m \cdot a_1 = \frac{W}{g} \times a_1$$

$$\therefore \quad a_1 = \frac{F_1 \times g}{W} = \frac{7350 \times 9.81}{15 \times 10^3} = 4.8 \text{ m/s}^2 \textbf{ Ans.}$$

The initial velocity ($u_1$), in this case, will be zero and the total distance travelled ($s_1$) by the truck during its downward movement will be $DB$, *i.e.*

$$s_1 = DB = BC + CD = 50 + 6.94 = 56.94 \text{ m}$$

Using the relation,
$$v_1^2 - u_1^2 = 2a_1 s_1$$

$$v_1^2 - 0 = 2 \times 4.8 \times 56.94 = 546.6$$

$$\therefore \quad v_1 = \sqrt{546.6} = 23.38 \text{ m/s } \textbf{Ans.}$$

**3. *Retardation of the truck on the level track***

Let $a_2$ be the retardation of the truck on the level track. A little consideration will show that on a level track, the total force acting on the truck is due to the track resistance which is equal to 150 N and it acts in the opposite direction of motion of the truck, *i.e.*

$$F_2 = -150 \text{ N}$$

Using the relation,
$$F_2 = m.a_2 = \frac{W}{g} \times a_2$$

$$\therefore \quad a_2 = \frac{F_2 \times g}{W} = \frac{-150 \times 9.81}{15 \times 10^3} = -0.098 \text{ m/s}^2 \textbf{ Ans.}$$

The –ve sign indicates retardation.

**4. *Distance travelled on the level track before coming to rest***

Let $s_2$ be the distance travelled on the level track before coming to rest. In this case, the initial velocity of the truck ($u_2$) will be equal to $v_1$, *i.e.* 23.38 m/s, and the final velocity ($v_2$) is zero. Using the relation,

$$v_2^2 - u_2^2 = 2a_2 s_2$$

$$0 - (23.38)^2 = -2 \times 0.098 s_2$$

$$\therefore \quad s_2 = \frac{(23.38)^2}{2 \times 0.098} = 2789 \text{ m } \textbf{Ans.}$$

## HIGHLIGHTS

1. Newton's first law of motion states that '*every body continues its state of rest or of uniform motion in a straight line, unless it is acted upon by some external force*'.
2. Newton's second law of motion states that '*the rate of change of momentum is directly proportional to the impressed force and takes place in the same direction in which the force acts*'.
3. Newton's third law of motion states that '*to every action, there is always an equal and opposite reaction*'.
4. The *mass* is the amount of matter contained in a body. It is expressed in kilograms (kg).
5. The *weight* of a body is a force by which it is attracted towards the centre of the earth. It is expressed in newtons (N). The relation between mass ($m$) and weight ($W$) is given by
$$W = m \, . \, g$$
where $W$ is in newtons, $m$ is in kg and g is the acceleration due to gravity in m/s².
6. The *momentum* is defined as the total motion possessed by a body. Mathematically,
$$\text{Momentum} = \text{Mass} \times \text{Velocity}$$

7. The *moment of momentum* or *angular momentum* is the product of mass moment of inertia ($I$) and the angular velocity ($\omega$) of the body. Mathematically,

   Moment of momentum or angular momentum $= I \cdot \omega$

8. The law of conservation of momentum states that '*the total momentum of a system of masses (i.e. moving bodies) in any one direction remains constant, if no external force acts on them*'. This law is applied to problems on impact *i.e.* collision of two bodies.

9. The *force* is defined as an agent, which produces or tends to produce, destroy or tends to destroy motion.

   According to Newton's second law of motion, the applied force or impressed force ($F$) is directly proportional to the rate of change of momentum. Mathematically,

   Rate of change of momentum,

   $$F = \frac{mv - mu}{t} = \frac{m(v-u)}{t} = m \cdot a \qquad \left(\because \frac{v-u}{t} = a\right)$$

   In S.I. system of units, the unit of force is newton (briefly written as N). A newton may be defined as the force, while acting upon a body of mass 1 kg produces an acceleration of 1 m/s$^2$ in the direction of which it acts. Thus

   $$1\text{ N} = 1\text{ kg} \times 1\text{ m/s}^2 = 1\text{ kg-m/s}^2$$

10. The *inertia force* is a force equal in magnitude but opposite in direction and collinear with the impressed force producing the acceleration. Mathematically,

    Inertia force, $F_I = -m \cdot a$

11. The *impulse* is the product of force ($F$) and time ($t$). Mathematically,

    $$\text{Impulse} = F \times t = \frac{m(v-u)}{t} \times t$$
    $$= m(v-u) = \text{Change of momentum}$$

    If a force acts for a very short time, it is then known as *impulsive force* or *blow*. It occurs in collisions, in explosions, in striking of a nail or a pile by a hammer.

12. When a pile of mass $m$ is driven into the ground by the impact of a pile hammer of mass $M$ and with a velocity $v$, then the common velocity ($V$) of the pile and pile hammer after impact is given by

    $$V = \frac{M \cdot v}{(M+m)}$$

    where $v = \sqrt{2gh}$, $h$ being the height through which the pile hammer falls before striking the pile. The average resistance of the ground to penetration ($R$) will be opposite to the retarding force. The retardation ($a$) is given by

    $$a = \frac{V^2}{2s}$$

    where $s$ is the distance through which the pile is driven into the ground.

    $$\therefore \quad R = (M+m)a + (M+m)g$$

13. D'Alembert's principle states that '*the resultant force acting on the body together with the reversed effective force or inertia force ($F_I$) are in equilibrium*'.

14. When a lift (elevator or cage) carrying some mass ($m$) and moving with uniform acceleration ($a$), then reaction of the lift or tension in the cable supporting the lift,

    $$R = m(g+a) \qquad \text{...(when the lift is moving upwards)}$$
    $$= m(g-a) \qquad \text{...(when the lift is moving downwards)}$$

**15. When a body of mass ($m$) is sliding down an inclined rough plane (see Fig. 16.9), then the acceleration ($a$) of the body is given by**

$$a = g(\sin\alpha - \mu\cos\alpha)$$

**If the body is to move up the inclined plane under the action of pull $P$, then force acting on the body for the motion up the plane**

$$= P - (W\sin\alpha + \mu W\cos\alpha) \quad \text{...}(W = mg)$$

**where** $\alpha$ = **Angle of inclination of the plane with the horizontal, and**

$\mu$ = **Coefficient of friction between the body and the plane.**

## NUMERICAL PROBLEMS

1. A body of mass 150 kg produces an acceleration of 2.5 m/s$^2$ in the direction of the force. Determine the force. [**Ans.** 375 N]
2. A car whose mass is 2500 kg moves at the rate of 60 kmph. After the engine is shut off, it is brought to rest by brakes in 250 m. Find the force exerted, assuming it to be uniform. [**Ans.** 1390 N]
3. A force of 10 N acts on a body having a mass of 2 kg for 15 seconds. If the body starts from rest, find the acceleration produced and the distance moved in this interval. [**Ans.** 5 m/s$^2$; 562.5 m]
4. A force of 200 N acts on a body having a mass of 300 kg for 90 seconds. If the initial velocity of the body is 20 m/s, determine the final velocity of the body when the force acts
(*a*) in the direction of motion; and (*b*) in the opposite direction of motion. [**Ans.** 80 m/s; –40 m/s]
5. A man of mass 65 kg dives into a swimming pool from a tower of height 20 m. He was found to go down in water by 2 m and then starts rising. Find the average resistance of water. Neglect the resistance of air. [**Ans.** 6370 N]
6. A bullet of mass 30 g is fired horizontally with a velocity of 400 m/s from a gun carried in a carriage, which together with the gun has a mass of 150 kg. The resistance to sliding of the carriage over the ice on which it rests is 40 N. Find: 1. the velocity with which the gun will recoil; 2. the distance, in which it comes to rest; and 3. the time taken by the gun in coming to rest. [**Ans.** 0.08 m/s; 0.012 m; 0.3 s]
7. A hammer of mass 1500 kg drops from a height of 0.6 m on a pile of mass 750 kg. Calculate:
1. The common velocity after impact, assuming plastic impact; and
2. The average resistance of the ground, if the pile comes to rest after penetrating 50 mm into the ground. [**Ans.** 2.287 m/s; 139 747.5 N]
8. A block of mass 100 kg rests on a rough horizontal surface and is acted upon by a push $P$ applied at an angle of 30° to the horizontal. Determine the magnitude of $P$ required to give the block an acceleration of 3 m/s$^2$ to the right. The coefficient of friction between the block and the plane is 0.25. [**Ans.** 736.2 N]
9. Two blocks of weights 25 N and 100 N are connected by a thread and move along a rough horizontal plane under the action of a force 50 N applied to the weight of 100 N, as shown in Fig. 16.13. The coefficient of friction between the sliding surfaces of the weights of the plane is 0.3. Determine the acceleration of the weights and tension in the thread using D'Alembert's principle. [**Ans.** 0.981 m/s$^2$; 10 N]

**Fig. 16.13**

10. An elevator of mass 100 kg is moving with uniform acceleration of 0.8 m/s$^2$. Determine the tension in the cable supporting the elevator, when (*a*) the elevator is moving upwards; and (*b*) when the elevator is moving downwards. Take $g$ = 9.8 m/s$^2$. [**Ans.** 1060 N; 900 N]
11. A cage is lowered down a mine shaft from rest. It carries 10 men each weighing 600 N. The cage attains a speed of 10 m/s is 15 m. Find the pressure exerted by each man on the floor of the cage. Take $g$ = 9.8 m/s$^2$. [**Ans.** 395.9 N]

12. A lift is moving up and down with an acceleration of 0.6 m/s$^2$. A man holds a spring balance from which a parcel weighing 15 N is hung. What will be the reading in the spring balance when (*a*) the lift is moving upward; and (*b*) the lift is moving downward. Take g = 9.8 m/s$^2$. [**Ans.** 15.92 N; 14.08 N]
13. A lift has an upward acceleration of 1 m/s$^2$. What pressure will a man, weighing 600 N, exert on the floor of the lift? What pressure would he exert if the lift had an acceleration of 1 m/s$^2$ downwards? What upward acceleration would cause his weight to exert a pressure of 700 N of the floor? [**Ans.** 661.22 N; 538.78 N; 1.63 m/s$^2$]
14. A body of weight 500 N is initially stationary on a 20° rough inclined plane. The coefficient of friction between the body and the plane is 0.25. What distance must the body slide along the inclined plane, before it reaches a speed of 15 m/s? Also find the time elapsed. [**Ans.** 107.14 m; 14.286 s]
15. A locomotive weighing 1 MN is moving up a slope of 1 in 100 and is pulling a train weighing 2.4 MN. The frictional resistance to the train amounts to 8 N per kN. If the speed of the train is 50 km/h and its acceleration is 0.2 m/s$^2$, find the power exerted by the engine. [**Ans.** 1814.2 kW]
16. At a mine, a train of small dump cars is first pulled along a level track from *A* to *B*, as shown in Fig. 16.14 and then up a 5% grade. At some point *C*, 40 m up the grade, the last car gets detached while the velocity is 24 km/h. If the car weighs 20 kN and the track resistance is 8 N per kN, find how far up the grade will the car continue before stopping at *D*? [**Ans.** 79 m]

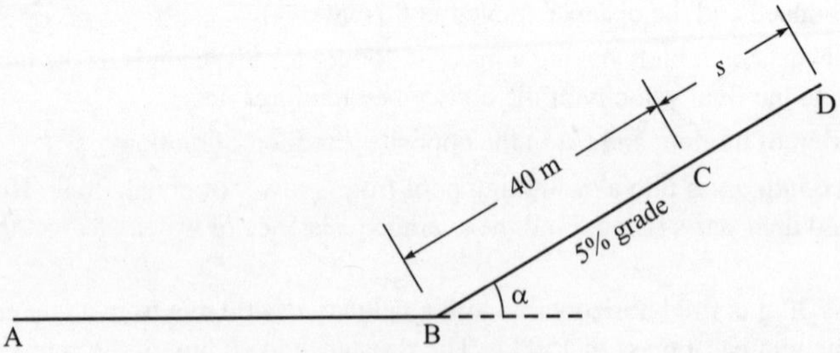


**Fig. 16.14**

## REVIEW QUESTIONS

1. State Newton's laws of motion. Which of these laws is called law of inertia?
2. Distinguish clearly between mass and weight.
3. Define momentum and moment of momentum.
4. Derive the relationship between impulse and linear momentum.
5. State and explain the law of conservation of momentum.
6. What is the difference between force and inertia force?
7. How the average resistance of the ground to penetration is determined when a pile is driven by a pile hammer into the ground?
8. Explain D'Alembert's principle. How is it applied in solving problems related to dynamics?
9. Derive an expression for the tension in the cable supporting a lift when (*a*) it is going up; and (*b*) it is coming down.
10. A body is moving along a rough inclined plane. Derive the expression for the acceleration of the body when

    (*a*) the body is sliding down the inclined plane; and

    (*b*) the body is moving up the inclined plane under the action of pull *P*.

## OBJECTIVE TYPE QUESTIONS

1. To every action, there is always an equal and opposite reaction. This statement is according to
   (*a*) Newton's first law of motion (*b*) Newton's second law of motion
   (*c*) Newton's third law of motion (*d*) none of these
2. A science teacher claimed that Newton's third law of motion is involved while studying the motion of rockets. His statement is
   (*a*) justified (*b*) not justified
3. The matter contained in a body is called
   (*a*) impulsive force (*b*) mass (*c*) weight (*d*) momentum
4. The force by which the body is attracted towards the centre of the earth is called
   (*a*) impulsive force (*b*) mass (*c*) weight (*d*) momentum
5. The total motion possessed by a body is called momentum.
   (*a*) True (*b*) False
6. The force applied on a body of mass 100 kg to produce an acceleration of 5 m/s$^2$, is
   (*a*) 20 N (*b*) 100 N (*c*) 500 N (*d*) none of these
7. The application of law of conservation of momentum is the firing of a shot from a gun.
   (*a*) Yes (*b*) No
8. D'Alembert's principle basically depends upon
   (*a*) Newton's first law of motion (*b*) Newton's second law of motion
   (*c*) Newton's third law of motion (*d*) none of these
9. The tension in the cable supporting a lift is more when the lift is moving .............. with an acceleration.
   (*a*) upwards (*b*) downwards
10. If the tension in the cable supporting a lift moving downwards is one-half the tension when it is moving upwards, then the acceleration of the lift is
   (*a*) $\frac{g}{2}$ (*b*) $\frac{g}{3}$ (*c*) $\frac{g}{4}$ (*d*) none of these

***Answers to Objective Type Questions***

| | | | | |
|---|---|---|---|---|
| **1.** (*c*) | **2.** (*a*) | **3.** (*b*) | **4.** (*c*) | **5.** (*a*) |
| **6.** (*c*) | **7.** (*a*) | **8.** (*b*) | **9.** (*a*) | **10.** (*d*) |

# 17

# MOTION OF CONNECTED BODIES



## 17.1 INTRODUCTION

This chapter deals with the motion of two bodies connected by an inextensible string and subjected to some external forces. It may be noted that the acceleration and velocity of both the bodies will be same. Though the two bodies may be connected in many ways, yet the following are important from the subject point of view:

1. Motion of two bodies connected by a string passing over a smooth pulley
2. Motion of two bodies connected by a string when one body is lying on a horizontal surface and the other is hanging free
3. Motion of two bodies connected by a string when one body is lying on an inclined plane and the other is hanging free
4. Motion of two bodies connected by a string over two inclined planes

These are discussed in detail in the following pages.

## 17.2 MOTION OF TWO BODIES CONNECTED BY A STRING PASSING OVER A SMOOTH PULLEY

Consider two bodies of weights $W_1$ and $W_2$ connected with a light inextensible string and passing over a smooth pulley, as shown in Fig. 17.1. Let $W_1$ is greater than $W_2$. Thus the weight $W_1$ will move downwards whereas the smaller weight $W_2$ will move upwards. Since the string is inextensible, the upward acceleration of $W_2$ and the downward acceleration of $W_1$, will be equal. As the string passes over a smooth pulley, the tension on both sides of the pulley will be equal.

Let $T$ = Tension in both sides of the string, and

$a$ = Acceleration of the bodies.

Consider the downward motion of weight $W_1$. As the weight $W_1$ is moving downwards, the tension $T$ acts upwards.

$\therefore$ Net force acting downwards $= W_1 - T$ ...(*i*)

Since the weight $W_1$ is moving downwards with an acceleration $a$, therefore force acting on the body

$$= \frac{W_1}{g} \times a \quad \text{...(ii)}$$

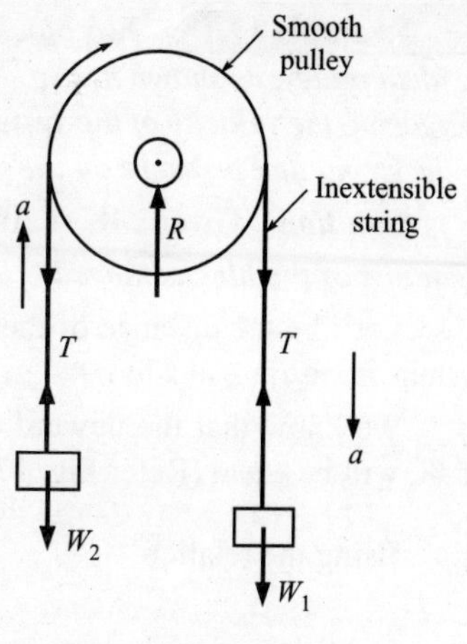


Fig. 17.1

From equations (*i*) and (*ii*), we have

$$W_1 - T = \frac{W_1}{g} \times a \quad \text{...(iii)}$$

Now considering the upward motion of weight $W_2$. As the weight $W_2$ is acting downwards, the tension $T$ acts upwards. Since the weight $W_2$ is moving upwards, therefore

Net force acting upwards $= T - W_2$ ...(*iv*)

The weight $W_2$ is moving upwards with an acceleration $a$, therefore force acting on the body

$$= \frac{W_2}{g} \times a \quad \text{...(v)}$$

From equations (*iv*) and (*v*), we have

$$T - W_2 = \frac{W_2}{g} \times a \quad \text{...(vi)}$$

Adding equations (*iii*) and (*vi*),

$$W_1 - W_2 = \left(\frac{W_1 + W_2}{g}\right) a$$

$$\therefore \quad a = \frac{(W_1 - W_2)\,g}{W_1 + W_2} \quad \text{...(vii)}$$

Substituting this value of $a$ in equation (*iii*), we have

$$W_1 - T = \frac{W_1}{g} \times \frac{(W_1 - W_2)g}{(W_1 + W_2)} = \frac{W_1(W_1 - W_2)}{W_1 + W_2}$$

$$\therefore \quad T = W_1 - \left[\frac{W_1(W_1 - W_2)}{W_1 + W_2}\right] = \frac{W_1(W_1 + W_2) - W_1(W_1 - W_2)}{W_1 + W_2}$$

$$= \frac{W_1^2 + W_1W_2 - W_1^2 + W_1W_2}{W_1 + W_2} = \frac{2\,W_1W_2}{W_1 + W_2} \quad \text{...(viii)}$$

**Notes: 1.** We know that $W_1 = m_1 \cdot g$, and $W_2 = m_2 \cdot g$, where $m_1$ and $m_2$ are the masses of the two bodies. Now, the equations (*vii*) and (*viii*) may be written as,

$$a = \frac{(m_1 - m_2)\,g}{m_1 + m_2}$$

and

$$T = \frac{2m_1m_2}{m_1 + m_2} \times g$$

**2.** Since the pulley is being pressed downwards by the tension ($T$) along the string on both of its sides, therefore the pressure on the pulley or reaction of the pulley,

$$R = 2T = 2 \times \frac{2W_1W_2}{W_1 + W_2} = \frac{4W_1W_2}{W_1 + W_2} = \frac{4m_1m_2}{m_1 + m_2} \times g$$

**Example 17.1** *Two blocks, weighing 200 N and 300 N, are hung to the ends of a rope passing over an ideal pulley, as shown in Fig. 17.2. How much distance the blocks will move in increasing the velocity of the system from 2 m/s to 4 m/s? How much is the tension in the string and pressure on the pulley?*

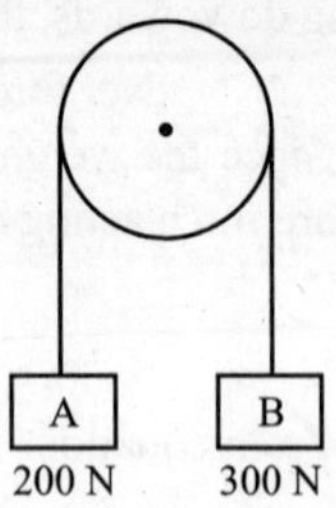


Fig. 17.2

***Solution*** Given : $W_2 = 200$ N ; $W_1 = 300$ N ; $u = 2$ m/s ; $v = 4$ m/s

***Distance of the blocks moved***

Let $s$ be the distance of the blocks moved in increasing the velocity of the system from $u = 2$ m/s to $v = 4$ m/s.

We know that the upward acceleration of $W_2$ and downward acceleration of $W_1$ will be equal (Refer Fig. 17.3).

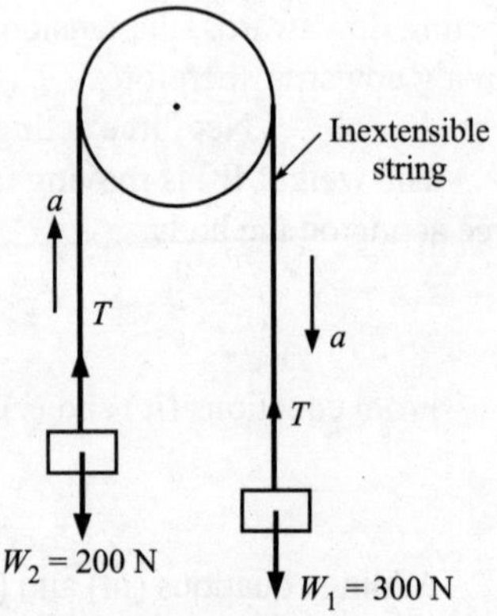


Fig. 17.3

Using the relation,
$$a = \frac{(W_1 - W_2)g}{W_1 + W_2}$$
$$= \frac{(300 - 200)9.81}{300 + 200} = \frac{981}{500}$$
$$= 1.962 \text{ m/s}^2$$

Again, using the relation,
$$v^2 - u^2 = 2as$$
$$4^2 - 2^2 = 2 \times 1.962\, s$$
$$\therefore \quad s = \frac{4^2 - 2^2}{2 \times 1.962} = \frac{16 - 4}{2 \times 1.962}$$
$$= 3.06 \text{ m} \quad \textbf{Ans.}$$

***Tension in the string***

We know that tension in the string,
$$T = \frac{2W_1W_2}{W_1 + W_2}$$
$$= \frac{2 \times 300 \times 200}{300 + 200} = 2.4 \text{ N} \quad \textbf{Ans.}$$

***Pressure on the pulley***

We know that pressure on the pulley,
$$R = 2T = 2 \times 2.4 = 4.8 \text{ N} \quad \textbf{Ans.}$$

**Example 17.2** *A pulley whose axis passes through the centre O carries loads, as shown in Fig. 17.4. Neglecting the inertia of the pulley and assuming that the cord is inextensible, determine the acceleration of the block A. How much weight should be added to or taken away from the block A if the acceleration of the block A is required to be $\frac{g}{3}$ downwards?*

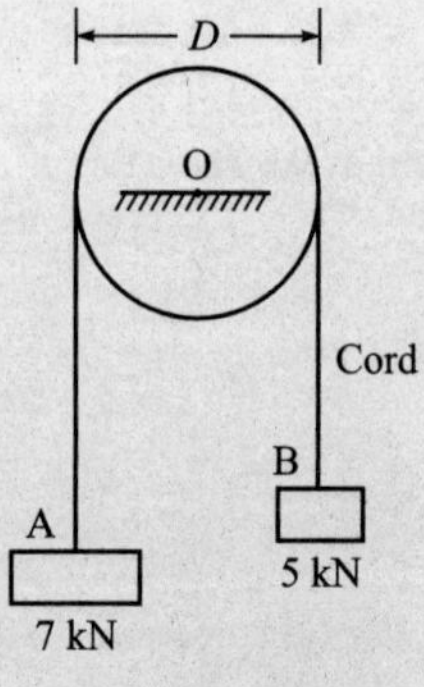


Fig. 17.4

***Solution*** Given : $W_1 = 7$ kN = 7000 N ; $W_2 = 5$ kN = 5000 N

***Acceleration of the block A***

Let $a$ = Acceleration of the block $A$ or acceleration of the system.

Using the relation,
$$a = \frac{(W_1 - W_2)g}{W_1 + W_2}$$

$$= \frac{(7000 - 5000)\,g}{7000 + 5000} = \frac{g}{6}$$

$$= \frac{9.81}{6} = 1.635 \text{ m/s}^2 \textbf{ Ans.}$$

***Weight added or taken away from block A (of weight 7000 N) to make the acceleration $\frac{g}{3}$ downwards***

Let $W_1'$ be the total weight of the block when acceleration ($a$) is $\frac{g}{3}$.

Using the relation, $$a = \frac{(W_1' \;\; W_2)g}{W_1' \;\; W_2}$$

or $$\frac{g}{3} = \frac{(W_1' - W_2)g}{W_1' + W_2}$$

$$\frac{1}{3} = \frac{W_1' - W_2}{W_1' + W_2}$$

$$W_1' + W_2 = 3(W_1' - W_2) = 3W_1' - 3W_2$$

$\therefore$ $$2W_1' = 4W_2 \quad \text{or} \quad W_1' = 2W_2 = 2 \times 5000 = 10\,000 \text{ N}$$

Thus, load to be added to block $A = W_1' - W_1 = 10\,000 - 7000 = 3000$ N **Ans.**

**Example 17.3** *A system of frictionless pulleys carries two weights $W_1$ and $W_2$, hung by the inextensible string, as shown in Fig. 17.5. Using D'Alembert's principle, find the acceleration of the weights $W_1$ and $W_2$ and tension $T_1$ in the string. Given that $W_1 = 300$ N, and $W_2 = 200$ N.*

***Solution*** Given : $W_1 = 300$ N ; $W_2 = 200$ N

***Acceleration of weights $W_1$ and $W_2$***

Let $T_1$ be the tension in the string. Assume that when weight $W_1$ travels a unit distance, then weight $W_2$ travels half the distance. Since acceleration is proportional to the distance, therefore acceleration of weight $W_2$ will be one-half of acceleration of weight $W_1$.

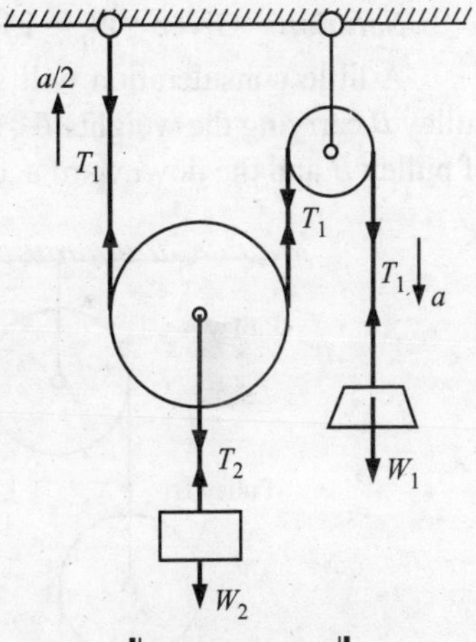

**Fig. 17.5**

Let $a$ = Acceleration of weight $W_1$.

$\therefore$ Acceleration of weight $W_2 = \frac{a}{2}$

Now using D'Alembert's principle for the downward motion of weight $W_1$, we have

$$W_1 - T_1 = \frac{W_1}{g} \times a$$

or $$300 - T_1 = \frac{300}{g} \times a \qquad ...(i)$$

Similarly, for upward motion of weight $W_2$,

$$T_2 - W_2 = \frac{W_2}{g} \times \frac{a}{2}$$

or $$2T_1 - 200 = \frac{200}{g} \times \frac{a}{2} = \frac{100}{g} \times a \qquad ...(\because T_2 = 2T_1) \quad ...(ii)$$

Multiplying equation (*i*) by 2, we have

$$600 - 2T_1 = \frac{600}{g} \times a \qquad ...(iii)$$

Adding equations (*ii*) and (*iii*), we have

$$400 = \frac{700}{g} \times a$$

$$\therefore \quad a = \frac{400 \times g}{700} = 5.6 \text{ m/s}^2$$

Thus, acceleration of weight $W_1$ $= a = 5.6 \text{ m/s}^2$ **Ans.**

and acceleration of weight $W_2$ $= \frac{a}{2} = \frac{5.6}{2} = 2.8 \text{ m/s}^2$ **Ans.**

***Tension in the string***

Let $T_1$ = Tension in the string.

Substituting the value of $a$ in equation (*i*), we have

$$300 - T_1 = \frac{300}{g} \times a = \frac{300}{9.81} \times 5.6 = 171.25$$

$$\therefore \quad T_1 = 300 - 171.25 = 128.75 \text{ N} \quad \textbf{Ans.}$$

**Example 17.4** *A system of weights connected by strings, passing over pulleys A and B is shown in Fig. 17.6. Find the accelerations of the three weights, assuming weightless strings and ideal condition for pulleys.*

***Solution*** Given : $W_1 = 150$ N ; $W_2 = 60$ N ; $W_3 = 40$ N

A little consideration will show that the weight of $W_1 = 150$ N will move downwards, whereas the pulley *B* carrying the weights $W_2$ and $W_3$ (*i.e.* 60 + 40 = 100 N) will move upwards. The upward acceleration of pulley *B* and the downward acceleration of weight $W_1 = 150$ N will be same. Let this acceleration is $a_1$.

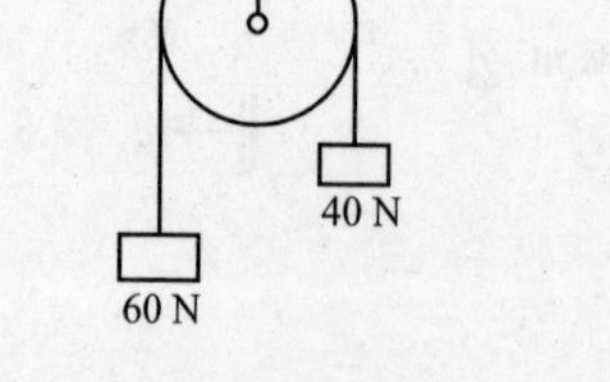


**Fig. 17.6**

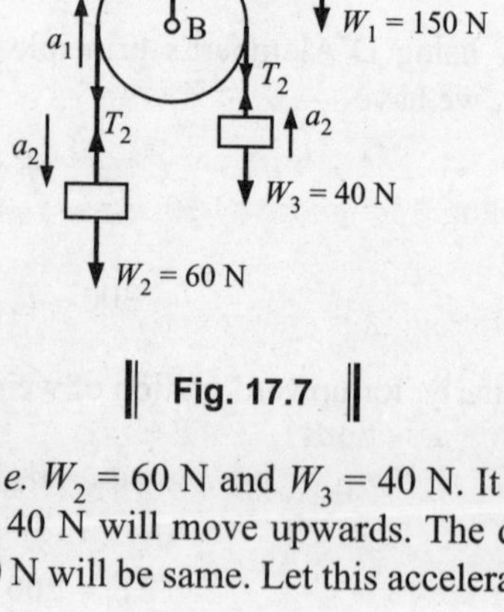


**Fig. 17.7**

The string passing over the pulley *B* carries two weights, *i.e.* $W_2 = 60$ N and $W_3 = 40$ N. It is ident that the weight 60 N will move downwards whereas the weight 40 N will move upwards. The downward acceleration of weight 60 N and upward acceleration of weight 40 N will be same. Let this acceleration is $a_2$. Thus,

Relative acceleration of weight 60 N with respect to pulley $B = a_2 - a_1$

and relative acceleration of weight 40 N with respect to pulley $B = a_2 + a_1$

Let $T_1$ and $T_2$ be the tension (in newtons) of the strings passing over the pulleys $A$ and $B$, respectively (Refer Fig. 17.7).

First of all, let us consider the motion of weight 150 N. The equation of motion is

$$150 - T_1 = \frac{150}{g} \times a_1 = \frac{150}{9.81} \times a_1 \quad ...(i)$$

Considering the motion of weight 60 N. It moves downward with an acceleration of $(a_2 - a_1)$. Thus, the equation of motion is

$$60 - T_2 = \frac{60}{9.81}(a_2 - a_1) \quad ...(ii)$$

Similarly, considering the motion of weight 40 N. It moves upwards with an acceleration of $(a_2 + a_1)$. Thus, the equation of motion is

$$T_2 - 40 = \frac{40}{9.81}(a_2 + a_1) \quad ...(iii)$$

Considering the motion of pulley $B$, we have

$$T_1 = 2T_2 \text{ or } T_2 = \frac{T_1}{2} \quad ...(iv)$$

Substituting the value of $T_2$ in equation (*ii*), we have

$$60 - \frac{T_1}{2} = \frac{60}{9.81}(a_2 - a_1)$$

or
$$120 - T_1 = \frac{120}{9.81}(a_2 - a_1) \quad ...(v)$$

Subtracting equation (*v*) from equation (*i*), we have

$$30 = \frac{150}{9.81} \times a_1 - \frac{120}{9.81}(a_2 - a_1)$$

$$= \frac{150a_1 - 120a_2 + 120a_1}{9.81} = \frac{270a_1 - 120a_2}{9.81}$$

or
$$30 \times 9.81 = 270a_1 - 120a_2 \quad ...(vi)$$

Now adding equations (*ii*) and (*iii*), we have

$$20 = \frac{60}{9.81}(a_2 - a_1) + \frac{40}{9.81}(a_2 + a_1)$$

$$= \frac{60a_2 - 60a_1 + 40a_2 + 40a_1}{9.81} = \frac{100a_2 - 20a_1}{9.81}$$

or
$$20 \times 9.81 = 100a_2 - 20a_1$$

Dividing throughout by 20, we have

$$9.81 = 5a_2 - a_1 \quad ...(vii)$$

Multiplying this equation by 270, we have

$$9.81 \times 270 = 270 \times 5a_2 - 270a_1 \quad ...(viii)$$

Adding equations (*vi*) and (*viii*), we have

$$30 \times 9.81 + 9.81 \times 270 = -120a_2 + 270 \times 5a_2$$

or
$$300 \times 9.81 = 1230a_2$$

$$\therefore \quad a_2 = \frac{300 \times 9.81}{1230} = 2.39 \text{ m/s}^2$$

Substituting the value of $a_2$ in equation (*vii*), we have

$$9.81 = 5 \times 2.39 - a_1$$

$$\therefore \quad a_1 = 5 \times 2.39 - 9.81 = 2.14 \text{ m/s}^2$$

From above, we find that

Acceleration of weight 150 N $= a_1 = 2.14 \text{ m/s}^2$ **Ans.**

Acceleration of weight 60 N $= a_2 - a_1 = 2.39 - 2.14 = 0.25 \text{ m/s}^2$ **Ans.**

and acceleration of weight 40 N $= a_2 + a_1 = 2.39 + 2.14 = 4.53 \text{ m/s}^2$ **Ans.**

## 17.3 MOTION OF TWO BODIES CONNECTED BY A STRING WHEN ONE BODY IS LYING ON A HORIZONTAL SURFACE AND THE OTHER IS HANGING FREE

We shall consider the following two cases:

**1. *When the horizontal surface is smooth***

Consider two bodies 1 and 2 of weights $W_1$ and $W_2$ connected by a light inextensible string, passing over a smooth pulley, as shown in Fig. 17.8. Let the body of weight $W_2$ is placed on a horizontal surface and the body of weight $W_1$ is hanging free. The weight $W_1$ is moving downwards whereas the weight $W_2$ moves on the smooth horizontal surface. The velocity and acceleration of the weight $W_1$ will be same as that of weight $W_2$. Since the string is inextensible and passing over a smooth pulley, therefore, the tension on the both sides of the pulley will be same.

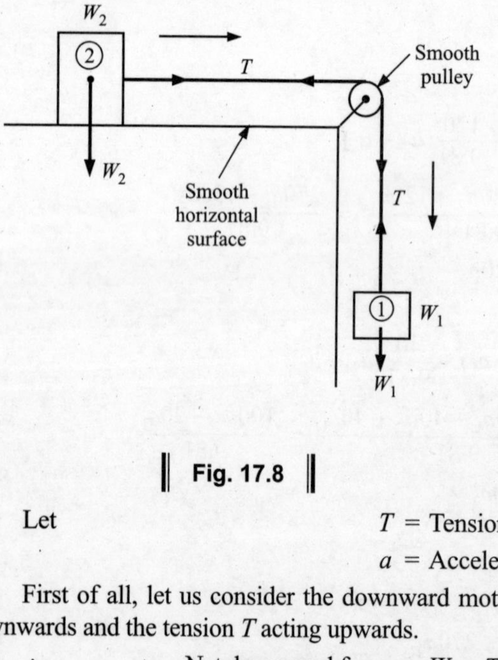


Fig. 17.8

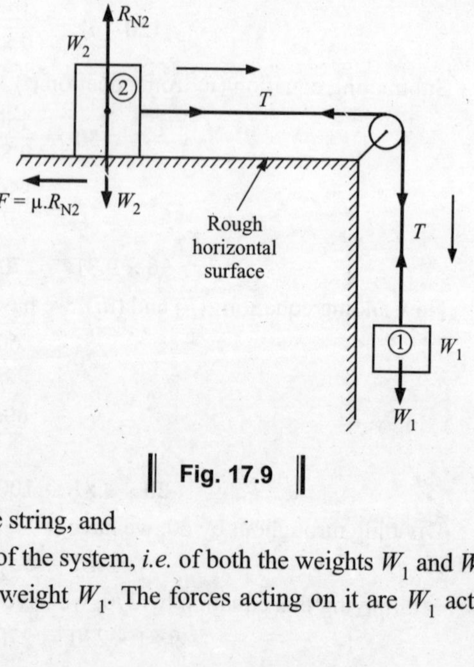


Fig. 17.9

Let $\quad T$ = Tension in the string, and

$a$ = Acceleration of the system, *i.e.* of both the weights $W_1$ and $W_2$.

First of all, let us consider the downward motion of weight $W_1$. The forces acting on it are $W_1$ acting downwards and the tension $T$ acting upwards.

$$\therefore \quad \text{Net downward force} = W_1 - T \qquad ...(i)$$

Since the weight $W_1$ is moving downward with an acceleration $a$, therefore force acting on $W_1$

$$= \text{Mass} \times \text{Acceleration} = \frac{W_1}{g} \times a \qquad ...(ii)$$

Equating equations (*i*) and (*ii*), we have

$$W_1 - T = \frac{W_1}{g} \times a \qquad ...(iii)$$

Now consider the motion of weight $W_2$ which moves on the horizontal surface with an acceleration $a$. The only force causing the motion of weight $W_2$ in the horizontal direction is tension $T$.

Since the weight $W_2$ is moving in the horizontal direction with an acceleration $a$, therefore the force acting on weight $W_2$

$$= \frac{W_2}{g} \times a$$

$\therefore$ Tension in the string, $T = \frac{W_2}{g} \times a \qquad ...(iv)$

Adding equations (*iii*) and (*iv*), we have

$$W_1 = \frac{W_1}{g} \times a + \frac{W_2}{g} \times a = (W_1 + W_2)\frac{a}{g}$$

$$\therefore \quad a = \frac{W_1 \cdot g}{(W_1 + W_2)}$$

and

$$T = \frac{W_2}{g} \times a = \frac{W_2}{g} \times \frac{W_1 \cdot g}{(W_1 + W_2)} = \frac{W_1 W_2}{(W_1 + W_2)}$$

**Note:** In terms of masses $m_1$ and $m_2$, $a = \frac{m_1 g}{m_1 + m_2}$ and $T = \frac{m_1 m_2 g}{m_1 + m_2}$

**2. *When the horizontal surface is rough***

When the horizontal surface is rough, the force of friction ($F$) will act in the opposite direction of motion of weight $W_2$, as shown in Fig. 17.9.

Let $\mu$ = Coefficient of friction between the weight $W_2$ and horizontal surface,

$R_{N2}$ = Normal reaction at the horizontal surface = $W_2$,

$F$ = Force of friction = $\mu \cdot R_{N2}$, ...($\because R_{N2} = W_2$)

$T$ = Tension in the string, and

$a$ = Acceleration of the system.

First of all, let us consider the downward motion of weight $W_1$ with an acceleration $a$. The net downward force acting on weight $W_1$

$$= W_1 - T \qquad ...(i)$$

We know that Force = Mass × Acceleration = $\frac{W_1}{g} \times a \qquad ...(ii)$

Equating equations (*i*) and (*ii*), we have

$$W_1 - T = \frac{W_1}{g} \times a \qquad ...(iii)$$

Now consider the motion of weight $W_2$ which moves towards right on the rough horizontal surface with an acceleration $a$. The forces acting on weight $W_2$ are the tension ($T$) towards right and the frictional force ($F$) = $\mu \cdot R_{N2}$, towards left.

∴ Net horizontal force acting towards right

$$= T - F = T - \mu . R_{N2} = T - \mu W_2 \qquad ...(iv)$$

We know that Force = Mass × Acceleration $= \dfrac{W_2}{g} \times a \qquad ...(v)$

Equating equations (*iv*) and (*v*), we have

$$T - \mu W_2 = \frac{W_2}{g} \times a \qquad ...(vi)$$

Adding equations (*iii*) and (*vi*), we have

$$W_1 - \mu W_2 = \frac{W_1}{g} \times a + \frac{W_2}{g} \times a = \frac{(W_1 + W_2)\,a}{g}$$

$$\therefore \quad a = \frac{(W_1 - \mu W_2)\,g}{W_1 + W_2} \qquad ...(vii)$$

Substituting this value of $a$ in equation (*iii*), we have

$$W_1 - T = \frac{W_1}{g} \times \frac{(W_1 - \mu W_2)\,g}{W_1 + W_2} = \frac{W_1 (W_1 - \mu W_2)}{W_1 + W_2}$$

$$\therefore \quad T = W_1 - \left[\frac{W_1 (W_1 - \mu W_2)}{W_1 + W_2}\right] = W_1 \left[1 - \frac{W_1 - \mu W_2}{W_1 + W_2}\right]$$

$$= W_1 \left[\frac{W_1 + W_2 - (W_1 - \mu W_2)}{W_1 + W_2}\right]$$

$$= \frac{W_1}{(W_1 + W_2)} \left[W_2 + \mu W_2\right] = \frac{W_1 W_2 (1 + \mu)}{W_1 + W_2}$$

**Note:** In terms of masses $m_1$ and $m_2$,

$$a = \frac{(m_1 - \mu m_2)\,g}{m_1 + m_2}$$

and

$$T = \frac{m_1 m_2 g (1 + \mu)}{m_1 + m_2}$$

**Example 17.5** *A body of weight 400 N lying on a horizontal plane is connected by a string with a weight of 900 N, which hangs over the edge of the surface. Calculate the acceleration of the system and tension in the string when*

1. *the horizontal surface is smooth ; and*
2. *the horizontal surface is rough. Take μ = 0.4.*

***Solution*** Given : $W_2 = 400$ N ; $W_1 = 900$ N ; $\mu = 0.4$

**1. *When the horizontal surface is smooth***

We know that acceleration of the system,

$$a = \frac{W_1 \cdot g}{W_1 + W_2} = \frac{900 \times 9.81}{900 + 400} = 6.79 \text{ m/s}^2 \quad \textbf{Ans.}$$

and tension in the string,

$$T = \frac{W_1 W_2}{W_1 + W_2} = \frac{900 \times 400}{900 + 400} = 276.9 \text{ N} \quad \textbf{Ans.}$$

**2. *When the horizontal surface is rough***

We know that acceleration of the system,

$$a = \frac{(W_1 - \mu W_2)g}{W_1 + W_2} = \frac{(900 - 0.4 \times 400)9.81}{900 + 400} = 5.584 \text{ m/s}^2 \textbf{ Ans.}$$

and tension in the string,

$$T = \frac{W_1 W_2 (1 + \mu)}{W_1 + W_2} = \frac{900 \times 400(1 + 0.4)}{900 + 400} = 387.7 \text{ N} \textbf{ Ans.}$$

**Example 17.6** *A string passing across a smooth table at right angle to two opposite edges has two masses $m_1$ and $m_2$ ($m_1 > m_2$) attached to its ends hanging vertically, as shown in Fig. 17.10. If a mass M be attached to the portion of the string which is on the table, find the acceleration of the system when left to itself.*

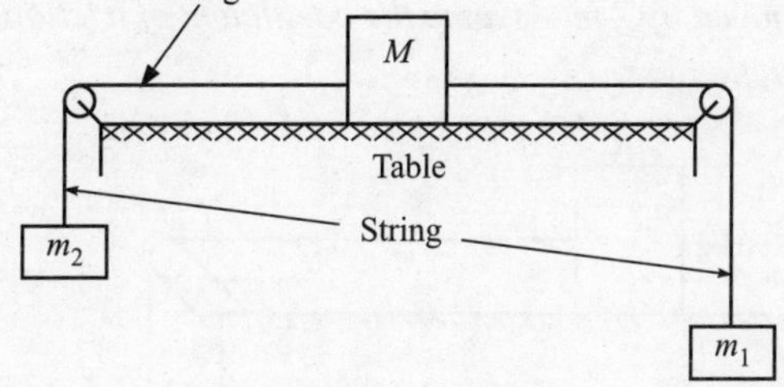


**Fig. 17.10**

***Solution*** Given : Two masses $m_1$ and $m_2$

Let $a$ be the acceleration of the system and $T_1$ and $T_2$ be the tensions in the string, as shown in Fig. 17.11. Since $m_1 > m_2$, therefore the mass $m_1$ will move downwards and mass $m_2$ will move upwards.

**Fig. 17.11**

The equations of motion for $m_1$ is

$$m_1 \cdot g - T_1 = m_1 \cdot a$$

or $$T_1 = m_1 \cdot g - m_1 \cdot a \qquad ...(i)$$

Similarly, the equation of motion for $m_2$ is

$$T_2 - m_2 \cdot g = m_2 \cdot a$$

or $$T_2 = m_2 \cdot g + m_2 \cdot a \qquad ...(ii)$$

and equation of motion for mass $M$ is

$$T_1 - T_2 = M \cdot a \quad ...(iii)$$

Substituting the values of $T_1$ and $T_2$ from equations (*i*) and (*ii*), we have

$$m_1 \cdot g - m_1 \cdot a - (m_2 \cdot g + m_2 \cdot a) = M \cdot a$$

$$m_1 \cdot g - m_2 \cdot g = m_1 a + m_2 a + M \cdot a$$

$$g\,(m_1 - m_2) = (m_1 + m_2 + M)a$$

$$\therefore \quad a = \frac{(m_1 - m_2)g}{m_1 + m_2 + M} \quad \textbf{Ans.}$$

**Example 17.7** *Two blocks A and B of mass 200 kg and 300 kg, respectively, are joined together by an inextensible weightless cable, as shown in Fig. 17.12. If the system is released from rest, determine the velocity of block A after it has moved by 2 m. Assume the coefficient of friction between block A and the plane as 0.25 and that the pulley is frictionless.*

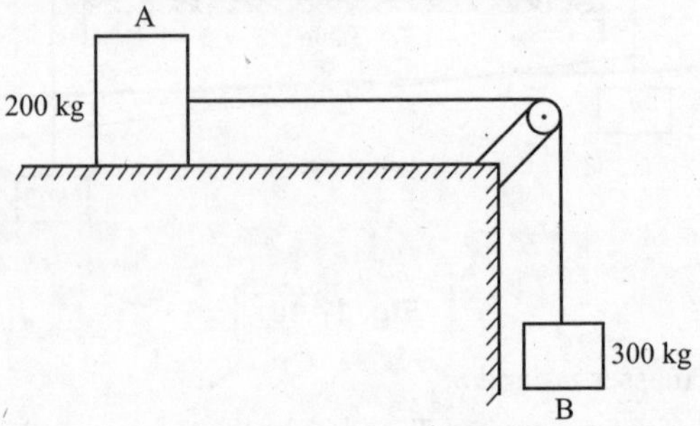


**Fig. 17.12**

***Solution*** Given : $m_A = 200$ kg ; $m_B = 300$ kg ; $s = 2$ m ; $\mu = 0.25$

Let $T$ = Tension in the cable in newtons,

$a$ = Acceleration of the blocks $A$ and $B$ in m/s$^2$,

$v$ = Velocity of the block $A$ in m/s.

First of all, let us consider the motion of the block $A$ in the horizontal direction. The forces acting on the block is weight $W_A = m_A.g$ acting downwards, tension $T$ causing the block to move towards right, and the frictional force $F = \mu.R_N$ acting towards left. The equation of motion for the block $A$ is

$$T - \mu \cdot R_N = m_A \cdot a \quad ...(\text{Refer Fig. 17.13})$$

or

$$T = \mu \cdot R_N + m_A \cdot a$$

$$= 0.25 \times 200 \times 9.81 + 200 \times a \quad ...(\because R_N = W_A = m_A.g)$$

$$= 490.5 + 200a \quad ...(i)$$

Now considering the downward motion of block $B$. The equation of motion for the block $B$ is

$$m_B \cdot g - T = m_B \cdot a$$

$$300 \times 9.81 - T = 300 \times a$$

or

$$T = 300 \times 9.81 - 300a = 2943 - 300a \quad ...(ii)$$

Equating equations (*i*) and (*ii*), we have

$$490.5 + 200a = 2943 - 300a$$

or

$$200a + 300a = 2943 - 490.5 = 2452.5$$

$$\therefore \quad a = \frac{2452.5}{200 + 300} = 4.905 \text{ m/s}^2$$

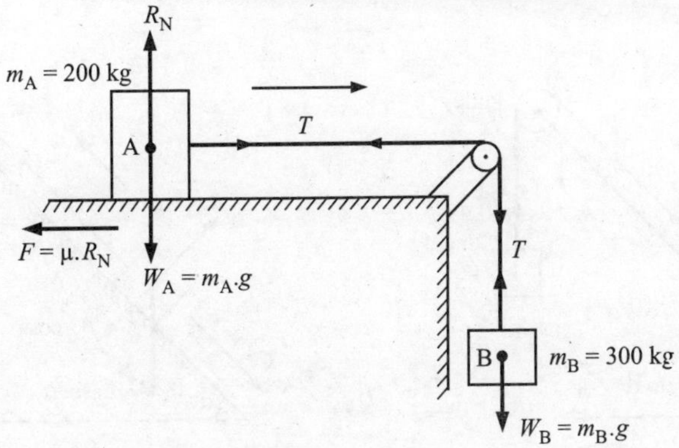


**Fig. 17.13**

Since the initial velocity of block *A* is zero (*i.e.* $u = 0$), therefore using the relation,

$$v^2 - u^2 = 2a \cdot s$$

$$v^2 - 0 = 2 \times 4.905 \times 2 = 19.62$$

$$\therefore \quad v = \sqrt{19.62} = 4.43 \text{ m/s} \quad \textbf{Ans.}$$

**Note:** In case the horizontal surface is smooth (*i.e.* $\mu = 0$), then acceleration of the system may be obtained by using the relation,

$$a = \frac{g \cdot W_B}{W_A + W_B} = \frac{9.81 \times 300}{200 + 300} = 16.35 \text{ m/s}^2$$

## 17.4 MOTION OF TWO BODIES CONNECTED BY A STRING WHEN ONE BODY IS LYING ON AN INCLINED PLANE AND THE OTHER IS HANGING FREE

We shall consider the following two cases:

**1. *When the inclined plane is smooth***

Consider two bodies 1 and 2 of weights $W_1$ and $W_2$ connected by a light inextensible string, passing over a smooth pulley, as shown in Fig. 17.14 (*a*). Let the body of weight $W_2$ is placed on a smooth inclined plane making an angle $\alpha$ with the horizontal, whereas the body of weight $W_1$ is hanging free. As the inclined plane is smooth, the acceleration of weight $W_1$ will be same as that of weight $W_2$. Since the string is inextensible and passing over a smooth pulley, therefore, the tension in both sides of the pulley will be same.

Let $T$ = Tension in the string, and

$a$ = Acceleration of the system, *i.e.* of both the weights $W_1$ and $W_2$.

First of all, let us consider the downward motion of weight $W_1$. The forces acting on it are $W_1$ acting downward and tension $T$ acting upwards.

$$\therefore \quad \text{Net downward force} = W_1 - T \qquad ...(i)$$

Since the weight $W_1$ is moving downwards with an acceleration $a$, therefore, the force acting on $W_1$

$$= \text{Mass} \times \text{Acceleration} = \frac{W_1}{g} \times a \qquad ...(ii)$$

Equating equations (*i*) and (*ii*), we have

$$W_1 - T = \frac{W_1}{g} \times a \qquad ...(iii)$$

Now consider the motion of weight $W_2$ which moves upwards on an inclined plane with an acceleration $a$. The forces acting on $W_2$ for the motion along the inclined plane are $W_2 \sin \alpha$ acting downwards and tension $T$ acting upwards.

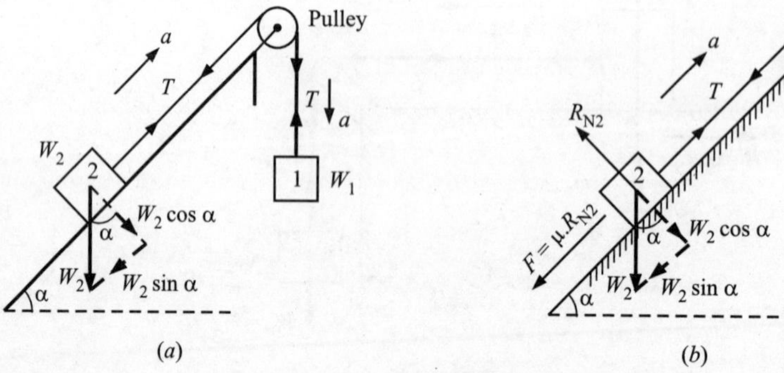


**Fig. 17.14**

∴ Net force acting on $W_2$ along the inclined plane acting upwards

$$= T - W_2 \sin \alpha \qquad ...(iv)$$

We know that net force $= \text{Mass} \times \text{Acceleration} = \dfrac{W_2}{g} \times a$ ...(*v*)

Equating equations (*iv*) and (*v*), we have

$$T - W_2 \sin \alpha = \frac{W_2}{g} \times a \qquad ...(vi)$$

Adding equations (*iii*) and (*vi*), we have

$$W_1 - W_2 \sin \alpha = \frac{W_1}{g} \times a + \frac{W_2}{g} \times a = \frac{a}{g}(W_1 + W_2)$$

$$\therefore \quad a = \frac{g(W_1 - W_2 \sin \alpha)}{W_1 + W_2} \qquad ...(vii)$$

Substituting this value of $a$ in equation (*iii*),

$$W_1 - T = \frac{W_1}{g} \times \frac{g(W_1 - W_2 \sin \alpha)}{W_1 + W_2} = \frac{W_1(W_1 - W_2 \sin \alpha)}{W_1 + W_2}$$

$$\therefore \quad T = W_1 - \left[\frac{W_1(W_1 - W_2 \sin \alpha)}{W_1 + W_2}\right] = W_1\left[1 - \frac{(W_1 - W_2 \sin \alpha)}{W_1 + W_2}\right]$$

$$= W_1\left[\frac{W_1 + W_2 - W_1 + W_2 \sin \alpha}{W_1 + W_2}\right] = \frac{W_1(W_2 + W_2 \sin \alpha)}{W_1 + W_2}$$

$$= \frac{W_1 W_2 (1 + \sin \alpha)}{W_1 + W_2} \qquad ...(viii)$$

**2. *When the inclined plane is rough***

When the inclined plane is rough, the force of friction ($F$) will act in the opposite direction of motion of $W_2$, as shown in Fig. 17.14 (*b*).

Let $\mu$ = Coefficient of friction between the weight $W_2$ and the inclined surface,

$R_{N2}$ = Normal reaction perpendicular to the inclined surface,

$F$ = Force of friction $= \mu . R_{N2} = \mu . W_2 \cos \alpha$ ...($\because R_{N2} = W_2 \cos \alpha$)

$T$ = Tension in the string, and

$a$ = Acceleration of the system.

First of all, let us consider the downward motion of weight $W_1$ with acceleration $a$. The net downward force acting on weight $W_1$ $= W_1 - T$ ...(*i*)

Since the weight $W_1$ is moving downward with acceleration $a$, therefore force acting on $W_1$

$$= \text{Mass} \times \text{Acceleration} = \frac{W_1}{g} \times a \quad ...(ii)$$

Equating equations (*i*) and (*ii*), we have

$$W_1 - T = \frac{W_1}{g} \times a \quad ...(iii)$$

Now consider the motion of weight $W_2$ which moves upwards on the rough inclined surface with an acceleration $a$. The forces acting on weight $W_2$ are the tension ($T$) acting upwards, component of weight $W_2$ along the inclined surface ($W_2 \sin \alpha$) acting downwards and frictional force ($F = \mu . R_{N2}$) acting downwards.

∴ Net upward force acting along the inclined plane

$$= T - W_2 \sin \alpha - \mu . R_{N2}$$

$$= T - W_2 \sin \alpha - \mu W_2 \cos \alpha \quad ...(\because R_{N2} = W_2 \cos \alpha) \quad ...(iv)$$

We know that net force $= \text{Mass} \times \text{acceleration} = \frac{W_2}{g} \times a$ ...(*v*)

Equating equations (*iv*) and (*v*), we have

$$T - W_2 \sin \alpha - \mu W_2 \cos \alpha = \frac{W_2}{g} \times a \quad ...(vi)$$

Adding equations (*iii*) and (*vi*), we have

$$W_1 - W_2 \sin \alpha - \mu W_2 \cos \alpha = \frac{W_1}{g} \times a + \frac{W_2}{g} \times a = \frac{W_1 + W_2}{g} \times a$$

$$\therefore \quad a = \frac{g(W_1 - W_2 \sin \alpha - \mu W_2 \cos \alpha)}{W_1 + W_2} \quad ...(vii)$$

Substituting this value of $a$ in equation (*iii*), we have

$$W_1 - T = \frac{W_1}{g} \times \frac{g(W_1 - W_2 \sin \alpha - \mu W_2 \cos \alpha)}{W_1 + W_2}$$

$$= \frac{W_1(W_1 - W_2 \sin \alpha - \mu W_2 \cos \alpha)}{W_1 + W_2}$$

$$\therefore \quad T = W_1 - \frac{W_1(W_1 - W_2 \sin \alpha - \mu W_2 \cos \alpha)}{W_1 + W_2}$$

$$= W_1\left[1 - \frac{(W_1 - W_2 \sin\alpha - \mu W_2 \cos\alpha)}{W_1 + W_2}\right]$$

$$= W_1\left[\frac{W_1 + W_2 - W_1 + W_2 \sin\alpha + \mu W_2 \cos\alpha}{W_1 + W_2}\right]$$

$$= W_1\left[\frac{W_2 + W_2 \sin\alpha + \mu W_2 \cos\alpha}{W_1 + W_2}\right]$$

$$= \frac{W_1 W_2}{W_1 + W_2}[1 + \sin\alpha + \mu \cos\alpha] \quad \text{...(viii)}$$

**Note:** For a smooth inclined plane, $\mu = 0$. Thus the equations (*vii*) and (*viii*) becomes,

$$a = \frac{g(W_1 - W_2 \sin\alpha)}{W_1 + W_2}$$

and

$$T = \frac{W_1 W_2 (1 + \sin\alpha)}{W_1 + W_2}$$

These expressions are same as already discussed.

**Example 17.8** *A body weighing 1500 N is placed on a smooth inclined plane inclined at 10° to the plane. It is pulled by a body weighing 800 N which hangs freely and it is connected to the first body by a light inextensible string passing over a frictionless pulley. Find : 1. the acceleration of the system ; and 2. the tension in the string.*

***Solution*** Given : $W_2 = 1500$ N ; $\alpha = 10°$ ; $W_1 = 800$ N

**1. *Acceleration of the system***

We know that acceleration of the system,

$$a = \frac{g(W_1 - W_2 \sin\alpha)}{W_1 + W_2} = \frac{9.81(800 - 1500 \sin 10°)}{800 + 1500}$$

$$= \frac{9.81(800 - 1500 \times 0.1736)}{2300} = \frac{9.81(800 - 260.4)}{2300}$$

$$= 2.3 \text{ m/s}^2 \quad \textbf{Ans.}$$

**2. *Tension in the string***

We know that tension in the string,

$$T = \frac{W_1 W_2 (1 + \sin\alpha)}{W_1 + W_2} = \frac{800 \times 1500 (1 + \sin 10°)}{800 + 1500}$$

$$= \frac{800 \times 1500 (1 + 0.1736)}{2300} = 612.3 \text{ N} \quad \textbf{Ans.}$$

**Example 17.9** *A body weighing 1200 N rests on a rough plane inclined at 12° to the horizontal. It is pulled up the plane by means of a light flexible rope running parallel to the plane and passing over a light frictionless pulley at the top of the plane. The portion of the rope beyond the pulley hangs vertically down and carries a weight of 800 N at its end. If the coefficient of friction for the plane and the body is 0.2, find : 1. tension in the rope ; 2. acceleration with which the body moves up the plane ; and 3. distance moved by the body in 3 seconds after starting from rest.*

***Solution*** Given : $W_2 = 1200$ N ; $\alpha = 12°$ ; $W_1 = 800$ N ; $\mu = 0.2$

**1.** ***Tension in the rope***

We know that tension in the rope, $T = \dfrac{W_1 W_2}{W_1 + W_2}\left[1 + \sin\alpha + \mu\cos\alpha\right]$

$$= \frac{800 \times 1200}{800 + 1200}\left[1 + \sin 12° + 0.2\cos 12°\right]$$

$$= 480[1 + 0.2079 + 0.2 \times 0.9781]$$

$$= 480 \times 1.4035 = 673.7 \text{ N} \quad \textbf{Ans.}$$

**2.** ***Acceleration with which the body moves up the plane***

We know that acceleration with which the body moves up the plane,

$$a = \frac{g(W_1 - W_2\sin\alpha - \mu W_2\cos\alpha)}{W_1 + W_2}$$

$$= \frac{9.81(800 - 1200\sin 12° - 0.2 \times 1200\cos 12°)}{800 + 1200}$$

$$= \frac{9.81(800 - 1200 \times 0.2079 - 0.2 \times 1200 \times 0.9781)}{800 + 1200}$$

$$= \frac{9.81(800 - 249.48 - 234.74)}{2000} = 1.549 \text{ m/s}^2 \quad \textbf{Ans.}$$

**3.** ***Distance moved by the body in 3 seconds after starting from rest***

Since the body starts from rest, therefore initial velocity, $u = 0$ and time ($t$) = 3 s. Thus, the distance moved by the body,

$$s = u.t + \frac{1}{2}a.t^2 = 0 + \frac{1}{2} \times 1.549 \times 3^2 = 6.9705 \text{ m} \quad \textbf{Ans.}$$

**Example 17.10** *Determine the resulting motion of the body B, assuming the pulleys to be smooth and weightless, as shown in Fig. 17.15.*

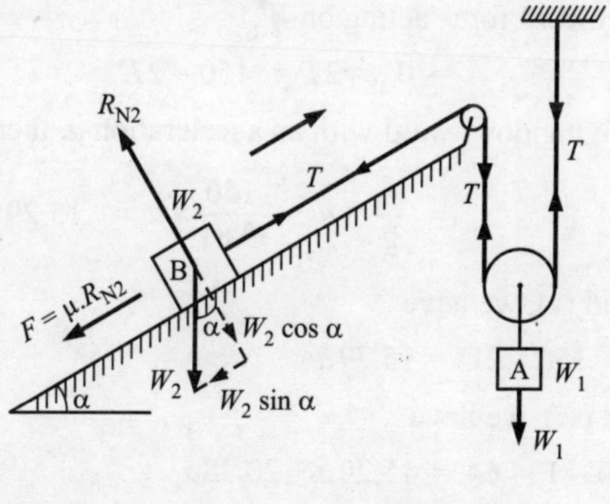


**Fig. 17.15**

*If the system starts from rest, find the velocity of body B after 8 seconds. Given that $W_1 = 150$ N ; $W_2 = 100$ N ; $\alpha = 30°$ and $\mu = 0.2$.*

***Solution*** Given : $W_1 = 150$ N ; $W_2 = 100$ N ; $\alpha = 30°$ ; $\mu = 0.2$

Let

$a$ = Acceleration of the body $B$,

$T$ = Tension in the string,

$R_{N2}$ = Normal reaction of weight $W_2$ perpendicular to the inclined surface = $W_2 \cos \alpha$, and

$F$ = Force of friction = $\mu \cdot R_{N2} = \mu W_2 \cos \alpha$ ...($\because R_{N2} = W_2 \cos \alpha$)

First of all, consider the motion of the body $B$ (*i.e.* weight $W_2$) which moves upwards on the rough inclined surface with acceleration $a$. The forces acting on weight $W_2$ are the tension ($T$) acting upwards, component of weight $W_2$ along the inclined surface ($W_2 \sin \alpha$) acting downwards and frictional force ($F = \mu . R_{N2}$) acting downwards.

$\therefore$ Net force acting upward along the inclined surface

$$= T - W_2 \sin \alpha - F = T - W_2 \sin \alpha - \mu . R_{N2}$$

$$= T - W_2 \sin \alpha - \mu W_2 \cos \alpha$$

$$= T - 100 \sin 30° - 0.2 \times 100 \cos 30°$$

$$= T - 100 \times 0.5 - 0.2 \times 100 \times 0.866$$

$$= T - 50 - 17.32 = T - 67.32 \quad ...(i)$$

Since the body $B$ is moving upwards with an acceleration $a$, therefore force acting on it

$$= \text{Mass} \times \text{Acceleration} = \frac{W_2}{g} \times a = \frac{100}{9.81} \times a = 10.19a \quad ...(ii)$$

Equating equations (*i*) and (*ii*),

$$T - 67.32 = 10.19a$$

Multiplying both sides by 2, we have

$$2T - 134.64 = 20.38a \quad ...(iii)$$

Now considering the motion of body $A$ (*i.e.* weight $W_1$) which moves downwards. It may be noted that acceleration of weight $W_1$ will be one-half of acceleration of weight $W_2$ $\left(i.e. \frac{a}{2}\right)$.

We know that the net downward force acting on $W_1$

$$= W_1 - 2T = 150 - 2T \quad ...(iv)$$

Since the weight $W_1$ is moving downward with an acceleration $a$, therefore force acting on $W_1$

$$= \frac{W_1}{g} \times a = \frac{150}{9.81} \times a = 15.29a \quad ...(v)$$

Equating equations (*iv*) and (*v*), we have

$$150 - 2T = 15.29a \quad ...(vi)$$

Adding equations (*iii*) and (*vi*), we have

$$150 - 134.64 = 15.29a + 20.38a$$

or

$$15.36 = 35.67a$$

$\therefore$

$$a = \frac{15.36}{35.67} = 0.43 \text{ m/s}^2$$

and velocity of body $B$ after 8 seconds, if the system starts from rest is given by the relation

$$v = u + a.t = 0 + 0.43 \times 8 = 3.44 \text{ m/s}$$ **Ans.**

## 17.5 MOTION OF TWO BODIES CONNECTED BY A STRING OVER TWO INCLINED PLANES

We shall consider the following two cases:

**1. *When the inclined planes are smooth***

Consider two bodies 1 and 2 of weights $W_1$ and $W_2$ connected by a light inextensible string, passing over a smooth pulley, as shown in Fig. 17.16 (*a*). Let the body of weight $W_2$ is placed on a smooth inclined plane *AB* making an angle $\alpha_2$ with the horizontal and the body of weight $W_1$ is placed on a smooth inclined plane *BC* making an angle $\alpha_1$ with the horizontal. As the inclined planes are smooth, the acceleration of both the weights $W_1$ and $W_2$ will be same. Since the string is inextensible and passing over a smooth pulley, therefore the tension in both the strings will be same.

Let $T$ = Tension in the string, and

$a$ = Acceleration of the system, *i.e.* of both the weights $W_1$ and $W_2$.

First of all, let us consider the downward motion of weight $W_1$ along the inclined plane *BC*. The forces acting on it are $W_1 \sin \alpha_1$ acting downwards and tension $T$ acting upwards.

$$\therefore \quad \text{Net downward force} = W_1 \sin \alpha_1 - T \qquad ...(i)$$

Since the weight $W_1$ is moving downwards with an acceleration $a$, therefore the force acting on $W_1$

$$= \text{Mass} \times \text{Acceleration} = \frac{W_1}{g} \times a \qquad ...(ii)$$

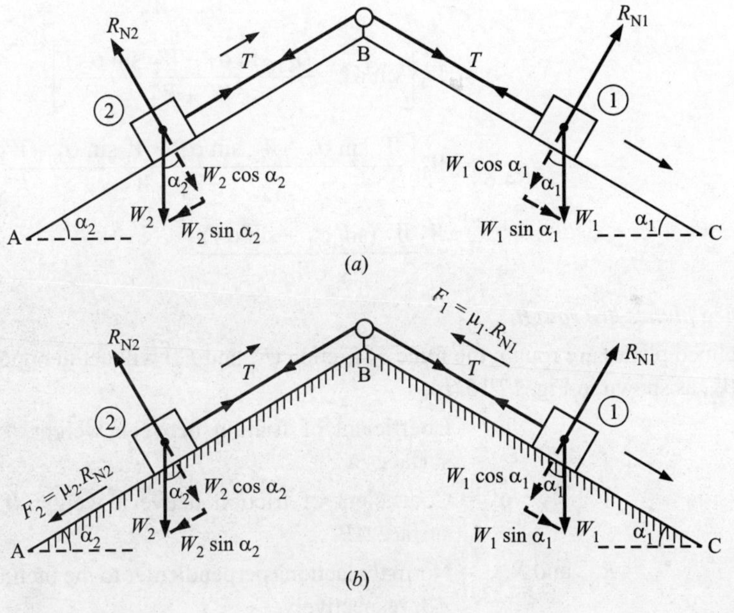


**Fig. 17.16**

Equating equations (*i*) and (*ii*), we have

$$W_1 \sin \alpha_1 - T = \frac{W_1}{g} \times a \qquad ...(iii)$$

Now consider the motion of weight $W_2$ which moves upwards on an inclined plane *AB* with an acceleration $a$. The forces acting on $W_2$ for the motion along the inclined plane *AB* are $W_2 \sin \alpha_2$ acting downwards and tension $T$ acting upwards.

∴ Net force acting on $W_2$ along the inclined plane $AB$ acting upwards

$$= T - W_2 \sin \alpha_2 \quad \text{...(iv)}$$

We know that net force

$$= \text{Mass} \times \text{Acceleration} = \frac{W_2}{g} \times a \quad \text{...(v)}$$

Equating equations (*iv*) and (*v*), we have

$$T - W_2 \sin \alpha_2 = \frac{W_2}{g} \times a \quad \text{...(vi)}$$

Adding equations (*iii*) and (*vi*), we have

$$W_1 \sin \alpha_1 - W_2 \sin \alpha_2 = \frac{W_1}{g} \times a + \frac{W_2}{g} \times a = \frac{W_1 + W_2}{g} \times a$$

$$\therefore \quad a = \frac{g(W_1 \sin \alpha_1 - W_2 \sin \alpha_2)}{W_1 + W_2} \quad \text{...(vii)}$$

Substituting this value of $a$ in equation (*iii*), we have

$$W_1 \sin \alpha_1 - T = \frac{W_1}{g} \times \frac{g(W_1 \sin \alpha_1 - W_2 \sin \alpha_2)}{W_1 + W_2}$$

$$\therefore \quad T = W_1 \sin \alpha_1 - \frac{W_1 (W_1 \sin \alpha_1 - W_2 \sin \alpha_2)}{W_1 + W_2}$$

$$= W_1 \left[ \sin \alpha_1 - \frac{(W_1 \sin \alpha_1 - W_2 \sin \alpha_2)}{W_1 + W_2} \right]$$

$$= W_1 \left[ \frac{W_1 \sin \alpha_1 + W_2 \sin \alpha_1 - W_1 \sin \alpha_1 + W_2 \sin \alpha_2}{W_1 + W_2} \right]$$

$$= \frac{W_1 W_2 (\sin \alpha_1 + \sin \alpha_2)}{W_1 + W_2} \quad \text{...(viii)}$$

**2. *When the inclined planes are rough***

When the inclined planes are rough, the force of friction ($F_1$ and $F_2$) will act in opposite direction of the motion of $W_1$ and $W_2$, as shown in Fig. 17.16 (*b*).

Let

$\mu_1$ = Coefficient of friction between weight $W_1$ and the inclined surface $BC$,

$\mu_2$ = Coefficient of friction between weight $W_2$ and the inclined surface $AB$,

$R_{N1}$ and $R_{N2}$ = Normal reactions perpendicular to the inclined surfaces $BC$ and $AB$, respectively,

$F_1$ and $F_2$ = Force of friction, such that $F_1 = \mu_1 \cdot R_{N1}$ and $F_2 = \mu_2 \cdot R_{N2}$,

$T$ = Tension in the string, and

$a$ = Acceleration of the system.

First of all, let us consider the downward motion of weight $W_1$ along the inclined plane $BC$ with an acceleration $a$. The forces acting on $W_1$ are the tension ($T$) acting upwards, frictional force ($F_1 = \mu_1 \cdot R_{N1}$) acting upwards and the component of weight $W_1$ along the inclined surface $BC$ ($W_1 \sin \alpha_1$) acting downwards.

∴ Net force acting on $W_1$ along the inclined plane $BC$ acting downwards

$$= W_1 \sin \alpha_1 - F_1 - T$$

$= W_1 \sin \alpha_1 - \mu_1 \cdot R_{N1} - T$

$= W_1 \sin \alpha_1 - \mu_1 . W_1 \cos \alpha_1 - T$ ...($\because$ $R_{N1} = W_1 \cos \alpha_1$) ...(*i*)

Since the weight $W_1$ is moving downwards with acceleration $a$, therefore net force acting on $W_1$

$$= \text{Mass} \times \text{Acceleration} = \frac{W_1}{g} \times a \quad ...(ii)$$

Equating equations (*i*) and (*ii*), we have

$$W_1 \sin \alpha_1 - \mu_1 W_1 \cos \alpha_1 - T = \frac{W_1}{g} \times a \quad ...(iii)$$

Now consider the motion of weight $W_2$ which moves upwards on the rough inclined plane $AB$ with an acceleration $a$. The forces acting on weight $W_2$ are the tension ($T$) acting upwards, component of weight $W_2$ along the inclined surface $AB$ ($W_2 \sin \alpha_2$) acting downwards and frictional force ($F_2 = \mu_2.R_{N2}$) acting downwards.

$\therefore$ Net upward force on $W_2$ acting along the inclined plane $AB$

$= T - W_2 \sin \alpha_2 - \mu_2.R_{N2}$

$= T - W_2 \sin \alpha_2 - \mu_2 \cdot W_2 \cos \alpha_2$ ...($\because$ $R_{N2} = W_2 \cos \alpha_2$) ...(*iv*)

Since the weight $W_2$ is moving upwards with acceleration $a$, therefore net force acting on $W_2$

$$= \text{Mass} \times \text{Acceleration} = \frac{W_2}{g} \times a \quad ...(v)$$

Equating equations (*iv*) and (*v*), we have

$$T - W_2 \sin \alpha_2 - \mu_2 W_2 \cos \alpha_2 = \frac{W_2}{g} \times a \quad ...(vi)$$

Adding equations (*iii*) and (*vi*), we have

$$W_1 \sin \alpha_1 - \mu_1 W_1 \cos \alpha_1 - W_2 \sin \alpha_2 - \mu_2 W_2 \cos \alpha_2$$

$$= \frac{W_1}{g} \times a + \frac{W_2}{g} \times a = \frac{W_1 + W_2}{g} \times a$$

$$\therefore \quad a = \frac{g(W_1 \sin \alpha_1 - W_2 \sin \alpha_2 - \mu_1 W_1 \cos \alpha_1 - \mu_2 W_2 \cos \alpha_2)}{W_1 + W_2} \quad ...(vii)$$

Substituting this value of $a$ in equation (*iii*),

$$W_1 \sin \alpha_1 - \mu_1 W_1 \cos \alpha_1 - T$$

$$= \frac{W_1}{g} \times \frac{g(W_1 \sin \alpha_1 - W_2 \sin \alpha_2 - \mu_1 W_1 \cos \alpha_1 - \mu_2 W_2 \cos \alpha_2)}{W_1 + W_2}$$

$$\therefore \quad T = (W_1 \sin \alpha_1 - \mu_1 W_1 \cos \alpha_1) - \frac{W_1(W_1 \sin \alpha_1 - W_2 \sin \alpha_2 - \mu_1 W_1 \cos \alpha_1 - \mu_2 W_2 \cos \alpha_2)}{W_1 + W_2}$$

$$= \frac{1}{W_1 + W_2}\Big[(W_1 + W_2)(W_1 \sin \alpha_1 - \mu_1 W_1 \cos \alpha_1)$$

$$- W_1 (W_1 \sin \alpha_1 - W_2 \sin \alpha_2 - \mu_1 W_1 \cos \alpha_1 - \mu_2 W_2 \cos \alpha_2)\Big]$$

$$= \frac{1}{W_1 + W_2}\Big[W_1^2 \sin \alpha_1 - \mu_1 W_1^2 \cos \alpha_1 + W_1 W_2 \sin \alpha_1 - \mu_1 W_1 W_2 \cos \alpha_1$$

$$- W_1^2 \sin \alpha_1 + W_1 W_2 \sin \alpha_2 + \mu_1 W_1^2 \cos \alpha_1 + \mu_2 W_1 W_2 \cos \alpha_2\Big]$$

$$= \frac{1}{W_1 + W_2}\left[W_1W_2 \sin\alpha_1 - \mu_1W_1W_2 \cos\alpha_1 + W_1W_2 \sin\alpha_2 + \mu_2W_1W_2 \cos\alpha_2\right]$$

$$= \frac{1}{W_1 + W_2}\left[W_1W_2(\sin\alpha_1 + \sin\alpha_2) - W_1W_2(\mu_1 \cos\alpha_1 - \mu_2 \cos\alpha_2)\right]$$

$$= \frac{W_1W_2}{W_1 + W_2}\left[(\sin\alpha_1 + \sin\alpha_2) - (\mu_1 \cos\alpha_1 - \mu_2 \cos\alpha_2)\right] \qquad ...(viii)$$

**Note:** For smooth inclined planes, the coefficient of friction is zero, *i.e.* $\mu_1 = 0$, and $\mu_2 = 0$. Thus the equations (*vii*) and (*viii*) become

$$a = \frac{g(W_1 \sin\alpha_1 - W_2 \sin\alpha_2)}{W_1 + W_2} \qquad ...(\text{same as before})$$

and

$$T = \frac{W_1W_2}{W_1 + W_2}(\sin\alpha_1 + \sin\alpha_2) \qquad ...(\text{same as before})$$

**Example 17.11** *Two bodies of weights 1500 N and 2000 N rest on two smooth inclined planes. The inclination of the planes are 25° and 30° with the horizontal, as shown in Fig. 17.17. Both the bodies are connected by a light inextensible string and passes over a smooth pulley. Find the acceleration of the system and tension in the string.*

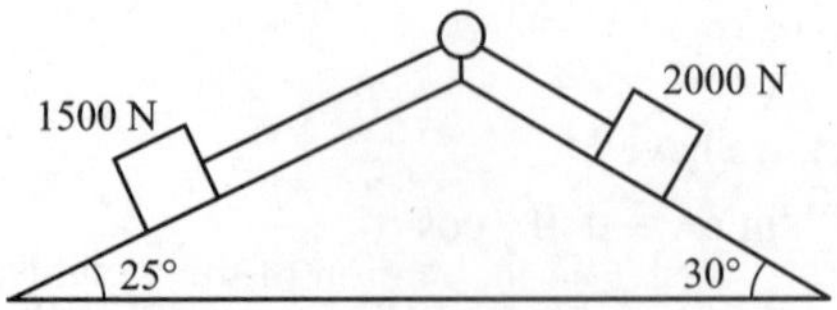


**Fig. 17.17**

***Solution*** Given : $W_2 = 1500$ N ; $\alpha_2 = 25°$ ; $W_1 = 2000$ N ; $\alpha_1 = 30°$

***Acceleration of the system***

We know that acceleration of the system,

$$a = \frac{g(W_1 \sin\alpha_1 - W_2 \sin\alpha_2)}{W_1 + W_2} = \frac{9.81(2000 \sin 30° - 1500 \sin 25°)}{2000 + 1500}$$

$$= \frac{9.81(2000 \times 0.5 - 1500 \times 0.4226)}{3500} = \frac{9.81(1000 - 634)}{3500}$$

$$= 1.026 \text{ m/s}^2 \quad \textbf{Ans.}$$

***Tension in the string***

We know that tension in the string,

$$T = \frac{W_1W_2(\sin\alpha_1 + \sin\alpha_2)}{W_1 + W_2} = \frac{2000 \times 1500(\sin 30° + \sin 25°)}{2000 + 1500}$$

$$= \frac{2000 \times 1500(0.5 + 0.4226)}{3500} = 790.8 \text{ N} \quad \textbf{Ans.}$$

**Example 17.12** *Two bodies A and B of weights 200 N and 300 N, respectively, are connected with the help of a cord, as shown in Fig. 17.18 (a).*

*The weight A is on inclined plane of 20° with the horizontal and the weight B is on inclined plane of 60° with the horizontal. If both the inclined planes are smooth, determine the acceleration of the weights and tension in the cord.*

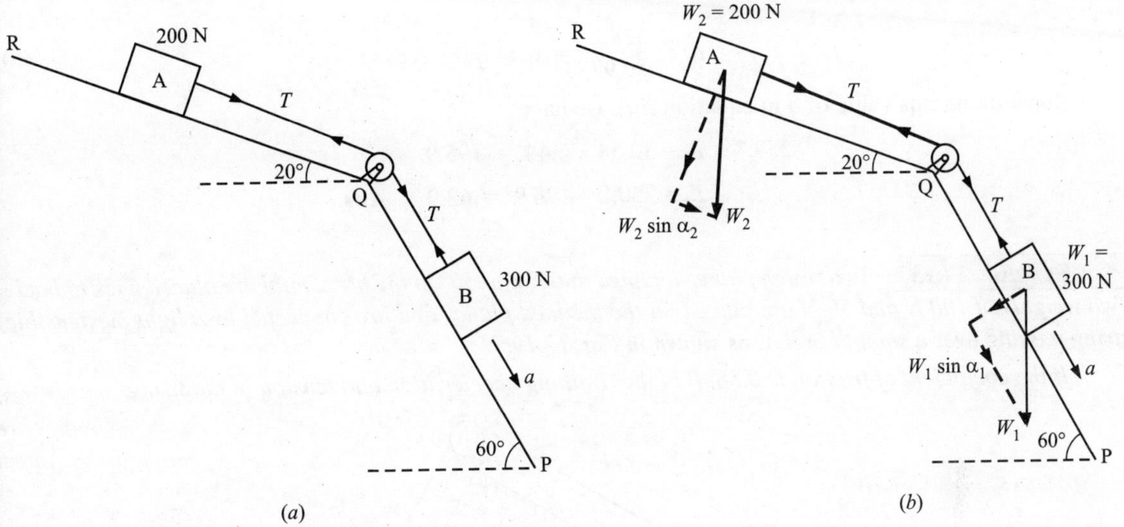


**Fig. 17.18**

***Solution*** Given : $W_2 = 200$ N ; $W_1 = 300$ N ; $\alpha_2 = 20°$ ; $\alpha_1 = 60°$

Let

$a$ = Acceleration of the weights, and

$T$ = Tension in the cord.

First of all, let us consider the downward motion of weight $W_1$ along the inclined plane $PQ$ as shown in Fig. 17.18 (*b*). The forces acting on it are $W_1 \sin \alpha_1$ acting downward and tension $T$ upwards.

$\therefore$ Net downward force $= W_1 \sin \alpha_1 - T$

$$= 300 \sin 60° - T = 300 \times 0.866 - T = 259.8 - T \quad ...(i)$$

Since the weight $W_1$ is acting downwards with acceleration $a$, therefore force acting on $W_1$

$$= \frac{W_1}{g} \times a = \frac{300}{9.81} \times a = 30.58a \quad ...(ii)$$

Equating equations (*i*) and (*ii*), we have

$$259.8 - T = 30.58a \quad ...(iii)$$

Now consider the motion of body $A$ which also moves downward along the plane $QR$. We know that the forces acting on it are $W_2 \sin \alpha_2$ acting downwards and the tension $T$ also acting downwards.

$\therefore$ Total force acting downward $= W_2 \sin a_2 + T = 200 \sin 20° + T$

$$= 200 \times 0.342 + T = 68.4 + T \quad ...(iv)$$

Since the weight $W_2$ is moving downward with acceleration $a$, therefore force acting on $W_2$

$$= \frac{W_2}{g} \times a = \frac{200}{9.81} \times a = 20.39a \quad ...(v)$$

Equating equations (*iv*) and (*v*),

$$68.4 + T = 20.39a \quad \text{...(vi)}$$

Adding equations (*iii*) and (*vi*), we have

$$259.8 + 68.4 = 30.58a + 20.39a$$

or

$$328.2 = 50.97a$$

$$\therefore \quad a = \frac{328.2}{50.97} = 6.44 \text{ m/s}^2 \quad \textbf{Ans.}$$

Substituting this value of $a$ in equation (*iii*), we have

$$259.8 - T = 30.58 \times 6.44 = 196.9$$

$$\therefore \quad T = 259.8 - 196.9 = 62.9 \text{ N} \quad \textbf{Ans.}$$

**Example 17.13** *Two rough planes inclined at 60° and 30° to the horizontal are placed back to back. Two weights of 100 N and 50 N are placed on the inclined planes and are connected by a light inextensible string passing over a smooth pulley, as shown in Fig. 17.19.*

*If the coefficient of friction is 0.58, find the resulting acceleration and tension in the string.*

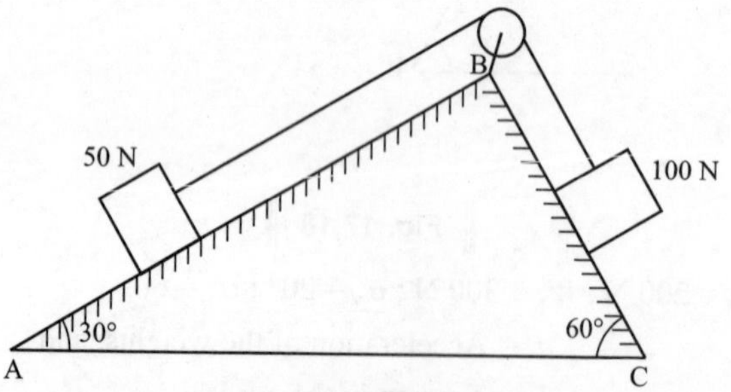


**Fig. 17.19**

***Solution*** Given : $\alpha_1 = 60°$ ; $\alpha_2 = 30°$ ; $W_1 = 100$ N ; $W_2 = 50$ N ; $\mu_1 = \mu_2 = 0.58$

***Resulting acceleration***

We know that the resulting acceleration,

$$a = \frac{g(W_1 \sin \alpha_1 - W_2 \sin \alpha_2 - \mu_1 W_1 \cos \alpha_1 - \mu_2 W_2 \cos \alpha_2)}{W_1 + W_2}$$

$$= \frac{9.81(100 \sin 60° - 50 \sin 30° - 0.58 \times 100 \cos 60° - 0.58 \times 50 \cos 30°)}{100 + 50}$$

$$= \frac{9.81(100 \times 0.866 - 50 \times 0.5 - 58 \times 0.5 - 29 \times 0.866)}{150}$$

$$= \frac{9.81(86.6 - 25 - 29 - 25.1)}{150} = \frac{9.81(86.6 - 79.1)}{150} = 0.4905 \text{ m/s}^2 \quad \textbf{Ans.}$$

***Tension in the string***

We know that tension in the string,

$$T = \frac{W_1 W_2}{W_1 + W_2}[(\sin \alpha_1 + \sin \alpha_2) - (\mu_1 \cos \alpha_1 - \mu_2 \cos \alpha_2)]$$

$$= \frac{100 \times 50}{100 + 50}[(\sin 60° + \sin 30°) - (0.58 \cos 60° - 0.58 \cos 30°)]$$

$$= 33[(0.866 + 0.5) - (0.58 \times 0.5 - 0.58 \times 0.866)]$$

$$= 33.33[1.366 + 0.21] = 52.53 \text{ N}$$ **Ans.**

## HIGHLIGHTS

1. Consider two bodies of weights $W_1$ and $W_2$ connected with a light inextensible string and passing over a smooth pulley, then acceleration of the system (*i.e.* of weights $W_1$ and $W_2$) is

$$a = \frac{(W_1 - W_2)g}{W_1 + W_2}$$

Tension in the string, $T = \dfrac{2W_1W_2}{W_1 + W_2}$

and pressure on the pulley or reaction of the pulley,

$$R = \frac{4W_1W_2}{W_1 + W_2}$$

2. Consider two bodies of weights $W_1$ and $W_2$ connected by a light inextensible string passing over a smooth pulley. Let one body of weight $W_2$ is lying on a horizontal surface and the other is hanging free. Following two cases are considered:

***(a) When the horizontal surface is smooth***

Acceleration of the system, $a = \dfrac{W_1 \cdot g}{W_1 + W_2}$

and tension in the string, $T = \dfrac{W_1W_2}{W_1 + W_2}$

***(b) When the horizontal surface is rough having coefficient of friction μ***

Acceleration of the system, $a = \dfrac{(W_1 - \mu W_2)g}{W_1 + W_2}$

and tension in the string, $T = \dfrac{W_1W_2(1 + \mu)}{W_1 + W_2}$

3. Consider two bodies of weights $W_1$ and $W_2$ connected by a light inextensible string passing over a smooth pulley. Let one body of weight $W_2$ is lying on an inclined plane and the other of weight $W_1$ is hanging free. Following two cases are considered:

***(a) When the inclined plane is smooth***

Acceleration of the system, $a = \dfrac{g(W_1 - W_2 \sin \alpha)}{W_1 + W_2}$

and tension in the string, $T = \dfrac{W_1W_2(1 + \sin \alpha)}{W_1 + W_2}$

where $\alpha$ is the angle of inclination of the inclined plane with the horizontal.

***(b) When the inclined plane is rough having coefficient of friction μ***

Acceleration of the system, $a = \dfrac{g(W_1 - W_2 \sin \alpha - \mu W_2 \cos \alpha)}{W_1 + W_2}$

and tension in the string, $T = \dfrac{W_1W_2}{W_1 + W_2}[1 + \sin \alpha + \mu \cos \alpha]$

4. Consider two bodies of weights $W_1$ and $W_2$ connected by a light inextensible string passing over a smooth pulley. Let both the weights $W_1$ and $W_2$ lying on the inclined planes having angle of inclination $\alpha_1$ and $\alpha_2$, respectively, with the horizontal. Following two cases are considered:

***(a) When both the inclined planes are smooth***

$$\text{Acceleration of the system, } a = \frac{g(W_1 \sin\alpha_1 - W_2 \sin\alpha_2)}{W_1 + W_2}$$

$$\text{and tension in the string, } \quad T = \frac{W_1 W_2 (\sin\alpha_1 + \sin\alpha_2)}{W_1 + W_2}$$

***(b) When both the inclined planes are rough having coefficient of friction $\mu_1$ and $\mu_2$***

$$\text{Acceleration of the system, } a = \frac{g(W_1 \sin\alpha_1 - W_2 \sin\alpha_2 - \mu_1 W_1 \cos\alpha_1 - \mu_2 W_2 \cos\alpha_2)}{W_1 + W_2}$$

$$\text{and tension in the string, } \quad T = \frac{W_1 W_2}{W_1 + W_2}[(\sin\alpha_1 + \sin\alpha_2) - (\mu_1 \cos\alpha_1 - \mu_2 \cos\alpha_2)]$$

**Note:** In all the above cases, $W_1 = m_1 g$, and $W_2 = m_2 g$, where $m_1$ and $m_2$ are the masses of the two bodies.

## NUMERICAL PROBLEMS

1. Two bodies weighing 600 N and 200 N are connected by means of an inextensible string passing over the frictionless pulley. Determine the magnitude of the acceleration of the weights and tension in the string.
[**Ans.** 4.905 m/s$^2$; 300 N]

2. A system of frictionless pulleys carries two weights $W_1$ and $W_2$, hung by the inextensible string, as shown in Fig. 17.20. Using D'Alembert's principle, find the acceleration of weights $W_1$ and $W_2$ and tension in the string. Given that $W_1 = 400$ N and $W_2 = 250$ N. [**Ans.** 5.14 m/s$^2$; 190.4 N]

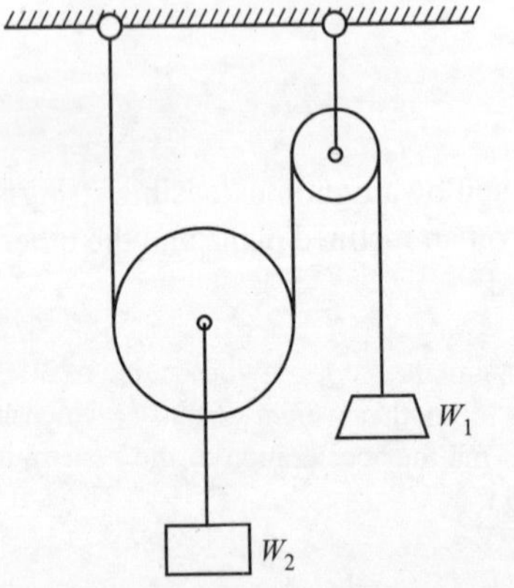


Fig. 17.20

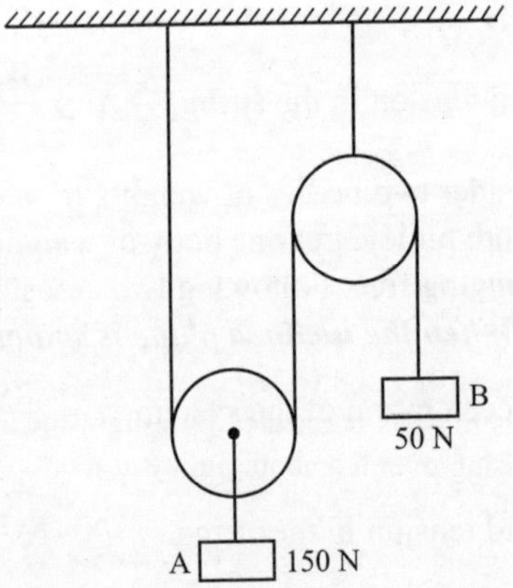


Fig. 17.21

3. Determine the acceleration of the blocks $A$ and $B$ of weights 150 N and 50 N, respectively, connected by a string and a frictionless and weightless pulley as shown in Fig. 17.21. Also, find the tension in the string.
[**Ans.** 2.8 m/s$^2$; 1.4 m/s$^2$; 64.27 N]

**4.** A fixed pulley carries a light inextensible string to one end of which is attached another pulley; the other end supports a mass of 30 kg. Over the second pulley passes another light string carrying masses 12 kg and 8 kg, respectively, attached to its ends. Determine the acceleration of the three masses. Take $g = 10$ m/s$^2$.

[**Ans.** 2.2 m/s$^2$; 0.24 m/s$^2$; 4.64 m/s$^2$]

**5.** A body of weight 100 N lying on a horizontal surface is connected by a string with a body of weight 250 N, which hangs over the edge of the surface. Find the acceleration of the system and tension in the string when

1. the horizontal surface is smooth; and
2. the horizontal surface is rough, the coefficient of friction being 0.25.

[**Ans.** 7 m/s$^2$, 71.43 N; 6.3 m/s$^2$, 89.3 N]

**6.** Two bodies *A* and *B* of weights 200 N and 100 N, respectively, are connected to the two ends of a light inextensible string passing over a smooth pulley. The body *A* is placed on a rough horizontal surface while the body *B* is hanging free in air. The coefficient of friction between the body *A* and the horizontal surface is 0.25. If the system is released from rest and the body *B* falls through a vertical distance of 2 m, find the acceleration of the system and the velocity attained by body *B*. [**Ans.** 1.635 m/s$^2$; 2.557 m/s]

**7.** A body *A* of weight 100 N is held on a rough horizontal table. An elastic string connected to the body *A* passes over a smooth pulley at the end of the table and then under a second smooth pulley carrying a body *B* of weight 50 N, as shown in Fig. 17.22. The other end of the string is fixed to a point above the second pulley. When the 100 N body is released, it moves with an acceleration of 1 m/s$^2$. Determine the value of coefficient of friction between the body *A* and the table. [**Ans.** 0.1225]

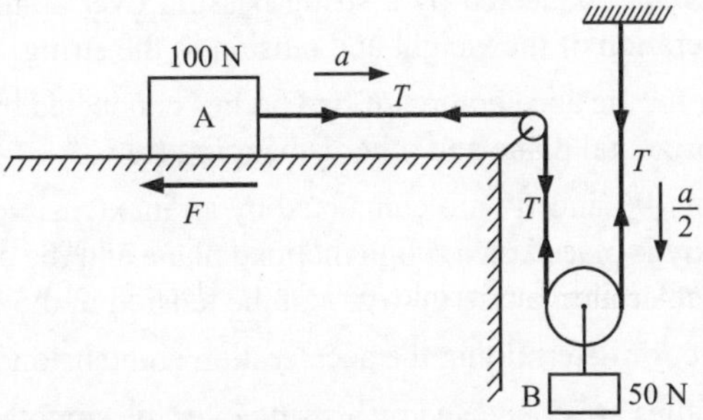


**Fig. 17.22**

**8.** A body of weight 100 N resting on a horizontal table is connected by a string passing over a smooth pulley at the edge of the table, to another body of weight 15 N and hanging vertically. Initially, the friction between the weight 100 N and the table is just sufficient to prevent motion. If an additional weight of 5 N is added to the weight of 15 N, determine the acceleration of the two weights and tension in the string.

[**Ans.** 0.408 m/s$^2$; 19.17 N]

**9.** A body of weight 1200 N is placed on a smooth inclined plane having an inclination of 30° to the horizontal. This mass is connected by a light inextensible string to another weight of 800 N which hangs freely after passing over a smooth pulley at the top of the plane. Find the acceleration of the system, and tension in the string. [**Ans.** 0.981 m/s$^2$; 720 N]

**10.** Two bodies of weights 400 N and 250 N are connected to the two ends of a light inextensible string, which passes over a smooth pulley. The weight 400 N is placed on a rough inclined plane whose angle of inclination is 15° to the horizontal. The weight 250 N is hanging free. If the coefficient of friction between the weight 400 N and inclined plane is 0.2, determine

1. the acceleration of the system;
2. the tension in the string; and
3. the distance moved by the weight 250 N in 2 seconds starting from the rest.

[**Ans.** 1.043 m/s$^2$; 223.4 N; 2.086 m]

11. Two smooth inclined planes, whose inclination, with the horizontal are 60° and 30°, are placed back to back. Two bodies of weight 200 N and 100 N are placed on them and are connected by a light inextensible string passing over a smooth pulley, as shown in Fig. 17.23.

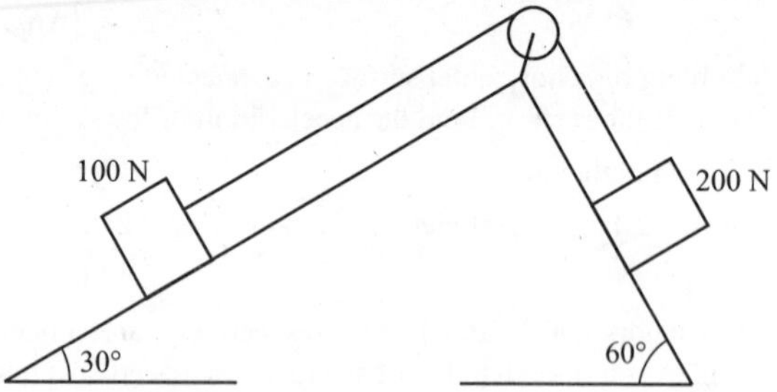


**Fig. 17.23**

Calculate the acceleration of the system and tension in the string. [**Ans.** 4.03 m/s²; 91 N]

12. In the above exercise, if the coefficient of friction between the planes and weights is 0.2, determine the acceleration of both the weights and tension in the string. [**Ans.** 2.81 m/s²; 96 N]

## REVIEW QUESTIONS

1. Two unequal weights are connected by a string passing over a small smooth pulley. Derive the relation for the acceleration of the weight and tension in the string.
2. Derive an expression for the acceleration of two bodies connected by a string, when one of them is lying on a smooth horizontal plane and other is hanging free.
3. Two bodies of weights $W_1$ and $W_2$ are connected by an inextensible string passing over a smooth pulley. One of the body is placed on a rough inclined plane and the other is hanging free. Obtain an expression for the acceleration of the system and the tension in the string.
4. Explain the procedure for determining the acceleration and tension in the following cases:
   (*a*) When the two bodies are connected by a string over the smooth inclined planes; and
   (*b*) When the two bodies are connected by a string over the rough inclined planes.

## OBJECTIVE TYPE QUESTIONS

1. Two bodies of weights $W_1$ and $W_2$ are hung from the ends of a rope passing over a frictionless pulley, as shown in Fig. 17.24. The acceleration of bodies will be

   (*a*) $\dfrac{(W_1 - W_2)g}{W_1 + W_2}$ (*b*) $\dfrac{2(W_1 - W_2)g}{W_1 + W_2}$

   (*c*) $\dfrac{(W_1 + W_2)g}{W_1 - W_2}$ (*d*) $\dfrac{2(W_1 + W_2)g}{W_1 - W_2}$

2. In the above question, as shown in Fig. 17.24, the tension ($T$) in the string will be

   (*a*) $\dfrac{W_1 W_2}{W_1 + W_2}$ (*b*) $\dfrac{2W_1 W_2}{W_1 + W_2}$

   (*c*) $\dfrac{W_1 + W_2}{W_1 W_2}$ (*d*) $\dfrac{W_1 + W_2}{2W_1 W_2}$

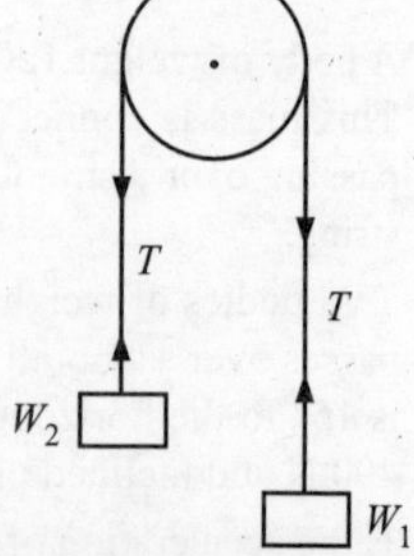


**Fig. 17.24**

**3.** A block of weight 200 N lying on a rough horizontal plane is connected by a light inextensible string passing over a smooth pulley to another weight of 50 N which hangs freely, as shown in Fig. 17.25. The tension in the string will ............ with the increase in coefficient of friction.

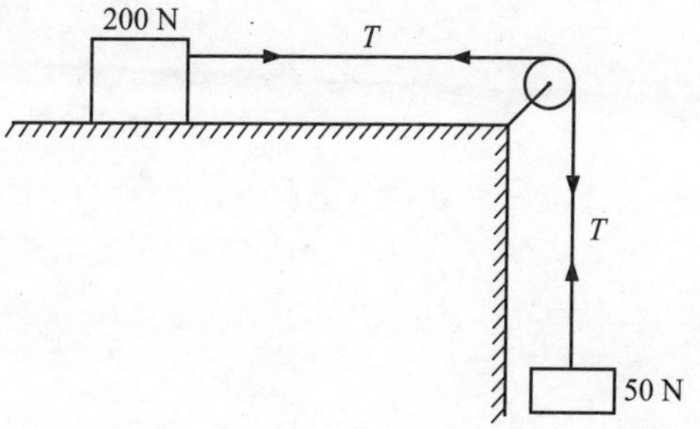


**Fig. 17.25**

(*a*) increase (*b*) decrease (*c*) not be effected

**4.** A body of weight $W_2$ placed on an inclined smooth plane is connected to a light inextensible string passing over a smooth pulley to a weight $W_1$ which moves vertically downwards, as shown in Fig. 17.26. The tension in the string is

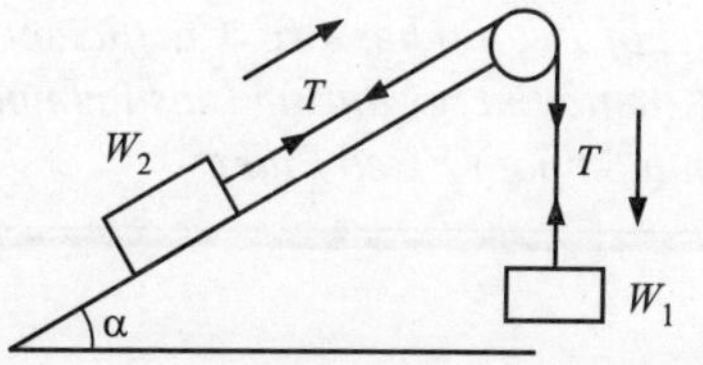


**Fig. 17.26**

(*a*) $\dfrac{W_2 \sin \alpha}{W_1}$ (*b*) $\dfrac{W_1 \sin \alpha}{W_2}$ (*c*) $\dfrac{(W_1 + W_2) \sin \alpha}{W_1 W_2}$ (*d*) $\dfrac{W_1 W_2 (1 + \sin \alpha)}{W_1 + W_2}$

**5.** In Fig. 17.26, if the weight $W_2$ is placed on a rough inclined plane having coefficient of friction μ, then the tension in the string will be

(*a*) $\dfrac{W_1 W_2}{W_1 + W_2}[1 + \mu \sin \alpha]$ (*b*) $\dfrac{W_1 W_2}{W_1 + W_2}[1 + \mu \cos \alpha]$

(*c*) $\dfrac{W_1 W_2}{W_1 + W_2}[1 + \sin \alpha + \mu \cos \alpha]$ (*d*) $\dfrac{W_1 W_2}{W_1 + W_2}[1 + \cos \alpha + \mu \sin \alpha]$

***Answers to Objective Type Questions***

**1.** (*a*) **2.** (*b*) **3.** (*a*) **4.** (*d*) **5.** (*c*)

18

# COLLISION OF ELASTIC BODIES



## 18.1 INTRODUCTION

When the balls of different materials (say a glass ball and an ivory ball) are allowed to fall on a marble floor, they will rebound to different heights due to their elasticity. The elasticity is that property of bodies, by virtue of which they rebound after impact. It may be noted that a body which rebounds to a greater height is said to be more elastic than a body which rebounds to a lesser height. If a body does not rebound at all, after its impact, then it is called an *inelastic body.*

When the two elastic bodies collide with each other, they are first deformed and then they separate apart due to the elastic properties and start moving with different velocities. The velocity with which the two bodies separate after impact or collision depends upon the properties of the material (*i.e.* shape, size, elastic property and line of impact) of which they are made.

## 18.2 IMPORTANT TERMS USED IN COLLISION

The following are the important terms used in collision.

1. ***Restitution*** When two elastic bodies collide with each other, they tend to compress each other. Immediately after this, the two bodies attempt to regain their original shapes, due to their elasticity. This process of regaining the original shape is called *restitution*.
2. ***Time of compression or compression period*** The time taken by the bodies in compression, after the instant of collision, is called *time of compression* or *compression period*.
3. ***Time of restitution or restitution period*** The period of time from the end of compression stage to the instant when the bodies separate (*i.e.* the time for which the restitution takes place) is called the *time of restitution* or *restitution period.*

4. ***Time of collision or period of impact*** The sum of the time of compression and the time of restitution is called *time of collision* or *period of impact.*
5. ***Line of impact*** It is the line joining the centres of the two bodies with the point of contact.

## 18.3 LAW OF CONSERVATION OF MOMENTUM

We have already discussed that the total momentum of a system of colliding bodies remains constant before and after the collision, if no external force acts on them.

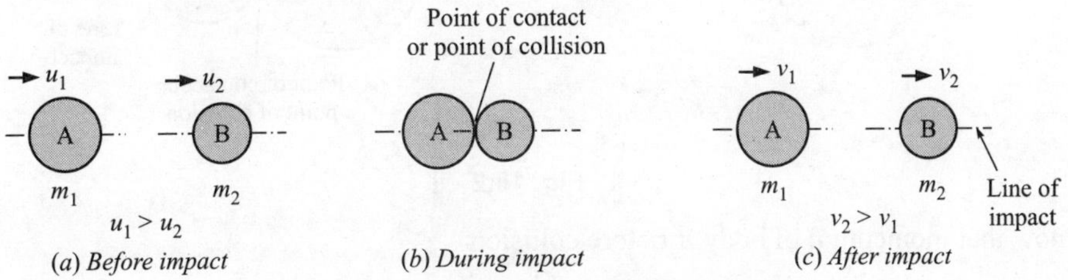


**Fig. 18.1**

Consider two bodies *A* and *B* colliding with each other, as shown in Fig. 18.1.

Let $m_1$ = Mass of the body *A*,

$u_1$ = Initial velocity of the body *A*,

$v_1$ = Final velocity of the body *A*,

$m_2$, $u_2$ and $v_2$ = Corresponding values for the body *B*.

We know that initial momentum of the body *A* before impact

$$= m_1u_1$$

and initial momentum of the body *B* before impact

$$= m_2u_2$$

∴ Total initial momentum before impact

$$= m_1u_1 + m_2u_2$$

Similarly, total final momentum after impact

$$= m_1v_1 + m_2v_2$$

Now according to the law of conservation of momentum,

Total initial momentum before impact = Total final momentum after impact

$$\therefore \quad m_1u_1 + m_2u_2 = m_1v_1 + m_2v_2$$

## 18.4 TYPES OF IMPACT

The following are the two types of *impact:

1. Direct impact; and 2. Indirect or oblique impact

These two types of impact are discussed, in detail, in the following pages:

## 18.5 DIRECT IMPACT OF TWO BODIES

The collision between the two bodies which are moving along the line of impact before the collision, is called *direct impact*. Consider two bodies *A* and *B* of mass $m_1$ and $m_2$ having a direct impact as shown in Fig. 18.2.

* When two bodies collide with each other, it is said to have an *impact.*

Let $u_1$ = Initial velocity of the first body *A*,

$v_1$ = Final velocity of the first body *A*,

$u_2$ and $v_2$ = Initial and final velocity of the second body *B*.

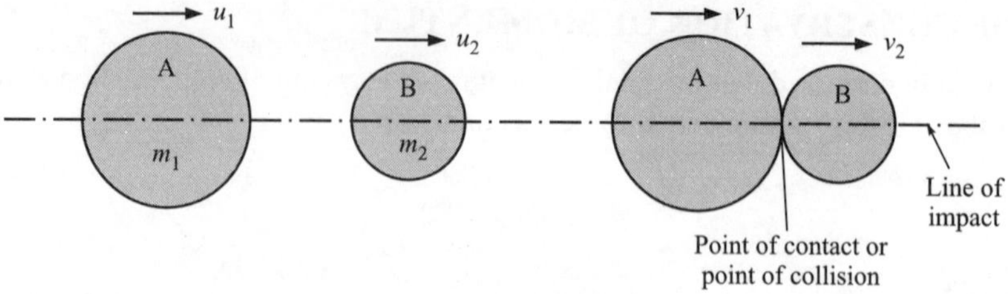


**Fig. 18.2**

We know that momentum of body *A* before collision

$$= m_1u_1$$

and momentum of body *B* before collision $= m_2u_2$

∴ Total initial momentum before collision

$$= m_1u_1 + m_2u_2$$

Similarly, total final momentum after collision

$$= m_1v_1 + m_2v_2$$

According to the principle of conservation of momentum,

Total initial momentum = Total final momentum

*i.e.* $$m_1u_1 + m_2u_2 = m_1v_1 + m_2v_2$$

**Note:** We know that the velocity is a vector quantity. Thus while solving problems, its direction should always be kept in view. If the velocity of a body is taken as *positive* in one direction, then the velocity in the opposite direction should be taken as *negative*.

## 18.6 COEFFICIENT OF RESTITUTION

Consider two bodies *A* and *B* moving along the same line with initial velocities $u_1$ and $u_2$, respectively, as shown in Fig. 18.3 (*a*). Let after the direct impact, their final velocities are $v_1$ and $v_2$, respectively, as shown in Fig. 18.3 (*c*).

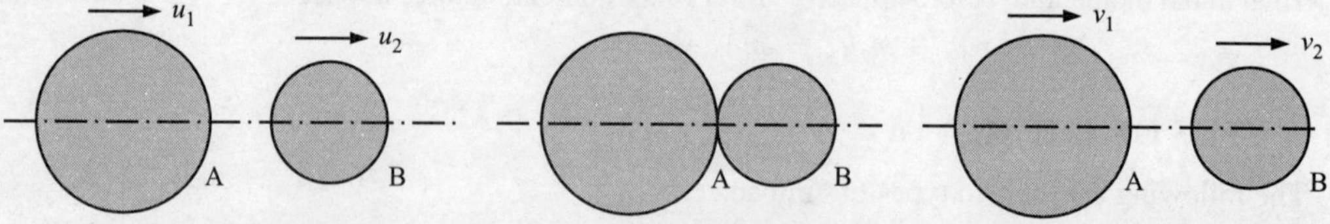


(*a*) *Before impact* (*b*) *During impact* (*c*) *After impact*

**Fig. 18.3**

It was observed by Sir Isaac Newton that when the collision of two bodies takes place, then the relative velocity of the two bodies after impact (or relative velocity of separation) bears a constant ratio to the relative velocity of the two bodies before impact (or relative velocity of approach) and is in the opposite direction. Mathematically,

$$\frac{v_1 - v_2}{-(u_1 - u_2)} = e, \text{ a constant}$$

or
$$\frac{v_2 - v_1}{u_1 - u_2} = e$$

This constant $e$ is known as *coefficient of restitution.* Thus, it may now be defined as the ratio of relative velocity after impact to the relative velocity before impact. It may also be defined as the ratio of relative velocity of separation to the relative velocity of approach.

It may be noted that

1. Since the physical bodies are not perfectly elastic, therefore relative velocity of the two bodies after impact is always less than the relative velocity of the two bodies before impact.
2. The value of $e = 0$ for perfectly inelastic bodies, *i.e.* the two bodies cling together and the relative velocity of separation is zero.
3. The value of $e = 1$ for perfectly elastic bodies, *i.e.* the relative velocity of the two bodies before and after impact is same.
4. In case the bodies are neither perfectly inelastic nor perfectly elastic, then the value of $e$ lies between 0 and 1.
5. The coefficient of restitution not only depends upon the material, but it also depends upon the shape and size of the body.

**Example 18.1** *Two balls A and B having the same mass move towards each other and collide. If the coefficient of restitution at the impact is e, derive the relation that must exist between their velocities before collision, so that the ball A comes to rest after collision.*

***Solution*** Let $u_1$ and $u_2$ be the velocities of balls $A$ and $B$ before collision (Refer Fig. 18.4) and $v_1$ and $v_2$ be the velocities after collision.

Let $m$ be the mass of $A$ and $B$.

We know that momentum before impact

= Momentum after impact.

or
$$mu_1 - mu_2 = mv_1 + mv_2$$

or
$$u_1 - u_2 = v_1 + v_2 = v_2 \quad ...(\because v_1 = 0) \quad ...(i)$$

By the law of collision,
$$v_2 - v_1 = -e\,(-u_2 - u_1)$$

or
$$v_2 - 0 = e(u_1 + u_2)$$
$$v_2 = e(u_1 + u_2) \quad ...(ii)$$

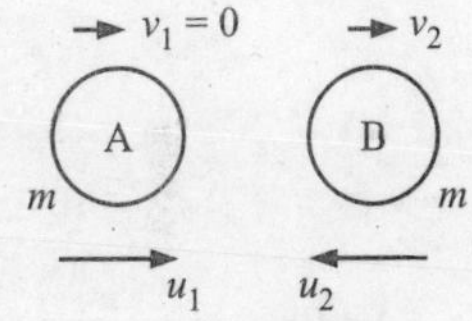


**Fig. 18.4**

From equations (*i*) and (*ii*), we have

$$u_1 - u_2 = e(u_1 + u_2) = eu_1 + eu_2$$

or
$$u_1 - eu_1 = u_2 + eu_2$$
$$u_1(1 - e) = u_2\,(1 + e)$$

$$\therefore \quad \frac{u_1}{u_2} = \frac{1+e}{1-e}$$

**Example 18.2** *A body of mass 10 kg is moving at a speed of 4 m/s towards right. It collides with another body of mass 5 kg moving at a speed of 2 m/s towards right. The 10 kg body has a final velocity of 3 m/s towards right. Find the final velocity of the 5 kg body and the coefficient of restitution.*

***Solution*** Given : $m_1 = 10$ kg ; $u_1 = 4$ m/s (towards right) ; $m_2 = 5$ kg ; $u_2 = 2$ m/s (towards right) ; $v_1 = 3$ m/s (towards right)

***Final velocity of the 5 kg body***

Let $v_2$ = Final velocity of the 5 kg body.

According to the principle of conservation of momentum,

$$m_1u_1 + m_2u_2 = m_1v_1 + m_2v_2$$

$$10 \times 4 + 5 \times 2 = 10 \times 3 + 5 \times v_2$$

$$50 = 30 + 5v_2$$

$$\therefore \quad v_2 = \frac{50 - 30}{5} = 4 \text{ m/s (towards right)} \quad \textbf{Ans.}$$

***Coefficient of restitution***

We know that coefficient of restitution,

$$e = \frac{v_2 - v_1}{u_1 - u_2} = \frac{4 - 3}{4 - 2} = 0.5 \quad \textbf{Ans.}$$

**Example 18.3** *Two bodies of mass 20 kg and 12 kg moving in opposite direction at initial velocities of 6 m/s and 8 m/s, respectively, collide with each other. Find their velocities after impact when the coefficient of restitution is 0.75.*

***Solution*** Given : $m_1 = 20$ kg ; $m_2 = 12$ kg ; $u_1 = 6$ m/s ; $u_2 = -8$ m/s (because of opposite direction) ; $e = 0.75$

Let $v_1$ and $v_2$ be the velocities of the two bodies after impact. We know that according to principle of conservation of momentum,

$$m_1u_1 + m_2u_2 = m_1v_1 + m_2v_2$$

$$20 \times 6 + 12\,(-8) = 20v_1 + 12v_2$$

$$24 = 20v_1 + 12v_2$$

or $$v_1 + 0.6v_2 = 1.2 \qquad ...(i)$$

We also know that coefficient of restitution,

$$e = \frac{v_2 - v_1}{u_1 - u_2}$$

$$0.75 = \frac{v_2 - v_1}{6 - (-8)} = \frac{v_2 - v_1}{14}$$

or $$v_2 - v_1 = 0.75 \times 14 = 10.5 \qquad ...(ii)$$

Adding equations (*i*) and (*ii*), we have

$$1.6v_2 = 11.7$$

$$\therefore \quad v_2 = \frac{11.7}{1.6} = 7.3 \text{ m/s} \quad \textbf{Ans.}$$

From equation (*i*), we have $v_1 = 1.2 - 0.6v_2$

$$= 1.2 - 0.6 \times 7.3 = 1.2 - 4.38 = -3.18 \text{ m/s} \quad \textbf{Ans.}$$

The negative sign shows that the direction of $v_1$ is opposite to that of $v_2$.

**Example 18.4** *Prove that the two perfect elastic bodies of equal masses exchange velocities in case of direct impact.*

***Solution*** Given : $m_1 = m_2$

Since the bodies are perfectly elastic, therefore coefficient of restitution, $e = 1$.

Let $m_1$ = Mass of the first body,

$u_1$ = Velocity of the first body before impact,

$v_1$ = Velocity of the first body after impact,

$m_2$, $u_2$ and $v_2$ = Corresponding values for the second body.

We know that coefficient of restitution,

$$e = \frac{v_2 - v_1}{u_1 - u_2} \quad \text{or} \quad 1 = \frac{v_2 - v_1}{u_1 - u_2}$$

$$\therefore \quad u_1 - u_2 = v_2 - v_1 \quad \text{...}(i)$$

According to the principle of conservation of momentum,

Total initial momentum before impact = Total final momentum after impact

or $$m_1u_1 + m_2u_2 = m_1v_1 + m_2v_2$$

$$u_1 + u_2 = v_1 + v_2 \quad \text{...}(\because m_1 = m_2) \text{...}(ii)$$

Adding equations (*i*) and (*ii*), we have

$$2u_1 = 2v_2 \quad \text{or} \quad u_1 = v_2 \quad \text{...}(iii)$$

From equation (*i*), $$u_1 - u_2 = u_1 - v_1 \quad \text{or} \quad u_2 = v_1 \quad \text{...}(iv)$$

From equation (*iii*), we see that velocity of the first body before impact is equal to velocity of the second body after impact.

From equation (*iv*), we see that velocity of the second body before impact is equal to velocity of the first body after impact.

Hence, the two perfect elastic bodies of equal masses exchange velocities after the impact.

## 18.7 DIRECT IMPACT OF A BODY ON A FIXED PLANE

Consider a body having a direct impact on a fixed plane, as shown in Fig. 18.5.

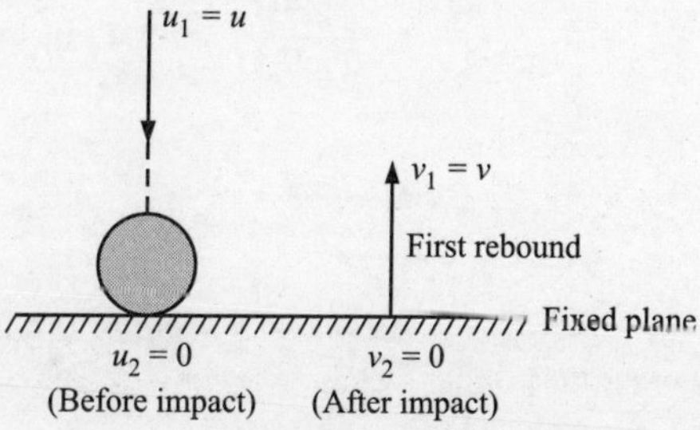


**Fig. 18.5**

Since the fixed plane remains at the same position before and after the impact, therefore the initial and final velocities of the fixed plane are zero.

Let $u$ = Initial velocity of the body before impact,

$v$ = Final velocity of the body after impact, and

$e$ = Coefficient of restitution.

We know that relative velocity of the body before impact (or velocity of approach)

$$= u_1 - u_2 = u_1 - 0 = u_1$$

and the relative velocity of the body after impact (or velocity of separation)

$$= v_2 - v_1 = 0 - v_1 = -v$$

$$\therefore \quad \text{Coefficient of restitution, } e = \frac{\text{Relative velocity of the body after impact}}{\text{Relative velocity of the body before impact}} = \frac{-v}{u}$$

or $$v = -eu$$

The –ve sign shows that the velocity after impact is opposite to the velocity before impact.

Hence, the body strikes the fixed plane with a velocity $v$ and it rebounds with a velocity $= e\,.\,u$

In order to find the height of rebound after impact, consider that a body falls freely from a height $h$ metres above the fixed plane. The velocity ($v$) with which it strikes the fixed plane is given by the relation

$$v^2 = u^2 + 2gh$$

or $$v = \sqrt{2gh} \qquad ...(\because\ u = 0)$$

After the impact, the body will move up with a velocity $v_1 = e\,.\,v = e\sqrt{2gh}$

Let the height of rebound is $h_1$ at which the velocity of body (say $v_2$) is zero.

$$\therefore \quad 0 = (e\sqrt{2gh})^2 - 2gh_1 \qquad ...(\text{–ve sign for upward motion})$$

or $$e^2 \times 2gh = 2gh_1$$

$$\therefore \quad h_1 = e^2\,.\,h$$

**Example 18.5** *A glass marble drops from a height of 3 m upon a horizontal floor. If the coefficient of restitution is 0.9, find the height to which it rises after impact.*

***Solution*** Given : $h = 3$ m ; $e = 0.9$

Let $h_1$ = Height to which a glass marble rises after impact.

We know that the velocity of the glass marble at which it strikes the horizontal floor,

$$v = \sqrt{2gh} = \sqrt{2 \times 9.81 \times 3} = 7.672 \text{ m/s}$$

and velocity after impact, $$v_1 = e\,.\,v = 0.9 \times 7.672 = 6.9 \text{ m/s}$$

$$\therefore \quad h_1 = \frac{v_1^2}{2g} = \frac{(6.9)^2}{2 \times 9.81} = 2.43 \text{ m} \quad \textbf{Ans.} \qquad ...(\because\ v_1 = \sqrt{2gh_1})$$

**Note:** We can also find $h_1$ by applying the relation,

$$h_1 = e^2h = (0.9)^2\,3 = 2.43 \text{ m} \quad \textbf{Ans.}$$

**Example 18.6** *A ball is dropped from a height of 9 m upon a horizontal slab. If it rebounds to a height of 6 m, find the coefficient of restitution.*

***Solution*** Given : $h = 9$ m ; $h_1 = 6$ m

Let $e$ = Coefficient of restitution.

We know that velocity of the ball at which it strikes the horizontal slab,

$$v = \sqrt{2gh} = \sqrt{2 \times 9.81 \times 9} = 13.3 \text{ m/s}$$

and height of rebound, $$h_1 = \frac{v_1^2}{2g} = \frac{(e.v)^2}{2g} \qquad ...(\because\ v_1 = e\,.\,v)$$

$$\therefore \quad e = \sqrt{\frac{2g\,h_1}{v^2}} = \sqrt{\frac{2 \times 9.81 \times 6}{(13.3)^2}} = \sqrt{\frac{117.72}{176.89}}$$

$$= \sqrt{0.6655} = 0.815 \quad \textbf{Ans.}$$

**Note:** The value of $e$ may also be obtained by applying the relation:

$$h_1 = e^2h$$

or $$e = \sqrt{\frac{h_1}{h}} = \sqrt{\frac{6}{9}} = \sqrt{0.6666} = 0.816 \quad \textbf{Ans.}$$

**Example 18.7** *A ball is dropped from a height of 1000 mm on a smooth floor. The height of the first bounce is 810 mm. Determine : 1. Coefficient of restitution ; and 2. Expected height of the second bounce.*

***Solution*** Given : $h = 1000$ mm = 1 m ; $h_1 = 810$ mm = 0.81 m

**1. *Coefficient of restitution***

Let $e$ = Coefficient of restitution.

We know that velocity of the ball at which it strikes the floor,

$$v = \sqrt{2gh} = \sqrt{2 \times 9.81 \times 1} = 4.43 \text{ m/s}$$

and height of rebound,

$$h_1 = \frac{v_1^2}{2g} = \frac{(e.v)^2}{2g} \qquad ...(\because v_1 = e.v)$$

$$\therefore \quad e = \sqrt{\frac{2gh_1}{v^2}} = \sqrt{\frac{2 \times 9.81 \times 0.81}{(4.43)^2}} = \sqrt{\frac{15.89}{19.62}} = \sqrt{0.81} = 0.9 \textbf{ Ans.}$$

**2. *Expected height of the second bounce***

Let $h_2$ = Expected height of the second bounce.

We know that $h_2 = e^2 h_1 = (0.9)^2\, 0.81 = 0.6561$ m **Ans.**

**Example 18.8** *Three perfectly elastic balls A, B and C of masses 2 kg, 6 kg and 12 kg are moving in the same direction with velocities of 12 m/s, 4 m/s, and 2 m/s, respectively. If the ball A strikes with the ball B, which in turn strikes with the ball C, prove that the balls A and B will be brought to rest by the impact.*

***Solution*** Given : $m_1 = 2$ kg ; $m_2 = 6$ kg ; $m_3 = 12$ kg ; $u_1 = 12$ m/s ; $u_2 = 4$ m/s ; $u_3 = 2$ m/s

Since the balls are perfectly elastic, therefore the coefficient of restitution, $e = 1$.

First of all, let us consider the impact of the first ball $A$ and second ball $B$. Let $v_1$ and $v_2$ be the final velocity of the first ball and the second ball, respectively, after impact. According to the principle of conservation of momentum,

$$m_1 u_1 + m_2 u_2 = m_1 v_1 + m_2 v_2$$

$$2 \times 12 + 6 \times 4 = 2v_1 + 6v_2$$

$$48 = 2v_1 + 6v_2$$

or

$$v_1 + 3v_2 = 24 \qquad ...(i)$$

We know that coefficient of restitution,

$$e = \frac{v_2 - v_1}{u_1 - u_2}$$

or

$$v_2 - v_1 = e(u_1 - u_2) = 1(12 - 4) = 8 \qquad ...(ii)$$

Adding equations (*i*) and (*ii*), $4v_2 = 32$ or $v_2 = 8$ m/s

Substituting the value of $v_2$ in equation (*ii*), we have

$$8 - v_1 = 8 \text{ or } v_1 = 0$$

Thus the velocity of the first ball after the impact of second ball is zero. Hence, the first ball will be brought to rest by the impact of first and second ball. **Ans.**

Now consider the impact of the second ball $B$ and the third ball $C$. The initial velocity of the second ball will be equal to the final velocity of the second ball after impact, *i.e.* $u_2 = v_2 = 8$ m/s.

Let $v_2$ and $v_3$ be the final velocity of the second and third ball, respectively, after impact. According to the principle of conservation of momentum,

$$m_2 u_2 + m_3 u_3 = m_2 v_2 + m_3 v_3$$

$$6 \times 8 + 12 \times 2 = 6v_2 + 12v_3$$

$$72 = 6v_2 + 12v_3$$

or $$v_2 + 2v_3 = 12 \quad ...(iii)$$

We know that coefficient of restitution,

$$e = \frac{v_3 - v_2}{u_2 - u_3}$$

or $$v_3 - v_2 = e(u_2 - u_3) = 1(8 - 2) = 6 \quad ...(iv)$$

Adding equations (*iii*) and (*iv*), we have

$$3v_3 = 18 \quad \text{or} \quad v_3 = 6 \text{ m/s}$$

Substituting the value of $v_3$ in equation (*iv*), we have

$$6 - v_2 = 6 \quad \text{or} \quad v_2 = 0$$

Thus the velocity of second ball after the impact with third ball is zero. Hence, the second ball will be brought to rest by the impact of second and third ball. **Ans.**

## 18.8 LOSS OF KINETIC ENERGY DURING ELASTIC IMPACT

The loss of kinetic energy during elastic impact is given by the difference of kinetic energy of the two bodies before and after the impact. Consider two bodies 1 and 2 having an elastic impact, as shown in Fig. 18.6.

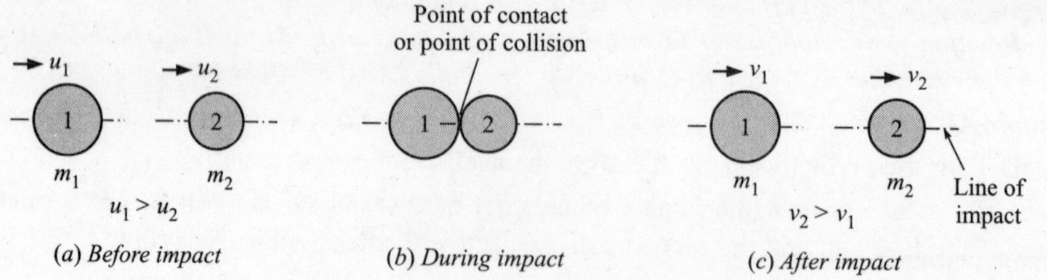


**Fig. 18.6**

Let $m_1$ = Mass of the first body,

$u_1$ = Velocity of the first body before impact,

$v_1$ = Velocity of the first body after impact,

$m_2$, $u_2$ and $v_2$ = Corresponding values for the second body,

$e$ = Coefficient of restitution, and

$E_L$ = Loss of kinetic energy during impact.

We know that the kinetic energy of the first body, before impact

$$= \frac{1}{2} m_1 \cdot u_1^2$$

Similarly, kinetic energy of the second body, before impact

$$= \frac{1}{2} m_2 \cdot u_2^2$$

∴ Total kinetic energy of the two bodies, before impact,

$$E_1 = \frac{1}{2} m_1 \cdot u_1^2 + \frac{1}{2} m_2 \cdot u_2^2 \quad ...(i)$$

Similarly, total kinetic energy of the two bodies, after impact

$$E_2 = \frac{1}{2} m_1 \cdot v_1^2 + \frac{1}{2} m_2 \cdot v_2^2 \quad ...(ii)$$

∴ Loss of kinetic energy during impact,

$$E_L = E_1 - E_2 = \left(\frac{1}{2}m_1 \cdot u_1^2 + \frac{1}{2}m_2 \cdot u_2^2\right) - \left(\frac{1}{2}m_1 \cdot v_1^2 + \frac{1}{2}m_2 \cdot v_2^2\right)$$

$$= \frac{1}{2}\left[(m_1 \cdot u_1^2 + m_2 \cdot u_2^2) - (m_1 \cdot v_1^2 + m_2 \cdot v_2^2)\right]$$

Multiplying the numerator and denominator by $(m_1 + m_2)$,

$$E_L = \frac{1}{2(m_1 + m_2)}\left[(m_1 + m_2)(m_1 \cdot u_1^2 + m_2 \cdot u_2^2) - (m_1 + m_2)(m_1 \cdot v_1^2 + m_2 \cdot v_2^2)\right]$$

$$= \frac{1}{2(m_1 + m_2)}\Big[(m_1^2 \cdot u_1^2 + m_1 \cdot m_2 \cdot u_2^2 + m_1 \cdot m_2 \cdot u_1^2 + m_2^2 \cdot u_2^2)$$
$$- (m_1^2 \cdot v_1^2 + m_1 \cdot m_2 \cdot v_2^2 + m_1 \cdot m_2 \cdot v_1^2 + m_2^2 \cdot v_2^2)\Big]$$

$$= \frac{1}{2(m_1 + m_2)}\Big[\{m_1^2 \cdot u_1^2 + m_2^2 \cdot u_2^2 + m_1 \cdot m_2 (u_1^2 + u_2^2)\}$$
$$- \{m_1^2 \cdot v_1^2 + m_2^2 \cdot v_2^2 + m_1 \cdot m_2 (v_1^2 + v_2^2)\}\Big]$$

$$= \frac{1}{2(m_1 + m_2)}\Big[\{(m_1 \cdot u_1 + m_2 \cdot u_2)^2 - (2m_1 m_2 u_1 u_2) + m_1 \cdot m_2 (u_1 - u_2)^2 + (2m_1 m_2 u_1 u_2)\}$$
$$-\{(m_1 \cdot v_1 + m_2 \cdot v_2)^2 - (2m_1 m_2 v_1 v_2) + m_1 \cdot m_2 (v_1 - v_2)^2 + (2m_1 m_2 v_1 v_2)\}\Big]$$

$$= \frac{1}{2(m_1 + m_2)}\Big[\{(m_1 \cdot u_1 + m_2 \cdot u_2)^2 + m_1 \cdot m_2 (u_1 - u_2)^2\}$$
$$- \{(m_1 \cdot v_1 + m_2 \cdot v_2)^2 + m_1 m_2 (v_1 - v_2)^2\}\Big]$$

We know that in an elastic impact,

Total momentum before impact = Total momentum after impact

*i.e.* $m_1 \cdot u_1 + m_2 \cdot u_2 = m_1 \cdot v_1 + m_2 \cdot v_2$

or $(m_1 \cdot u_1 + m_2 \cdot u_2)^2 = (m_1 \cdot v_1 + m_2 \cdot v_2)^2$ ...(Squaring both sides)

∴ Loss of kinetic energy due to impact,

$$E_L = \frac{1}{2(m_1 + m_2)}\left[m_1 \cdot m_2 (u_1 - u_2)^2 - m_1 \cdot m_2 (v_1 - v_2)^2\right]$$

Substituting $v_1 - v_2 = e(u_1 - u_2)$ in the above equation, we have

$$E_L = \frac{1}{2(m_1 + m_2)}\left[m_1 \cdot m_2 (u_1 - u_2)^2 - m_1 \cdot m_2 \cdot e^2 (u_1 - u_2)^2\right]$$

$$= \frac{m_1 \cdot m_2}{2(m_1 + m_2)}(u_1 - u_2)^2 (1 - e^2)$$

**Notes:** 1. The loss of kinetic energy may be found out by calculating the kinetic energy of the system before impact, and then by subtracting from it the kinetic energy of the system after impact.

2. For perfectly inelastic bodies, $e = 0$, therefore

$$E_L = \frac{m_1 \cdot m_2}{2(m_1 + m_2)}(u_1 - u_2)^2$$ ...(same as before)

3. For perfectly elastic bodies, $e = 1$, therefore $E_L = 0$.

4. If weights (instead of masses) of the two bodies are given, then the same may be used in all the relations.

**Example 18.9** *A body of mass 50 kg moving at 3 m/s strikes another body of mass 25 kg at 1.5 m/s in the same direction. Both the bodies adhere after impact. Find the common velocity with which the two bodies will move. Determine, also, the loss of kinetic energy due to the impact.*

***Solution*** Given : $m_1 = 50$ kg ; $u_1 = 3$ m/s ; $m_2 = 25$ kg ; $u_2 = 1.5$ m/s

***Common velocity of the two bodies after impact***

Let $v$ = Common velocity of the two bodies after impact.

We know that total momentum before impact

$$= m_1u_1 + m_2u_2 = 50 \times 3 + 25 \times 1.5$$

$$= 150 + 37.5 = 187.5 \text{ kg-m/s} \quad ...(i)$$

and total momentum after impact $= (m_1 + m_2)\, v = (50 + 25)\, v = 75v \quad ...(ii)$

According to the principle of conservation of momentum, the total momentum before impact is equal to the total momentum after impact. Therefore equating equations (*i*) and (*ii*), we have

$$187.5 = 75v$$

$$\therefore \quad v = \frac{187.5}{75} = 2.5 \text{ m/s } \textbf{Ans.}$$

***Loss of kinetic energy due to impact***

We know that kinetic energy before impact,

$$E_1 = \frac{1}{2} m_1 u_1^2 + \frac{1}{2} m_2 u_2^2$$

$$= \frac{1}{2} \times 50 \times 3^2 + \frac{1}{2} \times 25 (1.5)^2 = 225 + 28.12 = 253.12 \text{ N-m}$$

and kinetic energy after impact $E_2 = \frac{1}{2}(m_1 + m_2)v^2 = \frac{1}{2}(50 + 25)(2.5)^2 = 234.37$ N-m

∴ Loss of kinetic energy due to impact,

$$E_L = E_1 - E_2 = 253.12 - 234.37 = 18.75 \text{ N-m } \textbf{Ans.}$$

**Example 18.10** *A sphere of mass 1 kg moving at 3 m/s overtakes and collides with another sphere of mass 5 kg moving at 0.6 m/s in the same direction. Find the velocities of the two spheres after the impact and the loss of kinetic energy during impact. The coefficient of restitution between the two spheres is 0.75.*

***Solution*** Given : $m_1 = 1$ kg ; $u_1 = 3$ m/s ; $m_2 = 5$ kg ; $u_2 = 0.6$ m/s ; $e = 0.75$

***Velocities of the two spheres after the impact***

Let $v_1$ = Velocity of the first sphere after the impact, and

$v_2$ = Velocity of the second sphere after the impact.

According to the principle of conservation of momentum,

Total momentum before impact = Total momentum after impact

$$m_1u_1 + m_2u_2 = m_1v_1 + m_2v_2$$

$$1 \times 3 + 5 \times 0.6 = 1 \times v_1 + 5 \times v_2$$

$$3 + 3 = v_1 + 5v_2$$

or $$v_1 + 5v_2 = 6 \quad ...(i)$$

We know that coefficient of restitution,

$$e = \frac{v_2 - v_1}{u_1 - u_2}$$

$$0.75 = \frac{v_2 - v_1}{3 - 0.6} = \frac{v_2 - v_1}{2.4}$$

$$\therefore \quad v_2 - v_1 = 0.75 \times 2.4 = 1.8 \qquad ...(ii)$$

Adding equations (*i*) and (*ii*), we have

$$6v_2 = 7.8$$

$$\therefore \quad v_2 = \frac{7.8}{6} = 1.3 \text{ m/s} \quad \textbf{Ans.}$$

From equation (*ii*), $v_1 = v_2 - 1.8 = 1.3 - 1.8 = -0.5$ m/s **Ans.**

The –ve sign indicates that direction of first sphere after the impact is reversed.

***Loss of kinetic energy during impact***

We know that loss of kinetic energy during impact,

$$E_L = \frac{m_1 m_2}{2(m_1 + m_2)}(u_1 - u_2)^2 (1 - e^2)$$

$$= \frac{1 \times 5}{2(1 + 5)}(3 - 0.6)^2 (1 - 0.75^2)$$

$$= \frac{5}{12}(2.4)^2 (1 - 0.5625) = 1.05 \text{ N-m} \quad \textbf{Ans.}$$

**Note:** The loss of kinetic energy during impact may also be obtained as discussed below:

We know that total kinetic energy before impact,

$$E_1 = \frac{1}{2}m_1 u_1^2 + \frac{1}{2}m_2 u_2^2$$

$$= \frac{1}{2} \times 1 \times 3^2 + \frac{1}{2} \times 5 \times 0.6^2 = 4.5 + 0.9 = 5.4 \text{ N-m}$$

and total kinetic energy after impact,

$$E_2 = \frac{1}{2}m_1 v_1^2 + \frac{1}{2}m_2 v_2^2$$

$$= \frac{1}{2} \times 1(-0.5)^2 + \frac{1}{2} \times 5(1.3)^2 = 0.125 + 4.225 = 4.35 \text{ N-m}$$

$\therefore$ Loss of kinetic energy during impact, $E_L = E_1 - E_2 = 5.4 - 4.35 = 1.05$ N-m **Ans.**

## 18.9 INDIRECT OR OBLIQUE IMPACT OF TWO BODIES

When the two bodies are not moving along the line of impact but at some angle to the line of impact, then the collision is known as *indirect* or *oblique impact.* Fig. 18.7 shows two bodies 1 and 2 moving with a velocity $u_1$ and $u_2$ making an angle $\theta_1$ and $\theta_2$ with the horizontal. Since the two bodies 1 and 2 are not moving along the line of impact before collision, hence the collision is known as indirect or oblique impact.

Let

$m_1$ = Mass of body 1,

$u_1$ = Initial velocity of body 1 beforc collision,

$\theta_1$ = Angle made by body 1 with the horizontal before collision,

$v_1$ = Final velocity of body 1 after collision,

$\phi_1$ = Angle made by body 1 with the horizontal after collision,

$m_2, u_2, \theta_2, v_2$ and $\phi_2$ = Corresponding values of body 2.

We know that component of $u_1$ along the line of impact

$$= u_1 \cos \theta_1$$

Component of $u_2$ along the line of impact

$$= u_2 \cos \theta_2$$

Component of $v_1$ along the line of impact

$$= v_1 \cos \phi_1$$

and component of $v_2$ along the line of impact $= v_2 \cos \phi_2$

According to the principle of conservation of momentum, the total initial momentum in any direction must be equal to the total final momentum in that direction.

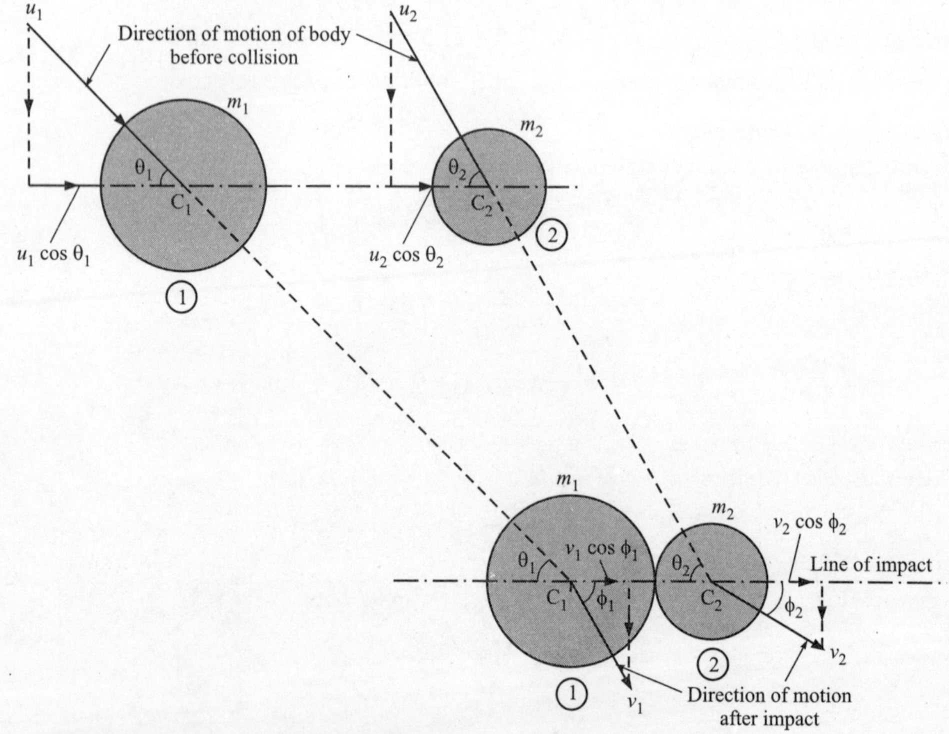


**Fig. 18.7**

We know that initial momentum of body 1 before collision in horizontal direction

$$= m_1 u_1 \cos \theta_1$$

and initial momentum of body 2 before collision in horizontal direction

$$= m_2 u_2 \cos \theta_2$$

∴ Total initial momentum in horizontal direction before collision

$$= m_1 u_1 \cos \theta_1 + m_2 u_2 \cos \theta_2 \qquad ...(i)$$

Similarly, total final momentum in horizontal direction after collision

$$= m_1 v_1 \cos \phi_1 + m_2 v_2 \cos \phi_2 \qquad ...(ii)$$

Equating equations (*i*) and (*ii*), we have

$$m_1 u_1 \cos \theta_1 + m_2 u_2 \cos \theta_2 = m_1 v_1 \cos \phi_1 + m_2 v_2 \cos \phi_2$$

**Note:** The coefficient of restitution ($e$) for the indirect or oblique impact is given by

$$e = \frac{v_2 \cos \phi_2 - v_1 \cos \phi_1}{u_1 \cos \theta_1 - u_2 \cos \theta_2}$$

**Example 18.11** *A ball of mass 1 kg moving with a velocity of 1 m/s impinges on a ball of mass 3 kg moving with a velocity of 3 m/s. The velocities of the two balls are parallel and inclined at 30° to the line joining their centres at the instant of impact. If the coefficient of restitution is 0.5, determine the velocity and direction of the balls immediately after the impact.*

***Solution*** Given : $m_1 = 1$ kg ; $u_1 = 1$ m/s ; $m_2 = 3$ kg ; $u_2 = 3$ m/s ; $\theta_1 = \theta_2 = 30°$ ; $e = 0.5$

***Velocity and direction of the balls immediately after the impact***

Let $v_1$ and $v_2$ be the velocity of the balls immediately after the impact, along the directions $\phi_1$ and $\phi_2$, respectively, with the line joining their centres, as shown in Fig. 18.8. Refer Fig. 18.9 also for clear understanding.

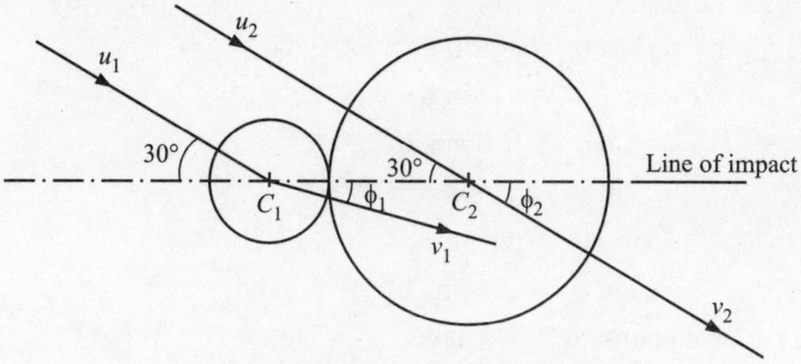


**Fig. 18.8**

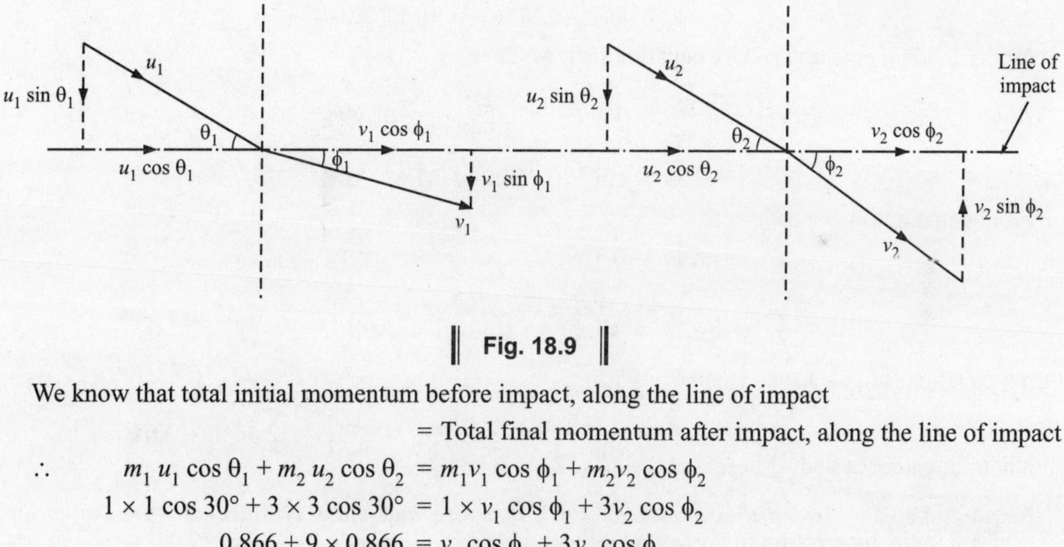


**Fig. 18.9**

We know that total initial momentum before impact, along the line of impact

= Total final momentum after impact, along the line of impact

$$\therefore \quad m_1 u_1 \cos \theta_1 + m_2 u_2 \cos \theta_2 = m_1 v_1 \cos \phi_1 + m_2 v_2 \cos \phi_2$$

$$1 \times 1 \cos 30° + 3 \times 3 \cos 30° = 1 \times v_1 \cos \phi_1 + 3 v_2 \cos \phi_2$$

$$0.866 + 9 \times 0.866 = v_1 \cos \phi_1 + 3 v_2 \cos \phi_2$$

or

$$v_1 \cos \phi_1 + 3 v_2 \cos \phi_2 = 0.866 + 9 \times 0.866 = 8.66 \quad ...(i)$$

We know that coefficient of restitution,

$$e = \frac{v_2 \cos \phi_2 - v_1 \cos \phi_1}{u_1 \cos \theta_1 - u_2 \cos \theta_2}$$

$$0.5 = \frac{v_2 \cos \phi_2 - v_1 \cos \phi_1}{1 \cos 30° - 3 \cos 30°} = \frac{v_2 \cos \phi_2 - v_1 \cos \phi_1}{0.866 - 3 \times 0.866}$$

$$= \frac{v_2 \cos \phi_2 - v_1 \cos \phi_1}{-1.732}$$

or $\quad v_1 \cos\phi_1 - v_2 \cos\phi_2 = 0.5 \times 1.732 = 0.866$ ...(ii)

Subtracting equation (*ii*) from equation (*i*), we have

$$4v_2 \cos\phi_2 = 8.66 - 0.866 = 7.794$$

$$\therefore \quad v_2 \cos\phi_2 = \frac{7.794}{4} = 1.9485 \quad ...(iii)$$

From equation (*ii*), $\quad v_1 \cos\phi_1 = v_2 \cos\phi_2 + 0.866$

$$= 1.9485 + 0.866 = 2.8145 \quad ...(iv)$$

Since the component of the velocity perpendicular to the line of impact remains same before and after impact, therefore

$$u_1 \cos\theta_1 = v_1 \sin\phi_1$$

or $\quad 1 \times \sin 30° = v_1 \sin\phi_1$

$$\therefore \quad v_1 \sin\phi_1 = 1 \times \sin 30° = 0.5 \quad ...(v)$$

Also $\quad u_2 \sin\theta_2 = v_2 \sin\phi_2$

$$3 \sin 30° = v_2 \sin\phi_2$$

$$\therefore \quad v_2 \sin\phi_2 = 3 \times \sin 30° = 3 \times 0.5 = 1.5 \quad ...(vi)$$

Dividing equation (*v*) by equation (*iv*), we have

$$\frac{v_1 \sin\phi_1}{v_1 \cos\phi_1} = \tan\phi_1 = \frac{0.5}{2.8145} = 0.1776$$

$$\therefore \quad \phi_1 = \tan^{-1}(0.1776) = 10.1° \quad \textbf{Ans.}$$

Again dividing equation (*vi*) by equation (*iii*), we have

$$\frac{v_2 \sin\phi_2}{v_2 \cos\phi_2} = \tan\phi_2 = \frac{1.5}{1.9485} = 0.7698$$

$$\therefore \quad \phi_2 = \tan^{-1}(0.7698) = 37.6° \quad \textbf{Ans.}$$

From equation (*v*), we have

$$v_1 \sin\phi_1 = 0.5$$

$$\therefore \quad v_1 = \frac{0.5}{\sin\phi_1} = \frac{0.5}{\sin 10.1°} = \frac{0.5}{0.1753} = 2.85 \text{ m/s} \quad \textbf{Ans.}$$

and from equation (*vi*), we have $\; v_2 \sin\phi_2 = 1.5$

$$\therefore \quad v_2 = \frac{1.5}{\sin\phi_2} = \frac{1.5}{\sin 37.6°} = \frac{1.5}{0.61} = 2.46 \text{ m/s} \quad \textbf{Ans.}$$

**Example 18.12** *Two billiard balls of the same size and mass collide with the velocities of approach 1.6 m/s and 3.5 m/s, respectively. The angle between their directions of motion is 45°. For a coefficient of restitution of 0.9, what are the final velocities of the balls immediately after the impact?*

***Solution*** Given : $m_1 = m_2 = m$ ; $u_1 = 1.6$ m/s ; $u_2 = 3.5$ m/s ; $\theta_1 = \theta_2 = 45°$ ; $e = 0.9$

***Final velocities of the balls immediately after the impact***

Let $v_1$ and $v_2$ be the final velocities of the balls after the impact, along the directions $\phi_1$ and $\phi_2$, respectively, with the line of impact, as shown in Fig. 18.10. Refer also Fig. 18.11, for the clear understanding.

We know that total initial momentum before impact, along the line of impact

= Total final momentum after impact, along the line of impact

$$\therefore \quad m_1 u_1 \cos\theta_1 + m_2 u_2 \cos\theta_2 = m_1 v_1 \cos\phi_1 + m_2 v_2 \cos\phi_2$$

Since $m_1 = m_2 = m$, therefore

$$u_1 \cos\theta_1 + u_2 \cos\theta_2 = v_1 \cos\phi_1 + v_2 \cos\phi_2$$

$$1.6 \cos 45° + 3.5 \cos 45° = v_1 \cos\phi_1 + v_2 \cos\phi_2$$

$$1.6 \times 0.707 + 3.5 \times 0.707 = v_1 \cos\phi_1 + v_2 \cos\phi_2$$

$$1.1312 + 2.4745 = v_1 \cos\phi_1 + v_2 \cos\phi_2$$

or $$v_1 \cos\phi_1 + v_2 \cos\phi_2 = 1.1312 + 2.4745 = 3.6057 \quad \text{...(i)}$$

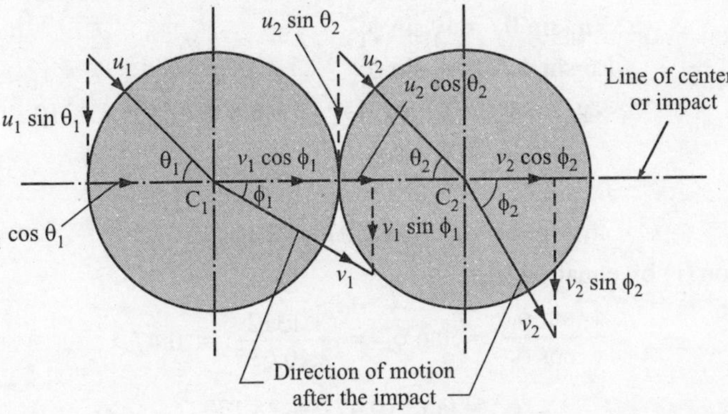


**Fig. 18.10**

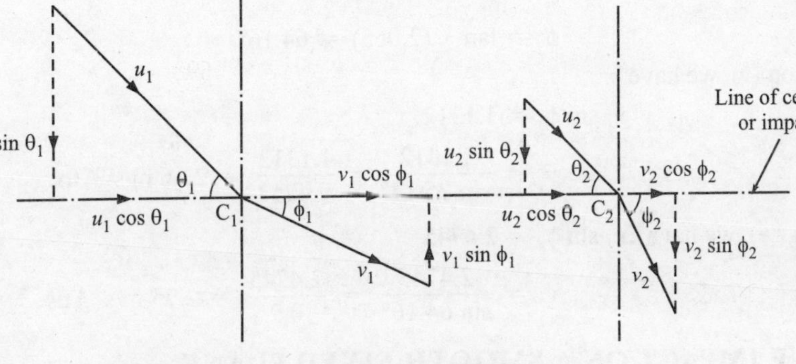


**Fig. 18.11**

We know that coefficient of restitution,

$$e = \frac{v_2 \cos\phi_2 - v_1 \cos\phi_1}{u_1 \cos\theta_1 - u_2 \cos\theta_2}$$

$$0.9 = \frac{v_2 \cos\phi_2 - v_1 \cos\phi_1}{1.6 \cos 45° - 3.5 \cos 45°} = \frac{v_2 \cos\phi_2 - v_1 \cos\phi_1}{1.6 \times 0.707 - 3.5 \times 0.707}$$

$$= \frac{v_2 \cos\phi_2 - v_1 \cos\phi_1}{1.1312 - 2.4745} = \frac{v_2 \cos\phi_2 - v_1 \cos\phi_1}{-1.3433}$$

or $$v_1 \cos\phi_1 - v_2 \cos\phi_2 = 0.9 \times 1.3433 = 1.209 \quad \text{...(ii)}$$

Adding equations (*i*) and (*ii*), we have

$$2v_1 \cos\phi_1 = 3.6057 + 1.209 = 4.8147$$

or $$v_1 \cos \phi_1 = \frac{4.8147}{2} = 2.407\,35 \qquad ...(iii)$$

From equation (*i*), $$v_2 \cos \phi_2 = 3.6057 - v_1 \cos \phi_1$$
$$= 3.6057 - 2.407\,35 = 1.198\,35 \qquad ...(iv)$$

Since the balls are smooth, therefore no force perpendicular to the line of centres is possible. Hence the velocities in the direction *i.e.* normal to the line of centres remain unchanged. In other words, the component of velocity perpendicular to the line of impact remain same before and after impact.

$$\therefore \quad u_1 \sin \theta_1 = v_1 \sin \phi_1$$
$$1.6 \sin 45° = v_1 \sin \phi_1$$

or $$v_1 \sin \phi_1 = 1.6 \sin 45° = 1.6 \times 0.707 = 1.1312 \qquad ...(v)$$

Also $$u_2 \sin \theta_2 = v_2 \sin \phi_2$$
$$3.5 \sin 45° = v_2 \sin \phi_2$$

or $$v_2 \sin \phi_2 = 3.5 \times 0.707 = 2.4745 \qquad ...(vi)$$

Dividing equation (*v*) by equation (*iii*), we have

$$\frac{v_1 \sin \phi_1}{v_1 \cos \phi_1} = \tan \phi_1 = \frac{1.1312}{2.407\,35} = 0.47$$

$$\therefore \quad \phi_1 = \tan^{-1}(0.47) = 25.17°$$

Again dividing equation (*vi*) by equation (*iv*), we have

$$\frac{v_2 \sin \phi_2}{v_2 \cos \phi_2} = \tan \phi_2 = \frac{2.4745}{1.198\,35} = 2.065$$

$$\therefore \quad \phi_2 = \tan^{-1}(2.065) = 64.16°$$

From equation (*v*), we have

$$v_1 \sin \phi_1 = 1.1312$$

$$\therefore \quad v_1 = \frac{1.1312}{\sin 25.17°} = \frac{1.1312}{0.4253} = 2.66 \text{ m/s} \quad \textbf{Ans.}$$

and from equation (*vi*), we have $v_2 \sin \phi_2 = 2.4745$

$$\therefore \quad v_2 = \frac{2.4745}{\sin 64.16°} = \frac{2.4745}{0.9} = 2.75 \text{ m/s} \quad \textbf{Ans.}$$

## 18.10 OBLIQUE IMPACT ON A SMOOTH FIXED PLANE

Consider a smooth sphere of mass $m$ impinges at point $C$ on a smooth horizontal plane. Let $u$ and $v$ be the velocities of the sphere before and after the impact, making an angle $\theta$ and $\phi$, respectively, with the normal to the plane, as shown in Fig. 18.12.

Since the plane is smooth, therefore there is no force parallel to the plane on the sphere. Thus the component of the velocity parallel to the plane remain unchanged

$$\therefore \quad v \sin \phi = u \sin \theta \qquad ...(i)$$

We know that coefficient of restitution,

$$e = \frac{v \cos \phi - 0}{u \cos \theta - 0} = \frac{v \cos \phi}{u \cos \theta}$$

$$= \frac{\text{Normal component of the velocity after impact}}{\text{Normal component of the velocity before impact}}$$

or $$v \cos \phi = e . u \cos \theta \qquad ...(ii)$$

Squaring equation (*i*) and (*ii*) and adding

$$v^2 \sin^2 \phi + v^2 \cos^2 \phi = u^2 \sin^2 \theta + e^2 u^2 \cos^2 \theta$$

$$v^2(\sin^2 \phi + \cos^2 \phi) = u^2 (\sin^2 \theta + e^2 \cos^2 \theta)$$

or
$$v^2 = u^2 (\sin^2 \theta + e^2 \cos^2 \theta) \quad ...(\because \sin^2 \phi + \cos^2 \phi = 1) \quad ...(iii)$$

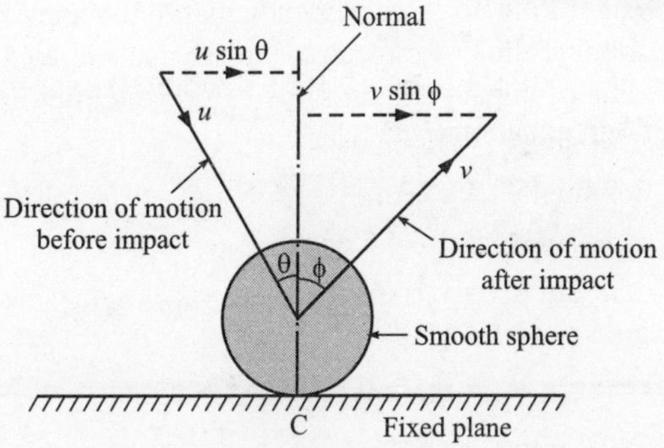


**Fig. 18.12**

Dividing equation (*ii*) by equation (*i*), we have

$$\frac{v \cos \phi}{v \sin \phi} = \frac{e.u \cos \theta}{u \sin \theta}$$

or
$$\cot \phi = e \cot \theta \quad ...(iv)$$

**Example 18.13** *A ball moving with a initial velocity of 2 m/s impinges on a fixed horizontal plane at an angle of 30° to the plane. Find the velocity and direction of motion of the ball after impact. The coefficient of restitution is 0.8.*

***Solution*** Given : $u = 2$ m/s ; $\theta = 90° - 30° = 60°$ ; $e = 0.8$

(Refer Fig. 18.13)

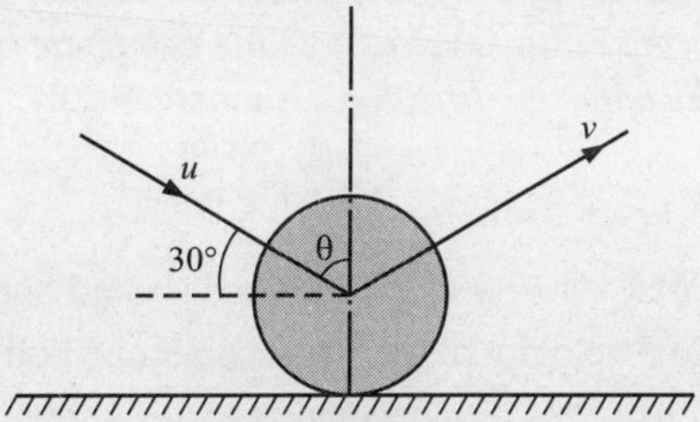


**Fig. 18.13**

Let $v$ = Velocity of the ball after impact in m/s, and

$\phi$ = Its angle of inclination with the normal to the plane.

Since the component of the velocity parallel to the plane remain unchanged, therefore

$$v \sin \phi = u \sin \theta = 2 \sin 60° = 2 \times \frac{\sqrt{3}}{2} = \sqrt{3} \quad ...(i)$$

We know that coefficient of restitution,

$$e = \frac{v \cos \phi}{u \cos \theta}$$

$\therefore \quad v \cos \phi = e.u \cos \theta = 0.8 \times 2 \cos 60° = 0.8 \times 2 \times \frac{1}{2} = 0.8 \quad ...(ii)$

Dividing equation (*i*) by equation (*ii*), we have

$$\frac{v \sin \phi}{v \cos \phi} = \frac{\sqrt{3}}{0.8} = 2.165$$

or $\tan \phi = 2.165$

$\therefore \quad \phi = \tan^{-1} (2.165) = 65.2°$ **Ans.**

Substituting this value of $\phi$ in equation (*i*),

$$v \sin 65.2° = \sqrt{3} = 1.732$$

$$v \times 0.9078 = 1.732$$

$\therefore \quad v = \frac{1.732}{0.9078} = 1.907 \text{ m/s}$ **Ans.**

## HIGHLIGHTS

1. When two elastic bodies collide with each other, they tend to compress each other. Immediately after this, the two bodies attempt to regain their original shapes, due to their elasticity. This process of regaining the original shape is called *restitution.*
2. The time taken by the bodies in compression, after the instant of collision, is called *time of compression* or *compression period.*
3. The period of time from the end of compression stage to the instant when the bodies separate, is called *time of restitution* or *restitution period.*
4. The sum of the time of compression and the time of restitution is called *time of collision* or *period of impact.*
5. The line joining the centres of the two bodies with the point of contact is called *line of impact.*
6. According to the principle of conservation of momentum, the total momentum of a system of colliding bodies remain constant before and after the collision, if no external force acts on them.
7. Newton's law of collision of elastic bodies, states that '*when two moving bodies collide with each other, then the relative velocity of the two bodies after impact (or relative velocity of separation) bears a constant ratio to the relative velocity of the two bodies before impact (or relative velocity of approach) and is in the opposite direction*'. Mathematically,

$$v_1 - v_2 = -e (u_1 - u_2)$$

or $v_2 - v_1 = e (u_1 - u_2)$

where $u_1$ and $u_2$ = Initial velocity of the first and second body, respectively, before impact,

$v_1$ and $v_2$ = Final velocity of the first and second body, respectively, after impact, and

$e$ = Constant of proportionality and is known as coefficient of restitution.

From above, coefficient of restitution,

$$e = \frac{v_2 - v_1}{u_1 - u_2}$$

The value of $e = 1$ for perfectly elastic bodies and $e = 0$ for perfectly inelastic bodies.

8. The collision between the two bodies which are moving along the line of impact before the collision is known as *direct impact.* For the direct impact between two bodies of mass $m_1$ and $m_2$,

Total initial momentum before impact

= Total final momentum after impact

*i.e.* $m_1 u_1 + m_2 u_2 = m_1 v_1 + m_2 v_2$

**9.** For the direct impact of a body on a fixed plane, coefficient of restitution,

$$e = \frac{v}{u}$$

where $u$ = Initial velocity of the body before impact, and

$v$ = Final velocity of the body after impact.

**10.** The loss of kinetic energy during impact is given by

$$E_L = \frac{m_1 m_2}{2(m_1 + m_2)} (u_1 - u_2)^2 (1 - e^2)$$

or

$$E_L = \frac{1}{2}\left[(m_1 u_1^2 + m_2 u_2^2) - (m_1 v_1^2 + m_2 v_2^2)\right]$$

**11.** The collision between the two bodies which are moving at some angles to the line of impact before the collision, is known as *indirect* or *oblique impact*.

**12.** For an indirect or oblique impact between two bodies,

$$m_1 u_1 \cos\theta_1 + m_2 u_2 \cos\theta_2 = m_1 v_1 \cos\phi_1 + m_2 v_2 \cos\phi_2$$

where $\theta_1$ and $\theta_2$ are the angles of $u_1$ and $u_2$, with the horizontal before collision and $\phi_1$ and $\phi_2$ are the angles of $v_1$ and $v_2$ with the horizontal after collision.

**13.** For an indirect or oblique impact on a fixed plane, coefficient of restitution,

$$e = \frac{v \cos\phi}{u \cos\theta}$$

## NUMERICAL PROBLEMS

**1.** Two bodies of mass 30 kg and 15 kg moving in the opposite direction at initial velocities of 6 m/s and 10 m/s, respectively, collide with each other. Find the velocity of each body after impact, if the coefficient of restitution is 0.8. [**Ans.** –3.6 m/s; 9.2 m/s]

**2.** A body of mass 8 kg is moving to the right at a speed of 3 m/s. It strikes a body of mass 1 kg which is moving to the left at a speed of 10 m/s. The final velocity of 1 kg body is 4 m/s to the right. Find the final velocity of 8 kg body and the coefficient of restitution. [**Ans.** 1.25 m/s; 0.2115]

**3.** A glass marble drops from a height of 3.5 m upon a horizontal floor. If the coefficient of restitution is 0.8, find the height to which it rises after impact. [**Ans.** 2.24 m]

**4.** A ball of mass 0.6 kg is dropped on a horizontal floor from a height of 10 m. The ball rebounds due to impact with the horizontal floor to the height of 6 m. Find the coefficient of restitution between the floor and the ball. [**Ans.** 0.7746]

**5.** A ball drops from the ceiling of a room and after rebounding twice from the floor, it reaches a height equal to one-fourth that of the ceiling. Show that the coefficient of restitution is 0.707.

**6.** A ball is dropped from a height of 2 m on a smooth floor. The height of the ball after the first bounce is 1.62 m. Find the coefficient of restitution between the ball and the floor. What will be the height after the second and the third bounce. [**Ans.** 0.9; 1.3122 m; 1.063 m]

**7.** Three perfectly elastic balls 1, 2 and 3 of masses 2 kg, 4 kg and 8 kg move in the same direction with velocities of 4 m/s, 1 m/s and 0.75 m/s, respectively. If the first ball strikes with the second ball, which in turn strikes the third ball, prove that the first and second ball will be brought to rest by the impact.

**8.** A sphere collides directly on an equal sphere which is at rest. If the coefficient of restitution is $e$, prove that the velocities of the sphere after impact are $(1 - e)$: $(1 + e)$.

**9.** A vehicle of mass 500 kg and moving with a velocity 15 m/s collides with a vehicle of mass 400 kg moving at 10 m/s. Both the vehicles get coupled together due to the impact. Find the common velocity with which the two vehicles will move and the loss of kinetic energy due to the impact. [**Ans.** 12.78 m/s; 2752 N-m]

**10.** A ball of mass 0.1 kg impinges directly on another ball of the same mass which is at rest. The first ball is reduced to rest by the impact. If the coefficient of restitution is 0.7, find the loss of kinetic energy.

[**Ans.** 0.052 $v_2^2$]

**11.** A ball of mass 4 kg moving with a velocity of 6 m/s strikes another body of mass 2 kg and velocity 3 m/s, both moving in the same direction. If the coefficient of restitution is 0.4, find:

(*a*) the velocity of the two balls; and

(*b*) the loss of kinetic energy during impact.

If both the balls move in the opposite direction, determine the values of (*a*) and (*b*).

[**Ans.** 4.6 m/s, 5.8 m/s; 5.04 N-m; 1.8 m/s, 5.4 m/s; 45.36 N-m]

**12.** A ball of mass 1 kg, moving with a velocity of 6 m/s strikes on a ball of mass 2 kg moving with a velocity of 2 m/s. At the instant of impact, the velocities of the two balls are parallel and inclined at 30° to the line joining their centres. If the coefficient of restitution is 0.5, determine the direction and velocity of the balls immediately after the impact. [**Ans.** 36.8°, 58.9°; 16.7 m/s, 30.4 m/s]

**13.** A ball moving with a velocity of 12 m/s strikes on a fixed plane making an angle of 60° with the fixed plane. If the angle made by the velocity of the ball after impact is 45°, with the fixed plane, find the velocity of the body after the impact and the coefficient of restitution. [**Ans.** $6\sqrt{2}$ m/s; 0.58]

**14.** A ball moving with an initial velocity of 4 m/s hits against a smooth horizontal plane, its direction being inclined at an angle of 30° to the plane. If the coefficient of restitution is 0.8, find the direction and velocity of the ball after hitting the plane. [**Ans.** 65.2°; 3.816 m/s]

## REVIEW QUESTIONS

1. What do you understand by an elastic body and an inelastic body?
2. Define the following terms:

   (*a*) Restitution; (*b*) Time of compression; (*c*) Time of restitution; (*d*) Time of collision; and (*e*) Line of impact.
3. Explain the term coefficient of restitution.
4. Find an expression for the coefficient of restitution in terms of initial and final velocities of the two bodies having the direct impact.
5. Prove that the loss of kinetic energy during the elastic impact of two bodies is given by

$$E_L = \frac{m_1 m_2}{2(m_1 + m_2)}(u_1 - u_2)^2 (1 - e^2)$$

where $m_1$ = Mass of the first body,

$u_1$ = Initial velocity of the first body before impact,

$v_1$ = Final velocity of the first body after impact,

$m_2$, $u_2$ and $v_2$ = Corresponding values of the second body, and

$e$ = Coefficient of restitution = $\dfrac{v_2 - v_1}{u_1 - u_2}$.

6. What is the difference between the impact of the two bodies and the impact of a body on a fixed plane?
7. Show that the coefficient of restitution in case of a body which is having oblique impact on a smooth fixed plane is given by

$$e = \frac{v \cos \phi}{u \cos \theta}$$

where $u$ and $v$ are the velocities of the body before and after the impact making an angle $\theta$ and $\phi$, respectively, with the normal to the plane.

## OBJECTIVE TYPE QUESTIONS

1. The bodies which rebound after impact are called
   (*a*) inelastic bodies (*b*) elastic bodies
2. The sum of time of compression and the time of restitution is called
   (*a*) time of collision (*b*) period of collision
   (*c*) period of impact (*d*) all of these
3. The collision between the two bodies which are moving along the line of impact before the collision, is called
   (*a*) restitution (*b*) indirect impact
   (*c*) direct impact (*d*) none of these
4. The coefficient of restitution is defined as the ratio of
   (*a*) relative velocity after impact to the relative velocity before impact
   (*b*) relative velocity of separation to the relative velocity of approach
   (*c*) relative velocity before impact to the relative velocity after impact
   (*d*) relative velocity of approach to the relative velocity of separation
5. The relative velocity of the two bodies after impact is always ............... the relative velocity of the two bodies before impact.
   (*a*) greater than (*b*) less than
6. The coefficient of restitution for perfectly inelastic bodies is
   (*a*) zero (*b*) between zero and one (*c*) one (*d*) more than one
7. The coefficient of restitution for perfectly elastic bodies is
   (*a*) zero (*b*) between zero and one (*c*) one (*d*) more than one
8. The kinetic energy of a body before impact is ............... the kinetic energy after impact.
   (*a*) equal to (*b*) less than (*c*) more than
9. A body of mass *m* moving with a constant velocity *v* strikes another body of the same mass *m* moving with same velocity but in opposite direction. The common velocity of both the bodies after collision is
   (*a*) $v$ (*b*) $2v$ (*c*) $4v$ (*d*) $8v$
10. A sphere of mass *m* impinges on a smooth horizontal plane. If *u* and *v* are the velocities of the sphere before and after the impact making an angle θ and ϕ with the normal to the plane, then the coefficient of restitution is given by
    (*a*) $e = \dfrac{v \cot \phi}{u \cot \phi}$ (*b*) $e = \dfrac{v \tan \phi}{u \tan \theta}$
    (*c*) $e = \dfrac{v \sin \phi}{u \sin \theta}$ (*d*) $e = \dfrac{v \cos \phi}{u \cos \theta}$

***Answers to Objective Type Questions***

| | | | | |
|---|---|---|---|---|
| **1.** (*b*) | **2.** (*d*) | **3.** (*c*) | **4.** (*a*), (*b*) | **5.** (*b*) |
| **6.** (*a*) | **7.** (*c*) | **8.** (*c*) | **9.** (*b*) | **10.** (*d*) |

# 19

# KINETICS OF ROTARY MOTION



## 19.1 INTRODUCTION

When a rigid body rotates about a fixed axis, then this motion is known as *motion of rotation* or *rotary motion.* The axis about which the body rotates is called *axis of rotation.* In the rotary motion, each particle of the body moves in a circle and centres of all these particles lies at the axis of rotation. In Chapter 14, on circular motion, we have only discussed the geometry of motion, but in this chapter, we shall deal with the forces which influence the motion of rotation.

## 19.2 TORQUE

When an external force acting on a body has a tendency to rotate the body about an axis, then the force is said to exert a torque about that axis. The torque or moment of a force about an axis of rotation is equal to the product of the magnitude of force and the perpendicular distance of the line of action of the force from the axis of rotation. The torque or the moment of force ($F$) acting at a distance $r$ from the axis of rotation, as shown in Fig. 19.1, is given by

$$T = F \times r$$

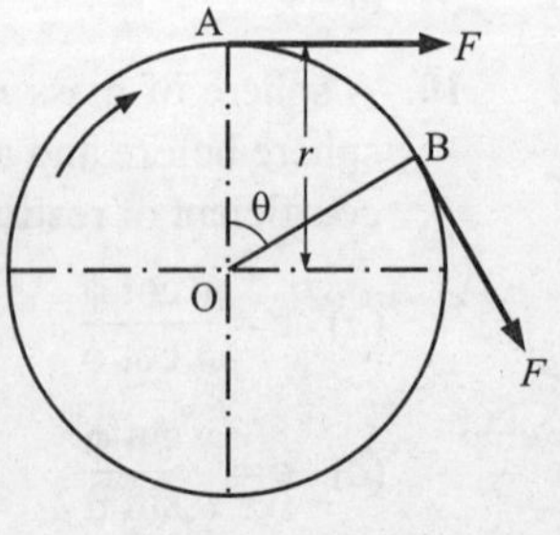


Fig. 19.1

When the force ($F$) is in newtons (N) and the distance ($r$) is in metres (m), then the unit of torque will be N-m.

## 19.3 WORK DONE BY A TORQUE

Consider that a body is pivoted at $O$ and rotates through a small angle ($\theta$) in radians such that the point $A$ shifts to point $B$, as shown in Fig. 19.1. We see that the distance moved by the force $F$

$$= \text{Length of arc } AB = r \cdot \theta$$

We know that work done by the force $F$ (acting tangentially) in rotating the body from $A$ to $B$

$$= \text{Force} \times \text{Distance moved by the force}$$

$$= F \times \text{Length of arc } AB = F \times r \cdot \theta$$

$$= (F \times r)\,\theta = T \cdot \theta \qquad ...(\because T = F \times r)$$

Thus, the work done by the torque is equal to the produce of torque ($T$) and the angular displacement ($\theta$) in radians. The practical unit of work is N-m. The work of 1 N-m is known as joule (briefly written as J) such that 1 N-m = 1 J.

## 19.4 LAWS OF ROTARY MOTION

The following are the three Newton's laws of rotary motion. These laws are similar to the three Newton's laws of linear motion as already discussed.

1. ***Newton's First Law of Rotary Motion*** It states that '*every body continues in its state of rest or of uniform motion of rotation about an axis, unless it is acted upon by some external torque*'.
2. ***Newton's Second Law of Rotary Motion*** It states that '*the rate of change of angular momentum of a body is directly proportional to the external torque and takes place in the same direction in which the torque acts*'.
3. ***Newton's Third Law of Rotary Motion*** It states that '*to every torque, there is always an equal and opposite torque*'.

## 19.5 MOMENT OF MOMENTUM OR ANGULAR MOMENTUM

We have already discussed that the linear momentum for a body moving along a straight line is the product of the mass of the body and its velocity (*i.e.* $m \cdot v$).

Let us now consider a body of total mass $m$ rotating with an angular velocity of $\omega$ rad/s, about the fixed axis $O$, as shown in Fig. 19.2. Since the body is composed of numerous small particles, therefore let us take one of these small particles having a mass $dm$ and at a distance $r$ from the axis of rotation. Let $v$ is its linear velocity acting tangentially at any instant. We know that momentum is the product of mass and velocity, therefore momentum of mass $dm$

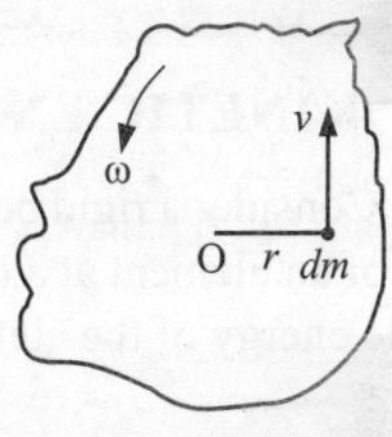


**Fig. 19.2**

$$= dm \times v$$

$$= dm \times \omega \times r \qquad ...(\because v = \omega \times r)$$

and moment of momentum of mass $dm$ about $O$

$$= dm \times \omega \times r \times r = dm \cdot \omega \cdot r^2$$

$$= I_m \cdot \omega$$

where $\qquad I_m$ = Mass moment of inertia of mass $dm$ about $O$ = $dm \cdot r^2$

∴ Moment of momentum or angular momentum of the whole body about $O$

$$= \int I_m \,.\omega = I.\omega$$

where $\qquad \int I_m = I = m \cdot k^2$

= *Mass moment of inertia of the whole body about $O$,

$m$ = Mass of the whole body, and

$k$ = Radius of gyration.

Thus, we see that the *moment of momentum* or *angular momentum* is the product of mass moment of inertia ($I$) and the angular velocity ($\omega$) of the body.

* The mass moment of inertia of various sections is given in Chapter 5 on Moment of Inertia.

## 19.6 RELATION BETWEEN TORQUE AND MOMENT OF INERTIA

We have already discussed that the product of mass moment of inertia ($I$) and the angular velocity ($\omega$) of a rotating body is called angular momentum. Thus

$$\text{Angular momentum} = I \cdot \omega$$

According to Newton's second law of rotary motion, the torque is directly proportional to the rate of change of angular momentum. Mathematically,

$$\text{Torque } (T) \propto \text{Rate of change of angular momentum}$$

or $$T \propto \frac{d(I \cdot \omega)}{dt}$$

Since $I$ is constant, therefore $$T \propto I \times \frac{d\omega}{dt} = kI \cdot \alpha$$

where $\frac{d\omega}{dt}$ = Angular acceleration ($\alpha$) and $k$ is a constant of proportionality.

Taking unit torque ($T = 1$) to produce unit angular acceleration ($\alpha = 1$) in a body of unit mass of moment of inertia ($I = 1$), we have $k = 1$. Thus,

$$T = I \cdot \alpha = I \times \frac{d\omega}{dt}$$

When $I$ is in kg-m$^2$ and $\alpha$ in rad/s$^2$, then the unit of torque is in N-m.

**Note:** When the torque is applied to the body for infinite small interval of time $dt$, then it is known as *impulsive torque* or *angular torque*.

$\therefore$ Impulsive torque, $T_I = I \, . \, d\omega$

## 19.7 KINETIC ENERGY DUE TO ROTATION

Consider a rigid body of mass $m$ rotating about point $O$ with constant angular velocity $\omega$. Let $dm$ be the mass of an element at a distance $r$ from the axis of rotation. If $v$ is the linear velocity of the element, then the kinetic energy of the elementary mass

$$= \frac{1}{2} \times \text{Mass (Velocity)}^2 = \frac{1}{2} dm \, v^2$$

$$= \frac{1}{2} dm \, (\omega . r)^2 = \frac{1}{2} \omega^2 r^2 \, dm \qquad ...(\because v = \omega . r)$$

Thus, kinetic energy of the whole body

$$= \int \frac{1}{2} \omega^2 r^2 \, dm = \frac{1}{2} \omega^2 \int r^2 \, dm$$

We know that $\int r^2 \cdot dm$ represents the mass moment of inertia ($I$) of the whole body about the axis of rotation.

$\therefore$ Kinetic energy of a rotating body $= \frac{1}{2} I \cdot \omega^2$

When mass is in kg, the kinetic energy will be in N-m.

**Note:** When a body has both linear and angular motion, *e.g.* in the locomotive driving wheels and wheels of a moving car, then the total kinetic energy of the body is equal to the sum of kinetic energies of translation and rotation.

$\therefore$ Total kinetic energy $= \frac{1}{2} mv^2 + \frac{1}{2} I\omega^2$

**Example 19.1** *A right circular disc which weighs 136 kg and is 762 mm in diameter is free to rotate about its geometric axis and is constantly accelerated from rest to 300 r.p.m. in 20 seconds. Determine the constant torque required to produce this acceleration.*

***Solution*** Given : $m = 136$ kg ; $D = 762$ mm or $R = 381$ mm $= 0.381$ m ; $N = 300$ r.p.m. or $\omega = \frac{2\pi \times 300}{60}$ $= 31.42$ rad/s ; $t = 20$ s

Let $\alpha$ = Angular acceleration of the disc in rad/s$^2$.

Since the disc is accelerated from rest, therefore initial angular velocity,

$$\omega_0 = 0$$

Using the relation,

$$\omega = \omega_0 + \alpha.t$$

$$31.42 = 0 + \alpha \times 20$$

$$\therefore \quad \alpha = \frac{31.42}{20} = 1.57 \text{ rad/s}^2$$

We know that mass moment of inertia of the disc,

$$I = 0.5\, mR^2 = 0.5 \times 136\, (0.381)^2 = 9.87 \text{ kg-m}^2$$

$\therefore$ Torque required, $T = I \cdot \alpha = 9.87 \times 1.57 = 15.5$ N-m **Ans.**

**Example 19.2** *The flywheel of a steam engine has a radius of gyration of 1 m and mass 2500 kg. The starting torque of the steam engine is 1500 N-m and may be assumed constant. Determine: 1. Angular acceleration of the flywheel ; and 2. Kinetic energy of the flywheel after 10 seconds from the start.*

***Solution*** Given : $k = 1$ m ; $m = 2500$ kg ; $T = 1500$ N-m

**1. *Angular acceleration of the flywheel***

Let $\alpha$ = Angular acceleration of the flywheel.

We know that mass moment of inertia of the flywheel,

$$I = m \cdot k^2 = 2500 \times 1^2 = 2500 \text{ kg-m}^2$$

We know that torque,

$$T = I \cdot \alpha$$

$$\therefore \quad \alpha = \frac{T}{I} = \frac{1500}{2500} = 0.6 \text{ rad/s}^2 \textbf{ Ans.}$$

**2. *Kinetic energy of the flywheel after 10 seconds from the start***

First of all, let us find the angular speed of the flywheel ($\omega$) after $t = 10$ seconds from the start (*i.e.* $\omega_0 = 0$).

We know that

$$\omega = \omega_0 + \alpha.t = 0 + 0.6 \times 10 = 6 \text{ rad/s}$$

$\therefore$ Kinetic energy of the flywheel, $E = \frac{1}{2} I \cdot \omega^2 = \frac{1}{2} \times 2500 \times 6^2 = 45\,000$ J $= 45$ kJ **Ans.**

**Example 19.3** *A uniform homogeneous cylinder rolls without slip along a horizontal level surface with a translational velocity of 200 mm/s. If its weight is 1 N and its radius is 100 mm, what is its total kinetic energy?*

***Solution*** Given : $v = 200$ mm/s $= 0.2$ m/s ; $w = 1$ N ; $R = 100$ mm $= 0.1$ m

Let $m$ = Mass of the cylinder in kg $= \frac{w}{g} = \frac{1}{9.81}$ kg

We know that kinetic energy due to translation

$$= \frac{1}{2} mv^2 = \frac{1}{2} \times \frac{1}{9.81} (0.2)^2 = 2.04 \times 10^{-3} \text{ N-m}$$

Angular velocity of the cylinder, $\omega = \frac{v}{R} = \frac{0.2}{0.1} = 2$ rad/s

and mass moment of inertia of a uniform cylinder

$$I = 0.5mR^2 = 0.5 \times \frac{1}{9.81}(0.1)^2 = 0.51 \times 10^{-3} \text{ kg-m}^2$$

$\therefore$ Kinetic energy due to rotation $= \frac{1}{2}I\omega^2 = \frac{1}{2} \times 0.51 \times 10^{-3} (2)^2 = 1.02 \times 10^{-3}$ N-m

We know that total kinetic energy = Kinetic energy due to translation + Kinetic energy due to rotation

$$= 2.04 \times 10^{-3} + 1.02 \times 10^{-3} = 3.06 \times 10^{-3} \text{ N-m}$$

$$= 3.06 \text{ N-mm } \textbf{Ans.}$$

**Example 19.4** *A flywheel of mass 5000 kg and having a radius of gyration 1 m, loses its speed from 400 r.p.m. to 280 r.p.m. in 2 minutes. Calculate : 1. The retarding torque acting on it ; 2. The change in kinetic energy during the above period ; and 3. The change in its angular momentum during the period.*

***Solution*** Given : $m = 5000$ kg ; $k = 1$ m ; $N_1 = 400$ r.p.m. ; $N_2 = 280$ r.p.m. ; $t = 2$ min = 120 s

We know that initial angular velocity,

$$\omega_0 = \frac{2\pi N_1}{60} = \frac{2\pi \times 400}{60} = 41.9 \text{ rad/s}$$

and final angular velocity, $\omega = \frac{2\pi N_2}{60} = \frac{2\pi \times 280}{60} = 29.3$ rad/s

Let $\alpha$ = Angular acceleration of the flywheel in rad/s².

Using the relation, $\omega = \omega_0 + \alpha . t$

$$\therefore \quad \alpha = \frac{\omega - \omega_0}{t} = \frac{29.3 - 41.9}{120} = -0.105 \text{ rad/s}^2$$

The –ve sign indicates retardation.

**1. *Retarding torque acting on the flywheel***

We know that mass moment of inertia of the flywheel,

$$I = m.k^2 = 5000(1)^2 = 5000 \text{ kg-m}^2$$

$\therefore$ Retarding torque, $T = I . \alpha = 5000 \times 0.105 = 525$ N-m **Ans.**

**2. *Change in kinetic energy***

We know that initial kinetic energy,

$$E_0 = \frac{1}{2}I(\omega_0)^2 = \frac{1}{2} \times 5000(41.9)^2 = 4.39 \times 10^6 \text{ N-m}$$

and final kinetic energy, $E = \frac{1}{2}I\omega^2 = \frac{1}{2} \times 5000(29.3)^2 = 2.14 \times 10^6$ N-m

$\therefore$ Change in kinetic energy $= E_0 - E = 4.39 \times 10^6 - 2.14 \times 10^6$

$$= 2.25 \times 10^6 \text{ N-m } \textbf{Ans.}$$

**3. *Change in angular momentum***

We know that initial angular momentum

$$= I.\omega_0 = 5000 \times 41.9 = 209.5 \times 10^3 \text{ N-m-s}$$

and final angular momentum $= I . \omega = 5000 \times 29.3 = 146.5 \times 10^3$ N-m-s

Change in angular momentum $= 209.5 \times 10^3 - 146.5 \times 10^3 = 63 \times 10^3$ N-m-s **Ans.**

**Example 19.5** *A train of cars weighing 60 kN is hauled up an incline of 1 in 30 by a rope coiled on a winding drum at the top of the incline. The drum is 1 m in diameter, weighs 10 kN and has a radius of gyration of 0.45 m. The resistance of the cars is 10 N/kN. The drum axle friction is negligible. If a torque of 3 kN-m is applied to the drum from a driving motor, determine the tension in the rope.*

***Solution*** Given : $W = 60$ kN $= 60 \times 10^3$ N ; $\sin\theta = \frac{1}{30}$ ; $d = 1$ m or $r = 0.5$ m ; $w = 10$ kN $= 10 \times 10^3$ N ; $k = 0.45$ m ; $F = 10$ N/kN $= 10 \times 60 = 600$ N ; $T = 3$ kN-m $= 3000$ N-m

Refer Fig. 19.3.

Let $P$ = Tension in the rope,

$\alpha$ = Angular acceleration at which the drum rotates, and

$a$ = Linear acceleration of the train of cars.

We know that mass moment of inertia of the drum,

$$I = \frac{w}{g} \times k^2$$

$$= \frac{10 \times 10^3}{9.81}(0.45)^2$$

$$= 206.4 \text{ kg-m}^2$$

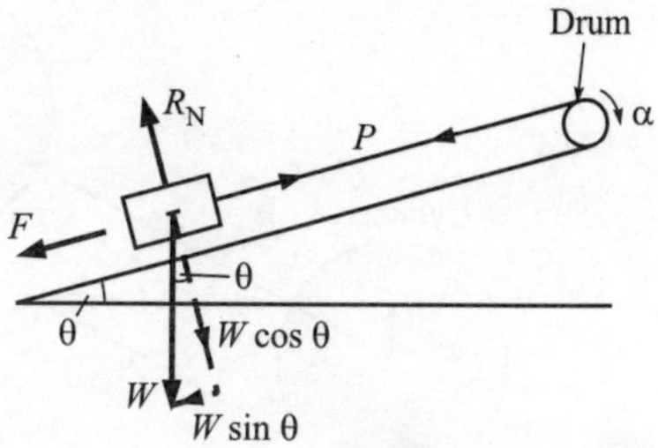


**Fig. 19.3**

and torque applied to the drum, $T = I \cdot \alpha$

$$\therefore \quad \alpha = \frac{T}{I} = \frac{3000}{206.4}$$

$$= 14.5 \text{ rad/s}^2$$

We know that linear acceleration of the train of cars,

$$a = r \cdot \alpha = 0.5 \times 14.5 = 7.25 \text{ m/s}^2$$

According to D'Alembert's principle for linear motion, we have

$$P - F - W\sin\theta = \frac{W}{g} \times a$$

$$P - 600 - 60 \times 10^3 \times \frac{1}{30} = \frac{60 \times 10^3}{9.81} \times 7.25$$

$$P - 2600 = 44\,342$$

$$\therefore \quad P = 44\,342 + 2600 = 46\,942 \text{ N} = 46.942 \text{ kN} \quad \textbf{Ans.}$$

**Example 19.6** *A homogeneous cylinder and a homogeneous sphere of equal weights, 80 N each and of equal radius R, are connected together by a weightless rod and roll without slip down the plane inclined at 28° with the horizontal, as shown in Fig. 19.4 (a). Determine the force in the rod. Assume the bearings to be frictionless.*

***Solution*** Given : $W_1 = W_2 = 80$ N ; $\theta = 28°$

Let $P$ = Force in the rod.

$R_{N1}$ and $R_{N2}$ = Normal reaction of the plane on the cylinder and sphere, respectively,

$F_1$ = Force of friction between the cylinder and the inclined plane, and

$F_2$ = Force of friction between the sphere and the inclined plane.

The weights $W_1$ and $W_2$ are resolved into two components, *i.e.* one parallel to the inclined plane and the other perpendicular to the inclined plane, as shown in Fig. 19.4 (*b*). The equilibrium equation may be written as

$$W_1 \sin\theta + W_2 \sin\theta = F_1 + F_2$$

or $$2W \sin\theta = 2F \qquad ...(\because W_1 = W_2 \text{ and } F_1 = F_2)$$

$\therefore$ $$F = W \sin\theta = 80 \times \sin 28° = 80 \times 0.4695 = 37.56 \text{ N}$$

Let $a$ = Linear acceleration of the system, and

$$\alpha = \text{Angular acceleration of the system} = \frac{a}{r}$$

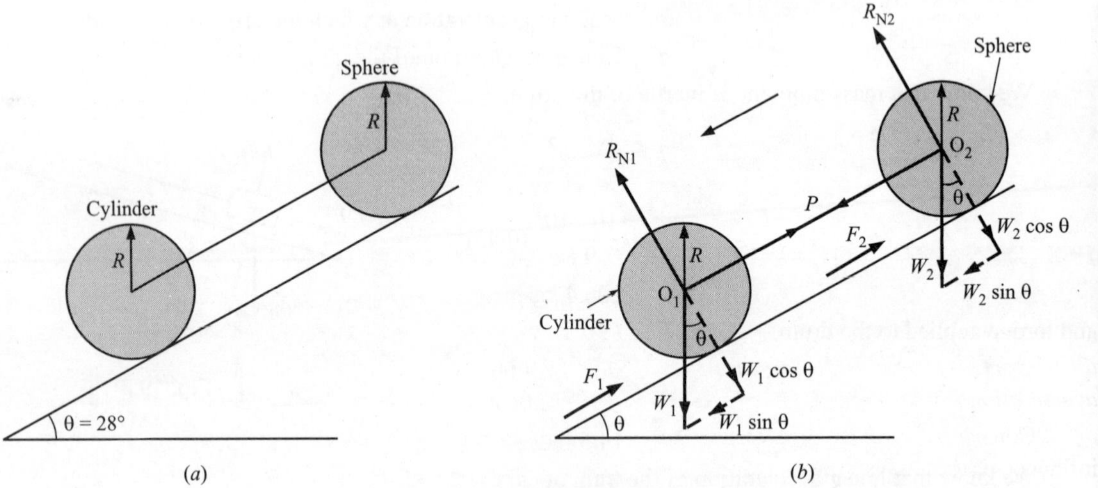


**Fig. 19.4**

We know that mass moment of inertia of the cylinder

$$I_1 = 0.5 \times \frac{W}{g} R^2$$

and mass moment of inertia of the sphere,

$$I_2 = 0.4 \times \frac{W}{g} R^2$$

Torque acting on the cylinder, $$T_1 = I_1 \cdot \alpha = I_1 \times \frac{a}{R}$$

and torque acting on the sphere, $$T_2 = I_2 \cdot \alpha = I_2 \times \frac{a}{R}$$

We also know that torque acting on the cylinder due to $F_1$

$$= F_1 \times R$$

and torque acting on the sphere due to $F_2$ $= F_2 \times R$

$\therefore$ $$T_1 + T_2 = F_1 \times R + F_2 \times R$$

$$I_1 \times \frac{a}{R} + I_2 \times \frac{a}{R} = F \times R + F \times R \qquad ...(\because F_1 = F_2 = F)$$

$$0.5 \times \frac{W}{g} R^2 \times \frac{a}{R} + 0.4 \times \frac{W}{g} R^2 \times \frac{a}{R} = 2F \times R$$

$$\frac{0.9W}{g} \times a = 2F$$

$$\therefore \quad a = \frac{2F \times g}{0.9W} = \frac{2 \times 37.56 \times 9.81}{0.9 \times 80} = 10.23 \text{ m/s}^2$$

Now consider the motion of the sphere parallel to the plane, the equation of motion is written as

$$P + W_2 \sin\theta - F_2 = \frac{W_2}{g} \times a$$

$$P + 80 \sin 28° - 37.56 = \frac{80}{9.81} \times 10.23 = 83.42$$

$$P + 80 \times 0.4695 - 37.56 = 83.42$$

$$\therefore \quad P = 83.42 + 37.56 - 80 \times 0.4695 = 83.42 \text{ N} \quad \textbf{Ans.}$$

**Note:** The value of $P$ may also be obtained by considering the motion of the cylinder parallel to the plane. In this case, the equation of motion is

$$P + F_1 - W_1 \sin\theta = \frac{W_1}{g} \times a$$

## 19.8 DYNAMIC EQUATION OF EQUILIBRIUM FOR ROTATING BODIES

We have already discussed that the torque causes the body to rotate. D'Alembert's principle when applied to rotating bodies, states that '*when torques act on a system having rotating motion, then the sum of all the torques acting on the system including the inertia torque taken in opposite direction of the angular acceleration is zero*'.

Consider a circular disc of moment of inertia ($I$) rotating at an angular acceleration $\alpha$ under the influence of torque $T$ acting in a clockwise direction. The inertia torque on the disc (equal to $I.\ \alpha$) will act in the anticlockwise direction. According to D'Alembert's principle,

$$T - I \cdot \alpha = 0$$

This is known as *dynamic equation of equilibrium* for rotating bodies.

## 19.9 MOTION OF ROTATION OF A BODY ATTACHED TO A STRING AND PASSING OVER A PULLEY

Consider a body of weight $W$ is attached to one end of an inextensible string passing over a pulley of weight ($W_P$). The other end of the string is tied to the periphery of the pulley, as shown in Fig. 19.5. Let the body moves downwards under the force of gravity.

Let
$r$ = Radius of the pulley,
$I$ = Mass moment of inertia of the pulley about the axis of rotation,
$k$ = Radius of gyration of the pulley,
$P$ = Tension in the string,
$a$ = Linear acceleration of the body,
$\alpha$ = Angular acceleration of the pulley.

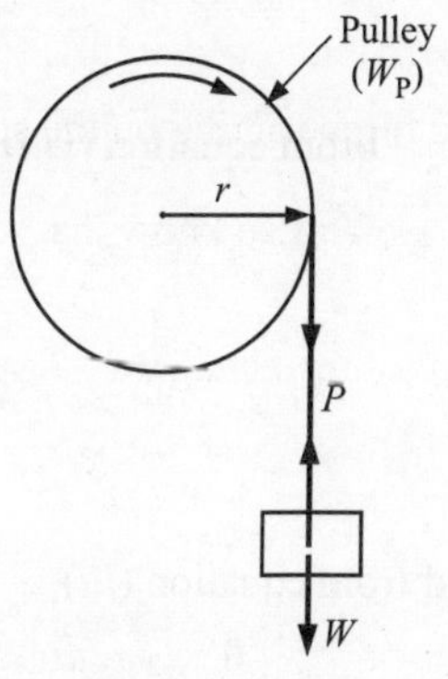

Fig. 19.5

First of all, consider the downward motion of the body. The forces acting on the body are $W$ acting downward and tension $P$ in the string acting upwards.

$\therefore$ Net force on the body acting downward

$$= W - P \quad \quad ...(i)$$

Since the body is moving downwards with an acceleration $a$, therefore net force

$$= \text{Mass} \times \text{Acceleration} = \frac{W}{g} \times a \qquad ...(ii)$$

Equating equations (*i*) and (*ii*), we have

$$W - P = \frac{W}{g} \times a \qquad ...(iii)$$

Now consider the motion of the pulley. We know that torque acting on the pulley,

$$T = I \times \alpha \qquad ...(iv)$$

We also know that torque, $T$ = Tension in the string × Radius of pulley = $P \times r$ ...(*v*)

Equating equations (*iv*) and (*v*), we have

$$I \times \alpha = P \times r$$

$$\therefore \quad P = \frac{I \times \alpha}{r} \qquad ...(vi)$$

We know that linear acceleration, $a$ = Angular acceleration × Radius = $\alpha . r$

$$\text{or} \quad \alpha = \frac{a}{r}$$

From equation (*vi*),

$$P = \frac{I}{r} \times \frac{a}{r} = \frac{I \times a}{r^2} \qquad ...(vii)$$

Substituting this value of $P$ in equation (*iii*), we have

$$W - \frac{I \times a}{r^2} = \frac{W}{g} \times a$$

$$\text{or} \quad W = \frac{W}{g} \times a + \frac{I \times a}{r^2} = \left(\frac{W}{g} + \frac{I}{r^2}\right) a$$

$$\therefore \quad a = \frac{W}{\dfrac{W}{g} + \dfrac{I}{r^2}} \qquad ...(viii)$$

Substituting the value of $a$ in equation (*vii*), we have

$$P = \frac{I}{r^2} \times \frac{W}{\dfrac{W}{g} + \dfrac{I}{r^2}} = \frac{I . W}{\dfrac{W . r^2}{g} + I} \qquad ...(ix)$$

If the pulley is a solid disc, then mass moment of inertia of the pulley is given by

$$I = 0.5 \times \text{Mass of the pulley} \times (\text{radius})^2 = 0.5 \times \frac{W_P}{g} \times r^2$$

From equation (*viii*),

$$a = \frac{W}{\dfrac{W}{g} + \dfrac{0.5 W_P \times r^2}{g . r^2}} = \frac{W}{\dfrac{W}{g} + \dfrac{0.5 W_P}{g}}$$

$$= \frac{W \cdot g}{W + \dfrac{W_P}{2}} = \frac{2W \cdot g}{2W + W_P} \qquad ...(x)$$

and from equation (*iii*),

$$W - P = \frac{W}{g} \times a = \frac{W}{g} \times \frac{2W \cdot g}{2W + W_P} = \frac{2W^2}{2W + W_P}$$

$$\therefore \quad P = W - \frac{2W^2}{2W + W_P} = \frac{2W^2 + W \times W_P - 2W^2}{2W + W_P} = \frac{W \times W_P}{2W + W_P} \qquad ...(xi)$$

**Example 19.7** *A pulley of weight 100 N has a radius of 0.5 m. A block of weight 250 N is suspended by a light rope wound round the pulley, and the other end of the rope being attached to the periphery of the pulley. If the block moves downwards, find : 1. Acceleration of the block ; and 2. Tension in the string.*

***Solution*** Given : $W_P = 100$ N ; $r = 0.5$ m ; $W = 250$ N

**1. *Acceleration of the block***

Considering the pulley to be a solid disc. We know that acceleration of the block,

$$a = \frac{2W \cdot g}{2W + W_P} = \frac{2 \times 250 \times 9.81}{2 \times 250 + 100} = 8.175 \text{ m/s}^2 \quad \textbf{Ans.}$$

**2. *Tension in the slope***

We know that tension in the rope, $P = \dfrac{W \times W_P}{2W + W_P} = \dfrac{250 \times 100}{2 \times 250 + 100} = 41.67 \text{ N}$ **Ans.**

**Example 19.8** *A body of weight 50 N is connected with an inextensible cord wrapped around a cylinder C of weight 90 N and radius 110 mm and at rest position, as shown in Fig. 19.6. If the weight is allowed to fall, determine the velocity after coming down a height of 70 mm.*

***Solution*** Given : $W = 50$ N ; $W_C = 90$ N ; $r = 110$ mm $= 0.11$ m ; $s = 70$ mm $= 0.07$ m

Let $v$ = Velocity of the body after coming down a distance of 70 mm (*i.e.* $s = 0.07$ m).

We know that acceleration of the body,

$$a = \frac{2W \cdot g}{2W + W_C} = \frac{2 \times 50 \times 9.81}{2 \times 50 + 90} = \frac{981}{190} = 5.163 \text{ m/s}^2$$

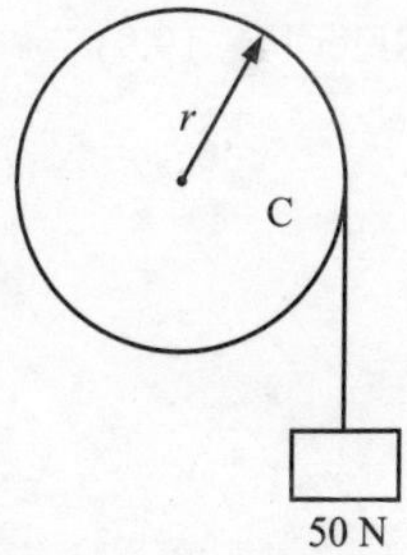


**Fig. 19.6**

Now using the relation, $v^2 - u^2 = 2a.s$

$$v^2 - 0 = 2 \times 5.163 \times 0.07 = 0.723 \qquad ...(\because u = 0)$$

$$\therefore \quad v = \sqrt{0.723} = 0.85 \text{ m/s}^2 \quad \textbf{Ans.}$$

**Example 19.9** *A drum has a radius of gyration 300 mm and weight 1850 N, is supported by small hubs which rests in bearings. A rope is wrapped around the drum carries a load of 1000 N on its free end, as shown in Fig. 19.7.*

*Neglecting friction in the bearings, determine the acceleration of the load, angular acceleration of the drum and tension in the rope. Take $g = 10$ m/s².*

***Solution*** Given : $k = 300$ mm $= 0.3$ m ; $W_D = 1850$ N ; $W = 1000$ N ; $r = 375$ mm $= 0.375$ m

***Acceleration of the load***

We know that mass moment of inertia of the drum,

$$I = \frac{W_D}{g} \times k^2 = \frac{1850}{10}(0.3)^2 = 16.65 \text{ kg-m}^2$$

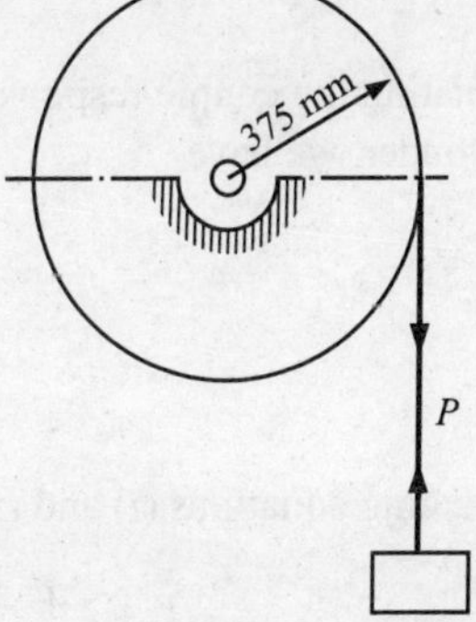


**Fig. 19.7**

and acceleration of the load,
$$a = \frac{W}{\frac{W}{g}+\frac{I}{r^2}} = \frac{1000}{\frac{1000}{10}+\frac{16.65}{(0.375)^2}} = \frac{1000}{100+118.9}$$
$$= 4.56 \text{ m/s}^2 \textbf{ Ans.}$$

***Angular acceleration of the drum***

We know that angular acceleration of the drum,
$$\alpha = \frac{a}{r} = \frac{4.56}{0.375} = 12.16 \text{ rad/s}^2 \textbf{ Ans.}$$

***Tension in the rope***

We know that tension in the rope, $P = \frac{I \times \alpha}{r} = \frac{16.65 \times 12.16}{0.375} = 540 \text{ N}$ **Ans.**

**Example 19.10** *A homogeneous solid circular cylinder of radius 0.25 m and weight 30 N is suspended by a cord with its one end held fixed at A, which is wrapped around the former, as shown in Fig. 19.8. If the cylinder is released from rest, what will be its velocity when it has dropped 3.3 m ? Determine also the tension in the string.*

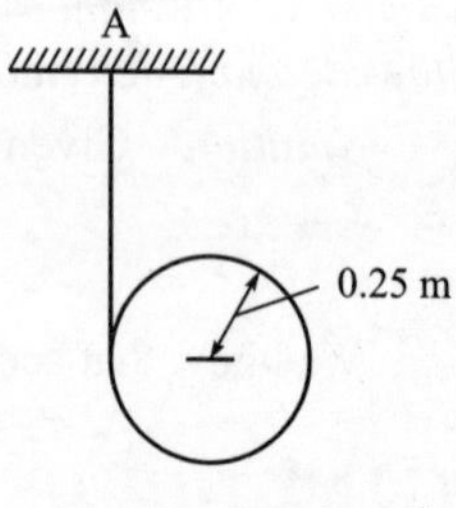

Fig. 19.8

***Solution*** Given : $r = 0.25$ m ; $W = 30$ N ; $s = 3.3$ m

(Refer Fig. 19.9)

Let
$P$ = Tension in the string,
$\alpha$ = Angular acceleration of the cylinder,
$a$ = Linear acceleration of the cylinder
$= \alpha \cdot r$, and
$k$ = Radius of gyration of the cylinder $= \frac{r}{\sqrt{2}}$.

First of all, consider the downward motion of the cylinder. The equation of motion is given by
$$W - P = \frac{W}{g} \times a$$
or
$$P = W - \frac{W}{g} \times a \quad ...(i)$$

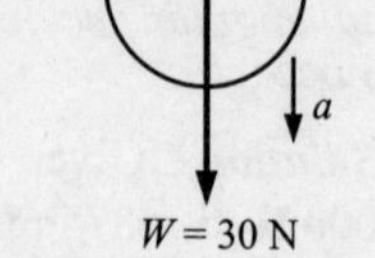

Fig. 19.9

Now consider the rotating motion due to a couple of $P \times r$. We know that torque acting on the cylinder,
$$T = I \cdot \alpha$$

Equating the couple responsible for rotating the cylinder and the torque on the cylinder, we have
$$P \times r = I \times \alpha = \frac{W}{g} \times k^2 \times \frac{a}{r}$$
or
$$P = \frac{W}{g} \times k^2 \times \frac{a}{r^2} \quad ...(ii)$$

Equating equations (*i*) and (*ii*), we have
$$W - \frac{W}{g} \times a = \frac{W}{g} \times k^2 \times \frac{a}{r^2}$$

or
$$W = \frac{Wa}{g}\left(1 + \frac{k^2}{r^2}\right)$$

$$\therefore \quad a = \frac{g}{1 + \frac{k^2}{r^2}} = \frac{r^2 \cdot g}{r^2 + k^2}$$

Since $k = \frac{r}{\sqrt{2}}$ or $k^2 = \frac{r^2}{2} = 0.5r^2$, therefore

$$a = \frac{r^2 \times g}{r^2 + 0.5r^2} = \frac{g}{1.5} = \frac{9.81}{1.5} = 6.54 \text{ m/s}^2$$

***Velocity of the cylinder when it has dropped 3.3 m***

Let $v$ = Velocity of the cylinder when it has dropped 3.3 m (*i.e.* when $s$ = 3.3 m).

Using the relation,
$$v^2 - u^2 = 2a.s$$
$$v^2 - 0 = 2 \times 6.54 \times 3.3 = 43.164 \quad \text{...}(\because u = 0)$$
$$\therefore \quad v = \sqrt{43.164} = 6.57 \text{ m/s} \quad \textbf{Ans.}$$

***Tension in the string***

Substituting the value of $a$ in equation (*i*), we have tension in the string,

$$P = W - \frac{W}{g} \times a = 30 - \frac{30}{9.81} \times 6.57 = 30 - 20 = 10 \text{ N} \quad \textbf{Ans.}$$

## 19.10 MOTION OF TWO BODIES CONNECTED BY A STRING AND PASSING OVER A PULLEY

Consider that two bodies $A$ and $B$ of weights $W_1$ and $W_2$, respectively, are attached to the two ends of a string passing over a pulley of weight $W_P$, as shown in Fig. 19.10.

Let $r$ = Radius of the pulley,

$I$ = Mass moment of inertia of the pulley about the axis of rotation,

$k$ = Radius of gyration of the pulley,

$a$ = Linear acceleration of the system, *i.e.* of the two bodies $A$ and $B$,

$\alpha$ = Angular acceleration of the pulley, and

$P_1$ and $P_2$ = Tension in the string to which the bodies $A$ and $B$ are attached.

When the weight of body $A$ ($W_1$) is greater than weight $W_2$, then the weight $W_1$ will move downwards while the weight of body $B$ ($W_2$) will move upwards with the same acceleration.

First of all, consider the downward motion of body $A$. We know that

$$\text{Net force acting on the body } A = W_1 - P_1 \quad \text{...}(i)$$

Since the body $A$ is moving downwards with an acceleration $a$, therefore force acting on it

$$= \text{Mass} \times \text{Acceleration}$$
$$= \frac{W_1}{g} \times a \quad \text{...}(ii)$$

Equating equations (*i*) and (*ii*), we have

$$W_1 - P_1 = \frac{W_1}{g} \times a \quad \text{...}(iii)$$

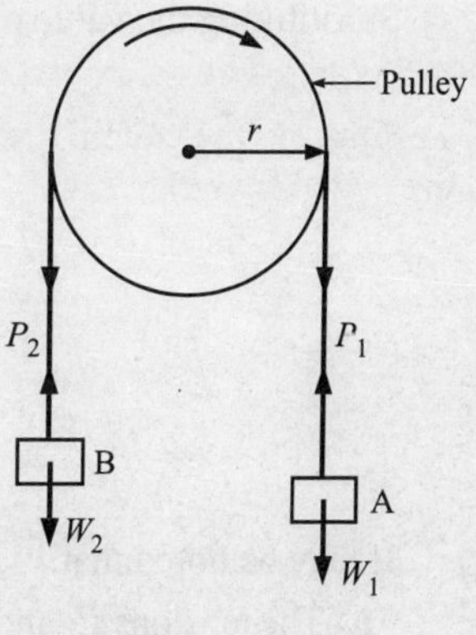


Fig. 19.10

Considering the upward motion of body $B$. We know that net force acting on body $B$

$$= P_2 - W_2 \qquad ...(\because P_2 > W_2) \quad ...(iv)$$

Since the body $B$ is moving upwards with an acceleration $a$, therefore force acting on it

$$= \frac{W_2}{g} \times a \qquad ...(v)$$

Equating equations (*iv*) and (*v*), we have

$$P_2 - W_2 = \frac{W_2}{g} \times a \qquad ...(vi)$$

Now consider the motion of the pulley which is rotating with an angular acceleration α. We know that angular acceleration,

$$\alpha = \frac{a}{r} \quad \text{or} \quad a = \alpha . r \qquad ...(vii)$$

and torque acting on the pulley,

$$T = I \times \alpha \qquad ...(viii)$$

We also know that torque acting on the pulley,

$$T = \text{Torque due to tension } P_1 - \text{Torque due to tension } P_2$$

$$= P_1 \times r - P_2 \times r = (P_1 - P_1)\, r \qquad ...(ix)$$

Equating equations (*viii*) and (*ix*),

$$(P_1 - P_2) r = I \times \alpha$$

or

$$P_1 - P_2 = \frac{I \times \alpha}{r} \qquad ...(x)$$

Adding equations (*iii*), (*vi*) and (*x*), we have

$$W_1 - W_2 = \frac{W_1}{g} \times a + \frac{W_2}{g} \times a + \frac{I \times \alpha}{r}$$

$$= \left(\frac{W_1}{g} + \frac{W_2}{g}\right) a + \frac{I \times a}{r^2} \qquad ...\left(\because \alpha = \frac{a}{r}\right) \quad ...(xi)$$

When the pulley is considered as a solid disc, then mass moment of inertia of the pulley is given by

$$I = 0.5 \times \text{Mass of the pulley} \times (\text{radius})^2$$

$$= 0.5 \times \frac{W_P}{g} \times r^2 = \frac{W_P}{2g} \times r^2$$

Substituting the value of $I$ in equation (*xi*), we have

$$W_1 - W_2 = \left(\frac{W_1}{g} + \frac{W_2}{g}\right) a + \frac{W_P}{2g} \times r^2 \times \frac{a}{r^2}$$

$$= \frac{(W_1 + W_2)\, a}{g} + \frac{W_P \times a}{2g} = \frac{a}{g}\left(W_1 + W_2 + \frac{W_P}{2}\right)$$

$$\therefore \quad a = \frac{(W_1 - W_2)\, g}{W_1 + W_2 + \frac{W_P}{2}} \qquad ...(xii)$$

It may be noted that

1. The tensions $P_1$ and $P_2$ in the string may be obtained by substituting the value of $a$ in equations (*iii*) and (*vi*), *i.e.*

$$W_1 - P_1 = \frac{W_1}{g} \times a \quad \text{or} \quad P_1 = W_1 - \frac{W_1}{g} \times a = \frac{W_1}{g}(g - a) \qquad ...(xiii)$$

and

$$P_2 - W_2 = \frac{W_2}{g} \times a \quad \text{or} \quad P_2 = W_2 + \frac{W_2}{g} \times a = \frac{W_2}{g}(g + a) \qquad ...(xiv)$$

**2.** The torque acting on the pulley due to difference of tensions in the two strings,

$$T_2 = (P_1 - P_2)r$$

and torque due to rotation of pulley, $T_1 = I \cdot \alpha$

∴ Net torque which must be applied to the pulley,

$$T = T_2 - T_1$$

**Example 19.11** *Two blocks weighing 100 N and 40 N are supported at the ends of a rope of negligible weight which is passing over the rough surface of a pulley mounted on a horizontal axle. The pulley may be assumed as a solid disc with a weight of 50 N. Friction in the bearings of the pulley may be neglected. Find the linear acceleration of the blocks and the tension on the two sides of the rope.*

***Solution*** Given : $W_1 = 100$ N ; $W_2 = 40$ N ; $W_P = 50$ N

The arrangement of the two blocks is shown in Fig. 19.11.

***Linear acceleration of the blocks***

We know that linear acceleration of the blocks,

$$a = \frac{(W_1 - W_2)g}{W_1 + W_2 + \frac{W_P}{2}}$$

$$= \frac{(100 - 40)9.81}{100 + 40 + \frac{50}{2}} = 3.567 \text{ m/s}^2 \textbf{ Ans.}$$

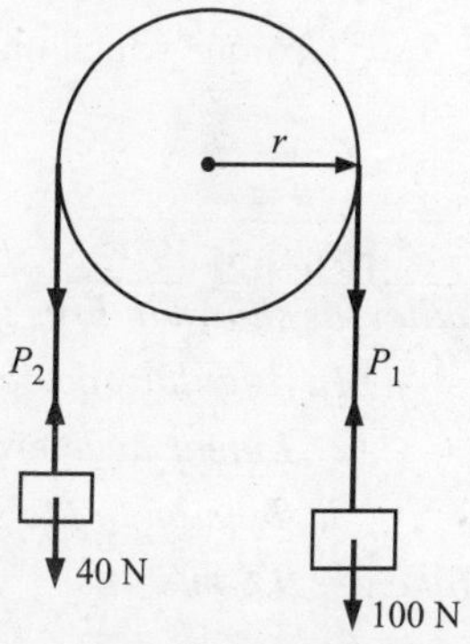


Fig. 19.11

***Tension on the two sides of the rope***

We know that tension,

$$P_1 = \frac{W_1}{g}(g - a)$$

$$= \frac{100}{9.81}(9.81 - 3.567) = 63.64 \text{ N} \textbf{ Ans.}$$

and tension,

$$P_2 = \frac{W_2}{g}(g + a)$$

$$= \frac{40}{9.81}(9.81 + 3.567) = 54.54 \text{ N} \textbf{ Ans.}$$

**Example 19.12** *Two bodies A and B of weights 8000 N and 6000 N are attached at the ends of a flexible rope. The rope passes over a pulley of 800 mm diameter. The pulley has a weight of 1000 N with a radius of gyration as 400 mm about its axis of rotation.*

*Determine the torque, which must be applied to the pulley to raise the body of 8000 N, with an acceleration of 1 m/s². Neglect friction of the spindle.*

***Solution*** Given : $W_1 = 8000$ N ; $W_2 = 6000$ N ; $d = 800$ mm or $r = 400$ mm = 0.4 m ; $W_P = 1000$ N ; $k = 400$ mm = 0.4 m

Since the body of 8000 N is being raised, therefore the acceleration ($a$) is taken as –ve, *i.e.* $a = -1$ m/s².

We know that tension in the rope carrying body $A$ of weight 8000 N,

$$P_1 = \frac{W_1}{g}(g - a) = \frac{8000}{9.81}[9.81 - (-1)] = 8815.5 \text{ N}$$

and tension in the rope carrying body $B$ of weight 6000 N,

$$P_2 = \frac{W_2}{g}(g + a) = \frac{6000}{9.81}(9.81 - 1) = 5388.4 \text{ N}$$

∴ Torque acting on the pulley due to difference of tensions,

$$T_2 = (P_1 - P_2)\, r = (8815.5 - 5388.4)\, 0.4 = 1371 \text{ N-m}$$

We know that mass moment of inertia of the pulley,

$$I = \frac{W_P}{g} \times k^2 = \frac{1000}{9.81} \times (0.4)^2 = 16.3 \text{ kg-m}^2$$

and angular acceleration of the pulley, $\alpha = \dfrac{a}{r} = \dfrac{-1}{0.4} = -2.5 \text{ rad/s}^2$

We know that torque acting on the pulley due to rotation,

$$T_1 = I \cdot \alpha = 16.3 \times (-2.5) = -40.75 \text{ N-m}$$

∴ Torque which must be applied to the pulley

$$= T_2 - T_1 = 1371 - (-40.75) = 1411.75 \text{ N-m}$$ **Ans.**

**Example 19.13** *Two blocks A and B of weights 150 N and 50 N are suspended from a compound pulley, as shown in Fig. 19.12. The weight of the pulley is 40 N and the radius of gyration is 70 mm. Find*

1. *Angular acceleration of the pulley ;*
2. *Linear acceleration of the blocks A and B ; and*
3. *Tension in the string.*

*Take g = 9.8 m/s².*

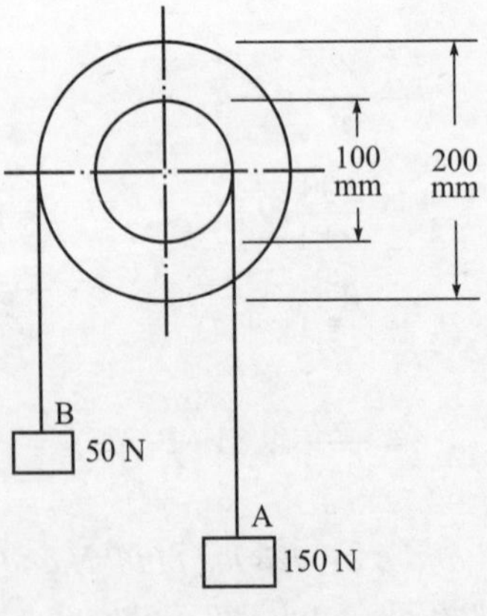


Fig. 19.12

***Solution*** Given : $W_1 = 150$ N ; $W_2 = 50$ N ; $W_P = 40$ N ; $k = 70$ mm $= 0.07$ m ; $d_1 = 100$ mm or $r_1 = 50$ mm $= 0.05$ m ; $d_2 = 200$ mm or $r_2 = 100$ mm $= 0.1$ m

**1. *Angular acceleration of the pulley***

Let $\alpha$ = Angular acceleration of the pulley;

$P_1$ and $P_2$ = Tension in the string carrying the weight of 150 N and 50 N, respectively (Refer Fig. 19.13).

We know that turning moment of weight 150 N

$$= W_1 \times r_1 = 150 \times 0.05 = 7.5 \text{ N-m}$$

and turning moment of weight 50 N $= W_2 \times r_2 = 50 \times 0.1 = 5$ N-m

Since the turning moment of 150 N is more than the turning moment of 50 N, therefore the weight 150 N will move downwards and the weight 50 N will move upwards, when the system is released.

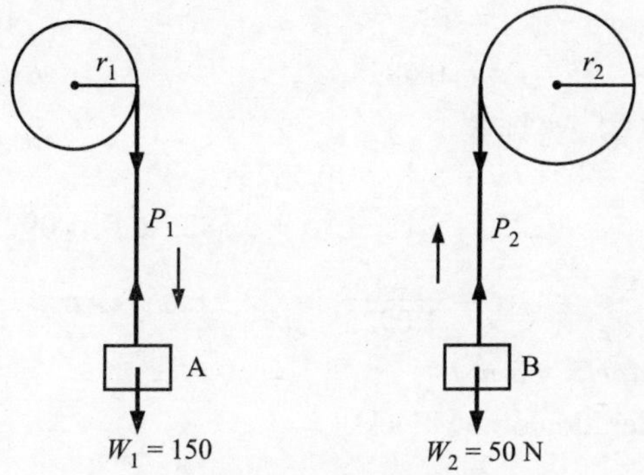


**Fig. 19.13**

Let $a_1$ = Acceleration of the weight 150 N, and

$a_2$ = Acceleration of the weight 50 N.

Considering the downward motion of block $A$ of weight $W_1 = 150$ N. The equation of motion is given by

$$W_1 - P_1 = \frac{W_1}{g} \times a_1$$

$$150 - P_1 = \frac{150}{9.8} \times a_1 = 15.3 a_1 \qquad ...(i)$$

Considering the upward motion of block $B$ of weight $W_2 = 50$ N. The equation of motion is given by

$$P_2 - W_2 = \frac{W_2}{g} \times a_2$$

$$P_2 - 50 = \frac{50}{9.8} \times a_2 = 5.1 a_2 \qquad ...(ii)$$

We know that linear acceleration of the body is equal to angular acceleration of the pulley ($\alpha$).

$$\therefore \quad a_1 = \alpha \cdot r_1 = \alpha \times 0.05 = 0.05\alpha$$

$$a_2 = \alpha \cdot r_2 = \alpha \times 0.1 = 0.1\alpha$$

Torque on the pulley due to difference of tensions

$$= P_1 \cdot r_1 - P_2 \cdot r_2 = P_1 \times 0.05 - P_2 \times 0.1 \qquad ...(iii)$$

We know that mass moment of inertia of the pulley,

$$I = \frac{W_P}{g} \times k^2 = \frac{40}{9.8}(0.07)^2 = 0.02 \text{ kg-m}^2$$

$\therefore$ Torque on the pulley due to rotation,

$$= I \cdot \alpha = 0.02\,\alpha \qquad ...(iv)$$

Equating equations (*iii*) and (*iv*), we have

$$P_1 \times 0.05 - P_2 \times 0.1 = 0.02\alpha$$

Dividing throughout by 0.05, we get

$$P_1 - 2P_2 = 0.4\alpha$$

or $$P_1 = 2P_2 + 0.4\alpha \quad ...(v)$$

Substituting this value of $P_1$ in equation (*i*), we have

$$150 - (2P_2 + 0.4\alpha) = 15.3a_1 = 15.3 \times 0.05\,\alpha$$

$$150 - 2P_2 - 0.4\alpha = 0.765\alpha$$

or $$150 - 2P_2 = 0.765\alpha + 0.4\alpha = 1.165\alpha$$

Dividing throughout by 2, we get

$$75 - P_2 = 0.5825\alpha \quad ...(vi)$$

Adding equation (*ii*) and (*vi*), we have

$$-50 + 75 = 5.1a_2 + 0.5825\,\alpha$$

$$25 = 5.1 \times 0.1\alpha + 0.5825\alpha = 1.0925\alpha$$

$$\therefore \quad \alpha = \frac{25}{1.0925} = 22.9 \text{ rad/s}^2 \quad \textbf{Ans.}$$

**2. *Linear acceleration of the blocks A and B***

We know that linear acceleration of the block *A*,

$$a_1 = \alpha \cdot r_1 = 22.9 \times 0.05 = 1.145 \text{ m/s}^2 \quad \textbf{Ans.}$$

and linear acceleration of the block *B*, $a_2 = \alpha \cdot r_2 = 22.9 \times 0.1 = 2.29 \text{ m/s}^2$ **Ans.**

**3. *Tension in the string***

From equation (*vi*), tension in the string carrying the weight of 50 N,

$$P_2 = 75 + 0.5825\alpha$$

$$= 75 + 0.5825 \times 22.9 = 75 + 13.34 = 88.34 \text{ N} \quad \textbf{Ans.}$$

Now from equation (*v*), $$P_1 = 2P_2 + 0.4\alpha$$

$$= 2 \times 88.34 + 0.4 \times 22.9 = 176.68 + 9.16$$

$$= 185.84 \text{ N} \quad \textbf{Ans.}$$

## 19.11 MOTION OF A CIRCULAR BODY ROLLING ON A ROUGH HORIZONTAL PLANE WITHOUT SLIPPING

Consider that a hanging body *A* of weight *w* and a circular body *B* of weight *W* are tied to a string passing over a smooth pulley, as shown in Fig. 19.14 (*a*).

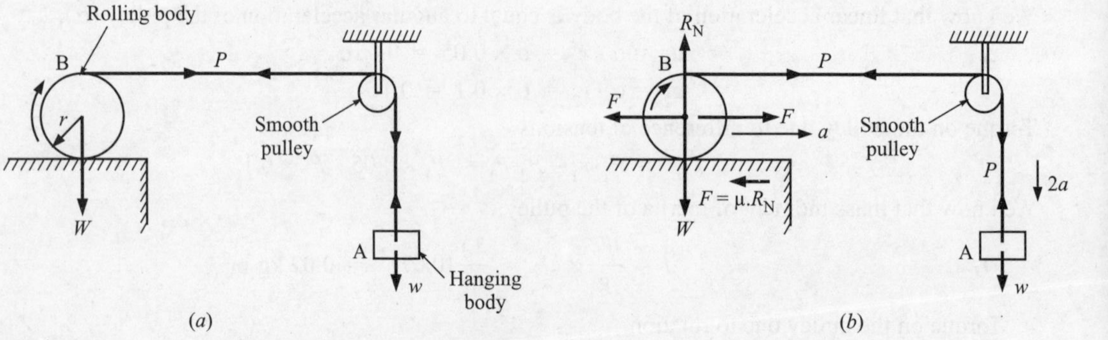


**Fig. 19.14**

The circular body *B* rolls on a rough horizontal plane without slipping.

Let $r$ = Radius of the rolling body $B$,

$k$ = Radius of gyration of the rolling body $B$,

$I$ = Mass moment of inertia of the rolling body $B = \frac{W}{g} \times k^2$,

$\alpha$ = Angular acceleration of the rolling body $B$,

$a$ = Linear acceleration of the rolling body $B = \alpha \cdot r$,

$\mu$ = Coefficient of friction between the rolling body $B$ and the horizontal surface, and

$P$ = Tension in the string.

A little consideration will show that when a rolling body $B$ makes one complete revolution, the displacement of the hanging body $A$ will be twice the circumference of the rolling body. Hence, the acceleration of the hanging body will always be double that of acceleration of the rolling body $A$.

∴ Acceleration of the hanging body $A = 2a$

First of all, let us consider the downward motion of the hanging body $A$. The equation of motion is given by

$$w - P = \frac{w}{g} \times 2a$$

or
$$P = w - \frac{2w \cdot a}{g} \qquad ...(i)$$

Now consider the motion of the rotating body $B$. Since it tends to rotate towards right, therefore the force of friction ($F = \mu . R_N$) will act towards left, as shown in Fig. 19.14 (*b*). Let us introduce two forces equal and opposite to the frictional force $F$ through the centre of the rolling body which does not affect the motion of the system. The rolling body is subjected to a force (equal to $P - F$) acting towards right which is responsible for the linear motion of the body. The rolling body is also subjected to a couple of magnitude $F \times r$, which is responsible for rolling of the body.

First of all, consider the linear motion of the rotating body $B$ due to the force ($P - F$). Since the body is moving towards right with an acceleration $a$, therefore the equation of motion is written as

$$P - F = \frac{W}{g} \times a$$

or
$$P = F + \frac{W}{g} \times a \qquad ...(ii)$$

Now considering the rolling motion of the rotating body due to the couple of $F \times r$, which is responsible for rolling of the body. We know that the torque acting on the rotating body $B$,

$$T = I \cdot \alpha$$

Equating the couple responsible for rolling and the torque on the rotating body $B$, we have

$$F \times r = I \times \alpha$$

Multiplying both sides by $r$, we have

$$F \times r^2 = I \times \alpha \times r = I . a \qquad ...(\because a = \alpha \cdot r)$$

∴
$$F = \frac{I \cdot a}{r^2} = \frac{W \cdot k^2}{g} \times \frac{a}{r^2}$$

Substituting this value of $F$ in equation (*ii*), we have

$$P = \frac{W \cdot k^2}{g} \times \frac{a}{r^2} + \frac{W}{g} \times a$$

Now substituting the value of $P$ from equation ($i$), we have

$$w - \frac{2w \cdot a}{g} = \frac{W \cdot k^2 \cdot a}{g r^2} + \frac{W \cdot a}{g}$$

or
$$w = \frac{2w \cdot a}{g} + \frac{Wk^2 a}{g r^2} + \frac{Wa}{g} = \frac{a}{g}\left(2w + \frac{Wk^2}{r^2} + W\right)$$

$\therefore$
$$a = \frac{w \cdot g}{W + 2w + \frac{Wk^2}{r^2}}$$

This expression shows that the linear acceleration of the rolling body is independent of coefficient of friction.

**Example 19.14** *A solid right circular roller B of weight 120 N is being pulled by another hanging body A of weight 60 N, along the horizontal plane, as shown in Fig. 19.14 (a). The body A is attached to the end of a string wound round the circumference of the roller B. Assuming that there is no slipping of the roller and the string is inextensible, determine the linear acceleration of the roller on the horizontal plane.*

***Solution*** Given : $W = 120$ N ; $w = 60$ N

We know that for a solid right circular roller of radius $r$, the radius of gyration,

$$k = \frac{r}{\sqrt{2}} \quad \text{or} \quad k^2 = \frac{r^2}{2} = 0.5 r^2$$

$\therefore$ Linear acceleration of the roller on the horizontal plane,

$$a = \frac{w \cdot g}{W + 2w + \frac{Wk^2}{r^2}}$$

$$= \frac{60 \times 9.81}{120 + 2 \times 60 + \frac{120 \times 0.5 r^2}{r^2}} = \frac{588.6}{300} = 1.962 \text{ m/s}^2 \quad \textbf{Ans.}$$

## 19.12 MOTION OF A CIRCULAR BODY ROLLING DOWN A ROUGH INCLINED PLANE WITHOUT SLIPPING

Consider that a circular body of weight $W$ is rolling down a rough inclined plane which makes an angle $\theta$ with the horizontal, as shown in Fig. 19.15.

Let $r$ = Radius of the rolling body,

$k$ = Radius of gyration of the rolling body,

$I$ = Moment of inertia of the rolling body = $\frac{W}{g} \times k^2$,

$\alpha$ = Angular acceleration of the rolling body,

$a$ = Linear acceleration of the rolling body,

$\mu$ = Coefficient of friction between the rolling body and the inclined plane.

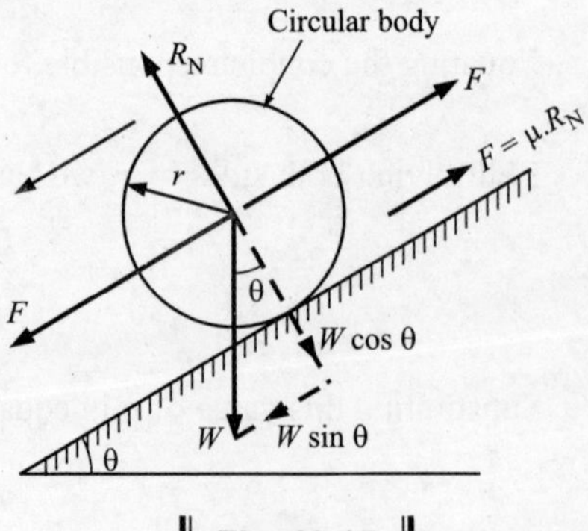


**Fig. 19.15**

The weight $W$ of the rolling body has two components: one parallel to the inclined plane of magnitude $W \sin \theta$ and the other perpendicular to the inclined plane of magnitude $W \cos \theta$. We know that the normal reaction,

$$R_N = W \cos \theta$$

and force of friction, $$F = \mu \cdot R_N = \mu \cdot W \cos \theta$$

Since the rolling body rolls down the inclined plane, therefore the force of friction ($F = \mu . R_N = \mu \cdot W \cos \theta$) will act upwards, as shown in Fig. 19.15. Let us introduce two forces equal and opposite to the frictional force $F$ through the centre of the body which does not affect the motion of the body. The rolling body is subjected to the following two forces:

***(a)*** Force ($W \sin \theta - F$) acting downwards, which is responsible for the linear motion of the body, and

***(b)*** A couple of magnitude $F \times r$, which is responsible for rolling down the body.

First of all, let us consider the linear motion of the rolling body, due to the force ($W \sin \theta - F$). Since the body is moving down with an acceleration $a$, therefore the equation of motion is written as

$$W \sin \theta - F = \frac{W}{g} \times a \qquad ...(i)$$

Now considering the rolling motion of the rotating body due to the couple of $F \times r$, which is responsible for rolling of the body. We know that torque acting on the body,

$$T = I \cdot \alpha$$

Equating the couple responsible for rolling and torque on the rotating body, we have

$$F \times r = I \times \alpha$$

Multiplying both sides by $r$, we have

$$F \times r^2 = I \times \alpha \times r = I \cdot a \qquad ...(\because a = \alpha \cdot r)$$

$$\therefore \quad F = \frac{I \cdot a}{r^2} = \frac{W \cdot k^2}{g} \times \frac{a}{r^2} \qquad ...(ii)$$

Substituting the value of $F$ in equation ($i$), we have

$$W \sin \theta - \frac{W \cdot k^2 \cdot a}{g r^2} = \frac{W}{g} \times a$$

or $$\sin \theta - \frac{k^2 \cdot a}{g r^2} = \frac{a}{g}$$

or $$\frac{a}{g} + \frac{k^2 \cdot a}{g r^2} = \sin \theta$$

$$\frac{a}{g}\left(1 + \frac{k^2}{r^2}\right) = \sin \theta$$

$$\frac{a}{g}\left(\frac{r^2 + k^2}{r^2}\right) = \sin \theta$$

$$\therefore \quad a = \frac{g \sin \theta}{\dfrac{r^2 + k^2}{r^2}} = \frac{r^2 \cdot g \sin \theta}{r^2 + k^2} \qquad ...(iii)$$

This expression shows that the linear acceleration of the rotating body is independent of the weight of the body.

Substituting the value of $a$ in equation (*ii*), we have

$$F = \frac{Wk^2}{g} \times \frac{a}{r^2} = \frac{Wk^2}{g \cdot r^2} \times \frac{r^2 \cdot g \sin\theta}{r^2 + k^2} = \frac{Wk^2 \sin\theta}{r^2 + k^2} \quad ...(iv)$$

For the circular body to roll down the plane without slipping, the applied force must be less than (or equal to) the available frictional force, *i.e.*

$$F \leq \mu \times W \cos\theta$$

or
$$\frac{Wk^2 \sin\theta}{r^2 + k^2} \leq \mu \times W \cos\theta$$

or
$$\frac{\sin\theta}{\cos\theta} \leq \mu \times W \times \frac{r^2 + k^2}{Wk^2}$$

$$\tan\theta \leq \mu\left(\frac{r^2 + k^2}{k^2}\right)$$

or
$$\mu \geq \frac{k^2 . \tan\theta}{r^2 + k^2}$$

For minimum value of $\mu$, the above equation may be written as

$$\mu = \frac{k^2 \cdot \tan\theta}{r^2 + k^2} \quad ...(v)$$

**Example 19.15** *A solid uniform circular disc having a diameter 750 mm and weight 2700 N, rolls without slipping down a plane inclined at an angle of 25° with the horizontal. Determine the frictional force F and acceleration of the mass centre.*

***Solution*** Given : $d = 750$ mm or $r = 375$ mm $= 0.375$ m ; $W = 2700$ N ; $\theta = 25°$

***Frictional force***

We know that for a uniform circular disc, radius of gyration,

$$k = \frac{r}{\sqrt{2}} \text{ or } k^2 = \frac{r^2}{2} = 0.5r^2 = 0.5(0.375)^2 = 0.0703 \text{ m}^2$$

$\therefore$ Frictional force,
$$F = \frac{Wk^2 \sin\theta}{r^2 + k^2}$$

$$= \frac{2700 \times 0.0703 \sin 25°}{(0.375)^2 + 0.0703} = \frac{189.81 \times 0.4226}{0.1406 + 0.0703} = \frac{80.214}{0.2109}$$

$$= 380.34 \text{ N} \quad \textbf{Ans.}$$

***Acceleration of the mass centre***

We know that acceleration of the mass centre,

$$a = \frac{r^2 g \sin\theta}{r^2 + k^2} = \frac{(0.375)^2\, 9.81 \times \sin 25°}{(0.375)^2 + 0.0703}$$

$$= \frac{0.1406 \times 9.81 \times 0.4226}{0.1406 + 0.0703} = \frac{0.583}{0.2109} = 2.765 \text{ m/s} \quad \textbf{Ans.}$$

**Example 19.16** *A solid uniformly thick wheel of radius 1 m and weight 400 N is released with no initial velocity at the top of an inclined plane, which makes an angle of 30° with the horizontal. It rolls down without slipping. Determine :*

1. *The minimum value of coefficient of friction ; and*
2. *The velocity of the centre of the wheel after it has travelled a distance of 4 m down the inclined plane.*

***Solution*** Given : $r = 1$ m ; $W = 400$ N ; $\alpha = 30°$ ; $s = 4$ m

**1. *Minimum value of coefficient of friction***

We know that for a uniformly thick wheel, radius of gyration,

$$k = \frac{r}{\sqrt{2}} \text{ or } k^2 = \frac{r^2}{2} = 0.5r^2$$

and minimum value of coefficient of friction,

$$\mu = \frac{k^2 \cdot \tan\theta}{r^2 + k^2} = \frac{0.5r^2 \times \tan 30°}{r^2 + 0.5r^2}$$

$$= \frac{0.5 \times 0.5773}{1 + 0.5} = 0.192 \textbf{ Ans.}$$

**2. *Velocity of the centre of the wheel***

Let $v$ = Velocity of the centre of the wheel.

We know that acceleration of the wheel when it rolls down the plane,

$$a = \frac{r^2 \cdot g \sin\theta}{r^2 + k^2} = \frac{r^2 \times 9.81 \sin 30°}{r^2 + 0.5r^2}$$

$$= \frac{9.81 \times 0.5}{1.5} = 3.27 \text{ m/s}^2$$

It is given that the wheel has travelled a distance of 4 m (*i.e.* $s = 4$ m) down the inclined plane. Now using the relation,

$$v^2 - u^2 = 2a \cdot s$$

$$v^2 - 0 = 2 \times 3.27 \times 4 = 26.16 \quad \text{...}(\because u = 0)$$

$$\therefore \quad v = \sqrt{26.16} = 5.11 \text{ m/s } \textbf{Ans.}$$

## HIGHLIGHTS

1. The motion of a rigid body rotating about a fixed axis, is known as *rotary motion* or *motion of rotation.* In the rotary motion, each particle of the body moves in a circle and centres of all these particles lie at the axis of rotation.
2. The torque is defined as the turning moment of the force on a body on which it acts. It is equal to the product of the magnitude of the force and the perpendicular distance of the line of action of the force from the axis of rotation.
3. The work done by the torque is equal to the product of the torque ($T$) and the angular displacement ($\theta$) in radians. Mathematically,

   Work done by the torque $= T \cdot \theta$

   The practical unit of work is N-m.

4. Newton's first law of rotary motion states that '*every body continues in its state of rest or of uniform motion of rotation about an axis, unless it is acted upon by some external torque*'.

5. Newton's second law of rotary motion states that '*the rate of change of angular momentum of a body is directly proportional to the external torque and takes place in the same direction in which the torque acts*'.

6. Newton's third law of rotary motion states that '*to every torque, there is always an equal and opposite torque*'.

7. The angular momentum is equal to the product of mass moment of inertia ($I$) and the angular velocity ($\omega$) of a rotating body. Mathematically,

$$\text{Angular momentum} = I \cdot \omega$$

8. The torque is directly proportional to the rate of change of angular momentum. Mathematically,

$$T = I \cdot \alpha$$

when mass moment of inertia ($I$) is in kg-m$^2$ and angular acceleration ($\alpha$) in rad/s$^2$, then torque will be in N-m.

9. The kinetic energy of a rotating body $= \frac{1}{2} I \cdot \omega^2$

10. When a body has both linear and angular motion, *e.g.* in the locomotive driving wheels and wheels of a moving car, then total kinetic energy of the body is equal to the sum of kinetic energy of translation and rotation. Mathematically,

$$\text{Total kinetic energy} = \frac{1}{2} mv^2 + \frac{1}{2} I\omega^2$$

11. The D'Alembert's principle when applied to rotating bodies states that '*when torques act on a system having rotating motion, then the sum of all the torques acting on the system including the inertia taken in opposite direction of the angular acceleration is zero*'. Mathematically,

$$T - I \cdot \alpha = 0$$

This is known as *dynamic equation of equilibrium* for rotating bodies.

12. When a body of weight ($W$) is attached to one end of an inextensible string passing over a pulley of weight ($W_P$) and the other end of the string is tied to the periphery of the pulley, then linear acceleration of the weight $W$ with which it moves downward is given by

$$a = \frac{W}{\frac{W}{g} + \frac{I}{r^2}}$$

and tension in the string ($P$) is given by

$$P = \frac{I \cdot W}{\frac{W \cdot r^2}{g} + I}$$

When the pulley is a solid disc, then mass moment of inertia of the pulley is given by

$$I = 0.5 \times \frac{W_P}{g} \times r^2$$

where $r$ is the radius of the pulley.

**13.** When two bodies of weights $W_1$ and $W_2$ ($W_1 > W_2$) are attached to the two ends of a string passing over a pulley of weight $W_P$, then the linear acceleration of the system, (assuming the pulley to be a solid disc) is given by

$$a = \frac{(W_1 - W_2)\,g}{W_1 + W_2 + \dfrac{W_P}{2}}$$

and tension in the two sides of the string are given by

$$P_1 = \frac{W_1}{g}(g - a); \text{ and } P_2 = \frac{W_2}{g}(g + a)$$

**14.** A circular body of weight $W_1$ is attached to a hanging body of weight $w$ by a string passing over a smooth pulley. The circular body rolls on a rough horizontal plane without slipping. The linear acceleration of the rolling body is given by

$$a = \frac{w \cdot g}{W + 2w + \dfrac{Wk^2}{r^2}}$$

where $k$ is the radius of gyration and $r$ is the radius of the rolling body.

**15.** When a circular body rolls on a rough inclined plane which makes an angle θ with the horizontal, then linear acceleration of the rolling body is given by

$$a = \frac{r^2 \cdot g \sin \theta}{r^2 + k^2}$$

## NUMERICAL PROBLEMS

**1.** A flywheel fitted on the crank shaft of a steam engine has a mass of 1 tonne and a radius of gyration 0.4 m. If the starting torque of the engine is 650 N-m and may be assumed constant. Find: 1. Angular acceleration of the flywheel; and 2. Kinetic energy of the flywheel after 10 seconds from the start.

[**Ans.** 4.06 rad/s$^2$; 131.87 N-m]

**2.** A flywheel of weight 24.5 kN and radius of gyration of 1.2 m loses its speed from 250 r.p.m. to 120 r.p.m. in an interval of 1.8 minutes. Find the retarding torque and change in kinetic energy.

[**Ans.** 445 N-m; 930.3 kN-m]

**3.** A uniform homogeneous cylinder rolls without slipping along a horizontal level surface with a translational velocity of 0.2 m/s. If its mass is 0.15 kg and its radius is 125 mm, determine its total kinetic energy.

[**Ans.** $4.5 \times 10^{-3}$ N-m]

**4.** A weight of 600 N suspended by a light rope wound round a pulley of weight 400 N and radius 600 mm. The other end of the rope is fixed to the periphery of the pulley. Find the acceleration of the descending weight and the pull in the rope. Find also the velocity after the weight has descended 10 m. Take $g = 9.8$ m/s$^2$.

[**Ans.** 7.357 m/s$^2$; 150 N; 12.13 m/s]

**5.** A solid cylindrical pulley of weight 800 N, having 800 mm radius of gyration and 2 m diameter, is rotated by an electric motor, which exerts a uniform torque of 6000 N-m. A body of weight 3000 N is to be lifted by a wire wrapped round the pulley. Find: 1. Acceleration of the body; and 2. Tension in the rope.

[**Ans.** 8.38 m/s$^2$; 5562 N]

**6.** Two bodies of weights 1000 N and 400 N are supported by a rope of negligible weight passing over a pulley, as shown in Fig. 19.16. The pulley may be taken as a solid disc. The weight of the pulley is 500 N. Find the linear acceleration of the bodies and tensions $P_1$ and $P_2$ in the two parts of the string.

[**Ans.** 3.567 m/s$^2$; 636.4 N; 545.4 N]

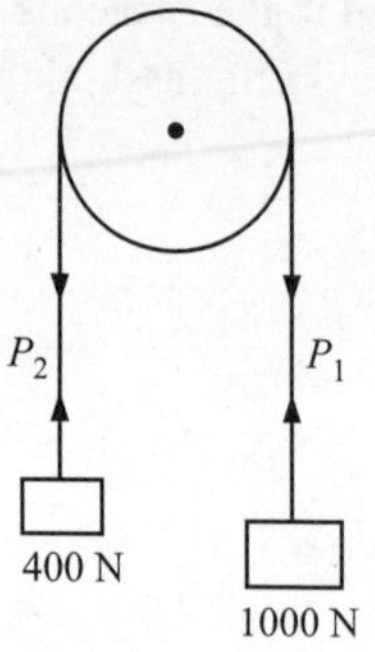


**Fig. 19.16**

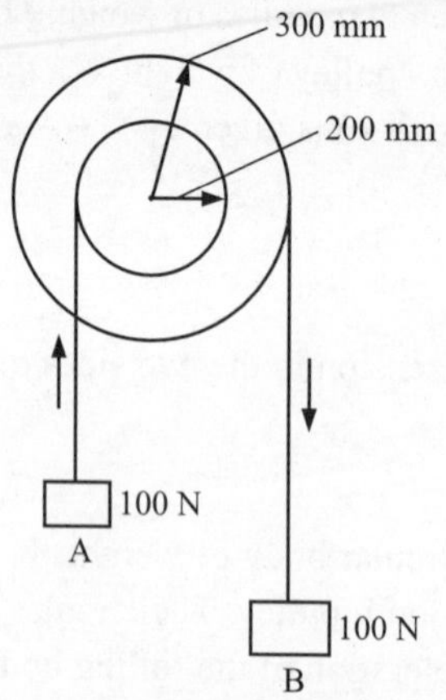


**Fig. 19.17**

7. Two blocks $A$ and $B$ each of 100 N are suspended from a compound pulley, as shown in Fig. 19.17. The weight of the pulley is 300 N and its radius of gyration may be taken as 0.25 m. Determine:
   1. Angular acceleration of the pulley;
   2. Linear acceleration of the blocks $A$ and $B$; and
   3. Tension in the string. [**Ans.** 3.09 rad/s$^2$; 0.927 m/s$^2$, 0.618 m/s$^2$; 106.31 N, 90.54 N]
8. A solid right circular roller $B$ of weight 150 N lies on a rough horizontal plane. It is being pulled by another hanging body $A$ of weight 150 N. Assuming that the string connecting the roller $B$ and body $A$ is inextensible and there is no slipping of the roller, find the linear acceleration of the roller. [**Ans.** 2.8 m/s$^2$]
9. A solid sphere of radius $r$ rolls down an inclined plane inclined at 30° with the horizontal. Find the acceleration of the sphere and the minimum coefficient of friction compatible with the rolling motion. [**Ans.** 3.5 m/s$^2$; 0.165]
   [**Hint:** Mass M.I. of solid sphere = $0.4\ mr^2$]
10. A uniform solid cylinder of radius $R$ and weight $W$ starts from rest and rolls down a rough inclined plane, with inclination $\alpha$, without slipping. Find the acceleration of the centre of the cylinder and the time taken to cover a distance $s$ along the plane. $\left[\textbf{Ans. } \frac{2}{3} g \sin\theta ; \sqrt{\frac{3s}{g \sin\theta}}\right]$

## REVIEW QUESTIONS

1. Define torque. Write an expression for the work done by a torque.
2. Explain the laws of motion for a body in rotational motion.
3. What do you understand by angular momentum? Prove that the angular momentum is the product of moment of inertia and angular velocity.
4. Derive a relation between torque and moment of inertia.
5. Show that the kinetic energy of a rotating body is $\frac{1}{2} I \cdot \omega^2$, where $I$ is the mass moment of inertia of the body about its axis of rotation and $\omega$ is the angular speed.
6. Derive a relation for the tension in a string to which a body is attached and passing over a pulley.
7. Two weights $W_1$ and $W_2$ are attached to the two ends of a string passing over a pulley of weight $W_P$. If $W_1$ is greater than $W_2$, then show that the linear acceleration of the system is given by

$$a = \frac{(W_1 - W_2)g}{W_1 + W_2 + \frac{W_P}{2}}$$

**8.** Derive a relation for the linear acceleration of a circular body rolling down a rough inclined plane without slipping.

## OBJECTIVE TYPE QUESTIONS

**1.** The product of mass moment of inertia and the angular velocity of the body is known as
(*a*) torque (*b*) linear momentum
(*c*) angular momentum (*d*) none of these

**2.** The product of mass moment of inertia and the angular acceleration of the body is known as
(*a*) torque (*b*) linear momentum
(*c*) angular momentum (*d*) none of these

**3.** When a body has both linear and angular motion, then total kinetic energy of the body is equal to the .................. of kinetic energy of translation and rotation.
(*a*) difference (*b*) sum

**4.** When two bodies of weights $W_1$ and $W_2$ ($W_1 > W_2$) are attached to the two ends of a string passing over a pulley of weight $W_P$, then the linear acceleration of the system is

(*a*) $\dfrac{W_1 + W_2}{W_1 + W_2 + W_P}$ (*b*) $\dfrac{W_1 + W_2 + 0.5W_P}{(W_1 - W_2)\,g}$

(*c*) $\dfrac{W_1 + W_2 + W_P}{W_1 + W_2}$ (*d*) $\dfrac{(W_1 - W_2)\,g}{W_1 + W_2 + 0.5W_P}$

**5.** When a circular body rolls on a rough inclined plane which makes an angle θ with the horizontal, then linear acceleration of the rolling body is

(*a*) $\dfrac{r^2 \cdot g \sin\theta}{r^2 + k^2}$ (*b*) $\dfrac{r^2 \cdot g \sin\theta}{r^2 - k^2}$

(*c*) $\dfrac{r^2 + k^2}{r^2 \cdot g \sin\theta}$ (*d*) $\dfrac{r^2 - k^2}{r^2 \cdot g \sin\theta}$

***Answers to Objective Type Questions***

**1.** (*c*) **2.** (*a*) **3.** (*b*) **4.** (*d*) **5.** (*a*)

# 20

# MECHANICAL VIBRATIONS



## 20.1 INTRODUCTION

A body is said to vibrate if it has a to-and-fro (or up and down) motion. When any elastic body such as spring, shaft or beam is displaced at the other end from its equilibrium position by the application of external forces and then released, it starts cyclic motion. Such cyclic motion due to elastic deformation under the action of external forces is known as *vibration*.

Consider a body which is held in equilibrium position by elastic constraints. Whenever the body is displaced from its equilibrium position by the external forces, work is done on the elastic constraints of the body, and thus the energy due to internal forces is stored as elastic or strain energy. It may be noted that when the external force is removed and the body is released, the internal forces causes the body to move towards the equilibrium position. Considering the motion to be frictionless, the strain energy stored in the body is converted into kinetic energy during the period when the body reaches equilibrium position. At equilibrium position, the body has the maximum kinetic energy and due to which the body continues to move in the opposite direction. Now, when the body passes through the equilibrium position, the whole kinetic energy is utilised to overcome the elastic forces and is stored in the form of strain energy. In this way, the vibratory motion is repeated indefinitely.

**Note:** A vibrating system involves the transfer of strain energy to kinetic energy and vice versa in alternating way.

## 20.2 TECHNICAL TERMS USED IN VIBRATIONS

The following technical terms related to vibratory motion are important from the subject point of view:

(*i*) ***Amplitude*** It is the maximum displacement of the body on either side of its mean position. It is measured in metres.

(*ii*) ***Period of vibration or time period*** It is the time taken by a motion to repeat itself or it is the time required to complete one cycle. It is measured in seconds.

***(iii)* Cycle** It is the motion completed during one time period.

***(iv)* Frequency** It is the number of cycles completed in one second. It is expressed in hertz (briefly written as Hz) which is equal to one cycle per second. It is inverse of the time period.

***(v)* Resonance** When the frequency of the external force is the same as that of the natural frequency of the system, it is said that the system has reached the state of resonance. Resonance results in large amplitudes of vibration and may be very dangerous.

## 20.3 DEGREE OF FREEDOM

The number of independent coordinates required to specify the vibratory motion of a system is known as its *degree of freedom*. A system is said to have *n* degrees of freedom, if it requires *n* independent coordinates to specify completely the configuration of the system at any instant. In general, it is equal to the number of independent displacements that are possible. It may be noted that, this number varies from zero to infinity (∞).

A system where only one independent coordinate is required to specify the configuration of system, then the system is said to have *single degree of freedom*. A single spring mass system as shown Fig. 20.1 (*a*) or simple pendulum oscillating in one plane as shown Fig. 20.1 (*b*) or crank slider mechanism as shown in Fig. 20.1 (*c*), respectively, are the examples of single degree of freedom systems. The single degree of freedom systems are also called as simple systems.

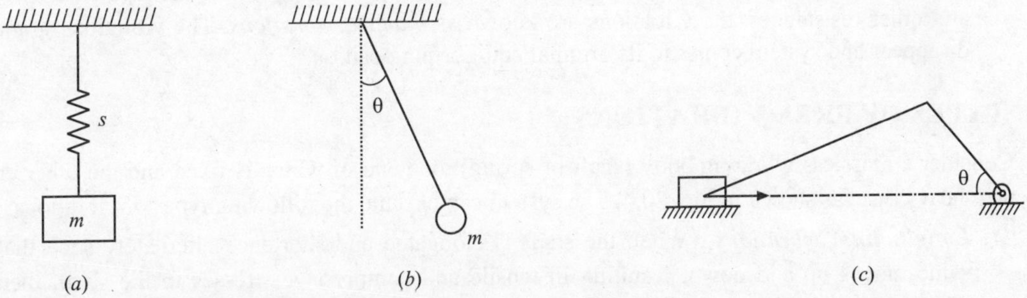


**Fig. 20.1** *Single degree of freedom*

A system where there are two independent coordinates which are required to specify the configuration of system, then the system is said to have *two degrees of freedom*. A two spring, two mass system constrained to move in one direction without rotation as shown in Fig 20.2 (*a*) or a double pendulum as shown in Fig. 20.2 (*b*) are the examples of two degrees of freedom systems.

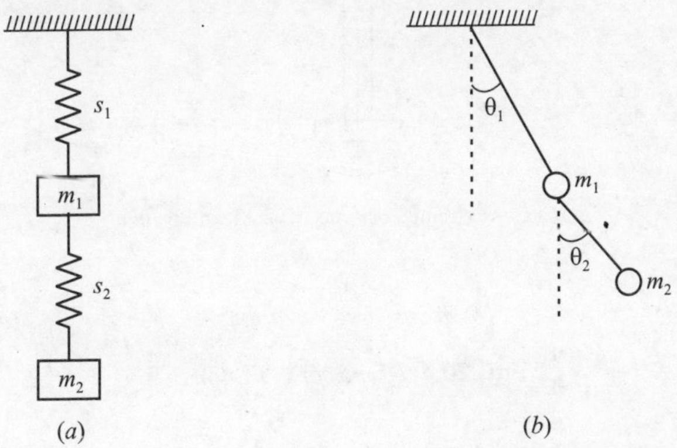


**Fig. 20.2** *Two degrees of freedom*

A system which has continuously distributed mass such as flexible beam stretched between two supports, as shown in Fig. 20.3, has *infinite degrees of freedom*. It may be noted that such system may be considered as equivalent to an infinite number of masses concentrated at different points.

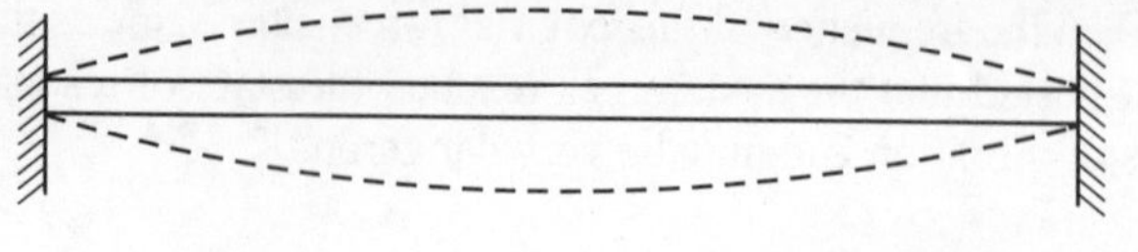

**Fig. 20.3** *Infinite degrees of freedom*

## 20.4 TYPES OF VIBRATIONS

The following types of vibrations are important from the subject point of view:

*(i)* ***Free or natural vibrations*** The elastic vibrations in which no external forces and friction acts on a body, after the initial displacement, are known as *free* or *natural vibrations.*

*(ii)* ***Forced vibrations*** When a repeated external force (*i.e.* periodic disturbing force) acts continuously on a body, the vibrations are known as *forced vibrations.* The frequency of the vibrations is same as that of the applied force and is independent of their own natural frequency of vibrations.

*(iii)* ***Damped vibrations*** When the energy of a vibrating system is gradually dissipated by friction and other resistances, the vibrations are known as *damped vibrations*. The vibrations gradually disappear and system comes to its original equilibrium position.

## 20.5 TYPES OF FREE VIBRATIONS

Consider a massless vibrating body (shaft or spring), one end of which is fixed and the other end is carrying a heavy disc, as shown in Fig. 20.4. The system can execute the following types of vibrations:

1. ***Longitudinal vibrations*** When the shaft is elongated and shortened alternately such that the same moves up and down, resulting in tensile and compressive stresses in the shaft, then the vibrations are known as *longitudinal vibrations*. In this case, the different particles of the shaft move parallel to the axis of shaft, as shown in Fig. 20.4 (*a*).

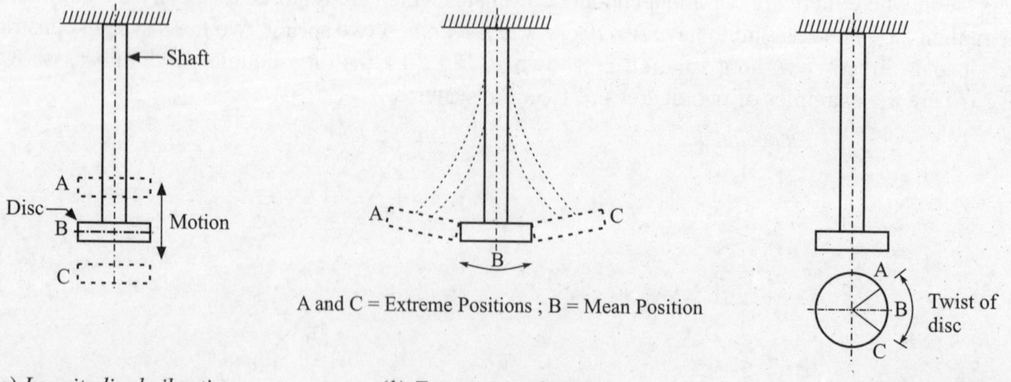


(*a*) *Longitudinal vibrations* (*b*) *Transverse vibrations* (*c*) *Torsional vibrations*

**Fig. 20.4** *Types of vibrations*

2. ***Transverse vibrations*** When the shaft is straight and bent alternately such that the tensile and compressive stresses are induced as a result of bending, then the vibrations are known as *transverse*

*vibrations*. In this case, the different particles of the shaft move perpendicular to the axis of shaft in the longitudinal direction, as shown in Fig. 20.4 (*b*).

3. ***Torsional vibrations*** When the shaft is twisted and untwisted alternately such that the torsional shear stresses are induced, then the vibrations are known as *torsional vibrations*. In this case, the different particles of the shaft move in a circle about the axis of the shaft, as shown in Fig. 20.4 (*c*).

## 20.6 NATURAL FREQUENCY OF FREE LONGITUDINAL VIBRATIONS

The natural frequency of the free longitudinal vibrations may be found out by any of the following methods:

**1. *Equilibrium method (Newton's method)***

Consider a helical spring of *negligible mass suspended vertically from a rigid support, as shown in Fig. 20.5 (*a*). The other end of the spring is free.

If a body of mass $m$ is suspended from the free end, the spring is stretched by a distance $\delta$, as shown in Fig. 20.5 (*b*). Thus, $\delta$ is the static deflection of the spring under the weight of mass $m$.

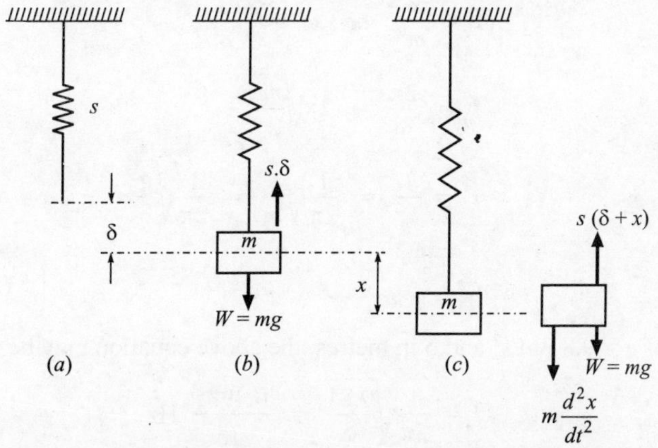


**Fig. 20.5** *Natural frequency of free longitudinal vibrations*

Let $s$ = Stiffness of the spring, *i.e.* force required to produce unit displacement in the direction of vibration.

In the static equilibrium position, as shown in Fig. 20.5 (*b*), the downward force due to the weight ($W = mg$) of the suspended mass is balanced by the upward force of the spring, such that

$$W = s \cdot \delta \quad \text{or} \quad mg = s \cdot \delta \qquad \text{....}(i)$$

Now, the mass is pulled in downward direction through a distance $x$, as shown in Fig. 20.5 (*c*), and is then released, therefore after some time $t$,

Restoring force = Downward Force – Upward force of spring

...(Direction sign for upward force from the equilibrium position is taken as negative)

$$= W - s(\delta + x) = mg - s \cdot \delta - s.x \qquad ...(\because W = mg)$$

$$= s \cdot \delta - s \cdot \delta - s \cdot x = -s \cdot x \qquad \text{...[ From equation } (i)] \quad ...(ii)$$

We know that according to Newton's second law of motion,

Accelerating force = Mass × Acceleration

$$= m \times \frac{d^2x}{dt^2} \qquad ...\left[\because \text{Acceleration} = \frac{dv}{dt} = \frac{d}{dt}\left(\frac{dx}{dt}\right) = \frac{d^2x}{dt^2}\right] \quad ...(iii)$$

* A spring is considered of negligible mass only if its mass is very less as compared to the suspended mass.

The equation of motion of the body of mass $m$ after time $t$ is given by equations (*ii*) and (*iii*). Thus,

$$m \times \frac{d^2x}{dt^2} = -s \cdot x \quad \text{or} \quad m \times \frac{d^2x}{dt^2} + s \cdot x = 0$$

$$\therefore \qquad \frac{d^2x}{dt^2} + \frac{s \cdot x}{m} = 0 \qquad \ldots(iv)$$

The above equation (*iv*) is the second order linear differential equation of motion for a single degree of freedom for a spring-mass system having free vibrations.

We know that the fundamental equation of simple harmonic motion is

$$\frac{d^2x}{dt^2} + \omega^2 \cdot x = 0 \qquad \ldots(v)$$

where $\omega$ = Angular velocity.

Comparing equation (*iv*) and (*v*), we have

$$\omega^2 = \frac{s}{m} \quad \text{or} \quad \omega = \sqrt{\frac{s}{m}} \qquad \ldots(vi)$$

$\therefore$ Time period,

$$t_p = \frac{2\pi}{\omega} = 2\pi\sqrt{\frac{m}{s}} \qquad \ldots(vii)$$

and natural frequency,

$$f_n = \frac{1}{t_p} = \frac{1}{2\pi}\sqrt{\frac{s}{m}} = \frac{1}{2\pi}\sqrt{\frac{g}{\delta}} \qquad \ldots(viii)$$

$$\ldots\left[\because \text{From equation } (i), \ \frac{s}{m} = \frac{g}{\delta}\right]$$

Taking the value of $g = 9.81$ m/s$^2$ and $\delta$ in metres, the above equation may be written as,

$$f_n = \frac{1}{2\pi}\sqrt{\frac{9.81}{\delta}} = \frac{0.4985}{\sqrt{\delta}} \text{ Hz} \qquad \ldots(ix)$$

It may be noted that the value of static deflection $\delta$ (static deflection of the spring produced by the load $W$) may be found out from the given conditions of the problem. For longitudinal vibrations, it may be obtained from relation,

$$E = \frac{\text{Stress}}{\text{Strain}} = \frac{\left(\frac{W}{A}\right)}{\left(\frac{\delta}{l}\right)} = \frac{W}{A} \times \frac{l}{\delta} \quad \text{or} \quad \delta = \frac{W.l}{E.A}$$

where $W$ = Weight suspended at the free end,

$l$ = Length of the spring or shaft,

$E$ = Young's modulus of the material of the spring or shaft, and

$A$ = Cross-sectional area.

**Note:** When the mass of the spring $m_1$ and the mass of the disc $m$ at the end is given, then the natural frequency of vibration,

$$f_n = \frac{1}{2\pi}\sqrt{\frac{s}{m + \frac{m_1}{3}}}$$

**2. *Energy method***

In the free vibrations, the total mechanical energy (*i.e.* sum of *kinetic energy and potential energy) remains constant. In other words,

$$K.E. + P.E. = \text{Constant} \quad \text{or} \quad \frac{d}{dt}(K.E. + P.E.) = 0 \qquad ...(i)$$

We know that kinetic energy,

$$K.E. = \frac{1}{2}mv^2 = \frac{1}{2}m\left(\frac{dx}{dt}\right)^2 \qquad ...(ii)$$

and potential energy,

$$P.E. = \text{Mean force} \times \text{Displacement}$$

$$= \left(\frac{0 + s.x}{2}\right)x = \frac{1}{2}s.x^2 \qquad ...(iii)$$

Substituting the value of kinetic energy. (*K.E.*) and potential energy (*P.E.*) in equation (*i*), we have

$$\frac{d}{dt}\left[\frac{1}{2}m\left(\frac{dx}{dt}\right)^2 + \frac{1}{2}s.x^2\right] = 0$$

$$\frac{1}{2}m \times 2 \times \frac{dx}{dt} \times \frac{d^2x}{dt^2} + \frac{1}{2}s \times 2x \times \frac{dx}{dt} = 0$$

or

$$m \times \frac{d^2x}{dt^2} + s \cdot x = 0 \quad \text{or} \quad \frac{d^2x}{dt^2} + \frac{s \times x}{m} = 0 \qquad ...(iv)$$

The above equation (*iv*) is same as obtained in the previous method. Therefore, the time period and the natural frequency may be calculated as discussed in the previous method.

**3. *Rayleigh's method***

In this method, the maximum kinetic energy at the mean position (where potential energy is zero) is equal to the maximum potential energy (or strain energy) at the extreme position (where kinetic energy is zero). Assuming the motion executed by the vibration to be simple harmonic, then

$$x = X \sin \omega \cdot t \qquad ...(i)$$

where $x$ = Displacement of the body having simple harmonic motion from the mean position after time $t$ seconds, and

$X$ = Maximum displacement of the body from the mean position to the extreme position.

Now, differentiating equation (*i*), we get

$$\frac{dx}{dt} = \omega \times X \cos(\omega \cdot t)$$

Since at the mean position, $t = 0$, therefore maximum velocity at the mean position,

$$v \frac{dx}{dt} = \omega \cdot X$$

Maximum kinetic energy at the mean position

$$= \frac{1}{2} \times m \cdot v^2 = \frac{1}{2} \times m \cdot \omega^2 \cdot X^2 \qquad ...(ii)$$

* *Kinetic energy* is due to the motion of the body.

*Potential energy* is taken with respect to a certain datum position, which is equal to the amount of work required to move the body from the datum position. In case of vibrations, the datum position is taken as equilibrium position at which the potential energy of the body is taken as zero.

and maximum potential energy at mean position,

$$= \left(\frac{0 + s \cdot X}{2}\right) X = \frac{1}{2} \times s.X^2 \qquad ...(iii)$$

Equating equations (*ii*) and (*iii*), we have

$$\frac{1}{2} \times m \cdot \omega^2 \cdot X^2 = \frac{1}{2} \times s.X^2 \quad \text{or} \quad \omega^2 = \frac{s}{m}\text{; and } \omega = \sqrt{\frac{s}{m}}$$

∴ Time period,

$$t_p = \frac{2\pi}{\omega} = 2\pi\sqrt{\frac{m}{s}} \qquad \text{...(same as before)}$$

and natural frequency,

$$f_n = \frac{1}{t_p} = \frac{\omega}{2\pi} = \frac{1}{2\pi}\sqrt{\frac{s}{m}} \qquad \text{...(same as before)}$$

It may be noted that in all the above expressions, $\omega$ is known as natural circular frequency and is generally denoted by $\omega_n$.

The following points must be noted:

1. In *vertical vibrating system* as discussed above, the system vibrates about the static equilibrium position, assumed by the mass after suspension.
2. *In horizontal vibrating system,* as shown in Fig. 20.6, the mass moves on a horizontal surface and the equilibrium position corresponds to the natural length of the spring. Thus, the gravity has no effect on its motion and the system vibrates about the original position. The equation of motion in this case is same as that of vertical system.

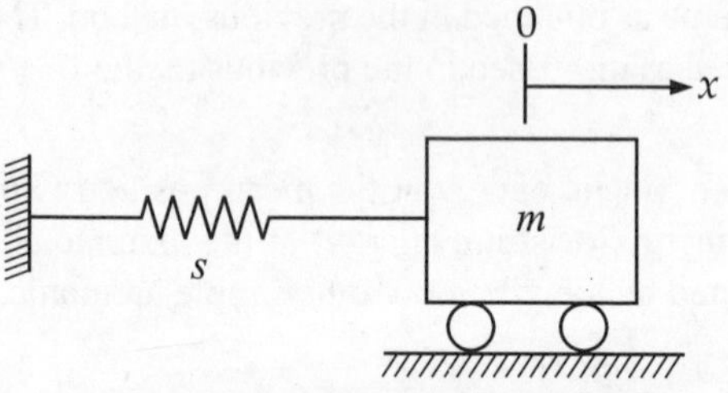


**Fig. 20.6**

**Example 20.1** *A mass of 0.5 kg is suspended by a spring having stiffness of 800 N/m. Calculate the static deflection of the spring and natural frequency of the longitudinal vibrations of the system.*

***Solution*** Given : $m = 0.5$ kg ; $s = 800$ N/m

***Static deflection of the spring***

Let $\delta$ = Static deflection of the spring.

We know that in static equilibrium position,

$$mg = s.\delta$$

$$\therefore \quad \delta = \frac{mg}{s} = \frac{0.5 \times 9.81}{800} = 6.131 \times 10^{-3}\ \text{m} = 6.131\ \text{mm} \quad \textbf{Ans.}$$

...(Taking $g = 9.81$ m/s$^2$)

***Natural frequency of the longitudinal vibrations***

We know that natural frequency of the longitudinal vibrations of the system,

$$f_n = \frac{0.4985}{\sqrt{\delta}} = \frac{0.4985}{\sqrt{6.131 \times 10^{-3}}} = \frac{0.4985}{0.0783} = 6.367\ \text{Hz} \quad \textbf{Ans.}$$

**Example 20.2** *A vertical shaft is fixed at the top and carries a disc of mass 500 kg at its free end. The diameter of the shaft is 100 mm and its length is 1 m. The modulus of elasticity is 200 GN/m². Neglecting the mass of the shaft, determine the frequency of longitudinal vibrations.*

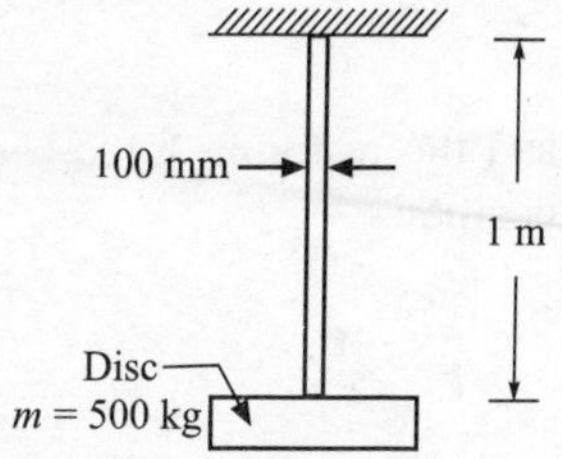


Fig. 20.7

***Solution*** Given : $m$ = 500 kg ; $d$ = 100 mm = 0.1 m ; $l$ = 1 m ; $E = 200 \text{ GN/m}^2 = 200 \times 10^9 \text{ N/m}^2$

The shaft system is shown in Fig. 20.7.

We know that cross-sectional area of the shaft,

$$A = \frac{\pi}{4} \times d^2 = \frac{\pi}{4}(0.1)^2$$

$$= 7.854 \times 10^{-3} \text{ m}^2$$

We know that static deflection the shaft,

$$\delta = \frac{Wl}{AE} = \frac{500 \times 9.81 \times 1}{7.854 \times 10^{-3} \times 200 \times 10^9} = 3.123 \times 10^{-6} \text{ m}$$

...($\because W = mg$)

$\therefore$ Natural frequency of longitudinal vibrations,

$$f_n = \frac{0.4985}{\sqrt{\delta}} = \frac{0.4985}{\sqrt{3.123 \times 10^{-6}}} = 282.1 \text{ Hz} \quad \textbf{Ans.}$$

**Example 20.3** *A vertical U-tube manometer open to atmosphere at both ends contains a column of length l of certain liquid of mass density ρ, as shown in Fig. 20.8. A sudden increase of pressure on one column forces the level of liquid down. When the pressure is released liquid starts vibrating. Determine the equation of vibration by equilibrium method and energy method.*

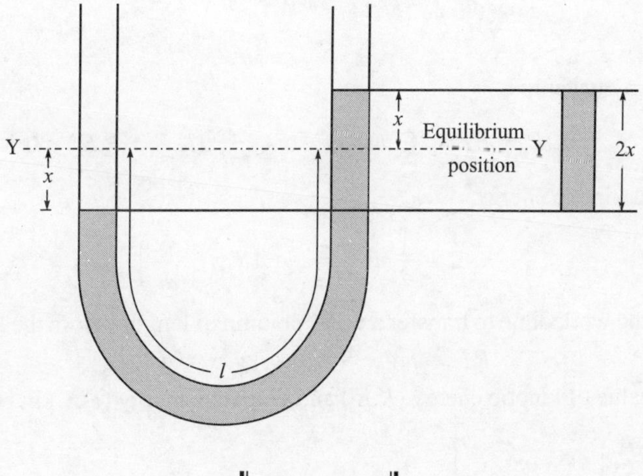


Fig. 20.8

***Solution***

Let $A$ = Cross-sectional area of tube in m²,

$\rho$ = Mass density of liquid in kg/m³, and

$l$ = Total length of liquid column in m.

***Equilibrium method***

The static equilibrium position of the liquid column is when these are in level, as shown in figure. It may be noted that at any instant during the vibration, one column rises by a distance $x$ while the other falls by the same distance.

We know that accelerating force (*i.e.* inertia force) of the mass of the total liquid

$$= \text{Mass} \times \text{Acceleration} = (A.l.\rho) \times \frac{d^2x}{dt^2} \quad ...(i)$$

...(∵ Mass = Volume × Density = Area × Length × Density)

and the restoring force (*i.e.* net force) is equal to the weight of the liquid column above the equilibrium position

= – Mass of the liquid column above *Y-Y* × Acceleration due to gravity

$$= -(A.2x.\ \rho)\, g \quad ...(ii)$$

Equating equations (*i*) and (*ii*), we have

$$(A.l.\rho)\frac{d^2x}{dt^2} = -(A.2x.\ \rho)\, g \quad \text{or} \quad (A.l.\rho)\frac{d^2x}{dt^2} + (A.2x.\ \rho)\, g = 0$$

$$\therefore \quad \frac{d^2x}{dt^2} + \frac{2g}{l} \times x = 0 \ \textbf{Ans.}$$

It may be noted that solving the above differential equation, as done in Sec. 20.7, gives,

$$\omega = \sqrt{\frac{2g}{l}}$$

Time period,
$$t_p = \frac{2\pi}{\omega} = 2\pi\sqrt{\frac{l}{2g}}$$

and natural frequency,
$$f_n = \frac{1}{t_p} = \frac{\omega}{2\pi} = \frac{1}{2\pi}\sqrt{\frac{2g}{l}}$$

***Energy method***

We know that at any instant,

$$K.E. + P.E. = \text{Constant} \quad \text{or} \quad \frac{d}{dt}(K.E. + P.E.) = 0 \quad ...(i)$$

We also know that kinetic energy,

$$K.E. = \frac{1}{2}mv^2 = \frac{1}{2}(A.l.\rho)\left(\frac{dx}{dt}\right)^2 \quad ...(ii)$$

and potential energy is the work done to transfer a water column of length *x* from the left side to the right side,

$$P.E. = mg.x = (A.x.\rho)\, g.\, x = (A.\rho.g)\, x^2 \quad ...(iii)$$

Substituting the value of kinetic energy (*K.E.*) and potential energy (*P.E.*) in equation (*i*), we get

$$\frac{d}{dt}\left[\frac{1}{2}A.l.\rho\left(\frac{dx}{dt}\right)^2 + A.\rho.g.\,x^2\right] = 0$$

$$\frac{1}{2}A.l.\rho \times 2\left(\frac{dx}{dt}\right)\left(\frac{d^2x}{dt^2}\right) + A.\rho.g.\,2x\left(\frac{dx}{dt}\right) = 0$$

$$\therefore \quad \frac{d^2x}{dt^2} + \frac{2g}{l} \times x = 0 \ \textbf{Ans.}$$

**Note:** From the expression of time period derived above, it may be seen that the time period is independent of the effects of density and the area of the tube (or diameter of the tube). Therefore, if the fluid is changed or diameter is increased or decreased, then there will be no effect on time period of the vibration.

## 20.7 NATURAL FREQUENCY OF FREE TRANSVERSE VIBRATIONS

Consider a shaft of negligible mass whose one end is fixed and on the other end carries a body of weight $W$ is attached, as shown in Fig. 20.9. The distance $\delta$ is the static deflection due to the weight of the body.

Let $\quad s$ = Stiffness of the shaft, and

$\quad x$ = Displacement of the body from mean position after time $t$.

We have already discussed in the previous article that,

$$\text{Restoring force} = -s\cdot x \qquad \text{...}(i)$$

and

$$\text{Accelerating force} = m\times\frac{d^2x}{dt^2} \qquad \text{...}(ii)$$

Equating equations (*i*) and (*ii*), the equation of motion becomes,

$$m\times\frac{d^2x}{dt^2} = -s\cdot x$$

or

$$m\times\frac{d^2x}{dt^2} + s\cdot x = 0$$

$\therefore$

$$\frac{d^2x}{dt^2} + \frac{s}{m}\times x = 0 \qquad \text{...}(iii)$$

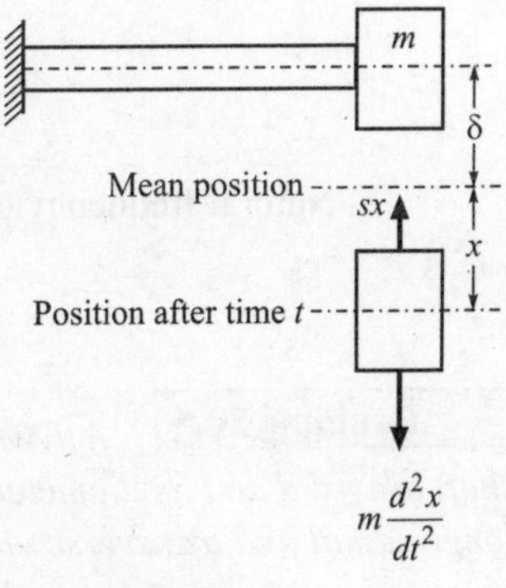


**Fig. 20.9** *Natural frequency of free transverse vibration*

The above equation (*iii*) is same as before.

Hence, the time period and natural frequency of transverse vibrations will be same as that of longitudinal vibrations. Therefore,

Time period, $$t_p = \frac{2\pi}{\omega} = 2\pi\sqrt{\frac{m}{s}}$$

and natural frequency, $$f_n = \frac{1}{t_p} = \frac{1}{2\pi}\sqrt{\frac{s}{m}} = \frac{1}{2\pi}\sqrt{\frac{g}{\delta}}$$

In the above equation, $\delta$ is the deflection of the shaft under the load $W$ when the shaft is treated as a cantilever. Thus, the following two situations are important from the subject point of view:

**(*a*)** The static deflection $\delta$ of the cantilever beam loaded at free end is given by

$$\delta = \frac{Wl^3}{3EI}$$

where $\quad l$ = Length of the shaft,

$E$ = Young's modulus for the material of the shaft, and

$I$ = Moment of inertia of the shaft.

**(*b*)** The static deflection $\delta$ of the fixed beam with an eccentric load is given by

$$\delta = \frac{Wa^3b^3}{3EIl^3} \quad \text{(at the point of load)}$$

where $\quad a$ and $b$ = Distances of the point load from the fixed end, respectively.

**Example 20.4** *A vertical shaft is fixed at the top and carries a disc of mass 100 kg at its free end. The diameter of the shaft is 100 mm and its length is 1 m. The modulus of elasticity is 200 GN/m². Neglecting the mass of the shaft, determine the natural frequency of transverse vibrations.*

***Solution*** Given : $m$ = 100 kg ; $d$ = 100 mm = 0.1 m ; $l$ = 1 m ; $E$ = 200 GN/m² = 200 × 10⁹ N/m²

The shaft system is shown in Fig. 20.10.

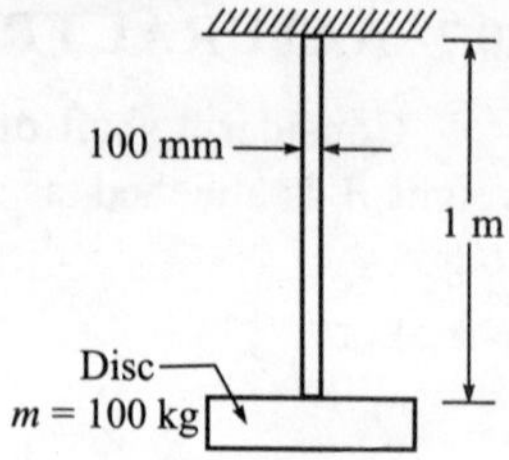


Fig. 20.10

We know that moment of inertia of the shaft,

$$I = \frac{\pi}{64} \times d^4 = \frac{\pi}{64}(0.1)^4$$

$$= 4.909 \times 10^{-6}\text{ m}^4$$

We know that static deflection of the shaft,

$$\delta = \frac{Wl^3}{3EI} = \frac{100 \times 9.81 \times 1^3}{3 \times 200 \times 10^9 \times 4.909 \times 10^{-6}}$$

$$= 0.333 \times 10^{-3}\text{ m}$$

∴ Natural frequency of transverse vibration,

$$f_n = \frac{0.4985}{\sqrt{\delta}} = \frac{0.4985}{\sqrt{0.333 \times 10^{-3}}} = 27.32 \text{ Hz } \textbf{Ans.}$$

**Example 20.5** *A flywheel is mounted on a vertical shaft, as shown in Fig. 20.11. The both ends of the shaft are fixed and its diameter is 50 mm. The flywheel has a mass of 500 kg. Find the natural frequencies of longitudinal and transverse vibrations. Take E = 200 GN/m².*

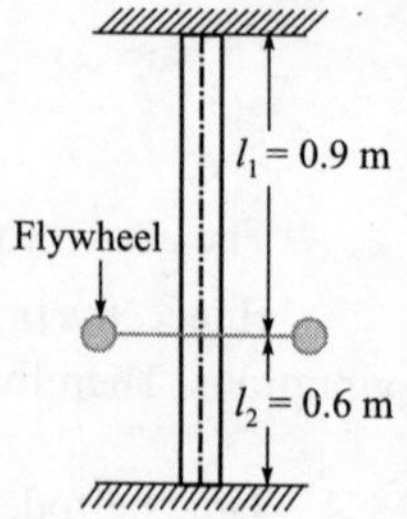


Fig. 20.11

***Solution*** Given : $d = 50$ mm $= 0.05$ m ; $m = 500$ kg ; $E = 200$ GN/m² $= 200 \times 10^9$ N/m²

We know that cross-sectional area of shaft,

$$A = \frac{\pi}{4} \times d^2 = \frac{\pi}{4}(0.05)^2 = 1.96 \times 10^{-3}\text{ m}^2$$

and moment of inertia of shaft,

$$I = \frac{\pi}{4} \times d^4 = \frac{\pi}{64}(0.05)^4 = 0.307 \times 10^{-6}\text{ m}^4$$

***Natural frequency of longitudinal vibration***

Let $m_1$ = Mass of flywheel carried by the length $l_1$.

∴ $m - m_1$ = Mass of flywheel carried by length $l_2$.

We know that extension of length $l_1$

$$= \frac{W_1 \cdot l_1}{A.E} = \frac{m_1 \cdot g \cdot l_1}{A.E.} \quad ...(i)$$

Similarly, compression of length $l_2$

$$= \frac{(W - W_1)l_2}{A.E} = \frac{(m - m_1)\, g.l_2}{A.E} \quad ...(ii)$$

Since extension of length $l_1$ must be equal to compression of length $l_2$, therefore equating equations (*i*) and (*ii*), we have

$$m_1 \cdot l_1 = (m - m_1)l_2$$

$$m_1 \times 0.9 = (500 - m_1)\,0.6 = 300 - 0.6\, m_1 \quad \text{or} \quad m_1 = 200 \text{ kg}$$

∴ Extension of length $l_1$,

$$\delta = \frac{m_1 \cdot g \cdot l_1}{A.E} = \frac{200 \times 9.81 \times 0.9}{1.96 \times 10^{-3} \times 200 \times 100^9} = 4.5 \times 10^{-6}\text{ m}$$

We know that natural frequency of longitudinal vibration,

$$f_n = \frac{0.4985}{\sqrt{\delta}} = \frac{0.4985}{\sqrt{4.5 \times 10^{-6}}} = 235 \text{ Hz } \textbf{Ans.}$$

***Natural frequency of transverse vibration***

We know that the static deflection for a shaft fixed at both ends and carrying a point load is given by

$$\delta = \frac{Wa^3b^3}{3EIl^3} = \frac{500 \times 9.81(0.9)^3(0.6)^3}{3 \times 200 \times 10^9 \times 0.307 \times 10^{-6}(1.5)^3} = 1.24 \times 10^{-3}\ \text{m}$$

...(Substituting $W = m.g$; $a = l_1$, and $b = l_2$)

We know that natural frequency of transverse vibration,

$$f_n = \frac{0.4985}{\sqrt{\delta}} = \frac{0.4985}{\sqrt{1.24 \times 10^{-3}}} = 14.24\ \text{Hz}\ \textbf{Ans.}$$

## 20.8 NATURAL FREQUENCY OF FREE TORSIONAL VIBRATIONS

Consider a shaft of length $l$ rigidly fixed at its upper end and carries a heavy disc of mass $m$ at its lower end, as shown in Fig. 20.12. The shaft is assumed to be of negligible mass. When the disc is displaced slightly in the angular manner about the axis of the shaft and then released, it will start oscillating about the axis and will execute *torsional vibrations.*

Let $\theta$ = Angular displacement of the disc from its mean position after time $t$,

$q$ = Torsional stiffness of the shaft or torque required to twist the shaft per radian within elastic limits,

$I$ = Mass moment of inertia of disc = $m.k^2$, and

$k$ = Radius of gyration.

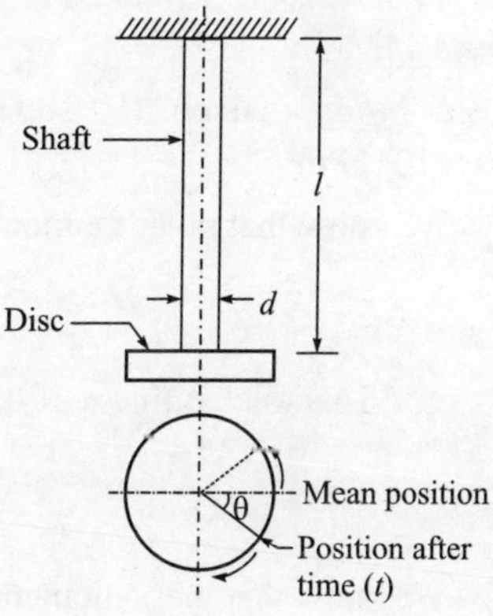


**Fig. 20.12** *Natural frequency of free torsional vibration*

$\therefore$ Restoring torque $= -q \cdot \theta$ ...(*i*)

(–ve sign is used as restoring torque acts opposite to the angular displacement)

and Accelerating torque $= I \times \dfrac{d^2\theta}{dt^2}$ ...(*ii*)

For equilibrium, equating equations (*i*) and (*ii*), we have

$$I \times \frac{d^2\theta}{dt^2} = -q \cdot \theta$$

or

$$I \times \frac{d^2\theta}{dt^2} + q.\theta = 0$$

$\therefore$

$$\frac{d^2\theta}{dt^2} + \frac{q}{I} \times \theta = 0 \quad \text{...(iii)}$$

The fundamental equation of simple harmonic motion is

$$\frac{d^2\theta}{dt^2} + \omega^2 \cdot x = 0 \quad \text{...(iv)}$$

where $\omega$ = Angular velocity.

Comparing equations (*iii*) and (*iv*), we have

$$\omega^2 = \frac{q}{I} \quad \text{or} \quad \omega = \sqrt{\frac{q}{I}} \quad \text{...(v)}$$

$\therefore$ Time period,

$$t_p = \frac{2\pi}{\omega} = 2\pi\sqrt{\frac{I}{q}} \quad \text{...(vi)}$$

and natural frequency,

$$f_n = \frac{1}{t_p} = \frac{1}{2\pi}\sqrt{\frac{q}{I}} \quad \text{...(vii)}$$

The value of torsional stiffness $q$ may also be obtained from the torsional equation,

$$\frac{T}{J} = \frac{C \cdot \theta}{l} \quad \text{or} \quad \frac{T}{\theta} = \frac{C \cdot J}{l}$$

$$\therefore \qquad q = \frac{C \cdot J}{l} \qquad \left( \because \frac{T}{\theta} = q \right)$$

where $C$ = Modulus of rigidity of the shaft material,

$J$ = Polar moment of inertia of shaft cross-section = $\frac{\pi}{32} d^4$,

where $d$ is the diameter of the shaft, and

$l$ = Length of the shaft.

**Note:** When the mass moment inertia of shaft $I_1$ and the mass moment of inertia of inertia of the disc $I$ at the end is given, then the natural frequency of torsional vibration,

$$f_n = \frac{1}{2\pi} \sqrt{\frac{q}{I + \frac{I_1}{3}}}$$

**Example 20.6** *A shaft of 100 mm diameter and 1 m long has its upper end fixed at the top. At the other end, it carries a disc of mass 500 kg at a radius of gyration of 450 mm. The modulus of rigidity for the shaft material is 80 GN/m². Neglecting the mass of the shaft, determine the frequency of torsional vibrations of the shaft.*

***Solution*** Given : $d$ = 100 mm = 0.1 m ; $l$ = 1 m ; $m$ = 500 kg ; $k$ = 450 mm = 0.45 m ; $C$ = 80 GN/m² = $80 \times 10^9$ N/m²

We know that polar moment of inertia of the shaft,

$$J = \frac{\pi}{32} \times d^4 = \frac{\pi}{32} (0.1)^4 = 9.82 \times 10^{-6} \text{ m}^4$$

$\therefore$ Torsional stiffness of the shaft,

$$q = \frac{C \cdot J}{l} = \frac{80 \times 10^9 \times 9.82 \times 10^{-6}}{1} = 785.6 \times 10^3 \text{ N-m}$$

We know that mass moment of inertia of the shaft,

$$I = m \cdot k^2 = 500 (0.45)^2 = 101.25 \text{ kg-m}^2$$

$\therefore$ Frequency of torsional vibrations,

$$f_n = \frac{1}{2\pi} \sqrt{\frac{q}{I}} = \frac{1}{2\pi} \sqrt{\frac{785.6 \times 10^3}{101.25}} = \frac{88.1}{2\pi} = 14 \text{ Hz} \quad \textbf{Ans.}$$

**Example 20.7** *A brass shaft of 10 mm diameter and 1 m long has its upper end fixed at the top. At the other end, at its centre it carries a steel disc of 300 mm diameter and 30 mm thickness. The density of steel is 7860 kg/m³ and the modulus of rigidity for the brass is 35 GN/m². Neglecting the mass of the shaft, determine the frequency of torsional vibrations of the shaft.*

***Solution*** Given : $d$ = 10 mm = 0.01 m ; $l$ = 1 m ; $D$ = 300 mm = 0.3 m or $R$ = 0.15 m ; $t$ = 30 mm = 0.03 m ; $\rho$ = 7860 kg/ m³ ; $C$ = 35 GN/m² = $35 \times 10^9$ N/m²

We know that polar moment of inertia of the shaft,

$$J = \frac{\pi}{32} \times d^4 = \frac{\pi}{32} (0.01)^4 = 9.82 \times 10^{-10} \text{ m}^4$$

∴ Torsional stiffness of the shaft,

$$q = \frac{C \cdot J}{l} = \frac{35 \times 10^9 \times 9.82 \times 10^{-10}}{1} = 34.37 \text{ N-m}$$

We know that mass of the disc,

$$m = \text{Volume} \times \text{Density} = \left(\frac{\pi}{4} \times D^2 \times t\right)\rho$$

$$= \left(\frac{\pi}{4} \times (0.3)^2 \times 0.03\right) 7860 = 16.67 \text{ kg}$$

∴ Mass moment of inertia of the disc about the axis of the shaft,

$$I = {}^*0.5mR^2 = 0.5 \times 16.67\,(0.15)^2 = 0.\,1875 \text{ kg-m}^2$$

We know that frequency of torsional vibrations,

$$f_n = \frac{1}{2\pi}\sqrt{\frac{q}{I}} = \frac{1}{2\pi}\sqrt{\frac{34.37}{0.1875}} = \frac{13.54}{2\pi} = 2.15 \text{ Hz} \quad \textbf{Ans.}$$

**Example 20.8** *A flywheel is mounted on a vertical shaft of 40 mm diameter, as shown in Fig. 20.13. The both ends of the shaft are fixed. The flywheel has mass of 96 kg and its radius of gyration is 0.4 m. Determine the natural frequency of torsional vibration if the modulus of rigidity for the shaft material is 85 GN/m².*

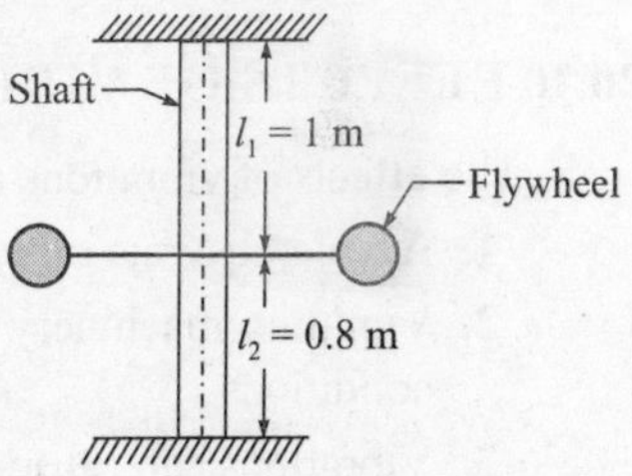


Fig. 20.13

***Solution*** Given : $d = 40$ mm $= 0.04$ m ; $m = 96$ kg ; $k = 0.4$ m ; $C = 85$ GN/m² $= 85 \times 10^9$ N/m² ; $l_1 = 1$ m; $l_2 = 0.8$ m

We know that polar moment of inertia of the shaft,

$$J = \frac{\pi}{32} \times d^4 = \frac{\pi}{32}(0.04)^4$$

$$= 0.25 \times 10^{-6} \text{ m}^4$$

∴ Torsional stiffness of the shaft for length $l_1$,

$$q_1 = \frac{C \cdot J}{l_1} = \frac{85 \times 10^9 \times 0.25 \times 10^{-6}}{1} = 21.25 \times 10^3 \text{ N-m}$$

Similarly, torsional stiffness of the shaft for the length $l_2$,

$$q_2 = \frac{C \cdot J}{l_2} = \frac{85 \times 10^9 \times 0.25 \times 10^{-6}}{0.8} = 26.56 \times 10^3 \text{ N-m}$$

∴ Total torsional stiffness of the shaft,

$$^{**}q = q_1 + q_2 = 21.25 \times 10^3 + 26.56 \times 10^3 = 47.81 \times 10^3 \text{ N-m}$$

We know that mass moment of inertia of the shaft,

$$I = m \cdot k^2 = 96\,(0.4)^2 = 15.36 \text{ kg-m}^2$$

∴ Frequency of torsional vibrations,

$$f_n = \frac{1}{2\pi}\sqrt{\frac{q}{1}} = \frac{1}{2\pi}\sqrt{\frac{47.81 \times 10^3}{15.36}} = \frac{55.8}{2\pi} = 8.88 \text{ Hz} \quad \textbf{Ans.}$$

* Mass moment of inertia of a disc has already been discussed in Chapter 5.

** When shafts are connected in parallel, then total stiffness (similar to spring system),

$$q = q_1 + q_2$$

## 20.9 CAUSES OF VIBRATIONS

The vibrations can result from a number of conditions, acting alone or in combination. The main causes of vibrations are as follows:

1. ***Misalignment/shaft run-out*** The vibrations can result when machine shafts are misaligned. Angular misalignment occurs when the axes of a motor and pump are not parallel. Misalignment can be caused during assembly or develop over time, due to thermal expansion, components shifting or improper reassembly after maintenance.
2. ***Wear*** The worn out moving components such as ball or roller bearings, drive belts or gears, might cause vibrations. A gear tooth that is heavily chipped or worn, or a drive belt that is breaking down, can also produce vibrations.
3. ***Looseness*** It can cause vibrations in both rotating and non-rotating machine components. Looseness is often due to excessive bearing clearances, loose mounting bolts, mismatched parts, corrosion and cracked structures.
4. ***Unbalance*** A heavy spot in a rotating component will cause vibrations when the unbalanced mass rotates around the machine's axis, creating a centrifugal force. Unbalance could be caused by manufacturing defects (machining errors, casting flaws) or maintenance issues. It may be noted that as machine speed increases the effects of unbalance becomes greater. Unbalance can severely reduce bearing life as well as cause undue machine vibrations.

## 20.10 EFFECTS OF VIBRATIONS

The effects of vibrations are as follows:

1. Accelerates rate of wear (*i.e.* reduce bearing life) and reduce equipment's functional life.
2. Vibrating machinery can create noise, cause safety problems and lead to degradation in working conditions.
3. Vibrations can cause machinery to consume excessive power and may damage product quality.
4. Vibration can cause human discomfort in the form of physical and mental strains.

**Note:** Though vibrations have harmful effects as listed above, the vibration phenomenon does some useful work in musical instruments, vibrating screens, shakers etc. The vibrations are also used to relive pain in physiotherapy.

## 20.11 ELIMINATION OF VIBRATIONS

The elimination or reduction of the undesirable vibration can be obtained as follows:

1. Correcting the misalignment, looseness in machine components.
2. Proper balancing of rotating components in machine.
3. Timely replacement of worn out moving parts with excessive clearance.
4. For controlling vibrations in machines or equipment install special mounts (*i.e.* vibration absorbers) designed to absorb the vibrations, in the machine foundation.
5. Isolate the vibrating component with a sub-frame or springs to reduce the energy of vibration from being transferred to the rest of system.

## HIGHLIGHTS

1. When any elastic body such as spring, beam or shaft is displaced from the equilibrium position by the application of external forces, and then released, they execute a vibratory motion.
2. The vibrations are of three types: free or natural vibrations; forced vibrations; and damped vibrations.

3. The number of cycles completed in one second is known as *frequency*. It is expressed in hertz (written as Hz) which is equal to one cycle per second.
4. When the frequency of external force is the same as the natural frequency of the system, resonance takes place.
5. Free vibrations are of three types: longitudinal vibrations; transverse vibrations; and torsional vibrations.
6. The natural frequency of free longitudinal vibrations may be found out by:
   **(*a*)** Equilibrium method; **(*b*)** Energy method; and **(*c*)** Rayleigh's method.
7. The natural frequency ($f_n$) of free longitudinal vibrations and transverse vibrations is

**(*i*)** $$f_n = \frac{1}{2\pi}\sqrt{\frac{s}{m}}$$ ...(in terms of stiffness and mass)

**(*ii*)** $$f_n = \frac{1}{2\pi}\sqrt{\frac{g}{\delta}} = \frac{0.4985}{\sqrt{\delta}}$$ ...(in terms of static deflection)

The value of $\delta$ may be obtained as

**(*a*)** $$\delta = \frac{W \cdot l}{E \cdot A}$$ ...(for longitudinal free vibrations)

**(*b*)** $$\delta = \frac{W\, l^3}{3EI}$$ ...(for transverse vibrations)

8. The natural frequency of torsional vibration is

$$f_n = \frac{1}{2\pi}\sqrt{\frac{q}{I}}$$

where $q$ = Torsional stiffness of the shaft $= \dfrac{T}{\theta} = \dfrac{C \cdot J}{l}$, and

$I$ = Mass moment of inertia of disc attached at the end of the shaft.

## NUMERICAL PROBLEMS

1. A helical spring of negligible mass extends 0.12 mm under a weight of 6 N. It supports a mass of 50 kg. The spring-mass system is displaced vertically through 12 mm and released. Find the natural frequency of longitudinal vibration of the system. **[Ans.** 5.03 Hz]
2. A vertical shaft is fixed at the top and carries a disc of mass 100 kg at its free end. The diameter of the shaft is 50 mm and its length is 300 mm. The modulus of elasticity is 200 GN/m$^2$. Neglecting the mass of the shaft, determine the frequency of longitudinal vibrations of the shaft. **[Ans.** 575 Hz]
3. A truck of mass 3.64 tonnes when empty is observed to settle 46 mm during the loading of 3 tonnes of cargo. What time period of vertical vibration will the truck have on its spring: 1. when loaded; and 2. when empty. **[Ans.** 0.64 s; 0.474 s]
4. A mass of 25 kg is attached to the free end of a spring of stiffness 2500 N/m. The other end of the spring is fixed. Find the natural frequency of vibration when the mass of the spring is 4 kg. **[Ans.** 1.55 Hz]
5. A V-tube manometer of constant cross-sectional area $A$ having liquid of length $l$ is filled with a liquid of density $\rho$, has both ends open to atmosphere as shown in Fig. 20.14. Each arm of manometer is inclined at an angle $\alpha$. Find the natural frequency of vibrations of liquid column when free surface of the liquid is slightly displaced in one arm.

$$\left[\textbf{Ans. } \frac{1}{2\pi}\sqrt{\frac{2g}{l\cos\alpha}}\ \text{Hz}\right]$$

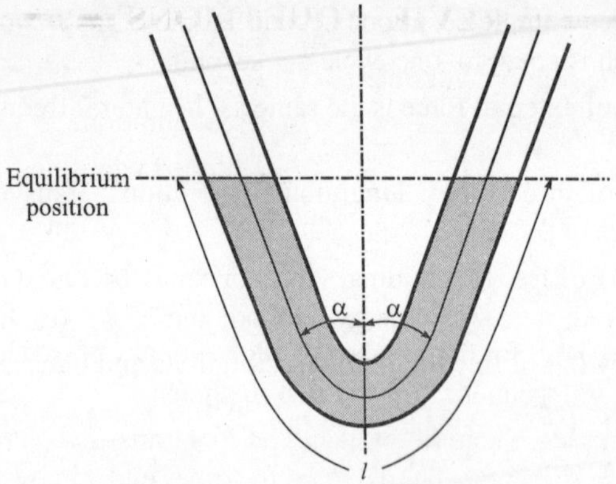


Fig. 20.14

6. A vertical shaft is fixed at the top and carries a disc of mass 200 kg at its free end. The diameter of the shaft is 100 mm and its length is 1 m. The modulus of elasticity is 200 GN/m$^2$. Neglecting the mass of the shaft, determine the frequency of transverse vibrations of the shaft. **[Ans.** 19.05 Hz]
7. A flywheel is mounted on vertical shaft, as shown in Fig. 20.15. The both ends of the shaft are fixed and its diameter is 50 mm. The flywheel has a mass of 500 kg. Find the natural frequencies of longitudinal and transverse vibrations. Take $E = 200$ GN/m$^2$. **[Ans.** 307.5 Hz; 31.74 Hz]

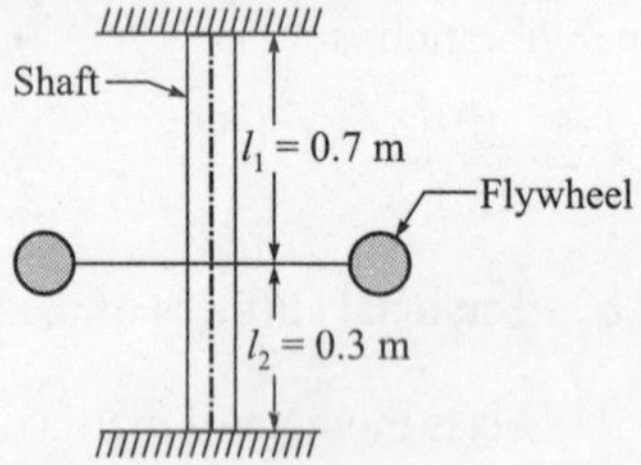


Fig. 20.15

8. A torsional pendulum has a natural frequency of 5 Hz. The inertia of mass fixed at free end is $9.8 \times 10^{-3}$ kg-m$^2$. Determine the length of steel wire of diameter 2 mm which should be used for this pendulum. **[Ans.** 13.4 mm]
9. A vertical shaft is fixed at the top and carries a flywheel of mass 500 kg, welded at its bottom. The radius of gyration of flywheel is 250 mm. The diameter of shaft is 100 mm and its length is 1 m. The modulus of rigidity for the shaft material is 82 GN/m$^2$. Neglecting the mass of the shaft, determine the frequency of torsional vibrations of the shaft. **[Ans.** 25.54 Hz]
10. A vertical shaft 5 mm in diameter and 1 m in length has its upper end fixed to the ceiling. At the lower end, it carries a rotor of diameter 200 mm and mass 2.4 kg. The modulus of rigidity for the material of rotor is 85 GN/m$^2$. Calculate the frequency of torsional vibrations for the system. **[Ans.** 3.32 Hz]
11. A flywheel is mounted on vertical shaft of 100 mm diameter, as shown in Fig. 20.16. The both ends of the shaft are fixed. The flywheel has a mass of 1200 kg and its radius of gyration is 0.65 m. Determine the natural frequency of torsional vibrations if the modulus of rigidity for the shaft material is 84 GN/m$^2$. **[Ans.** 9.26 Hz]




Fig. 20.16

## REVIEW QUESTIONS

**1.** Define and explain the following terms:
(*a*) Vibrations (*b*) Free vibrations
(*c*) Forced vibrations (*d*) Damped vibrations

**2.** Explain the following terms related to vibrations:
(*a*) Period (*b*) Cycle
(*c*) Frequency (*d*) Resonance

**3.** Name the different methods for finding the natural frequency of free longitudinal vibrations. Derive the expression of natural frequency by any two methods.

**4.** Derive the expression for the natural frequency of free transverse vibration for a cantilever shaft of negligible mass whose one end is fixed and on the other end a body of weight $W$ is attached.

**5.** Derive the expression for the natural frequency of free torsional vibrations of the shaft of negligible mass fixed at one end and carrying a disc of mass $m$ on the free end.

**6.** Describe briefly the causes of vibrations in machines, their effects and elimination methods.

## OBJECTIVE TYPE QUESTIONS

**1.** Resonance is a phenomenon in which the frequency of the external exciting force
(*a* ) is half the natural frequency of the system
(*b*) is double the natural frequency of the system
(*c*) coincides with the natural frequency of the system
(*d*) does not coincides with the natural frequency of the system

**2.** In forced vibration, the body vibrates under the influence of
(*a*) an externally applied force (*b*) perpendicular to the axis of the shaft
(*c*) damping force (*d*) its own mass

**3.** In longitudinal vibrations, the differenct particles of shaft moves
(*a*) parallel to the axis of shaft (*b*) perpendicular to the axis of the shaft
(*c*) in a circle about the axis of the shaft (*d*) in an ellipse about the axis of the shaft

**4.** In transverse vibrations, the different particles of shaft moves
(*a*) parallel to the axis of shaft (*b*) perpendicular to the axis of the shaft
(*c*) in a circle about the axis of the shaft (*d*) in an ellipse about the axis of the shaft

**5.** In torsional vibrations, the different particles of shaft moves
(*a*) parallel to the axis of shaft (*b*) perpendicular to the axis of the shaft
(*c*) in a circle about the axis of the shaft (*d*) in an ellipse about the axis of the shaft

**6.** The natural frequency (in Hz) of free longitudinal vibrations is equal to

(*a*) $\frac{1}{2\pi}\sqrt{\frac{s}{m}}$ (*b*) $\frac{1}{2\pi}\sqrt{\frac{g}{\delta}}$

(*c*) $\frac{0.4985}{\sqrt{\delta}}$ (*d*) any one of these

where $m$ = Mass of the body in kg,
$s$ = Stiffness of the body in N/m, and
$\delta$ = Static deflection of the body in metres.

**7.** The natural frequency (in Hz) of free longitudinal vibrations if mass of the spring is taking into account is equal to

(*a*) $\frac{1}{2\pi}\sqrt{\frac{s}{m + m_1}}$ (*b*) $\frac{1}{2\pi}\sqrt{\frac{s}{m + \frac{m_1}{3}}}$

(*c*) $\frac{1}{2\pi}\sqrt{\frac{s}{m + \frac{m_1}{2}}}$ (*d*) none of these

where $m_1$ = Mass of the spring in kg.

**8.** Two vibratory systems are shown in the Fig. 20.17.

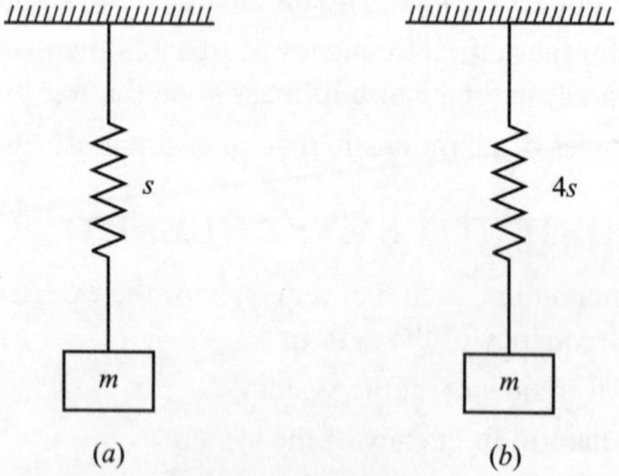


**Fig. 20.17**

The ratio of natural frequency of longitudinal vibration of the second system to that of the first is

(*a*) 4 (*b*) 2 (*c*) 0.5 (*d*) 0.25

**9.** A simple spring mass vibrating system has a natural frequency of *N*. If the spring stiffness is halved and the mass is doubled, then the natural frequency will become

(*a*) *N*/2 (*b*) 2*N* (*c*) 4*N* (*d*) 8*N*

**10.** The natural frequency of free torsional vibrations of a shaft is

(*a*) $2\pi\sqrt{\frac{q}{I}}$ (*b*) $2\pi\sqrt{q \cdot I}$ (*c*) $\frac{1}{2\pi}\sqrt{\frac{q}{I}}$ (*d*) $\frac{1}{2\pi}\sqrt{q.I}$

where $q$ = Torsional stiffness of the shaft, and

$I$ = Mass moment of inertia of the disc attached at the end of the shaft.

***Answers to Objective Type Questions***

| | | | | |
|---|---|---|---|---|
| **1.** (*c*) | **2.** (*a*) | **3.** (*a*) | **4.** (*b*) | **5.** (*c*) |
| **6.** (*d*) | **7.** (*b*) | **8.** (*b*) | **9.** (*a*) | **10.** (*c*) |

21

# FORCES IN SPACE (VECTOR APPROACH)

**1.** *Introduction* **2.** *Representation of a Vector* **3.** *Important Terms Used in Vector Algebra* **4.** *Magnitude of a Vector* **5.** *Multiplication of a Vector by a Scalar Quantity* **6.** *Rectangular Components of a Vector in Three Dimensions* **7.** *Unit Vector* **8.** *Important Vector Operations* **9.** *Vector Addition* **10.** *Properties of Vector Addition* **11.** *Scalar or Dot product* **12.** *Important Mathematical Results of Scalar Product* **13.** *Angle between Two Vectors in Terms of Scalar Product* **14.** *Geometrical Interpretation of Scalar Product* **15.** *Properties of Scalar Product* **16.** *Scalar Product in Terms of Rectangular Components* **17.** *Vector or Cross Product* **18.** *Important Mathematical Results of Vector Product* **19.** *Geometrical Interpretation of Vector Product* **20.** *Properties of Vector Product* **21.** *Vector Product in Terms of Rectangular Components* **22.** *Angle between Two Vectors in Terms of Vector Product* **23.** *Components of a Force in Space* **24.** *Components of a Force based on Two Given Points on Its Line of Action* **25.** *Position Vector of a Given Point* **26.** *Resultant Force of Concurrent Force System* **27.** *Condition for Equilibrium of Forces in Space* **28.** *Moment of a Force*

## 21.1 INTRODUCTION

The physical quantities which have only magnitude and no direction are called as *scalar quantities* or *scalars*. A scalar quantity can be specified by a number along with proper unit. For example: length, mass, time, distance, speed, volume, temperature, etc.

The physical quantities which have both magnitude and direction are called *vector quantities* or *vectors*. A vector quantity is specified by a number with a unit and its direction. For example: displacement, velocity, acceleration, force, moment, torque, etc.

The following points must be noted for vector quantities:

1. They possess both magnitude and direction.
2. They change if either magnitude or direction or both change.
3. They do not obey the ordinary laws of algebra but follow the special laws of vector algebra.

In this chapter, we will first study the fundamentals of vector algebra and then understand the methodology for solving the problems related to forces in space (*i.e.* forces lying in different planes) through vector algebra approach.

## 21.2 REPRESENTATION OF A VECTOR

A vector quantity is represented by a straight line with arrow at one of the ends, as shown in Fig. 21.1 The length of the line gives the magnitude and arrow head placed at one of the ends indicates the direction The direction of the vector may be measured by an angle θ from some known reference direction.

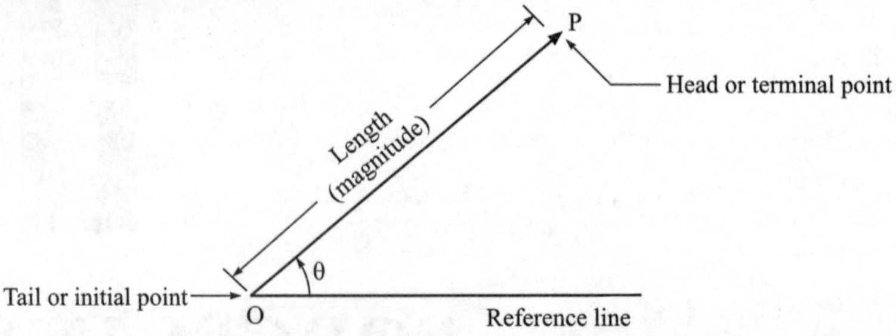


**Fig. 21.1** *Representation of a vector*

The point *P* is called as head or terminal point and point *O* is called as tail or initial point of the vecto $\overrightarrow{OP}$. The length *OP* represents the magnitude of vector $\overrightarrow{OP}$. It may be noted that in simple notation, the vector may also be represented by a single alphabet such as $\vec{A}, \vec{B}$ etc. Sometimes the vectors are represente by bold-faced letters **A, B** etc.

## 21.3 IMPORTANT TERMS USED IN VECTOR ALGEBRA

The following terms of vector quantities are important from the subject point of view:

(*i*) ***Equal vectors*** The vectors which are equal in magnitude and parallel to each other and have the same direction are called as *equal vectors*. In Fig. 21.2, the vectors $\vec{A}$ and $\vec{B}$ are equal vectors.

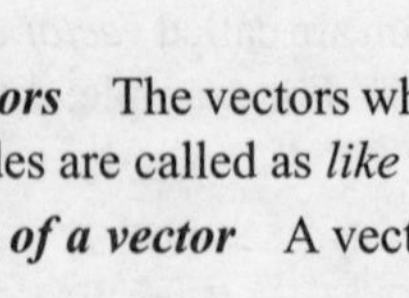


**Fig. 21.2** *Equal vectors*

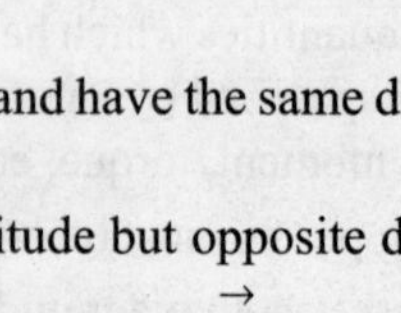


**Fig. 21.3** *Negative of a vector*

(*ii*) ***Like vectors*** The vectors which are parallel to each other and have the same direction but unequa magnitudes are called as *like vectors*.

(*iii*) ***Negative of a vector*** A vector which has the same magnitude but opposite direction is called as *negative of a vector*. In Fig. 21.3, the vector $\vec{B}$ is negative of vector $\vec{A}$.

(*iv*) ***Unit vector*** A vector of unit magnitude drawn in the direction of a given vector is called as a *uni vector*.

(*v*) ***Zero or null vector*** A vector of zero magnitude and an arbitrary direction is called as *zero* or *nul vector*.

**(vi) *Fixed vector*** A vector which has a fixed (specified) point of application is called as a *fixed vector*, as shown in Fig. 21.4. Thus, it is a vector whose point of application cannot be changed without changing the physical significance effect of the vector. For example: force vector.

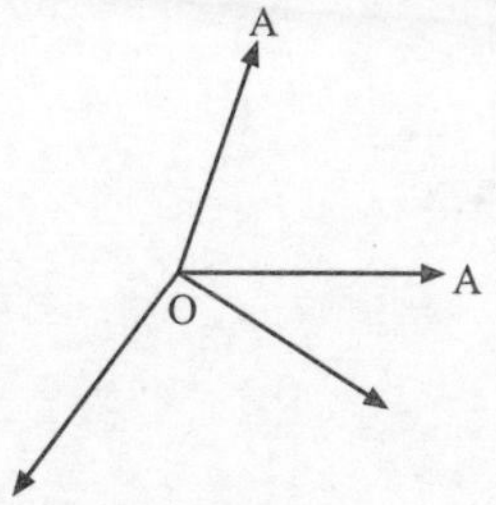


**Fig. 21.4** *Fixed vector*

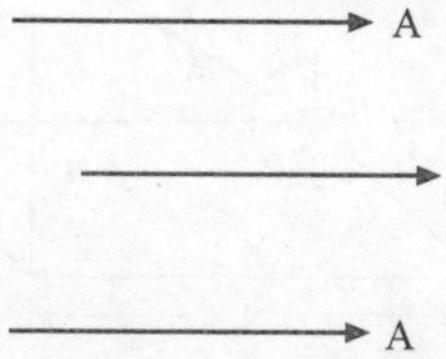


**Fig. 21.5** *Free vector*

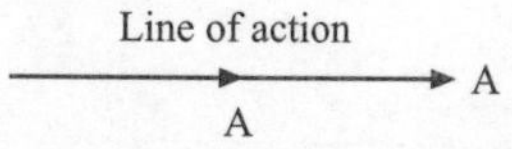


**Fig. 21.6** *Sliding vector*

**(vii) *Free vector*** A vector which can be moved anywhere in space, maintaining its magnitude and direction is called as a *free vector*, as shown in Fig. 21.5. Thus, it is a vector whose action is not confined to or associated with a unique line in space. For example: couple vector.

**(viii) *Sliding vector*** A vector which may be moved (applied) anywhere along the line of action maintaining its magnitude and direction is called as a *sliding vector*, as shown in Fig. 21.6. For example: velocity vector.

## 21.4 MAGNITUDE OF A VECTOR

The magnitude of a vector corresponds to the length of the vector. It is a positive scalar quantity. Mathematically,

$$\text{Magnitude of vector } \vec{A} = \left|\vec{A}\right| = A$$

**Note:** Bold-faced capital letter represents vector (as already discussed in Sec. 21.2) whereas normal capital letter represents the magnitude of vector.

## 21.5 MULTIPLICATION OF A VECTOR BY A SCALAR QUANTITY

When a vector $\vec{A}$ is multiplied by a scalar quantity (real number) $m$, we get another vector $m\vec{A}$. Mathematically,

$$\text{Vector } \vec{A} \text{ multiplied by scalar } m = m\vec{A}$$

**Notes:** **1.** If $m$ is positive, then the direction of $m\vec{A}$ is same as that of $\vec{A}$.

**2.** If $m$ is negative, then the direction of $m\vec{A}$ is opposite as that of $\vec{A}$.

## 21.6 RECTANGULAR COMPONENTS OF A VECTOR IN THREE DIMENSIONS

Consider a vector $\vec{A}$ represented by $\overrightarrow{OP}$, as shown in Fig. 21.7. Taking $O$ as origin, construct a rectangular box (parallelepiped) with its three edges along the three rectangular axes, *i.e.* $X$, $Y$ and $Z$ axes.

The vector $\vec{A}$ in terms of rectangular components may be written as

$$\vec{A} = \vec{A}_x + \vec{A}_y + \vec{A}_z \quad \text{...}(i)$$

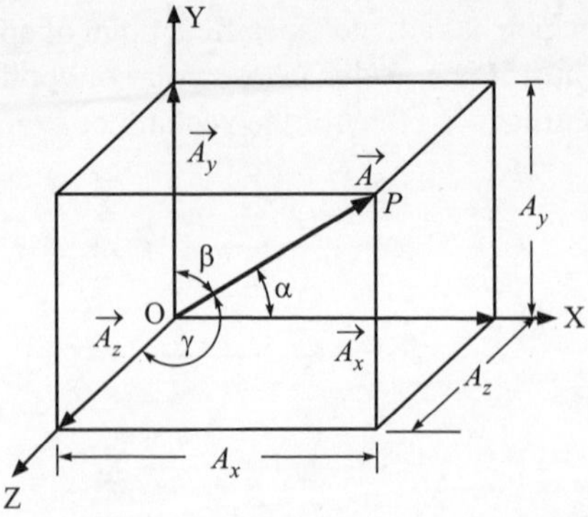


**Fig. 21.7** *Rectangular components of a vector in three dimensions*

Let $\hat{i}$ = Unit vector (vector of unit length) in the positive *X*-direction,

$\hat{j}$ = Unit vector (vector of unit length) in the positive *Y*-direction, and

$\hat{k}$ = Unit vector (vector of unit length) in the positive *Z*-direction.

It may be noted that based on the concept of unit vectors,

$$\vec{A_x} = A_x\hat{i}\,; \qquad \vec{A_y} = A_y\hat{j}\,; \qquad \vec{A_z} = A_z\hat{k}$$

Substituting the values of $\vec{A_x}$, $\vec{A_y}$ and $\vec{A_z}$ in equation (*i*), the vector $\vec{A}$ is given as

$$\vec{A} = A_x\hat{i} + A_y\hat{j} + A_z\hat{k} \qquad ...(ii)$$

where $A_x$, $A_y$ and $A_z$ are the magnitude of the three components of vector $\vec{A}$ in *X*, *Y* and *Z* axes, respectively.

The magnitude of the vector $\vec{A}$,

$$A = \left|\vec{A}\right| = \sqrt{A_x^2 + A_y^2 + A_z^2} \qquad ...(iii)$$

If the vector $\vec{A}$ makes an angle α with *X*-axis, angle β with *Y*-axis, and angle γ with *Z*-axis, then its components are given as

$$A_x = A\cos\alpha; \quad A_y = A\cos\beta; \text{ and } \quad A_z = A\cos\gamma$$

Substituting the values of $A_x$, $A_y$ and $A_z$ in equation (*ii*), the vector $\vec{A}$ is given as

$$\vec{A} = (A\cos\alpha)\hat{i} + (A\cos\beta)\hat{j} + (A\cos\gamma)\hat{k} \qquad ...(iv)$$

## 21.7 UNIT VECTOR

We have already discussed that the vector having unit magnitude drawn in the direction of a given vector is called as a *unit vector*. A unit vector in the direction of a given vector is found by dividing the given vector by its magnitude. It is denoted as $\hat{A}$ and is pronounced as 'A cap'.

Thus, unit vector corresponding to a vector $\vec{A}$ is given as,

$$\text{Unit vector} \quad \text{or} \quad \hat{A} = \frac{\vec{A}}{|\vec{A}|} = \frac{A_x\hat{i} + A_y\hat{j} + A_z\hat{k}}{\sqrt{A_x^2 + A_y^2 + A_z^2}}$$

The magnitude of unit vector is always 1 (unity).

**Example 21.1** *Calculate the magnitude of vector $10\hat{i} + 9\hat{j} + 12\hat{k}$. Also find the unit vector along this vector.*

***Solution*** Given : $\vec{A} = A_x\hat{i} + A_y\hat{j} + A_z\hat{k} = 10\hat{i} + 9\hat{j} + 12\hat{k}$

***Magnitude of vector***

We know that magnitude of vector,

$$|\vec{A}| = \sqrt{A_x^2 + A_y^2 + A_z^2}$$

$$= \sqrt{10^2 + 9^2 + 12^2} = \sqrt{325} = 18.03 \text{ or } 18 \textbf{ Ans.}$$

***Unit vector along this vector***

We know that unit vector along this vector,

$$\hat{A} = \frac{\vec{A}}{|\vec{A}|} = \frac{A_x\hat{i} + A_y\hat{j} + A_z\hat{k}}{|\vec{A}|}$$

$$= \frac{10\hat{i} + 9\hat{j} + 12\hat{k}}{18} = 0.55\hat{i} + 0.5\hat{j} + 0.67\hat{k} \textbf{ Ans.}$$

## 21.8 IMPORTANT VECTOR OPERATIONS

The following are the important vector operations:

*(i)* Vector addition; *(ii)* Scalar or dot product; and *(iii)* Vector or cross product

## 21.9 VECTOR ADDITION

Consider two vectors $\vec{A}$ and $\vec{B}$ in space which are required to be added.

Let $\vec{A} = A_x\hat{i} + A_y\hat{j} + A_z\hat{k}$, and

$$\vec{B} = B_x\hat{i} + B_y\hat{j} + B_z\hat{k}$$

Now, the sum of the two vectors (*i.e.* resultant vector, $\vec{R}$) is obtained by adding their corresponding components.

$$\therefore \quad \vec{R} = \vec{A} + \vec{B} = (A_x\hat{i} + A_y\hat{j} + A_z\hat{k}) + (B_x\hat{i} + B_y\hat{j} + B_z\hat{k})$$

$$= (A_x + B_x)\hat{i} + (A_y + B_y)\hat{j} + (A_z + B_z)\hat{k}$$

It may be noted that, in general

$$\vec{R} = \vec{A} + \vec{B} + ...$$

$$R_x\hat{i} + R_y\hat{j} + R_z\hat{k} = (A_x + B_x + ...)\hat{i} + (A_y + B_y + ...)\hat{j} + (A_z + B_z + ...)\hat{k}$$

## 21.10 PROPERTIES OF VECTOR ADDITION

The important properties of vector addition are as follows:

1. The vector addition is commutative *i.e.*,

$$\vec{A} + \vec{B} = \vec{B} + \vec{A}$$

2. The vector addition is associative *i.e.*,

$$\left(\vec{A} + \vec{B}\right) + \vec{C} = \vec{A} + \left(\vec{B} + \vec{C}\right)$$

## 21.11 SCALAR OR DOT PRODUCT

The scalar or dot product of two vectors $\vec{A}$ and $\vec{B}$ is defined as the magnitude of $\vec{A}$ and $\vec{B}$ and cosine of the angle θ between them. Mathematically,

$$\vec{A}.\vec{B} = \left|\vec{A}\right|\left|\vec{B}\right| \cos\theta = AB\cos\theta$$

**Notes: 1.** As $A$, $B$ and $\cos\theta$ are all scalars, so the dot product of $\vec{A}$ and $\vec{B}$ is scalar quantity.

**2.** Both $\vec{A}$ and $\vec{B}$ have direction but their dot product has no direction.

## 21.12 IMPORTANT MATHEMATICAL RESULTS OF SCALAR PRODUCT

The following mathematical results of scalar product (or dot product) are important from subject point of view:

1. If $\vec{A}$ and $\vec{B}$ are two vectors *perpendicular* or at *right angle* to each other (*i.e.* θ = 90°), then their scalar product is zero.

$$\vec{A}\cdot\vec{B} = AB\cos 90° = 0 \qquad ...(\because \cos 90° = 0)$$

2. If $\vec{A}$ and $\vec{B}$ are two vectors *parallel to each other* and *having same direction* (*i.e.* θ = 0°), then their scalar product has the maximum value.

$$\vec{A}\cdot\vec{B} = AB\cos 0° = AB \qquad ...(\because \cos 0° = 1)$$

3. If $\vec{A}$ and $\vec{B}$ are two vectors *parallel to each other* and *having opposite direction* (*i.e.* θ = 180°), then their scalar product has the maximum negative value.

$$\vec{A}\cdot\vec{B} = AB\cos 180° = AB \qquad ...(\because \cos 180° = -1)$$

## 20.13 ANGLE BETWEEN TWO VECTORS IN TERMS OF SCALAR PRODUCT

The cosine of the angle θ between the vectors $\vec{A}$ and $\vec{B}$ from the scalar product (or dot product) is given by

$$\cos\theta = \frac{\vec{A}\cdot\vec{B}}{|\vec{A}||\vec{B}|} = \frac{\vec{A}\cdot\vec{B}}{AB}$$

or

$$\theta = \cos^{-1}\left(\frac{\vec{A}\cdot\vec{B}}{AB}\right) = \cos^{-1}\left(\frac{A_x B_x + A_y B_y + A_z B_z}{\sqrt{(A_x^2 + A_y^2 + A_z^2)}\,\sqrt{(B_x^2 + B_y^2 + B_z^2)}}\right)^*$$

## 21.14 GEOMETRICAL INTERPRETATION OF SCALAR PRODUCT

We have already discussed in previous article that scalar product (or dot product) of two vectors is given as

$$\vec{A}\cdot\vec{B} = AB\cos\theta \qquad ...(i)$$

The above equation (*i*) may also be written as

$$\vec{A}\cdot\vec{B} = AB\cos\theta = A(B\cos\theta)$$

$$= A \times \text{Magnitude of component of } \vec{B} \text{ in the direction of } \vec{A} \qquad ...(ii)$$

This condition is shown in Fig. 21.8 (*a*).

(*a*) *Project of* $\vec{B}$ *on* $\vec{A}$ (*b*) *Project of* $\vec{A}$ *on* $\vec{B}$

**Fig. 21.8** *Geometrical interpretation of dot product*

Similarly, the equation (*i*) may also be written as

$$\vec{A}\cdot\vec{B} = AB\cos\theta = B(A\cos\theta)$$

$$= B \times \text{Magnitude of component of } \vec{A} \text{ in the direction of } \vec{B} \qquad ...(iii)$$

This, condition is shown in Fig. 21.8 (*b*).

Thus, the scalar product of two vectors is equal to the product of magnitude of one vector and the magnitude of component of other vector in the direction of first vector.

## 21.15 PROPERTIES OF SCALAR PRODUCT

The important properties of scalar product (or dot product) are as follows:

**1.** The scalar product is commutative *i.e.*,

$$\vec{A}\cdot\vec{B} = \vec{B}\cdot\vec{A}$$

---

* $\vec{A}\cdot\vec{B} = (A_x\hat{i} + A_y\hat{j} + A_z\hat{k})\cdot(B_x\hat{i} + B_y\hat{j} + B_z\hat{k}) = A_xB_x + A_yB_y + A_zB_z$

For details, refer Sec. 21.16.

2. The scalar product is distributive over addition *i.e.*,

$$\vec{A}\cdot(\vec{B}+\vec{C}) = \vec{A}\cdot\vec{B}+\vec{A}\cdot\vec{C}$$

2. The scalar product of a vector with itself is equal to the square of its magnitude.

$$\vec{A}\cdot\vec{A} = AA\cos 0° = A^2$$

3. The scalar product of two similar base vectors is unity and that of two different base vectors is zero.

$$\hat{i}\cdot\hat{i} = \hat{j}\cdot\hat{j} = \hat{k}\cdot\hat{k} = 1$$

and $$\hat{i}\cdot\hat{j} = \hat{j}\cdot\hat{k} = \hat{k}\cdot\hat{i} = 0$$

## 21.16 SCALAR PRODUCT IN TERMS OF RECTANGULAR COMPONENTS

The vectors $\vec{A}$ and $\vec{B}$ in terms of rectangular components is given as

$$\vec{A} = A_x\hat{i}+A_y\hat{j}+A_z\hat{k}$$

$$\vec{B} = B_x\hat{i}+B_y\hat{j}+B_z\hat{k}$$

The scalar product (or dot product) of these vectors in terms of rectangular components,

$$\begin{aligned}\vec{A}\cdot\vec{B} &= (A_x\hat{i}+A_y\hat{j}+A_z\hat{k})\cdot(B_x\hat{i}+B_y\hat{j}+B_z\hat{k})\\
&= A_x\hat{i}\cdot(B_x\hat{i}+B_y\hat{j}+B_z\hat{k})+A_y\hat{j}\cdot(B_x\hat{i}+B_y\hat{j}+B_z\hat{k})\\
&\quad + A_z\hat{k}\cdot(B_x\hat{i}+B_y\hat{j}+B_z\hat{k})\\
&= A_xB_x\hat{i}\cdot\hat{i}+A_xB_y\hat{i}\cdot\hat{j}+A_xB_z\hat{i}\cdot\hat{k}+A_yB_x\hat{j}\cdot\hat{i}+A_yB_y\hat{j}\cdot\hat{j}+A_yB_z\hat{j}\cdot\hat{k}\\
&\quad + A_zB_x\hat{k}\cdot\hat{i}+A_zB_y\hat{k}\cdot\hat{j}+A_zB_z\hat{k}\cdot\hat{k}\\
&= A_xB_x\times 1+A_xB_y\times 0+A_xB_z\times 0+A_yB_x\times 0+A_yB_y\times 1+A_yB_z\times 0\\
&\quad + A_zB_x\times 0+A_zB_y\times 0+A_zB_z\times 1\end{aligned}$$

or $$\vec{A}\cdot\vec{B} = A_xB_x+A_yB_y+A_zB_z$$

**Example 21.2** *The two vectors are given as*

$$\vec{A}=\hat{i}+2\hat{j}-\hat{k} \text{ and } \vec{B}=-\hat{i}+\hat{j}-2\hat{k}$$

*Determine: 1. Dot product of the vectors; and 2. Angle between the vectors.*

***Solution*** Given : $\vec{A} = A_x\hat{i}+A_y\hat{j}+A_z\hat{k}=\hat{i}+2\hat{j}-\hat{k}$ ; $\vec{B} = B_x\hat{i}+B_y\hat{j}+B_z\hat{k}=-\hat{i}+\hat{j}-2\hat{k}$

1. ***Dot product of the vectors***

We know that dot product of two vectors,

$$\begin{aligned}\vec{A}\cdot\vec{B} &= A_xB_x+A_yB_y+A_zB_z\\
&= [1\times(-1)]+[2\times 1]+[(-1)\times(-2)] = -1+2+2 = 3 \textbf{ Ans.}\end{aligned}$$

**2.** ***Angle between the vectors***

Let $\theta$ = Angle between the vectors.

We know that magnitude of vectors,

$$\left|\vec{A}\right| = \sqrt{A_x^2 + A_y^2 + A_z^2} = \sqrt{1^2 + 2^2 + (-1)^2} = \sqrt{6}$$

and
$$\left|\vec{B}\right| = \sqrt{B_x^2 + B_y^2 + B_z^2} = \sqrt{(-1)^2 + 1^2 + (-2)^2} = \sqrt{6}$$

We also know that the cos of angle $\theta$ between the vectors,

$$\cos\theta = \frac{\vec{A}.\vec{B}}{\left|\vec{A}\right|\left|\vec{B}\right|} = \frac{3}{\sqrt{6}\times\sqrt{6}} = \frac{3}{6} = \frac{1}{2}$$

$$\therefore \qquad \theta = \cos^{-1}\left(\frac{1}{2}\right) = 60° \textbf{ Ans.}$$

## 21.17 VECTOR OR CROSS PRODUCT

The vector or cross product of two vectors $\vec{A}$ and $\vec{B}$ is defined as the vector whose magnitude is equal to the product of the magnitudes of two vectors $\vec{A}$ and $\vec{B}$ and sine of the angle $\theta$ between them and whose direction is perpendicular to the plane of the two vectors $\vec{A}$ and $\vec{B}$. Mathematically,

$$\vec{A}\times\vec{B} = \left|\vec{A}\right|\left|\vec{B}\right| \sin\theta\,\hat{n} = AB\sin\theta\,\hat{n}$$

where $\hat{n}$ is a unit vector perpendicular to the plane of $\vec{A}$ and $\vec{}$ and its direction is given by right hand rule. Thus, the direction of $\vec{A}\times\vec{B}$ is same as that of unit vector $\hat{n}$.

It may be noted that magnitude of cross product of two vectors $\vec{A}$ and $\vec{B}$ is given as

$$\left|\vec{A}\times\vec{B}\right| = \left|\vec{A}\right|\left|\vec{B}\right| \sin\theta = AB\sin\theta$$

## 21.18 IMPORTANT MATHEMATICAL RESULTS OF VECTOR PRODUCT

The following mathematical results of vector product (or cross product) are important from subject point of view:

**1.** If $\vec{A}$ and $\vec{B}$ are two vectors *perpendicular* or *at right angle* to each other (*i.e.* $\theta = 90°$), then the magnitude of their vector product is equal to the product of their magnitudes.

$$\vec{A}\times\vec{B} = AB\sin 90°\,\hat{n} = AB\,\hat{n} \qquad ...(\because \sin 90° = 1)$$

$$\therefore \qquad \left|\vec{A}\times\vec{B}\right| = AB \qquad ...(\because |\hat{n}| = 1)$$

**2.** If $\vec{A}$ and $\vec{B}$ are two vectors *parallel to each other* and *having same* or *opposite direction* (*i.e.* $\theta = 0°$ or $180°$), then their vector product is a null vector.

$$\vec{A}\times\vec{B} = AB\sin(0° \text{ or } 180°)\,\hat{n} = \vec{0} \qquad ...(\because \sin 0° \text{ or } \sin 180° = 0)$$

## 21.19 GEOMETRICAL INTERPRETATION OF VECTOR PRODUCT

Consider two vectors $\vec{A}$ and $\vec{B}$ are represented by the sides *OP* and *OQ* of a parallelogram *OPRQ* and angle θ between them, as shown in Fig. 21.9. Draw *QN* perpendicular to *OP*.

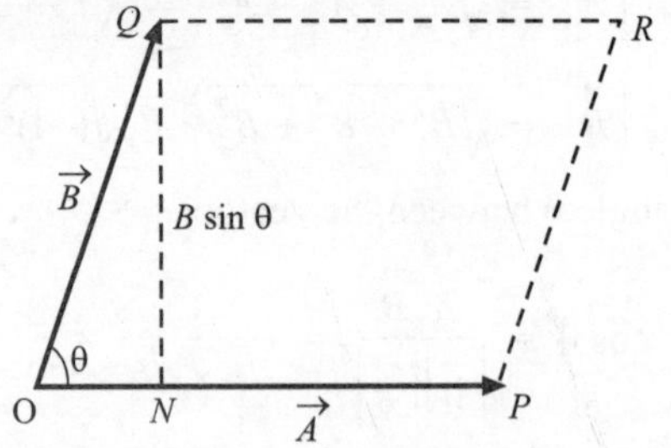


**Fig. 21.9** *Geometrical interpretation of cross product*

We have already discussed in previous article that magnitude of the vector product (or cross product) of two vectors is given as

$$\left|\vec{A} \times \vec{B}\right| = AB \sin\theta = OP.OQ \sin\theta$$

$$= OP.QN \qquad ...(QN = OQ \sin\theta)$$

$$= \text{Area of parallelogram } OPRQ$$

Thus, the magnitude of the vector product of two vectors is equal to area of the parallelogram formed by the two vectors as its adjacent sides.

## 21.20 PROPERTIES OF VECTOR PRODUCT

The important properties of vector product are as follows:

(*i*) The vector product is non-commutative *i.e.*,

$$\vec{A} \times \vec{B} \neq \vec{B} \times \vec{A}, \text{ rather } \vec{A} \times \vec{B} = -(\vec{B} \times \vec{A})$$

(*ii*) The vector product is distributive over addition *i.e.*,

$$\vec{A} \times (\vec{B} + \vec{C}) = \vec{A} \times \vec{B} + \vec{A} \times \vec{C}$$

(*iii*) The vector product of a vector with itself is a null vector.

$$\vec{A} \times \vec{A} = AA \sin 0° \, n = \vec{0}$$

(*iv*) The vector product of two unit vectors is given as

$$\hat{i} \times \hat{i} = \hat{j} \times \hat{j} = \hat{k} \times \hat{k} = \vec{0}$$

$$\hat{i} \times \hat{j} = \hat{k}\,;\ \hat{j} \times \hat{k} = \hat{i}\,;\ \hat{k} \times \hat{i} = \hat{j}$$

$$\hat{j} \times \hat{i} = -\hat{k}\,;\ \hat{k} \times \hat{j} = -\hat{i}\,;\ \hat{i} \times \hat{k} = -\hat{j}$$

## 21.21 VECTOR PRODUCT IN TERMS OF RECTANGULAR COMPONENTS

The vectors $\vec{A}$ and $\vec{B}$ in terms of rectangular components are given as

$$\vec{A} = A_x\hat{i} + A_y\hat{j} + A_z\hat{k}$$

$$\vec{B} = B_x\hat{i} + B_y\hat{j} + B_z\hat{k}$$

The vector product of these vectors in terms of rectangular components,

$$\begin{aligned}\vec{A}\times\vec{B} &= (A_x\hat{i} + A_y\hat{j} + A_z\hat{k})\times(B_x\hat{i} + B_y\hat{j} + B_z\hat{k})\\
&= A_x\hat{i}\times(B_x\hat{i} + B_y\hat{j} + B_z\hat{k}) + A_y\hat{j}\ (B_x\hat{i} + B_y\hat{j} + B_z\hat{k})\\
&\qquad + A_z\hat{k}\times(B_x\hat{i} + B_y\hat{j} + B_z\hat{k})\\
&= A_xB_x(\hat{i}\times\hat{i}) + A_xB_y(\hat{i}\times\hat{j}) + A_xB_z(\hat{i}\times\hat{k}) + A_yB_x(\hat{j}\times\hat{i}) + A_yB_y(\hat{j}\times\hat{j})\\
&\qquad + A_yB_z(\hat{j}\times\hat{k}) + A_zB_x(\hat{k}\times\hat{i}) + A_zB_y(\hat{k}\times\hat{j}) + A_zB_z(\hat{k}\times\hat{k})\\
&= A_xB_x\times 0 + A_xB_y\times\hat{k} + A_xB_z\times(-j) + A_yB_x\times(-\hat{k}) + A_yB_y\times 0\\
&\qquad + A_yB_z\times\hat{i} + A_zB_x\times\hat{j} + A_zB_y\times(-\hat{i}) + A_zB_z\times 0\\
&= (A_yB_z - A_zB_y)\hat{i} + (A_zB_x - A_xB_z)\hat{j} + (A_xB_y - A_yB_x)\hat{k}\end{aligned}$$

or

$$\vec{A}\times\vec{B} = \begin{vmatrix}\hat{i} & \hat{j} & \hat{k}\\ A_x & A_y & A_z\\ B_x & B_y & B_z\end{vmatrix}$$

The above equation expresses $A\times B$ in determinant form.

## 21.22 ANGLE BETWEEN TWO VECTORS IN TERMS OF VECTOR PRODUCT

The sine of the angle θ between the vectors $\vec{A}$ and $\vec{B}$ from the vector product (or cross product) is given by

$$\sin\theta = \frac{\left|\vec{A}\times\vec{B}\right|}{\left|\vec{A}\right|\left|\vec{B}\right|} = \frac{\left|\vec{A}\times\vec{B}\right|}{AB}$$

or

$$\theta = \sin^{-1}\left(\frac{\left|\vec{A}\times\vec{B}\right|}{AB}\right)$$

**Example 21.3** *The two vectors are given as*

$$\vec{A} = 2\hat{i} - 3\hat{j} + 6\hat{k} \quad and \quad \vec{B} = \hat{i} + \hat{j} - \hat{k}$$

*Determine : 1. Cross product of the vectors ; 2. Angle between the vectors ; and 3. Unit vector normal to both the vectors.*

***Solution*** Given : $\vec{A} = A_x\hat{i} + A_y\hat{j} + A_z\hat{k} = 2\hat{i} - 3\hat{j} + 6\hat{k}$ ; $\vec{B} = B_x\hat{i} + B_y\hat{j} + B_z\hat{k} = \hat{i} + \hat{j} - \hat{k}$

**1. *Cross product of the vectors***

We know that cross product of two vectors,

$$\vec{A} \times \vec{B} = \begin{vmatrix} \hat{i} & \hat{j} & \hat{k} \\ A_x & A_y & A_z \\ B_x & B_y & B_z \end{vmatrix} = \begin{vmatrix} \hat{i} & \hat{j} & \hat{k} \\ 2 & -3 & 6 \\ 1 & 1 & -1 \end{vmatrix}$$

$$= (A_y B_z - A_z B_y)\hat{i} + (A_z B_x - A_x B_z)\hat{j} + (A_x B_y - A_y B_x)\hat{k}$$

$$= [(-3)(-1) - (6)(1)]\hat{i} + [(6)(1) - (2)(-1)]\hat{j} + [(2)(1) - (-3)(1)]\hat{k}$$

$$= (3 - 6)\hat{i} + (6 + 2)\hat{j} + (2 + 3)\hat{k} = -3\hat{i} + 8\hat{j} + 5\hat{k} \quad \textbf{Ans.}$$

**2. *Angle between the vectors***

Let $\theta$ = Angle between the vectors.

We know that magnitude of vectors $\vec{A}$ and $\vec{B}$,

$$\left|\vec{A}\right| = \sqrt{A_x^2 + A_y^2 + A_z^2} = \sqrt{2^2 + (-3)^2 + 6^2} = \sqrt{49} = 7$$

and

$$\left|\vec{B}\right| = \sqrt{B_x^2 + B_y^2 + B_z^2} = \sqrt{1^2 + 1^2 + (-1)^2} = \sqrt{3} = 1.732$$

Similarly, magnitude of $\vec{A} \times \vec{B}$,

$$\left|\vec{A} \times \vec{B}\right| = \sqrt{(-3)^2 + 8^2 + 5^2} = \sqrt{98} = 9.9$$

We also know that the sin of angle $\theta$ between the vectors,

$$\sin\theta = \frac{\left|\vec{A} \times \vec{B}\right|}{\left|\vec{A}\right|\left|\vec{B}\right|} = \frac{9.9}{7 \times 1.732} = 0.816$$

$$\therefore \quad \theta = \sin^{-1}(0.816) = 54.69° \quad \textbf{Ans.}$$

**3. *Unit vectors normal to both the vectors***

We know that unit vector normal to both the vectors,

$$\hat{n} = \frac{\vec{A} \times \vec{B}}{\left|\vec{A} \times \vec{B}\right|} = \frac{-3\hat{i} + 8\hat{j} + 5\hat{k}}{9.9}$$

$$= -0.303\hat{i} + 0.808\hat{j} + 0.505\hat{k} \quad \textbf{Ans.}$$

## 21.23 COMPONENTS OF A FORCE IN SPACE

Consider a force $\vec{F}$, acting at the origin $O$ of the rectangular co-ordinate system. The force $\vec{F}$ is represented by the vector $\overrightarrow{OP}$, as shown in Fig. 21.10. Now, through the point $P$ draw planes parallel to co-ordinate planes, thus, completing a rectangular box (parallelepiped).

It may be noted that, the diagonal of the box is then represented by force $\vec{F}$ and its three components $\vec{F}_x, \vec{F}_y$ and $\vec{F}_z$ by its three edges along the three rectangular axes, *i.e.* $X$, $Y$ and $Z$ axes.

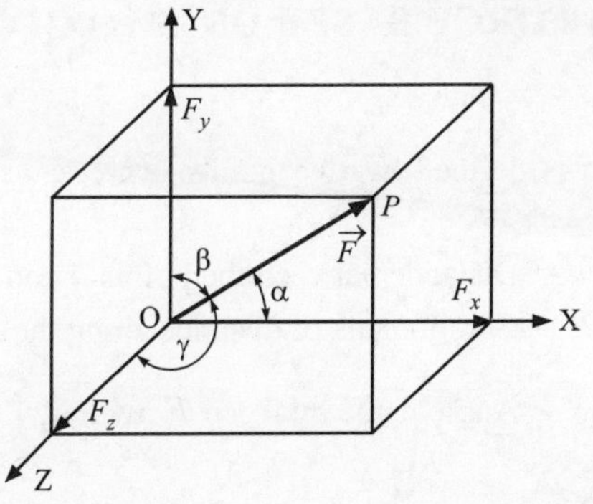


**Fig. 21.10** *Components of a force*

Thus, the force $\vec{F}$ in terms of rectangular components may be written as

$$\vec{F} = \vec{F}_x + \vec{F}_y + \vec{F}_z \qquad ...(i)$$

Let $\hat{i}$ = Unit vector in the positive *X*-axis,

$\hat{j}$ = Unit vector in the positive *Y*-axis, and

$\hat{k}$ = Unit vector in the positive *Z*-axis.

It may be noted that based on the concept of unit vectors,

$$\vec{F}_x = F_x\hat{i}\,; \qquad \vec{F}_y = F_y\hat{j}\,; \qquad \vec{F}_z = F_z\hat{k}$$

Substituting the values of $\vec{F}_x, \vec{F}_y$ and $\vec{F}_z$ in equation (*i*), the force $\vec{F}$ in terms of unit vectors may be given as

$$\vec{F} = F_x\hat{i} + F_y\hat{j} + F_z\hat{k} \qquad ...(ii)$$

where $F_x$, $F_y$ and $F_z$ are the magnitude of the three components of force $\vec{F}$ in *X*, *Y* and *Z* axes, respectively.

The magnitude of the force $\vec{F}$ is given as

$$F = \left|\vec{F}\right| = \sqrt{F_x^2 + F_y^2 + F_z^2} \qquad ...(iii)$$

If the force $\vec{F}$ makes an angle α with *X*-axis, angle β with *Y*-axis, and angle γ with *Z*-axis, then its components are given as

$$F_x = F\cos\alpha; \qquad F_y = F\cos\beta;\text{ and} \qquad F_z = F\cos\gamma$$

Substituting the values of $F_x$, $F_y$ and $F_z$ in the above equation (*ii*), the force $\vec{F}$ may be written as

$$\vec{F} = (F\cos\alpha)\hat{i} + (F\cos\beta)\,\hat{j} + (F\cos\gamma)\,\hat{k} \qquad ...(iv)$$

**Notes: 1.** The cosines of α, β and γ are known as *direction cosines* of the force $\vec{F}$ and are denoted as

$$l = \cos\alpha; \qquad m = \cos\beta;\text{ and} \qquad n = \cos\gamma$$

**2.** The three angles are related as

$$\cos^2\alpha + \cos^2\beta + \cos^2\gamma = 1 \quad \text{or} \quad l^2 + m^2 + n^2 = 1$$

## 21.24 COMPONENTS OF A FORCE BASED ON TWO GIVEN POINTS ON ITS LINE OF ACTION

Consider a force $\vec{F}$, which is defined by two points, $A(x_1, y_1, z_1)$ and $B(x_2, y_2, z_2)$, as shown in Fig. 21.11.

Let $d$ = Distance between the points $A$ and $B$,

$dx$, $dy$ and $dz$ = Components of distance along the $X$, $Y$ and $Z$ axes, respectively, and

$\alpha$, $\beta$ and $\gamma$ = Angle made by force $\vec{F}$ with $X$, $Y$ and $Z$ axes, respectively.

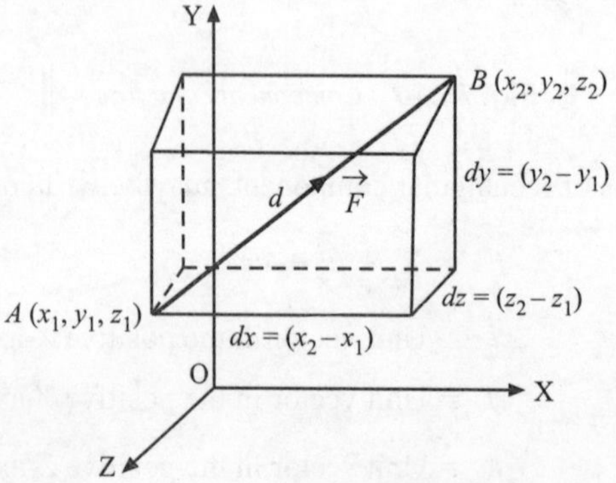


**Fig. 21.11** *Force defined by two points*

We know that from the geometry of the figure,

$$dx = (x_2 - x_1); \qquad dy = (y_2 - y_1); \text{ and } \qquad dz = (z_2 - z_1) \qquad ...(i)$$

Also, the components of the distance $AB$ along $X$, $Y$ and $Z$ axes in terms of distance and direction cosines is given as

$$dx = d\cos\alpha; \qquad dy = d\cos\beta; \text{ and } \qquad dz = d\cos\gamma \qquad ...(ii)$$

and distance $d$ in terms of co-ordinates

$$d = \sqrt{dx^2 + dy^2 + dz^2}$$

We know that force $\vec{F}$ in space may be written as

$$\vec{F} = F_x\hat{i} + F_y\hat{j} + F_z\hat{k}$$

Also, the components of force $\vec{F}$ along $X$, $Y$ and $Z$ axes in terms of magnitude of force and direction cosines is given as

$$F_x = F\cos\alpha; \qquad F_y = F\cos\beta; \text{ and } \qquad F_z = F\cos\gamma \qquad ...(iii)$$

From equations (*ii*) and (*iii*), we get

$$\frac{F_x}{dx} = \frac{F\cos\alpha}{d\cos\alpha} = \frac{F}{d}$$

Similarly,

$$\frac{F_y}{dy} = \frac{F\cos\beta}{d\cos\beta} = \frac{F}{d}; \text{ and } \frac{F_z}{dz} = \frac{F\cos\gamma}{d\cos\gamma} = \frac{F}{d}$$

$$\frac{F_x}{dx} = \frac{F_y}{dy} = \frac{F_z}{dz} = \frac{F}{d} \qquad ...(iv)$$

Substituting the value of *dx*, *dy*, and *dz* from equation (*i*) in above equation (*iv*), we get

$$\frac{F_x}{(x_2 - x_1)} = \frac{F_y}{(y_2 - y_1)} = \frac{F_z}{(z_2 - z_1)} = \frac{F}{d} \quad ...(v)$$

## 21.25 POSITION VECTOR OF A GIVEN POINT

Consider a point $A(x, y, z)$ in space whose position vector with respect to origin $O$ is given by vector $\overrightarrow{OA}$. It is denoted as $\vec{r}$ and is given as

$$\vec{r} = x\hat{i} + y\hat{j} + z\hat{k}$$

The position vector of point $A(x_1, y_1, z_1)$ with respect to point $B(x_2, y_2, z_2)$ is given as

$$\vec{r} = (x_2 - x_1)\hat{i} + (y_2 - y_1)\hat{j} + (z_2 - z_1)\hat{k}$$

**Example 21.4** *A force vector is given as*

$$\vec{F} = 250\hat{i} - 375\hat{j} + 500\hat{k}$$

*Determine : 1. Magnitude of force ; and 2. Direction with respect to co-ordinate axes.*

***Solution*** Given : $\vec{F} = F_x\hat{i} + F_y\hat{j} + F_z\hat{k} = 250\hat{i} - 375\hat{j} + 500\hat{k}$

**1. *Magnitude of force***

We know that magnitude of force,

$$\vec{F} = \sqrt{F_x^2 + F_y^2 + F_z^2} = \sqrt{250^2 + (-375)^2 + 500^2}$$

$$= 673.14 \text{ N} \quad \textbf{Ans.}$$

**2. *Direction with respect to co-ordinate axes***

Let $\alpha$ = Angle made by force $\vec{F}$ with $X$ axis,

$\beta$ = Angle made by force $\vec{F}$ with $Y$ axis, and

$\gamma$ = Angle made by force $\vec{F}$ with $Z$ axis.

We know that component of force along $X$-axis is given as

$$F_x = F\cos\alpha \quad \text{or} \quad \cos\alpha = \frac{F_x}{F}$$

$$\therefore \quad \alpha = \cos^{-1}\left(\frac{F_x}{F}\right) = \cos^{-1}\left(\frac{250}{673.14}\right) = 68.2° \quad \textbf{Ans.}$$

Similarly,

$$\beta = \cos^{-1}\left(\frac{F_y}{F}\right) = \cos^{-1}\left(\frac{-375}{673.14}\right) = 123.85° \quad \textbf{Ans.}$$

and

$$\gamma = \cos^{-1}\left(\frac{F_z}{F}\right) = \cos^{-1}\left(\frac{500}{673.14}\right) = 42.03° \quad \textbf{Ans.}$$

**Example 21.5** *A force of magnitude 748 N is directed from a point A (4, 1, 3) metres towards a point B (–2, 3, –1) metres. Determine the force vector F.*

***Solution*** Given : $|\vec{F}| = F = 748$ N ; Co-ordinates of point $A = (4, 1, 3)$ ; Co-ordinates of point $B = (-2, 3, -1)$

The position vector of point $A$ based on the co-ordinates of point $A$ (4, 1, 3) is given as

$$\overrightarrow{OA} = 4\hat{i} + 1\hat{j} + 3\hat{k}$$

Similarly, position vector of point $B$ based on the co-ordinates of point $B$ (–2, 3, –1) is given as

$$\overrightarrow{OB} = -2\hat{i} + 3\hat{j} - 1\hat{k}$$

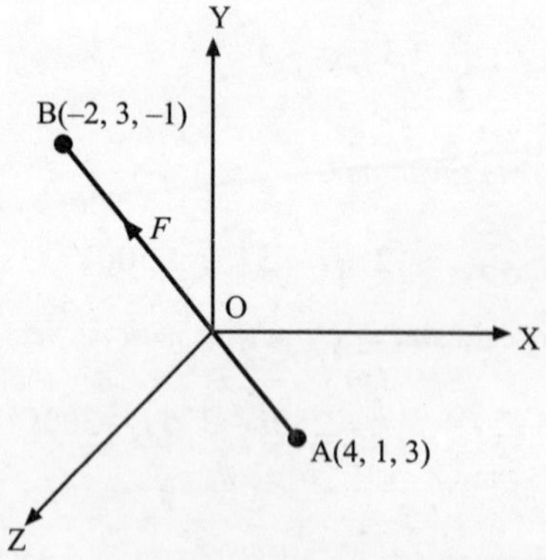


**Fig. 21.12** *Force defined by two points*

∴ Position vector $AB$ (*i.e.* vector joining the points $A$ and $B$) in the direction of force $\vec{F}$ as shown in Fig. 21.12 is given as

$$\overrightarrow{AB} = \text{Position vector of } B - \text{Position vector of } A$$

$$= \overrightarrow{OB} - \overrightarrow{OA}$$

$$= (-2\hat{i} + 3\hat{j} - 1\hat{k}) - (4\hat{i} + 1\hat{j} + 3\hat{k})$$

$$= (-2-4)\hat{i} + (3-1)\hat{j} + (-1-3)\hat{k}$$

$$= -6\hat{i} + 2\hat{j} - 4\hat{k}$$

∴ Unit vector in the direction of force $\vec{F}$

$$= \frac{\overrightarrow{AB}}{|\overrightarrow{AB}|} = \frac{-6\hat{i} + 2\hat{j} - 4\hat{k}}{\sqrt{(-6)^2 + 2^2 + (-4)^2}}$$

$$= \frac{-6\hat{i} + 2\hat{j} - 4\hat{k}}{\sqrt{56}} = \frac{-6\hat{i} + 2\hat{j} - 4\hat{k}}{7.48}$$

We know that force vector $\vec{F}$ is given as

$$= F \times \text{Unit vector in the direction of force}$$

$$= 748 \times \frac{-6\hat{i} + 2\hat{j} - 4\hat{k}}{7.48} = 100(-6\hat{i} + 2\hat{j} - 4\hat{k})$$

$$= -600\hat{i} + 200\hat{j} - 400\hat{k} \quad \textbf{Ans.}$$

## 21.26 RESULTANT FORCE OF CONCURRENT FORCE SYSTEM

If a number of forces are acting on a body are replaced by a single force which has the same effect as that of the forces together, then that single force is called as *resultant force*. It is denoted by $\vec{R}$.

Let $$\vec{R} = R_x\hat{i} + R_y\hat{j} + R_z\hat{k} = \text{Resultant force.}$$

Consider $\vec{F_1}, \vec{F_2}, \vec{F_3}, ..., \vec{F_n}$ are the forces acting on a body.

The resultant force on the body is given by

$$\vec{R} = \vec{F_1} + \vec{F_2} + \vec{F_3} + ... + \vec{F_n} \quad ...(i)$$

where
$$\vec{F_1} = F_{1x}\hat{i} + F_{1y}\hat{j} + F_{1z}\hat{k}$$

$$\vec{F_2} = F_{2x}\hat{i} + F_{2y}\hat{j} + F_{2z}\hat{k}$$

$$\vec{F_3} = F_{3x}\hat{i} + F_{3y}\hat{j} + F_{3z}\hat{k}\text{, and so on.}$$

Thus, substituting these values in equation (*i*), we get

$$\vec{R} = (F_{1x}\hat{i} + F_{1y}\hat{j} + F_{1z}\hat{k}) + (F_{2x}\hat{i} + F_{2y}\hat{j} + F_{2z}\hat{k}) + (F_{3x}\hat{i} + F_{3y}\hat{j} + F_{3z}\hat{k}) + ... + (F_{nx}\hat{i} + F_{ny}\hat{j} + F_{nz}\hat{k})$$

$$= (F_{1x} + F_{2x} + F_{3x} + ... + F_{nx})\hat{i} + (F_{1y} + F_{2y} + F_{3y} + ... + F_{ny})\hat{j} + (F_{1z} + F_{2z} + F_{3z} + ... + F_{nz})\hat{k}$$

$$= (\Sigma F_x)\hat{i} + (\Sigma F_y)\hat{j} + (\Sigma F_z)\hat{k} \quad ...(ii)$$

We know that,

$$\vec{R} = R_x\hat{i} + R_y\hat{j} + R_z\hat{k}$$

$$\therefore \quad R_x = \Sigma F_x; \quad R_y = \Sigma F_y; \text{ and } \quad R_z = \Sigma F_z \quad ...(iii)$$

Also, the magnitude of the resultant force,

$$R = \sqrt{R_x^2 + R_y^2 + R_z^2}$$

$$= \sqrt{(\Sigma F_x)^2 + (\Sigma F_y)^2 + (\Sigma F_z)^2} \quad ...(iv)$$

## 21.27 CONDITION FOR EQUILIBRIUM OF FORCES IN SPACE

A body is said to be in equilibrium if the resultant force due to the number of concurrent forces ($F_1$, $F_2$, $F_3$, ...) acting on it is zero.

$$\vec{R} = \vec{F_1} + \vec{F_2} + \vec{F_3} + ... = \Sigma F = 0$$

Since the body is in equilibrium, therefore the sum of forces in each direction of *X*, *Y* and *Z* axes is also zero, *i.e.*

$$R_x = F_{1x} + F_{2x} + F_{3x} + \ldots = \Sigma F_x = 0$$
$$R_y = F_{1y} + F_{2y} + F_{3y} + \ldots = \Sigma F_y = 0$$
$$R_z = F_{1z} + F_{2z} + F_{3z} + \ldots = \Sigma F_z = 0$$

**Example 21.6** *A pole AB of length 8 m is held in position by two cables BC and BD as shown in the Fig. 21.13. If the tensions in the cable BC is 500 N and in the cable BD is 800 N, determine the magnitude and direction of resultant of force exerted at B by two cables.*

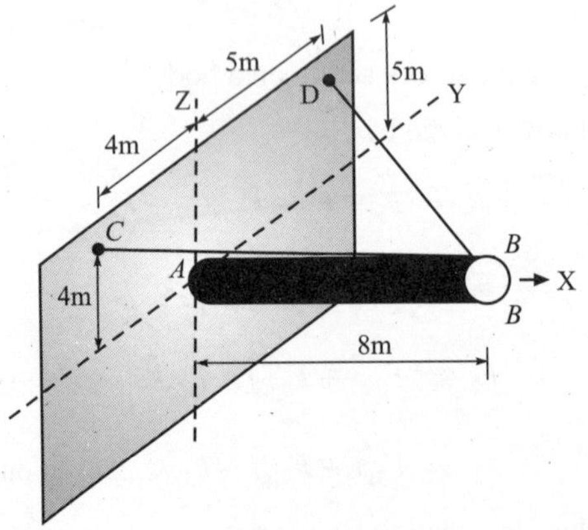


**Fig. 21.13**

***Solution*** Given : $F_1 = 500$ N ; $F_2 = 800$ N ; Co-ordinates of point $B = (8, 0, 0)$ ; Co-ordinates of point $C = (0, -4, 4)$ ; Co-ordinates of point $D = (0, 5, 5)$

Let $\vec{R} = R_x\hat{i} + R_y\hat{j} + R_z\hat{k}$ = Resultant force vector,

$\alpha$ = Angle made by resultant force $\vec{R}$ with *X*-axis,

$\beta$ = Angle made by resultant force $\vec{R}$ with *Y*-axis, and

$\gamma$ = Angle made by resultant force $\vec{R}$ with *Z*-axis.

We know that position vector *BC* (*i.e.* vector joining the points *B* and *C*) in the direction of force $\overrightarrow{F_{BC}}$ is given as

$$\overrightarrow{BC} = (0-8)\hat{i} + (-4-0)\hat{j} + (4-0)\hat{k} = -8\hat{i} - 4\hat{j} + 4\hat{k}$$

∴ Force vector in the direction *BC*,

$$\vec{F_1} = F_1 \times \text{Unit vector in the direction of force} = F_1 \times \frac{\overrightarrow{BC}}{\left|\overrightarrow{BC}\right|}$$

$$= 500 \times \frac{-8\hat{i} - 4\hat{j} + 4\hat{k}}{\sqrt{(-8)^2 + (-4)^2 + 4^2}} = 500 \times \frac{-8\hat{i} - 4\hat{j} + 4\hat{k}}{9.8}$$

$$= 51.02\,(-8\hat{i} - 4\hat{j} + 4\hat{k}) = -408.16\hat{i} - 204.08\hat{j} + 204.08\hat{k}$$

Similarly, we know that position vector *BD* (*i.e.* vector joining the points *B* and *D*) in the direction of force $\overrightarrow{F_{BD}}$ is given as

$$\overrightarrow{BD} = (0-8)\hat{i} + (5-0)\hat{j} + (5-0)\hat{k} = -8\hat{i} + 5\hat{j} + 5\hat{k}$$

∴ Force vector in the direction of *BD*,

$$\vec{F_2} = F_2 \times \text{Unit vector in the direction of force} = F_2 \times \frac{\overrightarrow{BD}}{|\overrightarrow{BD}|}$$

$$= 800 \times \frac{-8\hat{i} + 5\hat{j} + 5\hat{k}}{\sqrt{(-8)^2 + 5^2 + 5^2}} = 800 \times \frac{-8\hat{i} + 5\hat{j} + 5\hat{k}}{10.68}$$

$$= 74.91\,(-8i + 5\hat{j} + 5\hat{k}) = -599.28\hat{i} + 374.55\hat{j} + 374.55\hat{k}$$

We know that the resultant force vector,

$$\vec{R} = R_x\hat{i} + R_y\hat{j} + R_z\hat{k} = \vec{F_1} + \vec{F_2}$$

$$= (-408.16\hat{i} - 204.08\hat{j} + 204.08\hat{k}) + (-599.28\hat{i} + 374.55\hat{j} + 374.55\hat{k})$$

$$= (-408.16 - 599.28)\hat{i} + (-204.08 + 374.55)\hat{j} + (204.08 + 374.55)\hat{k}$$

$$= -1007.44\hat{i} + 170.47\hat{j} + 578.63\hat{k}$$

and magnitude of the resultant force,

$$R = \sqrt{R_x^2 + R_y^2 + R_z^2}$$

$$= \sqrt{(-1007.44)^2 + (170.47)^2 + (578.63)^2} = 1174.23 \text{ N}$$

We know that component of resultant force along *X*-axis is given as

$$R_x = R\cos\alpha \quad \text{or} \quad \cos\alpha = \frac{R_x}{R}$$

∴ $$\alpha = \cos^{-1}\left(\frac{R_x}{R}\right) = \cos^{-1}\left(\frac{-1007.44}{1174.23}\right) = 149.09° \quad \textbf{Ans.}$$

Similarly, $$\beta = \cos^{-1}\left(\frac{R_y}{R}\right) = \cos^{-1}\left(\frac{170.47}{1174.23}\right) = 81.65° \quad \textbf{Ans.}$$

and $$\gamma = \cos^{-1}\left(\frac{R_z}{R}\right) = \cos^{-1}\left(\frac{578.63}{1174.23}\right) = 60.48° \quad \textbf{Ans.}$$

**Example 21.7** *A load of 2 kN is supported by three cables AB, AC and AD and tied at a common point A 0.9 m below the ground, as shown in the Fig. 21.14. The cable AD lies in YZ plane. Determine the forces in the cables AB, AC and AD.*

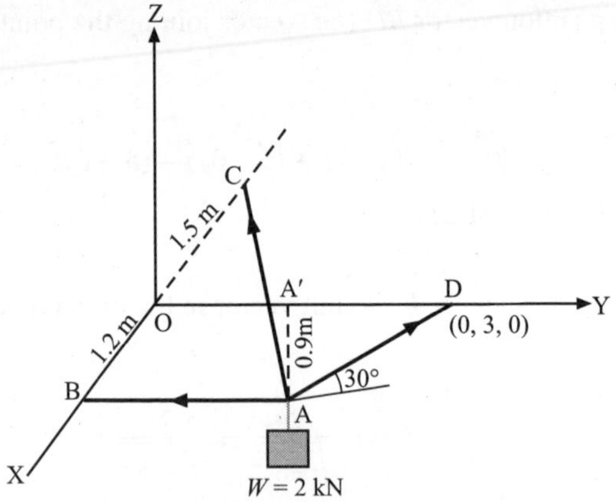


**Fig. 21.14**

***Solution*** Given : $W = 2$ kN = 2000 N (vertically downward) ; Co-ordinates of point $B = (1.2, 0, 0)$ ; Co-ordinates of point $C = (-1.5, 0, 0)$ ; Co-ordinates of point $D = (0, 3, 0)$

Let $F_1$ = Force in the cable *AB*,

$F_2$ = Force in the cable *AC*, and

$F_3$ = Force in the cable *AD*.

First of all, let us find the co-ordinates of point *A*. From the geometry of the figure, for triangle *AA'D* we find that $\angle D = 30°$.

$\therefore$ $Y$ co-ordinate of $A = OA' = OD - A'D = 3 - 0.9 \cot 30° = 3 - 1.56 = 1.44$ m

$$...(\because \cot 30° = \frac{A'D}{AA'} \quad \text{or} \quad AD' = AA' \cos 30°)$$

Thus,

Co-ordinates of point $A = (0, 1.44, -0.9)$

We know that position vector *AB* (*i.e.* vector joining the points *A* and *B*) in the direction of force $\vec{F_1}$ is given as

$$\overrightarrow{AB} = (1.2 - 0)\hat{i} + (0 - 1.44)\hat{j} + [0 - (-0.9)]\hat{k}$$
$$= 1.2\hat{i} - 1.44\hat{j} + 0.9\hat{k}$$

$\therefore$ Force vector in the direction *AB*,

$$\vec{F_1} = F_1 \times \text{Unit vector in the direction of force} = F_1 \times \frac{\overrightarrow{AB}}{|\overrightarrow{AB}|}$$

$$= F_1 \times \frac{1.2\hat{i} - 1.44\hat{j} + 0.9\hat{k}}{\sqrt{(1.2)^2 + (-1.44)^2 + (0.9)^2}}$$

$$= F_1 \times \frac{1.2\hat{i} - 0.44\hat{j} + 0.9\hat{k}}{2.08}$$

$$= F_1 (0.58\hat{i} - 0.69\hat{j} + 0.43\hat{k}) \qquad ...(i)$$

Similarly, we know that position vector *AC* (*i.e.* vector joining the points *A* and *C*) in the direction of force $\vec{F_2}$ is given as

$$\overrightarrow{AC} = (-1.5-0)\hat{i} + (0-1.44)\hat{j} + [0-(-0.9)]\hat{k}$$
$$= -1.5\hat{i} - 1.44\hat{j} + 0.9\hat{k}$$

∴ Force vector in the direction of $AC$,

$$\vec{F_2} = F_2 \times \text{Unit vector in the direction of force} = F_2 \times \frac{\overrightarrow{AC}}{|\overrightarrow{AC}|}$$
$$= F_2 \times \frac{-1.5\hat{i} - 1.44\hat{j} + 0.9\hat{k}}{\sqrt{(-1.5)^2 + (-1.44)^2 + (0.9)^2}}$$
$$= F_2 \times \frac{-1.5\hat{i} - 1.44\hat{j} + 0.9\hat{k}}{2.26}$$
$$= F_2\,(0.66\hat{i} - 0.64\hat{j} + 0.4\hat{k}) \qquad ...(ii)$$

and we know that position vector $AD$ (*i.e.* vector joining the points $A$ and $D$) in the direction of force $\vec{F_3}$ is given as

$$\overrightarrow{AD} = (0-0)\hat{i} + (3-1.44)\hat{j} + [0-(-0.9)]\hat{k}$$
$$= 0\hat{i} + 1.56\hat{j} + 0.9\hat{k}$$

∴ Force vector in the direction of $AD$,

$$\overrightarrow{F_3} = F_3 \times \text{Unit vector in the direction of force} = F_3 \times \frac{\overrightarrow{AD}}{|\overrightarrow{AD}|}$$
$$= F_3 \times \frac{0\hat{i} + 1.56\hat{j} + 0.9\hat{k}}{\sqrt{0^2 + (1.56)^2 + (0.9)^2}}$$
$$= F_3 \times \frac{0\hat{i} + 1.56\hat{j} + 0.9\hat{k}}{1.8}$$
$$= F_3\,(0\hat{i} + 0.87\hat{j} + 0.5\hat{k}) \qquad ...(iii)$$

It may be noted that the force acting on point $A$ are $W, F_1, F_2$ and $F_3$. Since the system is in equilibrium, hence the sum of forces in each direction of $X$, $Y$ and $Z$ axes is zero. Therefore, from equations (*i*), (*ii*) and (*iii*),

For $X$-axis, $\quad 0.58F_1 - 0.66F_2 = 0 \quad \text{or} \quad F_1 = 1.14F_2 \qquad ...(iv)$

For $Y$-axis,

$$-0.69F_1 - 0.64F_2 + 0.87F_3 = 0$$
$$-0.69\,(1.14F_2) - 0.64F_2 + 0.87F_3 = 0 \qquad \text{...[Substituting the value of } F_1 \text{ from equation } (iv)]$$
$$-0.79F_2 - 0.64F_2 + 0.87F_3 = 0$$
$$1.43F_2 = 0.87F_3 \quad \text{or} \quad F_2 = 0.61F_3 \qquad ...(v)$$

For $Z$-axis,

$$0.43F_1 + 0.4F_2 + 0.5F_3 - 2000 = 0$$

$$0.43(1.14 \times 0.61\ F_3) + 0.4(0.61F_3) + 0.5F_3 - 2000 = 0$$

...[Substituting the value of $F_1$ and $F_2$ form equations (*iv*) and (*v*), respectively]

$$0.3F_3 + 0.24F_3 + 0.5F_3 = 2000 \quad \text{or} \quad 1.04F_3 = 2000$$

$$\therefore \qquad F_3 = 1923.08 \text{ N } \textbf{Ans.}$$

From equations (*iv*) and (*v*), we get

$$F_2 = 0.61\ F_3 = 0.6 \times 1923.08 = 1173.08 \text{ N} \quad \textbf{Ans.}$$

and $$F_1 = 1.41\ F_2 = 1.41 \times 1173.08 = 1654.04 \text{ N} \quad \textbf{Ans.}$$

## 21.28 MOMENT OF A FORCE

The moment of force $\vec{F}$ about a point $O$ is obtained as the vector (cross) product of the position vector (perpendicular distance) of $O$ from the line of action of force and the force.

Let $\vec{r}$ = Position vector with respect to $O$ = $x\hat{i} + y\hat{j} + z\hat{k}$, and

$\vec{F}$ = Force vector = $F_x\hat{i} + F_y\hat{j} + F_z\hat{k}$

Thus, moment of a force $\vec{F}$ about a point $O$ is given as

$$\vec{M} = \vec{r} \times \vec{F} = \begin{vmatrix} \hat{i} & \hat{j} & \hat{k} \\ x & y & z \\ F_x & F_y & F_z \end{vmatrix} \qquad ...(i)$$

$$= (yF_z - zF_y)\hat{i} - (zF_x - xF_z)\hat{j} + (xF_y - yF_x)\hat{k} \qquad ...(ii)$$

The above equation (*ii*) in simple form may be written as

$$\vec{M} = M_x\hat{i} + M_y\hat{j} + M_z\hat{k} \qquad ...(iii)$$

where $M_x = yF_z - zF_y$ = Moment of $\vec{F}$ about $X$-axis,

$M_y = zF_x - xF_z$ = Moment of $\vec{F}$ about $Y$-axis, and

$M_z = xF_y - yF_x$ = Moment of $\vec{F}$ about $Z$-axis.

**Note:** The magnitude of moment $\vec{M}$ is given as

$$|\vec{M}| = |\vec{r}| \times |\vec{F}| \sin\theta = rF\sin\theta$$

where $\theta$ = Angle between force $F$ and position vector $r$.

**Example 21.8** *A force $F = 2\hat{i} + 4\hat{j} - 3\hat{k}$ is applied at a point P( 1, 1, –2). Find the moment of the force F about the point O (2, –1, 2).*

***Solution*** Given : $\vec{F} = F_x\hat{i} + F_y\hat{j} + F_z k = 2\hat{i} + 4\hat{j} - 3\hat{k}$ ; Co-ordinates of point $P$ = (1, 1, –2) ; Co-ordinates of point $O$ = (2, –1, 2)

Refer Fig. 21.15.

We know that position vector of point $P$ with co-ordinates (1, 1, –2)

$$= \hat{i} + \hat{j} - 2\hat{k}$$

Similarly, the position vector of point $O$ with co-ordinates ( 2, –1, 2)

$$= 2\hat{i} - \hat{j} + 2\hat{k}$$

The position vector $r$ of point $P$ w.r.t. $O$ is given as

= Position vector of point $P$ – Position vector of point $O$

$$= (\hat{i} + \hat{j} - 2\hat{k}) - (2\hat{i} - \hat{j} + 2\hat{k})$$

$$= (1-2)\hat{i} + (1+1)\hat{j} + [(-2)-(-2)]\hat{k} = -\hat{i} + 2\hat{j} - 4\hat{k}$$

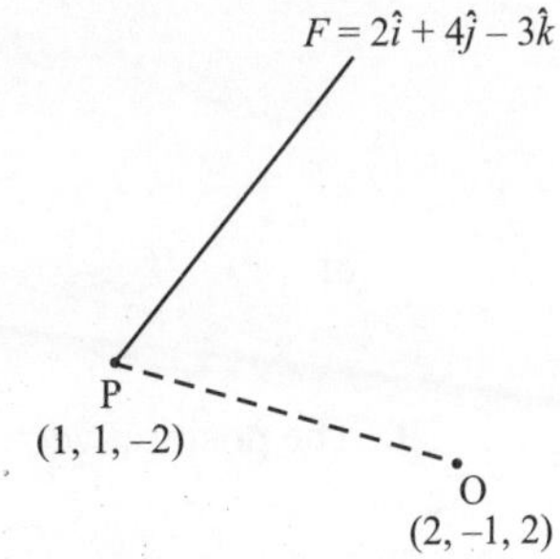

**Fig. 21.15**

We know that moment of force,

$$\vec{M} = \vec{r} \times \vec{F} = \begin{vmatrix} \hat{i} & \hat{j} & \hat{k} \\ x & y & z \\ F_x & F_y & F_z \end{vmatrix} = \begin{vmatrix} \hat{i} & \hat{j} & \hat{k} \\ -1 & 2 & -4 \\ 2 & 4 & -3 \end{vmatrix}$$

$$= (A_y B_z - A_z B_y)\hat{i} + (A_z B_x - A_x B_z)\hat{j} + (A_x B_y - A_y B_x)\hat{k}$$

$$= [(2)(-3) - (-4)(4)]\hat{i} + [(-4)(2) - (-1)(-3)]\hat{j} + [(-1)(4) - (2)(2)]\hat{k}$$

$$= (-6+16)\hat{i} + (-8-3)\hat{j} + (-4-4)\hat{k}$$

$$= 10\hat{i} - 11\hat{j} - 8\hat{k}$$ **Ans.**

## HIGHLIGHTS

1. The physical quantities which have only magnitude and no direction are called as *scalar quantities* or *scalars.*
2. The physical quantities which have both magnitude and direction are called as *vector quantities* or *vectors.*
3. The magnitude of vector corresponds to the length of the vector. It is represented as $|\vec{A}|$ or $A$.
4. If $A_x$, $A_y$, $A_z$ are the rectangular components of $A$ and $\hat{i}, \hat{j}, \hat{k}$ are the unit vectors along $X$, $Y$ and $Z$ axes, then

$$\vec{A} = A_x \hat{i} + A_y \hat{j} + A_z \hat{k}$$

5. The scalar or dot product of two vectors $\vec{A}$ and $\vec{B}$ is defined as

$$\vec{A}.\vec{B} = |\vec{A}||\vec{B}| \cos\theta = AB\cos\theta$$

where $\theta$ = Angle between the two vectors.

6. In terms of rectangular components

$$\vec{A}.\vec{B} = A_x B_x + A_y B_y + A_z B_z$$

7. The vector or cross product of two vectors $\vec{A}$ and $\vec{B}$ is defined as

$$\vec{A} \times \vec{B} = |\vec{A}||\vec{B}| \sin\theta\, \hat{n} = AB\sin\theta\, \hat{n}$$

where $\hat{n}$ = Unit vector perpendicular to the plane of $\vec{A}$ and $\vec{B}$

8. In terms of rectangular components

$$\vec{A} \times \vec{B} = (A_y B_z - A_z B_y)\hat{i} + (A_z B_x - A_x B_z)\hat{j} + (A_x B_y - A_y B_x)\hat{k}$$

or
$$\vec{A} \times \vec{B} = \begin{vmatrix} \hat{i} & \hat{j} & \hat{k} \\ A_x & A_y & A_z \\ B_x & B_y & B_z \end{vmatrix}$$

**9.** The position vector of point $A$ $(x_1, y_1, z_1)$ with respect to point $B$ $(x_2, y_2, z_2)$ is given as

$$\vec{r} = (x_2 - x_1)\hat{i} + (y_2 - y_1)\hat{j} + (z_2 - z_1)\hat{k}$$

**10.** If a number of force are acting on a body are replaced by a single force which has the same effect as that of the forces together, then that single force is called as *resultant force*. It is denoted by $\vec{R}$.

**11.** The moment of a force $\vec{F}$ about a point $O$ is given as

$$\vec{M} = \vec{r} \times \vec{F} = \begin{vmatrix} \hat{i} & \hat{j} & \hat{k} \\ x & y & z \\ F_x & F_y & F_z \end{vmatrix}$$

$$= (yF_z - zF_y)\hat{i} - (zF_x - xF_z)\hat{j} + (xF_y - yF_x)\hat{k}$$

## NUMERICAL PROBLEMS

**1.** A body moves 300 m to the east and then 400 m to the north. Calculate the total displacement of the body and the total distance travelled. [**Ans.** 500 m; 700 m]

**2.** Four forces $(2\hat{i} + 3\hat{j} - \hat{k}), (\hat{i} + 2\hat{j} - 3\hat{k}), (\hat{i} + 2\hat{j} + 3\hat{k})$ and $(2\hat{j} - 3\hat{j} + 4\hat{k})$, are acting at a point in space. Calculate the magnitude of the resultant force acting on the point. [**Ans.** 7.81]

**3.** Calculate the resultant of these three vectors

$$\vec{A} = 10\hat{i} + 5\hat{j} + 3\hat{k}\ ;\ \vec{B} = 2\hat{i} - 4\hat{j} + 5\hat{k}\ ;\ \vec{C} = 6\hat{i} - 5\hat{j} + 2\hat{k}$$

Also find the magnitude of the resultant vector and the unit vector along the resultant vector.

[**Ans.** $18\hat{i} - 3\hat{j} + 9\hat{k}$ ; 22 ; $0.82\hat{i} - 0.18\hat{j} + 0.54\hat{k}$ ]

**4.** A body is moving with a velocity of 5 m/s towards east. After 10 s, if velocity changes to 5 m/s towards north, calculate the acceleration of the body. [**Ans.** 0.707 m/s$^2$]

**5.** A force of 250 N makes angle of 45°, 60° and 120°, respectively, with $X$, $Y$ and $Z$ axes. Express the force in vector form. [**Ans.** $176.78\hat{i} + 125\hat{j} - 125\hat{k}$ ]

**6.** The co-ordinates of the initial and terminal points of a vector are (3, 1, –2) and (4, –7, 10), respectively. Express the vector and find its angles with the axes. [**Ans.** $\hat{i} + 8\hat{j} + 12\hat{k}$ ; 86.04° ; 123.57 ; 33.9°]

**7.** The two vectors are given as

$$\vec{A} = 2\hat{i} + 3\hat{j} + \hat{k} \text{ and } \vec{B} = 3\hat{i} - 3\hat{j} + 4\hat{k}$$

Determine: 1. Dot product of the vectors; 2. Cross product of the vector and the unit vector along it; and 3. Angle between the vector $\vec{A}$ and the vector resulting from their cross product.

[**Ans.** 1 ; $15\hat{i} - 5\hat{j} - 15\hat{k}$ , $0.69\hat{i} - 0.23\hat{j} - 0.69\hat{k}$ ; 90°]

**8.** The concurrent forces $F_1 = 810$ N; $F_2 = 1200$ N; $F_3 = 1600$ N act at point $A$, as shown in Fig. 21.16. The forces $F_1$ and $F_2$ are acting along the line $AB$ and $AC$, respectively. Find the direction of $F_3$ so that resultant of the three forces is parallel to $Y$-axis. Also find the magnitude of the resultant force.

**[Ans.** $788.8\hat{i} - 1391.5\hat{j} - 38.4\hat{k}$; 3094.5 N (downwards)]

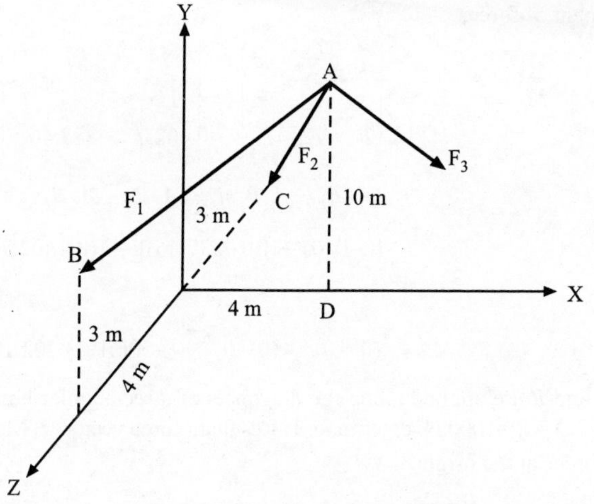


**Fig. 21.16**

**9.** A force $\vec{F} = 5\hat{i} + 2\hat{j} - 6\hat{k}$ is acting at a point $A$ whose position vector from origin $O$ is $6.5\hat{i} + 4.2j - 8\hat{k}$.

Calculate the moment of force about origin $O$ and also find its magnitude. **[Ans.** $-9.2\hat{i} - \hat{j} - 8\hat{k}$; 12.23 N-m ]

**10.** Determine the moment of force 525 N about point $A$, as shown in Fig. 20.17. **[Ans.** $-104.02\hat{k}$ ]

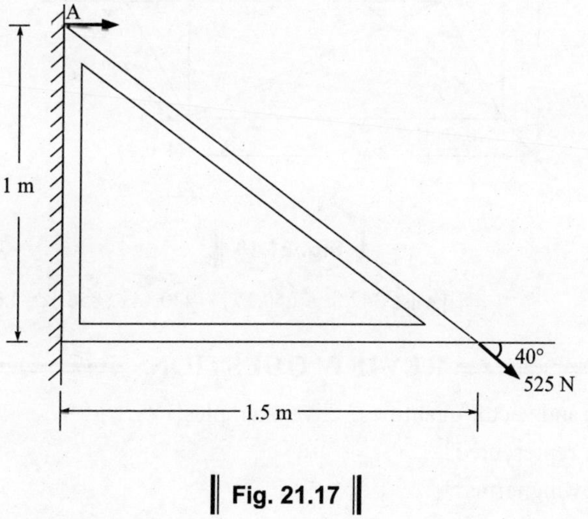


**Fig. 21.17**

**[Hint:** Given: $F = |\vec{F}| = 525$ N; $x = 1.5$; $y = -1$; $z = 0$

We know that position vector w.r.t. to origin $O$,

$$\vec{r} = x\hat{i} + y\hat{j} + z\hat{k} = 1.5\hat{i} - 1\hat{j} + 0\hat{k}$$

We also know that force vector $F$,

$$\vec{F} = F_x\hat{i} + F_y\hat{j} + F_z\hat{k} = 525 \cos 40^\circ\ \hat{i} + 252 \sin 40^\circ\ \hat{j} + 0\ \hat{k}$$

$$= 402.7\hat{i} - 337.46\hat{k} + 0\hat{k}$$

We know that moment of force,

$$\vec{M} = \vec{r} \times \vec{F} = \begin{vmatrix} \hat{i} & \hat{j} & \hat{k} \\ x & y & z \\ F_x & F_y & F_z \end{vmatrix} = \begin{vmatrix} \hat{i} & \hat{j} & \hat{k} \\ 1.5 & -1 & 0 \\ 402.7 & -337.46 & 0 \end{vmatrix}$$

$$= (A_y B_z - A_z B_y)\hat{i} + (A_z B_x - A_x B_z)\hat{j} + (A_x B_y - A_y B_x)\hat{k}$$

$$= [(-1)(0) - (0)(-337.46)]\hat{i} + [(0)(402.7) - (1.5)(0)]\hat{j}$$

$$+ [(1.5)(-337.46) - (-1)(402.7)]\hat{k}$$

$$= (0 + 0)\hat{i} + (0 - 0)\hat{j} + (-506.19 + 402.17)\hat{k} = -104.02\hat{k}$$ **Ans.]**

**11.** Three forces $P$, $Q$ and $R$ are applied along the diagonals of a rectangular box (parallelepiped), as shown in Fig. 21.18. If $2P = 2Q = R = 1000$ N, determine: 1. Resultant force vector; 2. Magnitude of resultant force; and 3. Moment of all forces at the origin.

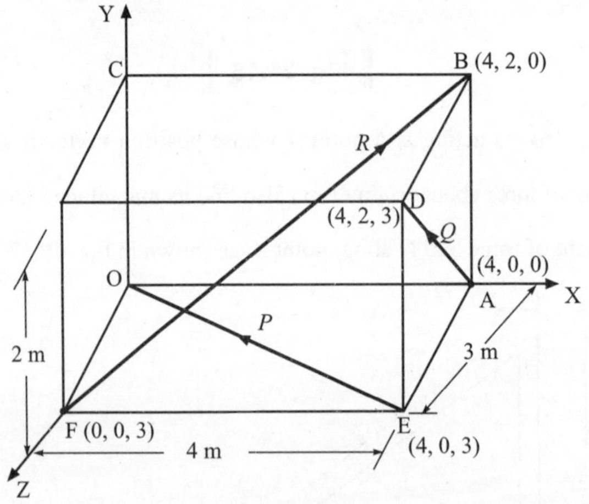


**Fig. 21.18**

[**Ans.** $342.8\hat{i} + 649.17\hat{j} - 440.44\hat{k}$; $856.1$; $-1114.2\hat{i} + 561.76\hat{j} + 1111.08\hat{k}$ ]

## REVIEW QUESTIONS

1. What are scalar and vector quantities? Give examples.
2. How is a vector represented?
3. Define the following terms:
   (*a*) Unit vector (*b*) Equal vector (*c*) Negative of a vector.
4. Explain how a vector can be resolved into its rectangular components in three dimensions.
5. What are the different ways in which a vector can be multiplied by another vector?
6. Define scalar or dot product of two vectors. Give its geometrical interpretation.
7. Define vector or cross product of two vectors. Give its geometrical interpretation.
8. What do you understand by moment of a force?

## OBJECTIVE TYPE QUESTIONS

**1.** Which of the following is a vector quantity?

(*a*) Mass (*b*) Distance (*c*) Force (*d*) Temperature

**2.** Which of the following is not a vector quantity?

(*a*) Mass (*b*) Velocity (*c*) Force (*d*) Moment

**3.** If a unit vector is represented by $0.5\hat{i} + 0.8\hat{j} + c\hat{k}$, then the value of *c* is

(*a*) 1 (*b*) $\sqrt{0.8}$ (*c*) $\sqrt{0.01}$ (*d*) $\sqrt{0.11}$

[**Hint:** For a unit vector, $\sqrt{0.5^2 + 0.8^2 + c^2} = 1$ or $c = \sqrt{1 - 0.89} = \sqrt{0.11}$ ]

**4.** If a vector *A* is multiplied by a scalar quantity *k*, then its magnitude becomes

(*a*) *k* times the vector (*b*) $k^2$ times the vector

(*c*) –*k* times the vector (*d*) $k^{-1}$ times the vector

**5.** If we add two vectors of equal magnitude but opposite direction, we get

(*a*) unit vector (*b*) unit scalar (*c*) zero vector (*d*) zero scalar

**6.** If $\vec{A} = \hat{i} + 2\hat{j} + \hat{k}$, $\vec{B} = 2\hat{i} + 3\hat{k}$, and $\vec{C} = -2\hat{i} + 4\hat{k}$, then the magnitude of the vector $A + B - C$ is

(*a*) $\sqrt{20}$ (*b*) $\sqrt{29}$ (*c*) $\sqrt{65}$ (*d*) $\sqrt{69}$

**7.** Two vectors are perpendicular if

(*a*) $\vec{A}.\vec{B} = 1$ (*b*) $\vec{A} \times \vec{B} = 0$ (*c*) $\vec{A}.\vec{B} = 0$ (*d*) $\vec{A}.\vec{B} = AB$

**8.** If $|\vec{A} \times \vec{B}| = |\vec{A}.\vec{B}|$, then angle between $\vec{A}$ and $\vec{B}$ will be

(*a*) 30° (*b*) 45° (*c*) 60° (*d*) 90°

**9.** The angle between the two vectors $\vec{A} = 3\hat{i} + 4\hat{j} + 5\hat{k}$ and $\vec{B} = 3\hat{i} + 4\hat{j} - 5\hat{k}$ will be

(*a*) zero (*b*) 45º (*c*) 90º (*d*) 180º

**10.** If $\vec{A} \times \vec{B} = \vec{B} \times \vec{A}$, then the angle between $\vec{A}$ and $\vec{B}$ is

(*a*) $\pi$ (*b*) $\frac{\pi}{2}$ (*c*) $\frac{\pi}{3}$ (*d*) $\frac{\pi}{4}$

***Answers to Objective Type Questions***

| | | | | |
|---|---|---|---|---|
| **1.** (*c*) | **2.** (*a*) | **3.** (*d*) | **4.** (*a*) | **5.** (*c*) |
| **6.** (*b*) | **7.** (*c*) | **8.** (*b*) | **9.** (*c*) | **10.** (*a*) |

# Index

## A



## B



## C



## D



## E

## N



## O



## P



## R

## S



## T



## U



## V



## W